Botany and History of Comfrey
Garden Uses of Comfrey

Taxonomy and Nomenclature of Comfrey
Borage Family, Symphytum Genus
Symphytum Species
Symphytum Species Classifications
Symphytum Genus Description
Symphytum Species Overview
Details about Symphytum Species
Details about Symphytum Species Hybrids

Prehistory, Ancient Times, Middle Ages and Comfrey
Renaissance and Comfrey: 1400s to 1600s
Age of Enlightenment and Comfrey: 1700s
Age of Revolution and Comfrey: 1800s
The 1900s and Comfrey

Garden Uses of Comfrey: Compost, Fertilizer, Potting Mix
Growing Vegetables and Fruit with Comfrey Fertilizer
Nutritional Value of Comfrey

Volume 1
Comfrey Book Series

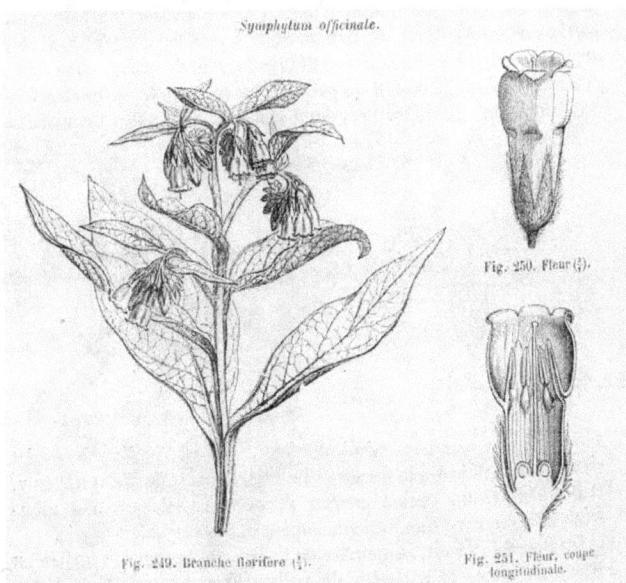

'Histoire des Plantes' by H. Baillon, Volume 10, Paris, France, 1891.
335 illustrations including Symphytum officinale page 346.

©Copyright 2022 Nancy Shirley. All rights reserved. Published in the United States.
Nantahala Farm and Garden www.nantahala-farm.com ncfarmgarden@gmail.com
ISBN 978-0-9890851-1-3

Basic Table of Contents for Volume 1

Preface for Volume 1 and 2 . . . . . . page 8

PART A: BOTANY OF COMFREY . . . . page 10
- Chapter 1: Taxonomy and Nomenclature of Comfrey . . . . . . page 11
- Chapter 2: Borage Family, Symphytum Genus . . . . page 23
- Chapter 3: Symphytum Species . . . . page 38
- Chapter 4: Symphytum Species Classifications . . . . page 53
- Chapter 5: Symphytum Genus Description . . . . page 83
- Chapter 6: Symphytum Species Overview . . . . page 110
- Chapter 7: Details about Symphytum Species: Asperum or Asperrimum (Prickly Comfrey) . . . . . . page 135
- Chapter 8: Details about Symphytum Species: Officinale (Common Comfrey) . . . . page 150
- Chapter 9: Details about Symphytum Species: No Asperum, No Officinale, No Hybrids . . . . page 182
- Chapter 10: Details about Symphytum Species Hybrid: Russian Comfrey . . . . page 263
- Chapter 11: Details about Symphytum Species Hybrids: Not Russian Comfrey . . . . page 300

PART B: HISTORY OF COMFREY . . . . page 309
- Chapter 12: Prehistory, Ancient Times, Middle Ages and Comfrey . . . . . . page 309
- Chapter 13: Renaissance and Comfrey 1400s-1600s . . . . page 322
- Chapter 14: Age of Enlightenment and Comfrey 1700s . . . . page 335
- Chapter 15: Age of Revolutions and Comfrey 1800s . . . . page 341
- Chapter 16: The 1900s and Comfrey . . . . page 355

PART C: GARDEN USES AND NUTRITIONAL VALUE . . . . page 369
- Chapter 17: Garden Uses of Comfrey: Compost, Fertilizer, Potting Mix . . . . page 369
- Chapter 18: Garden Uses of Comfrey: Growing Vegetables & Fruit with Comfrey Fertilizer . . . . . page 383
- Chapter 19: Nutritional Value of Comfrey . . . . page 407

Artwork

Symphytum armeniacum
    page 10

Symphytum asperum / asperrimum
    pages 8, 82, 149, 262, 334, 354, 382

Symphytum bulbosum
    page 334, 354

Symphytum caucasicum
    back cover

Symphytum cordatum
    page 340

Symphytum officinale
    front/back cover, pages 1, 37, 52, 299, 308, 354

Symphytum tauricum
    page 134

Symphytum tuberosum
    back cover; pages 37

Symphytum uplandicum
    page 149

Symphytum unknown
    pages 2, 7, 22, 109, 181, 321, 368, 439

Grand Consoude 1776

'Flora Parisiensis', also titled inside 'Introduction a la Flore des Environs de Paris', Volume 1, by Pierre Bulliard, published by P.F. Didot, Paris, France, 1776. Plate 87, Grand Consoude.

"Detail of the characteristic parts:
1. A life-size monopetalous flower.
2. An open flower, so as to show five stamens and five pointed scales which occupy the intervals left by the stamens.
3. The open chalice, at the bottom of which are four seeds.
4. State of the calyx when it envelops the seeds ready for their state of maturity."

Translated from French.

Book Cover by: Baker Vail Design

Expanded Table of Contents for Volume 1

Preface for Volume 1 and 2......page 8
Complex, Fascinating Comfrey
Definitions of Words
Finding Botanical Literature
Fair Use and Registration Marks
Legal Disclaimers

PART A: BOTANY OF COMFREY...p. 10

Chapter 1: Taxonomy & Nomenclature of Comfrey...p. 11
Classification of Life (Taxonomy of Comfrey)
Carl Linnaeus, Father of Taxonomy
Taxonomic Classifications Below Genus
    Cultivars
Two Basic Types of Taxonomy (Rank versus Group)
Methods of Determining Proper Taxonomic Relationships
    Morphological Taxonomy
    Chemical Taxonomy
    Cytological Taxonomy
Plant Nomenclature (Naming)
    Examples of Botanical Names
    More Plant Nomenclature Rules
    Illegitimate Botanical Names
Botanical Terminology and Finding Botanical Literature
    Recommended Books about Botany
    Common Botanical Words
    Standard Reference for Botanical Color Identification
    Free Websites to Find Botanical Literature
Organizations Involved in Naming/Cataloging Species
    Databases of Plants

Chapter 2: Borage Family, Symphytum Genus...p. 23
Boraginaceae Family (Borage)
    Plants in the Borage Family
    Chromosomes and Genome in the Borage Family

Symphytum (Comfrey) Genus
    Early Botanical History of Symphytum (Botanists)
        (With Examples of How Abbreviations are Used)
        Bauhin 1623, Parkinson 1640, Ray 1686,
        Tournefort 1700, Linnaeus 1753, Gaertner 1788,
        Schreber 1789, Jussieu 1789, Schkuhr 1791,
        Lamarck 1791, Sowerby 1791, Sibthorp 1794,
        Schmidt 1794, Lepechin 1797, Willdenow 1799.
    Later Botanical History of Symphytum (Botanists)
        Bieberstein 1808, Lehmann 1818,
        Schimper 1835, G. Don 1838, Fries 1839,
        De Candolle 1846, Ledebour 1847, Steven 1851,
        Bentham 1858, Kerner 1863, Schur 1866,
        Schultz 1875, Boissier 1879, Nyman 1884,
        Bucknall 1912, Tutin 1952, Komarov 1953,
        Popov 1953, Wade 1958, Perring 1962,
        Wickens 1969, Pawlowski 1972, Stearn 1986,
        Elci/Hilger/Erik 2008.

Chapter 3: Symphytum Species...p. 38
Comfrey Names (Ancient, Informal & Foreign Language)
Accepted Latin Names of Symphytum Species
List of All Symphytum Species
Symphytum Species at Herbariums (Dried)
Symphytum Species at Botanic Gardens (Living)
Symphytum Species at Gene Banks (Living)

Chapter 4: Symphytum Species Classifications...p. 53
Overview of Symphytum Species Classifications
1596 Bauhin
1623 Bauhin
1669 Ray
1846 De Candolle
1847 Ledebour
1879 Boissier
1881 Nyman
1887 Kuntze
1910 Kuznetsov
1913 Bucknall
1953 Popov
1967 Gadella and Kliphuis
1969 Wickens
1972 Pawlowski
1975 Perring
1985 Stearn
1990 Sandbrink
1991 Arnold Arboretum
2000 Slavik
2011 Hacioglu and Erik
2013 Gviniashvili
2019   (Accepted names of common species with
      synonyms and chromosome numbers.)
Taxonomy of Symphytum Species is Always Changing
Methods of Surveying and Collecting Symphytum in Field
Methods of Authentication of Comfrey Species

Chapter 5: Symphytum Genus Description...p. 83
Comfrey Confused with Foxglove
Symphytum Genus Description
Comfrey Leaf Description
    Comfrey Leaves and Pyrrolizidine Alkaloids
Comfrey Root Description
    Using Comfrey Roots to Tell Age of Plant
    Comfrey Root and Pyrrolizidine Alkaloids
Comfrey Flower
    Overview of Parts of a Flower
    Comfrey Flower Overview
    Comfrey Flower Buds, Bloom, Nectar & Pollination
    Comfrey Flower Perforations by Insects
    Comfrey Flower Color Changes Related to Soil Type
    Comfrey Flower and Pyrrolizidine Alkaloids
Comfrey Seed (Fruit)
Comfrey Seedling

Chapter 6: Symphytum Species Overview...p. 110
Comfrey Chromosomes Overview
Hybrids, Hybrid Swarms, Introgression
Overview of Crossbreeding Among Symphytum Species
Overview of Distribution of Symphytum Species
Most Frequently Grown Species of Comfrey
Overview Comparison of Symphytum Species Plus Keys
    Overview Russian Comfrey No.4 vs No.14 vs Officinale

Chapter 7: Details about Symphytum Species: Asperum or Asperrimum (Prickly Comfrey)...p. 135
Current Botanical Nomenclature
    S. asperum
    S. asperrimum
    S. armeniacum
    S. echinatum
    S. peregrinum Ledeb.
S. asperum is Correct Botanical Name
Subspecies and Varieties of S. asperum
    S. sepulcrale
    S. sylvaticum
History Prickly Comfrey in Great Britain, Ireland and USA
Distribution
Description
Large Forage Plant
Growth Patterns and Yield
Breeds with Other Symphytums
Chromosomes
Alkaloids

Chapter 8: Details about Symphytum Species: Officinale (Common Comfrey)....p. 150
Current Botanical Nomenclature
    S. bohemicum
    S. patens Sibth.
    S. tanaicense
    S. uliginosum
S. officinale Subspecies and Varieties
Herbalist's Shop
Distribution (Geographic Locations)
My Common Comfrey
Description
    More about Flowers and Leaf Wings
    Flower Colors
Flowers and Pollination
Hollow versus Solid Stem
More about Roots
Chromosomes
    S. officinale (2n=24 or 2n=48) vs (2n=40)
    S. officinale Chromosomes 2n = 40
    S. officinale Chromosomes 2n = 44
S. officinale Breeding with Itself Overview
S. officinale Breeds Between Its Purple & White 2n=48
S. officinale Breeds Between Its Own 2n=40, 2n=48
S. officinale Rarely Breeds Between Its 2n=24 & 2n=48
Constituents (Chemicals)
Alkaloids

Chapter 9: Details about Symphytum Species: No Asperum, No Officinale, No Hybrids....p. 182
anatolicum
    Current Botanical Nomenclature
    Description
    Chromosomes
angustifolium
armeniacum (see 'S. asperum')
bohemicum
    Current Botanical Nomenclature
    Description
    Chromosomes
bornmuelleri (see brachycalyx)
brachycalyx
    Current Botanical Nomenclature
    Description
bulbosum
    Current Botanical Nomenclature
    S. zeyheri
    Description
    Chromosomes
bullatum (see tauricum)
caucasicum
    Current Botanical Nomenclature
    Description
    Chromosomes
    Alkaloids
circinale
    Current Botanical Nomenclature
    Description
coccineum (see rubrum)
cordatum
    Current Botanical Nomenclature
    Description
    Chromosomes
creticum
cycladense
    Current Botanical Nomenclature
    Description
davisii
    Current Botanical Nomenclature
    Description
floribundum
    Current Botanical Nomenclature
    Description
Goldsmith (see grandiflorum)
grandiflorum
    Current Botanical Nomenclature
    S. goldsmith
    Description
    Chromosomes
    Alkaloids
gussonei
    Current Botanical Nomenclature
    Description
hajastanum
    Current Botanical Nomenclature
    Description

Hidcote (see 'Details about Symphytum Species Hybrids')
ibericum
    Current Botanical Nomenclature
    Description
    Chromosomes
    Alkaloids
kurdicum
    Current Botanical Nomenclature
    Distribution
    Description
leonhardtianum (see tuberosum)
mediterraneum (see floribundum)
naxicola
    Current Botanical Nomenclature
    Description
nodosum
norfolk (see 'Details about Symphytum Species Hybrids')
norvicense ('Details about Symphytum Species Hybrids')
orientale
    Current Botanical Nomenclature
        S. patens Fries
    Description
    Chromosomes
ottomanum
    Current Botanical Nomenclature
    Description
    Chromosomes
palaestinum
patens
    S. patens Fries (and see S. orientale)
    S. patens Sibth. (and see S. officinale)
peregrinum ('Details about Symphytum Species Hybrids')
podcumicum
    Current Botanical Nomenclature
    Description
pseudobulbosum
    Description
rubrum (Red)
savvalense
    Description
sepulcrale
    Current Botanical Nomenclature
    Description
sylvaticum
    Current Botanical Nomenclature
    Description
    Alkaloids
tanaicense
    Current Botanical Nomenclature
    Description
    Chromosomes
tauricum
    Current Botanical Nomenclature
    Description
    Chromosomes
tuberosum (Tuberous Comfrey)
    Current Botanical Nomenclature
    S. tuberosum Subspecies and Varieties
        S. angustifolium
        S. besseri
        S. bulbosum
        S. floribundum
        S. gussonei
        S. leonhardtianum
        S. mediterraneum
        S. nodosum
        S. popovii
        S. zeyheri and others
    S. tuberosum Distribution (Locations)
    Description
    S. tuberosum Breeds with Other Symphytum Species
    Chromosomes
    Alkaloids
uliginosum
    Current Botanical Nomenclature
    Description
    Chromosomes
uplandicum ('Details about Symphytum Species Hybrids')
zeyheri (see S. bulbosum)

Chapter 10: Details about Symphytum Species Hybrid: Russian Comfrey....p. 263

    (uplandicum, peregrinum auct. non Ledeb.)
Current Botanical Nomenclature
Hybrid Origin and Classification (peregrinum, uplandicum)
    Overview
    Hybrid Origin of Russian Comfrey
Russian Comfrey Distribution (Locations)
Description
Determining Hybrid Character: Chromosomes, Chemicals
F1 Hybrid
Flowers
High Yield of Russian Comfrey
    Yield of Russian Comfrey No. 4 versus No. 14
Preferred Soil Type and Roots
Strains (Cultivars) of Russian Comfrey
    Turner Strain (Bocking No. 1, 16, 17)
    Stephenson Strain (Bocking No. 2, 14)
        Bocking No. 14
    Webster Strain (Bocking No. 3-11, Bocking Mixture)
        Bocking No. 4
        Gibson Strain (improved Webster Strain)
    Other Russian Comfrey Cultivars and Varieties
        Bocking No. 12
        Moorland Heather
        Variegated Varieties
Russian Comfrey Chromosomes
    Overview
    Various Cytotypes
    Cytotype $2n = 34$
    Cytotype $2n = 36$
    Cytotype $2n = 40$
    Cytotype $2n = 42$
    Cytotype $2n = 44$
        $(2n = 36)$ vs $(2n = 40)$

Russian Comfrey Alkaloids
Creating New Hybrids of Russian Comfrey
Russian Comfrey Breeds with Other Symphytums

Chapter 11: Details about Symphytum Species Hybrids: Not Russian Comfrey....p. 300
Overview
Goldsmith (see S. grandiflorum)
Hidcote Blue and Pink
    Current Botanical Nomenclature
    Description
    Chromosomes
peregrinum (see 'Russian Comfrey')
Hybrids with S. officinale
    (excludes Hidcote and Russian Comfrey)
    S. officinale x S. asperum x S. tuberosum
    S. officinale x S. bohemicum
    S. officinale x S. cordatum
    S. officinale x S. peregrinum
    S. officinale x S. tuberosum
Other Hybrids:
    S. tuberosum x S. cordatum
    S. tuberosum x S. bulbosum
    S. asperum x S. caucasicum
    S. asperum x S. orientale = S. norvicense
    S. caucasicum x S. orientale
uplandicum (see 'Russian Comfrey')

PART B: HISTORY OF COMFREY

Chapter 12:
Prehistory, Ancient Times, Middle Ages and Comfrey
    ....p. 309
History of the Word 'Comfrey'
Prehistory of Comfrey
400 BC - 400 AD: Ancient Times
    Herodotus, 484-425 BC, Greek
    Nicander of Colophon, 200-101 BC, Greek
    Pliny the Elder, 23-79 AD, Roman
    Pedanius Dioscorides, 40-90 AD, Greek
    Galen of Pergamon, 130-200 AD, Greek
401 - 1400 AD: Middle Ages
    Johnson Papyrus, early 400s AD, Greek
    Aetios or Aetius, 500 AD, Greek
    Aegineta, 625-690 AD, Greek
    Anglo-Saxon Leech Books, 900-1000+ AD
    Christian Monasteries, 340-1400 AD, Europe
    Hildegard of Bingen, 1098-1179 AD, German
    Gilbert Anglicus, 1230-1250 AD, English

Chapter 13: Renaissance & Comfrey 1400-1600s....p. 322
Overview of Renaissance
1401 - 1600: Early Renaissance
    Paracelsus, 1493-1541, Swiss
    Otto Brunfels, 1488-1534, German
    Jean Ruel, 1536, 'De Natura Stirpium', French
    Jean Fernel, 1542, 'Naturali Parte Medicinae', French
    Hieronymus Bock 1552 'De Stirpium', German
    William Turner, 1551-1568, 'A New Herball', English
    Rembert Dodoens, 1578, 'A New Herball', Flemish
    John Gerard 1597 'Herball or General Historie', English
1600s: Renaissance
    John Parkinson 1640 'Theatrum Botanicum', English
    North American Colonists, 1649
    Johann Bauhin, 1651, 'Historia Plantarum', Swiss
    Nicholas Culpeper 1653 'Complete Herbal', English

Chapter 14:
Age of Enlightenment and Comfrey 1700s....p. 335
Joseph Tournefort 1700 French botanist, defined genus
William Salmon 1710 'Botanologia, The English Herbal'
Carl Linnaeus 1753 'Species Plantarum', Swedish
J. Busch 1771 sent Symphytum from Russia to England
I. Lepechin 1774-1802, described plant species, Russian
W. Withering 1787 'Systematic Arrangement British Plants'
Thomas Christy 1790 'Forage Plants', English
James Sowerby 1791-1814, 'English Botany' books
William Woodville 1794 'Medical Botany', English

Chapter 15:
Age of Revolutions and Comfrey 1800s....p. 341
William Curtis, early 1800s, 7 Symphytums in England
Prickly Comfrey becomes popular in England.
Many writings in Great Britain and USA about Comfrey
Thomas Christy promotes Prickly and Russian Comfrey
De Candolle 1846 classifies Symphytum species
Carl von Ledebour 1847 classifies Symphytum species
Henry Doubleday, late 1800s, developed Russian Comfrey

Chapter 16: The 1900s and Comfrey....p. 355
Webster Nurseries 1900 grows Russian Comfrey
Cedric Bucknall 1912 classifies Symphytum species
Vernon Stephenson 1942 grows Russian Comfrey
Lawrence Hills 1948 grows Russian Comfrey
Peggy Greer 1952 grows Russian Comfrey
Newman Turner 1952 Comfrey as veterinary medicine
Russian Comfrey goes to Canada and Africa 1953
Tom Tutin 1952-1980 wrote about British & Europe Flora
Lawrence Hills 1953 wrote book 'Russian Comfrey'
Henry Doubleday Research Association founded 1954
Russian Comfrey goes to Japan and Australia late 1950s
Alkaloids found in Common Comfrey 1968
Lawrence Hills 1975 wrote book 'Comfrey Report'
Lawrence Hills 1976 'Comfrey: Fodder, Food, Remedy'
Australia 1984 puts Comfrey on 'Poison Schedule'
Germany & European Union 1992 regulate Comfrey
United Kingdom 1993 bans some internal use of Comfrey
United States 1994 begins regulation of Comfrey

PART C:
GARDEN USES AND NUTRITIONAL VALUE

Chapter 17: Garden Uses of Comfrey:
   Compost, Fertilizer, Potting Mix....p. 369
Comfrey Roots Break Up Hard Soil
Comfrey As Mulch
Grow-It-Yourself Fertilizer
Minerals in Comfrey Overview (see 'Nutritional Value')
   Comfrey as Potassium Fertilizer
Comfrey as Liquid Manure (Liquid Fertilizer)
   How to Make Liquid Fertilizer with Water (Comfrey Tea)
   Make Liquid Fertilizer with No Water (Concentrate)
   How to Use Liquid Fertilizer
Making Compost with Comfrey
   How to Make Compost
   Deep Roots Reach Nutrients
   Mix Comfrey with High Carbon Material
   Comfrey is a Compost Activator
Potting Mixture with Leaf Mold
   How to Make Non-Comfrey Leaf Mold
   Make Potting Mixtures with Leaf Mold and Comfrey

Chapter 18: Garden Uses of Comfrey:
   Growing Vegetables & Fruit with Comfrey Fertilizer
                                                                 ....p. 383
Growing Potatoes with Comfrey as Fertilizer
   Putting Comfrey Compost in Soil for Potatoes
   Timing Planting of Potatoes with Comfrey Growth
   How Much Comfrey Per Row, and Type of Comfrey
   How to Use Comfrey in Potato Rows
   Digging Potatoes in the Fall
   Preparing Potato Patch in the Fall
Growing Vegetables with Comfrey as Fertilizer
   Cabbage, Bean, Fruit, Lettuce, Onion, Potato, Etc.
   Fermented Extracts from Comfrey Help Plants Grow
   Tomatoes and Comfrey
   Onions, Garlic and Comfrey
   Beans / Peas and Comfrey
   Wheat Seedlings Helped by Comfrey
   Vegetable Crops Not Helped
   Large Dose of Comfrey Reduces Growth of Lettuce
Fruit and Comfrey (Small Plant, Bush, Tree)
   Small Fruiting Plants and Bushes
   Fruit Trees and Other Trees
   Fruit Trees, Calcium and Comfrey
Comfrey Heals Other Plants
Protects Plants from Insects, Slugs and Snails
Comfrey as Plant Antifungal
Permaculture and Forest Gardens
   Permaculture Polycultures
Dynamic Accumulator and Soil Improver
Comfrey as Weed Barrier
Comfrey as Wind Barrier
Aquaponic Food Production
Beneficial Insects and Comfrey
   Bees and Comfrey

Chapter 19: Nutritional Value of Comfrey....p. 407
Dry Matter
Nitrogen-Phosphorus-Potassium (NPK)
Protein (Crude Protein)
   Amino Acids in Comfrey
Fiber (Crude Fiber)
Minerals (Ash) (see 'Garden Uses of Comfrey')
   Potassium
Vitamins in Comfrey
Vitamin B12
   Doubts About Vitamin B12 in Comfrey
Mucilage in Comfrey
Fat, Protein, Carbohydrate, Fiber, Ash, Miscellaneous
   Fructans (Fructose Polymers)
Other Constituents including Phenolic Acids
   Overview
   Rosmarinic Acid (Phenolic Acid)
   Caffeic Acid Polyether (Phenolic Acid)
   Choline
   Chlorphyll
Digestibility of Comfrey Leaf
   Definitions of Digestibility Terms
   Digestibiity Overview
   Nutrition and Digestibility of Leaves/Stems & Maturity
   Digestibility of Protein
   Digestible Dry Matter (DDM)
Energy: Gross and Metabolizable

1. Latin: **Symphytum flore pallido**
2. Latin: **Symphytum maius flore purpureo**
3. Latin: **Symphytum maius flore rubro**
Fol. 110 in Hortus Eystettensis (Garden of Eichstatt)
by Basilius Besler, **1640**, Nuremberg, Germany

Preface for Volume 1 and 2

Some references to 'Sections' and 'Subsections' are in Volume 2.

Complex, Fascinating Comfrey
When I started writing this book I had no idea there was so much information available about Comfrey.

Volume 1
The botany is complex and still has areas of uncertainty. Botanical research continues with updates about taxonomy, chromosomes and species.

The history section is an overview of medical references, taxonomical investigations, livestock feeding, garden use, and international trade. Comfrey has been used medicinally from at least 400 BC in Greece.

Comfrey can be used to help your garden grow. It is beneficial for compost, fertilization, and permaculture. The nutritional value is examined.

Volume 2
This volume shows how Comfrey is used as a forage plant for livestock. Also how to successfully make comfrey hay and silage.

The healing and potentially dangerous properties of Comfrey are investigated from both the scientific / research and the everyday / practical points of view.

Then how to take care of your Comfrey plants including propagation, fertilization and harvesting. Yields are given to help determine your expected productivity. Last is how to get rid of unwanted Comfrey plants.

Definitions of Words
I have included many definitions of words for those who are not botanical, historical, geographical, medical or livestock experts. As much as possible I have given the complete words for abbreviations various authors have used. It is surprising and sometimes frustrating how many abbreviations are used by expert writers. In many cases they are writing for other experts, so they assume everyone knows what the abbreviations mean.

At times I had to do a lot of searching to find out what they meant. At one point I could not figure out the meaning of a botanical phrase, so I contacted one of the well-known plant organizations about the meaning. I was put in contact with a botanical classification expert with a Ph.D. who works in the Department of Botany at Harvard University, Cambridge, Massachusetts.

I have included definitions of some words that may be unfamiliar to people who do not speak English as their first language. This book is relevant to people all over the world. This is also useful for clarifying the meaning of what the author is trying to say for those who are not experts in that field.

Many authors wrote research reports about a particular country or region and expected the only readers to be people in those areas. So towns, cities and other geographic regions are frequently mentioned with no country or other identifier for the reader to know where they are located. I defined as many as I could. Some ancient writings are harder to figure out.

In certain situations I include more detailed descriptions about the region because it is helpful to understand the botany about Comfrey both from a taxonomical and cultivation point of view. If an author uses measurements in metric, then I added the English/Imperial measurements, and vice-versa. This way people in all countries understand the length, weight or distance.

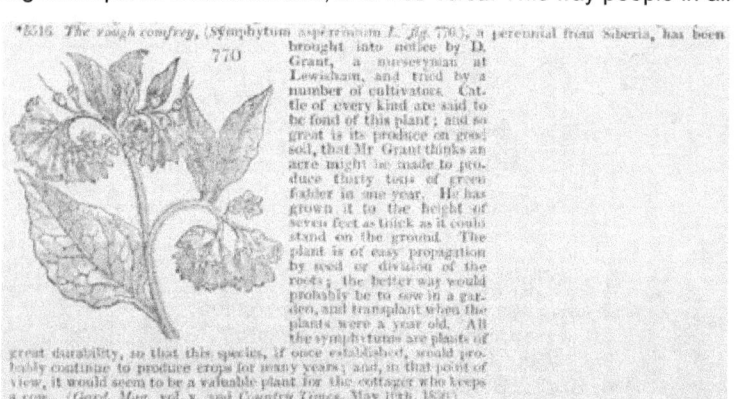

Rough Comfrey = Symphytum Asperrimum 1835

'An Encyclopaedia of Agriculture: Comprising the Theory and Practice of the Valuation, Transfer, Laying Out, Improvement and Management of Landed Property' by J.C. Loudon, FLGZ, HS, London, England, 1436 pages, **1835**. Comfrey page 870, image 770.

Finding Botanical Literature
It was a fun mystery to find some of the more obscure literature. In some cases, I contacted the authors who were able to send me pdfs of their research reports.

Some articles are written in languages other than English with no translations available. I tried online translators but in most cases they were not very helpful. For people who are able to translate reports into English, I am hoping you will send me translations. My goal is to write an update to this book with the new information I receive.

For all research articles in this book, some may not be relevant to the average person. However, I like to include them because there are researchers and other scientists who are interested in them. These scientists can then find these articles and read the entire report.

For an unknown reason, sometimes the search function for a particular term works better online at the original source such as www.biodiversitylibrary.org or https://www.hathitrust.org, than it does on a pdf that I downloaded from the same site.

Fair Use and Registration Marks
Fair use is a doctrine in United States copyright law that allows limited use of copyrighted material without requiring permission from the rights holders. It provides for the legal, non-licensed citation or incorporation of copyrighted material in another author's work.

"Title 17 United States Copyright Act, Section 107. Chapter 1 - Subject Matter and Scope of Copyright.
Limitations on exclusive rights: Fair use.
Notwithstanding the provisions of sections 106 and 106A, the fair use of a copyrighted work, including such use by reproduction in copies or phonorecords or by any other means specified by that section, for purposes such as criticism, comment, news reporting, teaching, scholarship, or research, is not an infringement of copyright."

All the contents of this book are under the protection of Section 107 of the United States Copyright Law. Under the fair use rule of copyright law, an author may make limited use of another author's work for purposes of criticism and comment such as quoting or excerpting a work in review or criticism for purposes of illustration or comment. There are also provisions for education, teaching and research, among others.

As much as possible I put '®' after a company name or registered product if they used it on their website or literature. The '®' symbol means it is federally registered with the United States Patent and Trademark Office, Alexandria, Virginia, either on the Principal Register or Supplemental Register. Please consider all company names and products as potentially trademarked or registered.

Legal Disclaimers
This book is for entertainment purposes only. This is especially true for the medical and medicinal information provided. Statements about health and healing for humans, pets or livestock are not a substitute for professional medical advice, diagnosis and treatment. If you, your family, pets or livestock have health issues, please contact your physician, veterinarian or other health care provider.

This book includes ancient, folk, traditional and various historical accounts of healing that may not be accurate or safe to follow. And there is great controversy about the safety of Comfrey for internal use. I have quoted all of the information I could find about that, but it is up to you and your health care provider to decide what is right. I can not make that decision, especially because each person or animal has unique health needs.

This book is the result of extensive investigation from many sources. Some quotes are from scientific research or years of practical experience. Others are opinions, folklore or grapevine/gossip. The information may not be accurate or appropriate for you to use in your situation. Therefore, I assume no responsibility if you decide to follow the quoted articles, reports, books, etc. Every quote is referenced with author, publication and year. It is up to you to do your own research to see if it applies to you.

Agricultural information may or may not be useful for your environment or type of farm/garden. Check with your local horticultural, agricultural or livestock department/agency to see what is right for you. Go online to various websites, forums and blogs to see what is good agriculture and ranching practice to achieve your goals. Some jurisdictions consider Comfrey to be invasive and unwanted.

Researchers, farmers and historical records sometimes have very different ideas about the safety of Comfrey as food for animals. Contact your veterinarian or other livestock specialist to see what is right for you and your animals.

I mention many companies, organizations, products, resources, services, journals, books, etc. Except for 'Nantahala Farm and Garden' and 'Western North Carolina Farm and Garden Calendar', I do not have a connection with any of them. I am not recommending or endorsing them. I am providing references so you can do your own research about them.

PART A

BOTANY OF COMFREY

*"**Comfrey is the common name given to the genus Symphytum, a member of the Boraginaceae (Borage) family.** It is native of Eastern Europe and Western Asia in the Caucasus area.*
Comfrey is a coarse perennial herb up to five feet (1.5 meter) high, leafy, branching, with hairy stems and leaves; the lower leaves are generally larger than the upper ones; extensive fleshy root system; flowers whitish, yellowish or dull purple; generally adapted to moist cool places."
-Comfrey Report: The Story of the World's Fastest Protein Builder and Herbal Healer, Conservation Gardening and Farming Series: Series C by Lawrence D. Hills. England: Henry Doubleday Research Association, 1975, page 26.

Comfrey is native to Europe and countries around the Mediteranean Sea (Spain, France, Italy, Sicily, Slovenia, Croatia, Serbia, Albania, Greece, Turkey, Syria, Lebanon, Israel, Egypt, Libya, Tunisia, Algeria, Morocco). **It is also native to the Caucasus.** The Caucasus Mountains are located at the border of Europe and Asia, between the Black Sea and Caspian Sea and occupied by Russia, Georgia, Azerbaijan, and Armenia.

*"**Russian Comfrey is a perennial fodder crop, in the lucerne (alfalfa) class for nutritional value but with a vastly greater yield.** This yield, in from six to eight cuts between early April and the end of November, totals forty tons an acre for a poor crop, and a hundred tons for a good one.*
It is a member of the order Boraginaceae (Borage-wort, Borage-root) and therefore avoids the galaxy of viruses and eelworms that beset so many modern crops.
***Its high average protein of 24 percent of the dry matter, and low average fibre at 10 percent when cut at the leafy stage,** makes it ideally suited for pigs and poultry."*
-Russian Comfrey: A Hundred Tons an Acre of Stock or Compost for Farm, Garden or Smallholding by Lawrence D. Hills. London England: Faber and Faber, Limited, 1953, page 15.

*"**The Comfrey plant has the quickest turnover of capital of all, to build cut after cut of leaves and stems,** any one of which is a season's work for most crops. Much of this leaf-won food from sun and air is hoarded in the great roots, and gives the starting stock the energy that allows a first cut before the soil is warmed and kindly for surface-rooting grasses.*
***The Russian Comfrey grows with semi-tropical speed in a cold climate,** and therefore there must be something rather unusual about the chloroplasts in its cells and its metabolism in general."*
-Russian Comfrey: A Hundred Tons an Acre of Stock or Compost for Farm, Garden or Smallholding by Lawrence D. Hills. London England: Faber and Faber, Limited, 1953, page 143.

*"**Comfrey (Symphytum spp.) is the fastest growing plant in the temperate zone.** Its taproot is extremely deep, and loaded with allantoin, double that of leaves. Its leaves are very important, due to content of allantoin.*
***Protein content is so great that Comfrey is the highest protein yielding plant known.** If the soil is good, protein content is 25% - 30%. It has a more balanced ratio of calcium (2) to phosphorus (1), than alfalfa has, as a forage for cows."*
-The Organic Method Primer: A Practical Explanation: The How and Why for the Beginner and the Experienced by Bargyla and Gylver Rateaver. San Diego, California: The Rateavers, 1993, page 163.
   (sp.= single species) (spp.= more than 1 species)

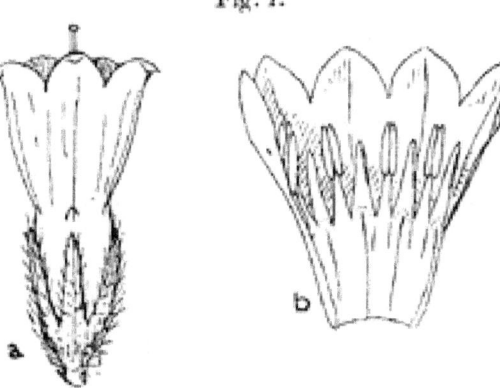

Fig. 1.

Symphytum armeniacum.
a. Flower.   b. Corolla, opened to show stamens and corolla-scales.   × 2.

Symphytum Armeniacum Flower

'A Revision of the Genus Symphytum, Tourn.' by Cedric Bucknall, Journal of the Linnean Society of London, England, Botanical Journal, Volume 41, Issue 284, December 1913.

a. Flower
b. Corolla, opened to show stamens and corolla-scales

Chapter 1

Taxonomy and Nomenclature of Comfrey

Classification of Life (Taxonomy of Comfrey)

Taxonomy is the science of defining and naming groups of biological organisms on the basis of shared characteristics. Organisms are grouped together into taxa (singular: taxon). A taxon is a taxonomic group of any rank, such as a class, family or species. **These groups are given a taxonomic rank.** Groups are combined to form a larger group of higher rank, thus creating a taxonomic hierarchy.

The taxonomic classifications of life are divided starting at the largest grouping:
Domain, Kingdom (Regnum), Division (Divisio or Phylum), Class (Classis), Order (Ordo), Family (Familia), Genus, Species.

"*Rank*   *Latin*
Kingdom   Plantae   Plantes, Planta, Vegetal, Plants
Subkingdom   Viridiplantae
Infrakingdom   Streptophyta   Land plants
Superdivision   Embryophyta
Division   Tracheophyta   Vascular plants, tracheophytes
Subdivision   Spermatophytina   Sermatophytes, seed plants, phanérogames
Class   Magnoliopsida
Superorder   Asteranae
Order   Boraginales
Family   **Boraginaceae**   **Borage, bourraches**
Genus   **Symphytum L.**   **Comfrey** "

-ITIS (Integrated Taxonomic Information System®) Report, www.itis.gov, 2018. The ITIS is the result of a partnership of United States Federal agencies formed to satisfy their mutual needs for scientifically credible taxonomic information.

"Kingdom   Plantae
Phylum   Tracheophyta
Class   Magnoliopsida
Order   Boraginales
Family   **Boraginaceae**
Genus   **Symphytum L.** "

-Global Biodiversity Information Facility® (GBIF), www.gbif.org, 2018. It is an international network and research infrastructure funded by the world's governments and aimed at providing open access to data about all types of life on Earth.

"Class: Equisetopsida C. Agardh
Subclass: Magnoliidae Novak ex Takht.
Superorder: Asteranae Takht.
Order: Boraginales Juss. ex Bercht. & J. Presl
Family: **Boraginaceae Juss.**
Genus: **Symphytum**
The Tropicos system follows the 2009 Angiosperm Phylogeny Group (APG III) and as updated by the Angiosperm Phylogeny Group in 2016 (APG IV). In this classification all green plants are included in the Class Equisetopsida."
-Tropicos®, www.tropicos.org, Missouri Botanical Garden, Saint Louis, Missouri; Plant database with 1.3 million scientific names and 4.4 million specimen records, 2019.

"*Rank*   *Latin*   *English*
Kingdom   Plantae   Plants
Subkingdom   Tracheobionta   Vascular plants
Superdivision   Spermatophyta   Seed plants
Division   Magnoliophyta   Flowering plants
Class   Magnoliopsida   Dicotyledons
Subclass   Asteridae
Order   Lamiales
Family   **Boraginaceae**   **Borage family**
Genus   **Symphytum L.**   **Comfrey** "

-United States Department of Agriculture, Natural Resources Conservation Service®, Plants Database, www.plants.usda.gov and https://plants.sc.egov.usda.gov/core/profile?symbol=SYMPH2, 2018. The 'Plants Database' provides standardized information about the vascular plants, mosses, liverworts, hornworts, and lichens of the United States and its territories.

"Family: Boraginaceae
Subfamily: Boraginoideae
Tribe: Boragineae
Subtribe: Boraginiae
Genus: Symphytum L."
-United States National Plant Germplasm System (NPGS), https://npgsweb.ars-grin.gov, United States Department of Agriculture (USDA), Agriculture Research Service, **2018**. NPGS is effort to safeguard genetic diversity of agriculturally important plants.

> **"Boraginaceae is a family of herbs, shrubs and trees with a cosmopolitan distribution.** *The family comprises about 130 genera and 2300 species (Mabberely 1997). Boraginaceae has for a long time provided a set of controversial phylogenetic problems at different taxonomic levels.*
> **Boraginoideae, the largest subfamily of Boraginaceae, has variously been divided into 4 to about 20 tribes, for example, Popov 1953, 12 tribes), even though most authors have accepted only 4 to 7 tribes.**
> **Symphytum L. genus is in Boragineae tribe** *according to De Candolle (1846), Bentham and Hooker (1873), Baillon (1888), Gurke (1897), Al-Shehbaz (1991), Riedl (1997), and Takhtajan (1997)."*
> -'Tribes of Boraginoideae (Boraginaceae) and Placement of Antiphytum, Echiochilon, Ogastemma and Sericostoma: A Phylogenetic Analysis Based on atpB Plastid DNA Sequence Data' by E. Langstrom (Sweden) and M.W. Chase (England), Plant Systematics and Evolution, Volume 234, pages 137-153, 2002.

Carl Linnaeus, Father of Taxonomy

Swedish botanist, zoologist and physician Carl (Carolus) Linnaeus (Carl von Linne) is the father of taxonomy because he developed a system known as 'Linnaean Taxonomy' for categorizing organisms and naming them using binomial nomenclature (two names). In botany, the abbreviation L. indicates Linnaeus as the authority for a species' name. In older publications, the abbreviation 'Linn.' is used.
He was Professor of Medicine and Botany at Uppsala University, Sweden. One of the most acclaimed scientists in Europe.

Linnaean Taxonomy is described in his 1735 book 'Systema Naturae' (System of Nature through the Three Kingdoms of Nature, According to Classes, Orders, Genera and Species, with Characters, Differences, Synonyms, Places).

> **"Systema Naturae** *is one of the major works of the Swedish botanist, zoologist and physician Carl Linnaeus (1707-1778) and introduced the Linnaean taxonomy. Although the system, now known as binomial nomenclature, was partially developed by the Bauhin brothers, Gaspard and Johann, 200 years earlier, Linnaeus was first to use it consistently throughout his book. The first edition was published in 1735.*
> *The tenth edition of this book (1758) is considered the starting point of zoological nomenclature. In 1766-1768 Linnaeus published the much enhanced 12th edition, the last under his authorship.*
> *Another again enhanced work in the same style and titled 'Systema Naturae' was published by Johann Friedrich Gmelin between 1788 and 1793.*
> **Systema Vegetabilium** *is a book published in 4 editions, following 12 earlier editions known as Systema Naturae. The first edition, published in 1774 and edited by Johan Andreas Murray is counted as edition 13 because it continues from the 12th edition of Systema Naturae. All names in it are attributed to Carl Linnaeus.*
> *The second edition, (counted as the 14th of Systema Naturae) published in 1784, includes plant species described by J.A. Murray and Carl Peter Thunberg.*
> *The third edition (counted as the 15th), was edited by Christiaan Hendrik Persoon.*
> *Although the fourth edition, purportedly the 16th edition of Linnaeus's work was published in five volumes between 1824 and 1828, and attributed to Kurt Sprengel, the International Plant Names Index suggests that it should be counted as edition 17.*
> *The 16th edition (abbreviated Syst. Veg., ed. 15 bis [Roemer & Schultes]) was written by Johann Jacob Roemer, Josef August Schultes and Julius Hermann Schultes. It was published in 7 volumes between 1817-1830 under the name Systema Vegetabilium: Secundum Classes, Ordines, Genera, Species. Cum characteribus differentiis et synonymis. Nova Editio, speciebus inde ab Editione XV. Detectis aucta et locupletata."*
> -Wikipedia®: The Free Encyclopedia, www.wikipedia.org, 2019.

See subsection 'Early Botanical History' in section 'Borage Family, Symphytum Genus' (Chapter 2).

In this binomial {two terms} nomenclature {name or designation} there are three kingdoms, divided into classes, that are then divided into orders, genera {singular: genus}, and species {singular: species}, with an additional rank lower than species.

'Genera Plantarum (Genera of Plants), Edition 1' by Swedish naturalist Carl Linnaeus, 1737. The first edition of 'Genera Plantarum' contains descriptions of the 935 plant genera known to Linnaeus.
Linnaeus also wrote the book 'Species Plantarum' (Species of Plants) published in 1753. It lists every species of plant known at the time, classified into genera (plural of genus). It was the first book to consistently apply binomial names.

Taxonomic Classifications Below Genus

The **'International Code of Nomenclature'** for algae, fungi, and plants (ICN) is the set of rules and recommendations that govern the scientific naming of all organisms. According to Article 3 of the ICN, the most important ranks of taxa are kingdom, division / phylum, class, order, family, genus, and species. **According to Article 4 the secondary ranks of taxa are tribe, section, series, variety and form.** There are an indefinite number of ranks.

Genus (gen.)
    subgenus (subgen.)
        Section (sectio, sect.)
            subsection (subsectio, subsect.)
                Series (ser.)
                      subseries (subser.)
Species (sp. = singular, spp. = plural)
    subspecies (subsp. = singular, subspp. = plural)
        Variety (varietas, var.)
            subvarietas (subvariety, subvar.)
                Form (forma, f.)
                      subform (subforma, subf.)

A section is a taxonomic rank below genus and above species. The subgenus, if present, is higher than the section, and the rank of series, if present, is below the section. Sections may in turn be divided into subsections.

Comfrey examples of Subgenus:
    subgenus Ramosa (Symphytum)
    subgenus Simplicia (Tuberosum)

> *"The genus Symphytum (Boraginaceae) contains 41 species, which are classified in the two subgenera, Symphytum with 29 and Tuberosum with 12 species (J. Sandbrink, unpublished). Subgenus Tuberosum is characterized by the presence of creeping rhizomes."*
> -'Hybridization in Symphytum: Pattern and Process' by T.W.J. Gadella, Sommerfeltia: The Journal of Natural History Museum and University of Oslo, Norway, Volume 11, pages 79-96, 1990.

Comfrey examples of Sections:
    Section Coerulea Bucknall
    Section Lingulata Pawlowski
    Section Symphytum Officinalis

Comfrey examples of Subsection:
    Subsection Ochroleuca Kusn.
    Subsection Cyanea Kusn.

Comfrey example of Series:
    Series Caucasica

Comfrey examples of Species:
    Symphytum asperum Lepech.
    Symphytum officinale L.
    Symphytum peregrinum Ledeb.

Comfrey examples of Subspecies:
    Symphytum tuberosum **subsp.** nodosum (genus, species, subspecies)
    Symphytum sylvaticum **subsp.** sepulcrale (genus, species, subspecies)

Comfrey examples of Variety:
    Symphytum officinale **var.** patens (genus, species, variety)
    Symphytum ibericum **var.** abchasicum (genus, species, variety)

Comfrey examples of Subvariety:
    S. officinale **subvar.** ochroleucum
    S. officinale **subvar.** purpureum

Comfrey examples of Form:
- Symphytum uplandicum Nyman **f.** coeruleum (Petitm. ex Thell.) P.D.Sell
- Symphytum uplandicum Nyman **f.** densiflorum (Buckn.) P.D.Sell (Fl. Gr. Brit. Ireland 3: 519. 2009)
- Symphytum uplandicum Nyman **f.** discolor (Buckn.) P.D.Sell (Fl. Gr. Brit. Ireland 3: 519. 2009)
- Symphytum uplandicum Nyman **f.** lilacinum (Buckn.) P.D.Sell (Fl. Gr. Brit. Ireland 3: 519. 2009)

Cultivars (Cultivated Plants)

Article 28 of the 'International Code of Nomenclature' (ICN) and the **'International Code of Nomenclature for Cultivated Plants'** (ICNCP or Cultivated Plant Code) define the rules for cultivars and cultivar groups. "Cultivar: The basic independent category used for organisms in agriculture, forestry, and horticulture."
A cultivar name is the scientific Latin botanical name followed by a cultivar epithet (descriptive phrase). This epithet is usually in common language

According to the ICNCP the word cultivar is used in two ways:
First, as a 'classification category' as defined in Article 2: *'The basic category of cultivated plants whose nomenclature is governed by this Code is the cultivar. There are 2 other classification categories for cultigens, grex and group. The Code then defines a cultivar as a taxonomic unit within the classification category of cultivar.'*
Second: *'A cultivar is an assemblage of plants that (a) has been selected for a particular character or combination of characters, (b) is distinct, uniform and stable in those characters, and (c) when propagated by appropriate means, retains those characters.'*

('Grex' is derived from Latin with gregis meaning flock. It is used to describe hybrids of orchids.)
Comfrey examples are:
- **Symphytum grandiflorum 'Hidcote Blue'** (genus, species, cultivar).
- **Symphytum ibericum 'Jubilee'** (genus, species, cultivar).
- **Symphytum x uplandicum 'Bocking No. 4'** (genus, hybrid species, cultivar).

Two Basic Types of Taxonony (Rank versus Group)

1. Linnaean Taxonomy (Rank)
Linnaean taxonomy is based on rank. Taxonomic rank is the level of a group of organisms (a taxon) in a taxonomic hierarchy. Examples of taxonomic ranks are species, genus, family, order, class, phylum, kingdom, domain. This is described above. It is the most frequent method of categorizing organisms.

2. Clade Taxonomy (Cladistics or Groups)
Organisms are categorized in groups (clades) based on the most recent common ancestor with shared derived characteristics. Rank is ignored.
Phylogenetics is a type of cladistics. Phylogentics is the study of the evolutionary history and relationships among organisms. Inherited traits such as DNA sequences or morphology are evaluated using a model of evolution (phylogenetic tree). This tree is a diagrammatic hypothesis about the history of evolution.

Methods of Determining Proper Taxonomic Relationships

See subsection 'Methods of Authentication of Comfrey Species' in section 'Symphytum Species Classifications'. It includes DNA barcoding plus other methods.

1. Morphological Taxonomy (Morphotaxonomy)
Morphology is the study of the form and structure of organisms. This includes the outside appearance (shape, structure, arrangement, color, pattern, size) and internal structure (anatomy).
Morphology is the primary method for classifying organisms into various taxonomic groups. Similarities in morphological characteristics are used to group plants together. For example, all flowering plants with ovules enclosed in an ovary are grouped together as Angiosperms. Then the angiosperms are further classified into Dicotyledons and Monocotyledons, based on differences in their roots, leaf veins, flower symmetry and number of cotyledons in the embryo.

2. Chemical Taxonomy (Chemotaxonomy)
Chemotaxonomy uses phytochemical (biologically active compounds in plants) data to determine the proper taxonomic category. So far 33 groups of chemicals in plants have been found to be of taxonomic significance. Examples in Comfrey include alkaloids and triterpenoids.

"Thirty members attended the Northeast Regional Meeting in York (England) on Saturday, 7th October, 1967. The morning session was held at St William's College, by kind permission of the Dean and Chapter, under the Chairmanship

of Dr F. H. Perring.
Dr G. A. Nelson, of the School of Medicine, University of Leeds gave an illuminating paper on **the problematical Symphytum spp. in Britain**, in which he described nine species and hybrids illustrating these with coloured slides. **He also showed how every species could be identified using the chromatography method frequently used by pharmacognosists."**
-'Proceedings of Botanical Society of the British Isles, Volume 7' edited by E.F. Greenwood, London, England, **1967-1969**, page 503.
> ('spp.' is the abbreviation for species. It refers to all species in that genus. Sometimes it is used when the specific species in that genus is not known.)
> (Chromatography is passing a mixture in solution, suspension or gas through a medium so that the parts separate.)
> (Pharmacognosy is the study of medicinal drugs obtained from plants and other natural sources.)

"For a consideration of the chemotaxonomical value of a specific marker it is necessary to investigate its variation between different individuals from single species.
From our survey of several cytotypes of Symphytum officinale (Common Comfrey) with 2n = 24, 40, and 48, and harvested at different times and locations, only slight qualitative variations in phytosterol and triterpenoid profiles can be demonstrated. The same applies to S. asperum (Prickly Comfrey) and S. x uplandicum (Russian Comfrey).
> **Because of their ubiquitous occurrence in extracts of the investigated taxa, in general, the phytosterols seem to be of no value as chemotaxonomical markers here.**

The triterpenoid isobauerenol, on the other hand, seems useful, for its presence in S. x uplandicum (2n = 36 or 40) readily demonstrates the hybrid character of this taxon which was artificially derived from S. officinale (2n = 40 or 48) plants containing isobauerenol and from S. asperum (2n = 32) plants which lack this typical compound.
> **However, this specific marker gives no information concerning the question, whether the S. officinale cytotype 2n = 40 belongs to a hybrid swarm between S. asperum and S. officinale (2n = 48)**, as was suggested by *Basler, or has to be considered as conspecific (same species) with S. officinale. For this question one has to use pyrrolizidine alkaloids as markers.

The data we have presented may also be useful for verifying the origin of Comfrey roots used in pharmaceutical preparations or for teas (**Roitman, 1981), especially when pyrrolizidine alkaloid patterns are also included in the phytochemical investigations as was outlined already by us (***Huizing et. al., 1982)."
-'**Chemotaxonomical Investigations of the Symphytum Officinale Polyploid Complex and S. Asperum (Boraginaceae): Phytosterols and Triterpenoids**' by H.J. Huizing, Th.M. Malingre, Th.W.J. Gadella, and E. Kliphuis, all from The Netherlands; Plant Systematics and Evolution, Volume 143, pages 285-292, **1983**.
(* -'Cytotaxonomic Studies on the Boraginaceen Genus Symphytum L.: Studies on Predominantly Northern German Plants of the Species S. Asperum Lepech., S. officinale L. and S. x Uplandicum Nym.' by Armin Basler, Botanische Jahrbucher fur Systematik: Pflanzengeschichte und Pflanzengeographie {Botanical Yearbooks for Systematics: Plant History and Plant Geography}, Volume 92, pages 508-553, 1972. All in German. I was unable to find this report. If you have an English translation, could you please send it to me.)
(** -'Comfrey and Liver Damage' by James N. Roitman, The Lancet, Volume 317, Issue 8226, page 944, April 25 1981.)
(*** -'Chemotaxonomical Investigations of the Symphytum Officinale Polyploid Complex and S. Asperum {Boraginaceae}: The Pyrrolizidine Alkaloids' by H.J. Huizing, Th.W.J. Gadella, and E. Kliphuis, all from The Netherlands; Plant Systematics and Evolution, Volume 140, pages 279-292, 1982.)
> (A cytotype is an individual of a species that has a different chromosome number/ploidy or structure than another in the same species. Karyotype is the number and appearance of chromosomes in the nucleus of an eukaryotic cell.)

"Chemotaxonomical studies on herbarium material (dried) have some disadvantages which are not met when using fresh material. The herbarium material is unique and therefore an experiment can not be repeated. The treatment of herbarium plant material and, therefore, also the range of decomposition in this material of pyrrolizidine alkaloids and triterpenes are unknown.
Fresh plant material should always be preferred to herbarium material for chemotaxonomical studies."
-'**Chemotaxonomy of the Symphytum Officinale Agg. (Boraginaceae)**' by Jaarsma, Lohmanns, Gadella and Malingre, Plant Systematics and Evolution, 167, pages 113-127, **1989**.

3. Cytological Taxonomy (Cytotaxonomy)
Cytotaxonomy investigates cellular structures such as somatic (body) chromosomes to classify organisms. The number, structure, and behaviour of chromosomes is of great value in taxonomy. Chromosome numbers are determined at mitosis (normal cell division) and can be diploid (2n = 2x) or polyploid (triploid or higher).
Body cells and individuals are described by the number of chromosome sets {ploidy level} they have: monoploid: 1 set; diploid: 2 sets = 2n = 2x, triploid: 3 sets = 2n = 3x, tetraploid: 4 sets = 2n = 4x, pentaploid: 5 sets = 2n = 5x, etc.

Cytotaxonomy methods include the 'genome and plasmon method' and the 'karyological method' (karyotaxonomy). Genome is the DNA of the nucleus. Plasmon is the DNA of cytoplasmic organelles. Karyotaxonomy classifies organisms by comparing the number, shape and size of chromosomes. The chromosome banding technique uses stains that create dark and light areas.

Plant Nomenclature (Naming)

A botanical name is a formal scientific name conforming to the 'International Code of Nomenclature'® for algae, fungi, and plants (ICN). The purpose of a formal botanical name is to have a single name that is accepted and used worldwide.

For the naming of cultivated plants there is a separate code, the 'International Code of Nomenclature for Cultivated Plants' that supplements the ICN.

The Code is periodically updated by the 'International Botanical Congress' with the support of the 'International Association for Plant Taxonomy'. Each new edition supersedes earlier editions and is retroactive back to 1753 (the publication of 'Species Plantarum' by Carl Linnaeus).

In botanical nomenclature the first publication of a name for a taxon is the correct one. The formal starting date is 1753.

Article 13 of 'International Code of Nomenclature' for algae, fungi, and plants states: *"Generic names that appear in Linnaeus'* **Species Plantarum** *edition 1 (1753), and edition 2 (1762–1763), are associated with the first subsequent description given under those names in Linnaeus'* **Genera Plantarum** *edition 5 (1754) and edition 6 (1764)."*

In a genus name, 'Gen. pl.' means 'Genera Plantarum' that was published by Swedish naturalist Carl Linnaeus. It is a complementary volume to his book 'Species Plantarum'. The first edition of 'Genera Plantarum' contains descriptions of the 935 plant genera known to Linnaeus at that time.

Examples of Botanical Names

Depending on rank, botanical names may be in one part (genus and above), two parts (below the rank of genus) or three parts (below the rank of species). Cultivated plant names may have another term added so there are four parts to the name. Except for cultivars, the name of a plant can never have more than three parts.

Examples of three part names:
- Symphytum tuberosum subsp. nodosum (genus, species, subspecies)
- Symphytum sylvaticum subsp. sepulcrale (genus, species, subspecies)
- Symphytum officinale var. patens (genus, species, variety)
- Symphytum ibericum var. abchasicum (genus, species, variety)
- Symphytum grandiflorum 'Hidcote Blue' (genus, species, cultivar)

Examples of four part names:
- Symphytum officinale subsp. officinale var. ochroleucum (genus, species, subspecies, variety)
- Symphytum officinale subsp. officinale var. purpureum
- Symphytum officinale var. rubra 'Raspberry Queen' (genus, species, variety, cultivar)
- I have seen this Symphytum name with 4 parts but do not think it is technically valid.

Example of hybrid names:
- Symphytum x uplandicum
- Symphytum officinale x Symphytum asperum x Symphytum grandiflorum
- Symphytum x hidcotense

Examples of names with author citation:

Author citation is the person who validly published a botanical name while fulfilling the formal requirements of the 'International Code of Nomenclature'.

Symphytum officinale L. (The 'L.' stands for Carl von Linnaeus.)

Symphytum officinale L. Sp. Pl. 1: 136 1753.
(Author 'Carl von Linnaeus' published in 'Species Plantarum', Volume 1, page 136 in 1753.)

Symphytum sylvaticum var.hordokopii(Kurtto)R.R.Mill Fl.Turkey&EAegeanIsl.10:188 1988
(Originally published by 'Kurtto', then revised publication by 'R.R.Mill'. Published in 'Flora of Turkey and the East Aegean Islands', Volume 10, page 188, in 1988.)

Symphytum uliginosum A. Kern. Oesterr. Bot. Z. 13(7): 227-228 1863.
(Author 'A. Kern.' published in 'Osterreichische Botanische Zeitschrift', Volume 13, No. 7, pages 227-288, in 1863.)

Symphytum officinale subsp. uliginosum (A. Kern.) Nyman Consp. Fl. Eur. 509 1881.
(Author 'A. Kern' was revised by 'Nyman' in 'Conspectus Florae Europaeae', page 509, in 1881.)

Symphytum uliginosum auct. non Kern.
(auct.= auctorum= various authors)
('auct. non' is the preferred syntax for a plant name that has been misapplied by a subsequent author

or authors {auct. or auctt.} such that it actually represents a different taxon from the one to which Kern's name correctly applies.)

Symphytum x uplandicum Nyman Syll. Fl. Eur. 80 1855.
(Author 'Carl Frederik Nyman' published in 'Sylloge Florae Europaeae', page 80, in 1855.)

Symphytum asperrimum Donn ex Sims
('ex' means an initial description did not satisfy the rules for valid publication, but that the same name was subsequently validly published by a second author or by same author in subsequent publication.)

Symphytum tauricum sensu Coste non Willd.
('sec. Coste' or 'sensu Coste' indicates that the intended meaning is the one defined by that author. The term 'sec.' is an abbreviation of 'secundum' which means 'following' or 'in accordance with'.)
(In biology, 'homonym' is a name for a taxon that is identical in spelling to another such name, that belongs to a different taxon. The term 'non' means that there is a homonym, and this indicates which taxon is correct. Or to put it another way: If an author wishes to indicate that a name has been used for a different taxon, then there is citation of the name and the author followed by the word 'non' or 'not' and the name of the author who first used the name.)

More Plant Nomenclature Rules

"Rules: The scientific, or Latin, names of plants, both wild and cultivated are formulated and written according to rules governed by the 'International Code of Botanical Nomenclature', July 2005 (Vienna Code).
Distinguishable groups of cultivated plants, whose origin or selection is due primarily to mankind, are given epithets (i.e., names, such as cultivar names) formed according to the rules and provisions of the **'International Code of Nomenclature for Cultivated Plants'**, February 2004.
The aim of these codes is to promote uniformity, accuracy and stability in formulating the scientific names of all plants (Botanical Code) and in formulating the cultivar names of agricultural, forestry, and horticultural plants (Cultivated Plant Code)."

"Genus and Species Names: Plant names may include a genus, specific epithet, a name rank below species (such as a subspecies and/or botanical variety), Latin name authorities, and the cultivar or release name."

"Hybrid Names: Validly published hybrid names are signified by the symbol 'x'.
Hybrids at the generic level are written with an 'x' immediately prior to the genus name, such as in the following example: xElyleymus colvillensis (Lepage) Barkworth.
For a hybrid at species level an 'x' is placed immediately prior to the specific epithet, as in this example: Quercus xdeamii Trel."

"Subspecific & Varietal Names: Terms 'subspecies' and 'variety' are used to designate first and second divisions of a species.
A **'subspecies'** is a grouping within a species used to describe geographically isolated variants, a category above 'variety', and is indicated by the abbreviation 'subsp.' in the scientific name.
A **'variety'** consists of more or less recognizable entities within species that are not genetically isolated from each other, below the level of subspecies, and are indicated by the abbreviation 'var.' in the scientific name.
When the subspecies or variety name is the same as the specific epithet (this is called a typical expression), then the authority is included only after the species name, as in the following example: Cornus sericea L. subsp. sericea.
When the subspecies name or variety is different than the species name, then both the species authority and the subspecies or variety authority are used, as in the following example: Cornus sericea L. subsp. occidentalis (Torr. & Gary) Fosberg."

"Plant Cultivars:
A **'cultivar' is a taxon that has been selected for a particular attribute or combination of attributes,** and this is clearly distinct, uniform, and stable in its characteristics that when propagated by appropriate means, retains those characteristics. The cultivated plants covered by the 'International Code of Nomenclature for Cultivated Plants' may arise by deliberate hybridization or by accidental hybridization in cultivation, by selection from existing cultivated stock, or may be a selection from variants within a wild population and maintained as a recognizable entity solely by continued propagation.
Cultivar names may be given to the following types of propagated materials: clones, graft-chimeras, seed (as long as the propagated material retains unique characteristics of the parents), line, multi-line, F1 hybrids, and genetically modified plants. **The words 'variety' and 'form' are not synonyms for the word cultivars according to the 'International Code of Nomenclature for Cultivated Plants'.** The Code considers these terms botanical classifications. **The 'Association of Official Seed Certifying Agencies'® (AOSCA) considers the terms 'cultivar' and 'variety' equivalent."**

-United States Department of Agriculture 'National Plant Materials Manual', 190-V-NPMM, Fourth Edition, Part 542 Acronyms, 542.2 Plant Nomenclature, July **2010**.

Illegitimate Botanical Names

International Code of Nomenclature for Algae, Fungi, and Plants: Chapter V.- Rejection of Names, Article 52

"**52.1. A name, unless conserved (Article 14) or sanctioned (Article 15), is illegitimate and is to be rejected if it was nomenclaturally superfluous when published**, i.e. if the taxon to which it was applied, as circumscribed by its author, definitely included the type (as qualified in Article 52.2) of a name which ought to have been adopted, or of which the epithet ought to have been adopted, under the rules (but see Article 52.3)."

'Nomen illegitimum', abbreviated 'nom. illeg.', is Latin for 'illegitimate name'. A 'superfluous name' is often an 'illegitimate name'. In Latin it is 'nomen superfluum', abbreviated 'nom. superfl.'

A 'nomen illegitimum' is a validly published name, but one that contravenes some of the articles laid down by the 'International Botanical Congress'. It could could be illegitimate because:
- Article 52: it was superfluous at its time of publication, i.e., the taxon (as represented by the type) already has a name, or
- Articles 53 and 54: the name has already been applied to another plant (a homonym).

International Code of Nomenclature for Algae, Fungi, and Plants: Chapter II.- Status, Typification and Priority of Names
Section 4, Limitation of the Principle of Priority, Article 14
"14.1. In order to avoid disadvantageous nomenclatural changes entailed by the strict application of the rules, and especially of the principle of priority in starting from the dates given in Article 13 and F.1, this Code provides, in App. II–IV, **lists of names of families, genera, and species that are conserved (nomina conservanda)** (see Rec. 50E.1).
Conserved names are legitimate even though initially they may have been illegitimate. The name of a subdivision of a genus or of an infraspecific taxon may be conserved with a conserved type and listed in App. III and IV, respectively, when it is the basionym or replaced synonym of a name of a genus or species that could not continue to be used in its current sense without conservation."

Botanical Terminology and Finding Botanical Literature

Recommended Books about Botany (I do not have any connection with these authors.)

A Botanist's Vocabulary: 1300 Terms Explained and Illustrated by Susan K. Pell and Bobbi Angell. Portland, Oregon: Timber Press, 2016.
Botanical Latin: History, Grammar, Syntax, Terminology and Vocabulary, Fourth Edition by William T. Stearn. Portland, Oregon: Timber Press, 1992.
Botany in a Day: The Patterns Method of Plant Identification by Thomas J. Elpel. Montana: Hops Press LLC, 2008.
Plant Identification Terminology: An Illustrated Glossary by James and Melinda Harris. Utah: Spring Lake Publishing, 2001.
The Cambridge Illustrated Glossary of Botanical Terms by Michael Hickey and Clive King. England: Cambridge University Press, 2001.
The Kew Plant Glossary: An Illustrated Dictionary of Plant Terms by Henk Beentje. Richmond, England: Kew Royal Botanic Gardens, 2016.

Common Botanical Words

Alternate: Leaves are arranged alternately along the stems.
Annual: A plant that completes its life cycle in one year or less.
Anthers: It carries the pollen of the flowers, located at the tip of the stamens (male reproductive organs).
Axil of a Leaf: Where the stem of a leaf (the petiole) joins the stem.
Biennial: A plant that completes its life cycle in two years.
Calyx: The part of the flower underneath or behind the corolla that consists of sepals.
Ciliate: Fine hairs along the leaf margin.
Cordate: The leaf is wide and heart-shaped.
Corolla (petals): The colorful part of the flower above the calyx.
Decurrent: The lower portion of a leaf that lies flat against the stem.
Ellipsoid: A 3-dimensional flattened structure that is more broad toward the middle than at its two outermost edges. The 2-dimensional counterpart of ellipsoid is 'elliptic.'
Fibrous: The root system consists of a loose collection of thin branching roots.
Filament: A long thread-like structure with an anther at its tip; the filament connects the anther to the base of the flower.
Fleshy: A fleshy root is a central taproot that can be nearly as wide as it is long. Sometimes fibrous roots are described as fleshy. This means that they are somewhat thicker than ordinary fibrous roots.
Fruit: Refers to the seeds and their surrounding covering.
Lanceolate: The leaf is broader at the base than at the apex (outer tip); the apex is pointed.
Leaf Nodes: Where the leaves join the stem.
Nutlets: Refers to seeds with a very hard coating.
Ovaries: The female reproductive organs of a flower that contain developing seeds (ovules).

Ovate: The leaf is rather wide, but narrows to a point at both ends, especially at the apex.
Ovoid: A seed or seed capsule that is oval or egg-shaped.
Perennial: A plant that lives for more than 2 years.
Petiole: The stem of a leaf.
Pinnate: A central vein along length of the leaf, from which side veins radiate outward from an acute angle.
Pistil: The female organ of a flower consisting of a stigma, style, and ovulum (ovary).
Pubescent: The surface of a leaf or stem that is densely covered with fine short hairs.
Raceme: An elongated cluster of flowers arranged individually along a central stem.
Rhizomatous: A rhizomatous root system has shallow underground runners (rhizomes) producing plantlets.
Rugose: The rough-wrinkly surface of a seed.
Sepal: A leaf-like segment at the base of the flower.
Sessile: A leaf that joins the stem without a petiole, or a flower that joins the stalk without a pedicel.
Stamens: Male reproductive organs of a flower; stamen consists of thread-like filament with a pollen-bearing anther at its apex.
Stigma: The tip of a pistil that receives the pollen.
Stoloniferous: A stoloniferous root system has above ground runners (stolons) that produce new plantlets.
Style: This is a long, thread-like structure that connects the stigma with the ovary.
Taproot: The root system has a central root growing vertically down that originates from the base of the plant.
Tuberous: A root system with a loose collection of coarse roots that occasionally thicken into fleshy underground tubers.

-'Definitions and Line Drawings of Botanical Terminology', Illinois Wildflowers, www.illinoiswildflowers.info. Contains descriptions, photographs, and range maps of many wildflowers and other plants in Illinois, Dr. John Hilty, 2018. And other sources.

Standard Reference for Botanical Color Identification

"Colour is an important criterion in the description and identification of cultivated plant varieties (cultivars). In fact it is frequently the only distinguishing feature between one cultivar and another; it is therefore a significant diagnostic character.
The RHS (Royal Horticultural Society) Colour Chart is a tool that horticulturalists can use to describe colour and it is, for horticulturalists, the standard reference for colour identifications.
Such descriptions are not only published in horticultural publications but importantly form part of the description that accompanies the herbarium specimen.
When consulting herbarium specimens, living colour is invariably lost, and it is the colour chart that forms a pivotal role in the descriptive label that accompanies the dried pressed plant.
The colour chart is invaluable when describing a range of plants in the same colour range. The colour chart has evolved and endeavoured to put colour description on a more objective, scientific and systematic basis."
-'**The Royal Horticultural Society's Colour Chart**: An Everyday Tool for Use in the Herbarium. Its Past, Present and Future' by Susan Grayer, Natural Sciences Collections Association, NatSCA News, Issue 18, pages 19-26, 2009.
(Horticulture is the art and science of garden cultivation and management.)

As of 2018, the Sixth Revised Edition of 'RHS Large Colour Chart' (2015) is available from the Royal Horticultural Society's online shop: www.rhsshop.co.uk. The chart has 920 colors. Every color has a code number.

The '**Royal Horticultural Society**' (RHS) was founded in 1804 as the 'Horticultural Society of London'. It is based in London, England. It helps train professional and amateur gardeners with formal courses up to the most complete one of 'Master of Horticulture'. It promotes horticulture through flower shows including the 'Chelsea Flower Show', 'Hampton Court Palace Flower Show', 'Tatton Park Flower Show' and 'Cardiff Flower Show'.

The 'Royal Horticultural Society' has four gardens in England: Wisley Garden in Surrey county, Rosemoor Garden in Devon county, Hyde Hall in Essex county, and Harlow Carr in North Yorkshire county. The RHS is custodian of the Lindley Library in London, England with branches at its four gardens. The library is based on the book collection of John Lindley. The RHS Herbarium has a collection of 3,300 watercolors, 30,000 color slides, and many digital images.

Munsell® Color Charts for Plant Tissue
Munsell® sells 17 'Color Charts' for analyzing plant tissue color. The color of plant tissues reflects the influence of light, temperature, soil composition, toxic substances and parasites. Sometimes plant color reveals genetic origins. Munsell Color Products, Grand Rapids, Michigan, https://munsell.com.

Recommended Free Websites to Find Botanical Literature (If you have others to recommend, please let me know.)

AgroAtlas®: Interactive Agricultural Ecological Atlas of Russia and Neighboring Countries
Economic plants and their diseases, pests and weeds. The Russian-English Agricultural Atlas is the world's most comprehensive information on the geographic distribution of plant agriculture in Russia and neighboring countries. It contains 1500 maps of the distribution of 100 crops, 560 wild crop relatives, 640 diseases, pests and weeds, and 200

environmental parameters. www.agroatlas.ru

Over 60 Russian scientists from 3 different institutes developed the maps, descriptions and literature citations on AgroAtlas. It is difficult to find Russian Comfrey literature translated into English. Please send me pdfs in English.

BASE®: Bielefeld Academic Search Engine, Germany
BASE is one of the world's largest search engines for academic web resources. It has 120 million documents from 6,000 sources. Full text is available for 60% of the documents for free. www.base-search.net

Biblioteca Digital: Real Jardin Botanico CSIC®, Spain
The Digital Library of the Royal Botanic Garden (Consejo Superior de Investigaciones Cientificas® = Spanish National Research Council) is an online resource about botanical bibliography. It is the botany digital library of publications from the Iberian Peninsula, Balearic Islands, Macaronesia, North Africa, Mediterranean region and Latin America. It provides journals and other botanical works (historical, floristic, taxonomic). http://bibdigital.rjb.csic.es

Biodiversity Heritage Library®
BHL operates as a worldwide consortium of natural history, botanical, research, and national libraries working together. As of 2018, BHL has 141,359 titles and 233,203 volumes with literature going back as far as the 1400s. It is the foundational literature component of 'Encyclopedia of Life' (www.eol.org). www.biodiversitylibrary.org

Through the 'Global Names Recognition and Discovery' (GNRD) service, BHL indexes taxonomic names throughout the collection, allowing researchers to locate publications about specific taxa, http://gnrd.globalnames.org.

Botanicus®, United States
Botanicus is a free portal to digitized historic (1480-1950 AD) botanical literature from the 'Missouri Botanical Garden Library' in Saint Louis, Missiouri. www.botanicus.org

At the top of the page, click on 'Names', then type in 'Symphytum' in the lower search box.

BSBI Archive (Botanical Society of Britain and Ireland®)
Botanical literature of Britain and Ireland are available in digital form for free:

'BSBI Database' where you can search for botanical literature by genus, with many publications from the 1950s to 2000. http://rbg-web2.rbge.org.uk/BSBI

'BSBI News' magazine from 1972 to 2014. http://archive.bsbi.org.uk/bsbi_news.html

'Botanical Exchange Clubs' 1839-1946. http://archive.bsbi.org.uk/bec_reports.html

'Journal of Botany' 1834-1940. http://archive.bsbi.org.uk/journal_of_botany.html

'Proceedings of the Botanical Society of the British Isles' 1954-1969. http://archive.bsbi.org.uk/proceedings.html

'Watsonia', the journal of the BSBI from 1949 to 2010. http://archive.bsbi.org.uk/watsonia.html

Chromosome Counts Database®
CCDB is a comprehensive community resource for plant chromosome numbers. It combines data resources into an extensive central database. http://ccdb.tau.ac.il

CORE®
It brings together millions of the world's 'open access' research papers. Open access refers to research materials which are distributed online, free of cost. https://core.ac.uk

Google Books®
It is the search engine with the most comprehensive index of full-text books. Sometimes there is a pdf of entire book. Sometimes just a few pages are available. https://books.google.com

Google Scholar®
Google Scholar is a search engine that indexes the full text or metadata of scholarly literature. It includes most peer-reviewed online academic journals and books, conference papers, theses and dissertations, preprints, abstracts, and technical reports. (Metadata is a set of data that describes information about other data.) https://scholar.google.com

HathiTrust Digital Library®
HathiTrust is a partnership of more than 140 academic and research institutions, offering a collection of millions of digitized titles from around the world (as of 2019, over 17 million volumes). Sometimes there is a pdf of entire article/book. Sometimes just a few pages are available. www.hathitrust.org, babel.hathitrust.org

International Plant Names Index® (IPNI)
IPNI is a database of the names and basic bibliographical details of seed plants, ferns and lycophytes (fern allies). It is a collaboration between the 'Royal Botanic Gardens, Kew', 'Harvard University Herbaria', and 'Australian National Herbarium'. www.ipni.org

You can do a **botanical publications search, by full title or abbreviation**:

www.ipni.org/ipni/publicationsearchpage.do
You can also do an **author search**: http://www.ipni.org/ipni/authorsearchpage.do

Internet Archive®
Internet Archive is a non-profit library of millions of free books, movies, software, music and more. It has 17 million books, articles and other printed references. https://archive.org

National Library of Australia® (Trove®)
Trove brings together content from libraries, museums, archives, repositories and other research and collecting organisations. It is a collaboration between the 'National Library of Australia', Australia's State and Territory libraries and hundreds of cultural and research institutions around Australia. Sometimes there is a pdf of entire article/book. Sometimes just a few pages are available. Good for more than just Australia.

Naturalis Biodiversity Center®, Netherlands
Guide to plant species descriptions published in seed lists from Botanic Gardens from 1800-1900. Botanic Gardens produced annual seed lists for exchange. In some from the 1800s they include diagnoses, descriptions and notes. Based in Netherlands. http://seedlists.naturalis.nl

PubMed®: United States National Library of Medicine, U.S. National Institues of Health
PubMed has 28 million citations for biomedical literature from MEDLINE, life science journals, and books. Citations sometimes include links to the full-text. www.ncbi.nlm.nih.gov/pubmed
PubMed Central®
PubMed Central® is a free full-text archive of 5 million biomedical and life sciences journal literature at the United States National Institutes of Health's 'National Library of Medicine' (NIH/NLM). www.ncbi.nlm.nih.gov/pmc

Soil and Health Library®
It has free downloadable e-books about radical agriculture, natural hygiene/nature, and homestead living. https://soilandhealth.org

Organizations Involved in Naming and Cataloging Plant Species

Australian National Herbarium (Canberra Act, Australia) www.anbg.gov.au/cpbr/herbarium
Harvard University Herbaria (Cambridge, Massachusetts) https://huh.harvard.edu
International Association for Plant Taxonomy: facilitates governance of the 'International Code of Nomenclature for
    algae, fungi and plants'. (Bratislava, Slovakia) www.iaptglobal.org
Missouri Botanical Garden (Saint Louis, Missouri) www.missouribotanicalgarden.org
Royal Botanic Gardens Kew (London, England) www.kew.org
United States Department of Agriculture (Washington, D.C.) www.usda.gov

Databases of Plants

Australian Plant Name Index® (APNI)
Maintained by 'Australian National Botanic Gardens' in collaboration with 'Centre for Australian National Biodiversity Research' and 'Australian Biological Resources Study'. APNI deals with plant names and their usage in the scientific literature, whether as a current name or synonym. www.anbg.gov.au/apni

'BSBI List of British & Irish Vascular Plants and Stoneworts'® is a database that includes accepted scientific names, good synonymy, vernacular names and Kent numbers. BSBI = Botanical Society of Britain and Ireland®. Based in London, England. www.nhm.ac.uk/our-science/data/uk-species/checklists/NHMSYS0000436459/index.html

'Catalogue of Life'® is an online database of the world's known species of animals, plants, fungi and micro-organisms. It consists of a single integrated species checklist and taxonomic hierarchy. Based in The Netherlands. www.catalogueoflife.org. Search: www.catalogueoflife.org/annual-checklist/2015/search/all

'Database of Vascular Plants of Canada'® (VASCAN) is a comprehensive and curated checklist of all vascular plants reported in Canada, Greenland (part of Denmark), and Saint Pierre / Miquelon (near Newfoundland, part of France). It was developed at the Universite de Montreal Biodiversity Centre, Quebec, Canada, http://data.canadensys.net/vascan

'Chromosome Counts Database'® (CCDB) is a comprehensive resource for plant chromosome numbers. CCDB combines existing data resources into an extensive central database that is updated regularly by the community. Every chromosome number includes the original database and the research article reference. http://ccdb.tau.ac.il

'**Encyclopedia of Life**'® (EOL) is supported by these institutions: 'Atlas of Living Australia', 'La Comision Nacional para el Conocimiento y Uso de la Biodiversidad' (Mexico), 'Harvard University Marine Biological Laboratory' (Massachusetts), 'New Library of Alexandria' (Bibliotheca Alexandrina, Egpyt) and 'Smithsonian Institution's National Museum of Natural History' (Washington, DC). www.eol.org

'**Global Biodiversity Information Facility**'® (GBIF), or 'Global Compositae Checklist', is an international network and research infrastructure funded by world governments to provide open access to data about all types of life. In Denmark. www.gbif.org

'**Integrated Taxonomic Information System**'® (ITIS) is a partnership of United States Federal agencies formed to meet needs for scientifically credible taxonomic information. It is a collaboration among 1 Canadian, 1 Mexican and 8 United States organizations. ITIS is a partner of 'Species 2000' and the 'Global Biodiversity Information Facility' (GBIF). ITIS and 'Species 2000 Catalogue of Life' (CoL) provide the taxonomic backbone to the 'Encyclopedia of Life' (EOL). www.itis.gov

'**International Plant Names Index**'® (IPNI) is a collaboration between 'Royal Botanic Gardens Kew' in England (Index Kewensis), 'Harvard University Herbaria' in Massachusetts (Gray Herbarium Index, https://kiki.huh.harvard.edu/databases), and 'Australian National Herbarium' (APNI). It is a database of the names and basic bibliographical details of seed plants, ferns and lycophytes (fern allies). www.ipni.org
You can also do a **botanical publications search**: www.ipni.org/ipni/publicationsearchpage.do
You can also do an **author search**: http://www.ipni.org/ipni/authorsearchpage.do

'**Plants Database**'®, 'United States Department of Agriculture', 'Natural Resources Conservation Service'®. It provides standardized information about vascular plants, mosses, liverworts, hornworts, and lichens in USA. www.plants.usda.gov

'**Plants of the World Online**'® is from Kew Royal Botanic Gardens, London, England. Its goal is to enable users to access information on all the world's known seed-bearing plants. With over 8.5 million items, Kew houses the largest and most diverse botanical and mycological collections in the world. www.plantsoftheworldonline.org

'**The Barcode of Life Data Systems**'® (BOLD) aids the acquisition, storage, analysis, and publication of DNA barcode records. It assembles molecular, morphological, and distributional data. All databases: www.boldsystems.org/index.php/databases
Public Data Portal: www.boldsystems.org/index.php/Public_BINSearch?searchtype=records

'**The Euro+Med PlantBase**'®: The Information Resource for Euro-Mediterranean Plant Diversity. It integrates and evaluates information from 'Flora Europaea', 'Med-Checklist', 'Flora of Macaronesia', and from regional and national floras and checklists. It provides access to the total European flora of vascular plants in 222 plant families, **2019**. www.emplantbase.org and http://ww2.bgbm.org/EuroPlusMed/query.asp

'**The Plant List**'® is a collaboration between 'Royal Botanic Gardens Kew' (England) and 'Missouri Botanical Garden' (United States), updating stopped in 2013. www.theplantlist.org. Now part of 'World Flora Online' (WFO). www.worldfloraonline.org

'**Tropicos**'® has all of the nomenclatural, bibliographic, and specimen data accumulated in the 'Missouri Botanical Garden' electronic databases during the past 30 years. www.tropicos.org
   '**Index to Plant Chromosome Numbers**'® (IPCN) www.tropicos.org/Project/IPCN
   It indexes plant chromosome numbers of naturally occurring and cultivated plants published throughout the world.
   IPCN is based at the 'Missouri Botanical Garden'.

'**United States National Plant Germplasm System**'® (NPGS) is a collaborative effort to safeguard the genetic diversity of agriculturally important plants. The NPGS is managed by the 'Agricultural Research Service'® (ARS), the in-house research agency of the United States 'Department of Agriculture'® (USDA). https://npgsweb.ars-grin.gov

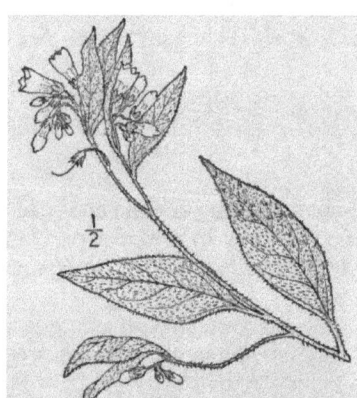

Symphytum 1913

'An Illustrated Flora of the Northern United
States and Canada, Volume 3:
Gentianceae to Compositae (Gentian to Thistle)'
by Nathaniel Lord Britton, PH.D., Sc.D., LL.D.
and Honorable Addison Brown, A.B., LL.D.,
New York Botanical Garden, 1913.

Chapter 2

Borage Family, Symphytum Genus

<u>Boraginaceae Family (Borage)</u>

The Boraginaceae or borage family (also called the forget-me-not family) includes a variety of shrubs, trees, and herbs, totaling around 2,000 species in 146 genera. Plants in the borage family are annual or perennial herbs, frequently with rough, stiff hairs.
Genus is a taxonomic rank that is above species and below family. In binomial nomenclature, the genus name is the first part of the binomial species name. Genera is the plural form of genus.

*"**The Boraginaceae is a family of dicotyledonous plants** of the order Tubiflorae of Engler and Gilg (1924), the order Polemoniales of Bentham and Hooker (1873) and the Boraginales of Hutchinson (1926).*
The family is world-wide in distribution. *There are various estimates of the size of the Boraginaceae. Gurke (1897) lists and describes 85 genera. Willis (1948) states that there are 100 genera and 1800 species in the family."*
-'Cytogenetic Studies on the Boraginaceae' by Donald M. Britton, Brittonia; The New York Botanical Garden, Volume 7, No. 4, pages 233-266, December 10 **1951**.

"Boraginaceae Family:
Herbs or dwarf shrubs, often hispid. Leaves alternate, exstipulate, simple. Flowers usually in scorpioid cymes, usually actinomorphic. Calyx 5-toothed or-lobed. Corolla 5-lobed, cylindrical, campanulate, hypocrateriform or rotate, usually with a distinct tube and limb; tube often with 5 scales, invaginations, or tufts or lines of hairs inside, sometimes with an annulus at the base. Stamens 5, inserted on the corolla and alternating with the lobes. Ovary superior, 2- or 4-locular; style usually simple, arising from between the 4 lobes of the ovary (gynobasic), rarely terminal. Fruit of 2 or 4 nutlets (rarely 1 or 3 by abortion)."
-Flora Europaea, Volume 3: Diapensiaceae to Myoporaceae, editors Tutin, Heywood, Burges, Moore, Valentine, Walters and Webb. United Kingdom: Cambridge University Press, **1972**, pages 83-84. A 5-volume encyclopedia of plants, published between 1964 and 1993.
> (This section includes a key to identification of the Borage family. Plant identification is the process of matching a specimen plant to a known taxon. One method is using a single-access key also called dichotomous key or sequential key. The sequence and structure of identification steps is fixed by the author of the key. At each point in the decision process, multiple alternatives are offered, each leading to a result or a further choice.)

"Boraginaceae: Borage Family:
Annual to perennial herbs, often hispid or scabrid; leaves alternate, simple, entire or so, exstipulate, sessile or petiolate.
Flowers in often spiralled cymes, *actinomorphic to weakly zygomorphic, bisexual, hypogynous; sepals 5, united into tube with 5 lobes or teeth; petals 5, fused into tube with distal limb, the latter with 5 lobes, mostly blue to pink, often with a knob, scale or hair-tuft at throat of tube; stamens 5, borne on corolla-tube; ovary 4-celled, deeply 4-lobed, with 1 ovule per cell; style 1, arising from base of ovary where the 4 cells meet; stigma 1, capitate or bilobed; fruit a cluster of four 1-seeded nutlets (schizocarp). Like Verbenaceae (Verbena family) and Lamiaceae (Mint family) in its 4-celled ovary with 1 ovule per cell and a fruit of 4 nutlets, but differing from both in usually alternate leaves, and spiralled cymose inflorescence.*
Much value is placed in many keys (for determining genus/species) on the presence or absence of folds, scales or bands of hairs at the throat of the corolla-tube; *these are often difficult to make out and very little use is made of them here. When they are well developed they may meet in the centre or around the style and the corolla-tube appears closed."*
-New Flora of the British Isles: Identification of Wild Vascular Plants of the British Isles edited by Clive Stace. Cambridge, England: Cambridge University Press, **1991**, page 640.

*"**The Old-World Boragineaceae tribe Boragineae consists of 16 genera and some 170 species** with a distribution range centering in the Mediterranean basin and Middle-East and extending through Europe and Tropical Africa, with a second minor center in the Cape region (south Africa).*
The group has been considered as an off-shoot of the tribe Lithospermeae which originated independently of the Eritrichieae-Cynoglosseae lineage (Johnston 1924; Riedl 1968).
Delimitation of the tribe and of some of its genera (e.g., Anchusa, Symphytum, Nonea) has been widely debated mainly due to a remarkable diversity of forms *with reticulate variation of morphological, palynological and karyological characters. Recent studies on the structure and micromorphology of reproductive organs (Bigazzi & Selvi 1998; Bigazzi et al. 1997, 1999; Bigazzi & Selvi 2000) revealed high inter- and intrageneric polymorphism, providing new insights on the systematic relationships within the group."*
-'Leaf Surface and Anatomy in Boraginaceae Tribe Boragineae with Respect to Ecology and Taxonomy' by Federico Selvi and Massimo Bigazzi, University Firenze, Italy; Flora: Morphology, Distribution, Functional Ecology of Plants, Volume 196, pages 269-285, **2001**.
> (Reticulate means resembling a network such as of veins or lines crossing a reticulate leaf.)

(Morphology studies the form of living organisms, and the relationships between their structures. This includes both the outward and internal appearance {shape, structure, color, pattern, size}.)
(Palynology is the study of pollen grains, especially in archaeological or geological deposits.)
(Karyology is the cytological characteristics of the cell nucleus especially the chromosomes of a single cell.)

"**Boragineae contains two large monophyletic groups. The first consists of three moderately to well-supported branches, Borago-Symphytum, Pulmonaria–Nonea, and Brunnera.**"
-'Molecular Systematics of Boraginaceae Tribe Boragineae Based on ITS1 and trnL Sequences, with Special Reference to Anchusa s.l.' by H.H. Hilger, F. Selvi, A. Papini and M. Bigazzi (Germany and Italy), Annals of Botany, Volume 94, No. 2, pages 201-212, **2004**.
(Monophyletic groups are descended from a common evolutionary ancestor or group, especially one not shared with other groups.)

"*The delimitation (boundaries) and tribal subdivision of the Boraginaceae (family) are discussed, and a synonymic (synonyms) survey of the genera accepted.*
The family Boraginaceae is a natural group already recognised by Caesalpinus (De Plantis, 1583), Jussieu (Gen. Pl., 1789) and Candolle & Candolle (Prodr. 9: 466-559. 1845; 10: 1-178. 1846).
It is formed by about 100 genera and almost 2000 species distributed mainly in temperate, cold and subtropical areas.
Candolle & Candolle (1845, 1846) divided Boraginaceae into four tribes: *Cordieae, Ehretieae, Heliotropieae and Boragineae, the latter subdivided into six subtribes: Cerinthinae, Echiinae, Lithosperminae, Craniosperminae, Anchusinae and Cynoglossinae.*
This tribal classification was adopted by Bentham & Hooker *(Gen. Pl. 2: 832-865. 1876) but recognising only four subtribes within Boragineae: Cynoglossinae, Eritrichinae (the Craniosperminae of the Candolles), Anchusinae and Lithosperminae, including here the Cerinthinae, Echiinae and Lithosperminae of the Candolles.*
Gurke *(in Engler & Prantl, Nat. Pflanzenfam. 4(3a): 71-131. 1893)* **divided the family into four subfamilies:** *Cordioideae, Ehretioideae, Heliotropioideae and Boraginoideae, the latter subdivided into seven tribes, which were reduced to four by Johnston (in Contr. Gray Herb. Harvard Univ. 73: 42-78. 1924) according to morphological and palynological characters: Lithospermeae, Anchuseae, Eritrichieae and Cynoglosseae.*"
-'The Euro+Med Treatment of Boraginaceae' by Benito Valdes, Universidad de Sevilla, Spain; Willdenowia: Annals of the Botanic Garden and Botanical Museum Berlin (BGBM), Germany, Volume 34, No. 1, pages 59-61, **2004**.

"**The plants of the Borage family are often rough and hairy, usually with simple, alternate leaves.**
The flowers are bisexual and mostly regular. They have 5 separate sepals and 5 united petals. There are 5 stamens; these are attached to the corolla tube, alternate with the petals. The ovary is positioned superior. It consists of 2 united carpels (bicarpellate, simple pistil) and produces 4 separate nutlets or sometimes achenes (dry seeds). False partitions may make the ovary appear 4-chambered.
Notice the variations in the nutlets among the examples. Some genera produce fewer than 4 nutlets due to abortion. You will usually be able to see the aborted nutlets around the developed ones.
Worldwide, there are approximately 100 genera, representing about 2,000 species. About 22 genera are native to North America.
The flower spikes often curl like a scorpion tail with the flowers blooming on the upper surface, similar to members of the Waterleaf family."
-Botany in a Day: The Patterns Method of Plant Identification by Thomas J. Elpel. Montana: Hops Press LLC, **2008**, page 144.
(In botany, bisexual means each flower has both female eggs and male sperm. Sepal is one of the individual leaves or parts of the calyx of a flower. Calyx is the usually green outer whorl of a flower. Nutlet is any of the segments of an ovary which splits into parts. Stamen is the pollen-bearing part of a flower, consisting of the filament and anther.)

"**Boraginaceae s. str., Borage Family:**
Herbs with stiff hairs. Leaves alternate, simple.
Inflorescence a scorpioid or helicoid cyme. Flowers sympetalous, actinomorphic, 5-merous.
Corolla often pink as young, then blue or purple.
Anthers attached to corolla. Ovary superior, 2-carpellate, 4 locules.
Style 1, attached to base of ovary, in center. Fruit a schizocarp with 4 nutlets.
Examples: *borage (Borago), forget-me-not (Myosotis),* **Comfrey (Symphytum),** *lungwort (Pulmonaria), viper's bugloss (Echium).*"
-'Field Identification of the 50 Most Common Plant Families in Temperate Regions Including Agricultural, Horticultural and Wild Species' by Lena Struwe, Rutgers: The State University of New Jersey, New Brunswick, New Jersey, **2009**.
(sensu stricto = s.s., s. str., sens. str., sens. strict. In the strict/narrow sense. It is added after a taxon to mean it is being used in the sense of the original author, or without taxa which may otherwise be associated with it.)

"**Boraginaceae:**
The Borage family comprises 142 genera representing approximately 2450 species.
The scientific name of the family refers to the genus Borago, *which consists of just three species. The best known of these is Borage (Borago officinalis). This plant is cultivated as a kitchen herb and bee plant.*
This family contains few other plants of economic value and is mostly limited to ornamental plants.

Most of the plants in this family are covered in coarse hairs.
*The flowers and fruits are usually arranged on a scorpioid cyme. A scorpioid cyme can be considered to be a formed cyme (monochasium) on which just one of the side axes develops. Boraginaceae have a short main axis with one flower; below this flower is a forked side axis with the branches positioned at right angles, each possessing a single flower at the terminus. Initially, the scorpioid cyme is small and furled. **During florescence, the scorpioid cyme unfurls,** and during ripening of the fruit the scorpioid cyme becomes longer. This elongation increases the distance between the flowers, in which the fruits finish ripening."*
-A Manual for the Identification of Plant Seeds and Fruits by R.T.J. Cappers and R.M. Bekker, Netherlands: Barkhuis Publishing, **2013**, page 67.
    (A cyme is an inflorescence in which each floral axis ends in a single flower.
    An inflorescence is a group of flowers arranged on a stem that is composed of a main branch.)

"**Summary of the infrafamilial classification of Boraginaceae proposed in this study:**
Subfamily: Boraginoideae. Tribe: Boragineae. Subtribe: Boraginiinae.
    Accepted genera: *(approximate total number of species per genus / number of species included in the analyses)*
    Anchusa (35 / 1), Anchusella, Borago (5 / 1), Brunnera (3 / 1), Cynoglottis, Gastrocotyle, Hormuzakia, Lycopsis (2 / 1), Melanortocarya, Nonea, Pentaglottis (1 / 1), Phyllocara, Pulmonaria (17 / 1), **Symphytum (35 / 2),** *Trachystemon (1 /1).*
    Approximate Number of Species: 140.
Tribe Boragineae Rchb., Fl. Germ. Excurs. 1: 340. 1831 - Type: Borago L.
    = Anchuseae W.D.J.Koch, Syn. Fl. Germ. Helv.: 497. 1837 - Type: Anchusa L.
    = Symphyteae D.Don in Edinburgh New Philos. J. 13: 240. 1832 - Type: Symphytum L.
Subtribe Boraginiinae G.Don, Gen. Hist. 4: 307, 309. 1837– 1838 - Type: Borago L.
    = Anchusinae Dumort., Fl. Belg.: 40. 1827 ('Anchuseae') - Type: Anchusa L.
    = Pulmonarinae Dumort., Fl. Belg.: 41. 1827 ('Pulmonarieae') – Type: Pulmonaria L.
    = Symphytinae D.Don in Sweet, Hort. Brit., ed. 3: 489. 1839 ('Symphyteae') - Type: Symphytum L.
Genera: Symphytum L. 1753 (incl. Procopiania Gusul. 1928, x Procopiphytum Pawl. 1971)."
-'The Borage Family (Boraginaceae s.str.): A Revised Infrafamilial Classification Based on New Phylogenetic Evidence, With Emphasis on the Placement of Some Enigmatic Genera' by Juliana Chacon, et. al from Chile, Czech Republic, Germany, Italy, Russia, United States; Taxon, Volume 65, No. 3, pages 523-546, June **2016**.
    (Infrafamily is below subfamily. Family: Boraginaceae. Subfamily: Boraginoideae. Tribe: Boragineae. Subtribe:
    Boraginiinae. Genus: Symphytum.)

"These plants **(Borage Family)** have alternately arranged leaves, or a combination of alternate and opposite leaves. The leaf blades usually have a narrow shape; many are linear or lance-shaped. They are smooth-edged or toothed, and some have petioles (stalk that attaches leaf to plant stem).
Most species have bisexual flowers, but some taxa (group of organisms) are dioecious (male and female). Most pollination is by hymenopterans (an order of insects), such as bees.
Most species have inflorescences that have a coiling shape, at least when new. The flower has a usually five-lobed calyx. The corolla (petals) varies in shape from rotate to bell-shaped to tubular, but it generally has five lobes. It can be green, white, yellow, orange, pink, purple, or blue. There are five stamens and one style with one or two stigmas. The fruit is a drupe, sometimes fleshy.
Most members of this family have hairy leaves. The coarse character of the hairs is due to cystoliths (mineral concretion) of silicon dioxide and calcium carbonate. These hairs can induce an adverse skin reaction, including itching and rash in some individuals."
-Wikipedia®: The Free Encyclopedia, www.wikipedia.org, **2018**.
    (Stamen is the pollen-bearing part of a flower, consisting of filament and anther. Style is a long, slender stalk that
    connects stigma and ovary. The stigma is the sticky stem of the pistil of the female part of a plant. A drupe is fruit with a
    soft, fleshy part covered by a skinlike outer layer surrounding an inner stone.)

Plants in the Borage Family

Commonly known plants in the Boraginaceae or borage family: alkanet (Alkanna tinctoria), bluebells- mountain (Mertensia ciliata), bluebells- tall (Mertensia paniculata), bluebells- small/long (Mertensia longiflora), blue stickseed (Hackelia micrantha), borage (Borago officinalis), catheweed (Aspergo procumbens), **Comfrey (Symphytum spp.)**, fiddleneck (Amsinckia spp.), forget-me-not (Myosotis spp.), geigertree (Cordia sebestena), green alkanet (Pentaglottis sempervirens), heliotrope (Heliotropium spp.), hound's tongue/tooth or beggar's ticks (Cynoglossum spp.), lungwort (Pulmonaria spp.), marble seed (Onosmodium molle), miner's candle (Cryptantha spp.), oysterplant (Mertensia maritima), puccoon/stoneseed (Lithospermum), purple viper's bugloss/Salvation Jane (Echium plantagineum), Scorpion Weed (Phacelia), Siberian bugloss (Brunnera macrophylla), viper's bugloss (Echium vulgare), Virginia bluebell (Mertensia virginica), yellow cryptantha (Cryptantha confertiflora) and others.
    ('spp.' is the abbreviation for species. It refers to all species in that genus. Sometimes it is used when the
    specific species in that genus is not known.)

Cynoglossum virginianum, also known as the Wild Comfrey or hounds-tongue, is in the Borage family in the genus Cynoglossum. **It is not in the Symphytum genus.** It is native to the eastern United States and most of Canada. This plant will not be discussed in this book except this section to reduce confusion.

> *"Cynoglossum virginianum:*
> ***Flower:*** *2 to 4 widely spreading clusters at the tip of the stem, each initially in a tight coil, unwinding and elongating with maturity. Clusters are mostly unbranched but occasionally fork.* **Flowers are bright blue-violet to nearly white,** *about 1/3 inch (0.8 cm) across, short-tubular with 5 spreading lobes, the lobes sometimes ragged around the edges. A white bead-like collar rings the mouth of the tube; stamens and styles are hidden inside the tube. The calyx behind the flower is densely hairy and about as long as the tube with 5 triangular lobes. Flower stalks are about 1/3 inch (0.84 cm) long and densely hairy.*
> ***Leaves and Stems:*** *Leaves are basal and alternate, oblong-elliptic, 2 1/2 to 8 inches (6.3-20.3 cm) long, 3/4 to 2 3/4 inches (1.9-6.9 cm) wide, toothless, stiff-hairy on both surfaces and around the edges. Basal and the lowest stem leaves are largest and stalked, becoming smaller as they ascend the stem; the middle and upper leaves are stalkless and clasp the stem. Stems are erect, unbranched and densely covered in stiff hairs, the hairs spreading in the lower plant and appressed in the flower clusters.*
> ***Fruit:*** *Fruit is a cluster of 4 nutlets, each rounded at the top, joined at the base, and covered in bristles. The nutlets separate and fall away when mature."*
> -Minnesota Wildflowers: A Field Guide to the Flora of Minnesota®, New Brighton, www.minnesotawildflowers.info/flower/wild-comfrey, 2018.

Chromosomes and Genome in the Borage Family

See subsection 'Comfrey Chromosomes Overview' in section 'Symphytum Species Overview' (Chapter 6).
For DNA barcoding, see subsection 'Methods of Authentication of Comfrey Species' in section 'Symphytum Species Classifications' (Chapter 4).

A **chromosome** is a DeoxyriboNucleic Acid (DNA) molecule with part of the genetic material (genome) of an organism. A chromosome is a package containing a part of a genome. It is a threadlike structure of nucleic acids and protein found in the nucleus of cells. In humans, each cell has 23 pairs of chromosomes, for a total of 46. Twenty-two pairs, called autosomes, are the same in both males and females. The 23rd pair is for the sex chromosome for male and female.

In molecular biology and genetics, a **genome** is the genetic material of an organism. It is the DNA. Each genome has all the information needed to build and maintain that organism. In humans, the genome is more than 3 billion DNA base pairs. It is found in all cells that have a nucleus.

"**Genome size** *is the total amount of DNA contained within one copy of a single genome. It is typically measured in terms of mass in* **picograms** *(trillionths of a gram, abbreviated* **pg**) *or less frequently in Daltons or as the total number of* **nucleotide base pairs typically in megabases** *(millions of base pairs, abbreviated* **Mb or Mbp**). *One picogram equals 978 megabases. In diploid organisms, genome size is used interchangeably with the term* **C-value**.
C-value *is the amount, in picograms, of DNA contained within a haploid nucleus (e.g., a gamete) or one half the amount in a diploid somatic cell of a eukaryotic organism."*
-Wikipedia®: The Free Encyclopedia, www.wikipedia.org, 2019.
> (Body cells and individuals are described by the number of chromosome sets {ploidy level} they have: monoploid: 1 set; diploid: 2 sets = 2n = 2x, triploid: 3 sets = 2n = 3x, tetraploid: 4 sets = 2n = 4x, pentaploid: 5 sets = 2n = 5x, etc.)
> (Haploid means a cell has a single set of unpaired chromosomes as found in a germ cell. It has half the number of chromosomes of a somatic/body cell. It can unite with a cell from the opposite sex to form a new individual; a gamete.)

"**Chromosome numbers in the Boraginaceae family** *are known for about 555 species (about 22 percent of the total) in 60 genera. Nearly 78 percent of the species investigated are either exclusively diploid or include diploid and polypoid populations. The lowest chromosome number (2n = 8) in the family has been reported for three species of Arnebia Forsskal and for Amsinckia lunaris Macbr. (see Bolkhovskikh et al.), whereas* **the highest chromosome number (2n = 144) was recorded for Symphytum cordatum L.** *(*Grau, 1968).*
Although the cytological data do not seem to support the infrafamilial classification of the family, they are somewhat useful in the delimitation and affinities of certain genera."
-'Journal of the Arnold Arboretum, Supplementary Series, Volume 1, Harvard University, Cambridge, Massachusetts, **1991**. This volume is a collection of contributions toward a 'Generic Flora of the Southeastern United States'. 'The Genera of Boraginaceae in the Southeastern United States' by Ihsan A. Al-Shehbaz, pages 1-40.
(* -'Cytologische Untersuchungen an Boraginaceae I' {Cytological Investigations on Boraginaceae} by J. Grau, Mitteilungen der Botanischen Staatssammlung Munchen {Communications of the Botanical State Collection Munich, Germany}, Volume 7, pages 277-294, 1968. I do not have an English translation of this report. If you have one, could you send it to me.)

"***The family Boraginaceae Juss. (Jussieu) is known for its considerable chromosome variation,*** *a consequence of various cytological processes, such as chromosome fusion or fragmentation, polyploidy or aneuploidy (Britton 1951, Coppi et al. 2006).*

These processes seem to be common in the family and play a crucial role in the evolution of many genera, such as Borago L. (Selvi et al. 2006), Cerinthe L. (Selvi et al. 2009), Myosotis L. (Stepankova 2001, 2006), Nonea Medik. (Selvi et al. 2002, Bigazzi & Selvi 2003), Onosma L. (Martonfi et al. 2008), Omphalodes Mill. (Grau 1967) and Pulmonaria L. (Sauer 1975).
In addition, occurrence of B chromosomes is quite common *(Gadella 1972, Sauer 1975, Bigazzi & Selvi 2003, Bedini et al. 2012).*
All these processes are also important for genome evolution in the genus Symphytum L. *(Grau 1968, *1971, Gadella & Kliphuis 1978, Murin & Majovsky 1982). Gadella & Kliphuis (1978) report a high frequency of polyploids in comparison with other genera of Boraginaceae with the occurrence of polyploidy, as in Onosma, Myosotis and Pulmonaria.*

This phenomenon is well illustrated by the four ploidy levels reported for the Symphytum officinale complex *(e.g. Markowa & Iwanowa 1970, Gadella & Kliphuis 1978)* **or, even more surprisingly, the eight ploidy levels reported in the Symphytum tuberosum complex** *(Murin & Majovsky 1982), which range from presumably diploid (2n = 2x = 18) up to octodecaploid cytotypes (2n = 18x = 144)."*

-'Symphytum Tuberosum Complex in Central Europe: Cytogeography, Morphology, Ecology and Taxonomy' by L. Kobrlova, M. Hrones, P. Koutecky, M. Stech, and B. Travnicek, Preslia: The Journal of the Czech Botanical Society, Czech Republic; Volume 88, pages 77-112, March **2016**.

(* -'Cytologische Untersuchungen an Boraginaceae II' {Cytological Investigations on Boraginaceae} by J. Grau, Mitteilungen der Botanischen Staatssammlung Munchen {Communications of the Botanical State Collection Munich, Germany}, Volume 9, pages 177-194, 1971.)

(In 1789 a French botanist, Antoine Laurent de Jussieu, published a plant classification system, 'Genera Plantarum', that included a description of 'Borragineae' as one of 100 orders, i.e., families. Many of his families are still used in modern classification. 'Borragineae' was based on the genus Borago. When Linnaeus created genus Borago, he used Latin name 'burra' that means 'hairy garment', due to the fine, short hairs of the plants.)

(A karyotype is the number and appearance of chromosomes in the nucleus of an eukaryotic cell. In addition to the normal karyotype, many animals and plants contain B chromosomes known as supernumerary, accessory, conditionally-dispensable, or lineage-specific chromosomes. These chromosomes are not essential for the life of a species, and are not found in most of the individuals. Some species of Symphytum have B chromosomes.)

"**We present the first study of Genome Size (GS) variation and evolution in the family Boraginaceae by analysing all native and naturalized taxa of Boraginaceae in the Czech Republic.**
Thirty-eight taxa (274 individuals) were analysed by flow cytometry. We found 60-fold overall variation in GS, with the lowest value in Myosotis sylvatica (2C = 0.56 pg) and the highest value in Lycopsis arvensis (2C = 33.63 pg).
Most of the analysed species possessed very small or small genomes (94.74% of all analysed taxa). *Monoploid GS varied 35-fold. We also focused on correlates of GS with several factors, such as length of life, residence status, selected environmental factors and phylogeny.*

Short-lived plants had a significantly smaller Genome Size than that of the perennial plants, and plants of ruderal habitats possessed a smaller GS than that of plants from natural habitats.
Moreover, members of Boraginoideae had a significantly higher GS than the members of Cynoglossoideae. **Genome Size was not correlated with the occurrence of polyploidy;** *therefore, GS variation in Boraginaceae is presumably driven by the proliferation of transposable elements."*

-'First Insights into the Evolution of Genome Size in the Borage Family: A Complete Data Set for Boraginaceae from the Czech Republic' by Lucie Kobrlova and Michal Hrones, Department of Botany, Palacky University, Olomouc, Czech Republic; Botanical Journal of the Linnean Society, Volume 189, No. 2, pages 115-131, February **2019**.

(Ruderal species are the first plants to colonize disturbed lands.)

Symphytum (Comfrey) Genus

Early Botanical History of Symphytum (Botanists) (in chronological order: **1623 to 1799**)

Chronological means arranged according to the order of time. In this book when 'in chronological order' is mentioned, it is from early years to later years. Dates in bold are when important works were written.

See the book '**A Biographical Index of British and Irish Botanists**' compiled by James Britten, F.L.S. (British Museum) and G.S. Boulger, F.L.S., F.G.S. (City of London College), London, England, **1893**. It includes the meaning of abbreviations of books and other sources used.

Bauhin (1623), Parkinson (1640), Ray (1686), Tournefort (1700), Linnaeus (1753), Gaertner (1788), Schreber (1789), Jussieu (1789), Schkuhr (1791), Lamarck (1791), Sowerby (1791), Sibthorp (1794), Schmidt (1794), Lepechin (1797), Willdenow (1799)

This subsection includes examples of how botanical abbreviations are used.
See subsection 'Plant Nomenclature' in section 'Taxonomy and Nomenclature of Comfrey' (Chapter 1).
See subsection 'History of the Word Comfrey' in section 'History' (Chapter 12).

Gaspard (Caspar) Bauhin, 1560-1624, was a Swiss botanist and medical doctor. He was Professor of Greek, and Chair of Anatomy / Botany at the University of Basel, Switzerland.
His 'Phytopinax' described thousands of plants and classified them in a way similar to the later binomial nomenclature of Carl Linnaeus. ('**Phytopinax**: Seu Enumeratio Plantarum' {The Enumeration of Plants} by Caspar Bauhin, Basel, Switzerland, 1596.)
See subsections '1596 Bauhin' and '1623 Bauhin' in section 'Symphytum Species Classifications'.
He wrote '**Prodromus Theatri Botanici**' in 1620 and '**Pinax Theatri Botanici**' in 1623. Both are in Latin. **In 'Pinax' he includes 6 categories of Symphytum on page 259: Symphytum Consolida Major, Symphytum Majus Tuberosa, Symphytum Minus Tuberosa, Consolida Major Amplexicaulis, Symphytum Minus Borraginis Facie, Symphyto Viribus Affinis.**

> *"The brothers Johann Bauhin (1541-1631) and Gaspard Bauhin (1560-1624) were Swiss botanists who worked separately but along similar lines. The most important book which they produced was the latter's **'Pinax Theatri Botanici' (1623)**. Its title (Pinax = register) indicates one of its most significant features- the listing not only of all the 6,000 or so species known to him but also all their synonyms, i.e., the various names given to each species by previous workers. Thus the chaotic state of nomenclature which existed was to a considerable degree brought to order. In addition G. Bauhin is notable for his recognition of genera as well as species as major taxonomic levels, and for using a binary nomenclature composed of the generic name followed by a single specific epithet to designate many of the species. Thus, in two major ways, **Bauhin's 'Pinax' foreshadowed Linnaeus' great works**."*
> -Plant Taxonomy and Biosystematics, Second Edition by Clive A. Stace, Professor of Plant Taxonomy, University of Leicester, England. London, England: Edward Arnold, 1989, page 20.

John Parkinson (1567-1650) was an English Master Herbalist and apothecary to King James I, England. In his treatise on plants '**Theatrum Botanicum: The Theater of Plants Or An Herball of Large Extent' (1640)**, Chapter 24, pages 522-524:
> "***Symphitum majus**, Great Comfrey:*
> ***Symphitum majus vulgare**, Common Great Comfrey,*
> ***Symphitum majus purpureo flore**, Great Comfrey with purple flowers,*
> ***Symphitum tuberosum**, Comfrey with knobbed rootes,*
> ***Symphitum angustifolium apulum**, Narrow Comfrey of Naples."*

John Ray (Joannis Raii), 1627-1705, was an English naturalist and Fellow of the Royal Society of London, England. He published 23 works on botany, zoology and natural theology from 1660 to 1703. Ray wrote 'Catalogus Plantarum Angliae' in 1670. His classification of plants in his **1686 'Historia Plantarum' (History of Plants)**, was an important step towards modern taxonomy that included creating a biological definition of species.
Historically, the letters i, y, and j were interchangeable, so Raii (Latin) = Rai's = Ray's.
'Raii Syn. 230' = **Ray's 'Synopsis Methodica Stirpium Britannicarum'**, edition 3, page 230, written in 1724:
> "***Symphytum magnum** J.B. III. 593. majus vulgare Park. 523.*
> *Symphytum, Confolida major C.B. Pin. 259. Confolida major Ger. 660. Comfrey."*

'**Institutiones Rei Herbariae**', 3 volumes, in Latin by Joseph Pitton de Tournefort, 1700. It includes 9000 species in 698 genera, which directly influenced Carl Linnaeus.
This book was originally published in French in 1694 as 'Elements de Botanique, ou Methode Pour Reconnaitre les Plantes' (Botanical Elements, or Method to Recognize Plants). It was translated into Latin in 1700 and 1719 as 'Aquisextiensis, Doctoris Medici Parisiensis, Academiae Regiae Scientiarum Socii, et in Horto Regio Botanices Professoris, Institutiones Rei Herbariae'.
> *Tourn. inst. t 56. (Tomus II and III are art.) (Tomus Primus = First Volume is the text.)*
> Tournefort, Institutiones, tab 56 = plate (art) 56, page 138, "**Genus Symphytum, Consoude**".
>> "***Symphytum Consolida Major** (5 flores: purpureo, purpureo-caeruleo, albo, luteo, variegato).*
>> ***Symphytum Majus Tuberosa***
>> ***Symphytum Echii Folio Ampliote** and **Symphytum Echii Folio Angustiore***
>> ***Symphytum a Congeneribus Differt Floris Forma**."

Tournefort was a French botanist who lived from 1656-1708. He was the first to make a concise definition of the concept of genus for plants. He was Professor of Botany at the Jardin des Plantes in Paris, France. His herbarium collection of 6,963 specimens is stored at the 'Museum National d'Histoire Naturelle' in Paris.

> *"**Tournefort's concept of the genus Symphytum in 1700 and Linnaeus's in 1753 were the same**."*
> -'The Greek Species of Symphytum Boraginaceae' by William T. Stearn, British Museum of Natural History, London, England; Annales Musei Goulandris, Greece, Volume 7, pages 175-220 (1985 or 1986).

'**Genera Plantarum, Edition 1**' by Swedish naturalist Carl Linnaeus, written 1737.
The first edition of 'Genera Plantarum' contains descriptions of the 935 plant genera known to Linnaeus.
'**Genera Plantarum, Edition 5**' by Swedish naturalist Carl Linnaeus, 1754.
'**Genera Plantarum, Edition 6**' by Swedish naturalist Carl Linnaeus, 1764.

*Lin. gen. 185. (Linnaeus 'Genera Plantarum', 1764, page 76, "**185. Symphytum**".)*
See subsection 'Carl Linnaeus' in section 'Taxonomy and Nomenclature of Comfrey' (Chapter 1).

'Species Plantarum, Edition 1' by Swedish naturalist Carl Linnaeus, written 1753.
The first edition of 'Species Plantarum' lists every species of plant known at the time, classified into genera (plural of genus). It was the first book to consistently apply binomial names. There were 2 volumes.
*Lin. Sp. Pl. 1: 136. 1753 (Linnaeus 'Species Plantarum', **Volume 1, Symphytum page 136.**)*
*"**Symphytum officinale, Symphytum tuberosum, Symphytum orientale.**"*

'Species Plantarum, Edition 2' by Swedish naturalist Carl Linnaeus, 1762-1763.
*Lin. Sp. Pl. 195. (Linnaeus 'Species Plantarum', **Symphytum page 195.**)*

'De Fructibus et Seminibus Plantarum: Vol. I (Of Fruits & Seedling Plants)' by Joseph Gaertner, 1788. *Gaertn. fruct. 1. p. 325. t. 67. f. 4. (Gaertner, de Fructibus, Volume 1, page 325. tab. 67.) In Latin. Tab. 67 = Plate/art LXVII).*
*"**CCCCXVII. Symphytum:**
Symphytum officinale, Symphytum majus,
Symphytum foliis ovato lanceolatis decurrentibus."*
'De Fructibus' has extremely accurate descriptions of plants with over a thousand species. This book introduced a new era in plant morphology. Gaertner was a German botanist who lived from 1732-1791. He was Professor of Anatomy in Tubingen, Germany, and Professor of Botany at St. Petersburg, Russia.
(Morphology studies the form of living organisms, and the relationships between their structures. This includes both the outward and internal appearance {shape, structure, color, pattern, size}.)

'Genera Plantarum: Volume I' edited by Johann Christian Daniel von Schreber, 1789.
*Schreb. no. 245. (**Schreber genus number 245**, page 101, "**245. Symphytum**".)*
Schreber was a German botanist who lived from 1739-1810. He was Professor of Materia Medica at the University of Erlangen, Germany. He was the President of the 'German Academy of Sciences Leopoldina' and was a member of the 'Royal Swedish Academy of Sciences' and a 'Fellow of the Royal Society' of London, England.

'Genera Plantarum Secundum Ordines Naturales Disposita' by Antoine Laurent de Jussieu, 1789.
*Juss. gen. 131. ed. (Jussieu Genera page 131, "**Symphytum, T.L., Consoude**".)*
Jussieu's 'Genera Plantarum' was a study of flowering plants based on a methodology using multiple characters to define groups. Many plant families are attributed to Jussieu.
Jussieu was a French botanist who lived from 1748-1836. He was the first to publish a natural classification of flowering plants, with much of his system still being used. He was Professor of Botany at the 'Jardin des Plantes' in Paris, France. He was a member of the 'Royal Swedish Academy of Sciences' in Stockholm, Sweden.

'Botanisches Handbuch der Mehresten Theils in Deutschland Wild Wachsenden' (Manual of Botany) by Christian Schkuhr, 1791.
*Schkuhr. Bot. Handb. tab. 30. (Schkuhr, Botanisches Handbuch, tab. 30, page 101, "LXXXIX Geschl. Tab. XXX. **Symphytum Beinwell**".) (geschlossen?=closed)*
Schkuhr was a German gardener, artist and botanist who lived from 1741-1811. For this manual he made his own drawings and engravings, many based on observations from his home-made microscope. Besides taxonomy, he wrote about the economic and agricultural uses of plants. He was a Physical Scientist at the University of Wittenberg, Germany. He was an expert on the flora of Wittenberg, Germany.

'Encyclopedie Methodique ou par Ordre de Matieres: Botanique' {Methodical Encyclopedia by Order of Subject Matter: Botany} by Jean Baptiste Pierre Antoine de Monet de Lamarck, 1783-1808.
'Encyclopedie Methodique' was published between 1782 and 1832 with 210-216 volumes. Botany was one of 26 divisions in the encyclopedia. It was 8 volumes with 12,002 pages and 1,000 plates/illustrations.
Symphytum is in Volume 2 under 'Consoude' on page 97, written 1786-1788.
*"**Consoude, Symphytum: Consoude officinale - Symphytum officinale, Consoude tubereuse - Symphytum tuberosum, Consoude du Levant - Symphytum orientale.**"*
Five Botany 'Supplements' were published from 1810-1817. Symphytum is in Volume 2 on page 345.
*"**Consoude, Symphytum - Symphytum officinale, Consoude en coeur - Symphytum cordatum, Consoude de la Tauride - Symphytum tauricum, Consoude herissee - Symphytum asperrimum.**"*
'L'Illustration des Genres (Genus Illustrations)', Volume I: 1791. Volume II: 1793, Volume III: 1800.
Lam. Illustr. gen. no. 261. tab. 93. (Lamarck Illustration; number 261, page 407; art 93, page 408.)
Also called: **'Tableau Encyclopedique et Methodique des Trois Regnes de la Nautre: Botanique. Premiere Livraison (Volume 1)', 1791**-1793. An illustrated encyclopedia of plants, animals and minerals published in Paris,

France. Jean Baptiste Pierre Antoine de Monet De Lamarck contributed plants and taxonomy.
Symphytum text is on page 407. Artwork is tab/plate 93.
Lamarck was a French naturalist who lived from 1744-1829. He was a member of the 'French Academy of Sciences'. He was Chair of Botany at 'Jardin des Plantes', Paris, France. He was Professor of Zoology at the 'Museum National d'Histoire Naturelle' in Paris. Lamarck named a large number of species, many of which have become synonyms.
French naturalist and zooloigist, Georges Cuvier (1769-1832) wrote:
"L'Illustration des Genres: The precision of the descriptions and of the definitions of Linnaeus is maintained, as in the institutions of Tournefort, with figures adapted to give body to these abstractions, and to appeal both to the eye and to the mind, and not only are the flowers and fruits represented, but often the entire plant."

'English Botany (Sowerby's); or Coloured Figures of British Plants with their Essential Characters, Synonyms and Places of Growth' by John T. Boswell {Syme} and **John Edward Sowerby, London, England.** Sowerby, 1757-1822, was a botanical illustrator and natural historian.
'English Botany' was a major publication of British plants comprising 36 volumes totaling 5,416 pages, issued in 267 monthly parts over 23 years from **1791** to 1814. It became the most comprehensive, illustrated flora of Great Britain published up to that time. It contained 2,592 hand-colored engravings so anyone could identify plants. The descriptions are scientifically accurate and use binomial nomenclature. Includes general discussions and cultivation for the average gardener. Abbreviated 'Eng. Bot.".
English Botany, Volume 7, 1867, third edition, pages 114 to 117:
*"**Symphytum Genus:***
Symphytum officinale var genuinum: *Corolla ochrcous or more or less stained with pale purple.*
Symphytum officinale var patens: *Corolla bright dark purple.*
Symphytum tuberosum *(tuberous Comfrey)."*

'Flora Oxoniensis Exhibens Plantas in Agro Oxoniensi' (Flora of Oxfordshire, England) by John Sibthorp, 1794, page 70.
*"**Symphytum.** Gen. Pl. 245. (Genera Plantarum by Schreber, plate 245.)*
Corolla limbus tubulato-ventricofus, fauce claufa radiis fubulatis.
***Symphytum officinale** (Common Comfrey), **Symphytum patens** (Red-flowered Comfrey)."*
John Sibthorp, 1758-1796, was an English botanist and Fellow of the 'Royal Society of London'. He was Professor of Botany at the University of Oxford, England. He was one of the founders of the Linnean Society.
His botanical author abbreviation is 'Sibth.'. He was the first to publish Symphytum patens Sibth. (Fl. Oxon. 70. 1794).

Franz Wilibald Schmidt, 1764-1796, Czech botanist, was Professor of Botany in Prague, Czechoslovakia. He wrote a number of writings on the flora of Bohemia and on physical-economic topics. From **1794** to 1795 he created his 4-volume work **'Flora Boemica Inchoata'** (Bohemia Plants Started/Found). Bohemia is the westernmost part of the Czech Republic.
His botanical author abbreviation is 'F.W.Schmidt'. A Comfrey species first published by Schmidt: Symphytum bohemicum F.W.Schmidt (Fl. Boem. iii. 13. t. 263). (t. 263 = plate / artwork 263)
Volume 3 contains Symphytum on pages 12 to 14: *"**Symphytum:***
Symphytum officinale, Symphytum bohemicum, Symphytum tuberosum."
The 'Flora Boemica Inchoata' plate numbers 1 to 244 were made, but did not get published. Most likely plates 245 and higher were never made.

*"**Franz Willibald Schmidt (1764-1796): He soon became a leading personality in botany in Bohemia; he described many new species and a number of them are generally accepted now.** His complete herbarium collection was deposited at PRC. (The herbarium collection of the Charles University in Prague, Czech Republic: Herbarium Universitatis Carolinae Pragensis.)*
Of ten generally accepted names based on European material, six can be readily typified with plants collected by F.W. Schmidt. *Published and unpublished figures drawn by F.W. Schmidt are also important elements of the original material of his names; they certainly serve as a good tool for the interpretation of his names (see also *Skalicky 1982)."*
-'Generally Accepted Plant Names Based on Material from the Czech Republic and Published in 1753-1820' by Jan Kirschner, Lida Kirschnerova and Jan Stepanek, Institute of Botany, Academy of Sciences, Pruhonice 1, Czech Republic; Preslia: The Journal of the Czech Botanical Society, Volume 79, pages 323-365, 2007.
(* -'Index Iconum Plantarum Vascularium Initio Botanicae Bohemicae 1' by Vladimir Skalicky; Folia Geobotanica et Phytotaxonomica, Volume 17, Issue 4, pages 393-420, 1982.)

Ivan Ivanovich Lepechin = Ivan Ivanovich Lepyokhin = Ivan Lepekhin, 1737 or 1740 to 1802.
"Ivan Ivanovich Lepechin was a Russian physician, naturalist and explorer. He was born and died in St. Petersburg (Russia). Lepechin studied natural sciences in Strasbourg (France), where he earned his doctorate before returning to St. Petersburg to work on the 'Dictionary of the Russian Academy' at the 'Imperial Academy of Sciences'.
In 1768 Lepechin embarked on an expedition to the Volga (central Russia) and Caspian Sea. He continued on in 1769 to the Ural mountains (western Russia), where he travelled for five years, making scientific collections and observa-

tions. He then explored Siberia in 1774-1775. Following his travels, Lepechin was appointed head of the 'Imperial Botanical Gardens' in St. Petersburg, which he led for 28 years, until his death. He developed an expert knowledge of medicinal plants and described 29 new species of flowering plants."
-'Lepechin, Ivan Ivanovich (1737-1802)', JSTOR® Global Plants, www.jstor.org, 2018. JSTOR is a digital library for scholars, researchers, and students. It provides access to more than 12 million academic journal articles, books, and primary sources in 75 disciplines.

'**Nova Acta Academiae Scientiarum Imperialis Petropolitanae**, Praecedit Historia Ejusdem Academiae' (Petersburg Imperial Academy of Sciences) by Imperatorskaia Akademia Nauk (Russia), **1797**-1798.

Volume 14, Part 2 includes article by Ivan Lepechin about Symphyti Asperi (Nova {new} species) in French, on pages 442 to 444. Tab 7 (Symphytum asperum artwork) is 12 pages after the last of the text on page 826 (the illustrations do not have page numbers.)

The standard author abbreviation 'Lepech.' indicates him as the author when citing a botanical name.
He was the first to publish this Symphytum species:
Symphytum asperum Lepech. (Nova Acta Acad. Sci. Imp. Petrop. Hist. Acad. 14(2): 442, t. 7. 1805)

Carl Ludwig Willdenow, 1765-1812, was a German botanist, taxonomist and pharmacist. He is one of the founders of phytogeography, the study of the geographic distribution of plants. His herbarium of more than 20,000 species is preserved in the 'Botanical Garden' in Berlin, Germany.
He wrote: '**Enumeratio Plantarum Horti Regii Berolinensis:** Continens descriptiones omnium vegetabilium in horto dicto cultorum' {An Enumeration of the Royal Garden Plants Berlin} by Carl Ludwig Willdenow. Berolini {Berlin}, Germany, 1809. Symphytum is in Part 1, pages 183-184.

"**Symphytum:** Symphytum officinale, Symphytum tuberosum, Symphytum asperrimum, Symphytum tauricum, Symphytum orientale, Symphytum cordatum."

The author botanical abbreviation is ' Willd.'. He was the first to publish these Comfrey species:
Symphytum cordatum Willd. (Neue Schriften Ges. Naturf. Freunde Berlin ii. **1799** 121.)
Symphytum tauricum Willd (Neue Schriften Ges. Naturf. Freunde Berlin ii. 1799 121. t. 5. f. 1.)

Later Botanical History of Symphytum (Botanists)
In chronological order: **1800 to present.** Dates in bold are when important works were written.

Bieberstein (1808), Lehmann (1818), Schimper (1835), G. Don (1838), Fries (1839), De Candolle (1846), Ledebour (1847), Steven (1851), Kerner (1863), Schur (1866), Schultz (1875), Boissier (1879), Nyman (1884), Bucknall (1912), Tutin (1952), Komarov (1953), Popov (1953), Wade (1958), Perring (1962), Wickens (1969), Pawlowski (1972)

Friedrich August Marschall von Bieberstein, 1768-1826, was a German botanist who was an early explorer of the flora and archaeology of the southern part of Imperial Russia that included the Caucasus and Novorossiya (New Russia).
He wrote the first flora catalog of the Crimeo-Caucasian region: '**Flora Taurico Caucasica:** Exhibens Stirpes Phaenogamas in Chersoneso Taurica et Regionibus Caucasicis Sponte Crescentes', **1808**-1819. The two volumes contain descriptions of 2,322 Spermatophyte (plants that produce seeds) species of the Caucasus.
In 1819, he published a supplementary volume. The books use the then new Linnaean classification system.
The 10,000 specimens of his plant collection was given to the 'Russian Academy of Sciences' and is now stored at the 'Komarov Botanical Institute' in St. Petersburg, Russia. **His botanical author citation abbreviation is 'M.Bieb.'.**
In 'Flora Taurico Caucasica', Volume 1 on pages 128 to 130 he lists:
Symphytum caucasicum, Symphytum orientale, Symphytum asperrimum, Symphytum cordatum.
He was the first to publish these Symphytum species:
Symphytum caucasicum M.Bieb. (Fl. Taur.-Caucas. 1: 128. 1808)
Symphytum cordatum M.Bieb. (Fl. Taur.-Caucas. 1: 130. 1808)

'**Plantae e Familia Asperifoliarum Nuciferae**' (Of a Plant from the Family of Asperifoliarum Nuciferae) by Johann Georg Christian Lehmann, **1818.** Written in Latin.
Lehm. asper, p. 3. and 343. (Lehmann, Asperifoliarum, page 3 and 343.)
Page 3 is a synopsis of the genus Symphytum. Pages 342 to 354 list: Symphytum cordatum, Symphytum tuberosum, Symphytum bullatum, Symphytum orientale, Symphytum caucasicum, Symphytum officinale, Symphytum asperrimum.
"**Symphytum (Genus):**
Linn. Gen. plant. ed. Schreb. no. 245.
Juss. Gen. plant. ed. Paris. pag. 131.
Lam. Illustr. gen. no. 261. tab. 93.
Gaertner. De fructib. et seminib. plant Vol. I. no. 417. tab.67. fig. 5.
Schkuhr. Bot. Handb. tab. 30.

Tournef. Inst. rei herb. tab. 56. "
Lehmann was a German botanist who lived from 1792-1860. He received a Doctorate in Medicine and a Doctorate in Philosophy from the University of Jena, Germany. He was Professor of Physics and Natural Sciences and head librarian at the 'Gymnasium Academicum in Hamburg', Germany. He was the founder of the 'Hamburg Botanical Garden'.

Karl (Carl) Friedrich Schimper, 1803-1867, was a German botanist and naturalist. He pioneered research in the field of plant morphology. He wrote '**Beschreibung des Symphytum Zeyheri** und Seiner Zwei Deutschen Verwandten der S. Bulbosum Schimper und S. Tuberosum Jacq.' (Description of the Symphytum Zeyheri and Two German Relatives of the S. Bulbosum Chimera and S. Tuberosum Jacq.), Mit 6 Steintafeln, Universitatsbuchhandlung von C. F. Winter, Heidelberg, Germany, **1835**. All in German. (If someone has an English translation of this, I would appreciate having a copy.)
His standard author abbreviation is 'K.F.Schimp.'. He was the first to publish this Symphytum species:
    Symphytum bulbosum K.F.Schimp. (Flora 8 (1 no. 2): 17. 1825)
        Flora oder Botanische Zeitung (Flora or Botanical Newspaper), Volume 8, No. 1, pages 17 to 24, year 1825.

"Carl Friedrich Schimper's recognition, naming and description of Symphytum bulbosum as a new species probably stands as his most lasting contribution to botany.
He was a keen field botanist who during his schooldays in Mannheim, Germany contributed substantially to F.W.L. Succow's 'Flora Manheminesis' (1822).
Botanizing in May 1824 at **Heidelberg, west Germany,** *where he was then a theological student, he found this Comfrey growing in vineyards there near the Krappfabrik, where it had become naturalized.*
It aroused his interest and so led to two scholarly publications on Symphytum; other botanists had confused S. bulbosum with S. tuberosum from which he carefully distinguished it."
-'The Greek Species of Symphytum Boraginaceae' by William T. Stearn, British Museum of Natural History, London, England; Annales Musei Goulandris, Greece, Volume 7, pages 175-220, (1985 or 1986, different sources give different dates, even the pdf has both dates). (I am not sure what 'Krappfabrik' is.)

'A General History of the Dichlamydeous Plants Comprising Complete Descriptions of the Different Orders: Volume IV: Corolliflorae' (A General System of Gardening and Botany) by George Don, 1838, pages 306, 307, 312 to 313.
    Symphytum ds. (ds = Don System = **G. Don** *= George Don, General System of Gardening and Botany)*
George Don was a Scottish botanist who lived from 1798-1856. He was a 'Fellow of the Linnean Society', London, England. In 1821 he went to Brazil (South America), the West Indies (Caribbean) and Sierra Leone (West Africa) to collect specimens for the 'Royal Horticultural Society' of London, England.
    "Order Boragineae: *this order contains plants agreeing with Borago in important characters.*
    Synopsis of the Genera:
        Tribe I: Boragieae.
        Tribe II: Symphyteae.
            Genera: Symphytum, *Colsmannia, Stomotechium, Onosma,*
            Onosmodium, Pulmonaria, Mertensia, Cerinthe.
        Tribe III: Lithospermeae. Tribe IV: Buglosseae. Tribe V: Heliotropeae.
"Tribe II: Symphyteae: This tribe agrees with Symphytum in the characters given.
D. Don, in edinb. phil. journ. July, Oct. 1832.
(D. Don = David Don, *professor of Botany, King's College, London. Edinburgh Philosophical Journal.)*
Corolla tubular, truncate, with very short lobes. Inflorescence revolute. Stamens inclosed. Nuts fixed to bottom of calyx.
Symphytum Genus *(from symphyo to make unite; and phyton, a plant; in reference to healing qualities of the plant.)*
    Tourn. inst. t 56. = Lin. gen. 185. =Schreb. no. 245. = Juss. gen. 131. ed.
    Usteri. p. 142. = Gaertn. fruct. 1. p. 325. t. 67. f. 4.
    Lehm. asper, p. 3. and 343. = Schkuhr, handb. t. 30. (tab.30)
Lin. syst. Pentandria, Monogynia. Calyx 5-parted. Corolla cylindrically campanulate; throat furnished with 5 subulate, vaulted processes, which connive into a cone. Nuts 4, 1-celled, ovate, fixed to the bottom of the calyx, imperforated at the base. Rough, herbaceous plants, with broad leaves and terminal, twin racemes of flowers.
Symphytum Species:
    Symphytum officinalis = Official Comfrey and also var patens.
    Symphytum bohemicum = Bohemian Comfrey.
    Symphytum tuberosum = Tuberous-Rooted Comfrey.
    Symphytum zeycheri = Zeycher's Comfrey.
    Symphytum bulbosum = Bulbous-Rooted Comfrey.
    Symphytum cordatum = Cordate-Leaved Comfrey (heart shaped).
    Symphytum orientale = Oriental Comfrey.
    Symphytum caucasicum = Caucasian Comfrey.
    Symphytum tauricum = Taurian Comfrey and also var bullatum.
    Symphytum asperrimum = Very Rough Comfrey and also var hybridum.
    Symphytum echinatum = Echinated Comfrey (with dense bristles).

Symphytum racemosum = Racemose-Flowered Comfrey (clustered)"

Swedish mycologist (fungus expert) and botanist **Elias Magnus Fries** was 'Fellow of the Royal Society of London', England, and 'Fellow of the Royal Society of Endinburgh', Scotland. He lived from 1794-1878. He wrote 'Systema Mycologicum', 'Elenchus Fungorum', 'Monographia Hymenomycetum Sueciae' and 'Hymenomycetes Europaei'.
Fries and Ponten wrote **'Novitiarum Florae Sueciae Mantissa Altera**, Quam Venia Ampliss. Facult. Philosoph., I.' **(Flora of Sweden), 1839.** It seems there were 2 volumes. I could not find volume 2. In Latin.
The standard botanical author abbreviation is 'Fr.'. He was the first to publish Symphytum patens Fr. (Novit. Fl. Suec. Mant. 2: 13. 1839). Symphytum patens is in Volume I on pages 13 and 14.

Augustin Pyramus De Candolle, 1778-1841, was a Swiss botanist.
Alphonse Pyramus De Candolle, 1806-1893, a French-Swiss botanist, was Augustin's son.
Augustin De Candolle documented hundreds of plant families and created a new plant classification system. He was interested in phytogeography (plant distribution), agronomy, paleontology (study of fossils), medical botany and economic botany.
'**Prodromus Systematis Naturalis Regni Vegetabilis**' is 17 volumes written from 1824 to 1873. Augustin Pyramus De Candolle wrote it as a summary of all known seed plants, including taxonomy, ecology, evolution and biogeography. He authored the first seven volumes. His son, Alphonse Pyramus De Candolle, wrote ten more volumes, with help from other authors.
> De Candolle further developed his concept of families. This system was published before there were internationally accepted rules for botanical nomenclature.
> Alphonse de Candolle wrote 'Origin of Cultivated Plants' book published by 'The International Scientific Series, Volume XLVIII', New York, New York, 1882.

Symphytum is in 'Prodromus Systematis Naturalis Regni Vegetabilis', **Volume 10,** pages 36 to 40.
> See subsection '1846 De Candolle' in section 'Symphytum Species Classifications'.

The abbreviation for Augustin Pyramus de Candolle is 'DC'. He was the first to publish these species:
> Symphytum donii DC. (Prodr. [A. P. de Candolle] 10: 37. **1846**)
> Symphytum grandiflorum DC. (Prodr. [A. P. de Candolle] 10: 40. 1846)
> Symphytum orientale Pinard ex DC. (Prodr. [A. P. de Candolle] 10: 40. 1846)
> Symphytum coeruleum Hort.Angl. ex DC. (Prodr. [A. P. de Candolle] 10: 40. 1846; nom. inval.)
> Symphytum armenum Gundelsh. ex DC. (Prodr. [A.P. de Candolle] 10: 66. 1846)
>> Accepted name: Species Onosma sericeum Willd. (Source: www.gbif.org)
> Symphytum fruticosum Moc. ex DC. (Prodr. [A. P. de Candolle] 10: 121. 1846)
>> Synonym: Antiphytum mexicanum DC. (Source: Swedish Wikipedia®.)

The standard botanical abbreviation for Alphonse Pyramus de Candolle is 'A.DC.'.

Carl (Karl) Friedrich von Ledebour, 1785-1851, was a German-Estonian (German/Swedish) botanist. He was born in Stralsund which was at that time a part of Sweden. He travelled extensively in Russia and wrote the first comprehensive book on Russian flora: '**Flora Rossica**; sive, Enumeratio plantarum in totius Imperii Rossici provinciis Europaeis, Asiaticis et Americanis hucusque observatarum' (Complete Flora of the Russian Empire), **1847**-1849.
See subsection '1847 Ledebour' in section 'Symphytum Species Classifications'.
The botanical abbreviation for author citation is 'Ledeb.'. He published these Symphytum species:
> Symphytum echinatum Ledeb. (Hort. Dorpat. Suppl. 1811, 5.)
> Symphytum majus Gueldenst. ex Ledeb. (Fl. Ross., Ledeb. 3 1,8: 115. 1847)
> Symphytum peregrinum Ledeb. (Ind. Sem. Hort. Dorpat. 1820, 4; et ex Spreng. Syst. i. 563.)
> Symphytum peregrinum Ledeb. (Index Seminum Horti Academici Dorpatensis. 1820)

Christian von Steven (Khristian Khristianovich Steven), 1781-1863, was a Finnish-born Russian botanist and entomologist (insect expert). He studied medicine at Saint Petersburg University, Russia. In 1815, he was elected a member of the 'Royal Swedish Academy of Sciences'.
He helped in the formation of the 'Nikitsky Botanical Garden' in the Crimea (eastern Europe) and was the director starting 1824. He wrote 'Verzeichnis der auf der Taurischen Halbinsel Wildwachsenden Pflanzen' (List of Plants that Grow Wild on the Taurian {Turkish} Peninsula) in 1856.
The botanical abbreviation for author citation is 'Steven'. Two Comfrey species he was the first to publish:
> **Symphytum ibericum** Steven (Bull. Soc. Imp. Naturalistes Moscou 24, 1: 579-580, **1851**)
> **Symphytum tanaicense** Steven (Bull. Soc. Imp. Naturalistes Moscou 24, 1: 577, 1851).
>> (Bulletin de la Societe Imperiale des Naturalistes Moscou, Volume 24, No.1, 1851.)

He also writes about: Symphytum officinale, Symphytum caucasicum, Symphytum asperrimum, Symphytum tauricum.

George Bentham, 1800-1884, was a British botanist known as 'the premier systematic botanist of the nineteenth century'. His greatest work was his classification of plants the '**Genera Plantarum:** Ad Exemplaria Imprimis in Herberiis Kewensibus Servata

Definita' in 3 volumes, begun in 1862, and finished in 1883 in collaboration with Joseph Dalton Hooker. Symphytum is on page 854 of Volume 2, Part 2, 1876.

His most famous work was '**Handbook of the British Flora:** A Description of the Flowering Plants and Ferns Indigenous to, or Naturalised in, the British Isles', first published in 1858. The second edition in 1863-1865 was illustrated. This was used for over a century, with many editions. After his death it was **edited by Joseph Dalton Hooker** (1817-1911), a British botanist, with editions as late as 1954.

The 'Handbook' 1858 edition lists 'The Borage Family: Comfrey- Symphytum' on pages 379-380:
> "Rough, hairy perennials. The genus contains but few species, nearly resembling each other, and extends over Europe and northern Asia. Common Comfrey, Symphytum officinale Linn.
> Tuberous Comfrey, Symphytum tuberosum Linn. (a luxuriant garden specimen)."

Anton Joseph Kerner = Anton Kerner Ritter von Marilaun, 1831-1898, was an Austrian botanist. In 1858 he was made Professor of Botany at the Polytechnic Institute at Buda (Budapest, Hungary), and in 1860 was made Professor of Natural History at the University of Innsbruck, Austria. In 1878 he was made Professor of Systematic Botany at the University of Vienna, Austria, as well as curator of their botanical garden.

Kerner was interested in phytogeography (geographic disbribution of plants) and phytosociology (plant communities). He wrote the 6 volume 'The Natural History of Plants, Their Forms, Growth, Reproduction, and Distribution' (Pflanzenleben = Plant Life or Plantscapes) in 1887. Later translated from German to English by F.W. Oliver.

His botanical author citation is 'A.Kern.'. He was the first to publish these Symphytum species:
> Symphytum angustifolium A.Kern. (Oesterr. Bot. Z. 13: 227. **1863**)
> Symphytum officinale L. subsp. uliginosum (A.Kern.)Nyman (Consp. Fl. Eur. 509. 1881, as uliginosum)
> Symphytum uliginosum A.Kern. (Oesterr. Bot. Z. 13: 227. 1863)
> > 'Descriptiones Plantarum Novarum Florae Hungaricae et Transsilvanicae' by A. Kerner, Oesterreichische Botanische Zeitschrift (Austrian Botanical Journal), Vienna, Austria, Volume 13, No. 7, pages 227-288, 1863.

Philipp Johann Ferdinand Schur, 1799-1878, German-Austrian botanist and pharmacist born in Konigsberg (a former German city that is now Kaliningrad, Russia). He wrote **'Enumeratio Plantarum Transsilvaniae'** about botany in Transylvania (central Romania) in **1866**.

His botanical author citation abbreviation is 'Schur'. He was the first to publish these Symphytum species:
> Symphytum bulbosum Schur (Enum. Pl. Transsilv. 468. 1866)
> Symphytum nodosum Schur (Enum. Pl. Transsilv. 468. 1866)

On pages 467 and 468 he wrote about Symphytum officinale (albiforum, ochroleucum, patens, angustifolium), and S. tuberosum.

Friedrich Wilhelm Schultz, 1804-1876, was a German pharmacist and botanist. He wrote 'Flora der Pfalz' about plants in southwest Germany. With Paul Constant Billot, he was co-author of 'Archives de la Flore de France et d'Allemagne' (Archives of the Flora of France and Germany). He was a founder of the scientific society Pollichia, that studied nature in the Rheinland-Pfalz region (western Germany).

His botanical author abbreviation is 'F.W.Schultz'. He was the first to publish these Symphytum species:
> Symphytum gussonei F.W.Schultz (in Arch. Fl. 1874 p. lviii; et in Flora, lviii. **1875** 218.)
> Symphytum mediterraneum Guss. ex F.W.Schultz (Flora 58: 218. 1875)
> > 'Flora oder Botanische Zeitung' {Flora or Botanical Newspaper}, Regensburg, Germany, Volume 58. In this 'Flora' article 'Beitrage zur Flora der Pflaz' by Dr. F. Schultz, written September 1874, pages 216 to 218, he also writes about Symphytum bulbosum, Symphytum tuberosum, and Symphytum floribundum. Pflaz is the Palatinate region in southwest Germany.

Pierre Edmond Boissier, 1810-1885, was a Swiss botanist, explorer and mathematician. He collected plants extensively in Europe, north Africa and western Asia. He wrote '**Flora Orientalis** Sive Enumeratio Plantarum in Oriente, a Graecia et Aegypto ad Indiae Fines' (Plants Flora, Flora East and the East, from Greece and Egypt to the Borders of India) published in Geneva, Switzerland. It is 5 volumes plus a supplement (1867-1888). **Symphytum is listed in Volume IV in 1879.**

His botanical author citation abbreviation is 'Boiss.'. He was the first to publish these Symphytum species:
> Symphytum anatolicum Boiss. (Diagn. Pl. Orient. ser. 1, 4: 43. 1844)
> > ('Diagnoses Plantarum Orientalium Novarum', Series 1, No. 4, page 43, by Pierre Edmond Boissier, Lipsiae {Leipzig}, Germany, 1844.)
> Symphytum brachycalyx Boiss. (Diagn. Pl. Orient. ser. 1, 4: 43. 1844)
> Symphytum kurdicum Boiss. & Hausskn. ex Boiss. (Pl. Or. Nov. Dec. ii. 5. 1875)
> > (Plantarum Orientalium Novarum, annotated as Kurdicum.)
> Symphytum palestinae Boiss. (Diagn. Pl. Orient. ser. 1, 11: 94. 1849)
> Symphytum sepulcrale Boiss. & Balansa ex Boiss. (Fl. Orient. [Boissier] 4(1): 174. 1875) (Flora Orientalis by Boissier)
> Symphytum sylvaticum Boiss. (Pl. Or. Nov. Dec. ii. 4. 1875)

Carl Fredrik Nyman, 1820-1893, was from Sweden. He was curator at the 'Swedish Museum of Natural History' in Stockholm from 1855 to 1889. With Heinrich Wilhelm Schott and Theodor Kotschy, he was editor of 'Analecta Botanica'. He wrote **'Conspectus Florae Europaeae' from 1878 to 1882.**
The botanical author abbreviation is 'Nyman'. He published these Symphytum species:
> Symphytum floribundum Shuttlew. ex Nyman (Consp. Fl. Eur. 3: 509. 1881)
> Symphytum officinale L. subsp. uliginosum (A.Kern.) Nyman (Consp. Fl. Eur. 509. 1881) as uliginosum
> Symphytum uplandicum Nyman (Syll. Fl. Eur. 80. 1855) (Sylloge Florae Europaeae)

In 'Conspectus Florae Europaeae' on pages 509 and 510 he also wrote about: Symphytum officinale, Symphytum mediterraneum, Symphytum molle, Symphytum orientale, Symphytum tauricum, Symphytum cordatum, Symphytum anatolicum, Symphytum tuberosum, Symphytum gussonei, Symphytum bulbosum, Symphytum ottomanum.

Cedric Bucknall, 1849-1921, was an English botanist and organist with a Bachelor of Music from Oxford University, England. He traveled across Europe ("Carinthia, the Apennines, Naples, Sicily, the Baleares, and Southern Spain") collecting plants, which he then catalogued. **In 1912 he wrote the report 'Some Hybrids of the Genus Symphytum' and in 1913 he wrote 'A Revision of the Genus Symphytum, Tourn.'**
The standard botanical author citation is 'Buckn.'. He was the first to publish these Symphytum species:
> Symphytum armeniacum Buckn. (J. Linn. Soc., Bot. 41: 520. 1913)
> Symphytum x bicknellii Buckn. (J. Linn. Soc., Bot. 41: 552. 1913)
> Symphytum bornmuelleri Buckn. (J. Linn. Soc., Bot. xli. 536 1913)
> Symphytum x densiflorum Buckn. (J. Bot. 50: 334. 1912)
> Symphytum x discolor Buckn. (J. Bot. 50: 333. 1912)
> Symphytum x lilacinum Buckn. (J. Bot. 50: 334. 1912)
> Symphytum x polonicum Blocki ex Buckn. (J. Linn. Soc., Bot. 41: 552. 1913)

See subsection '1913 Bucknall' in section 'Symphytum Species Classifications' (Chapter 4).

Thomas Gaskell Tutin (T.G. Tutin), 1908-1987, was born in Surrey county, England. He was Fellow of the 'Royal Society of London' and Professor of Botany at the University of Leicester, England. He was President of the 'Botanical Society of the British Isles' from 1957 to 1961.
With A.R. Clapham and E.F. Warburg, he wrote **'Flora of the British Isles'** in 1952. He was editor of **'Flora Europaea'**, a 5-volume encyclopedia of plants, published between 1964 and 1993. **He wrote about the Symphytum genus with works published from 1952 to 1980.** His research covered S. asperum, S. officinale, S. tuberosum, S. x uplandicum and others.

> *"The 'Flora Europea' is conservative in its format: no conspicuous technical innovations detract from its usefulness as a flora of the type that has been developed in the last one hundred and fifty years. The general system of the families follows that of Engler. The keys (to identification) seem to work smoothly: in the cases where the book was used by students in taxonomy at Utrecht University, the reports were favorable, and sometimes even very complimentary, in contrast to some of the older works that had also been used.*
> *In the treatment of the species, references are given to the original publication of the accepted names."*
> -'Taxonomic Literature: Flora Europaea: Review' by F.A.S.; Taxon: International Association for Plant Taxonomy, Volume 14, No. 3, pages 105-107, March 1965.

Vladimir Leontyevich Komarov, 1869-1945, was a Russian botanist and geographer. He was a member of the 'Russian Academy of Sciences' in St. Petersburg. He served as President of the 'Academy of Sciences of the USSR' from 1936-1945. He was senior editor of the **'Flora of the USSR'** (Flora SSSR), 30 volumes published in Leningrad by Botanicheskii Institut (Akademiia Nauk SSSR) between 1934-1960.
> **Symphytum is in Volume 19, written 1953** on pages 207-216 in English version and pages 279-291 in Russian version. This part was written by Mikhail Grigorevich Popov (see next listing).
> *"Tribe 5. Anchuseae DC. in Meisn. Comm. (1838) 189, Prodr, X (1846) 27. An ancient Mediterranean tribe (Caucasian flora), showing numerous features of affinity with Echieae, Lithospermeae and Myosotis and none with the Cynoglossinae or Eritricheae.*
>> ***Genus Symphytum L .:*** *This ancient Mediterranean genus consists mainly of rather tall herbs whose habitats are mainly in the wooded, relict islands of the ancient Mediterranean. Some of its species are found in Central Europe and one even in the taiga zone of the Boreal region. About 20-25 species."*

Komarov's most important publication was the study of flora of the Far East, Kamchatka (peninsula in eastern Russia), and nearby areas of north and eastern Asia ('Florae Peninsulae Kamtschatka', 1927.) He studied the evolution of vegetables and the theory of speciation. He wrote textbooks and popular books on botany. He was one of the founders of the 'Russian Botanical Society'. He wrote: 'De Gymnospermis Nonnullis Asiaticis I, II', 1923-1924. The botanical author abbreviation for him is 'Kom.'. The 'Komarov Botanical Institute' and the 'Komarov Botanical Garden' of the 'Russian Academy of Sciences' in Saint Petersburg, Russia are named after him. The Institute houses herbarium collections with over seven million specimens of plants and fungi. This collection is the largest in Russia, and one of the three largest in the world.

The botanist **Mikhail Grigorevich Popov**, 1893-1955, was born in Volsk, Saratov Oblast, Russia. He was the author of more than 200 scientific publications devoted to taxonomy and research on flora of different regions of Eurasia, in particular, Siberia, middle Asia, Caucasus, Carpathians, Sakhalin Peninsula (in north Pacific Ocean) and in the region of Lake Baikal (in Siberia). He published 300 new names of plants. **Popov's report about 'Family Boraginaceae' is in 'Flora of the U.S.S.R., Volume 19'** edited by V.L. Komarov, Leningrad, Russia, **1953**, pages 73 to 508 in English and pages 97 to 692 in Russian.
See subsection '1953 Popov' in section 'Symphytum Species Classifications'.
'Popov' is the botanical abbreviation for author citation.
Probably this Symphytum species was named in honor of Popov: Symphytum popovii Dobrocz. -- Ukrayins'k. Bot. Zhurn. (The Ukrainian Botanical Journal) 25(6): 60. 1968.
> (The Caucasus Mountains are located at the border of Europe and Asia, between the Black Sea and Caspian Sea and occupied by Russia, Georgia, Azerbaijan, and Armenia.)
> (The Carpathian Mountains or Carpathians form an arc across central and eastern Europe. It includes parts of Austria, Czech Republic, Hungary, Poland, Serbia, Slovakia, Ukraine and Romania.)
> (U.S.S.R. or Soviet Union was a state in Eurasia that existed from 1922 to 1991. It is now Russia and 12 nations.)

Arthur Edward Wade, 1895-1989, was born in Leicester, England. He was a 'Fellow of the Linnean Society', London, England. He worked at the Department of Botany at the 'National Museum of Wales' in Cardiff, Wales. He was interested in the Boraginaceae family and became an authority on Symphytum and Myosotis. In 1934 he published with H.A. Hyde 'Welsh Flowering Plants'. In 1970 he published 'The Flora of Monmouthshire'.
He wrote the article: **'The History of Symphytum Asperum Lepech. and S. x Uplandicum in Britain' in 1958.** (Watsonia, Volume 4, pages 117-118, 1958. Watsonia was the journal of the 'Botanical Society of Britain and Ireland' from 1949 to 2010. It is now called 'New Journal of Botany'.)

Franklyn Hugh Perring, 1927-2003, was an English botanist and nature conservationist. He received a Ph.D. in Ecology and Biogeography from Queens College, Cambridge, England. **His botanical author citation abbreviation is 'F.H.Perring'.**
Franklyn Perring and S. Max Walters wrote **'Atlas of the British Flora'**, published by the 'Botanical Society of the British Isles' (BSBI), first published **1962**. Perring started working on this book in 1954 by recruiting 1,500 volunteer botanists to map the locations of 1,700 species of flowering plants and ferns. Professor David Webb helped with the plant mapping in Ireland.
Perring wrote **'Symphytum Survey'** published in 'Watsonia: Journal and Proceedings of the Botanical Society of the British Isles', London, England, Volumes 8 and 10, 1970 and 1975.
He wrote the **Symphytum article in the book 'The Common Ground of Wild and Cultivated Plants**: B.S.B.I. Conference Report No. 22' published by the 'Botanical Society of the British Isles' in 1994.
Perring and T.W.J. Gadella wrote the **Symphytum article in Volume 6 of 'The European Garden Flora**: A Manual for the Identification of Plants Cultivated in Europe, Both Out-of-Doors and Under Glass' published by Cambridge University Press, England in 2000.
See subsection '1975 Perring' in section 'Symphytum Species Classifications'.
> ('Franklyn Hugh Perring Obituary', Watsonia: Journal and Proceedings of the Botanical Society of the British Isles, London, England, Volume 26, pages 197-211, 2006.)

Gerald Ernest Wickens, born 1927, is an English botanist. He worked on phanerogamous (spermatophyte = plants with seeds) taxonomy at Kew Gardens, Edinburgh, Scotland. In 2001 he wrote 'Economic Botany: Principles and Practices'. **His botanical author citation abbreviation is 'Wickens'. He was the first to publish these Symphytum species:**
> Symphytum aintabicum Hub.-Mor. & Wickens (Notes Roy. Bot. Gard. Edinburgh 29: 174. **1969**)
> Symphytum davisii Wickens (Notes Roy. Bot. Gard. Edinburgh 29: 168. 1969)
> Symphytum longipetiolatum Wickens (Notes Roy. Bot. Gard. Edinburgh 29: 179. 1969)
> Symphytum longisetum Hub.-Mor. & Wickens (Notes Roy. Bot. Gard. Edinburgh 29: 174. 1969)

See subsection '1969 Wickens' in section 'Symphytum Species Classifications' (Chapter 4).

Bogumil Pawlowski, 1898-1971, was a Polish botanist interested in plant taxonomy, phytogeography (plant distribution), and phytosociology (plant communities). He was a member of the 'Polish Academy of Sciences', and Professor at the Jagiellonian University in Krakow, Poland. He was Director of the 'Institute of Botany of the Academy of Sciences' in Krakow.
He was the author or co-author of over 100 papers about botany, such as 'Flora Polska' that describes plants in Poland. **He wrote the article 'Symphytum L.' in Flora Europaea, Volume 3**, pages 103-105, in **1972**. The 'Flora Europaea' is a 5-volume encyclopedia of plants, published between 1964 and 1993 by Cambridge University Press, England. It describes all the national floras of Europe to identify any wild or widely cultivated plant to the subspecies level.
The standard botanical author abbreviation is 'Pawl.'. See subsection '1972 Pawlowski' in section 'Symphytum Species Classifications'. **He was the first to publish these Symphytum species:**
> Symphytum sect. Graeca Pawl. (Fragm. Florist. Geobot. 17 1: 19. 1971)

Symphytum cycladense Pawl. (Fragm. Florist. Geobot. 17 1: 25. 1971)
Symphytum x hyerense Pawl. (Fragm. Florist. Geobot. 17 1: 31. 1971)
Symphytum icaricum Pawl. (Fragm. Florist. Geobot. 17 1: 19. 1971)
Symphytum naxicola Pawl. (Fragm. Florist. Geobot. 17 1: 21. 1971)
(Fragmenta Floristica et Geobotanica Polonica, W. Szafer Institute of Botany, Polish Academy of Sciences, Krakow, Poland, Volume 17, No. 1, 1971: 'Symphyta Mediterranea Nova vel Minus Cognita: Nowe lub malo znane srodziemnomorskie zywokosty' {**New or Little Known Mediterranean Comfrey**} by Bogumil Pawlowski, pages 17-37. If you have an English translation, I would appreciate a copy.)

'**The Greek Species of Symphytum Boraginaceae**' **by William T. Stearn**, Annales Musei Goulandris, Greece, Volume 7, pages 175-220, (1985 or **1986**, different sources give different dates).
"*Five species of Symphytum occur on Greek territory, namely S. bulbosum Schimper widespread in mainland Greece from the Peloponnese northward, S. ottomanum widespread in northern Greece, S. tuberosum subsp. angustifolium (A. Kerner) Nyman only in the extreme north-west, S. davisii subsp. davisii, naxicola, icaricum and cycladense endemic (native) to Aegean islands and S. anatolicum Boiss., a species of Western Asia Minor, on Lesvos, Kos and Samos.*"

'**Procopiania (Boraginaceae): Separate Genus or Part of Symphytum?**' **by Burcu Elci, Hartmut H. Hilger, Sadik Erik** from Ankara, Turkey and Berlin, Germany, April **2008**. 'Systematics 2008: Programme and Abstracts', 18th International Symposium, German Botanical Society, Gottingen, Germany.
"***The genus Symphytum L. (Boraginaceae) is a mesophytic genus of approximately 40 species, mainly distributed in the Euro-Siberian and Mediterranean regions.*** *With 19 species, Turkey holds the largest number of species.* ***Procopiania Gusul. is distributed in the Aegean archipelago, southern Greece and western Anatolia (Asian Turkey) with three species.***
The genus was described by ***Gusuleac*** *in 1928 with P. cretica (Willd.) Gusul., formerly belonging to Trachystemon. This division is based mainly on the presence or absence of bracts in the inflorescence and shape of fornices.*
Runemark *(1967) did not agree with Gusuleac's approach, included Procopiania into Symphytum and supported his hypothesis by morphological, palynological and chromosomal investigations.*
Wickens *(1969) in his treatment of Turkish taxa also synonymized Procopiania with Symphytum.*
However, ***Pawlowski*** *(1971) and* ***Stearn*** *(1986) followed Gusuleac and treated Procopiania as separate genus.*
In ***Hilger*** *et al. (2004) only one species of each genus was included.*
The affinities and differences of pollen structure of Procopiania and Symphytum is discussed in comparison with outgroups and compared with the findings of ***Harmata*** *(1977, 1981). We examined the relationship of Symphytum and Procopiania - sistergroup or inclusion - by adding palynological and molecular to morphological data. Species were chosen from the Boraginae and Symphytum in order to compare the similarities/dissimilarities between closely distributed species of both genera.*
In the tree yielded from the nuclear ITS 1 region, P. circinale and P. creticum cluster together. They are not sister to Symphytum but belong to the ingroup of Symphytum. Thus, the generic rank of Procopiania is not justified."
(I only have the abstract of the article. If you have all of it, could you please send it to me.)
(The Aegean Islands are in the Aegean Sea, with mainland Greece to the west and north, and Turkey to the east.)

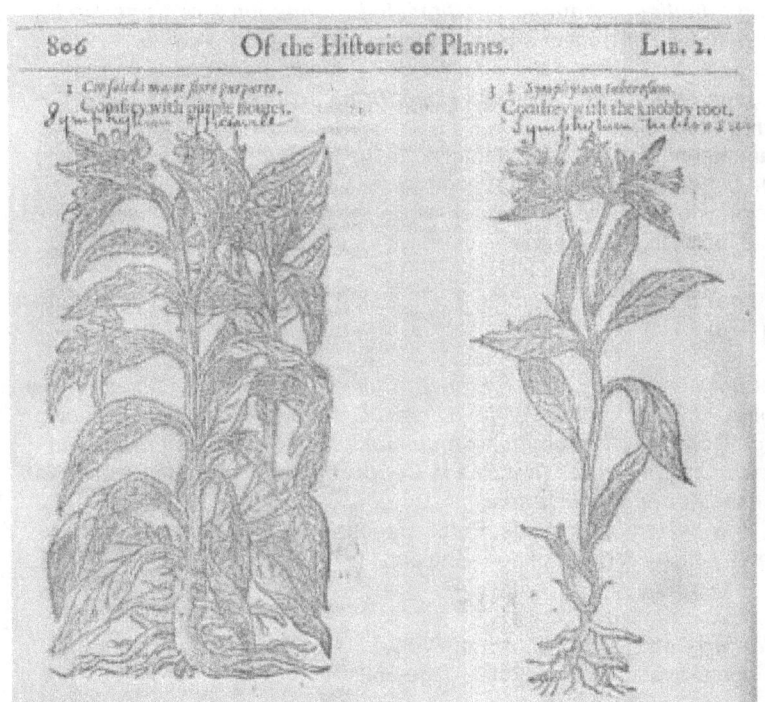

Symphytum officinale 1633
Symphytum tuberosoum

'The Herball, or, Generall Historie of Plantes Gathered by Gerarde of London, Master in Chirurgerie (Surgery)' by John Gerard, London, England.
First published in 1597 (1,484 pages with illustrations).

This art is from 1633 publication of 'The Herball' (1,730 pages): 'Of the Historie of Plants', page 806, Lib. 2.

Chapter 3

Symphytum Species

Comfrey Names (Ancient, Informal and Foreign Language)

Other names for Comfrey are: Abraham, all heal, asses-ears, backwort, beinwell, black root, blackwort, boneset, borraja, bruisewort, bourrache, buyuk, church bells, confirma, consolida, consolidae radix, consoude, consound, consoude, consuelda, cumfrey, gewone smeerworte, gum plant, healing herb, herbe aux charpentiers, grande consoude, herbe a la coupure, Issac-and-Jacob (from flower color variations), knitback, knitbone, langue-de-vache, liane chique, nipbone, oreille d'ane, salsify, saracen's root (desert nomadic tribe), slippery root, solidago, wallwort and yalluc (Saxon).

"*Latin:*
Symphytum consolida major. *Bauh. pin. 259.*
Consolida major flore purpureo. *Ger. Em. 806.*
Symphytum magnum. *Raii Syn. 230.*
Symphytum Consolida major, flore pallide luteo, quae femina. *Tourn. Paris. 306.*
Symphytum officinale. *Lin. Sp. Pl. 195. Eng. Bot. t. 817.* "
-'The British Flora Medica, or, History of the Medicinal plants of Great Britain, Volume 1' by Benjamin Herbert Barton and Thomas Castle, London, England, **1838**. (Symphytum officinale, pages 211-215.)
(Caspar Bauhin, Swiss botanist wrote 'Phytopinax' in 1596. Abbreviation: Bauh. pin. Page 259.)
(Ger. Em. = Gerard, emended {= altered, modified}. John Gerard published 'The Herball or General Historie of Plantes' in 1597. Gerard's book was revised by Thomas Johnson in 1636. Ger. Em. 806 = John Gerard's 'The Herball', edited 1636, page 806.)
(Raii Syn. = Ray's, Synopsis Methodica Stirpium Britannicarum'. Historically, letters i, y, and j were interchangeable, so Raii = Rai's = Ray's [Raii is Latin for Ray's]. Raii Syn. 230 = Ray's 'Synopsis', edition 3, page 230, written 1724. Joannis Raii = John Ray)
(Tourn. Paris. 306. Joseph Pitton de Tournefort wrote 'Histoire des Plantes qui Naissent aux Environs de Paris, avec leur Usage dans la Medecine' {History of Plants Born around Paris, with their Use in Medicine} in 1698. Symphytum is on page 306.)
(Lin. Sp. Pl. 195. = Linnaeus wrote 'Species Plantarum' in 1753. The second edition was written in 1762-1763 with Symphytum officinale on page 195.)
(Eng. Bot. t. 817. = 'English Botany (Sowerby's)', plate 817. In Volume 7, Third Edition, written 1867, plate MCXV, between pages 114 and 115, has written on it 'E.B. 817'. MCXV = 1115.)

*"**The botanic name of this plant is Symphytum, a word of Greek origin, signifying 'springing forth in company'**, which is descriptive of its growth, both root and herb, but not of its nature."*
-'The New England Farmer, and Horticultural Register' by Joseph Breck, Volume 23, New Series Volume 13, Boston, Massachusetts, **1845**. ('Symphytum or Comfrey, as Food for Men and Cattle' by Ezekiel Rich, Troy, New Hampshire, page 10, July 10 1844.)

"Comfrey L. confirma, from its supposed strengthening qualities, Symphytum officinale L., Consound, or Consoud L., consolida,* 'quia tanta praestantia est, ut carnes, dum coquuntur, conglutinet addita, unde nomen," ** Pliny, xxvii. 6.
A name given in the middle ages to several different plants, and among them to the daisy, Bellis perennis L. to the Comfrey, Symphytum officinale L. and to the Bugle, Ajuga reptans L., Knit-back L. confirma, from being used as a strengthener or restorative, the Comfrey, Symphytum officinale L."
-'On the Popular Names of British Plants, Being an Explanation of the Origin and Meaning of the Name of Our Indigenous and Most Commonly Cultivated Species' by R.C.A. Prior, M.D., London, England, **1863**, page 52.
(* Literal translation from Latin: 'because it is of such outstanding is to say, as the flesh, while the men cook, glueth were added, and from which the name of the'. Or 'because it works so exquisitely that, when it is done with the meat while roasting, the pieces stick together, hence the name'.)
(** -Pliny the Elder's 'Natural History'. 'Naturalis Historia', book 27, chapter 6, written 77-79 A.D. It is written in Latin by a Roman, about all of the natural world with 10 volumes, containing 37 books.)

*"Professor J.A. Barral, Perpetual Secretary of the 'Central Society of Agriculture in France' ('Societe Centrale de l'Agriculture en France') writes in the Journal de l'Agriculture, of 7th October, 1876 (No. 391): 'All the different Comfreys are favourably known, and peasants in the country know particularly well the large **Comfrey (Symphytum officinale) which they have christened amongst themselves, cows tongue (langue de vache) and cutting grass (herbe a la Coupure) besides other names. The name of Comfrey is derived from the property it possesses of curing wounds.**"*
-'Forage Plants and Their Economic Conservation by the New System of Ensilage: Part I: Caucasian Prickly Comfrey' by Thomas Christy, Jun., F.L.S. (Fellow of the Linnean Society), Christy & Co., London, England, **1877**, page 8.

*"**Symphytum: Grow-together-plant**.
Dioscorides' name: sumfuton, for healing plants, including Comfrey. Conferva of Pliny."*
-The Names of Plants by David Gledhill. England: Cambridge University Press, **2008**, page 368.

Albania: Kufilma Mjekesere, Kufilme
Arabia: Sanfitun, Azan hemar
China: Ju He Cao, Ju He Cao Shu
Croatia: Ljubicasti Gavez, Kostival Cesky
Czechoslovakia: Kostival Lekarsky
Denmark: Kulsukkerrod, Laege-Kulsukker
Estonia: Harilik Varemerohi
Finland: Rohtorauniovrtti, Raunioyrtti, Mustajuuri, tummarohtorauniovrtti
France: Consoude Officinale, Grande Consoude, Oreille de Vache, Conflee, Confee, Consyre, Herbe aux Charpentiers, Herbe a la Coupure, Langue de Vache, Oreilles de d'Ane
Germany: Arznei-Beinwell, Echter Beinwell, Echte Wallwurz, Schwarzwurz, Gemeiner Beinwell, Wilder Komfrey, Gewohnlicher Beinwell
Greece: Stekouli
Hungary: Fekete Nadalyto
India: Sankootun
Italy: Consolida Maggiore, Orecchia D'asino, Concuardie, Erba di San Lorenzo, Simfito
Japan: Hireharisou (Hirehariso), Konfuri
Korea: Keom Peu Li
Latvia: Arstniecibas Tauksakne
Lithuania: Vaistine Tauke
Mexico: Carqueja, Consuela mayor, Sinfito
Netherlands: Gewone Smeerwortel, Gewone Sweerwortel
Norway: Valurt
Persia: Hmawr kurdi and Gwsh khr
Poland: Zywokost Lekarski
Portugal: Confrei, Consolda-Maior, Consolida-Maior (Brazil), Grande-Consolda, Orelhas-de-Asno
Romania: Tataneasa
Russia: Okopnik, Okopnik Lekarstvennyj
Slovakia: Kostihoj Lekarsky
Slovenia: Navadni Gabez
Spain: Consuelda Mayor, Consuelda, Consolida, Oreja De Asno, Hierba de las Cortaduras, Ricasuelda, Sinfito Mayor, Suelda
Sweden: Vallort, Akta vallort
Turkey: Kafes Otu, Karakafes, Mayasilotu, Merkep Kulagi, Karakafesotu, Sinfit, Buyuk kara-kafes out, Ezmangag
Wales: Llysiau'r Cwlwm

<u>Accepted Latin Names of Symphytum Species</u>

"Species can be defined as distinguishable groups of genotypes that remain distinct in the face of potential or actual hybridization and gene flow. This is similar to Charles Darwin's usage of species to divide biodiversity by means of gaps or troughs in the distributions of phenotypes and genotypes."
-'Hybrid Speciation' by James Mallet, University College, London, England; Nature (multidisciplinary scientific journal published in London), Vol 446, No. 15, pages 279-283, March 2007.
    (Phenotype is the observable properties of an organism. Genotype is the genetic constitution of an organism.)

Below are the accepted Latin name for Symphytum (Comfrey) species. Many Comfrey names are unresolved indicating that the data sources included provided no evidence or view as to whether the name should be treated as accepted or not, or there were conflicting opinions that could not be readily resolved.

" 'The Plant List' includes 120 scientific plant names of species rank for the genus Symphytum.
Of these 7 are accepted species names:
Symphytum abchasicum Trautv.
Symphytum asperum Lepechin (Prickly Comfrey)
Symphytum besseri Zaver.
Symphytum caucasicum M.Bieb.
Symphytum officinale L. (Common Comfrey)
Symphytum popovii Dobrocz.
Symphytum x uplandicum Nyman (Russian Comfrey)."
-The Plant List®, www.theplantlist.org from the World Checklist Database®, 2018.

"5 accepted species of Symphytum:
Symphytum besleria Hanst.
Symphytum asperum Lepech. (Rough or Prickly Comfrey)

Symphytum officinale L. (Common Comfrey)
Symphytum tuberosum L. (Tuberous Comfrey)
Symphytum x uplandicum Nyman (Russian or Upland Comfrey)."
-**ITIS**® (Integrated Taxonomic Information System) Report, www.itis.gov, 2018. The ITIS is the result of a partnership of United States Federal agencies formed for their needs for scientifically credible taxonomic information.

"**5 accepted species of Symphytum:**
Symphytum asperum Lepechin (Prickly Comfrey)
    Symphytum asperrimum Donn ex Sims
Symphytum officinale L. (Common Comfrey)
    Symphytum officinale L. ssp. uliginosum auct. non (Kern.) Nyman
Symphytum uliginosum auct. non Kern.
Symphytum tuberosum L. (Tuberous Comfrey)
Symphytum x uplandicum Nyman (pro sp.) [asperum × officinale] (Russian Comfrey)
    Symphytum peregrinum auct. non Ledeb."
-U.S. Department of Agriculture, Natural Resources Conservation Service®, **Plants Database**®, www.plants.usda.gov, 2018.

"**25 accepted species of Symphytum:**
Symphytum aintabicum Hub.-Mor. & Wicken
Symphytum anatolicum Boiss.
Symphytum asperum Lepech.
Symphytum bornmuelleri Buckn.
Symphytum brachycalyx Boiss.
Symphytum bulbosum K.F.Schimp.
Symphytum caucasicum M.Bieb.
Symphytum circinale Runemark
Symphytum cordatum Waldst. & Kit. ex Willd.
Symphytum creticum (Willd.) Runemark
Symphytum davisii Wickens
Symphytum x floribundum Shuttlew. ex Buckn.
Symphytum grandiflorum DC.
Symphytum gussonei F.W.Schultz
Symphytum hajastanum Gvin.
Symphytum ibiricum Steven
Symphytum kurdicum Boiss. & Hausskn.
Symphytum longisetum Hub.-Mor. & Wickens
Symphytum officinale L.
Symphytum orientale L.
Symphytum ottomanum Friv.
Symphytum pseudobulbosum Azn.
Symphytum savvalense Kurtto
Symphytum sylvaticum Boiss.
Symphytum tauricum Willd.
Symphytum tuberosum L.
Symphytum x uplandicum Nyman
Symphytum x wettsteinii Sennholz"
-**Plants of the World Online**®, www.plantsoftheworldonline.org, Kew Science, Royal Botanic Gardens, London, England, Information on all the world's known seed-bearing plants, 2018.

"**40 accepted Symphytum species:**
Symphytum aintabicum Hub.-Mor. & Wickens
Symphytum anatolicum Boiss.
Symphytum asperum Lepech.
Symphytum asperum x officinale
Symphytum bornmuelleri Bucknall.
Symphytum bottii
Symphytum brachycalyx Boiss.
Symphytum bulbosum C.Schimper
Symphytum carpaticum Yu.M.Frolov
Symphytum caucasicum M.Bieb.
Symphytum caucasicum x uplandicum
Symphytum circinale Runemark
Symphytum cordatum Waldst. & Kit.
Symphytum creticum (Willd.) Greuter & Rech.fil.

Symphytum davisii Wickens
Symphytum x floribundum Shuttlew. ex Buckn.
Symphytum grandiflorum DC.
Symphytum gussonei F.W.Schultz
Symphytum hajastanum Gviniashvili
Symphytum x hidcotense P.D.Sell
Symphytum hyerense Pawl.
Symphytum ibiricum Steven
Symphytum kurdicum Boiss. & Hausskn. ex Boiss.
Symphytum longisetum Hub.-Mor. & Wickens
Symphytum microcalyx Opiz
Symphytum mosquense S.R.Majorov & D.D.Sokoloff
Symphytum officinale L.
Symphytum officinale x tuberosum subsp. angustifolium
Symphytum orientale L.
Symphytum ottomanum Friv.
Symphytum popovii Dobrocz.
Symphytum podcumicum Yu.M.Frolov
Symphytum pseudobulbosum Azn.
Symphytum savvalense A.Kurtto
Symphytum sylvaticum Boiss.
Symphytum tauricum Willd.
Symphytum tuberosum L.
Symphytum tuberosum x uplandicum
Symphytum uplandicum Nyman
Symphytum wettsteinii Sennh."

-**Global Biodiversity Information Facility** (GBIF), www.gbif.org, 2018. It is an international network and research infrastructure funded by the world's governments and aimed at providing anyone open access to data about all types of life on Earth.

List of All Symphytum Species

"**Symphytum Species:** *The list of species recognized in this study, including their synonyms, is presented:*
S. floribundum is not included as a separate species, since Kurtto (1981) showed that this species is most probably an interspecific (between species) hybrid of S. officinale and S. orientale.

1. Symphytum abchasicum Trautv.
      synonym: *S. grandiflorum DC. var. abchasicum (Trautv.) Kuz.*
                            S. ibericum Stev. var. abchasicum (Trautv.) Gviniasv.

2. Symphytum aintabicum Huber-Morath & Wickens

3. Symphytum anatolicum Boiss.
      synonym: *S. sicyosinum Candargy*

4. Symphytum armeniacum Buckn.
      synonym: *S. asperum Lepech. var. armeniacum (Buckn.) Kurtto*

5. Symphytum asperum Lepech.
      synonym: *S. echinatum Ledeb.*
                            S. patens Fries (includes S. asperum and S. x uplandicum)
                            S. asperrimum Sims in Curtis
                            S. orientale sensu Fries
                            S. majus Guldenst. ex Ledeb.
                            S. asperrimum Roem. & Schult.

6. Symphytum bornmuelleri Buckn.

7. Symphytum brachycalyx Boiss.
      synonym: *S. orientale L. var. angustior DC.*
                            S. palaestinum Boiss. var. dentatum Boiss.
                            S. palaestinum Boiss. var. majus Buckn.
                            S. palaestinum Boiss. var. strigosum Post in Post.

8. Symphytum bulbosum Schimp.
      synonym: *S. filipendulum Bischoff*
                            S. dusii Gmel.
                            S. punctatum Gend.
                            S. macrolepis Gay ex Reich.
                            S. brochum Bory & Chamn.
                            S. zeyheri Schimp.
                            S. tuberosum L. var. zeyheri (Schimp.) Fiori.

                              S. tuberosum L. var. bulbosum (Schimp.) Fiori.

9. *Symphytum caucasicum* Bieb.
       synonym: S. racemosum Steph. ex Roem. & Schult.
                       S. donii DC.

10. *Symphytum circinale* Runem.
       synonym: Procopiana circinnalis (Runem.) Pawlow.

11. *Symphytum ciscaucasicum* Gviniashv.

12. *Symphytum cordatum* Waldst. et Kitaib.
       synonym: S. pannonicum Pers.
                       S. cordifolium Baumg.

13. *Symphytum creticum* (Willd.) Runem.
       synonym: Borago cretica Willd.
                       Trachystemon creticum D. Don ex G. Don
                       Psilostemon creticum DC.
                       Procopiana cretica (Willd.) Gusul.

14. *Symphytum cycladense* Pawl.
       synonym: S. davisii Wick. ssp. cycladense (Pawl.) Stearn.

15. *Symphytum davisii* Wick.
       synonym: S. davisii Wick. ssp. davisii

16. *Symphytum euboicum* (Runem.) Runem.
       synonym: Procopiania euboica Runem. in Rech.

17. *Symphytum grandiflorum* DC.
       synonym: S. cordatum sensu Bieb.

18. *Symphytum gussonei* Schulz
       synonym: S. mediterraneum Guss.

19. *Symphytum hajastanum* Gviniashvili

20. *Symphytum ibericum* Stev.

21. *Symphytum icaricum* Pawl.
       synonym: S. davisii Wick. ssp. icaricum (Pawl.) Stearn

22. *Symphytum insulare* (Pawl.) Greuter & Burdet.
       synonym: Procopiania insularis Pawl.

23. *Symphytum kurdicum* Boiss. et Hausskn.

24. *Symphytum longipetiolatum* Wick.

25. *Symphytum longisetum* Huber-Morath & Wick.

26. *Symphytum mediterraneum* Koch

27. *Symphytum naxicola* Pawl.
       synonym: S. anatolicum Halacs.
                       S. davisii Wick. ssp. naxicola (Wick.) Stearn

28. *Symphytum nodosum* Schur.
       synonym: S. tuberosum L. ssp. angustifolium (Kern.) Buckn.
                       S. leonhardtianum Pugsl.
                       S. tuberosum L. var. nodosum Gusul.

29. *Symphytum officinale* L.
       synonym: S. bohemicum Schmidt (2n = 24, white flowers).

30. *Symphytum orientate* L.
       synonym: S. tauricum Sims.
                       S. jacquinianum Tausch.

31. *Symphytum ottoman um* Friv.
       synonym: S. bulbosum Schimp. var. ottomanum (Friv.) 0. Kuntze
                       Procopiana euboica Runem. in Rech.
                       S. euboicum (Runem.) Runem.

32. *Symphytum palaestinum* Boiss.

33. *Symphytum peregrinum* Ledeb.

34. *Symphytum pseudobulbosum* Azn.

35. *Symphytum savvalense* Kurtto

36. *Symphytum sepulcrale* Boiss. & Bal.
       synonym: S. longipetiolalum Wickens (synonym with S. sepulcrale Boiss. & Bal. var. sepulcrale)

37. *Symphytum sylvaticum* Boiss.
       synonym: S. sepulcrale Boiss. & Bal. ssp. sylvaticum (Boiss.) Kurtto

38. *Symphytum tanaicense* Stev.
       synonym: S. officinale L. ssp. uliginosum (Kern.) Nym.
                       S. vetteri Thellung
                       S. uliginosum Kern.

39. *Symphytum tauricum* Willd.

    synonym: S. bullatum Hornem.
                     S. orientate sensu Pallas non L.

40. Symphytum tuberosum L.
    synonym: S. leonhardtianum Pugsley
                     S. foliosum Rehm.
                     S. tuberosum L. ssp. tuberosum
                     S. tuberosum L. ssp. angustifolium (Kern.) Nym. = S. nodosum Schimp.

41. Symphytum zeyheri Schimp."

-'Phylogenetic Relationships in the Genus Symphytum L. (Boraginaceae)' by J.M. Sandbrink, J. Van Brederode and T.W.J. Gadella, Department of Genome Evolution, University of Utrecht, Netherlands; Proceedings: Koninklijke Nederlandse Akademie van Wetenschappen (Royal Netherlands Academy of Sciences), Series C, Volume 93, No. 3, pages 295-334, **1990**.

"The Old World genus Symphytum L. belongs to the tribe 'Boragineae Bercht. et J. Presl', a major monophyletic (common evolutionary ancestor) group within the family Boraginaceae (Hilger et al. 2004).
With approximately 40 species, it is one of the largest genera in this tribe (*Bucknall 1913, **Sandbrink et al. 1990). It includes perennial, roughly hirsute (hairy) plants, which are morphologically well characterized by creeping, mostly fleshy rhizomes, alternate leaves, double scorpioid cymes (= boragoids) with tubular flowers and five corolla appendages (= fornices) inside the flower.
The geographical range of the genus covers almost the whole of Europe and Asia Minor, as well as part of Western Asia and Siberia (Bucknall 1913). *The centre of its diversity is situated in the Pontic area and in the western parts of the Irano-Turanian region, primarily in the mountain ranges around the Black Sea (Gadella & Kliphuis 1978, Davis 1988).*
In central Europe, the following native species are recognized:
    S. cordatum Waldst. et Kit.;
    S. officinale complex *(S. bohemicum F.W. Schmidt, S. officinale s. str., S. tanaicense Steven);*
    S. tuberosum complex *(S. angustifolium A. Kern.; Kerner 1863, Murin & Majovsky 1982, Marhold & Hindak 1998; and S. tuberosum L.; Pawlowski 1963, Gams 1966, Smejkal 1978, Majovsky & Hegedusova 1993, Slavik 2000, Danihelka 2012)*
Five additional non-native species originating mostly from eastern Europe, the eastern Mediterranean and the Caucasus are also reported: S. asperum Lepech., S. bulbosum K. F. Schimp., S. caucasicum M. Bieb., S. orientale L. and S. tauricum Willd. (Pawloski 1963, Gams 1966, Smejkal 1978, Danihelka et al. 2012, Bomble & Schmitz 2013)."

-'Symphytum Tuberosum Complex in Central Europe: Cytogeography, Morphology, Ecology and Taxonomy' by L. Kobrlova, M. Hrones, P. Koutecky, M. Stech, and B. Travnicek, Preslia: The Journal of the Czech Botanical Society, Czech Republic; Volume 88, pages 77-112, March **2016**.

(* -'A Revision of the Genus Symphytum, Tourn.' by Cedric Bucknall, {Mus. Bac. Oxon= Bachelor of Music, Oxford University}, Journal of the Linnean Society of London, England, Botanical Journal, Volume 41, Issue 284, pages 491-556, December 1913.)
(** -'Phylogenetic Relationships in the Genus Symphytum L. {Boraginaceae}' by J.M. Sandbrink, J. Van Brederode and T.W.J. Gadella, Proceedings: Koninklijke Nederlandse Akademie van Wetenschappen {Royal Dutch Academy of Sciences}, Series C, Volume 93, No. 3, pages 295-334, 1990. I was unable to get this report. If you have one, could you please send it to me.)

    (The Irano-Turanian Region is located within the Tethyan Subkingdom of the Holarctic Kingdom. It is divided into 12 provinces according to Armen Takhtajan, a Soviet-Armenian botanist who created a classification system for flowering plants. This area has great diversity and an abundance of species.)

"74 species of Symphytum listed:
Symphytum abchasicum
Symphytum aggr.
Symphytum aintabicum
Symphytum anatolicum
Symphytum angustifolium
Symphytum armeniacum
Symphytum asperrimum
Symphytum asperum
Symphytum asperum aggr.
Symphytum asperum var. armeniacum
Symphytum besseri
Symphytum bohemicum
Symphytum bornmuelleri
Symphytum brachycalyx
Symphytum bulbosum
Symphytum carpaticum
Symphytum caucasicum
Symphytum circinale
Symphytum ciscaucasicum
Symphytum cordatum
Symphytum creticum
Symphytum cycladense

Symphytum davisii
Symphytum euboicum
Symphytum floribundum
Symphytum grandiflorum
Symphytum gussonei
Symphytum hajastanum
Symphytum ibericum
Symphytum ibericum var. abchasicum
Symphytum ibiricum
Symphytum icaricum
Symphytum incarnatum
Symphytum insulare
Symphytum kurdicum
Symphytum leonhardtianum
Symphytum longipetiolatum
Symphytum longisetum
Symphytum mediterraneum
Symphytum microcalyx
Symphytum molle
Symphytum naxicola
Symphytum nodosum
Symphytum officinale
Symphytum officinale aggr.
Symphytum officinale subsp. bohemicum
Symphytum officinale subsp. officinale
Symphytum officinale subsp. uliginosum
Symphytum officinale var. glabrescens
Symphytum orientale
Symphytum ottomanum
Symphytum palaestinum
Symphytum patens
Symphytum peregrinum
Symphytum podcumicum
Symphytum popovii
Symphytum pseudobulbosum
Symphytum savvalense
Symphytum sepulcrale
Symphytum sepulcrale subsp. sylvaticum
Symphytum sylvaticum
Symphytum sylvaticum subsp. sepulcrale
Symphytum sylvaticum subsp. sylvaticum
Symphytum tanaicense
Symphytum tauricum
Symphytum tuberosum
Symphytum tuberosum subsp. angustifolium
Symphytum tuberosum subsp. bulbosum
Symphytum tuberosum subsp. nodosum
Symphytum tuberosum subsp. tuberosum
Symphytum uliginosum
Symphytum x uplandicum
Symphytum zeyheri "

-**Global Plant Checklist® (GPC)** managed by 'International Organization for Plant Information' (IOPI), http://ww2.bgbm.org/IOPI/gpc/default.asp, **2018**.

"*147 species of Symphytum listed:*
   APNI= Australian Plant Names Index®: *Australian National Botanic Gardens, Canberra, Act.*
   GCI= Gray Card Index®: *Harvard University Herbaria, Cambridge, Massachusetts.*
   IK= Index Kewensis®: *Royal Botanic Gardens, Kew, registers all botanical names for seed plants at the rank of species and genera. Also names of taxonomic families and ranks below that of species.*

Boraginaceae Symphytum L. -- Sp. Pl. 1: 136. 1753 [1 May 1753] (GCI)
Boraginaceae Symphytum L. -- Species Plantarum (1 May 1753) 136, Genera Plantarum ed. 5 1754 (APNI)
Boraginaceae Symphytum L. -- Sp. Pl. 1: 136. 1753 [1 May 1753] (IK)
Boraginaceae Symphytum sect. Graeca Pawl. -- Fragm. Florist. Geobot. 17(1): 19. 1971 (IK)
Boraginaceae Symphytum abchasicum Trautv. -- Bull. Soc. Imp. Naturalistes Moscou xliii. (1870) I. 72. (IK)

Boraginaceae Symphytum aintabicum Hub.-Mor. & Wickens -- Roy. Bot. Gard. Edinburgh 29: 174. 1969 (IK)
Boraginaceae Symphytum album hort. ex Steud. -- Nomencl. Bot. [Steudel] 821. 1821 (IK)
Boraginaceae Symphytum ambiguum Pau. -- Not. Bot. Fl. Espan. iii. (1889) 32. (IK)
Boraginaceae Symphytum americanum W.Young -- Cat. Arbres, Arbustes & Pl. Herb. Am. 47 (1783). (IK)
Boraginaceae Symphytum anatolicum Boiss. -- Diagn. Pl. Orient. ser. 1, 4: 43. 1844 [Jun 1844] (IK)
Boraginaceae Symphytum angustifolium A.Kern. -- Oesterr. Bot. Z. 13: 227. 1863 (IK)
Boraginaceae Symphytum armeniacum Buckn. -- J. Linn. Soc., Bot. 41: 520. 1913 (IK)
Boraginaceae Symphytum armenum Gundelsh. ex DC. --Prodr.[A.P. deCandolle] 10:66. 1846[8 Apr 1846] (IK)
Boraginaceae Symphytum asperrimum d'Urv. -- Mém. Soc. Linn. Paris 1: 276. 1822 (IK)
Boraginaceae Symphytum asperrimum Donn -- Bot. Mag. 24: t. 929. 1806 (IK)
Boraginaceae Symphytum asperum Lepech. -- in Nov. Act. Acad. Petrop. xiv. (1805) 442. t. 7. (IK)
Boraginaceae Symphytum asperum Lepech. var. armeniacum(Buckn.)Kurtto-Ann.Bot.Fenn19(3):186.1982(IK)
Boraginaceae Symphytum azureum H.C.Hall -- in Tijdschr. Wetensch. ii. (1849) 128. (IK)
Boraginaceae Symphytum beckii Petr. -- Allg. Bot. Z. Syst. xiii. 146 (1907). (IK)
Boraginaceae Symphytum besseri Zaver. -- Ukrayins'k. Bot. Zhurn. 19(5): 56. 1962 (IK)
Boraginaceae Symphytum x bicknellii Buckn. -- J. Linn. Soc., Bot. 41: 552. 1913 (IK)
Boraginaceae Symphytum bohemicum F.W.Schmidt -- Fl. Boem. iii. 13. t. 263. (IK)
Boraginaceae Symphytum bornmuelleri Buckn. -- J. Linn. Soc., Bot. xli. 536 (1913). (IK)
Boraginaceae Symphytum borragineum Tausch -- Flora 19(2): 393. 1836 (IK)
Boraginaceae Symphytum brachycalyx Boiss. -- Diagn. Pl. Orient. ser. 1, 4: 43. 1844 [Jun 1844] (IK)
Boraginaceae Symphytum brochum Bory & Chaub. -- Nouv. Fl. Pelop. 13. t. 8. 1838 (IK)
Boraginaceae Symphytum bulbosum K.F.Schimp. -- Flora 8(1, no. 2): 17. 1825 [14 Jan 1825] (IK)
Boraginaceae Symphytum bulbosum Schur -- Enum. Pl. Transsilv. 468. 1866 [Apr-Jun 1866] (IK)
Boraginaceae Symphytum bullatum Hornem. -- Hort. Bot. Hafn. i. 179. 1813 (IK)
Boraginaceae Symphytum calcaratum E.D.Clarke -- Travels Eur. Asia & Africa ii. 651. (IK)
Boraginaceae Symphytum carmosina Hort.Angl. ex Steud. -- Nomencl. Bot. [Steudel], ed. 2. 2: 654. 1841 (IK)
Boraginaceae Symphytum carpaticum Yu.M.Frolov --Bot. Zhurn.(Moscow & Leningrad)72(10): 1396. 1987 (IK)
Boraginaceae Symphytum caucasicum M.Bieb. -- Fl. Taur.-Caucas. 1: 128. 1808 (IK)
Boraginaceae Symphytum caucasicum D.Don -- in Sweet, Brit. Flow. Gard. Ser. II. t. 294. (IK)
Boraginaceae Symphytum circinale Runemark -- Bot. Not. 120: 90. 1967 (IK)
Boraginaceae Symphytum ciscaucasium Gvin. -- Zametki Sist. Geogr. Rast. 29: 55. 1972 (IK)
Boraginaceae Symphytum clusii C.C.Gmel. -- Fl. Bad. iv. 144. (IK)
Boraginaceae Symphytum coccineum hort. ex Schltdl. -- Bot. Zeitung (Berlin) 7: 731. 1849 (IK)
Boraginaceae Symphytum coeruleum Hort.Angl. ex DC.-Prodr.[A. P. de Candolle]10:40.1846 [8 Apr 1846] (IK)
Boraginaceae Symphytum coeruleum Petitm. ex Thell --Vierteljahrsschr. Naturf. Ges.Zurich 52: 459. 1907 (IK)
Boraginaceae Symphytum commune Faegri--BergensMus.Arbok1931,Naturvidensk.Rekke,No.4,30(1932).(IK)
Boraginaceae Symphytum consolida Gueldenst.-Reis.Russland(Gueldenst.)i.420; exLedeb.Fl.Ross. iii.114(IK)
Boraginaceae Symphytum cordatum M.Bieb. -- Fl. Taur.-Caucas. 1: 130. 1808 (IK)
Boraginaceae Symphytum cordatum Waldst. & Kit. -- Descr. Icon. Pl. Hung. 1: 6, t. 7. [1799-1802] (IK)
Boraginaceae Symphytum cordatum Willd. -- Neue Schriften Ges. Naturf. Freunde Berlin ii. (1799) 121. (IK)
Boraginaceae Symphytum cordifolium Baumg. -- Enum. Stirp. Transsilv. 1: 126. 1816 (IK)
Boraginaceae Symphytum creticum (Willd.) Runemark -- Boissiera xiii. 100 (1967). (IK)
Boraginaceae Symphytum creticum (Willd.) Runemark var. squamulatum (Pawl.) R.L.Jahn -- in R. Jahn & P. Schonfelder, Exkursionsfl. Kreta 28 (1995), without exact basionym page:. (IK)
Boraginaceae Symphytum cycladense Pawl. -- Fragm. Florist. Geobot. 17(1): 25. 1971 (IK)
Boraginaceae Symphytum davisii Wickens -- Notes Roy. Bot. Gard. Edinburgh 29: 168. 1969 (IK)
Boraginaceae Symphytum davisii Wickens subsp.cycladense(Pawl)Stearn-AnnMusGoulandris7:196(1986)(IK)
Boraginaceae Symphytum davisii Wickens subsp.icaricum (Pawl.)Stearn-Ann.Mus.Goulandris7:198(1986)(IK)
Boraginaceae Symphytum davisii Wickens subsp.naxicola (Pawl.)Stearn- Ann.Mus.Goulandris7:198(1986)(IK)
Boraginaceae Symphytum x densiflorum Buckn. -- J. Bot. 50: 334. 1912 (IK)
Boraginaceae Symphytum dichroanthum Teyber -- Verh. K.K. Zool.-Bot. Ges. Wien 56: 73. 1906 (IK)
Boraginaceae Symphytum x discolor Buckn. -- J. Bot. 50: 333. 1912 (IK)
Boraginaceae Symphytum donii DC. -- Prodr. [A. P. de Candolle] 10: 37. 1846 [8 Apr 1846] (IK)
Boraginaceae Symphytum echinatum Ledeb. -- Hort. Dorpat. Suppl. (1811) 5. (IK)
Boraginaceae Symphytum elatum Tausch -- Flora 19(2): 393. 1836 (IK)
Boraginaceae Symphytum euboicum Runemark exWickens-RoyBotGardEdinburgh29:167.1969(IK)
Boraginaceae Symphytum ferrariense Massal. -- Bull. Soc. Bot. Ital. 1913, 78. hybr. (IK)
Boraginaceae Symphytum filipendulum Bisch. -- Flora 9(2): 562. 1826 (IK)
Boraginaceae Symphytum floribundum Shuttlew. ex Nyman -- Consp. Fl. Eur. 3: 509. 1881 [Jul 1881] (IK)
Boraginaceae Symphytum foliosum Rehmann -- Verh. K.K. Zool.-Bot. Ges. Wien 18: 495. 1868 (IK)
Boraginaceae Symphytum fruticosum Sesse & Moc. -- Naturaleza (Mexico City) ser. 2, 1: 21. 1888 (GCI)
Boraginaceae Symphytum fruticosum Moc., Sesse & Moc.- Pl. Nov.Hisp. 21(Naturaleza Mexico City);1888(IK)
Boraginaceae Symphytum fruticosum Moc. ex DC. -- Prodr.[A. P. de Candolle] 10:121. 1846 [8 Apr 1846] (IK)
Boraginaceae Symphytum grandiflorum DC. -- Prodr. [A. P. de Candolle] 10: 40. 1846 [8 Apr 1846] (IK)

Boraginaceae Symphytum gussonei F.W.Schultz -- in Arch. Fl. (1874) p. lviii; et in Flora, lviii. (1875) 218. (IK)
Boraginaceae Symphytum hajastanum Gvin. --Zametki Sist. Geogr. Rast. 26: 73. 1967 ; Not. Syst.(Tbilisi) (IK)
Boraginaceae Symphytum x hidcotense P.D.Sell -- Fl. Gr. Brit. Ireland 3: 520 (361). 2009
Boraginaceae Symphytum hirsutum Raf. -- Med. Fl. 2: 95. 1830 (IK)
Boraginaceae Symphytum x hyerense Pawl. -- Fragm. Florist. Geobot. 17(1): 31. 1971 (IK)
Boraginaceae Symphytum ibericum Steven -- Bull. Soc. Imp. Naturalistes Moscou 24(1): 579 (-580). 1851 (IK)
Boraginaceae Symphytum icaricum Pawl. -- Fragm. Florist. Geobot. 17(1): 19. 1971 (IK)
Boraginaceae Symphytum insulare (Pawl.) Greuter & Burdet -- Willdenowia 11(1): 39. 1981 (IK)
Boraginaceae Symphytum intermedium Fisch.-Cat.Jard.Pl.Gorenki ed.2, 27; Roem.&Schult.Syst4: 661812(IK)
Boraginaceae Symphytum jacquinianum Tausch -- Flora 19(2): 392, in obs. 1836 (IK)
Boraginaceae Symphytum kurdicum Boiss. & Hausskn. ex Boiss. -- Pl. Or. Nov. Dec. ii. 5. (IK)
Boraginaceae Symphytum laeve Besser -- Cat. Hort. Crem. Suppl. (1812) (Quidl). (IK)
Boraginaceae Symphytum leonhardtianum Pugsley -- J. Bot. 69: 95. 1931 (IK)
Boraginaceae Symphytum x lilacinum Buckn. -- J. Bot. 50: 334. 1912 (IK)
Boraginaceae Symphytum longipetiolatum Wickens -- Notes Roy. Bot. Gard. Edinburgh 29: 179. 1969 (IK)
Boraginaceae Symphytum longisetum Hub.-Mor. & Wickens --Roy. Bot. Gard. Edinburgh 29: 174. 1969 (IK)
Boraginaceae Symphytum macrolepis J.Gay ex Rchb. -- Fl. Germ. Excurs. 347. (IK)
Boraginaceae Symphytum majus Gueldenst. ex Ledeb. -- Fl. Ross. (Ledeb.) 3(1,8): 115. 1847 [Oct 1847] (IK)
Boraginaceae Symphytum majus Bubani -- Fl. Pyren. (Bubani) 1: 501. 1897 [Oct-Nov 1897] (IK)
Boraginaceae Symphytum mediterraneum W.D.J.Koch -- Syn. Fl. Germ. Helv. 1(2): 500. 1837 (IK)
Boraginaceae Symphytum mediterraneum Guss. ex F.W.Schultz -- Flora 58: 218. 1875 (IK)
Boraginaceae Symphytum microcalyx Opiz -- Seznam 94. 1852 (IK)
Boraginaceae Symphytum minus Bubani -- Fl. Pyren. (Bubani) 1: 502. 1897 [Oct-Nov 1897] (IK)
Boraginaceae Symphytum minus W.Young -- Cat. Arbres, Arbustes & Pl. Herb. Am. 47 (1783). nomen. (IK)
Boraginaceae Symphytum molle Janka -- Termesz. Fuzetek i. (1877) 29. (IK)
Boraginaceae Symphytum x mosquense S.R.Majorov&D.D.Sokoloff-Novosti Sist.Vyssh.Rast.31:241(1998)(IK)
Boraginaceae Symphytum multicaule Teyber -- Verh. K.K. Zool.-Bot. Ges. Wien 56: 71. 1906 (IK)
Boraginaceae Symphytum naxicola Pawl. -- Fragm. Florist. Geobot. 17(1): 21. 1971 (IK)
Boraginaceae Symphytum nodosum Schur -- Enum. Pl. Transsilv. 468. 1866 [Apr-Jun 1866] (IK)
Boraginaceae Symphytum x norvicense Leaney & C.L.O'Reilly -Watsonia 27(4): 373 (372-374). [Aug 2009]
Boraginaceae Symphytum officinale L. -- Sp. Pl. 1: 136. 1753 [1 May 1753] (IK)
Boraginaceae Symphytum officinaleLsubsp.uliginosum(A.Kern.)Nyman-Consp.Fl.Eur.509.1881*S. uliginosum
Boraginaceae Symphytum orientale Pall. -- Tabl. Phys. Topogr. Taur. 47; Bieb. Fl. Taur. Cauc. i. 129. (IK)
Boraginaceae Symphytum orientale Pinard ex DC. -- Prodr. [A. P. de Candolle] 10: 40. 1846 [8 Apr 1846] (IK)
Boraginaceae Symphytum orientale L. -- Sp. Pl. 1: 136. 1753 [1 May 1753] (IK)
Boraginaceae Symphytum ottomanum Friv. -- Flora 19(2): 439. 1836 (IK)
Boraginaceae Symphytum palaestinum var. violaceum Feinbrun -- Israel J. Bot. 25(1-2): 79. 1976 (IK)
Boraginaceae Symphytum palestinae Boiss. -- Diagn. Pl. Orient. ser. 1, 11: 94. 1849 [Mar-Apr 1849] (IK)
Boraginaceae Symphytum pannonicum Pers. -- Syn. Pl. [Persoon] 1: 161. 1805 [1 Apr-15 Jun 1805] (IK)
Boraginaceae Symphytum patens Sibth. -- Fl. Oxon. 70. 1794 (IK)
Boraginaceae Symphytum patens Fr. -- Novit. Fl. Suec. Mant. 2: 13. 1839 (IK)
Boraginaceae Symphytum peregrinum Ledeb. -Index Seminum Horti Academici Dorpatensis.1820 (APNI)
Boraginaceae Symphytum peregrinum Ledeb.-- Ind. Sem.Hort.Dorpat.(1820) 4; et ex Spreng. Syst. i. 563. (IK)
Boraginaceae Symphytum x perringianum P.H.Oswald & P.D.Sell -- Fl. Gr. Brit. Ireland 3: 520 (362-363). 2009
Boraginaceae Symphytum pictum hort. -- Gard. Chron. (1884) I. 827. (IK)
Boraginaceae Symphytum podcumicum Yu.M.Frolov --Bot.Zhurn.(Moscow & Leningrad) 70(4): 533.1985 (IK)
Boraginaceae Symphytum x polonicum Blocki ex Buckn. -- J. Linn. Soc., Bot. 41: 552. 1913 (IK)
Boraginaceae Symphytum popovii Dobrocz. -- Ukrayins'k. Bot. Zhurn. 25(6): 60. 1968 (IK)
Boraginaceae Symphytum pseudobulbosum Azn. -- Bull. Herb. Boissier Ser. II. iii. 588. (IK)
Boraginaceae Symphytum punctatum Gaudin -- Fl. Helv. ii. 41. (IK)
Boraginaceae Symphytum racemosum Steph. ex Roem. & Schult. -- Syst. Veg., ed. 15 bis 4: 752. 1819 (IK)
Boraginaceae Symphytum x rakosiense (Soo) Penzes -- Bot. Közlem. 38: 149, in obs. 1941 (IK)
Boraginaceae Symphytum regium S.G.Gmel. -- Reise Russland (S.G.Gmel.) 3: 363, t. 36. f. 1. 1774 (IK)
Boraginaceae Symphytum savvalense Kurtto -- Ann. Bot. Fenn. 19(3): 189. 1982 (IK)
Boraginaceae Symphytum secundum S.G.Gmel. -- Reise Russland (S.G.Gmel.) 3: 363, t. 36. f. 1. 1774 (IK)
Boraginaceae Symphytum sepulcrale Boiss. & Balansa ex Boiss.-Fl. Orient. [Boissier] 4(1): 174. 1875 (IK)
Boraginaceae Symphytum sepulcrale Boiss&Balansa exBoiss.var.hordokopiiKurtto-Ann.Bot.Fenn.1982(IK)
Boraginaceae Symphytum sepulcrale Boiss.&Balansa exBoiss.subsp.sylvaticumKurtto- Ann.BotFenn1982 (IK)
Boraginaceae Symphytum sicyosmum P.Candargy -- Bull. Soc. Bot. France 44: 150. 1897 (IK)
Boraginaceae Symphytum stenophyllum Beck -- Annal. Naturh. Hofmus. Wien ii. (1887) 132. (IK)
Boraginaceae Symphytum sylvaticum Boiss. -- Pl. Or. Nov. Dec. ii. 4. (IK)
Boraginaceae Symphytum sylvaticum Boiss. var. hordokopii (Kurtto) R.R.Mill -- Fl. Turkey 10: 188 (1988):(IK)
Boraginaceae Symphytum sylvaticum Boiss.subsp.sepulcrale(Balansa)GreuterBurdet-Willdenowia1984(IK)
Boraginaceae Symphytum tanaicense Steven -- Bull. Soc. Imp. Naturalistes Moscou xxiv. (1851) I. 577. (IK)

Boraginaceae Symphytum tauricum Willd. -Neue Schriften Ges.Naturf.FreundeBerlin ii.(1799) 121.t 5. f 1 (IK)
Boraginaceae Symphytum tuberosum L. -- Sp. Pl. 1: 136. 1753 [1 May 1753] (IK)
Boraginaceae Symphytum uliginosum A.Kern. -- Oesterr. Bot. Z. 13: 227. 1863 (IK)
Boraginaceae Symphytum x ullepitschii Wettst. -- in A. Kern., Sched. Fl. Austro-Hung. vi. (1893) 37. (IK)
Boraginaceae Symphytum uplandicum Nyman -- Syll. Fl. Eur. 80. 1855 (IK)
Boraginaceae Symphytum uplandicum Nyman f.coeruleum(Petitm.exThell.)PD.SellFlGrBritIreland3:519 2009
Boraginaceae Symphytum uplandicum Nyman f. densiflorum (Buckn.) P.D.Sell-Fl.Gr.Brit. Ireland 3: 519. 2009
Boraginaceae Symphytum uplandicum Nyman f. discolor (Buckn.) P.D.Sell -- Fl. Gr. Brit. Ireland 3: 519. 2009
Boraginaceae Symphytum uplandicum Nyman f. lilacinum (Buckn.) P.D.Sell -- Fl. Gr. Brit. Ireland 3: 519. 2009
Boraginaceae Symphytum variegatum hort. -- Gard. Chron. n.s. 13: 779, cum cit. falsa. 1880 [1880] (IK)
Boraginaceae Symphytum vetteri Thell. -- Vierteljahrsschr. Naturf. Ges. Zurich lii. 460 (1907). (IK)
Boraginaceae Symphytum violaceum Gaterau -Descr. Pl.Montauban48(1789);videRaynal inTaxon,(1968). (IK)
Boraginaceae Symphytum wettsteinii Sennholz-Sitzungsber-BotGesWien;BotCentral(1888);BeckNied(1893)
Boraginaceae Symphytum x zahlbruckneri G.Beck. -- Fl. Nied. Oest. ii. II. (1893) 964. (IK)
Boraginaceae Symphytum zeyheri Schimp. -- Flora 12(2): 418. 1829 (IK) "
-'**International Plant Names Index**'®. **(IPNI),** www.ipni.org, A database of the names and associated basic bibliographical details of seed plants, ferns and lycophytes, **2018**.

('Hort.' or hortulanorum is used for a name that saw significant use in horticultural literature, usually 19th century and earlier, but was never properly published. This term is used so that non-wild, cultivated plants can be examined by taxonomists so they can be established as species, and published.)

"*Symphytum asperum* Lepech.
Symphytum asperum x caucasicum
Symphytum brachycalyx Boiss.
Symphytum bulbosum K.F. Schimper
Symphytum caucasicum M. Bieb.
Symphytum grandiflorum DC. = *Symphytum ibericum* Steven
Symphytum officinale L.
   Symphytum officinale subsp. *bohemicum* (F.W. Schmidt) Pers. =*Symphytum bohemicum* F.W. Schmidt
   Symphytum officinale subsp. *officinale* L.
   Symphytum officinale subsp. *uliginosum* (A. Kern.) Nyman
Symphytum officinale x asperum = *S. x uplandicum* Nyman = *Symphytum x peregrinum* auct., non Ledeb.
Symphytum officinale x asperum x grandiflorum = *S.* 'Hidcote Blue' hort. ex G. Thomas = *Symphytum x tauricum* auct., non Willd.
Symphytum officinale x asperum x tuberosum
Symphytum orientale L.
Symphytum tauricum Willd. = *Symphytum orientale* Pall. non L.
Symphytum tuberosum L.
   Symphytum tuberosum subsp. *nodosum* (Schur) Soo
   Symphytum tuberosum subsp. *tuberosum* L."
-'**BSBI List of British & Irish Vascular Plants and Stoneworts**'® is a database that includes accepted scientific names, good synonymy, vernacular names and Kent numbers, **2018**.
www.nhm.ac.uk/our-science/data/uk-species/checklists/NHMSYS0000436459/index.html

Symphytum Species at Herbariums (Dried)
Herbarium is a collection of dried plant specimens mounted on sheets of paper. Plants are collected, identified by experts, pressed, and mounted to archival paper so all major morphological characteristics are visible. They are labeled with the scientific name, name of the collector, and information about where/how collected. Specimens are filed according to families and genera.

"**Family Boraginaceae with Genus Symphytum:**
   Symphytum anatolicum
   Symphytum armeniacum
   Symphytum asperrimum
   Symphytum besseri
   Symphytum bornmuelleri
   Symphytum brachycalyx
   Symphytum bulbosum
   Symphytum bullatum
   Symphytum caucasicum
   Symphytum circinale
   Symphytum ciscaucasium
   Symphytum coeruleum
   Symphytum cordatum
   Symphytum cretica

Symphytum cycladense
Symphytum davisii
Symphytum floribundum
Symphytum fruticosum
Symphytum hajastanum
Symphytum icaricum
Symphytum indet.
Symphytum kurdicum
Symphytum longipetiolatum
Symphytum naxicola
Symphytum nodosum
Symphytum officinale
Symphytum orientale
Symphytum ottomanum
Symphytum palestinae
Symphytum palestinum
Symphytum popovii
Symphytum pseudobulbosum
Symphytum racemosum
Symphytum savvalense
Symphytum sepulcrale
Symphytum sylvaticum
Symphytum tanaicense
Symphytum tauricum
Symphytum tuberosum
Symphytum tuberosum l. x cordatum waldst. & kit.
Symphytum uliginosum
Symphytum ullepitschii
Symphytum x polonicum
Symphytum x ullepitschii
Symphytum x uplandicum
Symphytum zeyheri

Herbarium: *(Herbarium Name, Herbarium Code, Country, Number of Specimens)*
  Academy of Natural Sciences (PH), United States (2)
  Agriculture and Agri-Food Canada (DAO) (3)
  Biozentrum Klein Flottbek und Botanischer Garten der Universitat Hamburg (HBG), Germany (1)
  Botanic Garden and Botanical Museum Berlin-Dahlem, Freie Universitat Berlin (B), Germany (23)
  Botanical Museum, Finnish Museum of Natural History - University of Helsinki (H), Finland (1)
  Botanische Staatssammlung Munchen (M), Germany (12)
  Conservatoire et Jardin botaniques de la Ville de Geneve (G), Switzerland (12)
  Field Museum of Natural History, Chicago (F), Illinois (1)
  Herbarium Senckenbergianum Frankfurt am Main (FR), Germany (7)
  Herbier de l'Universite Montpellier II (MPU), France (3)
  Institute of Botany of the National Academy of Sciences of Armenia (ERE) (1)
  Institute of Botany, Slovak Academy of Sciences Herbarium (SAV), Slovakia (6)
  Linnean Society of London Herbarium (LINN), England (13)
  Lund University Botanical Museum (LD), Sweden (4)
  Martin-Luther-Universitat (HAL), Germany (2)
  Museum Botanicum Hauniense, University of Copenhagen, Denmark (C) (2)
  Museum National d'Histoire Naturelle (P), France (13)
  National Botanic Garden of Belgium (BR) (2)
  National Herbarium of Georgia (TBI), country Georgia (2)
  National Herbarium of Ukraine, M.G. Kholodny Institute of Botany (KW) (16)
  Natural History Museum (BM), London, England (9)
  Naturalis Biodiversity Centre, formerly Leiden University (L), Netherlands (2)
  Naturalis Biodiversity Centre, formerly Wageningen University (WAG), Netherlands (1)
  Royal Botanic Garden Edinburgh (E), Scotland (7)
  Royal Botanic Gardens, Kew (K), London, England (4)
  Royal Horticultural Society (WSY), Surrey, England (2)
  Swedish Museum of Natural History Department of Botany (S), Sweden (6)
  The Gray Herbarium (GH), Massachusetts (3)
  The William and Lynda Steere Herbarium of the New York Botanical Garden (NYBG) (1)
  United States National Herbarium, Smithsonian Institution (US), Washington DC (1) "

-'**Global Plants'® by JSTOR®, 2019**. 'Global Plants' is a community-contributed database with two million high resolution plant

specimen images and other foundational materials from the collections of more than 330 herbaria from more than 70 countries. The largest of its kind, it is an essential resource for institutions supporting research and teaching in botany, ecology, and conservation studies. Contributing herbaria are partners of the 'The Global Plants Initiative' (GPI). https://plants.jstor.org
JSTOR® is part of ITHAKA, a not-for-profit organization helping the academic community use digital technologies to preserve the scholarly record and to advance research and teaching in sustainable ways.

"The list below gives the names for all the taxa of the herbarium Symphytum specimens held in the RHS Herbarium at Wisley, Surrey, England. Synonyms are not listed:
Symphytum asperum
Symphytum 'Belsay Gold'
Symphytum bohemicum
Symphytum bulbosum
Symphytum caucasicum
Symphytum cretica
Symphytum 'Goldsmith'
Symphytum grandiflorum
Symphytum 'Hidcote Blue'
Symphytum 'Hidcote Pink'
Symphytum ibericum
Symphytum officinale
Symphytum orientale
Symphytum ottomanum
Symphytum tauricum
Symphytum tuberosum
Symphytum tuberosum x symphytum 'Hidcote Blue'
Symphytum x uplandicum
Symphytum x uplandicum 'Lugh Samhoildanach'
Symphytum x uplandicum 'Variegatum' "
-'**Index to the Taxa Present in RHS Herbarium (WSY= Wisley)**' by the Royal Horticultural Society, London, England, **2009**. Names are according to the RHS Horticultural Database: http://apps.rhs.org.uk/horticulturaldatabase
Search RHS specimen database: http://rbg-web2.rbge.org.uk/rhs/herbspec.php

"Herbarium Symphytum Specimens (France):
Symphytum abchasicum Trautv.
Symphytum anatolicum Boiss.
Symphytum asperrimum Donn ex Sims
Symphytum asperum Lepech.
Symphytum asperum x officinale
Symphytum bohemicum F.W.Schmidt
Symphytum brachycalyx Boiss.
Symphytum bulbosum K.F.Schimp.
Symphytum caucasicum M.Bieb.
Symphytum cordatum Waldst. & Kit ex Willd.
Symphytum discolor Buckn.
Symphytum donii DC.
Symphytum echinatum Ledeb.
Symphytum floribundum Shuttlew. ex Nyman
Symphytum foliosum Rehm.
Symphytum grandiflorum DC.
Symphytum gussonei F.Schultz
Symphytum ibericum Stev.
Symphytum kurdicum Boiss. & Hausskn. ex Boiss.
Symphytum mediterraneum W.D.J.Koch
Symphytum molle Janka
Symphytum officinale L. "
-'**National Museum of Natural History**' in Paris, France (Museum National d'Histoire Naturelle, MNHN), has over 9,500,000 specimens. This major botanical collection is from over 300 years of exploration and study. Sources are global, with strong representation for France, **2018**. Vascular plant database: https://science.mnhn.fr/institution/mnhn/collection/p/item/search/form

120 records with most of them Symphytum officinale (New York):
"Symphytum angustifolium A.Kern.
Symphytum asperrimum Donn ex Sims
Symphytum asperum Lepechin
Symphytum officinale L.

Symphytum tauricum var. ibericum Kuntze
Symphytum tuberosum L.
Symphytum uliginosum A.Kern.
Symphytum x uplandicum Nyman"
-New York Botanical Garden, Bronx, New York. This list is from the 'C. V. Starr Virtual Herbarium', the digitized specimens of the William and Lynda Steere Herbarium. As of **2018**, three million specimen records and two million images are available. Every month 30,000 new records and images are added.   http://sycamore.nybg.org, Search: http://sycamore.nybg.org/science/vh

2 records of Symphytum (Germany):
"Symphytum podcumicum: Collected by Ju. Frolov, Russian Federation, Caucasus prov. Stavropol, July 1977.
Symphytum ottomanum: Collected by I. Frivaldszky, Balkan. Ex herb. Th. Waage, Berlin, Germany."
-Herbarium Hamburgense, Germany, holds 1 million sheets of dried seed plants which have been collected worldwide during the past 200 years. A main focus is in tropical regions as well as in northern Germany, **2019**. www.herbariumhamburgense.de

The 'Herbarium of Vascular Plants' of the 'Komarov Botanical Institute' (LE), 'Russian Academy of Sciences', St. Petersburg, Russia, is one of the largest herbariums in the world. The Komarov Botanical Institue was founded in 1715. It contains more than 6,000,000 dried specimens from all over the world. It has a library of over 500,000 botanical volumes. In 1931 they published a 30-volume study on the flora of the Soviet Union. The 54-acre botanic garden has 6,700 species of plants. www.binran.ru   www.binran.ru/collections/index_e.html
(I was unable to find a list of Symphytum specimens for this herbarium. If you have one, please send it to me.)

*"**Symphytum Genus: Specimen Depositories and Number of Specimens:**
    Mined from GenBank, NCBI, United States [16]
    Finnish Museum of Natural History, Finland [9]
    Royal Ontario Museum, Green Plant Herbarium, Canada (TRT) [7]
    University of British Columbia, Herbarium, Canada [6]
    University of Guelph, Ontario Agricultural College Herbarium, Canada [5]
    National Museum Wales [4]
    Royal Botanic Garden, Edinburgh, Scotland [3]
    Naturalis Biodiversity Centre, Netherlands [3]
    3 Others [4] "*
-'Taxonomy Browser: Genus Symphytum' by Pensoft®: Taxon Profile, http://ptp.pensoft.eu, Pensoft® is an independent academic publishing company with cutting-edge publishing tools and workflows, https://pensoft.net, **2019**.

Symphytum Species at Botanic Gardens (Living)

*"**Denver Botanic Gardens** strives to entertain and delight while spreading the collective wisdom of the Gardens through outreach, collaboration and education. Our conservation programs play a major role in saving species and protecting natural habitats for future generations. **Living Symphytum plants include:**
    Symphytum asperum, Symphytum grandiflorum, Symphytum grandiflorum Goldsmith,
    Symphytum grandiflorum Hidcote variegated, Symphytum officinale, Symphytum x uplandicum,
    Symphytum x uplandicum Axminster Gold."*
-Denver Botanic Gardens, www.botanicgardens.org and http://navigate.botanicgardens.org/FindPlant.html, Colorado, **2019**.
It has North America's largest collection of plants from cold temperate climates around the world.

*"**Founded in 1859, the Missouri Botanical Garden is the nation's oldest botanical garden.** The Garden is a center for botanical research and science education. It's mission is driven by the need to protect and conserve plants and their ecosystems. **Living Symphytum plants include:**
    Symphytum asperum, Symphytum Goldsmith, Symphytum grandiflorum,
    Symphytum Hidcote Blue, Symphytum officinale, Symphytum x uplandicum,
    Symphytum x uplandicum Axminster Gold, Symphytum x uplandicum Variegatum."*
-Missouri Botanical Garden, www.missouribotanicalgarden.org/plantfinder/plantfindersearch.aspx, Saint Louis, Missouri, **2019**.

*"**The Chicago Botanic Garden** opened more than 40 years ago as a beautiful place to visit, and it has matured into one of the world's great living museums and conservation science centers. **Living Symphytum plants include:**
    Symphytum asperum, Symphytum caucasicum, Symphytum grandiflorum Hidcote Blue,
    Symphytum rubrum, Symphytum x uplandicum Axminster Gold."*
-Chicago Botanic Garden, www.chicagobotanic.org/plantcollections/plantfinder, Glencoe, Illinois, **2019**.

*"**Millennium Seed Bank: Seed List:***

Serial No.	Plant Name	Location	Year collected	Wild	Germination	Viability
135263	Symphytum officinale	United Kingdom	1999	Yes	100%	100%
622657	Symphytum ottomanum	Bulgaria	2010	Yes	4%	9%

The Royal Botanic Gardens Kew 'Millennium Seed Bank' (MSB) is a growing collection of seeds from around the world, aiming to provide a safety net for species at risk of extinction.
Collecting seeds and preserving them ex situ (away from their natural habitat) offers an economical and effective way to save seeds and keep them for posterity. The rate of environmental change is happening so fast that it is not always possible to conserve plants within threatened habitats. In the future, if required, they can be germinated and reintroduced to the wild or used in scientific research. Seeds are available for use in research, restoration and re-introduction."
'Royal Botanic Gardens Kew, www.kew.org/science/engage/accessing-our-science/access-our-collections and http://apps.kew.org/seedlist/SeedlistServlet, London, England, **2019**.

Symphytum Species at Gene Banks (Living)

Gene banks are a repository which preserve genetic material. For plants, this is 'in vitro' storage, freezing cuttings from the plant, or stocking seeds (seedbank). It is possible to unfreeze material and grow it. Accession is an individual sample in a gene bank.

"In 'ex situ' field Gene Banks there are about 2,200 accessions stored at the MTT research stations around Finland. The material has been grouped into ligneous ornamentals, perennial ornamentals, herbs/ spices, fruits/berries and vegetables. **Decisions on long-term storage responsibility** have so far been taken for 406 accessions of vegetables and fruits and berries. In addition, a number of herbs, spices and medicinal plants will soon be accepted for **long-term conservation.**
>They belong to the following genera (number of accessions in parentheses): Acorus (11), Arnica (15), Artemisia (2), Carum (1), Chenopodium (10), Hypericum (40), Hyssopus (1), Inula (1), Leonurus (12), Levisticum (1), Mentha (32), Myrrhis (1), Nicotiana (1), Rhodiola (18), Salvia (1), Solidago (39), **Symphytum (1)** and Tanacetum (21)."

-'Fourth National Report on the Implementation of the Convention on Biological Diversity in Finland' by Ministry of the Environment, Helsinki, Finland, **2009**, page 113. ('Ex situ' conservation is preservation of biological diversity outside natural habitats.)

"Seeds from the collections provide genetic traits essential for confronting a wide range of agricultural and environmental challenges and the changing needs of United States agriculture for food, fiber, bioenergy, industrial uses, medicine, shelter, sustainable agriculture, and land restoration. Our collections are dynamic and widely used by public and private plant breeders and researchers worldwide.
>To maximize the benefit of these collections, information about them is stored in a data management system called the '**Germplasm Resources Information Network**' (GRIN Global). GRIN Global keeps track of all data we collect on the sample including information on plant characteristics and sample origin: www.ars-grin.gov/npgs.

'United States National Plant Germplasm System' Symphytum Accessions (living):

Plant ID	Plant Name	Taxonomy	Origin	Material	Maintained By
PI 233329		**Symphytum peregrinum**	Canada, Manitoba	Plant	NC7
Ames 26915	RGE 386	**Symphytum asperum**	United States, Illinois	Plant	NC7"

-'National Laboratory for Genetic Resources Preservation: Seed Brochure', United States Department of Agriculture, Agriculture Research Service, www.ars.usda.gov, Plant and Animal Genetic Resources Preservation, Fort Collins, Colorado, August **2013**. Symphytum accession data is from 2019. (NC7 is North Central Regional Plant Station, USDA, ARS, Iowa State University, Regional Plant Introduction Station, Ames, Iowa.)

*"**Ex-situ conservation of plant genetic resources:** If genetic resources are not within but outside their natural habitat preserved (eg., in a scale for the propagation field gene bank and into a freezing warehouse), one speaks of 'ex situ' - maintenance. The conservation is - where possible - take the form of seeds.*
'**In-situ' / on farm conservation:** The in-situ conservation is the conservation among the natural living conditions.
Symphytum conserved in Austria 'National Inventory':
>Symphytum officinale (Common)   Accession Name/Number: BVAL-901607
>Symphytum (officinale)   Accession Name/Number: WIES-D3
>Symphytum x uplandicum (Russian)   Accesison Name/Number: WIES-D43
>Symphytum caucasicum (Blue)   Accession Name/Number: WIES-D44"

-'AGES®: Plant Genetic Resources for Food and Agriculture', Pflanzengenetische Ressourcen, Database and National Inventory, www.genbank.at, Linz, Austria, **2019**.

*"The '**European Search Catalogue for Plant Genetic Resources**' (EURISCO) provides information about more than 2 million accessions of crop plants and their wild relatives, preserved 'ex situ' by almost 400 institutes.* Based on a network of 'National Inventories' of 43 member countries and represents an important effort for preservation of world's agrobiological diversity by providing information about the large genetic diversity kept by the collaborating institutions.
Symphytum species include: asperum, brachycalyx, carpaticum, caucasicum, grandiflorum, officinale, ottomanum, tuberosum, uplandicum."
-EURISCO®, https://eurisco.ipk-gatersleben.de, 'Leibniz Institute of Plant Genetics and Crop Plant Research' (IPK), Gatersleben, Germany. Part of 'European Cooperative Programme for Plant Genetic Resources', **2019**.

*"**Svalbard Global Seed Vault:** Way up north, in the permafrost, 1300 kilometers (807 miles) beyond the Arctic Circle, is the world's largest secure seed storage, Svalbard Global Seed Vault,* opened by the Norwegian Government in February

2008. From all across the globe, crates of seeds are sent here for safe and secure long-term storage in cold and dry rock vaults.

Total Symphytum Seeds:

SGSV Genus ID	Genus	Vernacular Name	Accessions	Seeds	Taxa	Genebanks
1876089	Symphytum	Comfrey	18	7,075	5	3

Individual Symphytum Accessions in Genebank:

Institute code	Accession number	Full scientific name	Country of collection
DEU146	SYM 25	**Symphytum asperum Lepech.**	Georgia
DEU146	SYM 22	Symphytum asperum Lepech.	Georgia

Example: Symphytum asperum: 2 accessions, 30 seeds, 1 genebank.

Institute code	Accession number	Full scientific name	Country of collection
AUT001	BVAL-901607	**Symphytum officinale L.**	Austria
DEU146	SYM 26	Symphytum officinale L.	Sweden
DEU146	SYM 17	Symphytum officinale L. subsp. officinale	Germany
DEU146	SYM 17	Symphytum officinale L. subsp. officinale	Germany
DEU146	SYM 1	Symphytum officinale L. subsp. officinale	Unknown
DEU146	SYM 24	Symphytum officinale L. subsp. officinale	Unknown
DEU146	SYM 23	Symphytum officinale L. subsp. officinale	Austria
DEU146	SYM 4	**Symphytum x uplandicum Nyman**	Unknown
DEU146	SYM 9	Symphytum x uplandicum Nyman	Unknown
DEU146	SYM 19	Symphytum x uplandicum Nyman	Germany
DEU146	SYM 15	Symphytum x uplandicum Nyman	Germany
DEU146	SYM 12	Symphytum x uplandicum Nyman	Germany
DEU146	SYM 11	Symphytum x uplandicum Nyman	Germany
DEU146	SYM 16	Symphytum x uplandicum Nyman	Germany
DEU146	SYM 20	Symphytum x uplandicum Nyman	Germany
DEU146	SYM 14	Symphytum x uplandicum Nyman	Germany "

-Svalbard Global Seed Vault®, Nordic Genetic Resource Center (NordGen), www.nordgen.org/sgsv; and Royal Norwegian Ministry of Agriculture and Food, www.regjeringen.no, **2019**. It is a long-term seed storage facility, built to stand the test of time, and the challenge of natural or man-made disasters. The Seed Vault represents the world's largest collection of crop diversity.

"Scientific name	Country of origin	Biological status	Holding institute
Symphytum officinale L.	Romania	Traditional cultivar	ROM020
Symphytum asperum Lepech.		Other	USA020
Symphytum officinale	Yugoslavia	Other	USA005
Symphytum peregrinum Ledeb.		Traditional cultivar	USA020
Symphytum x uplandicum Nyman	Germany		DEU146
Symphytum x uplandicum Nyman			DEU146
Symphytum officinale L. subsp. officinale			DEU146
Symphytum officinale L.	Germany	Wild	DEU626
Symphytum officinale L.	Ukraine	Wild	UKR019
Symphytum officinale L.	Croatia	Natural	HRV041
Symphytum asperum Lepech.	Georgia	Natural	GBR004
Symphytum brachycalyx Boiss.	Jordan	Natural	GBR004
Symphytum officinale L.	Italy	Wild	ITA389
Symphytum spp.		Other	USA020
Symphytum x uplandicum Nyman	Germany		DEU146
Symphytum asperum Lepech.	Russia		DEU146
Symphytum sp.	Austria		DEU146 "

-Genesys®: The Global Gateway to Genetic Resources, **2019**. Genesys is an online platform where you can find information about 'Plant Genetic Resources for Food and Agriculture' (PGRFA) conserved in genebanks worldwide. There are 4,017,707 accessions in 2,190 datasets from 462 institutes. Organizations that work with this project include CGIAR International Genebanks, ECPGR EURISCO network, and USDA ARS NPGS. www.genesys-pgr.org

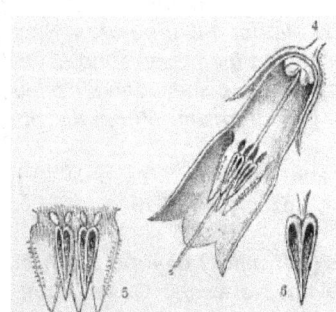

Symphytum Flower 1895

Figure 278: Sprinkling Apparatus
4. Longitudinal section through a flower of Symphytum officinale
5. Two stamens and three scales of same beset with prickles
6. Single stamen of Symphytum

'The Natural History of Plants, Their Forms, Growth, Reproduction, and Distribution', Volume II, by Anton Kerner von Marilaun, Austrian botanist. Published by Blackie & Son, London, England, 1895.

Chapter 4

Symphytum Species Classifications

Overview of Symphytum Species Classifications
For Botanical History of Symphytum (botanists), see subsection 'Symphytum Genus' in section 'Borage Family, Symphytum Genus' (Chapter 2).

"Divisions of the Symphytum genus have included those of De Candolle (1846), Boissier (1879), Kunze / Kuntze (1887), Kusnezow / Kuznetsov (1910) and Bucknall (1912-1913).
Symphytum officinale:
> **Bucknall** divided Symphytum officinale into the two forms alpha-ochroleum D.C. (white or yellowish corolla) and beta-purpureum Pers. (dull purple corolla); the leaves of both possess decurrent bases extending from one node to the next. In S. officinale var. lanceolatum Weinm. the leaves were only partially decurrent, the wings terminating between nodes; flowers were purple or violet.
> **Clapham, Tutin & Warburg (1962) and Bentham and Hooker (1954) list no varieties**, stating that flowers of S. officinale are yellowish white, sometimes purple or pink.

Symphytum x uplandicum:
> Those plants with partially decurrent petioles and blue or purple-blue flowers are described as 'Blue Comfrey', S. x uplandicum Nym. (Clapham, Tutin and Warburg, 1962)."

-'British Medicinal Species of the Genus Symphytum' by K.R. Fell and Janet M. Peck, Pharmacognosy Research Laboratories, University of Bradford, England; Planta Medica: Journal of Medicinal Plant and Natural Product Research, Volume 16, No. 2, pages 208-216, May **1968**.
(* -Handbook of British Flora: A Description of the Flowering Plants and Ferns Indigenous to the British Isles by George Bentham and revised by Joseph D. Hooker. Ashford, England: Reeve Ltd., 1954. A sixth revised edition was published in 1920.)

"Classification of the species of the genus Symphytum:
Linnaeus 1753: After Linnaeus distinguished three species in 1753, many new Symphytum species have been described in the eighteenth and nineteenth century.
Bucknall 1913: In the only revision of the entire genus, Bucknall (1913) recognized 25 species and some interspecific (between species) hybrids.
Symphytum in Geographical Areas: All other studies on Symphytum have been restricted to the species in a particular geographical area, e.g.: the Caucasus (Kuznetzow, 1910; Gviniashvili, 1976), Turkey and adjacent areas (Wickens 1969, 1978; Kurtto 1982, 1985), Europe -the Caucasus not included- (Pawlowski 1971, 1972), and Greece (Stearn, 1985).
Several authors have worked on the classification of the genus.
Subgeneric Classifications:
> **Boissier:** The first subgeneric (below genus, above species) classification was provided by Boissier (1879). He recognized two groups:
>> A. A group of 11 species sharing the characters fusiform roots and corolla scales not exserted from the corolla.
>> B. A group of 4 species, characterized by creeping (tuberous) roots and exserted corolla scales.
>
> This bifurcation in the subgeneric classification has been used by later authors as well.
> **Kuznetzow:** Kuznetzow (1910) recognized 21 species. He followed Boissiers classification and named the A. group: subgenus EuSymphytum (containing 19 species) and the B. group: Bulbosum (2 species).
> **Bucknall:** Bucknall (1913) also distinguished two groups in his revision of the genus: Ramosa with 18 species and Symplicia with 7 species.
>> The delimitation of the two subgenera was further specified by Bucknall (1913): the species in Ramosa have branched stems, fusiform roots and many-flowered inflorescences with white, blue and purple flowers and the species in Simplicia have unbranched stems, creeping roots and few-flowered inflorescences with light yellow flowers. Both groups were further subdivided into sections.
>
> Bucknall (1913) based the subdivision of the Ramosa on the colour of the corolla, the size of the plants, the decurrency of the leaves and the shape and surface characters of the nutlets and the subdivision of the second subgenus Simplicia on root characteristics and leaf shape.

Sections and Series:
> **Pawlowski** (1961) studied 16 species and paid special and meticulous attention to the corolla scales, their shape and marginal papillae, to the filaments and the root type. Instead of subgenera he recognized five sections, three of which were further split into series.
> Later Pawlowski (1971) added a sixth section, Graeca, to classify two recently described species from Greece.
> **Wickens** (1969) recognized 27 species and followed Bucknall's classification system. He created a new section to accommodate Procopiana and classified some recently described species.

Wickens, Pawlowski, Gviniashvili, Kurtto, and Stearn Classification of Symphytum Genus:

Subgenera	Sections	Species
1. Symphytum Wick (= Ramosa Buckn.)	a. Symphytum Buckn.	1. officinale
		2. tanaicense
	b. Coerulea Buckn.	1. armeniacum
		2. asperum
		3. hajastanum
		4. peregrinum
		5. savvalense
		6. sepulcrale
	c. Albida Buckn.	1. anatolicum
		2. cycladense
		3. davisii
		4. euboicum
		5. ottomanum
		6. sylvaticum
		7. tauricum
	d. Graeca Pawl.	1. icaricum
		2. naxicola
	e. Procopiania Wick.	1. circinale
		2. creticum
		3. insulare
	f. Orientalia Buckn.	1. caucasicum
		2. kurdicum
		3. orientale
		4. pseudobulbosum
	g. Suborientalia Buckn.	1. aintabicum
		2. bornmuelleri
		3. brachycalyx
		4. longisetum
		5. palaestinum
2. Simplicia Buckn.	a. Tuberosa Buckn.	1. bulbosum
		2. gussonei
		3. meditteraneum
		4. nodosum
		5. tuberosum
		6. zeyheri
	b. Cordata Buckn.	1. abchasicum
		2. ciscaucasicum
		3. cordatum
		4. grandiflorum
		5. ibericum
		6. longipetiolatum"

-'**Phylogenetic Relationships in the Genus Symphytum L. (Boraginaceae)' by J.M. Sandbrink**, J. Van Brederode and T.W.J. Gadella, Department of Genome Evolution, University of Utrecht, Netherlands; Proceedings: Koninklijke Nederlandse Akademie van Wetenschappen (Royal Netherlands Academy of Sciences), Series C, Volume 93, No. 3, pages 295-334, **1990**.

Symphytum Species Classifications 1596 by Bauhin

"**Symphytum:** *Dioscorides lib.4. cap. 9 & 10. Latinis and vulgo Consolida maior.*
    I. *Symphytum seu consolida maior.*
        Consolida maior mas and foemina Brunsel.
        Symphytum alum seu alus Lobel.
    II. *Symphytum maius tuberosa radice.*
        Symphytum tuberosum Lobel.
    III. *Consolida maior amplexicaulis fore purpureo Thalyo.*
    IV. *Symphytum maculosum seu Pulmonaria latifolia.*
    V. *Symphytum seu Pulmonaria angustifolia.*
    VI. *Symphytum seu Pulmonaria angustifolia altera.*
        An Consolida maior sylu.
        Symphytum syluestre Cordo histor.
    VII. *Symphytum Borraginis facie.*
        Symphytum pumilum repens Boraginis facie seu Borrago minima Herbariorum Lobel. "

-'**Phytopinax**: Seu Enumeratio Plantarum' (The Enumeration of Plants) by **Caspar (Gaspard) Bauhin**, Basel, Switzerland, **1596**, 'Liber VII., Sectio. II.', pages 499-501. All in Latin.
>   Gaspard Bauhin or Caspar Bauhin, 1560-1624, was a Swiss botanist and medical doctor. His 'Phytopinax' described thousands of plants and classified them in a way similar to the later binomial nomenclature of Carl Linnaeus. He was Professor of Greek, and Chair of Anatomy / Botany at the University of Basel, Switzerland. He also wrote 'Prodromus Theatri Botanici' in 1620 and 'Pinax Theatri Botanici' in 1623. The Symphytum listing is similar to this one.

See sub-subsection 'Bauhin 1623' in subsection 'Symphytum Genus' in section 'Borage Family, Symphytum Genus' (Chapter 2).

Symphytum Species Classifications 1623 by Bauhin

"**Symphytum:** *Dioscorides lib.4. cap. 9 & 10.*
- I. *Symphytum Consolida major.*
- II. *Symphytum majus tuberosa radice.*
- III. *Symphytum minus tubersoa radice.*
- IV. *Consolida major amplexicaulis flore purpureo.*
- V. *Symphytum minus Borraginis facie.*
- VI. *Symphyto viribus affinis.* "

-'**Pinax Theatri Botanici**' (Illustrated Exposition of Plants) by **Caspar (Gaspard) Bauhin**, Basel, Switzerland, **1623**, page 259.
>   'Pinax Theatri Botanici' has names, synonyms and references for 6,000 species. This book was a botanical landmark that was important for 100 years. When Carl Linnaeus cites 'Bauh. pin.', he is referring to 'Pinax Theatri Botanici'.

See sub-subsection 'Bauhin 1623' in subsection 'Symphytum Genus' in section 'Borage Family, Symphytum Genus' (Chapter 2).

Symphytum Species Classifications 1669 by Ray

"*The first work of *John Ray's (which contributed materially to the development of systematic botany) was his contribution of the 'Tables of Plants' to Dr John Wilkins's 'Real Character and a Philosophical Language', published in **1669**.*
The following is a summary of Ray's first attempt at a system of classification:
>   He begins by distinguishing Herbs, Shrubs, and Trees.
>   **Proceeding to the detailed classification of Herbs,** he divides them into Imperfect 'which either do want or seem to want some of the more essential parts of Plants, viz. (namely) either Root, Stalk, or Seed,' the Cryptogamia of Linnaeus; and Perfect 'having all the essential parts belonging to a Plant.'
>   **The Perfect Herbs** are arranged in three main groups according to (1) their leaves, (2) their flowers, (3) their seed-vessel, each group being subdivided in various ways.
>   **Herbs Considered According to their Leaves:**
>   With Long Leaves:
>   >   Frumentaceous, 'such whose seed is used by men for food, either Bread, Pudding, Broth, or Drink' (Cereals).
>   >   Non-Frumentaceous (other Grasses, Sedges, Reeds).
>   >   Gramineous Herbs of Bulbous Roots (Bulbous Monocotyledons).
>   >   Herbs of Affinity to Bulbous Roots (other Monocotyledons).
>   Herbs of Round Leaves (e.g. {for example} Petasites, Viola, Pinguicula, Drosera).
>   Herbs of Nervous Leaves (e.g. Veratrum, Plantago, Gentiana, Polygonum).
>   Succulent Herbs (Sedum, Saxifraga).
>   **Herbs considered according to Superficies (outward appearance) of their Leaves, or their Manner of Growing:**
>   >   More rough (e.g. Borago, Anchusa, Echium):
>   >   **Less rough (e.g. Pulmonaria, Symphytum, Heliotropium):**
>   >   Stellate leaves (e.g. Asparagus, Galium)."

-'Makers of British Botany: A Collection of Biographies by Living Botanists' edited by F.W. Oliver, Cambridge University, Press, England, 1913, page 28.
(* -'Tables Concerning Plants' or 'Tables of Plants' written by John Ray for inclusion in 'An Essay Towards a Real Character and a Philosophical Language' by Bishop Wilkins, 1669.)
See sub-subsection 'Ray 1686' in subsection 'Symphytum Genus' in section 'Borage Family, Symphytum Genus' (Chapter 2).

Symphytum Species Classifications 1846 by De Candolle

"**Symphytum.** *Tourn. inst. 138. t. 56. Linn. gen. n. 185. Lam. ill. t. 93. Goertn. fruct. 1. p. 325. t. 67. Lehm. asp. p. 343. Endl. gen. n. 3776. Spenn. in ic. fl. germ. fasc. 17.*" De Candolle lists these Symphytum species:
>   "**S. officinale** (Linn. sp. 195) (alpha ochroleucum, beta purpereum, gamma lanceolatum)
>   **S. Donii**
>   **S. peregriunum** (Ledeb. ex Spreng, syst. 1. p. 563)
>   **S. caucasicum** (Bieb. fl. taur. n. 326)
>   **S. asperrimum** (Sims bot. mag. t. 929)

  S. tuberosum (Linn. sp. 195)
  S. bulbosum (Schimp. in Flora 8. p. 1. p. 17. Koch syn. 500)
  S. ottomanum (Friv. in Flora 1836. p. 439)
  S. anatolicum (Boiss. diagn. 4. p. 43)
  S. tauricum (Willd. nov. act. nat. cur. berol. 5. p. 120. t. 6. f. 1. ex enum. h. ber. 183)
  S. orientale (Linn. sp. 195)
  S. brachycalyx (Boiss. diagn. 4. p.43)
  S. grandiflorum
  S. cordatum (Willd. act. soc. berol. 2. p. 121)
  S. echinatum (Ledeb. h. Dorp. suppl. 1811. p. 5)
  S. racemosum (Steph. in Willd. herb. ex Roem. et Seh. syst. 4. p. 752)
  S. mediterraneum (Koch syn. fl. germ. ed. 2. p. 575). "

-'**Prodromus Systematis Naturalis Regni Vegetabilis**, sive, Enumeratio contracta ordinum generum specierumque plantarum huc usque cognitarium, juxta methodi naturalis, normas digesta, Pars X' (An essay of the system of the natural of the Kingdom of Vegetable, or, An enumeration of the genera and species plant has thus far been contracted cognitarium of the orders, according to the method of the natural, to the norms of being digested, **Volume 10**) by **De Candolle, 1846**.
Standard botanical abbreviation: **Prodr. (DC.)**. Written in Latin. **Symphytum pages 36-40 and 587 appendix.**

  Prodromus Systematis Naturalis Regni Vegetabilis is 17 volumes written from 1824 to 1873. Augustin Pyramus De Candolle wrote it as a summary of all known seed plants, including taxonomy, ecology, evolution and biogeography. He authored the first 7 volumes. His son, Alphonse Pyramus De Candolle, wrote 10 more volumes, with help from others. See sub-subsection 'De Candolle 1846' in subsection 'Symphytum Genus' in section 'Borage Family, Symphytum Genus' (Chapter 2).

  *"**Classification of the Species: In De Candolle's 'Prodromus,' the first work in which all the known species of Symphytum were brought together, no attempt was made to divide the genus into groups or to classify the species.** The first author to do this was Boissier in the 'Flora Orientalis'."*

-'A Revision of the Genus Symphytum, Tourn.' by Cedric Bucknall, (Mus. Bac. Oxon= Bachelor of Music, Oxford University); Journal of the Linnean Society of London, England, Botanical Journal, Volume 41, Issue 284, pages 491-556, December 1913.

Symphytum Species Classifications 1847 by Ledebour

"605. **Symphytum. L.** *Dec. Prodr. X, p. 36.* *Endlicher gen. p. 649.* *Meisner gen. p. 280.*
  1. **S. officinale L.** (L. Sp. 195)
    S. officinale
    S. Consolida
    S. foliis ovato-lanceolatis
    Consolida major
    alpha ochroleucum
      S. officinale flor ochroleucis
      S. officinale flor albis
      S. officinale
      Consolida, Symphytum album.
    beta purpureum
      S. patens
    gamma lanceolutum
  2. **S. peregrinum** (Ledeb. Ind. sem. h. dorpat. p. a. 1820, p. 4)
  3. **S. caucasicum** (M. a Bieb. Fl. t. c. I, p. 128; III, p. 128)
  4. **S. asperrimum** (Sims bot. Mag. t. 929)
    S. asperrimum
    S. asperum
    S. majus
    S. echinatum
  5. **S. tuberosum** (L. Sp. 195)
  6. **S. tauricum** (Willd. in nov. Act. Soc. nat. scrut. berol. III, p. 120, t. 6, f. 1)
    S. tauricum
    S. orientale (non L.)
    S. bullatum
    S. orientale L.
  7. **S. cordatum** (Willd. in nov. Act. Soc. nat. scrut. berol. II, p. 121)
    S. cordatum
    S. pannonicum
    S. cordifolium "

-'**Flora Rossica**; sive, Enumeratio plantarum in totius Imperii Rossici provinciis Europaeis, Asiaticis et Americanis hucusque observatarum' (Complete Flora of the Russian Empire), Volume 3, by **Carl {Karl} Friedrich von Ledebour, 1847**-1849. Symphytum pages 113-116. All in Latin.

"Among the many German naturalists who labored under the Tzars and Tzarinas of Russia to make known the vegetation and fauna of that vast realm, Carl Friedrich von Ledebour holds an honored place for being the first man to complete a comprehensive flora, one with detailed descriptions and synonymy, covering northern Asia, the Caucasus, and Alaska, as well as European Russia."

-'Ledebour's Flora Rossica, Icones Plantarum Novarum, and Flora Altaica' review by William T. Stearn, Journal of the Arnold Arboretum, Harvard University, Massachusetts, Volume XXII, **1941**. Article on pages 225 to 230.

See sub-subsection 'Ledebour 1847' in subsection 'Symphytum Genus' in section 'Borage Family, Symphytum Genus' (Chapter 2).

Symphytum Species Classifications 1879 by Boissier

*"**Symphytum** (Tourn. Inst. 138, tab. 56)*
  *1. **S. officinale** (L Sp. 195) (Ic. Engl. bot. tab. 817) (Rchb. Ic. tab 102)*
  *2. **S. orientale** (L Sp. 195)*
  *3. **S. caucasicum** (M.B. Taur. Cauc. I, p. 128)   **S. racemosum***
  *4. **S. tauricum** (Willd. Nov. Act. Berol. III, p. 120, tab. 6)*
  *5. **S. sylvaticum** (Boiss. in Bourg. exs. Armen. 1862, Dec. 2, pg. 4)*
  *6. **S. anatolicum** (Boiss. Diagn. Ser. I, 4, p. 43)*
  *7. **S. brachycalyx** (Boiss. Diagn. Ser. I, 4, p. 43)*
  *8. **S. palaestinum** (Boiss. Diagn. Ser. I, II, p. 94)*
  *9. **S. kurdicum** (Boiss. et Haussk. Dec. II, p. 5)*
  *10. **S. sepulcrale** (Boiss. et Bal. in Bal. pl. Pont. 1866)*
  *11. **S. asperrimum** (Sims Bot. Mag. tab. 929)*
  *12. **S. tuberosum** (L. Sp. 195)*
  *13. **S. grandiflorum** (D.C. Prodr. X, p. 40)*
  *14. **S. bulbosum** (Schimp. Bot. Zeit. VIII, I, p. 17)*
  *15. **S. ottomanum** (Friv. Fl. 1836, p. 439)"*

-'**Flora Orientalis** Sive Enumeratio Plantarum in Oriente, a Graecia et Aegypto ad Indiae Fines (Plants Flora, Flora East and the East, from Greece and Egypt to the Borders of India), Volume IV' by **Edmond Boissier**, Society Physics Geneva, Society Linnean London; Geneva, Switzerland, **1879**. Symphytum pages 171-177. All in Latin, 5 volumes plus supplement (1867-1888). The volumes cover these countries and areas as they were then called: Greece, Macedonia and Thrace, Asia Minor and Armenia, Egypt and Arabia, Palestine, Syria and Mesopotamia, Crimea and Caucasus, Persia, Turkestan, Afghanistan and Baluchistan. It is a masterful piece of descriptive botany.

See sub-subsection 'Boissier 1879' in subsection 'Symphytum Genus' in section 'Borage Family, Symphytum Genus' (Chapter 2).

*"**The first author to do this (divide the Symphytum genus into groups or to classify the species) was Boissier,** whose arrangement in the 'Flora Orientalis' is given below:*
I. Corolla-scales included:
  1. Root fusiform or branched: S. officinale Linn., orientale Linn., caucasicum Bieb., tauricum Willd., sylvaticum Boiss., anatolicum Boiss., brachycalyx Boiss., palaestinum Boiss., kurdicum Boiss. & Hausskn., asperrimum Bieb., sepulcrale Boiss. & Bal.
  2. Root tuberous: S. tuberosum, Linn., grandiflorum, DC.
II. Corolla-scales exserted: S. bulbosum Schimp., ottomanum Friv."

-'A Revision of the Genus Symphytum, Tourn.' by Cedric Bucknall, (Mus. Bac. Oxon= Bachelor of Music, Oxford University), Journal of the Linnean Society of London, England, Botanical Journal, Volume 41, Issue 284, pages 491-556, December 1913.

  (Fusiform means tapering at both ends, in other words: spindle shaped.)
  (Exserted means protruding or projecting beyond the corolla of a flower.)

Symphytum Species Classifications 1881 by Nyman

*"**Symphytum L.:***
  *1. **S. officinale L.** -Exs. Fr. V. 3*
    S. uliginosum Kern. Suppl. 17. Hung.(r.).
    S. patens Sibith.
    S. bohemicum Schm.
    S. microcalyx Op.
  *2. **S. mediterraneum K.** -Exs. Sz. hb. norm. nov. ser. 106. Gall.mer. (Hieres, Toulon).*
    S. floribundum Shuttew. exs.
  *3. **S. molle Jka** in Termesz Fuzetek. I. (1877). -Hungar.*
  *4. **S. orientale L.** -Byzant. (ad rivulos primo vere : L.).*

      5. *S. tauricum W.* -Ross.mer. Dobr.
      6. *S. cordatum W.K. ap. W. (1799)*
          S. pannonicum P.
          S. cordifolium Bmg.
      7. *S. anatolicum Boiss.*
          S. asperrimum d'Urv -Arch. (ins. Cos.)
      8. *S. tuberosum L.* -Exs. Rchb. 26. Bill. 2713. Sz X. 917.
          S. angustifolium Kern. (1863) Hung.
      9. *S. gussonei* F.Sz arch. 1874 et in Flora 1875. 218.
          S. mediterraneum Guss. (non K.); Tod. sic. exs. 1393.
    10. *S. bulbosum schmp.* (1825).
          S. macrolepis Gay in Rchb. exc. 347 iet exs. 851
          S. clusii Gm. bad. (1826).
          S. punctatum Gaud. (1828).
          S. zeyheri Schmp. (1829). Syll. 81.
          S. brochum Bory. Chaub.
    11. *S. ottomanum Friv.*
          S. bullatum Guebh. pl. mold. 401."

-'**Conspectus Florae Europaeae:** seu Enumeratio Methodica Plantarum Phanerogamarum Europae Indigenarum, Indicatio Distributionis Geographicae Singularum etc.' by **Carl Fredrik Nyman**, Orebro Sueciae {Sweden}, 1878-1882. Symphytum was written about in **1881**, pages 509-510. All in Latin.
See sub-subsection 'Nyman 1884' in subsection 'Symphytum Genus' in section 'Borage Family, Symphytum Genus' (Chapter 2).

Symphytum Species Classifications 1887 by Kuntze / Kunze

"***Symphytum officinale L.*** *delta caucasicum O. Ktze. f. coeruleum O. Ktze. Kutais.*
S. tauricum W. *beta ibericum O. Ktze. Batum.*
In Symphytum, the species rearranged to date are uncertain, as they are found in A.DC. and Boisser.
    On heterostyly and the 'Kelchmerkmalen' recognized by several authors (the calyxes are soon farther, then with straight corners, now more narrowly with + - bent back tips, the lobes soon broader now narrower, soon longer soon shorter) and partly on the rather based on variable hair.
    Hairiness is constant only in some local races, and when it decreases it disappears first on the surface of the leaf; it is incorrect to distinguish only two kinds of hair, because medium-sized ones are not rare.
In addition, S. tauricum (and probably the S. silvaticum Boiss, which hardly remains of S. tuberosum, according to the description) of Boisser (who follows Nyman) is erroneously placed in the group: *Radix fusiformis vel ramosa, whereas it is the Tuberose group heard.*
I only differentiate between the following 5 types:
    1. S. bulbosum Schimper.
    2. S. tuberosum L.
    3. S. tauricum W.
    4. S. officinale L.
    5. S orientale L."

'**Planta Orientali-Rossicae, Volume 10**' by **Dr. Otto Kuntze (Kunze), 1887**. Published in Saint Petersburg, Russia. Written in German. (This was translated online from German to English. If you have a better translation, please send it to me.)
    (Heterostyly means having styles of different lengths relative to the stamens in the flowers of different individual plants, to reduce self-fertilization.) (I could not find a definition of 'Kelchmerkmalen'. An online translator: 'chalice features'.)
    German: "Symphytum officinale L. delta caucasicum O. Ktze. f. coeruleum O. Ktze. Kutais.
S. tauricum W. beta ibericum O. Ktze. Batum. Bei Symphytum sind die bisher aufgesteliten Arten unsicher, da sie bei A. DC. und Boisser z. Th. auf Heterostylie und den von mehreren Autoren als veranderlich erkannten Kelchmerkmalen (die Kelche sind bald weiter, dann mit graden Zipfeln, bald enger mit +- zuruckgebogenen Spitzen, die Zipfel bald breiter bald schmaler, bald langer bald kurzer) und theils auf der ziemlich veranderlichen Behaarung beruhen. Die Behaarung ist nur bei manchen Localrassen constant und wenn sie abnimmt, verschwindet sie auf der Blattoberflache zuerst; es ist unrichtig nur zweierlei Haare zu unterscheiden, da mittelgrosse nicht selten sind. Ausserdem wird S. tauricum (und wahrscheinlich auch das mir fremd gebliebene, der Beschreibung nach von S. tuberosum kaum abweichende S. silvaticum Boiss.) von Boisser (dem Nyman folgt) irrig in die Gruppe: Radix fusiformis vel ramosa gestellt, wahrend es zur tuberosen Gruppe gehort. Ich unterscheide nur folgende 5 Arten:"
Carl Ernst Otto Kuntze (1843-1907) was a German botanist. His revolutionary ideas about botanic nomenclature created conflicts between competing sets of 'Rules of Botanical Nomenclature', the precursors to the current 'International Code of Nomenclature for algae, fungi, and plants'. **The standard botanical author abbreviation is 'Kuntze'.**

    "***The next attempt to group the species was made by Otto Kunze*** *(Pl. orient.-rossic. in Act. Hort. Petropol. x. 1887, page 219).* ***He reduces the number of species to five, and apparently regards all the others as synonymous with these.*** *His 'species' are:*

1. *S. bulbosum* including *S. ottomanum*.
2. *S. tuberosum* including *S. sylvaticum*.
3. *S. tauricum* including *S. grandiflorum*
4. *S. officinale* including *S. uliginosum, caucasicum,* and *mediterraneum*.
5. *S. orientale* including *S. peregrinum, asperum, palaestinum, kurdicum, anatolicum, brachycalyx, sepucrale,* and *grandiflorum* hort. nec DC.

I altogether fail to see how any characters could justify such an incongruous (not in harmony) grouping of the species as is here displayed."
-'A Revision of the Genus Symphytum, Tourn.' by Cedric Bucknall, (Mus. Bac. Oxon= Bachelor of Music, Oxford University), Journal of the Linnean Society of London, England, Botanical Journal, Volume 41, Issue 284, pages 491-556, December 1913.

Symphytum Species Classifications 1910 by Kuznetsov / Kusnetsov / Kusnezow

"Traditionally, the Symphytum genus is divided into 2 to 9 sections, based on various infrageneric classifications (Boissier 1879, *Kuznetsov 1910, Bucknall 1913, Pawlowski 1961, Wickens 1969, Sandbrink et al. 1990)."
-'Symphytum Tuberosum Complex in Central Europe: Cytogeography, Morphology, Ecology and Taxonomy' by L. Kobrlova, M. Hrones, P. Koutecky, M. Stech, and B. Travnicek, Preslia: The Journal of the Czech Botanical Society, Czech Republic; Volume 88, pages 77-112, March 2016.
(* -'Kavkazkie Vidy Roda Symphytum (Tourn.) L.' **(Caucasian Species of the Genus Symphytum Tourn. L.) by N.I. Kuznetsov (Kusnetsov or Kusnezow)**, Memoires de l'Academie Imperiale des Sciences de Saint-Petersbourg, Serie. 8, Physiques et Mathematiques, Volume 25, No. 5, pages 1-94, **1910**. If you have an English translation of this, I would appreciate a copy.)
(Infrageneric is subclassification within a genus of organisms.)

"Kusnezow in the 'Caucasian Species of the Genus Symphytum' (1910) retains Boissier's main divisions, but carries the subdivision of these a step further. The following is an outline of his system:
I. Eusymphytum
    i. Cyanea.
        A. *Symphytum officinale, uliginosum, mediterraneum, molle, caucasicum.*
        B. *Symphytum orientale, brachycalyx, palaestinum, kurdicum, anatolicum,*
            sepulcrale, asperum, peregrinum.
    ii. Ochroleuca.
        Symphytum bulbosum, ottomanum.
The weak point in all these arrangements is that the main sections are founded on a comparatively unimportant character, viz. (that is), the relative length of the corolla-scales and the corolla. *From this it results that S. bulbosum and S. ottomanum, in which the scales are longer than the corolla, are associated together, and each is separated from the species to which it is naturally allied.*
To the characters of the groups adopted by Kusnezow the same objection applies, with the further disadvantage that they are not sufficiently definite and admit of numerous exceptions, *especially with regard to the branching of the stem and the colour of the flowers."*
-'A Revision of the Genus Symphytum, Tourn.' by Cedric Bucknall, (Mus. Bac. Oxon= Bachelor of Music, Oxford University), Journal of the Linnean Society of London, England, Botanical Journal, Volume 41, Issue 284, pages 491-556, December 1913.

*"Kusnetsov (1910) and Bucknall (1913) regarded **Symphytum asperum and Symphytum sepulcrale as distinct species**, mainly on account of differences in the calyx, indumentum and leaf characters.*
Asperum:
    *As a native plant **S. asperum var. asperum** grows in the Caucasus region **(Kusnetsov 1910)**. In Iran its distribution seems to be poorly known. Riedl (*1967) also cited specimens from the Talysh region, but all these very probably belong to the **closely allied species S. peregrinum Ledeb. (Kusnetsov 1910)**.*
*In the papers mentioned above **S. sylvaticum** has been placed near **S. anatolicum Boiss.**, an endemite of southwest Turkey (Pawlowski 1971), **S. tauricum Willd.**, a circumeuxine relic species (**Gviniasvili 1976) and **S. ottomanum Friv., a species of the east Balkans (see Kusnetsov 1910).***
All these are characterized by their slenderness and abundant branching.
S. ottomanum and S. tauricum always have yellowish white corollas; S.anatolicum has either white or violet corollas."
-'Taxonomy of the Symphytum Asperum Aggregate (Boraginaceae), Especially in Turkey' by Arto Kurtto, Botanical Museum, University of Helsinki, Finland; Annales Botanici Fennici (Finnish Botanical Annals), Helsinki, Volume 19, No. 3, pages 177-192, 1982.
(* -'Flora Iranica: Boraginaceae, Issue 48' by Harald Riedl. Edited by K.H. Rechinger. Graz, Austria: Naturhistorisches Museum (Natural History Museum), Akademische Druck und Verlagsanstalt, 1967.)
(** -'Kavkazkie Predstaviteli roda Symphytum L' {Caucasian Representatives of the Symphytum Clan} by C.N. Gviniasvili, 146 pages, Tbilisi, country of Georgia, 1976. I do not have this. If you have an English translation, please let me know.)
(Talysh Mountains are in southeastern Azerbaijan and northwestern Iran.)
(Endemism means a species is unique to a defined geographic, ecological location.)

(Species aggregate is a species group that is closely related and difficult to distinguish among them. Similar but not identical is a single species that displays large variations but can not strictly be subdivided: s.l. for 'sensu lato' {in a wider/broader sense} or s. ampl. for 'sensu amplo' {in a very wide/relaxed sense}.)

"**There are four notable 20th-century contributions to understanidng of the genus Symphytum in the Balkan Peninsula and Asia Minor (Kuznezow, Bucknall, Wickens, Pawlowski).**
The first of these is a 90-page revision of the Caucasian species by Nicolai Ivanovitch Kuznezow (1864-1932), entirely in Russian apart from Latin descriptions, published in 'Memoires de l'Academie Imperiale des Sciences de St. Petersbourg', Ser. VIII, Volume 25, No. 5, 1910.
Kuznezow dealt in great detail with the Caucsian species S. asperum, S. peregrinum, S. caucasicum and S. grandiflorum but also provided a classification of the genus Symphytum as a whole.
This he divided into two main groups, section Eusymphytum further divided into Cyanea (including S. officinale, S. orientale, S. anatolicum and others), and section Ochroleuca (including S. tuberosum and others), and section Bulbosum consisting of S. bulbosum and S. ottomanum.
For taxonomic purposes this revision has been superseded (replaced) by Ts. N. Gvinashvili's 'Kavkazskie Predstativeli Roda Symphytum L.', 1976."
-'The Greek Species of Symphytum Boraginaceae' by William T. Stearn, British Museum of Natural History, London, England; Annales Musei Goulandris, Greece, Volume 7, pages 175-220, (1985 or 1986, different sources give different dates, even the pdf has both dates).

(Asia Minor, Asian Turkey, Anatolia or Anatolian peninsula is the westernmost part of Asia. It includes most of Turkey.)

Symphytum Species Classifications 1913 by Bucknall

"**The geographical area of the Symphytum genus includes almost the whole of Europe (except Lapland = Finland), Asia Minor, a part of Siberia, and Persia.**
In altitude it ranges from sea-level to 8000 feet (2438 meters), the Caucasian species (S. caucasicum Bieb., and S. grandiflorum DC.) attaining 7000 feet (2133 meters), and S. asperum, Lepech., and S. peregrinum, Ledeb., 8000 feet.
S. officinale Linn., extending from the west of Europe into Siberia over 79 degrees of longitude and 24 degrees of latitude, has the widest range of all.
That of S. tuberosum Linn., comes next in extent, but falls far short of S. officinale, extending from Spain to the Black Sea and from central Germany to Turkey, with a small outlying area in the north of England and Scotland.
Reckoning from north-west to south-east, S. bulbosum Schimp., S. uliginosum Kern., S. Zeyheri Schimp., S. ottomanum Friv., S. cordatum Waldst. & Kit., S. tauricum Willd., S. anatolicum Boiss., S. orientale Linn., S. brachycalyx Boiss., and S. paloestinurn Boiss., occupy larger or smaller areas in central and southern Europe and in Asia Minor, sometimes distinct and sometimes overlapping.
Of these, S. paloestinum reaches farthest south, to about 31 degrees north latitude.
S. armeniacum Rucknall appears to extend from Armenia to the Caucasus.
S. arperum Lepech., S. caucasicum Bieb., and S. grandiflorum DC., are found in the Caucasian region.
And S. peregrinum Ledeb. in the extreme south-east of the Caucasus and in Persia.
S. mediterraneum Koch, and S. floribundum Shuttlw. occur only in a few spots in south-east France.
And S. Gussonei Schultz, is probably confined to Sicily.
S. pseudobulbosum Azn., S. bornmuelleri Bucknall, S. sylvaticum Boiss., S. sepulcrale Boiss. & Bal., and S. kurdicum Boiss. & Haussn. have as yet only been met with in small areas in Asia Minor.
This study has convinced me that the Symphytum genus is naturally divided into two sections, one containing the plants with branched stem and fusiform root, and the other those with a simple stem and a more or less creeping and tuberous root.
Of these sections the two British species of Symphytum are examples, the Ramosa being represented by S. officinale, and the Simplicia by S. tuberosum.
The Ramosa may be readily divided by the character of the calyx, as to whether its segments are longer or shorter than the calyx-tube. The groups thus formed may be subdivided by the leaves being decurrent or not, by the colour of the flowers, and by the corolla-scales being included or exserted.
By these means well defined sections are formed, in which the species are connected by important characters. There are some cases in which it is difficult to decide on the true relationship of the plant, but I believe that on the whole no species will be found to be far from its proper position in the arrangement I propose.
Conspectus (brief survey) of the Sections:
Division I. Ramosa.
  Subdivision i. Segmentata.
    Section 1. Officinalia: S. officinale Linn.; S. uliginosum Kern.
    Section 2. Caerulea: S. asperum Lepech.; S. peregrinum Ledeb.; S. sepulcrale Boiss. & Bal.; S. armeniacum Bucknall.
    Section 3. Albida: S. tauricum Willd.; S. sylvaticum Boiss.; S. anatolicum Boiss.; S. ottomanuum Friv.
  Subdivision ii. Dentata.
    Section 4. Orientalia: S. orientale Linn.; S. kurdicum Boiss. & Haussskn.; S. caucasicum Bieb.;
      S. floribundum Shuttlw.; S. pseudobulbosum Azn.

  Section 5. Suborientalia: *S. palaestinum Boiss.; S. brachycalyx Boiss.; S. bornmuelleri Bucknall.*
Division II. Simplicia.
  Section 6. Tuberosa: *S. tuberosum Linn.; S. mediterraneum Koch; S. gussonei Schultz;*
   S. bulbosum Schimp.; S. zeyheri Schimp.
  Section 7. Cordata: *S. cordatum Waldst. & Kit.; S. grandiflorum DC."*
-'A Revision of the Genus Symphytum, Tourn.' by Cedric Bucknall, (Mus. Bac. Oxon= Bachelor of Music, Oxford University), Journal of the Linnean Society of London, England, Botanical Journal, Volume 41, Issue 284, pages 491-556, December **1913**. See sub-subsection 'Bucknall 1912' in subsection 'Symphytum Genus' in section 'Borage Family, Symphytum Genus' (Chapter 2).

  "The genus Symphytum was divided into two species groups by Boissier (*1879) and Kusnetsov (**1910). They based their Classification on the length of the corolla scales (included or exserted) and on the colour of the corolla, respectively.
  Bucknall (1913) did not accept this rather arbitrary and artificial Classification. He based his own Classification on the stem type (branched or simple) and on the root type (fusiform and branched, or creeping and tuberous).
   In the group with branched stems and fusiform roots he distinguished five sections, two of which will be treated in this paper, with the sections Officinalia (with S. officinale L. and S. uliginosum Kern.) and the section Coerulea (with S. asperum Lepech. and S. peregrinum Ledeb., and two other species, left out of consideration within this paper)."
-'Cyto- and Chemotaxonomical Studies on the Sections Officinalia and Coerulea of the Genus Symphytum' by Th.W.J. Gadella, E. Kliphuis and H.J. Huizing; Botanica Helvetica (now called Alpine Botany), Volume 93, pages 169-192, 1983.
(* -'Flora Orientalis Sive Enumeratio Plantarum in Oriente, a Graecia et Aegypto ad Indiae Fines {Plants Flora, Flora East and the East, from Greece and Egypt to the Borders of India}, Volume IV' by Edmond Boissier, Society Physics Geneva, Society Linnean London; Geneva, Switzerland, 1879.)
(** -'Kavkazkie Vidy Roda Symphytum Tourn. L.' {Caucasian Species of the Genus Symphytum Tourn. L.} by N. Kuznetsov, Memoires de l'Academie Imperiale des Sciences de Saint-Petersbourg, Serie. 8, Physiques et Mathematiques, Volume 25, No. 5, pages 1-94, 1910. If you have an English translation of this, I would appreciate a copy.) (Fusiform means tapering at both ends, i.e., spindle shaped.)

Symphytum Species Classifications 1953 by Popov

*"***Genus 1199. Symphytum L.*** Sp. pl. (1753) 136; Gurke in Nat. Pflanzenf. IV, 3a, 112; Kuzn. in Mem. Acad. Sc. Petersb. VIII ser. XXV. 5 (1910) 22.*
This ancient Mediterranean genus consists mainly of rather tall herbs whose habitats are mainly in the wooded, relict islands of the ancient Mediterranean. Some of its species are found in Central Europe and one even in the taiga zone of the Boreal region. About 20-25 species.
Subsection 1. Ochroleuca *Kusn. l.c. (1910) 41, l.c. (1913) 236.*
Corolla pale yellow, neither blue nor violet, without anthocyanin (blue, violet or red color).
  1. S. cordatum *W. et K. Ic. Pl. rar. Hung. I (1802) 6;*
   S. cordifolium *Baumg. Enum. Stirp. Transsilv. I (1816) 126.*
   S. pannonicum *Pers. Syn. II (1805) 161.*
Shady mountain forests, often spruce, rarely broadleaved.
European part: U. Dns. (Upper Dniester), M. Dnp. (Middle Dnieper) (Podolia).
General Distribution: Central Europe (Germany, Poland, Czechoslovakia, Hungary, Austria, Switzerland) (Carpathians).
  2. S. tuberosum *L. Sp. pl. (1753) 136; DC. Prodr. X, 38;*
   S. nodosum *Schur, Eniun. Pl. Transsylv. (1866) 468.*
   S. leonhardtianum *Pugsley, Journ. of Bot. 69 (1931) 89.*
   S. foliosum *Rehm.*
Mainly in broadleaved, beech, hornbeam and oak forests.
European part: U. Dns. (Upper Dniester) (Carpathians and Subcarpathians), M. Dnp. (Middle Dnieper) (Podolia), Bes. (Bessarabia), L. Don (Lower Don) (very doubtful on the Mius River).
General Distribution: Central Europe, Balkan Peninsula and Asia Minor (Balkans).
This unique Central European species, endemic (native) to the Carpathians occurs mainly in the mountain-forests, occasionally it descends to the lowlands adjacent to these mountains; these localities are Kremenets (city in western Ukraine) and Pochaev (Pochaiv, city in western Ukraine) (all the others refer to Symphytum tauricum Willd.).
It is difficult to indicate to which species this one is related; perhaps the closest is S. cordatum with which it is sometimes sympatric, but hybridization apparently occurs rarely in spite of the wide distribution of these species in the Carpathians.
  3. S. grandiflorum *DC. Prodr. X (1846) 40;*
   S. ibericum *Stev. ex M.B. Fl. taur.-cauc. III (1819) 647, nomen;*
   S. abchasicum *Trautv. in Bull. Soc. Nat. Mosc. XLIII (1870) 72.*
   S. grandiflorum var. abchasicum *(Trautv.) Kusn. l. c. 46.*
Usually broadleaved forests, the lower mountain belt up to 1,000 meters (3280 feet).
Caucasus: Cisc. (Ciscaucasia= north Caucasus) (western part- Maikop), W. Transc. (Western Transcaucasia).

(Probably present in the adjacent Lazistan= southeastern shore of the Black Sea.)
This endemic (native), Colchic species is related to S. cordatum and S. tauricum, but sharply distinguished from them by the rhizome and the general habit.
It appears to me that these three species are not as closely related as thought by Kuznetsov.
The branches of the long rhizome are developed from the radix (root). The branch extends to the surface of the earth and produces there three shoots, one of them a fertile stem and the other two sterile ones, bearing only leaves. Later, towards the fall, these shoots become procumbent (lying on ground) just as branches of the rhizome; sometimes their leaves are preserved until the next year, usually not dying. These shoots are 10-20 (30) cm (3.9- 7.8- 11.8 inches) long.

4. S. tauricum Willd. *in Ges. Naturf. Fr. Neue Schr. II (1799) 120;*
      S. bullatum *Horn. Hort. Hafn. I (1813) 179.*
Apparently weedy in open forests.
European part: Bes. (Bessarabia), Bl. (Black Sea area), M. Dnp. (Middle Dnieper), L. Don (Lower Don), Crim. (Crimea); Caucasus: W. Transc. (West Transcaucasia) (extreme northern part along the Black Sea). Endemic (native). (Occurring possibly in the north Balkans).
Phylogenetically, this species may be regarded as the monocarpic (flowering only once) derivative of the typically perennial, forest species S. grandiflorum DC.

5. S. orientale L. *Sp. pl. (1753) 136; Prodr. X, 39;*
Wet places in forests.
European part: Bl. (Black Sea area) (near Odessa), M. Dnp. (Middle Dnieper) (S. Podolia, in Turchaninov herbarium). General distribution: Asia Minor, from Constantinople (Istanbul, Turkey) to Ankara (Turkey).
This is a very ambiguous species for which Linnaeus cited two different species (synonyms) by Tournefort, one with blue flowers, the other with white from Constantinople.
The first synonym probably refers to S. asperum Lepech. and the second to what De Candolle and Boissier regarded as S. orientale, though some older authors were of another opinion.
Roemer and Schultes (Syst. IV, 65), like Willdenow (*Enum. H. Berol. I, 183), referred Tournefort's plant with the white flowers to S. tauricum Willd.
Likewise, S. orientale in Hegis (p. 2223) is described as having leaves that taper gradually to petioles, without a cordate base.
I accept S. orientale as the specimen which was described by Boissier: the Wiedeman specimens from Byzantium (Istanbul, Turkey) and Ankara (Turkey) and the southern coast of Pontus (in Turkey) (according to Boissier from Bithynia). This S. orientale resembles S. caucasicum more than S. tauricum and in general it is intermediate between them.

Subsection 2. Cyanea Kusn. in Mem. Acad. Sc. Petersb. XXV, 5 (1910) 23; **Corolla with anthocyanin (red/purple), violet, blue, pink, sometimes white in albino plants.**

6. S. caucasicum M.B. *Fl. taur.-cauc. (1808) 128; DC. Prodr. X, 38;*
Edges and glades in the forests of the drying parts of the Caucasus.
Caucasus: Cisc (Ciscaucasia= north Caucasus), Dag. (Dagestan in north Caucasus, Russia), E. Transc. (Eastern Transcaucasia). Endemic (native).
By its decurrent upper leaves this species resembles S. officinale and is distinguished from all the other Caucasian species with anthocyanin corollas.
It differs noticeably from S. officinale by the calyx being cleft for only 1/3, its teeth obtuse, the corolla blue and not dingy violet, and by the pale, wrinkled nutlets (instead of black, smooth, luminous).

7. S. asperum Lepech. *in Nov. Acta Acad. Petrop. XIV (1805) 442;*
      S. asperrimum Sims, *Bot. Mag. (1806) tab. 929;*
      S. echinatum *Ldb. Ind. Sem. Hort. Dorp. Suppl. (1811) 5 et (1820) 4;*
Damp places in mountains, banks of small rivers and streams, edges of forest, meadows, from foothills to the subalpine belt.
European part: introduced into the Baltic area. Lad.-Ilm. (Ladoga-Il'men) (Ladoga is a lake in northwest Russia), Dv.-Pech. (Dvina-Pechora) (Dvina and Pechora are rivers in northwest Russia.), U. V. (Upper Volga), M. Dnp. (Middle Dnieper); Caucasus: everywhere (except for Talysh?).
General distribution: Central Europe (introduced and cultivated).
In the Caucasus there are individuals with densely bristly villous stems without decurrent leaves and with shorter calyx teeth; these may be hybrid forms of S. asperum x S. caucasicum. They are also often distinguished by their low habit.

8. S. peregrinum *Ldb. Ind. Sem. Hort. Dorp. (1820) 4; Sprengel, Syst. I, 563; DC. Prodr. X, 37; Ldb. Fl. Ross. III, 114; Kuzn. in. Mem. Acad. St. Petersb. XXV, 5 (1910) 31; Mat. Fl. Kavk. IV, 2, 228; Grossg. Fl. Kavk. III, 258.- Ic.: Kuzn., op. cit. (1910).*
Forests. Caucasus: Tal. (Talysh). Endemic (native).
I am retaining the independent status of this species only as a concession to Kuznetsov' s authority.
I did not observe the distinctions indicated, i. e., the size of the calyx compared with the corolla, the depth of its lobes, and the acuteness of the lobes (teeth).
I believe that Boissier was correct in including it in the synonymy of S. asperum (asperrimum) as a garden, hybrid form. As a matter of fact Ledebour did not indicate the locality of his species when he described it for the first time in 1820. It was only 29 years later (in 'Fl. Ross.') that he mentioned Talysh as such, adding that he had only seen cultivated specimens.

I find Kuznetsov' s claim, that this species 'maybe identified with S. asperum Lepech (= S. asperrimum Sims),' rather strange. I could not find any basis for Sprengel' s report on Podolia (southwest Ukraine and northeast Moldova) as a locality, which was repeated by De Candolle.
9. S. officinale L. *Sp. pl. (1753) 136; DC. Prodr. X, 37;*
    S. bohemicum Schmidt, *Fl. Boem. III (1795) 13, tab. 263.*
    S. tanaicense Stev. *in Bull. Soc. Nat. Mosc. XXIV (1851) 577.*
    S. uliginosum Kern, *in Oesterr. Bot. Zeitschr. XIII (1863) 227.*
Damp meadows, swampy places, rivers and streams in the taiga and the steppe zones.
European part: nearly everywhere except for the Arctic part, Crimea (rarely); Caucasus: Cisc. (Ciscaucasia) (rarely); West Siberia: Ob (Ob region from the eastern slopes of the Urals to the Yenisei River) (seldom), U. Tob. (Upper Tobol) (rarely); Central Asia: Dzu.-Tarb. (Dzungaria-Tarbatagai), Balkh. (Lake Balkhash area). General distribution: Central Europe.
Some varieties of this widespread European species have been distinguished according to the color of the corolla:
var. purpureum *Pers. (Ench. I, 161; DC. Prodr. X, 37) with red or pink corollas;*
var. ochroleucum *DC. l.c.=* **S. bohemicum Schmidt** *with straw-colored or white corollas, mainly in the western parts, others according to the general morphology which is dependent on the geographical zone.*
Hegi claimed accordingly two varieties:
    var. lanceolatum Weinm. *(in Bull. Soc. Nat. Mosc. VII (1837) 57; Ldb. Fl. Ross. III, 114; Rouy, l.c. 290.* **S. angustifolium** *(Opiz) G. Beck).*
    var. stenophyllum Cel. *-var. glabrescens Nickels apud Kirschleger*
        -S. uliginosum Kerner -S. vetteri Thellung.
Also known are hybrids with S. asperum, a species cultivated in the distribution area of S. officinale, under the name S. uplandicum Nym. *(Syll. 1854-55, 80; Hegi, FL Fl. V, 3, 2222)."*
-'**Flora of the U.S.S.R.**', 30 volumes, 1934-1960, edited by V.L. Komarov; Botanical Institute of the Academy of Sciences of the U.S.S.R, Leningrad, Russia, with **Boraginaceae article by Mikhail Grigorevich Popov, in Volume 19,** pages 279-291 Russian version, pages 207-216 English version, **1953**.
(* -'Enumeratio Plantarum Horti Regii Berolinensis: Continens descriptiones omnium vegetabilium in horto dicto cultorum, Part 1' {An Enumeration of the Royal Garden Plants Berlin} by Carl Ludwig Willdenow. Berolini {Berlin}, Germany, 1809, pages 183-184.)
See sub-subsection 'Popov 1953' in subsection 'Symphytum Genus' in section 'Borage Family, Symphytum Genus' (Chapter 2).
    (Relict means a surviving species of an otherwise extinct group, or a remnant of a formerly widespread species that lives in isolated areas.)
    (Taiga, or boreal or snow forest, is a plant community of coniferous forests with mostly pines, spruces and larches.)
    (The Black Sea is between far-southeastern Europe and far-western Asia and Turkey.)
    (The Dniester or Dnister River is in Eastern Europe. It starts in Ukraine and then goes through Moldova, finally flowing into the Black Sea in Ukraine again.)
    (The Dnieper River is the fourth longest river in Europe, starting in Valdai Hills near Smolensk, Russia, {near Europe}, and going through Russia, Belarus and Ukraine to the Black Sea.)
    (Podolia or Podilia is in west-central and southwest Ukraine and northeast Moldova.)
    (The Carpathian Mountains or Carpathians form an arc 1,500 kilometers {932 miles} long across central and eastern Europe. It includes parts of Austria, Czech Republic, Hungary, Poland, Serbia, Slovakia, Ukraine and Romania. It is the second-longest mountain range in Europe.)
    (Bessarabia is in Eastern Europe, bounded by the Dniester river on the east and the Prut river on the west. Most of it is in Moldova, with the rest in the Ukrainian Budjak region.)
    (The Don River is a major Eurasian river and fifth longest river in Europe. It starts in Novomoskovsk, Russia {120 killometers = 74 miles south of Moscow}, and flows 1,870 kilometers {1162 miles} to Sea of Azov in eastern Europe.)
    (Sympatric means occurring in the same geographic location; overlapping in distribution. For speciation it means taking place without geographical separation.)
    (The Euxine-Colchic deciduous forest region, with temperate broadleaf and mixed forests, is located along the southern shore of the Black Sea.)
    (Crimea is a peninsula on the northern coast of the Black Sea in Eastern Europe that is almost completely surrounded by the Black Sea and Sea of Azov to the northeast.)
    (The Caucasus Mountains are located at the border of Europe and Asia, between the Black Sea and Caspian Sea and occupied by Russia, Georgia, Azerbaijan, and Armenia.)
    (The Volga River is the longest and largest in Europe. It flows through central Russia and into the Caspian Sea. The Caspian Sea is between Europe and Asia. It is bounded by Kazakhstan to the northeast, Russia to the northwest, Azerbaijan to the west, Iran to the south, and Turkmenistan to the southeast.)
    (Talysh Mountains are in southeastern Azerbaijan and northwestern Iran.)
    (A steppe is montane grasslands/shrublands and temperate grasslands with no trees except for those near rivers and lakes. The world's largest steppe is in eastern Europe, central Asia, and neighbouring countries.)

"Families, Genera and Subdivision of Genera in 'Flora of Russia':
Detailed and interesting descriptions are provided for Chenopodiaceae (M. Iljin), Umbelliferae (B. Schischkin), **Boraginaceae (M. Popov),** Labiatae (S. Juzep-czuk), Campanulaceae (A. Fedorov) and some other families.
The system of families was rarely revised by the authors of the 'Flora'. Still, a new subfamily of the Labiatae was

established by A. Pojarkova; as many as 6 new tribes were described by A. Fedorov; **many new subtribes were distinguished among the Boraginaceae by M. Popov,** *etc. A certain reserve in introducing such changes is perfectly justifiable, since most families are inadequately represented in the U.S.S.R."*
-'The Flora of the USSR: Taxonomic Literature Review' by M.E. Kirpicznikov, Komarov Botanical Institute of the Academy of Sciences of the U.S.S.R., Leningrad, Russia; Taxon: International Association for Plant Taxonomy, Volume 18, No. 6, pages 685-708, December 1969.

Symphytum Species Classifications 1967 by Gadella and Kliphuis

*"***The genus Symphytum contains about 25 species (*Bucknall, 1913).** *In the Netherlands it is represented by one indigenous (native) species, S. officinale L.; two others, S. asperum Lepech. and S. bulbosum Schimp., have been introduced but have not become established.*
S. officinale is a variable species.
       Kerner (1863) described a segregate, S. uliginosum, which was reduced to subspecific rank by Nyman (1878): S. officinale L. ssp. (subspecies) uliginosum (Kern.) Nym.**
            The differences between subspecies officinale and uliginosum are:
            in ssp. officinale the stems and leaves are more or less densely hispid, in ssp. uliginosum they are provided with small white deciduous prickles with tubercular base. Furthermore in the first-mentioned ssp. the upper stem-leaves are entirely decurrent but in the last named they are not or partially decurrent.
            There is also a difference in the colour of the flowers: white, yellowish, or dark purple in ssp. officinale, and red-purple or violet in ssp. uliginosum.
S. asperum (Prickly Comfrey), which occurs in the Caucasus, Armenia, and Persia (Iran) hybridizes with S. officinale (*Lindman, 1911; Bucknall, 1912; ****Faegri, 1931; *****Tutin, 1956; ******Wade, 1958).**
       This hybrid, S. x uplandicum Nym., has also been recorded from the Netherlands (de Wever, 1932; Kloos, 1934; van Ooststroom and Reichgelt, 1961), but it is apparently not very common.
       In England the populations of hybrids are remarkably uniform, although great variations can be observed.
       Backcrossing with S. officinale may result in highly variable hybrid-swarms (Wade, 1958). Neither in England nor in the Netherlands is it likely that backcrossing in the direction of S. asperum occurs, in view of extreme rarity of this species.
In Western Europe S. officinale is most often characterized by white flowers (Tutin, 1956), whereas no white-flowering individuals have been reported from Russia (Steven, 1851; Popov, 1953).
       According to Tutin (l.c.) it is tempting to suggest that the purple-flowered form may be due to introgression between S. officinale and S. asperum.
In Britain S. x uplandicum is probably the commonest Symphytum (Tutin, l.c.).
Wade (1958) is of the opinion that originally most of the hybrids were introduced into Britain from Russia as fodder plants. The hybrids have purple flowers *(Tutin, l.c.)."*
-'Cytotaxonomic Studies in the Genus Symphytum I: Symphytum Officinale L. in the Netherlands' by Th.W.J. Gadella and E. Kliphuis, Botanical Museum and Herbarium of the State University of Utrecht, Netherlands, February 25 **1967**.
(* -'A Revision of the Genus Symphytum, Tourn.' by Cedric Bucknall, {Mus. Bac. Oxon= Bachelor of Music, Oxford University}, Journal of the Linnean Society of London, England, Botanical Journal, Volume 41, Issue 284, pages 491-556, December 1913.)
(** -'Descriptiones Plantarum Novarum Florae Hungaricae et Transsilvanicae' by A. Kerner, Oesterreichische Botanische Zeitschrift {Austrian Botanical Journal}, Vienna, Austria, Volume 13, No. 7, pages 227-288, 1863.)
(*** -'Ueber Symphytum Orientale L. und Symphytum Uplandicum Nym.' by C.A.M. Lindman; Botaniska Notiser {Botanical Notes}, Lund, Sweden, Nummer 2, pages 71-77, 1911. In German.)
(**** -'Uber in Skandinavien Gefundenen Symphytum-Arten: Nebst einigen Betrachtungen uber das Artproblem Innerhalb der Betreffenden Artgruppe' {About Symphytum Species Found in Scandinavia} by Knut Faegri, Bergens Museum Arbok, Norway, 47 pages, 1931. In German. If anyone has an English translation of it, I would appreciate having it.)
(***** -'The Genus Symphytum in Britain' by T.G. Tutin, University College of Leicester, England; Watsonia: Journal of the Botanical Society of the British Isles, Volume 3, pages 280-281, February 1956.)
(****** -'The History of Symphytum Asperum Lepech. and S. x Uplandicum in Britain' by A.E. Wade, Department of Botany, National Museum of Wales, Watsonia, Volume 4, pages 117-118, 1958. Watsonia was the journal of the 'Botanical Society of Britain and Ireland' from 1949 to 2010. It is now called 'New Journal of Botany'.)
      ('l.c.' is Latin for 'loco citado' which means 'locally cited'. In other words, the author's name was cited in the early part of the book or article with a date and publication name.)
(Introgression is the transfer of genetic information from one species to another as a result of hybridization between them and repeated backcrossing.)

      "The above mentioned results demonstrate that the 2n = 24 and the 2n = 48 chromosome type are morphologically very similar and can easily be distinguished from the 2n = 40 type.
      --- **2n = 40** --- (Symphytum officinale and others)
      The plants of the 2n = 40 type also prefer habitats in which diploid or tetraploid plants were never observed. It seems, therefore, that the 2n = 40 plants have a different ecological preference. *In general the 2n =40 type is represented on peat soils. These plants grow under very moist conditions and always in open vegetation.*
            After an examination of some specimens, quoted by Bucknall (1913), e.g. (for example) S. uliginosum Kern.,

(Herb. Gower, Caucasus, 1820, K), S. uliginosum Kern., (Veszto, Hungary, Borbas, 1877, K), we have come to the conclusion that the Dutch plants of the 2n = 40 type closely resemble these specimens.

The specimens examined by Bucknall are somewhat more slender, whereas the mediodorsal row of hairs of the outer surface of the sepals is less densely haired. **Moreover, the plants of the 2n = 40 type largely match the original description of S. uliginosum Kern.** Comparison with the type material, however, is necessary before further conclusions can be drawn.

An interesting problem is formed by the distribution area of S. uliginosum. According to Bucknall (l.c.) this species occurs in Hungary and in the south of Russia. Some forms which resemble S. uliginosum to a certain extent are found in Austria, France (Alsace) and in the west of Switzerland (Hegi, 1927). Probably these forms are not to be regarded as indigenous (native) but as naturalized (Hegi, l.c.). Rothmaler (1963) is of the opinion that S. officinale L. subsp. uliginosum (Kern.) Nym. occurs in the Rhine province of Western Germany.

Therefore, it is not impossible that the range of S. uliginosum extends to Western Europe. Herbarium studies are necessary to solve this problem.

--- 2n = 48 --- (Tetraploid Symphytum officinale and others)

Plants with the chromosome number 2n = 48 were never found on peaty soil. The plants occur in dryer places than plants with the chromosome number 2n = 40, but they sometimes were found in very moist places.

> Tetraploids occur at roadsides, in or near woods (but never completely shaded), on dikes (dams), in forelands of rivers, by pools and ditches (always on clayey and sandy soils), sometimes in places which are frequently flooded at high tide in the freshwater tidal deltas, de Biesbosch (Netherlands).

Tetraploid plants grew sometimes intermingled with diploid plants. Therefore it seems probable that diploid and tetraploid plants may have the same ecological requirements.

--- Flower Color & Flowering ---

All diploid plants are white-flowered. The 2n = 40 type sometimes has also white flowers, but the majority have light, dark or red purple corollas.

> There is an indication that in the wild that the flowering period of the 2n = 40 type is shorter than that of the 2n = 24 and 2n = 48 type.

--- Overview ---

The morphological and cytological characteristics as well as the ecological preferences of the 2n =40 type seem to support the treatment of this cytotype as a separate species. On the other hand, the crossability of the 2n = 40 and the 2n = 48 type indicate that these types are better regarded as conspecific (same species).

A study of the fertility of the hybrid, including its morphological characters, is necessary before it will be possible to arrive at more definite conclusions.

--- Genetic Mixing of Diploids and Tetraploids --- (Symphytum officinale and others)

The interfertility of plants with different chromosome numbers: In view of the fact that in populations with diploids and tetraploids no intermediate chromosome numbers could be observed, it seems probable that, **at least in the wild, no gene-exchange occurs. It is clear that mixed populations are rare. Generally within the same population all individuals have the same chromosome number."**

-'Cytotaxonomic Studies in the Genus Symphytum I: Symphytum Officinale L. in the Netherlands' by Th.W.J. **Gadella and E. Kliphuis**, Botanical Museum and Herbarium of the State University of Utrecht, Netherlands, February 25 **1967**.

> (Body cells and individuals are described by the number of chromosome sets {ploidy level} they have: monoploid: 1 set; diploid: 2 sets = 2n = 2x, triploid: 3 sets = 2n = 3x, tetraploid: 4 sets = 2n = 4x, pentaploid: 5 sets = 2n = 5x, etc. For more about chromosomes, see subsection 'Details about each Species of Symphytum'.)

Cytotaxonomic References for the above:
Britton, D. M., Brittonia, 7, 233-266 (1951).
Bucknall, C., Journal of Bot., 50, 332-337 (1912).
Bucknall, C., Journal Linn. Soc. Bot., 41, 491-556 (1913).
Datta, S., Journal Ind. Bot. Soc., 12, 131-152 (1933).
Faegri, K., Bergens Museum Arbok, 1-47 (1931).
Gadella, Th. W. J. and E. Kliphuis, Acta Bot. Neerl., 12, 195-230 (1963).
Gadella, Th. W. J. and E. Kliphuis, Proc. Roy. Noth. Acad. Sei., Ser. C70, 7-20 (1967).
Hegi, G., Illustrierte Flora von Mittel-Europa V (3), 2220-2229 (1927). Munchen.
Kerner, A., Osterr. Bot. Zeitschr., 13, 227-228 (1863).
Kloos, A. W., Ned. Kruidk. Arch., 44, 132 (1934).
Lindman, C. A. M., Bot. Not., 1911, 71-77 (1911).
Love, A. and D. Love, Acta Hort. Gothob., 20, 65-291 (1956).
Nyman, C. F., Conspectus Florae Europaeae (1878-1882). Orebro.
Ooststroom, S. J. van and Th. Rbichgelt, Flora Neerlandical IV(1), 107-110 (1961).
Popov, M. G., in Komarov, V. L. Flora U.S.S.R., 9, 279-291, (1953). Moskva and Leningrad.
Rothmaler, W., Excursionsflora von Deutschland, IV, 262 (1963). Berlin.
Steven, C., Bull. Soc. Imp. Nat. Mosc., 24, 577 (1851).
Strey, M., Planta 14, 682-730 (1931).
Tarnavschi, I. T., Bull. Jard. Mus. Bot. Univ. Cluj., 28, suppl., 1-130 (1948).
Tutin, T. G., Watsonia, 3, 280-281 (1956).

Wade, A. E., Watsonia, 4, 117-118 (1958).
Wever, A. db, Natuurhist. Maandbl., 21, 112-116 (1932).

Symphytum Species Classifications 1969 by Wickens

"The earlier classification of Boissier (1879) heavily weighted the corolla scales with which Bucknall quite rightly disagreed. He, in turn, emphasized calyx and flower colour.
 It is not proposed to challenge the primary divisions of Bucknall's classification at present, since insufficient material has been studied. *The few chromosome counts available (Darlington & Wylie 1955; Love & Love 1961) are certainly of interest and appear to support the present generic divisions.*
It is suggested, however, that Bucknall's heavy weight of calyx shape and corolla colour leads to an artificial classification that disguises such affinities *as S. palaestinum and S. aintabicum with S. anatolicum, and S. palaestinum var. majus with S. orientale and S. kurdicum. Their overall relationships cut right across Bucknall's sections, and in fact fit more closely with the earlier attempts of Boissier.*
Faegri *(*1931) united Bucknall's sections Symphytum (Officinalia) and Caerulea by creating Symphytum commune, to which he subordinated S. officinale, uliginosum, S. asperum and S. peregrinum as subspecies, while retaining S. sepulcrale and S. armeniacum as species. It is a concept that has found little support.*
Pawlowski *(**1961) has made a thorough study of the species available to and published a synopsis, with particularly valuable illustrations of the corolla scales (fornices), including a new classification. Unfortunately this only includes ten of the known species, but it is interesting to note that he associates S. tauricum, S. grandifiorum and S. orientale in one section (Lingulata), and S. bulbosum and S. ottomanum in another (Bulbosum).*
S. circinale and S. creticum are treated here as constituting a new section (Section Procopiania) *on account of their long corolla lobes and basal hairs on the filaments (cf. {compare to} Runemark, ***1967).*
The other two Aegean species, S. euboicum and S. davisii, *on account of their affinities with S. ottomanum and S. anatolicum, are placed for the present in Section Albida until such time as the sectional divisions can be fully revised.*
It must be borne in mind that the status of some of the species is open to question in the absence of sufficient material and biosystematic information."
-**'A Revision of Symphytum in Turkey and Adjacent Areas' by G.E. Wickens**, Royal Botanic Gardens: Kew; Notes from the Royal Botanic Garden Edinburgh, Scotland, Volume 29, pages 157-180, **1969**. Includes Turkey, Bulgaria, Greece, Aegean Islands and Caucasia.
(* -'Uber in Skandinavien Gefundenen Symphytum-Arten: Nebst einigen Betrachtungen uber das Artproblem Innerhalb der Betreffenden Artgruppe' {About Symphytum Species Found in Scandinavia} by Knut Faegri, Bergens Museum Arbok, Norway, 47 pages, 1931. In German. If anyone has an English translation of it, I would appreciate having it.)
(** -'Observations ad {on} Genus Symphytum L. Pertinentes' by Bogumilus Pawlowski, Fragmenta Floristica et Geobotanica, Poland, Volume 7, No. 2, pages 327-356, 1961. All in Polish. If you have an English translation, I would appreciate a copy.)
(*** -'Studies in the Aegean Flora XI: Procopiana {Boraginaceae} included into Symphytum' by Hans Runemark, Botaniska Notiser {Botanical Notes}, Sweden, Volume 120, pages 84-94, 1967. I was unable to get this report. If you have it, could you please send it to me.)
 (**Wickens describes these 27 Symphytum species:** S. asperum, S. peregrinum, S. hajastanum, S. sepulcrale, S. armeniacum, S. tauricum, S. sylvaticum, S. anatolicum, S. ottomanum, S. euboicum, S. davisii, S. circinale, S. creticum, S. caucasicum, S. orientale, S. kurdicum, S. pseudobulbosum, S. palaestinum, S. aintabicum, S. longisetum, S. brachycalyx, S. bornmuelleri, S. tuberosum, S. bulbosum, S. zeyheri, S. grandiflorum, S. longipetiolatum.
 He also includes a 'Key to the Species' for all 27 of these Symphytums. For keys see subsection 'Overview Comparison of Symphytum Species + Keys' in section 'Symphytum Species Overview'.)
See sub-subsection 'Wickens 1969' in subsection 'Symphytum Genus' in section 'Borage Family, Symphytum Genus' (Chapter 2).

 "Traditionally, the (Symphytum) genus is divided into 2-9 sections, based on various infrageneric classifications (Boissier 1879, Kuznetsov 1910, Bucknall 1913, Pawlowski 1961, *Wickens 1969, Sandbrink et al. 1990)."
-'Symphytum Tuberosum Complex in Central Europe: Cytogeography, Morphology, Ecology and Taxonomy' by L. Kobrlova, M. Hrones, P. Koutecky, M. Stech, and B. Travnicek, Preslia: The Journal of the Czech Botanical Society, Czech Republic; Volume 88, pages 77-112, March 2016.
(* -'A Revision of Symphytum in Turkey and Adjacent Areas' by G.E. Wickens, Royal Botanic Gardens: Kew; Notes from the Royal Botanic Garden Edinburgh, Scotland, Volume 29, pages 157-180, **1969**. Includes Turkey, Bulgaria, Greece, Aegean Islands and Caucasia.) (Infrageneric is subclassification within a genus of organisms.)

Symphytum Species Classifications 1972 by Pawlowski

"Symphytum L. (species in Europe):
 S. asperum
 S. bulbosum
 S. cordatum
 S. cycladense

S. davisii
S. floribundum
S. gussonei
S. ibiricum (ibericum)
S. naxicola
S. officinale, subspecies officinale and subspecies uliginosum.
S. orientale
S. ottomanum
S. tauricum
S. tuberosum, subspecies tuberosum and subspecies nodosum."

-'Symphytum L.' by B. **Pawlowski** with editors Tutin, Heywood, Burges, Moore, Valentine, Walters and Webb, **Flora Europaea, Volume 3,** pages 103-105, **1972**. The 'Flora Europaea' is a 5-volume encyclopedia of plants, published between 1964 and 1993 by Cambridge University Press, England. It describes all the national floras of Europe to identify any wild or widely cultivated plant to the subspecies level. It provides geographical distribution, habitat preference, and chromosome number.
See sub-subsection 'Pawlowski 1972' in subsection 'Symphytum Genus' in section 'Borage Family, Symphytum Genus'.

Clade 2: Symphytum icaricum and Symphytum naxicola form a group which is proposed to the Graeca section."
-'Symphyta Mediterranea Nova vel Minus Cognita: Nowe lub malo znane srodziemnomorskie zywokosty' {New or Little Known Mediterranean Comfrey} by Bogumil **Pawlowski**; Fragmenta Floristica et Geobotanica Polonica, W. Szafer Institute of Botany, Polish Academy of Sciences, Krakow, Poland, Volume 17, No. 1. pages 17-37, **1971**. If you have an English translation, I would appreciate a copy.)
(Symphytum icaricum Pawl.: Fragm. Florist. Geobot. 17(1): 19. 1971 in East Aegean Islands.)
(Symphytum naxicola Pawl.: Fragm. Florist. Geobot. 17(1): 21. 1971 in Greece.)

"The Polish botanist Bogumil Pawlowski (1896-1971) had made a very detailed and critical study of various European Symphytum species in which he paid special attention to the corolla scales (fornices), finding unsuspected diversity in their shape, length and marginal papillae when observed at a magnification of about x30. He illustrated their various types in 1961 and proposed a modification of Bucknall's classification largely based upon them.
He modified this classification yet further in 1971 as a consequence of the study of further material when preparing an account for the 'Flora Europaea', Volume 3, 1972.
    In this 1971 account he described three new Aegean species, S. icaricum (here separated from S. davisii), S. naxicola and S. cycladense. *The species then known were classified as follows:*
        Sectio (Section) Tuberosa Bucknall 1913: *S. tuberosum L.*
        Sectio Lingulata Pawl. 1961: *S. davisii Wickens, S. cycladense Pawl.,*
            S. anatolicum Bliss.
        Sectio Graeca Pawl. 1971: *S. icaricum Pawl, S. naxicola Pawl.*
        Sectio Bulbosum Kuzn. 1910: *S. bulbosum Schimper, S. ottomanum Frivald.*
-'The Greek Species of Symphytum Boraginaceae' by William T. Stearn, British Museum of Natural History, London, England; Annales Musei Goulandris, Greece, Volume 7, pages 175-220, (1985 or 1986, different sources give different dates, even the pdf has both dates).

"Traditionally, the Symphytum genus is divided into 2 to 9 sections, based on various infrageneric classifications (Boissier 1879, Kuznetsov 1910, Bucknall 1913, *Pawlowski 1961, Wickens 1969, Sandbrink et al. 1990)."
-'Symphytum Tuberosum Complex in Central Europe: Cytogeography, Morphology, Ecology and Taxonomy' by L. Kobrlova, M. Hrones, P. Koutecky, M. Stech, and B. Travnicek, Preslia: The Journal of the Czech Botanical Society, Czech Republic; Volume 88, pages 77-112, March 2016.
(* -'Observations ad Genus Symphytum L. Pertinentes' by Bogumilus Pawlowski, Fragmenta Floristica et Geobotanica, Poland, Volume 7, No. 2, pages 327-356, 1961. In Polish. If you have English translation, I would appreciate a copy.)

"The Eurasian genus Symphytum L., according to Pawlowski (1972), is represented by 14 species in Europe.
Five of them occur in peninsular Italy (*Pignatti 1982):
    S. officinale L., S. bulbosum Schimp. and S. tuberosum L. are native.
    Symphytum asperum Lepechin and S. orientale L. are naturalized aliens in north Italy and Tuscany, respectively. A sixth, S. gussonei F. W. Schultz, is endemic (native) to Sicily and considered as an insular (island) vicariant of S. tuberosum."
-'Symphytum Tanaicense (Boraginaceae) New for the Italian Flora' by Peruzzi, Garbari and Bottega, Willdenowia, Berlin, Germany, 31, pages 33-41, 2001.
(* -'Flora d'Italia' by Sandro Pignatti, Volume 2, Bologna, Italy, 1982. There are 4 volumes.)
(Vicariance is the separation of a group of organisms by a geographic barrier that results in differentiation of the original group into new varieties or species.)

Symphytum Species Classifications 1975 by Perring

"Symphytum Survey: The problems of the identification of taxa within the Symphytum officinale L. complex were previously discussed at the 1968 Annual General Meeting of the B.S.B.I. (Botanical Society of the British Isles) (*Proceedings Bot. Soc. Br. Isl, 7: 553-556, 1969), when the Symphytum survey was launched.
    Soon after the survey began, a collection of about 50 plants reflecting variation noted in the field was assembled in 'Cambridge University Botanic Garden' (England).
In August 1969 the collection was visited by **two Dutch botanists, W.J. Gadella and E. Kliphuis, who have made an extensive European study of the genus Symphytum.** They took living material of 26 plants from this collection and from a field excursion in south-western England back to Holland for cultivation and subsequent cytological examination.
The results of their survey disclose that the following taxa occur in Britain:
1. Symphytum asperum Lepech. *(Prickly Comfrey)* **Chromosomes 2n = 32**.
    Tall, up to 1 to 5 meters (3.28 to 16.40 feet). Stems and midrib of the underside of the leaves covered with stiff prickles. At least the lower and middle cauline leaves petiolate; leaves decurrent. Calyx very small in flower, usually one fifth as long as the corolla, enlarging rapidly in fruit. Corolla red in bud, becoming clear blue on opening; the limb widening gradually towards the apex. Anthers exceeding the filaments.
2. S. officinale L. *(Common Comfrey)*
    Variable in height. Stems and leaves rather softly hairy. Leaves strongly decurrent; length of decurrence usually exceeding one internode. Calyx half as long as the corolla, enlarging only slightly in fruit. Limb of corolla urceolate. Anthers equalling or exceeding the filaments. The two following variants occur:
        a. Diploid plants. Chromosomes 2n = 24.
        Short, usually less than 1 meter (3.28 feet). Corolla always with greenish-yellow buds becoming creamy-white on opening.
        b. Tetraploid plants. Chromosomes 2n = 48.
        Taller, up to 1 to 5 meters (3.28 to 16.40 feet). Corolla variable in colour between creamy white and red-purple. Populations of pure cream, pure carmine (vivid crimson) and mixed colours all occur.
Plants of S. officinale with cream-coloured flowers are scattered throughout the lowlands of England. *Evidence from chromosome counts and habitat data suggests that the diploids are almost confined to 100 kilometer square (38 square miles) 52 (TL).*
Plants of S. officinale with carmine-coloured (vivid crimson) flowers occur occasionally by themselves, particularly in the Midlands and northern England, whilst populations with mixed colours are almost confined to an area which includes the Thames Valley, and rivers and streams in Devon, Wiltshire, Dorset, Hampshire and West Sussex, England.
Overall, S. officinale sensu lato (in the broad sense) is much less widespread than previously realized; *it becomes a very rare plant in the north and west and is also almost absent from our three most eastern counties, Kent, Suffolk and Norfolk, England.*
Accounts in 'County Floras' for these areas which claim that S. officinale, particularly the purple-flowered form, is widespread may need revision.
3. S. x uplandicum Nyman *(Russian Comfrey)*
    A range of hybrids between S. asperum and S. officinale. Leaves generally lacking the decurrence of S. officinale but never petiolate. Calyx similar to that of S. officinale. Filaments equalling or exceeding the anthers. Flower colour varying from reddish-purple to violet, but never clear blue as in S. asperum or creamy-white as in some plants of S. officinale. The two following variants occur:
        a. 2n = 36. *Plants with leaves not or scarcely decurrent and with flower buds deep purple and opening to violet or violet-blue.*
        b. 2n = 40. *Plants with leaves frequently slightly decurrent and with flower buds varying from pink to magenta and opening to a wide range of colours from pink to violet.*
    The purple-budded S. x uplandicum is much less frequent than the pink-budded variant. *Both are scattered throughout the British Isles, occurring in the north and west in the absence of S. officinale.*
All the taxa described above can be separated on the basis of the measurements, descriptions and assessment of flower- and bud-colour which were asked for during the survey, with the exception of the diploid and tetraploid creamy-white-flowered S. officinale, which are morphologically indistinguishable.
4. Prickly Comfrey (Symphytum asperum):
    About 700 survey cards were received and, on the basis of these returns, it seems that **S. asperum is a very rare species of which only two populations are known.**"
-'**Symphytum Survey' by F.H. Perring,** Watsonia: Journal and Proceedings of the Botanical Society of the British Isles', London, England, Volume 10, Part 3, pages 233-323, January **1975**, page 296.
    (England is generally lower and flatter than the rest of the United Kingdom. It is divided into lowlands {south, east, and midlands}, and rougher terrain of upland areas of north and west. East Anglia is the lowest area of England, which includes the Fens. The highest area is northwest. In England, a 'mountain' is land over 600 meters {1968 feet} with most in northern England. This is useful for determining where different Comfrey species are more likely to grow.)
    ('52 TL' is the United Kingdom's 'Ordnance Survey National Grid' reference number.)
(* -'Network Research, Symphytum Survey' by Franklyn Perring, Biological Records Centre, Monks Wood, Huntingdon, Cambridgeshire County, England; Volume 7, No. 4, 1969. In 'Proceedings of Botanical Society of the British Isles, Volume 7' edited by E.F. Greenwood, London, England, 1967-1969, pages 553-556.)

See sub-subsection 'Perring 1962' in subsection 'Symphytum Genus' in section 'Borage Family, Symphytum Genus' (Chapter 2).

Symphytum Species Classifications 1985 by Stearn

"*The mainland Greek Symphytum species present no taxonomic problems, but the status of the taxa on the Aegean Islands described as S. davisii, S. naxicola, S. cycladense and S. icaricum is debatable,* because they grow on different islands and thus are geographically separated though very closely allied.
I have therefore, after much reluctance to differ from Pawlowski, treated them here as insular (island) vicariants of one species, S. davisii, namely subsp. davisii, naxicola, cycladense and icaricum.
These are low-growing plants, in habit rather like of Onosma and likewise mostly occurring among the rocks."
-'**The Greek Species of Symphytum Boraginaceae' by William T. Stearn**, British Museum of Natural History, London, England; Annales Musei Goulandris, Greece, Volume 7, pages 175-220, (**1985 or 1986**, different sources give different dates).
(Vicariance is the separation of a group of organisms by a geographic barrier that results in differentiation of the original group into new varieties or species.)
(Onosma is a genus of flowering plants in the Boraginaceae family. They are native to the Mediterranean and west Asia. They grow in dry, sunny habitats in rocky, sandy soil.)

Symphytum Species Classifications 1990 by Sandbrink

"*The genus Symphytum (Boraginaceae) is well defined, but infragenetic relationships are less clear and have often caused confusion in literature.* **Most earlier studies on the genus contain descriptions of species, comparisons of morphological variation and geographical distribution patterns. No attempts have been made to describe the evolutionary processes which have led to the present-day variation in the genus.**
In this report, evolutionary processes and phylogenetic relationships in the genus are investigated.
All characters used in the construction of this based on cpDNA (chloroplast DNA) were consistent with the most parsimonious tree found, indicating a high reliability of cpDNA as a tool to reconstruct the evolutionary history of Symphytum.
A bifurcation (2 branches) of two groups of four species was found in this phylogenetic tree. **The results of the analysis of these two datasets combined with literature data on morphology, geographicall distribution, chromosome numbers, chemotaxonomy and interspecific hybridization patterns are used to infer phylogenetic relationships between species of the genus Symphytum.**
It is concluded that hybridization must have played a major role in speciation (species creation) during the evolution of the genus Symphytum.
Almost all Symphytum species are clustered in one of the six groups clearly recognizable in all parsimonious trees. The monophyly (clade) of each of these six groups is in some cases exclusively defined by only one or two shared characters states, but quite often one or two species of a group do not have the character state predominantly present in the group, but do share other typical characters states with species of the same group. **This stresses that only a small part of the morphological variation is actually reflecting the evolutionary history of the genus. The recognized phylogenetic groups are:**
I. S. abchasicum, S. grandiflorum, S. ibericum and S. ciscaucasicum.
II. S. tuberosum (including S. nodosum), S. bulbosum, S. zeyheri, S. cordatum, S. gussonei, S. mediterraneum.
These two groups are tightly linked in a single group on a higher level, with S. tauricum and S. longisetum loosely attached.
III. S. officinale, S. asperum, S. tanaicense, S. armeniacum, S. savvalense, S. hajastanum,
S. sepulcrale (including S. longipetiolatum) and S. peregrinum.
IV. S. ottomanum, S. euboicum, S. pseudobulbosum, S. circinale, S. creticum and S. insulare.
V. S. cycladense, S. anatolicum, S. davisii, S. naxicola and S. icaricum.
VI. S. aintabicum, S. brachycalyx, S. bornmuelleri, S. palaestinum, S. caucasicum, S. kurdicum, S. orientate and S. sylvaticum."
-'**Phylogenetic Relationships in the Genus Symphytum L. (Boraginaceae)**' by J.M. Sandbrink, J. Van Brederode and T.W.J. Gadella, Department of Genome Evolution, University of Utrecht, Netherlands; Proceedings: Koninklijke Nederlandse Akademie van Wetenschappen (Royal Netherlands Academy of Sciences), Series C, Volume 93, No. 3, pages 295-334, **1990**.
(Infrageneric is subclassification within a genus of organisms.)
(A phylogenetic or evolutionary tree shows the relationships among species based on similarities and differences in physical or genetic characteristics.)
(Parsimony is the principle that the simplest explanation that can explain the data is preferred.)
(A monophyletic group or clade is a group of organisms of all descendants of a common ancestor or ancestral population.)

"***Traditionally, the Symphytum genus is divided into 2 to 9 sections, based on various infrageneric classifications*** (Boissier 1879, Kuznetsov 1910, Bucknall 1913, Pawlowski 1961, Wickens 1969, *Sandbrink et al. 1990)."
-'Symphytum Tuberosum Complex in Central Europe: Cytogeography, Morphology, Ecology and Taxonomy' by L. Kobrlova, M. Hrones, P. Koutecky, M. Stech, and B. Travnicek, Preslia: The Journal of the Czech Botanical Society, Czech Republic; Volume 88, pages 77-112, March 2016.

"As summarized by Sandbrink et al. (1990), the Symphytum genus consists of 9 sections; 7 of these were constructed by Bucknall (1913), one by Pawlowski (1971) and one by Wickens (1969).
Suborientalia Section:
    Sandbrink et al. *(1990) placed S. aintabicum to Bucknall's Suborientalia section. In Turkey, S. aintabicum grows in S. brachycalx distribution range. These two clustered with S. sylvaticum.*
Coerulea Section:
    S. asperum and S. armeniacum were proposed to Coerulea section by Bucknall (1913).
    S. armeniacum was first proposed as sub-species of S. asperum by Kurtto (1982) then Greuter et al. (*1984) regarded this taxon as synonym of S. asperum. Phylogenetical analyses supported this synonymy. **Sandbrink et al.** *(1990) placed S. savvalense in this Coerulea section as well.*
Cordata Section:
    S. grandiflorum was suggested for Cordata section by Bucknall (1913). According to **Sandbrink et al.** *(1990) S. ibericum belongs to Bucknall's Cordata section."*
-'Phylogeny of Symphytum L. (Boraginaceae) with Special Emphasis on Turkish Species' by B. Hacioglu and S. Erik, African Journal of Biotechnology, Volume 10, No. 69, pages 15483-15493, November 7 2011.
(*-Med-Checklist: A Critical Inventory of Vascular Plants of the Circum-Mediterranean Countries by W. Greuter, H.M. Burdet and G Long, Organization for the Phyto-Taxonomic Investigation of the Mediterranean Area. Geneva, Switzerland: Editions des Conservatoire et Jardin botaniques de la Ville de Geneve; and Berlin, Germany: Botanischer Garten and Botanisches Museum Berlin-Dahlem, 1984.)

Symphytum Species Classifications 1991 by Arnold Arboretum

*"**Symphytum is a well-defined Old World (Europe and Asia) genus of about 35 species distributed primarily in Turkey (20 species, nine endemic = native), the Caucasus (11 species, four endemic), and the Balkan Peninsula (southeast Europe) and Aegean Islands (11 species, four endemic).***
    Symphytum is poorly represented in Europe and Central Asia and apparently has no indigenous (native) taxa beyond those continents.
    The genus is represented in North America by three naturalized species *(Symphytum officinale, Symphytum asperum, Symphytum tuberosum) and in the southeastern United States by one (Symphytum officinale).*
Bucknall's division of Symphytum into two 'Subgenera' and seven Sections has since been amended by *Gviniashvili (or Ts.N. Gviniaschvili, Gviniaviliz) (1969, 1972), Pawlowski (**1961, 1971), and Wickens (***1969).
As many as ten Sections and several Series are now recognized."
-'**Journal of the Arnold Arboretum**, Supplementary Series, Volume 1, Harvard University, Cambridge, Massachusetts, **1991**. This volume is a collection of contributions toward a '**Generic Flora of the Southeastern United States**', by Ihsan A. Al-Shehbaz, Missouri Botanical Garden, Saint Louis, Missouri, Symphytum pages 149-158.
(* -'Review of the Caucasian Species of Genus Symphytum L.' by Tsiala Gviniashvili, Scientist, LEPL Ilia State University Institute of Botany, Tbilisi, Georgia; International Caucasian Foresty Symposium, Artvin Coruh University, Artvin, Turkey, October 24-26 2013.)
(** -'Observations ad (on) Genus Symphytum L. Pertinentes' by Bogumilus Pawlowski, Fragmenta Floristica et Geobotanica, Poland, Volume 7, No. 2, pages 327-356, 1961. All in Polish. If you have an English translation, could you please send it to me.)
(*** -'A Revision of Symphytum in Turkey and Adjacent Areas' by G.E. Wickens, Royal Botanic Gardens: Kew; Notes from the Royal Botanic Garden Edinburgh, Scotland, Volume 29, pages 157-180, 1969. Includes Turkey, Bulgaria, Greece, Aegean Islands and Caucasia.)
(The Aegean Islands are in the Aegean Sea, with mainland Greece to the west and north, and Turkey to the east.)
(Taxonomic ranks: Genus, Subgenus, Section, Subsection, Series, Species, Subspecies, Variety, Subvariety, Form, Subform.)

Symphytum Species Classifications 2000 by Slavik

"Symphytum L.:
Section 1, Symphytum Officinalia Buckn.
    Symphytum bohemicum F.W. Schmidt
    Symphytum Symphytum officinale L.
Section 2, Caerulea Bukn.
    Symphytum asperum Lepechin
    Symphytum x uplandicum Nyman
Section 3, Tuberosa Buckn.
    Symphytum tuberosum L."
-Kvetena Ceske Republiky (Flora of the Czech Republic), 9 Volumes, edited by B. Slavík, J. Jun. Chrtek and J. Stepankova. Praha (Prague), Czech Republic: Academia- Academy of Sciences of the Czech Republic, 1998-2010. 'Symphytum L. Kostival' in Boraginaceae Family, article by B. Slavik, Volume 6, pages 202-210, **2000**. All in Czech. (If you have an English translation, I would appreciate a copy.)

Symphytum Species Classifications 2011 by Hacioglu and Erik

"The genus Symphytum L. is a mesophytic genus with approximately 40 species, with a centre of origin in the Pontic province of the Euro-Siberian region. With 18 species Turkey holds the largest number (*Tarikahya, 2010).
The distribution area of the genus is mainly Euro-Siberian and Mediterranean regions."
-'**Phylogeny of Symphytum L. (Boraginaceae) with Special Emphasis on Turkish Species**' by B. Hacioglu and S. Erik (Central Research Institute for Field Crops, Ankara, Turkey; Hacettepe University, Ankara, Turkey; Berlin Free University Biology Institute, Germany); African Journal of Biotechnology, Nigeria, Volume 10, No. 69, pages 15483-15493, November 7 **2011**.
(* -'The Revision of Turkish Symphytum L. {Boraginaceae} Genus' by B. Tarikahya, Ph.D. thesis, Institute for the Graduate Studies in Science and Engineering, Department of Biology, Botany Section, Hacettepe University, Ankara, Turkey, (2010). All in Turkish. If you have an English translation, could you please send it to me.)
(Mesophyte is a plant growing under conditions of well-balanced moisture supply.)

"Morphology: There are two main clusters and four Symphytum species groups as follows:
Group 1:
    S. sylvaticum, S. savvalense and S. asperum belong to this group.
        These species are characteristic with having robust stem, pleiocormus under the ground and many blue flowers in the inflorescence.
    S. sylvaticum was proposed to Albida section by Bucknall (*1913). However this species is now combined with S. sepulcrale under the name S. sylvaticum (**Tarikahya and Erik, 2010).
    S. asperum was proposed to Coerulea Section by Bucknall and S. savvalense was proposed to the same section by Sandbrink et al. (1990).
        S. asperum has the widest distribution range. It grows naturally in Georgia, Caucasus and in North Iran.
In Group 1, two of the three species are endemic (native) to Turkey (S. savvalense and S. sylvaticum). This group of plants is very similar in habit but they also have some particular characters to be named as separate species.
S. sylvaticum: Lower stem leaves are asperous puberulous, corolla 10 to 14(-17) mm.
S. savvalense: Corolla about 10 mm, tube 4 to 5 mm, corolla distinctly concolor (one color).
S. asperum: Corolla longer than 10 mm, tube 5 to 9 mm, corolla discolor (changes color), lower stem leaves are asperous, corolla 14 to 20 mm.
Group 2:
    This group of species was proposed for Orientalia section by Bucknall *because of morphological similarities of calyx, stem and flowers (Bucknall, 1913). In Group 2 there are morphological similarities between* **S. orientale and S. kurdicum**. *However, distribution areas were far away from each other.*
Group 3:
    This group of plants grows in Mediterranean region or Mediterranean-like habitats.
    The characteristic properties of this group are having slender stem and branched roots.
    S. bornmuelleri, S. brachycalyx were proposed to Suborientalia section by Bucknall (1913).
    S. longisetum and S. aintabicum were proposed to the same section by Sandrink et al. (1990).
        Except for S. brachycalyx and S. tauricum, all species of this group are endemic.
        In Group 3, S. brachycalyx had the widest distribution area: from Taurus (mountain) range (southern Turkey) till Philistine (northwest Syria and southeast Turkey) in south.
        S. longisetum grows only in Mersin (south Turkey) on S. brachycalyx's distribution range, but morphologically, it is more closely related to S. anatolicum which grows in the Aegean region of Turkey.
        S. anatolicum can be considered as the transition species between Greek species and Turkish species.
        S. bornmuelleri is an endemic species that is distributed in the north of central Anatolia (Asian Turkey) that has morphological affinities with S. brachycalyx.
        S. aintabicum has a small distribution area in east of Mediterranean region.
        S. tauricum grows in Russia, Ukraine, Romania and Bulgaria, but only in Sinop in (north)Turkey.
    S. brachycalyx, S. anatolicum and S. tauricum seems to be the dominant species of this group. *Most probably other species of this group are variated from these species in particular habitats.*
Group 4:
    These species are characterized with exserted corolla scales and they grow in Marmara region in (northwest) Turkey.
    All species of this group were proposed to different sections by Bucknall (1913):
        S. pseudobulbosum to Orientalia, S. bulbosum to Tuberosa and
        S. ottomanum to Albida section.
    However, S. tuberosum, S. officinale, S. circinale and S. ibericum did not form a group with Turkish species. *In Group 4, only S. pseudobulbosum is endemic.*
    S. bulbosum has the widest distribution range in this group (very common in south Europe).
    Also according to phylogenetic tree, it seems to be the ancestral species of this group.
Bucknall's divisions (Ramosa and Simplicia) and subdivisions (Ramosa, Segmentata, Dentata) were not supported in molecular level. *The molecular data currently available do not coincide with the Bucknall's Orientalia, Suborientalia and Albida sections.*

Sections are revised due to molecular data, morphological investigations of several plant material and field observations. This investigation covered 24 species growing in Turkey and adjacent areas.

Tuberosa Buckn., Suborientalia Buckn., 'Cordata Pawl.', Graeca Pawl. and Procopiania Wick. sections were reserved (stay the same), as well as Coerulea Buckn. section. However in this section, **S. armeniacum is synonym of S. asperum** (Greuter et al., 1984) and **S. sepulcrale is synonym of S. sylvaticum** (Tarikahya and Erik, 2010).

> Hence, due to morphological affinities, S. sylvaticum should be transferred from section Albida to Coerulea, while S. pseudobulbosum should be transferred from the section Orientalia to section Albida.

Distribution of investigated species to sections is as follows:
- ***Officinalia Buckn.-*** S. officinale
- ***Coerulea Buckn.-*** S. asperum, S. sylvaticum, S. savvalense
- ***Albida Buckn.-*** S. tauricum, S. anatolicum, S. ottomanum, S. pseudobulbosum, S. longisetum
- ***Orientalia Buckn.-*** S. orientale, S. kurdicum
- ***Suborientalia Buckn.-*** S. brachycalyx, S. bornmuelleri, S. aintabicum
- ***Tuberosa Buckn.-*** S. tuberosum, S. bulbosum, S. gussonei
- ***Cordata Buckn.-*** S. ibericum, S. cordatum
- ***Procopiania Wick.-*** S. circinale, S. creticum, S. insulare
- ***Graeca Pawl.-*** S. icaricum, S. naxicola "

-'Phylogeny of Symphytum L. (Boraginaceae) with Special Emphasis on Turkish Species' by B. Hacioglu and S. Erik (Central Research Institute for Field Crops, Ankara, Turkey; Hacettepe University, Ankara, Turkey; Berlin Free University Biology Institute, Germany); African Journal of Biotechnology, Nigeria, Volume 10, No. 69, pages 15483-15493, November 7 **2011**.
(* -'A Revision of the Genus Symphytum, Tourn.' by Cedric Bucknall, {Mus. Bac. Oxon= Bachelor of Music, Oxford University}, Journal of the Linnean Society of London, England, Botanical Journal, Volume 41, Issue 284, pages 491-556, December 1913.)
(** -'The Revision of Turkish Symphytum L. {Boraginaceae} Genus' by B. Tarikahya, Ph.D. thesis, Institute for the Graduate Studies in Science and Engineering, Department of Biology, Botany Section, Hacettepe University, Ankara, Turkey, 2010. All in Turkish. If you have an English translation, could you please send it to me. In an online translation, only this sentence made sense: **"The Symphytum genus had 21 taxa according to 'Flora of Turkey', however, this number was reduced to 18 with our survey."**)

(I could not find the definition of 'pleiocormus'. However, 'pleio' = plio = pleo means more. 'Cormus' means an organism made up of a number of individuals, such as would be formed by a process of budding from a parent stalk where the buds remain attached.)

(A phylogenetic or evolutionary tree shows the relationships among species based on similarities and differences in physical or genetic characteristics.)

(They may have meant 'Cordata Buckn.' rather than 'Cordata Pawl.')

"The vegetation in the region of Kastamonu, Turkey is mostly part of the Pontic (Black Sea region of Turkey) province of Euro-Siberian territory, but also has similarities with the vegetation types of both the Mediterranean and the Balkans. In fact, although the vegetation within the hilly zone of the Kastamonu region is under Euxine (Black Sea) influence, the Mediterranean (Sea) influence is also quite important; and the montane zone is undoubtedly affected by Euro-Asiatic types.

There are 2 regions of Turkey where summer rainfall is heavier than in winter. The first is Kastamonu, North of Ankara, where the situation is rather confused because the vegetation here is Euxine in character and the region should be considered as being under the influence of a Continental type of Oceanic climate. The Kastamonu region is appreciably less cold and the minimum values of the coldest month are between -5 and -6 C (21-23 F)."

-'The Climate and Vegetation of Turkey' by Y. Akman and O. Ketenoglu, Department of Biology, Ankara University, Turkey, Proceedings of Royal Society of Edinburgh, Section B: Biological Sciences, Volume 89 B, pages 123-134, 1986.

(Kastamonu is the capital district of the Kastamonu Province, Turkey. Kastamonu has a humid continental climate using the 32 degree isotherm, with cold winters and warm summers. The average annual precipitation is evenly distributed throughout the year, with spring being the wettest season. This environmental information in Turkey is included because it gives insight into what Comfrey likes and how it developed.)

Symphytum Species Classifications 2013 by Gviniashvili

"***Caucasian species of genus Symphytum L., 10 in all, are represented in the Caucasus by 3 Sections:***
   1. Section Coerulea Buckn. (S. asperum Lepech., S. peregrinum Ledeb.,
   S. hajastanum Gviniaschvili, S. caucasicum Bieb.).

> In Section Coerulea, which we relate to sub-endemic Caucasian-Asia-Minor Section, together with S. asperum, three closely related species: S. peregrinum, S. hajastanum and S. caucasicum should be mentioned.
> **S. peregrinum** is a hircanic species, whose geographic distribution may be extended to northern part of Iran.
> **S. caucasicum**, a somewhat isolated, endemic (native) Caucasian species.
> Taxa remarcated morphologically, geographically and coenotically are also characterized by different chromosome numbers: **S. asperum 2n = 32, S. peregrinum 2n = 40, S. caucasicum 2n = 24.**

   2. Section Lingulata Pawl. (S. ibericum Stev., S. grandiflorum DC., S. abchasicum Trautv.,
   S. ciscaucasicum Gviniaschvili, S.tauricum Willd.).

S. ibericum and *S. grandiflorum* are vicarious species.
    S. ibericum is distributed in western part of Caucasus and north-eastern Anatolia (Asia Minor, Turkey).
    S. grandiflorum occupies a small area in eastern Transcaucasus (South Caucasus).
S. abchasicum, the local endemic species of Abchasia (northwest Georgia), is characteristic for mixed oak woods and endemic species of north-western part of the Caucasus.
S. ciscaucasicum lives in beech woods.
Along with coenotic similarity, these closely related species are characterized by geographical isolation and caryological differentiation: **S. ibericum 2n = 24;**
S. grandiflorum 2n = 60, S. ciscaucasicum 2n = 36. These endemic species represent ancient, tertiary period autochtonic (aboriginal, indigenous) groups of Caucasian flora.
S. tauricum can be found more or less sporadically. The characteristic disjuncted (not joined) area indicates that S. tauricum belongs to the group of relict circum-euxine (around Black Sea) species.
   3. **Section Symphytum Officinalis** *(S.officinale L.)*.
    Symphytum is related to typical elements of Eastern-Mediterranean flora. Polymorphic (several forms) species *S. officinale*, which is characteristic of deciduous shrublands, is broadly distributed in whole Europe, but it is not characteristic of Caucasian flora, and is known only from the North Caucasus.

The study of Caucasian species of Symphytum indicates the active process of speciation (creating new species) within old relict groups in different populations and coenosis."

-'**Review of the Caucasian Species of Genus Symphytum L.**' by **Tsiala Gviniashvili**, Scientist, LEPL Ilia State University Institute of Botany, Tbilisi, Georgia; International Caucasian Foresty Symposium, Artvin Coruh University, Artvin, Turkey, October 24-26 **2013**.

   (The Caucasus Mountains are located at the border of Europe and Asia, between the Black Sea and Caspian Sea and occupied by Russia, Georgia, Azerbaijan, and Armenia.)
   (Endemic or native species are from only one geographic area or zone such as an island, country, continent or habitat type. Habitat restriction occurs due to climate, physical barriers and biology. Sub-endemic species are mostly in one area but also grow in other areas.)
   (Hircanic means about Hyrcania, an ancient country of Asia, southeast of the Caspian {Hyrcanian} Sea.)
   (Coenospecies is a collection of related species able to hybridize.)
   (Vicarious species are closely related species distributed in various areas or found in the same area but in different ecological conditions.)
   (Caryological or karyological is the cytological characteristics of the cell nucleus especially the chromosomes.)
   (Relict means a surviving species of an otherwise extinct group, or a remnant of a formerly widespread species that lives in isolated areas.)

Symphytum hajastanum Gvin. {Ts.N. Gviniaschvili} Zametki Sist. Geogr. Rast. 26: 73. 1967 ; Not. Syst.; {Tbilisi}. Notes: U.S.S.R. {Caucas.}. Illus. = 'Zametki po Sistematike i Geografii Rastenii'. Published by 'Tbilisis Botanikuri Instituti', Georgia, Volume 26, 1967. In Russian.

Symphytum Species Classifications 2018-2019
    Names of common species with synonyms and chromosome numbers.
    This taxonomy changes over time as research continues.

S. asperrimum Sims = S. asperrimum Donn ex Sims = S. asperrimum Donn = S. asperum Lepech. $2n = 32$.

S. bohemicum F.W. Schmidt = S. officinale L. subsp. bohemicum (F.W. Schmidt) Celak. $2n = 24$.
(There is lack of agreement about the terminology for S. bohemicum, S. tanaicense, and S. uliginosum.)

S. bulbosum K.F. Schimp. = S. tuberosum L. subsp. bulbosum (K. F. Schimp.) P. Fourn.
    $2n = 48, 96, 104, 120$.

S. caucasicum M. Bieb. $2n = 48$.

S. grandiflorum DC. (Some say = S. ibericum Steven.) $2n = 60$.

S. ibericum Steven = S. ibericum Steven ex M. Bieb. $2n = 24, 60$.

S. officinale L. $2n = 24, 40, 48$ are the most common; others exist.
    S. officinale L. subsp. bohemicum (F.W. Schmidt) Celak.
    S. officinale L. subsp. tanaicense Steven
    S. officinale L. subsp. uliginosum (Kern.) Nyman
        S. uliginosum non Kern = S. officinale L. subsp. uliginosum non (Kern) Nyman

S. orientale L. = S. tauricum sensu H. J. Coste, non Willd. $2n = 32$.

S. peregrinum auct. non Ledeb. = S. x uplandicum Nyman (pro sp.)   (pro sp. = hybrid)   2n = 36, 40.
S. peregrinum Ledeb.
    "Symphytum peregrinum Ledeb. is an unresolved name."
    -The Plant List®, www.theplantlist.org from the World Checklist Database, 2018.
    (There is confusion among botanists and others about the correct meaning of 'S. peregrinum Ledeb.'.)

S. tanaicense Steven = S. officinale L. subsp. tanaicense Steven   2n = 40.
(Some say S. tanaicense Steven = S. uliginosum Kern.)
(There is lack of agreement about the terminology for S. bohemicum, S. tanaicense, and S. uliginosum.)

S. tauricum sensu H. J. Coste, non Willd. = S. orientale L.
S. tauricum Willd.   2n = 18.

S. tuberosum L.   2n = 32, 64, 96, 120, 128, 144.
S. tuberosum L. subsp. angustifolium (A. Kern.) Nyman = S. tuberosum L. subsp. nodosum (Schur) Soo
    2n = 18, 72, 96, 100.
S. tuberosum L. subsp. bulbosum (K. F. Schimp.) P. Fourn. = S. bulbosum K.F. Schimp.

S. uliginosum Kern. = S. officinale L. subsp. uliginosum (Kern.) Nyman   2n = 40.
    S. uliginosum non Kern = S. officinale L. subsp. uliginosum non (Kern) Nyman
(There is lack of agreement about the terminology for S. bohemicum, S. tanaicense, and S. uliginosum.)

S. x uplandicum Nyman (pro sp.) = S. peregrinum auct. non Ledeb.   (pro sp. = hybrid)   2n = 36, 40.

Taxonomy of Symphytum Species is Always Changing

*"**The chemotaxonomic hypothesis**, proposed by Gadella and collaborators, based on the presence of the triterpene isobauerenol in S. officinale and its absence in S. asperum Lepech, and the presence of the
pyrrolizidine alkaloid echimidine in S. asperum and its absence in S. officinale, can no longer be applied absolutely to the S. officinale species complex.*
The pyrrolizidine alkaloid and triterpene pattern of S. officinale (2n = 24) and S. bohemicum (2n = 24) is identical. S. bohemicum is morphologically, cytologically and phytochemically very similar to S. officinale.
Furthermore, it readily crosses with the white flowered western European diploids of S. officinale. Therefore it seems likely that these two taxa are conspecific."
-'Chemotaxonomy of the Symphytum Officinale Agg. (Boraginaceae)' by Jaarsma, Lohmanns, Gadella and Malingre, Plant Systematics and Evolution, 167, pages 113-127, **1989**.
    (Chemotaxonomy uses phytochemical {biologically active compounds in plants} data to determine the proper taxonomic category.) (Conspecific means belonging to the same species.)

    "Symphytum tanaicense shows a pyrrolizidine alkaloid and triterpene pattern similar to S. officinale (2n = 40). Also on morphological and cytological grounds they are very similar.
    It seems highly probable that S. tanaicense is conspecific with S. officinale (2n = 40) and represents an intraspecific variant only.
    S. officinale var. lanceolatum contained no pyrrolizidine alkaloids but did contain isobauerenol. This feature points to an origin from S. officinale."
    -'Chemotaxonomy of the Symphytum Officinale Agg. (Boraginaceae)' by Jaarsma, Lohmanns, Gadella and Malingre, Plant Systematics and Evolution, 167, pages 113-127, **1989**.
        (Intraspecific means occurring within a species or involving members of one species.)

"These Symphytum studies showed that experimentally based ecological and genetic studies in a large polymorphic plant population over several years may provide adequate information about the relationships between pattern and process. The processes underlying the patterns should be studied experimentally.
Biosystematics combines the study of pattern and process and is therefore of vital importance for the investigations of plant evolution. *The role of botanical gardens for making contributions to our knowledge of population variation and evolution continues to be invaluable."*
-'Hybridization in Symphytum: Pattern and Process' by T.W.J. Gadella, Sommerfeltia: The Journal of Natural History Museum and University of Oslo, Norway, Volume 11, pages 79-96, **1990**.
        (Morphology is a branch of biology that studies the form of living organisms, and the relationships between their structures. This includes both outward and internal appearance {shape, structure, color, pattern, size}.)
        (Polymorphism is two or more different morphs, forms or phenotypes in a species. Phenotype is the observable characteristics of an individual from the interaction of its genotype with the environment.)
        (Biosystematics is taxonomy based on the study of the genetic evolution of populations.)

"These changes are an attempt to reflect more accurately both the historical details of the naming of taxa (the nomenclature) and the increase in knowledge about the actual relationships between organisms (the taxonomy).
    Nomenclature: *Sometimes a species name will change as a result of nomenclatural research, for example, because someone has discovered that there is an older, perfectly valid name, for the same taxon.*
    Taxonomy: *Ever since Darwin, the taxonomy of organisms is required to attempt to reflect their phylogeny, in other words, how organisms are classified is supposed to represent their tree of descent.*
Sometimes it is discovered that a species needs to be moved to another genus, or even to a brand new genus. Sometimes a genus is broken up into many genera. Sometimes a family is broken up into many families.
Currently in biology, molecular research is leading to a lot of discoveries about taxa at every level, so a great number of taxonomic changes are resulting, and will continue to result from this."
-'Why do Scientific Names Change?' by Susan J. Hewitt, www.inaturalist.org, March 3, **2016**. Hewitt is a malacologist (studies mollusks) with 49 scientific papers published. Originally from Britain, now living in New York. At 'iNaturalist Network' citizen scientists share biological data with each other. It is made available to 'Global Biodiversity Information Facility', www.gbif.org.

"Plant taxonomy is well known for being turbulent, and traditionally not having any close agreement on circumscription and placement of taxa.
    Circumscription: *the definition of what does and does not belong to a given taxon, from a particular taxonomic viewpoint or taxonomic system.*
    Taxa *is the plural of taxon. Taxon is a group of one or more populations of an organism or organisms seen by taxonomists to form a unit."*
-Wikipedia®: The Free Encyclopedia, www.wikipedia.org, **2018**.

Symphytum Sections:
The number of Sections (varying from 2 to 10) and the species in each change with new research.
Even now there are disagreements about where some Symphytum species belong. A few species are listed:
    Section Albida Bucknall
        S. anatolicum, S. longisetum, S. ottomanum, S. pseudobulbosum, S. tauricum
    Section Coerulea Bucknall
        S. asperum Lepech., S. caucasicum Bieb., S. hajastanum Gviniaschvili,
        S. peregrinum Ledeb., S. savvalense, S. sylvaticum
    Section Cordata Bucknall
        S. cordatum, S. ibericum
    Section Graeca Pawlowski
        S. icaricum, S. naxicola
    Section Lingulata Pawlowski
        S. abchasicum Trautv., S. ciscaucasicum Gviniaschvili, S. grandiflorum DC.,
        S. ibericum Stev., S.tauricum Willd.
    Section Orientalia Bucknall
        S. kurdicum, S. orientale
    Section Procopiania Wickens
        S. circinale, S. creticum, S. insulare
    Section Suborientalia Bucknall
        S. aintabicum, S. bornmuelleri, S. brachycalyx, S. longisetum, S. palaestinum Boiss.
    Section Symphytum Officinalis
        S.officinale L., S. uliginosum Kern.
    Section Tuberosa Bucknall
        S. bulbosum, S. gussonei, S. tuberosum L. and others

Methods of Surveying and Collecting Symphytum in the Field

"If all three of these taxa (S. officinale, S. asperum, S. x uplandicum) hybridise freely, the origin of the Symphytum problem is not far to seek; recorders will have had great difficulty in applying names correctly to the plants which they found. It is for this reason that we are asking 'Botanical Society of the British Isles' members to provide not names, but a few simple measurements on 6 stems in a population on the record cards provided.
    For this purpose a population consists of a group of plants growing in the same locality and in the same
    habitat. All plants from one population should be within 50 yards (150 feet = 45.7 meters) of each other.
If the Symphytum population is variable, stems should be selected to reflect that variation. Six stems should be sufficient, but a second card could be used for the same locality if necessary.
Observations should be made when some flowers are fully open, and flowers which have just opened should be chosen for measuring. When buds and flowers differ in colour, this can be noted by putting b = bud and o = open flower in the appropriate box on the card.
Decurrence has to be based on an average for the stem; wings which protrude less than 1 mm, or lines on the stem should be ignored.

In general, measurements are not required from more than one population of apparently similar plants in each 10 kilometer square (3.8 square miles)."
-'Network Research, Symphytum Survey' by Franklyn Perring, Biological Records Centre, Monks Wood, Huntingdon, Cambridgeshire County, England; Volume 7, No. 4, **1969**. In 'Proceedings of Botanical Society of the British Isles, Volume 7' edited by E.F. Greenwood, London, England, 1967-1969, pages 553-556.

    (The 'Symphytum Survey' of the British Isles began in 1954. These surveys are ongoing, and you can volunteer by contacting the 'Botanical Society of Britain and Ireland', https://bsbi.org.)

"Symphytum is a difficult, critical genus, and it is necessary to make notes on the colour of both buds and mature flowers in the field, in order to make a confident determination."
-'Symphytum Asperum Lepechin: A New Irish Species' by J.A.N. Parnell, School of Botany, Trinity College, Dublin, Ireland; The Irish Naturalists' Journal Ltd., Holywood, County Down, Ireland, Volume 21, No. 11, pages 498-499, July **1985**.

"How to Collect Plants for a Voucher Specimen:
For herbaceous plants always collect at least part of the root system, stems, leaves, and, if possible, the flowers and/or fruit, and seeds. *Clean all soil from roots (rinse off mud, then brush when dry).*
Arrange your plant specimen on the newspaper carefully, so that it looks as natural as possible, without too many parts overlapping each other. Turn at least one leaf upside down so the undersurface can be seen when pressed.

    Sandwich herbs between two blotters and then two cardboard sheets. Drying can be done at room temperature or over a couple of heat lamps, but take care not to overheat the press which would discolour the plants.
    Arrange the plants carefully with a minimum of overlap on the herbarium page (acid-free mounting paper).
    Store the herbarium pages in a dry and dark place.

Voucher specimen label: *Scientific Name, Common Name, Location/ Habitat, Collector, Collection Date, Identification Number, Identified By."*
-'Good Practices for Plant Identification in the Herbal Industry' by Tim Brigham, Michelle Schroder and Wendy Cocksedge, Centre for Non-Timber Resources, Royal Roads University, Victoria, British Columbia, Canada;
'Saskatchewan Herb and Spice Association' and 'National Herb and Spice Coalition', Agriculture and Agri-Food Canada, February **2004**.

    (Herbaceous plants have little or no woody tissue. It can be an annual, biennial or perennial. The leaves or entire plant dies back every year when the weather gets cold.)

"What is a voucher and why is it important in research? As a preserved specimen of an identified taxon deposited in a permanent and accessible storage facility, the voucher serves as the supporting material for published studies of the taxon and ensures that the science is repeatable.
Vouchers are crucial in authenticating the taxonomy of an organism, as a tool for identifying localities of the taxon, and for additional taxonomic, genetic, ecological, and/or environmental research.

    A voucher can be broadly defined as a representative sample of an expertly identified organism that is deposited and stored at a facility from which researchers may later obtain the specimen for examination and further study.
    As such, a voucher is a specimen that has been specifically collected and accessioned to support a research project (e.g., genetic analysis of a taxon) or activity (e.g., a floristic survey of a park).
    For plants, a voucher typically consists of a herbarium specimen, a pressed and dried sample of an individual containing aboveground structures (leaves, stems, flowers, and/or fruits) and below ground structures when possible.

The voucher must include an identification label that ideally lists the recognized scientific name of the plant, its accepted taxonomic authority, the name of the person who identified the sampled plant, the collector's name, date of collection, habitat of the collection site, locality of the site (preferably consisting of GPS coordinates, with the corresponding datum and degree of accuracy), and perhaps a collection number assigned by the collector; a unique accession number is sometimes added later by the herbarium.
The specimen might also include annotations (notes) indicating taxonomic changes made by an expert familiar with the taxon. The voucher specimen should be deposited in an official herbarium that is located at a recognized institution and committed to the long term maintenance of its collection, such as those herbaria registered with 'Index Herbariorum' (http://sweetgum.nybg.org/ih)."
-Why Vouchers Matter in Botanical Research' by Theresa M. Culley, Editor-in-Chief, Applications in Plant Sciences, Department of Biological Sciences, University of Cincinnati, Ohio; Applications in Plant Sciences: Official Publication of the Botanical Society of America, Volume 1, No. 11, 5 pages, November **2013**.

    (A taxon is a taxonomic group of any rank, such as a species, family, or class.)
    (An voucher example: 'Voucher Specimen No. CANB 286704 Australian National Herbarium, Canberra'.)
    (Accession is an addition to a plant collection, usually in reference to an herbarium.)
    (Species of Comfrey are sometimes incorrectly identified in research reports. See sub-subsection 'Proper Botanical Identification is Needed' in subsection 'More Alkaloid Research is Needed' in section 'Alkaloids in Comfrey' in Volume 2.)

"The first 'Atlas of the British Flora', *published in 1962, pioneered the use of 'dot-maps' aligned to the OS (Ordnance Survey) grid and was hugely influential, leading to the first 'Red Data List for Great Britain' and establishing the methods used to map both plants and animals across Europe and North America.*

A repeat 'Atlas' was published in 2002 based on fieldwork carried out by BSBI (Botanical Society of Britain and Ireland) members and others from 1987-1999, with the results of this monumental work also having a far reaching impact. BSBI is now producing Atlas 2020. The scheme covers Britain, Ireland, the Isle of Man and the Channel Islands.
For more detail on what to record, including taxa which frequently cause confusion or are difficult to identify, see Stroh et al. (2015) 'Notes on Identification Works and Some Difficult and Under-Recorded Taxa'.
For basic information on field recording, see pdf by Ellis (2015): 'A Beginners Guide to Recording'."
-'Atlas of the British and Irish Flora 2020 Instruction Booklet' by 'Botanical Society of Britain and Ireland' and 'Cambridge University Botanic Gardens' (Peter A. Stroh), England, June **2015**.

> "The standard flora for the 'Atlas 2020' project is edition 3 of C.A. Stace's 'New Flora of the British Isles', Cambridge University Press, 2010.
> **Symphytum:** In addition to Stace, pages 555-558, see *Rich & Jermy, pages 235-236, and **Sell & Murrell, Volume 3, pages 358-363, which briefly describes some of the Symphytum x hidcotense cultivars.
> **Symphytum officinale subsp. officinale, subsp. bohemicum:**
> Subsp. officinale is less common than and over-recorded for S. x uplandicum, which itself is fertile and backcrosses to subsp. officinale forming intermediates.
> Subsp. bohemicum has only recently been discovered by British botanists and requires further study. Stace, page 557."

-'Notes on Identification Works and Some Difficult and Under-Recorded Taxa' by P.A. Stroh, D.A. Pearman, F.J. Rumsey and K.J. Walker; 'Botanical Society of Britain and Ireland', Bristol, England, and 'Biological Records Centre', Gifford, England, 43 pages, **2015**.
(* -Plant Crib: Handbook for Field Identification compiled by T.C.G. Rich and M.D.B. Rich with editorial assistance of F.H. Perring for the BSBI Monitoring Scheme. London, England: Botanical Society of the British Isles, 1988. There is also an updated 1998 edition by T.C.G. Rich and A.C. Jermy.)
(** -Flora of Great Britain and Ireland by Peter D. Sell and Gina Murrell. England: Cambridge University Press. Updates- Volume 1: 2018; Volume 2: 2014; Volume 3: 2009; Volume 4: 2006; Volume 5: 1997. Symphytum is in Volume 3.)

"**Specimens may need to be collected so that correct identifications can be made and a checkable voucher preserved for posterity. Specimens are also often valuable in especially difficult groups so they can be used for taxonomic study.**
As a general rule, collect only when you can take a good representative specimen without doing significant harm to the population, preserve it well so as to show the diagnostic features, label it properly, and if possible deposit it in a secure and well-curated herbarium.
Effective collecting is chiefly a matter of common sense, in that **one needs to collect a specimen that will show as many as possible of the diagnostic features of the genus or group.**
The more one knows about a genus, the better the specimens one is able to collect, and the more likely one is to get it identified. It is no accident that the best specimens are usually collected by the best authorities on the group in question.
Try to collect typical specimens, or several to show the range of variation; if in doubt, indicate the limits of variation on the label. Collect fruits (nutlets) as well as flowers if both are available. Remember that the lower leaves can differ from the upper ones, and do not just collect the top of the stem or the inflorescence.
For hybrids, as well as collecting the range, it is often helpful to collect the supposed parents from the same area.
Boraginaceae: Always record flower colour and size, and especially for Symphytum press a separate corolla opened out."
-'Collecting and Pressing Specimens' by Arthur O. Chater; Windover, Penyrangor, Aberystwyth, Dyfed, Wales, date unknown but around 2008-**2018**.

"**Drying of the Specimens Symphytum kurdicum and Symphytum tuberosum:**
> After the plant collection from the fields, the living plants were treated in Sulaimani Polytechnic University, Technical College of Applied Sciences Herbarium (SPUH), Iraq. The plants have been dried by utilizing newspapers and cardboard in a room temperature for **absorbing moisture specimens by newspapers. Repeating the process of drying for 7 days was done to make the herbarium specimens.**

Mounting and Labeling of Specimens:
> The specimens were held very carefully then adhered to the white rectangular boards that have standard size. **The labeling was done by giving them special numbers, herbarium name, district name, scientific name, common names, position, date of collection, the altitude, GPS coordination, type of soil and the name of collector with other required information.** Then, the samples were saved in the herbarium of Sulaimani Polytechnic University."

-'A Comparative Systematic Study of the Genus Symphytum L. (Boraginaceae) with New First Record of the Species Symphytum Tuberosum L. from Iraq' by Adel Mohan Aday Al-Zubaidy and Sherzad Rasul Abdalla Tobakari, Plant Production Dept, Technical College of Applied Sciences, Sulaimani Polytechnic University, Iraq; Plant Archives, Vol.18, No. 2, pages 2068-2076, Nov **2018**.

Methods of Authentication of Comfrey Species

Methods include morphology, overall habit and habitat of plant, standard microscopic analysis, molecular testing, chemical analysis such as chromatography, genetic analysis of DNA / chromosomes, and others. Morphology studies the form of living organisms, and relationships between their structures. This includes both the outward and internal appearance {shape, structure, color, pattern, size}. Roots, stems, leaves, flowers, inflorescences, pollen, fruits and nutlets are examined.

See subsection 'Methods of Determining Proper Taxonomic Relationships' in section 'Taxonomy and Nomenclature of Comfrey' (Chapter 1). See sections 'Comfrey Heals: Allantoin' and 'Alkaloids in Comfrey' for comfrey chemical analysis in Volume 2.

For help identifying a live Comfrey plant, see subsection 'Overview Comparison of Symphytum Species including Keys' in section 'Symphytum Species Overview' (Chapter 6).

"Testing Techniques:
    *1. **Macroscopic/Organoleptic (sense organs):*** *The unaided senses of sight, smell or taste. Included here, however, would be use of a hand lens with 4x to 20x magnification for visual identification. These methods are typically used for an herb or botanical ingredient in whole or uncut form.*
    *2. **Microscopic analysis:*** *Use of higher magnification than provided by a hand lens, and special light or staining techniques to examine powdered or chopped representative sample material. Analysis is based on observation of specific microscopic characteristics that have been established for the specific dietary supplement ingredient. This analysis also is used to identify some adulterants.*
    *3. **Chemical analysis:*** *Chromatography: Techniques that are based on the differential affinities of substances for a gas or liquid mobile medium and a stationary adsorbing medium. Analysis is based on observational comparison of a test pattern and a reference chromatogram for the dietary supplement ingredient of interest.*
Certificate of Authenticity:
    Scientific Name: *(Genus, Species), Common Name, Cultivated or Wildcrafted.*
    Plant Part: *Fruit/Seed, Inflorescence, Whole Plant, Aerial (Leaf, Stem), Bark, Root Bark, Root.*
    Country and Province/State of Origin*, Date of Harvest.*
    Stage of plant development at time of harvest: *Pre-bloom, In bloom, Post bloom, Dormant.*
    Batch/Lot/Shipment Identification Number: *Grower/Harvester Identification Number."*
    Certification: *I certify that this plant material is correctly identified as described, following the 'Recommended Practices for Verification of Plant Identification'."*
-'Good Practices for Plant Identification in the Herbal Industry' by Tim Brigham, Michelle Schroder and Wendy Cocksedge, Centre for Non-Timber Resources, Royal Roads University, Victoria, British Columbia, Canada; 'Saskatchewan Herb and Spice Association' and 'National Herb and Spice Coalition', Agriculture and Agri-Food Canada, February **2004**.

"We used DNA barcoding to conduct a blind test of the authenticity for:
    (i) 44 herbal products representing 12 companies and 30 different species of herbs, and
    (ii) 50 leaf samples collected from 42 herbal species.
We recovered DNA barcodes from most herbal products (91%) and all leaf samples (100%), with 95% species resolution using a tiered approach (rbcL + ITS2).
Most (59%) of the products tested contained DNA barcodes from plant species not listed on the labels.
    Although we were able to authenticate almost half (48%) of the products, one-third of these also contained contaminants and or fillers not listed on the label. Product substitution occurred in 30/44 of the products tested and only 2 out of 12 companies had products without any substitution, contamination or fillers. Some of the contaminants we found pose serious health risks to consumers.

Market Label or Herb Leaf	Sample	Barcode ID	ITS2 ID	rbcL ID
Symphytum officinale capsule	HP7C	S. officinale + Medicago sativaa	S. officinale + Medicago sativaa	Medicago sativaa
Symphytum officinale fresh	HP90L	S. officinale	S. officinale	Boraginaceae "

-'DNA Barcoding Detects Contamination and Substitution in North American Herbal Products' by Steven G. Newmaster, Meghan Grguric, Dhivya Shanmughanandhan, Sathishkumar Ramalingam and Subramanyam Ragupathy, University of Guelph, Ontario, Canada, and Bharathiar University, Tamil Nadu, India; BMC Medicine, Volume 11, No. 222, **2013**.
    (ID is identification. ITS2 is Internal Transcribed Spacer 2 region of nuclear ribosomal DNA.
    The region of chloroplast genes includes rbcL: Ribulose-1,5-bisphosphate carboxylase oxygenase, large subunit.)
    (Medicago sativaa is alfalfa/lucerne. So Comfrey leaf capsules were mixed with alfalfa but it was not listed on the label.)

"Authentication of botanicals by morphology, or the study of a plant's external form, requires no costly technology, but because plant species are defined by their morphological characteristics, it is in fact the most rigorous possible means of identifying a plant. *All other methods, as reliable as they can be, are no more than proxies or substitutes for morphological data; indeed, the reliability of other methods can be fully verified only by observing their performance in samples that have been botanically identified by morphology.*
Moreover, botanical identification of plants is not an exotic skill that only those with advanced degrees in botany can learn. Identifying plants by their appearance, combined with other sensory inputs such as odor, texture, and flavor, is an activity that human beings are naturally adapted to, having relied upon it for survival for hundreds of thousands of years."
-Botanicals: Methods and Techniques for Quality and Authenticity edited by Kurt A. Reynertson and Khalid Mahmood. Boca Raton, Flordia: CRC Press, **2015**. Chapter 4: 'Using Traditional Taxonomy and Vouchers in Authentication and Quality Control' by Wendy L. Applequist.

"Although there are a wide range of technologies developed for species authentication and adulterant detection, from microscopy and organolepsis (sense organs), high-performance thin layer chromatography (HPTLC) and thin layer chromatography (TLC), and Fourier transform infrared (FTIR) and near infrared (NIR) to name a few, still major challenges remain especially regarding differentiation between closely related species and identification of multiple species in a mixture. As a result, there is widespread adulteration in the natural products industry.
Using DNA for Authentication:
The first contract-testing laboratory in the United States to specialize in DNA-based (deoxyribonucleic acid) technologies for species identification and adulterant detection, AuthenTechnologies® LLC, was founded in 2010.
> Since that time DNA-based methods have quickly become a preferred test for routine quality control across a wide array of natural product ingredient suppliers and manufacturers. As a result, the '*National Institute of Standards and Technology' (NIST) launched the first line of DNA-authenticated 'Standard Reference Materials' ® for botanical species in 2013 in partnership with AuthenTechnologies®.
> Recently a more highly reproducible and informative analysis is the comparison of gene sequences from a specific stretch of DNA, or gene, referred to here as DNA sequencing, or barcoding; **DNA barcoding is the use of short DNA sequences for the identification of an organism.**"

-Botanicals: Methods and Techniques for Quality and Authenticity edited by Kurt A. Reynertson and Khalid Mahmood. Boca Raton, Flordia: CRC Press, **2015**. Chapter 5: 'The DNA Toolkit: A Practical User's Guide to Genetic Methods of Botanical Authentication' by Danica T. Harbaugh Reynaud.
(* -National Institute of Standards and Technology (NIST), United States Department of Commerce, Gaithersburg, Maryland, www.nist.gov/srm. NIST supports accurate and compatible measurements by certifying and providing over 1200 'Standard Reference Materials'® with well-characterized composition or properties, or both.)

"**This paper focuses on assessing the effectiveness of species identification using DNA barcoding standard loci (matK and rbcL) for a selected group of Polish flora.** PCR amplification success was 100% of all samples for the rbcL primer and 64% for matK primers giving in sum 87 fragments for sequencing.
The aligned sequences show that in the Vaccinium, Dryopters and Symphytum species groups barcoding is very useful and makes it possible to identify most species in those groups.
In the Symphytum genus only the rbcL locus shows differences between the species.
PCR success summary for loci, species and fragment lengths:

Species	Locus rbcL	Locus matK
Symphytum officinale	+(650bp)	+(800bp)
Symphytum tuberosum	+(650bp)	+(800bp)

In spite of very significant differences between these Symphytum species, the sequences of locus matK, believed to be very useful in distinguishing between taxa at the species level, showed no differentiation within the tested samples.
Also, in the nucleotide sequence of rbcL only a very slight difference was found between the tested Symphytum species. We found only one substitution occurring in the 385-site sequence.
> Symphytum officinale species is characterized by the presence of an 'A' nucleotide at this point while the S. tuberosum sample has a 'C' nucleotide. As a result, this locus can be used to distinguish these species from each other."

-'Estimating the Effectiveness of Species Identification by Sequencing of Two Chloroplast DNA Loci (matK and rbcL) in Selected Groups of Polish Flora' by M. Combik and Z. Mirek, Polish Academy of Sciences, Krakow, Poland; DNA Barcodes Journal, Volume 3, No. 1, pages 17-26, January **2015**.

"A DNA barcode is a standardized, short (< 1000 bp = base pairs) and highly variable segment of DNA which is compared to a reference database through a sequence alignment algorithm for species identification (Shinwari et al., 2014).
In Pakistan about 70% of the population is dependent on the plant derived traditional medicine for primary health care system. In developed countries like Germany, 80% of population have used herbal medicine at least once.
The very first step in the process of quality assurance is the authentic identification of the plant species.
The identification of herbs is traditionally carried out by morphological characters, but in case of cryptic species or phenotypically variable species the chances of misidentification are greater (Vijayan & Tsou, 2010).
> There can be serious consequences of use of a misidentified medicinal plant. For example, **Digitalis purpurea (foxglove) could be mistaken for Symphytum officinale** because of morphological similarity in their leaves."

-'DNA Barcoding: A Tool for Standardization of Herbal Medicinal Products (HMPS) of Lamiaceae from Pakistan' by Nadia Batool Zahra, Zabta Khan Shinwari and Muhammad Qaiser, Pakistan; Pakistan Journal of Botany, Volume 48, No. 5, pages 2167-2174, **2016**.

"In our annual review, NSF® AuthenTechnologies® found that more than 50% of the plant samples we tested contained DNA from other botanical species.
DNA testing is a good first step toward identifying the presence of other botanical species.
> However, chemical testing is needed to quantify the DNA in order to indicate how much of another botanical species is in the sample. Additional testing such as microscopy or organoleptic testing may also be warranted.

Traditional DNA testing techniques have their limitations, however. **DNA barcoding**, for instance, which uses Sanger Sequencing and universal primers, falls short in authenticating samples that are processed, including impure samples of commercial products.
Luckily, there's a next generation of DNA sequencing that can help identify numerous species in a wide range of

processed materials, and in only a few hours and for less, compared to many other traditional methods.
We call this technology **Target-Specific DNA Sequencing®** (TSDS). TSDS detects and identifies the target species and potential other species including contaminants such as allergens."
-'Next-Generation DNA Testing for Botanicals' by Danica Harbaugh Reynaud, Ph.D., Botanical Taxonomist and Geneticist, NSF® AuthenTechnologies®, Ann Arbor, Michigan; Nutritional Outlook, Volume 19, Issue 8, September 21, **2016**.
(Organoleptic testing is the analysis of products and materials with the sense organs.)

"***A number of approaches can be implemented to ensure plant-based material authentication for cosmetic applications.***
The choice of technique will depend on the data available in a given domain and on the more or less processed material to be assayed, as well as the kind of adulteration or quality defects to be detected.
> For example, molecular markers can help to distinguish between two species and thus can detect adulterations with a wrong plant, but they are unable to provide evidence of poor content in an active molecule, the latter requiring analytical chemistry methods.

The authentication methods described here require prior knowledge concerning the plants to be identified in order to be able, **based on comparisons of morphological criteria, molecular markers, or phytochemical profiles,** to affirm a sample's identity or its dissimilarity with the species to which it is supposed to belong.
> ***It is therefore important to have access to databases with botanical, genomic, and biochemical information.***"

-'A Critical View of Different Botanical, Molecular, and Chemical Techniques Used in Authentication of Plant Materials for Cosmetic Applications' by Samantha Drouet, et al, France and Thailand; Cosmetics Journal, Volume 5, No. 30, April **2018**.

"**The current study includes a comparative morphological study of the genus Symphytum L.** within the family Boraginaceae in relation to the phenotypic study, the study of the external manifestations of pollen, the environment and geographical distribution. The study included the study of the characteristics of roots, stems, leaves, flowers, inflorescence, fruits and nutlets.
Variation in characteristics was discussed and it was noted that **the characteristics of flowers were more important in taxonomic terms in identification and isolating studied species.**
The study also indicated that the pollen of all studied species varies in form and size and have characteristics of taxonomic significance that may be adopted in isolation and diagnosis of these species which studied for the first time in Iraq."
-'A Comparative Systematic Study of the Genus Symphytum L. (Boraginaceae) with New First Record of the Species Symphytum Tuberosum L. from Iraq' by Adel Mohan Aday Al-Zubaidy and Sherzad Rasul Abdalla Tobakari, Plant Production Department, Technical College of Applied Sciences, Sulaimani Polytechnic University, Iraq; Plant Archives, Volume 18, No. 2, pages 2068-2076, November **2018**.

"*DNA Meta-Barcoding:*
The most important methodological outcomes are the following:
> (1) the DNA extraction method greatly influences amplification success;
> (2) the main problem for the application of metabarcoding is DNA purity, not integrity or quantity;
> (3) the 'non-amplifiable' samples can be amplified with polymerases resistant to inhibitors.

Using this optimized workflow, **we analysed a broad set of plant products (teas, spices and herbal remedies)** using two NGS (Next Generation Sequencing) platforms.
The analysis revealed the problem of both the presence of extraneous (unwanted) components and the absence of labelled ones.
> Notably, for teas, no correlation was found between the price and either the absence of labelled components or presence of unlabelled ones. For spices, a negative correlation was found between the price and presence of unlabelled components.

A plant genus was considered 'found' in a sample if the reads corresponding to it were detected in more than 1% of all ITS (Internal Transcribed Spacer) reads in at least two of three replicates of each sequencing platform.
The products contained from 1 (sample S4) to 34 (sample D6) plant components according to the labels.
In sample D6, only 10 of 34 labelled plants were found by both platforms, among which Achillea, Tanacetum and Symphytum were found on the verge of the 1% threshold. Most of the labelled plants were not detected at all by both platforms in any of the replicates.
> The most frequently found unlabelled plants were common field weeds (Elymus = wild rye, Convolvulus = bindweed, Calystegia = false bindweed) that could be mixed with product components during the collection of raw plant material from fields or from the wild.

Economically motivated adulteration of food products is a known worldwide problem, as well as their contamination. These problems become especially topical with the globalization of food markets. High-throughput sequencing technologies offer a rapid and reliable method to analyse species composition in food."
-'Improved Protocols of ITS1-Based Metabarcoding and Their Application in the Analysis of Plant-Containing Products' by Denis O. Omelchenko, et al, Moscow, Russia; Genes: Journal of Genetics and Genomics by MDPI, Basel, Switzerland, Volume 10, No. 122, **2019**.

"***Please be aware that you may find companies offering 'DNA authentication' that has not been adequately tested to prove that the markers they use will always either accept the desired species or reject closely related species.***
> Barcoding is impossible in some species that hybridize readily or are very recently evolved, and has not been

rigorously tested in many other species.
*I am sure you can find a company to do this, but the results may then be of uncertain value. You'd want to **ask whether there is published information showing acceptable performance in distinguishing species**, and if not, what's the basis for their test. Since Comfrey species vary in their toxicity and chemical content, it is important to product safety and quality. ITS {Internal Transcribed Spacer} or cpDNA {chloroplast DNA} barcoding sequences do not show this. Therefore, I'd favor chemical authentication instead of DNA."*
-Plant Science Research Staff, Missouri Botanical Garden®, Saint Louis, Missouri, founded in 1859, the United States' oldest botanical garden, www.missouribotanicalgarden.org, **2019**.

*"The United States Department of Agriculture '**Germplasm Resources Information Network**' (GRIN, www.ars-grin.gov), provides '**National Genetic Resources Program**' (NGRP) personnel and germplasm users access to databases for the maintenance of passport, characterization, evaluation, inventory, and distribution data that is important for the effective management and utilization of national germplasm collections.*
It is the NGRP's responsibility to: acquire, characterize, preserve, document, and distribute to scientists, germplasm of all lifeforms important for food and agricultural production.
National Plant Germplasm System *(NPGS), National Agricultural Library, United States Department of Agriculture, National Agricultural Library, https://data.nal.usda.gov/dataset/national-plant-germplasm-system and https://npgsweb.ars-grin.gov/gringlobal. The database is available to the public. National Plant Germplasm System is a collaborative effort to safeguard genetic diversity of agriculturally important plants."*
-United States Department of Agriculture, Agricultural Research Service, Washington, DC and Beltsville, Maryland, **2019**.

*"**Flora Research Laboratories® LLC is one of the industry leaders in botanical testing.** We can perform identity by microscopy and 'High-Performance Thin-Layer Chromatography' (HPTLC) and screen for 'Foreign Organic Matter' (FOM).*
We offer a full suite of 'Current Good Manufacturing Practice' (cGMP) compliant testing for the dietary supplement industry including testing for heavy metals, pesticides, solvents, potency and more. We can perform Ash, 'Acid Insoluble Ash' (AIA) and 'Loss on Drying (LOD). While not a routine test, we can evaluate the presence of pyrrolizidine alkaloids.
*James Neal-Kababick, founder and director, developed many of the highly specialized detection techniques used today in our laboratory. This same innovation is applied to all of our **botanical authentication techniques as well as our various phytoforenisc screening methods used for examination of dietary supplement raw materials and finished products**."*
-Flora Research Laboratories® LLC, Grants Pass, Orgeon, **2019**. A 'Drug Enforcement Administration' registered and inspected botanical testing laboratory. (I do not have any connection with this company.)

*"**Our DNA testing services are appropriate for a wide range of materials, species, products and ingredients, including:** Raw ingredients, Dietary supplement capsules and tablets, Botanical drugs, Herbal tinctures, Botanical extracts, Food products, Pressed oils, Seasoning blends, Herbs and spices, Seeds and nuts, Meat and meat products, Fish and fish products, Grain and grain products, Fruit and vegetable purees and juices, Vegetable and meat protein powders, Mushrooms and mushroom blends.*
Species We Test:
    NSF® International currently has capabilities for identifying thousands of species of plants, animals, and fungi.
    Raw materials tested: *fresh, dried, whole, cut, or powdered ingredients or products that are not extracts.* **Includes Symphytum (Comfrey): raw.** *'Not raw' materials tested: botanical extracts, oils, finished dietary supplement and food products, as well as complex blends. Using next-generation sequencing (NGS®), our laboratory provides species-level identification for the organism that you need identified. Our database is filled with sequences from vouchered materials to compare your samples to, enabling us to provide the most accurate identification.*
Species Identification:
    Through the use of NGS®, we are able to identify every constituent of a sample, not just the target ingredient. For collaborators with product blends or powders, we can help validate the composition of your sample.
    If you have a single ingredient in your material, we can confirm the identity of that sample through targeted sequencing.
When used in conjunction with other traditional identification methods such as chemical testing, DNA testing is an important tool to ensure the identity of a sample."
-'NSF® International: Species and Materials We Test', www.nsf.org, Ann Arbor, Michigan, **2019**. An independent, accredited public health organization that facilitates the development of standards, and tests/certifies products and systems. Previously called AuthenTechnologies®. (I do not have any connection with this company. Other companies also do botanical DNA testing.)

*"**Symphytum officinale Transcriptome or Gene expression:***
RNA-Sequencing *of Symphytum officinale (Taxonomy ID: 278672). Accession: PRJNA415509.*
Data Type: Transcriptome or Gene expression. Scope: Monoisolate.
Registration date: October 23 2017, Nanjing University, China. www.ncbi.nlm.nih.gov/bioproject/415509
    SRA (Sequence Read Archive) Experiments:
        SRX3974888: SOLS. *1 ILLUMINA (Illumina HiSeq 4000).*
        Run: SRR7042966. Number of Spots: 31,452,246. Number of Bases: 9.4G. Size: 3.3Gb. Published: April 19 2018.
        SRX3974887: SORT. *1 ILLUMINA (Illumina HiSeq 4000).*
        Run: SRR7042967. Number of Spots: 33,419,657. Number of Bases: 10G. Size: 3.5Gb. Published: April 19 2018."
-National Center for Biotechnology Information (NCBI), www.ncbi.nlm.nih.gov. United States National Library of Medicine, Bethesda, Maryland, **2019**. Advances science and health by providing access to biomedical and genomic information.

(Transcriptome is all of the messenger RNA {mRNA} molecules expressed from the genes of an organism. Gene expression is the phenotype appearance of a characteristic from a particular gene. Phenotype is the observable characteristics of an individual from the interaction of its genotype with the environment.)

(Genome is the complete set of genes in a cell or organism. In other words, it is the haploid set of chromosomes in a gamete, microorganism, or cell of a multicellular organism.)

(Monoisolate is a single animal, cultured cell-line or inbred population.)

(SRA or 'Sequence Read Archive' makes biological sequence data for the research community to enhance reproducibility and allow discoveries by comparing data sets. SRA stores raw sequencing data and alignment information from high-throughput sequencing platforms, including Roche 454 GS System®, Illumina Genome Analyzer®, Applied Biosystems SOLiD System®, Helicos Heliscope®, Complete Genomics®, and Pacific Biosciences SMRT®. www.ncbi.nlm.nih.gov/sra.)

"**Medicinal Plants in Commerce: Genetic reference data for use in botanical product identity testing.**
Project funded by 'The Nature's Bounty Co.' as part of their **'Herbal Authenticity Program'** to publish genetic reference data for use by industry as standards in botanical product identity testing and in method development.

Accession: PRJNA515225. **Data Type: Genome sequencing and assembly.** Scope: Multispecies.
Registration date: January 15 2019. '**DNA4 Technologies LLC**' and 'Missiouri Botanical Garden'.
BioSample for BioProject: 191 species including: 157. Borago cretica leaf tissue from voucher Gamdogel 520:
Identifiers: BioSample: SAMN10997872; Sample name: DNA4-Tech_1961.
Organism: Symphytum creticum. Isolate: MO-2423524.
Package: Plant; version 1.0. Accession: SAMN10997872, ID: 10997872.
Development stage: adult. Tissue: leaf. Geographic location: missing.
Biomaterial provider: Missouri Botanical Garden. Collected by Gamdogel, May 10 1935."

-National Center for Biotechnology Information (NCBI), United States National Library of Medicine, Bethesda, Maryland, www.ncbi.nlm.nih.gov/bioproject?LinkName=biosample_bioproject&from_uid=10997872 and www.ncbi.nlm.nih.gov/biosample/10997872, **2019**.

(DNA4 Technologies LLC, www.dna4tech.com, Baltimore, Maryland. "DNA4 Technologies is a biotechnology company that offers full service DNA based tests for the identification of natural product identity and purity. We utilize the latest advances in genomics and bioinformatics to offer unbiased, quantitative estimates of the species content found in products derived from natural sources. Our extensive reference databases allow for the testing of plant, animal, and probiotic based products in addition to screening of impurities as well as GMO modification.")

"Since knowledge about Genome Size (GS) and its variation has markedly increased, GS has been considered an important biological character. For example, **intraspecific variation in Genome Size may indicate microevolutionary differentiation that could eventually be taxonomically significant** (Murray, 2005); therefore, GS could be a useful parameter for estimating cryptic taxonomic differentiation within species or aggregates (Greilhuber, 1998; Ohri, 1998). Several studies showed that GS allowed discrimination between closely related taxa (e.g. Zonneveld, 2001; Mahelka et al., 2005).

Similarly, we revealed Genome Size was helpful when distinguishing close relatives, such as in Buglossoides arvensis (L.) I.M.Johnst., **Symphytum officinale L., S. tuberosum** and Pulmonaria officinalis species groups."

-'First Insights into the Evolution of Genome Size in the Borage Family: A Complete Data Set for Boraginaceae from the Czech Republic' by Lucie Kobrlova and Michal Hrones, Department of Botany, Palacky University, Olomouc, Czech Republic; Botanical Journal of the Linnean Society, Volume 189, No. 2, pages 115-131, February **2019**.

Symphytum asperrimum (Prickly Comfrey) 1877

'Gartenflora', Volume 26, by Eduard August von Regel, Stuttgart, Germany, 1877. Page 151.

It was a monthly illustrated botanical magazine published 1852-1940. Described as a 'General Monthly Magazine for German, Russian and Swiss Horticulture and Botany'.

Regel was Director of the Russian Imperial Botanical Garden of Saint Petersburg, Russia from 1855-1892.

Chapter 5

Symphytum Genus Description

<u>Comfrey Confused with Foxglove</u>

Do not harvest Comfrey yourself unless you are sure what it is. If in doubt, do not use it.

"The identification of a species is based on morphological characteristics, whether one is a scientist or a herbalist.
However, the sets of characteristics used to delineate (define) species need not necessarily be the same in the value system of the herbalist and the value system of the botanist.
Furthermore, whereas Linnean names are the 'lingua franca' of botany, and are used worldwide, the plants intended to be indicated by a herbal name may differ from user to user or region to region in a similar manner to the way the common names of plants vary in English.
In addition to this cultural problem, there are, of course, also problems of simple misidentification by people collecting their own herbs. *Professional herbalists and curanderos (healers) are often extremely knowledgeable about the plants they use and can unerringly (with no error) select the plants they want.*
Less experienced people, however, in an attempt to collect the same material as the herbalist, may collect mixtures of species or totally different plants.
This explains tragedies, such as one reported from Washington state in 1977 in which two people who thought they were collecting Comfrey (Symphytum) in fact collected foxglove (Digitalis) and killed themselves by drinking a tea prepared from it*.*"
-'Herbal Teas and Toxins: Novel Aspects of Pyrrolizidine Poisoning in the United States' by Ryan J. Huxtable, Department of Pharmacology, University of Arizona, Tucson; Perspectives in Biology and Medicine, Johns Hopkins University Press, Volume 24, Number 1, pages 1-14, Autumn **1980**.
(* -'Poisoning Associated with Herbal Teas: Arizona, Washington', Morbidity and Mortality Weekly Report, Epidemiologic Notes and Reports, Centers for Disease Control and Prevention, Atlanta, Georgia, Volume 26, No. 32, pages 264-269, August 12 1977.)
> (Morphology is a branch of biology that studies the form of living organisms, and the relationships between their structures. This includes both the outward and internal appearance {shape, structure, color, pattern, size}.)
> ('Lingua Franca' is a language used as a common or commercial means of communication among people of various languages.)

> *"An elderly Chehalis, Washington state, couple attended a health spa that recommended Comfrey tea as an herbal remedy for their arthritis. The couple had experimented with various herbal teas, but the woman's knowledge of plants was limited. On Saturday, May 7, 1977,* **she picked what she believed to be Comfrey plants and made herbal tea, which she and her husband drank with their lunch.**
> **One hour later, they became incapacitated (disabled) with nausea, vomiting, dizziness, and sweating.**
> **Later in the afternoon, the husband discovered some foxglove plants in the refrigerator. Realizing that this herb - the leaves of which are similar to Comfrey - had mistakenly been substituted for Comfrey in their tea, he immediately called an ambulance. When the ambulance arrived at 4:30 PM, his wife already was dead."**
> -'Poisoning Associated with Herbal Teas: Arizona, Washington', Morbidity and Mortality Weekly Report, Epidemiologic Notes and Reports, Centers for Disease Control and Prevention, Atlanta, Georgia, Volume 26, No. 32, pages 264-269, August 12 **1977**.

"When it is not in bloom, Comfrey can be confused with foxglove, a deadly poisonous plant.
Foxglove (Digitalis purpurea L.) or Deadmen's Bells, Witch's Bells is in the Snapdragon family (Scrophulariaceae). *The name foxglove is derived from the shape of the blossoms, which bear a resemblance to glove fingers.*
Foxglove is extremely poisonous. *A leaf chewed and swallowed may cause paralysis and sudden heart failure. Its long green leaves are powdered into digitalis, the cardiac stimulant that keeps millions of heart patients alive.*
Foxglove Identification: *A biennial, foxglove forms a rosette of long-stalked leaves in its first year; in the second, stems grow 2-5 feet (0.6-1.5 meters) tall. The leaves are lance-shaped to oval. Spires of white to pinkish-lavender to red thimble-shaped flowers (June-September) are speckled inside with red dots."*
-'Magic and Medicine of Plants' by Reader's Digest. Pleasantville, New York: The Reader's Digest Association, Inc., **1986**.
> (Have a professional herbalist go with you when you gather Comfrey in the wild.)

*"****Identification:*** *Before using medicinal plants that have been collected from the wild, it is essential that they be correctly identified.* ***If in doubt, do not use the herb. The wrong identification of herbs has led to many cases of poisoning.***
Foxglove leaves (Digitalis purpurea), for example, are often mistaken for Comfrey."
-'Encyclopedia of Organic Gardening: The Complete Guide to Natural & Chemical-Free Gardening' by The Henry Doubleday Research Association, edited by Pauline Pears. London, England: DK (Dorling Kindersley), **2001**, page 290.

"Just because it looks like the herb in question doesn't mean it is the herb in question. Plants are trickier than you might think to identify in the wild.
For example, Comfrey and foxglove- the plants themselves as they grow in the wild- look quite alike. Moreover, the range for Comfrey and the range for foxglove overlap considerably; the two can grow side-by-side.
What would happen then if you decide to harvest your own Comfrey and end up with foxglove by accident? Brew some Comfrey tea, and it will probably clear your sinuses. Brew some foxglove tea, and it could very easily stop your heart."
-Western Herbs for Martial Artists and Contact Athletes: Effective Treatments for Common Sports Injuries by Susan Lynn Peterson, Ph.D. Wolfeboro, New Hampshire: YMAA Publication Center, **2010**.
    (The safest way to use Comfrey is to grow your own so you know for sure it's identity.)

Symphytum Genus Description (in chronological order: old to more recent)

"**Symphytum:** *The calyx of the Symphytum is an erect, acute; pentangular, permanent perianthium, divided into five parts at the edge: the corolla consists of a single petal, formed into a very short tube, and a tubulated, ventricose limb, somewhat thicker than the tube, and divided into five segments at the edge, obtuse and reflex. The opening is furnished with five subulated rays, shorter than the limb, and connivent, so as to form a cone. The stamina are five subulated filaments, placed alternately with the radii of the aperture: the antherae are acute, erect, and covered; the germina are four; the style is filiform, of the length of the corolla; the stigma is simple. There is no pericarpium, but the calyx grows larger, and contains four gibbous, pointed seeds, placed with their apices connivent.*
This genus comprehends the Symphytum of Tournefort, and the Consolida of *Rivinus."
-'**A History of Plants:** A General Natural History, or New and Accurate Descriptions of the Animals, Vegetables and Minerals of the Different Parts of the World' by John Hill, M.D., London, England, **1751**, page 254.
    (* Augustus Quirinus Rivinus, 1652-1723, also known as August Q. Bachmann, was a German physician and botanist. In the 1690 'Introductio Generalis in rem Herbariam' and other books, he introduced ideas later used by botanists such as Joseph Pitton de Tournefort and Carl Linnaeus. He classified plants according to the structure of the flower. Like John Ray he used dichotomous keys {2 choices to determine where a plant belongs, then another 2 choices, etc. down to species}.)

"**Symphytum. Dioscorides. Comfrey.** *Calyx 5-parted; corolla cylindrical, bellshape, mouth closed; arches awl shape, forming a cone; nuts 4, 1-celled, perforated at the base; gynobasis flat, small.*"
-'**A Natural Arrangement of British Plants**, According to their Relations to Each Other, as Pointed Out by Jussieu, De Candolle, Brown, etc., Volume II' by Samuel Frederick Gray, lecturer on botany, the materia medica and pharmaceutic chemistry, London, England, **1821**, page 355-356.

"**Symphytum;** *a genus of the class Pentandria, order Monogynia.*
Calix: *perianth five-parted, erect, five-cornered, acute, permanent.*
Corolla: *one-petalled, bell-shaped; tube very short; border tubular, bellying, a little thicker than the tube; mouth five-toothed, obtuse, reflexed; throat fenced by five lanceolate rays, spinulose at the edge, shorter than the border, converging into a cone.*
Stamina: *filamenta five, awl-shaped, alternate with the rays of the throat; antherae acute, erect, covered.* **Pistil:** *germina four; style filiform, length of the corolla; stigma simple.*
Pericarp: *none; calix larger, widened.*
Seeds: *four, gibbous, acuminate, converging at the tips.*"
-'**The Universal Herbal; or, Botanical, Medical, and Agricultural Dictionary;** Containing an Account of All the Known Plants in the World, Arranged According to the Linnean system. Specifying the uses to which they are or may be applied, whether as food, as medicine, or in the arts and manufactures, with the best methods of propagation, and the most recent agricultural improvements' by Thomas Green, London, England, **1824**, page 641.
    (Linnaean taxonomy is a biological classification system created by Swedish botanist, zoologist and physician Carolus Linnaeus in his book 'Systema Naturae' of 1735 and subsequent works. In this binomial {two terms} nomenclature {name or designation} there are three kingdoms, divided into classes, that are then divided into orders, genera {singular: genus}, and species {singular: species}, with an additional rank lower than species.)

"*The genus to which this plant belongs is well represented by our native Symphytum officinale, of which we find the following description in* **Dr. Symes new edition of *'Sowerby's English Botany'**, *which we here transcribe on account of its recent date.*
    '**Genus Symphytum**: *Calyx five cleft or five partite; corolla regular, or nearly so, cylindrical-clavate; throat with five lanceolate acute scales; limb erect, sub-campanulate, five toothed; stamens exserted beyond the corolla tube, but included in the limb; anthers with short filaments, not apendiculate, not connivent round the style. Nucules ovoid, smooth, sunk in and attached to the flat receptacle by a concave surface. Soft or bristly hispid herbs, with succulent stems. Flowers in scorpioid racemes arranged in pairs, usually with opposite leaves at their base. Corolla large, yellow, blue, purple, red or white.*'
Dr. Syme has described three native forms under the following names: Symphytum officinale var genuinam, Symphytum officinale var patens, Symphytum tuberosum."
-'**Forage Plants and Their Economic Conservation by the New System of Ensilage**: Part I: Caucasian Prickly Comfrey' by

Thomas Christy, Jun., F.L.S. (Fellow of the Linnean Society), Christy & Co., London, England, **1877**, page 3.
(* -'English Botany {Sowerby's}; or Coloured Figures of British Plants: Volume 7, Third Edition' by John T. Boswell {Syme} and John Edward Sowerby, London, England, 1867, 1880. It was a major publication of British plants comprising 36 volumes, issued in 267 monthly parts over 23 years from 1791 to 1814. This quote is page 114 in the 1867 edition.)

"**Genus VIII.: Symphytum. Tournef.**
Calyx 5-cleft or 5-partite. Corolla regular or nearly so, cylindrical-clavate; throat with 5 lanceolate acute scales; limb erect, subcampanulate, 5-toothed. Stamens exserted beyond the corolla tube, but included in the limb; anthers with short filaments, not apendiculate, not connivent round the style. Nucules ovoid, smooth, sunk in and attached to the flat receptacle by a concave surface.
Soft or bristly-hispid herbs, with succulent stems. Flowers in scorpioid racemes arranged in pairs, usually with opposite leaves at their base. **Corolla large, yellow, blue, purple, red, or white.**
The name of this genus of plants is derived from the Greek word 'to unite', because the species are supposed to agglutinate the lips of wounds."
-'**English Botany (Sowerby's)**; or Coloured Figures of British Plants: Volume 7' by John T. Boswell and John Edward Sowerby, London, England, **1880**, page 114.

"*Symphytum [Tourn.] L. Sp. Pl. 136. 1753.*
Erect coarse hairy perennial branching herbs, with thick mucilaginous roots, alternate entire leaves, those of the stem mostly clasping, the uppermost tending to be opposite, the lower long-petioled.
Flowers yellow, blue, or purple, in terminal simple or forked scorpioid racemes. Calyx deeply 5-cleft. Corolla tubular, slightly dilated above, 5-toothed or 5-lobed, the lobes short, the throat with 5 crests below the lobes. Stamens 5, included, inserted on the corolla-tube; filaments slender. Ovary 4-divided; style filiform.
Nutlets 4, obliquely ovoid, slightly incurved, wrinkled, inserted by their bases on the flat receptacle, the scar of the attachment broad, concave, dentate.
Greek, grow-together, from its supposed healing virtues. About 15 species, natives of the Old World."
-'**An Illustrated Flora of the Northern United States and Canada**, Volume 3: Gentianceae to Compositae (Gentian to Thistle)' by Nathaniel Lord Britton, PH.D., Sc.D., LL.D. and Hon. Addison Brown, A.B., LL.D., New York Botanical Garden, **1913**, page 92.

"**A modern botanical account of the Symphytums will be found in 'Flora of the British Isles'** by A.R. Clapham, T.G. Tutin and E.F. Warburg (Cambridge University Press, **1952**)."
-Comfrey Report No. 1 by H.D.R.A. (Henry Doubleday Research Association), 1955
(I was unable to find a copy of Comfrey Report No. 1. If anyone has it, please let me know.)

"**Symphytum L.:** Hispid perennial herbs. Radical leaves petioled, cauline usually sessile or decurrent. Flowers in terminal forked scorpioidal cymes, nodding. Calyx campanulate or tubular, 5-toothed. Corolla funnel-shaped or subcylindrical, shortly and broadly 5-lobed. Scales 5, linear or subulate, ciliate, connivent, included, rarely exserted. Stamens included.
Nutlets 4, ovoid, smooth or granulate, base annular, toothed, teeth clasping the receptacle.
About 25 species in Europe, Asia Minor (Asian Turkey), western Siberia (part of Russia) and Persia (Iran)."
-Flora of the British Isles, Second Edition by A.R. Clapham, T.G. Tutin and E.F. Warburg. England: Cambridge University Press, 1962, (first edition **1952**).

"**Symphytum L.:** Perennial, usually hispid herbs. Flowers in short ebracteate terminal cymes. Calyx lobed to 1/4 or almost to the base, accrescent. Corolla variously coloured, with cylindrical tube and tubular-campanulate limb, with triangular to semicircular lobes much shorter than the rest of the corolla; throat with 5 long scales, usually with marginal papillae. Stamens included, inserted at about the middle of the tube; filaments not more than 1 1/2 times as long as anthers. Style exserted; stigma very small, entire.
Nutlets ovoid, erect, sometimes curved, usually verruculose and often rugose, concave at base with a thickened collar-like ring. The number of the cauline leaves given in the descriptions does not include the pairs of leaves on the inflorescence-branches. The length of the filaments refers only to the part not concealed by the anther.
Literature:	C. Bucknall, Jour. Linn. Soc. London (Bot.) 41:491-556 (1913).
	B. Pawlowski, Fragm. Fl. Geobot. 7:327-356 (1961); op. cit. 17:17-37 (1971):
	G.E. Wickens, Notes Roy. Bot. Gard. Edinb. 29:157-180 (1969).

S. asperum
S. bulbosum
S. cordatum
S. cycladense
S. davisii
S. floribundum
S. gussonei
S. ibiricum (ibericum)
S. naxicola
S. officinale, Subspecies officinale and Subspecies uliginosum.

S. orientale
S. ottomanum
S. tauricum
S. tuberosum, Subspecies *tuberosum* and Subspecies *nodosum*.
S. x uplandicum"
-**Flora Europaea, Volume 3: Diapensiaceae to Myoporaceae,** editors Tutin, Heywood, Burges, Moore, Valentine, Walters and Webb. United Kingdom: Cambridge University Press, **1972**, page 103. (Symphytum L., pages 103-105 by B. Pawlowski.) 5-volume encyclopedia of plants, published between 1964 and 1993. It describes all national floras of Europe to identify any wild or widely cultivated plant to subspecies level. It provides geographical distribution, habitat preference, and chromosome number.

*"Detailed anatomical studies were made of the different organs of Sympytum officinale L. to determine the disposition of tissues. Some of the important characteristics of Comfrey are the presence of **glandular and non-glandular hairs; hooked trichomes** only on the abaxial side of the leaf; separate **vascular bundles** forming an arc in the leaf and petiole; and **large sac-like structures in the root**.*
*The anatomical aspect is important in studying the structure of plant parts and the major constituents of plants. This study was conducted to provide **morpho-anatomical description of the different organs of the plant as well as the tissues where the active principles are located**."*
-'**Anatomy of Symphytum Officinale L.**' by Ludivina S. De Padua, University of the Phillipines at Los Banos, Laguna; The Philippine Journal of Science, Volume 107, No. 1-2, pages 41-50, March-June **1978**.
(Trichome is a small hair or other outgrowth from the epidermis of a plant. Epidermis is the outer layer. Stomata is plural of stoma. Stoma is a tiny pore in the epidermis of a leaf that lets gases moves in and out.)
(Petiole is the stalk of a leaf that attaches to the stem.)
(Morpho-anatomy is study of anatomical forms and structures, especially characteristics that help distinguish species.)

"Symphytum L. - Comfreys:
Hispid perennials; leaves ovate-elliptic, subcordate to broadly cuneate at base, the basal long-petiolate. Flowers in rather dense cymes forming terminal panicle; calyx lobed about 1/5 to 9/10 to base; corolla actinomorphic, various colours, with limb little wider than tube and about as long, the limb with short lobes; stamens equal, included, alternating with 5 long corolla-scales; style simple, exserted.
Nutlets smooth to granulate, sometimes also ridged, with collar-like base."
-**New Flora of the British Isles:** Identification of Wild Vascular Plants of the British Isles edited by Clive Stace. Cambridge, England: Cambridge University Press, **1991**, page 645.

"After a historical synthesis on the taxonomic and nomenclatural framework proposed by several authors for the genus Symphytum, the results of micro-, macromorphological and cytological investigations on the Italian taxa are reported.
For each unit, besides the chromosome number, non glandular hairs and secretory structures of stem, leaves and flowers have been described.
As regards Symphytum taxonomy and phytogeography, *besides the specimens cultivated in the 'Botanic Garden of Pisa' (central Italy) and collected from several stands in Italy, the exsiccata from all major Italian and European Herbaria have been checked. Nomenclatural type, synonymies, distribution in Italy, ecology and other data from both bibliographical and original observations have been indicated for each species.*
An analytical key based on some morphological characters is proposed.
References on the most significant iconography (illustrations) are also given."
-'Il Genere Symphytum L. (Boraginaceae) in Italia: Revisione Biosistematica' (**The Genus Symphytum L. Boraginaceae in Italy: A Biosystematic Revision**) by Stefania Bottega and Fabio Garbari, Webbia: Journal of Plant Taxonomy and Geography, Volume 58, No. 2, pages 243-280, **2003**. (If you have an English translation of this article, I would appreciate having a copy.)

<u>Comfrey Leaf Description</u>

"Protection of Green Leaves (Boragineae family) Against Attacks of Animals:
Another form of weapon originating from the epidermal cells of leaves consists of stiff hairs or bristles, *with hard silicified cell-wall and sharp apex, which prick and wound like needles, though only unicellular; they are called pointed bristles. They usually project from the surface of the green leaves, closely crowded together, and their points are turned in the direction from which an attack might be expected. They appear gigantic in comparison with barbs, for even the smallest are much longer than these, and the largest resemble pins with their heads imbedded in the leaf-blade.*
This comparison becomes the more fitting since the pointed bristles are surrounded at their base by very regularly arranged cells which rise above the surface like a cushion, or often like a short white cone. The bristle itself on the end of this pedestal is formed of a single cell, which, when fully developed, loses its protoplasm and becomes filled with air.
The wall of this elongated cell is hardened by the deposition of silica, *and is usually unequally thickened by small knobs. **Although pointed bristles are developed in numerous groups of the vegetable kingdom, one group is especially so armed. This is the family of the Boragineae, which has been thus named, indeed, in consequence of its characteristic armour**. Examples of the equipment described are furnished in abundance particularly by species of the Viper's Bugloss (Echium), and of the genera Onosma, **Comfrey (Symphytum)**, and Borage (Borago)."*

-'The Natural History of Plants, Their Forms, Growth, Reproduction, and Distribution' (Pflanzenleben = Plant Life or Plantscapes, 1887) by Anton Kerner von Marilaun, Austrian botanist. Translated by F.W. Oliver from 6 volumes in German. Published by Blackie & Son, London, England, **1895**. Volume II, Symphytum pages 97-99, 191, 275, 441, 585, 744. Also Volume I, Symphytum pages 441 and 744, translated 1894.

(The word 'borage' derives from medieval Latin 'burra' which means rough-coated. It may also be derived from Latin 'borra' which means short wool or rough hair.)

"**Symphytum leaves** are sometimes bullate, and are entire or very occasionally partially dentate, but this is probably an abnormal condition. The margin, however, is often repand-sinuate and minutely eroso-denticulate.
In outline they vary from subrotund to oblong, ovate, or narrowly lanceolate, with the base cordate or rounded, suddenly contracted or gradually attenuated.
The blade of the leaf is more or less decurrent on the petiole, on each side of which it forms a membranous margin, and when the leaves are sessile it sometimes forms wings or decurrent lines on the stem.
The relative length and breadth of the leaves may vary considerably in the same species, and the autumnal leaves are often very large in comparison with those of the flowering season.
The clothing of the leaves, like that of the stem, is of two kinds, but the hairs are generally shorter, straighter and more regularly disposed than on the stem.

On the upper surface the hairs are, in the dried state, appressed, all pointing towards the apex of the leaf, and are often borne on tubercular bases, especially on the older leaves.

The lower surface is more softly hairy with longer hairs, but often bears tubercular setae on the midrib and veins."
-'A Revision of the Genus Symphytum, Tourn.' by Cedric Bucknall, (Mus. Bac. Oxon= Bachelor of Music, Oxford University), Journal of the Linnean Society of London, England, Botanical Journal, Volume 41, Issue 284, pages 491-556, December **1913**.

"**Symphyti Folium (Comfrey leaf, Symphytum officinale):**
The microscopical characters are stomata (pores) of the cruciferous type on both surfaces; the numerous thick-walled hairs, frequently one-celled, but sometimes with a short basal cell, very sharply pointed, and many, especially those on the lower surface, **strongly curved at the tip in the form of a hook**; the small, spherical-headed cystolith hairs with unicellular stalks; the slightly wavy-walled, epidermal cells."
-'The British Pharmaceutical Codex: An Imperial Dispensatory for the Use of Medical Practitioners and Pharmacists' by the Council of the Pharmaceutical Society of Great Britain, London, England, **1923**, pages 86-87, 1036.

(Cystolith is the outgrowth of the epidermal cell wall, usually of calcium carbonate.)

"The 'British Pharmaceutical Codex of 1934' (page 1036) introduced a monograph for Comfrey leaf, but current and recent textbooks refer to the drug only as an adulterant of the leaf of Digitalis purpurea (Perrot 1943, Trease 1957, Wallis 1960). The only detailed work on the leaf is that of Bider (*1935); this is not illustrated and deals only with macroscopical features together with histological (microscopic) epidermal features, particularly the trichomes. Kay (**1938) has briefly described and illustrated the covering trichomes.
The history and uses of the leaves of Symphytum officinale L. are given together with an illustrated account of the macroscopy of the upper and lower leaves of the plant and of the anatomical structure of the leaf. The diagnostic characters of the powdered leaves are also recorded and illustrated.
Macroscopy:

Symphytum officinale L. is an erect hispid perennial, producing erect stems 2-3 feet (0.6-0.9 meters) in height; both cauline and radical leaves are present. The plant grows in damp places especially near rivers and streams.

The leaf is simple, broadly lanceolate in shape, about 10-20 cm (3.9-7.8 inch) in length and 3-4 cm (1.1-1.5 inch) wide, with an acute apex and an entire or slightly wavy margin. The lateral veins anastomose towards the margin of the leaf; this feature shows more prominently on the under surface.
Lamina, interneural region (flat part of leaf):

The lower leaves are petiolate. The petiole is winged and the lateral veins of the leaf run parallel with the midrib, as far as the junction of the petiole with the stem.

The upper leaves are sessile, the base being decurrent with the stem. They are similar in shape and features to the lower leaves, but are about half their size.

The leaves have a brownish-green colour, no characteristic odour, a slightly astringent taste and surface is hispid.
The upper epidermis is covered with a thin smooth cuticle and consists of one layer of polygonal cells having sinuous anticlinal walls. Covering trichomes are numerous, may be of two types and arise from both veins and interneural epidermises.

The mesophyll is clearly differentiated. The lower epidermis has a thin smooth cuticle.
Midrib:

The midrib has a typically dicotyledonous structure, the only variation being the number of individual bundles present in transverse sections from apex to the base of the leaf. Serial sections showed that one bundle occurs at the apex and about 12 at the base, this increase being due to the entrance of the successive lateral veins into the central area of the meristele."
-'The Anatomy of the Leaf of Symphytum Officinale L.', by J.M. Peck and K.R. Fell, Pharmacognosy Research Laboratory, Bradford Institute of Technology, West Yorkshire, England; The Journal of Pharmacy and Pharmacology, Royal Pharmaceutical Society of Great Britain, Volume 13, pages 154-65, March **1961**.

(* -'Beitrage zur Pharmakognosie der Boraginaceen und Verbenaceen: Vergleichende Anatomie des Laubblattes' {Contributions to the Pharmacognosy of Boraginaceae and Verbenaceae: Comparative Anatomy of the Foliage Leaf} thesis by Jorg Bider, University Basel, Switzerland, Symphytum page 34, 1935. In German. If you have English translation, I would appreciate a copy.)
(** -'The Microscopical Study of Drugs' by Lilian A. Kay. London, England: Bailliere, Tindall and Cox, 1938. Symphytum pages 70, 72.)

> (This report has a lot of details and drawings about S. officinale leaf. I only gave a little here.)
> (The 'British Pharmaceutical Codex' of 1923 and 1934, page 1036, both have Symphytum.)
> (Petiole is the stalk of a leaf that attaches to the stem.)

"**Symphytum officinale L. Leaf:** *Both upper and lower epidermis consist of a single layer of cells, some of which are quite large, the outer wall of which are not cutinized (not covered with wax).*
One of the most striking features of Comfrey leaf is the presence of many thick-walled unicellular hairs which have a characteristic rough feeling when handled. *These hairs include glandular types which have short or long stalks and variously-shaped heads, and non-glandular ones.*
Numerous hooked or curved trichomes, smaller than those on the adaxial (upper) side, cover the abaxial (lower) side of the leaf.
The vascular bundles of the leaf are, as in the petiole, amphicribal. Two smaller vascular bundles are situated on both sides of a large median vascular bundle, forming what appears to be an arc of vascular bundles in the midrib portion."
-'Anatomy of Symphytum Officinale L.' by Ludivina S. De Padua, University of the Phillipines at Los Banos, Laguna; The Philippine Journal of Science, Volume 107, No. 1-2, pages 41-50, March-June **1978**.

> (Trichome is a small hair or other outgrowth from the epidermis of a plant. Epidermis is the outer layer.)
> (Xylem is one of 2 types of transport tissue in vascular plants, phloem being the other. Vascular rays extend radially across the stem, helping conduction from the vascular bundles to tissues alongside them. Amphicribral is the phloem surrounding the xylem.)
> (Morpho-anatomy is study of anatomical forms and structures, especially characteristics that help distinguish species.)

"**Tribe Boragineae:**
Mean thickness of the leaf blade, *measured in areas between the primary and the secondary veins, range from about 100-150 micrometer, in Brunnera, Pentaglottis and* **most of Symphytum**.
All the taxa possess anomocytic stomata oval in shape and variable in size, randomly oriented and not sunken in the epidermal tissue, as is typical of most dicots. **No stomata are found on the upper surface of Brunnera, Pentaglottis, Symphytum sp.pl. and Trachystemon.**

> *The absence of stomata on the adaxial (upper) surface of Brunnera, Trachystemon, Pentaglottis and most of Symphytum is possibly an ancestral feature retained by genera linked to their nemoral habitats.*

Type 5 Trichomes: Exclusive of Symphytum and consisting of unicellular hooked hairs, without a distinct base. In most species of this genus, hooked trichomes are prevalent on the abaxial (lower) surface, *while they are mixed trichomes of types 2 and 3 on the upper surface.* **The wall of hooked hairs mostly contain silica deposits, though calcium is present.**

> *Within the Boragineae tribe, hooked hairs are exclusive of Symphytum, although in different amounts and distribution from species to species (*Fell & Peck 1968).*
> *In Symphytum ibericum, for example, they are much denser on the adaxial surface, while they are exclusively found on the abaxial epidermis in S. asperum and S. armeniacum, as in the case of S. officinale (**De Padua 1980).*
> **Presence and distribution of hooked hairs may have a taxonomic relevance in Symphytum, although this should be tested on a wider range of species.**

Most of the Boragineae taxa investigated have dorsi ventral leaves with adaxial photosynthetic tissue. A single layer of typical palisade cells is found in Anehusella, Brunnera, part of Cynoglottis, Pentaglottis, Pulmonaria, most of Symphytum and Trachystemon.
Venation (arrangement of veins in leaf):
Venation is basically reticulate, *with conspicuous (visible) primary veins and a variable number of more or less anastomized veinlets terminating blindly in mesophyll islets. Veins are prominent in the taxa with thin blades, such as Symphytum, Brunnera, Pentaglottis and Trachystemon.*
The ovate-acuminate leaves *of Brunnera orientalis, Pentaglottis and of* **part of Symphytum** *tend to a camptodromelike venation pattern (sensu {in the sense of} Stearn 1995), with secondary and tertiary veins running outwards without reaching the margins, but arching upwards without formation of loops.*
In the megaphyllic leaves *of Brunnera macrophylla and Trachystemon orientalis, and in* **most of Symphytum**, *venation is instead reticulate-pinnate, with veinlets forming a network reaching the margins.*"
-'Leaf Surface and Anatomy in Boraginaceae Tribe Boragineae with Respect to Ecology and Taxonomy' by Federico Selvi and Massimo Bigazzi, University Firenze, Italy; Flora: Morphology, Distribution, Functional Ecology of Plants, Volume 196, pages 269-285, **2001**.
(* -'British Medicinal Species of the Genus Symphytum' by K.R. Fell and Janet M. Peck, Pharmacognosy Research Laboratories, University of Bradford, England; Planta Medica: Journal of Medicinal Plant and Natural Product Research, Volume 16, No. 2, pages 208-216, May 1968.)
(** -'Anatomy of Symphytum Officinale L.' by Ludivina S. De Padua, University of the Phillipines at Los Banos, Laguna; The Philippine Journal of Science, Volume 107, No. 1-2, pages 41-50, March-June 1978. Some sources say 1979 or 1980.)

> (Micrometre or micrometer or micron equals 10^{-6} meter = one millionth of a meter = one thousandth of a millimeter =

0.001 mm = 0.000039 inch.)
(Stomata is plural of stoma. Stoma is a tiny pore in the epidermis of a leaf that lets gases moves in and out.)
(Reticulate means resembling a net or network such as of veins, fibers, or lines crossing a reticulate leaf.)

"**Symphytum asperum:** *A transverse section of the lamina, midrib and both epidermises was studied. There is a single layered epidermis on the upper and lower surface of the leaf. Mesophyll consists of single layer of palisade parenchyma cells and 3-4 layers of spongy parenchyma cells.*"
-'Micromorphology and Anatomy of Three Symphytum (Boraginaceae) Taxa from Turkey' by Oznur Ergen Akcin and Hilal Baki, Department of Biology, Sciences & Arts Faculty, Ordu University, Ordu, Turkey; Bangladesh Journal of Botany, Volume 36, No. 2, pages 93-103, December **2007**.
(This report gives the results from an Electron Scanning Microscope used on the leaves, stems and roots of Symphytum asperum Lepechin, S. ibericum Steven and S. sylvaticum Boiss.)

"***Symphytum Officinale* L. Leaf (Folium):**
Surface view:
Upper epidermis *is composed of cells with wavy anticlinal walls and anisocytic (rarely anomocytic) stomata around 25 micrometer long, wavy, sometimes beaded anticlinal walls; covering trichomes of two types containing cystoliths occur:*
  1. *short unicellular, about 100 micrometer long, broad/circular base, tapering, straight apex.*
  2. *acute long unicellular, up to 700 micrometer long, slender.*
Epidermal cells are arranged in a rosette pattern around trichome base; glandular trichomes around 70 micrometer long, with unicellular stalk and unicellular spheroidal head.
Lower epidermal *cells have wavy anticlinal walls; anisocytic (rarely anomocytic) stomata are more frequent than on upper epidermis; covering trichomes of two types occur:*
  1. *short unicellular, up to 150 micrometer long, small base, slender and thick walled, with apex mostly hooked and generally without a cystolith.*
  2. *long unicellular, up to 2 mm long, straight, thick walled, usually without a cystolith.*
Glandular trichomes about 120 micrometer long, with multicellular, uniseriate stalk and unicellular spheroidal head.
Transverse section:
Bifacial; palisade cells in one layer; spongy mesophyll with large intercellular spaces.
Comfrey Leaf Powder:
Fragments of epidermal cells with anisocytic stomata, unicellular covering trichomes (some with a cystolith and/or hooked apex), and glandular trichomes."
-American Herbal Pharmacopoeia®, Botanical Pharmacognosy: Microscopic Characterization of Botanical Medicines edited by Upton, Graff, Jolliffe, Langer and Williamson. Boca Raton, Florida: CRC Press, **2010**, pages 627-628.

"**Plants living in different ecological habitats can show significant variability in their histological and phytochemical characters.** *The main histological features of various populations of three medicinal plants from the Boraginaceae family were studied.* **Stems, petioles and leaves were investigated** *by light microscopy in vertical and transverse sections. The outline of the epidermal cells, as well as the shape and cell number of trichomes was studied in leaf surface casts.*
Populations of Symphytum officinale showed variance in height of epidermal cells in leaves and stems, length of palisade cells and number of intercellular spaces in leaves, and the size of the central cavity in the stem.
Boraginaceae bristles were found to be longer in plants in windy/shady habitats as opposed to sunny habitats, *both in the leaves and stems of Pulmonaria officinalis and Symphytum officinale, which might be connected to varying levels of exposure to wind.*
Longer epidermal cells were detected in the leaves and stems of both Echium vulgare and Symphytum officinale plants living in shady habitats, compared with shorter cells in sunny habitats. *Leaf mesophyll cells were shorter in shady habitats as opposed to longer cells in sunny habitats, both in Echium vulgare and Symphytum officinale.*
This combination of histological characters may contribute to the plant's adaptation to various amounts of sunshine. The reported data prove the polymorphism of studied taxa, as well as their ability to adapt to various ecological circumstances."
-'Histological Study of Some Echium Vulgare, Pulmonaria Officinallis and Symphytum Officinale Populations' by Nora Papp, Timea Bencsik, Kitti Nemeth, Kinga Gyergyak, Alexandra Sulc and Agnes Farkas, University of Pecs, Hungary; Natural Product Communications: An International Journal for Communications and Reviews Covering All Aspects of Natural Products Research, Volume 6, No. 10, pages 1475-1478, October **2011**.
(Histology is the study of the microscopic structure of tissues.)
(Phytochemicals are biologically active compounds in plants.) (Petiole is the stalk of a leaf that attaches to the stem.)
(Trichome is a small hair or outgrowth from the plant epidermis. Epidermis is the outer layer.)
(Polymorphism is the occurrence of different forms among the members of a population.)

"*The paper presents the ontogenetic investigation of 16 herbaceous (annual, biennial and perennial) representatives of Boraginaceae. Special attention is paid to the pubescence of vegetative organs and to* **trichome micromorphology.**
Symphytum cordatum, Symphytum tuberosum, Symphytum caucasicum:
Trichome Type: Simple (pili simplices).
Trichome Shape: Subulate (pili subulati). Unicellular trichomes, swollen at base, with elongated and acuminate tips, with or without subtending 'rosettes'.

Description: Hooked (pili subulati uncinati).
Symphytum cordatum, S. x uplandicum:
Type of Ultrasculpture: Fine-grained. Small particles, grains, pellets, etc. on the trichome surface.
Symphytum cordatum:
Type of Ultrasculpure: Tuberculous fine-grained. Besides small particles, grains and pellets, bigger tubercules (protuberances) are found on the trichome surface.
Symphytum caucasicum, S. x uplandicum:
Type of Ultrasculpture: Tuberculous. Rounded or oblong medium-sized tubercules (knobs) are well visible on the smooth surface of the trichome."

-'Pubescence of Vegetative Organs and Trichome Micromorphology in Some Boraginaceae at Different Ontogenetic Stages' by Rimma P. Barykina and Vitaly Y. Alyonkin, Lomonosov Moscow State University, Russia; Wulfenia, Regional Museum of Carinthia, Austria, Volume 23, pages 1-29, **2016**.

(Ontogenesis is the development of an organism or anatomical or behavioral feature from the earliest stage to maturity.)

Comfrey Leaves and Pyrrolizidine Alkaloids

See sub-subsection 'Comfrey Root and Pyrrolizidine Alkaloids' in subsection 'Comfrey Root Description' in section 'Symphytum Genus Description' (Chapter 5).
See sub-subsection 'Comfrey Flower and Pyrrolizidine Alkaloids' in subsection 'Comfrey Flower' in section 'Symphytum Genus Description' (Chapter 5).
See sections 'Alkaloids in Comfrey' in Volume 2.

"**HomoSpermidine Synthase (HSS) is the first specific enzyme in Pyrrolizidine Alkaloid (PA) biosynthesis, a pathway involved in the plant's chemical defense.**
Here, we demonstrate that the **tissue-specific expression of HSS** in three boraginaceous species, Heliotropium indicum, Symphytum officinale, and Cynoglossum officinale, is unique with respect to plant organ, tissue, and cell type.
In Symphytum officinale, HSS expression has been detected in the cells of the root endodermis and in leaves directly underneath developing inflorescences.
Molecular approaches have shown that the cellular localization of the alkaloid pathways is remarkably diverse and complex, often including the translocation of intermediates between multiple cell types (Facchini and St-Pierre, 2005; Ziegler and Facchini, 2008). **The alkaloids are accumulated in cell types and tissues that are in most cases distinct from those that are involved in alkaloid biosynthesis.**
1. Tissue-Specific Expression of HSS in Symphytum officinale:
    PA in roots only when using Tracer test:
        For Symphytum officinale, **Tracer-feeding experiments identified the roots as the exclusive tissue of Pyrrolizidine Alkaloid biosynthesis,** with rootless shoot cultures showing no incorporation of the tracer (*Frolich et al., 2007).
    PA in roots, leaves and flowers using RT-PCR test:
        The RT-PCR experiments confirmed the presence of HSS transcripts in roots but also indicated the presence of the HSS transcript in the leaves and open flower.
    PA in leaves next to closed flower buds using Protein Extract test:
        In contrast to the results of RT-PCR, protein extracts of leaves of two developmental stages (4 and 12 cm long = 1.5 and 4.7 inch) and of the open flower showed no signal (no PA).
        Type I leaves lay directly beyond a terminal inflorescence with fully opened flowers. **Type II leaves occurred at the same position but with flower buds still closed.** Type III leaves were on stems without any inflorescences.
        The immunoblot confirmed the expression of HSS in leaves of type II, whereas leaves of types I and III were devoid of any label. **This result supports the idea that HSS expression in the shoot (leaf) of S. officinale depends on the developmental stage of the leaves.**
2. Immunolocalization of HSS in Roots of Symphytum officinale:
    Young white Comfrey roots that developed during the same year were cut. **In all such sections, HSS expression was found exclusively in cells of the endodermis.**
        Those cells of the endodermis that were affected by the emerging lateral root did not show the HSS label, suggesting the absence of HSS expression in cells displaced from their original position within the root."

-'Distinct Cell-Specific Expression of Homospermidine Synthase Involved in Pyrrolizidine Alkaloid Biosynthesis in Three Species of the Boraginales' by Daniel Niemuller, Andreas Reimann and Dietrich Ober; Botanisches Institut und Botanischer Garten, Universitat Kiel, Germany, and Institut fur Pharmazeutische Biologie, Technische Universitat Braunschweig, Germany; Plant Physiology: American Society of Plant Biologists, Volume 159, No. 3, pages 920-929, July **2012**.
(* -'Tissue Distribution, Core Biosynthesis and Diversification of Pyrrolizidine Alkaloids of the Lycopsamine Type in Three Boraginaceae Species' by Cordula Frolich, Dietrich Ober and Thomas Hartmann, Braunschweig and Kiel, Germany; Phytochemistry, Volume 68, pages 1026-1037, 2007.)
(Biosynthesis is an enzyme-catalyzed process where simple compounds are converted into more complex products in an organism.)

(Inflorescence is a group of flowers arranged on a stem of a main branch or complicated arrangement of branches.)

"This protocol (procedure) delivers a method to determine the biosynthetic production capability of Comfrey leaves for pyrrolizidine alkaloids independently from other organs like roots or flowers.
> *The protocol applies and combines radioactive tracer experiments with standard and modern techniques like thin layer chromatography (TLC), solid-phase extraction (SPE), high-performance liquid chromatography (HPLC) and gas chromatography-mass spectrometry (GC-MS).*

Comfrey roots are known to be able to synthesize pyrrolizidine alkaloids (*Frolich et al., 2007) and the key enzyme for biosynthesis, homospermidine synthase (HSS), was localized in the endodermis cells.

In addition to this site of synthesis, there have been hints that also leaves of a certain developmental stage might be able to produce pyrrolizidine alkaloids (**Niemuller et al., 2012).

Two young Comfrey (Symphytum officinale) leaves subtending an inflorescence with unopened flower buds are cut from a Comfrey plant.

The data resulting from measurement of radioactivity of the sample aliquots with the Tri-Carb LSC in Steps A1 to A8 allow calculation of the total uptake of [14C]putrescine as a tracer by the Comfrey leaves given in decays per minute (dpm).

A comparison with standards (literature and NIST database) is already the first evidence for a possible incorporation of putrescine into alkaloids.

Tracer Feeding Experiment: Recovery of Applied Radiation from Leaf:
> *Total radioactivity applied: 169 kBq (100%).*
> **Extract purified via SCX-SPE (Pyrrolizidine Alkaloids enrichment): 15 kBq (8%)."**

-'Radioactive Tracer Feeding Experiments and Product Analysis to Determine the Biosynthetic Capability of Comfrey (Symphytum Officinale) Leaves for Pyrrolizidine Alkaloids' by Thomas Stegemann, Lars H. Kruse and Dietrich Ober, Botanisches Institut und Botanischer Garten (Botanical Institute and Botanical Garden), Universitat Kiel, Germany; Bio-Protocol, California, Volume 8, Issue 3, 12 pages, February 5 **2018**.

(* -'Tissue Distribution, Core Biosynthesis and Diversification of Pyrrolizidine Alkaloids of the Lycopsamine Type in Three Boraginaceae Species' by Cordula Frolich, Dietrich Ober and Thomas Hartmann, Braunschweig and Kiel, Germany; Phytochemistry, Volume 68, pages 1026-1037, 2007.)

(** -'Distinct Cell-Specific Expression of Homospermidine Synthase Involved in Pyrrolizidine Alkaloid Biosynthesis in Three Species of the Boraginales' by Daniel Niemuller, Andreas Reimann and Dietrich Ober; Botanisches Institut und Botanischer Garten, Universitat Kiel, Germany, and Institut fur Pharmazeutische Biologie, Technische Universitat Braunschweig, Germany; Plant Physiology: American Society of Plant Biologists, Volume 159, No. 3, pages 920-929, July 2012.)

(Subtend is a leaf or bract that extends under a flower to support it. A bract is a specialized leaf at base of a flower.)

<u>Comfrey Root Description</u>

"At Kew Gardens, they found the roots of an old plant of Symphytum asperrimum (Prickly Comfrey) had penetrated nine feet (2.7 meters) down, when moved in 1875.

The roots sometimes globe, and hold half a pint to a pint (1 to 2 cups = 236 to 473 ml) of gummy water."

-'Prickly Comfrey: Its History, Cultivation, Extraordinary Production, and Uses: A Letter Addressed to His Excellency Sir Hercules Robinson, President of the Agricultural Society of New South Wales' by Arthur T. Holroyd, Sydney, Australia, **1876**, page 8.
> (Founded in 1840, Kew Gardens in London, England houses the 'largest and most diverse botanical and mycological collections in the world'.)

"The dried Comfrey root (Symphytum officinale Linne.), as found in commerce, is in pieces varying from 1 to 4 or 5 inches (2.54 to 10.16 or 12.70 cm) **long, black, and courrugated (ridges and grooves) externally, dark-whitish and corneous (hornlike, hard) internally,** *nearly odorless, viscid (sticky), and slightly astringent (shrinks body tissue).*

It contains some tannic acid, a trace of starch, some sugar, and a large amount of mucilage, which is readily extracted by water. Asparagine in small amount was obtained from it by Henry and Plisson in 1829."

-'Kings American Dispensatory, Volume 2, 19th Edition' by Harvey Wickes Felter, M.D. and John Uri Lloyd, Phr.M., Ph.D., The Eclectic Medical Institute, Cincinnati, Ohio, **1905**, page 1870.

"Description of root, Symphytum officinale L.:
> *Comfrey has a large, deep, spindle-shaped root, thick and fleshy at the top, white inside, and covered with a thin, blackish brown bark.* **The dried root is hard, black, and very deeply and roughly wrinkled, breaking with a smooth, white, waxy fracture.**
> *As it occurs in commerce it is in pieces ranging from about an inch (2.5 cm) to several inches (2 inches = 5 cm) in length, only about one-fourth of an inch (0.6 cm) in thickness, and usually considerably bent.*
> *It has a very mucilaginous, somewhat sweetish and astringent taste, but no odor.*

Collection of Root:
> *The root is dug in autumn, or sometimes in early spring.* **Comfrey root when first dug is very fleshy and juicy, but about four-fifths of its weight is lost in drying."**

-'American Root Drugs: Bulletin No. 107' by Alice Henkel, Drug-Plant Investigations, Bureau of Plant Industry, United States Department of Agriculture, Washington, DC, October 25 **1907**.
>(An astringent taste is slightly bitter or acidic.)

*"**The root in the greater number of the Symphytum species is fusiform**, either simple or branched, and sometimes very thick and fleshy. **In the remaining species it is more or less creeping,** with nodular, tubercular or cylindrical thickenings, being simple or branched according to the species.*
The stem is correlated with the root: in the species with a fusiform root it is branched, constituting the Ramosa, and in those with a creeping root it is normally simple- the Simplicia.
>*In the Ramosa the branches may be simple, bearing only one pair of leaves and racemes, or compound and similar to the main stem. Again, the branches may be much smaller than the stem and arranged in a racemose manner, or the stem may be subdichotomous, with the branches approaching it in size.*
>*In the Simplicia the stem is often bifid at the apex, and sometimes bears rudimentary branches in the axils of the leaves, or more rarely fully developed flowering branches."*

-'A Revision of the Genus Symphytum, Tourn.' by Cedric Bucknall, (Mus. Bac. Oxon= Bachelor of Music, Oxford University), Journal of the Linnean Society of London, England, Botanical Journal, Volume 41, Issue 284, pages 491-556, December **1913**.
>(Fusiform means tapering at both ends, in other words: spindle shaped.)

*"**Symphyti Radix, Comfrey Root, Symphytum officinale:**
Comfrey consists of the dried root and rhizome of Symphytum officinale Linn. (Family Boraginaceae). The drug consists mainly of segments of the dried root, and occurs in cylindrical pieces from about 10 to 40 millimetres in length and about 5 to 10 millimetres in diameter.*
Externally, it is nearly black in colour and exhibits glistening crystals on the surface; *it is strongly wrinkled longitudinally. The fracture is short, the fractured surface being greyish-white and horny.*
The smoothed, transverse surface *shows a narrow bark, separated by a dark cambium line from the radiate wood, which is composed of narrow bundles separated by wide medullary rays.*
It has a mucilaginous taste but is without odour.
Comfrey contains allantoin, 0.6 to 0.8 percent, gum, tannin, resin and a trace of starch."

-'The British Pharmaceutical Codex: An Imperial Dispensatory for the Use of Medical Practitioners and Pharmacists' by the Council of the Pharmaceutical Society of Great Britain, London, England, **1923**, pages 86-87, 1036.
>(Longitudinally = lengthwise. Transverse = crosswise.)
>('Fracture' is how the root breaks transversely and the character of the broken surface:
>>Short: fractured surface is smooth. Horny: has hard horn-like broken surface.)
>
>(Cambium is plant tissue from which phloem, xylem, or cork grows by division.)

*"**The detailed anatomy of the roots of Symphytum officinale L. is described and figured. The diagnostic microscopical characters which may be used to identify these roots when present in the crushed or powdered condition are:**
>large cork cells, irregular in outline, up to 250 micron in tangential width; abundant parenchyma of phelloderm, phloem, and xylem, including medullary rays, of sub-rectangular cells arranged in radial files and containing mucilage.*

Also starch granules both simple and 2- to 4- compound; xylem vessels solitary or in groups of 2 or 3 (rarely more than 3), with many rows of bordered pits; corrected vessel number per mm is 40-56-85.
Absence of calcium oxalate crystals, laticiferous tissue and lignified elements other than vessels."

-'Comfrey (Symphytum Officinale L.) Root: Its Anatomy and Its Detection in Admixture with Chicory in Dandelion Coffee' by J.M. Rowson, Museum of The Pharmaceutical Society of Great Britain, Transactions of the Society, Journal Royal Microscopical Society, Great Britain, Volume 75, No. 2, pages 119-128, 1955 and/or January **1956**.
>(Micron is a metric unit of length = 10^{-6} {10 to the minus 6} meter. As of 1967 that term is no longer officially used by the General Conference on Weights and Measures {French CGPM}, although some still use it.)
>(Parenchyma is the functional tissue of an organ as distinguished from connective and supporting tissue.)
>(Xylem is one of two types of transport tissue in vascular plants, phloem being the other. Vascular rays extend radially across the stem, helping conduction from the vascular bundles to tissues alongside them.)
>(Radial sections are longitudinal sections cut parallel to the rays. Tangential sections are cut perpendicular to the rays. Rays are the radial system in secondary xylem and phloem. They run parallel to radii that pass through the center of the axis. They are composed of parenchyma cells which are usually elongated perpendicular to the long axis.)
>(Lignin is a complex organic polymer deposited in cell walls of plants, making them rigid and woody. A laticifer is a secretory cell in the leaves and/or stems that produce latex and rubber as secondary metabolites.)

*"**Most species of Symphytum, including the type of the genus, S. officinale, have compact, erect rootstocks with fleshy, more or less fusiform roots serving as storage organ.***
>*Such species have branched stems and* **constitute *Bucknall's subgenus Ramosa (i.e. Symphytum sensu stricto).**
In other Symphytum species the rhizome (root) is horizontal and creeping; it may be underground and fleshy, thus serving as a storage organ, and then fairly evenly thick or else tuberously expanded at intervals, or it may be above ground and rather slender, putting out tufts of leaves.
>*The stems are unbranched and shorter than in the subgenus Symphytum (sensu stricto).* ***These species constitute Bucknall's subgenus Simplicia.'"***

-'A Revision of Symphytum in Turkey and Adjacent Areas' by G.E. Wickens, Royal Botanic Gardens: Kew; Notes from the Royal Botanic Garden Edinburgh, Scotland, Volume 29, pages 157-180, **1969**. Includes Turkey, Bulgaria, Greece, Aegean Islands and Caucasia.
(* -'A Revision of the Genus Symphytum, Tourn.' by Cedric Bucknall, {Mus. Bac. Oxon= Bachelor of Music, Oxford University}, Journal of the Linnean Society of London, England, Botanical Journal, Volume 41, Issue 284, pages 491-556, December 1913.)
  (Bucknall's main sections were Division I: Ramosa {for example, S. officinale} and Division II: Simplicia {for example, S. tuberosum}.)
  (Sensu stricto = s.s., s. str., sens. str., sens. strict. In the strict/narrow sense. It is added after a taxon to mean it is being used in the sense of the original author, or without taxa which may otherwise be associated with it.)

"*Symphytum officinale L. Root:*
Transverse section of the roots shows a thick outer wall of the epidermis. The epidermis is composed of a single layer of cells somewhat larger and irregularly-shaped than those of the leaf or the petiole. Both the epidermis and cortex are not sclerified.
Large, sac-like structures were observed in the longitudinal sections of the roots, containing mostly tannins, mucilage and alkaloids.
The vascular system of Comfrey root *consists of collateral bundles forming a circle (circle) about a parenchymatous pith. The phloem parenchyma contain an abundance of reserve material. The xylem consists mostly of reticulate tracheary elements. Some cells of the pith are sclerified, quite large and some are elongated.*"
-'Anatomy of Symphytum Officinale L.' by Ludivina S. De Padua, University of the Phillipines at Los Banos, Laguna; The Philippine Journal of Science, Volume 107, No. 1-2, pages 41-50, March-June **1978**.
  (Transverse section means made at right angles to the long axis of the body or object.)
  (Sclerenchyma is the strengthening tissue in a plant, formed from cells with thickened walls. Sclerify means being converted to sclerenchyma.) (Pith is soft or spongy tissue.)

"**The main roots of a Comfrey plant can reach a diameter of 3 inches (7 cm) at their base. They die off after about four years, but are continuously replaced during the plant's forty-year life-span;** they produce powerful secretions, and have root hairs that appear to make available large quantities of plant foods and trace elements.
A tree with the same type and depth of root system would lock up nearly all its mineral gatherings in wood, but Comfrey keeps it minerals in sappy leaves and thick stems that are so high in protein that they have a carbon:nitrogen ratio of 9.8 to 1 and can be used as a kind of 'instant compost'."
-Fertility Gardening: The Organic Way to Make Your Garden Grow by Lawrence D. Hills. London, England: Cameron & Tayleur Books Limited, **1980**.

"*Symphytum officinale Root Morphology:*
The root of this species is a thick fleshy secondary structure branching little and tapering from a width of 2 to 4 cm (0.78-1.57 inches) to a fibrous tip over a length of up to 30 cm (11.8 inches). The surface is smooth.
Anatomy:
Internal to an outer periderm is a region of parenchyma containing, close to the cambium, regions of phloem. Internal to a continuous cambium is a largely parenchymatous xylem. The width of the xylem and phloem are more or less equal.
Periderm:
5 to 10 layers of tangentially flattened cells alternately placed rather than in rows.
Phloem/Parenchyma:
The parenchymatous ground tissue progresses from rectangular, radially elongated and oriented in radial rows close to the cambium to tangentially elongated and tangentially oriented cells adjacent to the periderm. In logitudinal section these are rectangular and storied between 50 and 200 micrometer across. Phloem occurs in a contiuous layer adjacent to the cambium.
Xylem:
Cambium is continuous and rather sinuous. Internal to the cambium are radially oriented rows of vessels giving way to randomly placed vessels at the centre of the root. Vessels are polygonal in transverse section 20 to 75 micrometers across. The xylem parenchyma is oriented in radial rows and appears as storied parenchyma in longitudinal section 30 to 60 micrometers across."
-'The Morphological and Anatomical Interpretation and Identification of Charred Vegetative Parenchymatous Plant Remains' by Jonathan G. Hather, Thesis for Ph.D., Institute of Archaeology, University College London, England, **1988**.

"I have not found much information on **root systems of Symphytum species** though it has been used to put them in Sections.
  For **S. officinale (Common Comfrey)**, Pawlowski (*1961) gives 'Radix crassa, fusiformis, verticalis' (Root thick, spindle, vertical).
  For **S. asperum (Prickly Comfrey) and S. x uplandicum (Russian Comfrey)** 'Radix crassa, fusiformis, more or less verticalis'.
  More or less horizontal rhizomes are mentioned only in connection with other species."
-'Atlas 2000 and a Problem with Purple Flowered Comfrey (continued)' by Eric Chicken, East Yorkshire, England; Botanical Society of Britain and Ireland (BSBI) News, Hertfordshire, England, No. 82, pages 48-49, September **1999**.
(* -'Observations ad {on} Genus Symphytum L. Pertinentes' by Bogumilus Pawlowski, Fragmenta Floristica et Geobotanica, Poland, Volume 7, No. 2, pages 327-356, 1961. All in Polish. If you have an English translation, could you please send it to me.)
(Taxonomic ranks: Genus, Subgenus, Section, Subsection, Series, Species, Subspecies, Variety, Subvariety, Form, Subform.)

"**The dark brown to black root pieces** are longitudinally corrugated on the outside and have a short fracture. The cross section shows a thin, pale-colored cortex and a whitish to pale brown radiate xylem with broad medullary (mid-layer) rays.
Under magnification, wide vessels, single or in groups of 2-3, are discernible, scattered in the radiate xylem parenchyma. Fragments of rhizome with pith (soft, spongy tissue) may also be present."
-Herbal Drugs and Phytopharmaceuticals: A Handbook for Practice on a Scientific Basis edited by Max Wichtl. Boca Raton, Florida: CRC Press, **2004**. Symphytum Radix: Comfrey Root, pages 590-592.
    (In botany, the cortex is the outermost layer of the stem or root.)

"**Fibrous roots in general (not just Comfrey) divide into a cluster immediately upon leaving the root crown.** They are sometimes thin and numerous, **sometimes fleshy and thick,** often creating a dense root-ball. We don't usually eat the roots of these plants, unless the fibrous roots grow from a rootstock.
Fibrous-rooted plants include strawberries and rhubarb. **Some fibrous roots swell into thicker storage organs, as in Comfrey and daylilies."**
-Edible Forest Gardens: Volume One: Vision & Theory by Dave Jacke and Eric Toensmeier. White River Junction, Vermont: Chelsea Green, **2005**, page 205.

"**Symphytum Officinale L., Comfrey Root (Radix):**
Transverse section:
    Dark brown cork; inside the cork is a phelloderm consisting of a layer of tangentially elongated parenchyma cells; secondary phloem of spheroidal parenchyma; secondary xylem predominantly of parenchyma.
    Near the vascular cambium, small groups of vessels are found from which small radial strands of vessels, interrupted by parenchyma, project toward the center of the root; within these strands, vessels up to 100 micrometer diameter are found singly or in small groups.
    Primary xylem has vessels found singly or in small groups.
    Parenchyma contains mucilage that becomes stringy and gluey after preparation in chloral hydrate; fibers and crystals are absent.
Longitudinal section: Vessels with reticulate wall thickening or bordered pits.
Starch: Granules mostly simple, more or less spherical, up to 10 micrometer diameter.
Comfrey Root Powder: Fragments of parenchyma; few vessels with bordered pits or reticulate walls; mucilage; starch (water)."
-American Herbal Pharmacopoeia®, Botanical Pharmacognosy: Microscopic Characterization of Botanical Medicines edited by Upton, Graff, Jolliffe, Langer and Williamson. Boca Raton, Florida: CRC Press, **2010**, page 631.

Using Comfrey Roots to Tell Age of Plant

"**Thirty-five herbaceous dicotyledonous perennial plant species, with permanent root systems, from 16 families, were examined for the presence of growth rings in the secondary root xylem.** Most of the species surveyed showed ring zonations in the roots, and these could be verified as annual growth rings in the ten species for which plants of known age were available.
    Symphytum officinale L. Zonal Pattern:
        VL- differential vessel lumina (cavity in a plant cell enclosed by cell walls).
        VB- zonal branching of vessel rays.
    Demarcation (clarity of rings): Clear.
Investigations, in spring, of the species possessing growth rings invariably showed the secondary xylem to be bounded by wider vessels whereas, later in the growing season, an expanding zone of abruptly- or gradually-narrower vessels was added to the outside. This observation provided evidence that the growth rings were formed annually.
Where growth rings were observed they resulted, at least partly, from variation in cell- or vessel-diameter between earlywood and latewood, as found in ringporous woody plants (e.g. Metcalfe and Chalk, 1983).
The potential value of 'herbchronology' (herb chronology) as a tool in ecological investigations of species and stands of perennial herbs of temperate zones is discussed. The determination of stand age structure by 'herbchronology' could be used to **trace the past development of a plant stand,** providing valuable data for studies of succession and for testing population models (Bart, 1995).
    Possible applications of 'herbchronology' will be hampered by root decay radiating from the central axis or, in the case of stunted individuals, by tightly-packed indistinguishable growth rings.
Furthermore, the distinctness of the growth rings generally decreases in the distal (away from the center) parts of the roots, a pattern which is also known for tree roots (Fritts, 1976).
However, in our proximal (near the center) cuttings of main roots, the growth rings could be distinguished without difficulty in most cases."
-Age-Determination of Dicotyledonous Herbaceous Perennials by Means of Annual Rings Exception or Rule?' by Hansjorg Dietz and Isolde Ullmann, Julius-von-Sachs Institut fur Biowissenschaften {Institute of Life Sciences}, Wurzburg, Germany; Annals of Botany, Volume 80, pages 377-379, **1997**.)
    (Herbchronology is the analysis of annual growth rings in the secondary root xylem of perennial herbaceous plants. A new growth ring is added each year to persistent roots.)

"Determining the ages of long-lived perennials:
Annual rings can be counted in a cross section of a root. *Dietz and Ullmann published a list of wild plants for which ring counting is or is not a viable technique.*
For example, the growth rings found in Comfrey roots (Symphytum officinale) and horseradish are clear and easy to read."
-Planting the Future: Saving Our Medicinal Herbs by Rosemary Gladstar and Pamela Hirsch. Rochester, Vermont: Healing Arts Press, **2000**, page 32.

Comfrey Root and Pyrrolizidine Alkaloids

See sub-subsection 'Comfrey Leaves and Pyrrolizidine Alkaloids' in subsection 'Comfrey Leaf Description' in section 'Symphytum Genus Description' (Chapter 5). It has information on roots.
See sections 'Alkaloids in Comfrey' in Volume 2.

"Three species of the Boraginaceae were studied: *greenhouse-grown plants of 'Heliotropium indicum', and 'Agrobacterium rhizogenes' transformed* ***roots cultures (hairy roots)*** *of 'Cynoglossum officinale' and* ***'Symphytum officinale'.*** *The species-specific Pyrrolizidine Alkaloid (PA) profiles of the three systems were established by GC-MS (Gas Chromatography-Mass Spectrometry).* ***All PAs are genuinely present as N-oxides.***
The sites of PA biosynthesis vary among species. ***PAs are synthesized only in roots of S. officinale.***
Classical tracer studies with radioactively labelled precursor amines (e.g., putrescine, spermidine and homospermidine) and various necine bases (trachelanthamidine, supinidine, retronecine, heliotridine) and potential ester alkaloid intermediates (e.g., trachelanthamine, supinine) were performed to evaluate the ***biosynthetic sequences.***
It was relevant to perform these comparative studies since the key enzyme of the core pathway, ***Homo Spermidine Synthase (HSS),*** *evolved independently in the Boraginaceae and, for instance, in the Asteraceae. Studies showed that the core pathway for the formation of trachelanthamidine from putrescine and spermidine via homospermidine is common to the pathway in Senecio ssp. (Asteraceae). In both pathways homospermidine is further processed by a b-hydroxyethylhydrazine sensitive diamine oxidase. Further steps of PA biosynthesis starting with trachelanthamidine as common precursor occur in two successive stages."*
-'Tissue Distribution, Core Biosynthesis and Diversification of Pyrrolizidine Alkaloids of the Lycopsamine Type in Three Boraginaceae Species' by Cordula Frolich, Dietrich Ober and Thomas Hartmann, Braunschweig and Kiel, Germany; Phytochemistry, Volume 68, pages 1026-1037, **2007**.

Comfrey Flower

Overview of Parts of a Flower

Inflorescence is a group of flowers arranged on a stem composed of a main branch or complicated arrangement of branches.

Parts of a flower are petals and associated structures in the perianth, and the reproductive parts. There are 4 structures attached to the tip of a short stalk that are arranged in a whorl. Whorls starting from the base of the flower and going up:
    Perianth: The calyx and corolla form the perianth.
        Calyx: Outermost whorl of green sepals that enclose the flower in the bud stage.
        Corolla: Soft petals with colors to attract pollinators.
    Reproduction:
        Androecium (male): The next whorl is stamens: a stalk called a filament, topped by an anther where pollen is produced. Pollen is a powder made of pollen grains that produce male gametes (sperm).
        Gynoecium (female): The inner whorl is carpels that form a hollow structure called an ovary, which produces ovules. They produce megaspores which develop into female gametophytes that create egg cells. The innermost whorl (ovary, style and stigma) is called a pistil. The sticky tip of the pistil, the stigma, receives the pollen.

Comfrey Flower Overview

*"**In the curled inflorescences of the Comfrey,** Forget-me-not, and Viper's Bugloss (Symphytum, Myosotis, Echium), **and many other Boragineae, the inflorescence may be seen to unfold and fix itself, so that the flowers in turn are placed in the position in which they are best seen by and most accessible to flying insects.***
Meanwhile the older flowers, whose time is over, and to which insect-visits are of no further use, move out of the way of those which have just opened, and always choose their position so as not to obstruct the entrance to the new flowers of the same inflorescence.
In this process not only the flower-stalk but the rachis of the whole inflorescence takes part, and it is interesting to observe how even widely distant parts of the stem are sympathetically affected, so to speak, and how all the different parts of the system of axes are extended, raised, depressed, and curved exactly as required for the purpose of affording the most favourable position

to each flower in turn."
-'The Natural History of Plants, Their Forms, Growth, Reproduction, and Distribution' (Pflanzenleben = Plant Life or Plantscapes, 1887) by Anton Kerner von Marilaun, Austrian botanist. Translated by F.W. Oliver from 6 volumes in German. Published by Blackie & Son, London, England, **1895**. Volume II, Symphytum pages 97-99, 191, 275, 441, 585, 744. Also Volume I, Symphytum pages 441 and 744, translated 1894.   (Rachis is a stem of a plant with flower stalks at short intervals.)

"The Symphytum inflorescence consists of a pair of scorpioid racemes each with a leaf at the base, and the flowers are arranged in two rows on the upper side of the rhachis.
In the Ramosa Division the flowers are generally more numerous and more shortly pedicelled than in the Simplicia Division, and consequently form denser racemes.
    The calyx furnishes a good character for the Division of the Ramosa *in the relative length of its segments or teeth, which are sometimes longer and sometimes shorter than the tube. These vary from triangular or lanceolate to linear, with the apex acute or obtuse, and are of importance in the discrimination of many of the species.*
The margins of the segments are ciliate, in the Simplicia often conspicuously so, because the back of the segments is puberulous and almost destitute (almost no) of long hairs.
In fruit the calyx *is more or less accrescent and becomes strongly hispid or setose, with the segments at first connivent over the nutlets; but when the fruit is abortive the segments are widely spreading.*
The corolla when well developed *is infundibuliform or subcylindrical, with a more or less ventricose limb which is generally equal in length to the tube, but occasionally falls short of it. The margin is divided into five broadly ovate or triangular teeth with a short, obtuse, spreading apiculus.*
    Owing to unfavourable conditions of growth the corolla is sometimes narrowly clavate with the mouth unexpanded and teeth narrow, acute and erect. Flowers in this condition should not be taken as characteristic of the species.
The stamens *are included, the anthers oblong with the cells either obtuse or apiculate at one or both ends. The relative length of the anther and filament is sometimes a good specific character, and it is customary to estimate this, not from the entire length of the filament, but from that part which is not concealed by the anther.*
The style *is filiform with a minute capitate stigma and in the mature flower is shortly exserted. It is sometimes bent about one millimetre below the stigma, and this has been given as a specific character of* **Symphytum peregrinum**, *in which it is frequently but not invariably present. This peculiarity sometimes occurs in other species, and appears to be caused by the elongation of the style while the flower-bud is still tightly closed.*
All the species flower in the spring and early summer, but in some cases they continue flowering throughout the summer and autumn."
-'A Revision of the Genus Symphytum, Tourn.' by Cedric Bucknall, (Mus. Bac. Oxon= Bachelor of Music, Oxford University), Journal of the Linnean Society of London, England, Botanical Journal, Volume 41, Issue 284, pages 491-556, December **1913**.
    (Raceme is a flower cluster with separate flowers attached by short stalks at equal distances along a stem.
    The flowers at the base of the stem develop first.)

"The character of the calyx of Symphytum is of great significance; the relative length and shape of the segments or teeth and the length of the calyx in relation to that of the corolla tube are of diagnostic importance.
The calyx is slightly zygomorphic, although this fact does not appear to have been widely recognized.
The corolla is either funnel-shaped or subcylindrical, **its colour ranging from white or yellowish-white to pink, lilac (pale pinkish-violet) or blue;** *Symphytum anatolicum is unusual in having plants with either blue or white flowers.*
The corolla scales vary from narrowly triangular to subulate, either acute or obtuse at the apex, and in the latter case they are often slight emarginate. The sacles are generally as long as the stamens and may or may not slightly overtop them; less often they are exserted. The stamens alternate with the corolla scales."
-'A Revision of Symphytum in Turkey and Adjacent Areas' by G.E. Wickens, Royal Botanic Gardens: Kew; Notes from the Royal Botanic Garden Edinburgh, Scotland, Volume 29, pages 157-180, **1969**. Includes Turkey, Bulgaria, Greece, Aegean Islands and Caucasia.   (Zygomorphic means a flower can be divided into equal halves along only one line; bilateral symmetry.)

"Comfrey and Elecampane: Their scale provides a massive foliage accent. In spite of their leaf size, both have flower interest, with honors going to the pink-flowered Comfrey."
-'Landscaping with Herbs' by Elisabeth W. Morss, Arnoldia: The Arnold Arboretum of Harvard University, Cambridge, Massachusetts, Volume 39, No. 4, pages 238-269, July **1979**. arnoldia.arboretum.harvard.edu

"Much work on the Symphytum complex was undertaken by the Dutch botanists Th.W.J. Gadella and E. Kliphuis, and together with F.H. Perring they also looked at some British populations.
A relatively recent statement of the accepted position is given by Perring (*1994) in which the **purple flowered Comfreys are:**
    S. officinale, 2n = 48, with red buds opening to purple flowers. (Common Comfrey)
    S. x uplandicum, 2n = 36, with deep purple buds opening to a colour ranging from purple violet to violet blue.
A frequent form of S. x uplandicum, 2n = 40, with pink buds opening blue, and the cream coloured forms of S. officinale, 2n = 24 and 2n = 48 do not pose a problem here, nor does **purple flowered S. officinale, 2n = 40 since it is stated to occur only in Holland.**
Assuming Dr. G.E. Marks (of the John Innes Institute, Norwich, England) determination is correct, and I am not in a position to doubt it, how is one to determine the Woodhall type (southeast Yorkshire, vice county 61 at Woodhall, SE/695.320) in the field if petiole decurrency is thought to be insufficient?

> *This petiole or leaf decurrency is in fact quite difficult to assess overall since it is less pronounced the further one goes down the stem.*

Stace (**1991) states that the nutlets of S. officinale are shiny compared with dull and minutely tuberculate in the case of the hybrid (Russian Comfrey)."

-'Atlas 2000 and a Problem with Purple Flowered Comfrey' by Eric Chicken, East Yorkshire, England; Botanical Society of Britain and Ireland (BSBI) News, Hertfordshire, England, No. 76, pages 22-23, September **1997**.

-'Atlas 2000 and a Problem with Purple Flowered Comfrey (continued)' by Eric Chicken, East Yorkshire, England; Botanical Society of Britain and Ireland (BSBI) News, Hertfordshire, England, No. 82, pages 48-49, September **1999**.

(* -The Common Ground of Wild and Cultivated Plants: B.S.B.I. Conference Report No. 22 by A. Roy Perry and R.G. Ellis, {F.H. Perring, pages 64-70}. Cardiff, Wales: Department of Biology, National Museum of Wales, 1994.)

(** -New Flora of the British Isles: Identification of Wild Vascular Plants of the British Isles edited by Clive Stace. Cambridge, England: Cambridge University Press, 1991, pages 645-648.)

"While in the sub-tropics and tropics Comfrey plants rarely flower, as there is not enough winter chill, but they also don't die back so the leaves are available all year round."

-'Comfrey' by Penny Woodward, Edible and Useful Plants, www.pennywoodward.com.au, Australia, November 27 **2012**. Author of books such as 'Herbs for Australian Gardens' and 'Grow Your Own Herbal Remedies'.

(If you live in a tropical region, could you let me know if this is your experience. I have seen this idea repeated in general articles. See the below research articles from 1966, 1967 and 1996.)

(Chilling requirement {hours} for a fruit tree is the minimum period of cold weather needed for it to blossom in spring.)

"Symphytum officinale: Flower formation needs a day length of at least 12 hours.
Shoot growth needs a day length of at least 14 hours. The shorter the day length, the more leaves are formed."

-'Annual Developmental Cycle of Roots and Shoots in Symphytum Officinale L.' by Karin Staesche, Institut fur Spezielle Botanik und Pharmakognosie der Universitat Tubingen (Institute of Special Botany and Pharmacognosy of the University of Tubingen), Germany; Planta, Berlin, Germany, Volume 71, No. 3, pages 268-282, September **1966**. All in German except this abstract is also in English. If you have an English translation, I would appreciate a copy.

(At latitude 0 degrees which is the Equator, the day length is around 12 hours for the entire year.)

(See subsection 'Photoperiod' in section 'Planting, Soil, Fertilization, Water, Disease' in Volume 2.)

"Symphytum officinale: Flowers are always formed after 16-19 leaves, even at a day length of 12 hours at which 26-29 leaves usually appear before flowers are formed."

-'Development of Root-Layers of Symphytum Officinale L.' by Karin Staesche, Institut fur Spezielle Botanik und Pharmakognosie der Universitat Tubingen (Institute of Special Botany and Pharmacognosy of the University of Tubingen), Germany; Planta, Berlin, Germany, Volume 75, No. 4, pages 352-357, **1967**.

"For Symphytum tuberosum in Spain, peak flowering in the beech wood was May 22.
Number of light hours in May in the beech forest:
May 6 = 12.5 hours, May 16 = 11 hours, May 26 = 12 hours.
Some variations in light hours were due to the cloudiness."

-'Phenology of Hyacinthoides Non-Scripta L Chouard, Melittis Melissophyllum L and Symphytum Tuberosum L in Two Deciduous Forests in the Cantabrian Mountains, Northwest Spain' by G. Gonzalez Sierra, A. Penas Merino and E. Alonso Herrero, Universidad de Leon, Spain; Vegetatio (Plant Ecology), Volume 122, No. 1, pages 69-82, **1996**.

(Phenology is study of cycles and seasons of natural phenomena, especially in relation to climate and plant/animal life.)

(Daylight hours in Cantabria, Spain goes from 14 hours on May 1 to 15 hours on May 31.)

Comfrey Flower Buds, Bloom, Nectar and Pollination

See sub-subsection 'Flowers and Pollination' in subsection 'officinale' in section 'Details About Each Symphytum Species' (Chapter 8).

See sub-subsection 'Bees and Comfrey' in subsection 'Beneficial Insects and Comfrey' in section 'Garden Uses of Comfrey'(Chapter 18).

The nectar in Comfrey flowers refills every 45-60 minutes. Long-tongued pollinators such as bumblebees and hummingbirds go inside the flower from the opening in the usual manner and pollinate the flower. Short-tongued pollinators such as carpenter bees, mason bees, and honeybees get nectar from outside the flower at its base by chewing a hole, therefore they do not pollinate it. This is called nectar robbing. Once a hole is created, other pollinators use the same hole to rob the flower.

"The function of the minute pricklets which are to be found inside the corolla of certain Asperifolise, notably of some species of Symphytum, is slightly different. The pricklets are here placed neither on the inner surface of the corolla nor on the filaments, but on certain epiblastemes of the corolla.
These epiblastemes, which alternate with the stamens, are made of tough tissue, have an elongated triangular outline, and are beset round their edges with small sharp teeth, which may almost be compared with the processes on a swordfish's spear. (See Plate II. fig. 74).

They project into the corolla-tube, and are so closely apposed to each other as to form a hollow cone, the apex of which points towards the mouth of the flower, and has in its centre an opening through which the style is seen to protrude (Plate II. fig. 73, longitudinal section of a flower of Symphytum officinale).
It is only at their very tips that the epiblastemes or throat-scales are without teeth. But it is these tips that surround the small central hole in the apex of the cone, which gives issue to the style; and **through this hole such insects as have a proboscis long enough to reach the bottom of the flower** can therefore suck the nectar without hurting that sensitive organ.
It is, moreover, only by this method of suction that they will come in contact first with the stigma and then with the anthers, and **so will convey pollen from one flower to another, and promote intercrossing.**
They will never thrust their tender proboscis through the lateral chinks between the adjoining scales, defended as these are by pricklets. So far, therefore, the pricklets act as 'path-pointers' to invited guests. But inasmuch as they also shut out all such smaller insects with shorter proboscis, as would, in their absence, get at the nectar without first striking the stigma and then getting powdered with pollen, they are to be considered not only as path-pointers, but also as protective appliances against unbidden guests."
-'Flowers and Their Unbidden Guests' by Dr. Anton Kerner, Professor of Botany in the University of Innsbruck, Austria; Translated by Dr. W. Ogle; C. Kegan Paul & Co., London, England, **1878**. Comfrey text pages 78-79, Comfrey art between pages 154-155.

"***Pollen Grains:***
The dimensions of pollen-grains are very various in diffierent groups of plants. Thus, whilst in the Forget-me-not (Myosotis), **Borage (Borago), Comfrey (Symphytum), and Boragineae generally,** as also in Artocarpere (e.g. Fious), **the pollen-grains are very small,** in Cannacese, Malvaceae, Cucurbitaceae and Nyctaginese, they are relatively large. The number of the grooves in pollen-grains is constant for a given species, and even for whole families of plants. **Nine or ten grooves in pollen-grains of Sherardia, Borago, and Symphytum.**
Pollen and Insects:
In the flowers of several Boragineae- Comfrey (Symphytum) and Gerinthe, for example- **there are peculiar scales, furnished with sharp prickles,** alternating with the anthers, and placed in such a position that insects are afraid to insert their proboscees except at the apex of the cone of anthers, and **in consequence the head alone and not the abdomen is, in this case, besprinkled with pollen.**"
-'The Natural History of Plants, Their Forms, Growth, Reproduction, and Distribution' (Pflanzenleben = Plant Life or Plantscapes, 1887) by Anton Kerner von Marilaun, Austrian botanist. Translated by F.W. Oliver from 6 volumes in German. Published by Blackie & Son, London, England, **1895**. Volume II, Symphytum pages 97-99, 191, 275, 441, 585, 744. Also Volume I, Symphytum pages 441 and 744, translated 1894.

"***Symphytum Tourn.:***
Homogamous bee flowers; with nectar secreted by an annular ridge at the base of the ovary, and stored in the base of the corolla. According to *Kerner, the peduncle bends down in late anthesis, so that the flower assumes a nodding or pendulous position, and the stigma is brought into the line of fall of the pollen, **thus rendering automatic self-pollination inevitable.**"
-'Handbuch der Blutenbiologie' {Handbook of Flower Pollination} by P. Knuth, Volume 3, part 2. v + 601 pages, Leipzig, Germany, 1905. [Boraginaceae, pages 63-67.] English translation by J.R.A. Davis, Volume 3, iv + 644 pages, Oxford, England, **1909**. [Boraginaceae, pages 115-142; 23 genera, including Symphytum.]
(* -'The Natural History of Plants, Their Forms, Growth, Reproduction, and Distribution, Volume II' {Pflanzenleben = Plant Life or Plantscapes, 1887} by Anton Kerner von Marilaun, Austrian botanist. Translated by F.W. Oliver from 6 volumes in German. Published by Blackie & Son, London, England, 1895.)   (Homogamous means the flowers do not differ sexually.)

"***Flower Abnormalities:*** Phyllody is abnormal leaflike development of floral organs.
Phyllody (also bracteody, chloranthy, or virescence) has been reported in the following genera:
    Corolla (petals), carpels and ovules of Symphytum.*"
-'Flower Abnormalities' by Vesta G. Meyer, Delta Branch, Mississippi Agricultural Experiment Station, Stoneville, Mississippi; The Botanical Review: New York Botanical Garden, Volume 32, No. 2, pages 165-218, April-June **1966**.
    (* -Vegetable Teratology: An Account of the Principal Deviations from the Usual Construction of Plants' by Maxwell T. Masters, M.D., F.L.S.. London, England: Ray Society, Robert Hardwicke, 534 pages, 1869.)
    (Chloranthy is abnormal development where flowers look like leaf buds. Virescence is an abnormality where plant parts are green that should not be green.)
    (Carpel is the female reproductive organ of a flower, that includes the ovary, stigma and style. An ovule is the part of a plant that develops into a seed when fertilized.)

"***Symphytum officinale:***
1. Flowers are always formed after 16-19 leaves, even at a day length of 12 hours at which 26-29 leaves usually appear before flowers are formed.
2. In cultures kept at temperatures of at least +10 C (50 F) fructosans are stored in the young shoot-born roots, while the **amount of fructosans is reduced in the buds,** in the subterranean shoot parts and in the old root pieces."
-'Development of Root-Layers of Symphytum Officinale L.' by Karin Staesche, Institut fur Spezielle Botanik und Pharmakognosie der Universitat Tubingen (Institute of Special Botany and Pharmacognosy of the University of Tubingen), Germany; Planta, Berlin, Germany, Volume 75, No. 4, pages 352-357, **1967**.
    (See sub-subsection 'Fructans' in subsection 'Fat, Protein, Carbohydrate, Fiber, Ash, Miscellaneous' in section 'Nutri-

tional Value of Comfrey' {Chapter 19}.)

-'**Pollen Morphology and Taxonomy in the Genera Symphytum L.** and Procopiana Gusuleac' (Palinotaksonomia Rodzajow Symphytum L. i Procopiana Gusuleac) by Krystyna Harmata; Zeszyty Naukowe UJ, Prace Botaniczne / Botanical Papers 5, Poland, pages 28-29, **1977**. (I was unable to get this report. If you have a translation in English, could you please send it to me.)
    -'**A Supplement to the Pollen Morphology and Taxonomy of Genera Symphytum L.** and Procopiana Gusuleac' (Uzupelnienie do Palinotaksonomii Rodzajow Symphytum L. i Procopiana Gusuleac) by Krystyna Harmata, Prace Botaniczne / Botanical Papers 8, pages 7-10, 1981. (I was unable to get this report. If you have a translation in English, could you please send it to me.)

"When an insect forces its way into the flower, the anthers are forced apart, and the pollen sifts itself over the body of the insect.
    In some hanging flowers (e.g. Symphytum officinale) the whole flower forms a similar apparatus ('Streukegel') with (throat) scales acting as arresting organs as long as the flower is not visited.
The tube appears as a type in its own right, with no limb in Castilleja, Kentrosiphon (with laternal entrance; Vogel 1954), **Symphytum officinale,** etc."
-The Principles of Pollination Ecology by Knut Faegri (Bergen, Norway) and L. Van Der Pijl (Nijmegen, Netherlands). Oxford, England: Pergamon Press, Third edition, **1979**, pages 52, 92. First edition 1966.

"**Pollen grains of 18 species of 13 genera in the Korean Boraginaceae were investigated** by using light and scanning electron microscopes.
    The palynological result revealed that pollen grains of the family possess tricolporate aperture and other types derived from it and are primarily divided into two groups by having and **lacking pseudocolpi.** The latter included tricolpate Ehretia, **8-colporate (8 grooves) Symphytum (Symphytum officinale),** and 4,6-colporate Lithospermum.
Pollen Key to the Boraginaceae Species:
    1. Aperture tricolporate, then it is Ehretia ovalifolia.
    1. Aperture 4,6,8-colporate or 3-colporate with 3 pseudocolpi.
        2. Aperture 4,6,8-colporate:
            3. EV not-constricted. Aperture 6,8-colporate, isoplar:
                4. 8-colporate, circular oval prolate, gemmate, **then it is Symphyum officinale.**"
-'A Palynotaxonomic Study of the Korean Boraginaceae' by Young Mee Ahn and Sangtae Lee, Department of Biology, Sung Kyun Kwan University, Suwon, South Korea; Korean Journal of Plant Taxonony, Vol 16, No. 3, pages 199-215, December **1986**.
    (Colporate is pollen grain having apertures which combine a rounded pore and a colpus {groove}.)

"***Knuth listed numerous pollinators of the flowers of seven species of Symphytum.** The corolla throat in the genus is closed by a conical structure made up of the five anthers and the five alternating faucal appendages.
    Nectar accumulating at the base of the corolla is 'legitimately' accessible to insects with a proboscis longer than 1 cm (0.39 inch). However, several species of bees with a short proboscis have been observed 'stealing' nectar by piercing holes or slits at the base of the corolla tube.
The flowers of many species of Symphytum are drooping, and pollen falling from the anthers to the long-exserted stigmas ensures self-pollination.
Pollen of Symphytum is (7-)8-10(-11) zonocolporate, prolate or dumbbell shaped, rounded at the poles, and with short, narrow colpi and circular pores (Ahn & Lee; Clarke; Sahay).
Elaiosomes on nutlets of species of Symphytum have been suspected of aiding in its dispersal by ants (**Bresinsky)."
-'Journal of the Arnold Arboretum, Supplementary Series, Volume 1, Harvard University, Cambridge, Massachusetts, **1991**. This volume is a collection of contributions toward a 'Generic Flora of the Southeastern United States'. 'The Genera of Boraginaceae in the Southeastern United States' by Ihsan A. Al-Shehbaz, page 152.
(* -'Handbuch der Blutenbiologie' {Handbook of Flower Pollination} by P. Knuth, Volume 3, part 2. v + 601 pages, Leipzig, Germany, 1905. [Boraginaceae, pages 63-67.] English translation by J.R.A. Davis, Volume 3, iv + 644 pages, Oxford, England, 1909. [Boraginaceae, pages 115-142; 23 genera, including Symphytum.])
(** -'Bau, Entwicklungsgeschichte und Inhaltsstoffe der Elaiosomen: Studien zur Myrmekochoren Verbreitung von Samen und Fruchten' {Construction, History of Development and Ingredients of Elaiosomes: Studies on the Myrmecochoric Distribution of Seeds and Fruits} by Andreas Bresinsky; Bibliotheca Botanica, Germany, Heft 126, 1963. In German.)
    (Elaiosomes are fleshy structures attached to seeds. It is rich in lipids and proteins which attract ants who then take seed to their nest and feed the elaiosome to larvae.)
    (In insects the proboscis is an elongated sucking mouthpart that is tubular and flexible.)

    "**Ants are splendidly efficient planters,** especially when it comes to spring flowers such as crocuses, snowdrops and lungwort (Verbascum thapsus), not to mention the many varieties of wild flowers with small seeds which are already resident in the garden, for example, purslane (Portulaca) and **Comfrey.**"
    -'Companion Planting: Successful Gardening the Organic Way' by Gertrud Franck; Thorsons Publishers Inc, New York, 1983. Translated from German.

"**Vibratory pollen collection by bumble bees: The buzz foraging vibrations of Bombus terrestris and Bombus hortorum workers,** measured by an accelerometer attached to the flower stem, reached relatively high acceleration magnitudes of 212 m/

s² (meter per second squared) at frequencies of 374 Hz (Hertz).
Bombus hortorum workers were more effective than Bombus terrestris workers at vibrating the anthers of Symphytum officinale (Comfrey) because they inserted their heads into the corolla for nectar. The escape buzz produced by Bombus terrestris during capture was similar to that produced during pollen foraging on Actinidia deliciosa (kiwifruit) and required 0.3 W (work power) to generate. The power required to create the escape buzz is obtained by estimating the waste heat generated."
-'Buzz Foraging Mechanism of Bumble Bees' by Marcus J. King, Research Engineer, Industrial Research Ltd, Christchurch, New Zealand; Journal of Apicultural Research, Volume 32, No. 1, pages 41-49, **1993**.

"**We compared the flowering and fruiting phenology of Symphytum tuberosum L. (Boraginaceae),** Hyacinthoides non-scripta (L.) Chouard (Liliaceae), and Melittis melissophyllum L. (Labiatae) in a beech wood and an oak wood situated in the same valley in **Spain**, taking into account their different microclimatic and edaphic (soil) characteristics and carrying out correlation and simple regression analyses.

Differences observed in the phenological cycles of the species studied included earlier flowering and longer cycles in the oak wood, except for H. non-scripta. The most important climatic factors were soil moisture and the relative humidity, followed by temperature and the number of daylight hours and photosynthetically active radiation (P.A.R.). The strong correlation observed in the fruiting of the three species indicated clear synchronization.

Earlier Flowering:
All the species began flowering earlier in the oak forest than in the beech forest. **For Symphytum tuberosum, peak flowering in the beech wood coincided with peak fruiting in the oak wood (May 22).** This species flowered earlier than the others in both woods.

Humidity and Temperature:
Symphytum tuberosum initiated its flowering when relative humidity was low, but with similar temperatures in both woods. The flowering of this species was inversely related to temperature in the oak forest and increased with relative humidity. Temperature seems only to correlate with Symphytum tuberosum flowering.

No specific climatic parameter was found to affect Symphytum tuberosum. In this species, we cannot establish clear relations between climatic parameters and phenological cycles, although these cycles seem to depend mainly on temperature and relative humidity."
-'Phenology of Hyacinthoides Non-Scripta L Chouard, Melittis Melissophyllum L and Symphytum Tuberosum L in Two Deciduous Forests in the Cantabrian Mountains, Northwest Spain' by G. Gonzalez Sierra, A. Penas Merino and E. Alonso Herrero, Universidad de Leon, Spain; Vegetatio (Plant Ecology), Volume 122, No. 1, pages 69-82, **1996**.
(Phenology is study of cycles and seasons of natural phenomena, especially in relation to climate and plant/animal life.)

"**Flower structure, character of primary and secondary attractants, species of insects visiting flowers in five species of Symphytum L.: S. asperum Lepech., S. carpaticum Frolov, S. officinale L., S. tanaicense Stev., S. x uplandicum Nym.** have been studied in the condition of their introduction to Leningrad district, Russia.
It has been shown that flowers of the species provide high 'density of food' because of big inflorescences, produce sufficient quantity of nectar and pollen and are attractive for some insect species.
But peculiarities of flower structure (long carolla) make them available only for bumblebees with long proboscis. So, only two species of bumblebees: Bombus hortorum and B. lucorum are true pollinators of Symphytum. Two other species B. lapidarius and B. derhamellus and bees. Apis mellifera (Western or European Honey Bee) are thieves of nectar.
Dynamics of nectar secretion and changes in sugar productivity do not differ in samples of different origin. **It was shown that there in Symphytum flowers besides these attractants for pollinators there are mechanisms of outcrossing:** dichogamy (prothandry) and hercogamy (features of flower structure preventing self-pollination) and possibly partly self-incompatibility."
-'Biology of Flowering and Pollination of Symphytum L. Species Introduced in Leningrad District' by N.M. Naida and M.A. Vishnyakova; Rastitel'nye Resursy, Leningrad, Russia, Volume 33, No. 3, pages 52-61, **1997**. In Russian. (If you have this in English, could you please send it to me.)
(Symphytum carpaticum Yu.M.Frolov is an unresolved name. Distribution is in the Ukraine.)
(Dichogamy is the ripening of stamens and pistils of a flower at different times so it can not self-fertilize.
Hercogamy means self-pollination is impossible due to structural obstacles.)

"**The results of a comparative survey on pollen morphology in the Boragineae (tribe)** by means of light, scanning and transmission electron microscopy are presented and discussed in relation to the taxonomy of the tribe. The presence of supratectum gemmae is distinctive of some genera (Borago, Symphytum and Trachystemon).
Cluster no. 8: Symphytum bulbosum-Type:
Pollen (7)-8-9-10-11-zonocolporate, medium-sized, P = 27-42 gin, E = 22-32 gin.
Shape from subprolate to prolate, with P/E = 1.25-1.40. Outlines elliptic in equatorial view and circular in polar view. Ectoapertures 10-18 gm long, narrow, fusiform; typically the exine, or probably only the gemmate tectum, protrudes at the equator to form a bridge-like structure over the aperture. Endoapertures lalongate, 2-5 ×4-9 gm, with elliptical profile. Polar exine 0.7-1.2 micrometer, with endexine 0.10-0.15 gm and simple columellae; tectum punctate, covered by spheroidal gemmae 0.15-0.20 gm in diameter.
Their density is remarkably lower in S. gussonei and S. tuberosum, causing the separation of these two taxa into a separate subcluster.
This Symphytum group is variable in the number of apertures and consists of all the investigated taxa of this widespread Eurasiatic genus. **Our data largely agree with palynological features reported for the Asiatic (*Ahn &

Lee 1986) and northwestern European (**Clarke 1977) members of this Symphytum genus.

Key to Pollen Types:
Tectum gemmate. . . . 2.
2a - Exine forming a bridge over the apertures, endoapertures separate . . .. **Symphytum bulbosum**.

On the basis of tectum ornamentation, 2 major groups of pollen types can be distinguished within the Boragineae tribe.
The first includes all types with a nongemmate tectum.

The second includes the three gemmate types, Symphytum bulbosum, Trachystemon orientalis and Borago officinalis. This is distinct from all the other by the thick exine and the branched columellae.

The palynological affinity among the gemmate types supports the view by Gusuleac (1928) of a common evolutive line originated by an hypothetic 'Paleoborago' ancestor.

The two species Symphytum circinale and Symphytum creticum, included by Gusuleac (1928) in the genus Procopiania (compare to also Pawlowski 1971), do not show any differentiation from the other members of Symphytum, suggesting the maintainance of Procopiania into Symphytum (*Runemark 1967).**

A further aspect emerging from this survey is that the recognized pollen types correspond to the taxonomic delimitations currently accepted in the relevant literature only in seven out of the 12 examined genera: Anchusella, Borago, Brunnera, Elizaldia, Lithodora, **Symphytum** and Trachystemon. **The generally adopted circumscription of these genera is therefore matched by palynological data.**"

-'Pollen Morphology in the Boragineae (Boraginaceae) in Relation to the Taxonomy of the Tribe' by Massimo Bigazzi and Federico Selvi, Plant Systematics and Evolution, Volume 213, pages 121-151, **1998**.

(* -'A Palynotaxonomic Study of the Korean Boraginaceae' by Young Mee Ahn and Sangtae Lee, Department of Biology, Sung Kyun Kwan University, Suwon, South Korea; Korean Journal of Plant Taxonony, Volume 16, No. 3, pages 199-215, Dec 1986.)

(** -'Northwest European Pollen Flora, 10: Boraginaceae' by G.C.S. Clarke, Review of Palaeobotany and Palynology, Volume 24, pages 59-101, 1977.)

(*** -'Studies in the Aegean Flora XI: Procopiana {Boraginaceae} included into Symphytum' by Hans Runemark, Botaniska Notiser {Botanical Notes}, Sweden, Volume 120, pages 84-94, 1967. I was unable to get this report. If you have it, could you please send it to me.)

(Palynology is the study of pollen grains, especially in archaeological or geological deposits.)

(Family: Boraginaceae, Tribe: Boragineae, Genus: Symphytum L.)

(Supratectum gemmae: Supra means above. Tectum is a roof-like structure. The outer pollen wall is composed of two layers: tectum and foot layer. Gemmae is a mass of cells.)

"Honey bees, Apis mellifera, use short-lived repellent scent marks to distinguish and reject flowers that have recently been visited by themselves or by siblings, and so save time that would otherwise be spent in probing empty flowers.
Conversely, both honey bees and bumblebees, Bombus spp., can mark rewarding flowers with scent marks that promote probing by conspecifics.

We examined detection of recently visited flowers in a mixed community of bumblebees foraging on Comfrey, Symphytum officinale, in southern England.
When foraging among inflorescences on a plant, three abundant species of Bombus probed fewer inflorescences than would be expected from random foraging. Bees frequently encountered inflorescences but departed without probing them for nectar.

Examination of the incidence of such rejections in the two most common species, Bombus terrestris and Bombus pascuorum:
When presented with inflorescences of known history, bees selectively rejected those that had been recently visited by themselves or by conspecifics compared with randomly selected inflorescences. They were also able to distinguish inflorescences that had been visited by other Bombus species.

Bees were unable to distinguish and reject inflorescences from which the nectar had been removed artificially. We conclude that these Bombus species are probably using scent marks left by previous visitors."

-'Foraging Bumblebees Avoid Flowers Already Visited by Conspecifics or by Other Bumblebee Species' by Dave Goulson, Sadie A. Hawson and Jane C. Stout, Division of Biodiversity and Ecology, School of Biological Sciences, University of Southampton, England; Animal Behaviour, Volume 55, No. 1, pages 199-206, **1998**.

"**We have found that foraging bumblebees (Bombus hortorum, B. pascuorum, B. pratorum and B. terrestris) not only avoid flowers of Symphytum offcinale that have recently been visited by conspecifics (same species) but also those that have been recently visited by heterospecifics (different species).**

We found that flowers were repellent to other bumblebee foragers for approximately 20 minutes and also that after this time nectar levels in Symphytum officinale flowers had largely replenished. Thus bumblebees could forage more efficiently by avoiding flowers with low rewards.

Nectar build-up rate in Symphytum officinale from nectar removal time:

5 minutes	0.0040 ml/flower
10 minutes	0.0042 ml/flower
20 minutes	0.0055 ml/flower
30 minutes	0.0070 ml/flower
40 minutes	0.0075 ml/flower
50 minutes	0.0085 ml/flower
60 minutes	**0.0090 ml/flower = maximum nectar**

Bumblebees collect nectar from S. officinale in a conventional manner (probing for nectar from the tubular opening of the corolla), or they rob nectar (collecting nectar through a hole bitten in the base of the flower corolla; *Inouye 1983) or they collect pollen [which requires sonicating (buzzing) the anthers to release pollen; **King 1993]."
-'Repellent Scent-Marking of Flowers by a Guild of Foraging Bumblebees (Bombus spp.)' by Jane C. Stout, Dave Goulson and John A. Allen, Biodiversity and Ecology Division, School of Biological Sciences, University of Southampton, England; Behavioral Ecology and Sociobiology, Volume 43, pages 317-326, **1998**.
(* -The Biology of Nectaries edited by Barbara Bentley and Thomas Ellas. New York, New York: Columbia University Press, 1983. Article 'The Ecology of Nectar Robbing' by D.W. Inouye, pages 152-173.)
(** -'Buzz Foraging Mechanism of Bumble Bees' by Marcus J. King, Research Engineer, Industrial Research Ltd, Christchurch, New Zealand; Journal of Apicultural Research, Volume 32, No. 1, pages 41-49, 1993.)

*"**Nectar secretion and nectar chemistry in the flowers of Comfrey (Symphytum officinale L., Common Comfrey) were examined in the four stages of anthesis:***
>*I- large buds, before pollen exposure.*
>*II - freshy opened flowers with the beginning of anther dehiscence.*
>*III- completely opened flowers in the maximum of pollen exposure.*
>*IV- flowers at the final stage of anthesis, without pollen in the anthers."*

Individual flower stays fresh 2.5 to 3 days, on average.
Disc-shaped nectaries of Symphytum officinale are located at the base of the four-lobbed ovary. Nectar is released through the modified stomata.
Start of nectar secretion was noted at the bud stage. Nectar volume, mass of nectar and sugars differed in the examined stages of anthesis and the biggest values were noted at the final stages.
>*Dominant sugar in nectar was sucrose with smaller amounts of fructose and glucose. The presence of amino acids was also recorded."*

-'Nectar Secretion in the Flowers of Comfrey (Symphytum officinale L.) and Nectar Chemistry' by M. Stpiczynska, Department of Botany, Agricultural University, Lublin, Poland; Acta Agrobotanica published by Polish Botanical Society, Volume 56, No. 1/2, pages 27-36, **2003**.
>(Anthesis is the period when a flower is fully open and functional.)
>(Dehiscence is the splitting along a built-in line of weakness to release its contents.)

"Symphytum, Onosma and Cerinthe bear streukegel (scatter-cone or dispersion-cone) blossoms with nectar.
In Symphytum and Cerinthe a single anther needs 1-4 hours for opening.
Pollen portioning results through successive ripening and opening within 1-3 days of the five anthers of a flower."
-'Flowers of Boraginaceae (Symphytum, Onosma, Cerinthe) and Andrena symphyti (Hymenoptera-Andrenidae): Morphology, Pollen Portioning, Vibratory Pollen Collection, Nectar Robbing' by Herwig Teppner, Karl-Franzens University of Graz, Austria; Phyton: Annales rei Botanicae, Horn, Austria, Volume 50, No. 2, pages 145-180, **2011**.
>(Pollen portioning is the sucessive release of pollen in small amounts so it lasts longer.)

*"**A deeper understanding of determination and mechanisms of formation of complex biological patterns, such as exines of pollen and spores,** has been a recurrent theme in many investigations on exine pattern ontogeny.*
After detailing the exine ontogeny, our purpose was to find out whether the sequence of sporoderm developmental events corresponds to self-assembling micellar mesophases, initiated by genomically determined physicochemical parameters and induced by surfactant glycoproteins at increasing concentrations.
>***An unusual, 'hybrid' type of tapetum was observed. What is observed in Symphytum exine development allows us to obtain more evidence for the hypothesis of the participation of micellar self-assembly in sporoderm development*** *and to bring together the concepts of micelles and of Sporopollenin Acceptor Particles (SAPs)."*

-'Exine and Tapetum Development in Symphytum Officinale (Boraginaceae). Exine Substructure and Its Interpretation' by Nina Gabarayeva, Valentina Grigorjeva and Svetlana Polevova, Komarov Botanical Institute, Saint Petersburg, Russia and Moscow State University, Russia; Plant Systematics and Evolution, Volume 296, No. 1-2, pages 101-120, September **2011**.
>(Ontogenesis is the development of an organism or anatomical feature from the earliest stage to maturity.)
>(Exine is the decay-resistant outer coating of a pollen grain or spore.)
>(The tapetum is a layer of cells found in the anther of flowering plants. It is between the sporangenous tissue and anther wall. Tapetum is important for the nutrition and development of pollen grains and pollen coat.)

*"**Bombus borealis: Northern amber bumble bee:***
Select food plants: *Vicia (Vetches), Cirsium (Thistles), Asters, Prunella,* **Symphytum officinale (Comfrey),** *Eupatorium. Uncommon. Tongue length: long. Nests underground. Can be confused with Bombus fervidus.*
Females *(queens and workers, colors refer to pile or 'hair'):*
>*Face yellow, thorax and T1-4 extensively yellow, with a black band between the wing bases, lower sides of the thorax usually black, T5 black or yellow.*
>*Midleg basitarsus with the distal posterior corner sharply pointed. Cheek distinctly longer than broad, clypeus surface very smooth and shiny with only a few very small punctures near the center. Hair of the face usually gray-yellow and usually paler than on the rest of the body and only rarely with many black hairs extensively intermixed, sides of the thorax predominantly black with yellow only within the dorsal half. Hair of medium length and even."*

-'Bumble Bees of the Eastern United States' by Sheila Colla, Leif Richardson and Paul Williams, 'United States Department of Agriculture Forest Service', 'Pollinator Partnership' and 'National Fish and Wildlife Foundation', FS-972, Washington, DC, March **2011**, pages 66-68.

    (Bombus borealis is native to Canada, Alaska and the northern United States. This species lives in woodland habitat.)

"In Common Comfrey (Symphytum officinale), the flower buds and flowers in bloom point downward. The calyxes with the ripening fruits are located on the upper surface of the scorpioid cyme. The flowers are arranged in a double scorpioid cyme. Note hairs on the stem, leaves, pedicels, and sepals.
Flowering starts along the underside of the scorpioid cyme. The flowers that are the first to come into bloom are also the first to form ripe fruits.
> *If flowering takes place over a longer period of time, it can happen that the fruits that formed the earliest have already dispersed, while the fruits of the flowers along the middle of the scorpioid cyme are still present in the calyx and the flowers at the end of the scorpioid cyme are still flowering."*

-A Manual for the Identification of Plant Seeds and Fruits by R.T.J. Cappers and R.M. Bekker, Netherlands: Barkhuis Publishing, **2013**, page 69.   (A cyme is an inflorescence in which each floral axis ends in a single flower.)

*"**Insect Pollinators:** There is growing evidence that insect pollinators are declining globally and agricultural intensification has been identified as a major cause of this decline.*
Further analyses will be conducted to determine if habitat effects were solely due to differences in plant diversity. Observational evidence indicates that this may be partly, if not solely, the case.
> ***Pollinator presence was strongly linked to specific plant species and numbers of pollinators within a specific site fluctuated depending on what plant species were in flower at the time of sampling.***
> ***In general, raspberry (Rubus idaeus) and Russian Comfrey (Symphytum x uplandicum) were important plant species in June**, thistles (Cirsium avense, Cirsium vulgare and Cirsium palustre), woundworts (Stachys sylvatica and Stachys palustris) were important in July and knapweed (Centaurea nigra) and marsh woundwort (Stachys palustris) were important in August.*

Maintaining and enhancing plant diversity will increase the likelihood of providing a constant source of nectar and pollen throughout the pollinator season and thus of safeguarding pollinator populations in intensive agricultural landscapes."
-'Safe Guarding Pollinator Populations in an Intensive Grassland Landscape' by Lorna J. Cole, Billy Harrison, Duncan Robertson and David I. McCracken, Sustainable Ecosystems Team, Ayr, Scotland; The Glasgow Naturalist: Glasgow Natural History Society, Scotland, Volume 26, Part 1, pages 25-28, **2014**. Journal published since 1908.

"Variations in attractiveness traits are known to strongly impact pollinator visitation patterns and on a larger scale pollination service. Indeed, greater plant attractiveness can enhance the frequency or number of flower visits: most pollinators are preferentially attracted to plants producing numerous, large flowers and/or greater rewards (in quality or quantity).
Larger floral display size of Symphytum officinale can also influence the abundance of visiting pollinators.*"
-'Competition with Wind-Pollinated Plant Species Alters Floral Traits of Insect-Pollinated Plant Species' by Floriane Flacher, Xavier Raynaud, Amandine Hansart, Eric Motard and Isabelle Dajoz, Paris, France; Scientific Reports: A Nature Research Journal, Volume 5, Article number: 13345, 10 pages, September 3 **2015**.
(* -'Floral Display Size in Comfrey Symphytum Officinale L Boraginaceae: Relationships with Visitation by Three Bumblebee Species and Subsequent Seed Set' by D. Goulson, J.C. Stout, S.A. Hawson and J.A. Allen, University of Southhampton, School of Biology, England; Oecologia {Plant and Animal Ecology}, Volume 113, No. 4, pages 502-508, 1998. For this article, see subsubsection 'Flowers and Pollination' in subsection 'officinale' in section 'Details About Each Symphytum Species' {Chapter 8}.)

*"**Osmia pilicornis Mason Bee is distributed from western temperate Europe to western Siberia**, where it exclusively occurs in open-structured, mesophilous and mainly deciduous woodland below 1000 meters (3280 feet) above sea level. It is rare with low population densities over most of its range.*
Pollen composition of female pollen loads of Osmia pilicornis Mason Bee:
    Symphytum   2.8% Pollen Grain Volume   4 loads = 7.7% of all pollen loads
Confirmed pollen sources for Osmia pilicornis Mason Bee *other than Pulmonaria and species of Fabaceae and Lamiaceae are Lonicera (Caprifoliaceae), Polygonatum (Asparagaceae; Benoist 1931), Potentilla (Rosaceae; Westrich 1989), Rubus (Rosaceae; L.A. Nilsson personal communication),* **Symphytum** *(Boraginaceae; L.A. Nilsson personal communication), Taraxacum (Asteraceae) and Viola (Violaceae; Wallis 1886, Friese 1911, Benoist 1931, Chambers 1949, Elfving 1968).*
Most of these latter pollen host genera occur at woodland sites reflecting the bee's exclusive habitat."
-'Distribution, Biology and Habitat of the Rare European Osmiine Bee Species Osmia (Melanosmia) Pilicornis (Hymenoptera, Megachilidae, Osmiini)' by Rainer Prosi, Heinz Wiesbauer and Andreas Muller from Germany, Austria and Switzerland; Journal of Hymenoptera Research, Volume 52, pages 1-36, **2016**.

    (Hymenoptera is an order of insects that includes sawflies, wasps, bees, and ants.)
    (Osmia pilicornis is also called the Fringe-Horned or Hairy-Horned Mason Bee.)

"Relative Time of Abscission:
Shedding of petals accompanying fruit growth was observed in families in which petals are generally shed when still turgid (filled with water), e.g. in some Boraginaceae (e.g. Symphytum, *Pulmonaria) and Lamiaceae."*
-Plant Evolutionary Developmental Biology: The Evolvability of the Phenotype by Alessandro Minelli, Italy. Cambridge, England:

Cambridge University Press, **2018**. Chapter 9: Pheno-Evo-Devo, page 295.
(Abscission is the natural detachment of parts of a plant such as dead leaves and ripe fruit.)

Comfrey Flower Perforations by Insects (Nectar Robbing)

See sub-subsection 'Flowers and Pollination in S. officinale' in subsection 'officinale' in section 'Details about Each Symphytum Species' (Chapter 8).

"Pollination:
The full fertility of many plants, as Charles Darwin has shown, depends largely on cross-pollination. Insects do not commonly visit flowers unless they get nectar or pollen in return, so that, **when a flower is constantly robbed, the regular pollinators do not receive their due share of nectar or pollen, their visits are fewer, and consequently there is less chance for cross-pollination.**
> If the plant is capable of self-pollination seeds may be developed, and often in great abundance, yet Darwin has shown that the progeny (offspring) of self-fertilized flowers is less vigorous than from cross-fertilized flowers.
> If the structure of a flower is such that self-fertilization is prevented, and insects do not go to it in the regular way, sterility may result. But in most of the flowers perforated there is an abundance of nectar, and insects which perforate flowers are very hasty in their visits, and therefore always leave some nectar.

Pollen is protected in various ways.
> **In Symphytum officinale and other members of the Borragineae, scales are developed which close over the throat of the corolla** (petals of a flower, forming a whorl within the sepals and enclosing the reproductive organs).

Insects Make Holes in Flowers:
The opinion is current that perforated flowers are not as productive as unperforated ones. Delpino has shown, in the case of Symphytum tuberosum (Tuberous Comfrey) and Polygala Chumcebuxus, that the perforated flowers are absolutely sterile.
Perforated flowers are not necessarily sterile, but are often quite productive, as is well shown in the case of Symphytum officinale (Common Comfrey) and Phlomis tuberosa. The flowers of Phlomis, Symphytum, and Monarda, are regularly perforated in the Botanic Garden (Saint Louis, Missouri).
> **I doubt whether there are many flowers in which one can find more perforations than in Symphytum officinale.** In stocks which have several thousand flowers, hardly one can be found which is not perforated. Several stocks in the Botanic Garden gave me ample opportunity of seeing the results from perforated flowers. I did not undertake to count the ripened fruits, but **the greater number of flowers developed some nutlets.**

It has been shown that flowers with deep-seated nectar are often perforated, and that in most cases the perforations are made by insects which are unable to get at the nectar in a normal way.

Bees:
Bombus mastrucatus (= Bombus wurflenii, found in Alps and middle Europe) is more addicted to this perforation habit than any other European bumble-bee, and following this are Bombus terrestris (buff-tailed bumblebee or large earth bumblebee, common in Europe), Bombus pratorum (early-nesting bumblebee, found in southern Europe), and Xylocopa violacea (violet carpenter bee). From observations thus far published, **Bombus is the most frequent perforator of flowers.**
There is a certain correlation in the length of the tongues of the Apidae family and the flowers they visit in a normal way, but when this limit is reached flowers are often perforated. The insect uses considerable ingenuity in perforating flowers, attacking them in close proximity to the nectary.
> In Germany, **Bombus hortorum** (small garden bumblebee), **Bombus senilis**, and **Bombus fragrans** (large bumblebee) never perforate flowers. Tongues measure respectively: worker 18-19 mm, 11-15 mm, 15 mm.
> **Bombus lapidarius**, tongue of female 12-14 mm, male 8-10 mm, worker 10-12 mm.
> > It perforates **the flowers of Symphytum officinale (corolla from 9-10 mm to 14 mm long).**
> **Bombus mastrucatus** (= Bombus wurflenii), tongue of a female 10-12.5 mm, worker 9-10 mm.
> **Bombus muscorum**, tongue of female 13-15 mm, male 12-13 mm, worker 9-10 mm.
> **Bombus pratorum** (early-nesting bumblebee), tongue of a female 12-14.5 mm, male 8-10 mm, worker 8-12 mm.
> **Bombus rajellus**, tongue of female 13-14 mm, male 10-11 mm, worker 11-13 mm.
> **Bombus terrestris** (buff-tailed bumblebee or large earth bumblebee) tongue of female 9-11 mm, worker 8-9 mm.

Although the rule seems to be that honey-bees do not perforate flowers, there seem to be exceptions, for no less an authority than Hermann Muller* states that they perforate the flowers of Erica tetralix, using their mandibles to bite holes in the tube of the corolla. The tongue of the honey-bee is only 6 mm long. Muller records many cases in which it uses perforations made by other insects.
> It is not always an easy matter to tell whether an insect makes the perforations, especially when these are in the form of longitudinal slits, or whether it is merely looking for the perforations of some other insect.
> **In flowers where the tissue is firm, these slits close over quite effectively, and, as I have convinced myself in the case of Symphytum and Phlomis, are not readily seen.**

The Carpenter Bees, belonging to the Genus Xylocopa, do considerable injury to flowers in more southern European latitudes, where they abound. Delpino, Comes, and others, find that Xylocopa violacea, a native of southern Europe, perforates many flowers. The insect applies its sharp and wedge-shaped maxillae to the grooved surface of the tube and splits this open 3 or 4 mm from the base.

While wasps do not generally perforate flowers, they are not above using those perforated by species of Bombus and other insects; for these, in their rapid visits to flowers, are certain to leave some nectar. Insects much lower in the scale than wasps frequently use the perforations made by species of Bombus.

Ants are especially fond of saccharine matter, and are frequent visitors to flowers, but only for nectar.

Their visits are entirely injurious to the plant. They frequently gnaw parts of the flowers and make irregular holes, thus gaining an entrance, or they use the perforations made by other insects.

Beetles, although not high in the scale of development, and certainly low as far as the adaptation to flowers and their pollination is concerned, show, in a few cases, some ingenuity in getting at nectar.

The acute observer Sprengel** found that large numbers of the flowers of Symphytum officinale were perforated by one of the flower-beetles, and that ants used these perforations."

-'The Transactions of the Academy of Science of St. Louis (Missouri)', Volume V, No. 1 and 2, **1886 to 1891**. Includes: 'On the Pollination of Phlomis tuberosa L., & the Perforation of Flowers' by L.H. Pammel, Shaw School of Botany, No 1, page 241-277.

(* Heinrich Ludwig Hermann Muller, 1829-1883, a German botanist who provided important evidence for Darwin's theory of evolution. In 1873 he wrote book 'Die Befruchtung der Blumen durch Insekten', translated 1883 as 'The Fertilisation of Flowers'.)

(** Christian Konrad Sprengel, 1750-1816, was a German naturalist, theologian, and teacher. He was the first to realize that the function of flowers was to attract insects to have cross-pollination, i.e., plant sexuality.)

(Apidae is the largest family within the superfamily Apoidea, with 5700 species of bees. It includes bumblebees, honey bees, stingless bees, carpenter bees, orchid bees, cuckoo bees, and others.)

"Bees were observed with the naked eye or with the help of a lens. **The occurrence of Andrena symphyti Schmiedeknecht 1883 was proven** on the following localities by collecting vouchers or by the **registration of the characteristic bite slits on Symphytum corollas.** Graz, Austria: **Symphytum tuberosum (some had bite slits), Symphytum officinale.**"

-'Flowers of Boraginaceae (Symphytum, Onosma, Cerinthe) and Andrena symphyti (Hymenoptera-Andrenidae): Morphology, Pollen Portioning, Vibratory Pollen Collection, Nectar Robbing' by Herwig Teppner, Karl-Franzens University of Graz, Austria; Phyton: Annales rei Botanicae, Horn, Austria, Volume 50, No. 2, pages 145-180, **2011**.

(The genus Andrena are mining bees. It is the largest genus in the family Andrenidae, and is almost worldwide in distribution except for Oceania and South America.)

"**Bee species: Andrena symphyti Schmiedeknecht, 1883.** Andrena bothriorhina Perez, 1902. Andrena furcata Friese, 1921. Andrena furcata var amaniensis Friese, 1921.

North Mediterranean species, widely distributed in Slovenia.

Oligolectic, specialized on Symphytum (Boraginaceae). Nests in burrows in the ground, excavated by itself. Univoltine (one brood of offspring per year). Flies from April to June.

As she has too short a tongue to reach the nectar at the base of the flower from inside, female cuts a slit in the corolla tube with her mandibles and steals nectar from the outside."

-'Bee Fauna of Slovenia' by Andrej Gogala, Discover Life, www.discoverlife.org, 2014. We provide free on-line tools to identify species, teach and study nature's wonders, report findings, build maps, and contribute to and learn from a growing, interactive encyclopedia of life with 1,410,395 species pages and 809,345 maps.

(Oligolecty means bees have a specialized preference for pollen sources, usually a single genus of flowering plants.)

Comfrey Flower Color Changes Related to Soil Type

It is possible sometimes to get white flowers on Comfrey plants if planted in a soilless medium that contains no nutrients. Then when planted in soil, the new flowers will turn their normal color. I do not know if this is due to pH or minerals or some other reason. If you have any experience with this, please let me know.

In some varieties of bigleaf hydrangeas (Hydrangea macrophylla), soil pH determines the flower color. Blooms are blue in acidic soil with pH 5.0 to 5.5. The same cultivar has pink blossoms in more neutral pH 6.5 to 7.0.

"**S. peregrinum Ledeb. (fide C. Bucknall).**

When growing on the banks of streams, is a tall, luxuriant plant, with flowers rose-coloured in bud, then bright blue, the stem without wings, and bearing abundant fruit.

When growing in dry localities, the flowers remain rose-coloured or are only partially blue, and the entire plant is not so well developed as when growing in moister situations."

-'Watson Exchange Club Report: 1912-1913', The Journal of Botany British and Foreign, London, England, Volume 52, page 275, **1914**.   (fide = according to) Most plants do not have flower color changes based on soil pH.

Comfrey Flower and Pyrrolizidine Alkaloids

See sub-subsection 'Comfrey Leaves and Pyrrolizidine Alkaloids' in subsection 'Comfrey Leaf Description' in section 'Symphytum Genus Description' (Chapter 5). It has information on flowers.

See sections 'Alkaloids in Comfrey' in Volume 2.

"Pyrrolizidine alkaloids (PAs) are a typical class of plant secondary metabolites that are constitutively produced as part of the plant's chemical defense.
> *While roots are a well-established site of pyrrolizidine alkaloid biosynthesis, Comfrey plants (Symphytum officinale; Boraginaceae) have been shown to additionally activate alkaloid production in specialized leaves and* **accumulate PAs in flowers during a short developmental stage in inflorescence development.**

To gain a better understanding of the accumulation and role of PAs in Comfrey flowers and fruits (nutlets), we have dissected and analyzed their tissues for PA content and patterns.

PAs are almost exclusively accumulated in the ovaries, while petals, sepals, and pollen hardly contain PAs. High levels of PAs are detectable in the fruit, but the elaiosome was shown to be PA free.
> *The absence of 7-acetyllycopsamine in floral parts while present in leaves and roots suggests that the additional site of PA biosynthesis provides the pool of PAs for translocation to floral structures.*

Our data suggest that PA accumulation has to be understood as a highly dynamic system resulting from a combination of efficient transport and additional sites of synthesis that are only temporarily active.

Our findings are further discussed in the context of the ecological roles of PAs in Comfrey flowers."

-'Specific Distribution of Pyrrolizidine Alkaloids in Floral Parts of Comfrey (Symphytum Officinale) and Its Implications for Flower Ecology' by Thomas Stegemann, Lars H. Kruse, Moritz Brutt and Dietrich Ober (Kiel, Germany and Ithaca, New York); Journal of Chemical Ecology, e-publication before print, July 28 **2018**.

Comfrey Seed (Fruit)

*"**The Symphytum nutlets (seeds) are ovate or oblong,** more or less curved, in two pairs, with the contiguous sides of each pair slightly flattened so that the nutlet is obliquely keeled. The surface is more or less distinctly marked with a few large facets or areolae and is smooth and shining or granulated and opaque.*

The base is annular and tumid, and furnished with teeth which clasp the torus while the nutlet is still attached, but are generally inflexed in the dried state.

The strophiole is white, oblong and protrudes through the excavated base of the nutlet.

In certain Symphytum species some plants produce abundance of fruit while others are completely sterile. *In one species, S. floribundum, Shuttlw., the fruit appears to be always undeveloped."*

-'A Revision of the Genus Symphytum, Tourn.' by Cedric Bucknall, (Mus. Bac. Oxon= Bachelor of Music, Oxford University), Journal of the Linnean Society of London, England, Botanical Journal, Volume 41, Issue 284, pages 491-556, December **1913**.

*"**Elaiosomes are fleshy appendages or protuberances on seeds and seedlike fruits (diaspores) of many angiosperms, which promote seed dispersal by ants. This dispersal syndrome, usually termed myrmecochory,** involves the diaspores being carried away from the parent plants, often to or near the ant nest, where the ants feed upon the fat- and protein-rich elaiosomes while leaving the seeds intact.*

Seeds of plants naturalized in the United States were examined for the presence of elaiosomes.
> *Seeds of 47 species belonging to 13 families (Asteraceae, **Boraginaceae,** Dipsacaceae, Euphorbiaceae, Fabaceae, Fumariaceae, Lamiaceae, Liliaceae, Poaceae, Polygonaceae, Resedaceae, Rosaceae, and Solanaceae) were found to have elaiosomes, indicating that these species are probably mymecochorous, i.e., dispersed by ants. These include important rangeland weeds such as bull thistle, Canada thistle, musk thistle, diffuse knapweed, spotted knapweed, and leafy spurge.*

Myrmecochory in naturalized species may enhance their weediness in areas where they are established and assist their colonization of new and relatively closed communities.

	Elaiosome Site	Source of Seeds Examined
Symphytum officinale	Base	Montana State University Herbarium & Seed Collection, Bozeman"

-'Elaiosomes on Weed Seeds and the Potential for Myrmecochory in Naturalized Plants' by Robert W. Pemberton and Delilah W. Irving, United States Department of Agriculture, Albany, California; Weed Science, Volume 38, pages 615-619, **1990**.

*"**Morphological and anatomical studies were made on the fruits (seeds) of 6 species of Comfrey** from various countries and geographical regions, grown at an experiment station of the N.I. Vavilov VIR in Russia. **The 6 species are Symphytum officinale (Common Comfrey), S. carpaticum, S. tanaicense, S x uplandicum (Russian Comfrey), S. asperum (Prickly Comfrey), and S. caucasicum.***
> *The eremi (i.e., the four semi-mericarps forming the coenobium) differ in form, colour, and the teeth of the attachment ring, as well as in quantitative features. The structure of the pericarp and seed-coat is more conservative. The pericarp has an outer epidermis with a cuticle, mesocarp and inner epidermis.*

Considerable variation occurs in the cuticle and outer epidermis, and this may be used to distinguish the Symphytum species. *The spermoderm is not differentiated and is of no taxonomic value."*

-'Morphological and Anatomical Investigation of the Fruits of Some Species of Symphytum L.' by N.M. Naida, TSKhA, Russia; Izvestiya Timiryazevskoi Sel'skokhozyaistvennoi Akademii (News of the Timiryazev Agricultural Academy), Moscow, Russia, No. 3, pages 122-131, **1996**. (I do not have this report. If you have a copy in English, could you please send it to me.)
> (Seed-coat is the outer protective covering of a seed.)
> (Pericarp is the walls of a ripened ovary or fruit. The pericarp is three distinct layers: epicarp {exocarp} which is the

outermost layer; mesocarp which is the middle layer; and endocarp which is the inner layer surrounding the ovary or seeds. Cuticle is a protective and waxy layer covering the epidermis.)

"Stace (*1991) states that the nutlets of S. officinale (Common Comfrey) are shiny compared with dull and minutely tuberculate in the case of the hybrid (Russian Comfrey)."
-'Atlas 2000 and a Problem with Purple Flowered Comfrey' by Eric Chicken, East Yorkshire, England; Botanical Society of Britain and Ireland (BSBI) News, Hertfordshire, England, No. 76, pages 22-23, September **1997**.
(* -New Flora of the British Isles: Identification of Wild Vascular Plants of the British Isles edited by Clive Stace. Cambridge, England: Cambridge University Press, 1991, pages 645-648.)

*"**There are at least 15 species of Comfrey (Symphytum) in Russia** and many of these are used for forage, as a nectar source for honey bees and as decorative plants. Some species are used for medicinal purposes.*
The introduction of these species into cultivation is hampered for a number of reasons, the main one being low seed productivity.
The authors reported on the pollination conditions of the species Symphytum L. (Boraginacese) of the Russian north-west Region and considered a number of flower characteristics to determine low seed set.
    *The workers found that **five species Symphytum (S. asperum Lepech., S. carpaticum Frolov, S. officinale L., S. tanaecense Stev. and S. x uplandicum Nym.) had sufficient primary insect attractants (pollen and nectar) and bright flower colour.***
    However, the flowers were all found to contain a long corolla tube that prevented access to the pollen and nectar by many insects. Comfrey was also found to have a number of mechanisms that prevented self-pollination and self-fertilisation.
The specificity of Comfrey pollination by two species of bumblebee (Bombus hortorium and Bombus lucorum) explain why low seed set is prevalent and why Comfrey is typically propagated vegetatively."
-'Comfrey' by N.M. Naida (St. Petersburg State Agrarian University, Russia) and M.A. Vishnuakova (All-Union Research Institute of Plant Breeding, St. Petersburg, Russia); Global Newsletter on Underutilized Crops, International Center for Underutilized Crops, University of Southhampton, Highfield, England, December **1999**, pages 14-15.
    (Symphytum carpaticum Yu.M.Frolov is an unresolved name. Distribution is in the Ukraine.)

*"**Members of the Boraginaceae are known to ripen fewer than 100% of their seeds, and some are routinely as low as 25%.** Melser and Klinkhamer (2001) report 25% seed set (not pollen-limited) in Cynoglossum officinale, a species in the Boraginaceae with very similar phenology to H. venusta.*
Goulson* et al. (1998) measured seed set in Symphytum officinale (Comfrey) in its native range at 1.18 per flower out of a possible four nutlets, also not pollen-limited. *This is a robust, common plant native to Europe and Asia, used in herbal medicine and widely introduced in other parts of the world."*
-'Reproductive Biology of Hackelia Venusta (Piper) St John (Boraginaceae)' by Norma Jean Taylor, Thesis for Master of Science, College of Forest Resources, University of Washington, Seattle, Washington, **2008**.
(* -'Floral Display Size in Comfrey Symphytum Officinale L Boraginaceae: Relationships with Visitation by Three Bumblebee Species and Subsequent Seed Set' by D. Goulson, J.C. Stout, S.A. Hawson and J.A. Allen, University of Southampton, School of Biology, England; Oecologia {Plant and Animal Ecology}, Volume 113, No. 4, pages 502-508, 1998. For this article, see sub-subsection 'Flowers and Pollination' in subsection 'officinale' in section 'Details About Each Symphytum Species'.)

*"**Mature seed samples of twenty-four Boraginaceae taxa collected from their natural habitats in Turkey were analysed by GC (Gas Chromatography) for total oil content and fatty acid composition.*** *The range of total fat in the taxa varied between 7.0 and 35.7%.*
    A high level of gamma-linolenic acid occurred in the seed oils of Symphytum, *Anchusa, Echium, and T. orientalis as alternative sources of this fatty acid.*
Total oils and total amounts of fatty acids in the seeds:

Taxa	Total oil	Saturated	Mono-unsaturate	Poly-unsaturate	Unsaturate total
Symphytum officinale	23.10%	11.06%	29.17%	59.32%	88.49%
Symphytum asperum	25.97	11.55	27.48	60.49	87.97
Symphytum orientale	32.10	12.65	23.42	63.65	87.07
Symphytum tuberosum ssp. nodosum	7.00	16.87	14.32	67.30	81.62

*Although characteristic differences at generic levels and substantial similarities at infrageneric levels were observed for the quantities of fatty acids examined, **S. tuberosum ssp. nodosum** and A. azurea var. azurea **exhibited relatively distinct profiles compared with the others species of its genus (Symphytum).***
    S. tuberosum ssp. nodosum as Euro-Sib (Euro-Siberian) element has a very limited distributional area in northwest Turkey, but is scattered in the larger zones of Europe and Balkans. However, the other species of Symphytum examined here are also commonly distributed in Russia and Caucasia. It is morphologically different from the other Symphytum species because of its root nodules."
-'Analysis of the Total Oil and Fatty Acid Composition of Seeds of some Boraginaceae Taxa from Turkey' by Tamer Ozcan, Division of Botany, Istanbul University, Turkey; Plant Systematics and Evolution, Volume 274, Issue 3-4, pages 143-153, September **2008**.

(Infrageneric is subclassification within a genus of organisms.)
(The Euro-Siberian region extends from Iceland around most of Europe via Siberia to Kamchatka {far eastern Russia}. Siberia is in northern Russia.)
(The Balkan Peninsula is surrounded by the Adriatic Sea to the west, the Mediterranean Sea and Marmara Sea to the south, and the Black Sea to the east. Its northern boundary is the Danube, Sava and Kupa Rivers.)
(The Caucasus Mountains are located at the border of Europe and Asia, between the Black Sea and Caspian Sea and occupied by Russia, Georgia, Azerbaijan, and Armenia.)

"**Micro- and macro-morphological characters of fruit (seed) surface of seven species of Symphytum L. in Turkey were investigated.**
Fruits of S. asperum Lepechin (Prickly Comfrey), S. sylvaticum Boiss., S. brachycaylx Boiss, and S. kurdicum Boiss. & Hausskn have indistinct basal ring which in S. ibericum, S. orientale and S. bornmuelleri is distinct.
Based on the structure and ornamentation of the fruit surface, tuberculate and rugose types can be distinguished.
Colour of fruits varies from light brown to dark brown.
The fruits of Boraginaceae are characterized by one-seeded mericarpids (nutlets) with a sclerified exocarp protecting the seeds (Diane et al. 2002). The number of seeds produced per capsule can serve as a diagnostic character (Juan et al. 2000).
Symphytum species has one-seeded nutlet.
Micro-morphological characters could be useful in solving taxonomic problems of Symphytum asperum, S. kurdicum, S. ibericum, S. orientale, S. bornmuelleri, S. brachycalyx and S. sylvaticum."
-'Fruit Coat Patterns and Morphological Properties of Seven Species of Symphytum L (Boraginaceae) from Turkey' by Oznur Akcin and Hilal Baki, Ordu University, Turkey; Bangladesh Journal of Botany, Volume 38, no. 2, December **2009**.

"**'Random Amplified Polymorphic DNA' (RAPD) and fatty acid profiles, and alpha-tocoferol contents of the seeds of some Symphytum species were analyzed for their differentiation.**
The RAPD technique is a frequently used tool to **establish phylogenetic relationships and genetic variations**, even in closely related organisms and cultivars.
Discriminative patterns were found in the examined species. Genotypic characteristics of three Symphytum species (**S. pseudobulbosum, S. ottomanum, S. orientale**) were experienced using 20 decamer RAPD primers.
The majority of band positions varied between species. *The total amplified products of 20 RAPD primers was 247 (average of 12.35 bands per primer), of which 189 bands were polymorphic, corresponding to nearly 76.1% genetic diversity. The number of bands for each RAPD primer varied from 8 (UBC320) to 18 (OPA7). The percent of polymorphic bands ranged from 45.4% (OPJ20) to 93.3% (OPA16).*
Major unsaturated fatty acids in the seeds were linoleic, gamma-linolenic and oleic acids; alpha-linolenic and eicosenoic acids exhibited lower levels.
Variations in quantities and total ratios of fatty acid groups, and alpha-tocopherol contents were also observed between the species, and the developmental stages of the seeds.
Significant differences were found for the whole series of fatty acids between species, in addition to the calculated ratios ($p < 0.05$).
Differences were also significant for fatty acid traits between two developmental stages of the seeds of endemic (native) S. pseudobulbosum ($p < 0.05$). In some morphological descriptors of the seeds, difference was found at a significant level ($p < 0.05$).
Obtained data based on genetic and biochemical variations seem to be useful for molecular delimitation of Symphytum, in addition to selection of the genotypes expressing a high amount of GLA (Gamma-Linolenic Acid)."
-'Differentiation of Symphytum Species Using RAPD and Seed Fatty Acid Patterns' by Tamer Ozcan, Istanbul University, Division of Botany, Turkey; Natural Product Communications, Volume 5, No. 4, pages 587-596, **2010**.

"**Symphytum officinale is a perennial plant that produces dry schizocarp fruits that are separating into four mericarpids ('nutlets'). At the base of the nutlets elaiosomes are formed that attract ants as part of the zoochorous (animal) fruit dispersal (*Peters et al. 2003).**"
-'Specific Distribution of Pyrrolizidine Alkaloids in Floral Parts of Comfrey (Symphytum Officinale) and Its Implications for Flower Ecology' by Thomas Stegemann, Lars H. Kruse, Moritz Brutt and Dietrich Ober (Kiel, Germany and Ithaca, New York); Journal of Chemical Ecology, e-publication before print, July 28 **2018**.
(* -'Seed Dispersal by Ants: Are Seed Preferences Influenced by Foraging Strategies or Historical Constraints?' by M. Peters, R. Oberrath and K. Bohning-Gaese, Zoologisches Forschungsinstitut und Museum Alexander Koenig, Bonn, and Institut fur Zoologie/Tierphysiologie, Kopernikusstr, Aachen, and Institut fur Zoologie, Johannes-Gutenberg-Universität Mainz, Germany; Flora, Volume 198, pages 413-420, 2003.)   (Schizocarp is a dry fruit that splits into single-seeded parts when ripe.)

<u>Comfrey Seedlings</u>

"***Symphytum cordatum*:**
Germination bivariant (2 variations): epigeous hypocotylar-cotyledonar or hypogeous (*Barykina & Alyonkin 2008).
In the former case, cotyledon blades are brought out to the surface. They are oblong-elliptic in shape (8 mm long, 4 mm wide), fleshy, with glabrous petioles of 3.5-4 mm long, 1 mm wide, with indumentum only on the adaxial side."

-'Pubescence of Vegetative Organs and Trichome Micromorphology in Some Boraginaceae at Different Ontogenetic Stages' by Rimma P. Barykina and Vitaly Y. Alyonkin, Lomonosov Moscow State University, Russia; Wulfenia, Regional Museum of Carinthia, Austria, Volume 23, pages 1-29, **2016**.
(* -'A Comparative Morphological Ontogenetic Study of Some Representatives of the Genus Symphytum L.: Sections Coerulea Buchn., Symphytum Pawl. and Tuberosum Buchn.' by R.P. Barykina and V. Alyonkin, Byulleten Moskovskogo Obshchestva Ispytatelei Prirody, Otdel Biologicheskii, Moscow, Russia, Volume 113, No. 5, pages 47-57, 2008. In Russian.)
(A seed has epigeal germination when the cotyledons of the germinating seed expand, throw off the seed shell and become photosynthetic above ground.)

"**All representatives of Boraginaceae covered by the present study are found to possess protein-free seeds with large embryos.** Their own nutrient supply (fatty oils, aleurone) is stored predominantly in the cotyledons which are much greater in length than the embryo's axial part.
Syncotyly is reported. Its traces can sometimes be noticed as early as at the final stage of embryogenesis, indicating the early manifestation of geophily (Brachybotrys paridiformis, Omphalodes verna, O. linifolia, Anchusa pseudoochroleuca, Lindelofia stylosa, Macrotomia euchroma, **Symphytum x uplandicum**, Trichodesma incanum, Hackelia uncinata). It is, however, much more pronounced in the seedling and sprouts. Syncotyly results in the formation of a hollow cotyledon tube relatively small in length (2–5 mm).
Symphytum × uplandicum Nyman:
In its large (2.3-2.5 mm long, 1.8-2 mm wide) embryo, the axial part is 3 times shorter than the cotyledons (0.65 mm and 1.9 mm respectively). The seedling emergence pattern is cotyledonary.
Cotyledon blades are oval to elliptical, born on long petioles. The cotyledon tube is short (2-3 mm), formed by the fused basal parts of the petioles. The tube wall is covered by small-celled epidermis lacking trichomes or stomata. Large-celled parenchyma (6-7 layers) comprises up to 8 vascular bundles, out of them paired bundles are median veins and singular ones are lateral veins. When entering the hypocotyl stele, the lateral bundles unite with the median ones. The node is unilacunar, two-traced."
-'Syncotyly in Seedlings and Sprouts of Some Boraginaceae: Genesis, Structure and Function of the Cotyledon Tube' by Rimma P. Barykina and Vitaly Y. Alyonkin, Department of Higher Plants, Faculty of Biology, Lomonosov Moscow State University, Russia; Wulfenia, Austria, Volume 24, pages 11-28, **2017**.
(Cotyledon is an embryonic leaf in seed-bearing plants. They are the first leaves to appear in a germinating seed.)
(Syncotyly is the fusion of cotyledons at early embryogenetic stages.)
(Geophily is the bearing flowers and fruit at or below ground level.)

"**Symphytum officinale Young Plant:**
Cotyledons elliptic or ovate, rounded tip, more or less stalked. Leaves entire, elliptic or ovate, regularly dentate."
-'Symphytum Officinale' by Bayer Crop Science®, Crop Compendium, Bayer® is a 150-year Life Science company in health care and agriculture, global headquarters are in Leverkusen, Germany, **2018**.

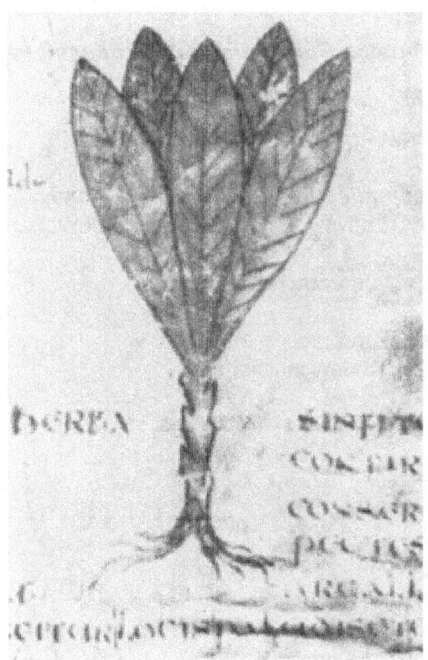

FIG. 28.—LEYDEN, Voss Q 9. SEVENTH CENTURY. 'SINFITOS.'

FIG. 29.—CASSEL, APULEIUS Fo. 14v. 'SINFITOS.'

Cassel (Kassel), Apuleius of the 7th and 9th century

"There is no figure of Symphyton in the Juliana Anicia manuscript, but our figure accords well with the Sinfitos of the Leyden (Leiden) Apuleius of the seventh century and better with that of the Cassel (Kassel) Apuleius of the ninth century."

-'The Herbal in Antiquity and Its Transmission to Later Ages' by Charles Singer; The Journal of Hellenic Studies: Society for the Promotion of Hellenic Studies, London, England, Volume 47, Part 1, pages 31-33, 1927.

Apuleius Barbarus also known as Apuleius Platonicus and Pseudo-Apuleius. He wrote the 4th-century Latin herbal known as 'Pseudo-Apuleius Herbarius' or 'Herbarium Apuleii Platonici'. Until the twelfth century it was the most influential herbal in Europe. A manuscript of the Apuleius herbal from the sixth or early seventh century is in Leiden, Netherlands.

Chapter 6

Symphytum Species Overview

Comfrey Chromosomes Overview (in chronological order from old to recent)

A **chromosome** is a DeoxyriboNucleic Acid (DNA) molecule with part of the genetic material (genome) of an organism. A chromosome is a package containing a piece of a genome. It is a threadlike structure of nucleic acids and protein found in the nucleus of cells.
In humans, each cell has 23 pairs of chromosomes, for a total of 46. Twenty-two pairs, called autosomes, are the same in both males and females. The 23rd pair is for the sex chromosomes for male and female.

In molecular biology and genetics, a **genome** is the genetic material of an organism. It is all of the DNA. Each genome has all the information needed to build and maintain that organism. In humans, the genome is more than 3 billion DNA base pairs.

"**Genome size** is the total amount of DNA contained within one copy of a single genome. It is typically measured in terms of mass in **picograms** (trillionths of a gram, abbreviated **pg**) or less frequently in Daltons or as the total number of **nucleotide base pairs typically in megabases** (millions of base pairs, abbreviated **Mb or Mbp**). One picogram equals 978 megabases. In diploid organisms, genome size is used interchangeably with the term **C-value**. C-value is the amount, in picograms, of DNA contained within a haploid nucleus (e.g., a gamete) or one half the amount in a diploid somatic cell of a eukaryotic organism."
-Wikipedia®: The Free Encyclopedia, www.wikipedia.org, 2019.

Cytotaxonomy investigates cellular structures such as somatic (body) chromosomes to classify organisms based on number, structure, and behaviour. Chromosome numbers are determined at mitosis (normal cell division) and can be diploid (2n = 2x) or polyploid (triploid or higher).
Body cells and species are described by the number of their chromosome sets {ploidy level}:
Monoploid: 1 set.   Diploid: 2 sets = 2n = 2x.   Triploid: 3 sets = 2n = 3x.
Tetraploid: 4 sets = 2n = 4x.   Pentaploid: 5 sets = 2n = 5x.   Etc.

See section 'Symphytum Species Classifications' (Chapter 4) for a list of accepted names of common species with synonyms and chromosome numbers. Also see individual species.
See subsection 'Chromosomes and Genome in the Borage Family' in section 'Borage Family, Symphytum Genus' (Chapter 2). For DNA barcoding and other methods, see subsection 'Methods of Authentication of Comfrey Species' in section 'Symphytum Species Classifications' (Chapter 4).

"*The genus Symphytum consists of 25 species found in Europe and adjacent Asia.*
The chromosomes are extremely difficult to fix without their ends adhering to one another. Consequently it is difficult to make accurate determinations of the chromosome numbers.
   The tendency for the ends of the chromosomes to stick together seems to reach its greatest intensity in this genus. Strey (*1931) considered the basic number in the genus to be 9. The species then fall into two groups: those with 36 chromosomes and those with 72 chromosomes.
This author found two species to have a gametic chromosome number of 20 and a somatic number of 40 respectively."
-'Cytogenetic Studies on the Boraginaceae' by Donald M. Britton, Brittonia (published by The New York Botanical Garden), Volume 7, No. 4, pages 233-266, December 10 **1951**.
(* -'Karyologische Studien an Borraginoideae' {Caryological studies on Borraginoideae'} by Martin Strey; Planta, Berlin, Germany, Volume 14, Issue 3-4, pages 682-730, October 1931. In German.)
   (Germplasm is the genetic material of germ cells. A germ cell has half the number of chromosomes of a
   somatic / body cell and can join with another germ cell from the opposite sex to form a new individual.)

"*Symphytum Chromosomes x = 9, 10.*

caucasicum	c. 36	Strey 1931	H	Caucasus
officinale	36	Suzuka 1950	HM	Europe, West Asia
uplandicum	36	Vaarama, L. & L. 1948	H	cultivated
peregrinum	36	Maude 1939	H	East Caucasus
asperum	40	Britton 1951	H	Russia, Persia (Iran)
tauricum	40	Britton 1951	H	South Russia
bulbosum	c. 72	Strey 1931	H	Europe
tuberosum	c.72		HM	Europe, Southwest Asia."

-Chromosome Atlas of Flowering Plants edited by C.D. Darlington and A.P. Wylie. London, England: G. Allen and Unwin Ltd., **1956**. First published 1953. Symphytum page 297.   ('c.' means around or about. I don't know what 'H' and 'HM' mean.)

"**Symphytum:** *Extensive cytological studies on the genus Symphytum exist only from recent times (Gadella & *Kliphuis 1967, **Runemark 1967).*
While the southeastern European Symphytum species, which were further studied by Runemark, form a separate group both palynologically (study of plant pollen) and cytologically (study of cells) **(2n = 28 in S. creticum*** and S. circinale****, 2n = 30 in S. anatolicum*****), S. officinale (according to Gadella & Kliphuis essentially with 2n = 24 and 2n = 48)** *and the clades around S. tuberosum may show tighter cytological relationships.*
Examined Plants:
    Symphytum asperum Lepechin
        Ordschonikidse, Kaukasus (Caucasus) (Wild material, Bot. Garten Halle) **2n = 32**
    Symphytum tuberosum L. ssp. nodosum (Schur) Soo (= S. leonhardtianum Pugsley)
        Bayern (Bavaria), Gauting bei Munchen (Munich) **2n = 96**
        Bayern, Allach, NW Munchen **2n = 96**
        Bayern, Weichenberg a. d. Donau **2n = 100**
        Jugoslawien (Yugoslavia), Slowenien (Slovenia), Cavn **2n = 96**
    Symphytum tuberosum L. s.l.
        Italien (Italy), Prov. Firenze, Vallombrosa **2n = 144**
The counts available so far for S. asperum (Strey around 2n = 36, Britton 2n = 40) seem to have not been made on pure material, *if one summarizes all previously available information (see Gadella & Kliphuis). The fact that both figures are from garden material could support this assumption. It would be to investigate whether plants with 2n = 40 chromosomes are hybrids between S. asperum and S. officinale."*
-'Cytologische Untersuchungen an Boraginaceae I' {**Cytological Investigations on Boraginaceae**} by J. Grau, Mitteilungen der Botanischen Staatssammlung Munchen {Communications of the Botanical State Collection Munich, Germany}, Volume 7, pages 277-294, **1968**. (I translated it online from the German.)
(* -'Cytotaxonomic Studies in the Genus Symphytum II: Crossing Experiments Between Symphytum Officinale L. and Symphytum Asperum Lepech' by Th.W.J. Gadella and E. Kliphuis, Botanical Museum and Herbarium of the State University of Utrecht, Netherlands, Acta Botanica Neerlandica, Volume 18, No. 4, pages 544-549, August 1969.)
(** -'Studies in the Aegean Flora XI: Procopiana {Boraginaceae} included into Symphytum' by Hans Runemark, Botaniska Notiser {Botanical Notes}, Sweden, Volume 120, pages 84-94, 1967. I was unable to get a copy of this report. If you have one, could you please send it to me.)
(*** Boraginaceae Symphytum circinale Runemark -- Bot. Not. 120: 90. 1967.)
(**** Boraginaceae Symphytum creticum (Willd.) Runemark -- Boissiera xiii. 100 1967.)
(***** Boraginaceae Symphytum anatolicum Boiss. -- Diagn. Pl. Orient. ser. 1, 4: 43. 1844)
    (Clade is a group of organisms evolved from a common ancestor.)
    (Sensu lato = s.l., sens. lat. In the broad sense. Used in taxonomy to clarify the scope of a taxon when it has
    been used to define more than one set of lower-level taxons.)

"**All results available show Symphytum to have a dysploid series of basic chromosome numbers**: *ascending 12-14-15-16-20 and descending 12-10.* **Probably x = 12 is its primary basic number.** *This number comprises two heterogenous (different genetic) groups and reaches with a dodekaploid (12-ploid) taxon its highest polyploid expression.*
    Symphytum bulbosum is hexaploid (6-ploid) with 2n = 84 *and is probably connected with the diploid species of 'Procopiana' which have the same basic number of x = 14.*
    More of the eastern species seem to have in common the basic number x = 15 (S. anatolicum, S. ibericum, S. cordatum).
    Symphytum asperum and S. Orientale show their strong relationship also in the same chromosome number (2n = 32)."
-'Cytologische Untersuchungen an Boraginaceae II' {**Cytological Investigations on Boraginaceae**} by J. Grau, Mitteilungen der Botanischen Staatssammlung Munchen {Communications of the Botanical State Collection Munich, Germany}, Volume 9, pages 177-194, **1971**.
(Dysploid is variation in chromosome numbers by less than a whole set of chromosomes, because of chromosomal rearrangements.)

"In our opinion in west Europe the following Symphytum 'pillars' occur in this highly intricate polyploid complex:
    a) S. officinale (2n = 24), native in west Europe, rare. (Common Comfrey)
    b) S. officinale (2n = 48), native, frequent.
    c) S. asperum (2n = 32), introduced, rare in west Europe. (Prickly Comfrey)
    d) S. x uplandicum (2n = 40), introduced, or escaped from cultivation; spontaneous origin highly improbable in western Europe, since S. asperum is very rare in west Europe. (Russian Comfrey)
    e) Dutch low-lying peat land plants (2n = 40); these plants belong either to S. uliginosum (or very near to this taxon), or they belong to a new, undescribed, taxon.
    f) Hybrids with 2n = 36. These plants originated by crossing (S. asperum 2n = 32) and (Dutch plant 2n = 40). They are common in some parts of west Europe and seem to be introduced. Spontaneous origin highly improbable in west Europe, since S. asperum is very rare in west Europe.
Hybridization between these 'pillars' regularly occurs, *not only in the experimental garden, but apparently also in nature. The various possibilities of intercrossing are described by Gadella & Kliphuis (1969) and by Gadella (1972)."*

-'**Cytotaxonomic Studies in the Genus Symphytum V**: Some Notes on W. European Plants with the Chromosome Number 2n=40' by Th.W.J. Gadella and E. Kliphuis, Botanische Jahrbucher fur Systematik (Botanical Yearbooks for Systematics), Volume 93, pages 530-538, **1973**.
(Polyploidy is a cell that has more than 2 paired {homologous} sets of chromosomes. Most species whose cells have nuclei {eukaryotes} are diploid, i.e., have 2 sets of chromosomes. One is inherited from each parent. Polyploidy is common in plants.)

"Most species of the genus Symphytum, showing most of the total variation share the basic number x = 12. This number should probably be regarded as the primary basic number from which the other numbers are derived.
The extent of the x = 12 series points to an ancient origin of this number. Other basic numbers are found (x = 14, 15, 16) and probably a fourth (x = 10). That is as long as the hybrid nature of the plants with 2n= 40 cannot be proved.
Symphytum is generally regarded as notoriously difficult in cytological respect; even so we usually obtained satisfactory results after fixation of root tips in Karpechenko's fixative and staining with Heidenhain's haematoxylin method. In many cases all chromosomes were clearly visible; their ends did not adhere to each other, thus making accurate counts possible. The best results were obtained by fixation of the material after a cold night."

-'**Cytotaxonomic Studies in the Genus Symphytum VIII**: Chromosome Numbers and Classification of Ten European Species' by T.W.J. Gadella and E. Kliphuis, Proceedings van de Koninklijke Nederlandse Akademie van Wetenschappen (Proceedings of Royal Netherlands Academy of Arts and Sciences): Series C, Biological & Medical Sciences, Volume 81, pages 162-172, **1978**.

"**Symphytum L. Chromosome Numbers:**

officinale L.	24 + 48	Italy	Gadella & Kliphuis 1970
	24 40 48	Netherlands	Gadella & Kliphuis 1967
	26 36		
	54	Bulgaria	*Markova & Ivanova 1970
(a) *officinale*			
(b) *uliginosum* (A. Kerner) Nyman			
------------			
asperum Lepechin			
------------			
x *uplandicum* Nyman	36	Finland	Vaarama 1948
------------			
tuberosum L.			
(a) *tuberosum*	44	Italy	Grau 1968
(b) *nodosum* (Schur) Soo	96	Germany	Grau 1968
	100	Yugoslavia	
	18	Romania	Tarnavschi 1948
	72	Bulgaria	Markova & Ivanova 1970
------------			
gussonei F.W. Schultz			
cordatum Waldst. & Kit. ex Willd.	18	Romania	Tarnavschi 1948
naxicola Pawl.	30	Greece	Runemark 1967
ibiricum Steven			
davisii Wickens			
cycladense Pawl.	30	Greece	Pawlowski, unpublished
tauricum Willd.	18	Romania	Tarnavschi 1948
orientale L.			
bulbosum C. Schimper			
ottomanum Friv.	20	Bulgaria	Markova & Ivanova 1970

This book has grown out of the requests by many users of 'Flora Europaea' (Tutin et al., 1964, 1968, 1972, 1976, 1980) for information on the sources of the chromosome numbers cited in that work."
-**Flora Europaea: Check-List and Chromosome Index** by D.M. Moore. Cambridge, England: Cambridge University Press, **1982**, page 179.
(* -'Karyologische Untersuchung der Vertreter der Fam. Boraginaceae, Labiatae und Scrophulariaceae in Bulgarien' {Caryological Examination of the Representatives of the Family Boraginaceae, Labiatae and Scrophulariaceae in Bulgaria} by Margarita Markova and P. Ivanova; Izvestiya na Boticeskiya Institut {Bulletin de l'Institut Botanique}, {also titled: Mitteilungen des Botanischen Instituts Sofia}, Bulgaria, Volume 20, pages 93-98, 1970. In German.)

"The present paper reports results of cytological studies of three samples of Comfrey introduced into China as forage plants. It is found that the chromosome numbers of these three different samples of Comfrey are the same, 2n = 40.
This number is consistent with that of the Symphytum peregrinum Ledeb. (Russian Comfrey) in literature, but different from that of the other species. Therefore from the view point of chromosome number we consider preliminarily that these three samples of Comfrey are identical with Symphytum peregrinum Ledeb."
-'**An Observation of Chromosome Numbers of Comfrey**' by Wang Guan-Lin, Li Zhen-Shan and Meng Qing-Xian, Acta Phytotaxonomica Sinica (now Journal of Systematics and Evolution), Beijing, China, Volume 21, No. 1, pages 55-59, **1983**.

"*Symphytum asperum* Lepech. (Prickly Comfrey), *S. uliginosum* Kern., *S. officinale* L. (Common Comfrey): *These three species hybridize and form hybrid swarms consisting of F1 and backcross hybrids (*, **, ***).*
    Symphytum officinale is cytologically heterogenous (diverse): 2n = 24, 2n = 40, 2n = 48, cytotypes occur in various parts of Europe.
    Symphytum asperum (2n = 32) does not hybridize with the diploid (2n = 24) form of S. officinale in Europe but produces hybrids (and backcrosses) with the 2n = 40 and 2n = 48 cytotypes of S. officinale.
    The primary hybrids, with 2n = 36, or 2n = 40, are collectively known under the name S. x uplandicum Nyman (Russian Comfrey).
In Europe the parental species are largely allopatric with a very small zone of overlap in the northwest Caucasus (Kusnetosov, 1910). In the zone of overlap, hyridization does not occur because S. officinale and S. asperum grow at different altitudes. **Apparently the hybrids arose outside the Caucasus** (****Tutin 1956; *****Wade 1958). In many parts of western, northwestern, or central Europe the hybrid swarms are more common than the parental species."

-'**Notes on Symphytum (Boraginaceae) in North America**' by T.W.J. Gadella, pages 1061-1067 in book: Annals of the Missouri Botanical Garden, Volume 71, Saint Louis, Missouri. Lawrence, Kansas: Allen Press, **1984**.

(* -'Cytological and Hybridization Studies in the Genus Symphytum' by Th.W.J. Gadella, Symposia Biologica Hungarica {Hungary Biological Symposium}, Volume 12, pages 189-199, 1972. I was unable to find a copy of this report. If you have it, could you please send it to me.)

(** -'Cytotaxonomic Studies in the Genus Symphytum V: Some Notes on W. European Plants with the Chromosome Number 2n=40' by Th.W.J. Gadella and E. Kliphuis, Botanische Jahrbücher fur Systematik {Botanical Yearbooks for Systematics}, Volume 93, pages 530-538, 1973.)

(*** -'Cytotaxonomic Studies on the Genus Symphytum VIII: Chromosome Numbers and Classification of Ten European Species' by Th.W.J. Gadella and E. Kliphuis, Proceedings van de Koninklijke Nederlandse Akademie van Wetenschappen {Proceedings of the Royal Netherlands Academy of Arts and Sciences} Section C, Volume 81, pages 162-172, 1978.)

(**** -'The Genus Symphytum in Britain' by T.G. Tutin, University College of Leicester, England; Watsonia: Journal of the Botanical Society of the British Isles, Volume 3, pages 280-281, February 1956.)

(***** -'The History of Symphytum Asperum Lepech. and S. x Uplandicum in Britain' by A.E. Wade, Department of Botany, National Museum of Wales, Watsonia, Volume 4, pages 117-118, 1958. Watsonia was the journal of the 'Botanical Society of Britain and Ireland' from 1949 to 2010. It is now called 'New Journal of Botany'.)

    (Hybrid or crossbreed is the result of combining the qualities of two organisms of different breeds, varieties, species or genera through sexual reproduction.)
    (A hybrid swarm is a group of hybrids that have survived past the first hybrid generation, with interbreeding taking place between hybrid individuals and also backcrossing with its parent type. These plants are highly variable, with genes and phenotypes {observable properties of an organism} of individuals ranging widely between the two parent types. Hybrid swarms blur the boundary between the parent taxa.)
    (A backcross is a cross of a hybrid with one of its parents or an individual genetically similar to its parent. This creates offspring with genetics closer to that of the parent.)
    (An F1 hybrid or 'Filial 1' hybrid is the first generation of offspring of distinctly different parental types that produce a new, uniform phenotype {observable properties of an organism} with a combination of characteristics from the parents. Mules are F1 hybrids between horse and donkey. This happens naturally, and includes hybrids between species, for example, peppermint is F1 hybrid of watermint and spearmint. In agronomy, these F1 hybrids are usually created by controlled pollination, sometimes hand-pollination. This is not GMO. A Genetically Modified Organism has been altered using genetic engineering techniques in a sophisticated laboratory, frequently using genes from unrelated species.)
    (Allopatric or geographic speciation occurs when populations of the same species become isolated from each other, and this prevents genetic exchange.)

"*Symphytum asperum and S. officinale, of which single individuals were studied, both had 2n = 45.*
Of 12 individuals of S. peregrinum studied, 11 also had 2n = 45, but one had 2n = 65.
Almost all chromosomes in all individuals had a submedian or subterminal centromere; chromosome length ranged from 0.8 micrometer to 2.0 micrometer and there was no SAT (Satellite) chromosome."

-'**Chromosome Numbers in Genus Symphytum**' by K. Shirato, T. Shintani, G. Nakanishi and A. Kamizyo, Chromosome Information Service, Japan, Volume 38, pages 21-23, **1985**.

    (This is not the usual number of chromosomes for these species. I could not find this report. If you have a copy in English, could you please send it to me.)
    (Micrometre or micrometer or micron equals 10^{-6} meter = one millionth of a meter = one thousandth of a millimeter = 0.001 mm = 0.000039 inch.)

"**Chromosome numbers known for species of Symphytum:**
    S. anatolicum   2n = 30 (Runemark 1967)
    S. asperum   2n = 32 (Grau 1968; Gadella & Kliphuis 1967; Gviniashvili 1972)
    S. bulbosum   2n = 72,84,96,120
                        (Grau 1971; Gadella & Kliphuis 1978; Strey 1931; Tarnayschi 1948)
    S. caucasicum   2n = 24 (Gviniashvili 1972)
    S. circinale   2n = 28 (Runemark 1967)

S. ciscaucasicum	2n = 36 (Gviniashvili 1972)
S. cordatum	2n = 60,120 (Grau 1971, *Wcislo 1972, Gadella & Kliphuis 1978)
S. creticum	2n = 28 (Runemark 1967)
S. cycladense	2n = 30 (Runemark 1967)
S. grandiflorum	2n = 60 (Gviniashvili 1969, Gadella & Kliphuis 1978)
S. ibericum	2n = 24 (Gviniashvili 1969, Gadella & Kliphuis 1978)
S. insulare	2n = 28 (Runemark 1967)
S. nodosum	2n = 32,64,96,144
	(**Van Loon & Oudemans 1982, Murin & Majovski 1982, Grau 1968, Wcislo 1972)
S. officinale	2n = 24,40,48 (Gadella & Kliphuis 1967, Basler 1972)
S. orientate	2n = 32 (Grau 1971, Gadella & Kliphuis 1978)
S. ottomanum	2n = 20,40
	(Markova&Ivanova1970, Tarnayschi1948,***Strid 1983, VanLoon&Oudemans1982)
S. peregrinum	2n = 40 (Gviniashvili 1972)
S. tauricum	2n = 18 (Tarnayschi 1948, Britton 1951)
S. tuberosum	2n = 64,72,84,96,144 (Murin & Majovsky 1982)"

-'**Phylogenetic Relationships in the Genus Symphytum L. (Boraginaceae)**' by J.M. Sandbrink, J. Van Brederode and T.W.J. Gadella, Department of Genome Evolution, University of Utrecht, Netherlands; Proceedings: Koninklijke Nederlandse Akademie van Wetenschappen (Royal Netherlands Academy of Sciences), Series C, Volume 93, No. 3, pages 295-334, **1990**.
(* -'Karyological Studies in Symphytum L.' by H. Wcislo; Acta Biologica Cracoviensia: Journal of Polish Academy of Sciences, Krakow, Poland, Series Botanica 15, pages 153-163, 1972.)
(** -'IOPB Chromosome Number Reports LXXV' edited by Askell Love; Taxon, Voume 31, No. 2, pages 342-368, May 1982. J. Chr. Van Loon and J.J.M.H. Oudemans, page 343.)
(*** -'IOPB Chromosome Number Reports LXXVIII' by A. Strid, Taxon, Volume 32, No. 1, page 139, 1983.)

"*Chromosome numbers are known for 16 species of Symphytum (about 46 percent of the total).*
Grau suggested that the primitive base number for the genus is 12, from which lower (x = 10) and higher (x = 14, 15, 16) base numbers were derived.
Aneuploidy and polyploidy apparently have played an important role in the evolution of the Symphytum genus.
   Decaploids based on 12 have been reported for **S. bulbosum** *Schimper and* **S. cordatum** *Waldst. & Kit. ex Willd. Dodecaploids (2n = 144) are known in* **S. tuberosum** *(Grau, 1968). This species also has hexaploid, heptaploid, and octoploid numbers based on 12, but Tarnauschi (1948; see Bolkhovskikh et al.) reported 2n = 18, a number that has not been observed by other workers.*
   Diploid and tetraploid counts (based on 12), as well as aneuploids (2n = 40-47) have been found in **S. officinale***. The 12 pairs of chromosomes of the diploid S. officinale can be distinguished morphologically by using the Feulgen-Giemsa banding technique (Mekki et al.)."*
-'**Journal of the Arnold Arboretum**, Supplementary Series, Volume 1, Harvard University, Cambridge, Massachusetts, **1991**. This volume is a collection of contributions toward a 'Generic Flora of the Southeastern United States'. 'The Genera of Boraginaceae in the Southeastern United States' by Ihsan A. Al-Shehbaz, page 151.
   (Aneuploidy is an abnormal number of chromosomes in a cell.)
   (Polyploidy is a cell that has more than two paired {homologous} sets of chromosomes. Most species whose cells have nuclei {eukaryotes} are diploid, i.e., they have two sets of chromosomes. One set is inherited from each parent. Polyploidy is common in plants.)

"
2n=	Cytology Reference	Kew Accession	Collector	Origin	Voucher
Symphytum bulbosum Schimp.					
48	87-461	1987-1948	Strid s.n.	Greece	K
48	88-28	none	Steam s.n.	cultivated	K
96	88-27	none	Stearn s.n.	cultivated	K
S. x Hidcote Blue ('c.' means around or about)					
c. 52	83-179	1983-242	Hidcote Garden	cultivated	K
S. ibericum Stev. ex Bieb.					
60	82-1389	1982-4913	Mclintock s.n.	cultivated	K

Symphytum bulbosum Schimp. 2n = 48. We also found different numbers (2n = c. 52, 60) in other Symphytum species. **These records, together with the wide array of numbers recorded by Fedorov (*1969), indicate a high level of chromosomal complexity in the Symphytum genus, in which the basic number is uncertain.**"
-'New Chromosome Numbers in Petaloid Monocotyledons and in Other Miscellaneous Angiosperms' by Margaret A.T. Johnson and P.E. Brandham, Jodrell Laboratory, Royal Botanic Gardens, Kew, Richmond, Surrey, England; Kew Bulletin, Volume 52, No. 1, pages 121-138, **1997**.
(* -'Chromosome Numbers of Flowering Plants' {Khromosomnye chisla tsvetkovykh rastenii} edited by A.A. Fedorov, authored by Z.V. Bolkhovskikh, V.G. Grip, O.I. Zakhar'eva, T.S. Matveeva, et al., in Russian, 927 pages. Leningrad, Russia: Nauka, 1969. If anyone has an English translation of the Symphytum section, I would appreciate a copy.)
   (s.n. means without number {Latin: sine numero}.)

"**Summary of the Genome Size (GS) estimations in the studied Czech Boraginaceae:**
For each taxon, the mean and standard deviation of the holoploid GS (2C value), the monoploid GS (1Cx value), an average chromosome size (C/n), the chromosome number and inferred ploidy level are given.

Taxon	Genome size (2C, pg; mean)	Genome Size (1Cx, pg)	Avg Chromosome Size (C/n, pg)	Chromosome # (2n)	Ploidy (x)
S. bohemicum	2.38 +- 0.006	1.1921	0.099	24+0-4B	2x
S. officinale	4.61 +- 0.110	1.1535	0.096	48	4x
S. tuberosum subsp. angustifolium	2.31 +- 0.067	0.5778	0.073	32	4x
S. tuberosum subsp. tuberosum	6.39 +- 0.363	0.5325	0.067	96	12x

More than one ploidy level within a species was detected only in Symphytum tuberosum L., with the tetraploid S. tuberosum subsp. angustifolium (A.Kern.) Nyman and the dodecaploid S. tuberosum subsp. tuberosum."
-'First Insights into the Evolution of Genome Size in the Borage Family: A Complete Data Set for Boraginaceae from the Czech Republic' by Lucie Kobrlova and Michal Hrones, Department of Botany, Palacky University, Olomouc, Czech Republic; Botanical Journal of the Linnean Society, Volume 189, No. 2, pages 115-131, February **2019**.

(Standard Deviation is a statistic that measures the dispersion of a dataset relative to its mean and is calculated as the square root of the variance. In other words, Standard Deviation is a measure of how spread out the numbers are.)

"**Genus Symphytum:**
  Specimen Records: 63.   **Specimens with Sequences: 29.**   **Specimens with Barcodes: 23.**
  Species: 11.   **Species With Barcodes: 6.**   Public Records: 46.
Specimen Depositories and Number of Specimens:
  Mined from GenBank®, NCBI, Bethesda, Maryland [16]
  Finnish Museum of Natural History, Finland [9]
  Royal Ontario Museum, Green Plant Herbarium (TRT), Canada [7]
  University of British Columbia, Herbarium, Canada [6]
  University of Guelph, OAC Herbarium, Canada [5]
  National Museum Wales, Great Britain [4]
  Royal Botanic Garden, Edinburgh, Scotland [3]
  Naturalis Biodiversity Centre, Leiden, Netherlands [3]
  3 Others [4]
Sequencing Labs and Number of Specimens:
  Mined from GenBank®, NCBI [16]
  National Botanic Garden of Wales [14]
  Biodiversity Institute of Ontario, Canada [6]
  Naturalis Biodiversity Centre [3].

Species	Specimens	Sequences	Barcodes >500bp
Symphytum armeniacum	2	2	2
Symphytum asperum	23	5	2
Symphytum bornmuelleri	1	1	0
Symphytum brachycalyx	1	1	0
Symphytum bulbosum	2	1	0
Symphytum officinale	18	13	10
Symphytum orientale	1	1	1
Symphytum tuberosum	9	14	14
Symphytum uplandicum	3	0	0
Symphytum x uplandicum	1	0	0
Symphytum x uplandicum	2	2	1"

-'Taxonomy Browser: Genus Symphytum' by Pensoft®: Taxon Profile, http://ptp.pensoft.eu, Pensoft® is an independent academic publishing company with cutting-edge publishing tools and workflows, https://pensoft.net, **2019**.

(NCBI GenBank® is a database with publicly available nucleotide sequences for more than 380,000 organisms at the genus level or lower. These are submissions from individual laboratories and batch submissions from large-scale sequencing projects, including 'Whole Genome Shotgun' {WGS} and environmental sampling projects. GenBank® is part of the 'International Nucleotide Sequence Database Collaboration', which also includes 'DNA DataBank of Japan' and 'European Nucleotide Archive'. www.ncbi.nlm.nih.gov/genbank)

"**Genbank: Symphytum L. with 240 results. Some examples:**
  Symphytum caucasicum isolate B9 tRNA-Leu (trnL-UAA) gene, partial sequence; trnL-trnF intergenic spacer, complete sequence; and tRNA-Phe (trnF-GAA) gene, partial sequence; chloroplast. 407 bp linear DNA. Accession: KX894532.1 GI: 1241192074
  Symphytum caucasicum isolate B9 18S ribosomal RNA gene, partial sequence; internal transcribed spacer 1, 5.8S ribosomal RNA gene, and internal transcribed spacer 2, complete sequence; and 26S ribosomal RNA gene, partial

sequence. 743 bp linear DNA. Accession: KX894530.1 GI: 1241192072.
Symphytum bornmuelleri tRNA-Leu (trnL) gene, partial sequence; trnL-trnF intergenic spacer, complete sequence; and tRNA-Phe (trnF) gene, partial sequence; chloroplast. 836 bp linear DNA. Accession: GQ285276.1 GI: 296144077.
Symphytum bulbosum tRNA-Leu (trnL) gene, partial sequence; trnL-trnF intergenic spacer, complete sequence; and tRNA-Phe (trnF) gene, partial sequence; chloroplast. 827 bp linear DNA. Accession: GQ285275.1 GI: 296144076."
-'Symphytum L., Indet. Comfrey: Sequences' by NBN Atlas, https://species.nbnatlas.org/species/NHMSYS0000464108, National Biodiversity Network®, **2019**. Provides database for the sharing of high quality biological data in the United Kingdom since 2000.

Hybrids, Hybrid Swarms, Introgression

Hybrid or Crossbreed
A hybrid or crossbreed is the result of combining the qualities of two organisms of different breeds, varieties, species or genera through sexual reproduction. Plant species that are capable of breeding with each other may not do so in nature due to geographical isolation, differences in flowering times, or differences in pollinators.

Chromosome Sets (Ploidy Number)
Body (somatic) cells, tissues and individuals are described by the number of chromosome sets (the ploidy level) they have:
- Monoploid: 1 set
- **Diploid: 2 sets = 2n = 2x**
- Triploid: 3 sets = 2n = 3x
- **Tetraploid: 4 sets = 2n = 4x**
- Pentaploid: 5 sets = 2n = 5x, etc.

Polyploidy in Plants
Polyploidy is a cell that has more than two paired (homologous) sets of chromosomes.
Examples include triploid, tetraploid and pentaploid.
Most species whose cells have nuclei (eukaryotes) are diploid, i.e., they have two sets of chromosomes. One set is inherited from each parent. Humans are diploid.
However, **polyploidy is common in plants.** Symphytum officinale has diploid (2n = 24) and tetraploid (2n = 48) varieties.

> *"Polyploidy or whole-genome duplication is now recognized as being present in almost all lineages of higher plants, with multiple rounds of polyploidy occurring in most extant (existing) species.*
> *The success of polyploidy, displacing the diploid ancestors of almost all plants, is well illustrated by the huge angiosperm diversity that is assumed to originate from recurrent polyploidization events.*
> *Strikingly, polyploidization often occurred prior to or simultaneously with major evolutionary transitions and adaptive radiation of species, supporting the concept that polyploidy plays a predominant role in bursts of adaptive speciation. Polyploidy results in immediate genetic redundancy and represents, with the emergence of new gene functions, an important source of novelty."*
> -'Polyploidy and Interspecific Hybridization: Partners for Adaptation, Speciation and Evolution in Plants' by Karine Alix, Pierre R. Gerard, Trude Schwarzacher and J.S. (Pat) Heslop-Harrison from University of Paris, France, or University of Leicester, England; Annals of Bontany, Volume 120, No. 2, pages 183-194, **2017**.

Polyploid Plant Hybrids
Polyploid plant hybrids duplicate chromosomes so that viable seed is produced. Hybrids that are not polyploid may be sterile because the chromosome numbers do not match.
If plant parents have different chromosome numbers, then the offspring have an odd number of chromosomes and can not produce chromosomally-balanced gametes (male or female germ cell).

F1 Hybrid
An F1 hybrid or 'Filial 1' hybrid is the first generation of offspring of distinctly different parental types that produce a new, uniform phenotype (observable properties of an organism) with a combination of characteristics from the parents.
Mules are F1 hybrids between horse and donkey. This happens naturally, and includes hybrids between species, for example, peppermint is an F1 hybrid of watermint and spearmint.
In agronomy (food crop production), these F1 hybrids are usually created by means of controlled pollination, sometimes by hand-pollination.
> This is not GMO. A Genetically Modified Organism has been altered using genetic engineering techniques in a sophisticated laboratory, frequently using genes from unrelated species.

Russian Comfrey Bocking No. 4 and No. 14 are F1 hybrids.

Backcross
A backcross is a cross of a hybrid with one of its parents or an individual genetically similar to its parent. This creates offspring with genetics closer to that of the parent.

Hybrid Swarm
A hybrid swarm is a group of hybrids that have survived past the first hybrid generation, with interbreeding taking place between hybrid individuals and also backcrossing with its parent type.
These plants are highly variable, with genes and phenotypes (observable properties of an organism) of individuals ranging widely between the two parent types. Hybrid swarms blur the boundary between the parent taxa.
A taxon is a taxonomic group of any rank, such as a class, family, or species. Taxa is plural of taxon.

Introgression
Introgression is the transfer of genetic information from one species to another as a result of hybridization between them and repeated backcrossing. The range of variation keeps increasing with each new backcross. It creates a complicated mixture of parental genes.

Species Complex
A species 'complex' is a group of closely related species that are extremely similar in appearance, so much so that the boundaries between them are frequently unclear.

*"**With reference to Dr. Bob Leaney's useful summary of his studies of a putative Symphytum hybrid,** I would like to explain why, as referee, I have refrained from determining the Norfolk Comfrey as S. x uplandicum x S. orientale.*
***There are no set or 'official' diagnostic criteria for the identification of hybrids.** Each case must be judged on its own merits (Lowe et al. 2004: Stace 1989). A taxonomist must review the available evidence and decide whether it is sufficient.*
Ultimately, 'it is a fundamental difficulty of a historical science like evolution that one can never establish the cause of a past event. It is only possible to show that certain causes such as a hybridization event are plausible or, at most, likely'. (Lewontin & Birch 1966 cited in Lowe et al. 2004 p.206).
The selection of the methods used to evidence hybridity will depend on resources and expertise available, as well as the characteristics of the plants to be investigated.
***The following is a summary of the evidence commonly used,** and ideally as many of the following criteria should be assessed as possible (Stace 1989):*
    ***1. Overall morphological intermediacy** which is often assessed by multivariate statistical analysis of morphometric data.*
    ***2. Cytology (the study of cell contents)** may be informative, in particular, a hybrid generally (but not always) will have a chromosome number intermediate between those of its putative parent taxa.*
    ***3. Reduced fertility** is also indicative of hybridity.*
    4. Molecular genetic analyses.
    5. Reciprocal artificial crosses.
    ***6. Distributional and ecological data** may also provide useful circumstantial evidence of putative parentage."*
-'Challenges when Determining a Putative Interspecific Hybrid' by Clare O'Reilly, Northumberland, England; Botanical Society of Britain and Ireland: BSBI News, Hertfordshire, England, No. 105, pages 9-11, April **2007**.
    (Putative means generally considered or reputed to be.)

*"**Both geneticists and taxonomists in Scandinavia have contributed greatly during past years to the discussion about, and the solution of, the species problem.** The geneticist starts with a supposedly homogeneous population and looks for differences, whereas the taxonomist starts with a supposedly heterogeneous population and looks for similarities.*
If the number of specific genes increases, the possibility of getting back the parents decreases very rapidly.
If, therefore, two natural species hybridise, and the hybrid is completely fertile and vital, in nature too, not merely in experimental gardens, the original species will very soon be lost in a rapidly increasing intermediate hybrid population.
    Such hybrids, however, are hardly to be expected except in crosses between closely related species which have differentiated from a common ancestor in comparatively recent times and have not become genetically stabilised as yet.
***The genus Symphytum represents such a case where the European S. officinale, the Persian S. asperum and the Caucasian S. peregrinum (perfectly 'good' species!) hybridise freely when brought together** as they are in Europe now.*
    The extreme forms, the 'species', are 'hybridised away', and the Symphytum population of Europe today consists to a great extent of intermediate types.
Under such circumstances I do not see any reason for keeping the old 'species' as species, even if they are still found in a pure state in many places (*Faegri 1931)."
-'Some Fundamental Problems of Taxonomy and Phylogenetics' by Knut Faegri, Bergen, Norway; The Botanical Review, Volume 3, Issue 8, pages 400-423, August **1937**. Scandinavia includes Denmark, Finland, Norway and Sweden, together with adjacent parts of northwest Russia- Kola and Karelia.
(* -'Uber in Skandinavien Gefundenen Symphytum-Arten: Nebst einigen Betrachtungen uber das Artproblem Innerhalb der Betreffenden Artgruppe' {About Symphytum Species Found in Scandinavia} by Knut Faegri, Bergens Museum Arbok, Norway, 47 pages, 1931. In German. If anyone has an English translation of it, I would appreciate having it.)

"Introgressive Hybridization:
***Most of the techniques presented here are comparatively simple ones that have been developed for analyzing interspecific (between different species) and intraspecific (between same species) variation.** Observation and measurement are used much as in traditional taxonomic work but refined to a point where they can be employed for analysis as*

well as for description.

By means of such techniques it is now possible for a trained observer to work intensively with a hybrid population in a region completely new to him and from it to deduce exact descriptions of the hybridizing species, *even when he has never seen that species.*

All the multiple-factor characters of an organism are linked with each other so strongly that in species crosses it would take scores of generations of directed breeding to break all the linkages.

Two criteria were pointed out specifically in 1939 (*Anderson, page 692):

> *'1. The intermediacy of separate characters will be correlated. Hybrids intermediate in one character will tend to be intermediate in others. Hybrids which are most like either parent in any one character will tend to resemble that parent in all other characters.*
>
> *2. Variation between individuals will lessen as parental character combinations are approached.'*

The first step in the analysis of any highly variable population is to discover at least two characters that are varying and to devise means for measuring this variation objectively.

The second step is to score a number of individual plants simultaneously for these two characters and then to plot the results as a scatter diagram."

-'Introgressive Hybridization' by Edgar Anderson, Geneticist; Missouri Botanical Garden, Professor of Botany, Washington University, Saint Louis, Missouri; book printed in New York, **1949**, pages vii, viii, 42-43.

(* -'Recombination in Species Crosses' by Edgar Anderson, Geneticist; Genetics, Volume, 24, pages 668-698, 1939.)

(This 1949 book goes into great detail about these methods. One of his methods is now called the 'Anderson Hybrid Index'.)

"Forms of hybrids:

A hybrid may arise by one of three ways in the wild; it may be the result of crossing between two or more different species; the resultant product may then be back-crossed by one of its parents, a process known as introgression. Provided the conditions be favourable these introgressants may, by natural selection, make use of a recombination of genes so that a hybrid swarm is formed.

Whilst the majority of hybrids are interspecific (between different species), crosses between genera are not unknown. In all cases it is the genes that control and exert their influence; they produce new combinations and bring about a gradual infiltration of the germ plasm.

Detection of hybrids:

Where hybrids take on characters intermediate between the putative (assumed) parents their detection in the field may be made with some confidence particularly in the cases of closely related parents.

> *They may differ in exhibiting more vigorous vegetative organs; they may flower earlier, longer and more profusely (abundantly); the flowers may be larger or smaller or the colour may be intermediate, and they may occur in intermediate habitats.*

In doubtful cases, the examination of pollen is useful. *It is necessary to find out the relative abundance of well-filled grains, a procedure facilitated by using a suitable chemical stain. If more than half are abortive and mis-shapen it is reasonable to assume that something has gone wrong with the meiotic (cell division) process to inhibit pollen-formation.*

> *Clues afforded by seed must always be suspect as there is such a long chain of events from bud stage to seed-setting involving pollination, fertilisation, development of the embryo and final seed-formation.*

Sometimes the absence of one of its parents may give rise to doubts concerning a presumed hybrid."

-'Transactions of the Norfolk and Norwich Naturalists Society' edited by E.A. Ellis, Cromer, Norfolk County, England; Volume 22, Part 4, pages 223-315, **1972**. 'Hybrids and Habitats' by E.L. Swann, pages 244-249.

> (Germplasm is the genetic material of germ cells. A germ cell has half the number of chromosomes of a somatic/body cell and can join with another germ cell from the opposite sex to form a gamete which is a new individual.)

"Hybrid Swarms and Introgression:

Sterile interspecific (between different species) hybrids are usually fairly easily recognized in the field, but fertile hybrids are often much less so, both because of the absence of the sterility criterion and because of the existence of a range of hybrids (multiform hybrid) rather than of one sort (uniform hybrid).

The two most obvious effects of hybrid fertility are the production of hybrid swarms, and the occurrence of introgression. *Any two species capable of interbreeding and producing fertile hybrids are, theoretically at least, able to show either or both of these phenomena, but, by strict definition of the terms, not both in the same population.*

> *Which of the two situations is the commoner is not know; there are probably more proven examples of hybrid swarms, but these are much easier to detect.*

Introgression *can be looked upon as taking place when the conditions are not conducive to the establishment of hybrid swarms.* ***The fertile F1 hybrids selectively interbreed with one or both parents rather than among themselves,*** *or the products of the former crosses survive selectively in preference to the products of the later.*

> *A typically introgressed population is one which is clearly referable to a particular species, but which possesses an extreme of variability outside that normally shown by that species and attributable to the influence of a second species."*

-Hybridization and the Flora of the British Isles edited by Clive A. Stace, University of Leicester, England in collaboration with 'Botanical Society of the British Isles' now called 'Botanical Society of Britain and Ireland'. London, England: Academic Press, **1975**, pages 48-49.

"If hybridization is caused by disturbance of equilibria, and two compatible species are brought together which were

previously (geographically) isolated, a common, non-discriminatory breeding system may lead to the extinction of the parent species (*Morley 1971), cf. {compare to} the well-known case of Symphytum officinale, S. asperum, and their invasive hybrid S. uplandicum.

Levin and Anderson (**1970) have analysed mathematically the situation when two plants flowering at the same time compete for the same pollinator, and deduce that this must lead to the elimination of the minority species 'because it suffers a larger percentage of heterospecific pollinations' (p. 465).

This is nothing but the old dictum that two species cannot occupy the same ecological niche. As niches are usually not absolutely identical, the competition pattern will be different in actual practice and will not lead to complete elimination."

-The Principles of Pollination Ecology by Knut Faegri (Bergen, Norway) and L. Van Der Pijl (Nijmegen, Netherlands). Oxford, England: Pergamon Press, Third edition, **1979**, pages 153-154. First edition 1966.

(* -'A Hybrid Swarm Between Two Hummingbird-Pollinated Species of Columnea {Gesneriaceae} in Jamaica' by B.D. Morley; The Botanical Journal of the Linnean Society, Volume 64, pages 81-96, 1971.)

(** -'Competition for Pollinators Between Simultaneously Flowering Species' by Donald A. Levin and Wyatt W. Anderson; The American Naturalist, Volume 104, No. 939, pages 455-467, September-October 1970.)

"***In studies where hybridization is suspected,*** *data frequently are presented as Anderson pictorialized scatter diagrams and Anderson hybrid indices. These two techniques, however, introduce inherent biases which influence the reader's interpretation of the data.*

In this article a new method based on a Euclidean distance coefficient is suggested *as a replacement for both the pictorialized scatter diagram and hybrid index. It is as simple to perform as the Andersonian techniques but relatively free of biases.*

The species concept as depicted in many older floras creates a static picture of organisms which is not borne out in nature *(Grant, 1957).* ***Hybridization between taxa often has been suggested to explain apparent discrepancies between species in concept and in nature*** *(Anderson and Schafer, 1931; Anderson and Hubricht, 1938; Heiser, 1947, 1949; Mason, 1949; Grant, 1952; Tucker, 1953; Anderson and Ander-son, 1954; Gottlieb, 1968; Keeley, 1976).*

The distance diagram requires that two reference points be formed. *In other words the taxa which are hypothesized to be hybridizing must be defined. This is analogous to the procedures of the hybrid index, discriminant analysis, and the Mahalanobis distance hybrid index.*

> *If geographical areas exist where individuals do not appear to be hybridizing, samples should be taken from these localities as well as the zone of suspected hybridization.*

The two reference points assigned, one to each taxon, are composed of the set of categories (characters) used in the study. *The characters must next be ranged between zero and one so that the different scales of measurement used on the various characters are weighted equally (Gower, 1971).*

The last step is to calculate the distance of each individual from the reference points *(dAh, dBh) and the distance between reference points (dAB), using the Euclidean distance equations."*

-'A Distance Coefficient as a Hybridization Index: An Example Using Mimulus Longiflorus and M. Flemingii (Scrophulariaceae) from Santa Cruz Island, California' by Harrington Wells, University of California; Taxon: International Association for Plant Taxonomy, Volume 29, No. 1, pages 53-65, February **1980**.

"*The following are known hybrid combinations in Symphytum:*
>   ***asperum x officinale*** *= S. x uplandicum Nyman (1854)*
>   ***bulbosum x tuberosum*** *= S. x bicknellii Bucknall (1913)*
>   ***caucasicum x orientale***
>   ***cordatum x officinale*** *= S. x polonicum Blocki ex Bucknall (1913)*
>   ***cordatum x tuberosum*** *= S. ullepitschii Wettst. (1893)*
>   ***grandiflorum (ibircum) x uplandicum*** *= S. 'Hidcote Blue' (1970)*
>   ***grandiforum (ibiricum) x tuberosum***
>   ***officinale x orientale*** *= S. x ferrariense Massal. (1913)*
>   *Regarding S. x ferrariense (S. floribundum Schuttlew. ex Bucknall),*
>   *cf. {compare to} A.Kurtto in Ann. Bot. Fennici 18:13-21 (1981).*
>   ***officinale x tuberosum*** *= S. foliosum Rehmann (1868)*
>   ***tuberosum x uplandicum***"

-'The Greek Species of Symphytum Boraginaceae' by William T. Stearn, British Museum of Natural History, London, England; Annales Musei Goulandris, Greece, Volume 7, pages 175-220, (**1985 or 1986**, different sources give different dates, even the pdf has both dates).

"***Symphytum hybridization studies have demonstrated that morphologically widely different species, belonging to different species groups or subgenera, can be crossed easily*** *(Bucknall 1913, Popov 1953, Gadella 1972, Smejkal 1978, Kurtto 1981, Stearn 1985).*

J. Sandbrink (unpublished) compared the chloroplast DNAs (CpDNA) of 8 Symphytum taxa belonging to different species groups and polyploid complexes. *He found two types of restriction patterns. The geographical distribution patterns of the species investigated in the CpDNA-tree are congruent (match) within the cpDNA divergences.*

>   ***The European species (including S. officinale) can be grouped together, the species of the other branch originate from the Caucasus and northwest Turkey and include S. asperum.***

Apparently the morphological evolution proceeded at a much faster rate than the molecular evolution. Likewise **hybridization studies have revealed that the morphological differentiation and sterility barriers did not evolve at the same rate.**"
-'Hybridization in Symphytum: Pattern and Process' by T.W.J. Gadella, Sommerfeltia: The Journal of Natural History Museum and University of Oslo, Norway, Volume 11, pages 79-96, **1990**.

"*A special, for drug manufacturing, topical preparation grown Comfrey variety is registered at 'European Plant Variety Office' under the name 'Symphytum x uplandicum Nyman, Harras'.*
The commonly used botanical name of Comfrey, Symphytum officinale L. (family Boraginaceae), is misleading insofar as Comfrey has no clearly defined species. 'Comfrey' is a complex of hybrids, which arise by crossing different Symphytum taxa.
The botanical denomination of Comfrey is therefore strictly speaking, Symphytum officinale L. sensu lato (in the broader sense). All hybrids are more or less proportion of Symphytum officinale sensu stricto, that is the actual plant in the narrower sense - a species that is in pure form, however rarely encountered.
Usually these are natural occurrences from Comfrey hybrids of Symphytum officinale sensu lato and Symphytum asperum Lepechin, the botanical according to today's nomenclature as Symphytum x uplandicum Nyman.
The registered variety denomination does not refer to another species, but documents that it is a botanically reproducible variety with clearly defined characteristics - therefore, properties, which for use in medicines are important.
'Symphytum x uplandicum Nyman, Harras' stands out in particular by the absence of pyrrolizidine alkaloids in the above-ground plant lines which makes the variety for medical use particularly suitable."
-'Comfrey Herbal Extract in Phytotherapy' by M. Schmidt, Zeitschrift fur Phytotherapie (Journal of Phytotherapy), Germany, Volume 33, No. 3, pages 114-117, **2012**.

(sensu stricto = s.s., s. str., sens. str., sens. strict. In the strict/narrow sense. It is added after a taxon to mean it is being used in the sense of the original author, or without taxa which may otherwise be associated with it.
sensu lato = s.l., sens. lat. In the broad sense. Used in taxonomy to clarify the scope of a taxon when it has been used to define more than one set of lower-level taxons. It includes all its subordinate taxa and/or other taxa that at other times are considered as distinct.)
(This was translated online from German to English. If you have a better translation, please send it to me.)

Overview of Crossbreeding Among Symphytum Species (in chronological order)

For more about crossbreeding, see the section 'Details About Symphytum Species Hybrids' (Chapter 11).

"**Symphytum officinale crossing with Symphytum peregrinum:**
In this paper I give the results of a study of the hybrids formed by Symphytum officinale L. (Common Comfrey) and Symphytum peregrinum Ledeb., *the latter being the plant which has been long established as a colonist in Britain, as well as in Scandinavia* **(S. uplandicum Nyman p.p.)** *(p.p.= pro parte = in part) and in other European countries.*
A native of south-east Caucasus and north Persia (Iran), **S. peregrinum** appears to have been introduced into Britain (together with Symphytum asperum) at the beginning of last century (early 1800s).
In the neighbourhood of Bath and Bristol (England) there are numerous stations (locations) for this plant, in some of which it is accompanied by **S. officinale, with which it hybridizes freely**.
At the end of April of this year, as soon as S. officinale was in flower, these localities were again examined, and **the hybrid plants were found to be in flower at the same time as that species (Common Comfrey) and earlier than S. peregrinum, of which only a few plants were in bloom.**
As the nutlets of the parent species are very distinct in character, and those of the hybrids mostly intermediate, it is well to gather specimens when the nutlets are fully formed and before all the flowers have fallen.
In order to show clearly the characters to be relied on in distinguishing them, I give concise descriptions of the parent species embodying the essential points of difference between them.
S. officinale L.:
Plant hispid, **sometimes sterile.** Stem 2-4 feet (0.6-1.2 meters) high, rarely asperous except when old, conspicuously winged from node to node; lower leaves elliptic- or ovate-lanceolate, attenuate into the petiole, upper narrowly lanceolate with a broad decurrent base.
Calyx with acuminate segments, hispid in fruit; **corolla white, yellowish, rose, purple, or striped with these colours**; anthers longer than filaments; **nutlets** even or minutely rugulose, faintly areolate, shining, black, not constricted above the broad base.
S. officinale L. alpha ochroleucum DC. Prod. x. p. 37 (1846); Oed. Fl. Dan. t. 664 (1777).
S. bohemicum Schm. Fl. Boem. Cent, tertia, p. 13 (1794).
Flowers white or yellowish. (S. officinale var. ochroleucum DC.)
S. officinale L. beta purpureum Pers. Ench. i. p. 161 (1805).
S. patens Sibth. Fl. Oxon. no. 70 (1794).
S. officinale, when sterile, has the segments of the calyx spreading after flowering, and it is not improbable that from this circumstance Sibthorp gave the name of patens to the 'red-flowered Comfrey'.
Flowers rose or purple.
S. peregrinum Ledeb.:
Plant hispid and setose, **very fertile.** Stem 4-7 feet (1.2-2.1 meters) high, stout, often asperous with tubercular-based setae, not

winged; lower leaves ovate-oblong, cordate, rounded or subattenuate at the base, upper ovate or ovate-lanceolate with narrow base, slightly or not at all decurrent.
Calyx-segments acute, tubercular-setose in fruit; **corolla crimson in bud, then pure blue, or shaded with rose and blue;** anthers as long or shorter than filaments; **nutlets** strongly striate and areolate, closely granulate, brownish-black, constricted above the narrow base."
-'**Some Hybrids of the Genus Symphytum**' by Cedric Bucknall, {Mus. Bac. Oxon= Bachelor of Music, Oxford University}, The Journal of Botany British and Foreign, London, England, Volume 50, pages 332-337, **1912**.
  (p.p. = pro parte {in Latin} = in part. This abbreviation is placed after the name of a synonym to indicate that some individuals identified as this species are synonymous with the accepted species.)

"**Hybrids formed by Symphytum officinale L. and Symphytum peregrinum Ledeb.:**
  x S. discolor mihi. (S. officinale alpha ochroleucum x < peregrinum)
    Habit and stature of S. peregrinum, sparingly fertile or quite sterile. Stem hispid, asperous, narrowly winged; lower leaves rather broadly ovate, upper lanceolate, narrowly decurrent.
    Calyx with acute or acuminate segments, slightly tubercular-setose in fruit; corolla whitish, or more or less tinted with pale rose and blue, turning to pale slaty blue when dry; anthers a little longer than filaments. Nutlets intermediate between those of the parents, faintly areolate, dotted with minute scattered points, shining, slightly constricted above the broad base.
  x S. lilacinum mihi. (S. officinale alpha ochroleucum x beta purpureum x < peregrinum)
    Habit and stature of S. peregrinum, sparingly fertile. Stem asperous, more conspicuously winged than in x S. discolor; lower leaves elliptic-lanceolate, rounded or attenuate at the base, narrower as a rule than in x S. discolor, upper lanceolate with broader decurrent base.
    Fruiting-calyx setose; corolla purplish with greenish-yellow tip when in bud, then pale purplish-rose, slaty blue tinged with purple when dry; anthers as long as the filaments; nutlets as in x S. discolor.
  x S. densiflorum mihi. (S. officinale beta purpureum x > peregrinum)
    Habit of S. officinale, but often taller, fertile or sterile. Stem asperous, narrowly winged; lower leaves narrowly oblong-lanceolate, shortly attenuate at the base, upper narrowly lanceolate, conspicuously decurrent.
    Flowers large, open, crowded; calyx with acute segments, hispid, rarely tubercular-setose in fruit; corolla reddish-violet, dark purple when dry; anthers as long or longer than filaments; nutlets as in x S. discolor, but often olivaceous-black.
  x S. coeruleum Petitmengin. (S. officinale alpha ochroleucum x peregrinum)
    Habit and stature of S. peregrinum, but almost entirely sterile. Stem hispid, asperous, partially and rather broadly winged; lower leaves oblong, attenuated into the petiole, upper lanceolate, semidecurrent from the rather broad base.
    Calyx with acute segments; corolla rose tipped with green when in bud, then bright blue or rose and blue, the tips of the lobes sometimes yellowish; anthers equal to filaments; nutlets (very seldom produced) closely granulated.
While S. peregrinum produces fruit abundantly, S. officinale, even when growing alone, and without any suspicion of hybridity, is often quite sterile.
In the case of the hybrids above described some plants are sterile, while others produce fruit sparingly, never in such abundance as S. peregrinum."
-'**Some Hybrids of the Genus Symphytum**' by Cedric Bucknall, {Mus. Bac. Oxon= Bachelor of Music, Oxford University}, The Journal of Botany British and Foreign, London, England, Volume 50, pages 332-337, **1912**.
  (**The symbol '<'** means the hybrid shows characters of S. peregrinum in lesser proportions.
  The symbol '>' means the hybrid shows characters of S. peregrinum in higher proportions.)
  (According to a current {2018} respected botanist: "The above names are valid, but incorrect. This is because, the hybrid names are at the rank of species, but their parents are of unequal rank {variety and species}.")
  (Olivaceous is a dusky yellowish green color; olive green.)
  (The last paragraph quoted in this report has the opposite fertility pattern of my Common Comfrey and Russian Comfrey. My Common Comfrey has very fertile seeds, and my Russian Comfrey does not. Apparently, my varieties are different from the varieties reported by Mr. Bucknall.)

"*So far as my experience goes a blue tint in the flowers is the result of hybridity with S. peregrinum Ledeb.*"
-'**A Revision of the Genus Symphytum, Tourn.**' by Cedric Bucknall, {Mus. Bac. Oxon= Bachelor of Music, Oxford University}, Journal of the Linnean Society of London, England, Botanical Journal, Volume 41, Issue 284, pages 491-556, December **1913**.

"**Symphytum officinale L. (Common Comfrey), and S. asperum Lepech. (Prickly Comfrey):**
The limits (geographic locations) of these species are disputed; it is probable that extensive hybridisation has occurred in Britain or that hybrid stock has been introduced.
For the time being it seems wisest to take a narrow view and only to class as species plants with the following combination of characters:
  S. officinale: *Flowers white or pale pink, anthers in front view longer than exposed portion of filaments; leaves strongly decurrent; pubescence soft, hairs not swollen at the base.*
  S. asperum: *Flowers blue, anthers in front view shorter than filaments; leaves not decurrent, with distinct petioles;*

pubescence stiff, hairs swollen at the base.
Any specimens belonging to this group without either of these two combinations should perhaps be placed under S. x uplandicum Nyman (Russian Comfrey).
    S. asperum is a very rare introduction.
    S. officinale occurs as a native in fens (wetlands) and marshes in eastern England.
    Symphytum x uplandicum Nyman (S. asperum Lepech. x S. officinale L.)
    This hybrid is widespread and often occurs, probably as an introduction, in the absence of either parent. Fertile."
-'**Hints on the Determination of Some Critical Species, Microspecies, Subspecies, Varieties and Hybrids in the British Flora**' compiled by Franklyn Perring, Proceedings of BSBI (Botanical Society of Britain and Ireland), Hertfordshire, England, Volume 4, **1962**, pages 365, 380.

"**The Symphytum Survey of the British Isles began early in 1968** as one of the new 'Network Research' projects. By the end of October over 400 completed cards had been returned. The exhibit was based on these and personal observations.
The initial results show that S. asperum Lepech. (Prickly Comfrey) is extremely rare, whilst S. x uplandicum Nyman (Russian Comfrey) is widespread and occurs throughout the British Isles.
The two forms of S. officinale L.: the cream var. ochroleucum DC., and the reddish-purple var. purpureum Pers., have different distribution patterns, the former mainly occurring in eastern England whilst the latter occurs in the south and west.
Introgression takes place between the two, and mixed colour populations are extremely abundant in the area of overlap, particularly in Berkshire County (southeast England), Hampshire County (south England) and Dorset County (southwest England).
Many of the populations measured are intermediate between S. x uplandicum and the various colour forms of S. officinale."
-'**Symphytum Survey: B.S.B.I.**' by F.H. Perring, Watsonia: Journal and Proceedings of the Botanical Society of the British Isles', London, England, Volume 8, **1970**, page 91.
    (Introgression is the transfer of genetic information from one species to another as a result of hybridization between them and repeated backcrossing.)

"**Natural hybrids between three or more species are fairly common in those genera which produce fertile binary hybrids. In the British flora there are fairly well substantiated examples** in x Dactyloglossum, Epilobium, Mentha, Rosa, Salix, **Symphytum** and Ulmus."
-'**Hybridization and the Flora of the British Isles**' edited by Clive A. Stace, University of Leicester, England in collaboration with 'Botanical Society of the British Isles' now called 'Botanical Society of Britain and Ireland'. London, England: Academic Press, **1975**, page 11.

"**Symphytum asperum Lepech. (Prickly Comfrey), S. uliginosum Kern., S. officinale L. (Common Comfrey): These three species hybridize and form hybrid swarms consisting of F1 and backcross hybrids**.
    Symphytum officinale is cytologically heterogenous (diverse): $2n = 24$, $2n = 40$, $2n = 48$, cytotypes occur in various parts of Europe.
    Symphytum asperum ($2n = 32$) does not hybridize with the diploid ($2n = 24$) form of S. officinale in Europe but produces hybrids (and backcrosses) with the $2n = 40$ and $2n = 48$ cytotypes of S. officinale.
The primary hybrids, with $2n = 36$, or $2n = 40$, are collectively known under the name S. x uplandicum Nyman (Russian Comfrey).
In Europe the parental species are largely allopatric with a very small zone of overlap in the northwest Caucasus (Kusnetosov, 1910). In the zone of overlap, hybridization does not occur because S. officinale and S. asperum grow at different altitudes.
Apparently the hybrids arose outside the Caucasus (Tutin 1956; Wade 1958). In many parts of western, northwestern, or central Europe the hybrid swarms are more common than the parental species."
-'**Notes on Symphytum (Boraginaceae) in North America**' by T.W.J. Gadella, page 1061-1067 in book: Annals of the Missouri Botanical Garden, Volume 71, Saint Louis, Missouri. Lawrence, Kansas: Allen Press, **1984**.
    (Allopatric or geographic speciation occurs when populations of the same species become isolated from each other, and this prevents genetic exchange.)

"**Natural interspecific (between different species) hybridization has been well documented in Symphytum.**
    Bucknall (1912, 1913) assigned names to hybrids between several pairs of species, but he gave different names to hybrids involving different varieties of a given species pair.
The hybrid origin of S. x uplandicum Nyman ($2n = 40$) from S. officinale ($2n = 48$) and S. asperum ($2n = 32$) was established by Gadella & Kliphuis (1973), and the natural and synthetic interspecific hybrids were remarkably similar.
    Symphytum x uplandicum derived from S. officinale with $2n = 40$ has a chromosome number of $2n = 36$.
The chromosome number in the hybrid is further complicated as a result of backcrossing to S. officinale, and five hybrid races with $2n = 34, 36, 40, 42,$ and 44 have been reported (Gadella & Kliphuis, 1975).
Both $2n = 36$ and $2n = 40$ cytological types of S. x uplandicum have been introduced into North America (Gadella, 1984).
Basler reported that although meiotic disturbance in S. x uplandicum is about 14 percent, pollen fertility is more than 95 percent.
    The hybrid is genetically stable, and it bridges, via backcrossing, the morphological and cytological gap between the parental species.
Interspecific hybrids between S. orientale and both S. officinale and S. ottomanum Frivald have been reported by Kurtto (1981,

1985, respectively).
Intraspecific (within same species) hybrids between diploid and tetraploid cytotypes of S. officinale are known (Basler; Gadella & Kliphuis, 1972).
The hybrid between S. icaricum Paw. (probably a subspecies of S. Davisii Wickens; see Stearn) and S. circinale Runemark was recognized by Pawlowski (1971) as an intergeneric (between different genus) one, since he placed the latter species in the segregate Procopiania."
-'**Journal of the Arnold Arboretum**, Supplementary Series, Volume 1, Harvard University, Cambridge, Massachusetts, **1991**. This volume is a collection of contributions toward a 'Generic Flora of the Southeastern United States'. 'The Genera of Boraginaceae in the Southeastern United States' by Ihsan A. Al-Shehbaz, pages 151-152.
>(Meiosis is a type of cell division that reduces the number of chromosomes in the parent cell by half and produces gamete cells. Meiotic disturbance means a problem with the cell division.)

*"***Neophytic Comfrey species and hybrids are more prevalent in the British Isles*** than observed in central Europe. This may be due to traditional British garden culture. But also because these clans are common in most German wild plants flora communities.*
Neophyte crossbreeding occur at different rates. *Sometimes there are only a few occurrences for many years, and only over decades does a change occur in colonization.*
Hybrid formation can significantly influence the evolution of the genus Symphytum in Europe.
Last but not least, it involves neophytic clans. The classic example is the established Symphytum x uplandicum.
Besides S. x hidcotense that is from Great Britain, there are other reported Symphytum hybrids. These too grow wild and start to establish.
>*As our cultural and settlement landscape offers new ecological niches, there may be contact between potential crossing partners caused by gardening that causes an evolutionary thrust and promotes hybrid species emergence."*
-'**Caucasian Comfrey (Symphytum caucasicum M. Bieb.) and Hidcote Comfrey (Symphytum x hidcotense P.D. Sell) in the Region of Aachen'** [Kaukasischer Beinwell {Symphytum caucasicum M. Bieb.} und Hidcote-Beinwell {Symphytum x hidcotense P.D. Sell} im Aachener Raum] by F. Wolfgang Bomble and Bruno G.A. Schmitz, Die Veroffentlichungen des Bochumer Botanischen Vereins e. V., Germany, Volume 4, No. 6, pages 50-54, **2012**. If you have an English translation of this, could you please send it to me.
>(In botany, a neophyte is a plant species that is not native to a geographical region, and was introduced in recent history. Plants that are long-established in an area are called archaeophytes. In Britain, neophytes are plant species introduced after 1492, when Christopher Columbus arrived in the Americas.)

*"How many hybrids (of all plants, not just Comfrey) are there in Britain and Ireland? There are 909 hybrids accepted from Britain and Ireland by Stace et al. (*2015).*
Most of them (744) are spontaneous hybrids, which are believed to have arisen here by hybridization in the wild; these totals include taxa with some spontaneous and some introduced populations. Three hybrids appear to be native but one parent is absent as a native, so they are treated separately from the spontaneous hybrids.
Ten hybrids of complex origin cannot easily be fitted into simple categories such as 'native x native' or 'native x neophyte'. They include the triple hybrids Symphytum asperum x S. officinale x S. tuberosum, which arises as a cross between the introduced hybrid Symphytum asperum x S. officinale and the neophyte S. tuberosum."
-'**Plant Hybrids in the Wild:** Evidence from Biological Recording' by Christopher D. Preston and David A. Pearman, Biological Journal of the Linnean Society of London, England, Volume 115, pages 555-572, **2015**.
(* -Hybrid Flora of the British Isles by Clive A. Stace, Christopher D. Preston and David A. Pearman, Hertfordshire, England: Botanical Society of Britain and Ireland, 2015.)

<u>**Overview of Distribution of Symphytum Species**</u> (in chronological order)
For details see each species in Chapters 7, 8, 9, 10, and 11.

"Roman hardihood (bold, daring men) conquered that island (Britain), and as the Symphytum officinale was in such repute as a vulnerary it was natural that they should carry the plant with them and naturalize it in the land of their conquest.
I deem Dr. Adams' hypothesis that the Romans introduced Symphytum officinale into Great Britain so plausible and reasonable that I unhesitatingly adopt it."
-'American Observer Medical Monthly: Devoted to Homoepathic Materia Medica, Surgery, Gynaecology, Obstetrics, Otology, Ophthalmology, Practice of Medicine', New Series Volume 17, Detroit, Michigan, 1880. Article: 'Materia Medica: The Empirical History of Symphytum Officinale' by Professor S.A. Jones, M.D., Ann Arbor, Michigan, August **1880**, pages 384-394.

*"**The geographical area of the Symphytum genus includes almost the whole of Europe (except Lapland= Finland), Asia Minor, a part of Siberia, and Persia (Iran).***
Sepulcrale, armeniacum, asperum and peregrinum:
>*It may be noted that, of the four members of this group, the distribution as far as is known:*
>***S. sepulcrale*** *is found furthest west, in the neighborhood of Trebizond, Turkey.*
>***S. armeniacum*** *comes next in Armenia and Georgia; here the area of **S. asperum** commences and extends northwards in the Caucasus.*

*While that of **S. peregrinum** extends southwards and eastwards from the south-east of the Caucasus into Persia."*
-'A Revision of the Genus Symphytum, Tourn.' by Cedric Bucknall, (Mus. Bac. Oxon= Bachelor of Music, Oxford University), Journal of the Linnean Society of London, England, Botanical Journal, Volume 41, Issue 284, pages 491-556, December **1913**.

"Symphytum L.: Wallwurz, Beinwurz, Beinwell, Beinheil in Central Europe, 1926:
Symphytum asperum Lepechin
Symphytum officinale L.
Symphytum tuberosum L.
Symphytum bulbosum Schimper"
-'Illustrierte Flora von Mittel-Europa: Mit Besonderer Berucksichtigung von Deutschland, Osterreich und der Schweiz' (**Illustrated Flora of Central Europe: With Special Consideration of Germany, Austria and Switzerland**), Volume 5, Part 3' by Gustav Hegi, Zurich, Switzerland; published by J.F. Lehmann, Munich, Germany, **1926**. Symphytum pages 2220-2230. In German, 13 volumes written from 1908 to 1931. If you have an English translation of the Symphytum pages, I would appreciate a copy.

*"**Two species of Symphytum are native to the British Isles, S. officinale L. the Common Comfrey, and S. tuberosum L. the Tuberous Comfrey.***
Three exotic (not native) species also occur, S. asperum Lepech. the Rough Comfrey of Caucasus and Persia (Iran), S. orientale L. of Turkey, and S. peregrinum Lebed. the Blue Comfrey, perhaps not identical with the wild species from the east Caucasus but with the hybrid S. x uplandicum Nyman.
All are found in waste places and hedge banks, while hybrids between S. officinale and S. peregrinum are of frequent occurence (Clapham, Tutin and Warburg, 1957)."
-'**Russian Comfrey**' by L.A. Willey and R.L. Knight, Journal of the National Institute of Agricultural Botany, Cambridge, England, Volume 9, No. 2, pages 139-144, **1962**.

"Of the 25 species of Symphytum occurring in Europe, Asia Minor, western Siberia and Persia, two species, both of medicinal interest, are indigenous (native) to Britain, viz. (namely), S. officinale L. and S. tuberosum L..
Others have been introduced, including S. asperum Lepech. and S. x uplandicum Nym. (S. peregrinum Ledeb.) (Clapham, Tutin and Warburg, 1962)."
-'**British Medicinal Species of the Genus Symphytum**' by K.R. Fell and Janet M. Peck, Pharmacognosy Research Laboratories, University of Bradford, England; Planta Medica: Journal of Medicinal Plant and Natural Product Research, Volume 16, No. 2, pages 208-216, May **1968**.

"Of the 33 species recognised for the Symphytum genus, 27 are included in this paper, of which 19 occur in Turkey (mainland), 9 being endemic (native); of the, remaining species 3 grow in Caucasia, 4 in the Aegean Islands, and 1 in Greece and Bulgaria.
S. officinale and S. uliginosum have not been included in this survey, although according to Bucknall the former is stated to occur in Turkey even though he cites no specimens, neither have I seen any; S. uliginosum is found in Hungary and south Russia.
Of the remaining excluded species, S. floribundum and S. mediterraneum occur in France and S. gussonei in Sicily and Galicia (northwest Spain).
The European species, S. bulbosum, has been included in this survey partly to emphasize its close affinities with Turkish species, and also to guard against the possibility of its eventual discovery in western Turkey.
It is remarkable how little material has been collected from Turkey-in-Europe since the days of *Aznavour, who collected extensively near the Bosphorus."
-'**A Revision of Symphytum in Turkey and Adjacent Areas**' by G.E. Wickens, Royal Botanic Gardens: Kew; Notes from the Royal Botanic Garden Edinburgh, Scotland, Volume 29, pages 157-180, **1969**. Includes Turkey, Bulgaria, Greece, Aegean Islands and Caucasia.
(* -'Un Symphytum Nouveau' {A New Symphytum} by G.V. Aznavour; Bulletin de l'Herbier Boissier, 2 Serie {Second Series}, Tome III {Volume 3}, p. 588-589, 1903. In French. Description of Symphytum pseudobulbosum Aznavour from the Asian side of the Bosphorus.)

>(The Caucasus Mountains are located at the border of Europe and Asia, between the Black Sea and Caspian Sea and occupied by Russia, Georgia, Azerbaijan, and Armenia.)
>(The Aegean Sea is a bay of the Mediterranean Sea located between the Greek and Anatolian peninsulas, i.e., between the mainlands of Greece and Turkey.)
>(The Bosphorus, also known as The Strait of Istanbul, is a narrow waterway located in northwestern Turkey.)

>"**Aznavour, Georges Vincent (1861-1920):**
>**Georges Aznavour was an amateur botanist of Armenian descent who made significant collections in the Istanbul, Turkey area from 1885 onwards.** His collections from the Bosphorus region comprise about 20,000 specimens, with the original herbarium deposited in Geneva, Switzerland.
>Aznavour published nearly 20 works, mostly in French language botanical journals, enumerating (listing) plants he had collected or received from Anatolia (Asian Turkey) and Syria, and describing new species.
>He also produced a substantial manuscript entitled 'Prodrome de la Flore de Constantinople', which was published posthumously as 'La Flore du Bosphore et des Environs', edited by Bertram Post.

In P.H. Davis' 'Flora of Turkey and the Aegean Islands' (1965-1985), 1,600 specimens collected by Aznavour are cited, including 20 new taxa. Aznavour is commemorated in at least five species with the epithet (name) aznavourii.
-'Aznavour, Georges Vincent (1861-1920)', JSTOR® Global Plants, www.jstor.org, 2018. JSTOR is a digital library for scholars, researchers, and students. It provides access to more than 12 million academic journal articles, books, and primary sources in 75 disciplines.

"Wickens (*1969) revised the species of the genus Symphytum in Turkey and adjacent areas. Of the 33 species recognized for the genus, 27 species occur in this area and represent all sections of the genus.
In fact in the Caucasus and Turkey the genus shows its greatest morphological diversity and is represented by the majority of its species. This may indicate that the centre of differentiation of the genus is situated in this area.
If this supposition is correct, the European species, and especially S. officinale, spread from this centre over the larger part of Europe."
-'**Cytotaxonomic Studies in the Genus Symphytum IV:** Cytogeographic Investigations in Symphytum Officinale L.' by Th.W.J. Gadella and E. Kliphuis, Instituut voor Systematische Plantkunde (Institute for Systematic Botany), Utrecht, Netherlands; Acta Botanica Neerlandica, Volume 21, Issue 2, pages 169-173, April **1972**.
(* -'A Revision of Symphytum in Turkey and Adjacent Areas' by G.E. Wickens, Royal Botanic Gardens: Kew; Notes from the Royal Botanic Garden Edinburgh, Scotland, Volume 29, pages 157-180, 1969. Includes Turkey, Bulgaria, Greece, Aegean Islands and Caucasia.)

*"**Five species of Symphytum occur on Greek territory**, namely **S. bulbosum Schimper** widespread in mainland Greece from the Peloponnese northward, **S. ottomanum** widespread in northern Greece, **S. tuberosum subsp. angustifolium (A. Kerner) Nyman** only in the extreme north-west, **S. davisii subsp. davisii, naxicola, icaricum and cycladense** endemic (native) to Aegean islands and **S. anatolicum Boiss.**, a species of Western Asia Minor, on Lesvos, Kos and Samos.*
Neither S. officinale nor S. orientale occurs in Greece.
*1. **S. tuberosum**, represented by the eastern subspecies angustifolium, has just managed to extend into the mountains in the extreme north west of modern Greece.*
*2. Of the other species occurring on Greek territory, **S. ottomanum** was described by *Frivaldszky in 1836 from Rumelia (now Turkey), the central part of the Balkan Peninsula, then under Turkish rule and thus with Ottoman Empire.*
*3. **S. bulbosum** by Carl Schimper in 1825 from plants naturalized near Heidelberg, west Germany.*
*4. **S. anatolicum** by Boissier in 1844 from western Asia Minor.*
*5. **S. davissi** by Wickens in 1969 from the Greek islands of Amorgos and Ikaria.*
*6. In 1973 Pawlowski distinguished from the last-named 3 other closely allied insular (island) Aegean species, **S. icaricum from Ikaria, S. naxicola from Naxos, and S. cycladense from Sikinos and Kardiotissa (all Greek islands)."***
-'**The Greek Species of Symphytum Boraginaceae**' by William T. Stearn, British Museum of Natural History, London, England; Annales Musei Goulandris, Greece, Volume 7, pages 175-220, (**1985 or 1986**, different sources give different dates).
(* -'Flora oder Botanische Zeitung' {Flora or Botanical Newspaper} or 'Flora oder Allgemeine Botanische Zeitung', Regensburg, Germany, Volume 19, No. 2. Symphytum ottomanum Frivaldszky by Dr. E. Frivaldszky, page 439, 1836.)
(The Peloponnese or Peloponnesus is a peninsula and region in southern Greece.)
(Asia Minor, Asian Turkey, Anatolia or Anatolian peninsula is the westernmost part of Asia. It includes most of Turkey.)
(Lesvos, Kos and Samos are Greek islands near Turkey.)

*"A phenetic analysis using 69 characters has shown that **Symphytum species can be divided into 6 geographically differentiated groups** (J. Sandbrink, unpublished)."*
-'**Hybridization in Symphytum: Pattern and Process**' by T.W.J. Gadella, Sommerfeltia: The Journal of Natural History Museum and University of Oslo, Norway, Volume 11, pages 79-96, **1990**.
(Phenetics classifies organisms based on overall similarity in morphology and other observed traits, regardless of evolution.)

"Flora of Slovakia: Symphytum L., Kostihoj (Slovak for Symphytum):
*   **Symphytum bohemicum** F.W. Schmidt (Kostihoj cesky)*
*   **Symphytum officinale** L. (Kostihoj lekarsky)*
*   **Symphytum tanaicense** Steven (Kostihoj mociarny)*
*   **Symphytum angustifolium** A. Kerner (Kostihoj uzkolisty)*
*   **Symphytum tuberosum** L. (Kostihoj hluznaty)*
*   **Symphytum cordatimi** Waldst. et Kit. in Willd. (Kostihoj srdcovitolisty)"*
-Flora Slovenska {Flora of Slovakia}, Volume 5/1 (V/1) or 5/2 (V/2), edited by L. Bertova and K. Goliasova. Bratislava, Slovakia: Veda, **1993**. Article: 'Symphytum L.' by J. Majovsky and Z. Hegedusova, pages 76-97. Some sources say 5/2 but I think 5/1 is correct. (If you have an English translation of the Symphytum pages, I would appreciate a copy.)
(Slovakia or Slovak Republic is in Central Europe. It is bordered by Poland to the north, Ukraine to the east, Hungary to the south, Austria to the west, and Czech Republic to the northwest.)

"The genus Symphytum L. (Boraginaceae) is a mesophytic genus of approximately 40 species, mainly distributed in the Euro-Siberian and Mediterranean regions.
With 19 species, Turkey holds the largest number of species."
-'**Procopiania (Boraginaceae): Separate Genus or Part of Symphytum?**' by Burcu Elci, Hartmut H. Hilger, Sadik Erik from

Ankara, Turkey and Berlin, Germany, April **2008**. 'Systematics 2008: Programme and Abstracts', 18th International Symposium, German Botanical Society, Gottingen, Germany.

"This paper presents a survey of high-altitude plant communities which occur in the Western Carpathians with an enumeration of the characteristic, transgressive and differential species of the individual alliances, orders and classes.
This study summarises the results of the syntaxonomical and nomenclatural (name) revisions of various types of high-altitude vegetation in the Western Carpathians and the longstanding research in the field.
Alpine regions show strong gradients in abiotic conditions and contain highly specialised biota. *The specific conditions in high mountains have given rise to a diverse mosaic of vegetation types, with an abundance of rare, relic and endemic (native) taxa. Rugged relief, heterogeneous bedrocks, and variable climatic and soil features all help to create an exceptional variety of habitats, including refugia that provide high-altitude plants with optimal conditions.*
Species of contact phytocoenoses with higher abundance in analysed syntaxa; naturally rare taxa:
    Symphytum tuberosum was found in:
        3 = Subthermophilous species in rich tall-grass vegetation on calcareous bedrock.
        4 = Calcareous meso-hygrophilous in tall-grass vegetation on gravelly soils.
        5 = Calcareous chionophilous hygrophilous in tall-grass vegetation of the Carpathians.
        7 = Tall-forb and fern-rich chionophilous communities on calcareous bedrock.
        8 = Natural nitrophilous tall-herb vegetation on alluviums and banks of montane streams in
          Carpathians and Hercynic mountains.
        9 = Subalpine communities of deciduous shrubs.
Carpathian sub-endemics (sub-native):
    Symphytum cordatum was found in:
        8 = Natural nitrophilous tall-herb vegetation on alluviums and banks of montane streams in Carpathians and Hercynic mountains."

-'High-Altitude Vegetation of the Western Carpathians: A Syntaxonomical Review' by Jan Kliment, Jozef Sibik, Ivana Svitkova, Ivan Jarolimek, Zuzana Dubravcova and Jana Uhlirova, from Blatnica and Bratislava, Slovakia; Biologia: Section Botany, Bratislava, Slovakia, Volume 65, No. 6, pages 965-989, November **2010**.

    (Carpathian Mountains are a mountain range that forms an arc 1,500 km {932 miles} across central and eastern Europe. They are the second-longest mountain range in Europe. **Western Carpathians include Austria, Czech Republic, Poland, Slovakia and Hungary.** Highest elevation is Gerlachovsky Stit, Slovakia at 2,655 meters or 8710 feet.)
    (In genetics, transgressive segregation is the formation of extreme phenotypes found in segregated hybrid populations compared to phenotypes in parental lines.)
    (Alpine regions are parts of mountains above the tree line, i.e., no trees grow.)
    (Abiotic factors include sunlight, temperature, wind, rain/snow, soil/rocks and altitude.)
    (Relict or relic means a surviving species of an otherwise extinct group, or a remnant of a formerly widespread species that lives in isolated areas.)
    (Refugia or refugium is an area where a population of organisms can survive through a period of unfavorable conditions such as glaciation. Glaciation is the process or condition of being covered by glaciers or ice sheets.)
    (Phytocoenosis or phytocenosis is a plant community in a specific geographical area that forms a uniform patch that is distinguishable from neighboring patches of different vegetation.)
    (Endemic or native species are from only one geographic area or zone such as an island, country, continent or habitat type. Habitat restriction occurs due to climate, physical barriers and biology. Sub-endemic species are mostly in one area but some also grow in other areas.)

"The Old World (Europe and Asia) genus Symphytum L. belongs to the tribe 'Boragineae Bercht. et J. Presl', a major monophyletic (common evolutionary ancestor) group within the family Boraginaceae (*Hilger et al. 2004).
With approximately 40 species, Symphytum is one of the largest genera in this tribe (Bucknall 1913, **Sandbrink et al. 1990). *It includes perennial, roughly hirsute (hairy) plants, which are morphologically well characterized by creeping, mostly fleshy rhizomes, alternate leaves, double scorpioid cymes (= boragoids) with tubular flowers and five corolla appendages (= fornices) inside the flower.*
The geographical range of the Symphytum genus covers almost the whole of Europe and Asia Minor (Asian Turkey), as well as part of Western Asia and Siberia (part of Russia) *(Bucknall 1913). The centre of its diversity is situated in the Pontic area and in the western parts of the Irano-Turanian region, primarily in the mountain ranges around the Black Sea (Gadella & Kliphuis 1978, ***Davis 1988).*
In central Europe, the following native species are recognized:
    S. cordatum Waldst. et Kit..
    S. officinale complex (S. bohemicum F.W. Schmidt, S. officinale s.str., S. tanaicense Steven).
    S. tuberosum complex
        (S. angustifolium A.Kern.: Kerner 1863, Murin & Majovsky 1982, Marhold & Hindak 1998.
        And S. tuberosum L.: Pawlowski 1963, Gams 1966, Smejkal 1978, Majovsky & Hegedusova 1993, Slavik 2000, Danihelka et al. 2012)."

-'**Symphytum Tuberosum Complex in Central Europe:** Cytogeography, Morphology, Ecology and Taxonomy' by L. Kobrlova, M. Hrones, P. Koutecky, M. Stech, and B. Travnicek, Preslia: The Journal of the Czech Botanical Society, Czech Republic; Volume 88, pages 77-112, March **2016**.

(* -'Molecular Systematics of Boraginaceae Tribe Boragineae Based on ITS1 and trnL Sequences, with Special Reference to Anchusa s.l.' by H.H. Hilger, F. Selvi, A. Papini and M. Bigazzi {Germany and Italy}, Annals of Botany, Volume 94, No. 2, pages 201-212, 2004.)
(** -'Phylogenetic Relationships in the Genus Symphytum L. {Boraginaceae}' by J.M. Sandbrink, J. Van Brederode and T.W.J. Gadella, Department of Genome Evolution, University of Utrecht, Netherlands; Proceedings: Koninklijke Nederlandse Akademie van Wetenschappen {Royal Netherlands Academy of Sciences}, Series C, Volume 93, No. 3, pages 295-334, 1990.)
(*** -Flora of Turkey and the East Aegean Islands, 11 Volumes by Peter H. Davis. Scotland: Edinburgh University Press, 1984-1988. 'Symphytum L.' in Supplement Volume 10, pages 186-189, 1988.)
> (Monophyletic is a group of organisms descended from a common evolutionary ancestor, especially one not shared with another group. The opposite is paraphyletic, a group of organisms descended from a common ancestor that does not include all the descendant groups.)
> (The Irano-Turanian Region is located within the Tethyan Subkingdom of the Holarctic Kingdom. It is divided into 12 provinces according to Armen Takhtajan, a Soviet-Armenian botanist who created a classification system for flowering plants. This area has great diversity and an abundance of species.)

"The last part of a revision of the genus Symphytum in the Czech Republic treat naturalised and cultivated species.
*Two naturalised taxa are traditionally recognised in the Czech flora, namely **Symphytum asperum** and **S. x uplandicum**.*
*In addition, the occurrence of three other alien species is discussed: **S. tauricum**, a Crimean species discovered in south-west Bohemia, **S. grandiflorum**, a Caucasian species found in Prague, and **S. x hidcotense**, an artificial hybrid taxon discovered in the vicinity of the town of Ivancice, Moravia. Symphytum grandiflorum and S. x hidcotense are reported from the Czech Republic for the first time.*
> *A detailed morphological description, ecology, distribution of these taxa and the history of their cultivation in our country (Czech Republic) are presented.*

In addition, several species cultivated in Central Europe (S. bulbosum, S. caucasicum, S. cordatum and S. orientale) are discussed."
-'**The Genus Symphytum L. (Comfrey) in the Czech Republic III: Introduced and Cultivated Species**' (Rod Symphytum Kostival v Ceske Republice III: Nepuvodni a Pestovane Druhy) by Lucie Kobrlova and Michal Hrones, Univerzity Palackeho v Olomouci, Czech Republic; Zpravy Ceskoslovenske Botanicke Spolecnosti (Czech Botanical Society News), Prague, Czech Reupublic, Volume 52, pages 225-248, December **2017**.
> (Naturalize means a plant or animal species establishes itself so that it lives successfully in the wild in a region where it is not native.)
> (A cultivated or domesticated species is one where people help and sometimes change its growth and reproduction.)
> (Crimea is a peninsula on the northern coast of the Black Sea in Eastern Europe that is almost completely surrounded by the Black Sea and Sea of Azov to the northeast.)
> (The Caucasus Mountains are located at the border of Europe and Asia, between the Black Sea and Caspian Sea and occupied by Russia, Georgia, Azerbaijan, and Armenia.)

<u>Most Frequently Grown Species of Comfrey</u>

Today (2019) the Comfrey plants you are most likely to find in the United States are Symphytum officinale (Common Comfrey) and Symphytum uplandicum (Russian Comfrey).

"Comfrey and the annual herb borage are the only agricultural crop plants in the Boraginaceae family.
> *Symphytum, the Comfrey genus, has about 25 species but only three are relevant to the crop known as Comfrey."*

-'**Comfrey: A Controversial Crop**' by Robert G. Robinson, University of Minnesota, Agricultural Experiment Station, Minnesota Report MR-191, Item No. AD-MR-2210, **1983**, page 2.
> (Borage, Borago officinalis also called starflower, is an annual herb in the family Boraginaceae. The leaves are edible, and the plant is grown in home gardens in Europe. It is commercially cultivated for 'borage seed oil' extract. The plant contains pyrrolizidine alkaloids.)

*"It appears that **S. officinale (Common Comfrey) and the two S. x uplandicum hybrids (Russian Comfrey) are the most common Symphytum taxa in North America.** The two cytotypes of S. x uplandicum (2n = 36 and 2n = 40 chromosomes) are present in almost equal proportions.*
Symphytum asperum (Prickly Comfrey) seems to be much rarer in North America."
-'**Notes on Symphytum (Boraginaceae) in North America**' by T.W.J. Gadella, page 1061-1067 in book: Annals of the Missouri Botanical Garden, Volume 71, Saint Louis, Missouri. Lawrence, Kansas: Allen Press, **1984**.

*"**Russian Comfrey (Symphytum uplandicum) is widely cultivated in England and Australia.***
Common Comfrey (Symphytum officinale) and Russian Comfrey are cultivated in Europe, Japan, North America, Australia, and doubtlessly elsewhere."
-**Toxicants of Plant Origin, Volume I: Alkaloids** edited by Peter R. Cheeke. Boca Raton, Florida: CRC Press, **1989**. Chapter 3: 'Human Health Implications of Pyrrolizidine Alkaloids and Herbs Containing Them' by Ryan J. Huxtable, University of Arizona.
(* -Flora of the British Isles, Second Edition by A.R. Clapham, T.G. Tutin and E.F. Warburg. England; Cambridge University

Press, 1962, first edition 1952.)

"Three plant species in the genus Symphytum are relevant to the crop known as Comfrey.
*Wild or **Common Comfrey, Symphytum officinale L.**, is native to England and extends throughout most of Europe into Central Asia and Western Siberia (part of Russia).*
***Prickly or Rough Comfrey** [S. asperum Lepechin (S. asperrimum Donn)], named for its bristly or hairy leaves, was brought to Britain from Russia about 1800.*
***Quaker or Russian Comfrey [S. x uplandicum Nyman (S. peregrinum Lebed.)] originated as a natural hybrid of S. officinale L. and S. asperum Lepechin.** This hybrid was called Russian or Caucasian Comfrey in reference to its country of origin.*

> *Cuttings of this hybrid were shipped to Canada in 1954, and it was named Quaker Comfrey, after the religion of Henry Doubleday, the British researcher responsible for promoting Comfrey as a food and forage. The majority of Comfrey grown in the United States can be traced to this introduction."*

-'**Comfrey**' by Teynor, Putnam, Doll, Kelling, Oelke, Undersander and Oplinger, 'Alternative Field Crops Manual', University of Wisconsin: Cooperative Extension, University of Minnesota: Center for Alternative Plant & Animal Products, and Minnesota Cooperative Extension, February 1992, updated November **1997**, page 1.

<u>**Overview Comparison of Symphytum Species including Keys**</u> (in chronological order)

'**Identification Keys' to finding a plant name is a 2-branched system. You start at the top and are given 2 descriptions. You choose the one that matches your plant.** That leads you to the next description where there are again 2 choices. Your choices lead you to the name of the plant.

*"**Symphytum Key (identify species):***
1. Upper leaves decurrent on stem, without petioles.....go to 2.
>   *or Upper leaves tapering to petiole, without decurrent bands on stem (not decurrent).....go to 5.*

2. Corolla blue-violet or violet- sky blue.....go to 3.
>   *or Corolla pale yellow, calyx dissected nearly to base. Upper leaves indistinctly decurrent. Rhizome nodose, horizontal. (Carpathians).....go to 4.*

3. Entire plant coarse-hirsute (nearly prickly). Calyx 5-parted nearly to base, half as long as corolla, with lanceolate long-acuminate lobes; corolla violet (throughout European part of USSR), **then it is S. officinale L.**
>   *or Entire plant finely grayish-hirsute (villous-downy). Calyx cleft for 1/3 into slightly unequal obtuse teeth, 1/3 as long as the sky blue corolla. (Caucasus),* **then it is S. caucasicum M.B.**

4. Cauline leaves elliptic or oblong, cuneately tapering to petiole; radical leaves absent, **then it is S. tuberosum L.**
>   *or Cauline leaves ovate, the lower two petioled, the upper 1-2 sessile, short-decurrent. Radical leaves with cordate-rounded, large blade, long petioled,* **then it is S. cordatum W. et K.**

5. Corolla sky blue. Stem clinging, beset with retrorse, prickly bristles.....go to 6.
>   *or Corolla pale yellow. Lower leaves more or less cordate, oblong.....go to 7.*

6. Calyx cleft for 3/4 or nearly to base, very small, 1/4 to 1/5 as long as corolla, its teeth very short, obtuse, **then it is S. asperum Lepech.**
>   *or Calyx parted up to 3/4, 1/3 to 1/2 as long as corolla, its teeth broader and longer, acute. Otherwise hardly distinguishable from the preceding species. (Talysh),* **then it is S. peregrinum Ldb.**

7. Stem strongly branching, entirely soft-villous. Rhizome short, tuberiform, not perennial. Leaves more or less regularly disposed along the erect stem, ovate, usually cordate at base, softly gray-villous but not tomentose, rather numerous. Calyx villous, cleft for 1/2 to 3/4, **then it is S. tauricum Willd.**
>   *or Stems not branching, simple or branching only at base. Rhizome horizontal, long, perennial..go to 8.*

8. Stems soft-villous with spreading short hairs. Leaves tomentose-downy beneath. Calyx spreading-bristly, teeth shorter than its tube, obtuse; style very long, about 5 mm, exserted from corolla. Resembling S. caucasicum but leaves not decurrent, and corolla whitish. Weedy or cultivated plants, **then it's S. orientale L.**
>   *or Stems not soft-villous. Cauline leaves few. Calyx glabrous, ciliate only at margins, rarely bristly.....go to 9.*

9. Radical leaves orbicular-cordate, large, 10-15 cm across; cauline leaves ovate, subglabrous, very short-hairy above. Stems erect, not branching, finely downy, not bristly. (Carpathians), **then S. cordatum W. et K.**
>   *or Radical leaves broadly ovate, smaller, 5-10 mm long, faintly cordate or rounded at base. Pubescence poor, without coarse thick prickly bristles. Stems usually elongate, rooting, ascending, branching at base. (Colchis),* **then it is S. grandiflorum DC."**

-'**Flora of the U.S.S.R., Volume 19**' edited by V.L. Komarov, Leningrad, Russia, **1953**. The Borage family part was written by Mikhail Grigorevich Popov, pages 73 to 508 in English version, and pages 97 to 692 in Russian version. Symphytum is pages 207-216 in English, and pages 279-291 in Russian.

(Plant identification is the process of matching a specimen plant to a known taxon. One method is using a single-access key also called dichotomous key or sequential key. The sequence and structure of identification steps is fixed by the author of the key. At each point in the decision process, multiple alternatives are offered, each leading to a result or a further choice.)

"***The Key we provide makes use only of characters that may be readily seen in the field;*** *the teeth of the calyx, which are of*

considerable diagnostic value, may be observed with the naked eye but the use of a hand lens is advised.
We must emphasize that it is necessary to look at the whole plant as general habit is important, *and the relative size of leaf on different parts of the stem also helps in identification.*
Colour of flower may help, but, being variable, should not be relied upon as a primary character.
Most of these Comfreys are roughly hairy but **the type of hairiness is to some extent different in each species,** *proving a useful guide.*
Identification Key for London, England:
Stems much branched; lowest leaves the longest:
 Teeth of calyx at least equal to its tube:
  Upper leaves continuing in a wing down the stem (decurrent), then it is **Symphytum officinale.**
  Upper leaves not (or very slightly) decurrent:
   Upper leaves sessile; calyx teeth acute, then it is **Symphytum peregrinum.**
   Upper leaves shortly stalked; calyx-teeth obtuse, then it is **Symphytum asperum.**
 Teeth of calyx half length of tube, then it is **Symphytum orientale.**
Stems more or less simple; lowest leaves shorter than mid-stem leaves:
 Tall upright plants; basal leaves shortly stalked, then it is **Symphytum tuberosum.**
 Short plants with creeping barren shoots; basal leaves on long stalks, then it is **S. grandiflorum.**"
-'**The Comfreys of the London Area**' by E.B. Bangerter and B. Welch, Department of Botany, British Natural History Museum, London, England; The London Naturalist: London Natural History Society, England, Volume 33, pages 55-58, **1954**.

"**The main morphological differences between the three taxa which have been commonly recognised:**

	S. officinale (Common)	S. peregrinum (Russian)	S. asperum (Prickly)
Flower colour	yellow-white (rarely purple)	purple	blue
Calyx-teeth	linear-subulate	variable	triangular
Calyx	not accrescent	somewhat accrescent	strongly accrescent
Leaf base	strongly decurrent	somewhat decurrent	not decurrent
Hairs	stiff	variable	almost prickly
Anthers	longer than filaments	equalling filaments	shorter than filaments."

-'**The Genus Symphytum in Britain**' by T.G. Tutin, University College of Leicester, England; Watsonia: Journal of the Botanical Society of the British Isles, Volume 3, pages 280-281, February **1956**. (Watsonia was the journal of the 'Botanical Society of Britain and Ireland' from 1949 to 2010. It is now called 'New Journal of Botany'.)
 (Morphology is a branch of biology that studies the form of living organisms, and the relationships between their structures. This includes both the outward and internal appearance {shape, structure, color, pattern, size}.)
 (Calyx is the sepals of a flower forming a whorl that encloses the petals and forms a protective layer around a flower bud. Anther is the part of a stamen that produces pollen and is on a stalk. Stamen is the male fertilizing organ of a flower, consisting of a pollen-containing anther and a filament. A filament is the slender part of a stamen that supports the anther.)
 (Accrescent means growing continuously; in particular, growing larger after flowering and frequently refers to the calyx.)
 (Decurrent means a plant part that extends downward; it is usually applied to leaf blades that partly wrap or have wings around the stem or petiole and extend down along the stem.)
 (Thomas Gaskell Tutin, Fellow of the Royal Society, lived from 1908-1987. He was Professor of Botany at the University of Leicester in England. He was co-author of the 'Flora of the British Isles' and 'Flora Europaea'. The 1591-page 'Flora of the British Isles', published in 1952, is regarded as the standard work on the subject. In 1954 at the eighth 'International Botanical Congress', a group of British botanists formed an editorial committee, with Tutin as chairman. They spent the next twenty years creating the 2392-page, 5-volume 'Flora Europaea'.)

"Describes and illustrates the species of Symphytum known to be in cultivation.
'**Key to the Cultivated Species of Symphytum**'*:
1. Stems essentially unbranched; roots tuberous; sterile shoots decumbent, then it is **S. grandiflorum.**
or
1. Stems much-branched; roots thick or spindle-shaped or with spindle-shaped divisions, then go to 2.
2. Calyx not lobed below the middle.   If Corollas (flowers) white, then it is **S. orientale.**
            If Corrolla blue, then it is **S. caucasicum.**
2. Calyx lobed below the middle and usually almost to the base.
 If Corollas white, cream-coloured or yellow:
  3. If cauline leaves strongly decurrent, ovate to oblong-ovate, cordate to rounded at the base, then it is **S. tauricum.**
  3. If cauline leaves strongly decurrent, often forming a wing down the stem to the next lower node, oblong-lanceolate to oblong-ovate, then it is **S. officinale.**
 If Corollas rose-purple, or rose-pink changing to blue or purple:
  4. If cauline leaves strongly decurrent, forming a wing down the stem; flowers dull purple or rose; calyx not enlarging in fruit; anthers shorter than filaments, then it is **S. officinale.**
  4. If cauline leaves not or only very shortly decurrent on the stem; calyx mostly enlarging in fruit; anthers longer than the filaments or almost equal to them in length:

*5. If expanded flowers purple; calyx-lobes linear-lanceolate, acuminate; anthers and filaments about equal in length, then it is **S. x uplandicum**.*

*5. If expanded flowers blue; calyx lobes triangular, obtuse; anthers longer than the filaments, then it is **S. asperum**.'"*

-'**Key to the Cultivated Species of Symphytum**' by J. Ingram, 'Proceedings of Botanical Society of the British Isles, Volume 4' edited by D.H. Kent, London, England, (**1960**-1962), pages 445-446.

(* -'Studies in the Cultivated Boraginaceae. 4 & 5. Symphytum' by J. Ingram; Baileya: A Quarterly Journal of Horticultural Taxonomy, Ithaca, New York, Volume 9, pages 1-12, 56, 92-99, 1961. A comprehensive key is provided to identification of the genera of Boraginaceae in cultivation. I was unable to get this report. If you have it, could you please send it to me. This was the only excerpt I could find.)

"Gross Morphology:

	S. officinale L. Common Comfrey	*S. x uplandicum Nym.* Russian Comfrey	*S. asperum Lepech.* Prickly Comfrey	*S. tuberosum L.* Tuberous
Plant	Erect and hispid	Scabrid - hispid	Scabrid	Hispid
Root	Thick & tuberous	Thick & tuberous	Thick & tuberous	Swollen, tuberous
Stem	Distinctly winged throughout entire internode.	Narrowly winged, wings ending between nodes.	Not winged.	Not winged.
Leaves Shape	Ovate lanceolate	Ovate lanceolate	Ovate, elliptical	Oval to spatulate
Base	Broadly decurrent	Narrowly decurrent	Not decurrent	Decurrent
Lower	Long petiole	Long petiole	Long petiole	Long petiole
Upper	Sessile	Sessile	Very short petiole	Sessile
Flowers	Yellowish white, rarely pink or purple.	Blue or purplish blue.	Pink, changing to blue.	Yellowish white.
Habitat	By rivers and streams.	Roadsides, hedge-banks, not by water.	Waste ground	Damp woods

Leaves of *S. officinale* and *S. x uplandicum* cannot be distinguished by histological characters, being accurately identified only by the nature of the decurrent petiole when still attached to the stem.

Leaves of *S. tuberosum* may be distinguished from those of the other two species in whole form, due to their smaller size and softer texture; microscopically the epidermal cells of the midrib and petiole may possess a striated cuticle or beaded anticlinal (at right angles) walls."

-'**British Medicinal Species of the Genus Symphytum**' by K.R. Fell and Janet M. Peck, Pharmacognosy Research Laboratories, University of Bradford, England; Planta Medica: Journal of Medicinal Plant and Natural Product Research, Volume 16, No. 2, pages 208-216, May **1968**.

(Histology is the study of the microscopic structure of tissues.)

(Hispid: rough or covered with bristles, stiff hairs. Scabrid: somewhat rough in texture.)

(Wing: a thin, flat margin bordering or extending from a structure; a thin, leaf-like membrane. Decurrent: a leaf that extends down the stem below the insertion.)

(Ovate: egg-shaped, with broader end at the base. / Lanceolate: shaped like a spear-head, longer than wide, broadest above base and narrowed to apex. / Elliptical: oblong with rounded ends. / Spatulate: narrowing downward from a rounded summit; spoon-shaped.)

(Petiole: the stalk of a leaf that attaches to the stem. Sessile: no footstalk of any kind.)

"The maps of *Symphytum officinale* L. and *S. asperum* Lepech. published in the 'Atlas of the British Flora' and that of *S. x uplandicum* Nyman in the 'Critical Supplement to the Atlas of the British Flora' were inadequate because observers have been trying to fit four taxa and several taxonomic situations into only three pigeon holes.

	Symphytum Officinale	*Symphytum Asperum*	*Symphytum x uplandicum*
Flower Colour	Yellowish-white or reddish-purple.	Red in bud, clear blue in full flower.	Purple to purple-violet.
Leaf Base	Strongly decurrent, running beyond one node.	Non-decurrent petiolate (leaves cordate).	Not, or only slightly decurrent.
Hairs: Leaf/Stem	Soft.	Asperous, rasping. Stem short, stout, hook bristles.	Hispid.
Anthers	Longer than filaments.	Shorter than filaments.	Equalling filaments.
Chromosomes	2n = 36.	2n = 40.	2n = 36 (fide Nelson*)
Habitat	Fens, streams, river banks,	Roadsides.	Hedgebanks and roadsides.

	wet ditches.		
Distribution	Lowland Britain; rare in Ireland- status uncertain.	Rare, always a relic of civilization.	Throughout lowland Britain; local in Ireland.

There are some other species of Symphytum in this country which might be recorded in error, and a summary of their characters is given as a guide and warning:

S. orientale L.	Pure white flowers. Flowers early (April). Calyx teeth only half length of tube.
S. tuberosum L.	Yellowish-white flowers. Calyx teeth 3 times as long as tube. MIddle leaves of stem largest.
S. bulbosum K.Schimper	Yellow flowers with exserted filaments. Flowers early (March to April).
S. grandiflorum DC.	Creamy-yellow flowers with red buds. Only species with long slender rhizomes.
S. caucasicum Bieb.	Flowers pale blue above, white below. Leaves softly hairy, not decurrent.
S. tauricum Willd.	Almost certainly extinct.
S. leonhardtianum Pugsl.	May never have existed in this country (fide Nelson, and McClintock 1968)."

-'**Network Research, Symphytum Survey**' by Franklyn Perring, Biological Records Centre, Monks Wood, Huntingdon, Cambridgeshire County, England; Volume 7, No. 4, **1969**. In 'Proceedings of Botanical Society of the British Isles, Volume 7' edited by E.F. Greenwood, London, England, 1967-1969, pages 553-556.
(* George A. Nelson, M.Sc, Ph.D., M.P.S., F.L.S., Department of Pharmacology, School of Medicine, University of Leeds, England. The article did not give any reference to his research papers.)   (Fide is Latin for 'by authority of'.)

"**Symphytum L. Identification Key:**
1. Nutlets smooth, shining; connective projecting beyond thecae; filament about as wide as anther, then it is **Symphytum officinale**.
1. Nutlets minutely verrucose and +- reticulate-rugose; connective usually not projecting beyond thecae; filament narrower than anther, then go to 2.
2. Scales of corolla exserted:
    3. Rhizome creeping, with tubers; corolla (excluding scales) (7-)8-11(-12) mm, then it is **Symphytum bulbosum**.
    3. Stock fusiform; corolla (excluding scales) 5-7 mm, then it is **Symphytum ottomanum**.
2. Scales of corolla included:
    4. Plant aculeolate at least on stems and on mid-vein of the leaves beneath:
        5. Leaves all petiolate, or the upper subsessile but not amplexicaul or decurrent, then it is **Symphytum asperum**.
        5. Upper leaves sessile and shortly decurrent or at least amplexicaul, then it is **Symphytum x uplandicum**.
    4. Plant pubescent to rigidly hispid, but not aculeolate:
        6. Scales of corolla ungulate, not or slightly widened in the lower part, rounded or +- emarginate at apex, the marginal papillae dense above, sparse below:
            7. Rhizome creeping, slender, producing ascending or procumbent non-flowering and flowering stems, then it is **Symphytum ibiricum**.
            7. Rhizome not creeping, or stock fusiform; non-flowering stems absent:
                8. Calyx lobed to 1/4 to 2/5, then it is **Symphytum orientale**.
                8. Calyx lobed to at least 3/5:
                    9. Leaves shallowly cordate, rounded or truncate at base; corolla pale yellow; plant asperous, then it is **Symphytum tauricum**.
                    9. Leaves mostly +- narrowed at base; corolla white or white tinged with pink; plant +- softly hispid:
                        10. Corolla (17-)18-22 mm; scales 5-6 mm, then it is **Symphytum davisii**.
                        10. Corolla 13-16.5 mm; scales 1.6-4 mm, then it is **Symphytum cycladense**.
        6. Scales of corolla lanceolate, gradually narrowed from the dilated base to the +- acute apex, the marginal papillae dense throughout:
            11. Corolla violet, or pink turning blue, then it is **Symphytum x uplandicum**.
            11. Corolla pale yellow or white:
                12. Corolla white; stock fusiform, then it is **Symphytum naxicola**.
                12. Corolla pale yellow; rhizome creeping:
                      13. Lower leaves cordate; rhizome of +- even thickness throughout, then it is **Symphytum cordatum**.
                    13. Leaves not cordate; rhizome with alternate thick tuberous and thin portions:
                          14. Connective not projecting beyond thecae, then it is **Symphytum tuberosum**.
                          14. Connective projecting slightly beyond thecae, then it is **Symphytum gussonei**."

-'**Symphytum L.**' by B. Pawlowski with editors Tutin, Heywood, Burges, Moore, Valentine, Walters and Webb, Flora Europaea,

Volume 3, pages 103-105, **1972**. The 'Flora Europaea' is a 5-volume encyclopedia of plants, published between 1964 and 1993 by Cambridge University Press, England. It describes all the national floras of Europe to identify any wild or widely cultivated plant to the subspecies level. It provides geographical distribution, habitat preference, and chromosome number.

"Faegri (*1931) regarded the relationship between the four species mentioned (S. officinale L., S. uliginosum Kern., S.asperum Lepech., S. peregrinum Ledeb.) as so very close that he decided to unite them as subspecies in one collective species; S. commune Faegri**.
> *Their close relationship was confirmed by hybridisation experiments (Gadella 1972, Gadella & Kliphuis 1973, 1978). Especially the production of somewhat fertile hybrids is indicative of a close affinity."*

-'**Cyto- and Chemotaxonomical Studies on the Sections Officinalia and Coerulea of the Genus Symphytum**' by Th.W.J. Gadella, E. Kliphuis and H.J. Huizing; Botanica Helvetica (now called Alpine Botany), Volume 93, pages 169-192, **1983**.
(* -'Uber in Skandinavien Gefundenen Symphytum-Arten: Nebst einigen Betrachtungen uber das Artproblem Innerhalb der Betreffenden Artgruppe' {About Symphytum Species Found in Scandinavia} by Knut Faegri, Bergens Museum Arbok, Norway, 47 pages, 1931. In German. If anyone has an English translation of it, I would appreciate having it.)
(** -Boraginaceae Symphytum Commune Faegri {Bergens Museum Arbok 1931, Naturvidensk. Rekke, No. 4, 30, 1932}.)

*"**Previous research described the use of pyrrolizidine alkaloids and triterpenes as chemotaxonomical markers within the Symphytum officinale (Common Comfrey) species complex and S. asperum (Prickly Comfrey).***
> *The alkaloid echidimine predominates in S. asperum and is present in the artificial hybrids between S. officinale and S. asperum known as S. x uplandicum Nym (Russian Comfrey).*
> *In 1/5 of the diploid plants (2n = 24) of S. officinale echimidine was found.*

The amount of echimidine found in S. officinale, however, is significantly lower then in S. x uplandicum and S. asperum where all plants contained echimidine *(data not published)."*
-'**Chemotaxonomy of the Symphytum Officinale Agg. (Boraginaceae)**' by Jaarsma, Lohmanns, Gadella and Malingre, Plant Systematics and Evolution, 167, pages 113-127, **1989**.
> (A cytotype is an individual of a species that has a different chromosomal factor to another.)

*"In a study of 16 forms belonging to five Symphytum species at the Pavlovsk Experiment Station in Leningrad province, Russia during 1985-1987, **green matter yield was highest in S. uplandicum (Russian Comfrey) form K16 (mean of 11.33 kg/m2). This species showed poor regrowth after the first cut; the best regrowth was shown by S. carpaticum and S. officinale. S. asperum (Prickly Comfrey) had the highest protein content. This species, together with S. uplandicum and S. carpaticum, appeared the most promising for breeding and growing.***"
-'Main Comfrey Species in the VIR Collection' by T.S. Selivanoa, Vavilov Institute of Plant Genetic Resources (VIR), Leningrad, USSR; Nauchno Tekhnicheski Byulleten (Scientific Technical Bullentin), No. 198, pages 49-51, **1990**.
> (This report says initial growth is better with Russian Comfrey but slower after cutting. I'm not sure why plant growth would differ between early spring growth and later growth after cuttings. I'm not sure what S. carpaticum is in English. It is mentioned in some foreign language sites.)

*"**One of the major problems to be faced in trying to understand Comfrey, either for the layman, the farmer, the specialist, the agriculturalist, the botanist or the chemist, lies in the lack of accurate scientific reports,** even in the various encyclopedias of the world.*
This is largely due to the fact that for nearly the first half of this century (1900-1950), very little research was done on Comfrey and the relationship between the various known types. It is not surprising then to find that many misleading and inaccurate statements have been published."
-Comfrey: Nature's Healing Herb & Health Food by Andrew Hughes. Japan: Sanyusha Publishing Co., Ltd, **1992**, page 132.

*"**Symphytum asperum / S. officinale Complex:***
The identification of plants in this complex (Perring 1994) is made difficult by the wide range of hybrids between these two species which have either arisen here or been introduced as forage plants and have subsequently introgressed with native populations.
For the purpose of recording, seven taxa or forms may usefully be recognised but some specimens will be difficult to place."
-'**Plant Crib: Handbook for Field Identification**' compiled by T.C.G. Rich and M.D.B. Rich with editorial assistance of F.H. Perring for the BSBI Monitoring Scheme. London, England: Botanical Society of the British Isles, 1988, page 74. There is an updated **1998** edition.
> (A species 'complex' is a group of closely related species that are extremely similar in appearance, so much so that the boundaries between them are frequently unclear.)
> (A taxon is a taxonomic group of any rank, such as a class, family, or species.)
> (Hybrid or crossbreed is the result of combining the qualities of two organisms of different breeds, varieties, species or genera through sexual reproduction.)
> (A hybrid swarm is a group of hybrids that have survived past the first hybrid generation, with interbreeding taking place between hybrid individuals and also backcrossing with its parent type. These plants are highly variable, with genes and phenotypes of individuals ranging widely between the two parent types. Hybrid swarms blur the boundary between the parent taxa.)
> (A backcross is a cross of a hybrid with one of its parents or an individual genetically similar to its parent. This creates

offspring with genetics closer to that of the parent.)
(Introgression is the transfer of genetic information from one species to another as a result of hybridization between them and repeated backcrossing.)

"**Symphytum asperum Lepechin (Prickly Comfrey), S. ibericum Steven, and S. sylvaticum Boiss. were examined morphologically, micromorphologically and anatomically.** Scanning electron microscopy was used to examine leaf surface and trichomes of these species.
These species had bifacial and hypostomatous leaf types. Epidermal cells of leaves were usually polygonal or irregular in form. The pattern of anticlinical cells may vary in different species and between the upper and lower epidermis of the same species. Stomata are anisocytic and anomocytic in three species. Stomata index is 27.5 for S. sylvaticum, 24.65 for S. ibericum and 21.86 for S. asperum.
Glandular trichomes are capitate in forms and more dense on the lower epidermis than upper epidermis. Eglandular trichomes are simple, short or long, unicellular or multicellular and thin or thick."
-'**Micromorphology and Anatomy of Three Symphytum (Boraginaceae) Taxa from Turkey'** by Oznur Ergen Akcin and Hilal Baki, Department of Biology, Sciences & Arts Faculty, Ordu University, Ordu, Turkey; Bangladesh Journal of Botany, Volume 36, No. 2, pages 93-103, December **2007**.
> (Trichome is a small hair or other outgrowth from the epidermis of a plant. Epidermis is the outer layer. Stomata is plural of stoma. Stoma is a tiny pore in the epidermis of a leaf that lets gases moves in and out.)

"**Most Comfreys are clump-forming herbaceous perennials that creep slowly via rhizomes.**
S. ibericum (to 40 cm = 1 foot 4 inches high) is evergreen and spreads faster, with stems rooting as well, making great low cover.
S. officinale (Common Comfrey) grows 1.2 to 1.5 meters (4 to 5 feet) high and 60 to 100 cm (2 feet to 3 feet 3 inches) wide.
S. orientale 90 cm (3 feet) high and 50 cm (1 foot 8 inches) wide, also evergreen.
S. tuberosum (Tuberous Comfrey) 40 to 60 cm (1 foot 4 inches to 2 feet) high and wide.
S. x uplandicum (Russian Comfrey) to 2 meters (6 feet 6 inches) high and 1 meter (3 feet 3 inches) wide."
-**Creating a Forest Garden:** Working with Nature to Grow Edible Crops by Martin Crawford. England: UIT Cambridge Ltd., **2010**, page 609.
> (I'm not sure what he means by 'evergreen'. I have seen S. ibericum listed as 'semi-evergreen' with hardiness zone 4-9. And S. orientale listed as 'semi-evergreen' with hardiness zone 3-9.)

"***Symphytum* L.**, *Sp. Pl.* 1: 136. 1753; *Gen. Pl.*, ed. 5: 66. 1754.
Key to Species:
1. Plant more than 50 cm (19.6 inches) high, with a fusiform, oblique or vertical rhizome; corolla white, red-purple, pink or pale blue, then go to 2.
or
1. Plant less than 35 cm (13.7 inches) high, with a horizontal, tuberous rhizome, more or less uniformly thickened or thin with spaced bulb-like thickenings; corolla pale yellow, then go to 5.
2. Leaf blade largely ovate; corolla infundibuliform, then it is **S. orientale**.
2. Leaf blade elliptic-lanceolate; corolla clavate, the limb campanulate or urceolate, then go to 3.
> **3.** Stem at least partially winged; calyx 6-10 mm long; nutlets smooth, then go to 4.
> **3.** Stem unwinged; calyx less than 4-5 mm long; nutlets rugose, then it is **S. asperum**.
>> **4.** Stem winged throughout, with wings 2-8 mm broad; corolla limb campanulate, then it is **S. officinale**.
>> **4.** Stem only partially winged, with wings less than 2 mm broad; corolla limb urceolate, then it is **S. tanaicense**.
>>> **5.** Corolla 10-15 mm long; faucal scale distinctly exerted, then it is **S. bulbosum**.
>>> **5.** Corolla 15-20 mm long; faucal scales included, then go to 6.
>>>> **6.** Rhizome slender, with distinctly spaced, tuberous thickenings; leaf blade contracted or attenuate at base, never decurrent; filament as long as the anther, then it is **S. gussonei**.
>>>> **6.** Rhizome stout, more or less uniformly enlarged; leaf blade decurrent on stem, sometimes prolonged in short wings; filament less than 1/2 of the anther length, then it is **S. tuberosum**."

-'**Synopsis of Boraginaceae subfam. Boraginoideae Tribe Boragineae in Italy'** is the same as 'Boraginaceae in Italy II' by L. Cecchi and F. Selvi, University of Florence, Firenze, Italy; Plant Biosystems: An International Journal Dealing with all Aspects of Plant Biology, Official Journal of the Societa Botanica Italiana, Vol 149, No. 4, pages 630-677, **2015**, Symphytum pages 636-646.

"**It can be very difficult to distinguish between different species of Comfrey** as all the species that grow in Finland share common habitats.
Generally speaking, the best marker is the length of the calyx lobes, and their form, the form and colour of the corolla, how decurrent the leaves are, and the surface of the carpels can also be helpful."
-'**Russian Comfrey: Symphytum x uplandicum**', NatureGate, University of Helsinki, Finland, www.luontoportti.com, **2019**. Eija and Jouko Lehmuskallio started documenting nature and developing identification services in the 1990s.

Overview of Russian Comfrey No. 4 versus No. 14 versus Officinale

"It is probably possible for a record yield of Comfrey to be secured from a single Bocking No. 4, No. 14 or No. 16, just as there are record yields of potatoes from a single seed tuber in contests organized by gardening periodicals that are won every year by the same people.
What counts on the farm is the ability of clones to stand up to cutting year after year and average a worthwhile yield in good seasons and bad."
-Comfrey: Fodder, Food and Remedy by Lawrence D. Hills. New York: Rizzoli Universe Books: 1976, page 75.

Russian Comfrey Bocking No. 4

Pros: Somewhat deeper roots than Russian Comfrey No. 14. Better tasting than No. 14 because it has less potash (potassium): average 4%. Slightly more protein than No. 14.
Cons: Starts growth in spring later than No. 14. Less allantoin (0.34%) than No. 14.
Thicker stems than No. 4 (a negative if using for compost, liquid fertilizer or drying).
Less potash than No. 14 (a negative when using as fertilizer).
More prone to rust than No. 14 but this problem is rare.
Neither Pro or Con: Very small wings on stems. The flower colour is Bishops Violet 34/3 (Royal Horticultural Society's colour chart, Great Britain). The leaves are broad and round tipped with unserrated edges. Stems are solid.
    (A plant fin or wing is a green, short, thin, paperlike structure extending along a stem. It is long along the stem but not tall away from the stem.)   (Potash/potassium varies with the season and soil.)

Russian Comfrey Bocking No. 14

Pros: Starts growing earlier in spring than Russian Comfrey No. 4. More rust resistant than No. 4. Highest in allantoin (0.44%). High in potash (potassium): average 7%.
Best for compost and liquid manure because high potash and thinner stems.
Better for drying than No. 4 because it has thinner stems.
Cons: Roots do not go as deep as No. 4 but both are very deep.
Does not taste as good as No. 4 because it has higher potassium. Slightly less protein.
Neither Pro or Con: No fins or wings on stems. Flowers are Imperial Purple 33/3 fading to Lilac Purple 031/3. The leaves are pointed, slightly serrated at the edges.

Symphytum officinale

Thick stems (most hollow). It has much larger fins/wings along the stalk than Russian Comfrey.
More prone to rust than Russian Comfrey. Lower yield than Russian Comfrey.
Lowest potash (potassium) at 3.09 percent so better tasting but not as good for fertilizer.

The Comfrey I currently sell on my website www.nantahala-farm.com is: Common Comfrey, Russian Comfrey No. 4, Russian Comfrey No. 14, Prickly Comfrey, and Symphytum Hidcote Blue.

Symphytum Tauricum 1740

'Plantarum Minus Cognitarum Centuria', Volume 5, by Johann C. Buxbaum, Petropoli (Saint Petersburg), Russia, 1740. Symphytum tauricum tab 68.

Buxbaum was born in Germany in 1693. He was a physician, botanist, and entomologist. In 1721 he moved to Russia to be a botanist in the Physical Garden, at Medical Collegium in Saint Petersburg.

Chapter 7

Details about Symphytum Species:
Asperum or Asperrimum (Prickly Comfrey)

Symphytum asperrimum (asperum) Species (Prickly or Rough Comfrey)
See subsection 'Prickly Comfrey as a Forage Crop' in section 'Livestock' in Volume 2.
See subsection 'General Cultivation of Prickly Comfrey' in section 'Care of Plant Overview and How to Propagate' in Volume 2.

Current Botanical Nomenclature

    Accepted Name or Synonym
        Symphytum asperum Lepech.
        Symphytum asperum var. armeniacum (Bucknall) Kurtto
        Symphytum asperrimum Donn ex Sims
        Symphytum armeniacum Bucknall.
        Symphytum echinatum Ledeb
        Symphytum peregrinum Ledeb.

    Not Accepted Name
        Symphytum asperrimum Donn
        Symphytum asperrimum Sims
        Symphytum asperrimum Urv. = Symphytum anatolicum Boiss.
        Symphytum majus Gueldenst.
        Symphytum majus Gueldenst. ex Ledeb.

"***Symphytum asperrimum d'Urv.*** *(Mem. Soc. Linn. Paris 1: 276. 1822)*
Symphytum asperrimum Donn *(Bot. Mag. 24: t. 929. 1806) (Curtis's Botanical Magazine)*
Symphytum asperum Lepech. *(Nova Acta Acad. Sci. Imp. Petrop. Hist. Acad. 14 2: 442, t. 7. 1805)*"
-'International Plant Names Index'® (IPNI), www.ipni.org, A database of the names and associated basic bibliographical details of seed plants, ferns and lycophytes, **2018**.
    (Lepech.= Ivan Ivanovich Lepechin from Russia, 1737-1802, was a naturalist, zoologist, explorer, and botanist. He was in charge of the Saint Petersburg Botanical Garden, Russia.)
    (Donn = James Donn from England, 1758-1813. In 1796 he published 'Hortus Cantabrigiensis', a catalog of plants grown in the 'Botanic Garden of Cambridge University', England, where he was Curator from 1790 until his death. He was a founder member of the Linnean Society, London, England.)

"***Symphytum asperum Lepech.*** *is an accepted name.* **Synonym: Symphytum armeniacum Bucknall.**
Symphytum asperrimum Donn ex Sims *is an accepted name. No synonyms.*"
-The Plant List®, www.theplantlist.org from the World Checklist Database, **2018**.
    (Sims = John Sims from England, 1749-1831. He was the editor of 'Curtis's Botanical Magazine', London, England from 1801 to 1807. He edited 'Annals of Botany' from 1805 to1806 with Charles Konig. He was a founder member of the Linnean Society, London, England.)
    ('ex' means an initial description did not satisfy the rules for valid publication, but the same name was subsequently validly published by a second author or authors.)

"*Boraginaceae* **Symphytum armeniacum Buckn.** *(J. Linn. Soc., Bot. 41: 520. 1913) basionym of: Boraginaceae Symphytum asperum Lepech. var. armeniacum (Buckn.) Kurtto (Ann. Bot. Fenn. 19 (3): 186. 1982).*"
-'International Plant Names Index'® (IPNI), www.ipni.org, A database of the names and associated basic bibliographical details of seed plants, ferns and lycophytes, **2018**.
    (Basionym or basyonym is the original name on which the new name is based.)

"***Symphytum armeniacum, sp. nov. (new species):***
Geographical Distribution: **Russia: Georgia, Caucasus. Turkey: Armenia.**
When I first recognised this as an undescribed species, and as distinct from the other members of the Caerulea genus, the specimen in the Cambridge, England Herbarium was the only one that I had seen.
This was collected in Armenia by Calvert (No. 222), at some time previous to 1866. As far as I was aware, no other specimen existed either in the British or Genevan, Switzerland, collections, until, on examination of the Fielding Herbarium, Oxford, England, I found several plants which were undoubtedly the same.
One of these is Calvert's No. 222; three others, collected in Georgia by the Hohenackers in 1831 and 1838, are named,

one of them S. asperrimum, and two others S. caucasicum.
As other plants so named by the same collectors are typical, this is an instance of the confusion which arises when two names have to serve for three different species.
S. armeniacum is most nearly allied to S. asperum and S. sepulcrale, *but differs from both in the sessile clasping leaves; from S. asperum it is clearly distinguished by the larger calyx with elongated linear segments, and from S. sepulcrale by the more setose and prickle-like clothing, and, by the more deeply divided calyx.*
It need not be confused with S. caucasicum*, which has the lower leaves gradually attenuated into the petiole, the upper partially decurrent and the calyx-segments shorter than the tube."*
-'A Revision of the Genus Symphytum, Tourn.' by Cedric Bucknall, {Mus. Bac. Oxon= Bachelor of Music, Oxford University}, Journal of the Linnean Society of London, England, Botanical Journal, Volume 41, Issue 284, pages 491-556, December **1913**.

"Symphytum asperum Lepech.
   Synonym: **Symphytum asperrimum Donn ex Sims***: Not accepted."*
-ITIS (Integrated Taxonomic Information System®) Report, www.itis.gov, **2018**. ITIS is a partnership of United States Federal agencies formed to satisfy their mutual needs for scientifically credible taxonomic information.

"Symphytum asperum Lepechin.
   Synonym: Symphytum armeniacum Bucknall.
   Czerepanov, S. K. 1981. Sosud. Rast. SSSR 509 pages. Nauka, Leningradskoe Otd-nie, Leningrad."
-Tropicos®, www.tropicos.org, Missouri Botanical Garden, Saint Louis, Missouri; Plant database with 1.3 million scientific names and 4.4 million specimen records, **2018**.

"Symphytum asperum Lepechin = Symphytum asperrimum Donn ex Sims."
-United States Department of Agriculture, Natural Resources Conservation Service®, Plants Database®, www.plants.usda.gov, **2018**.

"Symphytum asperum Lepechin*: Accepted scientific name:*
Synonyms:
   Symphytum asperrimum Sims (ambiguous synonym)
   Symphytum asperum var. armeniacum (Bucknall) Kurtto (synonym)
   Symphytum echinatum Ledeb. (synonym)
   Symphytum majus Gueldenst. ex Ledeb. (ambiguous synonym)
   Symphytum peregrinum Ledeb. (synonym) "
   Symphytum asperrimum Urv.: ambiguous synonym for Symphytum anatolicum Boiss."
-Catalogue of Life®: 2015 Annual Checklist, Online database of the world's known species of animals, plants, fungi and micro-organisms. It consists of a single integrated species checklist and taxonomic hierarchy., **2018**, www.catalogueoflife.org. Search: www.catalogueoflife.org/annual-checklist/2015/search/all

"Symphytum asperum Lepech.:
   = Symphytum armeniacum Buckn.
   = Symphytum asperrimum Sims
   = Symphytum asperum var. armeniacum (Bucknall) Kurtto
   = Symphytum echinatum Ledeb.
   = Symphytum majus Gueldenst.
   = Symphytum majus Gueldenst. ex Ledeb.
   = Symphytum peregrinum Ledeb."
-Global Biodiversity Information Facility® (GBIF), www.gbif.org, **2018**. International network and research infrastructure funded by the world's governments and aimed at providing anyone, anywhere, open access to data about all types of life on Earth.

Symphytum Asperum is Correct Botanical Name

"Symphyti Asperi Nova Species: *descripta ab Ioanne Lepechin. Conventui exhibita d. 20 Jan. 1802."*
(Symphytum Asperum New Species: *written by John Lepechin. The meeting will be presented February 20 1802.)*
-'Nova Acta Academiae Scientiarum Imperialis Petropolitanae, Praecedit Historia Ejusdem Academiae, Tomus 14' (Petersburg Imperial Academy of Sciences, Volume 14) by Imperatorskaia Akademia Nauk (Russia), Part 2, **1797**-1798. Includes article in Latin by **Ivan Lepechin** about Symphyti Asperi on pages 442-444.
(Boraginaceae Symphytum asperum Lepech.: Nova Acta Acad. Sci. Imp. Petrop. Hist. Acad. 14(2): 442, t. 7. 1805.)
(Ivan Ivanovich Lepyokhin = Ivan Lepekhin = Ivan Ivanovich Lepechin, 1737 or 1740 to 1802. The standard author abbreviation 'Lepech.' indicates him as the author when citing a botanical name.)
(See sub-subsection 'Lepechin 1797' in subsection 'Symphytum Genus' in section 'Borage Family, Symphytum Genus' {Chapter 2}.)

"In 1913 when dealing with the nomenclature of Symphytums it was decided to re-christen Symphytum asperrimum to

Symphytum asperum which is largely ignored by agriculturalists and so remains merely a synonym not used."
-Comfrey: Symphuo Symphytum: A Multi-Purpose Herb by Philip Clarke. Edinburgh, Scotland: Pentland Press, 1997, page 1.

*"In checking up the determinations and nomenclature (name) of the genus Symphytum in the 'Gray Herbarium' (Harvard University, Cambridge Massachusetts), it has come to my notice that **the rather generally introduced plant, that has gone under the name S. asperrimum Donn, must be known, for reasons of priority, as S. asperum Lepechin, as indicated by the following citations:***
    S. asperum Lepechin, Nov. Act. Acad. Petrop. xiv. 444, t. 7 (1805).
    S. asperrimum Donn in Sims, Bot. Mag. t. 129 (1806).
The work in which Lepechin published is in the library of the 'American Academy of Arts and Sciences' (Cambridge, Massachusetts). He gives a good description and also a fair plate (image), which show that there is no doubt as to the identity of his plant and that of Donn, published a year later.
Since Lepechin, the authority for the binomial which must be revived, *is a name not familiar to most American botanists, it may not be out of place to mention that* **he was a professor of botany and director of the 'Imperial Gardens at St. Petersburg' (Russia) during the last half of the eighteenth century (late 1700s).**
His name was connected with American botany, fourteen years after his death, by Willdenow's publication in 1816 of the genus Lepechinia, a group of Mexican mints."
-'The Correct Name of an Introduced Symphytum' by J. Francis Macbride; Rhodora Journal of the New England Botanical Club, Inc., Volume 18, No. 205, pages 23-25, January **1916**.

> (Herbarium: A collection of dried plant specimens mounted on sheets of paper. The plants are collected, identified by experts, pressed, and then mounted to archival paper so that all major morphological characteristics are visible. They are labeled with the scientific name, name of the collector, and information about where/how collected. The specimens are filed according to families and genera.)

"Symphytum asperum was formerly Symphytum asperrimum Donn, *and it is illustrated in color and described in 'Curtis's Botanical Magazine', Plate 929.*
Symphytum peregrinum (Russian Comfrey) is illustrated and described 73 years later (*Plate 6466)."
-Comfrey Report No. 1: The 1954 Research Results by Lawrence D. Hills, Henry Doubleday Research Association, Braintree, Essex, England, **1955**.
(* -S. peregrinum, Ledeb., Hooker in Journ. Bot. ix. p. 57, 1880, and Bot. Mag., t. {plate} 6466 1879.)

<u>Subspecies and Varieties of S. Asperum</u> (views about this taxonomy have changed over time)

S. asperum var. armeniacum (Buckn.) Kurtto
    S. sepulcrale Boiss. & Balansa ex Boiss. (Fl. Orient. [Boissier] 4(1): 174. 1875)
    S. sepulcrale Boiss. & Balansa ex Boiss. var. hordokopii Kurtto (Ann. Bot. Fenn. 19(3): 188. 1982)
    S. sepulcrale Boiss. & Balansa ex Boiss. subsp. sylvaticum (Boiss.) Kurtto (same as above)
S. sylvaticum Boiss. subsp. sepulcrale

See 'Symphytum sepulcrale' and 'Symphytum sylvaticum' in Chapter 9.

*"**Variegated Leaves:** From some correspondence in the 'Revue Horticole: Journal d'horticulture Practique', France, we condense the following remarks relating to the production of Varieties of Pelargonium. M. Lemoine asserts that his variegated Pelargoniums were all produced from seed; on the other hand he says that this variegation of the leaves is not perpetuated by dividing the root; plants so produced having always green leaves, even if the original stock were variegated.*
In illustration, he cites the variegated Symphytum asperrimum, which when propagated by parting the root, produces green leaves only."
-'The Gardeners Chronicle and Agricultural Gazette: A Newspaper of Rural Economy and General News', London, England, page 74, **1867**.
> (Variegated means leaves that are edged or patterned in a second color, especially white.)

"Symphytum asperum Lepechin was described in 1805 *from specimen(s) cultivated in the 'Botanical Garden of the Academy of Sciences in St. Petersbourg' (Leningrad, Russia). The seeds orginiated from the Caucasus ('Habitat in jugo montium Caucasi Rossici').*
A closely allied taxon was later described by Boissier (*1875) *from the collections of **Balansa from Lazistan in northeast Turkey under the name of* **S. sepulcrale Bliss. & Bal.**
> **Since then, S. sepulcrale has either been treated as an independent species or included in S. asperum without any taxonomic rank.**

Kusnetsov (1910) and Bucknall (1913) regarded S. asperum and S. sepulcrale as distinct species, mainly on account of differences in the calyx, indumentum (hairy bristles) and leaf characters.
Bucknall (1913) also described a new species, S. armeniacum Buckn., *from Caucasian and Armenian material. This species was closely related to the other two, but differed from S. asperum in the longer calyx, from S. sepulcrale in the 'more setose and prickle-like clothing and the more deeply divided calyx,' and from both species in the sessile, clasping leaves.*

> Wickens (1969) followed the treatment of Bucknall (1913). Later Wickens (1978) excluded from the combined description some characters earlier considered by him peculiar to S. sepulcrale, especially the softer indumentum,the calyx equalling the corolla tube, the violet corolla and the smooth or minutely tuberculate nutlets.
> **Gviniasvili (1976), who studied extensive material of S. asperum, chiefly from the Caucasus, included S. armeniacum in S. asperum and seemed to consider the species rank of S. sepulcrale doubtful."**

-'Taxonomy of the Symphytum Asperum Aggregate (Boraginaceae), Especially in Turkey' by Arto Kurtto, Botanical Museum, University of Helsinki, Finland; Annales Botanici Fennici (Finnish Botanical Annals), Helsinki, Vol 19, No. 3, page 177-192, **1982**.

(* -'Flora Orientalis Sive Enumeratio Plantarum in Oriente, a Graecia et Aegypto ad Indiae Fines {Plants Flora, Flora East and the East, from Greece and Egypt to the Borders of India}, Volume IV' by Edmond Boissier, Society Physics Geneva, Society Linnean London; Geneva, Switzerland, 1875, page 174. All in Latin.)

(** -Gaspard Joseph Benedict Balansa, also known as Benjamin Balansa or Benedict Balansa,1825-1891, was a French botanist. He collected specimens for the Museum National d'Histoire Naturelle, Paris, France.)

> ("Habitat in jugo montium Caucasi Rossici' = Lives in the yoke of the Caucasus Mountains Rossici.)
> (Species aggregate is a species group that is closely related and difficult to distinguish among them. Similar but not identical is a single species that displays large variations but can not strictly be subdivided: s.l. for 'sensu lato' {in a wider/broader sense} or s. ampl. for 'sensu amplo' {in a very wide/relaxed sense}.)

"**Symphytum armeniacum Buckn.** differs slightly from S. asperum in the indumentum, leaf shape and floral characters. It seems, however, to deserve only infraspecific rank and is treated as **S. asperum var. armeniacum (Buckn.) Kurtto, comb, et stat. nov.**
Symphytum armeniacum, S. sepulcrale and S. sylvaticum are lectotypified."

-'Taxonomy of the Symphytum Asperum Aggregate (Boraginaceae), Especially in Turkey' by Arto Kurtto, Botanical Museum, University of Helsinki, Finland; Annales Botanici Fennici (Finnish Botanical Annals), Helsinki, Vol 19, No. 3, page 177-192, **1982**.

> (Infraspecific is at a taxonomic level below that of species, e.g., subspecies, variety, cultivar, or form. Latin names at this level require the addition of a term denoting rank.)
> **(S. asperum var. armeniacum (Buckn.) Kurtto, comb, et stat. nov. = Symphytum asperum variety armeniacum, originally published by Bucknall, then revised publication by Kurtto, combination, and new status.)**
> (Lectotype is a specimen chosen as the type of a species if the author of the name fails to designate a type.)

"**Symphytum asperum Lepechin and S. sylvaticum Boiss. species are taxonomically very difficult to solve.** Many taxonomical arrangements were done by different researchers before with observing herbarium sheets.
In our research 25 populations were observed in field, and **we reached the decision that both of these species' infraspecific taxa should be synonymized (made the same)** because of the continuation of the characters used to distinguish taxa before.
S. asperum Lepechin var. armeniacum (Bucknall) Kurtto proposed by Kurtto was synonymed and included in S. asperum Lepechin.
S. sylvaticum Boiss subsp. sepulcrale var. sepulcrale and **S. sylvaticum Boiss subsp. sepulcrale (Boiss.&Bal.) Greuter&Burdet var. hordokopii (Kurtto) R.Mill.** were synonymed and included to S. sylvaticum.
The reasons of these taxonomical rearrangements were gathered from field observations, morphological measurements and scanning electron microscope images."

-'Taxonomy of Symphytum Asperum Lepechin and S. Sylvaticum Boiss. (Boraginaceae) Based on Macro- and Micro-Morphology' by Burcu Tarikahya and Sadik Erik, Hacettepe University, Department of Biology, Ankara, Turkey; Hacettepe Journal of Biology and Chemistry, Volume 38, No. 1, pages 47-61, **2010**.

History of Prickly Comfrey in Great Britain, Ireland and United States
See 'History' section, Chapters 12-16.

"*Caucasian Prickly Comfrey (Symphytum asperrimum):*
This perennial plant, introduced into England in the year 1790, has prominently attracted public attention during the last few years, as a forage plant, to be grown on a large scale, as a food for horned cattle, horses, sheep, pigs, and all kinds of poultry.
Its wonderfully quick growth, and fine cropping qualities, and its power of resisting drought, are making it a favourite with farmers and dairymen. With but little labour, beyond cutting the green leaves, Prickly Comfrey will yield, year after year, from 60 to 150 tons of forage per acre.
It comes up early in the spring, and may usually be cut for the first time in April, and will then continue yielding through the hottest and dryest summers, until the hard winter frosts cut down its foliage. It will grow well on clay, even the thickest and coldest, on loam, gravel, or pure sand."

-'New Commercial Plants with Directions How to Grow Them to the Best Advantage, No. 1 (1878) and No. 3 (1880)' by Thomas Christy, F.L.S. (Fellow of Linnaean Society), London, England. Number 1 includes Caucasian Prickly Comfrey pages 14-16. Number 3 includes Prickly Comfrey pages 11-14. Includes 'Caucasian Prickly Comfrey' and 'Russian Comfrey' advertisements.

"*S. asperum Lepechin (S. asperrimum Donn)* with $2n = 32$, *is native to the Caucasus.*
In Britain, Donn (*1831) recorded it as having been **first grown in the Cambridge Botanic Garden (England) in 1801** *and cultivated as a forage crop a few years later.*

*But according to Aiton (**1810) **it was introduced in 1799 by Conrad Loddiges from St. Petersburg in Russia."***
-'Comfrey Symphytum spp. {species} as a Forage Crop' by J.C. Forbes, A.D. McKelvie, and P.J.C. Saunders, North of Scotland College of Agriculture, Aberdeen, United Kingdom; Herbage Abstracts, Volume 49, No. 12, pages 523-539, 1979.
(* -'Hortus Cantabrigiensis: or An Accented Catalogue of Indigenous and Exotic Plants Cultivated in the Cambridge Botanic Garden, Thirteenth Edition' by James Donn, Curator, Fellow of the Linnean and Horticultural Societies, London, England, 1845. First published in 1796.)
(** -'Hortus Kewensis; or a Catalogue of the Plants Cultivated in The Royal Botanic Garden at Kew: Volume I' by William Townsend Aiton, London, England, 1810.)
  (The Caucasus Mountains are located at the border of Europe and Asia, between the Black Sea and Caspian
  Sea and occupied by Russia, Georgia, Azerbaijan, and Armenia.)

"The Prickly Comfrey, although first introduced into England, has not much engaged the attention of English agriculturists, but it appears to be extensively cultivated in several parts of Ireland.
It was extensively cultivated by the late Bishop of Kildare, Ireland, in a field at Glasnevin (suburb of Dublin, Ireland), where the plant is still found, growing up as a persistent weed, in spite of every attempt to eradicate it.
Many gentlemen, after the example of the Bishop of Kildare, who was a first-rate dairy-farmer, are reported to cultivate it in their villa or suburban farms around Dublin, and find it a very useful food for their dairy stock."
-'The Journal of the Royal Agricultural Society of England, Second Series, Volume 7', London, England, **1871**. Includes article 'XV: On the Composition and Nutritive Value of the Prickly Comfrey (Symphytum asperrimum)' by Dr. Augustus Voelcker, F.R.S. (Fellow of the Royal Society), pages 387-389.

"The latest novelty in Prickly Comfrey, is a new variety of the S. asperrimum, distinguished by the appellation (name) of 'Solid Stem'. Mr. Henry Doubleday, the introducer of this to the public, *claims for it that it is a much quicker grower, and a purer Symphytum than any hitherto obtainable, and that **its solid stem peculiarity makes it a heavier cropper, besides affording cuttings from the flower stem for propagation.***
In *'Forage Plants', Mr. Doubleday fully demonstrates the superiority of his new variety, and explains at length its difference from the ordinary S. asperrimum."
-'New Commercial Plants with Directions How to Grow Them to the Best Advantage, No. 1 (**1878**) and No. 3 (1880)' by Thomas Christy, F.L.S. (Fellow of Linnaean Society), London, England. Number 1 includes Caucasian Prickly Comfrey pages 14-16. Number 3 includes Prickly Comfrey pages 11-14. Includes 'Caucasian Prickly Comfrey' and 'Russian Comfrey' advertisements.
(* -'Forage Plants and Their Economic Conservation by the New System of Ensilage: Part I: Caucasian Prickly Comfrey' by Thomas Christy, Jun., F.L.S. {Fellow of the Linnean Society}, Christy & Co., London, England, 1877.)

  *"**Solid Stem Comfrey (Symphytum asperrimum). Christy's new variety.** Produced from root-cutting planted
  February, 1876. Sketched from nature July 31st same year. Height of plant: 3 feet (0.9 meters). Circumference:
  9 feet (2.7 meters). **Grown by Mr. H. Doubleday, Coggeshall, Kelvedon, Essex (England)."***
  -'Forage Plants and Their Economic Conservation by the New System of Ensilage: Part I:
  Caucasian Prickly Comfrey' by Thomas Christy, Christy & Co., London, England, 1877, page 12.

"Borage Family. Symphytum Asperrimum, Prickly Comfrey:
This plant, a native of Caucasus, was brought to England in 1811 as an ornamental plant. Later it was cultivated as a forage plant *to a limited extent, but was not esteemed by English farmers.*
In Ireland *it was grown more extensively and found quite valuable for dairy cattle. The Bishop of Kildare, Ireland, was especially conspicuous (noticeable) in its culture at Glassnevin (suburb of Dublin, Ireland).*
On the Carew Castle, Pembrokeshire County, Wales, farm, *the experiments showed a yield of forage reported at 82 tons per Irish acre: 28 1/2 tons in April, 31 tons in July, and 22 1/2 tons in September.*
A few years ago, Mr. Ashburner of Virginia introduced the best variety of Prickly Comfrey into America. *From him, as soon as possible, I obtained root cuttings and have grown it ever since. Some of the first plants obtained remain in full vigor, all perhaps that were not divided, nor permitted to seed.*
It is said to yield on rich land eighty tons per acre; and thirty tons is probably a medium or moderate crop. I do not doubt the estimate of the Carew Castle farm crop.
  On good land broad leaves may attain a length of 3 feet (0.9 meter) and form an immense conical pile for each plant. Like other things of value, it requires some work with attention and patience; and without these, one would better not touch it or any other crop. From one-fourth to one acre, on every farm properly managed, cannot be better occupied."
-'The Farmers Book of Grasses and Other Forage Plants for the Southern United States' by D.L. Phares, A.M. (Master of Arts), M.D., Professor of Biology, A&M College of Mississippi, Starkville, Misssissippi, **1881**, pages 21-23.
  (The Irish acre or plantation acre measured one Irish chain by one Irish furlong, or 4 Irish perches by 40, or 7840
  square yards, or 0.66 hectares or 1.62 statute acres.)

*"**Advertisement:**
Caucasian Prickly Comfrey will easily produce 100 tons to the acre. The new Russian variety is now ready, and surpasses all other kinds by its immense yield of fodder.*
New last season. Quite pure Symphytum asperrimum, solid-stem and root.
For export, cases of large roots, equal to 1000 crowns. From Thomas Christy and Co., London, England."

-'New Commercial Plants and Drugs, Issue 6' by Thomas Christy, F.L.S. (Fellow of the Linnaean Society), London, England, **1882**. Comfrey advertisement, page 110.
    (A Comfrey crown is a root section that includes part of the top of the plant. It has a leaf bud or buds at the top.)

"The plant (Prickly Comfrey) has been tested by a number of experiment stations, but has never come into much use in America. Yields have been reported by various American experiment stations in green matter to the acre as follows: Ontario Agricultural College, 9 3/4 tons in 4 cuttings; New York (Geneva), 14 to 16 tons; Vermont, 46 tons; North Carolina, 6 1/2 to 17 1/2 tons; Wisconsin, 33 1/2 tons.
Even with these large yields Comfrey can hardly compete with other forage crops. At the Wisconsin Experiment Station the yield of dry matter to the acre for red clover was 23 percent greater than that of Comfrey.
At the New York Experiment Station alfalfa yielded 16 tons of green matter as compared to 14 tons by Prickly Comfrey.
At the Pennsylvania Experiment Station the yield of digestible matter by Prickly Comfrey was considerably less than that produced by either Kafir corn or cowpeas.
The value of Prickly Comfrey would seem to be restricted entirely to that of a soilage crop where a large amount of green matter is to be grown on a limited acreage, but even in this respect it is surpassed by other crops."
-'Forage Plants and Their Culture: Rural Text-Book Series' by Charles V. Piper, M.S., Bureau of Plant Industry, Department of Agriculture, United States; edited by L.H. Bailey, New York, **1916**.
    (A 'soiling crop' is a crop cut green and fed to livestock immediately without further curing or processing.)

"Plants of the genus Symphytum are moderately common throughout Ireland: two species and a hybrid have been recorded (*Scannell and Synnott 1972). The commoner species is Symphytum officinale L. which occurs in 19 out of 40 Irish vice-counties.
The hybrid between it and Symphytum asperum Lepechin, Symphytum x uplandicum Nyman is the commonest Irish Symphytum and is recorded from 28 vice-counties.
Though Symphytum asperum itself has not been recorded from Ireland at all (Scannell and Synnott).
The other species, Symphytum tuberosum L. is far less common, only being recorded from six vice-counties.
Recently, whilst on field-work in the west of Ireland with E.R. and J.T.M. Parnell, I came across a plant of Symphytum asperum growing in a hedgebank beside the road leading to the Lake Isle of Innisfree, near Lough Gill, County Sligo, G7732 (northwest Ireland). **The identity of the specimen, now in TCD (Trinity College Dublin), has been kindly confirmed for me by Dr F. H. Perring.**
    Some evidence exists which suggests that this species (S. asperum) is the least persistent of all the symphyta in the British Isles and that it may, therefore, have been more common than it now is (**Wade 1958). It may well have declined due to improvements in agriculture and the elimination of hedgerows. This suggests that it is likely that more sites for S. asperum exist in Ireland in areas where agricultural improvement has not yet taken place."
-'Symphytum Asperum Lepechin: A New Irish Species' by J.A.N. Parnell, School of Botany, Trinity College, Dublin, Ireland; The Irish Naturalists' Journal Ltd., Holywood, County Down, Ireland, Volume 21, No. 11, pages 498-499, July **1985**.
(* -'Census Catalogue of the Flora of Ireland' by M.J. Scannell and D. Synnott; Stationery Office, Dublin, Ireland, 1972.)
(** -'The History of Symphytum Asperum Lepech. and S. x Uplandicum in Britain' by A.E. Wade, Department of Botany, National Museum of Wales, Watsonia, Volume 4, pages 117-118, 1958.)
    (A hedgerow is closely growing shrubs and trees, usually bordering a road or field.)

Distribution of Prickly Comfrey

"S. asperum Lepechin (S. asperrimum Donn) is native in the Caucasus but is now widely naturalised in Europe though very rare in Britain."
-'The Genus Symphytum in Britain' by T.G. Tutin, University College of Leicester, England; Watsonia: Journal of the Botanical Society of the British Isles, Volume 3, pages 280-281, February **1956**. (Watsonia was the journal of the 'Botanical Society of Britain and Ireland' from 1949 to 2010. It is now called 'New Journal of Botany'.)
(Naturalize means a plant or animal species establishes itself so it lives successfully in the wild in region where it is not native.)

"During the first half of the 19th century (early to mid 1800s), S. asperum was grown as a fodder plant, but to what degree is difficult to determine owing to confusion with S. x uplandicum.
As a naturalised plant in Britain, S. asperum is extremely rare, and I have been able to confirm only about 20 records extending over 120 years.
It is a less persistent species than most Comfreys and appears to have died out in most of its recorded localities in Britain. It seems to have disappeared by 1879. C. Bucknall searched for it without success during 1910-12."
-'The History of Symphytum Asperum Lepech. and S. x Uplandicum in Britain' by A.E. Wade, Department of Botany, National Museum of Wales, Watsonia, Volume 4, pages 117-118, **1958**. (Watsonia was the journal of the 'Botanical Society of Britain and Ireland' from 1949 to 2010. It is now called 'New Journal of Botany'.)
    (Arthur Edward Wade lived from 1895 to 1989. He was 'Fellow of the Linnean Society'. He worked at Department of Botany at 'National Museum of Wales' in Cardiff. Among flowering plants he was particularly interested in Boraginaceae. He was an authority on Symphytum and Myosotis. In 1934 he published with H. A. Hyde 'Welsh Flowering Plants'.)

"S. asperum appears at present to be the most widely grown species of Comfrey in the USSR (Russia and Eastern Europe)."
-'Comfrey Symphytum spp. as a Forage Crop' by J.C. Forbes, A.D. McKelvie, and P.J.C. Saunders, North of Scotland College of Agriculture, Aberdeen, United Kingdom; Herbage Abstracts, Volume 49, No. 12, pages 523-539, **1979**.

*"Symphytum asperum Lepech.: **104** verified records in England, Scotland and Wales.*
*Symphytum officinale L.: **7,334** verified records in England, Scotland and Wales.*
*S. x uplandicum Nyman: **9,335** verified records in England, Scotland and Wales."*
-National Biodiversity Network, NBN Atlas, United Kingdom, https://species.nbnatlas.org, **2018**.
    (These are the official records with this organisation. Of course, there are more of each species than just those verified. It is useful as a relative measure of the frequency of these species in that area.)

<u>Description of Prickly Comfrey</u> (in chronological order)

Symphytum asperrimum is a native of the Caucasus. The Caucasus Mountains are located at the border of Europe and Asia, between the Black Sea and Caspian Sea and occupied by Russia, Georgia, Azerbaijan, and Armenia.

The flower-stem leaves are opposite. **Asperriumum means roughest.** Its leaves and stems are hairy, covered with short, stiff bristles. **The flowers are pink or blue.** 'The Dictionary of Gardening' by the Royal Horticultural Society, 1952, recommends it as a plant good for a wild garden.

"Symphytum Asperrimum, Prickly Comfrey. Class and Order: Pentandria Monogynia.
This species of Symphytum, a native of Caucasus, is by far the largest of the genus, growing to the height of five feet (1.5 meters), and is really an ornamental, hardy perennial, which will thrive in any soil or situation.
It differs from Symphytum orientale not only in stature (height) and in the greater roughness of the leaves, but in the stems being not merely hispid, but covered with small curved prickles; the floral leaves are constantly opposite, which is seldom the case in orientale. The nectaries in both are flat, not fistulous (hollow).
According to Mr. Donn, it was introduced in 1801 (to England), we believe, by Mr. Loddiges, of Hackney, England.
Our drawing (plate) 929 was taken at the Botanic Garden at Brompton, London, where we have observed it some years in the greatest vigour.
Propagated by parting its roots or by seeds."
-'Curtis's Botanical Magazine; or, Flower-Garden Displayed in which the Most Ornamental Foreign Plants, Cultivated in the Open Ground, the Green-House, and the Stove, are Accurately Represented in their Natural Colours, Volumes 23-24' by John Sims, M.D., Fellow of the Linnean Society, London, England, **1806**.

"Symphytum Asperrimum; Prickly Comfrey.
Stems prickly; leaves oval, acute, stalked, the floral ones opposite; clusters in pairs. This species is the largest of the genus, rising to five feet (1.5 meter); and is an hardy ornamental perennial.
Flowers of a rich blue colour. Native of Caucasus, flowering through the summer."
-'The Universal Herbal; or, Botanical, Medical, and Agricultural Dictionary; Containing an Account of All the Known Plants in the World, Arranged According to the Linnean system. Specifying the uses to which they are or may be applied, whether as food, as medicine, or in the arts and manufactures, with the best methods of propagation, and the most recent agricultural improvements' by Thomas Green, London, England, **1824**, page 642.

"Symphytum asperrimum, Rough Comfrey:
This plant grows, in an apparently wild state, in a wet place far from any house, near the head of the valley leading from Oakford (Devon county, England) to The Rocks.
    If it had not been an Asiatic plant, I should have considered it wild; *but such being the case, it has no doubt been* ***accidentally naturalized***, *though I have not been able to trace its origin.*
Its specific character is, 'Symphitum foliis cordato-ovatis lanceolatisve, acuminatis, petiolatis, strigosis, asperrimis, summis oppositis subsessilibus; caule setis recurvis muricato; corolla limbo campanulaceo.' (Latin)
    *Roemer & Schultes, Syst. Veg. 4, 65. 'Symphitum asperrimum.'
    **Marschal Bieberstein, Fl. Tauro-Caucasica, i. p. 129, n. 328. Curtis, Bot. Mag., t. 929.
A native of the Caucasus, in wet places."
-'Flora Bathoniensis, Or, a Catalogue of the Plants Indigenous to the Vicinity of Bath' by Charles C. Babington, M.A., Fellow of the Linnean Society, London, England, **1834**, page 32.
(* -'Caroli a Linne Equitis Systema Vegetabilium: Secundum Classes, Ordines, Genera, Species. Cum Characteribus, Differentiis et Synonymiis', Volume 4 by Johann Jacob Roemer and Joseph August Schultes; Stuttgardtiae {Stuttgart, Germany}, **1819**. Symphytum pages 64-66. 'Symphitum asperrimum' page 65.)
(** -'Flora Taurico Caucasica: Exhibens Stirpes Phaenogamas in Chersoneso Taurica et Regionibus Caucasicis Sponte Crescentes' edited by Friedrich August Marschall von Bieberstein, Volumes 1 and 2 in 1808; Volume 3 and 4 in 1819.)

"Descripton of Prickly Comfrey: The seed stalks of Prickly Comfrey reach a height of 2 to 4 feet (0.60-1.21 meter) and are

surrounded by numerous long, heavy, rough leaves of a dark-green color somewhat mucilaginous in texture.
The bright-blue flowers are borne in nodding, one-sided clusters.
The roots are large and fleshy and in loose soil will reach a depth of 8 or 9 feet (2.43-2.74 meter).
The plant is hardy and will endure considerable cold or drought, making a very rapid growth when conditions are favorable."
-'Prickly Comfrey as a Forage Crop' by U.S. Department of Agriculture, Bureau of Plant Industry, Office of Forage Crop Investigations, Washington, DC, Circular No. 47, 10 pages, January 26 **1910**.

"**Symphytum asperimum Donn.** Sims, Bot. Mag. 24 : pi. 929. 1806. **Rough Comfrey.**
Similar to S. officinale, but the pubescence rougher, the hairs stiff and reflexed. Leaves ovate-lanceolate to oblong-lanceolate, long-acuminate at the apex, narrowed at the base, all but the uppermost petioled, slightly or not at all decurrent, the lower often 8 inches (20 cm) long.
Flower-clusters rather loose; calyx about half as long as the corolla-tube, its segments hispid; **corolla bluish-purple. Flowers June-August.**
Waste grounds, Massachusetts to Maryland. Adventive (introduced) or naturalized from Europe."
-'An Illustrated Flora of the Northern United States and Canada, Volume 3: Gentianceae to Compositae (Gentian to Thistle)' by Nathaniel Lord Britton, PH.D., Sc.D., LL.D. and Hon Addison Brown, A.B., L.L.D., New York Botanical Garden, **1913**, page 92.
>(A leaf wing or plant fin is a green, short, thin, paperlike structure extending along a stem. It is long along the stem but not tall away from the stem. Decurrent means a plant part that extends downward. It means leaf blades that partly wrap or have wings around stem or petiole and extend down along the stem. Petiole is the stalk that joins a leaf to a stem.)

"**Prickly Comfrey.** Botanical genus: Symphytum. **One species cultivated- Symphytum asperrimum.**
>'Russian Comfrey' is the commonest variety.
>There are two British wild varieties very similar to this: S. officinale and S. tuberosum.

The proper kind of Prickly Comfrey has very prickly leaves and stalks, is solid-stemmed, and has blue flowers.
Quantity of seed per acre for a Comfrey meadow is 6 pounds (2.7 kg); sown along with oats in March or April.
Usually propagated by sets or cuttings, planted a yard (3 feet = 0.9 meters) each way during spring; useful for growing in out-of-the-way corners.
Average produce per acre- 40 to 50 tons of forage in 6 or 8 cuttings, after first year. Comfrey is a perennial, and requires only to be kept clean and well manured. Requires deep, rich soils to grow on.
It (leaves) must be withered before eaten, but otherwise used fresh."
-'Notebook of Agricultural Facts and Figures for Farmers and Farm Students, 10th Edition' by Primrose McConnell, B.Sc., Fellow of Geological Society of London, Yeoman Farmer, North Wycke, Southminster, Essex, England, **1922**. Prickly Comfrey p. 236.
>(In the late 1800s and early 1900s Prickly Comfrey was sometimes called 'Russian' or Caucasian Comfrey. Prickly Comfrey is different from Henry Doubleday's and Lawrence Hill's Russian Comfrey which is Symphytum x uplandicum.)

"**The flowers of Prickly Comfrey (Symphytum asperrimum) are a pure cobalt blue.**"
-Compost, Comfrey and Green-Manure: 'Henry Doubleday Research Association' First Gardeners Report by Lawrence D. Hills. Braintree, Essex, England: Henry Doubleday Research Association, **1959**.
>(Cobalt blue is a deep blue to a strong greenish-blue color.)

"**S. asperum Lepech. (Rough Comfrey), S. asperrimum Donn ex Sims:**
A scabrid perennial up to about 150 cm (59 inches). Root thick, branched. Leaves ovate or elliptic, scabrid or with tuberclate bristles; lower 15-19 cm (5.9-7.4 inches), cordate or rounded at base, petioled; upper very shortly petioled, cuneate at base. Calyx 3-5 mm, accrescent, covered with short stout bristles; teeth linear-oblong, obtuse, becoming more or less triangular in fruit, 1-2 times as long as tube. Corolla 9-14 mm, at first pink, becoming clear blue. Scales lanceolate, about equalling stamens. Nutlets granulate. Flowers 6-7. 2n = 40 (chromosomes).
Hemicryptophyte Hs: Semi-rosette hemicryptophytes, with leafy stems but the lower leaves larger than the upper ones and the basal internodes shortened.
Introduced. Formerly cultivated, now occasionally naturalized in waste places. Caucasus and Persia (Iran)."
-Flora of the British Isles, Second Edition by A.R. Clapham, T.G. Tutin and E.F. Warburg. England: Cambridge University Press, **1962**, (first edition 1952).

>"**Note on Life Forms:** The Danish botanist Raunkiaer has classified plants, according to the position of the resting buds or persistent stem apices in relation to soil level, into a number of Life Forms, which have in turn been subdivided so as to convey more information about the plants included in the different groups.
>**Life Forms are a convenient method of indicating how a plant passes the unfavourable season (winter), and are also of interest because they show a correlation with climate.** The primary classes are:
>1. Phanerophytes: woody plants with buds more than 25 cm (9.8 inches), above soil level.
>2. Chamaephytes: woody or herbaceous plants with buds above soil surface but below 25 cm.
>3. **Hemicryptophytes: herbs (very rarely woody plants) with buds at soil level.**
>>Hp: Protohemicryptophytes, with uniformly leafy stems, but the basal leaves usually smaller than the rest.
>>**Hs: Semi-rosette hemicryptophytes, with leafy stems but the lower leaves larger than the upper ones and the basal inter-nodes shortened.**
>>Hr: Rosette hemicryptophytes, with leafless flowering stems and a basal rosette of leaves.

4. *Geophytes: herbs with buds below the soil surface.*
5. *Helophytes: marsh plants.*
6. *Hydrophytes: water plants.*
7. *Therophytes: plants which pass the unfavourable season as seeds."*
-Flora of the British Isles, Second Edition by A.R. Clapham, T.G. Tutin and E.F. Warburg. England: Cambridge University Press, **1962**, (first edition 1952).

>(Christen Christensen Raunkiaer, 1860-1938, was a Danish botanist, who was a pioneer of plant ecology. He discovered plant strategies of how to survive an unfavourable season that he called 'Life Forms', and he showed these strategies corresponded to climate zones. This scheme is still used today and is a forerunner of modern plant strategies such as J. Philip Grime's CSR Triangle theory.)

*"***S. asperum Lepechin** *in Nova Acta Acad. Sci. Petrop. 14 :442 (1805).*
    Ic.: Curtis, Bot. Mag. 24: t.929, 1806 (sub S. asperrimum).
Synonym: *S. orientate folio subrotundo aspero flare coeruleo. Tournef., Corell. 7 (1703).*
    S. orientale L, Sp. Pl. 136 (1753) p.p.
    S. asperrimum Sims in Curtis, Bot. Mag. 24:t.929 (1806).
    S. echinatum Ledeb. Index Sem. Hort. Dorpat. Suppl. 5 (1811).
    S. patens Fries, Nov. Fl. Suecica Mant. 2:13 (1839) p.p. cf. Fries, Mant.3:18 (1842).
    S. orientate sensu Fries, Nov. Fl Suecicae Mant. 3 :18 (1842) p.p. non L.
    S. majus Guldenst. ex Ledeb., Fl. Ross. 3 :115 (1847).
Picea (spruce and fir) forests, meadows and stream banks, 700-2000 meters (2296-6561 feet).
This appears to be a rather variable species as regards habit, leaf shape and leaf texture.
Linnaeus:
>*By including under his S. orientale both the blue- and white flowered Turkish 'Comfries' mentioned by Tournefort,* **Linnaeus failed to distinguish S. asperum Lepechin (of which he had seen neither an illustration nor a specimen) from S. orientale."*

-'A Revision of Symphytum in Turkey and Adjacent Areas' by G.E. Wickens, Royal Botanic Gardens: Kew; Notes from the Royal Botanic Garden Edinburgh, Scotland, Volume 29, pages 157-180, **1969**. Includes Turkey, Bulgaria, Greece, Aegean Islands and Caucasia.

*"***S. asperum Lepechin***, Nova Acta Acad. Sci. Petrop. {Acta Petersburg Russia Imperial Academy of Sciences} 14:442 (1805). Stock vertical. Stem up to 180 cm (70 inches = 5.9 feet), branched, aculeolate and with scattered hairs.*
Leaves ovate to oblong, cuneate to rounded or subcordate at base, petiolate or the uppermost sessile, not decurrent and not amplexicaul, densely setose and aculeolate at least on the mid-vein beneath.
Calyx 3-4 mm, lobed to about 2/3 to 3/4, the lobes rounded. **Corolla 11-17 mm, pink at first, turning blue.** *Scales linguate, slightly widened at the base; marginal papillae dense, narrowly cylindric-conical. Stamens with connective not projecting beyond thecae; filament narrower than and about as long as anther.*
Nutlets reticulate-rugose and finely verrucose.
Formerly cultivated for fodder, and naturalized in several parts of Europe: Austria, Belgium, Luxembourg, Britain, Denmark, Finland, France, Switzerland, Norway, Sweden. Also USSR {Baltic region, Central region, South-Western region} and southwest Asia."
-Flora Europaea, Volume 3, editors Tutin, Heywood, Burges, Moore, Valentine, Walters and Webb. Article 'Symphytum L.' by B. Pawlowski, pages 103-105, **1972**.
('Flora Europaea' is a 5-volume encyclopedia of plants, published between 1964 and 1993 by Cambridge University Press. It describes all the national floras of Europe to identify any wild or widely cultivated plant to the subspecies level. It provides geographical distribution, habitat preference, and chromosome number.)

*"***Symphytum asperum Lepechin** *in Nova Acta Acad Sci Imp Petrop Hist Acad 14:442, 1850.*
Synonyms: **S. orientale L., *Sp Pl 136, 1753,* pro parte excl. typ.**
    (In part, excludes types which other authors have subsequently included.)
    S. asperrimum Donn ex Sims, *in Bot Mag 24: t. 929, 1806.*
    S. echinatum Ledebour, *Index Sem Hort Dorpat Suppl 5, 1811.*
    S. patens Fries, *Novit Fl Suec Mant 2: 13, 1839,* **pro parte. (in part)**
    S. majus Guldenst. ex Ledebour, *Fl Ross 3: 115, 1847.*
Stem to 200 cm (78 inches), never winged, very scabrid with aculeate curved subretrorse hairs; the hairs with a tubercular base; basal leaves ovate-elliptic, with an acuminate apex and a rounded cordate base; lamina 15-19 cm long, 7-12 cm wide; petiole to 10 cm long; stem leaves gradually smaller, 10-20 cm long, 4-10 cm wide, ovate or elliptic, acuminate at the apex and cuneate at the base; leaves not decurrent; adaxial side of the leaf very scabrid with short more or less appressed hairs with a small tubercular base and smaller shorer hairs without a tubercular base; abaxial side of the leaves with shorter uncinate hairs and with setae on the veins.
Calyx to 3 mm long, divided to 3/4 of its length, calyx lobes linear oblong and obtuse in flower, becoming triangular in fruit; stiff marginal setae irregularly distributed; setae with a small tubercular base; corolla campanulate, 9-14 mm long, **red in bud, sky-blue in flower**, stamens 4-5 mm long; corolla scales shorter than stamens; anther longer than filament, connective not projecting beyond thecae; squamae of corolla lingulate, 6 mm long, 1 mm wide, with a broad rotundate apex; marginal papillae

fewer in number, longer and narrower than in all cytotypes of S. officinale, papillae acute, regularly distributed along margin.
Reproduction strictly allogamous (must cross fertilize), plants self-incompatible.
Fruit brown and dull, urceolate-granulate, 3-4 mm long and 3 mm wide at the base.
> The production of ripe nutlets varies considerably in different plants, even fertile plants may have many flowers which produce only 1, 2 or 3 (and often 0) nutlets which are able to germinate."

-'Notes on Symphytum (Boraginaceae) in North America' by T.W.J. Gadella, pages 1061-1067 in book: Annals of the Missouri Botanical Garden, Volume 71, Saint Louis, Missouri. Lawrence, Kansas: Allen Press, **1984**.

> (Global Biodiversity Information Facility®, www.gbif.org, 2018, lists these as synonyms of S. asperum Lepechin: S. armeniacum Buckn., S. asperrimum Sims, S. asperum var. armeniacum {Bucknall} Kurtto, S. echinatum Ledeb., S. majus Gueldenst., S. majus Gueldenst. ex Ledeb., S. peregrinum Ledeb.)

"S. asperum Lepechin - Rough Comfrey:
Stems erect, well-branched to 1.5 meter (4.9 feet), from thick, vertical root; stem-leaves shortly petiolate, not decurrent. Calyx divided about 2/3 to 4/5 to base, enlarging somewhat in fruit, with less sharply acute lobes than in S. officinale: **corolla sky blue when open.**
The shortly petiolate upper stem-leaves and white, sub-spiny bristles, are diagnostic.
Introduced; naturalized in rough and waste ground; formerly occasional, **now very rare, scattered over Britain, County Sligo (Ireland),** much over-recorded for S. x uplandicum: southwest Asia."

-New Flora of the British Isles: Identification of Wild Vascular Plants of the British Isles edited by Clive Stace. Cambridge, England: Cambridge University Press, **1991**, page 647.

> ('Over-recorded' means botanists thought they found a particular species but actually it was another species, so they acccidentally put in their records this incorrect geographical distribution of a plant.)

"S. asperum Lepech.:
Tall, up to 1.8 meters (5.9 feet). Stems and midrib of the undersides of the leaves covered in short, stout, hooked bristles. At least the lower and middle cauline leaves cordate and with long unwinged petioles.
Calyx tiny, 3-5 mm long in bud, with obtuse segments only 1/5 as long as the corolla but enlarging rapidly in fruit up to 8 mm long. Corolla 11-17 mm, red in bud, becoming sky blue on opening, the limb widening gradually towards the apex.
A very rare introduction, probably over-recorded in the past in error for the 2n = 36 form of S. x uplandicum (Russian Comfrey)."

-Plant Crib: Handbook for Field Identification compiled by T.C.G. Rich and M.D.B. Rich with editorial assistance of F.H. Perring for the BSBI Monitoring Scheme. London, England: Botanical Society of the British Isles, 1988, page 74. There is also an updated **1998** edition.

"Symphytum Asperum:
Morphological Properties:
> Perennial. Stem 30-90 cm (11.8-35.4 inches), hairy. Leaves ovate to elliptic-lanceolate. Basal leaves petiolate, cauline leaves shortly petiolate. Inflorescence helicoid cymes. Calyx 3-5 mm, hairy. Corolla pink at first, turning blue or lilac (pale pinkish-violet), 11-12 mm. Nutlets 3-4 mm, base constricted, brown-dark brown.

Anatomical Properties:
> **A transverse section taken from the root** was observed and the anatomical features were as follows: Periderm is multilayered. Phelloderm is clear. Cortex is 10-13 layered and parenchymatic. Parenchymatic cells are 29.25 +- 1.82 x 9.5 +- 1.04. Cambium cells are undistinguishable. Xylem is composed of sclerenchymatic cells and tracheary elements. Pith rays are present. The pith consists of primary xylem elements.
> **A transverse section taken from the middle part of the stem** was observed. Epidermal cells consist of a single layer and orbicular or rectangular. Trichomes are present on the cuticle. Collenchyma is 3-4 layered. Cortex cells are 78.25 +- 5.43 x 51.25 +- 5.88. and multilayered. Endodermis is between cortex and vascular tissue. The vascular tissues are collateral. Bundles are of different sizes. Xylem and phloem elements are clear. Cambium is distinguishable. Diameter of vessel members are 35.5 +- 6.4. The pith consists of large parenchymatic cells."

-'Micromorphology and Anatomy of Three Symphytum (Boraginaceae) Taxa from Turkey' by Oznur Ergen Akcin and Hilal Baki, Department of Biology, Sciences & Arts Faculty, Ordu University, Ordu, Turkey; Bangladesh Journal of Botany, Volume 36, No. 2, pages 93-103, December **2007**.   (Transverse section means made at right angles to the long axis of the body or object.)

"Symphytum asperum:
Ecology: A tall perennial herb, naturalised in rough and waste ground. Generally lowland, but reaching 315 meter (1033 feet) altitude at Sheldon, Derbyshire, England.
Status: Neophyte in Great Britain.
Trends: This species, introduced into Britain in 1799, was quite widely grown as an ornamental in the 19th and early 20th centuries, sometimes escaping or outcast from cultivation. It was recorded from the wild by 1862 (North Essex, Essex County, England), but was greatly over-recorded in the past for the 2n = 36 (chromosome) cytotype of S. x uplandicum (Russian Comfrey) (*Perring, 1994), and **the few confirmed post-1930 records reflect its present rarity in cultivation in Great Britain.**
World Distribution: Native of northeast Turkey, the Caucasus and northern Iran. Formerly cultivated and now naturalised in western and central Europe."

-'Online Atlas of the British and Irish Flora', https://www.brc.ac.uk/plantatlas, Distribution and ecology with photographs, **2018**.

(* -The Common Ground of Wild and Cultivated Plants: B.S.B.I. Conference Report No. 22 by A. Roy Perry and R.G. Ellis, {F.H. Perring, Symphytum-Comfrey, pages 64-70}. Cardiff, Wales: Department of Biology, National Museum of Wales, 1994. I was unable to get a copy of the Symphytum section in 'The Common Ground' book. If you have it, could you please send it to me.)
(In botany, a neophyte is a plant species that is not native to a geographical region, and was introduced in recent history. Plants that are long-established in an area are called archaeophytes. In Britain, neophytes are plant species introduced after 1492, when Christopher Columbus arrived in the Americas.)

Symphytum Asperum (Prickly Comfrey): Large Forage Plant
See sub-section 'Prickly Comfrey as a Forage Crop' in section 'Comfrey as Food and Medicine for Livestock' in Volume 2.

"**A variety with flowers of a rich blue colour Symphytum Asperimum, Prickly Comfrey, was introduced into this country (England) from the Caucasus in 1811 as a fodder (food for animals) plant.**
This species is the largest of the Symphytum genus, rising to 5 feet (1.5 meters) and more, with prickly stems and bold foliage, the leaves very large and oval, the hairs on them having bulbous (bulb shape) bases.
It was extensively recommended as a green food for most animals, it being claimed for it that it contained a considerable amount of flesh-forming substances, and was, moreover, both preventative and curative of 'foot and mouth disease' in cattle.
It has the advantage of producing large crops, two at least in a season, if cut before the flowers quite expand, and in favourable circumstances even more, so that 40 to 50 tons of green food per acre might be reckoned on. At the time of its introduction, a number of farmers and smallholders planted it."
-'A Modern Herbal: The Medicinal, Culinary, Cosmetic and Economic Properties, Cultivation and Folk-Lore of Herbs, Grasses, Fungi, Shrubs and Trees with their Scientific Uses' by Mrs. M. Grieve. New York: Dover Publications, 1971. First published **1931**.
(The Caucasus Mountains are located at the border of Europe and Asia, between the Black Sea and Caspian Sea and occupied by Russia, Georgia, Azerbaijan, and Armenia.)

"The merits of the Caucasian Prickly Comfrey are recorded from early writings, but a warning against the use of other species of the Symphytum genus which grow wild in parts of England (S. tuberosum and S. officinale) is included.
S. asperrimum has been recommended for its quick growth, fine cropping qualities and resistance to drought. With little labour, beyond cutting the green leaves, the crop will yield, year after year, from 50 to 60 tons of forage per acre."
-'Symphytum Asperrimum as a Fodder Plant' by G.W. Robinson, Curator, Chelsea Physic Garden (Apothecaries' Garden, London, England created 1673); Gardeners' Chronicle, Volume 111, page 131, **1942**. (I was unable to get a copy of this report. If you have it, could you please send it to me.)

"**In the 1870's and 1880's Comfrey enjoyed a wave of popularity** mainly through the work of four men whose experiments and writings form part of our basic knowledge of the crop.
The most famous of the four was not a Comfrey grower, he was **Dr. Auguste Voelcker, D.Sc., F.R.S. (Fellow of the Royal Society of London)**, Consulting Agricultural Chemist to the Royal Agricultural Society of England.
His article in the 'Journal' of that body (**1871**, Volume 7, second series, pages 387-9) entitled '**On the Composition and Nutritive Value of the Prickly Comfrey (Symphytum asperrimum)**', gave a very good account of it, including the Carnew Castle, County Wicklow, Ireland yields.
Even more important, it contains the first analysis ever made of Symphytum asperrimum, from plants grown in Oxfordshire county, England, which he carried out in 1869.
The general analysis, stems and leaves together, was as follows: In natural state (fresh): Water 90.66%, Nitrogenous organic compounds (flesh-forming matter) 2.72%, Non-nitrogenous compounds (heat- and fat-producing matter) 4.78%, Mineral matter (ash) 1.84%. Calculated Dry: Nitrogenous 29.12%, Non-nitrogenous 51.28%, Mineral 19.60%."
-Comfrey: Fodder, Food and Remedy by Lawrence D. Hills. New York: Rizzoli Universe Books: **1976**, page 22.
(Mr. Voelcker was known for his methodical and precise analytical practices in agricultural chemistry. This included his testing of concentrated feedstuffs such as Peruvian guano and artificial fertilizers.)

Symphytum Asperum (Prickly Comfrey) Growth Patterns and Yield

"As to produce, our obliging correspondent Dr. Derenzy, states, that **on the farm of the Reverend Henry Moore, of Carnew Castle, County Wicklow, Ireland it produced, in 1835, the enormous quantity of 82 tons to the Irish acre,** in three cuttings, viz. (that is) the first cutting the middle of April, 28 1/2 tons; second cutting the middle of July, 31 tons; third cutting, the middle of September, 22 1/2 tons. Total 82 tons."
-'A Cyclopaedia of Practical Husbandry and Rural Affairs in General' by Martin Doyle, London, England, **1851**, page 152.

"**Symphytum asperum is usually propagated from seed, which it produces fairly abundantly** (*Manlov 1970, Moissev and Votinova 1970).
Medvedev (**1974) established plots from six seed provenances of S. asperum and measured their yields from the second year onwards. **There were some consistent differences between provenances, but the yield of a provenance in the second year of growth was not necessarily indicative of its long-term productivity.**
He also studied the variation within a single provenance from the Caucasus and found a range in average yields over

seven years from nine seed slections of 59.4 to 97.8 tons green herbage/hectare."
-'Comfrey Symphytum spp. as a Forage Crop' by J.C. Forbes, A.D. McKelvie, and P.J.C. Saunders, North of Scotland College of Agriculture, Aberdeen, United Kingdom; Herbage Abstracts, Volume 49, No. 12, pages 523-539, **1979**.
(* -'Seed Production of Plants in Some Plant Communities of the Greater Caucasus', 1970. In Russian.)
(** -'The Duration of Economic Utilisation and Yields of 5 Species of Comfrey', 1974. In Russian.)
    (Provenance is a record of ownership, or the documented place of origin or earliest known history.)

"The biology of Symphytum asperum flowering and fruit bearing was studied in the Komi SSSR, Russian automomous republic. *The effect of soil type, plant density, mineral fertilizers, time of fertilization and leaf feeding with microelements on S. asperum seed productivity was studied.*
S. asperum seed growing in the central taiga subzone of the Komi SSSR should be based on obtaining seeds from thyrses. **Central inflorescences produced the best seeds.**
Seed plots should be located on mineral soils, but not on turf soils. Optimal area of plant nutrition is 70x50 cm (27.5 x 19.6 inch). **Total mineral fertilization should be conducted in seed plots not in early spring, but after the plants started growing when the soil gets dry.** *It is necessary to carry out leaf feeding with Mo (molybdenum), Cu (copper), Zn (zinc) and Basalts or a mixture of these microelements at the beginning of plant budding before plant rows close."*
-'Effect of Growing Conditions on Seed Production by Symphytum Asperum Lepech' or 'Effect of Cultivation Conditions on Symphytum Asperum Seed Productivity' by Y.M. Frolova and N.P. Frolova, Rastitel'nye Resursy, Leningrad, Russia, Volume 19, No. 3, pages 336-341, **1983**. (I could not find this article. If you have an English translation, could you please send it to me.)
    (Taiga, or boreal or snow forest, is a plant community of coniferous forests with mostly pines, spruces and larches.)
    (A thyrse is a type of inflorescence where the main axis grows indeterminately, and branches have determinate growth.)
    (A mineral soil is derived from minerals or rocks with little humus or organic matter.)
    (Basalt is an igneous rock formed from rapid cooling of magnesium-rich and iron-rich lava.)

"Prickly Comfrey is propagated by seed in the Soviet Union."
-'Comfrey' by Teynor, Putnam, Doll, Kelling, Oelke, Undersander and Oplinger, 'Alternative Field Crops Manual', University of Wisconsin: Cooperative Extension, University of Minnesota: Center for Alternative Plant & Animal Products, and Minnesota Cooperative Extension, February **1992**, updated November 1997, page 3.

"Annual and interannual (from one year to the next) phenomena and canopy behavior of Prickly Comfrey (Symphytum asperum Lep.) were studied in a 10-year experiment in Finland with 25 measurement sessions during growing season.
The results confirm the importance of long-term experiments in studying plant phenomena, biometrics and behavior. Prickly Comfrey produced a green canopy each year and growth started very early in spring. **Maximum plant height was less than 160 cm (5.2 feet).**
Annual phenomena (growth initiation, seedling phase, flower phase, seed phase, senescent / old age phase), interannual phenomena (initiation and youth, reproduction, new generation formation, plant death) and two population cycles (colonization and expansion) were measured. The duration of annual development up to canopy death can be expressed as x+2x+3x+2x, where x is initial growth.
The differences between the organs in the upper and lower parts are very considerable and should be taken into account in morphological descriptions of this species. *The upper and lower stems and leaves showed differential growth. Both stem and leaves were densely setose (bristles).*
Life Strategy:
The genetic structure and activity of Prickly Comfrey promotes generative (reproductive) development of the species. Its age can be measured over a single and several vegetation generations. The ability to change the angle of vertical stem growth after 9 weeks can be considered a functional behavior of Prickly Comfrey and part of its life strategy.
Old leaves were 3.8 times longer, 4 times broader and 2.4 times thicker than young leaves. Hairs were on average 3 times longer on old than on young leaves.
Flowers had contact with pollinators making relatively long visits to them."
-'Phenomena, Biometrics and Canopy Behavior of Prickly Comfrey (Symphytum Asperum Lep., Boraginaceae in a Long-Term Experiment' by Tadeusz Aniszewski, Research and Teaching Laboratory of Applied Botany, University of Eastern Finland, Acta Biologica Cracoviensia, Series Botanica (Biological Association Cracoviensia, Botanical Series) 54/1, pages 121-128, **2012**.
    (Senescence is biological aging, i.e., the gradual decrease in health and functioning.)

"Prickly Comfrey differs morphologically, genetically and chemotaxonomically from the widely distributed Common Comfrey (Symphytum officinale L.), less widely distributed Russian Comfrey (Symphytum x uplandicum Nym.), and Tuber Comfrey (Symphytum tuberosum L.), *a rare newcomer in southern Finland.*
    Prickly Comfrey (Symphytum asperum Lep.) differs from other Comfrey species in flower color and the sizes of organs, more vigorous growth (*Mossberg and Stenberg, 2005), DNA, ploidy level and chemotaxonomic markers (Gadella and Kliphuis, 1975, 1978; Gadella et al., 1983; Jaarsma et al., 1989, 1990; Ozcan, 2010; Barbakadze et al., 2011).
Among the Comfreys mentioned, Prickly Comfrey has the smallest chromosome number (2n = 32) and contains echimidine alkaloid but not isobauerenol, which is typically found in Common Comfrey.
    Russian Comfrey contains both of these alkaloids and is a cross hybrid between Prickly and Common Comfrey (Jaarsma et al., 1989).
Prickly Comfrey (Symphytum asperum Lep.) is native to the Caucasus and Iran (Mossberg and Stenberg, 2005) but has been

naturalized in many parts of Asia, Europe, America, Africa and Australia, especially in former British colonies (**Tutin, 1956; ***Wade, 1958)."
-'Phenomena, Biometrics and Canopy Behavior of Prickly Comfrey (Symphytum Asperum Lep., Boraginaceae in a Long-Term Experiment' by Tadeusz Aniszewski, Research and Teaching Laboratory of Applied Botany, University of Eastern Finland, Acta Biologica Cracoviensia, Series Botanica (Biological Association Cracoviensia, Botanical Series) 54/1, pages 121-128, **2012**.
(* -'Grand Guide to Northern Plants' or 'Extensive Manual of the North Flora' by B. Mossberg and L. Stenberg; Tammi, Helsinki, Finland, 2005, page 928. In Finnish.)
(** -'The Genus Symphytum in Britain' by T.G. Tutin, University College of Leicester, England; Watsonia: Journal of the Botanical Society of the British Isles, Volume 3, pages 280-281, February 1956.)
(*** -'The History of Symphytum Asperum Lepech. and S. x Uplandicum in Britain' by A.E. Wade, Department of Botany, National Museum of Wales, Watsonia, Volume 4, pages 117-118, 1958. Watsonia was the journal of the 'Botanical Society of Britain and Ireland' from 1949 to 2010. It is now called 'New Journal of Botany'.)

"In field trials near Minsk in Belarus (in eastern Europe), **S. asperum (Prickly Comfrey) yielded from 2 cuts 850, 960 and 920 hkg air-dry matter/ha (hectare) in the 2nd, 3rd and 4th year of growth, respectively, when the 1st cut was taken at the stage of bud formation;** *higher yields than these were obtained when the 1st cut was taken at flowering.*
CP (Crude Protein) content in the leaves ranged from 20.4 to 29.6% in the air-dry matter, depending on the year and growth stage. The stems contained 15.6% soluble carbohydrates and the leaves 8.8%. The contents of CP and soluble carbohydrates are also given for S. tauricum, S. officinale and S. caucasicum."
-'On the Introduction of Symphytum Asperum into Belorussia" by N.V. Smol'skii and I.I. Chekalinskaya, Tsentr Bot Sad Akademiia Nauk Belorus, SSR, Minsk, Belarus; Rastitel'nye Resursy (Vegetable Resources), Volume 6, No. 2, pages 223-237, 1970.
(Total yield is higher if several or many cuts are taken throughout the season.) (I'm not sure what 'hkg' is.)

Prickly Comfrey Breeds with Other Symphytum Species
See section 'Details about Symphytum Species Hybrids' (Chapter 11).

***"Symphytum asperum Lepech. (Prickly Comfrey), S. uliginosum Kern., S. officinale L. (Common Comfrey): These three species hybridize and form hybrid swarms consisting of F1 and backcross hybrids** (*, **, ***).*
Symphytum officinale is cytologically heterogenous: (chromosomes) 2n = 24, 2n = 40, 2n = 48, cytotypes occur in various parts of Europe.
Symphytum asperum (2n = 32) does not hybridize with the diploid (2n = 24) form of S. officinale in Europe but produces hybrids and backcrosses with the 2n = 40 and 2n = 48 cytotypes of S. officinale.
The primary hybrids, with 2n = 36, or 2n = 40, are collectively known under the name S. x uplandicum Nyman (Russian Comfrey).
In Europe the parental species are largely allopatric with a very small zone of overlap in the northwest Caucasus (Kusnetosov, 1910). In the zone of overlap, hybridization does not occur because S. officinale and S. asperum grow at different altitudes.
Apparently the hybrids arose outside the Caucasus** (*Tutin 1956; *****Wade 1958). In many parts of western, northwestern, or central Europe the hybrid swarms are more common than the parental species."*
-'**Notes on Symphytum (Boraginaceae) in North America**' by T.W.J. Gadella, page 1061-1067 in book: Annals of the Missouri Botanical Garden, Volume 71, Saint Louis, Missouri. Lawrence, Kansas: Allen Press, **1984**.
(* -'Cytological and Hybridization Studies in the Genus Symphytum' by Th.W.J. Gadella, Symposia Biologica Hungarica {Hungary Biological Symposium}, Volume 12, pages 189-199, 1972. I was unable to find a copy of this report. If you have it, could you please send it to me.)
(** -'Cytotaxonomic Studies in the Genus Symphytum V: Some Notes on W. European Plants with the Chromosome Number 2n=40' by Th.W.J. Gadella and E. Kliphuis, Botanische Jahrbücher fur Systematik {Botanical Yearbooks for Systematics}, Volume 93, pages 530-538, 1973.)
(*** -'Cytotaxonomic Studies on the Genus Symphytum VIII: Chromosome Numbers and Classification of Ten European Species' by Th.W.J. Gadella and E. Kliphuis, Proceedings van de Koninklijke Nederlandse Akademie van Wetenschappen {Proceedings of the Royal Netherlands Academy of Arts and Sciences} Section C, Volume 81, pages 162-172, 1978.)
(***** -'The Genus Symphytum in Britain' by T.G. Tutin, University College of Leicester, England; Watsonia: Journal of the Botanical Society of the British Isles, Volume 3, pages 280-281, February 1956.)
(****** -'The History of Symphytum Asperum Lepech. and S. x Uplandicum in Britain' by A.E. Wade, Department of Botany, National Museum of Wales, Watsonia, Volume 4, pages 117-118, 1958. Watsonia was the journal of the 'Botanical Society of Britain and Ireland' from 1949 to 2010. It is now called 'New Journal of Botany'.)
(Hybrid or crossbreed is the result of combining the qualities of two organisms of different breeds, varieties, species or genera through sexual reproduction.)
(A hybrid swarm is a group of hybrids that have survived past the first hybrid generation, with interbreeding taking place between hybrid individuals and also backcrossing with its parent type. These plants are highly variable, with genes and phenotypes of individuals ranging widely between the two parent types. Hybrid swarms blur the boundary between the parent taxa.)
(A backcross is a cross of a hybrid with one of its parents or an individual genetically similar to its parent. This creates offspring with genetics closer to that of the parent.)
(An F1 hybrid or 'Filial 1' hybrid is the first generation of offspring of distinctly different parental types that produce a new, uniform phenotype {observable properties of an organism} with a combination of characteristics from the parents. Mules are F1

hybrids between horse and donkey. This can happen naturally, and includes hybrids between species, for example, peppermint is an F1 hybrid of watermint and spearmint. In agronomy, these F1 hybrids are usually created by means of controlled pollination, sometimes by hand-pollination. This is not GMO. A Genetically Modified Organism has been altered using genetic engineering techniques in a sophisticated laboratory, frequently using genes from unrelated species.)
(Allopatric or geographic speciation occurs when populations of the same species become isolated from each other, and this prevents genetic exchange.)

Prickly Comfrey Chromosomes

"*Symphytum asperum Lepech.:*
*Coimbra, Portugal, **n = 20** (Britton), **2n = about 36** (*Strey 1931)."*
-'Cytogenetic Studies on the Boraginaceae' by Donald M. Britton, Brittonia: The New York Botanical Garden, Volume 7, No. 4, pages 233-266, December 10 **1951**.
(* -'Karyologische Studien an Borraginoideae' {Caryological studies on Borraginoideae'} by Martin Strey; Planta, Berlin, Germany, Volume 14, Issue 3-4, pages 682-730, October 1931. In German.)

"According to the literature cytological (cellular) studies Symphytum asperum yielded following results: **Symphytum asperum:**
*   **2n = + or - 36** (plus or minus 36) (Strey, 1931; cult. mat.= cultivated material);*
*   **2n = 40** (Britton, 1951; cult. mat.)."*
-'Cytotaxonomic Studies in the Genus Symphytum I: Symphytum Officinale L. in the Netherlands' by Th.W.J. Gadella and E. Kliphuis, Botanical Museum and Herbarium of the State University of Utrecht, Netherlands, February 25 **1967**.

*"**The species Symphytum asperum** was studied by Strey (1931) and Britton (1951), who found in material of garden origin the numbers 2n = 36 and 2n = 40 respectively.*
*   This could neither be confirmed by Grau (1968), nor by the present authors, who found number **2n = 32. Flowers blue**."*
-'Cytotaxonomic Studies in the Genus Symphytum II: Crossing Experiments Between Symphytum Officinale L. and Symphytum Asperum Lepech' by Th.W.J. Gadella and E. Kliphuis, Botanical Museum and Herbarium of the State University of Utrecht, Netherlands, Acta Botanica Neerlandica, Volume 18, No. 4, pages 544-549, August **1969**.

*"**Prickly Comfrey is a cytotype of (2n = 32)**."*
-'Comfrey: A Controversial Crop' by Robert G. Robinson, University of Minnesota, Agricultural Experiment Station, Minnesota Report MR-191, Item No. AD-MR-2210, **1983**, page 2.
(A cytotype is an individual of a species that has a different chromosomal factor to another.)

*"**S. asperum: 2n = 32**."*
-'Chemotaxonomy of the Symphytum Officinale Agg. (Boraginaceae)' by Jaarsma, Lohmanns, Gadella and Malingre, Plant Systematics and Evolution, 167, pages 113-127, **1989**.

Prickly Comfrey and Alkaloids
See sections 'Alkaloids in Comfrey' in Volume 2.

*"**Symphytum asperum Lepech, Prickly Comfrey,** Consound, (German: Rauher Beinwell, French: Consoud epineux), growing in Caucasia is cultivated and used as medicinal and useful plant in the southern parts of Russia and in central Europe like S. officinale and S. x uplandicum.*
In the dried leaves the total alkaloid content amounts up to 0.13%, in the dried roots up to 0.14 to 0.37%.
In these the alkaloids the N-oxides of intermedine and lycopsamine, the 7-acetyl derivatives of the latter two alkaloids symlandine, symviridine, myoscorpine, symphytine as well as the major alkaloid echimidine typical of S. asperum were detected.
The presence of the alkaloids asperumine, lasiocarpine, echiumine, makrotomine, and rousorine found in the older studies of *Manko et al by means of paper chromatography could not be confirmed.
Prickly Comfrey should no longer be used for medicinal purposes."
-'Medicinal Plants in Europe Containing Pyrrolizidine Alkaloids' by Erhard Thomas Roeder, Pharmazeutisches Institut (Pharmaceutical Institute) der Rheinischen Friedrichs-Wilhelms, University of Bonn, Germany, Pharmazie (Pharmacy) 50, pages 83-98, March **1995**.
(* -'Pyrrolizidine Alkaloids and Pyrrolizidine-N-oxide Alkaloids from Boraginaceous Plants' by I.V. Manko and Y.G. Borisyuk; Ukraine Khim. Zhur., Volume 23, page 362, 1957. I was unable to find this article.)

*"**S. asperrimum Lepech, Prickly Comfrey, contains up to 1300 mg/kg total alkaloids in the leaf, and 1400 to 3700 mg/kg in the root**."*
-'Using Herbs that Contain Pyrrolizidine Alkaloids' by Alison Denham, B.A., MNIMH (National Institute of Medical Herbalists), University of Central Lancashire, England; The European Journal of Herbal Medicine, Volume 2, No. 3, pages 27-38, **1996**.

"Comfrey (Symphytum asperum Lepech) is widely used in Kenya (east Africa) and other countries as a vegetable, herbal medicine and animal feed. It is known to have high nutritional and medicinal value.

In Kenya, no study has been done to ascertain the levels of these pyrrolizidine alkaloids (PAs) despite the fact that the plant is widely grown and consumed in central Kenya and other parts.

There is need to establish whether these PAs are also present in this species here in Kenya or whether the levels have been affected by genetic drift or epigenetic factors.

This study was aimed at determining the levels of the pyrrolizidine alkaloids (PAs) in the roots and leaves of Comfrey from two agro-ecological zones in Kenya, Kiambu and Kisii counties during wet (April 2012) and the dry (September 2012) seasons.

From the root samples, echimidine, 7-acetyllycopsamine, 3-acetyllycopsamine, triangularine and heliosupine were identified.

7-Acetyllycopsamine showed significantly higher levels during the wet season than in the dry season *(p=0.033, alpha=0.05, t-test).*

The other pyrrolizidine alkaloids did not vary significantly between the two seasons *(p>0.05, alpha=0.05, t-test).*

All the compounds reported in the root samples are associated with toxicity, and their values are above the tolerable levels as recommended by relevant regulatory bodies in various countries.

Echimidine, one of the most toxic PAs, was found in all the root samples. The leaf samples however, had levels of PAs below detectable limits *using GC-MS (Gas Chromatography - Mass Spectrometer).*

Pyrrolizidine Alkaloid	Number	Dry (Mean+-SE) ppm	Wet (Mean+-SE) ppm	p-value
Echimidine	12	3.55 +- 0.23	3.67 +- 0.56	0.841
Heliosupine	10	3.52 +- 0.38	4.27 +- 0.80	0.408
7-Acetyllycopsamine	6	4.08 +- 0.64	6.58 +- 0.85	0.033
3-Acetyllycopsamine	4	7.00 +- 2.06	6.53 +- 0.73	0.840
Triangularine	8	5.50 +- 0.71	3.61 +- 0.60	0.070

-'Determination of Levels of Pyrrolizidine Alkaloids in Symphytum Asperum Lepech Growing in Selected Parts of Kenya' thesis by Shylock O. Onduso, Kenyatta University, Kenya, 119 pages, October **2014**.

(Genetic drift is variation in different genotypes in a small population, owing to disappearance of particular genes as individuals die or do not reproduce. Epigenetic means nongenetic influences on gene expression.)

(Statistical 'mean' is the average used to derive the central tendency of the data. SE or Standard Error is the standard deviation of a statistical population. It measures the accuracy of the sample for representing a population. The p-value is the level of marginal significance in a statistical hypothesis representing the probability of the occurrence.)

(ppm= parts per million.)

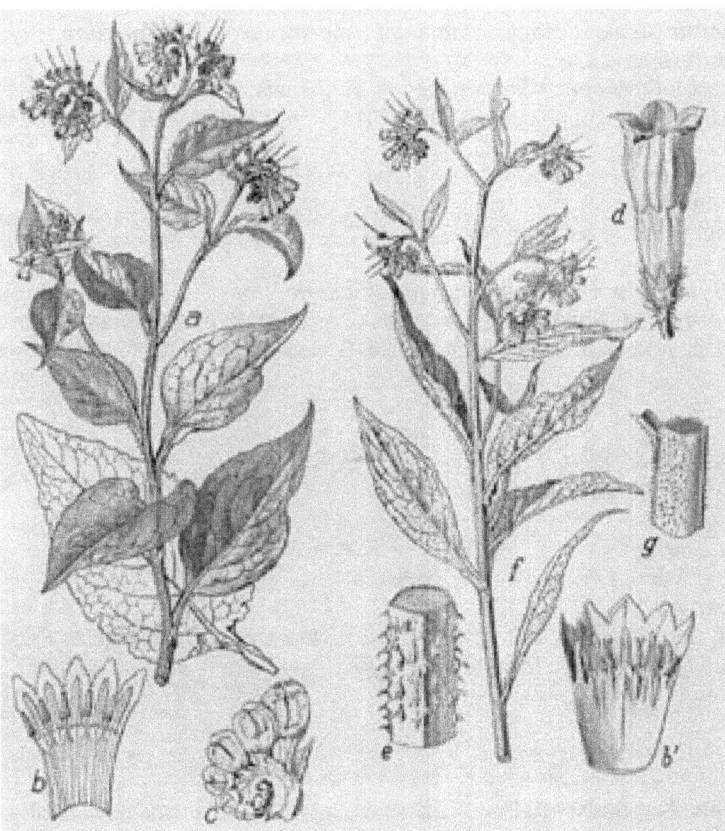

Symphytum Asperum Lepechin 1926
Symphytum Uplandicum Nyman

'Illustrierte Flora von Mittel-Europa: Mit Besonderer Berucksichtigung von Deutschland, Osterreich und der Schweiz' (Illustrated Flora of Central Europe: With Special Consideration of Germany, Austria and Switzerland), Volume 5, Part 3' by Gustav Hegi, Zurich, Switzerland. Published by J.F. Lehmann, Munich, Germany, 1926.

Symphytum pages 2220-2230. In German, 13 volumes written from 1908 to 1931.

Gustav Hegi (1876-1932) was a Swiss botanist. Illustrierte Flora von Mittel-Europa is one of the most comprehensive floras in the world. It contains morphological, ecological and phytogeographical information about all plant species in Central Europe.

Chapter 8

Details about Symphytum Species:
Officinale (Common Comfrey)

Symphytum officinale (Common Comfrey)
See 'Symphytum bohemicum', 'Symphytum patens', 'Symphytum tanaicense' and 'Symphytum uliginosum' in Chapter 9.

Current Botanical Nomenclature

    Accepted
        Symphytum officinale L.
        Symphytum officinale subsp. bohemicum (F. W. Schmidt) Celak.
        Symphytum officinale subsp. officinale
        Symphytum officinale subsp. officinale Sibth.
        Symphytum officinale subsp. uliginosum (A.Kern.) Nyman
            = Symphytum tanaicense Stev.

    Not Accepted
        Symphytum patens Sibth.
        Symphytum officinale ssp. uliginosum auct. non (Kern.) Nyman
        Symphytum officinale var. officinale L.

"Symphytum officinale L. (Sp. Pl. 1: 136. 1753)"
-'International Plant Names Index'® (IPNI), www.ipni.org, A database of the names and associated basic bibliographical details of seed plants, ferns and lycophytes, **2018**.   (L.= Carl Linnaeus who created the modern system of naming organisms.)

"Symphytum officinale: accepted name.
 Symphytum officinale subsp. bohemicum (F. W. Schmidt) Celak.**: accepted name. Infraspecific taxon.* ***Symphytum officinale subsp. officinale*: accepted name. Infraspecific taxon."*
-Catalogue of Life®: 2015 Annual Checklist, Online database of the world's known species of animals, plants, fungi and micro-organisms. It consists of a single integrated species checklist and taxonomic hierarchy., **2018**, www.catalogueoflife.org. Search: www.catalogueoflife.org/annual-checklist/2015/search/all
    (Infraspecific is at a taxonomic level below that of species, e.g., subspecies, variety, cultivar, or form. Latin names at this level require the addition of a term denoting rank.)
    (sp.= single species)   (spp.= more than 1 species)   (ssp. = subspecies)

*"**Symphytum patens Sibth.: ambiguous synonym for Symphytum officinale subsp. officinale Sibth.**"*
-Catalogue of Life®: 2015 Annual Checklist, Online database of the world's known species of animals, plants, fungi and micro-organisms. It consists of a single integrated species checklist and taxonomic hierarchy., **2018**, www.catalogueoflife.org. Search: www.catalogueoflife.org/annual-checklist/2015/search/all

"Symphytum officinale L.: Accepted.
 Symphytum officinale subsp. uliginosum (A.Kern.) Nyman*: Accepted."*
-The Plant List®, www.theplantlist.org from the World Checklist Database, **2018**.

"Symphytum officinale L. - name accepted (Common Comfrey) =
Symphytum officinale ssp. uliginosum auct. non (Kern.) Nyman - name not accepted =
Symphytum officinale var. officinale L. - name not accepted."
-ITIS (Integrated Taxonomic Information System®) Report, www.itis.gov, **2018**. The ITIS is a partnership of United States Federal agencies formed to satisfy their needs for scientifically credible taxonomic information.   (auct.= auctorum= various authors)

"Symphytum officinale L.
    Symphytum officinale L. ssp. uliginosum auct. non (Kern) Nyman =
    Symphytum uliginosum auct. non Kern."
-U.S. Department of Agriculture, Natural Resources Conservation Service®, Plants Database®, www.plants.usda.gov, **2018**.

*"**Symphytum officinale L. Sp. pl. 1:136. 1753** ('Species Plantarum', Volume 1, page 136, in year 1753.)*
    = Consolida major Gilib.

= *Symphytum officinale var. ochroleucum DC., 1846*
= *Symphytum officinale var. rectiflorum Touss. & Hoschede, 1898*
Immediate Children Subspecies:
Symphytum officinale subsp. *bohemicum* (F.W.Schmidt) Celak.
Symphytum officinale subsp. *officinale* L., 1753
Symphytum officinale subsp. *uliginosum* (A.Kern.) Nyman
Immediate Children Variety:
Symphytum officinale var. *borysthenicum* L.
Symphytum officinale var. *lanceolatum* A.DC.
Symphytum officinale var. *officinale* L."
-Global Biodiversity Information Facility® (GBIF), www.gbif.org, **2018**. International network and research infrastructure funded by the world's governments and aimed at providing anyone, anywhere, open access to data about all types of life on Earth.

S. officinale Subspecies and Varieties (in chronological order)

The taxonomic ranks: Genus, Subgenus, Section, Subsection, Series, Species, Subspecies, Variety, Subvariety, Form, Subform.

A **'subspecies'** is a grouping within a species used to describe geographically isolated variants, a category above 'variety'. It is indicated by the abbreviation 'subsp.' in the scientific name.
A **'variety'** consists of more or less recognizable entities within species that are not genetically isolated from each other, below the level of subspecies. It is indicated by the abbreviation 'var.' in the scientific name.

*"220. **Symphytum patens,** foliis ovato-lanceolatis rentibus, calyce patente tubo corollae breviori. **Red flowered Comfrey.** Ditches. Banks of Rivers. Banks of the Thames by Caversham. Fl. Maio. {Flowers in May.}"*
-'Flora Oxoniensis Exhibens Plantas in Agro Oxoniensi' (Flora of Oxfordshire, England) by John Sibthorp, **1794**, page 70.

"Reports of the Floral Committee. The subjects exhibited were as follows:
Symphytum officinale variegatum sulphureum: *From Mr. Salter. A distinct hardy perennial, with the **leaves broadly margined with yellowish-green.** Though a pretty plant it was not thought equal to the variety of S. tuberosum already noticed."*
-'Proceedings of the Royal Horticultural Society, Volume 1', London, England, pages 564 and 565, June 1 1859 to December 31 **1861**. April 9 1861, page 565.   (Variegated means leaves that are edged or patterned in a second color, especially white.)

*"**Symphytum officinale sulphureum variegatum** struck us as being a particularly useful subject, on account of the **fine bold character of its golden leaves and its almost perpetual blooming habit.**"*
-'The Gardeners Chronicle: A Weekly Illustrated Journal of Horticulture and Allied Subjects', New Series, January to June 1877: Volume VII, July to December **1877**: Volume VIII. London, England. This excerpt is from October 13, 1877, page 467.

"Two forms of S. officinale L.:
    *S. officinale L. var. ochroleucum DC. **(cream-colored flower)***
    *S. officinale L. var. purpureum Pers. **(reddish-purple flower)**."*
-'Symphytum Survey: B.S.B.I.' by F.H. Perring, Watsonia: Journal and Proceedings of the Botanical Society of the British Isles', London, England, Volume 8, **1970**, page 91.

*"In 'A Revision of the Genus Symphytum' by Bucknall (*1913) three taxa are recognized in the **'Section Officinalia':***
    a. Symphytum officinale L.
    b. Symphytum officinale L. var. lanceolatum Weinm.
    c. Symphytum uliginosum Kern. (= S. officinale L. subsp. uliginosum (Kern.) Nym.).
The plants of these taxa differ in the decurrence of the leaves, in the indument of the stems, leaves and sepals and in the colour of the flowers.
Also the distribution shows marked differences:
    The typical variety of S. officinale occurs throughout Europe, except for the Caucasus and the extreme north and south.
    The variety lanceolatum in Bessarabia (parts of Moldova and Ukraine in eastern Europe), the Crimea and the Caucasus.
    And finally S. uliginosum in Hungary and south Russia.
The (low-lying peat land) Dutch (Netherlands) plants with 2n = 40 have many features in common with S. uliginosum, but are *not entirely identical with the material collected in the type locality by Kerner.*
The variety lanceolatum Weinm. is intermediate between S. officinale and S. uliginosum.
    The width of the leaves are narrowly lanceolate in the variety lanceolatum. A distinctive appearance of the variety lanceolatum is said to be given by the uppermost leaves, which are supported in a nearly erect position and generally overtop the flowers (Bucknall, 1913).
    This, however, can also be found in many other plants with the chromosome numbers 2n = 24 and 2n = 48, which belong to the typical variety of S. officinale. Therefore, this character does not seem to be distinctive.
These facts led us to the conclusion that the Dutch plants with 2n = 40 agree largely but not entirely with S. uliginosum

and differ from S. officinale L. var. lanceolatum Weinm.
At any rate, the whole Section Officinalia seems to be a comparium in the sense of *Danser, to which also S. asperum can be added.
> *These experiments made clear that S. asperum and S. officinale form a very complicated species complex, not under natural conditions, but by the influence of man."*

-'Cytotaxonomic Studies in the Genus Symphytum III: Some Symphytum Hybrids in Belgium and the Netherlands' by Th.W.J. Gadella and E. Kliphuis, Biologisch Jaarboek (Biological Yearbook), **1971**.
(* -'A Revision of the Genus Symphytum, Tourn.' by Cedric Bucknall, {Mus. Bac. Oxon= Bachelor of Music, Oxford University}, Journal of the Linnean Society of London, England, Botanical Journal, Volume 41, Issue 284, pages 491-556, December 1913.)
(** -'Ueber die Begriffe Komparium, Kommiskuum und Konvivium und Ueber die Entstehungsweise der Konvivien' {On the Concepts of the Comparium, the Commiscuum and the Convivium and on the Origin of the Convivia} by B.H. Danser, Amsterdam, Netherlands; Genetica: An International Journal of Genetics and Evolution, Volume 11, Issue 5, pages 399-450, September 1929.)

> (Crimea is a peninsula on the northern coast of the Black Sea in Eastern Europe that is almost completely surrounded by the Black Sea and Sea of Azov to the northeast.)
> (The Caucasus Mountains are located at the border of Europe and Asia, between the Black Sea and Caspian Sea and occupied by Russia, Georgia, Azerbaijan, and Armenia.)
> (Comparium is a group of organisms that can interbreed and are equivalent in scope to a taxonomic genus. For Danser in 1929 comparium is a group comprising all coenospecies which may participate in mutual hybridization, whether direct or indirect. Coenospecies is a collection of related species able to hybridize.)
> (A species 'complex' is a group of closely related species that are extremely similar in appearance, so much so that the boundaries between them are frequently unclear.)

"Symphytum officinale L. subsp. (subspecies) officinale:
Middle and upper leaves deeply and broadly decurrent, and **the stem distinctly winged.** *Stem, leaves, pedicels and calyces more or less densely hairy and setose. Throughout the range of the species."*
-Flora Europaea, Volume 3: Diapensiaceae to Myoporaceae, editors Tutin, Heywood, Burges, Moore, Valentine, Walters and Webb. United Kingdom: Cambridge University Press, 1972. (5-volume encyclopedia of plants, published between 1964 and 1993), page 104, **1972**. (Symphytum L., pages 103-105 by B. Pawlowski.) ('Flora Europaea' describes all the national floras of Europe to identify any wild or widely cultivated plant to the subspecies level. It provides geographical distribution, habitat preference, and chromosome number.)

"Symphytum officinale L.:
Subsp. bohemicum (F. W. Schmidt) Celak:
> *Short, usually less than 1 meter (3.2 feet) tall. Corolla with greenish-yellow buds becoming creamy-white on opening.* **This form seems to be almost confined to relict fenland (marsh) in V.C. 29 & 31 (Vice-counties Cambridgeshire and Huntingdonshire, England).** *It is reproductively isolated from other cytotypes and never occurs in mixed colour populations in Britain.* **2n = 24, diploid.**

Subsp. officinale:
> *Taller, up to 1.5 meters (4.9 feet). Corolla variable in colour from creamy-white (var. ochroleucum DC.) to carmine (var. purpureum Pers.). Populations including pure cream, pure carmine (vivid crimson) and mixed colours, often with alternating vertical dark and light bands ('peppermint stripe') are frequent, especially in ditches and on stream and river banks in *V.C. 9, 11-13, 22-24 and 33-37.*
> **Three forms may be recorded: cream, carmine and 'peppermint stripe'. 2n = 48, tetraploid. "**

-Plant Crib: Handbook for Field Identification compiled by T.C.G. Rich and M.D.B. Rich with editorial assistance of F.H. Perring for the BSBI Monitoring Scheme. London, England: Botanical Society of the British Isles, 1988, page 74. There is also an updated **1998** edition.

> (*V.C. or Vice-County or biological Vice-County is a geographical division of the British Isles for biological recording and other scientific data-gathering. There are 112 vice-county boundaries for England, Scotland and Wales. There are 40 vice-counties in Ireland.
> V.C. 9, 11-13, 22-24 & 33-37= Vice-Counties Dorset, South Hampshire, North Hampshire, West Sussex, Berkshire, Oxfordshire, Buckinghamshire, East Gloucestershire, West Gloucestershire, Herefordshire and Worcestershire- all in England, and Monmouthshire in Wales.)
> (Relict means a surviving species of an otherwise extinct group, or a remnant of a formerly widespread species that lives in isolated areas.)

"Plants of the Symphytum officinale complex *were sampled at natural sites in southern and central Hesse (Germany) and investigated for 53 external characteristics.* **The position of each plant, whether belonging to the diploid (2n = 24) Symphytum bohemicum or to the tetraploid (2n = 48) Symphytum officinale, was defined by determining its chromosome number.**
> *Subsequently the morphological differences between the two taxa were examined using characteristics that can be easily collected in the field. The ranges of variation of most characteristics overlap considerably. Differences exist in the leaf-shape and the flower-structure.*

The two taxa are especially differentiated in two characteristics: in the ratio of the apical aperture of the corolla to the

maximum diameter of the tube, and in the length of the distal style segment protruding from the corolla.
The statistical analysis of several combinations of characteristics revealed a clear differentiation between Symphytum officinale and Symphytum bohemicum.
Finally the results are interpreted in the light of previously published work. The classification of the taxa within the complex is discussed.
Symphytum bohemicum fulfills the criteria of a biospecies and thus should be treated separately, even though not every individual can be unequivocally identified by its morphological characteristics."
-'Symphytum Officinale (Boraginaceae) in Southern and Central Hesse: Cytological and Morphological Investigation of the Discrimination of Taxa' (Symphytum Officinale Boraginaceae in Sud- und Mittelhessen: Cytologisch-Morphologische Untersuchungen zur Abgrenzung der Sippen) by Irith Wille, Botanik und Naturschutz in Hessen, Frankfurt am Main, Germany, Volume 10, pages 87-119, **1998**. All in German except abstract is also in English. (If you have an English translation, could you please send it to me.)
    (Biospecies is a group of interbreeding individuals isolated reproductively from other groups)

Three subspecies of Symphytum officinale are recognised by Gadella & Perring (*2000) as subsp. officinale, subsp. bohemicum, and subsp. uliginosum.
S. officinale subsp. uliginosum (A. Kern.) Nyman has not been recorded from the British Isles since before 1930 (**Clement & Foster 1994) although it may occur here (Perring 1994). It is distinguished from the other subspecies by the outer surface of the sepals being more sparsely pubescent and on chromosome number being **2n = 40**. It also nearly always has purple buds and flowers (some cytotypes of S. officinale and S. x uplandicum also have purple buds and flowers) and Gadella & Perring (2000) provide further details on its identification.
S. officinale subsp. bohemicum occurs in the Netherlands, and Eastern Europe, particularly in the Czech Republic and Hungary. In the British Isles, it is recorded from only three English vice counties: Cambridgeshire v.c. 29 and Huntingdonshire v.c. 31 (***Stace et al. 2003) and South Lincolnshire v.c. 53 (Perring 1994), with all records being from fens."
-'What is Symphytum Officinale Subsp. Bohemicum (Schmidt) Celak: A Taxon on the Red Data Waiting List' by Clare Coleman O'Reilly, County Durham, England; Botanical Society of Britain and Ireland (BSBI) News, Hertfordshire, England, No. 102, pages 46-48, April **2006**.
(* -The European Garden Flora: A Manual for the Identification of Plants Cultivated in Europe, Both Out-of-Doors and Under Glass, 6 Volumes, edited by James Cullen, Sabina G. Knees, and H. Suzanne Cubey. England: Cambridge University Press, 2000. Volume 6: 'Symphytum' by T.W.J. Gadella and F.H. Perring, pages 138-141.)
(** -Alien Plants of the British Isles: Provisional Catalogue of Vascular Plants Excluding Grasses by E.J. Clement and M.C. Foster. London, England: Botanical Socieity of Britain and Ireland, 1995.)
(*** -New Flora of the British Isles: Identification of Wild Vascular Plants of the British Isles edited by Clive Stace. Cambridge, England: Cambridge University Press, 2003.)

"**Symphytum officinale subsp. officinale is less common than and over-recorded for S. x uplandicum**, which itself is fertile and backcrosses to subsp. officinale forming intermediates.
Symphytum officinale subsp. bohemicum has only recently been discovered by British botanists and requires further study."
-'Atlas 2020 Notes on Identification Works and Difficult and Underrecorded Taxa' by P.A. Stroh, D.A. Pearman, F.J. Rumsey and K.J. Walker; 'Botanical Society of Britain and Ireland' and 'Biological Records Centre', Bristol, England, June **2015**, page 28.
    ('Over-recorded' means botanists thought they found a particular species but actually it was another species,
    so they put in their records this incorrect geographical distribution of a plant.)

"**Symphytum officinale L.**, Sp. Pl. 1: 136. 1753
    S. patens Sibth., Fl. Oxon.: 70. 1794;
        = **S. officinale L. var. purpureum Pers.**, Syn. Pl. 1: 161. 1805;
        = **S. officinale var. patens (Sibth.) Nyman**, Consp. Fl. Eur.: 509. 1881.
    S. bohemicum F.W.Schmidt, Fl. Boem. 3: 13. 1794 ;
        = **S. officinale var. bohemicum (F.W.Schmidt) Nyman**, Consp. Fl. Eur.: 509. 1881;
        = **S. officinale L. subsp. bohemicum (F.K.Schmidt.) Celak.**
            in Sitzungsber. Konigl. Bohm. Ges. Wiss., Math.-Naturwiss. Cl. 1891(1): 29. 1891;
        = **S. officinale L. var. typicum f. bohemicum (F.W.Schmidt)**
            Fiori in Fiori & Beg., Fl. Italia 2(3): 377. 1902.
    S. officinale L. var. ochroleucum DC., Prodr. 10: 37. 1846.
    S. tanaicense Steven in Bull. Soc. Imp. Nat. Moscou 24: 1851.
        = **S. officinale L. subsp. tanaicense (Steven) Soo** in Bot. Kozlem. 28: 127. 1931.
    S. uliginosum A.Kern. in Oesterr. Bot. Z. 13: 227. 1863;
        = **S. officinale L. subsp. uliginosum Nyman**, Consp. Fl. Eur.: 509. 1881;
        = **S. officinale L. f. uliginosum** Fiori in Fiori & Beg., Fl. Italia 2(3): 377. 1902."
-'Synopsis of Boraginaceae subfam. Boraginoideae Tribe Boragineae in Italy' is the same as 'Boraginaceae in Italy II' by L. Cecchi and F. Selvi, University of Florence, Firenze, Italy; Plant Biosystems: An International Journal Dealing with all Aspects of Plant Biology, Official Journal of the Societa Botanica Italiana, Vol 149, No. 4, page 630-677, **2015**, Symphytum page 636-646.

*"**Symphytum officinale: This species is widespread and somewhat difficult taxonomically.**
Several cytotypes have been reported (e.g. Gadella & Kliphuis 1969, 1972, Gadella 1972), mainly diploids (2n = 24), hypotetraploids (2n = 40) and tetraploids (2n = 48).*
In our opinion, only the tetraploid populations should be considered as S. officinale, whereas diploid populations correspond to S. bohemicum and hypotetraploids to S. tanaicense (synonym S. officinale subsp. uliginosum; *Gadella & Kliphuis 1969, *Majovsky & Hegedusova 1993).*
Symphytum officinale s.str. is distributed almost throughout the whole of Europe (***Hulten & Fries 1986). It was introduced to China (**Zhu et al. 1995) and North America (Gadella 1984), mostly as green forage for livestock and due to its use in traditional medicine.
It grows on wet meadows, along rivers and in humid ruderal habitats such as damp ditches or road edges. In the Czech Republic it is common from the lowlands to the mountains, being less frequent or under-recorded only in western Bohemia."
-'Distributions of Vascular Plants in the Czech Republic, Part 3' by Z. Kaplan, J. Danihelka, M. Lepsi, P. Lepsi, L. Ekrt, J. Chrtek Jr, J Kocian, J Prancl, L. Kobrlova, M. Hrones and V. Sulc, Czech Republic; Preslia: The Journal of the Czech Botanical Society, Volume 88, pages 459-544, **2016**. The third part of the publication series on the distributions of vascular plants in the Czech Republic includes grid maps.
(* -Flora Slovenska {Flora of Slovakia}, Volume 5/1 (V/1) or 5/2 (V/2), edited by L. Bertova and K. Goliasova. Bratislava, Slovakia: Veda, 1993. Article: 'Symphytum L.' by J. Majovsky and Z. Hegedusova, pages 76-97. Some sources say 5/2 but I think 5/1 is correct.)
(** -Flora of China, Volume 16 {Gentianaceae through Boraginaceae}, edited by Z.Y. Wu, P.H. Raven and D.Y. Hong. Science Press {Beijing, China} and Missouri Botanical Garden Press {Saint Louis, Missouri}, 1995. Article: 'Boraginaceae' by G.L. Zhu, H. Riedl and R.V. Kamelin, pages 329-427. Symphytum page 359. With 22 volumes it is the first modern English-language description of 31,000 species of vascular plants of China.)
(*** -Atlas of North European Vascular Plants North of the Tropic of Cancer, Volumes 1-3 by E. Hulten and M. Fries. Konigstein, Germany: Koeltz Scientific Books, 1986.)
    (sensu stricto = s.s., s. str., sens. str., sens. strict. In the strict/narrow sense. It is added after a taxon to mean it is being used in the sense of the original author, or without taxa which may otherwise be associated with it.)

Symphytum officinale 'Alba' or Symphytum officinale var. Alba:
It is the same as regular Common Comfrey except it has white flowers. In the late 1800s it was called Symphytum officinale fol. var Alba. This is not the same as 'White Comfrey' which is Symphytum orientale.

Herbalist's Shop

'Officinale' means 'of the Herbalist's shop'. 'Officinale' means it was the official medicinal plant sold in apothecaries and pharmacopoeias in the middle ages (5th- 15th centuries AD). 'Officinalis' or 'officinale' are Medieval Latin words that mean a plant was used in medicine and herbalism. The word frequently occurs as the second term of a two-part botanical name.
An apothecary is a medical professional who formulates and dispenses 'Materia Medica' to physicians, surgeons, and patients. Today a pharmacist (chemist in Britain) does the same job. Apothecary is also a shop or store where these medicines are sold. 'Materia Medica' is the scientific study of medicinal drugs and their sources, preparation, and use.

*"The term 'Officinal' first came into use in its application to medicines, according to the Oxford Dictionary, **in 1693; the Latin Officina being applied to the storeroom of a monastery in which medicines, etc., were kept, and so herbs, plants, drugs, etc., kept in stock in an apothecary's shop became 'Officinal' or as being of recognized utility.**"*
-The Medicinal Uses of Comfrey by Dr. Charles MacAlister, M.D., F.R.C.P., 1935.
(In Comfrey: Fodder, Food & Remedy, and Comfrey: An Ancient Medicinal Remedy.)

*"**Officinale:
The treasure house of knowledge, and the source of physical and spiritual healing in Europe for many centuries was the monastery and such religious institutions of learning;*** *the store room where the medical herbs were kept, the plants and drugs, etc. was the Officina, and the name of Common Comfrey comes from this interesting and significant fact of history. And for those interested in tracing its medical history, we can add that the Comfreys and their healing powers are recorded in the **Pharmacopoeias of various countries** as follows: Belgium, Radix (root) Symphyte; France, Consonde; Mexico, Sinfito; Portugal, Consolida Major; Spain, Sinfito Major.*
All names suggest making sound or well, binding or healing."
-Comfrey: Nature's Healing Herb & Health Food by Andrew Hughes. Japan: Sanyusha Publishing Co., Ltd, **1992**, page 122.

*"**The officina was the building, usually an out-building, in medieval monasteries where medical monks prepared medicaments and pharmaceutical preparations to heal the sick.*** *Dried extracts, infusions, decoctions, tinctures and distillates were prepared therein. Often the officina was attached to the medicinal or herbal gardens, also enclosed within the monastery precinct.*
When Linnaeus invented the binomial system of nomenclature, he gave the specific name 'officinalis' to dozens of herbs and plants whose medical use had been established in preceding millennia.
    In the 1735 (first edition) of his 'Systema Naturae', he acknowledged the historical traditions of healing by naming

scores of plants with the species designator, 'officinalis', as a generic qualifier.
Literally 'from the officina', the species name 'officinalis' thus embodied the history of many centuries of medicinal use and health lore.
Plants Officinalis: In 1753 Linnaeus published 'Plants Officinals' (or 'Plants Officinalis'). In this work he catalogued many examples of plants and herbs with long-established medical use.
> He wrote, 'Cetainly there can be no doubt that no plant is given, that cannot have great usefulness in sustaining and restoring people's health when that has been lost. Those that have been chosen to treat disease are the strongest, now called officinals, that one may make use of them more than others.' "

-'On Officinalis: The Names of Plants as One Enduring History of Therapeutic Medicine' by John Pearn, Queensland, Australia; Vesalius: International Society of the History of Medicine, Belgium; Vesalius Congress Supplement, pages 24-29, Dec **2010**.

Distribution (Geographic Locations) of S. officinale (in chronological order)

"The Common Comfrey (Symphytum officinale) is abundantly met with in England, but is rare in Scotland; the Tuberous Comfrey is commonly found in Scotland, but is seldom met with in England, the northern counties of England and North Wales being its extreme southern limit, so that except in the narrow zone of country common to both, there will be no possibility of mistaking the one species for the other."
-A Modern Herbal: The Medicinal, Culinary, Cosmetic and Economic Properties, Cultivation and Folk-Lore of Herbs, Grasses, Fungi, Shrubs and Trees with their Scientific Uses by Mrs. M. Grieve. New York: Dover Publications, 1971. First published **1931**.

"Comfrey (Symphytum officinale), Boraginaceae:
*This is a plant of moist and watery places, ditch-sides and by water taps, although **I have found a form of it growing upon sun-baked banks in many parts of Algeria (North Africa)**."*
-The Complete Herbal Handbook for Farm and Stable by Juliette de Bairacli Levy. London, England: Faber & Faber, Inc., **1952**.

*"S. officinale L. is a rather local plant of river banks and wet ditches **occurring in most of Europe, except the arctic, eastwards to the Caucasus, west Siberia and central Asia.***
> In western Europe it most commonly has yellowish-white flowers, but purple-flowered forms were recorded by *Ray in the 17th Century and still earlier by **Gerarde.
> It is interesting to note that Popov (***1953) does not mention yellowish-white form at all as a Russian or Asiatic plant."

-'The Genus Symphytum in Britain' by T.G. Tutin, University College of Leicester, England; Watsonia: Journal of the Botanical Society of the British Isles, Volume 3, pages 280-281, February **1956**. (Watsonia was the journal of the 'Botanical Society of Britain and Ireland' from 1949 to 2010. It is now called 'New Journal of Botany'.)
(* -J. Ray, Catalogus Plantarum Angliae, 79. London, 1670. John Ray, Fellow of the Royal Society, 1627-1705, was an English naturalist. He published important works on botany, zoology, and natural theology. His classification of plants in his 'Historia Plantarum', was an important step towards modern taxonomy.)
(** -John Gerard or Gerarde, 1545-1612, was an English botanist. He was the author of the 1,484-page illustrated 'Herball, or Generall Historie of Plantes', first published in 1597.)
(*** -M.G. Popov, in Komarov, V.L., Flora U.S.S.R., 9, Leningrad, 1953. The Russian botanist Mikhail Grigorievich Popov, 1893-1955, was the author of more than 200 scientific publications devoted to taxonomy and research on the flora of different regions of Eurasia, in particular, of Siberia, Middle Asia, Caucasus, the Carpathians, the Sakhalin Peninsula and in the region of Lake Baikal. He published about 300 new names of plants.)
> (The Caucasus Mountains are located at the border of Europe and Asia, between the Black Sea and Caspian Sea and occupied by Russia, Georgia, Azerbaijan, and Armenia.)

"Symphytum officinale:
*Still used in country districts in Britain as a poultice. Native. In damp places, especially beside rivers and streams. **Generally distributed throughout Great Britain, though less common in the north and probably not native there; common but probably nowhere native in Ireland. Central Scandinavia to Spain and eastwards to western Siberia and Turkey**."*
-Flora of the British Isles, Second Edition by A.R. Clapham, T.G. Tutin and E.F. Warburg. England: Cambridge University Press, **1962**, (first edition 1952).

Symphytum officinale Distribution:
Much of Europe, but rare in the extreme south and only as a naturalized alien in much of the north (Austria, Belgium/ Luxembourg, Britain, Bulgaria, Czechoslovakia, France, Germany, Switzerland, Netherlands, Hungary, Italy, Yugoslavia, maybe Portugal, Poland, Romania, USSR {Baltic region, Central region, Northern region, Southwestern region, Crimea, Southeastern region}, Sardinia, Sicily, Turkey-European part, Denmark, Finland, Ireland, Norway, Sweden)."
-Flora Europaea, Volume 3: Diapensiaceae to Myoporaceae, editors Tutin, Heywood, Burges, Moore, Valentine, Walters and Webb. United Kingdom: Cambridge University Press, **1972**, page 104. (Symphytum L., pages 103-105 by B. Pawlowski.) (5-volume encyclopedia of plants, published between 1964 and 1993). ('Flora Europaea' describes all the national floras of Europe to identify any wild or widely cultivated plant to the subspecies level. It provides geographical distribution, habitat preference, and chromosome number.)   (Naturalized plants become established and survive in a region other than their place of origin.)

"*Symphytum officinale* Linnaeus, Sp. Pl. 1: 136. 1753. In Chinese: Ju He Cao. Found in forests.
>**China: Fujian (province southeast China), Hebei (province north China), Liaoning (province northeast China), Taiwan, Xinjiang (territory northwest China).**
>Also in Kazakhstan, Kyrgyzstan, Russia, Tajikistan, Turkmenistan, Uzbekistan, Europe.

This species was introduced in China in 1963 as green forage for livestock."
-Flora of China, Volume 16 {Gentianaceae through Boraginaceae}, edited by Z.Y. Wu, P.H. Raven and D.Y. Hong. Science Press {Beijing, China} and Missouri Botanical Garden Press {Saint Louis, Missouri}, **1995**. Article: 'Boraginaceae' by G.L. Zhu, H. Riedl and R.V. Kamelin, pages 329-427. Symphytum page 359. With 22 volumes it is the first modern English-language description of 31,000 species of vascular plants of China.

"**Two species of Comfrey are plentiful in the United Kingdom, the Common *Symphytum officinale* and Tuberous *Symphytum tuberosum*** while a third, Soft Comfrey *Symphytum orientale* is less abundant.
The two commoner species can be separated by a number of means:
>**The Common Comfrey occurs mainly in southern regions, is found almost exclusively in damp places, has variable coloured flowers that can be white, cream, pink or purple and is a tall plant which can reach a height of 120 cm (3.9 feet).**
>*Tuberous Comfrey, on the other hand, has a more northerly distribution, can be found in drier areas, has yellow flowers and attains a maximum height of only 50 cm (1.6 feet).*

Both flower between May and July with odd plants hanging on a little later."
-'Growing and Collecting Wild and Cultivated Greenfoods and Seeds' by Dave Coles, Goring, Reading, England, **2003**.

"***Symphytum officinale*, Phytoclimatic Group 1 (PCG1):**

Temperature	StandardDev	Precipitation	Max1	Max2	Max3	Min1	Min2	Min3
9.48 C = 49 F	5.64	823 mm	21.33	12.80	33.65	-0.13	5.95	-17.02

Phytoclimate: Four phytoclimatic groups, i.e., **groups of species with similar responses to climatic variables**, were identified based on the species' climate optima (optimum ranges) within the study area.
>**1. PCG1 group contained 124 species, which mainly occurred in the northern part of France, and in Belgium and the Netherlands.** Few sample locations in the southern part of France had more than 30 percent of the occurring species belonging to PCG1.
>**2. PCG2** was a smaller group, containing 24 species having their optimum in the central part of France. The frequency of PCG2 species was lower both at northern locations (Belgium, the Netherlands) and at southern (Mediterranean) locations.
>**3.** This is in contrast with the frequency pattern of the species belonging to **PCG3** (containing 64 species) for which the frequency was high in southern and central France, but low in Belgium. Moreover, these species were almost totally absent in the Netherlands.
>**4.** The 24 species of the **PCG4** are clearly more adapted to the climatic conditions in the southern part of France, and did not occur in the northern part of the study area.

Precipitation (Rain, Snow): Variation in precipitation was represented by the total annual precipitation, expressed in mm.
Temperature- Mean, Standard Deviation: With regard to temperature, mean, minimum and maximum month temperatures were available. The mean annual temperature was calculated as the average mean temperature over the 12 months. Variation in temperature throughout the year was expressed by the standard deviation on the mean annual temperature (StDev), as a measure for continentality.
Temperature- Maximum and Minimum:
As survival and fitness of plant species often depends rather on extreme than on average climate values, some values representing maximum and minimum temperature throughout the year were calculated:
>***Max1:*** Mean temperature of the warmest month.
>***Max2:*** Average maximum month temperature.
>***Max3:*** Highest maximum temperature.
>***Min1:*** Mean temperature of the coldest month.
>***Min2:*** Average minimum month temperature.
>***Min3:*** Lowest minimum temperature."

-'Climate Gradients Explain Changes in Plant Community: Composition of the Forest Understorey' by Sebastiaan Van Der Veken, Beatrijs Bossuyt and Martin Hermy, Laboratory for Forest, Nature and Landscape Research, Catholic University of Leuven, Belgium; Belgian Journal of Botany, Volume 137, No. 1, pages 55-69, **2004**.
(Phytoclimate is the climate of a small area such as a plant community or a specific wooded area. It can be different than the climate of the larger region.)
(Continentality is difference between land and marine climates. There is a greater temperation range over land than water.)
>(See Symphytum tuberosum section in Chapter 9 for similar analysis by same authors.)

"Sustainable land management requires scientists to provide reliable data on diversity distribution patterns. Mean species richness per site in the three land cover types surveyed. **The most common species for each land cover type:**
>**Arable (suitable for growing crops): *Symphytum officinale*.**
>*Grassland (pastures and meadows): (no Symphytum)*
>*Forest: (no Symphytum)*"

-'Developing Robust Field Survey Protocols in Landscape Ecology: A Case Study on Birds, Plants and Butterflies' by Jacqueline Loos, Jan Hanspach, Henrik von Wehrden, Monica Beldean, Cosmin Ioan Moga and Joern Fischer from Germany and Romania; Biodiversity and Conservation, Volume 24, Issue 1, pages 33-46, January **2015**.

"*Native dominant species were defined 'a priori' from expert judgement, as being species with mainly or wholly competitor growth strategies (sensu Grime 1974*) that also commonly form mono-dominant stands alongside rivers in Britain.*
   These comprised *Angelica sylvestris, Carex acutiformis, Carex aquatilis, Carex riparia, Carex rostrata, Epilobium hirsutum, Filipendula ulmaria, Glyceria maxima, Oenanthe crocata, Petasites hybridus, Phalaris arundinacea, Phragmites australis, Sparganium erectum,* **Symphytum officinale** *and Typha latifolia.*
Significant indicator species for the nine invaded and uninvaded rivers sites:
   Symphytum officinale: *Very highly associated with this type of site, based on Monte Carlo permutation tests.*"
-'Effects of Invasive Alien Plants on Riparian Vegetation and their Response to Environmental Factors' by Zarah Pattison for Degreee of Doctor of Philosophy, Biological and Environmental Sciences, School of Natural Sciences, University of Stirling, Scotland, September **2016**.
(* -'Vegetation Classification by Reference to Strategies' by J.P. Grime, The University, Sheffield, England; Nature, Volume 250, pages 26-31, July 5 1974.)
   ('A priori' means based on theoretical deduction rather than empirical observation.) ('Sensu' is Latin for 'in the sense of'.)

"*The second part of a revision of the genus Symphytum in the Czech Republic deals with Symphytum officinale agg. (aggregate). In our flora, this complex is represented by two native species: S. officinale s.str. and S. bohemicum.*
Symphytum officinale occurs throughout the Czech Republic *and grows from the lowlands to the mountains. It is the most common and widespread species of the genus Symphytum.*
In contrast, S. bohemicum is classified as endangered and is confined to east, central and north Bohemia *(especially to the river basins of the Labe, Ohre, Metuje and Cidlina).*
An overview of morphological characters, habitat preferences and distribution maps are provided for both taxa."
-'The Genus Symphytum L. (Comfrey) in the Czech Republic II: S. Officinale Agg.' (Rod Symphytum Kostival v Ceske Republice II: S. Officinale Agg.) by Lucie Kobrlova, Univerzity Palackeho v Olomouci, Czech Republic; Zpravy Ceskoslovenske Botanicke Spolecnosti (Czech Botanical Society News), Prague, Czech Reupublic, Volume 52, pages 175-223, December **2017**. All in Czech except abstract is also in English.
   (Species aggregate is a species group that is closely related and difficult to distinguish among them. Similar but not identical is a single species that displays large variations but can not strictly be subdivided: s.l. for 'sensu lato' {in a wider/broader sense} or s. ampl. for 'sensu amplo' {in a very wide/relaxed sense}.)
   (sensu stricto = s.s., s. str., sens. str., sens. strict. In the strict/narrow sense. It is added after a taxon to mean it is being used in the sense of the original author, or without taxa which may otherwise be associated with it.)

"**Symphytum Officinale Distributional Range: Native**
  Asia-Temperate
    Western Asia: *Turkey*
    Caucasus: *Russian Federation-Ciscaucasia (Ciscaucasia)*
    Siberia: *Russian Federation-Western Siberia (Western Siberia)*
    Middle Asia: *Kazakhstan*
  Europe
    Northern Europe: *United Kingdom*
    Middle Europe: *Austria, Belgium, Czech Republic, Germany, Hungary, Netherlands,*
     Poland, Slovakia, Switzerland
    Eastern Europe: *Belarus, Moldova, Russian Federation-European part,*
     Ukraine (including Krym)
    Southeastern Europe: *Bulgaria, Croatia, Italy (including Sardinia, Sicily), Romania,*
     Serbia, Slovenia
    Southwestern Europe: *France, Spain.*"
-U.S. National Plant Germplasm System® (NPGS), https://npgsweb.ars-grin.gov, United States Department of Agriculture (USDA), Agriculture Research Service, **2018**. NPGS is collaborative effort to safeguard genetic diversity of agriculturally important plants.

"**Symphytum officinale:**
Ecology: *This tall perennial herb occurs on the banks of streams and rivers, in ditches, fens (wetland) and marshes, and on damp road verges (curb strip/lawn). Generally lowland, reaching 320 meters (1049 feet) altitude near Buxton (Derbyshire County, England).*
Status: *Native in Great Britain.*
Trends: *S. officinale was over-recorded for S. x uplandicum in the 1962 *'Atlas of the British Flora' and this confusion still obscures its true distribution. It may well be alien in much of northern and western Britain and in Ireland but* **it is impossible to determine the native range with any certainty** *and all records are mapped as if they are native. Both diploid and tetraploid (chromosome) cytotypes occur and are morphologically separable.*

S. x uplandicum sometimes back-crosses with this parent (**Perring, 1994).
World Distribution: *European Temperate element; also in central Asia and widely naturalised outside its native range."*
-'Online Atlas of the British and Irish Flora', https://www.brc.ac.uk/plantatlas, Distribution and ecology with photographs, **2018**.
(* -Atlas of the British Flora by Franklyn Perring and S. Max Walters. Harpenden, Hertfordshire County, England: Botanical Society of the British Isles. First published 1962, second edition 1976, third edition 1982.)
(** -The Common Ground of Wild and Cultivated Plants: B.S.B.I. Conference Report No. 22 by A. Roy Perry and R.G. Ellis, {F.H. Perring, pages 64-70}. Cardiff, Wales: Department of Biology, National Museum of Wales, 1994.)

Description of S. Officinale (in chronological order)

"Comfrey (Symphtym officinale): Description:
The common great Comfrey hath divers (various) very large and hairy green leaves, lying on the ground, **so hairy or prickly, that if they touch any tender part of the hands, face, or body, it will cause it to itch.** *The stalk that riseth up from among them, being two or three feet (0.6-0.9 meters) high, hollowed, and cornered, as also very hairy, having many suchlike leaves as grow below, but runs less and less up to the top.*
At the joints of the stalks it is divided into many branches, with some leaves thereon; and at the ends stand many flowers in order one above another, which are somewhat long and hollow like the finger of a glove, of a **pale whitish colour,** *after which come small black seed.*
The roots are great and long, spreading great thick branches under ground, black on the outside and whitish within, short or easy to break, and full of a glutinous (sticky) or clammy juice, of little or no taste.
There is another sort in all things like this, save only it is somewhat less, and beareth **flowers of a pale purple colour.**
Place: *They grow by ditches and water sides, and in divers fields that are moist,* *for therein they chiefly delight to grow.*
Time: *They flower in June and July, and give their seed in August."*
-'**Culpeper's English Physician and Complete Herbal** to Which are Now First Added Upwards of One Hundred Additional Herbs' by Nicholas Culpeper, English herbalist, London, England, **1652-1653**, pages 131-132.

"Common Comfrey: Symphytum foliis ovato-lanceolatis. The Symphytum, with ovato-lanceolate leaves.
The root is oblong, irregularly shaped, an inch (2.5 cm) thick, black on the outside, white within, and extremely viscous.
The radical leaves are a foot (0.3 meter) long, three or four inches (7.6-10.1 cm) broad in the middle, even at the edges, rough to the touch, of a pale green, and terminate in a point. The stalk grows to two feet (0.6 meter), or more, in height; the shape is round, but it is edged with membranes from the bases of the leaves; it is as thick as a finger, green, hairy, and rough to the touch; toward the top it is divided into several branches, which are covered with long series of flowers, moderately large, and **of a white colour, sometimes purple, sometimes yellow: this variation of the colour on the flower, has by some been made the distinction of two different species, but erroneously.**
The plant is a native of England, and is common in watery places.
C. Bauhine calls it Symphytum consolida major; others, Consolida major, and Symphytum majus, or Vulgare.
The root of this is a powerful agglutinant, good in the fluor albus."
-'**A History of Plants:** A General Natural History, or New and Accurate Descriptions of the Animals, Vegetables and Minerals of the Different Parts of the World' by John Hill, M.D., London, England, **1751**, page 254.
(For more about C. Bauhine, see '1596, 1623 Bauhin' in section 'Symphytum Species Classifications'.)
(Fluor albus is a white discharge from the uterus or vagina, i.e., leukorrhea.)

"Symphytum Officinale; Common Comfrey:
Leaves ovate-lanceolate, decurrent. Root perennial, fleshy, mucilaginous, externally black; stem two or three feet high (0.6-0.9 meters), upright, leafy, winged, hispid with deflexed hairs.
Clusters of flowers in pairs on a common stalk, with an odd flower between them, recurved, dense, hairy; **corolla (petals) yellowish white, sometimes purple,** *the rays down at each edge.* **The variety with a red or purple flower is more common in many parts of the continent than in England.**
Mr. Miller asserts, that the difference in colour is permanent in the plants raised from seeds; and that the purple and whitish-yellow flowers are never found mixed, where the plants grow wild. It is a native of Europe, and also of Siberia: common in watery places on the banks of rivers and ditches, **flowering from the end of May to September.**
Propagation:
This and the following species may be cultivated by sowing the seeds in the spring, or by parting their roots in the autumn, when almost every piece of a root will grow. They should be planted about two feet and a half feet (0.76 meters) asunder (apart), and will require no further care but to keep them clear from weeds; for they are hardy enough for any soil or situation."
-'**The Universal Herbal; or, Botanical, Medical, and Agricultural Dictionary;** Containing an Account of All the Known Plants in the World, Arranged According to the Linnean system. Specifying the uses to which they are or may be applied, whether as food, as medicine, or in the arts and manufactures, with the best methods of propagation, and the most recent agricultural improvements' by Thomas Green, London, England, **1824**, page 641.
(A leaf wing or plant fin is a green, short, thin, paperlike structure extending along a stem. It is long along the stem but not tall away from the stem. Decurrent means a plant part that extends downward. It means leaf blades that partly wrap or have wings around the stem or petiole and extend down along the stem. Petiole is stalk that joins a leaf to a stem.)

"**Comfrey: Symphytum officinale.** *Plate XIII. Fig. 155.*
Specific Characteristics: Leaves ovate-lanceolate, decurrent. Stem winged on the upper part. Root fleshy.
This plant is abundant in many places where the soil is damp or watery, especially by the side of rivers and ditches.
It grows from two to three or even four feet (0.60 to 0.91 to 1.21 meters) high. The whole plant is covered with short stiff hairs, especially the leaves, which are very large and broadly ovate-lanceolate in form; the margins of the upper ones are decurrent, or continued down the stem in wing-like ridges.
The flowers are in forked curled cymes; they are of a dingy yellow, or pale rust colour, or rarely purple, and open all the summer.
A smaller plant called the Tuberous Comfrey, found in the south of Scotland, appears to be only a variety of this, and possesses the same properties."
-'**The Useful Plants of Great Britain:** Part I, August' by John E. Sowerby (Illustrator) and C. Pierpoint Johnson (Description), London, England, **1862**, page 182.
  (A cyme is a flower cluster with a central stem. The flower at the end opens first. The other flowers develop as buds of the side stems.)

"**Species I: Symphytum Officinale Linn.**
  Plates (images) MCXV. MCXVI.
  Reich. Ic. Fl. Germ. et. Helv. Volume XVIII. Tab. (image) MCCCIII. Fig. 1.
    (Heinrich Gustav Reichenbach, *Icones Florae Germanicae et Helveticae, 1856-1858.)
  Billot, Fl. Gall. et Germ. Exsicc. No. 2887.
Rootstock vertical, passing insensibly into the thick fleshy root, which divides into large branches. Stem very thick, branched, strongly winged above. Leaves ovate or lanceolate, the upper ones, especially the pair at the base of the racemes, strongly decurrent and lanceolate.
Calyx segments triangular-lanceolate, acuminated, divided nearly to the base, in fruit submuricated on the central line of each segment, with stiff prickle-like bristles seated on large tubercles. Corolla about twice as long as the calyx; scales included. Plant clothed with short pubescence, intermixed with harsh bristly hairs, few of them gland-tipped.
  S. officinale Var. (variety) genuinum.
  S. officinale Var. patens: Corolla ochreous, or more or less stained with pale purple.
    (S. patens, Sibth. = John Sibthorp, Flora Oxoniensis. 1794.)
    (Reich. Fl. Germ. Excurs. p. 347. = Ludwig Reichenbach, Flora Germanica Excursoria. 1830-1832.)"
-'**English Botany (Sowerby's)**; or Coloured Figures of British Plants: Volume 7' by John T. Boswell and John Edward Sowerby, London, England, **1880**, page 114.
(* -'Icones Florae Germanicae et Helveticae, Volume XVIII' by Heinrich Gottlieb Ludwig Reichenbach and Heinrich Gustav Reichenbach; Lipsiae {Leipzig}, Germany, 1858. In Latin. Symphytum pages 57-58.)

"*Symphytum officinale L. Sp. Pl. 136. 1753.* **Comfrey. Healing-herb.**
Roots thick, deep; stem erect, branched, 2-3 feet (0.6-0.9 meters) high. Leaves lanceolate, ovate-lanceolate, or the lower ovate, pinnately veined, 3-10 inches (7.6-25.4 cm) long, acute or acuminate at the apex, narrowed into margined petioles, or the uppermost smaller and sessile, decurrent on the stem; petioles of the basal leaves sometimes 12 inches (30.4 cm) long.
Flowers numerous, *in dense racemes or clusters; calyx-segments ovate or ovate-lanceolate, acute or acuminate, much shorter than the corolla; corolla yellowish or purplish; nutlets brown, shining, slightly wrinkled.*
In waste places, *Newfoundland (Canada) to Minnesota, south to Virginia and North Carolina. Naturalized or adventive (introduced) from Europe. Native also of Asia.*
Flowers June to August. *Back- or black-wort. Bruisewort. Knitback. Boneset. Consound. Gum-plant."*
-'**An Illustrated Flora of the Northern United States and Canada**, Volume 3: Gentianaceae to Compositae (Gentian to Thistle)' by Nathaniel Lord Britton, PH.D., Sc.D., LL.D. and Hon Addison Brown, A.B., LL.D., New York Botanical Garden, **1913**, page 92.

"This well-known showy plant (*Symphytum officinale*) is a member of the Borage and Forget-me-not tribe, **Boraginaceae.** The plant is erect in habit and rough and hairy all over. There is a branched rootstock, the roots are fibrous and fleshy spindle-shaped, an inch (2.5 cm) or less in diameter and up to a foot (30.5 cm) long, smooth, blackish externally, and internally white, fleshy and juicy.
The leafy stem, 2 to 3 feet (0.60 to 0.91 meter) high, is stout, angular and hollow, broadly winged at the top and covered with bristly hairs. The lower, radical (near the roots) leaves are very large, up to 10 inches (25.4 cm) long, ovate (egg) in shape and covered with rough hairs which promote itching when touched.
The stem-leaves are decurrent, i.e. a portion of them runs down the stem, the body of the leaf being continued beyond its base and point of attachment with the stem. They decrease in size the higher they grow up the stem, which is much branched above and terminated by one-sided **clusters of drooping flowers, either creamy yellow, or purple,** growing on short stalks.
These racemes of flowers are given off in pairs, and are what is known as scorpoid in form, the curve they always assume suggesting, as the word implies, the curve of a scorpion's tail, the flowers being all placed on one side of the stem, gradually tapering from the fully-expanded blossom to the final and almost imperceptible bud at the extremity of the curve, as in the Forget-Me-Not. The corollas (all of the petals) are bell-shaped, the calyx (all of the sepals) deeply five-cleft, narrow to lance-shaped, spreading, more downy in the purple flowered type.
The fruit consists of four shining nutlets, perforated at the base, and adhering to the receptacle by their base.
Comfrey is in bloom throughout the greater part of summer, the first flowers opening at the end of April or early May."

-**A Modern Herbal:** The Medicinal, Culinary, Cosmetic and Economic Properties, Cultivation and Folk-Lore of Herbs, Grasses, Fungi, Shrubs and Trees with their Scientific Uses by Mrs. M. Grieve. New York: Dover Publications, 1971. First published **1931**.
>(Raceme is an unbranched, elongated inflorescence with pedicellate flowers maturing from the bottom upwards. Inforescence is the flowering part of a plant; a flower cluster. Nutlet is one of the lobes or sections of the mature ovary of some members of the Boraginaceae, Verbenaceae and Labiatae/Lamiaceae families.)

"Broadly speaking, Symphytum officinale have paler green foliage, a higher fibre content, and lesser growth speed than Russian Comfrey; even with increased vigour from good manurial treatment and cultivation, they cannot compare with the imported species (Russian Comfrey)."
-**Russian Comfrey: A Hundred Tons an Acre of Stock or Compost for Farm, Garden or Smallholding** by Lawrence D. Hills. London England: Faber and Faber, Limited, **1953**, page 21.

-'Determination of Allantoin in Comfrey Root (Symphytum officinale L.)' by F. Kaczmarek and A. Walicka, Biuletyn Instytutu Roslin Leczniczych, 4, Jan **1958**. pages 273-280. I am unable to find this. If you have a copy, could you please send it to me.

"S. officinale L.: An erect hispid perennial 30-120 cm (11.8-47.24 inches). Root thick, fleshy, fusiform, branched. Stem clothed with long deflexed conical hairs, branched, winged with decurrent leaf-bases. Leaves sparsely pilose, rarely with tuberclate bristles; the lower 15-25 cm (5.9-9.8 inches), ovate-lanceolate, petioled; upper oblong-lanceolate, broadly decurrent at base. Calyx 7-8 mm (0.27-0.31 inch), whitish, yellowish-white, purplish or pink. Scales triangular-subulate, scarcely longer than the stamens. Nutlets shining, black. Flower 5-6.   2n = 36 (chromosomes).
Hemicryptophyte Hs: Semi-rosette hemicryptophytes, with leafy stems but the lower leaves larger than the upper ones and the basal internodes shortened."
-**Flora of the British Isles**, Second Edition by A.R. Clapham, T.G. Tutin and E.F. Warburg. England: Cambridge University Press, **1962**, (first edition 1952).

>*"Note on Life Forms: The Danish botanist Raunkiaer has classified plants, according to the position of the resting buds or persistent stem apices in relation to soil level,* into a number of Life Forms, which have in turn been subdivided so as to convey more information about the plants included in the different groups.
>Life Forms are a convenient method of indicating how a plant passes the unfavourable season (winter), and are also of interest because they show a correlation with climate. The primary classes are:
>1. Phanerophytes: woody plants with buds more than 25 cm (9.8 inches), above soil level.
>2. Chamaephytes: woody or herbaceous plants with buds above soil surface but below 25 cm.
>3. **Hemicryptophytes: herbs (very rarely woody plants) with buds at soil level.**
>>Hp: Protohemicryptophytes, with uniformly leafy stems, but the basal leaves usually smaller than the rest.
>>**Hs: Semi-rosette hemicryptophytes, with leafy stems but the lower leaves larger than the upper ones and the basal internodes shortened.**
>>Hr: Rosette hemicryptophytes, with leafless flowering stems and a basal rosette of leaves.
>4. Geophytes: herbs with buds below the soil surface.
>5. Helophytes: marsh plants.
>6. Hydrophytes: water plants.
>7. Therophytes: plants which pass the unfavourable season as seeds."
>-**Flora of the British Isles**, Second Edition by A.R. Clapham, T.G. Tutin and E.F. Warburg. England: Cambridge University Press, **1962**, (first edition 1952).
>>(Christen Christensen Raunkiaer, 1860-1938, was a Danish botanist, who was a pioneer of plant ecology. He discovered plant strategies of how to survive an unfavourable season that he called 'Life Forms', and he showed these strategies corresponded to climate zones. This scheme is still used today and is a forerunner of modern plant strategies such as J. Philip Grime's CSR Triangle theory.)

"S. officinale L., Sp. Pl. 136 (1753).
Stock stout, vertical, branched. Stem 30-50 to 120 cm (11.8-19.6 to 47.2 inch), stout, erect, often branched. Leaves large, ovate-lanceolate to lanceolate, acuminate; the middle and upper sessile, often decurrent.
Cymes many-flowered. Calyx lobed to 3/4 to 4/5, with lanceolate lobes. Corolla 12-18 mm, purple-violet or dirty pink or white, with deflexed lobes. Scales broadly triangular-lanceolate, the lower marginal papillae shortly cylindric-conical, the upper much smaller and shorter, all dense. Stamens with connective projecting beyond thecae; filaments as wide as anther.
Nutlets 5-6 mm, black, very smooth, shining. River-banks and damp grassland.
-**Flora Europaea, Volume 3: Diapensiaceae to Myoporaceae,** editors Tutin, Heywood, Burges, Moore, Valentine, Walters and Webb. United Kingdom: Cambridge University Press, **1972**, page 104. (Symphytum L., pages 103-105 by B. Pawlowski.) (5-volume encyclopedia of plants, published between 1964 and 1993 describes all national floras of Europe to identify any wild or widely cultivated plant to subspecies level. It provides geographical distribution, habitat preference, and chromosome number.)

'Studies on Symphytum Species: HPLC Determination of Allantoin' by R. Dennis, C. Dezelak and J. Grime, Acta pharmaceutica Hungarica, Vol 57, No. 6, pages 267-274, Nov **1987**. I could not find this. If you have a copy, could you please send it to me.

"S. officinale L.- Common Comfrey:

Stems erect, well-branched, to 1.5 meter (4.9 feet), from thick, vertical root; stem-leaves sessile, long-decurrent; calyx divided about 2/3 to 4/5 way to base; corolla purplish or pale creamy-yellow.
Native; by streams and rivers, in fens and marshy places, also roadsides and rough ground; **locally frequent in British Isles, but less common than and over-recorded for S. x uplandicum.**
The flowers are often wrongly described as white; except for very rare albinos they are pale creamy-yellow (or purplish)."
-**New Flora of the British Isles:** Identification of Wild Vascular Plants of the British Isles edited by Clive Stace. Cambridge, England: Cambridge University Press, **1991**, page 647.
    ('Over-recorded' means botanists thought they found a particular species but actually it was another species, so they put in their records this incorrect geographical distribution of a plant.)

"*Symphytum officinale* L.:
Variable in height, up to 1.5 meters (4.9 feet). Stems and leaves covered in rather soft, long, deflexed, conical hairs. Leaves strongly and broadly decurrent, **the 'wings' exceeding one internode so that the stems show two pairs of 'wings' in any cross-section.**
Calyx 8-13 mm, about 1/2 as long as the corolla with lanceolate, acute teeth 2-3 times as long as tube, enlarging only slightly in fruit. Limb of the corolla urceolate. Stamens with anthers longer than filaments."
-Plant Crib: Handbook for Field Identification compiled by T.C.G. Rich and M.D.B. Rich with editorial assistance of F.H. Perring for the BSBI Monitoring Scheme. London, England: Botanical Society of the British Isles, 1988, page 74. There is also an updated **1998** edition.

"*Symphytum officinale* (Comfrey):
Hardiness Zone: 4-8; Light: full sun to partial sun. Soil pH: Acid / Neutral / Alakaline.
Clumping herb. Height: 3 to 5 feet (0.9-1.5 meters); Width: 3 to 5 feet.
Growth rate: fast. Native region: Eurasia.
Habitats: disturbed, old fields, edges, other habitats. Areas recently disturbed, whether by humans or natural causes. These plants can be observed as urban or garden 'weeds' or roadside plants. Old fields that are in the process of becoming forest. Edges are boundaries between two habitats, here generally indicating the boundary between forest and more open fields.
Medicinal: excellent. Some or all parts of the plant are poisonous.
Dynamic accumulator, invertebrate shelter, nectary.
Roots are fibrous and fleshy: Dividing into a large number of fleshy (thick or swollen) roots immediately upon leaving the crown.
Nuisance: persistent. Very difficult to remove once planted, often resprouting from even tiny pieces of root.
Nuisance: dispersive. Annoyingly successful at spreading by seed, whether in the garden or in the neighborhood."
-Edible Forest Gardens: Volume Two: Design & Practice by Dave Jacke and Eric Toensmeier. White River Junction, Vermont: Chelsea Green, **2005**, page 490.

"***Symphytum officinale* L. (Boraginaceae) or Common Comfrey is a perennial native of Europe and Asia** and has been naturalised throughout North America**. It is very common in all of Europe, especially in damp soils (*Bruneton, 1999).
There are about 25 species of the Symphytum genus, including further medicinal plants apart from Common Comfrey (e.g., S. asperum Lepechin, S. tuberosum L., and S. x uplandicum Nyman (synonym S. peregrinum, S. asperum x officinale), according to (**, ***,****).
Comfrey grows well in rich, moist, low meadows, or along ponds and river banks, where it may reach a height of 1.2 meters (3.9 feet).
Comfrey root is large, branching, and black on the outside with a creamy white interior containing slimy mucilage. The root is slimy and horn-like when dried.
Hollow, erect stems, also containing mucilage, are covered with bristly hairs that cause itching when in contact with the skin. The thick, somewhat succulent, veined leaves are covered with rough hairs. They are alternate and lance shaped, with lower leaves as large as 25 cm (9.8 inch) in length, and dark green on top and light green underneath. The lower ones and the basal ones are ovate-lanceolate and pulled together in the petiole; the upper ones are lanceolate and broad.
Small, bell shaped flowers grow from the axils of the smaller, upper leaves on red? stalks. Flowers are mauve to violet and form in dense, hanging clusters, blooming in summer. They are arranged in crowded, apical, 2-fayed hanging cymes. The calyx is fused and has 5 tips. The corolla is also fused and is cylindrical-campanulate with a pentangular tube and 5-tipped border. The tips are revolute and there are 5 awl-shaped scales in the mouth of the tube. The scales are close together in a clavate form and have a glandular tipped margin. There are 5 stamens and 1 style.
The ovary is 4-valved. The fruit consists of 4 smooth, glossy nutlets (**, *****)."
-'**European Medicines Agency- Assessment Report on Symphytum Officinale L., Radix**' by European Medicines Agency, Committee on Herbal Medicinal Products (HMPC), London, England, 27 pages, May 5 **2015**.
(* -Pharmacognosy, Phytochemistry, Medicinal Plants, 2nd Edition by J. Bruneton. Paris, France: Lavoisier Publishing, 1999.)
(** -'The Gale Encyclopedia of Alternative Medicine, Volume 1' edited by J.L. Longe, Thomson Gale, Michigan: Gale Group, pages 526-527, 2005.)
(*** -'Adverse Effects of Herbal Drugs' by De Smet, Keller, Hansel and Chandler, Editors, Springer-Verlag, Berlin-Heidelberg, Germany, pages 194-205, 220-222, published 1992.)
(**** -'Flora Europaea' by Tutin, Heywood, Burges, Moore, Valentine, Walters and Webb, University Press, Cambridge, United

Kingdom, 1992.)
(***** -'PDR for Herbal Medicines, 4th Edition' edited by J. Gruenwald, T. Brendler and C. Jaenicke, Thomson Healthcare Inc., Montvale, New Jersey, pages 212-214, published 2000.)   (Mauve is a pale purple color named after the mallow flower.)

"**Symphytum Officinale:** *Robust, taprooted perennial 30-120 cm (11.81 - 47.24 inch) tall, often with several clustered stems. Stem and flower cluster with spreading or bent back, stiff hairs.*
Young Plant: *Cotyledons elliptic or ovate, rounded tip, more or less stalked. Leaves entire, elliptic or ovate, regularly dentate.*
Stems: *50 - 100 cm (19.68 - 39.37 inch), simple to branched, sharp-bristly.*
Leaves: *Large, the basal (base) ones stalked, with ovate or lance-ovate blade mostly 15 - 30 cm (5.9 - 11.8 inch) long and 7 - 12 cm (2.7 - 4.7 inch) wide.*
Stem leaves alternate, gradually reduced and with shorter stalks but still ample, the upper leaves are commonly stalkless. **Stem evidently winged by the conspicuously downward-extended bases of the leaves.**"
-'**Symphytum Officinale**' by Bayer Crop Science®, Crop Compendium, Bayer® is a 150-year Life Science company in health care and agriculture, global headquarters are in Leverkusen, Germany, **2018**.
    (Cotyledon is the primary or rudimentary leaf of the embryo of seed plants.)

More about S. Officinale Flowers and Leaf Wings

A leaf wing or plant fin is a green, short, thin, paperlike structure extending along a stem. It is long along the stem but not tall away from the stem.
In botany, decurrent means a plant part that extends downward. It means leaf blades that partly wrap or have wings around the stem or petiole and extend down along the stem. Petiole is the stalk that joins a leaf to a stem.

"*Symphytum officinale flowers are 3/4 inch long (1.9 cm) and cream-yellow. They can be purple but it can be identified by the wide wings that continue right down the flower stems from leaf to leaf, and the pointed, long, narrow leaves.*"
-Comfrey: Fodder, Food and Remedy by Lawrence D. Hills. New York: Rizzoli Universe Books: **1976**, page 62.

"*In summary, one can say of Symphytum officinale that while it has good medical properties, it is far less valuable than Symphytum peregrinum.*
It has paler green leaves than peregrinum; it has a higher fiber content, lower speed of growth and even with the best manuring and location, it cannot compare with its later offspring.
Symphytum officinale can be identified by its flowers: white to cream and red, with one variety dull purple.
The leaf wings continue right down the flower stems from leaf to leaf; *the leaves are narrower and sharply pointed with thin stems, and the yield is low.*
It is also subject to rust (Melampsorella symphyti), which does not affect either asperrimum (Prickly Comfrey) or peregrinum."
-Comfrey: Nature's Healing Herb & Health Food by Andrew Hughes. Japan: Sanyusha Publishing Co., Ltd, **1992**, page 141.

"**Symphytum officinale L.: Common Comfrey.**
    Symphytum officinale L. ssp. uliginosum (Kern.) Nyman.
    Symphytum tanaicense Steven.
    Symphytum uliginosum Kern.
At present there appears to be two cytotypes of Symphytum officinale that may warrant recognition at some level.
    The 2n = 24 or 48 type (diploid and tretraploid, respectively):
        Appears to be the common form in New England (northeast United States). It shows hispid stems that are not harsh to the touch, **leaves with prominent decurrent wings,** *marginal setae of the sepals distributed in an irregular pattern, and* **white (usually diploid) or purple (usually tetraploid) corollas.**
    The 2n = 40 type:
        Known from North America but not yet documented in New England, has tuberculate-based hairs on the stem that are harsh to the touch, **leaves with shorter decurrent wings,** *marginal setae of the sepals distributed in a uniform manner, and* **purple corollas. This latter cytotype has been named ssp. uliginosum (or S. tanaicense at the rank of species).**"
-New England Wild Flower Society's Flora Novae Angliae: A Manual for the Identification of Native and Naturalized Higher Vascular Plants of New England by Arthur Haines. New Haven: Connecticut: Yale University Press, **2011**.
    (A cytotype is an individual of a species that has a different chromosomal factor to another.)
    (Body cells and individuals are described by the number of chromosome sets {ploidy level} they have: monoploid: 1 set; diploid: 2 sets = 2n = 2x, triploid: 3 sets = 2n = 3x, tetraploid: 4 sets = 2n = 4x, pentaploid: 5 sets = 2n = 5x, etc.)

"*Shoot structure of Symphytum officinale was studied in terms of its architecture, vascular anatomy and morphogenesis.*
Stems of Symphytum officinale also bear green wings of leafy nature. They run downwards along the stem pairwisely from each leaf base and bear vascular bundles which originate from the veins of leaf blade of second to fourth order.
*The stem wings of S. officinale are believed to be decurrent parts of leaf bases (*Sokoloff 2009).*
    Morphologically, stem wings of S. officinale are parts of leaf bases and receive their vascular supply from leaf veins."
-'Shoot Structure of Symphytum Officinale L. (Boraginaceae) in Relation to the Nature of its Axially Shifted Lateral Branches' by

Ksenia V. Kotelnikova, Daria J. Tretjakova and Maxim S. Nuraliev (all from Moscow State Univeristy, Russia); Wulfenia Journal: Multidisciplinary ISI Journal in All Fields of Sciences, Klagenfurt, Austria, Volume 18, pages 63-79, **2011**.
(* -Botany: Systematics of Higher Plants, Volume 4(2), edited by A.K. Timonin, D.D. Sokoloff and A.B. Shipunov. Moscow, Russia: ITS Akademiya, 2009. In Russian. Article: 'Lamiales' by D.D. Sokoloff, pages 284-288.)

"Symphytum officinale: Stem evidently winged by the conspicuously downward-extended bases of the leaves."
-'Symphytum Officinale' by Bayer Crop Science®, Crop Compendium, Bayer® is a 150-year Life Science company in health care and agriculture, global headquarters are in Leverkusen, Germany, **2018**.

Common Comfrey Flower Colors

"Symphytum officinale 'Purpureum' has purple flowers; 'Ochroleucum' has white or yellowish-white flowers, and 'Coccineum' has crimson flowers; 'Argenteum' has creamy marginal variegation on the leaves and produces purple flowers."*
-'Proceedings of Botanical Society of British Isles, Volume 7' edited by E.F. Greenwood, London, England, 1967-1969, page 433.
(* -'Cultivars of British Wild Plants' by E. Knowles, The Gardeners' Chronicle: A Weekly Illustrated Journal of Horticulture and Allied Subjects, London, England, Volume 163, No. 11, page 9, **1968**.)
(Variegated means leaves that are edged or patterned in a second color, especially white.)

"S. officinale occurs in two colour forms: yellowish-white: var. (variety) ochroleucum DC., and reddish-purple: var. purpureum Pers. (var. patens Sibth.). The former occurs mainly in south and east England whilst the latter occurs in the south and west.
Var. purpureum is not so frequent in natural habitats but if it was introduced, it must have been over 300 years ago as it was known to Ray and Gerarde (Tutin, 1956).
Where the distributions of the two colour varieties overlap, populations exhibiting all intermediate shades occur."
-'Network Research, Symphytum Survey' by Franklyn Perring, Biological Records Centre, Monks Wood, Huntingdon, Cambridgeshire County, England; Volume 7, No. 4, **1969**. In 'Proceedings of Botanical Society of the British Isles, Volume 7' edited by E.F. Greenwood, London, England, 1967-1969, pages 553-556.

"Except for the purple-flowered botanical variety patens (var. patens), Common Comfrey has cream-yellow flowers. Other cytotypes of this species found in continental Europe have white, red, or purple flowers."
-'Comfrey: A Controversial Crop' by Robert G. Robinson, University of Minnesota, Agricultural Experiment Station, Minnesota Report MR-191, Item No. AD-MR-2210, **1983**, page 2.

"S. x uplandicum Nyman was first described as **Symphytum patens E.M. Fries (non S. patens Sibth.)** *in 1839 from the province of Uppland in Sweden, where it must have resulted from the crossing of native S. officinale with introduced S. asperum. This invalid name was later replaced by the name S. uplandicum Nyman.*
The range of variation in the hybrid is great and plants may be found showing all combinations of the characters of the parents.
The fact that the parents maintain their distinctness would appear to be due mainly to their ecological isolation; *S. asperum (Prickly Comfrey) is a plant of dry places, while S. officinale (Common Comfrey), occurs only by water.*
Further investigation may show, nevertheless, that these two taxa should be regarded as subspecies in order to emphasise what appears to be their close genetical similarity.
On the scanty evidence available about the distribution of the two colour forms of S. officinale, it is perhaps unwise to draw conclusions, but it is tempting to suggest that the purple-flowered form may be due to introgression (repeated backcrossing) between S. officinale and S. asperum."
-'The Genus Symphytum in Britain' by T.G. Tutin, University College of Leicester, England; Watsonia: Journal of the Botanical Society of the British Isles, Volume 3, pages 280-281, February **1956**. (Watsonia was the journal of the 'Botanical Society of Britain and Ireland' from 1949 to 2010. It is now called 'New Journal of Botany'.)
(Introgression is the transfer of genetic information from one species to another as a result of hybridization between them and repeated backcrossing.)

"Much work on the Symphytum complex was undertaken by the Dutch botanists Th.W.J. Gadella and E. Kliphuis, and together with F.H. Perring they also looked at some British populations.
*A relatively recent statement of the accepted position is given by Perring (*1994) in which the* **purple flowered Comfreys are:**
S. officinale, 2n = 48, *with red buds opening to purple flowers. (Common Comfrey)*
S. x uplandicum, 2n = 36, with deep purple buds opening to a colour ranging from purple violet to violet blue. (Russian Comfrey)
A frequent form of S. x uplandicum, 2n = 40, with pink buds opening blue, and the **cream coloured forms of S. officinale, 2n = 24 and 2n = 48** *do not pose a problem here, nor does* **purple flowered S. officinale, 2n = 40** *since it is stated to occur only in Holland."*
-'Atlas 2000 and a Problem with Purple Flowered Comfrey' by Eric Chicken, East Yorkshire, England; Botanical Society of Britain and Ireland (BSBI) News, Hertfordshire, England, No. 76, pages 22-23, September **1997**.

(* -'The Common Ground of Wild and Cultivated Plants: B.S.B.I. Conference Report No. 22' by A. Roy Perry and R.G. Ellis, {F.H. Perring, pages 64-70}. Cardiff, Wales: Department of Biology, National Museum of Wales, 1994.)

Flowers and Pollination in S. officinale
See sub-subsection 'Comfrey Flower Perforations by Insects' in subsection 'Comfrey Flower' in section 'Symphytum Genus Description' (Chapter 5).
See sub-subsection 'Comfrey Flower Buds, Bloom, Nectar and Pollination' in subsection 'Comfrey Flower' in section 'Symphytum Genus Description' (Chapter 5).
See sub-subsection 'Bees and Comfrey' in subsection 'Beneficial Insects and Comfrey' in section 'Garden Uses of Comfrey' (Chapter 18).

"*S. officinale L.:*
The drooping flowers of this species are white or violet-purple in colour, *and their mechanism resembles that of Borago. The bell-shaped corolla is 14 mm long, and it is contracted above for a distance of 8 mm, so that **only long-tongued insects can suck the nectar legitimately.***
At the junction of the narrow and broader portions of the corolla there are triangular hollow scales, alternating with the filaments and covering the spaces between them. The spiny edges of these appendages prevent visitors from probing for nectar between the filaments, and **they are obliged to insert their proboscis (sucking mouthpart) in such a way that it must get dusted with pollen.**
The anthers converge to form a hollow cone surrounding the style, and they 'dehisce introrsely' in the bud, some of the pollen falling into the apex of the cone and some remaining clinging to them.
When an insect probes for nectar with its proboscis the anthers are displaced and some pollen falls out. The projecting stigma is the first part of the flower to be touched by a visitor, after which it is dusted by pollen. The arrangement is favourable to crossing by insect-visits, but should these fail automatic self-pollination apparently takes place.
Kerner states that the flower is at first horizontal, but comes to droop in late anthesis owing to a bending of the peduncle, so that the stigma is brought into the line of fall of the pollen, when autogamy (self fertilization) results.
Mouthpart of Insects:
A proboscis of at least 11 mm in length is required to reach the nectar by probing between the anthers, but one of 8 mm would be able to get at it between the filaments. As already explained, however, the latter way is barred by means of hollow scales, the edges of which are beset with minute prickles.
>**Insects with a proboscis less than 11 mm long can therefore only secure the nectar by perforating the corolla (thereby not getting nectar 'legitimately'). This is done extremely often by three kinds of bumble-bee,** i.e., Bombus terrester L. (proboscis 7-9 mm), Bombus pratorum L. (proboscis 8-9 mm), and Bombus lapidarius L. (proboscis 9-10 mm).
>**The honey-bee also sucks nectar through the holes thus made, but Loew says that but little harm is done to the flowers in this way.**

Pollen: *Warnstorf describes the pollen-grains as white in colour, ellipsoidal, smooth, on an average 33 micrometers long and 27 micrometers broad.
Visitors: Hermann Muller gives the following list of those insects that suck legitimately and effect pollination:
>*Order Diptera (flies).   Syrphidae (hoverfly):*
>>*Rhingia rostrata (small hoverfly)*
>
>*Order Hymenoptera (sawflies, wasps, bees, and ants). Apidae (bees):*
>>*Anthophora personata (solitary bees)*
>>*Apiomerus pilipes (bee assassin)*
>>*Bombus rajellus (bee)*
>>*Bombus sylvarum (carder bee)*
>>*Osmia aenea = Osmia caerulescens (blue mason bee)*
>>*Xylocopa violacea (violet carpenter bee)."*

-'Handbuch der Blutenbiologie' {Handbook of Flower Pollination} by P. Knuth, Volume 3, part 2. v + 601 pages, Leipzig, Germany, 1905. [Boraginaceae, pages 63-67.] English translation by J.R.A. Davis, Volume 3, iv + 644 pages, Oxford, England, **1909**. [Boraginaceae, pages 115-142; 23 genera, including Symphytum.]
(* -'Verhandlungen des Botanischen Vereins der v Provinz Brandenburg', Volumes 37-39 by Professor Dr. P. Ascherson, R. Beyer and Dr. M. Gurke, Berlin, Germany, 1896. Article by C. Warnstorf: 'Blutenbiologische Beobachtungen aus der Ruppiner Flora im Jahre 1895' {Biological Observations}, Abhandlungen {Treatises} page 44.)
>(Dehiscence means splitting along a line of weakness in a plant structure in order to release its contents. Anther dehiscence causes the release of pollen grains. If the pollen is released from the anther through a split on the outer side, it is extrorse dehiscence. If released from the inner side, it is introrse dehiscence.)
>(In insects the proboscis is an elongated sucking mouthpart that is tubular and flexible.)
>(Micrometre or micrometer or micron equals 10^{-6} meter = one millionth of a meter = one thousandth of a millimeter = 0.001 mm = 0.000039 inch.)

"Although blueberry flowers are nectar and pollen sources for native bees, blueberry blooms for approximately one month. *In contrast, adults of many of the species of native bees that pollinate lowbush blueberry in Maine are active prior to*

blueberry bloom and often for a considerable period after blueberry bloom.
To maintain their populations, appropriate and accessible alternative food sources must be provided before, perhaps during, and after blueberry bloom.
The literature survey results indicate that 28 species of native bees associated with blueberry in Maine collect Vaccinium spp. pollen. Coupled with our new pollen records from the pollen analysis work, **a total of 38 native bee species pollinate Vaccinium spp. and all are polylectic, meaning they forage on a diversity of plant species for nectar and pollen.**
Native bees that visit Symphytum officinale flowers:
    Bombus bimaculatus Cresson (Two-Spotted Bumblebee)
    Bombus perplexus Cresson (Confusing Bumblebee)
    Bombus ternarius Say (Orange-Belted Bumblebee or Tricolored Bumblebee)
    Bombus terricola terricola Kirby (Yellow-Banded Bumblebee)."
-'Alternative Forage Plants for Native (Wild) Bees Associated with Lowbush Blueberry, Vaccinium spp., in Maine' by C.S. Stubbs, H.A. Jacobson, E.A. Osgood and F.A. Drummond, Maine Agricultural Experiment Station, University of Maine, Orono, Maine; Technical Bulletin 148, February **1992**.

"**The fecundity (reproduction) of insect-pollinated Symphytum officinale plants may not be linearly related to the number of flowers produced, since floral display will influence pollinator foraging patterns.**
Bumblebee species differed in their response to the size of floral display.
    More individuals of Bombus pratorum (early-nesting bumblebee) and the nectar-robbing Bombus terrestris (buff-tailed or large earth bumblebee) were attracted to plants with larger floral displays, but Bombus pascuorum (common carder bee) exhibited no increase in recruitment according to display size.
Once attracted, all bee species visited more inflorescences (flowers) per plant on plants with more inflorescences.
Overall the visitation rate per inflorescence and seed set per Comfrey flower was independent of the number of inflorescences per plant. Variation in seed set was not explained by the numbers of bumblebees attracted or by the number of inflorescences they visited for any bee species.
    However, the mean seed set per flower (1.18) was far below the maximum possible (4 per flower). We suggest that in this system seed set is not limited by pollination but by other factors, possibly nutritional resources."
-'Floral Display Size in Comfrey Symphytum Officinale L Boraginaceae: Relationships with Visitation by Three Bumblebee Species and Subsequent Seed Set' by D. Goulson, J.C. Stout, S.A. Hawson and J.A. Allen, University of Southhampton, School of Biology, England; Oecologia (Plant and Animal Ecology), Volume 113, No. 4, pages 502-508, **1998**.

"**Timing and duration of stigma (female part of plant) receptivity in Comfrey (Symphytum officinale L., Boraginaceae),** a species with a dry-type stigma, were studied using two methods based on enzymatic activity (Peroxtesmo Ko and Perex tests) as well as on observations of pollen germination in hand-pollinated flowers of various stages. Moreover, structural changes of stigmatic tissue in relation to stigma receptivity were noted.
In Comfrey flowers the receptivity period lasted about 4-5 days. It started at the time of flower opening but full receptivity was concomitant (at same time) with pollen exposure. Flowers at the final life stage did not show restricted receptivity.
    The stigma was covered with short, unicellular papillae. With the beginning of receptivity, papillae arrangement loosened. Receptivity only occurred at the basal part of the papillae.
Germination of pollen grains started 3 hours after pollination. The results obtained with the Peroxtesmo Ko receptivity test were in line with the test for pollen germination. The Perex test was not applicable for Comfrey flowers due to the dark color of pistil tissues, which hides the test reaction."
-'Stigma Receptivity in Comfrey (Symphytum officinale L.) During the Course of Anthesis' by Marzena Masierowska and Malgorzata Stpiczynska, Department of Botany, Agricultural University of Lublin, Poland; Israel Journal of Plant Sciences, Volume 53, No. 1, pages 41-46, July **2005**.

"**Observations and measurements of Symphytum officinale L.** were taken in meadows of the Lublin, Poland area, in the district of Slawin in the years 2006-2007.
Symphytum officinale L. is a self incompatible species. The surface of the corolla is densely covered with non-glandular and a few glandular trichomes with lipids. The scales situated between stamen heads are densely covered by one-cell trichomes. The apical part of the hair cells is rich in lipids.
The inflorescence of Symphytum officinale consists of 14.65 flowers on average, with 2-7 open at any one time. The mean mass of pollen delivered was 9.87 mg per 100 anthers. The mean mass of pollen produced in anthers averaged 27.15% of the total dry anther weight. The amount of pollen delivered per inflorescence was 9.26 mg.
    The estimated magnitude of pollen flow in successive flowering stages differs and is 0.17 kg (5.99 ounces) per 1 ha (hectare) in the initial phase of blooming. At full bloom Symphytum officinale yield from 6.50 to 17.64 kg (229-622 ounce) per 1 ha of the community. Towards the end of the flowering period estimated pollen flow is 2.95 kg (104 ounce) per 1 ha, on average.
The dominant pollinators of Symphytum officinale were different Bombus spp (bumblebee species). The taxon is the spring source of nectar and pollen. Besides sugars and proteins, it is highly probable that lipids are important attractants.
Symphytum officinale should be also considered as a supplementary source of food for Apis mellifera (honey bee). Honeybees mainly collect nectar using the holes made by bumblebees in the base of the corolla. A direct access to the flower, however, is often used by Apis mellifera or Bombus spp."
-'Flowers Ecology and Pollen Output of Symphytum Officinale L' by Bozena Denisow, Department of Botany, University of Life

Sciences in Lublin, Poland; Journal of Apicultural (Beekeeping) Science, Volume 52, No. 2, pages 81-89, **2008**.
(Self-incompatibility is a genetic mechanism in flowering plants which prevent self-fertilization and encourage cross-fertilization.)

*"**Comfrey (Symphytum officinale) is a common perennial plant in pastures and meadows along the river Rhine in Germany**, where it is frequently visited by local bumblebees for nectar and pollen.*
Bumblebees leave traces of cuticular hydrocarbons on flowers they visit, with the amount deposited being positively related to the number of visits.
On 29th July 2007, we recorded insect visits to flowers of 63 individual plants in 10-minute intervals distributed evenly over the day from 800 to 1600 hours (8 am to 4 pm) (on average 5.5 intervals per plant, or 55 minutes of observation).
Bumblebees were the only regular flower visitors (99% of visits)*, and the occasional visits by other insects (unidentified syrphid flies {hover flies} and solitary bees) were excluded from further analysis.*
Workers of Bombus pascuorum (common carder bee) were the most abundant visitors of Comfrey at the time and contributed roughly 80% of all observed flower visits.
The remaining 20% of visits were by Bombus hortorum (small garden bumblebee) (13%), B. terrestris (buff-tailed or large earth bumblebee) (4%), and B. pratorum (early-nesting bumblebee) (3%).
*The extracts of visited Comfrey corollae contained alkenes of chain lengths from 21 to 31, corresponding well with that found in tarsal (part of foot) extracts of the visiting species of bumblebees. Together, these data suggest that **flowers retain a long-term quantitative record of bumblebee visitation**."*
-'Hydrocarbon Footprints as Record of Bumblebee Flower Visitation' by S. Witjes and T. Eltz, Department of Neurobiology, University of Dusseldorf, Germany; Journal of Chemical Ecology, Volume 35, No. 11, pages 1320-1325, **2009**.

*"**Comfrey (S. officinale) is a polycarpous (producing fruit many times)**, perennial herb that grows in moist habitats. In Germany it is a common plant in pastures and meadows, especially along rivers (Dull and Kutzelnigg 2005; Hegi 1966).*
Its flowers are open mainly from May to July, and are frequently visited by bumblebees.
Comfrey has been reported to be self-incompatible (Goulson et al. 1998), but self-pollination may occasionally occur if pollinators are absent or rare (*Dull and Kutzelnigg 2005; **Hegi 1926).
S. officinale plants set a maximum of four relatively large seeds per flower (Dull and Kutzelnigg 2005; Goulson et al. 1998; Hegi 1926), *and the proportion of developed seeds can be assessed accurately for individual flowers.*
The measurement of insect visits to flowers is essential for basic and applied pollination ecology.
In three consecutive years we recorded bumblebee visitation to wild plants of Comfrey, Symphytum officinale, *and later used gas chromatography/mass spectrometry (GC/MS) to quantify bumblebee-derived unsaturated hydrocarbons (UHCs) extracted from flowers. The UHCs washed from corollae were most similar to the tarsal UHC profile of **the most abundant bumblebee species, Bombus pascuorum (common carder bee)**, in all 3 years. Comfrey plants were visited by six different bumblebee species.*
Seed set of Comfrey plants was positively correlated with overall bumblebee visitation and the total amount of UHCs on flowers."
-'Reconstructing the Pollinator Community and Predicting Seed Set from Hydrocarbon Footprints on Flowers' by S. Witjes and K. Witsch, Sensory Ecology Group, University of Dusseldorf, Germany; Oecologia (Plant and Animal Ecology), Volume 165, No. 4, pages 1017-1029, **2011**.
(* -'Taschenlexikon der Pflanzen Deutschlands' {Pocket Dictionary of the Plants of Germany} by R. Dull and H. Kutzelnigg; Quelle & Meyer Verlag, Wiebelsheim, Germany, 2005.)
(** -'Illustrierte Flora von Mittel-Europa: Mit Besonderer Berucksichtigung von Deutschland, Osterreich und der Schweiz' {Illustrated Flora of Central Europe: With Special Consideration of Germany, Austria and Switzerland}, Volume 5, Part 3' by Gustav Hegi, Zurich, Switzerland; published by J.F. Lehmann, Munich, Germany, 1926. Symphytum pages 2220-2230. In German, 13 volumes written from 1908 to 1931.)
(Polycarpous means consisting of two or more carpels or ovaries, especially with the ovaries remaining free or distinct. This means the plant produces seeds many times throughout the season.)

*"**Competition for pollinators occurs when, in a community of flowering plants, several simultaneously flowering plant species depend on the same pollinator.** Competition for pollinators increases interspecific (between different species) pollen transfer rates, thereby reducing the number of viable offspring.*
In order to decrease interspecific pollen transfer, plant species can distinguish themselves from competitors by having a divergent phenotype. Floral colour is an important signalling cue to attract potential pollinators and thus a major aspect of the flower phenotype.
Plant species that were visited by the same pollinator were considered specialist and competing for that pollinator, whereas plant species visited by a broad array of pollinators were considered non-competing generalists.
Seven Pollinator Guilds: *Generalists, Bibio, Musca, Eristalis, Rhingia campestris, Bombus pascuorum, Bombus terrestris.*
Bombus pascuorum (Common Carder Bumblebee) Guild: *Galeopsis tetrahit (52%); Linaria vulgaris (43%); Lotus corniculatus (59%); Stachys palustris (50%);* ***Symphytum officinale (62% degree of specialization);*** *Trifolium pratense (55%); Trifolium repens (51%); Vicia cracca (93%); Vicia sativa (82%).*
The degree of specialization is the percent (%) visited by one pollinator.
We expected that in plant species that heavily depended on only a few pollinator species, competition for pollination would be stronger than between generalist species, and that therefore flowers of specialist plant species would exhibit stronger spectral differences.

Clearly, for specialists, spectral dissimilarity is an efficient signalling cue, as it can be perceived by insects from far longer distances than is the case with morphological traits such as corolla length (Schemske 1976).
The present investigation of the spectral properties of co-flowering plants within pollinator guilds lends support to the hypothesis that flower communities are, at least in part, structured by plant-pollinator interactions."
-'Competition for Pollinators and Intra-Communal Spectral Dissimilarity of Flowers' by C.J. van der Kooi, I. Pen, M. Staal, D.G. Stavenga and J.T.M. Elzenga, University of Groningen, The Netherlands; Plant Biology: German Botanical Society and The Royal Botanical Society of the Netherlands, Volume 18, No. 1, pages 56-62, January **2016**.

Hollow versus Solid Stem in S. officinale
Usually Symphytum officinale is described as having a hollow stem. However, in some varieties (or perhaps within a variety) the stem is solid or some of the stems are solid.

"Symphytum officinale: The stalk that riseth up from among them (the leaves), being two or three feet (0.6-0.9 meter) high, hollowed, and cornered."
-'Culpeper's English Physician and Complete Herbal to Which are Now First Added Upwards of One Hundred Additional Herbs' by Nicholas Culpeper, written **1652-1653**, page 132

"Dr. Symes new edition of 'Sowerbys English Botany':
Symphytum officinale varies in several points of structure, namely: in hollowness and solidity of stem- *as we have found them both in our own meadows (in England) in a more or less hairy and decurrent (extending downward along the stem), or scarcely decurrent leaf, and in more or less striking floral recemes (simple flowers on short stalks of about equal length at equal distances).*
It is highly probable then that the species may be prone to run into varieties, *a matter of no little importance in an agricultural point of view, as one form may be more valuable than another."*
-'Forage Plants and Their Economic Conservation by the New System of Ensilage: Part I: Caucasian Prickly Comfrey' by Thomas Christy, Jun., F.L.S. (Fellow of the Linnean Society), Christy & Co., London, England, **1877**, pages 3, 4.

"We have since observed the same in our wild forms of Symphytum officinale; sometimes the stems are solid, both in the typical S. officinale and its variety, S. patens, which confirms us in our former conclusions as the result of experiment that they are only varieties, *though from all the evidence we can collect upon the subject, we are bound to conclude that in this, as in other cultivated plants, a variety is often of greater consequence than an admitted species."*
-'Forage Plants and Their Economic Conservation by the New System of Ensilage: Part I: Caucasian Prickly Comfrey' by Thomas Christy, Jun., F.L.S. (Fellow of the Linnean Society), Christy & Co., London, England, **1877**, page 6.

"The leafy stem, 2 to 3 feet (0.6 to 0.9 meter) high, is stout, angular and hollow, broadly winged at the top and covered with bristly hairs."
-A Modern Herbal: The Medicinal, Culinary, Cosmetic and Economic Properties, Cultivation and Folk-Lore of Herbs, Grasses, Fungi, Shrubs and Trees with their Scientific Uses by Mrs. M. Grieve. New York: Dover Publications, 1971. First published **1931**.

"Hollow, erect stems, *also containing mucilage, are covered with bristly hairs that cause itching when in contact with the skin."*
-'European Medicines Agency- Assessment Report on Symphytum Officinale L., Radix' by European Medicines Agency, Committee on Herbal Medicinal Products (HMPC), London, England, 27 pages, May 5 **2015**.

More about S. Officinale Roots
See subsection 'Comfrey Root Description' in section 'Symphytum Genus Description' (Chapter 5).

"Symphytum officinale L. (Boraginaceae) is a perennial herb known as Comfrey, *gum plant or boneset, and it is employed topically (on skin) as anti-inflammatory, emollient and mild anesthetic in phytotherapy (plant therapy), due to allantoin found in the underground organs and leaf.*
In order to contribute to the medicinal plant and vegetal drug identification, morpho-anatomical investigations of the root and rhizome, in secondary growth, were carried out. *The botanical material was fixed and prepared according to usual microtechniques.*
The underground organs are alike, slender and yellow to black coloured. *The root and the rhizome show similar periderm (outer layer) and secondary vascular tissues, yet they differ in that the root shows exarch primary xylem, while the rhizome presents endarch primary xylem and parenchymatous pith. Numerous parenchymatic cells which contain mucilage and amyloplasts are observed in the root and rhizome."*
-'Morpho-Anatomical Characterization of the Root and Rihizome of Symphytum Officinale L. (Boraginaceae)' by A.C.O. Toledo, M.R. Duarte and T. Nakashima, Universidade Estadual de Ponta Grossa (State University of Ponta Grossa), Parana, Brazil; Revista Brasileira de Farmacognosia (Brazilian Journal of Pharmacognosy), Volume 16, No. 2, pages 185-191, **2006**.
(Rhizome is a continuously growing horizontal underground stem that sends out lateral shoots at intervals.)
(Xylem is one of two types of transport tissue in vascular plants, phloem being the other. It transports water and nutrients from roots to shoots and leaves. Endarch is where the protoxylem is directed towards the center and metaxy-

lem elements towards the periphery. The exarch is where the protoxylem is directed towards the periphery and metaxylem towards the center.)

"The present study provides a detailed summary of pharmacognostical characters of Symphytum officinale roots to give clear standards for identification of the drug.
The findings of macroscopic studies unveiled that the color of pieces of roots were internally buff to light brown and dark brown to blackish brown externally with longitudinal wrinkles.
Transverse section of root showed dark brown cork next to the cork is a phelloderm consisting of layer of tangentially elongated parenchyma cells; endodermis; small groups of vessels and starch grains are found singly or in groups.
Powder microscopy revealed the presence of fragments of parenchyma, few vessels with bordered pits or reticulate walls; calcium oxalate crystals and rounded and oval starch grains.
The roots of Symphytum officinale showed the characteristic physicochemical (physical chemistry) values like total ash (11.1%), acid insoluble ash (0.74%), water soluble ash (0.83%), alcohol extractive value (18.3%), water soluble extractive value (22.3%) and loss on drying (13.5%).
Various pharmacognostic characters observed in the above study helps in standardization, identification and establishments of quality parameters of plant."
-'Pharmacognostical Studies of Roots of Symphytum Officinale' by Manpreet Kaur and Hayat M. Mukhtar, Punjab, India, International Journal of Pharmaceutical Sciences Review and Research, Vol 45, No 2, Article No 27, pages 146-148, July-August **2017**.
(Pharmacognosy is the study of medicinal drugs obtained from plants and other natural sources.)

Common Comfrey Chromosomes
See subsection 'Comfrey Chromosomes Overview' in section 'Symphytum Species Overview' (Chapter 6).

For more information about the complex relation between hybrids of Common Comfrey and Russian Comfrey, see subsections 'Hybrid Origin and Classification' and 'Russian Comfrey Chromosomes' in section 'Details about Symphytum Species Hybrid: Russian Comfrey' (Chapter 10).

*"**Symphytum officinale L. Chromosome Numbers 2n:***

Rhijnauwen near Bunnik (Utrecht, Netherlands)	48
Molenpolder near Westbroek (Utrecht)	40
near IJsselstein (Utrecht)	26
Betw. Jutphaas and IJsselstein (Utrecht)	48
Betw. Jutphaas and IJsselstein (Utrecht)	40
Fort Hoofddijk near Utrecht (Utrecht)	48

Species with both intraspecific (within same species) euploidy and aneuploidy:
Galium aparine L.; Iris pseudacorus L.; Poa nemoralis L.; Polygonatum multiflorum (L.) All.; Polygonatum odoratum (Mill.) Druce; Ranunculus acris L.; **Symphytum officinale L.***; Urtica urens L.; Veronica officinalis L.; Vicia cracca L.*
Symphytum officinale L.:
Three cytotypes of this species, with 2n = 26, 40 and 48 respectively, were found in a limited area. *Morphological and ecological differences are not known."*
-'Chromosome Numbers of Flowering Plants in the Netherlands' by Th.W.J. Gadella and E. Kliphuis, Botanical Museum and Herbarium, Utrecht, Netherlands; Acta Botanica Neerlandica, Volume 12, pages 195-230, **1963**. Symphytum pages 214, 224.
(Aneuploidy is an abnormal number of chromosomes in a cell. Euploidy is when a cell has a chromosome number that is an exact multiple of the monoploid/haploid number.)
(A cytotype is an individual of a species that has a different chromosomal factor to another.)

*"**Symphytum officinale L.***
In a previous paper three different chromosome numbers (2n = 26, 40 and 48) were reported (Gadella and Kliphuis, 1963).
The plants with 2n = 26 were collected near IJsselstein (province of Utrecht, Netherlands).
*During 1964 and 1965, many plants were sampled in this locality. **Cytological studies revealed the number 2n = 24 in most plants, whereas in others the occurrence of B-chromosomes could be demonstrated: 2n = 24 + 1 - 4B.***
The investigated plants with 2n = 24 are always white-flowering.
Further research on this population as well as on populations with the numbers 2n = 40 and 2n = 48 are in progress."
-'Chromosome Numbers of Flowering Plants in the Netherlands: III' by Th.W.J. Gadella and E. Kliphuis, Botanical Museum and Herbarium, Utrecht, Netherlands; Mededelingen van het Botanisch Museum en Herbarium van de Rijksuniversiteit te Utrecht (Announcements from the Botanic Museum and Herbarium of the University of Utrecht), Netherlands, Volume 271, Issue 1, pages 7-20, **1967**.
(A karyotype is the number and appearance of chromosomes in nucleus of an eukaryotic cell. In addition to normal karyotype, many animals and plants contain B chromosomes known as supernumerary, accessory, conditionally-dispensable, or lineage-specific chromosomes. These chromosomes are not essential for the life of a species, and are not found in most of the individuals.)

"According to the literature cytological studies in Symphytum officinale yielded the following results: **Symphytum officinale:**

$2n = 36$ (*Strey, 1931; cult, mat.= cultivated material);
$2n = 42$ (**Datta, 1933; cult, mat.);
$2n = 48$ (Tarnavschi, 1948; Roumania);
$2n = +$ or $- 40$ **(plus or minus 40)** (Love and Love, 1956; Iceland);
$2n = 26, 40, 48$ (Gadella and Kliphuis, 1963, 1967; the Netherlands).
The basic number of S. officinale seems to be x = 12.
The number 2n = 40 does not fit into the euploid series."
-'Cytotaxonomic Studies in the Genus Symphytum I: Symphytum Officinale L. in the Netherlands' by Th.W.J. Gadella and E. Kliphuis, Botanical Museum and Herbarium of the State University of Utrecht, Netherlands, February 25 **1967**.
(* -'Karyologische Untersuchungen an Boraginaceen I.' by M. Strey, Mitteilungen der Botanischen Staatssammlung Munchen, Volume 7, pages 277-294, 1931.)
(** -S. Datta; The Journal of the Indian Botanical Society, India, Volume 12, pages 131-152, 1933. Title of article was not given.)
(Cytology is the science that deals with the formation, structure, and function of cells.)
(Body cells and individuals are described by the number of chromosome sets {ploidy level} they have: monoploid: 1 set; diploid: 2 sets = $2n = 2x$, triploid: 3 sets = $2n = 3x$, tetraploid: 4 sets = $2n = 4x$, pentaploid: 5 sets = $2n = 5x$, etc.)

"**Symphytum officinale, 2n = 24; flowers white.**
Symphytum officinale, 2n = 40; flowers purple.
Symphytum officinale, 2n = 48; flowers white, red and purple.
It is clear that Symphytum officinale is a very complex species consisting of at least three cytological races.
The plants with 2n = 40 differ slightly from the diploids and tetraploids, (Gadella & Kliphuis, 1967), and may be assigned to the subspecies uliginosum (Kern.) Nym.
The diploids and tetraploids are very similar morphologically. According to our results diploid and tetraploid plants seem to be crossable to an extremely limited extent. In the experimental garden a hybrid could be obtained only twice. In nature, in a mixed population, hybrids are never met with.
Contrary, the plants with 2n = 40 and 2n = 48 chromosomes are fully interfertile. Their hybrids are also fertile and intercrossable. Crossing of these hybrids ($2n = 44$) with their parents ($2n = 40$ and $2n = 48$) is possible."
-'Cytotaxonomic Studies in the Genus Symphytum II: Crossing Experiments Between Symphytum Officinale L. and Symphytum Asperum Lepech' by Th.W.J. Gadella and E. Kliphuis, Botanical Museum and Herbarium of the State University of Utrecht, Netherlands; Acta Botanica Neerlandica, Volume 18, No. 4, pages 544-549, August **1969**.

"**Symphytum officinale L.:**
In the Austrian material **diploid plants** were not represented. These are, however, in east Germany, Czechoslovakia and Hungary, so that their appearance in Austria is not excluded.
The Dutch diploids and tetraploids are morphologically indistinguishable and do not cross.
The subspecies uliginosum occurs especially in eastern Europe, from Hungary to southern Russia.
In Austria it should be on unshaded and very swampy spots."
-'Cytotaxonomic Investigations on Flowering Plants from the East of Austria' (Zytotaxonomische Untersuchungen an Blutenpflanzen aus dem Osten Osterreichs) by Von Th.W.J. Gadella, E. Kliphuis and K.U. Kramer, Institut fur Systematische Botanik der Universitat Utrecht, Netherlands; Die Wissenschaftlichen Arbeiten aus dem Burgenland, Eisenstadt, Osterr, Austria, Vol 44, pages 187-195, **1970**. In German. (If you have English translation, I would appreciate a copy of parts about Symphytum.)

"**Boraginaceae: Symphytum officinale L.:**
Collection number: G.5463 and 5464. $2n = 24 + 4B$
Origin: In a moist meadow between La Sarre and Chateau de Sarre in Aosta Valley, Italy.
Symphytum officinale L. is a very variable species (Gadella and Kliphuis 1967, 1969; Gadella, Kliphuis and Kramer 1970, in press; Tutin 1956; Wade 1958).
Three cytotypes are known: 2n = 24, 2n = 40 and 2n = 48.
$2n = 24$:
    In the Netherlands diploids (2n = 24) seem to have a restricted distribution.
        Hitherto only two small populations in a small area (osier bed = willow tree area) were found (Gadella and
        Kliphuis 1967). This cytotype is also known from eastern Germany (near Gotha) and Hungary (near Dabas).
In the present study, plants with $2n = 24$ chromosomes have been reported from the Valley of Aosta, Italy. Dutch and Italian material investigated showed in the diploids the occurrence of **B chromosomes**. In the diploid plants from Germany and Hungary these additional chromosomes were never met with.
$2n = 48$:
    Many populations in western Germany, the Netherlands, Austria (Burgenland) and the north of Yugoslavia consist entirely of tetraploid plants (2n = 48). It seems to be the most common type in Europe. In the Netherlands it occurs frequently on dikes and along roads.
$2n = 40$:
    Plants with 2n = 40 chromosomes were found in the Netherlands only. They are very common in very moist places on low moor peat.
        Bucknall (*1913) treats in his revision of the genus Symphytum the species '**Symphytum uliginosum Kern.**'.
        The Dutch material with $2n = 40$ chromosomes closely matches Bucknall's description. According to Bucknall

(1913) this species occurs in Hungary and the south of Russia.
It is a remarkable fact that, as far as is known, all diploid plants (2n = 24) are white flowered. Tetraploids (2n = 48) have also purple and red flowers. *White flowered tetraploids and white flowered diploids are morphologically indistinguishable.*
The 2n = 40 type is nearly always purple flowered. *Morphologically this type is somewhat different from the 2n = 48 type.*
Crossing experiments showed that the diploids (2n = 24) are reproductively isolated from the 2n = 40 type and to a very large extent also from the tetraploid plants (2n = 48). *Only twice a triploid hybrid could be produced after many unsuccessful attempts.*
The 2n = 40 type is capable of exchanging genes with the tetraploid (2n = 48). Their hybrids (2n = 44) are fully fertile *(Gadella and Kliphuis 1969)."*
-'Cytotaxonomic Investigations in Some Angiosperms Collected in the Valley of Aosta and in the National Park Gran Paradiso' by Th.W.J. Gadella and E. Kliphuis, Institute of Systematic Botany, Department of Experimental Taxonomy, State University of Utrecht, Netherlands; Caryologia: International Journal of Cytology, Cytosystematics and Cytogenetics, London, England, Volume 23, No. 3, **1970**.
(* -'A Revision of the Genus Symphytum' by C. Bucknall, Botanical Journal of the Linnean Society, 41, pages 491-556, 1913.)
(Aosta is in the Italian Alps mountain range. Gran Paradiso National Park is an Italian park in the Graian Alps, between Aosta Valley and Piedmont regions.)

"Strey (*1931) and Laane (**1969) were able to demonstrate the occurrence of Common Comfrey plants with the number 2n = 36. Strey, who studied plants of garden origin, made the remark that S. officinale presents very serious difficulties with regard to the correct identification of its chromosome number. The plants investigated by Laane originated from Norway.
In spite of many chromosome counts neither the present authors nor Skaliisiska et al. (***1971) were able to confirm the results obtained by Strey and Laane. **The number 2n = 36 could not be found in Common Comfrey plants collected in their natural habitats, but experimental hybridization between plants with 2n = 24 and 2n = 48 resulted in the formation of hybrids with 2n = 36** (****Gadella & Kliphuis 1969).
Also the hybrid S. x uplandicum (Russian Comfrey), which is rather common in southern Scandinavia, is often characterized by the number 2n = 36.
The number 2n = 40 in Common Comfrey was counted repeatedly in Dutch plants by the present authors (Gadella & Kliphuis 1963, 1967, 1971) and in Icelandic plants by Love and Love (****1956).
This cytotype, however, will be left out of consideration in this paper, since it differs, at least in the Netherlands, both in morphology and ecological preference from diploid (2n = 24) and tetraploid (2n = 48) plants (Gadella & Kliphuis 1967).
Tetraploid plants (2n = 48) in Common Comfrey have been reported from Romania (Tarnavschi 1948), from Poland by Skalinska et al. (1971) and from various other European countries (Gadella & Kliphuis 1963, 1967; Gadella, Kliphuis & Kramer, 1970; Gadella 1972)."
-'Cytotaxonomic Studies in the Genus Symphytum IV: Cytogeographic Investigations in Symphytum Officinale L.' by Th.W.J. Gadella and E. Kliphuis, Instituut voor Systematische Plantkunde (Institute for Systematic Botany), Utrecht, Netherlands; Acta Botanica Neerlandica, Netherlands; Volume 21, Issue 2, pages 169-173, April **1972**.
(* -'Karyologische Untersuchungen an Boraginaceen I.' by M. Strey, Mitteilungen der Botanischen Staatssammlung Munchen, Volume 7, pages 277-294, 1931.)
(** -'Meiosis and Structural Hybridity in Some Norwegian Plant Species' by M.M. Laane, Blyttia, Norwegian Botanical Association, Norway, Volume 27, pages 141-173, 1969.)
(*** -'Studies in Chromosome Numbers of Polish Angiosperms, Eighth Contribution', by M. Skalinska, A. Jankun, H. Wcislo, et al., Acta Biologica Cracoviensia, Polish Academy of Sciences, Poland, Volume 14, pages 55-102, 1971.)
(**** -'Cytotaxonomic Studies in the Genus Symphytum II: Crossing Experiments Between Symphytum Officinale L. and Symphytum Asperum Lepech' by Th.W.J. Gadella and E. Kliphuis, Botanical Museum and Herbarium of the State University of Utrecht, Netherlands; Acta Botanica Neerlandica, Netherlands, Volume 18, No. 4, pages 544-549, August 1969.)
(**** -Cytotaxonomical Conspectus of the Icelandic Flora: Acta Horti Gotoburgensis Series, Volume 20, No. 4, pages 65-291 by Askell Love and Doris Love. Goteborg {Gothenburg}, Sweden: Goteborgs Botaniska Tradgard, 1956.)

"Chromosomes: 2n = 24, 24 + 4B, 26, c. (about) 36, 40, 48, 54."
-Flora Europaea, Volume 3: Diapensiaceae to Myoporaceae, editors Tutin, Heywood, Burges, Moore, Valentine, Walters and Webb. United Kingdom: Cambridge University Press, 1972. (5-volume encyclopedia of plants, published between 1964 and 1993), page 104, **1972**.
('Flora Europaea' describes all the national floras of Europe to identify any wild or widely cultivated plant to the subspecies level. It provides geographical distribution, habitat preference, and chromosome number.)

"Symphytum L. Chromosome Numbers:

officinale L.	24 + 48	Italy	Gadella & Kliphuis 1970
	24 40 48	Netherlands	Gadella & Kliphuis 1967
	26 36		
	54	Bulgaria	Markova & Ivanova 1970

(a) officinale
(b) uliginosum (A. Kerner) Nyman
This book has grown out of the requests by many users of 'Flora Europaea' (Tutin et al., 1964, 1968, 1972, 1976, 1980) for information on the sources of the chromosome numbers cited in that work."

-Flora Europaea: Check-List and Chromosome Index by D.M. Moore. Cambridge, England: Cambridge University Press, 1982, page 179.

*"In Symphytum officinale L. intraspecific (within a species) cytological variation could be demonstrated by various authors (Gadella & Kliphuis 1967, 1971, 1972; Grau 1971; Skalinska et al. 1971; *Basler 1972).* **There is no difference of opinion with regard to the systematic position of the diploid (2n = 24) and the tetraploid (2n = 48) plants. All authors assign these plants, collected in different parts of Europe, to the species S. officinale L."**
-'Cytotaxonomic Studies in the Genus Symphytum V: Some Notes on W. European Plants with the Chromosome Number 2n=40' by Th.W.J. Gadella and E. Kliphuis, Botanische Jahrbucher fur Systematik (Botanical Yearbooks for Systematics), Volume 93, pages 530-538, **1973**.
(* -'Cytotaxonomic Studies on the Boraginaceen Genus Symphytum L.: Studies on Predominantly Northern German Plants of the Species S. Asperum Lepech., S. officinale L. and S. x Uplandicum Nym.' by Armin Basler, Botanische Jahrbucher fur Systematik: Pflanzengeschichte und Pflanzengeographie {Botanical Yearbooks for Systematics: Plant History and Plant Geography}, Volume 92, pages 508-553, 1972. All in German. I was unable to find a copy of this. If you have an English translation, could you please send it to me.)

"S. officinale L.:
This species was studied in more detail by Gadella and Kliphuis (1967) and by Basler (1972). **They agree that the most common chromosome number for this species is 2n = 48**, *but that scattered diploid (2n = 24) populations occur in west, central and east Europe.*
The origin of the 2n = 40 form in S. officinale is again very difficult to explain.
Both in this species and in S. peregrinum, a hybrid derivation seems numerically the most plausible explanation (2n = 32) x (2n = 48) or reciprocal, *but experimental evidence is either lacking (S. peregrinum) or conflicting (the 2n = 40 form of S. officinale).*
We therefore provisionally propose the basic number x = 10 for the genus Symphytum."
-'Cytotaxonomic Studies in the Genus Symphytum VIII: Chromosome Numbers and Classification of Ten European Species' by T.W.J. Gadella and E. Kliphuis, Proceedings van de Koninklijke Nederlandse Akademie van Wetenschappen (Proceedings of the Royal Netherlands Academy of Arts and Sciences): Series C, Biological and Medical Sciences, Vol 81, pages 162-172, **1978**.
(A reciprocal cross shows the role of parental sex on a given inheritance pattern. In one cross, a male expressing the trait is crossed with a female not expressing the trait. In the other, a female expressing the trait is crossed with a male not expressing the trait.)

*"**The most common chromosome number of S. officinale (Common Comfrey) is 2n = 48 (tetraploid). Scattered diploid (2n = 24) populations occur in western, central and eastern Europe; they are white-flowered throughout. The tetraploids are white- or purple-flowered in western Europe and purple in eastern Europe.***
Populations in which purple- and white-flowered individuals occur intermingled are very common in western Europe. In eastern Europe, mixed populations (which seem to be rare) consist of white-flowered diploid plants with purple-flowered tetraploid plants. We did not find comparable situations in western Europe.
In one population in the Netherlands, white-flowered dipolids and tetraploids grew intermingled with purple flowered tetraploids.
The cytotype 2n = 56 *was found only once in a population from the Tatra mountains in Czechoslovakia.*
The cytotype 2n = 40 *is very common in low-lying peat regions in the Netherlands. Judging from the descriptions given by Basler (1972) these plants seem to occur also in Schleswig-Holstein (northern Germany). The plants are nearly always purple-flowered, only very exceptionally white.*
We differ from Basler, however, on the taxonomical position of the plants with 2n = 40. *Basler refers all plants with 2n = 40 to S. x uplandicum Nym. (Russian Comfrey), the hybrid between S. asperum (Prickly Comfrey) (2n = 32) and S. officinale (2n = 48)."*
-'Chemotaxonomical Investigations of the Symphytum Officinale Polyploid Complex and S. Asperum (Boraginaceae): The Pyrrolizidine Alkaloids' by H.J. Huizing, Th.W.J. Gadella, and E. Kliphuis, The Netherlands; Plant Systematics and Evolution, Volume 140, pages 279-292, **1982**.

*"**The 2n = 40 cytotype of S. officinale and S. x uplandicum (2n = 40) are not identical at all.***
They differ morphologically (Gadella & Kliphuis 1973) and chemically (Huizing, Gadella & Kliphuis 1982, Huizing, Malingre, Gadella & Kliphuis, submitted).
The 2n = 24 and 2n = 48 cytotypes of S. officinale are inseparable on morphological and chemical grounds as far as they have been studied. They do not hybridise in nature and with great difficulty in the experimental garden, giving rise to sterile triploid hybrids (2n = 36). *Such hybrids could be produced only between white-flowered parents."*
-'Cyto- and Chemotaxonomical Studies on the Sections Officinalia and Coerulea of the Genus Symphytum' by Th.W.J. Gadella, E. Kliphuis and H.J. Huizing; Botanica Helvetica (now called Alpine Botany), Swiss Botanical Society, Basel, Switzerland, Volume 93, pages 169-192, **1983**.

"Common Comfrey (S. officinale) has cytotypes of (2n = 40) and (2n = 48)."
-'Comfrey: A Controversial Crop' by Robert G. Robinson, University of Minnesota, Agricultural Experiment Station, Minnesota Report MR-191, Item No. AD-MR-2210, **1983**, page 2.

"At least 2 cytotypes of S. officinale occur in North America:
2n = 24 (diploid) or 2n = 48 (tetraploid) (these taxa are indistinguishable if the flower color of the 2n = 48 is white; diploids are always white-flowered), and 2n = 40.
    2n = 24 has white flowers; 2n = 48 has white or purple flowers.
    Plants with 2n = 40 are usually purple-flowered and usually occur in very moist habitats."
-'Notes on Symphytum (Boraginaceae) in North America' by T.W.J. Gadella, page 1061-1067 in book: Annals of the Missouri Botanical Garden, Volume 71, Saint Louis, Missouri. Lawrence, Kansas: Allen Press, **1984**.

"***Symphytum officinale* L.: 2n = 24 or 2n = 48**
    S. bohemicum Schmidt, Fl. Boem. 3: 13. tab 263. 1795.
    S. tanaicense Steven, Bull.Soc.Imp. Naturalistes Moscou 24:577. 1851.
    S. uliginosum Kerner, Oesterr. Bot. Z. 13: 227, 1863.
White or cream flowers in diploids (2n = 24) and tetraploids (2n = 48).
Purple or red (or various intermediate colors between white and dark purple) in tetraploids (2n = 48).
The diploid cytotype was assigned to Symphytum bohemicum Schmidt by A. Murin and J. Majovsky* (Acta Facultatis Rerum Naturalium Universitatis Comenianae Botanica, Volume 29, 1982.)
    The exact status of S. bohemicum, S. tanaicense, and S. uliginosum appears to require further investigation."
-'Notes on Symphytum (Boraginaceae) in North America' by T.W.J. Gadella, page 1061-1067 in book: Annals of the Missouri Botanical Garden, Volume 71, Saint Louis, Missouri. Lawrence, Kansas: Allen Press, **1984**.
(* -'Die Bedeutung der Polyploidie in der Entwicklung der in der Slowakei Wachsenden Arten der Gattung Symphytum L.' {The Importance of Polyploidy in the Development of the Growing Slovakia Species of the Genus Symphytum L.} by A. Murin and J. Majovsky, Acta Facultatis Rerum Naturalium Universitatis Comenianae Botanica, Slovakia, Volume 29, 1982, pages 1-25. I was unable to find a copy of this report. If you have an English translation, could you please send it to me.)

"**For *Symphytum officinale* L. the chromosome numbers 2n = 24, 40, 48 and 56 have been reported** (Gadella & Kliphuis 1967, 1978).
    In addition all the chromosome numbers between 2n = 40 and 2n = 48 occur occasionally in nature due to hybridization of the two tetraploid cytotypes (Gadella 1972, Gadella & Kliphuis 1971).
The diploid (2n = 24) cytotype is morphologically indistinguishable from the tetraploid plants with the chromosome number 2n = 48, except for the colour of the flowers which is always white in the diploid plants whereas it is white or purple in the tetraploid plants (l.c.).
Tetraploids:
The tetraploid cytotypes are fairly common and widely distributed throughout Europe, but the diploid plants have only been found in a small number of scattered populations in western, central and eastern Europe. They have been reported from Britain, France, Germany, Hungary, Italy and The Netherlands (Gadella & Kliphuis 1978, Gadella et al. 1983).
The results indicate that the Feulgen-Giemsa banding technique can be used in further studies about the origin of the karyotypes of the tetraploid cytotypes of S. officinale (2n = 40 and 48).
Diploids:
Morphologically the diploid plants of S. officinale are rather uniform, though plants from different parts of Europe may differ somewhat in the length of the flowering stems, the indumentum and the shape of the corolla (Kliphuis, personal communication).
    The twelve pairs of chromosomes of the diploid cytotype of Symphytum officinale L. (2n = 24) can be distinguished individually with the use of their Feulgen-Giemsa banding pattern in combination with the relative length of the chromosomes and the position of the centromeres.
The karyotypes of plants from France, Hungary and The Netherlands do not differ significantly, except for the number of satellites."
-'The Giemsa C-Banded Karyotype of Diploid Symphytum Officinale (Boraginaceae)' by L. Mekki, H. Hart, N.Z. El-Alfy, A. Dewedar, and T.W.J. Gadella (The Netherlands and Egypt); Acta Botanica Neerlandica, Netherlands, Volume 36, No. 1, pages 33-37, February **1987**.
    (A karyotype is the number and appearance of chromosomes in the nucleus of an eukaryotic cell. A eukaryotic cell contains a nucleus surrounded by a membrane and whose DNA is bound together by proteins into chromosomes.)
    ('l.c.' is Latin for 'loco citado' which means 'locally cited'. In other words, the author's name was cited in the early part of the book or article with a date and publication name.)

"**The pyrrolizidine alkaloid and triterpene pattern of S. officinale (2n = 24) and S. bohemicum (2n = 24) is identical. The species S. officinale includes cytotypes with 2n = 24, 40, 48, 54, and 56 chromosomes and is widespread across Europe.**
    White flowered diploid (2n = 24) populations occur in western, central and eastern Europe. This diploid S. officinale resembles the cream flowered S. bohemicum (2n = 24) from eastern Europe very closely in morphological respect (Murin and Majowski/Majovsky 1982, *Holub 1963).
    The 2n = 40 cytotype of S. officinale occurs in the Netherlands and northwest Germany. This type morphologically resembles plants from east Hungary and south Russia. These plants are referred to S. uliginosum Kern. by Soo (1926, 1931) and to S. tanaicense Steven by Degen (1930) and Dobrochaeya (1968) and may have a much wider distribution than S. officinale (2n = 40) from western Europe. Love and Love (1956) and Fernandes

and Leitao (1972) report the chromosome number 2n = 40 for plants from Iceland and Portugal, respectively. Love and Love (1975) refer this material to S. uliginosum (= S. tanaicense). The plants from western and eastern Europe may be identical and belong to the same taxon.

Plants with the higher chromosome numbers (2n = 48, 54, 56) show different distribution patterns. The tetraploids are widely distributed in Europe at lower altitudes. In western and central Europe they are white- or purple-flowered, in eastern Europe purple-flowered.

In south Poland, Czechoslovakia, Yugoslavia, and Bulgaria plants with 56 chromosomes have been found (sometimes plants occur within the same population with 50, 52, 53, and 54 chromosomes). These plants are usually purple flowered. The extra chromosomes are B chromosomes (Hart and Mekki, unpublished)."

-'Chemotaxonomy of the Symphytum Officinale Agg. (Boraginaceae)' by Jaarsma, Lohmanns, Gadella and Malingre, Plant Systematics and Evolution, 167, pages 113-127, **1989**.

(* -Acta Horti Botanici Pragensis. Praga {Prague}, Czech Republic: Universitas Carolina, Facultas Rerum Naturalium, 1962-1963. Article: 'Miscellanea ad Floram Cechoslovacum Pertinentia' by J. Holub, 1-17, pages 47- 59.)

"Gadella & Kliphuis (1967) identified **three main cytotypes in the Symphytum officinale group: 2n = 24, 40, and 48.**
    The tetraploid (2n = 48) is the most widespread in Europe and may have white or purplish flowers.
    The white-flowered diploid (2n = 24) may show, in the southern parts of its distributional area, the presence of B chromosomes or not (2n = 24 + 0-4 B); it is sporadic and uncommon and, in the above quoted author's opinion, virtually indistinguishable from the tetraploid biotype on a morphological basis."

-'Symphytum Tanaicense (Boraginaceae) New for the Italian Flora' by Peruzzi, Garbari and Bottega, Willdenowia, Berlin, Germany, 31, pages 33-41, **2001**. ('Cytotaxonomic Studies in the Genus Symphytum L. Symphytum Officinale L. in the Netherlands' by T.W.J. Gadella and E. Kliphuis, Proceedings of the Koninklijke Nederlandse Akademie van Wetenschappen {Royal Dutch Academy of Sciences}, C, 70, pages 378-391, published 1967.)

"**Symphytum officinale L.: Common Comfrey.**
    Symphytum officinale L. ssp. uliginosum (Kern.) Nyman (2n = 40)
    S. tanaicense Steven (species, 2n = 40)
    S. uliginosum Kern
At present there appears to be two cytotypes of Symphytum officinale that may warrant recognition at some level.
    The 2n = 24 or 48 type (diploid and tretraploid, respectively) appears to be the common form in New England (northeast United States). It shows hispid stems that are not harsh to the touch, **leaves with prominent decurrent wings,** marginal setae of the sepals distributed in an irregular pattern, and white (usually diploid) or purple (usually tetraploid) corollas.
    The 2n = 40 type, known from North America but yet documented in New England, has tuberculate-based hairs on the stem that are harsh to the touch, leaves with shorter decurrent wings, marginal setae of the sepals distributed in a uniform manner, and purple corollas.
    This latter cytotype has been named ssp. uliginosum (or S. tanaicense at the rank of species)."

-New England Wild Flower Society's Flora Novae Angliae: A Manual for the Identification of Native and Naturalized Higher Vascular Plants of New England by Arthur Haines. New Haven: Connecticut: Yale University Press, **2011**.

"**The family Boraginaceae Juss. is known for its considerable chromosome variation, which is a consequence of various cytological processes, such as chromosome fusion or fragmentation, polyploidy or aneuploidy** (Britton 1951, Coppi et al. 2006). These processes seem to be common in the family and play a crucial role in the evolution of many genera. In addition, the occurrence of B chromosomes is quite common (Gadella 1972, Sauer 1975, Bigazzi & Selvi 2003, Bedini et al. 2012).

All these processes are also important for genome evolution in the genus Symphytum L. (Grau 1968, *1971, Gadella & Kliphuis 1978, Murin & Majovsky/Majowsky 1982). Gadella & Kliphuis (1978) **report a high frequency of polyploids** in comparison with other genera of Boraginaceae with the occurrence of polyploidy, as in Onosma, Myosotis and Pulmonaria.
    This phenomenon is well illustrated by the four ploidy levels reported for the Symphytum officinale complex (e.g. **Markowa & Iwanowa 1970, Gadella & Kliphuis 1978)."

-'Symphytum Tuberosum Complex in Central Europe: Cytogeography, Morphology, Ecology and Taxonomy' by L. Kobrlova, M. Hrones, P. Koutecky, M. Stech, and B. Travnicek, Preslia: The Journal of the Czech Botanical Society, Czech Republic; Volume 88, pages 77-112, March **2016**.

(* -'Cytologische Untersuchungen an Boraginaceae II' {Cytological Investigations on Boraginaceae} by J. Grau, Mitteilungen der Botanischen Staatssammlung Munchen {Communications of the Botanical State Collection Munich, Germany}, Volume 9, pages 177-194, 1971.)

(** -'Karyologische Untersuchung der Vertreter der Fam. Boraginaceae, Labiatae und Scrophulariaceae in Bulgarien' {Caryological Examination of the Representatives of the Family Boraginaceae, Labiatae and Scrophulariaceae in Bulgaria} by Margarita Markova/Markowa and P. Ivanova/Iwanowa; Izvestiya na Botaniceskiya Institut {Bulletin de l'Institut Botanique}, {also titled: Mitteilungen des Botanischen Instituts Sofia}, Bulgaria, Volume 20, pages 93-98, 1970. In German.)

"**Symphytum officinale: Both diploid and tetraploid (chromosome) cytotypes occur and are morphologically separable. S. x uplandicum (Russian Comfrey) sometimes back-crosses with this parent** (* Perring, 1994)."

-'Online Atlas of the British and Irish Flora', https://www.brc.ac.uk/plantatlas, Distribution and ecology with photographs, **2018**.

(* -The Common Ground of Wild and Cultivated Plants: B.S.B.I. Conference Report No. 22 by A. Roy Perry and R.G. Ellis, {F.H. Perring, pages 64-70}. Cardiff, Wales: Department of Biology, National Museum of Wales, 1994.)

Symphytum officinale Chromosomes (2n=24 or 2n=48) vs (2n=40)

"**S. officinale 2n = 24 or 2n = 48:**
Stem to 120 cm (47 inches) long, distinctly winged, hispid; the indument renders the stems soft to the touch.
Basal leaves lanceolate or ovate, to 60 cm (23 inches) long, acute at the apex, acuminate and attenuate at the base; lamina 10-40 cm (3.9-15.7 inches) long, 2-12 cm (0.7-4.7 inches) wide; petiole 2-20 cm (0.7-7.8 inches) long; middle and upper stem leaves of the same type, but much smaller, adaxial side of the leaves with many short and long hairs, which are never scabrid; sometimes these hairs have a tubercular base that is not deciduous; abaxial side of the leaves with long appressed hairs along the veins and many shorter hairs between the veins; the leaf base is decurrent from node to node.
Calyx to 8 mm long, divided to 3/4 of its length, calyx lobes triangular lanceolate and acute, marginal stiff setae with an irregular distribution pattern; these marginal setae lack a tubercular base; corolla urceolate, 15-17 mm long.
White or cream flowers in diploids (2n = 24) and tetraploids (2n = 48).
Purple or red (or various intermediate colors between white and dark purple) in the tetraploids (2n = 48).
Stamens to 7 mm long, anthers longer than the filament and shorter than the corolla scales with which they alternate; connective projecting beyond the thecae; squamae of the corolla triangular-lanceolate, 7-7.5 mm long and 2 mm wide at the base; apex mucronate, papillae obtuse, papillae more densely crowded at the tip of the scale margin.
Reproduction: both cytotypes are obligate allogamous (must cross fertilize), they are strictly self-incompatible.
Fruit black and shiny, 4-5 mm long, 2-2.5 mm wide. The production of fruits (nutlets) varies considerably among different populations. Even plants with normal fertility have many flowers which produce only 3, 2, 1 or even 0 viable nutlets."
-'Notes on Symphytum (Boraginaceae) in North America' by T.W.J. Gadella, pages 1061-1067 in book: Annals of the Missouri Botanical Garden, Volume 71, Saint Louis, Missouri. Lawrence, Kansas: Allen Press, **1984**.

"**S. officinale 2n = 40:**
Stem to 70 cm (27 inch) long, distinctly winged, prickly and asperous, harsh to the touch; decurrence of the stem usually less pronounced than in the 2n = 24 and 2n = 48 cytotypes, at least in the upper leaves, but still distinctly present.
Shape of the leaf the same as in S. officinale (2n = 24, 48). Indument of the leaf on adaxial side of the lamina very scabrous with many short tubercular based prickly setae that are deciduous; between the hairs many short curved or uncinate hairs with or without a tubercular base.
Calyx to 9 mm long, divided to 3/4 of its length: calyx lobes triangular-lanceolate with an acute tip; stiff marginal setae in a very regular distribution pattern; some of the marginal and dorso-median hairs with a tubercular base; corolla urceolate, 16-19 mm, **usually dark or light purple, very occasionally white**; stamens as in the 24 or 48 cytotype of S. officinale, but stamens somewhat longer than squamae; papillae of the corolla scales more densely crowded in the middle of the scale margin, otherwise the same as in S. officinale 2n = 24, 48.
Reproduction obligate allogamous (must cross fertilize), plants strictly self-incompatible.
Fruit as in the 2n = 24 or 48 cytotypes of S. officinale. Even plants with normal fertility may have flowers which produce only 3, 2, 1 or even (and not occasionally) 0 viable nutlets."
-'Notes on Symphytum (Boraginaceae) in North America' by T.W.J. Gadella, pages 1061-1067 in book: Annals of the Missouri Botanical Garden, Volume 71, Saint Louis, Missouri. Lawrence, Kansas: Allen Press, **1984**.

Symphytum officinale Chromosomes 2n = 40

"**Symphytum*: The Symphytum officinale plants with 2n = 40, showed distinct morphological, ecological and phenological differences to the other two races (S. officinale 2n = 24 = diploid, 2n = 48 = tetraploid).**
S. officinale 2n = 40 was less tall, with stems and leaves prickly (the diploids and tetraploids had densely hispid, never scabrous, stems and leaves), the hairs were also found to be deciduous (persistent in the other two races), and upper stem leaves not, or hardly, decurrent (usually decurrent in the diploids and tetraploids).
S. officinale 2n = 40 invariably grew under very moist conditions on peaty soils; and always in the open, while the other two races were found in drier places, e.g. roadsides, near woods, ditches, river and pond sides, etc., but never on peaty soils, they were also found to be tolerant of light shade.
The flowering period was also noted to be shorter, and it was invariably in fruit in September when the other two cytotypes were still in flower.
It is concluded that the Dutch plant with 2n = 40 closely matches S. uliginosum Kern., a native of southern Russia and Hungary, which is naturalised in Austria, France and Switzerland.
    It is suggested that the species may in fact be native in the latter countries, in Holland, and possibly elsewhere in western Europe."
-'Proceedings of Botanical Society of the British Isles, Volume 7' edited by E.F. Greenwood, London, England, **1967**-

1969, page 432.
(* -'Cytotaxonomic Studies in the Genus Symphytum I: Symphytum Officinale L. in the Netherlands' by Th.W.J. Gadella and E. Kliphuis, Botanical Museum and Herbarium of the State University of Utrecht, Netherlands, February 25 1967.)
(Ecology is the interrelationship of organisms and their environment.)
(Phenology is relation between climate and periodic biological phenomena such as flowering, leafing and breeding.)

"A third cytotype of Symphytum offcinale L., 2n = 40, occurs in Western Europe, where it was found in the Netherlands (Dutch) and in Schleswig-Holstein (northern Germany).
Gadella and Kliphuis (1967, 1971) are of the opinion that many of these plants should be assigned to S. officinale.
Basler (*1972), on the other hand, regards all plants from northwest Germany as hybrids between S. asperum (Prickly Comfrey) and S. officinale, i.e., S. x uplandicum Nym. (Russian Comfrey).

Since S. asperum Lepech. has the chromosome number 2n = 32 (Gadella & Kliphuis 1969; Grau, 1971; Basler, l.c.; Gviniashvili 1972), the hybrid between this species and S. officinale may be charactetized by 2n = 40, but also other chromosome numbers can be expected.

In order to obtain a better insight into the way in which the cytotype 2n = 40 originated, the present authors compared the west European cytotype with artificially produced hybrids.

By crossing S. asperum and S. officinale (2n = 48) it proved impossible to produce plants that resemble those of the Dutch low-lying peat lands. The artificially produced hybrid can be assigned without any doubt to S. x uplandicum Nym. Plants from the Dutch low-lying peat lands, however, present serious taxonomic problems.

Therefore, two possibilities exist: the Dutch cytotype 2n = 40 is either regarded as a taxon of its own, or it may be assigned to S. ulginosum Kern.

These taxonomic problems are further complicated by the frequently occurring products of backcrossing to either S. officinale or S. asperum. From such crosses highly variable hybrid swarms may arise, which seem to connect S. asperum and S. officinale through a large chain of intermediates.

Never the Dutch low-lying peat land plants could be made artificially, which led us to the conclusion that these plants do not belong to S. uplandicum. *Neither their morphological characters, nor the results of crossing experiments support Basler's view."*
-'Cytotaxonomic Studies in the Genus Symphytum V: Some Notes on W. European Plants with the Chromosome Number 2n=40' by Th.W.J. Gadella and E. Kliphuis, Botanische Jahrbucher fur Systematik (Botanical Yearbooks for Systematics), Volume 93, pages 530-538, **1973**.
(* -'Cytotaxonomic Studies on the Boraginaceen Genus Symphytum L.: Studies on Predominantly Northern German Plants of the Species S. Asperum Lepech., S. officinale L. and S. x Uplandicum Nym.' by Armin Basler, Botanische Jahrbucher fur Systematik: Pflanzengeschichte und Pflanzengeographie {Botanical Yearbooks for Systematics: Plant History and Plant Geography}, Volume 92, pages 508-553, 1972. All in German. I was unable to find a copy of this report. If you have an English translation, could you please send it to me.)

('l.c.' is Latin for 'loco citado' which means 'locally cited'. In other words, the author's name was cited in the early part of the book or article with a date and publication name.)

(A backcross is a cross of a hybrid with one of its parents or an individual genetically similar to its parent. This creates offspring with genetics closer to that of the parent.)

(A hybrid swarm is a group of hybrids that have survived past the first hybrid generation, with interbreeding taking place between hybrid individuals and also backcrossing with its parent type. These plants are highly variable, with genes and phenotypes of individuals ranging widely between the two parent types. Hybrid swarms blur the boundary between the parent taxa.)

"The 2n = 40 cytotype of S. officinale differs both from the diploid (2n = 24) and tetraploid (2n = 48) cytotype (of S. officinale) in morphological aspect, but not in chemical respect.
It contains the same pyrrolizidine alkaloids as S. officinale (2n = 24, 48). For that reason and also because this cytotype is interfertile (can breed) with the 2n = 48 cytotype, ***it is regarded as conspecific (same species) with S. officinale (2n = 48).***

The lack of information of the exact identity of the Western 2n = 40 cytotype of S. officinale and the Eastern morphotype of S. tanaicense Stev. does, for the moment, not permit to give a taxonomic recognition and an assignment of the appropriate rank to these taxa."
-'Cyto- and Chemotaxonomical Studies on the Sections Officinalia and Coerulea of the Genus Symphytum' by Th.W.J. Gadella, E. Kliphuis and H.J. Huizing; Botanica Helvetica (now called Alpine Botany), Volume 93, pages 169-192, **1983**.
(Conspecific means belonging to the same species.)

"The origin of the 2n = 40 cytotype of S. officinale may be explained by a 5 step hypothesis: *auto-polyploidization followed by 4 steps of centric fusion."*
-'Hybridization in Symphytum: Pattern and Process' by T.W.J. Gadella, Sommerfeltia: The Journal of Natural History Museum and University of Oslo, Norway, Volume 11, pages 79-96, **1990**.

(Autoploidy is when an individual's chromosomes are more than 2 copies of the genome of a single ancestral species. An individual has more than 2 sets of chromosomes {4n} that are all derived from an original species {2n}.)

(Centric-fusion translocation occurs when chromosomes are rearranged where there is fusion of an entire long arm of one acrocentric chromosome with another intact long arm of acrocentric chromosome.)

S. officinale Chromosomes 2n = 44

"The Symphytum plants with 2n = 44 are hybrids between the subspecies uliginosum and officinale.
In spite of having different chromosome numbers, the plants of these subspecies are fully interfertile in the Netherlands. They are usually kept apart by the fact that the subspecies uliginosum grows in low moor peat and the subspecies officinale in clayey or sandy soil."
-'Cytotaxonomic Studies in the Genus Symphytum III: Some Symphytum Hybrids in Belgium and the Netherlands' by Th.W.J. Gadella and E. Kliphuis, Biologisch Jaarboek (Biological Yearbook), **1971**.

*"**Symphytum officinale L.** (with hypotetraploid chromosome number)*
    Ipswich (Vice County 25 E. Suffolk, England), FHP 69/43; G & K 10.287-10.288: **2n = 44**
    Twyford Forest (Vice County 53 S. Lincolnshire, England), G & K 10.191-10.192: **2n = 44**
The plants from Twyford Forest, Lincolnshire, with the chromosome number 2n = 44, have the same characters as the plants with 2n = 48 of S. officinale. *They are white-flowered and their leaves are strongly decurrent.*
    The plants are different from the hybrids (2n = 44) between S. x uplandicum (2n = 40) and S. officinale (2n = 48) described by Gadella and Kliphuis (1971). This hybrid is much taller (up to 1.7 meter = 5.5 feet) than the plants from Twyford Forest and has pinkish flowerbuds and pink flowers (which turn pinkish-blue when the flowers age).
The origin of the Twyford Forest plants may also be explained by assuming that the cytotypes 2n = 40 and 2n = 48 of S. officinale hybridized. The fact, however, that S. officinale (2n = 40) has not yet been found in Britain forms a serious objection against this theory. Moreover, the S. officinale plants with 2n = 40 are usually purple-flowered and exceptionally white-flowered. **Since the Twyford Forest plants are white-flowered, it seems highly probable that only white-flowered parents are involved in the formation of the plants concerned.**
For the time being we are unable to give a satisfactory explanation for the origin of the Twyford Forest plants, but we intend to continue our crossing experiments (including these cytologically deviating plants).
The same holds true for the plants from Ipswich, which also have the chromosome number **2n = 44. These plants, however, have purple flower buds and light-purple corollas.***"*
-'Cytotaxonomic Studies in the Genus Symphytum VI: Some Notes on Symphytum in Britain' by Th.W.J. Gadella, E. Kliphuis and F.H. Perring, Instituut voor Systematische Plantkunde (Institute for Systematic Botany), Utrecht, Netherlands, and Institute of Terrestrial Ecology, Abbots Ripton, England; Acta Botanica Neerlandica, Volume 23, No. 4, pages 433-437, August **1974**.
    (Hypotetraploid means having fewer than four times the haploid number of chromosomes in a cell nucleus. Another definition is having several chromosomes less than the tetraploid multiple of the basic genome.)

S. officinale Breeding with Itself Overview

My writeup of online translation from Dutch to English of below article about S. officinale:
    a. Purple 2n = 48 breeds with white 2n = 48.
    b. Purple 2n = 40 breeds with both purple and white 2n = 48, creating 2n = 44.
    c. White 2n = 40 only breeds with 2n = 40.
    d. White 2n = 24 only breeds with 2n = 24 except experimentally and then it is sterile.
"Symphytum:
a. Purple and white-flowering individuals of S. officinale (2n = 48) cross very easily and deliver a fertile F1 and F2.
b. Purple-flowered plants with 2n = 40 of S. officinale cross very easily with purple and white-flowered plants of S. officinale (2n = 48), resulting in fertile hybrids with 2n = 44.
c. White-flowered plants with 2n = 40 of S. officinale cross fertile with same cytotype, but because of the rarity of the white-flowered ones only used as pollen suppliers.
d. The white-flowered plants of S. officinale with 2n = 24 do not cross with the others cytotypes of S. officinale. In principle they are reproductively isolated. In an orchard at IJsselstein, Netherlands, white plants with 2n = 24 and 2n = 48 are mixed, mixed with purple flowering with 2n = 48. They do not cross there.
    After very much failed attempts succeeded in creating two (completely sterile) hybrids (2n = 36) in the experimental garden obtain after crossing white-flowered with 2n = 24 and 2n = 48. One artificial hybrid (2n = 32) was obtained, which is also sterile, between white-flowered plants with 2n = 24 and 2n = 40.
e. The bright blue flowering plants of S. asperum (2n = 32) provided after crossing with S. officinale two types of hybrids: with 2n = 36 (2n = 32 x 2n = 40 and reciprocity) and with 2n = 40 (2n = 32 x 2n = 48 and reciprocity; with the exception of S. officinale 2n = 48 male, white flowering)."
-'Variation and Hybridization in Some Taxa of the Genus Symphytum' (Variatie en Hybridisatie bij enkele Taxa van het Genus Symphytum) by Th.W.J. Gadella, Department of Population and Evolutionary Biology (Vakgroep Populatie en Evolutiebiologie, Utrecht, The Netherlands); Gorteria: Tijdschrift voor de Floristiek, de Plantenoecologie en het Vegetatie-Onderzoek van Nederland (Journal for Floristics, Plant Ecology and Vegetation Research in the Netherlands), Volume 9, No. 4, pages 88-93,

1978. All in Dutch except abstract was also in English.
>(An F1 hybrid or filial 1 hybrid is the first generation of offspring of distinctly different parental types that produce a new, uniform phenotype {observable properties of an organism} with a combination of characteristics from the parents. Mules are F1 hybrids between horse and donkey. This can happen naturally, and includes hybrids between species, for example, peppermint is an F1 hybrid of watermint and spearmint.)
>(F2 hybrids or second generation hybrids are self or cross pollination of F1 hybrids. They have more variability than F1s, but can have some desirable traits. They are not true to form.)

***"Triploid (2n = 36) hybrids between diploid (2n = 24) and tetraploid (2n = 48) plants of S. officinale are lacking in mixed populations**, possibly because homoploid pollen grows faster than heteroploid."*
-'Hybridization in Symphytum: Pattern and Process' by T.W.J. Gadella, Sommerfeltia: The Journal of Natural History Museum and University of Oslo, Norway, Volume 11, pages 79-96, **1990**.
>(Hybridization with no changes in chromosome number is homoploid hybrid speciation.
>Heteroploids have a chromosome number that is not the haploid or diploid number normal in the species.)

Example of S. officinale breeding with itself:
*"**Besides the well known species S. bohemicum and S. officinale s. str. (sensu stricto), an intermediate, partly independent taxon is grouped to the hybrid S. rakosiense.**
Symphytum officinale Group: Of the clans of the Symphytum officinale group in central Europe, two species can be detected in North Rhine-Westphalia (western Germany): **the diploid (2n = 24) Symphytum bohemicum and the tetraploid (2n = 48) S. officinale s. st. (sensu stricto).**
>In the Aachen urban area and adjoining areas (in North Rhine-Westphalia), **the author could in addition to typical yellowish white flowering Symphytum bohemicum (2n = 24) and S. officinale s. st. (2n = 48) with flowers that have so far only been medium to dark purple in color, he could also find pink flowering plants in which the idea of hybrids is obvious.**

The hybrid between S. bohemicum and S. officinale s. st. can according to *Buch & al. (2007) as S. x rakosiense (Soo) Penzes."*
-'Symphytum Bohemicum, S. Officinale s. str. (sensu stricto), S. x Rakosiense und S. Uplandicum s.l. (sensu lato) im Aachener Stadtgebiet' by F. Wolfgang Bomble, Die Veroffentlichungen des Bochumer Botanischen Vereins e. V. (The Publications of the Bochum Botanical Association), Germany, Volume 5, No. 5, pages 44-60, 2013. In German except abstract is also in English. If you have an English translation, I would appreciate a copy of it.)
(* -'Aspekte der Flora und Vegetation des NSG Rheinaue Friemersheim in Duisburg' {Aspects of the Flora and Vegetation of the Naturschutzgebiet Rhine Friemersheim in Duisburg, Germany} by C. Buch, G.H. Loos and P. Keil; Decheniana: Natural History Association of the Rhineland and Westphalia, Germany, Volume 160, pages 133-153, 2007.)
>(sensu stricto = s.s., s. str., sens. str., sens. strict. In the strict/narrow sense. It is added after a taxon to mean it is being used in the sense of the original author, or without taxa which may otherwise be associated with it.
>sensu lato = s.l., sens. lat. In the broad sense. Used in taxonomy to clarify the scope of a taxon when it has been used to define more than one set of lower-level taxons. It includes all its subordinate taxa and/or other taxa that at other times are considered as distinct.)

*"**Symphytum x rakosiense (Soo) Penzes** (Bot. Kozlem. 38: 149, in obs. 1941)"*
-'International Plant Names Index'® (IPNI), www.ipni.org, A database of the names and associated basic bibliographical details of seed plants, ferns and lycophytes, **2018**.

S. officinale Breeds Between Its Own Purple and White Flower 2n = 48 (Tetraploid)

*"**Symphytum officinale: From crossing experiments it became clear that white- and purple-flowered tetraploids (2n = 48) are interfertile (able to breed together)** (*Gadella & Kliphuis 1969).*
>*In this connection it is interesting to note that crosses between two light purple-flowered tetraploid individuals (originating from two different localities in the Netherlands) yielded the following results: from 27 nutlets harvested, **24 F1 plants turned out to be purple-flowered and 3 F1 plants white-flowered.**"*
-'Cytotaxonomic Studies in the Genus Symphytum IV: Cytogeographic Investigations in Symphytum Officinale L.' by Th.W.J. Gadella and E. Kliphuis, Instituut voor Systematische Plantkunde (Institute for Systematic Botany), Utrecht, Netherlands; Acta Botanica Neerlandica, Volume 21, Issue 2, pages 169-173, April **1972**.
(* -'Cytotaxonomic Studies in the Genus Symphytum II: Crossing Experiments Between Symphytum Officinale L. and Symphytum Asperum Lepech' by Th.W.J. Gadella and E. Kliphuis, Botanical Museum and Herbarium of the State University of Utrecht, Netherlands, Acta Botanica Neerlandica, Volume 18, No. 4, pages 544-549, August 1969.)

S. officinale Breeds Between Its Own 2n = 40 and 2n = 48

*"**Plants with 2n = 40 and 2n = 48 of Symphytum officinale hybridize easily. The hybrids are interfertile (breed with each other). Hybridization, followed by repeated backcrossing to one of the parents may occur. Environmental factors decide whether hybrid swarms or unidirectional (one direction) introgression develops.**

Sixteen populations were studied in the Netherlands where hybridization between the cytotypes 2n = 40 and 2n = 48 had taken place. Two large populations (Kinselmeer and Nijerk) were studied in depth, using frequency histograms and hybridization index as proposed by *Anderson and the distance diagrams by **Wells.
In the first population introgression is in the direction of the 2n = 40 parent, in the second in that of the 2n = 48 parent.
The chromosome number provides an extra marker for the study of successive backcrosses."
-'Population Variability, Hybridization and Introgression in Symphytum Officinale in the Netherlands' by T.W.J. Gadella and E. Kliphuis, Botanische Jahrbuecher fuer Systematik Pflanzengeschichte und Pflanzengeographie (Botanical Yearbooks for Systematics Plant History and Plant Geography), Volume 104, No. 4, pages 519-536, **1984**. (I was unable to get this report. If you have it, could you please send it to me.)
(* -'Introgressive Hybridization' by Edgar Anderson, Geneticist, Missouri Botanical Garden, Professor of Botany, Washington University, Saint Louis, Missouri; book printed in New York, 1949.)
(** -'A Distance Coefficient as a Hybridization Index: An Example Using Mimulus Longiflorus and M. Flemingii {Scrophulariaceae} from Santa Cruz Island, California' by Harrington Wells, University of California; Taxon: International Association for Plant Taxonomy, Volume 29, No. 1, pages 53-65, February 1980.)

*"**Introgressive hybridization between the 2n = 40 and 2n = 48 cytotypes of S. officinale was studied** in a large population along the border of Lake Kinselmeer, The Netherlands.*
*Introgression works in two different directions: **towards the 2n = 40 cytotype in the moist bank zone and towards the 2n = 48 cytotype in the dry clayey dike (dam) zone.***
Anderson's hybrid index and Well's distance diagram revealed a close correlation between cytological and morphological variation.
A flower colour marker and chromosome number variation indicated that gene flow is strongly restricted by distance."
-'Hybridization in Symphytum: Pattern and Process' by T.W.J. Gadella, Sommerfeltia: The Journal of Natural History Museum and University of Oslo, Norway, Volume 11, pages 79-96, **1990**.

S. officinale Rarely Breeds Between Its Own Diploid (2n = 24) and Tetraploid (2n = 48)

*"**A strong barrier between Common Comfrey diploids (2n = 24) and tetraploids (2n = 48) exists.***
Hybrids from crosses between diploid plants and purple-flowered tetraploid plants could never be obtained.
*****Only twice a hybrid with the chromosome number 2n = 36 (triploid) was obtained by crossing white-flowered tetraploids and diploids.*****
At IJsselstein, the Netherlands, a large population consisting of diploids and tetraploids was found: 46 plants were studied cytologically, with the following results:
    16 white-flowered individuals (2n = 24); 18 white-flowered individuals (2n = 48);
    *12 purple-flowered individuals (2n = 48). **No triploids were found.***
In Poland, Skalinska et al. (*1971) arrived at the same conclusion after studying a population near Biezanow.
    In a meadow a mixed population of diploids and tetraploids was found, but no triploids proved to occur. This population, however, differed from that near Usselstein in the absence of white-flowered tetraploids."
-'Cytotaxonomic Studies in the Genus Symphytum IV: Cytogeographic Investigations in Symphytum Officinale L.' by Th.W.J. Gadella and E. Kliphuis, Instituut voor Systematische Plantkunde (Institute for Systematic Botany), Utrecht, Netherlands; Acta Botanica Neerlandica, Volume 21, Issue 2, pages 169-173, April **1972**.
(* -'Studies in Chromosome Numbers of Polish Angiosperms, Eighth Contribution', by M. Skalinska, A. Jankun, H. Wcislo, et al., Acta Biologica Cracoviensia: Series Botanica, published by Polish Academy of Sciences, Volume 14, pages 55-102, 1971. I could not find this report.)

Constituents (Chemicals) of Common Comfrey
See section 'Nutritional Value of Comfrey' (Chapter 19). See sections 'Alkaloids in Comfrey' in Volume 2.

"Symphytum Officinale:

	Low PPM	*High PPM*	*Low Percent*	*High Percent*	*Standard Deviation*
Allantoin Leaf	*1,100*	*20,000*	*0.11%*	*2.00%*	*1.0*
Allantoin Root	*6,000*	*25,500*	*0.60%*	*2.55%*	*1.41 "*

-Handbook of Phytochemical Constituents of GRAS Herbs and Other Economic Plants by Dr. James A. Duke. Boca Raton, Florida: CRC Press, **2000**. (GRAS= Generally Regarded As Safe) This information is also at: Dr. Duke's Phytochemical and Ethnobotanical Databases, https://phytochem.nal.usda.gov/phytochem/search.　　(PPM is Parts Per Million.)
    (Standard Deviation is a statistic that measures the dispersion of a dataset relative to its mean and is calculated as the square root of the variance. In other words, Standard Deviation is a measure of how spread out the numbers are.)

*"**The Symphytum officinalis plant roots contain allantoin (0.6-2%), pyrrolizidine alkaloids (0.02–0.07%), polyphenolic acids, triterpenic saponosides, proteins, caffeic acid, chlorogenic acid, rosmarinic acid, tannins (2.4%), carotene (0.63%), choline, asparagine, coniferin, mucopolysaccharides, starch, gumiresins, phytosterols, carotenoids, and vitamins A / C / E, riboflavin (B2) and B12.***

Moreover, they contain an antigonadotropic principle- lithospermic acid (antioxidant) and immuno-stimulant polyosides, as well as **high amounts of mineral substances (Calcium, Potassium, Phosphorus, Magnesium, Iron, Manganese, Sodium, Zinc).**
The quantitative determinations of tannins, amino-acids, terpenoides, sterols, triterpenes, flavonoids, reducing agents, saponins, alkaloids were performed in the obtained extracts.
A quantitative determination of allantoin- a compound characteristic for the Comfrey- was made. **The highest allantoin quantity- 436.5 microgram/mL- being found in the concentrate of 50% methanolic extract.**
The obtained hydro-alcoholic Symphytum officinalis extracts, 10% mass concentration in 50% EtOH and 50% MeOH were concentrated through membranous procedures (ultrafiltration) on regenerated cellulose membranes with 5.000 Da cut-off.
The studied bioactive compounds were present in initial extracts, *while their presence diminished or even missed in permeates and become abundant in the concentrates."*
-'Phytochemical Study of Some Symphytum Officinalis Extracts Concentrated by Membranous Procedures' by Elena Neagu, Gabriela Paun and Lucian Gabriel Radu, Bucharest, Romania; UPB (University of Politehnica) Scientific Bulletin, Series B: Chemistry and Materials Science, Volume 73, Issue 3, pages 65-74, January **2011**.

(Antigonadotropic means it inhibits the physiological activity of gonadotropic hormones.)
(A permeate is a liquid that has passed through a filtration system.)

"The constituents of Comfrey (Symphytum officinale) root include 0.6–4.7% allantoin (Dennis et al., 1987); abundant mucilage polysaccharides (about 29%) composed of fructose and glucose units (Franz, 1969); phenolic acids such as rosmarinic acid (up to 0.2%), chlorogenic acid (0.012%) as well as caffeic acid (0.004%) and a-hydroxy caffeic acid (Andres, 1991; Grabias and Swiatek, 1998; Teuscher et al., 2009); glycopeptides and amino acids (Hiermann and Writzel, 1998); and triterpene saponins in the form of monodesmosidic and bidesmosidic glycosides based on the aglycones hederagenin (e.g. symphytoxide A), oleanolic acid (Aftab et al., 1996) and lithospermic acid (Wagner et al., 1970)."
-'Comfrey: A Clinical Overview' by Christiane Staiger in Germany, Phytotherapy Research, Volume 26, Issue 10, pages 1441-1448, February **2012**.

Common Comfrey Alkaloids
See the sections 'Alkaloids in Comfrey' in Volume 2.

*"**Two pyrrolizidine alkaloids, symphytine, a new compound, and echimidine have been isolated from the dried roots of Symphytum officinale (Common Comfrey).**"*
-'Studies on Constituents of Crude Drugs. I. Alkaloids of Symphytum Officinale Linn.' by T. Furuya and K. Araki from Kitasato University in Japan; Chemical & Pharmaceutical Bulletin, Volume 16, No. 12, pages 2512-2516, **1968**.

(**This was the first time research was done to see if Comfrey contains alkaloids.**
For more details about this report, see number 2 in the subsection 'Scientific Studies Showing Dangers of Alkaloids in Comfrey' in the section 'Alkaloids in Comfrey' in Volume 2.)

*"**Alkaloids, symphytine and echimidine, triterpenoids, isobauerenol, and phytosterols, beta-sitosterol, have heen isolated from the roots of Symphytum oficinale (Boraginaceae).**
The structure of symphytine was also determined as '7-tiglylretronecine viridiflorate (I), and is the first isolated pyrrolizidine alkaloids with tiglic acid."*
-'Alkaloids and Triterpenoids of Symphytum Officinale' by Tsutomu Furuya and Manabu Hikichi, Kitasato University, Tokyo, Japan; Phytochemistry, Volume 10, pages 2217-2220, **1971**.

*"**S. officinale s.s. (sensu stricto) plants, however, do not contain echimidine.** As a consequence, the presence of echimidine in alkaloid extracts from the S. x uplandicurn (Russian Comfrey) plants, clearly indicates their hybrid character.
Yet, further evidence from other chemotaxonomical markers, which are not related to pyrrolizidine alkaloids, should be of value in a further study of the interrelationships of taxa within the S. officinale species complex and S. asperum.*
In a communication by Furuya and Hikichi (*1971) in which the occurrence of echimidine in S. officinale s.l. (sensu lato) has been reported, *also the presence of a typical triterpenoid, isobauerenol and of phytosterols, beta-sitosterol and stigmasterol was mentioned.*
From our previous work, however, we have assumed that **Furuya and Hikichi might have used S. x uplandicum plants for their phytochemical survey instead of S. officinale plants.**
Although phytosterols are mostly considered to be of little value as chemotaxonomical markers, the triterpenoid isobauerenol might become of considerable interest in further taxonomical studies."
-'Chemotaxonomical Investigations of the Symphytum Officinale Polyploid Complex and S. Asperum (Boraginaceae): Phytosterols and Triterpenoids' by H.J. Huizing, Th.M. Malingre, Th.W.J. Gadella, and E. Kliphuis, all from The Netherlands; Plant Systematics and Evolution, Volume 143, pages 285-292, **1983**.
(* -'Alkaloids and Triterpenoids of Symphytum Officinale' by Tsutomu Furuya and Manabu Hikichi, Kitasato University, Tokyo, Japan; Phytochemistry, Volume 10, pages 2217-2220, 1971.)

(sensu stricto = s.s., s. str., sens. str., sens. strict. In the strict/narrow sense. It is added after a taxon to mean it is being used in the sense of the original author, or without taxa which may otherwise be associated with it.
sensu lato = s.l., sens. lat. In the broad sense. Used in taxonomy to clarify the scope of a taxon when it has been used

to define more than one set of lower-level taxons. It includes all its subordinate taxa and/or other taxa that at other times are considered as distinct.

More about this: After a species name has been first established by the author originally publishing the name, then specialist or later taxonomists may do more research on it and make changes based on new information. 'Taxonomic circumspection' is needed to classify properly. Circumscription means the taxonomist must decide which specimens are included in the species described, and which are excluded. It is in this process of species description that the question of the 'sense / sensu' arises, because the taxonomist must explain his view of the proper circumscription.)
(Chemotaxonomy is the classification of plants and animals based on similarities and differences in biochemical composition.)

"Within more than 300 samples of Symphytum officinale plants coming from over 150 different natural habitats, no alkaloid-free root was found. The alkaloid concentrations varied from 0.599% to 0.045%.
It was shown that pyrrolizidine alkaloids (PA) are distributed not uniformly within the plants.
In the underground parts of the plants, the PAs are especially localized at the extreme exodermis, in the center of the rhizome, in light young roots, and in fine, thin, as hairy roots indicated lateral branches.
> **The PA concentrations of the roots were a 100-fold higher than the PA concentrations of the aereal (aerial= above ground) parts.**

Since 'Symphyti herba (herb)' and 'Symphyti folium (leaf)' as well as 'Symphyti radix (root)' were positively evaluated for the external use in case of contusions, strains and sprainings, we suggest to use the aereal parts of the plants as an alternative for the roots."
-'Investigations Concerning the Content and the Pattern of Pyrrolizidine Alkaloids in Symphytum Officinale L.' by R. Mutterlein and C.G. Arnold, Pharmazeutische Zeitung Wissenschaft (Pharmaceutial Science Newspaper), Germany, Volume 138, No. 5/6, pages 119-125, **1993**. (I was unable to get this except for the abstract. If you have it in English, could you please send it to me.)
> (Exodermis is a layer in a root beneath the epidermis or velamen.)

"Symphytum officinale L. *(synonym S. consolida L.)*:
Both leaves and roots are used externally in cases of fractures, contused injuries, sprainings, contusions, strains, thrombophlebitis, mastitis, hematoma in the form of extracts, ointments, compress pastes, etc., internally as infusions and extracts in cases of gastro-intestinal diseases and respiratory tract diseases.
For vegetarians numerous recipes are offered for the preparation of Comfrey salad, spinach, souffles, soups, bread, rolls, and root beverages.
In dried leaves 0.02 to 0.18% and in the roots 0.25 to 0.29% alkaloids respectively their N-oxides were detected. The alkaloids include intermedine, lycopsamine, their 7-acetyl derivatives, symlandine, symviridine, myoscorpine, and symphytine.
> The presence of the alkaloids (echinatine, heliosupine N-oxide, heliotrine, lasiocarpine, and viridiflorine) isolated and characterized by a Russian and Polish working group by means of paper chromatography could not be confirmed.

A possible risk associated with the consumption of Symphytum in humans was repeatedly reported and in medical literature several cases of intoxication attributed to Symphytum and Comfrey are described.
In animal tests acute toxic effects were detected in rats and goats. Rats given for a longer period root drug or a mixture of the alkaloids intermedine and lycopsamine, that are also contained in the Comfrey, showed insuloma tumors of the pancreas, liver adenomas, hemangioendothelial sarcomata, and tumors of the bladder.
In the long-term test with the root drug carcinogenicity could be attributed to the main alkaloid symphytine. Moreover, administration of the total alkaloid extract resulted in a mutagenic effect.
In view of the risks associated with the alkaloid content neither the leaves nor the roots should be used internally. However, under certain conditions, there are no objections to an external use provided the skin is intact.
> In a study on the percutaneous absorption of alkaloids from an alcoholic plant extract that had been applied to the skin of rats 0.08 to 0.41% alkaloid N-oxides were detected in the urine even after two days."

-'Medicinal Plants in Europe Containing Pyrrolizidine Alkaloids' by Erhard Thomas Roeder, Pharmazeutisches Institut (Pharmaceutical Institute) der Rheinischen Friedrichs-Wilhelms, University of Bonn, Germany, Pharmazie (Pharmacy) 50, pages 83-98, March **1995**.
> (The above metioned tests are discussed in subsection 'Scientific Studies Showing Dangers of Alkaloids in Comfrey' in the section 'Alkaloids in Comfrey' in Volume 2. There is controversy on this topic.)

"Pyrrolizidine Alkaloid (PA) Concentration in Symphytum officinale (Common Comfrey):
Leaves:
> Roeder (1995) quotes concentrations of 200 mg/kg to 1800 mg/kg in dried leaves of S. officinale. **Awang* et al analysed 5 samples and found 200 to 2000 mg/kg total PAs. This last figure of 2000 mg/kg corresponds with the figure given by Roeder for S. x uplandicum leaf and may not be of S. officinale.**
> Mutterlein and Arnold** (1993) analysed 9 samples of large leaves of S. officinale and found an average concentration of 3 mg/kg.
> **Their research confirms earlier findings that PAs are most concentrated in the young small leaves. Two samples of young leaves yielded 87 mg/kg PAs in 5 cm (1.96 inch) long leaves collected in March 1991, and 16 mg/kg in leaves up to 15 cm (5.9 inch) long collected in April 1992.**
> This change in concentration ties in with the proposal that the role of PAs in plants is to protect young leaves against

insects and slug attack.

Roots:

Mutterlein and Arnold analysed 300 samples of S. officinale roots collected from 150 habitats in Germany and found 450 to 5990 mg/kg PAs, average 1700 mg/kg. *Roeder quotes from 2500 to 2900 mg/kg. Awang et al anlysed 5 samples yielding 700-1700 mg/kg."*

-'Using Herbs that Contain Pyrrolizidine Alkaloids' by Alison Denham, B.A., MNIMH (National Institute of Medical Herbalists), University of Central Lancashire, England; The European Journal of Herbal Medicine, Volume 2, No. 3, pages 27-38, **1996**.
(* -'Echimidine Content of Commercial Comfrey {Symphytum spp.- Boraginaceae}' by D.V.C Awang, Brian A. Dawson, Julie Fillion, Michel Girad and Daryl Kindack, Bureau of Drug Research, Health and Welfare Canada, Ottawa, Ontario, Canada; Journal of Herbs, Spices and Medicinal Plants, Volume 2, No. 1, pages 21-34, 1993.)
(** -'Investigations Concerning the Content and the Pattern of Pyrrolizidine Alkaloids in Symphytum Officinale L.' by R. Mutterlein and C.G. Arnold, Pharmazeutische Zeitung Wissenschaft {Pharmaceutial Science Newspaper}, Germany, Volume 138, No. 5/6, pages 119-125, **1993**. I was unable to get this report except for the abstract. If you have it in English, could you send it to me.)

*"***The alkaloids of S. officinale are the isomeric monoesters*** *lycopsamine and intermedine,* ***the isomeric diesters*** *7-acetyl lycopsamine and 7-acetyl intermedine, the isomeric diesters symphytine, symlandine and symviridine, and rarely, echimidine.* ***They are all based on retronecine.****"*

-'Using Herbs that Contain Pyrrolizidine Alkaloids' by Alison Denham, B.A., MNIMH (National Institute of Medical Herbalists), University of Central Lancashire, England; The European Journal of Herbal Medicine, Volume 2, No. 3, pages 27-38, **1996**.

(In chemistry, isomers are ions or molecules with identical formulas but different structures. Isomers do not necessarily have similar properties.)
(Retronecine is a pyrrolizidine alkaloid found in some plants in the genera Senecio and Crotalaria, and the family Boraginaceae. It is the most common central core for other pyrrolizidine alkaloids.)

*"****Comfrey consists of the dried root and rhizome of Symphytum officinale (Boraginaceae)****; the leaf has also been used. It contains about 0.7% of allantoin, large quantities of mucilage, and some tannin.*
It also contains different hepatotoxic (liver toxic) pyrrolizidine alkaloids.
The total content of pyrrolizidine alkaloids is approximately 0.3% of the dry weight of the root, usually lower in the leaves. The amount of pyrrollizidine alkaloids in the fresh plant may not be very high, but the ready-to-use preparations often have high levels, e.g. 270-2900 mg/kg *(Martindale, *Toxnet, **Council of Europe 2008, Cornell University)."*

-'Risk Profile Symphytum Officinale Extracts', CAS No. 84696-05-9, www.mattilsynet.no, Statens tilsyn for planter, fisk, dyr og naeringsmidler (State supervision of plants, fish, animals and food), Brumunddal, Norway, March 11 **2013**.
(* -Toxnet®: Toxicology Data Network, United States National Library of Medicine, Bethesda, Maryland, http://toxnet.nlm.nih.gov. Resource for searching databases on toxicology, hazardous chemicals, environmental health, and toxic releases.)
(** -'Symphytum Officinale Extracts, Monograph No. 38. Active Ingredients Used in Cosmetics: Safety Survey'; Council of Europe's Committee of Experts on Cosmetic Products, Council of Europe Publishing, France, pages 369-374, 2008. I could not find this monograph.)

Symphytum Consolida Major 1737

'A Curious Herbal: Containing Five Hundred Cuts, of the Most Useful Plants, Which are Now Used in the Practice of Physick Engraved on Folio Copper Plates, After Drawings Taken from the Life', Volume 1 by Elizabeth Blackwell, London, England, 1737.

Symphytum Plate 252.

Chapter 9

Details about Symphytum Species:
No Asperum, No Officinale, No Hybrids

Symphytum anatolicum

Current Botanical Nomenclature of S. anatolicum

Symphytum anatolicum Boiss. (*Diagn. Pl. Orient. ser. 1, 4: 43. 1844)
-'International Plant Names Index'® (IPNI), www.ipni.org, A database of the names and associated basic bibliographical details of seed plants, ferns and lycophytes, **2018**.
(* -'Diagnoses Plantarum Orientalium Novarum', Series 1, No. 4, page 43, by Pierre Edmond Boissier, Lipsiae {Leipzig}, Germany, 1844. Series 1: Volume 1: No. 1-7, Volume 2: No. 8-13. Series 2: Volume 3: No. 1-6. Published 1842-1859. In Latin.)

"Family: Boraginaceae Juss.
Genus: Symphytum Tourn. ex L.
Species: Symphytum anatolicum Boiss.
        This species is accepted, and its native range is east Aegean Islands to west and southwest Turkey.
Synonym: Symphytum sicyosmum Candargy"
-Plants of the World Online®, www.plantsoftheworldonline.org, Kew Science, Royal Botanic Gardens, London, England, Information on all the world's known seed-bearing plants, **2018**.
        (The Aegean Sea is a bay of the Mediterranean Sea located between the Greek and Anatolian peninsulas, i.e., between the mainlands of Greece and Turkey.)

        Boraginaceae **Symphytum sicyosmum P.Candargy** (Bull. Soc. Bot. France 44: 150. 1897)
        -'International Plant Names Index' (IPNI), www.ipni.org, A database of the names and associated basic bibliographical details of seed plants, ferns and lycophytes, **2018**.

Description of S. anatolicum

"***S. anatolicum Boiss.***, Diagn. ser. I, 4, p. 43 (1844).
Geographical Distribution: **Turkey: Asia Minor.**
This is a small, slender species, distinct and easily recognised. De Candolle mentions a 'specimen simplex subpedale', but all those which I have seen are copiously branched with numerous, mostly many-flowered, branches.
The leaves are rough with tubercular setae, and are generally attenuated into the petiole, but are occasionally rounded at the base, as in Bornmuller's plant from Mount Sipylos (Sipylus, Turkey).
The racemes often bear as many as 36 flowers, as in Boissier's Smyrna (Izmir, Turkey) plant, but in his plant from Tralles (Aydin, Turkey) they are as few as 12.
Flowers of two distinct colours are rare in this Symphytum genus, only occurring, in pure species, in this and in S. officinale. In the latter, two colours are often combined in the same flower, but in S. anatolicum, as far as can be seen in dried specimens, **they are either pure violet or white.** The corolla is generally narrow, and the tube is nearly twice as long as limb.
In fruit, the calyx-segments are broadly ovate at the base, rather suddenly attenuated to the apex, and become nearly glabrous."
-'A Revision of the Genus Symphytum, Tourn.' by Cedric Bucknall, (Mus. Bac. Oxon= Bachelor of Music, Oxford University), Journal of the Linnean Society of London, England, Botanical Journal, Volume 41, Issue 284, pages 491-556, December **1913**.

"***S. anatolicum Boiss.***, Diagn. ser. I, 4:43 (1844).
Synonym: S. sicyosinum Candargy in Bull. Soc. Bot. Fr. 44: 150 (1897).
Small, slender perennial herb, 15-45 cm (5.9-17.7 inch) tall. Stem puberulous, with curved tuberculate-based hairs. Lower leaves 6.5 cm (2.5 inch) long, 2.5 cm (0.9 inch) broad, oblong-ovate, shortly petiolate, subrepandous; upper leaves 3 cm (1.1 inch) long, 2 cm (0.7 inch) broad, sessile shortly decurrent. Leaves shortly tuberculate-setose.
Inflorescence 25- or more flowered. Calyx 5-7 mm long, divided nearly to base, segments linear-lanceolate, subacute; accrescent in fruit, enlarging to 13 mm. Corolla 14-15 mm long, **white or violet;** it is unusual for all the flowers on one plant to be either white or coloured in Symphytum. Corolla tube narrow, 2 mm wide, twice as long as calyx, limb spreading; corolla scales 4 mm long, linear, obtuse, scarcely exceeding the stamens. Stamen filaments 2.5 mm long; anthers 2 mm long.
Nutlets 4 x 1.5 mm, curved, reticulately veined, tuberculate. Flowers 4-9.
Montane (mountainous) woodland, 350-1390 meters (1148-4560 feet)."
-'A Revision of Symphytum in Turkey and Adjacent Areas' by G.E. Wickens, Royal Botanic Gardens: Kew; Notes from the Royal Botanic Garden Edinburgh, Scotland, Vol 29, page 157-180, **1969**. Includes Turkey, Bulgaria, Greece, Aegean Islands & Caucasia.

"***Symphytum anatolicum has a wide distribution in western Asia Minor and extends on to the Greek islands of Samos,***

Kos and Lesvos (if S. sicyosmum is conspecific)."
-'The Greek Species of Symphytum Boraginaceae' by William T. Stearn, British Museum of Natural History, London, England; Annales Musei Goulandris, Greece, Volume 7, pages 175-220, (**1985 or 1986**, different sources give different dates).
    (Samos, Kos and Lesvos are Greek islands near Turkey.) (Conspecific means same species.)

*"**Symphytum anatolicum Boiss. is an endemic (native) species in the flora of Turkey and the east Aegean Islands.***
*It was first collected by d'Urville on the island of Cos (Kos, Greek island) in 1820 and named **Symphytum asperrimum d'Urv.** (Mem. Soc. Linn. Paris 1: 276. 1822)*
*The plant was again collected by **Boissier** near Izmir (Smyrna, Greek city on Aegean coast of Anatolia) in 1842, and described by him as **S. anatolicum in 1844. It is a renaming for the (invalid, later homonym) S. asperrimum d'Urv., non S. asperrimum M.Bieb.***
*There are 4 collections mentioned in the protologue in Boissier's 'Flora Orientalis' (*1879). The herbarium specimen 'Hab. in umbrosis montium circa Smyrnam', housed in G-Boiss. Herbarium (Geneva Herbarium: Boissier's Flora Orientalis), was **chosen as lectotype as it best represents the species description.***
Symphytum anatolicum Boiss., Diagn. Ser. 1(4):43 (1844). =
Symphytum sicyosmum Cand. Ic: Pawlowski in Fragm. Fl. Geobot. Ann. 17: Pars 1, p. 29. (1971)."
-'Lectotypification of Symphytum Anatolicum (Boraginaceae)' by Burcu Tarikahya Hacioglu (Cankaya, Turkey) and Sadik Erik (Berlin, Germany), Turkish Journal of Botany, Volume 36, pages 101-102, **2012**.
(* -'Flora Orientalis Sive Enumeratio Plantarum in Oriente, a Graecia et Aegypto ad Indiae Fines {Plants Flora, Flora East and the East, from Greece and Egypt to the Borders of India}, Volume IV' by Edmond Boissier, Society Physics Geneva, Society Linnean London; Geneva, Switzerland, 1879.)
(** -'A Revision of Symphytum in Turkey and Adjacent Areas' by G.E. Wickens, Royal Botanic Gardens: Kew; Notes from the Royal Botanic Garden Edinburgh, Scotland, Volume 29, pages 157-180, 1969. Includes Turkey, Bulgaria, Greece, Aegean Islands and Caucasia.)
(*** -'Symphyta Mediterranea Nova vel Minus Cognita' by Bogumilus Pawlowski, Fragmenta Floristica et Geobotanica, Volume 17, No. 1, pages 17-37, 1971. I could not get a copy of this report. If you have an English translation, could you send it to me.)
    (In biology, 'homonym' is a name for a taxon that is identical in spelling to another such name, that belongs to a different taxon. The term 'non' means that there is a homonym, and this indicates which taxon is correct. Or to put it another way: If an author wishes to indicate that a name has been used for a different taxon, then there is citation of the name and the author followed by the word 'non' or 'not' and the name of the author who first used the name.)
    (Lectotype is a specimen chosen as the type of a species if the author of the name fails to designate a type.)
    (In taxonomy, protologue is the original material associated with a newly published name, comprising its description and informaiton such as illustrations, synonymy, etc.)

*"**Symphytum Anatolicum: Ana Kafesotu:***
Slender puberulous perennial, 12-45 cm (4.7-17.7 inch). Leaves oblong-ovate, lower shortly petiolate, subrepand, upper sessile, shortly decurrent.
Flowers 8-36. Calyx 4.5-8 mm, accrescent to 13 mm in fruit, divided nearly to base, lobes linear lanceolate, subacute.
***Corolla white or violet,** 8-16 mm, tube 2x calyx; scales 3.5-5 mm, linear, obtuse, scarcely exceeding stamens. Nutlets 3-4 mm, curved, reticulately veined, tuberculate.*
Fl. 4-9. Montane (mountain) woods, 50-1390 meter (164-4560 feet).
Endemic (native). East Mediterranean element."
-'Symphytum Anatolicum: Ana Kafesotu' published by Turkiye Bitkileri (Plants of Turkey), **2019**. Turkey has a great botanical richness with more than 9600 plant species. This website has been constructed to gather as much information as possible about native plants of Turkey. www.turkiyebitkileri.com

Chromosomes of S. anatolicum

"S. anatolicum Boiss.:
Both S. ottomanum and S. bulbosum have exserted corolla scales, which are absent in S. anatolicum.
The chromosome numbers of S. anatolicum (2n = 30) and S. ottomanum (2n = 48) are very different and do not support the inclusion of these species in the same Section.
On the basis of this information we hestiate to assign S. anatolicum to a subgeneric taxon."
-'Cytotaxonomic Studies in the Genus Symphytum VIII: Chromosome Numbers and Classification of Ten European Species' by T.W.J. Gadella and E. Kliphuis, Proceedings van de Koninklijke Nederlandse Akademie van Wetenschappen (Proceedings of the Royal Netherlands Academy of Arts and Sciences): Series C, Biological and Medical Sciences, Vol 81, page 162-172, **1978**.

Symphytum angustifolium
See 'Symphytum tuberosum' in this section.

*"**Symphytum angustifolium A.Kern.** (Oesterr. Bot. Z. 13: 227. 1863)"*
-'International Plant Names Index'® (IPNI), www.ipni.org, A database of the names and associated basic bibliographical details of seed plants, ferns and lycophytes, **2018**.

('Descriptiones Plantarum Novarum Florae Hungaricae et Transsilvanicae' by A. Kerner, Oesterreichische Botanische Zeitschrift (Austrian Botanical Journal), Vienna, Austria, Volume 13, No. 7, page 227, 1863.)

"Symphytum angustifolium A. Kern.:
> *synonym for Symphytum tuberosum subsp. angustifolium (A. Kerner) Nyman*
> **Symphytum tuberosum subsp. angustifolium (A. Kerner) Nyman:** *accepted name. Infraspecific taxon."*

-Catalogue of Life®: 2015 Annual Checklist, Online database of the world's known species of animals, plants, fungi and micro-organisms. It consists of a single integrated species checklist and taxonomic hierarchy., **2018**, www.catalogueoflife.org. Search: www.catalogueoflife.org/annual-checklist/2015/search/all
> (Infraspecific is at a taxonomic level below that of species, e.g., subspecies, variety, cultivar, or form. Latin names at this level require the addition of a term denoting rank.)

Symphytum armeniacum
See subsection 'Varieties of S. Asperum', in section 'Details about Symphytum Species: Asperum or Asperrimum' (Chapter 7).

Symphytum bohemicum
See 'Details about Symphytum Species: Officinale' (Chapter 8).

Current Botanical Nomenclature of S. bohemicum

"Symphytum officinale subspecies bohemicum (F.W. Schmidt)."
-National Biodiversity Network®, NBN Atlas, United Kingdom, https://species.nbnatlas.org, **2018**.

"Symphytum bohemicum F.W.Schmidt (Fl. Boem. iii. 13. t. 263.)"
-'International Plant Names Index'® (IPNI), www.ipni.org, A database of the names and associated basic bibliographical details of seed plants, ferns and lycophytes, **2018**.
(-'Flora Boemica Inchoata' {Bohemia Plants Started/Found}, Volume 3, page 13, plate/image 263, by Franz Wilibald Schmidt, Czechoslovakia, 1794.)

"Symphytum bohemicum F. W. Schmidt.
> *Synonym for Symphytum officinale subsp. bohemicum (F. W. Schmidt) Celak.*
> *Infraspecific taxon. Accepted name."*

-Catalogue of Life®: 2015 Annual Checklist, Online database of the world's known species of animals, plants, fungi and micro-organisms. It consists of a single integrated species checklist and taxonomic hierarchy., **2018**, www.catalogueoflife.org. Search: www.catalogueoflife.org/annual-checklist/2015/search/all
> (Infraspecific means at a taxonomic level below that of species, e.g., subspecies, variety, cultivar, or form.)

Description of S. bohemicum

*"**The pyrrolizidine alkaloid and triterpene pattern of S. officinale (2n = 24) and S. bohemicum (2n = 24) is identical. Symphytum bohemicum is morphologically, cytologically and phytochemically very similar to S. officinale.** Furthermore, it readily crosses with the white flowered west European diploids (2n = 24) of S. officinale. Therefore it seems likely that these two taxa are conspecific (same species)."*
-'Chemotaxonomy of the Symphytum Officinale Agg. {Boraginaceae}' by Jaarsma, Lohmanns, Gadella and Malingre, Plant Systematics and Evolution, 167, pages 113-127, **1989**.

*"**Symphytum officinale subsp. bohemicum** is on the 'Waiting List' as data deficient in 'The Vascular Plant Red Data List for Great Britain' (Cheffings & Farrell 2005), therefore it is a plant to look out for, yet **it may seem rather enigmatic: subsp. bohemicum is not in Stace (*1997) or Flora Europaea (**Pawlowski 1972).***
This taxon originated from chemotaxonomic research (***Jaarsma et al. 1989) which established that S. bohemicum Schmidt and S. officinale are best regarded as conspecific (same species).
Three subspecies of Symphytum officinale are recognised by Gadella & Perring (****2000) as subsp. officinale, subsp. bohemicum, and subsp. uliginosum.
> *S. officinale subsp. uliginosum (A. Kern.) Nyman has not been recorded from the British Isles since before 1930 (Clement & Foster 1994) although it may occur here (Perring 1994). It is distinguished from the other subspecies by the outer surface of the sepals being more sparsely pubescent and on chromosome number being 2n = 40. It also nearly always has purple buds and flowers (some cytotypes of S. officinale and S. x uplandicum also have purple buds and flowers) and Gadella & Perring (2000) provide further details on its identification.*

S. officinale subsp. bohemicum *occurs in the Netherlands, and Eastern Europe, particularly in the Czech Republic and Hungary. In the British Isles, it is recorded from only three English vice counties: Cambridgeshire v.c. 29 and Huntingdonshire v.c. 31 (Stace et al. 2003) and South Lincolnshire v.c. 53 (Perring 1994), with all records being from fens (marshes)."*
-'What is Symphytum Officinale Subsp. Bohemicum (Schmidt) Celak: A Taxon on the Red Data Waiting List' by Clare Coleman

O'Reilly, County Durham, England; Botanical Society of Britain and Ireland (BSBI) News, Hertfordshire, England, No. 102, pages 46-48, April **2006**.
(* -New Flora of the British Isles: Identification of Wild Vascular Plants of the British Isles edited by Clive Stace. Cambridge, England: Cambridge University Press, 1991, 1997.)
(** -Flora Europaea, Volume 3: Diapensiaceae to Myoporaceae, editors Tutin, Heywood, Burges, Moore, Valentine, Walters and Webb. United Kingdom: Cambridge University Press, 1972. 5-volume encyclopedia of plants, published between 1964 and 1993. Symphytum L., pages 103-105 by B. Pawlowski.)
(*** -'Chemotaxonomy of the Symphytum Officinale Agg. {Boraginaceae}' by Jaarsma, Lohmanns, Gadella and Malingre, Plant Systematics and Evolution, 167, pages 113-127, 1989.)
(**** -The European Garden Flora: A Manual for the Identification of Plants Cultivated in Europe, Both Out-of-Doors and Under Glass, 6 Volumes, edited by James Cullen, Sabina G. Knees, and H. Suzanne Cubey. England: Cambridge University Press, 2000. Volume 6: 'Symphytum' by T.W.J. Gadella and F.H. Perring, pages 138-141.)

"***Symphytum bohemicum F. W. Schmidt***, Fl. Boem. 3: 13 (1794):
Locality: *The protologue refers to a group of localities in the lowlands in the northern part of* **central Bohemia**, *by the Labe River near Melnik, Bohemia.*
Original material:
There is a single plant collected by F.W. Schmidt and corresponding to the protologue; it must be considered a syntype and is deposited in PRC (The herbarium collection of the Charles University in Prague, Czech Republic: Herbarium Universitatis Carolinae Pragensis.) Lectotype, designated here: PRC.
Taxonomic note:
Although the species is not always accepted in Floras (books), **it represents a taxon characterized by a peculiar ecology (mineral rich to subsaline alluvial meadows) and karyology (a diploid with 2n=24,** **Majovsky 1978). It is not to be confused with a pale flowered form of the tetraploid (2n = 48) S. officinale.*
Czech Republic: *Only in the northern part of Bohemia, mostly along the Labe River.*
Conservation note: it has become quite rare."
-'Generally Accepted Plant Names Based on Material from the Czech Republic and Published in 1753-1820' by Jan Kirschner, Lida Kirschnerova and Jan Stepanek, Institute of Botany, Academy of Sciences, Pruhonice 1, Czech Republic; Preslia: The Journal of the Czech Botanical Society, Volume 79, pages 323-365, **2007**.
(* -'Index of Chromosome Numbers of Slovakian Flora, Part 6' by J. Majovsky, Acta Facultatis Rerum Naturalium Universitatis Comenianae Botanica, Volume 26, pages 1-42, 1978. Chromosome numbers of 530 spermatophyte taxa are listed. I do not have a copy of this. If you have an English translation of the Symphytum sections, could you please send it to me.)
(In taxonomy, protologue is the original material associated with a newly published name, comprising its description and other informaiton such as illustrations, synonymy, etc.)
(Bohemia is the westernmost region of the Czech Republic. In a broader meaning, Bohemia refers to the entire Czech territory, including Moravia and Czech Silesia.)
(A syntype is each of a set of type specimens of equal status, upon which description and name of a new species is based. Lectotype is a specimen chosen as the type of a species if the author of the name fails to designate a type.)

"*The third part of the publication series on the distributions of vascular plants in the Czech Republic includes grid maps. The plants studied include 53 taxa classified in the 'Red List' of vascular plants of the Czech Republic, some of which have shown remarkable declines.*
Symphytum bohemicum, distributed mainly in central Europe, is confined to calcareous fens (wetlands) in the lowlands. The species is classified as endangered (**Grulich 2012).*
Symphytum bohemicum is a diploid member of the Symphytum officinale group.
It is quite well defined morphologically by its greenish or yellowish white flowers and only shortly decurrent leaves. It was described from central Bohemia by F.W. Schmidt as early as the late 18th century (**Kirschner et al. 2007).
Further records of the diploid white-flowered 'S. officinale' that we consider to be Symphytum bohemicum are from eastern England, the Netherlands, Germany, southern Poland, south-eastern Slovakia, northern Hungary, southwestern Slovenia and northern Italy (Gadella & Kliphuis 1969, 1972, Majovsky & Hegedusova 1993, Jogan et al. 2001 {Materials for the Atlas of Flora of Slovenia}, Stace 2010).
Even so, Symphytum bohemicum remains neglected in most national floras despite its morphological distinctiveness and strong reproductive isolation from S. officinale *(Gadella & Kliphuis 1969, 1972).*
Location: *In the Czech Republic S. bohemicum is found in calcareous fens in the lowlands along the middle and lower stretches of the Labe, Ohre,Metuje and Cidlina rivers in eastern, central and northern Bohemia.*"
-'Distributions of Vascular Plants in the Czech Republic, Part 3' by Z. Kaplan, J. Danihelka, M. Lepsi, P. Lepsi, L. Ekrt, J. Chrtek Jr, J Kocian, J Prancl, L. Kobrlova, M. Hrones and V. Sulc, Czech Republic; Preslia: The Journal of the Czech Botanical Society, Volume 88, pages 459-544, **2016**.
(* -'Red List of Vascular Plants of the Czech Republic, Third Edition' by V. Grulich; Preslia: Journal of the Czech Botanical Society, Volume 84, pages 631-645, 2012.)
(** -'Generally Accepted Plant Names Based on Material from the Czech Republic and Published in 1753-1820' by Jan Kirschner, Lida Kirschnerova and Jan Stepanek, Institute of Botany, Academy of Sciences, Pruhonice 1, Czech Republic; Preslia: The Journal of the Czech Botanical Society, Volume 79, pages 323-365, 2007.)

Chromosomes in S. bohemicum

"*Symphytum officinale L.: 2n = 24 or 2n = 48*
      ***S. bohemicum Schmidt**, Fl. Boem. 3: 13. tab 263. 1795.*
      S. tanaicense Steven, Bull.Soc.Imp. Naturalistes Moscou 24:577. 1851.
      S. uliginosum Kerner, Oesterr. Bot. Z. 13: 227, 1863.
White or cream flowers in diploids (2n = 24) and tetraploids (2n = 48).
Purple or red (or various intermediate colors between white and dark purple) in the tetraploids (2n = 48).
*The diploid cytotype was assigned to Symphytum bohemicum Schmidt by A. Murin and J. Majovsky (Acta Facultatis Rerum Naturalium Universitatis Comenianae Botanica, Volume 29, 1982.) **The exact status of S. bohemicum, S. tanaicense, and S. uliginosum appears to require further investigation.***"
-'Notes on Symphytum (Boraginaceae) in North America' by T.W.J. Gadella, page 1061-1067 in book: Annals of the Missouri Botanical Garden, Volume 71, Saint Louis, Missouri. Lawrence, Kansas: Allen Press, **1984**.

"*The pyrrolizidine alkaloid and triterpene pattern of S. officinale (2n = 24 chromosomes) and **S. bohemicum (2n = 24)** is identical.*"
-'Chemotaxonomy of the Symphytum Officinale Agg. (Boraginaceae)' by Jaarsma, Lohmanns, Gadella and Malingre, Plant Systematics and Evolution, 167, pages 113-127, **1989**.

"***Symphytum bohemicum F.W. Schmidt: Chromosome Count is 2n = 24.***"
-Tropicos®, www.tropicos.org, Missouri Botanical Garden, Saint Louis, Missouri; Plant database with 1.3 million scientific names and 4.4 million specimen records, **2018**.

Symphytum bornmuelleri (see 'S. brachycalyx' in this section.)

Symphytum brachycalyx (includes S. bornmuelleri)
See 'Symphytum palaestinum' in this section.

Current Botanical Nomenclature of S. brachycalyx

"***Symphytum brachycalyx Boiss.**, *Diagn. Pl. Orient. ser. 1, 4: 43 (1844). Remarks: Kurdistan.*"
-'International Plant Names Index'® (IPNI), www.ipni.org, A database of the names and associated basic bibliographical details of seed plants, ferns and lycophytes, **2019**.
(* -'Diagnoses Plantarum Orientalium Novarum', Series 1, No. 4, page 43, by Pierre Edmond Boissier, Lipsiae {Leipzig}, Germany, 1844.)

"***Symphytum brachycalyx Boiss.: species accepted.***
    = Symphytum orientale Pinard
    = Symphytum orientale Pinard ex DC.
    = Symphytum palaestinum Boiss.
    = Symphytum palaestinum var. violaceum N.Feinbrun."
-Global Biodiversity Information Facility® (GBIF), www.gbif.org, **2019**. International network and research infrastructure funded by the world's governments and aimed at providing anyone, anywhere, open access to data about all types of life on Earth.

"***Symphytum bornmuelleri Bucknall.: species accepted.** Published in: *J. Linn. Soc., Bot., 41: 536, 1913.*"
-Global Biodiversity Information Facility® (GBIF), www.gbif.org, **2019**. International network and research infrastructure funded by the world's governments and aimed at providing anyone, anywhere, open access to data about all types of life on Earth.
(* -'A Revision of the Genus Symphytum, Tourn.' by Cedric Bucknall, (Mus. Bac. Oxon= Bachelor of Music, Oxford University), Journal of the Linnean Society of London, England, Botanical Journal, Volume 41, Issue 284, page 536, December 1913.)
    (Joseph Friedrich Nicolaus Bornmueller or Bornmuller, 1862-1948, was a botanist born in Hildburghausen, Thuringia,
    Germany. He collected plants in Balkans, Greece, Middle East, Asia Minor, North Africa, Madeira and Canary Islands.)

Description of S. brachycalyx and S. bornmuelleri

"***Symphytum brachycalyx Boiss.**, Diagn. ser. I, iv. p. 43 (1844).*
*Geographical Distribution: **Turkey: Caria.***
The distinguishing characters of this species are the leaves gradually attenuated into the petiole and the campanulate calyx only slightly enlarged in fruit.
    From S. bornmuelleri it differs in the dichotomous branching of the stem and the more deeply divided, campanulate, not
    cylindrical calyx.
***It is doubtful whether this plant has been found outside the ancient province of Caria, in the south-west corner of Asia Minor.** All other plants referred to it are S. palaestinum or S. bornmuelleri.*"

-'A Revision of the Genus Symphytum, Tourn.' by Cedric Bucknall, (Mus. Bac. Oxon= Bachelor of Music, Oxford University), Journal of the Linnean Society of London, England, Botanical Journal, Volume 41, Issue 284, pages 491-556, December **1913**.
    (Caria was a region of western Anatolia extending along the coast from mid-Ionia {Mycale} south to Lycia and east to Phrygia. Today it is southwestern Anatolia {Asia Minor, Asia Turkey}.)

"***Symphytum bornmuelleri*** *sp. nova {new species}*:
Geographical Distribution: **Turkey: northern Asia Minor.**
This species, having been determined as S. brachycalyx Boiss., by Haussknecht, occasioned much uncertainty in my mind with regard to the limits of variation, both in that species and in S. palaestinum Boiss., until, by careful comparison of authentic specimens, **I realized that Bornmueller's plant could not be assigned to either of these.**
Boissier's description of the stem of S. brachycalyx is, in the 'Diagnoses', 'caulibus a basi ramosis', and in 'Flora Orientalis', 'caulibus subdichotome paniculato-corymbosis'. In actual specimens the stem is generally forked, with the branches comparable in size with the main stem.
In S. bornmuelleri the branches are arranged in a racemose manner in the axils of leaves and are small in comparison with the stem, and it is only at the top of mature plants that they are corymbose. The calyx also is very different from that of S. brachycalyx, which Boissier describes in the 'Diagnoses' as 'breviter campanulato laciniis triangulari-lanceolatis obtusiusculis tubo calycino brevioribus', and in 'Flora Orientalis' as 'calyce fructifero parum aucto breviter campanulato ad tertiani partem in dentes triangulares fisso."
    In S. bornmuelleri it is tubular, cylindric, much enlarged in fruit, with broadly ovate, obtuse teeth, which are scarcely one sixth of the length of the tube: shorter in proportion than in any other described species.
From S. palaestinum it differs in the same way, but in that species the calyx is still more enlarged in fruit. It also differs from both in being very shortly and sparingly pilose.
Having only been found in the north of Asia Minor, its area is, as far as is known, quite distinct from those of the related species, being about 200 miles (321 kilometer) distant from that of S. palaestinum and 600 miles (965 kilometer) from that of S. brachycalyx."
-'A Revision of the Genus Symphytum, Tourn.' by Cedric Bucknall, (Mus. Bac. Oxon= Bachelor of Music, Oxford University), Journal of the Linnean Society of London, England, Botanical Journal, Volume 41, Issue 284, pages 491-556, December **1913**.
    (Heinrich Carl Haussknecht, 1838-1903, was a pharmacist and botanical collector who was a native of Bennungen, Sachsen-Anhalt, Germany. He collected plants in Thuringia, lower Saxony, Greece and the Middle East {present-day Turkey, Syria, Iraq, Iran}. His herbarium is now at University of Jena in Thuringia, Germany.)

"***Symphytum Brachycalyx: Dere Kafesotu:***
Pubescent perennial, 20-60 cm (7.8-23.6 inch). Leaves linear-oblong or ovate to narrowly ovate or linear-lanceolate, lower petiolate, upper sessile.
Flowers about 9-15. Calyx 7-9 mm, accrescent to 15 mm in fruit, divided to 1/4-1/3, lobes linear-lanceolate. **Corolla white**, 12-15 mm; scales linear, equalling or exceeding stamens.
Nutlets 2.5-3 mm, constricted at base, areolate, tuberculate.
Flowers 4-7. Abies forest, shady places, often near streams, 800-2100 meter (2624-6889 feet).
West Syria. East Mediterranean element. **Related to S. bornmuelleri and S. orientale.**"
-'Symphytum Brachycalyx: Dere Kafesotu' published by Turkiye Bitkileri (Plants of Turkey), **2019**. Turkey has a great botanical richness with more than 9600 plant species. This website has been constructed to gather as much information as possible about native plants of Turkey. www.turkiyebitkileri.com
    ('Abies' are a genus of evergreen trees of temperate regions of the northern hemisphere that comprise the true firs.)

"***Symphytum Bornmuelleri: Kayin Kafesotu:***
Perennial, 15-60 cm (5.9-23.6 inch). Leaves oblong-ovate to ovate-lanceolate, asperous, upper sessile. Flowers 20 or more. Calyx 7 mm, accrescent to 12 mm in fruit, divided to 1/2 or less, lobes ovate, obtuse. **Corolla white,** 12-15 mm, scales linear, obtuse, +- equalling stamens.
Nutlets 3 mm, reticulate, minutely tuberculate.
Flowers 4-8. Shaded banks, Fagus woods, among boulders, 15-1900 meter (49-6233 feet).
Endemic (native). Euxine (Black Sea) element? **Related to S. brachycalyx.**"
-'Symphytum Bornmuelleri: Kayin Kafesotu' published by Turkiye Bitkileri (Plants of Turkey), **2019**. Turkey has a great botanical richness with more than 9600 plant species. This website has been constructed to gather as much information as possible about native plants of Turkey. www.turkiyebitkileri.com
    (Fagus or Beech is a genus of deciduous trees in family Fagaceae. Native to temperate Europe, Asia and North America.)

Symphytum bulbosum (Bulbous Comfrey, Symphytum zeyheri)
See 'Symphytum pseudobulbosum' in this section.
See 'Symphytum tuberosum' in this section.

Current Botanical Nomenclature of S. bulbosum

Karl (Carl) Friedrich Schimper =

C.Schimper or C.Schimp. (MOSS abbreviation) or K.F.Schimp (Missouri Botanical Garden abbreviation).

"Symphytum zeyheri C.Schimper: synonym for Symphytum bulbosum C.Schimper.
Symphytum bulbosum C.Schimper: accepted name.
Symphytum bulbosum Schur: ambiguous synonym for Symphytum tuberosum subsp. tuberosum Schur.
Symphytum tuberosum subsp. bulbosum (C.Schimper) P.Fourn.: synonym S. bulbosum C. Schimper.Infraspecific taxon."
-Catalogue of Life®: 2015 Annual Checklist, Online database of the world's known species of animals, plants, fungi and micro-organisms. It consists of a single integrated species checklist and taxonomic hierarchy., **2018**, www.catalogueoflife.org. Search: www.catalogueoflife.org/annual-checklist/2015/search/all

(Infraspecific is at a taxonomic level below that of species, e.g., subspecies, variety, cultivar, or form. Latin names at this level require the addition of a term denoting rank.)

"Symphytum zeyheri Schimp. (*Flora 12(2): 418. 1829)
Symphytum bulbosum K.F.Schimp. (**Flora 8 (1, no. 2): 17. 1825)
Symphytum bulbosum Schur (***Enum. Pl. Transsilv. 468. 1866)"
-'International Plant Names Index'® (IPNI), www.ipni.org, A database of the names and associated basic bibliographical details of seed plants, ferns and lycophytes, **2018**.
(* -'Flora oder Botanische Zeitung' {Flora or Botanical Newspaper}, Volume 12, No. 2, Regensburg, Germany, 1829. Symphytum page 418.)
(** -'Flora oder Botanische Zeitung' {Flora or Botanical Newspaper}, Volume 8, No. 1, Regensburg, Germany, 1825. Symphytum pages 17 to 24.)
(*** -'Enumeratio Plantarum Transsilvaniae, exhibens stirpes phanerogamas sponte crescentes atque frequentius cultas, cryptogamas vasculares, characeas, etiam muscos hepaticasque' (Flora of Transylvania) by Ferdinand Schur, Romania, 1866. Symphytum page 468.)

"Symphytum bulbosum K.F.Schimp. in Flora 8(1): 17. 1825.
   = S. tuberosum L. subsp. bulbosum (K.F.Schimp.) P.Fourn., Quatre Fl. France: 747. 1937
   = S. tuberosum L. var. bulbosum (K.F.Schimp.) Fiori in Fiori & Beg., Fl. Italia Nicotra, Syll. Fl. Sic.: 39. 1893."
-'Synopsis of Boraginaceae subfam. Boraginoideae Tribe Boragineae in Italy' is the same as 'Boraginaceae in Italy II' by L. Cecchi and F. Selvi, University of Florence, Firenze, Italy; Plant Biosystems: An International Journal Dealing with all Aspects of Plant Biology, Official Journal of the Societa Botanica Italiana, Vol 149, No. 4, pages 630-677, **2015**, Symphytum pages 636-646.

Symphytum zeyheri C.Schimper: synonym for Symphytum bulbosum C.Schimper

"S. zeyheri Schimp. in *Flora xii. (1829) p. 418.
*Geographical Distribution: **Southern Italy: Sicily, Sardinia, Corsica. Greece.***
This differs from S. bulbosum much in the same way that S. mediterraneum differs from S. tuberosum.
The plant is generally less tall and more compact, the lower leaves are more broadly ovate, rounded or subcordate at the base, and nearly as large as the succeeding ones, and the corolla is larger and infundibuliform. The stem often bears rudimentary branches in the lower axils and is occasionally branched at the base.
***Schimper states that it is easily distinguished at first sight and from afar,** and that he had seen many complete flowering and fruiting specimens in Herb. Zeyheriano (Zeyher's Herbarium). He describes the teeth of the fruiting-calyx as narrower than those of S. bulbosum, but in Sicilian plants containing well developed fruit, I find that the teeth are at least as broad as in that species.*
Beguinot,** on the label of an excellent specimen from Augusta, Sicily, quotes Gussone (**Fl. Sic. Syn. i. p. 226), who says that the characters of S. zeyheri are constant, and (Enum. Pl. Vasc. Inar. p. 218) that:*
   *'**S. zeyheri is not a southern form of S. bulbosum,** for that species also occurs with us, but in the mountain region, never in the lowlands, and careful comparison shows that they do not pass into each other.'*
   Beguinot continues that he can confirm this conclusion with regard to a plant from Ischia (island in Italy), erroneously recorded as S. bulbosum, and that the same form is found in Capri and Nisida (islands in Italy).
However this may be, the plants named S. zyheri by Beguinot, from Avellino, southern Italy, are scarcely distinguishable. at least when dry (herbarium sample), from S. bulbosum.
The plant gathered by Mr. Clarence Bicknell in Sardinia is similar to Sicilian (Italian islands) specimens."
-'A Revision of the Genus Symphytum, Tourn.' by Cedric Bucknall, (Mus. Bac. Oxon= Bachelor of Music, Oxford University), Journal of the Linnean Society of London, England, Botanical Journal, Volume 41, Issue 284, pages 491-556, December **1913**.
(* -'Flora oder Botanische Zeitung' {Flora or Botanical Newspaper}, Volume 12, No. 2, Regensburg, Germany, 1829. Symphytum page 418.)
(** -'Florae Siculae Synopsis, Exhibens Plantas Vasculares in Sicilia Insulisque' by Giovanni Gussone; Neapoli {Naples}, Italy, Volume 1, page 226, 1842. In Latin.)
(*** -'Enumeratio Plantarum Vascularium in Insula Inarime' by Giovanni Gussone; Neapoli {Naples}, Italy, page 218, 1855. In Latin.)

*"**S. zeyheri Schimper** in Flora (Regensburg) 12 :418 (1829).*

Synonym: S. tuberosum sensu Ucria Hort. Reg. Pan. 83 (1789) non L.
S. bulbosum sensu Guss., Fl. Sic. Prod. 1 :219 (1827) non Schimper.
S. brochum Bory & Chaub., Exp. Moree 65 (1832).

Roots either immediately tuberous, 20 mm diameter, or fine rhizome, 2 mm diameter, eventually forming a tuber. Perennial herb 20 cm (7.8 inch) tall; stem simple, pubescent and with tuberculate-based setae. Leaves sparsely pubescent, ovate or oblong-ovate, base subcordate or round; lower leaves 7-11 x 5-7 cm (2.7-4.3 x 1.9-2.7 inches), attenuated into winged, decurrent petiole, petiole sometimes long as lamina; upper leaves sessile, decurrent. Inflorescence about 10-flowered. Calyx 7 mm long, divided almost to base, segments linear-lanceolate, sub-acute. **Corolla white,** *infundibuliform, 10 mm long; corolla scales 4 mm long, broadly linear-lanceolate, subacute exserted 1 mm, exceeding stamens by 1.5 mm. Stamen filaments 1.5 mm long; anther 3 mm long. Style persistent, exserted 2 mm.*
Nutlets erect, 3.5 x 4 mm, constricted at base, reticulate-rugose, minutely tuberculate.
Range: Sardinia, Corsica, south Italy, Sicily, Greece, Turkey (northwest Anatolia)."
-'A Revision of Symphytum in Turkey and Adjacent Areas' by G.E. Wickens, Royal Botanic Gardens: Kew; Notes from the Royal Botanic Garden Edinburgh, Scotland, Volume 29, pages 157-180, **1969**. Includes Turkey, Bulgaria, Greece, Aegean Islands and Caucasia.

(Sardinia is an island in the Mediterranean Sea located west of the Italian Peninsula. Corsica is an island in the Mediterranean Sea southeast of mainland France.)

(Anatolia or Asian Turkey, the Anatolian peninsula or Anatolian plateau, is the westernmost part of Asia, that makes up the majority of modern-day Turkey. It is bounded by the Black Sea to the north, the Mediterranean Sea to the south, and the Aegean Sea to the west.)

Description of S. bulbosum

"**Symphytum bulbosum** Schimp. in Flora, viii. (1825) p.17:
Geographical Distribution: **From Germany (native?) and Switzerland to Sicily and Greece, and from southeast France to Bulgaria.**
The creeping root, simple stem, and exserted corolla-scales clearly separate this species from all others except S. zeyheri, *and from this it is distinguished by the smaller basal leaves, which are less abruptly contracted into the petiole, and by the rather smaller corolla.*

The root is very slender and far creeping (in *Bischoff's figure it is 4.5 dcm {= decimeter = 17.7 inch long) unbranched, with a few rootlets at the base of the stem and on the tubercles.

The leaves are more often rounded at the base than in S. tuberosum, but are sometimes gradually attenuated into the petiole.

De Candolle states that a Swiss specimen in his possession has small flowers with the corolla-scales and style scarcely exceeding the corolla, but that it must undoubtedly be referred to this species.
The plant from Solduno, Locarno, Switzerland, has similar flowers.
In the olive groves on the Italian Riviera, S. bulbosum grows to a large size. In Mr. Clarence Bicknell's herbarium in England there are plants with leaves measuring 17 cm (6.6 inch) long and 9 cm (3.5 inch) broad.
Schimper's description is taken from plants growing in the vineyards at Heidelberg, Germany. If S. bulbosum is native here, this must be the most northerly extension of its area."
-'A Revision of the Genus Symphytum, Tourn.' by Cedric Bucknall, (Mus. Bac. Oxon= Bachelor of Music, Oxford University), Journal of the Linnean Society of London, England, Botanical Journal, Volume 41, Issue 284, pages 491-556, December **1913**.
(* -'Flora oder Botanische Zeitung' issued by K. Bayerische Botanische Gesellschaft, Regensburg, Germany, Volume 9, No. 2, pages 561-562, 1826. Includes Dr. Bischoff {from Heidelberg, Germany} figure {tab/image 1} of S. filipendulum = S. bulbosum Schimp. on last page.)

"**Symphytum bulbosum** Schimp., 1825, in Flora, 8, 17-22. 9, Dorset; in great abundance by stream, Abbotsbury (Dorset county, England), April 17 1938, J.E. Lousley and A.W. Graveson.
(Hb. Lousley = Herbarium of J.E. Lousley, part of Reading University, Berkshire, England)
(Hb. Kew = Herbarium, Royal Botanic Gardens, Kew, London, England)
A sparsely hispid perennial. Rhizome slender, elongate with sub-rotund tubers. Stem 30-50 cm (11.8-19.6 inch) tall, simple or bifid, slender, flexuous, not winged, rough to the touch.
Lower leaves ovate, 7-10 cm (2.7-3.9 inch) long, rounded at the base, with petioles about equalling the blade; middle cauline leaves ovate-lanceolate, 12-17 cm (4.7-6.6 inch) long, acute, narrowed at the base to a winged petiole 6-8 cm (2.3-3.1 inch) long; upper leaves smaller, acute, sessile, slightly decurrent.
Calyx divided for 3/4 of its length into narrow, lanceolate, acute lobes. **Corolla yellowish-white,** 8-10 mm long, with small erect lobes much exceeded by the narrow lanceolate scales.
Distribution:
Native from Greece and the western Balkans through Istria (peninsula in Adriatic Sea), southern Switzerland and Italy to south France, Sicily and Corsica (island in Mediterranean Sea); naturalised in Germany and Austria.
This species has been much confused with S. tuberosum L., but may be distinguished at a glance by the exserted lanceolate corolla scales. The slender widely spreading rhizome bears brown tubers about the size of hazel nuts at intervals, whereas the rhizome of S. tuberosum is irregularly knobbly, and swollen at the base of the stem.

*The specimens on which this note is based were collected in April 1938, when Mr. A.W. Graveson took me to see what we assumed to be a puzzling variant of S. tuberosum, growing in abundance by the stream near Abbotsbury Swannery (colony of Mute swans). **Its origin here is clearly that of a planted garden plant, but it has spread outside cultivation and maintained itself in competition with native species for at least 32, and probably 68 years.***"
-'Symphytum Bulbosum' by J.E. Lousey, 'Proceedings of Botanical Society of the British Isles, Volume 4' edited by D.H. Kent, London, England, **1960**-1962, pages 43-44.

(The Adriatic Sea separates the Italian Peninsula from the Balkan peninsula. It is the northernmost part of Mediterranean Sea, extending from Strait of Otranto to northwest and Po Valley.)

"*S. bulbosum Schimper in Flora (Regensburg) 8 :17 (1825).*
Synonym: *S. filipendulum Bischoff in Flora (Regensburg) 9 :561 (1826).*
S. clusii C.C.Gmel., Fl. Bad. 4:144 (1826).
S. punctatum Gaudin, Fl. Helv. 2 :41 (1828).
S. tuberosum beta exsertum Loisel, Fl. Gall. ed. 2, 1 :152 (1828).
S. macrolepis Gay in Reich., Fl. Exc. 1 :347 (1832).
S. tuberosum beta clusii Caruel in Parl. Fl. Ital. 4 :879 (1884).
Range: *South Europe (excluding Iberian peninsula but including the Balkans). Although occurring in Greece and Bulgaria, this species apparently does not extend into modern European Turkey."*
-'A Revision of Symphytum in Turkey and Adjacent Areas' by G.E. Wickens, Royal Botanic Gardens: Kew; Notes from the Royal Botanic Garden Edinburgh, Scotland, Volume 29, pages 157-180, **1969**. Includes Turkey, Bulgaria, Greece, Aegean Islands and Caucasia.

"***S. bulbosum C. Schimper**, Flora (Regensb.) 8:17 (1825) **(includes S. zeyheri C. Schimper).***
Rhizome slender, creeping *producing subglobose tubers. Stems 15-50 cm (5.9-19.6 inches), simple or little-branched; stem and leaves with dense, very small hooked hairs and scattered setae up to 1.5-2.0 mm.*
Lower leaves ovate to elliptic-lanceolate, gradually attenuate or abruptly contracted into a long petiole, the uppermost sessile, slightly decurrent.
Calyx lobed to 1/3 to 6/7. Corolla (excluding scales) 7-8 to 11-12 mm, ***pale yellow****, with erect lobes which are 1/4-1/3 as long as the tube. Scales 5-5.5 to 9-10 mm, exserted for 1 to 4-5 mm, acute, lanceolate-subulate, rarely triangular-lanceolate; marginal papillae dense, about as long as wide. Stamens with filament 1/3 to 1/2 as long as anther; anthers 2.4 to 4 mm, minutely apiculate.*
Southern Europe, eastwards from Corsica (Albania, Bulgaria, Corsica, France, Germany, Switzerland, Italy, Yugoslavia, Sardinia, Sicily)."
-Flora Europaea, Volume 3: Diapensiaceae to Myoporaceae, editors Tutin, Heywood, Burges, Moore, Valentine, Walters and Webb. United Kingdom: Cambridge University Press, **1972**, page 105. (Symphytum L., pages 103-105 by B. Pawlowski.) (5-volume encyclopedia of plants, published between 1964 and 1993. 'Flora Europaea' describes all the national floras of Europe to identify any wild or widely cultivated plant to the subspecies level. It provides geographical distribution, habitat preference, and chromosome number.)

"**Distribution: Symphytum bulbosum, with which S. zeyheri is conspecific (same species), extends from** *southern France over southern Switzerland to the Balkan Peninsula and southward overy Italy to Sicily and eastward over Jugoslavia (Yugoslavia) and Bulgaria to southern Greece and north-east Turkey."*
-'The Greek Species of Symphytum Boraginaceae' by William T. Stearn, British Museum of Natural History, London, England; Annales Musei Goulandris, Greece, Volume 7, pages 175-220, (**1985 or 1986**, different sources give different dates).

"*S. bulbosum C. Schimper - Bulbous Comfrey:*
Stems simple, erect, to 50 cm (19.6 inch), from rhizome with subglobose tubers; stem-leaves petiolate to sessile, somewhat decurrent; calyx divided about 1/2 to 3/4 to base; ***corolla pale yellow.***
Introduced; naturalized in woods and by streams; ***very scattered in central and south Britain; Scotland and southeast Europe."***
-New Flora of the British Isles: Identification of Wild Vascular Plants of the British Isles edited by Clive Stace. Cambridge, England: Cambridge University Press, **1991**, page 648.

Chromosomes of S. bulbosum

"*S. bulbosum Schimp.:*
S. bulbosum was counted by Strey (1931): 2n = 72, by Tarnavschi (1948): 2n = 78, by Grau (1971): 2n = 84, and by us (Gadella and Kliphuis, 1978): 2n = 120. *In our opinion, Grau rightly discarded the observations by Strey and Tarnavschi.*
Wickens (1969) placed S. bulbosum in the section Tuberosum. Symphytum bulbosum and Symphytum tuberosum agree fairly well morphologically. *They share the thick tuberous rhizomes, which are creeping and alternately thick tuberous and thin in S. tuberosum, and creeping but more slender with subglobose tubers in S. bulbosum. The latter species (S. bulbosum) differs from the former (S. tuberosum) primarily in the exserted corlla-scales.*
We therefore agree with Wickens in placing S. bulbosum in the large polyploid complex of the Section Tuberosum."
-'Cytotaxonomic Studies in the Genus Symphytum VIII: Chromosome Numbers and Classification of Ten European Species' by

T.W.J. Gadella and E. Kliphuis, Proceedings van de Koninklijke Nederlandse Akademie van Wetenschappen (Proceedings of the Royal Netherlands Academy of Arts and Sciences): Series C, Biological and Medical Sciences, Vol 81, pages 162-172, **1978**.
    (Subglobose means imperfectly or nearly globose. Globose means having the shape of a sphere or ball.)
    (Polyploidy is a cell that has more than two paired {homologous} sets of chromosomes. Most species whose cells have nuclei {eukaryotes} are diploid, i.e., they have two sets of chromosomes. One set is inherited from each parent. Polyploidy is common in plants.)
    (A species 'complex' is a group of closely related species that are extremely similar in appearance, so much so that the boundaries between them are frequently unclear.)

"From Heidelberg, Germany, **Symphytum bulbosum** was transplanted by Schimper into the Schlosspark at Schwetzingen near Heidelberg, and here it survived, thus **enabling Grau to record its chromosome number as 2n = 84** in *Mitteil. Bot. Staatssamml. Munchen 9:185 (1971).
It remains to be ascertained whether populations elsewhere of so widespread a species are likewise hexaploid.
    A plant collected by me in June 1985 on Mount Parnassus, Greece at 1200 meters (3937 feet) has 2n = 48, accourding to Margaret Johson at Kew, London, England."
-'The Greek Species of Symphytum Boraginaceae' by William T. Stearn, British Museum of Natural History, London, England; Annales Musei Goulandris, Greece, Volume 7, pages 175-220, (**1985 or 1986**, different sources give different dates).
    (* -'Cytologische Untersuchungen an Boraginaceae II' {Cytological Investigations on Boraginaceae} by J. Grau, Mitteilungen der Botanischen Staatssammlung Munchen {Communications of the Botanical State Collection Munich, Germany}, Volume 9, pages 177-194, 1971.)

"**Symphytum: Two of our counts for S. bulbosum (2n = 48 and 96) were published by Stearn (*1986).** *The counts obtained here are a continuation of this work, recording the same numbers in other accessions:*
Symphytum bulbosum Schimp.

2n=	Cytology Reference	Kew Accession	Collector	Origin
48	87-461	1987-1948	Strid s.n.	Greece
48	88-28	none	Stearn s.n.	cultivated
96	88-27	none	Stearn s.n.	cultivated

We also found different numbers (2n= about 52, 60) in other species. These records, together with the wide array of numbers recorded by Fedorov (**1969), indicate a high level of chromosomal complexity in the Symphytum genus, in which the basic number is uncertain."
-'New Chromosome Numbers in Petaloid Monocotyledons and in Other Miscellaneous Angiosperms' by Margaret A.T. Johnson and P.E. Brandham, Jodrell Laboratory, Royal Botanic Gardens, Kew, Richmond, Surrey, England; Kew Bulletin, Volume 52, No. 1, pages 121-138, **1997**.
(* -'The Greek Species of Symphytum Boraginaceae' by W.T. Stearn, Annales Musei Goulandris, Greece, Volume 7, pages 175-220, 1986. If you have an English translation, could you please send it to me.)
(** -'Chromosome Numbers of Flowering Plants' {Khromosomnye chisla tsvetkovykh rastenii} edited by A.A. Fedorov, authored by Z.V. Bolkhovskikh, V.G. Grip, O.I. Zakhar'eva, T.S. Matveeva, et al., in Russian, 927 pages. Leningrad, Russia: Nauka, 1969. If anyone has an English translation of the Symphytum section, I would appreciate a copy.)
    (Accession is an addition to a plant collection, usually in reference to an herbarium.)

"**Symphytum bulbosum Schimp.: 2n = 104.**
    Italy: *Tuscany, Cascine Park of Florence, in shady habitats, April 1999, Bottega and Peruzzi, cultivated Herbarium Horti Botanici Pisani, Pisa, Italy.*
This species is quite common in southern Europe with few naturalized stations (locations) in Austria, England and Germany (Stearn 1986).
Previous chromosome number reports are 2n = 41 (*Tarnavschi 1935), 2n = 72 (Strey 1931), 2n = 84 (Grau 1971), 2n = 120 (Gadella and Kliphuis 1978). Our present count 2n = 104 is reported for the first time.
Due to the chromosome size (1.0-2.0 micrometer), the karyotype characters can not be represented."
-'Mediterranean Chromosome Number Reports 11 {Report 1229: Symphytum gussonei}' by S. Bottega, F. Garbari and L. Peruzzi; Flora Mediterrana, edited by G. Kamari, C. Blanche and F. Garbari, Volume 11, pages 436-439, **2001**.
(* -Chromosome Atlas of Flowering Plants edited by C.D. Darlington and A.P. Wylie. London, England: G. Allen and Unwin Ltd., 1961. Article by J.T. Tarnavschi.)

"**Symphytum bulbosum Schimp.: Chromosome Count is 2n = 48, 96, 104, 120.**"
-Tropicos®, www.tropicos.org, Missouri Botanical Garden, Saint Louis, Missouri; Plant database with 1.3 million scientific names and 4.4 million specimen records, **2018**.

<u>Symphytum bullatum</u> (see S. tauricum in this section)

<u>**Symphytum caucasicum (Caucasian Comfrey)**</u>

Current Botanical Nomenclature of S. caucasicum

"***Symphytum caucasicum M.Bieb.*** *(*Fl. Taur.-Caucas. 1: 128. 1808)*
Symphytum caucasicum D.Don *(**in Sweet, Brit. Flow. Gard. Ser. II. t. 294.)*"
-'International Plant Names Index'® (IPNI), www.ipni.org, A database of the names and associated basic bibliographical details of seed plants, ferns and lycophytes, **2018**.
(* -'Flora Taurico Caucasica: Exhibens Stirpes Phaenogamas in Chersoneso Taurica et Regionibus Caucasicis Sponte Crescentes', Volume 1 and 2, edited by Friedrich August Marschall von Bieberstein, Volumes 1 and 2 in 1808; Volume 3 and 4 in 1819. Symphytum page 128.)
(** -The British Flower Garden: Containing Coloured Figures and Descriptions of the Most Ornamental and Curious Hardy Flowering Plants, Series II, by Robert Sweet, F.L.S. London, England: James Ridgway and Sons, 1838. Seven volumes with Symphytum caucasian in volume 6, plate 294.)
> (M.Bieb. = Baron Friedrich August Marschall von Bieberstein, 1768-1826, was an early explorer of the flora and archaeology of the southern portion of Imperial Russia, including the Caucasus and Novorossiya. He compiled the first comprehensive flora catalogue of the Crimeo-Caucasian region.)

"***Symphytum caucasicum D.Don****: species doubtful.*"
-Global Biodiversity Information Facility® (GBIF), www.gbif.org, **2018**. International network and research infrastructure funded by the world's governments and aimed at providing anyone, anywhere, open access to data about all types of life on Earth.

"***Symphytum caucasicum M.Bieb.*** *is an accepted name. No synonyms are recorded for this name.*"
-The Plant List®, www.theplantlist.org from the World Checklist Database®, **2018**.

"***Symphytum caucasicum M. Bieb.****: accepted name.*"
-Catalogue of Life: 2015 Annual Checklist, Online database of the world's known species of animals, plants, fungi and micro-organisms. It consists of a single integrated species checklist and taxonomic hierarchy., **2018**, www.catalogueoflife.org. Search: www.catalogueoflife.org/annual-checklist/2015/search/all

Description of S. caucasicum

"***Symphytum caucasicum (Caucasian Comfrey) Description:***
Stem (two feet high = 0.6 meter) hairy near the bottom, higher up pubescent and viscous, slightly winged, flexuose, branched. Leaves ovato-lanceolate, hairy on both sides, but less harshly on the upper, and there, when young, subviscid, half-decurrent, the lower ones attenuated at the base, the upper pair oblique, sessile, and alternate.
Flowers:
Racemes terminal, geminate, many-flowered, secund, and involute, common peduncle and pedicels glanduloso-pubescent. Calyx angled, the angles and blunt teeth ciliated; when in fruit, distichous. **Corolla at first red-purple, but losing this colour as soon as it expands, and acquiring a lively azure hue.**
Tube longer than the calyx, sparingly and minutely pubescent on the outside, having a white, fleshy, narrow edge projecting internally from its base over the disk, teeth of the limb blunt and revolute in their edges, teeth of the throat erect, blunt, and having short, chrystalline ciliae on their edges. Stamens included, about as long as the teeth; filaments purplish; anthers yellow, rather shorter than the free portion of the filaments, bifid at both extremities. Pistil rather longer than the stamens; stigma bilobular, rounded; style slightly tapering, glabrous, lilac; germen light yellowish-green, seated on a white disk.
Seeds:
The unripe achenia are rough, irregularly depressed over their surface; and each is raised on a sandglass-shaped portion of the disk, the upper lobe of which projects from its lower side a simple row of short, dependent, subulate hairs.
The seeds of this plant were received at the Royal Botanic Garden, Edinburgh, Scotland, from Dr. Fischer, under the name here adopted, in 1830, and they blossomed, for the first time, in May, 1832.
> ***The profusion of lively-coloured flowers in this kind of Comfrey, which is less deformed by coarseness of herbage than others, makes it one of the most desirable for cultivation.*** "
-'Curtis's Botanical Magazine', Volume 6, New Series, by Samuel Curtis, F.L.S. and William Jackson Hooker, L.L.D., London, England, **1832**. Symphytum caucasicum: plate 3188.
> (Azure is a light shade of blue between blue and cyan. Cyan is greenish-blue. Azure is the color of a bright clear sky.)

"***Symphytum caucasicum, Caucasian Comfrey:***
Stems erect, angular, branched, scarcely 2 feet (0.6 meter) high, and clothed like the rest of the plant with recumbent bristly hairs. Leaves ovate-lanceolate, acute, wrinkled, entire, from 3 to 7 inches long (7.6-17.7 cm); the radical ones on long, simple, channelled footstalks; those of the stem sessile, and unequally decurrent at the base. Racemes forked, many flowered, furnished at the base with a single leafy bracte. Pedicels angular, bristly, about 3 lines long.
Calyx tubular, five-angled, ribbed, as long as the pedicels, with 5, ovate, blunt, connivent teeth. Corolla funnel-shaped, twice longer than the calyx, **before expansion of a rich pink, afterwards changing to an azure blue**, *the mouth ventricose, with 5, shallow, rounded, revolute lobes. Filaments glabrous, slightly compressed, blue. Anthers cream-coloured, as long as the filaments, composed of two distinct parallel cells, opening*
lengthways. Appendages ligulate, blunt, papillose, longer than the stamens. Style longer than the corolla. Stigma small, capitate.

Attractive Plant:
***Symphytum caucasicum* is a highly ornamental plant, having the brilliant blossoms of *Symphytum asperrimum* united to the more delicate habit of *Symphytum orientale*,** between which one might imagine it to be a spontaneous hybrid, so completely does it combine the characters of these two species.
From its dwarf habit and the beauty of its flowers it is admirably adapted to ornament front borders, for which purpose the large and coarse habit of S. asperrimum renders it unsuited.
    It is growing in the Chelsea Botanic Garden, London, England, where the plant had been introduced from the Imperial Botanic Garden at St. Petersburg, Russia.
Location:
It is found abundantly in moist shady places on the banks of the River Terek between Mosdok and Kisljar in Russia, and on the borders of woods throughout the greater part of the northern promontory of Caucasus, flowering in May and June."
-The British Flower Garden: Containing Coloured Figures and Descriptions of the Most Ornamental and Curious Hardy Flowering Plants, Series II, by Robert Sweet, F.L.S. London, England: James Ridgway and Sons, **1838**. Seven volumes with Symphytum caucasian in volume 6, plate 294.
    (The Caucasus Mountains are located at the border of Europe and Asia, between the Black Sea and Caspian Sea and occupied by Russia, Georgia, Azerbaijan, and Armenia.)

"Report on the Open Air Vegetation at the Royal Botanic Garden, Edinburgh, **Scotland** by Mr. M. Nab.
Recording the dates of their flowering:
    Symphytum caucasicurn   March 21 1871   April 10 1970   January 10 1869."
-'The Gardeners Chronicle and Agricultural Gazette for 1871' published in Covent Garden, WC, London, England, 1269 pages, January 7 to December 30 **1871**. Article on July 15 1871, page 910.
    (The dates varied depending on how long the cold weather lasted.)

'Forage Plants and Their Economic Conservation by the New System of Ensilage: Part I: **Caucasian Prickly Comfrey**' by Thomas Christy, Jun., F.L.S. (Fellow of the Linnean Society), Christy & Co., London, England, **1877** uses the name 'Caucasian Prickly Comfrey' for Prickly Comfrey.
Today 'Caucasian Comfrey' and 'Prickly Comfrey' are considered 2 different species. Mr. Christy's book is about Prickly Comfrey (Symphytum asperrimum).

"Notes on Open-Air Vegetation for April by Edinburgh Botanical Society, **Scotland**:
The following spring plants are annually recorded to show their **periods of flowering:**
    Symphytum caucasicum: March 15, 1876. April 5, 1877."
-'The Gardeners Chronicle: A Weekly Illustrated Journal of Horticulture and Allied Subjects', New Series, January to June 1877: Volume VII, July to December **1877**: Volume VIII. London, England. This excerpt is from May 19, 1877, page 631.

"**Symphytum caucasicum:**
This is the best plant I know of for growing under dense shade of trees; the flowers, too, are very beautiful. But the plant requires plenty of ground to ramble about in: it is not suitable for association with smaller kinds, as it would soon overrun them."
-'The Gardeners Chronicle: A Weekly Illustrated Journal of Horticulture and Allied Subjects', New Series, January to June 1877: Volume VII, July to December **1877**: Volume VIII. London, England. This excerpt is from December 1, 1877, page 694.

"**S. caucasicum, Bieb.**, Fl. Taur.-Cauc. iii. p. 128 (1808).
Geographical Distribution: Russia: **Caucasus and Transcaucasus.**
S. caucasicum is distinguished from the other blue-flowered species by the softly hairy stem and leaves, by the lower leaves, especially the radical ones, being gradually attenuated into the petiole, and by the tubular shortly toothed calyx.
The leaves vary from broadly ovate, as in the Cartalina Garden (Abastuman: government of Tiflis {Tbilisi}, Georgia) and Chelsea Physick Garden (London, England) plants, and in the figure t. 3188 of the *'Botanical Magazine', to oblong-lanceolate, as in the Nosdoe Taurasi (Italy?) and Georgia plants, and in those collected in Oxfordshire, England, by Mr. G.C. Druce.
    Some autumnal radical leaves of the Chelsea Garden plant which flowered at Clifton, England, in 1912, attained a large size (6 dcm. {decimeter} x 1.5 dcm.) (1.96-0.49 feet) with the blade of the leaf gradually attenuated nearly to the base of the petiole.
The leaves are wrinkled and repand, as shown in t. 294 of the **'British Flower Garden'. **The flowers are generally as described in the 'Botanical Magazine', 'red-purple then a lively azure', but in the Chelsea plant they were very pale blue,** and of the same colour when the plant flowered at Clifton in the succeeding year, during several months of variable weather. This plant was completely sterile.
S. caucasicum appears to have been first introduced into Britain in 1830, when Dr. Fischer sent seeds from St. Petersburg, Russia to Edinburgh, Scotland; the plants which sprung from these flowered in 1832. Seeds were also received at Chelsea from Dr. Fischer, probably at the same time, as Don's figure in 'British Flower Garden' was taken from the resulting plants in 1835.
On Don's figure of this plant, De Clandolle founded a new species, which he described in the 'Prodromus' as S. Donii.
The calyx-teeth in the figure are subulate, acuminate, almost equal to the tube, and in this it differs, as De Candolle says, from S. caucasicum.
    There is nothing in the drawing which separates it from S. caucasicum, and the description of the corolla-scales, 'ligulate, blunt', exactly fits those of that species. There is no known specimen of S. Donii, and no botanist appears to

*have even seen one, although the name has occasionally been suggested for other species. **It is clear, therefore, that S. Donii, being founded on erroneous drawing, does not exist as a species.**"*
-'A Revision of the Genus Symphytum, Tourn.' by Cedric Bucknall, (Mus. Bac. Oxon= Bachelor of Music, Oxford University), Journal of the Linnean Society of London, England, Botanical Journal, Volume 41, Issue 284, pages 491-556, December **1913**.
(* -'Curtis's Botanical Magazine', Volume 6, New Series, by Samuel Curtis, F.L.S. and William Jackson Hooker, L.L.D., London, England, 1832. Symphytum caucasicum: plate 3188.)
(** -The British Flower Garden: Containing Coloured Figures and Descriptions of the Most Ornamental and Curious Hardy Flowering Plants, Series II, by Robert Sweet, F.L.S. London, England: James Ridgway and Sons, 1838. Seven volumes with Symphytum caucasian in volume 6, plate 294.)

*"Two other introduced species to Britain are recorded as established locally: **S. caucasicum M. Bieb., which is like S. orientale but has clear blue flowers and sessile, shortly decurrent upper leaves.**"*
-'The Genus Symphytum in Britain' by T.G. Tutin, University College of Leicester, England; Watsonia: Journal of the Botanical Society of the British Isles, Volume 3, pages 280-281, February **1956**. (Watsonia was the journal of the 'Botanical Society of Britain and Ireland' from 1949 to 2010. It is now called 'New Journal of Botany'.)

"S. caucasicum M. Bieb. - Caucasian Comfrey:
*Stems erect, to 60 cm (23.6 inch), from thick, branched roots; stem-leaves sessile, shortly decurrent; calyx divided about 1/4 to 1/2 to base; **corolla blue.***
*Introduced; naturalized in hedgerows and other shady places; very scattered, and **often not persistent, in southeast England, Flintshire County (northeast Wales); Caucasus.**"*
-New Flora of the British Isles: Identification of Wild Vascular Plants of the British Isles edited by Clive Stace. Cambridge, England: Cambridge University Press, **1991**, page 648.
  (A hedgerow is closely growing shrubs and trees, usually bordering a road or field.)

"Caucasian Comfrey (Symphytum caucasicum):
***This coarse herbaceous perennial is native to the Caucasus Mountains** where it grows in waste places, along roadsides, in scrub, and along streams.*
Slender rhizomes with long thick roots form a rosette of leaves and then erect sparsely branched stems that become sprawling. The ovate (egg shaped) to lanceolate (tapering to a point) leaves are grayish green and softly hairy. The basal (bottom) leaves are up to eight inch (20 cm) long while the upper leaves are six inches (15 cm) long and attached to the stem for a short distance.
***The bell-shaped flowers** are drooping, 3/4 inches (1.9 cm) long, and are carried in terminal, arching, paired, scorpiod cymes in spring to early summer. **They open pink, aging to sky blue** and the petals are considerably longer than the sepals. Plants grow quickly and can be invasive."*
-'Plant Profile: Caucasian Comfrey (Symphytum caucasicum)', karensgardentips.com, from North Carolina and California, around **2009**.
  (Herbaceous plants do not have a persistent woody stem above ground, meaning all leaves / stems die back in winter.)
  (A rhizome is an elongated, usually horizontal underground plant stem.)
  (Rosette shape is similar to the arrangement of rose petals.)
  (Scorpiod cymes means the axis is curved and the flowers arise two-ranked and on alternate sides of the axis as in the forget-me-not flowers.)
  (Sepal is one of the individual leaves or parts of the calyx of a flower. Calyx is the usually green outer whorl of a flower consisting of separate or fused sepals.)

*"Stace (*2010) mentions as an essential feature to distinguish the Symphytum caucasicum versus the S. officinale s.l. (sensu lato) and S. x uplandicum, the less (1/4 to 1/2 instead of 2/3 to 4/5) deeply incised calyx."*
-'Caucasian Comfrey (Symphytum caucasicum M. Bieb.) and Hidcote Comfrey (Symphytum x hidcotense P.D. Sell) in the Region of Aachen' [Kaukasischer Beinwell {Symphytum caucasicum M. Bieb.} und Hidcote-Beinwell {Symphytum x hidcotense P.D. Sell} im Aachener Raum] by F. Wolfgang Bomble and Bruno G.A. Schmitz, Die Veroffentlichungen des Bochumer Botanischen Vereins e. V., Germany, Volume 4, No. 6, pages 50-54, **2012**.
(If you have an English translation of this, could you please send it to me.)
(* -New Flora of the British Isles: Identification of Wild Vascular Plants of the British Isles edited by Clive Stace. Cambridge, England: Cambridge University Press, 1991, 1997, 2010.)
  (sensu stricto = s.s., s. str., sens. str., sens. strict. In the strict/narrow sense. It is added after a taxon to mean it is being used in the sense of the original author, or without taxa which may otherwise be associated with it.
  sensu lato = s.l., sens. lat. In the broad sense. Used in taxonomy to clarify the scope of a taxon when it has been used to define more than one set of lower-level taxons. It includes all its subordinate taxa and/or other taxa that at other times are considered as distinct.)

*"**Symphytum caucasicum Bieb.:** One-headed, tap-rooted, herbaceous hemicryptophyte with semi-rosette, dicyclic, monocarpic shoots, is a typical representative of the Section Caerulea.*
***Natural distribution area of this species is restricted to the Caucasus, except for southeastern Azerbaijan.** The plant grows in meadows and forest margins and is cultivated as forage crop and bee plant (*Gviniashvili 1976; Frolov **1982, ***1989,*

****1997; *****Tikhomirov et al. 1999).
Germination *epigeous, hypocotylar. Cotyledons rounded or oval (14 to 16 mm long, 11 mm wide); petioles 4 to 5 mm in length, pubescent and widened at base. Blades have indumentums on both sides, pubescence more or less uniform, mixed, dense.*
Two types of hairs are present: *simple (sessile or subtended by rosettes) and glandular. Subtended (rosette) hairs long, one-celled, subulate (straight), swollen at base, with tuberculous surface. Pronounced rosettes are composed of 6 to 10 epidermal cells. Sessile hairs 2 to 2.5 times shorter than the former, unicellular, hooked. Glandular hairs two- to three-celled, with pronounced stalks and clavate heads.*
Juvenile leaves *of rosette shoots possess oblong-ovate petiolate leaves with entire margins and acuminate apex.*
Immature and mature vegetating plants *retain the same leaf shape, whereas their sizes increase (up to 45 cm in length and 22 cm in width) together with the petiole length reduction to 12 to 13 mm, until sessile leaves are formed.*
All leaves are covered by trichomes of two types.
> *Sessile leaves unicellular, short, subulate (straight or hooked, mostly located along the midrib), thin-walled, with greenish-brown contents near the tip and tuberculous surface.*
> *Subtended (rosette) hairs unicellular, long, subulate (straight or slightly bent), with thicker walls than in Symphytum cordatum or Symphytum tuberosum.*

Their cell cavities permanently possess lumpy dark green contents; surface tuberculous. Rosettes consist of 5 to 10 epidermal cells. Two- to three celled glandular hairs are also filled with green or greenish-brown contents. Glandular hairs are located along large veins and blade margins."
-'Pubescence of Vegetative Organs and Trichome Micromorphology in Some Boraginaceae at Different Ontogenetic Stages' by Rimma P. Barykina and Vitaly Y. Alyonkin, Lomonosov Moscow State University, Russia; Wulfenia, Regional Museum of Carinthia, Austria, Volume 23, pages 1-29, **2016**.
(* -'Kavkazskie Predstaviteli Roda Symphytum L. Boraginaceae Jus' book by T.N. {Ciala Nikolevna} Gviniashvili {Gviniasvili}, pages 130-135, Tbilisi, Georgia, 1976. In Russian. If you have the Symphytum pages in English, I would appreciate a copy.)
(** -'Common Comfrey in the Conditions of the North' or 'Symphytum Under Northern Conditions' by Y.M. {Yu.M.} Frolov; Nauka, Leningrad, Russia, **1982**. In Russian. Author botanical abbreviation: Yu.M.Frolov.)
(*** -'System of the Genus Symphytum L. (Flora SSSR)' by Y.M. Frolov; Syktyvkar, Komi Republic, Russia, **1989**. In Russian.)
(**** -'The Ontogeny of Caucasian Comfrey' by Y.M. Frolov; Proceedings of Komi Scientific Centre, Ural Department of Russian Academy of Sciences, Saint Petersburg, Russia, Volume 150, pages 83-102, **1997**. In Russian.)
(***** -'The Genus Symphytum L. {Boraginaceae} in Central Russia' by V.N. Tikhomirov, S.R. Majorov and D.D. Sokoloff; Novosti Sistematiki Vysshikh Rastenii {News of Higher Plants Systematics}, Moscow and Leningrad {St. Petersburg}, Russia, Volume 31, pages 231-245, 1999. In Russian. If you have the Symphytum pages in English, I would appreciate a copy.)
> (The nation Azerbaijan is bounded by the Caspian Sea and Caucasus Mountains.)
> (Pubescent means the surface of a leaf or stem is covered with fine short hairs.)
> (Ontogeny is the development of an organism from fertilization of the egg to maturity.)
> (Trichome is a small hair or other outgrowth from the epidermis of a plant. Epidermis is the outer layer.)

"Symphytum caucasicum **M. Bieb. Boraginaceae:**
Botany and Ecology: *Perennial, perhaps monocarpic (flowering only once), sometimes rosettes.*
Rhizome *reduced, short-fusiform, with long thick roots.*
Stem *usually single, 40-60 cm (15.7-23.6 inch) high, rather thick, shortly grayish soft-villous or downy-villous, with few short lateral branches.*
Radical leaves *withering at flowering; cauline leaves fairly numerous, ovate or oblong; lower leaves with relatively long petiole, obtuse, truncate or rounded at base; all the leaves 5-10 cm (1.9-3.9 inch) long, 2-4 cm (0.78-1.57 inch) wide, with apex acute, base cuneate, sparsely gray velutinous above, more dense-gray-hairy and velutinous below, venation subreticulate.*
Inflorescence *in cymes apical on stem and upper lateral branches, few-flowered, leafless, twisted to one side; pedicels drooping, 3-5 mm long, grayish-downy and sometimes with prickly short bristles. Flowers with calyx narrowly campanulate, 4-6 mm long, in fruit elongating up to 8-15 mm, grayish-glandular-downy, sometimes coarse with curved filiform bristles, incised for 1/3 into unequal lanceolate rounded-obtuse erect teeth; corolla blue, tube distinctly longer than calyx, about 7-8 mm long, entire corolla about 15 mm long, campanulate, with narrow triangular short teeth; style slightly exserted from corolla.*
Nutlets *pale, oblong, nearly straight, oblique, at large side finely but sharply netted-wrinkled, finely tuberculate within this net, 3-3.5 mm long.*
Endemic (native) in edges and glades in the forests of the drying parts of the Caucasus (Ciscaucasia, Dagestan and Eastern Transcaucasia).
> *Symphytum caucasicum grows in shrubberies, glades, damp forest fringes, near ravines and at watersides. It is a winter-hardy plant, but not very tolerant to moisture."*

-Ethnobotany of Caucasus by Ketevan Batsatsashvili, et. al. (Chapter: Symphytum caucasicum M. Bieb. Boraginaceae, pages 683-688). Cham, Switzerland: Springer, **2017**.   (A glade is an open area in a forest. They are frequently grassy meadows.)

"Symphytum caucasicum **M. Bieb.:**
Armenia: *In humid habitats, near water streams, ruderal places. Up to upper mountain belt, on an elevation 400-2000 meters (1312-6561 feet). Flowers from May to July, fruits (nutlets) from June to August.*
North Caucasus, East and South Transcaucasia. In Armenia in Upper Akhuryan, Lori, Idjevan, Sevan, Darelegis, Zangezur, Meghri floristic regions (*Takhtadjan 1954-2009).
Azerbaijan: *Endemic (native) of Caucasus. Distributed from lowland to subalpine belt almost throughout Azerbaijan. Grows on*

edges, in forests, among shrubs, in gardens, mountain meadows, along riverbanks, roads and in weedy areas. Flowering in May-June, fruiting in June-July (**Flora of Azerbaijan 1950-1961).
Georgia (country): Mostly in meadows and wetlands of lower and middle montane (mountainous) zones (***Ketskhoveli et al. 1971-2011). Distributed in Apkhazeti (Abkhazia), Racha-Lechkhumi, Imereti South Oseti, Kartli, Mtiuleti, Kakheti, Kiziki, Gare Kakheti, Trialeti, Meskheti (Ketskhoveli et al. 1971-2011)."
-Ethnobotany of Caucasus by Ketevan Batsatsashvili, et. al. (Chapter: Symphytum caucasicum M. Bieb. Boraginaceae, pages 683-688). Cham, Switzerland: Springer, **2017**.
(* -Flora of Armenia {Flora Armenii}, Volumes 1-11, by A.L. Takhtadjan. Yerevan, Armenia: Russian Academy of Sciences, 1954-2009. Volume 7: Boraginaceae family. In Russian.)
(** -Flora of Azerbaijan, Volume 1-VIII by I. Karyagin. Baku, Azerbaijan: Academy of Sciences of Azerbaijani Soviet Socialist Republic Press; 1950–1961. In Russian.)
(*** -Flora of Georgia, 16 Volumes, by N. Ketskhoveli, A. Kharadze and R. Gagnidze. Tbilisi, Georgia: Metsniereba, 1971-2011. In Georgian.)
      (Ruderal species are the first plants to colonize disturbed lands.)

"Favorite Early-Flowering Herbs:
Blue Comfrey (Symphytum caucasicum)- *Growing up to 3-4 feet (0.9-1.2 meter) tall,* **this is the showiest Comfrey,** *with the same properties in its foliage to heal bruises and broken bones as the dull reddish-colored one.*
Its masses of dangling blue bells attract hummingbirds and bees. Chickens gobble up stalks and spent flowers, and when the plant is thrown on the compost heap, it decomposes rapidly, releasing valuable nutrients."
-'Farming Magazine: People, Land, Community, www.farmingmagazine.net, Volume 17, Issue 1, No. 65, Mount Hope, Ohio, Spring **2017**. Includes article 'Flowering Herbs, Part I' by Jo Ann Gardner.

"S. caucasicum is a rhizomatous perennial eventually forming an extensive colony of **erect stems to 60 cm (23 inches) in height,** *bearing large lance-shaped leaves and* **nodding tubular bright blue flowers in summer.**
Plant range: Caucasus, Iran.*"*
-Royal Horticultural Society, London, England, www.rhs.org.uk, **2018**.

*"****Symphytum caucasicum superficially resembles S. asperum (Prickly Comfrey) and these species have been confused.*** *Both are erect, non-stoloniferous perennials with sky-blue corollas.*
In the former (caucasicum) the calyx is only shallowly divided at flowering (always less than half way to base) whereas in the latter (asperum) the calyx is always divided more than half way to base. This character is much less obvious in the fruiting stage. In addition, nutlets are minutely tuberculate in Symphytum caucasicum and more or less papillose with angular markings in S. asperum.
Finally, in 'pure' Symphytum asperum, leaves are not decurrent at all (the upper often shortly stalked) while in S. caucasicum leaves are always shortly decurrent along stem.
A putative (generally regarded as such) hybrid between these species has been recorded in the British Isles (*Learmonth & Perring 1995)."
-Manual of the Alien Plants of Belgium, http://alienplantsbelgium.be, Descriptions to the introduced plants that grow wild in Belgium, **2018**.
(* -'A Symphytum Comfrey Hybrid New to Britain' by R.W.C. Learmonth and F.H. Perring, Botanical Society of the British Isles News, Volume 69, No. 74, 1995.)

S. caucasicum Chromosomes

*"****Symphytum caucasicum, n = around 18*** *(*Strey 1931)."* (2n was not listed.)
-'Cytogenetic Studies on the Boraginaceae' by Donald M. Britton, Brittonia (published by The New York Botanical Garden), Volume 7, No. 4, pages 233-266, December 10 **1951**.
(* -'Karyologische Studien an Borraginoideae' {Caryological studies on Borraginoideae} by Martin Strey; Planta, Berlin, Germany, Volume 14, Issue 3-4, pages 682-730, October 1931. In German.)

"The possibility that S. caucasicum, S. peregrinum and S. asperum are somewhere sympatric in the Caucasus cannot be excluded; hence hybridization (if at all possibe; this remains to be tested experimentally), cannot be ruled out.
The present authors found 2n = 48 in material of S. caucasicum M. Bieb. of garden origin, *the exact origin of which could not be traced (oral communcation by Dr. Nelson, Leeds, England, who kindly put material from his garden at our disposal).*
Gviniashvili (*1972) on the other hand, counted 2n = 24 in authentic wild Caucasian material.*"*
-'Cytotaxonomic Studies in the Genus Symphytum VIII: Chromosome Numbers and Classification of Ten European Species' by T.W.J. Gadella and E. Kliphuis, Proceedings van de Koninklijke Nederlandse Akademie van Wetenschappen {Proceedings of the Royal Netherlands Academy of Arts and Sciences}: Series C, Biological and Medical Sciences, Vol 81, page 162-172, **1978**.
(* -'Some Data on the Karyology of Caucasic Species of Symphytum L. with Respect to their Taxonomy' by TS.N. Gviniashvili, Botaniceskij Zurnal, Lennigrad, Russia, Volume 57, pages 1120-1126, 1972 in Russian.
If you have a translation of this in English, I would appreciate a copy.)
    (Sympatric means occurring in the same geographical location; overlapping in distribution. For speciation it means taking place without geographical separation.)

"*Symphytum caucasicum* M. Bieb.: Chromosome count is 2n = 48."
-Tropicos®, www.tropicos.org, Missouri Botanical Garden, Saint Louis, Missouri; Plant database with 1.3 million scientific names and 4.4 million specimen records, **2018**.

"*Symphytum caucasicum* M. Bieb.: 2n = 24.
Source: TS.N. Gviniashvili, 1972. 'Some Data on the Karyology of Caucasic Species of Symphytum L. with Respect to their Taxonomy', (in Russian). Botaniceskij Zurnal, Russia, 57: 1120-1126."
-'Chromosome Counts Database'® (CCDB) is a comprehensive resource for plant chromosome numbers. Every chromosome number includes the original database and the research article reference, **2018**.   http://ccdb.tau.ac.il

S. caucasicum and Alkaloids

"*S. caucasicum* Bieb, Caucasian Comfrey, root was found to have 4800 mg/kg PAs (Pyrrolizidine Alkaloids). S. asperrimum is the parent of S. x uplandicum (Russian Comfrey) which contains echimidine.
It also has high PA levels and should not be used medicinally. **S. caucasicum also has high PA levels** and should not be used.
-'Using Herbs that Contain Pyrrolizidine Alkaloids' by Alison Denham, B.A., MNIMH (National Institute of Medical Herbalists), University of Central Lancashire, England; The European Journal of Herbal Medicine, Volume 2, No. 3, pages 27-38, **1996**.

"*Symphytum caucasicum* Bieb.:
Caucasian Comfrey (German: Kaukasischer Beinwell, French: Consoud de Caucase) **grows in Caucasia and is used there and in the European part of Russia as folk medicine.** In these areas it plays nearly the same important role as S. officinale in Central Europe.
In the 1970s it was repeatedly studied by *Manko et al. who found a total content of 0.48% of PAs. Among these the following alkaloids were detected by means of paper and thin-layer chromatography: the N-oxide of echimidine, asperumine, echinatine, and lasiocarpine.
Since the latter three alkaloids have not been found so far in other species of the genus Symphytum, further investigations of S. caucasicum are required.
Owing to its **high alkaloid content** an internal intake of this Comfrey as a medicinal plant is not recommendable."
-'Medicinal Plants in Europe Containing Pyrrolizidine Alkaloids' by Erhard Thomas Roeder, Pharmazeutisches Institut (Pharmaceutical Institute) der Rheinischen Friedrichs-Wilhelms, University of Bonn, Germany, Pharmazie (Pharmacy) 50, pages 83-98, March **1995**.
(* -I.V. Manko, Z.V. Melkumova and V.F. Malysheva; Rastitel'nye Resursy, Volume 8, page 538, Leningrad, Russia, 1972. It did not list the title of the article.)
(Also by Manko: 'Pyrrolizidine Alkaloids and Pyrrolizidine-N-oxide Alkaloids from Boraginaceous Plants' by I.V. Manko and Y.G. Borisyuk; Ukraine Khim. Zhur., Volume 23, page 362, 1957. I was unable to find this article.)

Symphytum circinale (Rhodian Comfrey)
See Symphytum creticum in this section.

Current Botanical Nomenclature of S. circinale

"*Symphytum circinale* Runemark (*Bot. Not. 120: 90. 1967)"
-'International Plant Names Index'® (IPNI), www.ipni.org, A database of the names and associated basic bibliographical details of seed plants, ferns and lycophytes, **2018**.
(* -'Studies in the Aegean Flora XI: Procopiana {Boraginaceae} included into Symphytum' by Hans Runemark, Botaniska Notiser (Botanical Notes), Sweden, Volume 120, page 90, 1967. I was unable to get this. If you have it, could you send it to me.)
  (P. Hans B. Runemark, 1927-2014, was a Swedish botanist and lichenologist, professor at Lund University, Sweden.
  After receiving his Ph.D. in lichenology, he studied vascular flora of the Aegean Islands in Greece with special interest in their native plants.)

"*Symphytum circinale* Runemark is an accepted name. Bot. Not. 120: 90 1967.
  Synonym: Procopiania circinalis (Runemark) Pawl. World Checklist Database (WCSP) in review."
-The Plant List®, www.theplantlist.org from the World Checklist Database®, **2019**.

"*Symphytum circinale* Runemark:
This species is accepted, and its native range is Greece (Evvoia) to southwest Turkey.
Synonym: Procopiania circinalis (Runemark) Pawl."
-'Plants of the World Online'® is from Kew Royal Botanic Gardens, London, England. It has information on all the world's known seed-bearing plants. With over 8.5 million items, Kew houses the largest and most diverse botanical and mycological collections in the world, **2019**. www.plantsoftheworldonline.org
  (Evvoia, Euboea or Evia is a Greek island.)

Description of S. circinale

"*S. circinale* Runem. in Bot. Notiser 120:90 (1967).
Slender perennial herb. Rhizome branched. Stem lax, 10-40 cm (3.9-15.7 inches) high, branched, tuberculate villose. Leaves tuberculate villose, scarcely dentate; lower leaves ovate-elliptic, 9 cm (3.5 inch) long, 5 cm (1.9 inch) broad, decurrent; petiole winged, as long as lamina; upper leaves parabolical, 3 cm (1.1 inch) long, 2 cm (0.7 inch) broad, sub-sessile or broadly winged, short petiole.
Inflorescence lax, about 10-flowered. Calyx 6-8 mm long, funnel-shaped, divided to lower quarter, segments lanceolate, acute, villose, enlarging in fruit to 9-10 mm. **Corolla white**, urceolate, tube 4-5 mm long, lobes about 8 mm long, at 45 degrees to stamens, upper half usually revolute; corolla scales 6 mm long, subulate, acute, base broader than filaments, margin with unicellular prickles. Stamen filaments about 11 mm from long, inserted 2 mm below base of corolla scales, hairy cuff at 1.5 mm from base of filament; anthers 2 mm long. Style 16 mm long, stigma capitate.
Nutlets 3-4 mm, obovate, erect, reticulate. Flowers 3-5.
Stream banks, rock crevices, shade, altitude 0-250 meters (0-820 feet)."
-'A Revision of Symphytum in Turkey and Adjacent Areas' by G.E. Wickens, Royal Botanic Gardens: Kew; Notes from the Royal Botanic Garden Edinburgh, Scotland, Volume 29, pages 157-180, **1969**. Includes Turkey, Bulgaria, Greece, Aegean Islands and Caucasia.

"**The phytogeography (geographic distribution of plants) of the Greek island Rodhos is briefly sketched.** The flora of Rodhos holds a dual position having affinities with the South Aegean islands as well as with the East Aegean islands and Southwest Anatolia.
Differentiation has apparently taken place in several taxa since the separation of Rodhos from the South Aegean islands and Southwest Anatolia, which is shown by the occurrence of **many allopatric taxa**.
> The occurrence of many local variants without taxonomic rank shows that this process of speciation is still not complete. Both these areas of ophiolithic rock and the limestone cliffs have apparently provided refugia for taxa with formerly wider distributions.

Evolution:
It is probable that genetic differences were already initiated between populations of the Pliocene mountain range, and that the subsequent formation of an island arc only accelerated the speciation process.
The sea between the islands has apparently formed a barrier to dispersal with ensuing **geographical isolation**. During periods of climatic fluctuation and edaphic (soil) changes in the Pleistocene, the populations may have been much more restricted than they are now, **enabling genetic drift to exert an influence over differentiation,** and bringing about more rapid evolutionary change.
The following groups of allopatric taxa demonstrate the differentiation that has probably occurred in the South Aegean islands:
> **Symphytum circinale** (Rodhos, Marmaris peninsula, the East Aegean islands).
> **S. insularis** (Karpathos, South and Central Aegean).
> **S. creticum** (Crete, Peloponnisos, Zakinthos)."

-'The Phytogeographical Position of Rodhos' by Annette Carlstrom, Department of Systematic Botany, Lund University, Sweden; Proceedings of the Royal Society of Edinburgh, Scotland, Volume 89 B, pages 79-88, **1986**.
> (Rhodhos or Rhodes is a Greek island in the southeast Aegean Sea that is 10 miles off the coast of Turkey.)
> (The Aegean Sea is a bay of the Mediterranean Sea located between the Greek and Anatolian peninsulas, i.e., between the mainlands of Greece and Turkey.)
> (Allopatric or geographic speciation occurs when populations of the same species become isolated from each other, and this prevents genetic exchange.)
> (Pliocene or Pleiocene is the epoch in the geologic timescale that extends from 5.3 million to 2.5 million years ago.)

"**The two species Symphytum circinale and Symphytum creticum, included by Gusuleac (*1928) in the genus Procopiania (compare to also Pawlowski 1971), do not show any pollen differentiation from the other members of Symphytum, suggesting the maintainance of Procopiania into Symphytum (**Runemark 1967).**"
-'Pollen Morphology in the Boragineae (Boraginaceae) in Relation to the Taxonomy of the Tribe' by Massimo Bigazzi and Federico Selvi, Plant Systematics and Evolution, Volume 213, pages 121-151, **1998**.
(* -'Die Monotypischen und Artenarmen Gattungen der Anchuseae {Caryolopha, Brunnera, Hormuzakia, Gastrocotyle, Phyllocara, Trachystemon, Procopiania und Borago}' by M. Gusuleac; Buletinul Facultatii de Stiinte din Cernauti, Romania, Volume 2, pages 394-461, 1928.)
(** -'Studies in the Aegean Flora XI: Procopiana {Boraginaceae} included into Symphytum' by Hans Runemark, Botaniska Notiser {Botanical Notes}, Sweden, Volume 120, pages 84-94, 1967. I was unable to get a copy of this report. If you have one, could you please send it to me.)
> (For more of this article, see subsection 'Comfrey Flower Buds, Bloom, Nectar and Pollination' in section 'Symphytum Genus Description'.)
> (Palynology is the study of pollen grains, especially in archaeological or geological deposits.)
> (Family: Boraginaceae, Tribe: Boragineae, Genus: Symphytum L.)

"**The island of Chalki, located west of Rhodos island, belongs to the East Aegean Islands and is situated at the east part of the South Aegean Island Arc.** We show that Chalki has the second highest percentage of Greek endemics in the phytogeo-

graphical region of the East Aegean Islands (the highest is Ilkaria, and Rhodos is fourth).
> *The flora of Chalki hosts 103 East Mediterranean elements, 13 of which are considered as range-restricted taxa.*
> ***Chalki constitutes the southwesternmost distributional border for at least five Anatolian taxa:*** *Gladiolus anatolicus (Boiss.) Stapf, Ophrys speculum subsp. regisferdinandii (Acht. & Kellerer ex Renz) Soo, Quercus aucheri Jaub. & Spach,* ***Symphytum circinale Runemark*** *and Verbascum propontideum Murb.*

Campanula rhodensis A.DC., Dianthus fruticosus subsp. rhodius (Rech. f.) Runemark, Nigella arvensis subsp. brevifolia Strid and ***Symphytum circinale may have resulted from random genetic drift during the climatic fluctuations and sea-level oscillations of the Pleistocene*** *(*Carlstrom 1986; Comes & al. 2008).*
Chalki also seems to have close phytogeographical affinities with the Mugla province of Turkey*, and more specifically, with the Marmaris (Datca) peninsula, since virtually all (apart from Ophrys speculum subsp. regisferdinandii) of its range-restricted East Mediterranean taxa occur also there.*
> *This phenomenon can be attributed to the cliff flora of the outer Marmaris (Datca) peninsula mainly resembling a slightly impoverished Aegean flora, since* ***it hosts several typical Aegean species,*** *such as Campanula hagielia,* ***Symphytum circinale*** *and Verbascum propontideum (Carlstrom 1986)."*

-'Contribution to the Vascular Flora of Chalki Island (East Aegean, Greece) and Biomonitoring of a Local Endemic Taxon' by Maria Tsakiri, Konstantinos Kougioumoutzis and Gregoris Iatrou from Greece; Willdenowia: Annals of the Botanic Garden and Botanical Museum, Berlin, Germany, Volume 46, No. 1, pages 175-190, **2016**.
(* -'The Phytogeographical Position of Rodhos' by Annette Carlstrom, Department of Systematic Botany, Lund University, Sweden; Proceedings of the Royal Society of Edinburgh, Scotland, Volume 89 B, pages 79-88, 1986.)
> (Chalki island is next to Rodos island. Chalki is much smaller than Rodos {Rhodos} but both are smaller than the largest East Aegean islands.)

*"****Symphytum Circinale: Akrep Kafesotu:***
Slender villous perennial, 10-40 cm (3.9-15.7 inch). Lower leaves ovate-elliptic, petiole winged, equalling lamina; upper ovate, subsessile or with short, broadly winged petiole.
Flowers about 10. Calyx 6-8 mm, accrescent to 10 mm in fruit, divided to 3/4, lobes lanceolate, acute. ***Corolla white,*** *tube 4,5-6 mm, lobes 1 1/2 to 2 times tube, spirally contorted above; scales subulate, about 6 mm; filaments about 11 mm, with hairy collar about 1.5 mm from base. Nutlets 3-4 mm, erect, reticulate.*
2n=28. *Fl. 3-5. Stream banks, rocks, crevices, shade, sea level to 1000 meter (3280 feet).*
Islands. East Mediterranean element."
-'Symphytum Circinale: Akrep Kafesotu' published by Turkiye Bitkileri (Plants of Turkey), **2019**. Turkey has a great botanical richness with more than 9600 plant species. This website has been constructed to gather as much information as possible about native plants of Turkey. www.turkiyebitkileri.com

<u>Symphytum coccineum</u> (See S. Rubrum in this section.)

<u>Symphytum cordatum</u>
See 'Symphytum grandiflorum' in this section.

Current Botanical Nomenclature of S. cordatum

*"****Symphytum cordatum Waldst. & Kit.*** *(*Descr. Icon. Pl. Hung. 1: 6, t. 7. 1799-1802)*
 Symphytum cordatum Willd. *(**Neue Schriften Ges. Naturf. Freunde Berlin ii. (1799) 121.)*
 Symphytum cordatum M.Bieb. *(***Fl. Taur.-Caucas. 1: 130. 1808)"*
-'International Plant Names Index'® (IPNI), www.ipni.org, A database of the names and associated basic bibliographical details of seed plants, ferns and lycophytes, **2018**.
(* -'Descriptiones et Icones Plantarum Rariorum Hungariae' {Descriptions and Pictures of the Rare Plants of Hungary} or {Plants and Descriptions Icons Engravings Hungary} by Franz de Paula Adam von Waldstein-Wartenbur and Pal Kitaibel; Vienna, Austria, Volume 1, page 6, plate 7, 1799-1802. In Latin.)
(** -'Der Gesellschaft Naturforschender Freunde zu Berlin, Neue Schriften' {The Society of Friends of Nature for Berlin, New Writings}, Volume 2, page 121, 1799. In German.)
(*** -'Flora Taurico Caucasica: Exhibens Stirpes Phaenogamas in Chersoneso Taurica et Regionibus Caucasicis Sponte Crescentes' edited by Friedrich August Marschall von Bieberstein, Volume 1, page 130, 1808.)
> (Franz de Paula Adam Norbert Wenzel Ludwig Valentin von Waldstein, 1759-1823, was an Austrian soldier, explorer and naturalist. From 1789 he studied the botany of Hungary with Pal Kitaibel. His herbarium is stored in Prague, Czech Republic. Pal Kitaibel, 1757-1817, was a Hungarian botanist / chemist and professor at University of Buda, Hungary.)

*"****Symphytum cordatum Bieb.****: ambiguous synonym for Symphytum grandiflorum DC.*
 Symphytum cordatum Waldst. & Kit.*: accepted name."*
-Catalogue of Life®: 2015 Annual Checklist, Online database of the world's known species of animals, plants, fungi and micro-organisms. It consists of a single integrated species checklist and taxonomic hierarchy., **2018**, www.catalogueoflife.org.
Search: www.catalogueoflife.org/annual-checklist/2015/search/all

"Symphytum cordatum Waldst. & Kit ex Willd. is an accepted name. No synonyms are recorded for this name."
-The Plant List®, www.theplantlist.org from the World Checklist Database, **2018**.

Description of S. cordatum

*"**The Symphytum cordatum W.K. plant is 700-900 meters (2296-2952 feet) above sea level and loves preferably shady wet forest spots.***
Symphytum cordatum differs at first glance from S. tuberosum. The former is higher, softly nodding, its broad, rather flattened leaves cover a large area, mostly touching the nearest plant."
-'Symphytum Cordatum W.K.' by J. Ullepitsch; Oesterreichische Botanische Zeitschrift (Austrian Botanical Journal), Volume 36, pages 298-299, Vienna, Austria, **1886**. In German.

*"**Symphytum cordatum Waldst. & Kit.** apud Willd. in Act. Soc. Berol. ii. (1799) p. 121:*
*Geographical Distribution: **Hungary: Banat, Transylvania (central Romania). Galicia (Spain). Moldavia. Russia: Volhynia.***
The large, subglabrous, deeply cordate leaves distinguish this from all other species.
The description of the root and manner of growth is taken from Waldstein & Kitaibel's *'Pl. rar. Hung'. The figure there given shows a root with two branches, one bearing a flowering stem and the other a tuft of radical leaves.
S. grandiflorum differs in the decumbent caulescent barren shoots."
-'A Revision of the Genus Symphytum, Tourn.' by Cedric Bucknall, (Mus. Bac. Oxon= Bachelor of Music, Oxford University), Journal of the Linnean Society of London, England, Botanical Journal, Volume 41, Issue 284, pages 491-556, December **1913**.
(* -'Descriptiones et Icones Plantarum Rariorum Hungariae' {Plants and Descriptions Icons Engravings Hungary} by Franz de Paula Adam von Waldstein-Wartenbur and Pal Kitaibel; Vienna, Austria, Volume 1, page 6, plate 7, 1799-1802. In Latin. Illustrations are end of the book.)

> (Banat is a region between central and eastern Europe that is now divided among three countries: the eastern part is in western Romania, the western part is in northeastern Serbia, and a small northern part is in southeastern Hungary.)
> (Moldavia is territory between Eastern Carpathians and Dniester River. The western half of Moldavia is part of Romania. The eastern side belongs to Republic of Moldova, and northern and southeastern parts are territories of Ukraine.)
> (Volhynia is between southeastern Poland, southwestern Belarus, and western Ukraine.)

*"**S. cordatum Waldst. & Kit. ex Willd.**, Ges. Naturf. Freunde Berlin Neue Schr. 2: 121 (1799):*
*Rhizome creeping, uniformly stout. Stem 15-35 (to 50) cm {5.9-13.7 (to 19.6)}, simple, sparsely hairy; hairs up to 1.1 mm. Cauline leaves 2 to 4 (to 5), the lower large, cordate, long-petiolate, the upper rounded at the base, shortly petiolate or sessile. Inflorescence with up to 20 flowers. Calyx deeply lobed. Corolla 13-18 mm, **pale yellow**, 1 1/2 to 2 times as long as calyx. Scales triangular-lanceolate, the lower marginal papillae 2 1/2 to 3 1/2 times as long as wide, the upper smaller but distinctly longer than wide. Stamens with filament 4/7 to 3/4 as long as anther.*
***2n = 18**. Deciduous woods.*
Carpathians (Carpathian mountains), central Romania, west Ukraine. Czechoslovakia, Poland, Romania, Russia (southwestern region)."
-Flora Europaea, Volume 3: Diapensiaceae to Myoporaceae, editors Tutin, Heywood, Burges, Moore, Valentine, Walters and Webb. United Kingdom: Cambridge University Press, **1972**, page 104. (Symphytum L., pages 103-105 by B. Pawlowski.) (5-volume encyclopedia of plants, published between 1964 and 1993. 'Flora Europaea' describes all the national floras of Europe to identify any wild or widely cultivated plant to the subspecies level. It provides geographical distribution, habitat preference, and chromosome number.)

> (The Carpathian Mountains are a mountain range that forms an arc 1,500 km {932 miles} across central and eastern Europe. They are the second-longest mountain range in Europe. Western Carpathians include Austria, Czech Republic, Poland, Slovakia and Hungary. Eastern Carpathians include southeastern Poland, eastern Slovakia, Ukraine and Romania. Southern Carpathians include Serbia and Romania.)

*"**S. cordatum Waldst. et Kit.** (Loew, *'Blutenbiol. Floristik,' page 280.)*
***This Hungarian species bears yellowish-white flowers**, which are shorter than those of S. officinale, but with longer prickles on the triangular scales, according to a description given by Loew of plants cultivated in the Berlin Botanic Garden, Germany."*
-'Handbuch der Blutenbiologie' {Handbook of Flower Pollination} by P. Knuth, Volume 3, part 2. v + 601 pages, Leipzig, Germany, 1905. [Boraginaceae, pages 63-67.] English translation by J.R.A. Davis, Volume 3, iv + 644 pages, Oxford, England, **1909**. [Boraginaceae, pages 115-142; 23 genera, including Symphytum.]
(* -'Blutenbiologische Floristik des Mitteleren und Nordlichen Europa sowie Gronlands' {Floral Biology Floristry of Central and Northern Europe and Greenland} by Dr. Ernst Loew, Stuttgart, Germany, page 280, 1894. In German.)

*"**Symphytum cordatum:***
A tuber- and rhizome-forming geophyte with a chain-like shortened rhizome, dicyclic, semi-rosette, monocarpic shoots.
***It can be found in Ukraine, Romania, Slovakia, Poland (in the Carpathians),** most often growing in montane (mountainous) fir forests and less often in broadleaf forests (Pawlowski 1961; Popov 1953).*
***Germination bivariant (2 variations)** - epigeous hypocotylar-cotyledonar or hypogeous (*Barykina and Alyonkin 2008). In the former case, cotyledon blades are brought out to the surface. They are oblong-elliptic in shape (8 mm long, 4 mm wide), fleshy,*

with glabrous petioles of 3.5 to 4 mm long, 1 mm wide, with indumentum only on the adaxial side.
Pubescence mixed, composed of two types of trichomes: simple sessile and glandular. Among sessile trichomes, two morphological subtypes can be identified: relatively long, subulate straight hairs and short hooked hairs; surface fine-grained, tuberculous. Glandular hairs one- or two-celled, with clavate heads and very short stalks.
In plants with hypogeous **germination,** the cotyledons remain covered by the nutlet's coat (pericarp) or partially released. In case of hypogeous germination, pubescence is the same as in aerial (epigeous) cotyledons, i.e., restricted to the adaxial side of the blade, yet it is scarcer and the hairs are very short, ephemerous.
Juvenile, immature and mature (vegetative and reproductive) plants are characterized by cordate petiolate leaves; near the inflorescence they are ovate or oblong, with short petioles. Leaf blades have indumentums on both adaxial and abaxial (mainly along protruding veins) sides and along the margins.
Trichomes simple (sessile or subtended by 'rosettes' and glandular).
> Sessile trichomes short, unicellular, subulate (hooked or sickle-shaped), thin-walled, with brown content concentrated near the hair's tip; surface fine-grained or tuberculous fine-grained.
> Subtended ('rosette') hairs long, unicellular, subulate (straight, sometimes sickle-shaped), with thicker walls and dark green content near the top; surface tuberculous fine-grained.

Solitary glandular hairs, composed of rounded to oblong heads and unicellular stalks, occur primarily along the veins and leaf margins. The entire petiole surface is covered by simple subulate transparent hairs, spaced apart from each other."
-'Pubescence of Vegetative Organs and Trichome Micromorphology in Some Boraginaceae at Different Ontogenetic Stages' by Rimma P. Barykina and Vitaly Y. Alyonkin, Lomonosov Moscow State University, Russia; Wulfenia, Regional Museum of Carinthia, Austria, Volume 23, pages 1-29, **2016**.
(* -'A Comparative Morphological Ontogenetic Study of Some Representatives of the Genus Symphytum L.: Sections Coerulea Buchn., Symphytum Pawl. and Tuberosum Buchn.' by R.P. Barykina and V. Alyonkin, Byulleten Moskovskogo Obshchestva Ispytatelei Prirody, Otdel Biologicheskii, Moscow, Russia, Volume 113, No. 5, pages 47-57, 2008. In Russian.)
> (A seed has epigeal germination when the cotyledons of the germinating seed expand, throw off the seed shell and become photosynthetic above ground.)
> (Pubescent means the surface of a leaf or stem is covered with fine short hairs.)
> (Trichome is a small hair or other outgrowth from the epidermis of a plant. Epidermis is the outer layer.)

Chromosomes of S. cordatum

"S. cordatum Waldst. & Kit. ex Willd.: 2n = 18."
-Flora Europaea, Volume 3: Diapensiaceae to Myoporaceae, editors Tutin, Heywood, Burges, Moore, Valentine, Walters and Webb. United Kingdom: Cambridge University Press, **1972**, page 104. (Symphytum L., pages 103-105 by B. Pawlowski.) (5-volume encyclopedia of plants, published between 1964 and 1993. 'Flora Europaea' describes all the national floras of Europe to identify any wild or widely cultivated plant to the subspecies level.)

"S. cordatum Waldst et Kitaib. ex Willd.: S. cordatum, a subendemic species of the Carpathians, was studied by Tarnavschi (*1948), who counted n = 13. This could neither be confirmed by Wcislo (**1972), who counted 2n = 120, nor by Grau (***1971), who found **2n = 60** in material of garden origin.
**The number 2n = 120 was confirmed by the present authors. This number represents the decaploid level in the x = 12 series."*
-'Cytotaxonomic Studies in the Genus Symphytum VIII: Chromosome Numbers and Classification of Ten European Species' by T.W.J. Gadella and E. Kliphuis, Proceedings van de Koninklijke Nederlandse Akademie van Wetenschappen (Proceedings of the Royal Netherlands Academy of Arts and Sciences): Series C, Biological and Medical Sciences, Vol 81, pages 162-172, **1978**.
(* -'Die Chromosomenzahlender Anthophyten-Florav on Rumanien mit Einem Ausblick auf das Polyploidie-Problem' {The Number of Chromosomes The Anthophyten-Florav in Romania with an Outlook on the Polyploidy Problem} by I.T. Tarnavschi; Bulletin du Jardin et du Musee Botaniques de Universite de Cluj, Roumanie, Volume 28, pages 1-130, 1948. In German.)
(** -'Karyological Studies in Symphytum L.' by H. Wcislo; Acta Biologica Cracoviensia: Journal of Polish Academy of Sciences, Krakow, Poland, Series Botanica 15, pages 153-163, 1972.)
(*** -'Cytologische Untersuchungen an Boraginaceae II' {Cytological Investigations on Boraginaceae} by J. Grau, Mitteilungen der Botanischen Staatssammlung Munchen {Communications of the Botanical State Collection Munich, Germany}, Volume 9, pages 177-194, 1971.)
> (Subendemic means it is largely localized in one natural area. Partially endemic or native.)
> (Body cells and individuals are described by the number of chromosome sets {ploidy level} they have: monoploid: 1 set; diploid: 2 sets = 2n = 2x, triploid: 3 sets = 2n = 3x, tetraploid: 4 sets = 2n = 4x, pentaploid: 5 sets = 2n = 5x, etc. Decaploid is ten complete sets of chromosomes in a single cell: 2n = 10x.)

"Symphytum cordatum Waldst. & Kit.: 2n = 72, 24, 40, 48, 56, 120."
-Tropicos®, www.tropicos.org, Missouri Botanical Garden, Saint Louis, Missouri; Plant database with 1.3 million scientific names and 4.4 million specimen records, **2018**.

Symphytum creticum
See Symphytum circinale in this section.

"Symphytum creticum (Willd.) Runemark (*Boissiera xiii. 100 (1967).)
Symphytum creticum (Willd.) Runemark var. squamulatum (Pawl.) R.L.Jahn"
(in R. Jahn & P. Schonfelder, Exkursionsfl. Kreta 28 (1995), without exact basionym page.)
-'International Plant Names Index'® (IPNI), www.ipni.org, A database of the names and associated basic bibliographical details of seed plants, ferns and lycophytes, **2018**.
(* -Boissiera: Conservatoire et Jardin Botaniques de la Ville de Geneve, Universite de Geneve, Institut de Botanique Systematique, Geneva, Switzerland, Volume 13, page 100, 1967.)

"S. creticum (Willd.) Runem. in Bot. Notiser 120 :89 (1967).
Synonym: Borago cretica Willd., Sp. Pl. 1 :778 (1797).
Trachystemon creticum D. Don ex G. Don, Gen. Syst. 4:309 (1838).
Psilostemon creticum DC., Prod. IO :36 (1846).
Procopiania cretica (Willd.) Gusul. in Bui. Fae. Sti. Cernauti 2 :435 (1928).
Slender perennial herb. Rhizome branched. Stem lax, 10-40 cm (3.9-15.7 inches) high, branched, tuberculate villose, rarely almost glabrous. Lower leaves elliptic, 2-8 cm (0.7-3.1 inches) long, winged petiole 1-5 cm (0.3-1.9 inches) long; upper leaves elliptic to broadly lanceolate, sessile.
Calyx 6-8 mm long, divided to lower quarter, segments lanceolate, acuminate; accrescent in fruit. **Corolla bluish-violet, rarely white**, *tube urceolate, about 5 mm long, lobes about 10 mm long, slightly recurved and at 45-75 degrees to stamens; corolla scales about 5 mm long, narrow-linear, subulate, base narrower than filaments, margins with unicellular prickles. Stamen filaments about 9 mm long, with 2 hairy lateral projections at 0.5 mm from base; anthers 2 mm long. Style 14 mm long, stigma capitate.*
Nutlets 3.5 x 1.5 mm, obovate, erect, minutely tuberculate. Fl. 3-4. Maritime shady rock crevices."
-'A Revision of Symphytum in Turkey and Adjacent Areas' by G.E. Wickens, Royal Botanic Gardens: Kew; Notes from the Royal Botanic Garden Edinburgh, Scotland, Volume 29, pages 157-180, **1969**. Includes Turkey, Bulgaria, Greece, Aegean Islands and Caucasia.

"The two species Symphytum circinale and Symphytum creticum, included by Gusuleac (*1928) in the genus Procopiania (compare to also Pawlowski 1971), do not show any pollen differentiation from the other members of Symphytum, suggesting the maintainance of Procopiania into Symphytum (**Runemark 1967)."
-'Pollen Morphology in the Boragineae (Boraginaceae) in Relation to the Taxonomy of the Tribe' by Massimo Bigazzi and Federico Selvi, Plant Systematics and Evolution, Volume 213, pages 121-151, **1998**.
(* -'Die Monotypischen und Artenarmen Gattungen der Anchuseae {Caryolopha, Brunnera, Hormuzakia, Gastrocotyle, Phyllocara, Trachystemon, Procopiania und Borago}' by M. Gusuleac; Buletinul Facultatii de Stiinte din Cernauti, Romania, Volume 2, pages 394-461, 1928.)
(** -'Studies in the Aegean Flora XI: Procopiana {Boraginaceae} included into Symphytum' by Hans Runemark, Botaniska Notiser {Botanical Notes}, Sweden, Vol 120, page 84-94, 1967. I was unable to get this. If you have one, could you send it.)
(For more of this article, see subsection 'Comfrey Flower Buds, Bloom, Nectar and Pollination' in section 'Symphytum Genus Description' {Chapter 5}.)
(Palynology is the study of pollen grains, especially in archaeological or geological deposits.)
(Family: Boraginaceae, Tribe: Boragineae, Genus: Symphytum L.)

"The Procopiania genus was instituted by Gusuleac (1928) to accommodate Symphytum creticum (Procopiania cretica), a south Aegean species with floral morphology intermediate between Borago and Symphytum *due to the corolla lobes being longer than the tube and the exserted stamens.*
In more recent times, Procopiania was accepted by some authors (Riedl, 1963; Pawlowski, 1971; Chater, 1972; Stearn, 1986) but not by others (Runemark, 1967; Wickens, 1969) who included it in Symphytum.
The data presented here showed that S. creticum is nearly identical to S. tuberosum in both ITS1 and trnL sequences."
-'Molecular Systematics of Boraginaceae Tribe Boragineae Based on ITS1 and trnL Sequences, with Special Reference to Anchusa s.l.' by H.H. Hilger, F. Selvi, A. Papini and M. Bigazzi (Germany and Italy), Annals of Botany, Volume 94, No. 2, pages 201-212, **2004**.   (Aegean Sea is a bay of Mediterranean Sea located between the Greek and Anatolian peninsulas, i.e., between mainlands of Greece and Turkey.)

Symphytum cycladense
See 'Symphytum davisii' in this section.

Current Botanical Nomenclature of S. cycladense

*"**Symphytum cycladense Pawl.** (*Fragm. Florist. Geobot. 17(1): 25. 1971)"*
-'International Plant Names Index'® (IPNI), www.ipni.org, A database of the names and associated basic bibliographical details of seed plants, ferns and lycophytes, **2018**.
(* -'Symphyta Mediterranea Nova vel Minus Cognita: Nowe lub malo znane srodziemnomorskie zywokosty' {New or Little Known Mediterranean Comfrey} by Bogumil Pawlowski; Fragmenta Floristica et Geobotanica Polonica, W. Szafer Institute of Botany,

Polish Academy of Sciences, Krakow, Poland, Vol 17, No. 1, page 25, 1971. If you have English translation, I would like a copy.)

"Symphytum cycladense Pawl.: synonym for Symphytum davisii subsp. cycladense (Pawl.) W.T. Stearn.
Symphytum davisii subsp. cycladense (Pawl.) W.T. Stearn: accepted name. Infraspecific taxon.
-Catalogue of Life®: 2015 Annual Checklist, Online database of the world's known species of animals, plants, fungi and micro-organisms. It consists of a single integrated species checklist and taxonomic hierarchy., **2018**, www.catalogueoflife.org. Search: www.catalogueoflife.org/annual-checklist/2015/search/all
(Infraspecific means at a taxonomic level below that of species, e.g., subspecies, variety, cultivar, or form.)

Description of S. cycladense

"S. cycladense Pawl., Fragm. Fl. Geobot. 17: 25 (1971):
Plant with numerous basal leaves in a rosette, densely and rather softly villous-pubescent, with setae and hooked hairs. Stem 5-20 cm (1.9-7.8 inch), simple or scarcely branched. Lower leaves ovate to ovate-oblong, shortly petiolate, greyish beneath. Calyx 6 to 7 (to 10) mm, lobed to 2/3 to 7/8. Corolla 13 to 16.5 mm, ?white. Scales 1.6 to 4 mm, lingulate, marginal papillae not more than 1 1/2 times as long as wide, the lower more or less scattered. Stamens 2.6 to 4.3 mm; anthers 2-3 mm, 4-6 times as long as wide; filament 1/3 to 1/2 as long as anther.
2n = 30. *South Aegean region (Sikinos island). Greece."*
-Flora Europaea, Volume 3: Diapensiaceae to Myoporaceae, editors Tutin, Heywood, Burges, Moore, Valentine, Walters and Webb. United Kingdom: Cambridge University Press, **1972**, page 104. (Symphytum L., pages 103-105 by B. Pawlowski.) (5-volume encyclopedia of plants, published between 1964 and 1993. 'Flora Europaea' describes all the national floras of Europe to identify any wild or widely cultivated plant to the subspecies level.)
(The Aegean Sea is a bay of the Mediterranean Sea located between the Greek and Anatolian peninsulas, i.e., between the mainlands of Greece and Turkey.)

*"**Taxon Name: Symphytum cycladense Pawl.**:*
Synonym: Symphytum davisii subspecies cycladense (Pawl.) Stearn.
Taxonomic Notes: Symphytum cycladense Pawl. is treated under the name Symphytum davisii Wickens ssp. cycladense (Pawl.) Stearn. in the 'Greek Red Data Book' (*Snogerup and Snogerup in: Phitos et al. 2009).
Geographic Distribution: Symphytum cycladense is a Greek endemic (native) with a very restricted range on the islands of Sikinos and the neighboring islet of Kardiotissa (Snogerup and Snogerup in: Phitos et al. 2009). The two known subpopulations are very small. The extent of occurrence and area of occupancy are very small as Sikinos itself is only 42 square km (16.2 square miles).
Vulnerability: There are no major threats known as the plant grows in an area that is difficult to access. However, the population trend is unknown too and until more information is available, it is precautionary assessed as 'Vulnerable D2'.
Population: The total number of individuals is not known but the number of mature individuals in each of the two subpopulations is rather small, up to a few hundred individuals (Snogerup and Snogerup in: Phitos et al. 2009). The population trend is unknown.
Habitat and Ecology: It grows in shady, stony and rocky places on limestone at 90-300 meters (295-984 feet) altitude.
Conservation Actions: Symphytum cycladense is listed as priority species on 'Annex II of the 'Habitats Directive' and under Appendix I of the 'Convention on the Conservation of European Wildlife and Natural Habitats' (Bern Convention).
It is listed in the 'National Red Data Book' as Vulnerable D' (Snogerup and Snogerup in: Phitos et al. 2009) and is protected by the Greek 'Presidential Decree 67/81'."
-'Symphytum Cycladense' assessment by S. Snogerup and B. Snogerup, The IUCN Red List of Threatened Species®, www.iucnredlist.org, 7 pages, **2013**.
(* -'The Red Data Book of Rare and Threatened Plants of Greece', Volume Two: E-Z, edited by D. Phitos, T. Constantinidis and G. Kamari. Patras, Greece: Hellenic Botanical Society, 2009. Symphytum article by Snogerup and Snogerup.)

"S. cycladense Pawl.: 2n = 30, Greece, Pawlowski, unpublished."
-Flora Europaea: Check-List and Chromosome Index by D.M. Moore. Cambridge, England: Cambridge University Press, **1982**, page 179.

Symphytum davisii
See 'Symphytum cycladense' in this section.
See 'Symphytum naxicola' in this section.

Current Botanical Nomenclature of S. davisii

"Symphytum davisii Wickens (*Notes Roy. Bot. Gard. Edinburgh 29: 168. 1969)."
-'International Plant Names Index'® (IPNI), www.ipni.org, A database of the names and associated basic bibliographical details of seed plants, ferns and lycophytes, **2018**.
(* -Notes from the Royal Botanic Garden, Edinburgh, Scotland, Volume 29, page 168, 1969.)

"*Symphytum davisii:* accepted name.
Symphytum davisii subsp. cycladense (Pawl.) W.T. Stearn: *accepted name. Infraspecific taxon.*
Symphytum davisii subsp. davisii: *accepted name. Infraspecific taxon.*
Symphytum davisii subsp. icaricum (Pawl.) W.T. Stearn: *accepted name. Infraspecific taxon.*
Symphytum davisii subsp. naxicola (Pawl.) W.T. Stearn: *accepted name. Infraspecific taxon.*"
-Catalogue of Life®: 2015 Annual Checklist, Online database of the world's known species of animals, plants, fungi and micro-organisms. It consists of a single integrated species checklist and taxonomic hierarchy., **2018**, www.catalogueoflife.org. Search: www.catalogueoflife.org/annual-checklist/2015/search/all
    (Infraspecific means at a taxonomic level below that of species, e.g., subspecies, variety, cultivar, or form.)

"*Family: Boraginaceae Juss. Genus: Symphytum Tourn. ex L.*
Symphytum davisii Wickens: *This species is accepted, and its native range is Greece (Amorgos).*
Accepted Infraspecifics:
   Symphytum davisii subsp. cycladense (Pawl.) Stearn
   Symphytum davisii subsp. icaricum (Pawl.) Stearn
   Symphytum davisii subsp. naxicola (Pawl.) Stearn"
-'Plants of the World Online'® is from Kew Royal Botanic Gardens, London, England. It has information on all the world's known seed-bearing plants. With over 8.5 million items, Kew houses the largest and most diverse botanical and mycological collections in the world, **2019**. www.plantsoftheworldonline.org
    (Amorgos is the easternmost island of the Cyclades island group and the nearest island to the neighboring Dodecanese island group in Greece.)

Description of S. davisii

"*S. davisii Wickens, sp. nov. (new species):*
This is a new and distinct species, very similar in vegetative habit to S. ottomanun Friv. from central Europe and Balkans. *It differs in having large flowers and corolla scales not exserted; floristically it is similar to a large-flowered S. anatolicum Boiss.*
Runemark (*1967) records S. creticum (Willd.) Runem. from the Cyclades (Greek islands), including Amorgos, but no specimens of it have been seen from that island.
S. creticum and the neighbouring S. circinale Runem. are very similar in habit to S. davisii but with major floral differences, having long, revolute corolla lobes and hairs at the base of the filaments."
-'A Revision of Symphytum in Turkey and Adjacent Areas' by G.E. Wickens, Royal Botanic Gardens: Kew; Notes from the Royal Botanic Garden Edinburgh, Scotland, Volume 29, pages 157-180, **1969**. Includes Turkey, Bulgaria, Greece, Aegean Islands and Caucasia.
(* -'Studies in the Aegean Flora XI: Procopiana {Boraginaceae} included into Symphytum' by Hans Runemark, Botaniska Notiser {Botanical Notes}, Sweden, Vol 120, pages 84-94, 1967. I was unable to get this. If you have it, could you please send it to me.)

"**S. davisii Wickens**, *Notes Roy. Bot. Gard. Edinb. 29: 168 (1969);*
Stock fusiform. Plant covered with rather soft hairs up to 4 mm and short setiform straight and hooked hairs, often more or less sericeous-villous. Stem 10-30 cm (3.9-11.8 inch), unwinged. Lower leaves rather long-petiolate, the upper sessile, not decurrent. Calyx 9-13 mm, lobed to 2/3 to 7/8. Corolla (17 to) 18 to 22 mm, **white or white tinged with pink.** *Scales 5 to 6 (to 6.5) mm, lingulate, the marginal papillae not more than 1 1/2 times as long as wide, the lower scattered. Stamens 5 to 6 (to 6.5) mm; anthers (3 to) 3.5 to 4.2 mm, filament 1/2 as long as or equalling anther.*
Shady rocks. South Aegean region (Amorgos island). Greece."
-Flora Europaea, Volume 3: Diapensiaceae to Myoporaceae, editors Tutin, Heywood, Burges, Moore, Valentine, Walters and Webb. United Kingdom: Cambridge University Press, **1972**, page 104. (Symphytum L., pages 103-105 by B. Pawlowski.) (5-volume encyclopedia of plants, published between 1964 and 1993. 'Flora Europaea' describes all the national floras of Europe to identify any wild or widely cultivated plant to the subspecies level. It provides geographical distribution, habitat preference, and chromosome number.)
    (The Aegean Sea is a bay of the Mediterranean Sea located between the Greek and Anatolian peninsulas, i.e., between the mainlands of Greece and Turkey.)

Symphytum floribundum (includes Symphytum mediterraneum)

Current Botanical Nomenclature of S. floribundum

"***Symphytum floribundum Shuttlew. ex Nyman*** *(*Consp. Fl. Eur. 3: 509. 1881)*"
-'International Plant Names Index'® (IPNI), www.ipni.org, A database of the names and associated basic bibliographical details of seed plants, ferns and lycophytes, **2018**.
(* -'Conspectus Florae Europaeae: seu Enumeratio Methodica Plantarum Phanerogamarum Europae Indigenarum, Indicatio Distributionis Geographicae Singularum etc.' by C.F. Nyman, Orebro Sueciae {Sweden}, Volume 3, page 509, 1881.)

*"**Symphytum x floribundum Shuttlew. ex Buckn.** is an accepted name.*
* **Symphytum x mediterraneum Guss. ex F.W.Schultz**: Illegitimate Synonym."*
-The Plant List®, www.theplantlist.org from the World Checklist Database, **2018**.

*"**Symphytum mediterraneum W.D.J.Koch** =*
Symphytum tuberosum subsp. mediterraneum (W.D.J.Koch) P.Fourn., 1937."
-Global Biodiversity Information Facility® (GBIF), www.gbif.org, **2019**. International network and research infrastructure funded by the world's governments and aimed at providing anyone, anywhere, open access to data about all types of life on Earth.

> *"**Symphytum mediterraneum W.D.J.Koch** (*Syn. Fl. Germ. Helv. 1(2): 500. 1837)*
> ***Symphytum mediterraneum Guss. ex F.W.Schultz** (Flora 58: 218. 1875)"*

-'International Plant Names Index'® (IPNI), www.ipni.org, A database of the names and associated basic bibliographical details of seed plants, ferns and lycophytes, **2018**.
(* -'Synopsis Florae Germanicae et Helveticae' by Wilhelm Daniel Joseph Koch, Frankfurt, Germany, edition 1, Volume 1, Section 2, page 500, 1837.)

International Code of Nomenclature for Algae, Fungi, and Plants®:
Chapter V.- Rejection of Names, Article 52
"**52.1. A name, unless conserved (Article 14) or sanctioned (Article 15), is illegitimate and is to be rejected if it was nomenclaturally superfluous when published,** i.e. if the taxon to which it was applied, as circumscribed by its author, definitely included the type (as qualified in Article 52.2) of a name which ought to have been adopted, or of which the epithet ought to have been adopted, under the rules (but see Article 52.3)."

'Nomen illegitimum', abbreviated 'nom. illeg.', is Latin for 'illegitimate name'. A 'superfluous name' is often an 'illegitimate name'. In Latin it is 'nomen superfluum', abbreviated 'nom. superfl.'
A 'nomen illegitimum' is a validly published name, but one that contravenes (goes against) some of the articles laid down by the 'International Botanical Congress'. It could be illegitimate because:
> Article 52: it was superfluous at its time of publication, i.e., the taxon (as represented by the type) already has a name, or
> Articles 53 and 54: the name has already been applied to another plant (a homonym).
> (In biology, 'homonym' is a name for a taxon that is identical in spelling to another such name, that belongs to a different taxon. The term 'non' means that there is a homonym, and this indicates which taxon is correct. Or to put it another way: If an author wishes to indicate that a name has been used for a different taxon, then there is citation of the name and the author followed by the word 'non' or 'not' and the name of the author who first used the name.)

"S. floribundum Shuttlew. ex Buckn. in J. Linn. Soc., Bot. 41: 531. 1913."
-'Synopsis of Boraginaceae subfam. Boraginoideae Tribe Boragineae in Italy' is the same as 'Boraginaceae in Italy II' by L. Cecchi and F. Selvi, University of Florence, Firenze, Italy; Plant Biosystems: An International Journal Dealing with all Aspects of Plant Biology, Official Journal of the Societa Botanica Italiana, Vol 149, No. 4, pages 630-677, **2015**, Symphytum page 636-646.

*"**Symphytum x floribundum Shuttlew. ex Buckn.:***
> *This is a synonym of Symphytum x ferrariense C.Massal.*
> *First published in *J. Linn. Soc., Bot. 41: 531 (1913)."*

-'Plants of the World Online'® is from Kew Royal Botanic Gardens, London, England. It has information on all the world's known seed-bearing plants. With over 8.5 million items, Kew houses the largest and most diverse botanical and mycological collections in the world, **2019**. www.plantsoftheworldonline.org
(* -'A Revision of the Genus Symphytum, Tourn.' by Cedric Bucknall, (Mus. Bac. Oxon= Bachelor of Music, Oxford University), Journal of the Linnean Society of London, England, Botanical Journal, Volume 41, Issue 284, page 531, December 1913.)

Description of S. floribundum

*"**Symphytum floribundum, Shuttlw. in scheda (1871)**: Geographical Distribution: **France: Var. (southeast France).***
Judging from herbarium specimens, this species may attain a height of 12 dcm. (decimeters) (3.9 feet) or more. Its distinguishing characters are the partially decurrent leaves, the dense many-flowered racemes, the short pedicels of the flowers, and the campanulate calyx not divided to the middle.
Shuttleworth described flowers as yellowish white, but Schultz states that, in plants he cultivated, they were white.
In certain Symphytum species some plants produce abundance of fruit while others are completely sterile.
In one species, S. floribundum, Shuttlw., the fruit appears to be always undeveloped.
> *It is remarkable that the plant appears to be always sterile. Schultz only found very small abortive fruits, and in all the specimens which I have seen, some of them well past flowering, the fruits were in the same condition.*

I must here give my reasons for concluding that this species, being a branched plant of the Ramosa division, cannot be, as Schultz supposed, S. mediterranean Koch, *because as I hope to prove, the latter is an unbranched plant allied to S. tuberosum. Unfortunately, Koch's type cannot be found, and we therefore have to rely on his description. An analysis of the*

description will show that it does in fact belong to the Tuberosum and not, like S. floribundum, to the Ramosa.
These three species (floribundum, mediterranean, gussonei) have always been confused by Italian and French authors.
Grenier and Godron give Koch's description with the addition of that of the root of S. gussonei. Coste describes S. floribundum, but attributes to it the same kind of root, and Rouy & Foucaud combine the characters and citations of all three species. It is clear that none of these authors had seen a complete specimen, if any, of S. floribundum.
So far as I have been able to ascertain, this plant has not been gathered at Hyeres since 1875, and has only been found since at Aups in 1882, and at Ampus (Albert in Jahandiez Cat. Pl. du Var). I have not seen a specimen from the latter locality."
-'A Revision of the Genus Symphytum, Tourn.' by Cedric Bucknall, (Mus. Bac. Oxon= Bachelor of Music, Oxford University), Journal of the Linnean Society of London, England, Botanical Journal, Volume 41, Issue 284, pages 491-556, December **1913**.
(* -Flore de France; ou, Description des Plantes qui Croissent Naturellement en France et en Corse' by M. Charles Grenier and M. Dominique Alexandre Godron; Paris, France. Three volumes, 1848-1856. Symphytum officinale, tuberosum, mediterraneum and bulbosum, Volume 2, 1850, pages 511-512.)

"**Symphytum mediterraneum Koch,** Syn. Fl. Germ. ed. 1, p. 500. (1837) non Schultz nec Guss.:
Geographic Distribution: **France: Var. (southeastern France); Bouches du Rhone (southern France); Alpes-Maritimes (southeast France near Italy)?**
In Koch's type specimen, which cannot now be found, the root was wanting; but Italian and French authors have attributed to it a root similar to that of S. bulbosum, viz. slender and creeping, with tubercles at intervals, because S. mediterraneum Gussone (Symphytum Gussonei Schultz), has such a root.
The comparison of plants named S. mediterraneum Koch, from Aubagne (southern France) with authentic Sicilian specimens of **S. Gussonei Schultz,** which I had tried in vain to reconcile with each other and with descriptions, shows that **there are two distinct forms, allied to S. tuberosum Linn.,** but varying from it in different directions, and occupying different areas.
The Aubagne plants are distinguished from S. tuberosum by the lower and more compact growth, by the flexuous stem which is bent in alternate directions at each node, and by the comparatively large cordate or rounded lower leaves which are broader than, but not so long as, the succeeding ones.
The size of the flowers is only a little less than is usual in S. tuberosum, and in this point the Aubagne plants do not correspond with Koch's statement that they are only half the size of those of that species. This has been copied by succeeding authors as a character of S. mediterraneum, but, as **the size of the flower is often variable in the Symphytum genus**, this need not be an obstacle to the acceptance of these plants as Koch's species.
The winged amplexicaul petioles form a well-marked feature of the Aubagne plants, and in this Koch states that S. mediterraneum differs from S. tuberosum. This character is, however, also found in the latter species, as well as in other members of the group, more especially in the primordial leaves, but it is not generally so conspicuous as in S. mediterraneum."
-'A Revision of the Genus Symphytum, Tourn.' by Cedric Bucknall, (Mus. Bac. Oxon= Bachelor of Music, Oxford University), Journal of the Linnean Society of London, England, Botanical Journal, Volume 41, Issue 284, pages 491-556, December **1913**.

"**Symphytum floribundum R.J. Shuttlew. ex Buckn.,** Jour. Linn. Soc. London (Bot.) 41: 531 (1913).
(**Symphytum mediterraneum** F.W. Schultz, non Koch):
An asperous-hispid plant with the calyx lobed to 1/4 to 1/2, **dirty white corolla** with the lobes not recurved, scales with all the marginal papillae similar, about 1 1/2 times as long as wide, the connective not projecting beyond the thecae and abortive pollen, is **probably a hybrid of Symphytum officinale and some other species**; it was known from south France (Var Department in southeast France), but has not been found for 75 years."
-Flora Europaea, Volume 3: Diapensiaceae to Myoporaceae, editors Tutin, Heywood, Burges, Moore, Valentine, Walters and Webb. United Kingdom: Cambridge University Press, **1972**, page 104. (Symphytum L., pages 103-105 by B. Pawlowski.) (5-volume encyclopedia of plants, published between 1964 and 1993. 'Flora Europaea' describes all the national floras of Europe to identify any wild or widely cultivated plant to the subspecies level. It provides geographical distribution, habitat preference, and chromosome number.)

"A comparison is made of the main morphological characters of Symphytum floribundum R. J. Shuttlew. ex Buckn., S. x ferrariense C. Massal., S. x hyerense Pawl., S. orientale L. and S. officinale L., in order to elucidate (clarify) the status of the three first-mentioned taxa. The cytology and distribution of the taxa are also briefly discussed.
Symphytum floribundum and S. x ferrariense are both more or less intermediate between S. orientale and S. officinale.
S. x hyerense is intermediate between S. officinale and S. floribundum.
Thus these three taxa seem to belong to the same **hybrid (swarm) S. officinale x orientale. The correct binary name of the hybrid is S x ferrariense C. Massal.
Symphytum floribundum is lectotypified. The name S. mediterraneum Koch, often adopted for S. floribundum, probably belongs to the synonymy (synonym) of S. tuberosum L.**"
-'Taxonomical Status of Symphytum Floribundum and S. x Ferrariense (Boraginaceae)' by Arto Kurtto, Botanical Museum, University of Helsinki, Finland; Annales Botanici Fennici (Finnish Botanical Annals), Helsinki, Volume 18, No. 1, pages 13-21, **1981**. (I was unable to get a copy of this report. If you have one, could you please send it to me.)
(Morphology is a branch of biology that studies the form of living organisms, and the relationships between their structures. This includes both the outward and internal appearance {shape, structure, color, pattern, size}.)
(A hybrid swarm is a group of hybrids that have survived past the first hybrid generation, with interbreeding taking place between hybrid individuals and also backcrossing with its parent type. These plants are highly variable, with genes and

phenotypes of individuals ranging widely between 2 parent types. Hybrid swarms blur boundary between parent taxa.)
(Lectotype is a specimen chosen as the type of a species if the author of the name fails to designate a type.)

Goldsmith (See Symphytum grandiflorum in this section.)

Symphytum grandiflorum
For 'Hidcote' see section 'Details about Symphytum Species Hybrids' (Chapter 11).
See 'Symphytum cordatum' in this section.
See 'Symphytum ibericum' in this section.

Current Botanical Nomenclature of S. grandiflorum

"Symphytum grandiflorum DC. (*Prodr. [A. P. de Candolle] 10: 40. 1846)"
-'International Plant Names Index'® (IPNI), www.ipni.org, A database of the names and associated basic bibliographical details of seed plants, ferns and lycophytes, **2018**.
(* -'Prodromus Systematis Naturalis Regni Vegetabilis' by Alphonse Pyramus De Candolle, Volume 10, page 40, 1846. In Latin.)
  ('DC.' = Augustin Pyramus {Pyrame} De Candolle, 1778-1841, was a Swiss botanist. He documented hundreds of plant families and created a new natural plant classification system, 'De Candolle System'. For more about De Candolle, see sub-subsection 'De Candolle 1846' in subsection 'Symphytum Genus' in section 'Borage Family, Symphytum Genus'.)

"Symphytum grandiflorum DC.: accepted name."
-Catalogue of Life®: 2015 Annual Checklist, Online database of the world's known species of animals, plants, fungi and micro-organisms. It consists of a single integrated species checklist and taxonomic hierarchy., **2018**, www.catalogueoflife.org.
Search: www.catalogueoflife.org/annual-checklist/2015/search/all

"Symphytum grandiflorum DC. is an unresolved name."
-The Plant List®, www.theplantlist.org from the World Checklist Database®, **2018**.

"Symphytum grandiflorum DC.: Species accepted.
  = Symphytum abchasicum Trautv.
  = Symphytum ciscaucasicum Gviniaschvili
  = Symphytum cordatum Bieb.
  = Symphytum ibericum Stev.
  = Symphytum ibericum Stev. ex Bieb.
  = Symphytum ibericum var. abchasicum (Trautv.) Gviniaschvili"
-Global Biodiversity Information Facility® (GBIF), www.gbif.org, **2019**. International network and research infrastructure funded by the world's governments and aimed at providing anyone, anywhere, open access to data about all types of life on Earth.

Symphytum 'Goldsmith' (Symphytum grandiflorum 'Goldsmith', Symphytum ibericum 'Jubilee')

"Symphytum 'Goldsmith' has handsome, yellow-edged leaves and is much less vigorous than Symphytum ibericum.
Unlike some of their green-leaved relations, variegated Comfrey plants are vigorous but not invasive."
-Right Place, Right Plant: Over 1,400 Plants for Every Situation in the Garden by Nicola Ferguson. New York: A Fireside Book, Simon & Schuster, **2005**, page 123.
(Variegated means leaves that are edged or patterned in a second color, usually white, cream or yellow.)

"Symphytum 'Goldsmith':
USDA Zone 4-8. Sun: Full sun to part shade.
Height: 0.75 to 1.00 feet (0.22 to 0.30 meter). Spread: 1.00 to 1.50 feet (0.30 to 0.45 meter).
'Goldsmith' is generally rather restrained and can be easily controlled.
'Goldsmith' is a variegated hybrid Comfrey cultivar (parentage unknown) *that typically grows in a low-spreading clump. It features light green crinkled leaves to 4 inch (10 cm) long, that are edged with golden yellow.*
Bell-shaped, bluebell-like, pink to blue to white flowers *appear in drooping clusters (scorpiod cymes) in mid-spring to early summer (generally May to June).*
'Goldsmith' is sometimes listed and sold as a cultivar of Symphytum grandiflorum.
'Goldsmith' is also synonymous with Symphytum ibericum 'Jubilee'.
Garden Uses: Borders. Naturalize in woodland gardens, shade gardens, cottage gardens or wildflower meadows."
-Missouri Botanical Garden, www.missouribotanicalgarden.org, Saint Louis, Missouri, founded in 1859, the United States' oldest botanical garden. 'Plant Finder' lists over 7,500 plants, **2018**.

*A **'variety'** consists of more or less recognizable entities within species that are not genetically isolated from each other, below the level of subspecies, and are indicated by the abbreviation 'var.' in the scientific name.*

*The words **'variety'** and 'form' are not synonyms for the word cultivars according to the 'International Code of Nomenclature for Cultivated Plants'. The Code considers these terms botanical classifications. The 'Association of Official Seed Certifying Agencies' (AOSCA) considers the terms **'cultivar' and 'variety'** equivalent."*
-USDA National Plant Materials Manual, 190-V-NPMM, Fourth Edition, Part 542 Acronyms, 542.2 Plant Nomenclature, July 2010.

"Symphytum 'Goldsmith' (V): Synonyms:
  Symphytum grandiflorum 'Goldsmith', Symphytum ibericum 'Jubilee',
  Symphytum ibericum 'Variegatum'.
'Goldsmith' is a clump-forming, variegated cultivar with coarse, hairy foliage marked, with green, gold and cream. Short tubular flowers in pale blue and cream are borne in spring.
Ultimate height: 0.1-0.5 metres (3.9-19.6 inch). Ultimate spread: 0.1-0.5 metres.
Time to ultimate height: 2-5 years.
  Hardiness: H7. Hardy in the severest European continental climates (less than -20 C = -4 F)."
-Royal Horticultural Society, London, England, www.rhs.org.uk. Inspiring everyone to grow, **2018**.

Description of S. grandiflorum

*"**Symphytum grandiflorum DC.** Prodr. x. p. 40. (1846):*
*Geographical Distribution: **Russia: West Transcaucasia. Turkey: Armenia.***
S. grandiflorum is easily recognised by the decumbent barren shoots with broadly ovate, cordate leaves.
  In the second year the main axis of the shoot is prolonged into an erect flowering stem, while an axillary branch remaining decumbent, causes the main stem to appear lateral.
*The root is irregularly cylindrical and tortuous (not simple as stated by Kusnezow), but copiously branched and far creeping. Its branches give rise to numerous stems which form large tufts (clumps, bunches). **In the autumn the stems attain 3-5 dcm (decimeter = 11.8 to 19.6 inch) in length, and the leaves increase greatly in size.***
  Stevens states that he had a specimen with a subrotund radical leaf half a foot (= 6 inches = 15 cm) long and broad which one would scarcely believe could belong to the same plant.
The corolla is very variable in size. *De Candolle described it as 18-20 mm long, but Steven states that the largest flower he had seen was only 12 mm long. In the plant from Erzerum, Turkey at Kew, London, England they are 20-22 mm, and in those from the Bristol and Cambridge Gardens, England, they are about 18 mm long.*
  The Bristol plant was already in fruit at the end of May 1912, and continued flowering throughout the year.
The var. (variety) abchasicum *differs from the type in the cauline leaves being rounded or gradually attenuated at the base, and in the calyx-segments being nearly as long as the corolla-tube. I have seen no specimens."*
-'A Revision of the Genus Symphytum, Tourn.' by Cedric Bucknall, (Mus. Bac. Oxon= Bachelor of Music, Oxford University), Journal of the Linnean Society of London, England, Botanical Journal, Volume 41, Issue 284, pages 491-556, December **1913**.
  (Decumbent means a plant or part of a plant lying along the ground with the end curving up.)

*"**Symphytum grandiflorum:***
*Occasionally found as a garden throw-out, **this species differs most conspicuously in its habit. It is low-growing (up to 12 inches = 30 cm high) with extensively creeping roots and decumbent barren shoots.***
The leaves are small, up to about 2 inches (5 cm) long, round-based on long stalks, but as the plant gets older the leaves may become three or four times this length. It is roughly hairy.
It flowers in April and May, the buds being brick-red and the opened corolla cream. *The calyx-teeth are three times the length of the tube but are not so sharply pointed as Symphytum tuberosum.*
We have seen specimens from Abbey Wood in Kent (southeast England), but it is recorded from Surrey (southeast England) localities not far outside our area (London, England)."
-'The Comfreys of the London Area' by E.B. Bangerter and B. Welch, Department of Botany, British Natural History Museum, London, England; The London Naturalist: London Natural History Society, England, Volume 33, pages 55-58, **1954**.

*"**S. grandiflorum DC. is another early-flowering species** which is often grown in gardens in Britain and has become naturalised in woods and hedgebanks.*
It is sometimes confused with S. tuberosum but may be readily distinguished by its spring flowering (with often a second crop of flowers in October and November), and by the long creeping sterile shoots which root at intervals."
-'The Genus Symphytum in Britain' by T.G. Tutin, University College of Leicester, England; Watsonia: Journal of the Botanical Society of the British Isles, Volume 3, pages 280-281, February **1956**. (Watsonia was the journal of the 'Botanical Society of Britain and Ireland' from 1949 to 2010. It is now called 'New Journal of Botany'.)

*"**S. grandiflorum DC.:** Hispid perennial with long, slender rhizomes. Lower leaves elliptic to ovate, long-petioled; blade 5-10 cm (1.9-3.9 inches); cauline leaves smaller, uppermost sessile.*
Calyx 4-5 mm, accrescent, divided nearly to base, teeth linear-lanceolate, obtuse.
*Corolla 10-15 mm, **yellowish-white**. Style about 20 mm, persistent. Fruiting calyx about 10 mm. Flowers 4-5 and sporadically later.*
Hemicryptophyte Hs: *Semi-rosette hemicryptophytes, with leafy stems but the lower leaves larger than the upper ones and the*

basal internodes shortened.
Introduced. Not infrequently naturalized in hedges and woods in south England and the Midlands.
Sometimes mistaken for S. tuberosum. *Native of the Caucasus."*
-Flora of the British Isles, Second Edition by A.R. Clapham, T.G. Tutin and E.F. Warburg. England: Cambridge University Press, **1962**, (first edition 1952).
(Midlands is a cultural and geographic area across central England that corresponds to early medieval Kingdom of Mercia.)
(The Caucasus Mountains are located at the border of Europe and Asia, between the Black Sea and Caspian Sea and occupied by Russia, Georgia, Azerbaijan, and Armenia.)

"*S. grandiflorum DC., Prod. 10 :40 (1846).*
var. grandiflorum *Ic.: Mem. Acad. Imp. Sci. St. Peters. ser. 8, 25(5):t. 1, fig. 12 (1910).*
      Synonym: *S. cordatum sensu Bieb., FL Taur.-Cauc. 1 :130 (1808) non Willd.*
                        S. ibericum Steven in Bieb., Fl. Taur.-Cauc. 3 :647 (1819).
      Range: Caucasia, Turkey (northeast Anatolia).
var. abchasicum *(Trautv.) Kusn. in Mem. Acad. Imp. Sci. St. Peters. ser. 8, 25(5): 46 (1910). Ic.: l.c. t.1, figs. 10, 13 & 14.*
      Synonym: *S. abchasicum Trautv. in Bull. Soc. Nat. Mosc. 43:72 (1870).*
                        S. iberIcum var., Steven, Observ. Asperif 579 (1851).
      Range: Caucasia."
-'A Revision of Symphytum in Turkey and Adjacent Areas' by G.E. Wickens, Royal Botanic Gardens: Kew; Notes from the Royal Botanic Garden Edinburgh, Scotland, Volume 29, pages 157-180, **1969**. Includes Turkey, Bulgaria, Greece, Aegean Islands and Caucasia.

"*Polygonum affine (fleece flower)* **grows best in partial shade**, *like Lamium maculatum (spotted dead-nettle), Oxalis oregana (redwood sorrel) and* **Sympthytum grandiflorum:**
These are the very cream of rapid covers for growing under large shrubs and trees, and save all work.
The dense carpets of broad, hairy leaves of the Symphytum, a small-growing Comfrey, are covered with cream bells in spring.
Beautiful, thoroughly weed-proof, and vigorous are **Symphytum grandiflorum and its Hidcote hybrids.**
      The whole subject of ground-cover plants is so wrapped up with nature and naturalness that one is apt to forget their uses in formal design. As a savings in labour certain ground-cover could be of immense benefit.
      The most efficient ground-cover plants could be planted in large formal areas and provide the answer beautifully. In the outlying areas Symphytum grandiflorum would be more suitable.
Leaf height is 7 inches (17 cm). Plant them 2 feet (0.6 meters) apart. Leaves are rich green, broad, pointed, hairy. From the Caucasus."
-Plants for Ground-Cover by Graham Stuart Thomas, V.M.H., Gardens Advisor to the National Trust. London, England: J.M. Dent & Sons Ltd, **1970**, pages 128, 129, 194, 219, 226.

"**The only really good flowering species among the sixteen in this Symphytum genus of borage and anchusa family, the Boraginaceae is S. grandiflorum, which is a ground cover plant, tolerant of shade, with stems that hug the soil.**
From March to June, it carries **pale yellow, bell-like flowers** *1/2 to 3/4 inch (10-15 mm) long on branched stalks rising about 7 inches (17 cm) above the mat of small, deeply veined leaves."*
-Fertility Gardening: The Organic Way to Make Your Garden Grow by Lawrence D. Hills. London, England: Cameron & Tayleur Books Limited, **1980**.

"*Fresh material of S. tuberosum contained the triterpene isobauerenol, but in herbarium material isobauerenol was lacking.*
In S. grandiflorum, neither fresh nor dried material contains isobauerenol. *In herbarium material of S. ibericum also no isobauerenol could be found."*
-'Chemo- and Karyotaxonomic Studies on Some Rhizomatous Species of the Genus Symphytum (Boraginaceae)' by T.A. Jaarsma, E. Lohmanns, H. Hendriks, T.W.J. Gadella and T.M. Malingre, Plant Systematics and Evolution, Volume 169, pages 31-39, **1990**.   (Triterpenes are made by animals, plants and fungi. It is the precursor to all steroids.)

"*S. grandiflorum DC. (S. ibiricum Steven) - Creeping Comfrey:*
Differs from S. 'Hidcote Blue' in unbranched stems to 40 cm (15.7 inch); **corolla pale yellow when open, pinkish-red earlier.**
Introduced; common in gardens and well naturalized in woods and hedges; scattered in central and south Britain, south Northumberland County (northeast England); Caucasus."
-New Flora of the British Isles: Identification of Wild Vascular Plants of the British Isles edited by Clive Stace. Cambridge, England: Cambridge University Press, **1991**, page 648.

"*Symphytum tuberosum versus S. grandiflorum:*
These two yellowish-flowered, patch-forming species may be confused when the above ground leafy stolons in S. grandiflorum DC. are mistaken for the underground rhizomes in S. tuberosum L. *(cf. {compare to} Stace's 'New Flora of the British Isles'; this can be quite difficult in practice).*
In flower, S. grandiflorum has the calyx divided nearly to the base with obtuse teeth, whilst S. tuberosum has acute teeth cut to about 3/4 way."

-Plant Crib: Handbook for Field Identification compiled by T.C.G. Rich and M.D.B. Rich with editorial assistance of F.H. Perring for the BSBI Monitoring Scheme. London, England: Botanical Society of the British Isles, 1988, page 74. There is also an updated **1998** edition.
      (A stolon is a creeping horizontal plant stem that takes root at points along its length, forming new plants.)

*"**Symphytum grandiflorum:** It is a ground-covering species that forms a low mound. It will spread vegetatively and to a certain degree by seed. **Large-flowered Comfrey is a fantastic ground cover for partial shade.**"*
-Edible Forest Gardens: Volume One: Vision & Theory by Dave Jacke and Eric Toensmeier. White River Junction, Vermont: Chelsea Green, **2005**, page 329.

*"**Symphytum grandiflorum (large-flowered Comfrey):**
Hardiness Zone: 4-8; Light: full sun to partial sun; Soil pH: Acid / Neutral / Alakaline.
Clumping herb. Height: 8-12 inches (20-30 cm); Width: 18 inches (45 cm) or more.
Growth rate: fast. Native region: Eurasia.
Habitats: disturbed, old fields, edges, other habitats.
      Areas recently disturbed, whether by humans or natural causes. These plants can be observed as urban or garden 'weeds' or roadside plants. Old fields that are in the process of becoming forest. Edges are boundaries between two habitats, here generally indicating the boundary between forest and more open fields.
Medicinal: excellent. Some or all parts of the plant are poisonous.
Dynamic accumulator, invertebrate shelter, nectary, ground cover.
Roots are rhizomatous: Underground stems send out shoots and roots periodically along their length. They can travel great distances, or stay close to the crown.
Nuisance: persistent. Very difficult to remove once planted, often resprouting from even tiny pieces of root."*
-Edible Forest Gardens: Volume Two: Design & Practice by Dave Jacke and Eric Toensmeier. White River Junction, Vermont: Chelsea Green, **2005**, page 490.

*"**Suggested Ground Covers for Georgia Landscapes:**
Comfrey Symphytum grandiflorum:
Sun or partial shade. **USDA Hardiness Zones 6, 7, 8.**
Mature height: 12-18 inches (0.30-0.45 meter). Mature spread: 3 feet (0.9 meter).
Growth rate: Medium. **Creamy white flowers in spring.**"*
-'Ground Covers for Georgia Landscapes: Circular 928' by Gary L. Wade, Extension Horticulturist; The University of Georgia Cooperative Extension, College of Agricultural and Environmental Sciences, Athens, Georgia, August **2011**.

*"**Symphytum grandiflorum DC (De Candolle) (Creeping Comfrey)** is a terrestrial perennial herbaceous species with large flowers belonging to Boraginaceae family. It is endemic (native) to the Caucasus region, particularly Georgia (country).
S. grandiflorum was described for the first time by Swiss botanists Augustin Pyramus De Candolle and published by his son Alphonse De Candolle later in 1846.
According to some published works, the main component of mucilage of S. grandiflorum are glucofructans (67%). Cellulose, uric acid, ketoses, aldoses, saccharose, starch and dextrins in minor extent are also reported."*
-'Investigation of Water-Soluble High Molecular Preparation of Symphytum Grandiflorum DC (Boraginaceae)' by Sopio Gokadze, et. al., from Tbilisi State Medical University, Georgia (country) or Madrid, Spain; Bulletin of the Georgian Natonal Academy of Sciences, Volume 11, No. 1, pages 115-121, **2017**.
(For more about De Candolle, see subsection '1824 De Candolle' in section 'Symphytum Species Classification' (Chapter 4).)

*"**Symphytum grandiflorum:**
Hardiness Zone: 5 to 8, Height: 1.00 to 1.50 feet (0.30-0.45 meter).
Spread: 1.50 to 2.00 feet (0.45-0.60 meter).
Sun: Full sun to part shade. **Tolerates close to full shade.**
This Comfrey can spread aggressively by creeping rhizomes. It is generally a coarse, hairy, rhizomatous perennial that is typically grown in borders and shade gardens for its dense attractive foliage and its spring flowers.
It typically forms a low-spreading foliage clump to 18 inches (0.45 meter) tall consisting of both ascending flowering stems and decumbent sterile stems. Crinkled, elliptic to ovate, medium to dark green leaves grow to 7 inches (17.78 cm) long on decumbent stems but to only 2 inches (5.08 cm) long on flowering stems.
Tubular, bluebell-like, creamy yellow to white flowers appear in drooping clusters (scorpiod cymes) in mid-spring to early summer.
Bloom Time: May to June. Cutting back stems promptly after flowering may encourage a rebloom."*
-Missouri Botanical Garden, Saint Louis, Missouri, founded in 1859, the United States' oldest botanical garden. 'Plant Finder'® lists over 7,500 plants, **2018**.

*"**Symphytum grandiflorum 'Sky-Blue-Pink':** This unusual form of the old cottage garden plant (Symphytum grandiflorum) is **one of the first spring flowers to appear in March.** Short, rough stems bearing hairy leaves carry terminal bunches of **long pink buds opening to blue and white tubular flowers.**"*
-Plant World Seeds: 3500 Seed Varieties, www.plant-world-seeds.com, Newton Abbot, Devon, England, **2018**.

(I do not have any connection with this company.)

S. grandiflorum Chromosomes

"The chromosome number 2n = 60, first counted by Gviniashvili (*1969) was confirmed.
Wickens (1969) placed S. grandiflorum and S. cordatum (2n = 60, 120) in the Section Cordata; this seems justified from the cytological point of view.
On the other hand, a morphological relationship between S. grandiflorum and the species of the section Lingulata cannot be denied."
-'Cytotaxonomic Studies in the Genus Symphytum VIII: Chromosome Numbers and Classification of Ten European Species' by T.W.J. Gadella and E. Kliphuis, Proceedings van de Koninklijke Nederlandse Akademie van Wetenschappen (Proceedings of the Royal Netherlands Academy of Arts and Sciences): Series C, Biological and Medical Sciences, Vol 81, pages 162-172, **1978**.
(* -'De Speciebus Caucasis Nonnulis Generis Symphytum L. {Sect. Lingulata Pawl.} Non Critica' by T.N. Gviniashvili; Notulae Systematicae ac Geographicae Instituti Botanici Thbilissiensis {Zametki po Sistematike i Geografii Rastenii}, Tbilisi, Georgia, Volume 27, pages 87-95, 1969. In Russian. If you have this in English, could you please send it to me.)

*"Cytologically **S. grandiflorum** (2n = 60) and S. ibericum (2n = 24) differ considerably."*
-'Chemo- and Karyotaxonomic Studies on Some Rhizomatous Species of the Genus Symphytum (Boraginaceae)' by T.A. Jaarsma, E. Lohmanns, H. Hendriks, T.W.J. Gadella and T.M. Malingre, Plant Systematics and Evolution, Volume 169, pages 31-39, **1990**.

*"**Symphytum grandiflorum DC.**: Chromosome Count is 2n = 60."*
-Tropicos®, www.tropicos.org, Missouri Botanical Garden, Saint Louis, Missouri; Plant database with 1.3 million scientific names and 4.4 million specimen records, **2018**.

S. grandiflorum and Alkaloids

*"In S. tuberosum subspp. {subspecies}, tuberosum and nodosum, **S. grandiflorum** and S. ibericum, the **presence of the pyrrolizidine alkaloids lycopsamine, echimidine and symphytine could be demonstrated.**"*
-'Chemo- and Karyotaxonomic Studies on Some Rhizomatous Species of the Genus Symphytum (Boraginaceae)' by T.A. Jaarsma, E. Lohmanns, H. Hendriks, T.W.J. Gadella and T.M. Malingre, Plant Systematics and Evolution, Volume 169, pages 31-39, **1990**.

"Only few papers have been published concerning chemical composition of S. grandiflorum.
It was documented to synthesis pyrrolizidine alkaloids of retronecine type lycopsamine, echimidine and symphytine. A methanolic alkaloids extract and a hexane extract containing triterpenes and phytosterols were obtained."
-'Investigation of Water-Soluble High Molecular Preparation of Symphytum Grandiflorum DC (Boraginaceae)' by Sopio Gokadze, et. al., from Tbilisi State Medical University, Georgia (country) or Madrid, Spain; Bulletin of the Georgian Natonal Academy of Sciences, Volume 11, No. 1, pages 115-121, **2017**.

Symphytum gussonei
See 'Symphytum tuberosum' in this section.

Current Botanical Nomenclature of S. gussonei

*"**Symphytum gussonei F.W.Schultz** (in *Arch. Fl. (1874) p. lviii; et in **Flora, lviii. (1875))"*
-'International Plant Names Index'® (IPNI), www.ipni.org, A database of the names and associated basic bibliographical details of seed plants, ferns and lycophytes, **2018**.
(* -'Archives de Flore {Archives de la Flore d'Europe}: Journal Botanique and Recueil Botanique, 2' by Friedrich Wilhelm Schultz, Wissembourg, France, page 27, 1874. In French.)
(** -'Flora oder Botanische Zeitung' {Flora or Botanical Newspaper}, Regensburg, Germany, Volume 58, page 218. Article 'Beitrage zur Flora der Pfalz', 1875. Pflaz is the Palatinate region in southwest Germany.)

*"**Symphytum gussonei F. Schultz**: accepted name."*
-Catalogue of Life®: 2015 Annual Checklist, Online database of the world's known species of animals, plants, fungi and micro-organisms. It consists of a single integrated species checklist and taxonomic hierarchy., **2018**, www.catalogueoflife.org.
Search: www.catalogueoflife.org/annual-checklist/2015/search/all

*"**Symphytum gussonei F.W.Schultz**: This species is accepted, and its native range is Sicilia."*
-'Plants of the World Online'® is from Kew Royal Botanic Gardens, London, England. It has information on all the world's known seed-bearing plants. With over 8.5 million items, Kew houses the largest and most diverse botanical and mycological collections in the world, **2019**. www.plantsoftheworldonline.org
(Sicilia or Sicily is the largest Mediterranean island. It is near the 'toe' part of Italy's 'boot'.)

Description of S. gussonei

"***Symphytum gussonei** Schultz*, in Arch. Fl. (1874) p. lviii:
Geographical Distribution: **Sicily. Galicia (northwest Spain).**
The creeping root, which is like that of S. bulbosum and S. zeyheri, alone distinguishes this from S. tuberosum.
> From S. mediterraneum it differs in the lower leaves being smaller than the succeeding ones, and from S. bulbosum and S. zeyheri in the included corolla-scales.

*Lojacono describes S. gussonei as a small plant with the lower leaves appressed to the ground, and few stem-leaves. He also states that the root is that of S. officinale, but, in my case, this must be an error. His plant is probably a small form of S. gussonei.
> S. nodosum Schur. is a plant 16-20 cm (6.2-7.8 inch) high, growing in rock-fissures at Surul, Gotzenburg, Transylvania, Romania. It resembles S. tuberosum except in the root, which is like that of S. bulbosum. Schur remarks 'An var. rupestris S. tuberosi insignis?'.

It is remarkable that plants apparently so similar as this and S. gussonei should occur in such different and widely separated localities as the lowlands of Sicily and the rocks of the Transylvanian mountains.
More complete examples are necessary in order to form an opinion as to the position of Schur's plant.
**Dr. Woloszczak's plant named S. nodosum from Skole, Ukraine is like S. tuberosum, but the root is wanting."
-'A Revision of the Genus Symphytum, Tourn.' by Cedric Bucknall, (Mus. Bac. Oxon= Bachelor of Music, Oxford University), Journal of the Linnean Society of London, England, Botanical Journal, Volume 41, Issue 284, pages 491-556, December **1913**.
(* -'Flora Sicula, o Descrizione delle Piante Vascolari Spontanee o Indigenate in Sicilia', 3 Volumes, by Pojero M. Lojacono. Palermo, Sicily, Italy, 1888-1909.)
(** -'Botanisches Zentralblatt', Volume 90, Leiden, Holland {Netherlands}, page 77, 1902. Paragraph about Eustachius Woloszczak.)

"***Symphytum gussonei** F.W. Schultz*, Arch. Fl. Eur. 58 (1874):
Like S. tuberosum but rhizome very slender, the tuberous portions more or less globose and very distant; stem slender; cauline leaves, 3-8.
Inflorescence with 4 to 10 (to 20) flowers; corolla 16-22 mm; scales 6-8.5 mm, up to 2 mm wide; stamens with filament 1/5 to 1/3 as long as anther; anthers (3.5) to 4 to 5 to (6) mm; connective projecting slightly beyond thecae. Sicily."
-Flora Europaea, Volume 3: Diapensiaceae to Myoporaceae, editors Tutin, Heywood, Burges, Moore, Valentine, Walters and Webb. United Kingdom: Cambridge University Press, **1972**, page 104. (Symphytum L., pages 103-105 by B. Pawlowski.) (5-volume encyclopedia of plants, published between 1964 and 1993. 'Flora Europaea' describes all the national floras of Europe to identify any wild or widely cultivated plant to the subspecies level.)

"***Symphytum gussonei** Schultz*: 2n = 96.
> **Sicily:** Vallone Cerasa, Mezzojuso (Province of Palermo), about 300 meter (984 feet), March 2000, Bottega, cultivated Herbarium Horti Botanici Pisani, Pisa, Italy.

S. gussonei is considred endemic (native) of Sicily, linked to S. tuberosum L. of the European continental areas.
The chromosme number 2n = 96 appears recorded for the first time.
The size of chromosomes ranges from 1.0 to 2.0 micrometer."
-'Mediterranean Chromosome Number Reports 11 {Report 1229: Symphytum gussonei}' by S. Bottega, F. Garbari and L. Peruzzi; Flora Mediterrana, edited by G. Kamari, C. Blanche and F. Garbari, Volume 11, pages 436-439, **2001**.

"***Symphytum gussonei** F.W.Schultz* in Arch. Fl.: 27. 1874, in Flora 58: 218. 1875.
= **S. tuberosum L. var. gussonei** (F.W.Schultz) Fiori in Fiori & Beg., Fl. Italia 2(3): 378. 1902."
-'Synopsis of Boraginaceae subfam. Boraginoideae Tribe Boragineae in Italy' is the same as 'Boraginaceae in Italy II' by L. Cecchi and F. Selvi, University of Florence, Firenze, Italy; Plant Biosystems: An International Journal Dealing with all Aspects of Plant Biology, Official Journal of the Societa Botanica Italiana, Vol 149, No. 4, pages 630-677, **2015**, Symphytum pages 636-646.

"***Symphytum gussonei** Schultz*, Arch. Fl.: 27. 1874
> = **Symphytum mediterraneum Gussone non Koch**, Florae Siculae Synopsis: 792. 1844
> = **Symphytum tuberosum var. australis Strobl**, Flora: 624-625. 1884

Chromosome Number: 2n = 96 (*Bottega et al., 2001). Obtained on material from two stations (locations) in Sicily: Madonie, Mongiarrati (Palermo), along a dry stream 166/2000 HBP; Vallone Cerasa, Mezzojuso (Palermo) 167/2000 HBP.
Ecology: Lives in the woods, on acid, mature soils, from 300 to 1000 meters (984-3280 feet) of altitude.
Organic Form: Rhizomatous geophyte (perennial).
Phenology: It blooms from March to April, with cases of late flowering in the month of May.
General Distribution: It is a Sicilian endemic (native) mainly spread over the Madonie, locus Classicus, and the Nebrodi."
-'Il Genere Symphytum L. (Boraginaceae) in Italia: Revisione Biosistematica' (The Genus Symphytum L. Boraginaceae in Italy: A Biosystematic Revision) by Stefania Bottega and Fabio Garbari, Webbia: Journal of Plant Taxonomy and Geography, Volume 58, No. 2, pages 243-280, **2003**. (If you have an English translation of this article, I would appreciate having a copy.)
(* -'Mediterranean Chromosome Number Reports 11 {Report 1229: Symphytum gussonei}' by S. Bottega, F. Garbari and L. Peruzzi; Flora Mediterrana, edited by G. Kamari, C. Blanche and F. Garbari, Volume 11, page 436, 2001.)

(Phenology is the relation between climate and periodic biological phenomena such as plant flowering, leafing and breeding.)

Symphytum hajastanum (Hajastanian Comfrey)

Current Botanical Nomenclature of S. hajastanum

"Symphytum hajastanum Gviniashvili: accepted name."
-Catalogue of Life®: 2015 Annual Checklist, Online database of the world's known species of animals, plants, fungi and micro-organisms. It consists of a single integrated species checklist and taxonomic hierarchy., **2018**, www.catalogueoflife.org. Search: www.catalogueoflife.org/annual-checklist/2015/search/all

"Symphytum hajastanum Gvin. (*Zametki Sist. Geogr. Rast. 26: 73. 1967 ; Not. Syst. (Tbilisi))"
-'International Plant Names Index'® (IPNI), www.ipni.org, A database of the names and associated basic bibliographical details of seed plants, ferns and lycophytes, **2018**.
  (* -Zametki Sist. Geogr. Rast. 26: 73. 1967 ; Not. Syst.; {Tbilisi}. Notes: U.S.S.R. {Caucas.}. Illus. = 'Zametki po Sistematike i Geografii Rastenii'. Published by 'Tbilisis Botanikuri Instituti', Georgia, Vol 26, 1967. Russian & Georgian.)

Description of S. hajastanum

"S. hajastanum Gviniaschvili in Notul. Syst. Geogr. Inst. Bot. Tbilisi fasc. 26:73 (1967). Ic: l.c. t. 1 & 2.
Perennial, taproot fusiform. Plant greenish-grey, softly hispid. Stem solitary, 40-75 cm (15.7-29.5 inch) tall, erect somewhat branched. Lower leaves lanceolate, 9-15 cm (3.5-5.9 inches) long, 2.5-4 cm (0.9-1.5 inches) broad, apex strongly acuminate, base cuneate, attenuated into long petiole.
Terminal racemes many-flowered; rachis and pedicels with curved hairs, densely puberulous and beset with long setae; pedicels 5-8 mm long, slender. Calyx 7-11 mm long, puberulous, with curved hairs and dense setae; calyx divided to base, segments linear, acute, subequal, accrescent in fruit, enlarging to 15 mm.
Corolla 14-19 mm long, **blue,** *tubular, limb narrow-infundibuliform, twice as long as calyx; corolla scales ligulate, apex obtuse, shorter than stamens. Anthers almost equalling filaments or somewhat shorter. Style exserted.*
Nutlets dark, 4.5-5 mm, with dense minute tuberculae and coarsely venose-areolate.
Near S. peregrinum Ledeb., S. hajastanum differs in the considerably narrower, lanceolate and strongly cuneate leaves, *the calyx divided almost to the base into linear segments, and the corolla scales shorter than the stamens.*
It is well distinguished from S. asperum *by its pubescence and larger calyces with acute linear segments (*Gviniaschvili 1967)."*
-'A Revision of Symphytum in Turkey and Adjacent Areas' by G.E. Wickens, Royal Botanic Gardens: Kew; Notes from the Royal Botanic Garden Edinburgh, Scotland, Volume 29, pages 157-180, **1969**. Includes Turkey, Bulgaria, Greece, Aegean Islands and Caucasia.
(* -'Species Nova Generis Symphytum L. ex Armenia' by TS.N. Gviniaschvili; Notulae Systematicae ac Geographicae Instituti Botanici Thbilissiensis {Zametki po Sistematike i Geografii Rastenii}, Tbilsi, Georgia, Volume 26, pages 73-75, (1967). In Russian. If you have this in English, could you please send it to me.)

"Taxon Name: Symphytum hajastanum Gvin.:
Common Name: *Hajastanian Comfrey.*
Taxonomic Notes: *Symphytum hajastanum Gvin. differs from* **closely related S. asperum Lepechin** *by morphology of pollen grains, ratio between size of calyx and corolla.*
Red List Category & Criteria: *Endangered B2ab(i,ii,iii,iv).*
Threats: *This species is listed as Endangered in view of its Area of Occupancy (AOO) estimated to be 8 square km (3 square miles), its existence in only two locations and an inferred and projected continuing decline in the EOO (Extent of Occurrence), AOO, the area extent and quality of habitat, and the number of locations and subpopulations, caused by habitat loss and degradation related to shifting agriculture and nomadic livestock farming, as both localities fall in an area of intensive agricultural activities. A part of the population grows in the buffer zone of the Sevan National Park, Armenia.*
Geographic Range Description: *This species is endemic (native) to Armenia. It occurs in Semyonovka in the Sevan floristic region and in Garni in the Yerevan floristic region.*
Population: *It is very rare. There are two subpopulations separated by a distance of 30 km (18.6 miles).*
Current Population Trend: *Unknown.*
Habitat and Ecology: *It grows on subalpine meadows and in shrublands, between 1,400-2,200 meter (4593-7217 feet) asl (above sea level).*
Conservation Actions: *No conservation measures are in place. The species occurs in the buffer zone of the Sevan National Park. Research on population numbers and range, population trends monitoring and improved management of protected areas are needed."*
-'Symphytum Hajastanum' assessment by K. Tamanyan, The IUCN Red List of Threatened Species®, www.iucnredlist.org, 8 pages, **2014**.

Symphytum Hidcote Blue (See section 'Details about Symphytum Species Hybrids' {Chapter 11}.)
Symphytum Hidcote Pink (See section 'Details about Symphytum Species Hybrids' {Chapter 11}.)

Symphytum ibericum (Creeping Comfrey, Dwarf Comfrey, Iberian Comfrey, Georgian Comfrey)
See 'Symphytum grandiflorum' in this section.

Symphytum ibericum 'Jubilee' = Symphytum ibericum 'Variegatum' = Symphytum grandiflorum 'Goldsmith'.

Current Botanical Nomenclature of S. ibericum

*"**Symphytum ibericum Stev.**: ambiguous synonym for Symphytum grandiflorum DC.*
***Symphytum ibericum Stev. ex Bieb.**: ambiguous synonym for Symphytum grandiflorum DC.*
***Symphytum ibericum var. abchasicum (Trautv.) Gviniaschvili**: synonym for Symphytum grandiflorum DC.*
Infraspecific taxon."
-Catalogue of Life®: 2015 Annual Checklist, Online database of the world's known species of animals, plants, fungi and micro-organisms. It consists of a single integrated species checklist and taxonomic hierarchy., **2018**, www.catalogueoflife.org.
Search: www.catalogueoflife.org/annual-checklist/2015/search/all
(Infraspecific is at a taxonomic level below that of species, e.g., subspecies, variety, cultivar, or form. Latin names at this level require the addition of a term denoting rank.)

*"**Symphytum ibericum Steven** (Bull. Soc. Imp. Naturalistes Moscou 24(1): 579 (-580). 1851)"*
-'International Plant Names Index'® (IPNI), www.ipni.org, A database of the names and associated basic bibliographical details of seed plants, ferns and lycophytes, **2018**.

*"**Symphytum ibericum Stev.** and **Symphytum ibericum Stev. ex Bieb.**, Creeping Comfrey:*
*Species is heterotypic synonym. **Synonym of Symphytum grandiflorum DC.**"*
-Global Biodiversity Information Facility® (GBIF), www.gbif.org, **2019**. Iternational network and research infrastructure funded by the world's governments and aimed at providing anyone, anywhere, open access to data about all types of life on Earth.
(In botanical nomenclature, a heterotypic synonym {taxonomic synonym} exists when a taxon is reduced in status and becomes part of a different taxon.)

Description of S. ibericum

*"**Hardy Herbaceous Plants for a Border Shaded by Trees:***
***Symphytum ibericum**, a plant with cream coloured drooping flowers.*
From 9 to 12 inches (22.8-30.4 cm) high. Grows very freely."
-'The Gardeners Chronicle: A Weekly Illustrated Journal of Horticulture and Allied Subjects', New Series, January to June 1877: Volume VII, July to December **1877**: Volume VIII. London, England. This excerpt is from December 1, 1877, page 694.

*"**S. ibericum Steven**, Bull. Soc. Nat. Moscou 24(1):579 (1851).*
***Rhizome creeping, rather slender,** funiculiform, producing more or less ascending non-flowering and flowering stems 12-40 cm (4.7-15.7 inches). Stems and leaves with very short hairs and setae up to 3 mm. Leaves ovate or elliptical, slightly cordate or rounded at base, the lower long-, the upper short-petiolate.*
*Calyx 4-7 mm, lobed to 2/3 to 5/6. Corolla 14-15 to 18-19 mm, **pale yellow**. Scales lingulate, rounded or sometimes emarginate at apex, the marginal papillae about as long as wide, the lower scattered, the upper dense. Stamens with filament 2/3 as long as the anther or more.*
***Naturalized in hedges and grassland in southern and eastern England (Britain) (Caucasian region).**"*
-Flora Europaea, Volume 3: Diapensiaceae to Myoporaceae, editors Tutin, Heywood, Burges, Moore, Valentine, Walters and Webb. United Kingdom: Cambridge University Press, **1972**, page 105. (Symphytum L., pages 103-105 by B. Pawlowski.) (5-volume encyclopedia of plants, published between 1964 and 1993. 'Flora Europaea' describes all the national floras of Europe to identify any wild or widely cultivated plant to the subspecies level. It provides geographical distribution, habitat preference, and chromosome number.)

*"**S. grandiflorum DC. (S. ibiricum Steven) - Creeping Comfrey:***
Differs from S. 'Hidcote Blue' in unbranched stems to 40 cm (15.7 inch); corolla pale yellow when open, pinkish-red earlier. Introduced; common in gardens and well naturalized in woods and hedges; scattered in central and south Britain, south Northumberland County (northeast England); Caucasus."
-New Flora of the British Isles: Identification of Wild Vascular Plants of the British Isles edited by Clive Stace. Cambridge, England: Cambridge University Press, **1991**, page 648.

*"**Symphytum ibericum** (Comfrey): Hardy perennial, **USDA Zone 5-9**. Good in heavy, clay soils.*
Height: 9-12 inch (23-30 cm).
Flowering time: late spring to early summer. Flower color: pale cream.

Its 1/2 inch (1 cm)-long flowers are carried in branched heads.
Hardly an exquisite embellishment to a garden, Symphytum ibericum is useful in certain places, including the ground beneath shrubs or trees.
The plant tolerates dry shade and, as long as the soil is not very dry, **it will creep densely and rapidly, soon forming a carpet of coarse, veined leaves,** *each leaf up to 10 inch (25 cm) long. It spreads at least 24 inch (60 cm) wide and is potentially invasive, especially in rich, moist soil."*
-Right Place, Right Plant: Over 1,400 Plants for Every Situation in the Garden by Nicola Ferguson. New York: A Fireside Book, Simon & Schuster, **2005**, page 123.

"***Symphytum ibericum* Morphological Properties:**
Perennial. Stem 27-38 cm (10.6-14.9 inch). Leaves ovate to ovate-lanceolate. Basal leaves petiolate, cauline leaves shortly petiolate. Inflorescence cymes. Calyx 4-5 mm, hairy. Corolla cream, 14-15 mm. Nutlets 2-3 mm, brown and minutely tuberculate.
Anatomical Properties:
A transverse section taken from the root *was observed as follows: Periderm is multilayered. Secondary cortex is multilayered. Cortex cells are 31.75 +- 4.75 x 23.75 +- 2.48. Cambium cells are distinguishable and 1-3 layers. Xylem is composed sclerenchymatic cells and tracheary elements. The pith consists of primary xylem elements.*
A transverse section taken from the middle part of the stem *was studied. Cuticle layer is thin and trichomes are sparse. Epidermal cells consist of a single layer and orbicular or rectangular. These cells are 22.5 +- 4.53 x 16.5 +- 1.24. Two to three layers of collenchyma located under the epidermis. Cortex cells are 57.25 +- 7.66 x 35.75 +- 5.90 and multilayered. Endodermis is between cortex and vascular tissue. Endoderm cells with starch are one layer and distinguishable. Vascular bundles are of different sizes. Xylem and phloem elements are clear. Cambium is distinguishable. Phloem cells are 12 +- 1.33 x 8.25 +- 0.99. The pith consists of large parenchymatic cells."*
-'Micromorphology and Anatomy of Three Symphytum (Boraginaceae) Taxa from Turkey' by Oznur Ergen Akcin and Hilal Baki, Department of Biology, Sciences & Arts Faculty, Ordu University, Ordu, Turkey; Bangladesh Journal of Botany, Volume 36, No. 2, pages 93-103, December **2007**.

"***Symphytum ibericum:*** *12 inch (0.3 meter) tall, leaves shiny green and crinkled;* **early spring flowers pale yellow;** *good groundcover.* **USDA Zone 4-9.**"
-'The Sandy Mush Herb Nursery Handbook 9': Herbs, perennials, wildflowers, flowering plants, shrubs and trees, www.sandymushherbs.com, Leicester, North Carolina, July 27 **2004**.

"***Symphytum ibericum* Steven ex M.Bieb.:**
***Symphytum ibericum* Steven is often confused with *S. grandiflorum* DC.,** *which differs principally in having larger flowers: calyx 6-8 mm, corolla 20-24 mm.*
Taxonomic Sources:
*Ketskhoveli, N., Kharadze, A., and Gagnidze, R. 2001. Flora of Georgia. Vol. XIII. Tbilisi.
**Czerepanov, S.K. 1995. Vascular Plants of Russia & Adjacent States (in former USSR).*
Range Description: *Symphytum ibericum is widespread in Georgia (country) occurring in Abkhazeti, Svaneti, Racha-Lechkhumi, Samegrelo, Imereti, Adjara and Kartli floristic regions.
In Turkey the species is recorded in Trabzon, Coruh, Rize and Artvin regions.*
Country Occurrence: *Native in Georgia; Turkey.*
Habitat and Ecology: *Symphytum ibericum can be found in rhododendron scrub on shaded coastline areas as well as Colchic rainforest from the lower through middle montane (mountainous) zone."*
-'Symphytum Ibericum, Georgian Comfrey' assessment by S. Shetekauri and T. Ekim, The IUCN Red List of Threatened Species®, www.iucnredlist.org, 7 pages, **2014**.
(* -Flora of Georgia {country}, 16 Volumes, by N. Ketskhoveli, A. Kharadze and R. Gagnidze. Tbilisi, Georgia: Metsniereba, 1971-2011. In Georgian.)
(** -Vascular Plants of Russia and Adjacent States (the Former USSR) by S. K. Czerepanov. New York: Cambridge University Press, 1995.)
(Euxine-Colchic deciduous forest, with temperate broadleaf and mixed forests, is located along southern shore of Black Sea.)

"***Symphytum ibericum* Plant:**
Early rosettes of bright golden leaves *brighten up the spring garden.* **Later appear the pale blue and pink flowers.** *This form comes true from seed,* **viable seeds being very few and very difficult to collect!**
***Symphytum ibericum* Seeds:**
We sow most seeds in an unheated greenhouse and wait for natural germination, as many seeds wait for spring before emerging regardless of when they are sown. Spring sowing will obviously give them a full season of growth.
These seeds can sometimes exhibit very deep dormancy so do not discard the seed tray until more than a year has passed."
-Plant World Seeds: 3500 Seed Varieties, www.plant-world-seeds.com, Newton Abbot, Devon, England, **2018**.
(Rosette shape is similar to the arrangement of rose petals.)   (I do not have any connection with this company.)

S. ibericum Chromosomes

"S. ibericum Stev. ex. Bieb.:
Gviniashvili's earlier (*1969) report (2n = 24) could be confirmed by us on the basis of material kindly sent to us by Gviniashvili from west Georgia (country).
Grau (**1971) counted 2n = 60 in material obtained from Munchen Botanical Garden (Munich, Bavaria, Germany).
Wickens (***1969) regarded S. ibericum as a variety of S. grandiflorum (2n = 60), but with Gviniashvili we think that Wickens opinion should be rejected on morphological, cytological and geographical grounds."
-'Cytotaxonomic Studies in the Genus Symphytum VIII: Chromosome Numbers and Classification of Ten European Species' by T.W.J. Gadella and E. Kliphuis, Proceedings van de Koninklijke Nederlandse Akademie van Wetenschappen (Proceedings of the Royal Netherlands Academy of Arts and Sciences): Series C, Biological and Medical Sciences, Vol 81, page 162-172, **1978**.
(* -'De Speciebus Caucasis Nonnulis Generis Symphytum L. {Sect. Lingulata Pawl.} Non Critica' by T.N. Gviniashvili; Notulae Systematicae ac Geographicae Instituti Botanici Thbilisiensis {Zametki po Sistematike i Geografii Rastenii}, Tbilisi, Georgia, Volume 27, pages 87-95, 1969. In Russian. If you have this in English, could you please send it to me.)
(** -'Cytologische Untersuchungen an Boraginaceae II' (Cytological Investigations on Boraginaceae) by J. Grau, Mitteilungen der Botanischen Staatssammlung Munchen (Communications of the Botanical State Collection Munich, Germany), Volume 9, pages 177-194, 1971.)
(*** -'A Revision of Symphytum in Turkey and Adjacent Areas' by G.E. Wickens, Royal Botanic Gardens: Kew; Notes from the Royal Botanic Garden Edinburgh, Scotland, Volume 29, pages 157-180, 1969. Includes Turkey, Bulgaria, Greece, Aegean Islands and Caucasia.)

"S. ibericum Stev. ex Bieb.

2n=	*Cytology Reference*	*Kew Accession*	*Collector*	*Origin*	*Voucher*
60	82-1389	1982-4913	Mclintock s.n.	cultivated	K "

-'New Chromosome Numbers in Petaloid Monocotyledons and in Other Miscellaneous Angiosperms' by Margaret A.T. Johnson and P.E. Brandham, Jodrell Laboratory, Royal Botanic Gardens, Kew, Richmond, Surrey, England; Kew Bulletin, Volume 52, No. 1, pages 121-138, **1997**.

"Symphytum ibericum Steven: Chromosome Count is 2n = 24, 60."
-Tropicos®, www.tropicos.org, Missouri Botanical Garden, Saint Louis, Missouri; Plant database with 1.3 million scientific names and 4.4 million specimen records, **2018**.

S. ibericum and Alkaloids

*"In S. tuberosum subspp, tuberosum and nodosum, S. grandiflorum and **S. ibericum the presence of the pyrrolizidine alkaloids lycopsamine, echimidine and symphytine could be demonstrated.***
Fresh material of S. tuberosum contained the triterpene isobauerenol, but in herbarium material isobauerenol was lacking. In S. grandiflorum, neither fresh nor dried material contains isobauerenol.
In herbarium material of S. ibericum also no isobauerenol could be found."
-'Chemo- and Karyotaxonomic Studies on Some Rhizomatous Species of the Genus Symphytum (Boraginaceae)' by T.A. Jaarsma, E. Lohmanns, H. Hendriks, T.W.J. Gadella and T.M. Malingre, Plant Systematics and Evolution, Volume 169, pages 31-39, 1990.

Symphytum kurdicum

Current Botanical Nomenclature of S. kurdicum

"Authors: Pierre Edmond Boissier and Heinrich Carl Haussknecht.
Published In: *Plantarum Orientalium Novarum 2: 5. 1875. (Pl. Orient. Nov.).   **Annotation: as 'Kurdicum'** "
-Tropicos®, www.tropicos.org, Missouri Botanical Garden, Saint Louis, Missouri; Plant database with 1.3 million scientific names and 4.4 million specimen records, **2019**.
(* -'Plantarum Orientalium Novarum Ex Flora Orientalis' {Plants, Flowers Represent the New East Asia} by Pierre Edmond Boissier, Geneva, Switzerland, Part 2, **1875**. Symphytum sylvaticum and Symphytum kurdicum, Part 2, pages 4-5. In Latin.)

"Symphytum kurdicum Boiss. & Hausskn.:
This species is accepted, and its native range is east Turkey."
-'Plants of the World Online'® is from Kew Royal Botanic Gardens, London, England. It has information on all the world's known seed-bearing plants. With over 8.5 million items, Kew houses the largest and most diverse botanical and mycological collections in the world, **2019**. www.plantsoftheworldonline.org

"Symphytum kurdicum Boiss. & Hausskn. ex Boiss.: Species accepted."
-Global Biodiversity Information Facility® (GBIF), www.gbif.org, **2019**. International network and research infrastructure funded by the world's governments and aimed at providing anyone, anywhere, open access to data about all types of life on Earth.

*"**Symphytum kurdicum Boiss. & Hausskn. is an unresolved name.** The record derives from WCSP (in review) (data supplied*

on 2012-03-23) which does not establish this name either as an accepted name or as a synonym with original publication details: Pl. Orient. Nov. 2: 5 1875."
-The Plant List®, www.theplantlist.org from the World Checklist Database, **2019**.

"***Symphytum kurdicum* Boiss. & Hausskn.** in Boissier, Pl. Orient. Nov. 2: 5. 1875.
Rank: Species. **Status: Accepted.**
Occurrence:

Region	Status	Source
Turkey	native	Calculated
Asiatic Turkey	native	Reference

Reference: *Greuter, W., Burdet, H.M. & Long, G. - Med-Checklist 1 Geneve & Berlin 1984."
-The Euro+Med PlantBase®: The Information Resource for Euro-Mediterranean Plant Diversity. It integrates and evaluates information from 'Flora Europaea', 'Med-Checklist', 'Flora of Macaronesia', and from regional and national floras and checklists. It provides access to the total European flora of vascular plants in 222 plant families, **2019**. www.emplantbase.org and http://ww2.bgbm.org/EuroPlusMed/query.asp
(*-Med-Checklist: A Critical Inventory of Vascular Plants of the Circum-Mediterranean Countries by W. Greuter, H.M. Burdet and G Long, Organization for the Phyto-Taxonomic Investigation of the Mediterranean Area. Geneva, Switzerland: Editions des Conservatoire et Jardin botaniques de la Ville de Geneve; and Berlin, Germany: Botanischer Garten and Botanisches Museum Berlin-Dahlem, 1984.)

"***Boraginaceae Symphytum kurdicum* Boiss. & Hausskn. ex Boiss.** Pl. Or. Nov. Dec. ii. 5. Notes: Kurdist."
-'International Plant Names Index'® (IPNI), www.ipni.org, A database of the names and associated basic bibliographical details of seed plants, ferns and lycophytes, **2019**.

Distribution of S. kurdicum

"*Symphytum kurdicum*:
Potential Conservation Status: **Regional Endemic (native).**
Found in the following Iraq sites:
S6 (Sulaimani. Peramagroon w Homer Qawm & Shadala Valley),
S11 (Sulaimani. Qara Dagh), D5 (Dohuk. Garagu),
S33 (Sulaimani. Gmo Mountain), D2A (Dohuk. Ser Amadia).
Citation: *Fl Iranica 48. (1967)."
-'Key Biodiversity Survey of Iraq: 2010 Site Review' by 'Nature Iraq' and 'Iraq Ministry of the Environment Report', Sulaimani city in Kurdistan region, Iraq, NI-0311-01P, **2011**. This report has been prepared to summarize and inform partner agencies on the status and progress of the biodiversity initiatives.
(* -Flora Iranica: Boraginaceae, Issue 48 by Harald Riedl. Edited by K.H. Rechinger. Graz, Austria: Naturhistorisches Museum {Natural History Museum}, 1967. If you have an English translation, I would appreciate a copy. www.iranicaonline.org/articles/flora-iranica-)

"***Symphytum kurdicum* Boiss. & Hausskn in Boiss. was found to be widely distributed in Iraq,** while *Symphytum tuberosum* L. was limited."
-'A Comparative Systematic Study of the Genus Symphytum L. (Boraginaceae) with New First Record of the Species Symphytum Tuberosum L. from Iraq' by Adel Mohan Aday Al-Zubaidy and Sherzad Rasul Abdalla Tobakari, Plant Production Department, Technical College of Applied Sciences, Sulaimani Polytechnic University, Iraq; Plant Archives, Volume 18, No. 2, pages 2068-2076, November **2018**.

Description of S. kurdicum

"***Symphytum kurdicum* Boiss. & Hausskn,** in *Boiss. Fl. Orient. iv. p. 174, (1879):
Geographical Distribution: **Turkey: Kurdistan.**
Nearly related to *S. orientale* Linn., from which it differs in being more asperous, in the membranaceous leaves, the upper of which are petiolate, not sessile.
According to Boissier's description the calyx is divided to the middle when in flower, but in all the specimens I have seen it is divided to one-quarter or scarcely more than one-third.
In the plants collected by **P. Sintenis in 1888 the calyx-teeth are triangular, acute, one-third as long as the calyx, and in those collected by Bornmuller in 1893 they are lanceolate, subobtuse, one-quarter to one-third as long as the calyx."
-'A Revision of the Genus Symphytum, Tourn.' by Cedric Bucknall, (Mus. Bac. Oxon= Bachelor of Music, Oxford University), Journal of the Linnean Society of London, England, Botanical Journal, Volume 41, Issue 284, pages 491-556, December **1913**.
(* -'Flora Orientalis Sive Enumeratio Plantarum in Oriente, a Graecia et Aegypto ad Indiae Fines (Plants Flora, Flora East and the East, from Greece and Egypt to the Borders of India), Volume IV' by Edmond Boissier, Society Physics Geneva, Society Linnean London; Geneva, Switzerland, 1879. All in Latin.)
(** -Paul Ernst Emil Sintenis, born 1847 in Seidenberg, Oberlausitz, Prussia, lived to 1907. He was a German botanist and pharmacist. He collected plants in Rhodes, Cyprus, Italy, Istria, Turkey, Syria, Iraq, Turkmenistan, Iran, and Greece. His

herbarium is stored by Lund University in Sweden.)

"Flowers and Flowering Calyx of S. kurdicum:
The flowers are perfect, bisexual, determinate inflorescence and actinomorphic symmetry.
Sepals divided near the middle of calyx have teeth in the species S. kurdicum.
The average length of tube is 2.5 mm, with the calyx teeth dimensions average 3.5 x 1 mm.
Symphytum kurdicum has longer flowering period which extended from March to August.*"*
-'A Comparative Systematic Study of the Genus Symphytum L. (Boraginaceae) with New First Record of the Species Symphytum Tuberosum L. from Iraq' by Adel Mohan Aday Al-Zubaidy and Sherzad Rasul Abdalla Tobakari, Plant Production Department, Technical College of Applied Sciences, Sulaimani Polytechnic University, Iraq; Plant Archives, Volume 18, No. 2, pages 2068-2076, November **2018**.

*"**Symphytum kurdicum:***
Vegetatively similar to Symphytum orientale *but more asperous and leaves more membranous.*
Related to Symphytum orientale.
Flowers about 10, cymes more compact than S. orientale. Calyx 7-9 mm, accrescent to 12 mm in fruit, divided to about 1/4 to 1/2, lobes lanceolate, acute to subobtuse. ***Corolla white,*** *16-19 mm; scales 6 mm, broadly subulate, obtuse, equalling stamens. Nutlets 4 mm, obliquely curved, areolate, tuberculate.*
Fl. 4-5. Abies forest, shady rocks, cliffs and stream banks, 1200-2000 meter (3937-6561 feet) altitude.
North Iraq, West Iran. Ir.-Tur. (Irano-Turanian) *element."*
-'Symphytum Kurdicum: Kurt Kafesotu' published by Turkiye Bitkileri (Plants of Turkey), **2019**. Turkey has a great botanical richness with more than 9600 plant species. This website has been constructed to gather as much information as possible about native plants of Turkey. www.turkiyebitkileri.com
  ('Fl. 4-5' probably means flowering April to May.)
  ('Abies' are a genus of evergreen trees of temperate regions of the northern hemisphere that comprise the true firs.)

Symphytum leonhardtianum (See 'Symphytum tuberosum' in this section.)

Symphytum mediterraneum (See 'Symphytum floribundum' in this section.)

Symphytum naxicola
See 'Symphytum davisii' in this section.

Current Botanical Nomenclature of S. naxicola

*"**Symphytum naxicola Pawl.** (*Fragm. Florist. Geobot. 17(1): 21. 1971)"*
-'International Plant Names Index'® (IPNI), www.ipni.org, A database of the names and associated basic bibliographical details of seed plants, ferns and lycophytes, **2018**.
(* -Fragmenta Floristica et Geobotanica Polonica, W. Szafer Institute of Botany, Polish Academy of Sciences, Krakow, Poland, Volume 17, No. 1, 1971: 'Symphyta Mediterranea Nova vel Minus Cognita: Nowe lub malo znane srodziemnomorskie zywokosty' {New or Little Known Mediterranean Comfrey} by Bogumil Pawlowski, pages 17-37. If you have an English translation, I would appreciate a copy.)

*"**Symphytum naxicola Pawl.**: synonym for Symphytum davisii subsp. naxicola (Pawl.) W.T. Stearn.*
 Symphytum davisii subsp. naxicola (Pawl.) W.T. Stearn*: accepted name. Infraspecific taxon."*
-Catalogue of Life®: 2015 Annual Checklist, Online database of the world's known species of animals, plants, fungi and microorganisms. It consists of a single integrated species checklist and taxonomic hierarchy., **2018**, www.catalogueoflife.org.
Search: www.catalogueoflife.org/annual-checklist/2015/search/all
  (Infraspecific is at a taxonomic level below that of species, e.g., subspecies, variety, cultivar, or form. Latin names at this level require the addition of a term denoting rank.)

Description of S. naxicola

*"**S. naxicola Pawl.***, *Fragm. Fl. Geobot. 17: 21 (1971):*
Stock fusiform. Plant often more or less sericeous-villous, with rather soft hairs up to 4 mm and shorter straight and hooked hairs and setae. Stem 12-45 cm (4.7-17.7 inches), unwinged. Lower leaves rather long-petiolate, the upper sessile, not or scarcely decurrent.
Calyx 7.5-12 mm, lobed to 3/4 to 7/8. Corolla (14 to) 15 to 17 (to 18) mm, ***white****. Scales (3) to 3.5 to 4 to (5) mm, triangular-lanceolate, gradually narrowed to the apex, the margin densely papillose throughout, the papillae not more than 1 1/2 times as long as wide. Stamens 4-5 mm; anthers 2.5 to 4 mm; filament mostly shorter than anther.*
2n = 30. *South Aegean region (Naxos, a Greek island). Greece."*

-Flora Europaea, Volume 3: Diapensiaceae to Myoporaceae, editors Tutin, Heywood, Burges, Moore, Valentine, Walters and Webb. United Kingdom: Cambridge University Press, **1972**, page 104. (Symphytum L., pages 103-105 by B. Pawlowski.) (5-volume encyclopedia of plants, published between 1964 and 1993.)
> (The Aegean Sea is a bay of the Mediterranean Sea located between the Greek and Anatolian peninsulas, i.e., between the mainlands of Greece and Turkey.)

"***S. naxicola Pawl.: 2n = 30***, *Greece, Runemark, 1967.*"
-Flora Europaea: Check-List and Chromosome Index by D.M. Moore. Cambridge, England: Cambridge University Press, **1982**, page 179.

Symphytum nodosum
See 'Symphytum tuberosum' in this section.

"***Symphytum nodosum Schur*** *(Enum. Pl. Transsilv. 468. 1866)*"
-'International Plant Names Index' (IPNI), www.ipni.org, A database of the names and associated basic bibliographical details of seed plants, ferns and lycophytes, **2018**.
(* -'Enumeratio Plantarum Transsilvaniae, exhibens stirpes phanerogamas sponte crescentes atque frequentius cultas, cryptogamas vasculares, characeas, etiam muscos hepaticasque' (Flora of Transylvania) by Ferdinand Schur, Romania, 1866.)

"***Symphytum tuberosum subsp. nodosum*** *(Schur) So, *Acta Geobot. Hung. 4: 192 (1941):*
Rhizome slender, the tuberous portions irregular oblong and separated. Cauline leaves usually 3-7.
Inflorescence mostly with 1-9(-20) flowers. Corolla-scales 4.5-6(-6.5) mm, up to 2.5 mm wide. Stamens with filament 1/4-1/3(-1/2) as long as anther. Nutlets 2.5-3.5 mm.
2n = 18, 72, 96, 100. *Central and southeast Europe.*"
-'Symphytum L.' by B. Pawlowski with editors Tutin, Heywood, Burges, Moore, Valentine, Walters and Webb, Flora Europaea, Volume 3, pages 103-105, **1972**. The 'Flora Europaea' is a 5-volume encyclopedia of plants, published between 1964 and 1993 by Cambridge University Press, England. It describes all the national floras of Europe to identify any wild or widely cultivated plant to the subspecies level. It provides geographical distribution, habitat preference, and chromosome number.
(* - Acta Geobotanica Hungarica, Editio Instituti Botanici Universitatis Debrecen, Hungary, Volume 4, page 192, 1941.)

"***Symphytum nodosum Schur***:
> *synonym for Symphytum tuberosum subsp. angustifolium (A. Kerner) Nyman.*
Symphytum tuberosum subsp. nodosum (Schur) Soo:
> *synonym for Symphytum tuberosum subsp. angustifolium (A. Kerner) Nyman. Infraspecific taxon.*"
-Catalogue of Life: 2015 Annual Checklist, Online database of the world's known species of animals, plants, fungi and micro-organisms. It consists of a single integrated species checklist and taxonomic hierarchy., **2018**, www.catalogueoflife.org. Search: www.catalogueoflife.org/annual-checklist/2015/search/all
> (Infraspecific is at a taxonomic level below that of species, e.g., subspecies, variety, cultivar, or form. Latin names at this level require the addition of a term denoting rank.)

Norfolk Comfrey = S. asperum x S. orientale (see 'Details about Sympytum Species Hybrids' {Chapter 11}.)

Symphytum norvicense = S. asperum x S. orientale (see 'Details about Sympytum Species Hybrids' {Chapter 11}.)

Symphytum orientale (White or Soft Comfrey)

Current Botanical Nomenclature of S. orientale

"***Symphytum orientale Pall.*** *(Tabl. Phys. Topogr. Taur. 47; Bieb. Fl. Taur. Cauc. i. 129.)*
Symphytum orientale Pinard ex DC. *(Prodr. [A. P. de Candolle] 10: 40. 1846)*
Symphytum orientale L. *(*Sp. Pl. 1: 136. 1753)*"
-'International Plant Names Index'® (IPNI), www.ipni.org, A database of the names and associated basic bibliographical details of seed plants, ferns and lycophytes, **2018**.
(* -'Species Plantarum, Volume 1' by Swedish naturalist Carol {Carl} Linnaeus, page 136, 1753.)

"***Symphytum orientale L.***: *accepted name.*
Symphytum orientale Pall.: *ambiguous synonym for Symphytum tauricum Willd.*
Symphytum orientale Pinard ex DC.: *ambiguous synonym for Symphytum brachycalyx Boiss.*
Symphytum jacquinianumTausch: *synonym.*
Symphytum patens Fries: *ambiguous synonym.*

Symphytum violaceum Gaterau: synonym."
-Catalogue of Life®: 2015 Annual Checklist, Online database of the world's known species of animals, plants, fungi and micro-organisms. It consists of a single integrated species checklist and taxonomic hierarchy., **2018**, www.catalogueoflife.org. Search: www.catalogueoflife.org/annual-checklist/2015/search/all

*"**Symphytum orientale L.**
= Symphytum jacquinianum Tausch
= **Symphytum patens Fries***
= Symphytum violaceum Gaterau*
    Varieties:
        *Symphytum orientale var. angustior DC.
        Symphytum orientale var. orientale "*
-Global Biodiversity Information Facility® (GBIF), www.gbif.org, **2018**. International network and research infrastructure funded by the world's governments and aimed at providing anyone, anywhere, open access to data about all types of life on Earth.
(* -Symphytum patens Fr.: Novit. Fl. Suec. Mant. 2: 13. 1839. = 'Novitiarum Florae Sueciae Mantissae Altera, Additis Plantis in Norvegia Recentius Detectis, II.' {Flora of Sweden} by E.M. Fries, Supplement page 13, Uppsala, Sweden, 1839. All in Latin. Includes Symphytum patens and Symphytum officinale.)

Description of S. orientale

*"**Symphytum Orientale; Eastern Comfrey:** Leaves ovate, subpetioled. Root perennial; stalks two feet (0.6 meters) high; flowers in bunches, blue. They appear in March, but seldom produce seeds in England. Found growing by the sides of rivulets (small streams) **near Constantinopole (Istanbul, Turkey)."***
-'The Universal Herbal; or, Botanical, Medical, and Agricultural Dictionary; Containing an Account of All the Known Plants in the World, Arranged According to the Linnean system. Specifying the uses to which they are or may be applied, whether as food, as medicine, or in the arts and manufactures, with the best methods of propagation, and the most recent agricultural improvements' by Thomas Green, London, England, **1824**, page 641.

*"**Symphytum orientale Linn.** Sp. Pl. ed. I, p. 136. 1753), pro parte {in part}
Geographical Distribution: **Neighborhood of Constantinople (Istanbul, Turkey).**
This has been confused with S. tauricum, Willd. It greatly resembles that species, but is distinguished by the shortly dentate tubular calyx and the larger flowers.
From S. palaestinum Boiss., S. brachycalyx Boiss., and S. Bornmuelleri Bucknall, it differs in the larger, thicker, less asperous leaves which are often cordate at the base, and from the first two by the racemose, not dichotomous branching of the stem. De Candolle's var. Beta angustior belongs partly or entirely to S. palaestinum Boiss.*
Linnaeus combined this with Tournefort's blue-flowered plant, which probably included more than one species, under the name of S. orientale. *But Fries was of opinion that the blue-flowered plants which are found in Sweden, and which also contain several distinct forms, are the true S. orientale Linn.*
    *As Tournefort's excellent description and figure prove that his **S. constantinopolitanum is the white-flowered plant which is now accepted as S. orientale,** Fries was perhaps justified in his opinion, and it would have been better if the name of S. constantinopolitanum had been retained for this, and S. orientale for one of the blue-flowered species."*
-'A Revision of the Genus Symphytum, Tourn.' by Cedric Bucknall, (Mus. Bac. Oxon= Bachelor of Music, Oxford University), Journal of the Linnean Society of London, England, Botanical Journal, Volume 41, Issue 284, pages 491-556, December **1913**.

S. orientale L. *Sp. pl. (1753) 136; Prodr. X, 39; Wet places in forests.
European part: Bl. (Black Sea area) (near Odessa, Ukraine), M. Dnp. (Middle Dnieper) (S. Podolia, in Turchaninov herbarium). General distribution: Asia Minor, from Constantinople (Istanbul, Turkey) to Ankara (Turkey).*
This is a very ambiguous species for which Linnaeus cited two different species (synonyms) by Tournefort, one with blue flowers, the other with white from Constantinople. The first synonym probably refers to S. asperum Lepech. and the second to what De Candolle and Boissier regarded as S. orientale, though some older authors were of another opinion.
    *Roemer and Schultes (Syst. IV, 65), like Willdenow (Enum. H. Berol. I, 183), referred Tournefort's plant with the white flowers to S. tauricum Willd.
    Likewise, S. orientale in Hegis (*page 2223) is described as having leaves that taper gradually to petioles, without a cordate base.*
I accept S. orientale as the specimen which was described by Boissier: *the Wiedeman specimens from Byzantium (Istanbul, Turkey) and Ankara (Turkey) and the southern coast of Pontus (in Turkey) (according to Boissier from Bithynia).* **This S. orientale resembles S. caucasicum more than S. tauricum and in general it is intermediate between them."**
-'Flora of the U.S.S.R.', 30 volumes, 1934-1960, edited by V.L. Komarov; Botanical Institute of the Academy of Sciences of the U.S.S.R, Leningrad, Russia, with article by Mikhail Grigorevich Popov, in Volume 19, pages 279-291 Russian version, pages 207-216 English version, **1953**.
(* -'Illustrierte Flora von Mittel-Europa: Mit Besonderer Berucksichtigung von Deutschland, Osterreich und der Schweiz' {Illustrated Flora of Central Europe: With Special Consideration of Germany, Austria and Switzerland}, Volume 5, Part 3' by Gustav Hegi or Hegis, Zurich, Switzerland; published by J.F. Lehmann, Munich, Germany, 1926. Symphytum pages 2220-2230.

In German, 13 volumes written from 1908 to 1931.)
>(The Black Sea is between far-southeastern Europe and the far-western edges of Asia and Turkey.)
>(The Dnieper River is the fourth longest river in Europe, starting in Valdai Hills near Smolensk, Russia, {near Europe}, and going through Russia, Belarus and Ukraine to the Black Sea.) (Podolia or Podilia is in west-central and southwest parts of Ukraine and northeast Moldova.)

"S. orientale L. is locally naturalised in Britain. It may be recognised by its early flowering (April-May), soft hairs, broad leaves and short calyx-teeth, which are about half as long as the tube."
-'The Genus Symphytum in Britain' by T.G. Tutin, University College of Leicester, England; Watsonia: Journal of the Botanical Society of the British Isles, Volume 3, pages 280-281, February **1956**. (Watsonia was the journal of the 'Botanical Society of Britain and Ireland' from 1949 to 2010. It is now called 'New Journal of Botany'.)

*"**S. orientale L.:** A softly pubescent perennial up to about 70 cm (27.5 inches). Root fusiform, branched. Stems sparsely puberulent and pilose, branched. Leaves softly pubescent, ovate or oblong, subacute, base cordate, truncate or rounded; lower up to 14 cm (5.5 inches), often less, petioled, petiole narrowly winged at top; upper subsessile.*
Calyx 7-9 mm, tubular, teeth about 1/2 length of tube, ovate or oblong, obtuse. Corolla 15-17 mm, white. Scales broad-subulate, slightly exceeding the stamens. Nutlets tuberclate, dark brown. Flowers 4-5.
Introduced. Naturalized in hedgebanks and grassy places, not uncommon in some districts of the British Isles. Turkey."
-Flora of the British Isles, Second Edition by A.R. Clapham, T.G. Tutin and E.F. Warburg. England: Cambridge University Press, **1962**, (first edition 1952). (Pubescent means the surface of a leaf or stem is covered with fine short hairs.)

"**S. orientale L.**, *Sp. Pl.* 136 (1753) p.p.
Synonym: *S. constantinopolitanum boraginis folio et facie flore albo* Tournef. Cor. 7 (1703).
S. tauricum Sims, Bot. Mag. t. 1921 (1817).
S. jacquinianum Tausch. in Flora (Regensburg) 19 :393 (1836).
Range: Northwest and west Turkey. Its limited distribution ensures that S. orientale will not be confused with the more membranous-leaved *S. palaestinum* var. *majus* from the Cilician Plain, Anti-Taurus and Amanus.
Linnaeus wrongly combined at least two of Tournefort's species so that his original description also includes S. asperum Lepechin."
-'A Revision of Symphytum in Turkey and Adjacent Areas' by G.E. Wickens, Royal Botanic Gardens: Kew; Notes from the Royal Botanic Garden Edinburgh, Scotland, Volume 29, pages 157-180, **1969**. Includes Turkey, Bulgaria, Greece, Aegean Islands and Caucasia.
>(The Cilician Plains are on the Mediterranean coast of Turkey. It includes the pine forests of the Taurus Mountains and the flat land of the plains near the Seyhan and Ceyhan Rivers.)
>(The Nur Mountains, also known as Alma-Dag and the ancient Amanus, is a mountain range in the Hatay Province of south-central Turkey.)

"**S. orientale L.**, Sp. Pl. 136 (1753) *(S. tauricum sensu Coste, non Willd.):*
Stock fusiform. Stem up to 70 cm (27 inch) or more, much branched. Leaves ovate, subcordate or rounded or truncated at base, densely hairy on both surfaces, rather softly hispid, often subtomentose beneath; the lower long-, the upper short-petiolate, the uppermost sessile, not decurrent.
Calyx 6 to 9-12 mm, lobed to 1/4-2/5. Corolla 13-14 to 18-19 mm, **white**; lobes not recurved. Scales lingulate the marginal papillae up to 2 1/2 times as long as wide. Stamens with filiament about 3/5 as long as anther; anthers 2.5 to 3.5-4.0 mm, 2 1/2-3 to 5- 5 1/2 times as long as wide.
Damp, shady places.
Around Istanbul (Turkey); southwestern Ukraine; locally naturalized elsewhere (Russia {Southwestern region}, Turkey-European part, Britain, France, maybe Switzerland, Italy). Northwestern Anatolia (Turkey)."
-Flora Europaea, Volume 3: Diapensiaceae to Myoporaceae, editors Tutin, Heywood, Burges, Moore, Valentine, Walters and Webb. United Kingdom: Cambridge University Press, **1972**, page 154. (Symphytum L., pages 103-105 by B. Pawlowski.)

*"Medvedev (*1970) explored the breeding possibilities of **the less cold-tolerant S. orientale L. and S. tauricum Willd.**"*
-'Comfrey Symphytum spp. as a Forage Crop' by J.C. Forbes, A.D. McKelvie, and P.J.C. Saunders, North of Scotland College of Agriculture, Aberdeen, United Kingdom; Herbage Abstracts, Volume 49, No. 12, pages 523-539, **1979**.
(* -'Basic Material and Methods for Selection in Comfrey, Part 1' {Iskhodnyi Material i Metody Selektsii Okopnika} by P.F. Medvedev; 5th Symposium on New Silage Plants, Leningrad, Russia, pages 68-69, 1970. If you have this in English, I would appreciate a copy.)

"Linnaeus' name S. orientale clearly comprises at least two species.
>**One of these is the true white-flowered S. orientale,** as now understood, endemic to northwest Turkey (e.g., *Wickens 1978).
>In addition the description includes S. asperum and perhaps other blue-flowered species (see the thorough discussions of **Lindman 1911 and Bucknall 1913).
>The name S. orientale has also been used for S. x uplandicum and S. tauricum.
Symphytum patens:

The name S. patens was discussed by Lindman (1911), Bucknall (1913) and Faegri (1931).
__S. patens Fries seems to include S. asperum and S. x uplandcium.__ The brief diagnosis of S. patens Sibth. clearly refers to S. officnale, although Faegri (1931) disagreed with this."
-'Taxonomy of the Symphytum Asperum Aggregate (Boraginaceae), Especially in Turkey' by Arto Kurtto, Botanical Museum, University of Helsinki, Finland; Annales Botanici Fennici (Finnish Botanical Annals), Helsinki, Vol 19, No. 3, page 177-192, **1982**.
(* -Flora of Turkey and the East Aegean Islands, 11 Volumes by Peter H. Davis. Scotland: Edinburgh University Press, 1984-1988. 'Symphytum L.' in Volume 6, pages 378-386 by G.E. Wickens, 1978.)
(** -'Ueber {About} Symphytum Orientale L. und Symphytum Uplandicum Nym.' by C.A.M. Lindman; Botaniska Notiser {Botanical Notes}, Lund, Sweden, Nummer 2, pages 71-77, 1911. In German.)

"S. orientale L. - White Comfrey:
Stems erect, little-branched, to 70 cm (27.5 inch), from thick branched roots; stem-leaves petiole to sessile but not decurrent; calyx divided about 1/4 to 2/5 to base; __corolla white.__
Introduced; naturalized in hedgerows and other shady places, often self-sown; frequent in east and south England and central Scotland, very scattered elsewhere; west Russia and Turkey."
-New Flora of the British Isles: Identification of Wild Vascular Plants of the British Isles edited by Clive Stace. Cambridge, England: Cambridge University Press, **1991**, page 648.
   (A hedgerow is closely growing shrubs and trees, usually bordering a road or field.)

"Symphytum orientale:
Ecology: *This perennial herb is found as an escape or outcast in hedgerows and copses, on lanesides, by roads and railways, and on waste ground. It is often naturalised, and sometimes regenerates from seed. Lowland.*
Status: *Neophyte (in Britain and Ireland).*
Trends: *__S. orientale was introduced to British gardens by 1752,__ when it was known to have been grown in Cambridge, England, and was known from the wild by 1849.*
Its distribution has increased greatly since the 1962 *'Atlas of the British Flora' *owing to a genuine spread, although part of the increase is also due to better recording of alien species.*
World Distribution: *Native of southern Russia, northwest Turkey and the Caucasus."*
-'Online Atlas of the British and Irish Flora', https://www.brc.ac.uk/plantatlas, Distribution and ecology with photographs, **2018**.
(* -Atlas of the British Flora by Franklyn Perring and S. Max Walters. Harpenden, Hertfordshire County, England: Botanical Society of the British Isles. First published 1962, second edition 1976, third edition 1982.)
   (Copse is a thicket of small trees and bushes.)
   (In botany, a neophyte is a plant species that is not native to a geographical region, and was introduced in recent history. Plants that are long-established in an area are called archaeophytes. In Britain, neophytes are plant species introduced after 1492, when Christopher Columbus arrived in the Americas.)

"Symphytum Orientale: Esek Kafesotu:
Robust perennial, 50-70 cm (19.6-27.5 inch), shortly villous. Leaves ovate to oblong-ovate, subcordate, rounded or truncate at base, lower long-petiolate, uppermost sessile.
Flowers about 20, in well developed forked scorpioid cymes. Calyx 6-9 mm, accrescent to 19 mm in fruit, divided to 1/4 to about 1/2; lobes ovate-oblong, obtuse. __Corolla white__, 15-17 mm; scales 3.5-5 mm, triangular-lanceolate, shortly exceeding stamens. Nutlets 3 mm, erect to slightly curved, areolate, tuberculate.
Fl. 4-6. Shady stream banks, Pinus nigra forest, near sea level to 1500 meter (4921 feet).
South Russia, Caucasus. Euro-Siberian element? __Related to S. kurdicum and S. brachycalyx.__"
-'Symphytum Orientale: Esek Kafesotu' published by Turkiye Bitkileri (Plants of Turkey), **2019**. Turkey has a great botanical richness with more than 9600 plant species. This website has been constructed to gather as much information as possible about native plants of Turkey. www.turkiyebitkileri.com
   (Pinus nigra, Austrian pine, or black pine grows across southern Mediterranean Europe from Spain to the eastern Mediterranean on Anatolian peninsula of Turkey and on Corsica/Cyprus, including Crimea, and in the high mountains of the Maghreb in North Africa.)

Chromosomes of S. orientale

*"__S. orientale L.: We found 2n = 32 (five counts)__, the same number as reported by Grau (*1971) for material of garden origin. Tarnavschi (**1948) counted 2n = 62, 63. These numbers require confirmation."*
-'Cytotaxonomic Studies in the Genus Symphytum VIII: Chromosome Numbers and Classification of Ten European Species' by T.W.J. Gadella and E. Kliphuis, Proceedings van de Koninklijke Nederlandse Akademie van Wetenschappen (Proceedings of the Royal Netherlands Academy of Arts and Sciences): Series C, Biological and Medical Sciences, Vol 81, pages 162-172, **1978**.
(* -'Cytologische Untersuchungen an Boraginaceae II' {Cytological Investigations on Boraginaceae} by J. Grau, Mitteilungen der Botanischen Staatssammlung Munchen {Communications of the Botanical State Collection Munich, Germany}, Volume 9, pages 177-194, 1971.)
(** -'Die Chromosomenzahlender Anthophyten-Florav on Rumanien mit Einem Ausblick auf das Polyploidie-Problem' {The Number of Chromosomes The Anthophyten-Florav in Romania with an Outlook on the Polyploidy Problem} by I.T. Tarnavschi; Bulletin du Jardin et du Musee Botaniques de Universite de Cluj, Roumanie, Volume 28, pages 1-130, 1948. In German. If you

have an English translation of the Symphytum pages, I would appreciate a copy.)

"Symphytum orientale L. 2n = 32.
Bulgaria: *Southern Black Sea coast, along the river Ropotamo, 42 degrees, 17 minutes north; 27 degrees, 53 minutes east; damp scrub, 1975, Markova L487 (SOM= Herbarium of the Institute of Botany, Bulgaria).*
The chromosome number 2n = 32 was established by the study of the same population L487 (*Markova 1983) and agrees with the results of Gadella & Kliphuis (1978) and Grau (1971) from material of garden origin, but not with the counts **2n = 62 and 2n = 63** (see **Fedorov 1969).
The karyotype consists of 2n = 4x = 14m + 16sm + 2sm - SAT = 32."
-'Mediterranean Chromosome Number Reports: 5' edited by G. Kamari (Greece), F. Felber (Switzerland), and F. Garbari (Italy); Flora Mediterranea (biogeography, floristics and systematic botany in the Mediterranean area), Volume 5, pages 261-373, **1995**, page 300.
(* -'IOPB Chromosome Number Reports LXXX' by Askell Love; Taxon, Volume 32, No. 3, pages 67-68, 504-511, August 1983. Includes Symphytum orientale and Symphytum tuberosum by Margarita Markova. Results are published in 'Chromosome Counts Database'® (CCDB) which is a comprehensive resource for plant chromosome numbers, http://ccdb.tau.ac.il)
(** -Chromosome Numbers of Flowering Plants (Khromosomnye chisla tsvetkovykh rastenii) (Hromosomnye cisla cvetkovyh rasteniy) edited by A.A. Fedorov, authored by Z.V. Bolkhovskikh, V.G. Grip, O.I. Zakhar'eva, T.S. Matveeva, et al., in Russian. Leningrad, Russia: Nauka, 1969. If anyone has an English translation of the Symphtum section, I would appreciate a copy.)
    (A karyotype is the number and appearance of chromosomes in the nucleus of an eukaryotic cell.)

"Symphytum orientale L.: 2n = 32.
Italy: *Tuscany, Cascine Park of Florence, in front of the Police barracks, on rich and shady soil, April 1999, Bottega and Peruzzi, cultivated Herbarium Horti Botanici Pisani, Pisa, Italy.*
S. orientale is native of Istanbul, Turkey, environs and of southwest Ukraine, but locally naturalized elsewhere *(Pawlowski 1972).*
*In Italy the species is recorded for the outskirts of Florence and some Botanical Gardens (*Viegi et al, 1974)."*
-'Mediterranean Chromosome Number Reports 11 {Report 1229: Symphytum gussonei}' by S. Bottega, F. Garbari and L. Peruzzi; Flora Mediterrana, edited by G. Kamari, C. Blanche and F. Garbari, Volume 11, pages 436-439, **2001**.
(* -'Flora Esotica d'Italia' by Lucia Viegi, Giovanna Cela Renzoni and Fabio Garbari, University of Pisa, Italy; Biogeographia: The Journal of Integrative Biogeography, Volume 4, 1973-1974. All in Italian.)

"Symphytum orientale L., *Sp. Pl. ed. I: 136. 1753, pro parte (in part).*
    = *S. constantinopolitanum boraginis folio et facie, flare alba, Tournefort Cor.: 7. 1703*
    = *S. tauricum Sims, Bot. Mag. t. 1912. 1817*
    = *S. jacquinianumTausch, Flora 19:393.1836*
    = *S. orientale var. normale O. Kuntze, Acta Horti Petrop. 10(1): 220. 1887*
    = *S. tauricum Coste, non Willd., Fl. Fr. 2: 582. 1903*
Chromosome Number: 2n = 32. *It was obtained on collected samples from the present population at Cascine di Firenze (344-345 / 1999 HBP) (Florence, Italy).*
Databank: *2n = 32 (Gadella & Kliphuis, 1978; Grau, 1971). Obtained on material from the surroundings of Istanbul, Turkey:* **2n = 62, 63** *(Tarnavschi, 1948).*
Ecology: *It lives in woods uncultivated and shady, from 0 to 1500 meters (0-4921 feet) of altitude.*
Biological Form: *Scaposa hemicriptophyte.*
Phenology: *It blooms from April to May*
General Distribution: *Euro-Siberian, endemic (native) to north-western Turkey, with sub-Mediterranean natural or semi-natural populations and Mediterranean (*Kurtto, 1985)."*
-'Il Genere Symphytum L. (Boraginaceae) in Italia: Revisione Biosistematica' (The Genus Symphytum L. Boraginaceae in Italy: A Biosystematic Revision) by Stefania Bottega and Fabio Garbari, Webbia: Journal of Plant Taxonomy and Geography, Volume 58, No. 2, pages 243-280, **2003**. (If you have an English translation of this article, I would appreciate having a copy.)
(* -'Taxonomy of Symphytum Ottomanum, S. Pseudobulbosum and S. Orientale (Boraginaceae)' by Arto Kurtto, Botanical Museum, University of Helsinki, Finland; Annales Botanici Fennici (Finnish Botanical Annals), Helsinki, Volume 22, No. 4, pages 319-331, 1985. I was unable to get this report. If you have it in English, could you please send it to me.)
    (Hemicryptophytes: herbs {very rarely woody plants} with buds at soil level.
    Hs: Semi-rosette hemicryptophytes, with leafy stems but the lower leaves larger than the upper ones and the basal internodes shortened.)
    (Scapose means consisting of a scape. Scape is a long, leafless flower stalk coming directly from a root.)
    (Phenology is relation between climate and periodic biological phenomena such as plant flowering, leafing, breeding.)

"Symphytum orientale L. 2n = 32.
in 'Cytotaxonomic Studies on the Genus Symphytum VIII: Chromosome Numbers and Classification of Ten European Species' by Th.W.J. Gadella and E. Kliphuis, Proceedings van de Koninklijke Nederlandse Akademie van Wetenschappen {Proceedings of the Royal Netherlands Academy of Arts and Sciences} Section C, Volume 81, pages 162-172, 1978."
-'Chromosome Counts Database'® (CCDB) is a comprehensive resource for plant chromosome numbers. Every chromosome number includes the original database and the research article reference, **2018**. http://ccdb.tau.ac.il

"Symphytum orientale L.: Chromosome Count is 2n = 32."
-Tropicos®, www.tropicos.org, Missouri Botanical Garden, Saint Louis, Missouri; Plant database with 1.3 million scientific names and 4.4 million specimen records, **2018**.

Symphytum ottomanum
See 'Symphytum pseudobulbosum' in this section.

Current Botanical Nomenclature of S. ottomanum

"Symphytum ottomanum Friv. (*Flora 19(2): 439. 1836)"
-'International Plant Names Index'® (IPNI), www.ipni.org, A database of the names and associated basic bibliographical details of seed plants, ferns and lycophytes, **2018**.
> (* -'Flora oder Botanische Zeitung' {Flora or Botanical Newspaper}, Regensburg, Germany, Volume 19, No. 2, 1836. Symphytum ottomanum page 439.)

"Symphytum ottomanum Friv.: accepted name."
-Catalogue of Life®: 2015 Annual Checklist, Online database of the world's known species of animals, plants, fungi and micro-organisms. It consists of a single integrated species checklist and taxonomic hierarchy., **2018**, www.catalogueoflife.org. Search: www.catalogueoflife.org/annual-checklist/2015/search/all

*"**Symphytum ottomanum Friv. is an accepted name.** This name is the accepted name of a species in the genus Symphytum (family Boraginaceae). The record derives from WCSP {World Checklist Database} (in review) which reports it as an accepted name with original publication details: Flora 19: 439 1836."*
-The Plant List®, www.theplantlist.org from the World Checklist Database, **2018**.

Description of S. ottomanum

"S. ottomanum Friv., in Flora, xix, p. 439 (1936):
*Geographical Distribution: **Hungary: Banat. Servia, Bulgaria. Greece: Thessaly, Euboea.***
This has hitherto been placed with S. bulbosum, Schimp. in a section characterised by the exserted corolla-scales, but with that species it has no other affinity. It differs from it fundamentally in the fusiform root, the branched stem, and the many-flowered racemes.
From S. pseudobulbosum, Azn., which has slightly exserted corolla-scales, it is distinguished by the more slender habit, the wingless stem, and the deeply divided calyx."
-'A Revision of the Genus Symphytum, Tourn.' by Cedric Bucknall, (Mus. Bac. Oxon= Bachelor of Music, Oxford University), Journal of the Linnean Society of London, England, Botanical Journal, Volume 41, Issue 284, pages 491-556, December **1913**.
> (Banat is a region between central and eastern Europe that is now divided among three countries: the eastern part is in western Romania, the western part is in northeastern Serbia, and a small northern part is in southeastern Hungary.)

"S. ottomanum Friv., Flora (Regensb.) 19: 439 (1936):
Stock fusiform. Stem 30-80 cm (0.98-2.6 feet), branched. Leaves ovate or ovate-lanceolate, more or less rounded-cuneate at base, the lower long-petiolate, the upper sessile and slightly decurrent.
*Flowers small. Calyx 3-5 mm, lobed to 3/5 to 4/5. Corolla (excluding scales) 5-7 mm, **pale yellow**; lobes about 1/3 as long as tube, more or less erect. Scales 5 to 9.5 mm, exserted for 2 to 5.5 mm, linear-lanceolate, more or less acute; marginal papillae dense, about as long as wide. Stamens with filament 1/3 to 3/5 as long as anther; anthers 2 to 2.5 mm, minutely apiculate.*
2n = 20.
*Woods. **Balkan peninsula, Romania. Albania, Bulgaria, Greece, Yugoslavia, Romania, Turkey (European part).**"*
-Flora Europaea, Volume 3: Diapensiaceae to Myoporaceae, editors Tutin, Heywood, Burges, Moore, Valentine, Walters and Webb. United Kingdom: Cambridge University Press, **1972**, page 104. (Symphytum L., pages 103-105 by B. Pawlowski.)
> (The Balkan Peninsula is surrounded by the Adriatic Sea to the west, the Mediterranean Sea and Marmara Sea to the south, and the Black Sea to the east. Its northern boundary is the Danube, Sava and Kupa Rivers.)

*"**A comparison is made of the main morphological characters of Symphytum ottomanum Friv., S. pseudobulbosum Aznavour and S. orientale L.,** in order to elucidate their ranges of variation and the taxonomical status of S. pseudobulbosum. The pollen morphology, cytology, distribution and ecology of the species are also discussed.*
In many characters S. pseudobulbosum is intermediate between S. ottomanum and S. orientale, but in some others it is much closer to one or other of the species. It seems to be a stabilized hybrid derivative of S. ottomanum and S. orientale.
The distribution area of S. pseudobulbosum is situated between the main areas of the putative (supposed) parents, and includes Turkey-in-Europe.
The three species are lectotypified and their affinities with other species of the genus are briefly discussed."
-'Taxonomy of Symphytum Ottomanum, S. Pseudobulbosum and S. Orientale (Boraginaceae)' by Arto Kurtto, Botanical Museum, University of Helsinki, Finland; Annales Botanici Fennici (Finnish Botanical Annals), Helsinki, Volume 22, No. 4, pages 319-331,

1985. (I was unable to get this report. If you have it, could you please send it to me.)

*"**Symphytum ottomanum is essentially a Balkan species** occuring in Romania, Jugoslavia (Yugoslavia), Albania, Bulgaria, the northern half of Greece and north-east Turkey, i.e, mostly in territory once included in the Ottoman Empire.*
Of the other species occurring on Greek territory, S. ottomanum was described by *Frivaldszky in 1836 from Rumelia (now Turkey), the central part of the Balkan Peninsula, then under Turkish rule and thus with the Ottoman Empire."
-'The Greek Species of Symphytum Boraginaceae' by William T. Stearn, British Museum of Natural History, London, England; Annales Musei Goulandris, Greece, Volume 7, pages 175-220, (**1985 or 1986**, different sources give different dates).
(* -'Flora oder Botanische Zeitung' {Flora or Botanical Newspaper}, Regensburg, Germany, Volume 19, No. 2, page 439, 1836.)

*"**Symphytum Ottomanum: Koru Kafesotu:***
Slender perennial, 30-80 cm (11.8-31.4 inch). Leaves ovate to ovate-lanceolate, lower narrowed into winged petiole equalling lamina, upper cuneate, sessile, slightly decurrent.
*Flowers about 20. Calyx 3-5 mm, accrescent to 10 mm in fruit, divided to below middle, lobes ovate-lanceolate. **Corolla whitish to pale yellow**, 5-7 mm; scales linear-lanceolate, exserted, 2-5.5 mm, exceeding stamens by 3.5 mm. Nutlets 2 mm, suberect, areolate, minutely tuberculate.*
*Fl. 5. Woods, about 500 meter (1640 feet). **Romania, Balkans.** East Mediterranean element?"*
-'Symphytum Ottomanum: Koru Kafesotu' published by Turkiye Bitkileri (Plants of Turkey), **2019**. Turkey has a great botanical richness with more than 9600 plant species. This website has been constructed to gather as much information as possible about native plants of Turkey. www.turkiyebitkileri.com

S. ottomanum Chromosomes

"Both S. ottomanum and S. bulbosum have exserted corolla scales, which are absent in S. anatolicum.
The chromosome numbers of S. anatolicum (2n = 30) and S. ottomanum (2n = 48) are very different and do not support the inclusion of these species in the same section."
-'Cytotaxonomic Studies in the Genus Symphytum VIII: Chromosome Numbers and Classification of Ten European Species' by T.W.J. Gadella and E. Kliphuis, Proceedings van de Koninklijke Nederlandse Akademie van Wetenschappen (Proceedings of the Royal Netherlands Academy of Arts and Sciences): Series C, Biological and Medical Sciences, Vol 81, pages 162-172, **1978**.

*"**S. ottomanum Friv.: 2n = 20**, Bulgaria, *Markova and Ivanova, 1970."*
-Flora Europaea: Check-List and Chromosome Index by D.M. Moore. Cambridge, England: Cambridge University Press, **1982**, page 179.
(* -'Karyologische Untersuchung der Vertreter der Fam. Boraginaceae, Labiatae und Scrophulariaceae in Bulgarien' {Caryological Examination of the Representatives of the Family Boraginaceae, Labiatae and Scrophulariaceae in Bulgaria} by Margarita Markova/Markowa and P. Ivanova/Iwanowa; Izvestiya na Botaniceskiya Institut {Bulletin de l'Institut Botanique}, {also titled: Mitteilungen des Botanischen Instituts Sofia}, Bulgaria, Volume 20, pages 93-98, 1970. In German.)

*"**Symphytum ottomanum Friv. 2n = 18, 20, 22.***
Bulgaria: *Pirin Mountain, above the town of Sandanski, damp ruderal places, 1986, Markova L1580 (SOM= Herbarium of the Institute of Botany, Bulgaria).*
    Western Rodopi, at locality Slivov dol, near the town of Backovo, woody damp places, 1982, Markova L1241 (SOM).
    Eastern Rodopi, around the village of Momeilgrad, scrubby places, 1980, Markova L1057 (SOM).
    Stara Planina Mountain, in the Steneto forest reserve, 1982, Markova L1316 (SOM).
    North-eastern Bulgaria, at the locality Pop Cair near the town of Sumen, scrub, 1978, Markova L959 (SOM).
    North-eastern Bulgaria, the hill above the village of Cestemensko, scrub, 1975, Markova L509 (SOM).
The number 2n = 20 for this taxon *was reported for the first time by Markova and Ivanova (1970). The same chromosome number, as well as **2n = 40**, has also been reported from elsewhere (*Loon & Oudemans 1982, Strid & Franzen 1983, **Strid & Andersson 1985, van Loon 1987). The latter number was found in different individuals from the same populations (Strid & Andersson 1987).*
The data here do not agree with the number 2n = 48, *found in material from the Botanical Gardens of Sofia (Bulgaria) and Palermo (Sicily, Italy) (Gadella & Kliphuis 1978).*
The chromosome number 2n = 18 (probably a descending aneuploid) *agrees with the data from a Romanian population (Tarnavsehi 1948 after Strid & Andersson 1987).*
The chromosome number 2n = 22 is here reported for the first time, and seems to be an ascending aneuploid. *The karyotype of population:*
    L1580 consists of 2n = 2x = 6m + 10sm + 2 sm - SAT + 2st = 20 chromosomes,
    L1316 and 1241 , 2n = 2x = 4m + 14 sm + 2st = 20 , and
    L1057, 2n = 2x = 4m + 12sm + 2sm - SAT +2st = 20."
-'Mediterranean Chromosome Number Reports: 5' edited by G. Kamari (Greece), F. Felber (Switzerland), and F. Garbari (Italy); Flora Mediterranea (Biogeography, Floristics and Systematic Botany in the Mediterranean Area), Vol 5, pages 261-373, **1995**.
(* -'IOPB Chromosome Number Reports LXXV' edited by Askell Love; Taxon, Voume 31, No. 2, pages 342-368, May 1982. J. Chr. Van Loon and J.J.M.H. Oudemans, page 343.)
(** -'Chromosome Numbers of Greek Mountain Plants: An annotated List of 115 Species' by A. Strid and A. Andersson; Bota-

nische Jahrbucher fur Systematik: Pflanzengeschichte und Pflanzengeographie {Botanical Yearbooks for Systematics: Plant History and Plant Geography}, Stuttgart, Germany, Volume 107, pages 203-228, 1985. I was not able to find this report.)
 (Ruderal species are first plants to colonize disturbed lands.) (Aneuploidy is an abnormal number of chromosomes in a cell.)

"**Symphytum ottomanum Friv.:**
Origin: Greece, Xanthi, near village Echinos, 400 meter (1312 feet), wood margins.
All plants showed the diploid number 2n = 2x = 20.
Karyotype consisted of eight pairs of metacentrics, one of submetacentrics and one of subtelocentrics.
Accordingly, intrachromosomal karyotype asymmetry was relatively low (A1 = 0.29; A2 = 0.26).
Our finding is in line with other previous reports from Bulgaria and Greece (Markova & Ivanova 1970; *Markova & Goranova 1995; **Strid 1983).
The occurrence of tetraploid plants with **2n = 40** (Strid & Andersson 1985) and of cytotypes with **2n = 48** (Gadella & Kliphuis 1978) may indicate infraspecific variation which should be better investigated.
  Somatic numbers (2n), base numbers (x), ploidy levels, karyotype formulas and mean chromosome length (L) of the investigated taxa:
    2n = 20, Base Number: 10, Ploidy Level: (2x)
    Karotype Formula: 16m + 2sm + 2 st
    Chromosome Length: 1.5 micro-meter."
-'Chromosome Studies in Mediterranean Species of Boraginaceae' by A. Coppi, F. Selvi and M. Bigazzi, Flora Mediterranea, Volume 16, pages 253-274, **2006**.
(* -'Mediterranean Chromosome Number Reports: 5' edited by G. Kamari {Greece}, F. Felber {Switzerland}, and F. Garbari {Italy}; Flora Mediterranea {Bbiogeography, Floristics and Systematic Botany in the Mediterranean Area}, Volume 5, pages 261-373, 1995. Includes research by Markova and Goranova.)
(** -'IOPB Chromosome Number Reports LXXVIII' by A. Strid, Taxon, Volume 32, No. 1, page 139, 1983.)
  (Infraspecific is at a taxonomic level below that of species, e.g., subspecies, variety, cultivar, or form. Latin names at this level require the addition of a term denoting rank.)

Symphytum palaestinum
See Symphytum brachycalyx in this section.

"**Symphytum palestinae Boiss. is an unresolved name.**
The record derives from WCSP (World Checklist Database) (in review) (data supplied on 2012-03-23) which does not establish this name either as an accepted name or as a synonym with original publication details: *Diagn. Pl. Orient. 11: 94 1849."
-The Plant List®, www.theplantlist.org from the World Checklist Database, **2018**.
(* -'Diagnoses Plantarum Orientalium Novarum, No. 11' by Pierre Edmond Boissier, Paris, France, page 94, 1849.)

"**Symphytum palaestinum Boiss.**
  Synonym of Symphytum palaestinum var. violaceum N.Feinbrun.
  Synonym of Symphytum brachycalyx Boiss.: accepted name."
-Global Biodiversity Information Facility® (GBIF), www.gbif.org, **2019**. International network and research infrastructure funded by the world's governments and aimed at providing anyone, anywhere, open access to data about all types of life on Earth.

"**S. palaestinum, Boiss.** *Diagn. ser. I. xi. p. 94 (1844).
Geographical Distribution: **Turkey: Palestine, Syria, southern Asia Minor.**
S. palaestinum forms, with S. brachycalyx and S. bornmuelleri, a group of closely allied species. These must be carefully distinguished from each other, but are, nevertheless, separated by well defined characters, and occupy areas which do not overlap.
  From S. brachycalyx, S. palaestinum differs in the lower leaves being more often rounded or subcordate at the base, although they are sometimes attenuated into the petiole as in that species, and in the calyx, which is much more accrescent in fruit.
  From S. bornmuelleri it is easily distinguished, even in a young stage, by the subdichotomous branching of the stem, and by the more deeply divided calyx with linear or triangular teeth, which are generally one-third as long as the calyx itself.
In **'Flora Orientalis'* Boissier states that Symphytum palaestinum leaves are not truly decurrent, but that the petioles are produced into the angles of the stem; this is not easily detected in the dry plant. In the 'Diagnoses', however, he describes the upper leaves as shortly decurrent, and this is seen to be the case in Boisaier's own specimens.
The area of S. palaestinum extends further south than that of any known species; commencing in the neighbourhood of Jerusalem in the Middle East, it includes a zone passing through Syria round the eastern shores of the Mediterrranean, and along the Taurus mountain range (southern Turkey) through Cilicia to Lycia."
-'A Revision of the Genus Symphytum, Tourn.' by Cedric Bucknall, (Mus. Bac. Oxon= Bachelor of Music, Oxford University), Journal of the Linnean Society of London, England, Botanical Journal, Volume 41, Issue 284, pages 491-556, December **1913**.
(* -'Diagnoses Plantarum Orientalium Novarum', Series 1, No. 11, page 94, by Pierre Edmond Boissier, Lipsiae {Leipzig}, Germany, 1844. Series 1: Volume 1: No. 1-7, Volume 2: No. 8-13. Series 2: Volume 3: No. 1-6. Published 1842-1859.)

(** -'Flora Orientalis Sive Enumeratio Plantarum in Oriente, a Graecia et Aegypto ad Indiae Fines {Plants Flora, Flora East and the East, from Greece and Egypt to the Borders of India}, Volume IV' by Edmond Boissier, Society Physics Geneva, Society Linnean London; Geneva, Switzerland, 1875 and 1879. All in Latin. Symphytum palaestinum, page 173-174.)
       (In antiquity, Cilicia was the south coastal region of Asia Minor. Extending inland from the southeastern coast of Turkey, Cilicia is north and northeast of the island of Cyprus.)
       (Lycia was a region in Anatolia in what are now the provinces of Antalya and Mugla on the southern coast of Turkey, bordering the Mediterranean Sea, and Burdur Province inland.)

Symphytum patens
See 'Details about Symphytum Species: Officinale' (Chapter 8) for **S. patens Sibth**.
See 'Symphytum orientale' in this section for **S. patens Fries**.

"Symphytum patens Sibth. (*Fl. Oxon. 70. 1794)
Symphytum patens Fr. (**Novit. Fl. Suec. Mant. 2: 13. 1839)"
-'International Plant Names Index'® (IPNI), www.ipni.org, A database of the names and associated basic bibliographical details of seed plants, ferns and lycophytes, **2018**.
(* -'Flora Oxoniensis Exhibens Plantas in Agro Oxoniensi' (Flora of Oxfordshire, England) by John Sibthorp, 1794, page 70.)
(** -'Novitiarum Florae Sueciae Mantissae Altera, Additis Plantis in Norvegia Recentius Detectis, II.' {Flora of Sweden} by E.M. Fries, Supplement page 13, Uppsala, Sweden, **1839**. All in Latin. Includes Symphytum patens and Symphytum officinale.)

"Symphytum patens Fries: ambiguous synonym for Symphytum orientale L.
Symphytum patens Sibth.: ambiguous synonym for Symphytum officinale subsp. officinale Sibth."
-Catalogue of Life®: 2015 Annual Checklist, Online database of the world's known species of animals, plants, fungi and micro-organisms. It consists of a single integrated species checklist and taxonomic hierarchy., **2018**, www.catalogueoflife.org. Search: www.catalogueoflife.org/annual-checklist/2015/search/all

"Symphytum patens Fries: synonym = Symphytum orientale L.: accepted name."
-Global Biodiversity Information Facility® (GBIF), www.gbif.org, **2018**. International network and research infrastructure funded by the world's governments and aimed at providing anyone, anywhere, open access to data about all types of life on Earth.

"Symphytum patens Sibth. is an unresolved name."
-The Plant List®, www.theplantlist.org from the World Checklist Database, **2018**.

Symphytum peregrinum
See section 'Details about Symphytum Species Hybrids: Russian Comfrey' (Chapter 10).

Symphytum podcumicum

Current Botanical Nomenclature of S. podcumicum

"Symphytum podcumicum Yu.M.Frolov (*Bot. Zhurn. (Moscow & Leningrad) 70(4): 533. 1985)."
-'International Plant Names Index'® (IPNI), www.ipni.org, A database of the names and associated basic bibliographical details of seed plants, ferns and lycophytes, **2018**.
(* -'A New Species of the Genus Symphytum {Boraginaceae} from the Caucasus' by Yu.M. Frolov, Botanicheskii Zhurnal {Botanical Journal}, Moscow and Leningrad, Russia, Volume 70, No. 4, pages 533, 1985.)

"Symphytum podcumicum Yu.M. Frolov: accepted name."
-Catalogue of Life®: 2015 Annual Checklist, Online database of the world's known species of animals, plants, fungi and micro-organisms. It consists of a single integrated species checklist and taxonomic hierarchy., **2018**, www.catalogueoflife.org. Search: www.catalogueoflife.org/annual-checklist/2015/search/all

"Symphytum podcumicum Yu.M.Frolov is an unresolved name.
The record derives from WCSP (World Checklist Database) (in review) (data supplied on 2012-03-23) which does not establish this name either as an accepted name or as a synonym with original publication details: Bot. Zhurn. (Moscow & Leningrad) 70: 533 1985."
-The Plant List®, www.theplantlist.org from the World Checklist Database, **2018**.

Description of S. podcumicum

"Taxon Name: Symphytum podcumicum Yu.M.Frolov.
Common Name: Podkumian Comfrey.

Taxonomic Notes: *Symphytum podcumicum Yu.M.Frolov is an unresolved name according to 'The Plant List' (2010) but accepted according to *Czerepanov (1995) and Takhtajan (manuscript).*
References:
    1. A. L. Takhtajan (editor). **Caucasus Flora Conspectus. Komarov Botanical Institute, St. Petersburg. Unpublished manuscript (in Russian).
    2. A. L. Chernogorov (Chair of the editorial board) 2002. Red Data Book of Stavropol Kray.
    Rare and threatened plant / animal species. Plants. Stavropol: 'Polygraphservis'. Volume I., page 432, (in Russian).
Red List Category & Criteria: *Endangered B1ab(iii,v)+2ab(iii,v); C2a(i).*
Threats:
This species is listed as Endangered in view of the EOO (Extent of Occurrence) estimated to be less than 5,000 square kilometers (1930 square miles), its Area of Occupancy (AOO) estimated to be less than 500 square km (193 square miles), its existence at less than five locations and an observed continuing decline in at least the area, extent and quality of habitat and the number of mature individuals caused by destruction of riparian forest ecosystem in Stavropol Kray (north Caucasus region in southern Russia).
The human impact on the population is strengthened by intrinsic factors such small population size, estimated to number fewer than 2,500 mature individuals, and no subpopulation estimated to contain more than 250 mature individuals. The species is not recorded in protected areas. Forest clear cutting, river degradation and recreation are the main threats to this species.
Geographic Range Description:
This species is endemic (native) to the Russian Federation. It is found in the Verkhne-Kumskiy floristic region in the surroundings of the Podkumok railway station and the distribution range covers a narrow riverside line from the Kislovodsk to Goryachevodsk - floristic regions according to Menitsky (***1991).
Current Population Trend: *Decreasing.*
Habitat and Ecology:
It occurs on alluvial deposits on riverbanks in willow riparian forests in the lower montane (mountain) zone.
Conservation Actions: *It is in the 'Red Data Book' of Stavropol Kray (2002). Research on the species biology and ecology, population numbers and range, for example recording all of the subpopulations, is needed, as well as population monitoring."*
-'Symphytum Podcumicum' Assessment by S. Litvinskaya, The IUCN Red List of Threatened Species®, www.iucnredlist.org, 9 pages, **2014**.
(* -Vascular Plants of Russia and Adjacent States (the Former USSR) by S. K. Czerepanov. New York: Cambridge University Press, 1995.)
(** -Caucasus Flora Conspectus (Konspekt Flory Kavkaza), 3 Volumes, edited by A.L. Takhtajan. Saint Petersburg, Russia: Saint Petersburg University Press, 2003-2008. In Russian.)
(*** -'Project Conspectus of the Flora of the Caucasus: A Map of the Floristic Regions' by U.L. Menitsky, Botanicheskiy Zhurnal, Volume 76, No. 11, pages 1513-1521, 1991. In Russian.)
    (Podkumok River is in Stavropol Krai, Russia, right tributary of the Kuma River.)
    (Alluvium is loose soil or sediment that has been eroded by water, and then deposited on dry land.)
    (Riparian is wetlands next to rivers and streams.)

Symphytum pseudobulbosum
See 'Symphytum bulbosum' in this section.

"Symphytum pseudobulbosum Azn. (*Bull. Herb. Boissier Ser. II. iii. 588. 1903)"
-'International Plant Names Index'® (IPNI), www.ipni.org, A database of the names and associated basic bibliographical details of seed plants, ferns and lycophytes, **2018**.
(* -'Un Symphytum Nouveau' {A New Symphytum} by G.V. Aznavour; Bulletin de l'Herbier Boissier, 2 Serie {Second Series}, Tome III {Volume 3}, p. 588-589, 1903. In French.)
(For a brief writeup about Aznavour, see subsection 'Overview of Distribution of Symphytum Species" in section 'Symphytum Species Overview' {Chapter 6}.)

"Symphytum pseudobulbosum Aznav.: *accepted name."*
-Catalogue of Life®: 2015 Annual Checklist, Online database of the world's known species of animals, plants, fungi and micro-organisms. It consists of a single integrated species checklist and taxonomic hierarchy., **2018**, www.catalogueoflife.org.
Search: www.catalogueoflife.org/annual-checklist/2015/search/all

Description of S. pseudobulbosum

"Symphytum pseudobulbosum Azn. (sp. nov.) {new species}:
Habitat*: Shady places, near homes and gardens.*
In Ak-baba, Hunkiar-iskelessi, Beicos, Gueuk-souyou (not far from Anadolou-hissari). Localities all on the Asian coast of the Bosphorus. Flowers April and May."
-'Un Symphytum Nouveau' {A New Symphytum} by G.V. Aznavour; Bulletin de l'Herbier Boissier, 2 Serie {Second Series}, Tome III {Volume 3}, p. 588-589, **1903**. In French. Description of Symphytum pseudobulbosum Aznavour from the Asian side of the Bosphorus. (Translated online from French to English. If you have a better translation, please send it to me.)

(The Bosphorus, also known as The Strait of Istanbul, is a narrow waterway located in northwestern Turkey.)

*"**S. pseudobulbosum Aznavour** in Bull. Herb. Boiss. Ser. ii. iii. p. 588 (1903).*
*Geographical Distribution: **Turkey: Asiatic side of the Bosphorus** in shady places.*
This, the most recently discovered species of the Symphytum genus, is well characterised by the stem being narrowly winged below, by the small white corolla with its scales shortly exserted, and by the calyx being divided scarcely to the middle.
It is easily distinguished from S. ottomanum Friv., which is a shorter, more slender plant, with denser, longer hairs, smaller leaves, deeply divided calyx, and longer corolla-scales."
-'A Revision of the Genus Symphytum, Tourn.' by Cedric Bucknall, (Mus. Bac. Oxon= Bachelor of Music, Oxford University), Journal of the Linnean Society of London, England, Botanical Journal, Volume 41, Issue 284, pages 491-556, December **1913**.

*"**S. pseudobulbosum Aznavour** in Bull. Herb. Boiss. Ser. 2, 3 :588 (1903).*
Root thick and fleshy. Stem branched, 45-65 cm (17.7-25.5 inches) tall, narrowly winged below, pubescent to scabrid, also with tuberculate-based hairs. Leaves slightly pubescent and with short tuberculate-based hairs; basal leaves 14 cm (5.5 inch) long, 7 cm (2.7 inch) broad, ovate, apex acute, base rounded with long, winged petiole; middle leaves oblong-lanceolate, shortly petioled; upper leaves ovate-lanceolate, sessile or shortly decurrent.
*Inflorescence about 15-flowered. Calyx 6-7 mm long, divided almost to middle, segments lanceolate, acute; calyx accrescent in fruit, enlarging to 13 mm. **Corolla yellowish-white**, 10 mm long, tube as long as calyx; corolla scales 5.5 mm long, triangular-lanceolate, acute, exserted 1 mm beyond corolla lobes. Stamen filaments 1.5 mm long, anthers 2.5 mm long; stamens shortly exserted. Style exserted 3 mm.*
Nutlets 2 mm, erect, areolate, tuberculate. Fl. 4-6. Shady places.
***This is the easiest of all the Turkish species to recognise** because of the slightly exserted corolla scales and stamens. Vegetatively, it is rather similar to the less robust S. ottomanum Friv., which has smaller leaves, wingless stem, deeply divided calyx, and corolla scales fully exserted. The latter is distributed throughout the Balkan Peninsula."*
-'A Revision of Symphytum in Turkey and Adjacent Areas' by G.E. Wickens, Royal Botanic Gardens: Kew; Notes from the Royal Botanic Garden Edinburgh, Scotland, Volume 29, pages 157-180, **1969**. Includes Turkey, Bulgaria, Greece, Aegean Islands and Caucasia.   (The Balkans or Balkan Peninsula is in southeastern Europe.)

*"**Symphytum pseudobulbosum Aznavour is one of the rarest species in its genus.** Thus far it has been considered endemic (native) to the Asiatic side of the Bosphorus in Turkey (e.g., Bucknall 1913; Wickens 1969, 1978).*
In 1981 in the 'Herbarium of the Faculty of Pharmacy', University of Istanbul (Turkey), I saw two specimens of Symphytum collected in 1975 and 1977 by Professor Dr. A. Baytop, from the province of Kirklareli in Turkey-in-Europe, about 200 kilometers (124 miles) northwest of the Bosphrous. Only after careful comparsion with the type material of S. pseudobulbosum did I realize they they represent that species.
***A comparison is made of the main morphological characters of Symphytum ottomanum Friv., S. pseudobulbosum Aznavour, and S. orientale L.,** in order to elucidate their ranges of variation and the taxonomical status of S. **pseudobulbosum**. The pollen morphology, cytology, distribution and ecology of the species are also discussed.*
In many characters S. pseudobulbosum is intermediate between S. ottomanum and S. orientale, but in some others it is much closer to one or other of the species. It seems to be a stabilized hybrid derivative of S. ottomanum and S. orientale.
The distribution area of S. pseudobulbosum is situated between the main areas of the putative (generally considered to be the) parents, and includes Turkey-in-Europe.
The three species are lectotypified and their affinities with other species of the genus are briefly discussed."
-'Taxonomy of Symphytum Ottomanum, S. Pseudobulbosum and S. Orientale (Boraginaceae)' by Arto Kurtto, Botanical Museum, University of Helsinki, Finland; Annales Botanici Fennici (Finnish Botanical Annals), Helsinki, Vol 22, No. 4, page 319-331, **1985**.
(I was unable to get this report. If you have it in English, could you please send it to me.)
  (Morphology is a branch of biology that studies the form of living organisms, and the relationships between their
   structures. This includes both the outward and internal appearance {shape, structure, color, pattern, size}.)
  (Lectotype is a specimen chosen as the type of a species if the author of the name fails to designate a type.)

*"**Symphytum Pseudobulbosum: Yalan Kafesotu:***
Perennial, 45-65 cm (17.7-25.5 inch), stems narrowly winged below. Leaves ovate, oblong-lanceolate to ovate-lanceolate; lower with long winged petioles, uppermost sessile or shortly decurrent.
*Flowers about 15. Calyx 6-7 mm, accrescent to 13 mm in fruit, divided almost to middle, lobes lanceolate, acute. **Corolla cream**, 10 mm; scales 5.5 mm, triangular-lanceolate, exserted for 1 mm; stamens shorter than scales, shortly exserted. Nutlets 2 mm, erect, areolate, tuberculate.*
Fl. 4-6. Shady places, near sea level. Endemic. East Mediterranean element?"
-'Symphytum Pseudobulbosum: Yalan Kafesotu' published by Turkiye Bitkileri (Plants of Turkey), **2019**. Turkey has a great botanical richness with more than 9600 plant species. This website has been constructed to gather as much information as possible about native plants of Turkey.   www.turkiyebitkileri.com

<u>Symphytum rubrum</u> (Red Comfrey)
See 'Symphytum grandiflorum' in this section.

See section 'Details about Symphytum Species: Officinale' (Chapter 8).

Also called **Symphytum x rubrum**.
This is an ornamental Comfrey that is a good groundcover. Plants form a low mound of hairy leaves that are 10 inches (25.4 cm) long. It is 1 1/2 to 2 feet (0.45-0.60 meter) tall. The flowers are deep, dark red. It is not as hardy as other types of Comfrey.

*"Symphytum rubrum: This is a more unusual type said to be a **hybrid between Symphytum officinale 'coccineum' and Symphytum grandiflorum**. It spreads but not unduly; is an admirable ground cover for cool places.*
Deep crimson tubular flowers *hang in little croziers (shephard's crook) above hairy green leaves.*
46 x 60 cm (18 x 24 inches)."
-Comfrey: Symphuo Symphytum: A Multi-Purpose Herb by Philip Clarke. Edinburgh, Scotland: Pentland Press, **1997**, page 20.

> *"Child of Symphytum officinale:*
> ***Symphytum coccineum hort. ex Schltdl.*** *is similar to Symphytum coccineum hort."*
> -Global Biodiversity Information Facility® (GBIF), www.gbif.org, **2018**. International network and research infrastructure funded by the world's governments and aimed at providing anyone, anywhere, open access to data about all types of life on Earth.   (In Latin 'coccineum' means scarlet. Scarlet is brilliant red.)

> *"**Symphytum coccineum Schltdl. is an unresolved name.***
> *The record derives from WCSP (World Checklist Database) (in review) (data supplied on 2012-03-23) which does not establish this name either as an accepted name or as a synonym with original publication details: *Bot. Zeitung (Berlin) 7: 731 1849."*
> -The Plant List®, www.theplantlist.org from the World Checklist Database, **2018**.
> (* -Botanische Zeitung (Botanical Newspaper) by Hugo Von Mohl and Diedrich Franz Leonhard Von Schlechtendal, Berlin, Germany, Jahrg. {Volume} 7, column 731, 1849. Symphytum coccineum Hort.)

"Symphytum rubrum (Red):
12 inches (0.3 meter) tall; red-flowered form; ***blooms very early in spring. USDA Zone 4-9.****"*
-'The Sandy Mush Herb Nursery Handbook 9': Herbs, perennials, wildflowers, flowering plants, shrubs and trees, www.sandymushherbs.com, Leicester, North Carolina, July 27 **2004**.

*"**Red Comfrey (Symphytum rubrum) is a hybrid and not very robust*** *and should be grown in soil containing plenty of humus so that it does not dry out. It needs moisture so a position in semi-shade is best.*
It seldom grows to more than 30 cm (11.8 inches) and ***comes into flower later than the other Comfreys, from July onwards.***
It is not invasive and will flower on into the autumn.*"*
-'Herbalpedia®: Comfrey' by Maureen Rogers, The Herb Growing & Marketing Network, Silver Spring, Pennsylvania, www.herbalpedia.com, **2006**.

"Symphytum x rubrum:
The hybrid Comfrey (Symphytum x rubrum) makes ***dark red, attractive flowers from June to July****.*
The flowers are tubular. Symphytum x rubrum grows bushy, and usually reaches a height of 20-40 cm (7.8-15.7 inch) and is up to 20-40 cm wide. Its egg-shaped leaves are dark green."
-Pflanzen Fuer Dich (Plants for You), Westerstede, Germany, **2018**, www.pflanzen-fuer-dich.de

Symphytum savvalense

"Symphytum savvalense Kurtto (*Ann. Bot. Fenn. 19(3): 189. 1982)"
-'International Plant Names Index'® (IPNI), www.ipni.org, A database of the names and associated basic bibliographical details of seed plants, ferns and lycophytes, **2018**.
(* -'Taxonomy of the Symphytum Asperum Aggregate (Boraginaceae), Especially in Turkey' by Arto Kurtto, Botanical Museum, University of Helsinki, Finland; Annales Botanici Fennici (Finnish Botanical Annals), Helsinki, Volume 19, No. 3, page 189, 1982.)

"Symphytum savvalense Kurtto: accepted name."
-Catalogue of Life®: 2015 Annual Checklist, Online database of the world's known species of animals, plants, fungi and micro-organisms. It consists of a single integrated species checklist and taxonomic hierarchy., **2018**, www.catalogueoflife.org.
Search: www.catalogueoflife.org/annual-checklist/2015/search/all

Description of S. savvalense

*"**Symphytum savvalense Kurtto, spec, nova (new species)**, is described from northeastern Turkey.*
In some characters Symphytum savvalense is +- intermediate between S. asperum var. asperum and S. sepulcrale,
especially in the indumentum, basal leaves, and surface structure and basal teeth of the mericarps. The length of the calyx and the form and colour of the mericarps are similar to those of S. asperum var. asperum, while the fornicles, stamens and form of

the calyx resemble those of S. sepulcrale.

The broad cauline leaves with distinct petioles and the obtuse ridges on the mericarps are features only exceptionally seen in S. asperum var. asperum and S. sepulcrale. Nor do the mericarp ridges form a distinct network as in S. asperum var. asperum and S. sepulcrale.

S. savvalense is completely fertile: *the pollen is excellent and well-developed mericarps are abundant. The size and form of the pollen grains do not differ significantly from those of S. asperum.*

There are at least three possible interpretations of this kind of character combination:

1. S. savvalense may be a hybrid derivative of S. asperum var asperum and S. sepulcrale, combining intermediate characters and characters peculiar to each of the parents.

2. It may also represent an intermediate form, through which S. sepulcrale has diverged from S. asperum var. asperum or vice versa.

3. Nor can the possiblity be entirely ruled out, that it is part of the common ancestral stock of both these taxa."

-'Taxonomy of the Symphytum Asperum Aggregate (Boraginaceae), Especially in Turkey' by Arto Kurtto, Botanical Museum, University of Helsinki, Finland; Annales Botanici Fennici (Finnish Botanical Annals), Helsinki, Vol 19, No. 3, page 177-192, **1982**.

"**Taxon Name: Symphytum savvalense Kurtto.**
Common Name: *Savvalian Comfrey.*
Taxonomic Notes: Symphytum savvalense Kurtto is known only from the type specimen.
It is intermediate in some characters between S. asperum Lepechin and S. sylvaticum Boiss.
Red List Category & Criteria: Critically Endangered B1ab(i,ii,iii)+2ab(i,ii,iii).
Threats:
This species is listed as Critically Endangered in view of its small Extent of Occurrence (EOO) and Area of Occupancy (AOO), both of which are estimated to be 4 square km (1.5 square miles), its existence at a single location (it is known only from the type locality) and an observed, as well as projected, continuing decline in EOO, AOO, and the area, extent and quality of habitat caused by copper mining and air pollution. There are no conservation measures in place.
Geographic Range Description:
This species is endemic (native) to northeast Anatolia in Turkey. It is an Euxine (Black Sea) element.
Current Population Trend: Decreasing.
Habitat and Ecology: It was collected in a forest clearing, at about 1,300 meters (4265 feet) asl (above sea level)."
-'Symphytum Savvalense, Savvalian Comfrey' assessment by T. Ekim, M. Vural, H. Duman, Z. Ayta and N. Adiguzel, The IUCN Red List of Threatened Species®, www.iucnredlist.org, 7 pages, **2014**.

Symphytum sepulcrale

See subsection 'Subspecies and Varieties of S. Asperum', in section 'Details about Symphytum Species: Asperum or Asperrimum' (Chapter 7).
See 'Symphytum savvalense' and 'Symphytum sylvaticum' in this section.

Current Botanical Nomenclature of S. sepulcrale

"***Symphytum sepulcrale*** **Boiss. & Balansa ex Boiss.** *(*Fl. Orient. [Boissier] 4(1): 174. 1875)"*
-'International Plant Names Index'® (IPNI), www.ipni.org, A database of the names and associated basic bibliographical details of seed plants, ferns and lycophytes, **2018**.
(* -'Flora Orientalis Sive Enumeratio Plantarum in Oriente, a Graecia et Aegypto ad Indiae Fines {Plants Flora, Flora East and the East, from Greece and Egypt to the Borders of India}, Volume IV' by Edmond Boissier, Society Physics Geneva, Society Linnean London; Geneva, Switzerland, 1875 and 1879. In Latin. Symphytum sepulcrale, page 174, Boiss. et Bal. in Bal. pl. Pont. 1866.)

"***Symphytum sepulcrale* Boiss. & Bal.:**
synonym for Symphytum sylvaticum subsp. sepulcrale (Boiss. & Bal.) Greuter & Burdet.
***Symphytum sepulcrale* subsp. *sylvaticum* (Boiss.) Kurtto:**
synonym for Symphytum sylvaticum subsp. sylvaticum (Boiss.) Kurtto. Infraspecific taxon.
***Symphytum sepulcrale* var. *hordokopii* A. Kurtto:**
synonym for Symphytum sylvaticum var. hordokopii (Kurtto) R.R. Mill. Infraspecific taxon.
***Symphytum sylvaticum* subsp. *sepulcrale* (Boiss. & Bal.) Greuter & Burdet:** accepted name.
Infraspecific taxon."
-Catalogue of Life®: 2015 Annual Checklist, Online database of the world's known species of animals, plants, fungi and micro-organisms. It consists of a single integrated species checklist and taxonomic hierarchy., **2018**, www.catalogueoflife.org.
Search: www.catalogueoflife.org/annual-checklist/2015/search/all
(Infraspecific is at a taxonomic level below that of species, e.g., subspecies, variety, cultivar, or form.
Latin names at this level require the addition of a term denoting rank.)

"***Symphytum sepulcrale* Boiss. & Balansa** is an unresolved name."
-The Plant List®, www.theplantlist.org from the World Checklist Database, **2018**.

Description of S. sepulcrale

"***S. sepulcrale, Boiss. & Bal.**, in Boiss. Fl. Orient. iv. p. 174 (1879):*
Geographical Distribution: Turkey: Lazistan.
Easily recognised *by oblong, obtuse calyx-segments which are almost truncate at apex and ciliate with long spreading hairs. In the 'Cimitiere Turc de Djimil', Lazistan, specimen the **nutlets** are smooth and shining, as described by Boissier, but in some at least of the 'Boejukdere supra Artabir', Armenia, plants they are venoso-reticulate and minutely tuberculated.*
It differs from S. armeniacum *in the leaves being petiolate and rounded or cordate at the base, by the slender hairs of the stem, and by the more deeply divided calyx, the segments of which bear long cilia on the margin.*
The specimen in the Sherard Herbarium, Oxford University, England, anticipates its collection by Balansa by more than 150 years. It consists only of a branch with a pair of leaves and racemes, but agrees in all respects with Balansa's plant."
-'A Revision of the Genus Symphytum, Tourn.' by Cedric Bucknall, {Mus. Bac. Oxon= Bachelor of Music, Oxford University}, Journal of the Linnean Society of London, England, Botanical Journal, Volume 41, Issue 284, pages 491-556, December **1913**.
    (Lazistan was an administrative region of Ottoman Empire, 1551-1925. It included Laz or Lazuri-speaking population on the southeastern shore of the Black Sea. It is now Rize Province and the shoreline of Artvin Province, both in Turkey.)

"***S. sepulcrale Boiss. & Bal.*** *in Boiss., Fl. Orient. 4:174 (1879).*
Villose perennial herb. Stem sulcate, branched. Leaves petiolate, membranous, ovate to broadly lanceolate, acute or acuminate, rounded or cordate at base; lower leaves 12 cm (4.7 inch) long, 5.5 cm (2.1 inch) broad; upper leaves 7 cm (2.7 inch), 2.5 cm (0.9 inch) broad.
Inflorescence 10-12 flowered. Calyx 6 mm long, divided nearly to base, segments lanceolate-obtuse. **Corolla violet,** *15 mm long, tube equalling calyx, 2-3 mm broad, corolla scales 4 mm long, broadly linear, equalling stamens. Stamen filaments 2 mm long; anthers 2 mm long. Style not exserted.*
Nutlets not seen, smooth or minutely tuberculate (Bucknall) Fl. 5. Shady places, meadows."
-'A Revision of Symphytum in Turkey and Adjacent Areas' by G.E. Wickens, Royal Botanic Gardens: Kew; Notes from the Royal Botanic Garden Edinburgh, Scotland, Volume 29, pages 157-180, **1969**. Includes Turkey, Bulgaria, Greece, Aegean Islands and Caucasia.

"***Symphytum sepulcrale Boiss. & Bal. differs from S. asperum Lepechin*** *in many morphological characters and the two taxa seem to be allopatric.*
S. sepulcrale is regarded as a distinct species endemic (native) to Turkey.
A local variant of S. sepulcrale with longer and softer indumentum and some long-petiolate cauline leaves is described as **var. hordokopii Kurtto, var. nova***.*
S. longipetiolatum Wickens is reduced to synonymy with S. sepulcrale.
S. sylvaticum Boiss. is included in S. sepulcrale as subsp. sylvaticum (Boiss.) Kurtto, comb, et stat. nov.
Symphytum armeniacum, S. sepulcrale and S. sylvaticum are lectotypified."
-'Taxonomy of the Symphytum Asperum Aggregate (Boraginaceae), Especially in Turkey' by Arto Kurtto, Botanical Museum, University of Helsinki, Finland; Annales Botanici Fennici (Finnish Botanical Annals), Helsinki, Vol 19, No. 3, page 177-192, **1982**.
    (Allopatric or geographic speciation occurs when populations of the same species become isolated from each other, and this prevents genetic exchange.)
    (subsp. sylvaticum {Boiss.} Kurtto, comb, et stat. nov. = subspecies sylvaticum, originally published by Boiss., then revised publication by Kurtto, combination, and new status.)
    (Lectotype is a specimen chosen as the type of a species if the author of the name fails to designate a type.)

<u>Symphytum sylvaticum</u>
See 'Symphytum sepulcrale' in this section.

Current Botancial Nomenclature of S. sylvaticum

"***Symphytum sylvaticum Boiss.*** *(*Pl. Or. Nov. Dec. ii. 4. 1875.)*
Basionym of Symphytum sepulcrale subsp. sylvaticum (Boiss.) Kurtto, Ann. Bot. Fenn. 19(3): 188 (1982). "
-'International Plant Names Index'® (IPNI), www.ipni.org, A database of the names and associated basic bibliographical details of seed plants, ferns and lycophytes, **2018**.
(* -'Plantarum Orientalium Novarum Ex Flora Orientalis' {Plants, Flowers Represent the New East Asia} by Pierre Edmond Boissier, Geneva, Switzerland, Part 2, **1875**. Symphytum sylvaticum and Symphytum kurdicum, Part 2, pages 4-5. In Latin.)
    (Basionym or basyonym is the original name on which the new name is based.)

"***Symphytum sylvaticum****: accepted name.*
Symphytum sylvaticum subsp. sepulcrale (Boiss. & Bal.) Greuter & Burdet*: accepted name.*
    Infraspecific taxon.
Symphytum sylvaticum subsp. sylvaticum*: accepted name. Infraspecific taxon.*
Symphytum sylvaticum var. hordokopii (Kurtto) R.R. Mill*: accepted name. Infraspecific taxon.*

***Symphytum sepulcrale* subsp. *sylvaticum* (Boiss.) Kurtto:**
synonym for Symphytum sylvaticum subsp. sylvaticum (Boiss.) Kurtto. Infraspecific taxon."
-Catalogue of Life®: 2015 Annual Checklist, Online database of the world's known species of animals, plants, fungi and micro-organisms. It consists of a single integrated species checklist and taxonomic hierarchy., **2018**, www.catalogueoflife.org. Search: www.catalogueoflife.org/annual-checklist/2015/search/all
(Infraspecific is at a taxonomic level below that of species, e.g., subspecies, variety, cultivar, or form.
Latin names at this level require the addition of a term denoting rank.)

"Symphytum sylvaticum Boiss. is an unresolved name."
-The Plant List®, www.theplantlist.org from the World Checklist Database, **2018**.

Description of S. sylvaticum

"S. sylvaticum, Boiss., *Fl. Orient. iv. p. 172 (1879).
Geographical Distribution: **Turkey: northern Asia Minor.**
Leaves thin, conspicuously tuberculato-setose; corolla-scales broad, longer than the stamens; nutlets strongly reticulato-rugose, shining but minutely tuberculated.
This is one of the species whose range is, as far as is known, extremely limited.
The above specimen is the only one contained in the herbaria which I have examined. It exactly corresponds with Boissier's description."
-'A Revision of the Genus Symphytum, Tourn.' by Cedric Bucknall, (Mus. Bac. Oxon= Bachelor of Music, Oxford University), Journal of the Linnean Society of London, England, Botanical Journal, Volume 41, Issue 284, pages 491-556, December **1913**.
(* -'Flora Orientalis Sive Enumeratio Plantarum in Oriente, a Graecia et Aegypto ad Indiae Fines (Plants Flora, Flora East and the East, from Greece and Egypt to the Borders of India), Volume IV' by Edmond Boissier, Society Physics Geneva, Society Linnean London; Geneva, Switzerland, 1879. Symphytum pages 171-177. All in Latin, 5 volumes plus supplement, 1867-1888.)

*"**S. sylvaticum Boiss. is included in S. sepulcrale as subsp. sylvaticum (Boiss.) Kurtto, comb, et stat. nov.***
Symphytum armeniacum, S. sepulcrale and S. sylvaticum are lectotypified."
-'Taxonomy of the Symphytum Asperum Aggregate (Boraginaceae), Especially in Turkey' by Arto Kurtto, Botanical Museum, University of Helsinki, Finland; Annales Botanici Fennici (Finnish Botanical Annals), Helsinki, Vol 19, No. 3, page 177-192, **1982**.
(subsp. sylvaticum {Boiss.} Kurtto, comb, et stat. nov. = subspecies sylvaticum, originally published by Boiss.,
then revised publication by Kurtto, combination, and new status.)
(Lectotype is a specimen chosen as the type of a species if the author of the name fails to designate a type.)

"*Symphytum sylvaticum* Morphological Properties: *Perennial. Stem 36-48 cm (14.1-18.8 inches). Leaves oblong-lanceolate, acute or acuminate. Basal leaves petiolate, cauline leaves sessile.*
Inflorescence scorpoid cymes to helicoid cymes. Pedicels 5-7 mm. Calyx 5-7 in flowers, 6-11 in fruits. Corolla white, 11-13 mm. Nutlets 3-4.5 mm, subglobose, minutely tuberculate.
Anatomical Properties:
A transverse section taken from the root *was observed as follows: Periderm is multilayered. Secondary cortex is multilayered and occupies 30% of the root. Cortex consists of parenchymatic cells and secondary phloem. Parenchymatic cells are 26.25 +- 3.10 x 4.73 +- 2.34. Cambium cells are distinguishable and 1-3 layers. Xylem is composed of sclerenchymatic cells and tracheary elements. The pith consists of primary xylem elements.*
A transverse section taken from the middle of the stem *was observed: Epidermal cells consist of a single layer and orbicular or rectangular. Trichomes are present on the cuticle. Collenchyma is three-four layered. Cortex cells are 78.25 +- 5.43 x 51.25 +- 5.88. and multilayered. Endodermis is between cortex and vascular tissue. The vascular tissues are collateral type. Bundles are of different sizes. Xylem and phloem elements are clear. Cambium is distinguishable. Diameter of vessel members are 35.5 +- 6.4. The pith consists of large parenchymatic cells."*
-'Micromorphology and Anatomy of Three Symphytum (Boraginaceae) Taxa from Turkey' by Oznur Ergen Akcin and Hilal Baki, Department of Biology, Sciences & Arts Faculty, Ordu University, Ordu, Turkey; Bangladesh Journal of Botany, Volume 36, No. 2, pages 93-103, December **2007**.

"*Symphytum asperum* Lepechin and *S. sylvaticum* Boiss. species are taxonomically very difficult to solve. *Many taxonomical arrangements were done by different researchers before with observing herbarium sheets.*
In our research 25 populations were observed in field, and **we reached the decision that both of these species' infraspecific taxa should be synonymized (made the same)** *because of the continuation of the characters used to distinguish taxa before.*
S. asperum Lepechin var. armeniacum (Bucknall) Kurtto proposed by Kurtto was synonymed and included in S. asperum Lepechin.
S. sylvaticum Boiss subsp. sepulcrale var. sepulcrale and **S. sylvaticum Boiss subsp. sepulcrale (Boiss.&Bal.) Greuter&Burdet var. hordokopii (Kurtto) R.Mill.** were synonymed and included to S. sylvaticum.
The reasons of these taxonomical rearrangements were gathered from field observations, morphological measurements and scanning electron microscope images."
-'Taxonomy of Symphytum Asperum Lepechin and S. Sylvaticum Boiss. (Boraginaceae) Based on Macro- and Micro-Morphology' by Burcu Tarikahya and Sadik Erik, Hacettepe University, Department of Biology, Ankara, Turkey; Hacettepe

Journal of Biology and Chemistry, Volume 38, No. 1, pages 47-61, **2010**.
>(Herbarium: A collection of dried plant specimens mounted on sheets of paper. The plants are collected, identified by experts, pressed, and then mounted to archival paper so that all major morphological characteristics are visible. They are labeled with the scientific name, name of the collector, and information about where/how collected. The specimens are filed according to families and genera.)
>(Infraspecific is at a taxonomic level below that of species, e.g., subspecies, variety, cultivar, or form. Latin names at this level require the addition of a term denoting rank.)

"Symphytum sylvaticum: Distribution: Turkey (northeast Anatolia)."
-World Plants, http://www.catalogueoflife.org, Netherlands, **2015**. The Catalogue holds essential information on the names, relationships and distributions of over 1.8 million species. Search: www.catalogueoflife.org/annual-checklist/2015/search/all
>(Anatolia or Asian Turkey, the Anatolian peninsula or Anatolian plateau, is the westernmost part of Asia, that makes up the majority of modern-day Turkey. It is bounded by the Black Sea to the north, the Mediterranean Sea to the south, and the Aegean Sea to the west.)

*"**Symphytum Sylvaticum: Tomara:** Sin (synonym): Symphytum longipetiolatum.*
Perennial with stout rhizome, stem 25-50 cm (9.8-19.6 inch), hispid. Leaves broadly ovate, asperous, cordate, lower long-petiolate, upper subsessile, attenuate.
*Flowers 15-18. Calyx 5.5-6 mm, accrescent to 10 mm in fruit, divided at least to 3/4, lobes linear-lanceolate, obtuse. **Corolla pink or blue**, 11-12 mm, scales 4-5 mm, lanceolate, subacute, exceeding stamens by 1-2 mm. Nutlets 3 mm, curved, areolate, minutely tuberculate.*
Fl. 5. Meadows and by stream with Alnus, Corylus, Fagus and Ulmus, 1000-1600 meter (3280-5249 feet).
Endemic. Euxine (Black Sea) element.
Intermediate between S. asperum, which it resembles in floral characters, and S. ibericum."
-'Symphytum Sylvaticum: Tomara' published by Turkiye Bitkileri (Plants of Turkey), **2019**. Turkey has a great botanical richness with more than 9600 plant species. This website has been constructed to gather as much information as possible about native plants of Turkey. www.turkiyebitkileri.com
>(Alnus genus, also called Alder, is in birch family Betulaceae, distributed throughout temperate northern hemisphere. Corylus genus or hazel are deciduous trees and large shrubs native to the temperate northern hemisphere. It is the birch family Betulaceae. Fagus genus or beech are deciduous trees in the family Fagaceae that are native to temperate Europe, Asia, and North America. Ulmus genus or elms are deciduous and semi-deciduous trees in family Ulmaceae.)

Symphytum sylvaticum Alkaloids

*"**Dried samples of Symphytum sylvaticum Boiss. subsp. sepulcrale (Boiss. & Bal.) Greuter & Burdet var. sepulcrale plant gave a yield of total alkaloids of 0.199% in the roots and 0.093% in the aerial parts.***
The major alkaloid in the roots was Echimidine-N-oxide (ENO). The ENO content was higher in the roots (0.1078%) than in the aerial parts of the plant (0.0061%).
The yield of ENO was 55.96-57.54% in the root and 6.66-6.86% in the aerial part crude alkaloid fraction. This explains why the alkaloid fraction from the root was more active than the alkaloid fraction from the aerial part compared with the fungal cultures used. Consequently, the activity was found to be mainly due to Echimidine-N-Oxide (ENO)."
-'Antifungal Activities of Different Extracts and Echimidine-N-oxide from Symphytum Sylvaticum Boiss. subsp. sepulcrale (Boiss. & Bal.) Greuter & Burdet var. Sepulcrale' by Murat Kartal (Turkey), Semra Kurucu (Turkey) and M. Iqbal Choudary (Pakistan), Turkish Journal of Medical Sciences, Ankara, Turkey, Volume 31, pages 487-492, 2001.

*"**There are 17 Symphytum species (Boraginaceae) growing in Turkey and 8 of these are endemic (native)**. This is the first phytochemical investigation carried out on pyrrolizidine alkaloids of these two endemic Symphytum species.*
>***Pyrrolizidine alkaloid (Echimidine-N-oxide) was isolated from Symphytum sylvaticum Boiss.*** *subsp. sepulcrale (Boiss. & Bal.) Greuter & Burdet var. sepulcrale.*
>***Pyrrolizidine alkaloid (Echimidine) was isolated from Symphytum aintabicum*** *Hub. - Mor. & Wickens.*
The structures of the isolated compounds were elucidated based on IR, EIMS, 1H, and 13C NMR analysis and also on 2D NMR (COSY, HMBC, HMQC) experiments."
-'Pyrrolizidine Alkaloids from Symphytum Sylvaticum Boiss. subsp. sepulcrale. (Boiss. & Bal.) Greuter & Burdet var. sepulcrale and Symphytum aintabicum Hub. Mor. & Wickens' by S. Kurucu, M. Kartal, M.I. Choudary and G. Topcu (University of Ankara, Turkey) (University of Karachi, Pakistan)(Istanbul University, Turkey); Turkish Journal of Chemistry, Vol 26 , page 195-199, **2002**.

Symphytum tanaicense (Symphytum officinale subsp. tanaicense Steven)

See section 'Details about Symphytum Species: Symphytum officinale' (Chapter 8).
See 'Symphytum uliginosum' in this section.

Current Botanical Nomenclature of S. tanaicense

"Symphytum tanaicense Steven (*Bull. Soc. Imp. Naturalistes Moscou xxiv. (1851) I. 577.)"
-'International Plant Names Index'® (IPNI), www.ipni.org, A database of the names and associated basic bibliographical details of seed plants, ferns and lycophytes, **2018**.
(* -'Bulletin de la Societe Imperiale des Naturalistes Moscou' {Bulletin of the Imperial Society of Naturalists of Moscow}, Russia, Volume 24, No.1, page 577, **1851**.)
(See sub-subsection 'Steven 1851' in subsection 'Symphytum Genus' in section 'Borage Family, Symphytum Genus' {Chapter 2}.)

"Symphytum tanaicense Stev.: accepted name."
-Catalogue of Life®: 2015 Annual Checklist, Online database of the world's known species of animals, plants, fungi and micro-organisms. It consists of a single integrated species checklist and taxonomic hierarchy., **2018**, www.catalogueoflife.org.
Search: www.catalogueoflife.org/annual-checklist/2015/search/all

"Symphytum tanaicense Stev.: synonym of Symphytum officinale subsp. uliginosum (A.Kern.) Nyman: accepted name."
-Global Biodiversity Information Facility® (GBIF), www.gbif.org, **2018**. International network and research infrastructure funded by the world's governments and aimed at providing anyone, anywhere, open access to data about all types of life on Earth.

Description of S. tanaicense

"*S. officinale is usually fairly hairy and often densely so above, but forms are found, especially in Eastern Europe, in which the plant loses much of its hairiness and becomes asperous with thinly scattered tubercular setae.*
    *When this is the case and the leaves are also narrowly lanceolate and only partially decurrent, the plant belongs to the Symphytum officinale variety lanceolatum Weinm., which no doubt consists of **a series of forms intermediate between typical S. officinale and the very distinct S. uliginosum Kerner.***
S. tanaicense Steven, which was gathered in the region of the river Don (Tanais) in 1817 was described in 1851. It is clear that this is either one of the above-mentioned intermediate forms, or S. uliginosum itself.
    That S. uliginosum is found in south Russia is proved by an excellent specimen in the Fielding Herbarium (Oxford, England) gathered at Novgorod-Sieversk in 1824 as S. officinale, of which there is also a typical specimen on the same sheet. It is therefore quite possible that S. uliginosum is identical with S. tanaicense, but as I have seen no authentic specimen of the latter this must remain doubtful."
-'A Revision of the Genus Symphytum, Tourn.' by Cedric Bucknall, {Mus. Bac. Oxon= Bachelor of Music, Oxford University}, Journal of the Linnean Society of London, England, Botanical Journal, Volume 41, Issue 284, pages 491-556, December **1913**.
    (The Don River is a major Eurasian river and the fifth longest river in Europe. It starts in Novomoskovsk, Russia {120 killometers = 74 miles south of Moscow}, and flows 1,870 kilometers {1162 miles} to the Sea of Azov in eastern Europe. From the 11th century BC to the 2nd century AD, Scythians called the river Silys. At that time Greeks called it Tanais.)
    (The Principality of Novgorod-Seversk was a medieval Rus principality centered on the town now called Novhorod-Siverskyi. Now that region is part of Russia, Ukraine and Belarus.)

"**There is some confusion with regard to the correct name of the taxon described by Kerner (*1863).**
Von Degen (**1930) remarked that Steven's (1851) description of S. tanaicense completely agrees with that of Kerner, so that the correct name is S. tanaicense Steven.
Both Nyman (***1878-1882) and R.V. Soo (****1931) reduced this taxon to subspecific rank:
    S. officinale L. subsp. uliginosum (Kern.) Nym. and
    S. officinale L. subsp. tanaicense (Stev.) Soo respectively.
According to the rules of botanical nomenclature S. officinale L. subsp. uliginosum (Kern.) Nym. is the correct name.
For final conclusions the type material of Steven must be consulted for comparison."
-'Cytotaxonomic Studies in the Genus Symphytum III: Some Symphytum Hybrids in Belgium and the Netherlands' by Th.W.J. Gadella and E. Kliphuis, Biologisch Jaarboek (Biological Yearbook), **1971**.
(* -'Descriptiones Plantarum Novarum Florae Hungaricae et Transsilvanicae' by A. Kerner, Oesterreichische Botanische Zeitschrift (Austrian Botanical Journal), Vienna, Austria, Volume 13, No. 7, pages 227-288, 1863.)
(** -'Bemerkungen uber einige orientalische Pflanzenarten LXXXIX. Uber Symphytum uliginosum Kern.' {Remarks about some Oriental Plant Species LXXXIX. About Symphytum uliginosum Kern.} by A. Von Degen, Magyar Botanikai Lapok {Hungarian Botanical Pages}, Volume 29, pages 144-148, 1930. If you have an English translation, could you send it to me.)
(*** -'Conspectus Florae Europaeae: seu Enumeratio Methodica Plantarum Phanerogamarum Europae Indigenarum, Indicatio Distributionis Geographicae Singularum etc.' by C.F. Nyman, Orebro Sueciae {Sweden}, 1878-1882.)
(**** -'Zur Kenntnis von Symphytum Uliginosum / Tanaicense' (To the Knowledge of Symphytum) by R. Von Soo, 'Kritische Bemerkungen und neue Beitrage zur Kenntnis der ungarischen Flora IV', Botanikai Kozlemenyek, XVIII, Volume 5, pages 125-133, 1931. If anyone has an English translation of this, I would appreciate having it.)

"*S. tanaicense Steven: The west European plants share a number of characters with east European plants of S. tanaicense Steven. However, the latter species has neither been studied cytologically nor chemically.*
If the material from the low-lying peat lands of The Netherlands is identical with Hungarian and south Russian plants, they should be assigned taxonomically to S. tanaicense."
-'Cyto- and Chemotaxonomical Studies on the Sections Officinalia and Coerulea of the Genus Symphytum' by Th.W.J. Gadella, E. Kliphuis and H.J. Huizing; Botanica Helvetica (now called Alpine Botany), Volume 93, pages 169-192, **1983**.

(For more of this article and another article that mentions S. tanaicense, see subsection 'Hybrid Origin and Classification {peregrinum, uplandicum}' in section 'Details about Symphytum Species Hybrids: Russian Comfrey' {Chapter 10}.)

"In the Netherlands **the 2n = 40 cytotype** is very common in the low lying peat lands of Noord Holland, Utrecht, and Friesland. Populations with 2n = 40 may have a much wider European distribution because some of the characters of this cytotype closely match those of plants from Hungary and southern Russia.
These later plants were referred to S. uliginosum Kern. by R.V. de Soo (1926) and to S. tanaicense Steven by Degen (1930). Symphytum tanaicense is the correct name."
-'Notes on Symphytum (Boraginaceae) in North America' by T.W.J. Gadella, page 1061-1067 in book: Annals of the Missouri Botanical Garden, Volume 71, Saint Louis, Missouri. Lawrence, Kansas: Allen Press, **1984**.

"*Symphytum officinale* L.: 2n = 24 or 2n = 48
    S. *bohemicum* Schmidt, Fl. Boem. 3: 13. tab 263. 1795.
    S. *tanaicense* Steven, Bull.Soc.Imp. Naturalistes Moscou 24:577. 1851
    S. *uliginosum* Kerner, Oesterr. Bot. Z. 13: 227, 1863.
White or cream flowers in diploids (2n = 24) and tetraploids (2n = 48).
Purple or red (or various intermediate colors between white and dark purple) in the tetraploids (2n = 48).
The diploid cytotype was assigned to Symphytum bohemicum Schmidt by A. Murin and J. Majovsky (Acta Facultatis Rerum Naturalium Universitatis Comenianae Botanica, Volume 29, 1982.)
The exact status of S. bohemicum, S. tanaicense, and S. uliginosum appears to require further investigation."
-'Notes on Symphytum (Boraginaceae) in North America' by T.W.J. Gadella, page 1061-1067 in book: Annals of the Missouri Botanical Garden, Volume 71, Saint Louis, Missouri. Lawrence, Kansas: Allen Press, **1984**.

"S. tanaicense shows a pyrrolizidine alkaloid and triterpene pattern similar to S. officinale (2n = 40). Also on morphological and cytological grounds they are very similar.
It seems highly probable that S. tanaicense is conspecific (same species) with S. officinale (2n = 40) and represents an intraspecific variant only.
S. officinale var. lanceolatum contained no pyrrolizidine alkaloids but did contain isobauerenol. This feature points to an origin from S. officinale."
-'Chemotaxonomy of the Symphytum Officinale Agg. {Boraginaceae}' by Jaarsma, Lohmanns, Gadella and Malingre, Plant Systematics and Evolution, 167, pages 113-127, **1989**.
    (Intraspecific means occurring within a species or involving members of one species.)

"In his revision of Symphytum, Bucknall (*1913) recognised **2 species in Symphytum 'section Officinalia'** (i.e., S. Section Symphytum):
    1. The widespread S. officinale L.:
        The whitish-flowered plants were assigned to S. officinale 'subvar. ochroleucum'.
    2. S. uliginosum A. Kern., known from Hungary and south Russia.
        The purple-flowered (purplish violet) ones to subvar. purpureum (Pers.) Buckn. (i.e., subvar. officinale), of which S. patens Sibth. was considered a synonym.
A taxon showing intermediate features between S. officinale and S. uliginosum was named S. officinale var. lanceolatum Weinm., with S. tanaicense as a tentative synonym.
On the basis of Steven's and Kerner's original descriptions and after study of the exsiccata in FI and PI, the identity of the Symphytum from the Massaciuccoli Lake (Tuscany, Italy) can be established as follows:
    Symphytum tanaicense Steven = Symphytum officinale subsp. tanaicense (Steven) =
    Symphytum uliginosum Kern = Symphytum officinale subsp. uliginosum Nym."
-'Symphytum Tanaicense (Boraginaceae) New for the Italian Flora' by Peruzzi, Garbari and Bottega, Willdenowia, Berlin, Germany, 31, pages 33-41, **2001**.
(* -'A Revision of the Genus Symphytum, Tourn.' by Cedric Bucknall, {Mus. Bac. Oxon= Bachelor of Music, Oxford University}, Journal of the Linnean Society of London, England, Botanical Journal, Volume 41, Issue 284, pages 491-556, December 1913.)
    (subvar.=subvariety, var.=variety) (Taxon is a group of organisms seen by taxonomists to form a unit.)

"**Species: Symphytum tanaicense Steven**
    Synonyms: Symphytum uliginosum A.Kern.
                     S. officinale L. subsp. uliginosum (A.Kern.) Nyman
                     S. uliginosum Kern. var. pseudopterum Borbas
    Family: Boraginaceae
    Common Name: Consolida del Don
Ecology: Symphytum tanaicense is a species linked to environments of marsh planes, such as **marshes, alluvial meadows, coastal wetlands, glades in riparian forests and banks of water courses.**
According to the data we detected, the species is heliophilous (likes sunlight) and appears ubiquitous (everywhere) as far as concerns the edaphic (soil) aspects, **growing in soils medium rich in nutrients, with neutral-sub-alkaline pH, with tolerance even at high concentrations of limestone.**
The texture of the two investigated sites varies from clayey-silt (coltano) with a prevalent skeleton (Massaciuccoli). Participates

in hygrophilous (likes moisture) plant communities."
-'Symphytum Tanaicense Steven: Schede per una Lista Rossa della Flora Vascolare e Crittogamica Italiana (Cards for a Red List of Italian Vascular and Cryptographic Flora)' by M. D'Antraccoli, F. Aiello and L. Peruzzi, Dipartimento di Biologia, Universita di Pisa, Italy; Informatore Botanico Italiano, Volume 47, No. 2, pages 245-289, **2015** / 2016. In Italian. (If you have an English translation, I would appreciate a copy.)   (Riparian is wetlands next to rivers and streams. Glade is a open area in a forest.)

S. tanaicense Chromosomes

"Symphytum officinale L. ssp. uliginosum (Kern.) Nyman (2n = 40)
S. tanaicense Steven (species, 2n = 40)
S. uliginosum Kern
At present there appears to be two cytotypes of Symphytum officinale that may warrant recognition at some level:
  1. The 2n = 24 or 48 type (diploid and tretraploid, respectively) *appears to be the common form in New England (northeastern United States). It shows hispid stems that are not harsh to the touch, leaves with prominent decurrent wings, marginal setae of the sepals distributed in an irregular pattern, and white (usually diploid) or purple (usually tetraploid) corollas.*
  2. The 2n = 40 type, *known from North America but yet documented in New England, has tuberculate-based hairs on the stem that are harsh to the touch, leaves with shorter decurrent wings, marginal setae of the sepals distributed in a uniform manner, and* **purple corollas.**
  This latter cytotype has been named ssp. uliginosum (or S. tanaicense at the rank of species)."
-New England Wild Flower Society's Flora Novae Angliae: A Manual for the Identification of Native and Naturalized Higher Vascular Plants of New England by Arthur Haines. New Haven: Connecticut: Yale University Press, **2011**.
  (A cytotype is an individual of a species that has a different chromosomal factor to another.)
  (Body cells and individuals are described by the number of chromosome sets {ploidy level} they have: monoploid: 1 set; diploid: 2 sets = 2n = 2x, triploid: 3 sets = 2n = 3x, tetraploid: 4 sets = 2n = 4x, pentaploid: 5 sets = 2n = 5x, etc.)

*"**Symphytum tanaicense Steven**, Bull. Soc. Imp. Nat. Mosc. 24. 1851*
  *= **S. uliginosum Kern. var. pseudopterum Borbas**, Flora von Budapest und Umgebung, 1879.*
Chromosome Number: 2n = 40 *(*Peruzzi & al., 2001). Obtained from the only one living population found in Tuscany, near Massaciuccoli, in the locality la Piaggetta (387-388/1999 HBP) (central Italy).*
Data Sheets: 2n = 40 *(Love & Love, 1956; Gadella & Kliphuis, 1967, 1969, 1973, 1974, 1978; Gadella et al., 1983; **Jaarsma et al., 1989; Sandbrink et al., 1990).*
Ecology: *Prefers humid, peaty environments, at 0 meters s.l.m. (sopra il livello del mare = above sea level).*
Biological Form: *Scaposa Hemicryptophyte.*
Phenology: *It blooms from April to June.*
General Distribution: *South of Russia, Hungary, Romania, Holland (Steven 1851, Bucknall 1913, ***Dobrochaeva 1967, Gadella et al. 1983), Slovakia, south-east of Poland (Sandbrink et al., 1990) and central-northern Italy."*
-'Il Genere Symphytum L. (Boraginaceae) in Italia: Revisione Biosistematica' (The Genus Symphytum L. Boraginaceae in Italy: A Biosystematic Revision) by Stefania Bottega and Fabio Garbari, Webbia: Journal of Plant Taxonomy and Geography, Volume 58, No. 2, pages 243-280, **2003**. (If you have an English translation of this article, I would appreciate having a copy.)
(* -'Symphytum Tanaicense {Boraginaceae} New for the Italian Flora' by Peruzzi, Garbari and Bottega, Willdenowia, Berlin, Germany, 31, pages 33-41, 2001.)
(** -'Chemotaxonomy of the Symphytum Officinale Agg. {Boraginaceae}' by Jaarsma, Lohmanns, Gadella and Malingre, Plant Systematics and Evolution, 167, pages 113-127, 1989.)
(*** -'On Taxonomy of the Genus Symphytum 1: Symphytum Section' by D.N. Dobrochaeva; Ukrayins'k Botanicnyi Zhurnal {Ukrainian Botanical Journal}, Volume 25, No. 5, pages 33-39, 1967 or 1968. In Russian. I could not find this article. If you have an English translation, I would appreciate a copy.)
  (Phenology is relation between climate and periodic biological phenomena such as plant flowering, leafing and breeding.)
  (Hemicryptophytes: herbs {very rarely woody plants} with buds at soil level.
  Hs: Semi-rosette hemicryptophytes, with leafy stems but the lower leaves larger than the upper ones and the basal internodes shortened.)
  (Scapose means consisting of a scape. Scape is a long, leafless flower stalk coming directly from a root.)

*"**Symphytum tanaicense Steven: Chromosome Count is 2n = 40.**"*
-Tropicos®, www.tropicos.org, Missouri Botanical Garden, Saint Louis, Missouri; Plant database with 1.3 million scientific names and 4.4 million specimen records, **2018**.

Symphytum tauricum (Crimean Comfrey)
Crimea is a peninsula on the northern coast of the Black Sea in Eastern Europe that is almost completely surrounded by the Black Sea and Sea of Azov to the northeast.

See 'Symphytum bulllatum' in this section.
See 'Symphytum ibericum' in this section.

See 'Symphytum orientale' in this section.

Current Botanical Nomenclature of S. tauricum

"Symphytum tauricum Willd. (*Neue Schriften Ges. Naturf. Freunde Berlin ii. (1799) 121. t. 5. f. 1.)"
-'International Plant Names Index'® (IPNI), www.ipni.org, A database of the names and associated basic bibliographical details of seed plants, ferns and lycophytes, **2018**.
(* -'Der Gesellschaft Naturforschender Freunde zu Berlin, Neue Schriften' {The Society of Friends of Nature for Berlin, New Writings}, Volume 2, page 121, 1799. Tab./image 5, figure 1. In German.)

"Symphytum tauricum Willd.: accepted name."
-Catalogue of Life®: 2015 Annual Checklist, Online database of the world's known species of animals, plants, fungi and micro-organisms. It consists of a single integrated species checklist and taxonomic hierarchy., **2018**, www.catalogueoflife.org. Search: www.catalogueoflife.org/annual-checklist/2015/search/all

"Symphytum tauricum Willd.
   = *Symphytum borragineum Tausch*
   = *Symphytum bullatum Hornem.*
   = *Symphytum orientale Pall.*
Varieties:
   Symphytum tauricum var. ibericum Kuntze
   Symphytum tauricum var. tauricum "
-Global Biodiversity Information Facility® (GBIF), www.gbif.org, **2018**. International network and research infrastructure funded by the world's governments and aimed at providing anyone, anywhere, open access to data about all types of life on Earth.

"Symphytum bullatum Hornem. is an unresolved name.
The record derives from WCSP (World Checklist Database) (in review) (data supplied on 2012-03-23) which does not establish this name either as an accepted name or as a synonym with original publication details:
*Hort. Bot. Hafn. 1: 179 1813."
-The Plant List®, www.theplantlist.org from the World Checklist Database, **2018**.
(* -'Hortus Regius Botanicus Hafniensis, In Usum Tyronum et Botanophilorum, Hauniae' by J.W. Hornemann, Professor of Botany, Copenhagen, Denmark, Volume 1, page 179, 1813. Three volumes.)
   (Jens Wilken Hornemann, 1770-1841, was a Danish botanist. He published 'Flora Danica' from 1801 to 1840. He was lecturer and director of the University of Copenhagen Botanical Garden, Denmark.)

Description of S. tauricum

"Symphytum tauricum (Taurian Comfrey) Description:
Stem herbaceous, scarcely a foot (0.3 meter) high, hispid and spotted with small reddish dots. Branches very divaricate. Leaves on very short winged footstalks, ovate, acute, with a broad base frequently cordate, rugose, margin finely undulated, so as to give the appearance of being crenulate, villous and hairy along the veins: floral leaves sessile, opposite, spreading.
   Flowers: *Racemes terminal, always two together, nodding. Calyx five-cleft: segments subulate. Corolla funnel-shaped, white: tube shorter than calyx: limb cup shaped: laciniae very obtuse. Nectaries five barren filaments, lance-shaped between the stamens, and a little longer than them. Style oblique, the length of the corolla.*
This species differs in many respects from Symphytum orientale, *in which the leaves are more approaching to lance shaped, not rugose, nor undulate at the margin; limb of the corolla spreading gradually from the tube to the mouth, not suddenly in form of a cup; style considerably exserted, which in this only equals the corolla.*
A hardy perennial. Flowers in April and May. **Native of Southern Tauria.**
Communicated by Mr. Joseph Knight, of the Exotic Nursery, King's-Road, Chelsea, London, England, who **raised it from seeds sent from Russia, under the name of Symphytum bullatum.**
It appears to have been described by Willdenow, in his *'Enumeration of the Plants' cultivated in the Berlin Garden, Germany, under the appellation which we have adopted."
-'Curtis's Botanical Magazine, Or, Flower-Garden Displayed', Volume 43, First of the New Series, by John Sims, M.D., Fellow of the Royal and Linnean Societies, London, England, **1816**. Symphytum tauricum, plate 1787 with text.
(* -'Enumeratio Plantarum Horti Regii Berolinensis: Continens descriptiones omnium vegetabilium in horto dicto cultorum' {An Enumeration of the Royal Garden Plants Berlin} by Carl Ludwig Willdenow. Berolini {Berlin}, Germany, 1809. Symphytum is in Part 1, pages 183-184.)
   (I'm not sure what region 'Tauria' is. Perhaps it means Crimea or the Tauric Peninsula, as it was called in antiquity. Crimea is a peninsula on the northern coast of the Black Sea in Eastern Europe that is almost completely surrounded by the Black Sea and Sea of Azov to the northeast.)

*"****Contrast of colour is also frequently produced by the corollas changing their colour at various stages of development.***
Especially remarkable in this respect are *the Bitter Vetches (Orobus vernus and Venetus), and several Boragineous plants belonging to widely different genera (e.g. Pulmonaria officinalis, Mertensia Sibirica,* **Symphytum tauricum***)."*

-'The Natural History of Plants, Their Forms, Growth, Reproduction, and Distribution' (Pflanzenleben = Plant Life or Plantscapes, 1887) by Anton Kerner von Marilaun, Austrian botanist. Translated by F.W. Oliver from 6 volumes in German. Published by Blackie & Son, London, England, **1895**. Volume II, Symphytum pages 97-99, 191, 275, 441, 585, 744.

"S. tauricum, Willd. in Neue Schrift. Nat. Berl. ii. p. 120, t. 6. fig. 1, (1799):
Geographical Distribution: **Russia: from Podolia to the Crimea.** *This is distinguished from S. orientale, which it resembles in the cordate leaves, by the deeply divided calyx with lanceolate segments.*
       The plant from Allesley, Warwick, England was mentioned as a garden escape by Syme in 'English Botany' in *1867, and he states that it had also been found on a hedge-bank near the Observatory at Cambridge, England, but specimens from this locality belong to S. orientale.
S. grandiflorum DC. has been taken for S. tauricum, *but differs greatly in the creeping root with decumbent leafy shoots and simple flowering stem with few-flowered racemes.*
S. bullatum, Hornem., *is said to differ in the smaller leaves of a darker green, the more hispid stem, the hairs with purple glands at the base, and the more inflated, pale corolla."*
-'A Revision of the Genus Symphytum, Tourn.' by Cedric Bucknall, (Mus. Bac. Oxon= Bachelor of Music, Oxford University), Journal of the Linnean Society of London, England, Botanical Journal, Volume 41, Issue 284, pages 491-556, December **1913**.
(* -'English Botany {Sowerby's}; or Coloured Figures of British Plants: Volume 7, Third Edition' by John T. Boswell {Syme} and John Edward Sowerby, London, England, 1867, 1880.)
    (Podolia or Podilia is in west-central and southwest parts of Ukraine and northeast Moldova.)
    (Decumbent means a plant or part of a plant lying along the ground with the end curving up.)

"Two other introduced species (to Britain) are recorded as established locally: **S. tauricum Willd. which differs from S. orientale in having bullate leaves with undulate margins and lanceolate acute calyx-teeth.***"*
-'The Genus Symphytum in Britain' by T.G. Tutin, University College of Leicester, England; Watsonia: Journal of the Botanical Society of the British Isles, Volume 3, pages 280-281, February **1956**. (Watsonia was the journal of the 'Botanical Society of Britain and Ireland' from 1949 to 2010. It is now called 'New Journal of Botany'.)
    (Willd. = Carl Ludwig Willdenow, 1765-1812, was a German botanist, pharmacist, and plant taxonomist. He is a founder of phytogeography which is the study of the geographic distribution of plants. His herbarium of more than 20,000 species is preserved in the 'Botanical Garden' in Berlin, Germany. See 'Willdenow 1799' in subsection 'Symphytum Genus' in section 'Borage Family, Symphytum Genus' {Chapter 2}.)

"S. tauricum Willd., Ges. Naturf. Freunde Berlin Neue Schr. 2: 120 (1799).
Stock fusiform. Stem 20-60 cm (7.8-23.6 inch), rather stout, much-branched, densely hairy with very short hooked hairs and setae up to 2-3 mm. All cauline leaves except the uppermost petiolate, slightly cordate to truncate or rounded at the base, not decurrent, rather densely hairy on both surfaces, asperous.
Calyx 4-7 mm, lobed to 3/5-5/6. Corolla 8-9 to 12-15 mm, **pale yellow***; lobes not recurved. Scales lingulate, rounded or sometimes emarginate at the apex, the marginal papillae up to twice as long as wide, scattered towards the base. Stamens with filament mostly shorter than anther; anthers 2-3 mm, 2 1/2 to 4-5 times as long as wide.*
Chromosomes: 2n=18. Woods.
Southeastern Europe: Bulgaria, Romania; Russia (Central region, Southwestern region, Crimea, Southeastern region).*"*
-Flora Europaea, Volume 3: Diapensiaceae to Myoporaceae, editors Tutin, Heywood, Burges, Moore, Valentine, Walters and Webb. United Kingdom: Cambridge University Press, **1972**, page 104. (Symphytum L., pages 103-105 by B. Pawlowski.) (5-volume encyclopedia of plants, published between 1964 and 1993.)

"Medvedev (*1970) explored the breeding possibilities of **the less cold-tolerant S. orientale L. and S. tauricum Willd.***"*
-'Comfrey Symphytum spp. as a Forage Crop' by J.C. Forbes, A.D. McKelvie, and P.J.C. Saunders, North of Scotland College of Agriculture, Aberdeen, United Kingdom; Herbage Abstracts, Volume 49, No. 12, pages 523-539, **1979**.
(* -'Basic Material and Methods for Selection in Comfrey, Part 1' {Iskhodnyi Material i Metody Selektsii Okopnika} by P.F. Medvedev; 5th Symposium on New Silage Plants, Leningrad, Russia, pages 68-69, 1970. If you have this in English, I would appreciate a copy.)

"S. tauricum Willd. - Crimean Comfrey:
Stems erect, well-branched, to 60 cm (23.6 inch), from thick, vertical root; stem-leaves petiolate to sessile but not decurrent; calyx divided about 3/5 to 5/6 to base; **corolla pale yellow.**
Introduced; naturalized on hedge-bank; one place in Cambridgeshire (east England) since 1973; southeast Europe."
-New Flora of the British Isles: Identification of Wild Vascular Plants of the British Isles edited by Clive Stace. Cambridge, England: Cambridge University Press, **1991**, page 648.

*"***Symphytum tauricum Willd. is European nemoral species distributed in Ukraine mostly in the northern and western regions of the Forest-Steppe and Steppe zones.**
This species has not been indicated in the last generalized monograph about the flora of the Left Bank of Dnipro/Dnieper River Area (Russia, Belarus, Ukraine) (*Bayrak, 1997). The only literature source attributed to the distribution of Symphytum tauricum in Poltava, Ukraine region is an old find of it in the vicinity of Poltava town (**Illichevsky, 1927): in the clay zone on the right bank of Vorskla river, rare and only to the south of Poltava town.

Based on this indication of S.O. Illichevsky **we could find this species in broadleaved forests of Rozsoshentsi forestry near Poltava town during our field studies in 2002-2005.**
> *Firstly the one plant of Symphytum tauricum we found in 2002 in oak forest near pine plantation not far from Velykyi Trostianets village of Poltava district. Now this location probably is extinct.*
>
> *In 2003 near Tiutiunnyky village a little population (8 plants in general, 2 from them were juvenile and 3 were senile {old}) was found; it is preserved now.*

The most numerous location of Symphytum tauricum now is situated in the southern part of Poltava district between Bulanove and Sapozhyne villages; it is allocated near broadleaved forests on areas with rough country. The current state of this population is good. The number of juvenile plants reaches 35% from all individuals in the population.

This species has no especial concurrents because it grows not only on grey forest soils but on clay loess soils too distributed on slopes of the right bank of Vorskla river. So probably the quantity of this population is able to increase in future *based on fact that only several nemoral species were found on areas with these ecological features."*

-'Symphytum Tauricum Willd. in Poltava Region' {Symphytum Tauricum Willd. na Poltavshchyni} by Denis A. Davydov and L.M. Gomlya; M.G. Kholodny Institute of Botany, NAS of Ukraine, Kyiv, Ukraine; Conference: Regional Problems of Nature Management & Plant and Animal Kingdoms Protection: Proceedings of Second International Scientific-Practical Conference, Kryvyi Rih, Ukraine, page 88, 2006. In Ukranian and English.

(* -'The Conspectus of the Flora of the Left Bank of Dnieper River Area' {Konspekt Flory Livoberezhnoho Prydnoprovya Sudynni Roslyny} by O.M. Bayrak; Poltava, Ukraine: Verstka, page 162, 1997. In Ukrainian.)

(** -'Flora of the Vicinities of Poltava Town with Full List of the Wild Vegetation' {Illichevsky S.O. Flora Okolyts Poltavy z Povnym Spyskom Dykoyi Roslynnosti}, Poltava, Ukraine: Poltavapoligraf, page 32, 1927. In Ukrainian.)

> ('Nemoral' refers to groves or woodland biome zones according to the Walter classification system. They have a moderate temperate climate with short frost periods. They are frost-resistant broadleaved deciduous forests.)
>
> (A steppe is montane grasslands/shrublands and temperate grasslands with no trees except for those near rivers and lakes. The world's largest steppe is in eastern Europe, central Asia, and neighbouring countries.)
>
> (The Dnieper River is the fourth longest river in Europe, starting in Valdai Hills near Smolensk, Russia, {near Europe}, and going through Russia, Belarus and Ukraine to the Black Sea. Podolia or Podilia is in west-central and southwest parts of Ukraine and northeast Moldova.)

"Calcium (Ca) in Plants: Calcium participates in the formation and development of the living organisms, accumulates in the protoplasm, vacuoles, chloroplasts, mitochondria. One of the most important properties of calcium is to set negative charges on the surface of protoplasm. Calcium is a basic indicator of soil fertility.

The analysis of calcium in forest herbaceous species: *Most studied species have Ca2+ within 1-2%.*
> ***Symphytum tauricum Willd.*** *fit in the limits of 2-3% in 'Forest B'* = *Durmast oak with linden and ash forest on brown clay soils over deeply gleyed clay (in central Moldova).*
>
> ***Symphytum tauricum Willd. Ca2+: June: 2.5%, September: 1.98%.***

Magnesium (Mg) in Plants: *Magnesium is at the border between macro and micronutrients. Its predominant role is catalytic; Mg participates in biomass formation, photosynthesis, activation of enzyme systems, enters into constitution of chlorophyll.*

Mg content in studied herbaceous plants ranged mostly within the limits of up to 1%, most plants had values less than 0.5%
> ***Symphytum tauricum Willd. in 'Forest B' Mg2+: June: 0.38%, September: 0.31%"***

-'Calcium and Magnesium in Plants and Soil, Codri Reservation' by Tamara Cojuhari, Tatiana Vrabie, S. Pana, and P. Koterniak, 'Journal of Botany', Academy of Sciences of Moldova, Botanical Garden Institute, Chisinau, Republic of Moldova, Volume VIII, NR. 1 (12), pages 121-130, **2016**.

> (Gley is a sticky waterlogged soil lacking oxygen; usually gray to blue.)
>
> (Codri or Codru are forests in the hilly part of central Moldova. The Republic of Moldova is a landlocked country in Eastern Europe, bordered by Romania to the west and Ukraine to the north, east, and south.)

*"**Symphytum tauricum is native around the Black Sea, i.e. in southern Ukraine, southern European Russia, Anatolia (Asian Turkey), Romania and Bulgaria** (*Smejkal 1978, **Wickens 1978, ***Fedorov 2001).*

In the Czech Republic the occurrence of S. tauricum was first reported by Smejkal (1978), based on a herbarium specimen collected in the town of Cernosice near Prague in 1912.
> *On the sheet one specimen of S. tauricum is mounted together with three specimens of S. tuberosum subsp. tuberosum. Unfortunately, it is not clear whether these plants originated from cultivation or not. They also may have been mixed accidentally in herbaria.*

In the 1980s, S. tauricum was repeatedly recorded from the vicinity of villages Miretice and Ptakova Lhota in south-western Bohemia, Czech Republic, where it was found on garden waste and road verge (curb strip/lawn). It is not clear if it was only an ephemeral (short-lived) occurrence or if it still grows on any of these localities."

-'Distributions of Vascular Plants in the Czech Republic, Part 3' by Z. Kaplan, J. Danihelka, M. Lepsi, P. Lepsi, L. Ekrt, J. Chrtek Jr, J Kocian, J Prancl, L. Kobrlova, M. Hrones and V. Sulc, Czech Republic; Preslia: The Journal of the Czech Botanical Society, Volume 88, pages 459-544, **2016**. The third part of the publication series on the distributions of vascular plants in the Czech Republic includes grid maps.

(* -'The Genus Symphytum in Czechoslovakia' {Rod Symphytum v Ceskoslovensku} by M. Smejkal; Zpravy Ceskoslovenske Botanicke Spolecnosti {Czech Botanical Society News}, Prague, Czech Reupublic, Volume 13, pages 145-161, 1978. In Czech. If you have an English translation, I would appreciate a copy.)

(** -Flora of Turkey and the East Aegean Islands, 11 Volumes by Peter H. Davis. Scotland: Edinburgh University Press, 1984-

1988. {'Symphytum L.' in Supplement Volume 10, pages 186-189, 1988.} {'Symphytum L.' in Volume 6, pages 378-386 by G.E. Wickens, 1978.})
(*** -Flora of Russia: The European Part & Bordering Regions, Volume 5 edited by An. A. Fedorov. Boca Raton, Florida: CRC Press, 2001.)

S. tauricum Chromosomes

"**Symphytum tauricum Willd.**: Hortus Botanicus, Amsterdam, Holland, **2n = 40**, Britton."
-'Cytogenetic Studies on the Boraginaceae' by Donald M. Britton, Brittonia (published by The New York Botanical Garden), Volume 7, No. 4, pages 233-266, December 10 **1951**.

"**Symphytum L. Chromosome Numbers:**
    tauricum Willd.   18   Romania   *Tarnavschi   1948
This book has grown out of the requests by many users of 'Flora Europaea' (Tutin et al., 1964, 1968, 1972, 1976, 1980) for information on the sources of the chromosome numbers cited in that work."
-Flora Europaea: Check-List and Chromosome Index by D.M. Moore. Cambridge, England: Cambridge University Press, **1982**, page 179.
(* -'Die Chromosomenzahlender Anthophyten-Florav on Rumanien mit Einem Ausblick auf das Polyploidie-Problem' {The Number of Chromosomes The Anthophyten-Florav in Romania with an Outlook on the Polyploidy Problem} by I.T. Tarnavschi; Bulletin du Jardin et du Musee Botaniques de Universite de Cluj, Roumanie, Volume 28, pages 1-130, 1948. In German.)

"**Symphytum tauricum Willd. 2n =:**
    18   Flora Europaea   Tarnavschi 1948
    18   Fedorov 1974   Tarnavschi 1948
    40   Fedorov 1974   Britton 1951
    40   Darlington 1955 Britton 1951 "
-'Chromosome Counts Database'® (CCDB) is a comprehensive resource for plant chromosome numbers. Every chromosome number includes the original database and the research article reference, **2018**. http://ccdb.tau.ac.il

Symphytum tuberosum (Tuberous Comfrey)
See 'Symphytum angustifolium' in this section.
See 'Symphytum bulbosum' in this section.
See 'Symphytum nodosum' in this section.
See 'Symphytum zeyheri' in this section.

Current Botanical Nomenclature of S. tuberosum

"**Symphytum tuberosum L.** (*Sp. Pl. 1: 136. 1753)"
-'International Plant Names Index'® (IPNI), www.ipni.org, A database of the names and associated basic bibliographical details of seed plants, ferns and lycophytes, **2018**.
(* -'Species Plantarum, Volume 1' by Swedish naturalist Carol {Carl} Linnaeus, page 136, 1753.)

"**Symphytum tuberosum**: accepted name.
Symphytum tuberosum subsp. angustifolium (A. Kerner) Nyman: accepted name. Infraspecific taxon. **Symphytum tuberosum subsp. bulbosum (C. Schimper) P.Fourn.**:
    synonym for Symphytum bulbosum C. Schimper. Infraspecific taxon.
Symphytum tuberosum subsp. nodosum (Schur) Soo:
    synonym for Symphytum tuberosum subsp. angustifolium (A. Kerner) Nyman. Infraspecific taxon.
Symphytum tuberosum subsp. tuberosum: accepted name. Infraspecific taxon."
-Catalogue of Life®: 2015 Annual Checklist, Online database of the world's known species of animals, plants, fungi and micro-organisms. It consists of a single integrated species checklist and taxonomic hierarchy., **2018**, www.catalogueoflife.org. Search: www.catalogueoflife.org/annual-checklist/2015/search/all
    (Infraspecific is at a taxonomic level below that of species, e.g., subspecies, variety, cultivar, or form.
    Latin names at this level require the addition of a term denoting rank.)

"**Symphytum tuberosum L.**: accepted name.
Subspecies:
    Symphytum tuberosum subsp. angustifolium (A.Kerner) Nyman
    Symphytum tuberosum subsp. tuberosum L., 1753
        = Symphytum bulbosum Schur
        = Symphytum minus Bub."
-Global Biodiversity Information Facility® (GBIF), www.gbif.org, **2018**. International network and research infrastructure funded by the world's governments and aimed at providing anyone, anywhere, open access to data about all types of life on Earth.

"Symphytum tuberosum L.: Accepted.
Symphytum tuberosum subsp. nodosum (Schur) Soo: Accepted."
-The Plant List®, www.theplantlist.org from the World Checklist Database, **2018**.

"Symphytum tuberosum L. (Sp. Pl. 1: 136 1753)
Symphytum tuberosum subsp. angustifolium Nyman
Symphytum tuberosum subsp. nodosum Soo"
-Tropicos®, www.tropicos.org, Missouri Botanical Garden, Saint Louis, Missouri; Plant database with 1.3 million scientific names and 4.4 million specimen records, **2018**.

S. tuberosum Subspecies and Varieties (Tuberosum Complex)

S. tuberosum subspecies tuberosum.
S. angustifolium, S. besseri, S. bulbosum, S. floribundum, S. foliosum, S. gussonei, S. leonhardtianum,
S. mediterraneum, S. nodosum, S. popovii, S. zeyheri.

A species 'complex' is a group of closely related species that are extremely similar in appearance, so much so that the boundaries between them are frequently unclear.

The taxonomic ranks: Genus, Subgenus, Section, Subsection, Series, Species, Subspecies, Variety, Subvariety, Form, Subform. A taxon is a taxonomic group of any rank, such as a class, family, or species. Taxa is plural of taxon.
A 'Section' is a taxonomic rank below genus and above species. The subgenus, if present, is higher than the section, and the rank of series, if present, is below the section. Sections may in turn be divided into subsections.

Symphytum Sections:
The number of Sections (varying from 2 to 10) and the species in each change with new research.
Even now there are disagreements about where some Symphytum species belong. A few species are listed:
    Section Albida Bucknall
        S. anatolicum, S. longisetum, S. ottomanum, S. pseudobulbosum, S. tauricum
    Section Coerulea Bucknall
        S. asperum Lepech., S. caucasicum Bieb., S. hajastanum Gviniaschvili,
        S. peregrinum Ledeb., S. savvalense, S. sylvaticum
    Section Cordata Bucknall
        S. cordatum, S. ibericum
    Section Graeca Pawlowski
        S. icaricum, S. naxicola
    Section Lingulata Pawlowski
        S. abchasicum Trautv., S. ciscaucasicum Gviniaschvili, S. grandiflorum DC.,
        S. ibericum Stev., S.tauricum Willd.
    Section Orientalia Bucknall
        S. kurdicum, S. orientale
    Section Procopiania Wickens
        S. circinale, S. creticum, S. insulare
    Section Suborientalia Bucknall
        S. aintabicum, S. bornmuelleri, S. brachycalyx, S. longisetum, S. palaestinum Boiss.
    Section Symphytum Officinalis
        S.officinale L., S. uliginosum Kern.
    Section Tuberosa Bucknall
        S. bulbosum, S. gussonei, S. tuberosum L. and others

"Reports of the Floral Committee. The subjects exhibited were as follows: **Symphytum tuberosum variegatum superbum:** *From Mr. Salter, F.R.H.S., Versailles Nursery, Hammersmith. A handsome variety of a hardy herbaceous perennial, in which* **the leaves were broadly margined with cream-colour.** *It was 'Commended'."*
-'Proceedings of the Royal Horticultural Society, Volume 1', London, England, pages 564 and 565, June 1 1859 to December 31 **1861**. April 9 1861, page 564.   (Variegated means leaves that are edged or patterned in a second color, especially white.)

    "Mr. Bull showed **Symphytum tuberosum variegatum superbum,** *with rather pretty greyish-green leaves bordered with yellow."*
    -'The Gardeners Chronicle and Agricultural Gazette: A Newspaper of Rural Economy and General News',
    London, England, page 74, **1867**.

"The plants of the Tuberosum Section are more nearly related than those of most of the other Sections of the Symphytum genus, and opinions have differed greatly as to their rank. *Teodoro Caruel states that, having studied these* *plants in the neighbourhood of Florence (Italy), where all the forms grow promiscuously (interbreeding) together, he has formed the opinion (*Fl. Ital. vi. p. 879) that **S. mediterraneum, S. gussonei, S. bulbosum, and S. zeyheri are all different forms or states of S. tuberosum,** and that this species varies in the following manner:*

    The root is thickened in different ways, sometimes for a considerable distance as in **Reich. Icon. t. 103, sometimes becoming large and nodose as in ***Jacq. t. 63, and sometimes it is thickened at intervals into lateral, sessile or stipitate tubercles as in ****Bischoffs figure of S. filipendulum (S. bulbosum Schimp.) in 'Flora,' ix. p. 561.

    The flowers vary in size, sometimes the entire flower, sometimes only the corolla being smaller, and in the latter case the corolla-scales and style are necessarily exserted.

*But it should be borne in mind, **firstly, that where S. tuberosum and S. bulbosum grow together, the circumstances are not favourable for forming an opinion as to the relationship of the members of this group, on account of the probability of hybridization taking place between them** and producing a series of forms which may be confused with independent species. **Secondly,** that other characters must be taken into consideration besides those of the root and flowers on which Caruel relies. **Thirdly, that so far as is known, the plants mentioned above do not appear to be hybrids, but stable forms occupying separate geographical areas.***

    In the case of S. tuberosum and S. bulbosum the areas are partly distinct and partly overlapping.

    That of S. mediterraneum is within the area of S. tuberosum, and just outside that of S. bulbosum.

    S. gussonei and S. zeyheri are within the area of S. bulbosum, and outside that of S. tuberosum.

After much consideration and careful comparison of numerous specimens, I have decided to treat these plants as species, *in accordance with the views of those authors who, having studied them in their natural habitats, or in other ways, have had the opportunity of forming a definite opinion, have arrived at the conclusion that they are distinct entities, and not chance variations of a single species."*

-'A Revision of the Genus Symphytum, Tourn.' by Cedric Bucknall, (Mus. Bac. Oxon= Bachelor of Music, Oxford University), Journal of the Linnean Society of London, England, Botanical Journal, Volume 41, Issue 284, pages 491-556, December **1913**.

(* -'Flora Italiana' by Teodoro Caruel, Florence, Italy, Volume 6, Part 3, page 879, 1885.)

(** -'Icones Florae Germanicae et Helveticae, Volume XVIII' by Heinrich Gottlieb Ludwig Reichenbach and Heinrich Gustav Reichenbach; Lipsiae (Leipzig), Germany, **1858**. In Latin. Symphytum tuberosum page 58.)

(*** -'Observationum Botanicarum, Iconibus ab Auctore Delineatis Illustratarum', Pars/Part 3, by Nikolaus Joseph Von Jacquin {Nicolai Josephi Jacquin}, Vindobonae {Vienna}, Austria, 1768. Symphytum tuberosum, page 12, tab. 63. Illustrations are at the end of the book. In Latin.)

(**** -'Flora oder Botanische Zeitung' issued by K. Bayerische Botanische Gesellschaft, Regensburg, Germany, Volume 9, No. 2, pages 561-562, 1826. Includes Dr. Bischoff {from Heidelberg, Germany} figure {tab/image 1} of S. filipendulum = S. bulbosum Schimp. on last page.)

    (Theodore {Teodoro} Caruel, 1830-1898, was an Italian botanist of French-English parentage who specialized in flora of Tuscany, Italy. His herbarium was donated to the botanical institute at the University of Pisa, Italy.)

*"**Symphytum tuberosum var. (variety) angustifolium Kern.** is a very distinct looking plant, of which I have only seen the specimens cited.*

In these the primordial leaves are very small and narrow, measuring with the petiole 3 cm x 0.3 cm (1.18 x 0.11 inch); the largest stem-leaves are 10 cm x 1.4 cm (3.9 x 0.55 inch), the flowers 15 mm long, the calyx 8 mm long, with nearly filiform segments only 0.5 mm wide at the base."

-'A Revision of the Genus Symphytum, Tourn.' by Cedric Bucknall, (Mus. Bac. Oxon= Bachelor of Music, Oxford University), Journal of the Linnean Society of London, England, Botanical Journal, Volume 41, Issue 284, pages 491-556, December **1913**.

*"**Forms of Symphytum Tuberosum L.:***

Symphytum leonhardtianum in Austria, Switzerland and Croatia:

Towards the end of May 1926 I was staying at **Salzburg, in Austria,** and one day, while ascending the **Gaissberg (Gaisberg Mountain)** there, I was struck by the appearance of a Symphytum which grew all over the mountain and was then in full beauty on the upper slopes, under the beech trees just bursting into leaf.

The plant could only be the Austrian Symphytum tuberosum.

    My previous experience of S. tuberosum was limited to a single gathering in Scotland, another in the Pyrenees, where one or two individuals only were seen, and a few sheets of British examples received from other botanists.

    But I quickly recognised that the Gaissberg Comfrey looked distinct, being a dwarfer and much more ornamental plant, with broader leaves and larger, brighter flowers.

In the following April 1927 **this dwarf Symphytum came under my notice in two localities near Lugano (southern Switzerland)**. In the spring of 1930 I again met with a similar form in small quantity a little south of Ragusa, in Dalmatia (part of Croatia).

The Lugano plant is a weak grower, while the British one becomes very luxuriant. I learn that Dr. Stapf, who is familiar with the Austrian plant, has observed independently that it is not identical with the S. tuberosum of British botanists.

It will now be seen that the common Austrian form cannot be confidently referred to any described species other than S. tuberosum L., and as **the Symphytum type in Herb. Linn. (Linnean Herbarium) is the narrow-leaved western plant, which has been shown to be distinct, a new name becomes necessary.**

    The name S. Leonhardtianum is therefore proposed, in honour of Professor Otto Leonhardt, of Nossen in Saxony (Germany), through whom my attention was originally directed to the plant.

Symphytum tuberosum:

S. tuberosum is mainly a plant of south-west Europe, and some doubt attaches to records from more eastern stations (locations). It usually grows in damp situations, in shady woods or copses (group of trees), in moist meadows and along ditches, and sometimes in hedgerows and wastes.

Symphytum leonhardtianum:
S. Leonhardtianum differs from S. tuberosum in its slender rootstock, with fewer and longer branches and consequently less tufted growth; in its shorter and less branched stem; in its fewer and broader leaves, which increase in size upwards nearly to the top of the stem; in its more conspicuous flowers, with shorter and more strongly ciliate calyx-lobes, and broader, more brightly coloured corolla; and in its rather smaller and paler nutlets.
It is pre-eminently a Central European species, its range extending westward from the French Alps to the Pyrenees, and on the east into Russia and the Balkan Peninsula.
It is commonly found in mountain woods, in drier situations than those affected by S. tuberosum."
-'The Forms of Symphytum Tuberosum L.' by Herbert William Pugsley, B.A., F.L.S., Journal of Botany: British and Foreign, London, England, Volume 69, pages 89-97, April **1931**.

(Pyrenees is a range of mountains in southwest Europe between Spain and France.)
('Linnean Herbarium' of the Linnean Society of London, England, has the specimens from the 'Herbarium of Carl Linnaeus', 1707-1778.)
(The Balkan Peninsula is surrounded by the Adriatic Sea to the west, the Mediterranean Sea and Marmara Sea to the south, and the Black Sea to the east. Its northern boundary is the Danube, Sava and Kupa Rivers.)
(Herbert William Pugsley, 1868-1947, English botanist who contributed 30,000 specimens to the British Natural History Museum, London, England. Pugsley is one of the greatest British amateur botanists. He published revisionary and detailed articles about British and European genera.)

"In the 'Flora of the North-East of Ireland', edition 2 (1938) all the Tuberous Comfrey records are attributed to S. leonhardtianum, although S. tuberosum is reported from elsewhere in Ireland.
The author gathered and examined material from three localities in north-east Ireland, and considered that they come within the range of variation shown within S. tuberosum.
A subsequent examination of herbarium material of S. tuberosum and **S. leonhardtianum** from Europe has led him to the conclusion that **the latter should be treated as a synonym of S. tuberosum.**"
-'Proceedings of Botanical Society of British Isles, Volume 7' edited by EF Greenwood, London, England, 1967-1969, page 432.
(* -'A Flora of the North-East of Ireland, Edition 2' by Samuel Alexander Stewart. Belfast, Ireland: The Quota Press, 1938.)

"**S. tuberosum L.** Sp. pl. (1753) 136; DC. Prodr. X, 38;
S. nodosum Schur, Enum. Pl. Transsylv. (1866) 468.
S. leonhardtianum Pugsley, Journ. of Bot. 69 (1931) 89.
S. leonhardtianum is distinguished from S. tuberosum by the following:
leaves 3-6, elliptic, ovate, rarely lanceolate; cymes with 5-20 flowers; calyx lobes nearly half as long as corolla tube; corolla 12-20 mm long, wider than in S. tuberosum; nutlets 2.5-3.5 mm long.
Occurs in the USSR in the Carpathians and near Pochaiv (Pochaiv, city in western Ukraine).
S. foliosum Rehm.
S. foliosum Rehm. (Verh. zool. -bot. Gesellsch. Wien, XVIII (1868) 495; Exs.: Fl. pol. exs. No. 851; Fl. exs. Austro-Hung. No. 3 709) is very close to S. tuberosum, but apparently there is some effect of hybridization with S. officinale which is manifested in the larger size of the plant, the decurrence even of the median leaves, the variability of the calyx (length of teeth). This view was also expressed by Pugsley (Journ. of Bot. 69 1931 92). The race is known from broadleaved forests along the Dniester River (Chernelitsa) from which the type was described in Lvov (Lviv in western Ukraine).
In S. tuberosum there are 6-12 cauline leaves, leaves oblong; cymes with 10 flowers; calyx lobes as long as corolla tube; corolla not as wide, 12-16 mm long; nutlets about 4 mm long.
Distribution:
Extreme west Europe. Mainly in broadleaved, beech, hornbeam and oak forests.
European part: U. Dns. (Upper Dniester) (Carpathians and Subcarpathians), M. Dnp. (Middle Dnieper) (Podolia), Bes. (Bessarabia), L. Don (Lower Don) (very doubtful on the Mius River).
General Distribution: Central Europe, Balkan Peninsula and Asia Minor (Balkans).
This unique Central European species, endemic (native) to the Carpathians occurs mainly in the mountain-forests, occasionally it descends to the lowlands adjacent to these mountains; these localities are Kremenets (city in western Ukraine) and Pochaev (Pochaiv, city in western Ukraine) (all the others refer to S. tauricum Willd.).
It is difficult to indicate to which species this one is related; perhaps the closest is S. cordatum with which it is sometimes sympatric, but hybridization apparently occurs rarely in spite of the wide distribution of these species in the Carpathians.
-'Flora of the U.S.S.R.', 30 volumes, 1934-1960, edited by V.L. Komarov; Botanical Institute of the Academy of Sciences of the U.S.S.R, Leningrad, Russia, with article by Mikhail Grigorevich Popov, in Volume 19, pages 279-291 Russian version, pages 207-216 English version, **1953**.
(The Black Sea is between far-southeastern Europe and the far-western edges of Asia and Turkey.)
(The Dniester or Dnister River is in Eastern Europe. It starts in Ukraine and then goes through Moldova, finally flowing into the Black Sea on Ukrainian territory again.)

(The Dnieper River is the fourth longest river in Europe, starting in Valdai Hills near Smolensk, Russia, {near Europe}, and going through Russia, Belarus and Ukraine to the Black Sea.
(Podolia or Podilia is in west-central and southwest parts of Ukraine and northeast Moldova.)
(The Carpathian Mountains or Carpathians form an arc 1,500 kilometers {932 miles} long across central and eastern Europe. It includes parts of Austria, Czech Republic, Hungary, Poland, Serbia, Slovakia, Ukraine and Romania. It is the second-longest mountain range in Europe. Western Carpathians include Austria, Czech Republic, Poland, Slovakia and Hungary. Eastern Carpathians include southeastern Poland, eastern Slovakia, Ukraine and Romania. Southern Carpathians include Serbia and Romania.)
(Bessarabia is in Eastern Europe, bounded by the Dniester river on the east and the Prut river on the west. Most of it is in Moldova, with the rest in the Ukrainian Budjak region.)
(The Don River is a major Eurasian river and the fifth longest river in Europe. It starts in Novomoskovsk, Russia {120 killometers = 74 miles south of Moscow}, and flows 1,870 kilometers {1162 miles} to the Sea of Azov in eastern Europe.)
(Sympatric means occurring in the same geographical location; overlapping in distribution. For speciation it means taking place without geographical separation.)

"**All the Tuberous Comfrey records in the 'Flora of the North-East of Ireland', 2nd Edition (1938, page 151) are attributed to S. leonhardtianum.** In it Praeger said true Tuberosum could not be quoted for the north, although known elsewhere in Ireland.
This split of Tuberosum was described by H.W. Pugsley from central Europe in the 'Journal of Botany (British and Foreign)', 1931 (69: 89-97, *'The Forms of Symphytum Tuberosum L.'). He said it differed in its slender rootstock with fewer and longer branches and consequently less tufted growth; in its shorter and less branched stems; in its fewer and broader leaves, which increase in size upwards, nearer to the top of the stem; in its more conspicuous flowers, with shorter and more strongly ciliate calyx lobes and broader, more brightly coloured, corolla and in its rather smaller and paler nutlets.
    The 'Flora' recorded it from Clandeboye and Conlig in County Down and from Antrim Castle in County Antrim (Ireland). Praeger (**'Proceedings of Royal Irish Academy', 1939, 243) added it from near Navan, County Meath vice county H 22 (eastern Ireland).
I collected specimens from the last three localities (Clandeboye, Conlig village, Slieve Donard Nurseries: Newcastle, County Down, northern Ireland). I exhibited them at the Meeting of the 'Botanical Society of the British Isles' in November, 1967, **but failed to see how they really corresponded with leonhardtianum or differed from tuberosum.** The quoted characters seemed to appear indiscriminately and there was no difference in general appearance.
    A search with Mr. E. Bangerter in the 'European Herbarium of the British Museum' (London, England) showed Pugsley had divided the sheets among the two species. But he and I could never be sure which would be found in which folder. There was an implication that Mr. A.J. Wilmott also considered leonhardtianum inseparable from tuberosum. This may account for there apparently being no note or reference to this species in any other book I have come across.
In the absence of clearer distinctions or further information, it seems best to treat S. leonhardtianum, as a synonym of S. tuberosum."
-'Symphytum Leonhardtianum Pugsley and S. Tuberosum L.' by D. McClintock, The Irish Naturalists' Journal Ltd., Holywood, County Down, Ireland, Volume 16, No. 1, pages 21-22, January **1968**.
(* -'The Forms of Symphytum Tuberosum L.' by H.W. Pugsley, Journal of Botany: British and Foreign, London, England, Volume 69, pages 89-97, 1931.)
(** -'A Further Contribution to the Flora of Ireland' by Robert Lloyd Praeger, Proceedings of the Royal Irish Academy, Dublin, Ireland, Volume 45, {Section B: Biological, Geological, and Chemical Science}, pages 231-254, 1939.)

"**S. tuberosum L. subsp. (subspecies) tuberosum:**
Rhizome stout, up to 12 mm in diameter, the tuberous portions of irregular shape and close together.
Cauline leaves usually 6-12.
Inflorescence with 8 to 16 (to 40) flowers. Corolla-scales 5.5-7.5 mm, not more than 2 mm wide. Stamens with filament 1/3 to 1/2 as long as anther. Nutlets up to 4 mm.
2n = 144 (chromosomes). West Europe."
-Flora Europaea, Volume 3: Diapensiaceae to Myoporaceae, editors Tutin, Heywood, Burges, Moore, Valentine, Walters and Webb. United Kingdom: Cambridge University Press, **1972**, page 104. (Symphytum L., pages 103-105 by B. Pawlowski.) (5-volume encyclopedia of plants, published between 1964 and 1993. 'Flora Europaea' describes all the national floras of Europe to identify any wild or widely cultivated plant to the subspecies level. It provides geographical distribution, habitat preference, and chromosome number.)

"**S. tuberosum subsp. nodosum** (Schur) Soo, Acta Geobot. Hung. 4: 192 (1941):
Rhizome slender, the tuberous portions irregular-oblong and separated. Cauline leaves usually 3-7.
Inflorescence mostly with 1 to 9 (to 20) flowers. Corolla-scales 4.5 to 6 (to 6.5) mm, up to 2.5 mm wide. Stamens with filament 1/4 to 1/3 (to 1/2) as long as anther. Nutlets 2.5 to 3.5 mm.
2n = 18, 72, 96, 100. Central and southeast Europe."
-Flora Europaea, Volume 3: Diapensiaceae to Myoporaceae, editors Tutin, Heywood, Burges, Moore, Valentine, Walters and Webb. United Kingdom: Cambridge University Press, **1972**, page 104. (Symphytum L., pages 103-105 by B. Pawlowski.) (5-volume encyclopedia of plants, published between 1964 and 1993. 'Flora Europaea' describes all the national floras of Europe to identify any wild or widely cultivated plant to the subspecies level. It provides geographical distribution, habitat preference, and

chromosome number.)

"*Symphytum tuberosum L.,* Sp. Pl. 1: 136. 1753.
    S. tuberosum L. subsp. *angustifolium (A.Kern.) Nyman,* Consp. Fl. Eur.: 510. 1881.
        = *S. angustifolium A.Kern.* in Ost. Bot. Z. 13: 227. 1863.
    S. mediterraneum W.D.J.Koch, Syn. Fl. Germ. Helv. 1: 500. 1837
        = *S. tuberosum L.* var. *mediterraneum (W.D.J.Koch)* Fiori & Beg., Fl. Italia 2(3):378. 1902.
    S. nodosum Schur, Enum. Pl. Transsilvaniae: 468. 1866.
        = *S. tuberosum L.* subsp. *nodosum (Schur) Soo* in Acta Geobot. Hung. 4: 192. 1941."

-'Synopsis of Boraginaceae subfam. Boraginoideae Tribe Boragineae in Italy' is the same as 'Boraginaceae in Italy II' by L. Cecchi and F. Selvi, University of Florence, Firenze, Italy; Plant Biosystems: An International Journal Dealing with all Aspects of Plant Biology, Official Journal of the Societa Botanica Italiana, Vol 149, No. 4, pages 630-677, **2015**, Symphytum page 636-646.

"In total, there are up to 10 taxa described within the *S. tuberosum* complex (in chronological order):
    S. tuberosum L. s. str.,
    S. mediterraneum W. D. J. Koch,
    S. angustifolium A. Kern.,
    S. nodosum Schur,
    S. foliosum Rehmann,
    S. gussonei F.W. Schultz,
    S. floribundum Shuttlew. ex Nyman,
    S. leonhardtianum Pugsley,
    S. besseri Zaver.,
    S. popovii Dobrocz.
    (Koch 1837, Kerner 1863, Schur 1866, *Rehmann 1868, Schultz 1872, 1875, Nyman 1884, Pugsley 1931, **Zaverucha 1962, ***Dobroczajeva 1968).

However, the treatments of Symphytum in 'Flora Europaea' and the 'Euro+Med Checklist' recognize only two species:
    1. Sicilian endemic (native) *S. gussonei.*
    2. The widespread *S. tuberosum,* the latter comprising the western European subsp. *tuberosum* and central-
    and eastern-European subsp. *angustifolium/nodosum* (Pawlowski 1972, ****Valdes 2011)."

-'Symphytum Tuberosum Complex in Central Europe: Cytogeography, Morphology, Ecology and Taxonomy' by L. Kobrlova, M. Hrones, P. Koutecky, M. Stech, and B. Travnicek, Preslia: The Journal of the Czech Botanical Society, Czech Republic; Volume 88, pages 77-112, March **2016**.

(* -'Botanische Fragmente aus Galizien' {Botanical Fragments from Galicia} by Dr. A. Rehmann in Verhandlungen der K.K. Zoologisch-Botanischen Gesellschaft in Wien, Vienna, Austria, Volume 18, page 495, 1868. Written in German. Galicia is in Spain.)
(** -'New Species of Plants from the Environs of Kremenets' {Novi Vidi Roslin z Okolic Kremencja} by B.V. Zaverucha; Ukrayins'kyi Botanichnyi Zhurnal, Volume 19, pages 49-63, 1962. Kremenets is in western Ukraine.)
(*** -'On Taxonomy of Genus Symphytum L. II. Section Tuberosa Buckn.' by D.N. Dobrochaeva / Dobroczajeva; Ukrayins'k Botanicnyi Zhurnal {Ukrainian Botanical Journal}, Volume 25, pages 58-62, 1968. In Russian. I was not able to find this article.)
(**** -'Boraginaceae' by B. Valdes; Euro+Med Plantbase: The Information Resource for Euro-Mediterranean Plant Diversity, http://ww2.bgbm.org/EuroPlusMed, 2011.)

"**The taxonomy of Symphytum tuberosum** in central Europe was revised recently by Kobrlova et al. (2016). They showed that two subspecies of *S. tuberosum* occur in the Czech Republic.
    The taxonomy within this group is quite intricate, especially due to the occurrence of high polyploids and considerable morphological variability.
In its broad circumscription, *S. tuberosum* is distributed all over Europe except for Scandinavia, the Netherlands, Belgium, north-western Germany, southernmost Spain and Portugal (Bucknall 1913, Murin & Majovsky 1982).
Symphytum tuberosum subsp. angustifolium is tetraploid (2n = 32), and has an obvious affinity to the Pannonian basin (Carpathian basin in central Europe) (Kobrlova et al. 2016).
    Until recently, this taxon was known only from northern Hungary and the southern part of Slovakia (Majovsky & Hegedusova 1993, *Marhold & Hindak 1998, both as **S. angustifolium**), but it has been omitted from flora accounts of the former country.
    It was recently discovered in south-eastern Moravia in the Czech Republic and confirmed for northern Hungary (Kobrlova et al. 2016).
In comparison with the type subspecies, S. tuberosum subsp. angustifolium is more thermophilous (warmth-loving), occurring mainly in the lowlands. It occurs rarely at higher altitudes, reaching them through warmer valleys.
    It grows in drier habitats than the type subspecies, such as thermophilous broad-leaved forests and semi-dry grasslands.
In the Czech Republic it is confined to the westernmost Carpathians in south-eastern Moravia, mainly to the Bile (Biele) Karpaty Mts (White Carpathian Mountains), Litencicke Vrchy hills, Chriby hills and Zdanicky les hills. Its northern distribution limit runs through central Moravia, its western limit west of the city of Brno, Czech Republic.
    The map is based only on revised herbarium specimens and our own field records as no earlier records exist."

-'Distributions of Vascular Plants in the Czech Republic, Part 3' by Z. Kaplan, J. Danihelka, M. Lepsi, P. Lepsi, L. Ekrt, J. Chrtek Jr, J Kocian, J Prancl, L. Kobrlova, M. Hrones and V. Sulc, Czech Republic; Preslia: The Journal of the Czech Botanical Society, Volume 88, pages 459-544, **2016**. The third part of the publication series on the distributions of vascular plants in the Czech Republic includes grid maps.
(* -Checklist of Non-Vascular and Vascular Plants of Slovakia edited by Karol Marhold and Frantisek Hindak. Bratislava, Slovakia: Veda- Publishing House of the Slovak Academy of Sciences, 1998.)

(In biological taxonomy, circumscription is the definition of a taxon {a group of organisms}. One goal is to define a stable circumscription for every taxon. This has not been achieved in most taxa of all organisms, not just Symphytum.)
(Polyploidy is a cell that has more than two paired {homologous} sets of chromosomes. Most species whose cells have nuclei {eukaryotes} are diploid, i.e., they have two sets of chromosomes. One set is inherited from each parent. Polyploidy is common in plants.)
(The Carpathian Basin or Pannonian Basin is in the southeastern part of central Europe. It is surrounded by the Carpathian Mountains and the Alps.)

"**Symphytum tuberosum subsp. tuberosum is dodecaploid (2n = 96) and it is the most widespread member of the S. tuberosum group** (Kobrlova et al. 2016).
In central Europe it is found in Austria, Germany (mostly in the south and along the lower stretches of the Elbe river), southern Poland, northern Slovakia, and in southern and western Hungary (Kobrlova et al. 2016).
In the Czech Republic Symphytum tuberosum subsp. tuberosum prefers shady, moist and also nutrient-rich habitats. It inhabits the banks of rivers or streams, forests in deep river valleys, the fringes of wet meadows, alder carrs, and alluvial, ravine and mesophilous forests. It was recorded from ruderal or disturbed places (e.g. roadsides and abandoned wet meadows) and parks.
It occurs mainly in southern and central Bohemia and in northern Moravia (both Czech Republic) and Silesia. The distribution map is based solely on revised herbarium specimens and our own field records."
-'Distributions of Vascular Plants in the Czech Republic, Part 3' by Z. Kaplan, J. Danihelka, M. Lepsi, P. Lepsi, L. Ekrt, J. Chrtek Jr, J Kocian, J Prancl, L. Kobrlova, M. Hrones and V. Sulc, Czech Republic; Preslia: The Journal of the Czech Botanical Society, Volume 88, pages 459-544, **2016**.

(Alder or Alnus is a genus of flowering plants in the birch family Betulaceae of trees and shrubs.)
(A 'carr' is a waterlogged wooded terrain.)
(Alluvium is loose soil or sediment eroded by water, and then deposited on dry land.)
(A ravine is narrower than a canyon and smaller than a valley. It is created by stream erosion.)
(Mesophilous is an area with a moderate amount of rain or snow.)
(Silesia is a historical region of central Europe located mostly in Poland, with small parts in Czech Republic and Germany.)

"**The Symphytum tuberosum complex belongs to one of the most complicated groups within the genus Symphytum Linnaeus (1753: 136) in Europe, mainly due to an occurrence of polyploidy and associated extensive morphological variability** (Gadella & Kliphuis 1978, Murín & Majovsky 1982, Kobrlova et al. 2016).
Despite current progress, the taxonomy of S. tuberosum is still not satisfactorily resolved. The members of this complex are distributed across Europe and Asia Minor (Bucknall 1913, Murin & Majovsky 1982, Kobrlova et al. 2016) and **a total of ten taxa have been described within this complex, three of them from Central Europe:**
  Symphytum tuberosum Linnaeus (*1753: 136)
  Symphytum angustifolium A.Kerner (**1863: 227)
  Symphytum leonhardtianum Pugsley (***1931: 95)."
-'Taxonomic Status and Typification of a Neglected Name Symphytum Leonhardtianum from the Symphytum Tuberosum Complex (Boraginaceae)" by Lucie Kobrlova, Terezie Mandakova and Michal Hrones (all from Czech Republic); Phytotaxa, Volume 349, No. 3, pages 225-236, **2018**.
(* -'Species Plantarum' by Linnaeus, Volume 1, Symphytum tuberosum page 136.)
(** -'Descriptiones Plantarum Novarum Florae Hungaricae et Transsilvanicae' by A. Kerner, Oesterreichische Botanische Zeitschrift {Austrian Botanical Journal}, Vienna, Austria, Volume 13, No. 7, page 227, 1863.)
(*** -'The Forms of Symphytum Tuberosum L.' by H.W. Pugsley, B.A., F.L.S., Journal of Botany: British and Foreign, London, England, Volume 69, page 95, April 1931.)

"**Symphytum angustifolium** was described from the plant material collected in the Pilis Mountains in northern Hungary as a narrow-leaved morph of S. tuberosum (Kerner 1863). Later, it was also discovered in Slovakia and in the south-eastern part of the Czech Republic.
It has been shown to have a tetraploid chromosome number (2n = 32; Murin & Majovsky 1982, Kobrlova et al. 2016).
Nevertheless, there has been much confusion surrounding this name, and it has been often synonymised with S. nodosum Schur (1866: 468) or applied to all populations of the S. tuberosum complex from East and Central Europe (cf. {compare to} Pawlowski 1972, *Smejkal 1978, Valdes 2011)."
-'Taxonomic Status and Typification of a Neglected Name Symphytum Leonhardtianum from the Symphytum Tuberosum Complex (Boraginaceae)' by Lucie Kobrlova, Terezie Mandakova and Michal Hrones (all from Czech Republic); Phytotaxa, Volume 349, No. 3, pages 225-236, **2018**.
(* -'The Genus Symphytum in Czechoslovakia' {Rod Symphytum v Ceskoslovensku} by M. Smejkal; Zpravy Ceskoslovenske Botanicke Spolecnosti {Czech Botanical Society News}, Prague, Czech Reupublic, Volume 13, pages 145-161, 1978. In Czech.

If you have an English translation, I would appreciate a copy.)

"Symphytum leonhardtianum, a member of the S. tuberosum complex, is investigated. This taxon was described by Pugsley in 1931, from the vicinity of Vienna, Austria. Nevertheless, it is generally not accepted in European floras.
In this study, we conducted an evaluation of this taxon using flow cytometry, karyology and morphological analysis. Flow cytometric and karyological investigations of **plants from the type locality of S. leonhardtianum revealed only dodecaploids (2n = 12x = 96), a ploidy level corresponding to the S. tuberosum subsp. tuberosum.**
The chromosome number of the S. tuberosum from Austria is here recorded for the first time. Morphological comparison of central European populations of S. tuberosum complex showed that S. leonhardtianum did not differ significantly from S. tuberosum subsp. tuberosum.
**Based on our findings, we propose treating the name S. leonhardtianum as a heterotypic synonym of S. tuberosum subsp. tuberosum. The lectotype of S. leonhardtianum is designated."*
-'Taxonomic Status and Typification of a Neglected Name Symphytum Leonhardtianum from the Symphytum Tuberosum Complex (Boraginaceae)" by Lucie Kobrlova, Terezie Mandakova and Michal Hrones (all from Czech Republic); Phytotaxa, Volume 349, No. 3, pages 225-236, **2018**.
　　　(Karyology is the cytological characteristics of the cell nucleus especially the chromosomes of a single cell.)
　　　(In botanical nomenclature, a heterotypic synonym {taxonomic synonym} exists when a taxon is reduced in status and becomes part of a different taxon.)
　　　(Lectotype is a specimen chosen as the type of a species if the author of the name fails to designate a type.)
End of S. tuberosum Subspecies and Varieties (Tuberosum Complex)

S. tuberosum Distribution (Geographical Location)

"The Common Comfrey is abundantly met with in England, but is rare in Scotland; **the Tuberous Comfrey is commonly found in Scotland, but is seldom met with in England, the northern counties of England and north Wales being its extreme southern limit,** *so that except in the narrow zone of country common to both, there will be no possibility of mistaking the one species for the other."*
-A Modern Herbal: The Medicinal, Culinary, Cosmetic and Economic Properties, Cultivation and Folk-Lore of Herbs, Grasses, Fungi, Shrubs and Trees with their Scientific Uses by Mrs. M. Grieve. New York: Dover Publications, 1971. First published **1931**.

*"**S. tuberosum L., common to the north of England,** is readily distinguished from S. officinale (Common Comfrey) by its smaller, softer leaves, white flowers and tuberous roots.*
Although possessing similar medicinal properties to S. officinale, it has been used less frequently."
-'British Medicinal Species of the Genus Symphytum' by K.R. Fell and Janet M. Peck, Pharmacognosy Research Laboratories, University of Bradford, England; Planta Medica: Journal of Medicinal Plant and Natural Product Research, Volume 16, No. 2, pages 208-216, May **1968**.

*"**Two species of Comfrey are plentiful in the United Kingdom, the Common Symphytum officinale and Tuberous Symphytum tuberosum** while a third, Soft Comfrey Symphytum orientale is less abundant.*
The two commoner species can be separated by a number of means:
　　　The Common Comfrey occurs mainly in southern regions, is found almost exclusively in damp places, has variable coloured flowers that can be white, cream, pink or purple and is a tall plant which can reach height of 120 cm (3.9 feet).
　　　Tuberous Comfrey, on the other hand, has a more northerly distribution, can be found in drier areas, has yellow flowers and attains a maximum height of only 50 cm (1.6 feet).
Both flower between May and July with odd plants hanging on a little later."
-'Growing and Collecting Wild and Cultivated Greenfoods and Seeds' by Dave Coles, Goring, Reading, England, **2003**.

*"**Symphytum tuberosum, Phytoclimatic Group 4 (PCG4):***

Temperature	St Dev	Precipitation	Max1	Max2	Max3	Min1	Min2	Min3
13.21 C	5.47	908 mm/year	25.20 C	17.35	38.55	2.42	7.99	-12.22
=55.7 F		=35 inch	=77 F	=63 F	=101 F	=36 F	=46 F	=10 F

Phytoclimate:
*Four phytoclimatic groups, i.e., **groups of species with similar responses to climatic variables**, were identified based on the species' climate optima (optimum ranges) within the study area.*
　　　***1. PCG1** group contained 124 species, which mainly occurred in the northern part of France, and in Belgium and the Netherlands.*
　　　***2. PCG2** was a smaller group, containing 24 species having their optimum in the central part of France. The frequency of PCG2 species was lower both at northern locations (Belgium, the Netherlands) and at southern (Mediterranean) locations.*
　　　*3. This is in contrast with the frequency pattern of the species belonging to **PCG3** (containing 64 species) for which the frequency was high in southern and central France, but low in Belgium. Moreover, these species were almost totally absent in the Netherlands.*
　　　4. The 24 species of the PCG4 are clearly more adapted to the climatic conditions in the southern part of France, and did not occur in the northern part of the study area.

Precipitation (Rain, Snow): Variation in precipitation was represented by total annual precipitation, expressed in mm.
Temperature- Mean, Standard Deviation:
>   *The mean annual temperature was calculated as the average mean temperature over the 12 months.*
>   *Variation in temperature throughout the year was expressed by the standard deviation on the mean annual temperature (StDev), as a measure for continentality.*

Temperature- Maximum and Minimum:
>   **As survival and fitness of plant species often depends rather on extreme than on average climate values,** *some values representing maximum and minimum temperature throughout the year were calculated:*
>   >   **Max1:** *Mean temperature of the warmest month.*
>   >   **Max2:** *Average maximum month temperature.*
>   >   **Max3:** *Highest maximum temperature.*
>   >   **Min1:** *Mean temperature of the coldest month.*
>   >   **Min2:** *Average minimum month temperature.*
>   >   **Min3:** *Lowest minimum temperature."*

-'Climate Gradients Explain Changes in Plant Community: Composition of the Forest Understorey' by Sebastiaan Van Der Veken, Beatrijs Bossuyt and Martin Hermy, Laboratory for Forest, Nature and Landscape Research, Catholic University of Leuven, Belgium; Belgian Journal of Botany, Volume 137, No. 1, pages 55-69, **2004**.

>   (Phytoclimate is the climate of a small area such as a plant community or a specific wooded area. It can be different than the climate of the larger region. Continentality is the difference between land and marine climates. There is a greater range of temperatures over land than water.)
>   (See section 'Details about Symphytum Species: Officinale' {Chapter 8} for similar analysis by same authors.)

"Assessing the Status of Doubtfully Native Species in the Flora of the British Isles:
Symphytum tuberosum in British Isles:
>   *In cultivation in 1596 A.D., with the first record in wild 1777 (Edinburgh, Scotland, 'Water of Leith' river).*
>   *It is not included in the *' British Plant Communities' (1991-2000) but it seems to be a part of woodland vegetation.* **It is widely distributed, dynamic, increasing fast, and persistent.** *It is occasionally grown in gardens.*
>   ***Watson (1847-1859) treats it as a native, but with caveats (limitations) regarding England and part of Scotland; it is not covered in ***Dunn (1905).*
>   **Recent floras all treat Symphytum tuberosum as native in British Isles.**

Germany, Spain, France, Belgium, Netherlands, and Luxembourg:
>   *The European picture is very interesting. Meusel ('Vergleichende Chorologie der Zentraleuropaischen Flora' {Comparative Chorology of Central European Flora}, Germany, 1965-1992) shows its range from northern Spain, across southern France to the Black Sea.*
>   *The German Atlas (Haeupler & Schonfelder, 'Atlas der Farn- und Blutenpflanzen der Bundersrepublik Deutschland', 1988) shows it as present only in the farthest south.*
>   *It is absent from the Low Countries, and present in France only in the centre and south - elsewhere it occurs as an adventive (new to the area) (e.g. des Abbayes 1971, 'Flora et Vegetation du Massif Armoricain').*

Perring & Walters (1962, 'Atlas of the British Flora') treated **Symphytum tuberosum probably native only in Scotland**, and we have followed this, although the recent Atlas (Preston et al., 2002, 'New Atlas of the British and Irish Flora) contains so many more records, including many more in Scotland too, though some of these may be due to past under-recording.
None of the Scottish floras cover this species in any depth.

Symphytum tuberosum is not native to Britain:
>   **In view of this and Symphytum tuberosum rapid and recent spread in England and Wales, the hallmark of an alien species, together with its continued spread in Scotland, it seems very likely to have been an introduction in the mid-seventeenth century,** *a conclusion echoed in Braithwaite et al. (****2006, 'Change in the British Flora').*
>   **Status Decision: Alien (neophyte). A species treated as a native, but about which I now have strong reservations."**

-Watsonia: Journal and Proceedings of the Botanical Society of the British Isles, London, England, Volume 26, **2007**. Includes 'Far Away from Any House: Assessing the Status of Doubtfully Native Species in the Flora of the British Isles' by D.A. Pearman, Cornwall, England, pages 271-290, Symphytum Tuberosum page 284.
(* -British Plant Communities, Volumes 1-5 edited by J.S. Rodwell. Cambridge, England: Cambridge University Press, 1991-2000.)
(** -'Cybele Britannica, Or British Plants and Their Geographical Relations', 4 Volumes by Hewett C. Watson, Longman & Co., London, England, 1847-1859.)
(*** -'Alien Flora of Britain' by Stephen T. Dunn, BA, FLS. West, Newman & Co., London, England, 1905.)
(**** -Change in the British Flora, 1987-2004 by M.E. Braithwaite, R.W. Ellis and C.D. Preston. London, England: Botanical Society of Britain and Ireland, 2006.)

>   (Chorology is the causal relations between geographical phenomena in a particular region, i.e., the study of the spatial distribution of organisms.)
>   ('Low Countries' or Benelux countries are the coast of northwestern Europe which includes Belgium, the Netherlands, and Luxembourg.)

"We experimentally assessed the survival and performance of four understorey forest plants (Omithogalum pyrenaicum, Scilla bifolia, Iris foetidissima and **Symphytum tuberosum**) **transplanted beyond their natural range limit over a 7-year**

period. All four species are characteristic of alluvial alder-ash forest understorey communities in Europe. The survival probability of plants transplanted beyond their natural geographic range is expected to decrease with the home-away distance.
Symphytum tuberosum ssp. tuberosum L.:
    It is a 20-60 cm (7.8-23.6 inch) tall hemicryptophyte flowering from April to June. It is a pale yellow-flowered Comfrey with tuberous rhizomes.
    Symphytum tuberosum is naturally restricted to northern Spain, across the sub-Mediterranean regions in France to the Black Sea; the species is regarded as alien in Britain *(*Pearman 2007).*
    S. tuberosum exhibited no signs of success at the experimental sites about 500 km (310 miles) away from the range edge. *Adults in the plots with natural vegetation cover died within two years of the initiation of this experiment, and adults in the cleared plots were extinct after five years.*
    *S. tuberosum adults only flowered in the first year of the experiment, and then showed a decline in performance over the following years. The poor performance and survival of this species, even when all above ground competition was removed, strongly suggests that **its occurrence in deciduous forests is controlled by environmental factors such as the local climate.***
    Surprisingly, S. tuberosum continues to spread rapidly throughout Britain after being introduced in the mid-17th century *(Pearman 2007)."*
-'Experimental Assessment of the Survival and Performance of Forest Herbs Transplanted Beyond Their Range Limit' by Sebastiaan Van der Veken, Pieter De Frenne, Lander Baeten, Eric Van Beek, Kris Verheyen and Martin Hermy, Belgium; Basic and Applied Ecology, Volume 13, No. 1, pages 10-19, February **2012**.
(* -Watsonia: Journal and Proceedings of the Botanical Society of the British Isles, London, England, Volume 26, (2007). Includes 'Far Away from Any House: Assessing the Status of Doubtfully Native Species in the Flora of the British Isles' by D.A. Pearman, Cornwall, England, pages 271-290, Symphytum Tuberosum page 284.)
    (Alluvium is loose, unconsolidated soil on land that has been eroded and shaped by water.)
    (Hemicryptophytes are herbs with buds at soil level.)

"United Kingdom: Lowland Mixed Deciduous Woodland:
This priority habitat brings together a wide range of lowland woodland types, on well-drained basic to acidic soils on steeply sloping to more or less level ground in the southern and eastern Scottish lowlands.
Given the range of soil conditions it is not surprising that the plant species composition of this priority habitat is very varied.
On the most base-rich soils (NVC W8) *the tree canopy is typically made up mainly of ash Fraxinus excelsior, wych elm Ulmus glabra and sycamore Acer pseudoplatanus, mixed with other species such as wild cherry Prunus avium, goat willow Salix caprea, elder Sambucus nigra and hawthorn Crataegus monogyna.*
Extensive carpets of Tuberous Comfrey (Symphytum tuberosum) are a feature of some W8 woods in southern and eastern Scotland. *In Scotland, W8 and W10 are the most common NVC communities in this priority habitat."*
-'Lowland Mixed Deciduous Woodland: UK BAP Priority Habitat' for United Kingdom Biodiversity Action Plan, Joint Nature Conservation Committee, jncc.defra.gov.uk, statutory advisor to United Kingdom government, **2014**. UK BAP is the government's response to the 'Convention on Biological Diversity' (CBD).
    (NVC is 'National Vegetation Classification' that divides vegetation types, e.g. woodland, calcareous grassland, mires, that are then divided into communities designated by a number and name.)
    (W8: Fraxinus excelsior / Acer campestre / Mercurialis perennis woodland. It is a community most abundant in the relatively warm, dry, lowlands of southern and eastern Britain. It occurs on various types of calcareous soils in areas where the effects of leaching are limited.)

*"**Distribution areas for Symphytum cordatum and Symphytum tuberosum largely overlap.***
S. tuberosum is primarily a mountain forest Central European species, *commonly found in the Carpathians, sometimes reaching the adjacent plains; unlike S. cordatum, it grows mainly in shady moist beech, horn beech and oak forests (*Hege 1972; **Tikhomirov et al. 1999)."*
-'Pubescence of Vegetative Organs and Trichome Micromorphology in Some Boraginaceae at Different Ontogenetic Stages' by Rimma P. Barykina and Vitaly Y. Alyonkin, Lomonosov Moscow State University, Russia; Wulfenia, Regional Museum of Carinthia, Austria, Volume 23, pages 1-29, **2016**.
(* -Flora Europaea, Volume 3, edited by T.G. Tutin. Cambridge, England: Cambridge University Press, 1972. Article: 'Boraginaceae Juss.' by J.C. Hege, pages 86-120.)
(** -'The Genus Symphytum L. {Boraginaceae} in Central Russia' by V.N. Tikhomirov, S.R. Majorov and D.D. Sokoloff; Novosti Sistematiki Vysshikh Rastenii {News of Higher Plants Systematics}, Moscow and Leningrad {St. Petersburg}, Russia, Volume 31, pages 231-245, 1999. In Russian. If you have the Symphytum pages in English, I would appreciate a copy.)

*"**A review of genus Symphytum in the Czech Republic is presented,** including an identification key and remarks on its distribution and ecology.*
The first part of this review is focused on the Symphytum tuberosum agg. (aggregate). *This group is taxonomically difficult and includes high polyploids. Two dominant ploidy levels were found in the Czech Republic: tetraploid (S. tuberosum subsp. angustifolium) and dodecaploid (S. tuberosum subsp. tuberosum).*
S. tuberosum subsp. angustifolium shows clear affinity to the margins of the Pannonian basin and was found only in Moravia (eastern Czech Republic).
The more common and widespread S. tuberosum subsp. tuberosum was demonstrated to occur in the entire country of

the Czech Republic, *but is almost absent from the Czech part of the Czech part of the Carpathian Mountains.*
A detailed overview of morphological characters useful for identification of these taxa, their ecological differentiation and distribution maps are presented."
-'The Genus Symphytum L. (Comfrey) in the Czech Republic I: S. Tuberosum Agg.' (Rod Symphytum Kostival v Ceske Republice I: S. Tuberosum Agg.) by Lucie Kobrlova, Michal Hrones and Bohumil Travnicek, Univerzity Palackeho v Olomouci, Czech Republic; Zpravy Ceskoslovenske Botanicke Spolecnosti (Czech Botanical Society News), Prague, Czech Reupublic, Volume 51, pages 221-256, December **2016**. All in Czech except abstract is also in English. If you have an English translation, I would appreciate a copy.

(Species aggregate is a species group that is closely related and difficult to distinguish among them. Similar but not identical is a single species that displays large variations but can not strictly be subdivided: s.l. for 'sensu lato' {in a wider/broader sense} or s. ampl. for 'sensu amplo' {in a very wide/relaxed sense}.)

(Tetraploid: 2n = 32. Dodecaploid: 2n = 96.)

S. tuberosum Description

*"**Symphytum foliis oblongis anguslioribus. The oblong, narrow-leaved Symphytum.***
The root is tuberous, long, branched, and brittle, black on the surface, white within.
The radical leaves are eight inches (20.3 cm) long, and two and a half (6.3 cm) in breadth, of a pale bright green colour, and not so rugose or rough to the touch, as those of the Common Comfrey.
The stalk is green, succulent, and edged with membranes from the bases of the leaves, so that it looks pentangular: the leaves stand at distances on it; they are oblong, narrow, pointed, and of a bright green.
The flowers are of a whitish colour, *longer, but not thicker than those of the Common Comfrey.*
This is frequent in the woods of Germany, and other places. ***It varies much in size, and has been described by C. Bauhine, in these states, as two species, under the names of Symphytum majus tuberosa radice, and Minus."***
-'A History of Plants: A General Natural History, or New and Accurate Descriptions of the Animals, Vegetables and Minerals of the Different Parts of the World' by John Hill, M.D., London, England, **1751**, page 254.

(For more about C. Bauhine, see '1596, 1623 Bauhin' in section 'Symphytum Species Classifications' {Chapter 4}.)

*"**Symphytum Tuberosum; Tuberous-rooted Comfrey.***
Leaves ovate, semi-decurrent, the uppermost opposite. This is suspected to be a variety of the preceding species (Symphytum officinale). It is a lower plant, with the root white on the outside; clusters in pairs, terminal; ***flowers yellowish or greenish-white,*** *cylindrical. It flowers from May to October.*
Native of Germany, Austria, France, Spain, and Italy, and observed in various parts of Scotland."
-'The Universal Herbal; or, Botanical, Medical, and Agricultural Dictionary; Containing an Account of All the Known Plants in the World, Arranged According to the Linnean system. Specifying the uses to which they are or may be applied, whether as food, as medicine, or in the arts and manufactures, with the best methods of propagation, and the most recent agricultural improvements' by Thomas Green, London, England, **1824**, page 641.

*"**Species II.: Symphytum Tuberosum. Linn.** Plate MCXVII.*
  *Reich. Ic. Fl. Germ. et. Helv. Vol. XVIII. Tab. MCCCIV.
  **Billot, Fl. Gall. et Germ. Exsicc. No. 2713.
Roots: *Rootstock horizontal, tuberous, knotted, fleshy, praemorse, branched, with slender root fibres.*
Rootstock fleshy, branching, the divisions somewhat resembling the tubers of the Jerusalem artichoke (but smaller), pale brown.
Stem and Leaves: *Stem rather thick, simple or nearly so, very slightly winged above. Leaves all oval or elliptical-oval, the upper ones slightly decurrent, especially the pair at the base of the racemes.*
Plant clothed with minute pubescence, intermixed with rather harsh bristly hairs, many of them gland-tipped.
Producing at the apex stems, but no tufts of radical leaves, as in Symphytum officinale.
The stems are 1 to 2 feet (0.3-0.6 meter) high, flexuous, ***much less winged and less hairy than in Symphytum officinale;*** *the leaves taper towards the base as well as the apex, and are more rugose, much less rough, and with the hairs on the under side of the veins much fewer and shorter.*
The plant is of a paler and yellower green, and the lower leaves have turned brown or withered before the flowers expand.
Flowers and Nutlets: *Calyx segments strap shaped, divided nearly to the base, in fruit not muricated, the hairs being seated on inconspicuous tubercles. Corolla about twice as long as the calyx; scales included.*
The calyx segments are longer, narrower, and less bristly. ***Corolla about 3/4 inch (1.9 cm) long, ochreous (light brownish-yellow), but rather deeper in colour than in Symphytum officinale.***
The mature fruit (nutlets) I have not seen, but, according to ***M.Godron, it is tubercular and contracted above the base."
-'English Botany (Sowerby's); or Coloured Figures of British Plants: Volume 7' by John T. Boswell and John Edward Sowerby, London, England, **1880**, page 116.

(* -'Icones Florae Germanicae et Helveticae, Volume XVIII' by Heinrich Gustav Reichenbach, Tab/Plate MCCCIV, 1856-1858. {Reich. Ic. Fl. Germ.})

(** -Paul Constant Billot, 1796-1863, was a French botanist born in Rambervillers. With botanist Friedrich Wilhelm Schultz, 1804-1876, he co-authored 'Archives de la Flore de France et d'Allemagne'. Billot's 'Annotations a la Flore de France et d'Allemagne', 1855, was printed with 'Flora Galliae et Germaniae Exsiccata' in 1856. It was continued by other botanists as 'Billotia'.)

(*** -Flore de France; ou, Description des Plantes qui Croissent Naturellement en France et en Corse' by M. Charles Grenier and M. Dominique Alexandre Godron; Paris, France. Three volumes, 1848-1856. Symphytum officinale, tuberosum, mediterraneum and bulbosum, Volume 2, 1850, pages 511-512.)

"**Symphytum tuberosum L.,** *with thickened tuberous roots, the nutlets granular-tuberculate, not shining, has been found in* **sandy meadows in Connecticut.**"
-'An Illustrated Flora of the Northern United States and Canada, Volume 3: Gentianceae to Compositae (Gentian to Thistle)' by Nathaniel Lord Britton, PH.D., Sc.D., LL.D. and Honorable Addison Brown, A.B., LL.D., NY Botanical Garden, **1913**, page 92.

"**Symphytum tuberosum Linn.** *Sp. Pl. ed. I. p. 136 (1753)*
Geographical Distribution: **From Britain, France and Spain to Turkey,**
    and from Germany and southwest Russia to Italy and Greece.
S. tuberosum is distinguished from allied species by the thick, branched, tuberous root, by the lower leaves which are smaller than the succeeding ones, and by the included stamens."
-'A Revision of the Genus Symphytum, Tourn.' by Cedric Bucknall, (Mus. Bac. Oxon= Bachelor of Music, Oxford University), Journal of the Linnean Society of London, England, Botanical Journal, Volume 41, Issue 284, pages 491-556, December **1913**.

"*There is another species,* **S. tuberosum,** *found in wet places from north Wales, Stafford and Lincoln (England), northwards into Scotland, and most common in the south of Scotland, though absent from Ireland.*
The stem is scarcely branched and but slightly winged, the bases of the leaves being hardly at all continued down the stem. Though also covered with hairs, the latter are not so bristly. The root-stock is short and horizontal with slender root fibres.
This is a much smaller plant than Symphytum officinale, the stem rarely more than a foot (0.3 meter) high, rather slender and leafy. *The lower radical (near the root) leaves are much as in S. officinale in form, but with longer footstalks.*
The flowers, creamy-yellow in colour though about the same size as those of S. officinale, are in much smaller masses."
-A Modern Herbal: The Medicinal, Culinary, Cosmetic and Economic Properties, Cultivation and Folk-Lore of Herbs, Grasses, Fungi, Shrubs and Trees with their Scientific Uses by Mrs. M. Grieve. New York: Dover Publications, 1971. First published **1931**.

"*As Gerard (1597, 'Herball, or Generall Historie of Plantes') says, we have one other native Comfrey, rare except in Scotland, 'with floures of an* **overworne yellow colour.** *The rootes are thicke, shorte, black without and tuberous.'*
This is S. tuberosum, possibly the 'Trottel', which grows only 9 inches to 21 inches (0.22-0.53 meter) high, **a smaller plant of no agricultural or medicinal value, with 72 chromosomes,** *a number which luckily keeps it out of the Comfrey (taxonomy) confusion.*"
-Russian Comfrey: A Hundred Tons an Acre of Stock or Compost for Farm, Garden or Smallholding by Lawrence D. Hills. London England: Faber and Faber, Limited, **1953**, page 21.
    (In the mid 1800s Trottel was sometimes thought to be Symphytum asperum.)

"**S. tuberosum L. is the only other native species (in Britain).** *(The other is S. officinale.) It is a local woodland plant found chiefly in northern England and in Scotland.*
It is not known to hybridise with any other species and is easily recognised by its little-branched stems, tuberous stock and by having the middle stem leaves larger than the basal ones."
-'The Genus Symphytum in Britain' by T.G. Tutin, University College of Leicester, England; Watsonia: Journal of the Botanical Society of the British Isles, Volume 3, pages 280-281, February **1956**. (Watsonia was the journal of the 'Botanical Society of Britain and Ireland' from 1949 to 2010. It is now called 'New Journal of Botany'.)
    (Basal means located at or near the base of a plant stem, or at the base of any other plant part.)

"**S. tuberosum L., Tuberous Comfrey:**
A hispid perennial 20-50 cm (7.8-19.6 inches). Root fibrous; rhizome stout, tuberous. Stems covered with reflexed bristles, simple or with one or two short branches near the top. Leaves puberulent and densely hispid; lower small, ovate or spathulate, narrowed at base, petioled; middle cauline 10-14 cm (3.9-5.5 inches), considerably larger than the lowest, ovate-lanceolate or elliptic, shortly petioled; upper sessile.
Calyx 7-8 mm, teeth lanceolate-acute, 3 times as long as tube. Corolla 12-16 mm (0.47-0.62 inch), **yellowish-white.** *Scales broadly triangular-subulate, acuminate, somewhat exceeding the stamens. Nutlets minutely tuberclate. Flowers 6-7.*
2n = about 72 (chromosomes).
Native. In damp woods and hedgebanks. Scattered throughout Great Britain, local but commoner in the north; an escape from cultivation in Ireland. Germany to Spain and eastwards to southwest Russia and Turkey."
-Flora of the British Isles, Second Edition by A.R. Clapham, T.G. Tutin and E.F. Warburg. England: Cambridge University Press, **1962**, (first edition 1952).

"**S. tuberosum L. subsp. nodosum (Schur) Soo** *in Acta Geobot. Hung. 4 :182 (1941). Ic.:*
S. tuberosum L. in Hegi, Illustr. Fl. Mittel Eur. 5(3):2226, t. 219, fig. 5 & 3162 (1906).
    Synonym: *S. nodosum Schur, Enum. Pl. Transyl. 468 (1866).*
            S. foliosum Rehm. in Verh. zool.-bot. Ges. Wien 18:495 (1868).
            S. tuberosum f. longifolium G. Beck in Ann. Naturh. Hofmus. Wien 2 :132 (1887).
            S. tuberosum alpha latifolium G. Beck, Fl. Nied.-Osterr. 2 :963 (1892).

S. leonhardtianum Pugsley in *J. Bot. (London)* 69 :95 (1931).
S. leonhardtianum var. *longifolium* Pugsley in *J. Bot. (London)* 69 :96 (1931).
Range: South Europe (including Balkans), south Russia, Turkey (Bosphorus).
Pugsley (*1931) noted that under the name S. tuberosum plants belonging to two taxa could be distinguished.
He carefully studied pre-Linnaean specimens in the Linnaean literature and found two sheets in the Linnaean Herbarium (London, England) labelled 'tuberosum 2' and '2'. These belong to the pre-1753 collection and can be accepted as typifying **S. tuberosum L. sensu stricto (i.e., subspecies tuberosum)**. This is a narrow leaved plant characteristic of **southwest Europe (United Kingdom, France and Spain)** and is illustrated by **Ross-Craig, Drawings of Brit. Pl. 21: t.4 (1965).
From it Pugsley distinguished the east European plants passing as S. tuberosum by giving them specific rank under the name S. Leonhardtianum Pugsley.
Schur had, however, earlier used the name S. nodosum for Transylvanian (central Romania) material and this epithet (name) must be adopted whether the taxon concerned is given specific rank, following Pugsley, or subspecific rank following Soo and Pawlowski.
Subsp. nodosum has a more slender rootstock than subsp. tuberosum and is less tufted in habit, with shorter and less branched stems bearing fewer (6-8) and broader leaves. The flowers are more conspicuous, with shorter and more strongly ciliate calyx lobes and a broader more brightly coloured corolla. The nutlets are also smaller and paler. A number of forms have been described by Pawlowski (***1961)."
-'A Revision of Symphytum in Turkey and Adjacent Areas' by G.E. Wickens, Royal Botanic Gardens: Kew; Notes from the Royal Botanic Garden Edinburgh, Scotland, Volume 29, pages 157-180, **1969**. Includes Turkey, Bulgaria, Greece, Aegean Islands and Caucasia.
(* -'The Forms of Symphytum Tuberosum L.' by H.W. Pugsley, Journal of Botany: British and Foreign, London, England, Volume 69, pages 89-97, 1931. I was unable to get a copy of this report. If you have a copy, could you please send it to me.)
(** -Drawings of British Plants: Being Illustrations of the Species of Flowering Plants Growing Naturally in the British Isles, Part 21, by Stella Ross-Craig, Illustrator. London, England: G. Bell & Sons, 1965. There were 31 parts, published from 1948 to 1973 that included 1300 lithographic plates.)
(*** -'Observations ad (on) Genus Symphytum L. Pertinentes' by Bogumilus Pawlowski, Fragmenta Floristica et Geobotanica, Poland, Volume 7, No. 2, pages 327-356, 1961. All in Polish. If you have an English translation, I would appreciate a copy.)
(Stella Ross-Craig, 1906-2006, was an English illustrator best known as a prolific illustrator of native flora. In 1929, she began work as a botanical illustrator and taxonomist at Kew Gardens, London, and was a contributor to 'Curtis's Botanical Magazine' and 'Icones Plantarum' of William Jackson Hooker. In 1999 she became the sixth person to receive the 'Kew International Medal'. She was a Fellow of the Linnean Society.)

"*S. tuberosum* L., *Sp. Pl.* 136 (1753).
Rhizome creeping with alternate thick tuberous and thin portions. Stem 10-15 to 40-60 cm (3.9-5.9 to 15.7-23.6 inch), simple or little-branched; stem and leaves more or less densely hairy; hairs up to 1.5 mm. Basal leaves (which disappear at flowering time) long-petiolate; the lower cauline leaves elliptical to lanceolate, the upper sessile, shortly and narrowly decurrent. Calyx 5-8 mm, lobed to 1/3 to 9/10. Corolla 13-19 mm, **pale yellow**; lobes with deflexed apex. Scales triangular-lanceolate; marginal papillae dense, the lower not more than 1 1/2 times as long as wide, the upper very small. Anthers 3 to 4.5 to 5 mm; connective not projecting beyond thecae.
Woods and other damp or shady places.
Western, central and south Europe, northwards to England, central Germany and northern Ukraine (Albania, Austria, Balearic Islands, Britain, Bulgaria, Corsica, Czechoslovakia, France, Germany, maybe Greece, Switzerland, Spain, Hungary, Italy, Yugoslavia, Poland, Romania, USSR (central region, southwestern region), Turkey- European part."
-Flora Europaea, Volume 3: Diapensiaceae to Myoporaceae, editors Tutin, Heywood, Burges, Moore, Valentine, Walters and Webb. United Kingdom: Cambridge University Press, **1972**, page 104. (Symphytum L., pages 103-105 by B. Pawlowski.) (5-volume encyclopedia of plants, published between 1964 and 1993. 'Flora Europaea' describes all the national floras of Europe to identify any wild or widely cultivated plant to the subspecies level. It provides geographical distribution, habitat preference, and chromosome number.)

"**S. tuberosum L. - Tuberous Comfrey:** Stems erect, little or not branched, to 60 cm (23.6 inch), from rhizomes with thick swollen regions; stem-leaves sessile, shortly decurrent. Calx divided about 3/4 to 9/10 to base; **corolla pale yellow.**
Native; damp woods, ditches and river banks; frequent in lowland Scotland, scattered in England, Scotland and Ireland, and perhaps only introduced."
-New Flora of the British Isles: Identification of Wild Vascular Plants of the British Isles edited by Clive Stace. Cambridge, England: Cambridge University Press, **1991**, page 647.

"Symphytum tuberosum versus S. grandiflorum:
These two yellowish-flowered, patch-forming species may be confused when the above-ground leafy stolons in S. grandiflorum DC. are mistaken for the underground rhizomes in S. tuberosum L. (cf. {compare to} Stace's 'New Flora'; this can be quite difficult in practice). In flower, S. grandiflorum has the calyx divided nearly to the base with obtuse teeth, whilst S. tuberosum has acute teeth cut to about 3/4 way."
-Plant Crib: Handbook for Field Identification compiled by T.C.G. Rich and M.D.B. Rich with editorial assistance of F.H. Perring for the BSBI Monitoring Scheme. London, England: Botanical Society of the British Isles, 1988, page 74. There is also an

updated **1998** edition. (A stolon is a creeping horizontal plant stem that takes root at points along its length, forming new plants.)

"Cluster no. 8: Symphytum bulbosum-Type:
Pollen (7)-8-9-10-11-zonocolporate, medium-sized, P = 27-42 gin, E = 22-32 gin. Shape from subprolate to prolate, with P/E = 1.25-1.40. **The pollen density is remarkably lower in S. gussonei and S. tuberosum, causing the separation of these two taxa into a separate subcluster.***"*
-'Pollen Morphology in the Boragineae (Boraginaceae) in Relation to the Taxonomy of the Tribe' by Massimo Bigazzi and Federico Selvi, Plant Systematics and Evolution, Volume 213, pages 121-151, **1998**.
(See subsection 'Comfrey Flower Buds, Bloom, Nectar and Pollination' in section 'Symphytum Genus Description' {Chapter 5}.)

"Total oils and total amounts of fatty acids in the seeds of the examined taxa:

Taxa	Total oil	Saturated	Mono-unsaturate	Poly-unsaturate
Symphytum officinale	23.10%	11.06%	29.17%	59.32%
Symphytum asperum	25.97	11.55	27.48	60.49
Symphytum orientale	32.10	12.65	23.42	63.65
Symphytum tuberosum ssp. nodosum	**7.00%**	**16.87%**	**14.32%**	**67.30%**

Although characteristic differences at generic levels and substantial similarities at infrageneric levels were observed for the quantities of fatty acids examined, **S. tuberosum ssp. nodosum** *and A. azurea var. azurea* **exhibited relatively distinct profiles compared with the others species of its genus (Symphytum).*

    S. tuberosum ssp. nodosum as Euro-Sib (Euro-Siberian) element has a very limited distributional area in northwest Turkey, but is scattered in the larger zones of Europe and Balkans. However, the other species of Symphytum examined here are also commonly distributed in Russia and Caucasia.
    S. tuberosum is morphologically different from the other Symphytum species because of its root nodules."
-'Analysis of the Total Oil and Fatty Acid Composition of Seeds of some Boraginaceae Taxa from Turkey' by Tamer Ozcan, Division of Botany, Istanbul University, Turkey; Plant Systematics and Evolution, Volume 274, Issue 3-4, pages 143-153, September **2008**.
    (The Euro-Siberian region extends from Iceland around most of Europe via Siberia to Kamchatka {far eastern Russia}. Siberia is in northern Russia.)
    (The Balkan Peninsula is surrounded by the Adriatic Sea to the west, the Mediterranean Sea and Marmara Sea to the south, and the Black Sea to the east. Its northern boundary is the Danube, Sava and Kupa Rivers.)
    (The Caucasus Mountains are located at the border of Europe and Asia, between the Black Sea and Caspian Sea and occupied by Russia, Georgia, Azerbaijan, and Armenia.)

"Symphytum tuberosum:
A tuber- and rhizome-forming geophyte with a chain-like shortened rhizome, dicyclic, semi-rosette, monocarpic shoots.
Often capable of forming hybrids with Symphytum cordatum, *e.g., Symphytum ullepitschii Wettst. (Popov 1953).*
Early ontogenetic stages and structure *of vegetative organs in S. tuberosum are similar to those of S. cordatum. Germination unstable: can be hypogeous or epigeous hypocotylar-cotyledonar. Cotyledons petiolate, rounded (5 to 8 mm long, 3 to 5 mm wide).*
Adaxial leaf surface and petioles pubescent. Pubescence moderate, mixed, uniform along the entire adaxial surface.
Trichomes of two types *can be reported: simple (subtended or on 'rosettes' and sessile) and glandular.*
The former elongated and acuminate, widened at base; 5 to 10-celled rosettes. Sessile hairs unicellular, hooked, 3 to 4 times shorter than the rosette ones. Glandular hairs are exactly the same as in S. cordatum.
Juvenile leaves *rounded-ovate or elliptic (1.5 to 2.3 cm long, 1.5 to 2 cm wide) (0.59 to 0.90 inch long, 0.59 to 0.78 inch wide), slightly acuminate at top, with rounded to cuneate leaf blade base. Petiole long (3 to 4 cm = 1.18 to 1.57 inch).*
    Leaf shape changes with the plants age.
    Immature individuals have rounded-ovate leaves (4 to 5 cm long, 3 to 4 cm wide), with acuminate apex and cordate blade base; petioles 4 to 5 cm (1.57 to 1.96 inch) long.
    Mature reproductive plants show pronounced heterophylly. Two to three lower leaves of a flowering annual shoot have small underdeveloped blades, middle three to five leaves, sessile or semi-sessile, are located close together, cuneate at base and with acuminate apex (length 5 to 10 cm, width 1.5 to 4 cm) (length 1.95 to 3.93 inch, width 0.59 to 1.57 inch) small bracts with decurrent base.
Leaf pubescence *in juvenile, immature and mature reproductive plants is similar to that of S. cordatum; some differences are related only to qualitative characteristics and the presence of scabrous-plicate surface of hairs, subtended by the rosettes."*
-'Pubescence of Vegetative Organs and Trichome Micromorphology in Some Boraginaceae at Different Ontogenetic Stages' by Rimma P. Barykina and Vitaly Y. Alyonkin, Lomonosov Moscow State University, Russia; Wulfenia, Regional Museum of Carinthia, Austria, Volume 23, pages 1-29, **2016**.
    (Pubescent means the surface of a leaf or stem is covered with fine short hairs.)
    (Trichome is a small hair or other outgrowth from the epidermis of a plant. Epidermis is the outer layer.)
    (Ontogeny is the development of an organism from fertilization of the egg to maturity.)
    (Heterophylly is when the same plant has different leaf forms.)

"Symphytum tuberosum:

Ecology: *The native habitats of this perennial herb are damp woodland, ditches, stream and river banks, where it occurs in both shaded and open situations. As an alien (not native), it occurs on roadside verges (curb strip/lawn), waste ground and other disturbed sites. Generally lowland, but reaching 335 meter (1099 feet) in Mid Perth (central Scotland).*
Status: *Native.*
Trends: ***This species has long been considered introduced to Ireland, Wales and most of England.***
In Scotland too its native status is sometimes questioned, partly in view of its late discovery *(1777), and the limit of any native range has been obscured by escapes. There is no significant change in its presumed native distribution since the 1962 *'Atlas of the British Flora', though alien records are much more frequent in England.*
World Distribution: *European Temperate element."*
-'Online Atlas of the British and Irish Flora', https://www.brc.ac.uk/plantatlas, Distribution and ecology with photographs, **2018**.
(* -Atlas of the British Flora by Franklyn Perring and S. Max Walters. Harpenden, Hertfordshire County, England: Botanical Society of the British Isles. First published 1962, second edition 1976, third edition 1982.)

"Tuberous Comfrey: Plants form extensive patches, spreading by means of a creeping tuberous rhizome.
Height x Width: 0.6 x 0.6 metre (1.9 x 1.9 feet).
Light: Semi-shade or full sun, but ***prefers semi-shade***. *Prefers moist soil.*
USDA Hardiness Zone: 4 - 8. ***It is in flower from May to June.***
Plants can be grown as a ground cover *when planted about 1.2 metres (3.9 feet) apart each way. A good, and sometimes rampant, ground cover plant for a shady border or woodland."*
-'Tuberous Comfrey, Symphytum Tuberosum' by Balkep: The Balkan Ecology Project: A Permaculture-Inspired, Grassroots Project, www.balkep.org, southeastern Europe, Bulgaria, **2019**.

S. tuberosum Breeds with Other Symphytum Species
See section 'Details about Symphytum Species Hybrids' for information about:
- S. officinale x S. asperum x S. tuberosum = S. x uplandicum x S. tuberosum
- S. officinale x S. tuberosum
- S. bulbosum x S. tuberosum
- S. cordatum x S. tuberosum

"Symphytum tuberosum, the Tuberous Comfrey *with large black roots rather like those of a Dahlia, and entirely different flowers and foliage,* ***has 72 chromosomes; it does not, therefore, interbreed (breed with other species)*** *and is of no agricultural or medicinal value, though it is sometimes tested by Research Stations as 'Comfrey'. "*
-Comfrey Report No. 1: The 1954 Research Results by Lawrence D. Hills, Henry Doubleday Research Association, Braintree, Essex, England, published **1955**.
> (This statement about not breeding with other species of Comfrey is not correct. Tuberosum will form hybrids with other species of Comfrey depending on their chromosome count.)

"(S. x uplandicum) x (S. tuberosum):
This cross occurs near the parents in very scattered localities in England and Scotland; it is intermediate in all characters, with ***yellow corollas tinged with blue or purple,*** *and tuberous rhizomes, and is at least partially sterile; endemic (native).*
Some plants might be S. officinale x S. tuberosum."
-New Flora of the British Isles: Identification of Wild Vascular Plants of the British Isles edited by Clive Stace. Cambridge, England: Cambridge University Press, **1991**, page 647.

"Symphytum tuberosum x uplandicum: species accepted.
Source: Dyntaxa Svensk taxonomisk database."
-Global Biodiversity Information Facility® (GBIF), www.gbif.org, **2018**. International network and research infrastructure funded by the world's governments and aimed at providing anyone, anywhere, open access to data about all types of life on Earth.
> (Dyntaxa is a Swedish taxonomic database of organisms occurring in Sweden.)

S. tuberosum Chromosomes

"S. tuberosum L. ssp. {subspecies} nodosum (Schur.) Soo and ssp. tuberosum:
Grau (*1968) found 2n = 96 in plants from Germany and north Yugoslavia. We found this same number for ssp. nodosum in plants from Czechoslovakia and north Yugoslavia, and 2n = 64 from Turkey in Europe.
Scottish material of the ssp. tuberosum also investigated by us turned out to have 2n = 84.
From these data it is very difficult to determine the basic number:
> *2n = 72, 84, 96, 144 point to x =12. And 2n = 64 and 2n = 96 point to x = 16.*
> *The numbers 2n = 64, 72 and 96 refer to the ssp. nodosum, and 2n = 84 and 144 to the ssp. tuberosum.*
> *Other related species are S. cordatum and S. bulbosum, which both show the chromosome number 2n = 120.*

If we classify these three species (tuberosum, cordatum, bulbosum) in one polyploid complex, *the numbers of the components are: 2n = 60, 64, 72, 84, 96, 120, 144.*
If this classification is correct, ***the most probable basic number is x = 12; only the number 2n = 64 does not fit this series."***
-'Cytotaxonomic Studies in the Genus Symphytum VIII: Chromosome Numbers and Classification of Ten European Species' by

T.W.J. Gadella and E. Kliphuis, Proceedings van de Koninklijke Nederlandse Akademie van Wetenschappen (Proceedings of the Royal Netherlands Academy of Arts and Sciences): Series C, Biological and Medical Sciences, Vol 81, pages 162-172, **1978**.
(* -'Cytologische Untersuchungen an Boraginaceae I' {Cytological Investigations on Boraginaceae} by J. Grau, Mitteilungen der Botanischen Staatssammlung Munchen {Communications of the Botanical State Collection Munich, Germany}, Volume 7, pages 277-294, 1968. I do not have an English translation of this report. If you have one, could you send it to me.)

(Yugoslavia was created in 1918 by uniting the Kingdoms of Serbs, Croats and Slovenes. It is now 8 nations including Croatia, Macedonia, Slovenia, Serbia and Bosnia/Herzegovina.)

(Polyploidy is a cell that has more than two paired {homologous} sets of chromosomes. Most species whose cells have nuclei {eukaryotes} are diploid, i.e., they have two sets of chromosomes. One set is inherited from each parent. Polyploidy is common in plants.)

"***Symphytum* L. Chromosome Numbers:**
tuberosum L.

(a) *tuberosum*	44	Italy	Grau 1968
(b) *nodosum (Schur) Soo*	96	Germany	Grau 1968
	100	Yugoslavia	
	18	Romania	*Tarnavschi 1948
	72	Bulgaria	**Markova & Ivanova 1970

This book has grown out of the requests by many users of 'Flora Europaea' (Tutin et al., 1964, 1968, 1972, 1976, 1980) for information on the sources of the chromosome numbers cited in that work."
-Flora Europaea: Check-List and Chromosome Index by D.M. Moore. Cambridge, England: Cambridge University Press, **1982**, page 179.
(* -'Die Chromosomenzahlender Anthophyten-Florav on Rumanien mit Einem Ausblick auf das Polyploidie-Problem' {The Number of Chromosomes The Anthophyten-Florav in Romania with an Outlook on the Polyploidy Problem} by I.T. Tarnavschi; Bulletin du Jardin et du Musee Botaniques de Universite de Cluj, Roumanie, Volume 28, pages 1-130, 1948. In German.)
(** -'Karyologische Untersuchung der Vertreter der Fam. Boraginaceae, Labiatae und Scrophulariaceae in Bulgarien' {Caryological Examination of the Representatives of the Family Boraginaceae, Labiatae and Scrophulariaceae in Bulgaria} by Margarita Markova and P. Ivanova; Izvestiya na Botaniceskiya Institut {Bulletin de l'Institut Botanique}, {also titled: Mitteilungen des Botanischen Instituts Sofia}, Bulgaria, Volume 20, pages 93-98, 1970. In German.)

"*Species, origins and chromosome numbers (2n) of studied materials in present study:*
Symphytum tuberosum L. subsp. *nodosum (Schur) Soo*
Locality A7, Trabzon in Turkey: Arakh, Ayvadere village, near fields, meadow areas, under Corylus spp, road sides. 250 meter (820 feet) above sea level. Collected living plants. **Chromosome number: 2n = 24.**
A great variation of chromosome numbers have been reported for this species:
2n = 18, 72, 96, 100 from Europe (*Pawlowski 1972).
2n = 64 from Turkey (**Davis 1988).
2n = 36 from Spain (***Luque 1989)."
-'Chromosome Numbers of the Twenty Two Turkish Plant Species' by Huseyin Inceer, Sema Hayirlioglu-Ayaz and Melahat Ozcan, Karadeniz Technical University, Trabzon, Turkey; Caryologia: International Journal of Cytology, Cytosystematics and Cytogenetics, Volume 60, No. 4, pages 349-357, **2007**.
(* -Flora Europaea, Volume 3: Diapensiaceae to Myoporaceae, editors Tutin, Heywood, Burges, Moore, Valentine, Walters and Webb. United Kingdom: Cambridge University Press, 1972. Symphytum L., pages 103-105 by B. Pawlowski.)
(** -Flora of Turkey and the East Aegean Islands, 11 Volumes by Peter H. Davis. Scotland: Edinburgh University Press, 1984-1988. 'Symphytum L.' in Supplement Volume 10, pages 186-189, 1988.)
(*** -'Estudio Cariologico de Boraginaceae Espanolas IV' by T. Luque; Boletim da Sociedade Broteriana, Coimbra, Portugal, Volume 2, No. 62, pages 211-220, 1989. If you have an English translation of this, I would appreciate a copy.)

"***Chromosome number (2n) and genome size (2C-value, in gigabase pairs)*** *of herbaceous dicot perennials on five sampling plots in valley Dolina Smrti, Slovenia:* ***Symphytum tuberosum* L. 2n = 64 2C = 5.73 Gbp.**"
-'Selective Significance of Genome Size in a Plant Community with Heavy Metal Pollution' by T. Vidic, J. Greilhuber, B. Vilhar and M. Dermastia, Kranj and Ljubljana, Slovenia, and Vienna, Austria; Ecological Applications: Ecological Society of America, Volume 19, No. 6, pages 1515-1521, **2009**.
('C' is the mass in picograms of DNA present in a haploid {gamete} chromosome set.)
(A base pair {bp} is a unit of two nucleobases bound to each other by hydrogen bonds. They form the building blocks of the DNA double helix. 1 Gbp {giga base pairs} = 1,000,000,000 bp.)

"***Genome size*** *is the total amount of DNA contained within one copy of a single genome. It is typically measured in terms of mass in* ***picograms*** *(trillionths (10^{-12}) of a gram, abbreviated* ***pg***) *or less frequently in Daltons or as the total number of* ***nucleotide base pairs typically in megabases*** *(millions of base pairs, abbreviated* ***Mb or Mbp***). *One picogram equals 978 megabases. In diploid organisms, genome size is used interchangeably with the term* ***C-value***. ***C-value*** *is the amount, in picograms, of DNA contained within a haploid nucleus (e.g., a gamete) or one half the amount in a diploid somatic cell of a eukaryotic organism.*"
-Wikipedia®: The Free Encyclopedia, www.wikipedia.org, **2019**.

"**Chromosome numbers, ploidy levels, and mean C-values (Gbp and pg)** *from Vidic et al.* compared to the mean C- and Cx-values of the present study, and the locality code.*

Taxon	2n	Ploidy	2C (Gbp)	1C (pg)	Mean values. 1C (pg)	1Cx (pg)	Accession data. Locality	1C (pg)	SD (pg)
S. tuberosum	64	8x	5.73	2.93	2.745	0.686	43	2.745	0.014

C-value or holoploid genome size *(Greilhuber et al.), that is, the DNA amount contained in the chromosome complement of an organism, is directly and positively correlated with cell size (Bennett, Knight and Beaulieu).*
Cx-value or monoploid genome size, *the DNA amount contained in the single basic chromosome set with chromosome number x has a strong positive influence on cell cycle duration (Bennett, Francis et al.)."*
-'Heavy Metal Pollution, Selection, and Genome Size: The Species of the Zerjav Study Revisited with Flow Cytometry' by Eva M. Temsch, Wilhelm Temsch, Luise Ehrendorfer-Schratt and Johann Greilhuber, University of Vienna, Austria; Hindawi Publishing Corporation, Journal of Botany, Volume 2010, Article ID 596542, May **2010**.
(* -'Selective Significance of Genome Size in a Plant Community with Heavy Metal Pollution' by T. Vidic, J. Greilhuber, B. Vilhar, and M. Dermastia; Ecological Applications, Volume 19, No. 6, pages 1515-1521, 2009.)

    (Body cells and individuals are described by the number of chromosome sets {ploidy level} they have: monoploid: 1 set; diploid: 2 sets = 2n = 2x, triploid: 3 sets = 2n = 3x, tetraploid: 4 sets = 2n = 4x, pentaploid: 5 sets = 2n = 5x, etc.)
    (Locality 43: District: Vienna, Austria. Province: Vienna. Village: Vienna/Lainzer Tiergarten. Altitude: 295 meters {967 feet} above sea level.)

"**The Symphytum tuberosum complex is a highly polyploid and taxonomically intriguing group.**
The S. tuberosum complex belongs to the widely accepted section Tuberosa Buckn., *which is characterized by:*
    (i) mostly tuberous rhizomes;
    (ii) triangular, densely papillose fornices that do not protrude from corollas;
    (iii) yellow flowers (Bucknall 1913, Pawlowski 1961, Wickens 1969, Sandbrink et al. 1990).
The section is a taxonomically difficult and still unresolved group with high-level polyploids and considerable morphological variation *(Gadella & Kliphuis 1978, *Murin & Majovsky 1982).*
    At least eight ploidy levels were recorded previously within this complex.
Based on flow cytometric screening of 271 central-European populations, **two dominant ploidy levels were revealed: tetraploid (2n = 4x = 32) and widespread dodecaploid (2n = 12x = 96).**
    The tetraploid cytotype *is mainly distributed along the southern and south-western margins of the west Carpathians where they abut (common boundary) the Pannonian basin, and found only in Slovakia, the Czech Republic (south-eastern Moravia) and Hungary; our findings represent the first records of this ploidy level for the latter two countries.*
    In contrast, the dodecaploid cytotype *occurs throughout the whole area studied.*
In addition to their geographic distributions, differences between the cytotypes in morphology and habitat requirements were detected using a multivariate morphometric analysis and analysis of a phytosociological database, respectively.
Based on this information and taking certain overlaps in morphological traits and habitat requirements into account, **we propose treating the dominant cytotypes as subspecies:**
    S. tuberosum subsp. tuberosum (dodecaploids)
    S. tuberosum subsp. angustifolium (tetraploids).
In some populations, aneuploids and several minority ploidy levels were also detected, including DNA-hexaploids (only within populations of tetraploids), DNA-decaploids and DNA-tetradecaploids (both only within populations of dodecaploids)."
-'Symphytum Tuberosum Complex in Central Europe: Cytogeography, Morphology, Ecology and Taxonomy' by L. Kobrlova, M. Hrones, P. Koutecky, M. Stech, and B. Travnicek, Preslia: The Journal of the Czech Botanical Society, Czech Republic; Volume 88, pages 77-112, March **2016**.
(* -'Die Bedeutung der Polyploidie in der Entwicklung der in der Slowakei Wachsenden Arten der Gattung Symphytum L.' (The Importance of Polyploidy in the Development of the Growing Slovakia Species of the Genus Symphytum L.) by A. Murin and J. Majovsky, Acta Facultatis Rerum Naturalium Universitatis Comenianae Botanica, Slovakia, Volume 29, 1982, pages 1-25. I was unable to find a copy of this report. If you have an English translation, could you please send it to me.)
    (A cytotype is an individual of a species that has a different chromosomal factor to another.)
    (Aneuploidy is an abnormal number of chromosomes in a cell.)

"**All these processes are also important for genome evolution in the genus Symphytum L.** *(Grau 1968, 1971, Gadella & Kliphuis 1978, Murin & Majovsky 1982).*
Gadella & Kliphuis (1978) **report a high frequency of polyploids in comparison with other genera of Boraginaceae family** *with the occurrence of polyploidy, as in Onosma, Myosotis and Pulmonaria genuses.*
    This phenomenon is well illustrated by the four ploidy levels reported for the Symphytum officinale complex (e.g.
    Markowa & Iwanowa 1970, Gadella & Kliphuis 1978) **or, even more surprisingly, the eight ploidy levels reported in the Symphytum tuberosum complex (Murin & Majovsky 1982), which range from presumably diploid (2n = 2x =18) up to octodecaploid cytotypes (2n = 18x = 144)."**
-'Symphytum Tuberosum Complex in Central Europe: Cytogeography, Morphology, Ecology and Taxonomy' by L. Kobrlova, M. Hrones, P. Koutecky, M. Stech, and B. Travnicek, Preslia: The Journal of the Czech Botanical Society, Czech Republic; Volume 88, pages 77-112, March **2016**.

"Symphytum tuberosum L.: Chromosome count is 2n = 32, 64, 96, 120, 128, 144."
-Tropicos®, www.tropicos.org, Missouri Botanical Garden, Saint Louis, Missouri; Plant database with 1.3 million scientific names and 4.4 million specimen records, **2018**.

"Two previous estimations of GS (Genome Size) were compared with our data for Symphytum tuberosum (1C = 2.75 pg, *Temsch et al., 2010; 1C = 2.93 pg, **Vidic et al., 2009 vs. 1C = 3.20 pg, this study).
Both of these values show small discrepancies, suggesting the possible detection of aneuploidy or other minor cytological differences, which have both been detected in populations of S. tuberosum s.l. to date *(***Kobrlova et al., 2016).*
> *In general, instrumental or methodological errors, potential differences among laboratories, use of a different reference standard (Dolezel et al., 1998), intraspecific GS variation (Smarda & Bures, 2010) and the effect of secondary metabolites should also be considered (e.g. Loureiro et al., 2006; Kolarcik et al., 2018)."*

-'First Insights into the Evolution of Genome Size in the Borage Family: A Complete Data Set for Boraginaceae from the Czech Republic' by Lucie Kobrlova and Michal Hrones, Department of Botany, Palacky University, Olomouc, Czech Republic; Botanical Journal of the Linnean Society, Volume 189, No. 2, pages 115-131, February **2019**.
(* -'Heavy Metal Pollution, Selection, and Genome Size: The Species of the Zerjav Study Revisited with Flow Cytometry' by Eva M. Temsch, Wilhelm Temsch, Luise Ehrendorfer-Schratt and Johann Greilhuber, University of Vienna, Austria; Hindawi Publishing Corporation, Journal of Botany, Volume 2010, Article ID 596542, May 2010.)
(** -'Selective Significance of Genome Size in a Plant Community with Heavy Metal Pollution' by T. Vidic, J. Greilhuber, B. Vilhar and M. Dermastia, Kranj and Ljubljana, Slovenia, and Vienna, Austria; Ecological Applications: Ecological Society of America, Volume 19, No. 6, pages 1515-1521, 2009.)
(*** -'Symphytum Tuberosum Complex in Central Europe: Cytogeography, Morphology, Ecology and Taxonomy' by L. Kobrlova, M. Hrones, P. Koutecky, M. Stech, and B. Travnicek, Preslia: The Journal of the Czech Botanical Society, Czech Republic; Volume 88, pages 77-112, March 2016.)
> (Sensu lato = s.l., sens. lat. In the broad sense. Used in taxonomy to clarify the scope of a taxon when it has been used to define more than one set of lower-level taxons. It includes all its subordinate taxa and/or other taxa that at other times are considered as distinct.)

S. tuberosum and Alkaloids (Low in Alkaloids, High in Allantoin)

*"**The alcoholic extract of Symphytum tuberosum also yielded six alkaloids, two of them identified as anadoline and echimidine.** In the aqueous (water) extract of the plant the following amino acids were detected: aspartic acid glycine, leucine, serine, valine, alanine, glutamic acid, proline, methionine, isoleucine, phenylalanine, histidine and lysine."*
-'Alkalloids and Other Compounds of Symphytum Tuberosum' by A. Ulubelen and F. Ocal, Faculty of Pharmacy, University of Istanbul, Turkey, Phytochemistry, Volume 16, Issue 4, pages 499-500, **1977**.

"In a previous comnunication (*Gray et al 1983) we reported the isolation of two pyrrolizidine alkaloids, symlandine and echimidine together with a large quantity of allantoin from the roots and rhizomes of Symphytum tuberosum L. (Tuberous Comfrey, family Boraginaceae).
> *Allantoin is used in the treatment of psoriasis and other skin diseases and is a component of many cosmetics (Nakao et al 1982). The occurrence of pyrrolizidine alkaloids in Symphytum spp. (species) is of considerable importance owing to their hepatotoxicity and carcinogenicity (**Schoental 1982).*

The dried leaves (2.5 kg = 5.5 pounds) of S. tuberosum (collected May 1984 at the National Botanic Garden, Glasnevin, Dublin, Ireland) were percolated with CHCl3 and then MeOH. Work up (cf. Gray et al 1983) led to the isolation of **two pyrrolizidine N-oxides, symlandine-N-oxide (1, 0.0032% w/w {weight by weight}) and echimidine-N-oxide (2, 0.004% w/w).**
The structures were determined on the basis of physico-chemical studies using high resolution 1H-NMR with spin decoupling, 13C-NMR and MS, by comparison of spectral data with those of the free bases (3 & 4, respectively), which were also found in the CHCl3 extract, and by conversion into the free bases.
Once again (cf. Gray et al 1983) the yield of allantoin was high (0.98% w/w from the MeOH extract) and that of the pyrrolizidine alkaloids low in comparison with S. officinale L. (*Tittel et al 1979), the species used in herbal medicine. This confirms that S. tuberosum is a useful alternative source of allantoin."*
-'Pyrrolizidine Alkaloid N-Oxides from Symphytum Tuberosum' by P. Bhandari and A.I. Gray, Department of Pharmacognosy, School of Pharmacy, Trinity College, Ireland, Journal of Pharmacy and Pharmacology, Volume 37, Issue S12, December **1985**.
(* -'Hepatotoxic Alkaloids and Allantoin in Symphytum Tuberosum' by Gray, A.I. et al, Journal of Pharmacy and Pharmacology, Volume 35, 13P, 1983. I was unable to get a copy of this report. If you have one, could you send it to me.)
(** -'Health Hazards of Pyrrolizidine Alkaloids: A Short Review' by R. Schoental; Toxicology Letters, Amsterdam, Netherlands, Volume 10, Issue 4, pages 323-326, 1982.)
(*** -'Quantitative Estimation of the Pyrrolizidinalkaloids of Radix Symphyti' by HPLC' by G. Tittel, H. Hinz and H. Wagner; Planta Medica: Journal of Medicinal Plant Research, Volume 37, Issue 9, pages 1-8, 1979. In German except abstract is in English.)
> (cf. = compare to other literature)

*"**In S. tuberosum subspp., tuberosum and nodosum, S. grandiflorum and S. ibericum the presence of the pyrrolizidine alkaloids lycopsamine, echimidine and symphytine could be demonstrated.**
The taxon S. tuberosum contains an unknown compound that seems to be specific for this taxon. This compound is not

the pyrrolizidine alkaloid anadoline which has previously been reported for this species. It is possibly represented by a peak on GC/MS (Gas Chromatography/Mass Spectrometry) with a molecular ion peak at m/z 623 (as TMS derivative) and can be used as a chemotaxonomic marker for the species S. tuberosum.
The pyrrolizidine alkaloid pattern of the two subspecies of S. tuberosum reinforces the close relationship. *Fresh material of S. tuberosum contained the triterpene isobauerenol, but in herbarium material isobauerenol was lacking."*
-'Chemo- and Karyotaxonomic Studies on Some Rhizomatous Species of the Genus Symphytum (Boraginaceae)' by T.A. Jaarsma, E. Lohmanns, H. Hendriks, T.W.J. Gadella and T.M. Malingre, Plant Systematics and Evolution, Volume 169, pages 31-39, **1990**.

"**Symphytum tuberosum L., Tuberous Comfrey** *(German: Knoten-Beinwell, French: Consoud tubereuse), grows in south-east Europe.* **Its allantoin content is very high; besides, it contains symlandine, echimidine and anadoline respectively the N-oxides of these alkaloids in a total concentration of only 0.02%*.**
Tuberous Comfrey is recommended for medicinal purposes as an alternative to the other Comfrey species**."
-'Medicinal Plants in Europe Containing Pyrrolizidine Alkaloids' by Erhard Thomas Roeder, Pharmazeutisches Institut (Pharmaceutical Institute) der Rheinischen Friedrichs-Wilhelms, University of Bonn, Germany, Pharmazie (Pharmacy) 50, pages 83-98, March **1995**.
(* -'Alkalloids and Other Compounds of Symphytum Tuberosum' by A. Ulubelen and F. Ocal, Faculty of Pharmacy, University of Istanbul, Turkey, Phytochemistry, Volume 16, Issue 4, pages 499-500, 1977.)
(** -'Pyrrolizidine Alkaloid N-Oxides from Symphytum Tuberosum' by P. Bhandari and A.I. Gray, Department of Pharmacognosy, School of Pharmacy, Trinity College, Ireland, Journal of Pharmacy and Pharmacology, Volume 37, Issue S12, December 1985.)

"**S. tuberosum L. has been shown to contain 72 mg/kg PA (Pyrrolizidine Alkaloid) N-oxides in the leaf, and 180 mg/kg in the root. S. tuberosum contains PA levels significantly lower than other Symphytum spp. (species) and relatively high levels of allantoin, 0.96% in the root and 0.98% in the leaf.**
These findings are only on single samples, but it may become the preferred species for medicinal use."
-'Using Herbs that Contain Pyrrolizidine Alkaloids' by Alison Denham, B.A., MNIMH (National Institute of Medical Herbalists), University of Central Lancashire, England; The European Journal of Herbal Medicine, Volume 2, No. 3, pages 27-38, **1996**.

"**To the extent that some members of the Comfrey plant family do not contain PAs,** *then the concerns FDA (Food and Drug Administration of the United States) has about a Comfrey-containing dietary supplement product or ingredient depends upon the exact species identified.*
The FDA ruling banned the internal use of Symphytum officinale (Common Comfrey), S. asperum (Prickly Comfrey), and S. x uplandicum (Russian Comfrey), as well as any other plant/substance containing pyrrolizidine alkaloids.
While FDA did not examine the safety of other Comfrey species such as **Symphytum tuberosum L. (Tuberous Comfrey), which is suggested to contain negligible amounts of pyrrolizidine alkaloids,** *FDA would rely on the presence of the pyrrolizidine alkaloid to determine whether this species was permitted to be used as a dietary ingredient."*
-Bioactive Compounds in Food: Edited by John Gilbert (United Kingdom) and Hamide Z. Senyuva (Turkey), United Kingdom, Oxford: Blackwell Publishing Ltd, **2008**, page 619.

Symphytum uliginosum
See section 'Details about Symphytum Species: Officinale' (Chapter 8}.
See 'Symphytum tanaicense' in this section.

Current Botanical Nomenclature of S. uliginosum

"*Symphytum tanaicense Steven, Bull. Soc. Imp. Nat. Mosc. 24. 1851*
        = S. uliginosum Kern. var. pseudopterum Borbas, Flora von Budapest und Umgebung, 1879."
-'Il Genere Symphytum L. (Boraginaceae) in Italia: Revisione Biosistematica' (The Genus Symphytum L. Boraginaceae in Italy: A Biosystematic Revision) by Stefania Bottega and Fabio Garbari, Webbia: Journal of Plant Taxonomy and Geography, Volume 58, No. 2, pages 243-280, **2003**. (If you have an English translation of this article, I would appreciate having a copy.)
(* -Flora von Budapest und Umgebung {Flora of Budapest, Hungary, and Surroundings}, 1879 is all the information I could find. Barbos wrote: 'Vasvarmegye Novenyfoldrajza es Floraja: Geographia atque Enumeratio Plantarum Comitatus Castferrei in Hungaria' {Plant Geography and Flora in Vas County, Hungary} by Vincze von Borbas, Veszto, Hungary, 1887.)

"**Symphytum uliginosum A.Kern.** *(*Oesterr. Bot. Z. 13: 227. 1863).*
Symphytum officinale subsp. uliginosum (A.Kern.) Nyman *(**Consp. Fl. Eur. 509. 1881)."*
-'International Plant Names Index'® (IPNI), www.ipni.org, A database of the names and associated basic bibliographical details of seed plants, ferns and lycophytes, **2018**.
(* -'Descriptiones Plantarum Novarum Florae Hungaricae et Transsilvanicae' by A. Kerner, Oesterreichische Botanische Zeitschrift {Austrian Botanical Journal}, Vienna, Austria, Volume 13, No. 7, page 227, 1863.)
(** -'Conspectus Florae Europaeae: seu Enumeratio Methodica Plantarum Phanerogamarum Europae Indigenarum, Indicatio Distributionis Geographicae Singularum etc.' by Carl Fredrik Nyman, Orebro Sueciae {Sweden}, page 509, 1881.)

***"Symphytum uliginosum Kern.**: synonym for Symphytum tanaicense Stev."*
-Catalogue of Life®: 2015 Annual Checklist, Online database of the world's known species of animals, plants, fungi and micro-organisms. It consists of a single integrated species checklist and taxonomic hierarchy., **2018**, www.catalogueoflife.org. Search: www.catalogueoflife.org/annual-checklist/2015/search/all

***"Symphytum uliginosum A.Kern.** is synonym of Symphytum officinale subsp. uliginosum (A.Kern.) Nyman."*
-The Plant List®, www.theplantlist.org from the World Checklist Database, **2018**.

***"Symphytum uliginosum Kern.** Published in: Oesterr. Bot. Z. 13:227. 1863*
*Synonym of **Symphytum officinale subsp. uliginosum (A.Kern.) Nyman**"*
-Global Biodiversity Information Facility® (GBIF), www.gbif.org, **2018**. International network and research infrastructure funded by the world's governments and aimed at providing anyone, anywhere, open access to data about all types of life on Earth.

Description of S. uliginosum

*"**Symphytum uliginosum:**
Differt a vicino S. officinali L, foliis scabris (nec hirsutis) caulinis in petiolum basi subdilatata amplexicaulem contractis (nec decurrentibus). In pratis uliginosis prope Pest cum S. officinali, Cirsio brachycephalo el Senecio paludoso."*
(Online translation: It is different from the neighbor S. officinale. Leaves are hairy....In meadows near Budapest....Senecio marshlands.)
-'Descriptiones Plantarum Novarum Florae Hungaricae et Transsilvanicae' (Descriptions of New Plants in Hungary and Transylvania/Romania) by A. Kerner, Oesterreichische Botanische Zeitschrift (Austrian Botanical Journal), Vienna, Austria, Volume 13, No. 7, pages 227-288, **1863**. All in Latin.

*"**S. uliginosum, Kern.** in Oestr. Bot. Zeitschr. xiii. (1863) p. 227.:
Geographical Distribution: **Hungary, south Russia**. The entire plant is glabrous except for small white deciduous prickles with tubercular bases, which render it scabrid and harsh to the touch.
The calyx segments are long and gradually atttenuated to an acute point, ciliate with small, strong, white prickles, and purplish in colour. **The flowers appear to be rose-coloured.** The nutlets are similar to those of S. officinale.
Kerner describes the leaves as not decurrent, but forms often occur in which the stem is partially winged, and this constitutes the **variety pseudopterum Borb**."*
-'A Revision of the Genus Symphytum, Tourn.' by Cedric Bucknall, {Mus. Bac. Oxon= Bachelor of Music, Oxford University}, Journal of the Linnean Society of London, England, Botanical Journal, Volume 41, Issue 284, pages 491-556, December **1913**.

*"**S. officinale L. subsp. uliginosum (A. Kerner) Nyman**, Consp. 509 (1881):
Leaves, even the uppermost, not or shortly and narrowly decurrent. Stem and leaves only sparsely covered with very short, stiff setae and densely verrucose-hispid. Sepals with stiff, apical setae, otherwise more or less glabrous. **East central Europe.**"*
-Flora Europaea, Volume 3: Diapensiaceae to Myoporaceae, editors Tutin, Heywood, Burges, Moore, Valentine, Walters and Webb. United Kingdom: Cambridge University Press, **1972**, page 104. (Symphytum L., pages 103-105 by B. Pawlowski.)

*"**Symphytum officinale L.: 2n = 24 or 2n = 48**
S. bohemicum Schmidt, Fl. Boem. 3: 13. tab 263. 1795.
S. tanaicense Steven, Bull.Soc.Imp. Naturalistes Moscou 24:577. 1851.
S. uliginosum Kerner, Oesterr. Bot. Z. 13: 227, 1863.
White or cream flowers in diploids (2n = 24) and tetraploids (2n = 48).
Purple or red (or various intermediate colors between white and dark purple) in the tetraploids (2n = 48).
The diploid cytotype was assigned to Symphytum bohemicum Schmidt by A. Murin and J. Majovsky (Acta Facultatis Rerum Naturalium Universitatis Comenianae Botanica, Volume 29, 1982.)
The exact status of S. bohemicum, S. tanaicense, and S. uliginosum appears to require further investigation."*
-'Notes on Symphytum (Boraginaceae) in North America' by T.W.J. Gadella, page 1061-1067 in book: Annals of the Missouri Botanical Garden, Volume 71, Saint Louis, Missouri. Lawrence, Kansas: Allen Press, **1984**.

*"**S. uliginosum var. pseudopterum Borbas is also characterized by decurrent stem leaves.** These data suggest a very close affinity of S. officinale (2n = 40) and S. tanaicense."*
-'Chemotaxonomy of the Symphytum Officinale Agg. (Boraginaceae)' by Jaarsma, Lohmanns, Gadella and Malingre, Plant Systematics and Evolution, Volume 167, pages 113-127, **1989**.

Three subspecies of Symphytum officinale are recognised by Gadella & Perring (*2000) as subsp. officinale, subsp. bohemicum, and subsp. uliginosum.
S. officinale subsp. uliginosum (A. Kern.) Nyman has not been recorded from the British Isles since before 1930 (**Clement & Foster 1994) although it may occur here (Perring 1994).
It is distinguished from the other subspecies by the outer surface of the sepals being more sparsely pubescent and on chromosome number being **2n = 40**.
It also nearly always has **purple buds and flowers** (some cytotypes of S. officinale and S. x uplandicum also

have purple buds and flowers) and Gadella & Perring (2000) provide further details on its identification.
S. officinale* subsp. *bohemicum *occurs in the Netherlands, and eastern Europe, particularly in the Czech Republic and Hungary. In the British Isles, it is recorded from only 3 English vice counties: Cambridgeshire v.c. 29 and Huntingdonshire v.c. 31 (Stace et al. 2003) and South Lincolnshire v.c. 53 (Perring 1994), with all records being from fens (marshes)."*
-'What is Symphytum Officinale Subsp. Bohemicum (Schmidt) Celak: A Taxon on the Red Data Waiting List' by Clare Coleman O'Reilly, County Durham, England; Botanical Society of Britain and Ireland (BSBI) News, Hertfordshire, England, No. 102, pages 46-48, April 2006.
(* -The European Garden Flora: A Manual for the Identification of Plants Cultivated in Europe, Both Out-of-Doors and Under Glass, 6 Volumes, edited by James Cullen, Sabina G. Knees, and H. Suzanne Cubey. England: Cambridge University Press, **2000**. Volume 6: 'Symphytum' by T.W.J. Gadella and F.H. Perring, pages 138-141.)
(** -Alien Plants of the British Isles: Provisional Catalogue of Vascular Plants Excluding Grasses by E.J. Clement and M.C. Foster. London, England: Botanical Socieity of Britain and Ireland, 1995.)

"*In his revision of Symphytum, Bucknall (*1913) recognised* **two species in Symphytum 'sect. Officinalia'** *(i.e., S. section Symphytum):*
    1. The widespread S. officinale L.: The whitish-flowered plants were assigned to S. officinale 'subvar. ochroleucum',
    2. ***S. uliginosum* A. Kern.**, *known from Hungary and south Russia: The* **purple-flowered (purplish violet) ones** *to*
    **subvar. purpureum (Pers.) Buckn. *(i.e., subvar. officinale), of which S. patens Sibth. was considered a synonym.*
A taxon showing intermediate features between S. officinale and S. uliginosum was named S. officinale var. lanceolatum Weinm., with S. tanaicense as a tentative synonym.
On the basis of Steven's and Kerner's original descriptions and after study of the exsiccata in Fl and Pl, the identity of the Symphytum from the Massaciuccoli Lake (Tuscany, Italy) can be established as follows:
    ***Symphytum tanaicense* Steven =**
    ***Symphytum officinale* subsp. *tanaicense* (Steven) =**
    ***Symphytum uliginosum* A. Kern =**
    ***Symphytum officinale* subsp. *uliginosum* Nym.**"
-'Symphytum Tanaicense (Boraginaceae) New for the Italian Flora' by Peruzzi, Garbari and Bottega, Willdenowia, Berlin, Germany, 31, pages 33-41, **2001**.
(* -'A Revision of the Genus Symphytum' by C. Bucknall, Botanical Journal of the Linnean Society, 41, pages 491-556, 1913.)
(** -Symphytum officinale L. Sp. 136, var. purpureum Pers. Syn. i. 161 = Synopsis Plantarum, Volume 1, page 161 by Christiaan Hendrik Persoon, 1805.)
    (sect.=section, subvar.=subvariety, var.=variety) (Taxon is a group of organisms seen by taxonomists to form a unit.)
    (Exsiccata are dried herbarium specimens, usually a set of identified specimens all belonging to the same taxon and distributed among various herbaria {plural for herbarium}.)

"***Symphytum tanaicense* Steven** *in Bull. Soc. Imp. Nat. Moscou 24: 1851.*
= *S. officinale* L. subsp. *tanaicense* (Steven) Soo *in Bot. Kozlem. 28: 127. 1931.*
    Locus classicus: Russia- 'ad Tanain inferiorem' {south part of Don River Valley}.
    Holotype (see Bottega & Garbari 2003: 258): 06.1817, Steven (H 1535841).
= *S. uliginosum* A.Kern. *in Oesterr. Bot. Z. 13: 227. 1863.*
= *S. officinale* L. subsp. *uliginosum* Nyman, *Consp. Fl. Eur.: 509. 1881;*
S. officinale L. f. uliginosum Fiori in Fiori & Beg., Fl. Italia 2(3): 377. 1902.
    Locus classicus: Hungary- 'In pratis uliginosis prope Pest' {meadows near Budapest, Hungary}.
    Lectotype (here designated): Hungary- 'Sumpfwiesen am Rakos bei Pest' {Marsh meadows...}, s.d., Kerner (WU 69901; iso-WU 69902, 69904; 'Sumpfige Wiesen langs d. Rakosbrache bei Pest' {Swampy meadows...}, WU 69903).
A rare species in Italy, *found near Pisa in Tuscany."*
-'Synopsis of Boraginaceae subfam. Boraginoideae Tribe Boragineae in Italy' is the same as 'Boraginaceae in Italy II' by L. Cecchi and F. Selvi, University of Florence, Firenze, Italy; Plant Biosystems: An International Journal Dealing with all Aspects of Plant Biology, Official Journal of the Societa Botanica Italiana, Volume 149, No. 4, pages 630-677, **2015**.
    ('Locus Classicus' is text that is the best known or most authoritative on a particular subject.)
    (Holotype is a single type specimen upon which the description and name of a new species is based.)
    (Lectotype is a specimen chosen as the type of a species if the author of the name fails to designate a type.)
    ('f.' = forma, which is an infraspecific rank. Infraspecific is at a taxonomic level below that of species, e.g., subspecies, variety, cultivar, or form. Latin names at this level require the addition of a term denoting rank.)

S. uliginosum Chromosomes

"*In the Netherlands* **the 2n = 40 cytotype** *is very common in the low lying peat lands of Noord Holland, Utrecht, and Friesland.*
    Populations with 2n = 40 may have a much wider European distribution because some of the characters of this cytotype closely match those of plants from Hungary and southern Russia.
These later plants (from Hungary and Russia) were referred to *S. uliginosum* Kern. by R.Von de Soo (*1926) and to *S. tanaicense* Steven by Degen (**1930). *Symphytum tanaicense* is the correct name. *The name is used by Fernald (***1950) in Gray's 'Manual of Botany'.*"

-'Notes on Symphytum (Boraginaceae) in North America' by T.W.J. Gadella, page 1061-1067 in book: Annals of the Missouri Botanical Garden, Volume 71, Saint Louis, Missouri. Lawrence, Kansas: Allen Press, **1984**.
(* -'Diagnoses Plantarum Novarum et Revisio Formarum Specierum Nonnularum' by R. Von de Soo, Repertorium Specierum Novarum Regni Vegetabilis {Repertoire of New Types of Vegetables}, Volume 27, pages 316-322, 1926. In Latin.)
(** -'Bemerkungen uber einige orientalische Pflanzenarten LXXXIX. Uber Symphytum uliginosum Kern.' {Remarks about some Oriental Plant Species LXXXIX. About Symphytum uliginosum Kern.} by A. Von Degen, Magyar Botanikai Lapok {Hungarian Botanical Pages}, Volume 29, pages 144-148, 1930. If you have an English translation, could you send it to me.)
(*** -'Gray's Manual of Botany: A Handbook of the Flowering Plants and Ferns of the Central and Northeastern United States and Adjacent Canada' by Asa Gray and M.L. Fernald. New York: American Book Company, 1950.)

"Symphytum officinale L. ssp. uliginosum (Kern.) Nyman (2n = 40)
S. tanaicense Steven (species, 2n = 40)
S. uliginosum Kern
At present there appears to be two cytotypes of Symphytum officinale that may warrant recognition at some level:
   1. The 2n = 24 or 48 type (diploid and tetraploid, respectively) appears to be the common form in New England (northeastern United States). It shows hispid stems that are not harsh to the touch, leaves with prominent decurrent wings, marginal setae of the sepals distributed in an irregular pattern, and white (usually diploid) or purple (usually tetraploid) corollas.
   *2. The 2n = 40 type, known from North America but yet documented in New England, has tuberculate-based hairs on the stem that are harsh to the touch, leaves with shorter decurrent wings, marginal setae of the sepals distributed in a uniform manner, and **purple corollas**.*
   This latter cytotype has been named ssp. uliginosum (or S. tanaicense at the rank of species)."
-New England Wild Flower Society's Flora Novae Angliae: A Manual for the Identification of Native and Naturalized Higher Vascular Plants of New England by Arthur Haines. New Haven: Connecticut: Yale University Press, **2011**.
   (A cytotype is an individual of a species that has a different chromosomal factor to another.)
   (Body cells and individuals are described by the number of chromosome sets {ploidy level} they have: monoploid: 1 set; diploid: 2 sets = 2n = 2x, triploid: 3 sets = 2n = 3x, tetraploid: 4 sets = 2n = 4x, pentaploid: 5 sets = 2n = 5x, etc.)

Symphytum uplandicum (See section 'Details about Symphytum Species Hybrids: Russian Comfrey' {Chapter 10}.)

Symphytum zeyheri (See S. bulbosum in this section.)

Fig. 421. Prickly comfrey (*Symphytum asperrimum*).

Prickly Comfrey (Symphytum asperrimum) 1911

'Cyclopedia of American Agriculture, Volume II: Crops: A Popular Survey of Agricultural Conditions, Practices and Ideals in the United States and Canada', 4 Volumes, edited by Liberty Hyde Bailey, New York, 1911.

PRICKLY COMFREY.

Prickly Comfrey 1879

"The picture of Prickly Comfrey given is from a cut kindly furnished us by Mr. James J. H. Gregory, the eminent seedsman of Marblehead, Massachussetts. This new forage plant is extensively grown in Europe for the feeding of stock. It is a deep rooted plant, and even in the hottest seasons will yield several cuttings of forage. It comes in earlier and lasts longer than almost any forage crop. In well plowed and well manured ground plant the cuttings three feet apart each way, giving them a liberal dressing of manure the first Winter."

'The Farmer's Magazine, A Monthly Journal for Planters and for Country Homes'; John Duncan, Editor. Louisville, Kentucky, April 21 1879.

Chapter 10

Details about Symphytum Species Hybrid: Russian Comfrey

Russian Comfrey, Quaker Comfrey, Upland Comfrey.

Russian Comfrey: S. officinale x S. asperum = S. x uplandicum
 (includes S. peregrinum which some consider not a hybrid)

See subsection 'Hybrids, Hybrid Swarms, Introgression' in section 'Symphytum Species Overview' (Chapter 6).

Current Botanical Nomenclature of Russian Comfrey

Symphytum x uplandicum Nyman = S. asperum x S. officinale = S. peregrinum auct. non Ledeb

See subsection below 'Hybrid Origin and Classification (peregrinum Ledeb. versus uplandicum)'.

The below plant databases state (2018):
 S. peregrinum Ledeb. is uncertain or no information: NPGS, Tropicos
 S. peregrinum Ledeb. is synonym for S. officinale L.: The Plant List, Plants of the World Online
 S. peregrinum Ledeb. is synonym for S. asperum Lepechin: Catalogue of Life, GBIF
 S. peregrinum auct. non Ledeb. is synonym for S. x uplandicum Nyman: ITIS, USDA Plants Database

"Symphytum x uplandicum Nyman	*Accepted*
Symphytum x uplandicum f. coeruleum (Petitm. ex Thell.) P.D.Sell	*Synonym*
Symphytum x uplandicum f. densiflorum (Buckn.) P.D.Sell	*Synonym*
Symphytum x uplandicum f. discolor (Buckn.) P.D.Sell	*Synonym*
Symphytum x uplandicum f. lilacinum (Buckn.) P.D.Sell	*Synonym"*

-The Plant List®, www.theplantlist.org from the World Checklist Database, **2018**.
 ('Nyman' is Carl Fredrik Nyman from Sweden, 1820-1893. He was a curator at the 'Swedish Museum of Natural History' in Stockholm from 1855 to 1889. With Heinrich Wilhelm Schott and Theodor Kotschy, he was editor of 'Analecta Botanica'. See 'Nyman 1884' in subsection 'Symphytum Genus' in section 'Borage Family, Symphytum Genus'.)
 ('f.' = forma, which is an infraspecific rank. Infraspecific is at a taxonomic level below that of species, e.g., subspecies, variety, cultivar, or form. Latin names at this level require the addition of a term denoting rank.)

"Symphytum uplandicum Nyman
 = Symphytum x peregrinum
 = Symphytum x uplandicum
 = Symphytum x coeruleum Petitm.
 = Symphytum x coeruleum Petitm. ex Thell.
 = Symphytum x densiflorum Buckn.
 = Symphytum x discolor Buckn.
 = Symphytum x lilacinum Buckn.
 = Symphytum x uplandicum f. coeruleum f. coeruleum (Petitm. ex Thell.) P.D.Sell
 = Symphytum x uplandicum f. densiflorum f. densiflorum (Buckn.) P.D.Sell
 = Symphytum x uplandicum f. discolor f. discolor (Buckn.) P.D.Sell
 = Symphytum x uplandicum f. lilacinum f. lilacinum (Buckn.) P.D.Sell"

-Global Biodiversity Information Facility® (GBIF), www.gbif.org, **2018**. International network and research infrastructure funded by the world's governments and aimed at providing anyone, anywhere, open access to data about all types of life on Earth.

"Symphytum peregrinum Ledeb.
  An uncertain taxon. *Publication: *Index sem. hort. dorpat. 1820:4. 1820.*
 Non Symphytum peregrinum auct. (= S. x uplandicum Nyman)"
-U.S. National Plant Germplasm System® (NPGS), https://npgsweb.ars-grin.gov, United States Department of Agriculture, Agriculture Research Service, **2018**. NPGS is collaborative effort to safeguard genetic diversity of agriculturally important plants.
(* -'Index Seminum Horti Academici Dorpatensis' by Carl {Karl} Friedrich von Ledebour, Dorpat {Tartu}, Estonia, page 4, 1820. In Latin. Includes Symphytum asperrimum, echinatum, peregrinum and caucasicum.)
 ('Ledeb' is Carl/Karl Friedrich von Ledebour from Germany/Sweden, 1785-1851. He was born in Stralsund which was at that time a part of Sweden. He travelled extensively in Russia and wrote the first comprehensive book on Russian flora. 'Flora Rossica; sive, Enumeratio plantarum in totius Imperii Rossici provinciis Europaeis, Asiaticis et Americanis

hucusque observatarum' {Complete Flora of the Russian Empire}, 1847-1849 with Symphytum in Volume 3. See 'Ledebour 1847' in subsection 'Symphytum Genus' in section 'Borage Family, Symphytum Genus'.)
(auct.= auctorum= various authors)

"*Boraginaceae Symphytum peregrinum Ledeb. (Index Seminum Horti Academici Dorpatensis. 1820)*
Boraginaceae Symphytum peregrinum Ledeb. (Ind. Sem. Hort. Dorpat. 1820 4; et ex Spreng. Syst. i. 563.)
Boraginaceae Symphytum uplandicum Nyman (*Syll. Fl. Eur. 80. 1855)"
-'International Plant Names Index'® (IPNI), www.ipni.org, A database of the names and associated basic bibliographical details of seed plants, ferns and lycophytes, **2018**.
(* -'Sylloge Florae Europaeae' Oerebroae {Oerebroe}, Sweden, by Carl Fredrik Nyman, page 80, 1854-1855.)

"*Symphytum peregrinum Ledeb. (no publication information).*
Symphytum x uplandicum Nyman. Published In: Sylloge Florae Europaeae 80. 1855."
-Tropicos®, www.tropicos.org, Missouri Botanical Garden, Saint Louis, Missouri; Plant database with 1.3 million scientific names and 4.4 million specimen records, **2018**.

"*Symphytum peregrinum Ledeb. is a synonym of Symphytum officinale L.*"
-The Plant List®, www.theplantlist.org from the World Checklist Database, **2018**.

"*Symphytum peregrinum auct. non Ledeb. Not accepted.*
It is synonym for: Symphytum x uplandicum Nyman (pro sp.) Accepted name."
-ITIS (Integrated Taxonomic Information System®) Report, www.itis.gov, **2018**. The ITIS is the result of a partnership of United States Federal agencies formed to satisfy their mutual needs for scientifically credible taxonomic information.
  (pro sp.= hybrid status)

"*Symphytum peregrinum Ledeb. synonym for Symphytum asperum Lepechin.*
Symphytum uplandicum Nym. accepted name."
-Catalogue of Life®: 2015 Annual Checklist, Online database of the world's known species of animals, plants, fungi and micro-organisms. It consists of a single integrated species checklist and taxonomic hierarchy., **2018**, www.catalogueoflife.org. Search: www.catalogueoflife.org/annual-checklist/2015/search/all

"*Symphytum peregrinum auct. non Ledeb*
= Symphytum x uplandicum Nyman (pro sp.) (asperum x officinale)."
-United States Department of Agriculture, Natural Resources Conservation Service, Plants Database®, www.plants.usda.gov, **2018**.   (sp.= single species) (spp.= more than 1 species)

> **International Code of Nomenclature for Algae, Fungi, and Plants®:**
> **Author Citations, Article 50:** "*When a taxon at the rank of species or below is transferred from the non-hybrid category to the hybrid category of the same rank, or vice versa, the author citation remains unchanged but may be followed by an indication in parentheses of the original category.*
> **Example 1:** *Stachys ambigua Sm. (1809) was published as the name of a species. If regarded as applying to a* **hybrid**, *it may be cited as S. x ambigua Sm.* **(pro sp.).**
> **Example 2:** *Salix x glaucops Andersson (1868) was published as the name of a hybrid. Later, Rydberg considered the taxon to be a species. If this view is accepted, the name may be cited as S. glaucops Andersson (pro hybr.).*"
> (The 'International Code of Nomenclature for algae, fungi, and plants'® {ICN} is the set of rules and recommendations that govern the scientific naming of all organisms.)
> (In botanical nomenclature, 'author citation' refers to the person or group who validly publish a botanical name, i.e., first publishing that fulfills the formal requirements of the 'International Code of Nomenclature'® of plants.)

"*The standard author abbreviation (citation) Nyman (Symphytum x uplandicum Nyman) is used to indicate Carl Fredrik Nyman (1820-1893), a Swedish botanist.*
Morphological and chemotaxonomic studies indicated that Symphytum x uplandicum is a hybrid between Symphytum officinale (Common Comfrey) and Symphytum asperum (Prickly Comfrey).
It interbreeds with Symphytum officinale.
Compared to Symphytum officinale, Symphytum x uplandicum is generally more bristly and has flowers which tend to be more blue or violet."
-www.flowersinsweden.com, Flowers in Sweden®, **2018**. No other information available about the site.

Hybrid Origin and Classification of Russian Comfrey (overview, then in chronological order)
See subsection 'Hybrids, Hybrid Swarms, Introgression' in section 'Symphytum Species Overview' (Chapter 6).

Overview: 'S. peregrinum Ledeb.' versus 'S. x uplandicum Nyman'

Some botanists think S. peregrinum Ledeb. is a pure species from the Caucasus (not a hybrid):
Ledebour 1820, Kuznetsov 1910, Bucknall 1913, Gviniashvili 1972,
Gadella/Kliphuis 1978, Gadella/Kliphuis/Huizing 1982
Bucknall believes S. peregrinum Ledeb. is a pure species that can hybridize with S. asperum or with S. officinale.

Some botanists think that if S. peregrinum is not a pure species, then it may be S. asperum x S. caucasicum from the Caucasus: Gadellea/Kliphuis 1978

Some botanists think S. peregrinum Ledeb. is a hybrid between S. officinale x S. asperum (with no mention of uplandicum): Fries 1839, Baker 1877, Boissier 1879, Popov 1953

Some botanists think S. peregrinum Ledeb. is the same as S. x uplandicum Nyman (hybrid between S. officinale and S. asperum): Lawrence Hills 1955, 1976

Some botanists think S. peregrinum Ledeb is part of the hybrid swarm of S. x uplandicum Nyman:
Clapham/Tutin/Warburg 1962, Gadella/Kliphuis 1971

Some botanists think S. x uplandicum Nyman is a stable hybrid species of S. officinale x S. asperum from the Caucasus: Wade 1958, Perring 1969

Some botanists think S. x uplandicum Nyman is a new hybrid species of S. officinale x S. asperum first formed in Britain: Tutin 1956 (hybrid swarm), Gadella/Kliphuis/Perring 1974, Gadella/Kliphuis/Huizing 1983

The articles below explain the above list.

Overview Article from 1968 about Hybrid Origin:

"Those plants with partially decurrent petioles and blue or purple-blue flowers are described as 'Blue Comfrey', S. x uplandicum Nyman (###Clapham, Tutin and Warburg, 1962).
The origin of this hybrid is confused:
Fries (*1839):
Hybridisation of S. asperum with S. officinale was observed in Uppland (Sweden) in 1839 where the specific name S. patens E.M. Fries was applied, but this was subsequently replaced by S. x uplandicum Nym. (****Tutin, 1956).
Baker (**1877-1878):
Considered that S. peregrinum might be the hybrid S. asperum x S. officinale.
Bucknall (***1912-1913):
Referred to S. peregrinum Ledeb. as the largest species of the Symphytum genus, being distinguished from S. officinale (Common Comfrey) by a non-winged stem and blue flowers.
He noted its resemblance to S. asperum Lepech. (Prickly Comfrey) and its tendency to hybridise with S. officinale; *in consequence it had frequently been misnamed.*
Tutin (****1956):
Stated that hybrids of S. officinale had arisen in Britain after the introduction of S. asperum as a fodder crop.
Agreed that the range of variation in hybrids was great; plants may be found showing all combinations of the parental characteristics. The parents appeared to maintain their distinctiveness mainly as a result of their ecological isolation; S. asperum grows on dry land and S. officinale by water.
Wade (#1958):
Considered that S. x uplandicum Nym. was introduced into Britain as a separate species from Russia for use as fodder, being first cultivated here in 1827 and confirmed that S. asperum (Prickly Comfrey) was rare in Britain, [cf. (compare to other literature) Bucknall (1912-1913)], being distinguished from S. x uplandicum Nym. with difficulty.
Wade regarded the latter (S. x uplandicum Nym.) to be a fixed hybrid, populations showing remarkable uniformity unless growing in close proximity to S. officinale, where they may show variation due to back crossing, which is relatively rare with S. asperum.
Ingram (##1961):
Confirmed that the hybrid possessed the characteristics of both parents, but was more likely to be confused with S. asperum (Prickly Comfrey).
Clapham, Tutin and Warburg (###1962):
Included the hybrid swarm under S. x uplandicum Nym. giving S. peregrinum auct. Ledeb. as the central form. "
-'British Medicinal Species of the Genus Symphytum' by K.R. Fell and Janet M. Peck, Pharmacognosy Research Laboratories, University of Bradford, England; Planta Medica: Journal of Medicinal Plant and Natural Product Research, Volume 16, No. 2, pages 208-216, May **1968**.
(* -'Novitiarum Florae Sueciae Mantissae Altera, Quam Venia Ampliss. Facult. Philosoph., I.' {Flora of Sweden} by E.M. Fries and Jonas Otto Ponten, pages 13-14, Uppsala, Sweden, 1839. All in Latin. His writings are below.)
(** -'Report of the Plants Gathered in 1878' edited by J.G. Baker for 'The Botanical Exchange Club Report for 1877-8' also called

'Report of the Curators for the London Botanical Exchange Club', published 1879, page 17. In botanical literature I found it listed as: "Baker, J. G.: Report of the Botanical Exchange Club, London, 17 1877-8". His report is below.)
(*** -'Some Hybrids of the Genus Symphytum' by Cedric Bucknall, {Mus. Bac. Oxon= Bachelor of Music, Oxford University}, The Journal of Botany British and Foreign, London, England, Volume 50, pages 332-337, 1912.)
(*** -'A Revision of the Genus Symphytum, Tourn.' by Cedric Bucknall, {Mus. Bac. Oxon= Bachelor of Music, Oxford University}, Journal of the Linnean Society of London, England, Botanical Journal, Volume 41, Issue 284, pages 491-556, December 1913.)
(**** -'The Genus Symphytum in Britain' by T.G. Tutin, University College of Leicester, England; Watsonia: Journal of the Botanical Society of the British Isles, Volume 3, pages 280-281, February 1956. Watsonia was the journal of the 'Botanical Society of Britain and Ireland' from 1949 to 2010. It is now called 'New Journal of Botany'.)
(# -'The History of Symphytum Asperum Lepech. and S. x Uplandicum in Britain' by A.E. Wade, Department of Botany, National Museum of Wales, Watsonia, Volume 4, pages 117-118, 1958.)
(## -'Studies in the Cultivated Boraginaceae. 5. Symphytum' by J. Ingram; Baileya: A Quarterly Journal of Horticultural Taxonomy, Ithaca, New York, Volume 9, pages 92-99, 1961. I was unable to get a copy of this report. If you have one, could you please send it to me.)
(### -Flora of the British Isles, Second Edition by A.R. Clapham, T.G. Tutin and E.F. Warburg. England; Cambridge University Press, 1962, pages 643-644.)

(A hybrid swarm is a group of hybrids that have survived past the first hybrid generation, with interbreeding taking place between hybrid individuals and also backcrossing with its parent type. These plants are highly variable, with genes and phenotypes of individuals ranging widely between the two parent types. Hybrid swarms blur the boundary between the parent taxa. See subsection 'Hybrids, Hybrid Swarms, Introgression' in section 'Symphytum Species Overview'.)
(A backcross is a cross of a hybrid with one of its parents or an individual genetically similar to its parent. This creates offspring with genetics closer to that of the parent.)

Hybrid Origin of Russian Comfrey (in chronological order from old to recent)

"Boraginaceae Symphytum peregrinum Ledeb.
(Index Seminum Horti Academici Dorpatensis. 1820, page 4, C.F. Ledebour.)
Latin: *S. caule hirsuto, foliis mollibus, radicalibus oblongis, caulinis ovatis acutis, calyce quinquepartite tubum corollae parum superante, laciniis erectis, corollae limbo campanulato, laciniis margine revolutis, stylo infracto.*
Online translation to English: *S. hairy stems, leaves soft, oblong radical, ovate, acute, calyx five parts of the corolla tube a little overwhelming, putting fringes corolla limb camp rings, lobes margin revolution, broken pen."*
-Listing found at 'Naturalis Biodiversity Center'®: Guide to plant species descriptions published in seed lists from Botanic Gardens from 1800-1900. http://seedlists.naturalis.nl   (If you have a better English translation, please let me know.)

*"17. **Symphytum patens,** caule solido tereti retrorsum hirto, foliis omnibus alternis ovato-oblongis leviter decurrentibus, radicalibus cordato-ovatis, corolllae limbo cylindrico glabro quinquefido tubum superante, antheris filamento brevioribus. Herb. Norm. V. n. 4. (Herbarium Normale)*
***S. patens. Sibth. et Auctt.?** nulli vero characterres distinctivi afferuntur; e calyce prorsus falsi.*
***In Suecia (Sweden) media orientali (Middle East), ut Upsaliae (and Uppsala)** in horrtis juxta amnem, **Enkoping (in Uppsala county, Sweden),** etc.; in Suecia meridiomali et occidentali numquam vidimus (**Symphytum officinale** vero etiam in Suecia media adest; Upsaliae, necentius introductum, in antiquo horto Linnaeano {antique garden Linnaeus}.)*
Facies potius Symphyti asperrimi, quam S. officinalis. (Make more Symphytum asperrimum, and Symphytum officinale = Symphytum patens.)"
-'Novitiarum Florae Sueciae Mantissae Altera, Quam Venia Ampliss. Facult. Philosoph., I.' {Flora of Sweden} by **E.M. Fries** and Jonas Otto Ponten, pages 13-14, Uppsala, Sweden, **1839**. All in Latin. (If you have an English translation, please send it to me.)

Russian Comfrey is mistakenly identified as Prickly Comfrey:
*"**Symphytum asperrimum:** The introduced British plant which has been so called by Babington in **'Flora Bathonensis', and Dr. Boswell in **'English Botany', of which Mr. Flower sends us a good supply this year from the long-known station (location) in the neighbourhood of Bath (southwest England), and Reverend W.H. Purchas from Grange Mill, near Wirksworth, Derbyshire (east midlands of England), **is evidently not the true wild S. asperrimum M.B. (M. Bieb.), of the Caucasus, but a garden hybrid between that species and officinale,** which is often planted for forage, and **which is most likely S. peregrinum, Ledeb.,** ***Fl. Ross., vol. iii., p. 114.*

S. asperrimum is a plant that grows five or six feet (1.5-1.8 meters) high, with stems densely clothed with very short, rigid, bristly pubescence, many of the bristles springing from white calcareous tubercles, leaves rough over the face with bristle-pointed white tubercles, like Anchusa italica (Italian Alkanet or bugloss), lower leaves of the flowering branches ovate and contracted suddenly at the base, and a flower-calyx not more than one-eighth of an inch (0.3 cm) long, with linear-oblong obtuse teeth not longer than the tube.
***The naturalised hybrid** has much less bristly stems, leaves without white tubercles on the face, lower leaves of the flowering branches both absolutely narrower and narrowed more gradually at the base, and a flower-calyx like that of Symphytum officinale, with acute linear teeth twice as long as the tube.*
Mr. Flower tells me that the Bath plant grows sometimes to the height of a man, so that it is not inferior to the true (wild) asperrimum in stature, although in its leaves and flowers it seems much nearer to officinale.

We have the true asperrimum in the Kew herbarium (London, England) from the neighbourhood of Stirling (central Scotland), gathered by G. Thomson."
-'Report of the Plants Gathered in **1878**' edited by J.G. Baker for 'The Botanical Exchange Club Report for 1877-8' also called 'Report of the Curators for the London Botanical Exchange Club', published 1879, page 17. In botanical literature I found it listed as: 'Baker, J.G.: Report of the Botanical Exchange Club, London, 17 1877-1878'.
(* -'Flora Bathoniensis, Or, a Catalogue of the Plants Indigenous to the Vicinity of Bath' by Charles C. Babington, M.A., Fellow of the Linnean Society, London, England, 1834.)
(** -'English Botany {Sowerby's}; or Coloured Figures of British Plants: Volume 7, Third Edition' by John T. Boswell {Syme} and John Edward Sowerby, London, England, 1867, 1880.)
(*** -'Flora Rossica; sive, Enumeratio plantarum in totius Imperii Rossici provinciis Europaeis, Asiaticis et Americanis hucusque observatarum' {Complete Flora of the Russian Empire}, Volume 3, by Carl {Karl} Friedrich von Ledebour, 1847-1849. Symphytum pages 113-114.)

"**Symphytum peregrinum.** *Native of the Caucasus. Nat. Ord. {Natural Order} Boragineae. Tribe Borageae.*
      Genus Symphytum, Linn.; (*Benth. et Hook. f. Gen. Pl. vol. ii. p. 854. 1876.)
S. peregrinum, Ledebour *Ind. Sem. Hort. Dorpat. 1820, p. 4;*
      Fl. Ross. vol. iii. p. 114; ('Flora Rossica' in Latin, 1847-1849)
      DC. Prodr. vol. x. p. 37; ('Prodromus Systematis Naturalis Regni Vegetabilis' in Latin, 1824-1873)
      Briggs in Report of Bot. Exchange Club for 1877-78, p. 17.
      S. asperrimum, Bab. Fl. Bathon. 32. ('Flora Bathoniensis', 1832)
The history of this plant, which is now well known under the erroneous (incorrect) name of Symphytum asperrimum is still obscure (unknown).
That it is not the true S. asperrimum of Donn, *figured by Sims in this work (**t. 929), is obvious from a comparison of that plate, in which the calyx is correctly represented as short, and shortly 5-cleft to the middle only, with obtuse lobes, and which has curved prickles on the stem, arising from conspicuous white tubercles.*
      It agrees well with the character of S. peregrinum given in Ledebour, except that the appendages between the stamens are rather shorter *(than longer) than the anthers, and the style is not always bent below the top (stylo infra apicem infracto), though it is sometimes so above the middle.*
From S. caucasicum it differs *in the stem not being hirsute (hairy), nor the leaves softly hoary (grayish white), and in the calyx being deeply divided.*
In the 'Report of the Botanical Exchange Club', *cited above, in which work I find the plant for the first time referred though doubtfully to S. peregrinum, it is suspected to be a garden hybrid between S. asperrimum and S. officinale, which latter is said to be often planted for forage.*
      This may be so, but there is no evidence of its hybridity, and **Ledebour gives a habitat** *for the indigenous (native) S. peregrinum, namely, Sawunt in Talysch province of Caucasus, at a height of 4000 feet (1219 meters) above the sea.*
And I have seen excellent dried specimens in the 'Kew Herbarium' (England), *collected by Besser (under the erroneous name of S. caucasicum, Bieb.), and by Wilhelms, collected in Iberia in 1824, and sent under the name of S. asperrimum to the late J. Gay, who has attached to the specimen the note, 'Je crois que c'est le Symphytum caucasicum M. B. et nullement le S. asperrimum' (I believe it is Symphytum caucasicum M. B. and not the S. asperrimum).*
Boissier in his ***'Flora Orientalis' *(vol. iii. p. 175) indeed says of S. peregrinum and another, 'formae hortenses forsan hybridae'.*
Lastly, for my own part, I see very little reason to regard it as other than a very large form of S. officinale, with the stem fistular below, probably originating from cultivation, and not from hybridization."
-'Curtis's Botanical Magazine, Comprising the Plants of the Royal Gardens of Kew, and of Other Botanical Establishments in Great Britain, Volume 105 of Whole Work, Or Volume 35 of Third Series' by Sir Joseph Dalton Hooker, M.D., C.B., K.C.S.I., F.R.S., F.L.S., London, England, **1879**. December 1: Symphytum peregrinum on Tab./plate 6466.
(* -Genera Plantarum: Ad Exemplaria Imprimis in Herberiis Kewensibus Servata Definita' in 3 volumes, begun in 1862, and finished in 1883 in collaboration with Joseph Dalton Hooker. Symphytum is on page 854 of Volume 2, Part 2, 1876.)
(** -'Curtis's Botanical Magazine; or, Flower-Garden Displayed in which the Most Ornamental Foreign Plants, Cultivated in the Open Ground, the Green-House, and the Stove, are Accurately Represented in their Natural Colours, Volume 23-24' by John Sims, M.D., Fellow of the Linnean Society, London, England, 1806. Symphytum asperrimum, plate 929.)
      (Hooker wrote above: 'Briggs in Report of Bot. Exchange Club for 1877-78, p. 17' but I think he meant 'Baker' instead of 'Briggs'. See the article above this one.)
      (***Hooker wrote: 'Flora Orientalis vol. iii. p. 175' but I could not find any reference to Symphytum in Volume 3. Symphytum is listed in Volume 4 written in 1879 on pages 171 to 175. Symphytum peregrinum is mentioned in the S. asperrimum listing. It is in Latin.)
      (Boissier wrote: 'S. peregrinum et S. echinatum Ledeb. Sem. Dorp.: 'formae hortenses forsan hybridae'. That means 'forms may hybridize'. I don't know what 'hortenses' means, though 'Hortense' is a French feminine given name that comes from Latin meaning gardener.)
      (The Iberian Peninsula or Iberia is in the southwest part of Europe. It is divided between Spain and Portugal, comprising most of their territory. It also includes Andorra, small parts of France, and Gibraltar.)

"**Symphytum Peregrinum, Ledeb.:**
In the 'Curtis's Botanical Magazine' for December 1879 (t. 6466), Sir Joseph Hooker adopts Mr. Baker's suggestion in the last

'Report of the Botanical Exchange Club' ('Report of the Plants Gathered in 1878') (quoted in *'Journal of Botany, British and Foreign' 1879, page 250), as to **the identity of the 'Symphytum asperrimum' of cultivation with 'S. peregrinum Ledeb.'.**"
-'Journal of Botany, British and Foreign', London, England, New Series Volume 9, Volume 18 of entire work, **1880**. Article: 'Short Notes: Symphytum Peregrinum, Ledeb.', pages 57-58.
(* -'Journal of Botany, British and Foreign', London, England, New Series Volume 8, Volume 17 of entire work, 1879. 'Notices of Books and Memoirs: Report of the Plants Gathered in 1878' edited by J.G. Baker', Symphytum asperrimum, pages 250-251.)
(Then the article quotes the above "Curtis's Botanical Magazine" article from 1879.)

"**Symphytum peregrinum Ledeb.?: S. Donii DC.?:**
Dr. Boswell {wrote 'English Botany'} sends cultivated specimens of this plant from Balmuto Castle, Scotland, July 1879, with the following remarks:
> '**Mr. Baker** supposes this to be a garden hybrid between S. asperrimum and S. officinale, between which it is intermediate, though much nearer S. asperrimum in everything except the shape of the limb of the corolla, which is that of S. officinale.
> It is no doubt from this that **Professor Babington** thinks it may be a luxuriant form of S. officinale. It is quite as tall and robust as S. asperrimum (or even more so), and young plants of it come up like weeds in the garden.
> The calyx is but little more deeply divided than that of S. asperrimum, though the divisions are more acute. In both plants they elongate after flowering. The bristles on the stem are less like prickles, but the so-called white tubercles, on which they are seated, become apparent only when dry, and are smaller than in S. asperrimum, in which they are equally green while the plant is alive.
> **It seems to agree better with DeCandole's description of S. Donii than with his or Ledebour's description of S. peregrinum**, as the stem leaves, especially the upper ones, are shortly decurrent.' "

-'Journal of Botany, British and Foreign', London, England, New Series Volume 9, Volume 18 of entire work, **1880**. Article: 'Notices of Books and Memoirs: Extracts from the Report of the Botanical Exchange Club of the British Isles for 1879', various Symphytums, page 381.

"**S. peregrinum, Ledeb:**
> On account of its similarity to S. asperum, its tendency to hybridize with S. officinale, incomplete descriptions and the absence of authentic specimens, it has been much misunderstood and misnamed.

Ledebour, although he does not describe the root, the colour of the flowers or the nutlets, either in the 'Index Seminum' or in the 'Flora Rossica', sufficiently well distinguishes this species from S. asperum.
> **S. peregrinum, Ledeb.**, *DC. Prod. x. p. 37 (1846); Ledeb. Fl. Ross. iii. p. 114 (1846-51).
> **S. officinale v. patens**, Syme, Eng. Bot. t. 1516 (1867).
> **S. uplandicum, Nym.**, Syll. Fl. Eur. p. 80 (1854-5) p.p.
> **S. asperum x officinale (S. peregrinum, Ledeb. ?)**, Baker in Rep. Bot. Exc. Club, p. 17, (1878).
>> (See Jour. Bot. viii. p. 250. (1879)).
> **S. peregrinum, Ledeb.**, Hooker in Journ. Bot. ix. p. 57 (1880), and Bot. Mag. t. 6466 (1879);
>> **Boiss. Fl. Orient. iv. p. 175 (1879) p.p.
> **S. Donii, DC.?**, Boswell in ***Jour. Bot. ix. p. 381 (1880).
> **S. peregrinum, Ldeb.**, ****Aschers. & Grsb. Fl. Nordostd. Flachl. p. 577 (1898-99);
>> *****Kusnez. Cauc. Sp. Gen. Symph. p. 31, t. i. f. A, 4 & 8 (1910).

Symphytum peregrinum is native on the banks of streams in the province of Talysch, in the extreme south-east of Transcaucasia (south Caucasus), where it ascends to 8000 feet (2438 meters), and also, according to Kusnezow/Kuznetsov, in Persia (Iran) and Turcomania (Turkmenistan).
> **S. peregrinum was described by Ledebour (******Ind. Sem. Hort. Dorp., 1820), presumably from plants from the province of Talysch in the extreme southeast of the Caucasus.**

Specimens exist which were grown in the 'Geneva Botanic Gardens' (Switzerland) between 1820 and 1830, and this species was subsequently introduced into Britain and Scandinavia as a fodder-plant. It is now extensively naturalised in these and other countries, and has even been cultivated with more or less success in India and Australia.
These discrepancies... (in this report Bucknall gave various descriptions of S. peregrinum by past botanists such as Hooker, Sprengel, De Candolle)**, together with the other difficulties which I have mentioned, have served to perplex botanists, and it is not surprising that they should have hesitated to accept the naturalised plant as a true species, or as that described by Ledebour, but that they should have looked upon it as a variable hybrid between S. asperum and S. officinale, or as a cultivated form of one of these species.**
About the year 1861 S. peregrinum began to attract attention as a naturalised plant in England, and was then and for some years after variously named S. asperrimum, S. officinale var. patens, and S. uplandicum Nym.
> **Boissier** in the 'Flora Orientalis,' iv. p. 175 (1879) states his opinion that this and also S. echinatum, Ledeb., are 'formae hortenses forsan hybridae' (forms may hybridize).
> **Sir Joseph Hooker** in the following year (Bot. Mag. 1879, t. 6466) quotes Ledebour's description and says that he considers that there is no evidence of hybridity, but that he regards it as a large form of S. officinale probably originating from cultivation and not from hybridity.
> **Dr. Boswell Syme** writes in the ******'Journal of Botany,' 1880, p. 381: 'It seems to agree better with De Candolle's description of S. Donii than with his or Ledebour's description of S. peregrinum, as the leaves, especially the upper ones, are shortly decurrent'.

Having studied all available specimens, both living and dried, and carefully compared them with, I believe, all the published descriptions, I am convinced that we have Ledebour's plant in Britain, and that it is a true species, which, when pure bred, produces fruit abundantly.
As I have shown in 'Journal of Botany,' ******* volume 51, 1912, page 332, **when it grows in company with S. officinale, it hybridises with it** and produces a series of forms which are more or less similar to one or other of the parents.
> One of these, with semidecurrent leaves and blue flowers, which I refer to **S. coeruleum Petitmengin**, has been widely distributed by root division, and has occasioned much of the difficulty in deciding on the status of this plant.

Lastly, there are intermediate plants between S. asperum and S. peregrinum, possibly hybrid, but more probably the result of difference of climate, soil and situation.
The appearance of Kusnezow's work, with a figure and an excellent and complete description of this species, has confirmed me in the opinion that our plant, when pure bred, is that described by Ledebour."
-'A Revision of the Genus Symphytum, Tourn.' by Cedric Bucknall, (Mus. Bac. Oxon= Bachelor of Music, Oxford University), Journal of the Linnean Society of London, England, Botanical Journal, Volume 41, Issue 284, pages 491-556, December **1913**.
(* -'Prodromus Systematis Naturalis Regni Vegetabilis, Pars X' {An essay of the system of the natural of the Kingdom of Vegetable, Part 10} by A.P. De Candolle, **1846**. Written in Latin. Symphytum pages 36-40, 587.)
(** 'Flora Orientalis Sive Enumeratio Plantarum in Oriente, a Graecia et Aegypto ad Indiae Fines {Plants Flora, Flora East and the East, from Greece and Egypt to the Borders of India}, Volume IV' by Edmond Boissier, Society Physics Geneva, Society Linnean London; Geneva, Switzerland, 1879. All in Latin.)
(*** -'Journal of Botany, British and Foreign', London, England, New Series Volume 9, Volume 18 of entire work, 1880. Article: 'Notices of Books and Memoirs: Extracts from the Report of the Botanical Exchange Club of the British Isles for 1879', various Symphytums, page 381.)
(**** -'Flora des Nordostdeutschen Flachlandes' {Flora of the North-East German Lowland} by Paul Friedrich August Ascherson and Karl Otto Robert Peter Paul Graebner; Berlin, Germany, page 577, 1898-1899. In German.)
(***** -'Kavkazkie Vidy Roda Symphytum (Tourn.) L.' {Caucasian Species of the Genus Symphytum Tourn. L.} by N. Kuznetsov/ Kusnezow, Memoires de l'Academie Imperiale des Sciences de Saint-Petersbourg, Serie. 8, Physiques et Mathematiques, Volume 25, No. 5, page 31, 1910. If you have an English translation of this, I would appreciate a copy.)
(****** -Boraginaceae Symphytum peregrinum Ledeb. --Index Seminum Horti Academici Dorpatensis. 1820)
(******* -'Some Hybrids of the Genus Symphytum' by Cedric Bucknall, {Mus. Bac. Oxon= Bachelor of Music, Oxford University}, The Journal of Botany British and Foreign, London, England, Volume 50, pages 332-337, 1912. Volume 51, year 1913, has a passing reference to this article, but no useful information.)
> (Talysh Mountains are in southeastern Azerbaijan and northwestern Iran.)
> (Symphytum coeruleum Petitm. ex Thell. --Vierteljahrsschr. Naturf. Ges. Zurich 52: 459. 1907)
> (Symphytum coeruleum Hort.Angl. ex DC. --Prodr. [A.P. de Candolle] 10: 40. 1846; nom. inval.)

"S. peregrinum Ledeb. (fide {according to} C. Bucknall):
This species forms a series of hybrids with the white and purple-flowered varieties of S. officinale, which have been described by the writer (Bucknall) in 'The Journal of Botany'*, Volume 51, page 332, (1912).
These are distinguished by the more or less winged stem, by the colour of the flowers, which are white, rose-coloured, bluish or purple, always changing to a cinereous (ash-gray) blue in the dried plant, and by the fruit being sparingly produced.
Typical S. peregrinum, as well as some of its hybrids, has often been named S. patens Sibth., but the latter is probably only S. officinale var. purpureum with undeveloped fruit, and the calyx-lobes, in consequence, spreading after the flowering instead of being connivent over the nutlets, as is the case when they are well developed.
S. peregrinum has also been confused with S. asperum Lepech. (S. asperrimum Donn and M.B.), which, in Britain, is a much rarer plant. It is distinguished by the small calyx with obtuse segments, the calyx in S. peregrinum being generally considerably larger, with acute lanceolate segments.
With regard to the clothing of hairs and prickles, and in other characters, both species are variable, and they are often difficult to separate except by the above-mentioned characters of the calyx; and when, owing to conditions of climate or situation, the flowers are imperfectly developed, even these characters are liable to be deceptive.
It is probable that intermediates, and possibly hybrids, occur, and that they are sometimes the cause of the difficulty in the accurate determination of these plants.
Ledebour, in the 'Flora Rossica' **, has well distinguished the two species, and complete descriptions, with remarks on the forms which occur both in the wild and naturalised state, will be found in the writer's 'Revision of the Genus Symphytum' in the 'Journal of the Linnean Society (Botany)' *** xli., December 1913. By Cedric Bucknall."
-'Watson Exchange Club Report: 1912-1913', The Journal of Botany British and Foreign, London, England, Volume 52, page 275, **1914**. This section by Cedric Bucknall.
(* -'Some Hybrids of the Genus Symphytum' by Cedric Bucknall, {Mus. Bac. Oxon= Bachelor of Music, Oxford University}, The Journal of Botany British and Foreign, London, England, Volume 50, pages 332-337, 1912.)
(** -'Flora Rossica; sive, Enumeratio plantarum in totius Imperii Rossici provinciis Europaeis, Asiaticis et Americanis hucusque observatarum' {Complete Flora of the Russian Empire}, Volume 3, by Carl {Karl} Friedrich von Ledebour, 1847-1849. Symphytum page 113.)
(*** -'A Revision of the Genus Symphytum, Tourn.' by Cedric Bucknall, {Mus. Bac. Oxon= Bachelor of Music, Oxford University}, Journal of the Linnean Society of London, England, Botanical Journal, Volume 41, Issue 284, pages 491-556, December 1913.)

"S. peregrinum Ldb.

Ind. Sem. Hort. Dorp. (1820) 4;   *Sprengel, Syst. I, 563;   DC. Prodr. X, 37;   Ldb. Fl. Ross. III, 114; Kuzn. in. Mem. Acad. St. Petersb. XXV, 5 (1910) 31;
**Mat. Fl. Kavk. IV, 2, 228;
***Grossg. Fl. Kavk. III, 258.- Ic.: Kuzn., op. cit. {in work already cited} (1910).
Forests. Caucasus: Tal. (Talysh). Endemic (native).
I am retaining the independent status of this species only as a concession to Kuznetsov's authority. I did not observe the distinctions indicated, i.e., the size of the calyx compared with the corolla, the depth of its lobes, and the acuteness of the lobes.
I believe that Boissier was correct in including it in the synonymy (synonym) of S. asperum (asperrimum) as a garden, hybrid form.
As a matter of fact Ledebour did not indicate the locality of his species when he described it for the first time in 1820. It was only 29 years later (in 'Fl. Ross.') that he mentioned Talysh as such, adding that he had only seen cultivated specimens.
I find Kuznetsov's claim, that this species 'may be identified with S. asperum Lepech (= S. asperrimum Sims)', rather strange.
I could not find any basis for Sprengel's report on Podolia (southwest Ukraine and northeast Moldova) as a locality, which was repeated by De Candolle."
-'Flora of the U.S.S.R.', 30 volumes, 1934-1960, edited by V.L. Komarov; Botanical Institute of the Academy of Sciences of the U.S.S.R, Leningrad, Russia, with article by Mikhail Grigorevich Popov, in Volume 19, pages 279-291 Russian version, pages 207-216 English version, **1953**.
(* -'Systema Vegetabilium, Volume I, Class 1-5, 16th Edition' {System of Vegetables} by Kurt Polycarp Joachim Sprengel, Gottingen, Germany, 1824-1825. In Latin. Symphytum pages 562-563.)
(** -'Flora Caucasica Critica' = 'Materialy dlia flory Kavkaza : Kriticheskoe sistematichesko-geograficheskoe izsliedovanie' by Nikolai Ivanovich Kuznetsov; Iurev, Novgorod, Russia, 1901-1913.)
(*** -'Flora Kavkaza' {Flora of the Caucasus}, Volumes 1-7 by Aleksandr Alfonsovic Grossgeim. Trudy Botaniceskogo sada SSR Armenii, Nauka, Baku, Azerbaijan, 1939-1967.)   (Talysh Mountains are in southeastern Azerbaijan and northwestern Iran.)

"The work of Professor K.V. Braid at the West of Scotland Agricultural College, and Dr. P.J. Watson of the Scottish Plant Breeding Station, has established that **the plant botanically described as Symphytum peregrinum is a true hybrid. Its name should be 'x S. uplandicum'**.
The parents are Symphytum asperrimum and Symphytum officinale, both of which have 36 chromosomes like their more productive offspring."
-Comfrey Report No. 1: The 1954 Research Results by Lawrence D. Hills, Henry Doubleday Research Association, Braintree, Essex, England, published **1955**.
   (A 'true' hybrid is one that has offspring that breed true. In other words, the offspring are identical to the parents.)

"**The identification of S. peregrinum** has caused much trouble to British botanists but a re-examination of the S. officinale-peregrinum-asperum complex suggests that **there are in fact only two species and a hybrid swarm involved.**
The genus Symphytum in Britain is therefore represented by the following :
    Native species: S. officinale L, S. tuberosum L.
    Naturalised species: S. asperum Lepechin, S. orientale L., S. grandiflorum DC,
                                  S. caucasicum M. Bieb., S. tauricum Willd.
    Hybrid swarm: S. x uplandicum Nyman."
-'The Genus Symphytum in Britain' by T.G. Tutin, University College of Leicester, England; Watsonia: Journal of the Botanical Society of the British Isles, Volume 3, pages 280-281, February **1956**. (Watsonia was the journal of the 'Botanical Society of Britain and Ireland' from 1949 to 2010. It is now called 'New Journal of Botany'.)
   (A hybrid swarm is a group of hybrids that have survived past the first hybrid generation, with interbreeding taking place
   between hybrid individuals and also backcrossing with its parent type. These plants are highly variable, with genes and
   phenotypes of individuals ranging widely between the two parent types. Hybrid swarms blur the boundary between the
   parent taxa.)

"**It is presumably from this introduction that the hybrids arose, some of which have gone under the name of 'S. peregrinum'; the whole hybrid swarm should in fact be known as 'S. x uplandicum Nyman*'.**
S. x uplandicum Nyman was first described as S. patens E.M. Fries** (non S. patens Sibth.) in 1839 from the province of Uppland in Sweden, where it must have resulted from the crossing of native S. officinale with introduced S. asperum. This invalid name was later replaced by the name S. uplandicum Nyman.
   It should be noted that the original, and presumably correct, spelling of the name is uplandicum in spite of the fact that
   the province after which it is named is called Uppland.
The hybrid is also well known from the Caucasus, where the geographical areas of the parents appear to overlap.
The range of variation in the hybrid is great and plants may be found showing all combinations of the characters of the parents.
The fact that the parents maintain their distinctness would appear to be due mainly to their ecological isolation; S. asperum is a plant of dry places, while S. officinale, occurs only by water.
Further investigation may show, nevertheless, that these two taxa should be regarded as subspecies in order to emphasize what appears to be their close genetical similarity.
On the scanty evidence available about the distribution of the two colour forms of S. officinale, it is perhaps unwise to draw
   conclusions, but it is tempting to suggest that the purple-flowered form may be due to introgression (repeated backcrossing)

between S. officinale and S. asperum."
-'The Genus Symphytum in Britain' by T.G. Tutin, University College of Leicester, England; Watsonia: Journal of the Botanical Society of the British Isles, Volume 3, pages 280-281, February **1956**. (Watsonia was the journal of the 'Botanical Society of Britain and Ireland' from 1949 to 2010. It is now called 'New Journal of Botany'.)
(* -C.F. Nyman, Sylloge Florae Europaeae, 80. Orebro, 1854-1855. Carl Fredrik Nyman was born in Stockholm. He was curator at the 'Swedish Museum of Natural History' in Stockholm. He was editor of 'Analecta Botanica' {Analect Botanical} published 1854 with Heinrich Wilhelm Schott and Theodor Kotschy.)
(** -E.M. Fries, Novitiarum Florae Sueciae Mantissae, 2, 13. Uppsala, 1839. Swedish mycologist and botanist Elias Magnus Fries was 'Fellow of the Royal Society of London' and 'Fellow of the Royal Society of Endinburgh'. He lived from 1794-1878. He wrote 'Systema Mycologicum', 'Elenchus Fungorum', 'Monographia Hymenomycetum Sueciae' and 'Hymenomycetes Europaei'.)
    (Introgression is the transfer of genetic information from one species to another as a result of hybridization between them and repeated backcrossing.)

"Tutin (1956) suggests that the introduction of S. asperum and its hybridisation with the native S. officinale gave rise to S. x uplandicum in this country. That such a hybridisation may have taken place in Britain is not disputed.
The cultivation of Comfrey seems to have become very popular from 1870 onwards, when Henry Doubleday of Coggeshall, Essex, England, imported S. x uplandicum from Leningrad (then St. Petersburg, Russia).
The figure (artwork) in 'Curtis Botanical Magazine', plate 6466 (1879) under the name of S. peregrinum was drawn from plants presented to Kew Gardens in 1875 by Thomas Christy, who was associated with Doubleday in his efforts to popularise its cultivation.
    Several seedsmen took up the distribution of the plant, including Messrs. Sutton of Reading, Berkshire, England, who continued to supply it until 1896.
    In 1900 Messrs. Webster of Stock, Essex, England imported it from Russia, and they and other nurserymen still carry stocks (as of 1958).
These and other introductions from Russia and elsewhere are undoubtedly the chief, if not the sole, origin of S. x uplandicum in Britain."
-'The History of Symphytum Asperum Lepech. and S. x Uplandicum in Britain' by A.E. Wade, Department of Botany, National Museum of Wales, Watsonia, Volume 4, pages 117-118, **1958**. (Watsonia was the journal of the 'Botanical Society of Britain and Ireland' from 1949 to 2010. It is now called 'New Journal of Botany'.)
    (Kew Gardens is a botanical garden in London, England that houses the 'largest and most diverse botanical and mycological collections in the world'. The library has more than 750,000 volumes, and the illustrations collection has more than 175,000 prints and drawings of plants.)

"S. x uplandicum is thought to be a hybrid between S. officinale and S. asperum.
It is possible that it originated in this country (Britain), but Wade (1958) could find no evidence for this.
    As S. asperum is very rare, whilst S. x uplandicum is very widespread, it seems that almost all the populations of S. x uplandicum have originated from material, introduced from Russia as a forage crop in the middle of the nineteenth century, known as Russian Comfrey.
The purple flower colour suggests that the S. officinale parent was var. purpureum rather than var. ochroleucum.
    Tutin (1956) reports that the 'Flora of the U.S.S.R.' does not include a yellowish-white form as a Russian or Asiatic plant, so that hybrids of Russian origin might be expected to have purple flowers.
Wade (1958) reports that S. x uplandicum breeds true from seed and produces uniform populations unless growing in proximity to S. officinale when backcrossing takes place. In such situations we can expect to find variation in flower-colour, leaf decurrence, etc. within a population.
Backcrossing can presumably occur between S. x uplandicum and either of the colour varieties of S. officinale or the intermediates between them."
-'Network Research, Symphytum Survey' by Franklyn Perring, Biological Records Centre, Monks Wood, Huntingdon, Cambridgeshire County, England; Volume 7, No. 4, **1969**. In 'Proceedings of Botanical Society of the British Isles, Volume 7' edited by E.F. Greenwood, London, England, 1967-1969, pages 553-556.

"S. peregrinum Ledebour Index Sem. Hort. Dorpat. 4 (1820).
No native specimens were seen by myself or by Bucknall, from whom the above description is taken; his description agrees with that of Kuznetsov (1910), who has monographed the Caucasian species of Symphytum.
Gviniaschvili/Gviniashvili (*1967) has confirmed that the species occurs in Caucasia.
Ledebour, Fl. Ross. 3:114 (1847), states that the original material was collected by Hohenacker from Swant in Talysh at 1200 meters (3937 feet), adding 'vidi cult.' ('I saw cultivated.')
Hohenacker in his Enum. Talisch, 77 (1837) only refers to his collecting trip in 1834. I have found no reference to an earlier visit to the type locality.
Since S. peregrinum had been considered by some botanists as synonymous with S. asperum, or as a hybrid between S. asperum and S. officinale, Kuznetsov compared authenticated cultivated material from Dorpat (Tartu in eastern Estonia) with Talysh and Iranian material and concluded that they were identical. i.e. (that is) no hybridisation had occurred in cultivation before the species was described by Ledebour. Kuznetsov's excellent distribution map shows S. peregrinum to be geographically distinct from the alleged parents.

A.E. Wade (in litt. = unpublished correspondence) is of the opinion that S. peregrinum is not synonymous with S. uplandicum as suggested by Bucknall, although he considers S. uplandicum may be of hybrid origin, with S. caucasicum as one parent.
As Kuznetsov has pointed out, S. peregrinum appears to be intermediate between S. caucasicum and S. asperum; the former occurs further to the north, the latter to the south.
Its exact status appears to require further investigation."
-'A Revision of Symphytum in Turkey and Adjacent Areas' by G.E. Wickens, Royal Botanic Gardens: Kew; Notes from the Royal Botanic Garden Edinburgh, Scotland, Volume 29, pages 157-180, **1969**. Includes Turkey, Bulgaria, Greece, Aegean Islands and Caucasia.
(* -'Species Nova Generis Symphytum L. ex Armenia' by TS.N. Gviniaschvili; Notulae Systematicae ac Geographicae Instituti Botanici Thbilissiensis {Zametki po Sistematike i Geografii Rastenii}, Tbilsi, Georgia, Volume 26, pages 73-75, 1967. In Russian. If you have this in English, could you please send it to me.)
(A monograph is a detailed written study of a single specialized subject.)

"Symphytum x uplandicum Nym. (2n = 36).
Gadella and Kliphuis (1969) showed that the species S. asperum Lepech. and S. officinale L. are able to intercross. Their hybrid, S. x uplandicum, is fairly common in various regions of Europe, notably in England, Ireland and southern Scandinavia. In the Netherlands the hybrid is rare.
The plants from Terziet (Netherlands) clearly showed intermediate characters: the size of the plants rather large, the leaves not decurrent and the colour of the corolla bluish (purple in bud).
Experimentally produced hybrids between S. officinale (2n = 40) and S. asperum (2n = 32) completely agree with the plants from Terziet.
Hybrids between S. officinale (2n = 48) and S. asperum (2n = 32), which are characterized by the chromosome number 2n = 40 and by pink flower buds, have not been found in the Netherlands up till now."*
-'Chromosome Numbers of Flowering Plants in the Netherlands: V' by Th.W.J. Gadella and E. Kliphuis, Botanical Museum and Herbarium, Utrecht, Netherlands; Mededelingen van het Botanisch Museum en Herbarium van de Rijksuniversiteit te Utrecht (Announcements from the Botanic Museum and Herbarium of the University of Utrecht), Netherlands, Volume 357, Issue 1, pages 335-343, **1971**.

*"The hybrid between S. officinale and S. asperum, **S. x uplandicum**, gave rise to some taxonomic confusion.*
In view of its presumed true-breeding, Bucknall (1913) considered the hybrid as a good species: S. peregrinum Ledeb. Kuznetsov (*1910), on the other hand, was able to demonstrate that S. peregrinum of Ledebour is not identical with S. x uplandicum Nym. The true S. peregrinum occurs in the Transcaucasian province of Talysh and differs morphologically from S. x uplandicum.
*Other authors, i.e. Gams in Hegi (**1927), Faegri (***1931), Tutin (1956), Wade (1958) and Gadella and Kliphuis (1969) are of the opinion that **S. x uplandicum is a hybrid, introduced into west Europe as a fodder plant.***
S. x uplandicum should be regarded as a collective name which covers a series of hybrids between S. asperum and S. officinale."
-'Cytotaxonomic Studies in the Genus Symphytum III: Some Symphytum Hybrids in Belgium and the Netherlands' by Th.W.J. Gadella and E. Kliphuis, Biologisch Jaarboek (Biological Yearbook), **1971**.
(* -'Kavkazkie Vidy Roda Symphytum Tourn. L.' {Caucasian Species of the Genus Symphytum Tourn. L.} by N. Kuznetsov, Memoires de l'Academie Imperiale des Sciences de Saint-Petersbourg, Serie. 8, Physiques et Mathematiques, Volume 25, No. 5, pages 1-94, 1910. If you have an English translation of this, I would appreciate a copy.)
(** -'Illustrierte Flora von Mittel-Europa: Mit Besonderer Berucksichtigung von Deutschland, Osterreich und der Schweiz' (Illustrated Flora of Central Europe: With Special Consideration of Germany, Austria and Switzerland), Volume 5, Part 3' by Gustav Hegi, Zurich, Switzerland; published by J.F. Lehmann, Munich, Germany, 1926. Symphytum pages 2220-2230. In German, 13 volumes written from 1908 to 1931. If you have an English translation of the Symphytum pages, I would appreciate a copy.)
(*** -'Uber in Skandinavien Gefundenen Symphytum-Arten: Nebst einigen Betrachtungen uber das Artproblem Innerhalb der Betreffenden Artgruppe' {About Symphytum Species Found in Scandinavia} by Knut Faegri, Bergens Museum Arbok, Norway, 47 pages, 1931. In German. If anyone has an English translation of it, I would appreciate having it.)

"Variation in colour sometimes indicates hybridity, and this is very apparent in the plant we call the Blue Comfrey, Symphytum x uplandicum, reputed to be the cross between the Rough Comfrey (S. asperum) and the cream-coloured plant of dyke- (dike, dam) and river-sides (S. officinale).
Authorities do not agree on the limits of this fertile plant. *So far as the writer is aware, S. asperum does not occur in Norfolk (eastern England), and it is very rare elsewhere in the British Isles.*
However, a walk along the banks of the river Wissey near Wretten Fen (Norfolk) is instructive so far as colour range is concerned; both on the wash-plain and by several dykes, **there are many thousands of plants in every shade of red and blue."**
-'Transactions of the Norfolk and Norwich Naturalists Society' edited by E.A. Ellis, Cromer, Norfolk County, England; Volume 22, Part 4, pages 223-315, **1972**. 'Hybrids and Habitats' by E.L. Swann, pages 244-249.

"S. x uplandicum 2n = 36:
*The Comfrey plants with the chromosome number 2n = 36 belong to S. x uplandicum Nym.. They probably originated as: **S. asperum (2n = 32) x S. officinale (2n = 40) creates the hybrid (2n = 36)**. This has been duplicated*

experimentally (Gadella & Kliphuis 1971).
In the Netherlands the 2n = 40 cytotype of S. officinale is very common in fens (marsh), but it seems to be absent from Britain. It was described by Gadella & Kliphuis (**1967, 1971).
It seems likely that the hybrid (S. x uplandicum, 2n = 36) was introduced into Great Britain, in view of the fact that hybridisation between the 2n = 40 form of S. officinale and S. asperum (2n = 32) is impossible in Britain, since the parental forms are either absent (S. officinale, 2n = 40) or extremely rare (S. asperum).
This hybrid (2n = 36) is rather tall (up to 1.5 meters = 4.9 feet), rough and prickly. The flower-buds are dark purple, **the corolla is purple or blue-purple.** The leaves are not or only slightly and very shortly decurrent.

S. x uplandicum 2n = 40:
British plants with 2n = 40 are also regarded as belonging to S. x uplandicum but they differ in a number of respects from the 2n = 36 hybrid.
In the first place the probable way in which they originated differs: **one of the parents of S. x uplandicum (2n = 40) is the tetraploid form (2n = 48) of S. officinale** instead of S. officinale with the chromosome number 2n = 40, the other parent is S. asperum. This has also been duplicated experimentally.
This S. x uplandicum hybrid (2n = 40) is rather tall (up to 1.3 meters = 4.2 feet) and is less rough and prickly than the S. x uplandicum (2n = 36) plants, at least in the majority of cases. The buds are usually pink, but sometimes purplish- or red-pink, e.g. in the plants from Stoke Ferry (Norfolk county, England) and Downham (Lancashire county, England).
The corolla is pink, turning bluish when the flowers age. They flower during a considerably shorter period than the 2n = 36 hybrids, at any rate in the experimental garden in Utrecht (Netherlands).
The 2n = 40 hybrid more frequently shows leaves decurrent than does the 2n = 36 hybrid: sometimes the leaves are decurrent up to half the distance to the next lower leaf.

Both the 2n = 36 and 2n = 40 hybrids have corollas that are gradually widened towards the top and have the portion above the constriction longer than the portion below, whereas the plants with 2n = 24 and 2n = 48 chromosomes have urceolate flowers in which the portions above and below the constriction are more or less equal in length.
Both the 2n = 36 and 2n = 40 hybrids are fertile and interfertile (can breed with each other), but the artificially made hybrid (2n = 38) between these hybrids has not yet been found in nature.
The two types of S. x uplandicum are usually clearly distinguishable by the characters mentioned above, but sometimes puzzling intermediate plants are found, perhaps indicating that segregation of characters occurs. The British S. x uplandicum hybrids completely agree with the artificially made corresponding hybrids cultivated in the botanical garden of Utrecht (Netherlands).
With regard to the origin of the S. x uplandicum hybrids, we agree with Tutin that they were introduced into Britain. In the majority of cases the local populations seem to be naturalized, but the possibility exists that **some plants may be regarded as of spontaneous local origin. At any rate, one of the parents, S. asperum, is rare in Britain."**
-'Cytotaxonomic Studies in the Genus Symphytum VI: Some Notes on Symphytum in Britain' by Th.W.J. Gadella, E. Kliphuis and F.H. Perring, Instituut voor Systematische Plantkunde (Institute for Systematic Botany), Utrecht, Netherlands, and Institute of Terrestrial Ecology, Abbots Ripton, England; Acta Botanica Neerlandica, Volume 23, No. 4, pages 433-437, August **1974**.
(* -'Cytotaxonomic Studies in the Genus Symphytum I: Symphytum Officinale L. in the Netherlands' by Th.W.J. Gadella and E. Kliphuis, Botanical Museum and Herbarium of the State University of Utrecht, Netherlands, February 25 1967.)

"**The Russian Comfrey, Symphytum peregrinum, introduced by Henry Doubleday in the 1870s, was, as far as we know, a first cross hybrid between the Common and Prickley Comfreys.**
In 'Curtis Botanical Magazine', it is called Symphytum peregrinum Ledeb., for it was regarded as the species described by Karl Von Ledebour (1785-1851), an Estonian botanist, from specimens he collected 4,000 feet (1219 meters) above sea level in the Caucasus.
The Royal Horticultural Society 'Dictionary of Gardening' now (1976) agrees that S. uplandicum may be the correct name, and to anyone who has grown Comfrey, **the fact that the rarely set seed produces mixed seedlings, instead of breeding true like all species, is clear proof that S. peregrinum was and is a hybrid.**
I shall continue however to refer to the cultivated Comfreys as Symphytum peregrinum."
-Comfrey: Fodder, Food and Remedy by Lawrence D. Hills. New York: Rizzoli Universe Books: **1976**, pages 29, 30.

"**S. asperum Crossing with S. officinale:**
The bright blue flowering plants of S. asperum (2n = 32) provided after crossing with S. officinale **two types of hybrids: with 2n = 36 (2n = 32 x 2n = 40 and reciprocity) and with 2n = 40 (2n = 32 x 2n = 48 and reciprocity; with the exception of S. officinale 2n = 48 male, white flowering)."**
-'Variation and Hybridization in Some Taxa of the Genus Symphytum' (Variatie en Hybridisatie bij enkele Taxa van het Genus Symphytum) by Th.W.J. Gadella, Department of Population and Evolutionary Biology (Vakgroep Populatie en Evolutiebiologie, Utrecht, The Netherlands); Gorteria: Tijdschrift voor de Floristiek, de Plantenoecologie en het Vegetatie-Onderzoek van Nederland (Journal for Floristics, Plant Ecology and Vegetation Research in the Netherlands), Volume 9, No. 4, pages 88-93, **1978**. All in Dutch except abstract was also in English.

(A reciprocal cross shows the role of parental sex on a given inheritance pattern. In one cross, a male expressing the trait is crossed with a female not expressing the trait. In the other, a female expressing the trait is crossed with a male not expressing the trait.)

"**Several authors (Bucknall 1913, Boissier 1879, Popov 1953) are of the opinion that S. peregrinum is not a separate**

species.
Bucknall suggested that S. peregrinum is synonymous with S. x uplandicum; he regards the later as of hybrid origin, with S. caucasicum as one of the parents.
Kuznetsow (1910), however, regarded S. peregrinum as a good species, distinct both morphologically and ecologically (S. peregrinum grows indeed at lower elevations than S. asperum). Kuznetsow holds that S. peregrinum is closely related with S. asperum and S. caucasicum.
Gviniashvili (*,**1972) seems to agree with him by placing S. caucasicum in the same section Coerulea but in the series Caucasica. The species indeed share the blue colour of the corolla. **Gviniashvili (1972) regards S. peregrinum as a good species; a conclusion we share.**
We are therefore of the opinion that the origin of the chromosome number 2n = 40 is still unexplained.
The possibility that S. caucasicum, S. peregrinum and S. asperum are somewhere sympatric in the Caucasus cannot be excluded; hence hybridization cannot be ruled out.
In that case the origin of the chromosome number of S. peregrinum (2n = 40) might be explained by assuming that the tetraploid form of S. caucasicum (2n = 48) crossed with S. asperum (2n = 32)."
-'Cytotaxonomic Studies in the Genus Symphytum VIII: Chromosome Numbers and Classification of Ten European Species' by T.W.J. Gadella and E. Kliphuis, Proceedings van de Koninklijke Nederlandse Akademie van Wetenschappen (Proceedings of the Royal Netherlands Academy of Arts and Sciences): Series C, Biological and Medical Sciences, Vol 81, page 162-172, **1978**.
(* -'Some Data on the Karyology of Caucasic Species of Symphytum L. with Respect to their Taxonomy' by TS.N. Gviniashvili, Botaniceskij Zurnal, Lennigrad, Russia, Volume 57, pages 1120-1126, 1972 in Russian. If you have a translation of this in English, I would appreciate a copy.)
(** -'Species Nova Generis Symphytum L. e Caucaso Boreali-Occidentale' by TS.N. Gviniashvili, Notulae Syst. ac Geograph Inst Bot Thblissiense, Volume 29, pages 55-60, 1972, in Russian. If you have an English translation, I would appreciate a copy.)
  (In the above excerpt from 'A Revision of the Genus Symphytum, Tourn.' by Cedric Bucknall, 1913, Bucknall
   wrote: *"I am convinced that we have Ledebour's plant in Britain, and that it is a true species, which, when pure bred, produces fruit abundantly."*)
  (Sympatric means occurring in the same geographical location; overlapping in distribution. For speciation it means taking place without geographical separation.)

"Two different hybrids in the genus Symphytum were collected in the Dutch province of Limburg, the first near Terziet (2n = 36), the second in Gulpen and on the Dolsberg between Gulpen and Wijlre (2n = 44).
The first mentioned plants (2n = 36) were indistinguishable from artificial hybrids between the 2n = 40 form of S. officinale and S. asperum (2n = 32).
The other hybrids (2n = 44) were identical with the backcross of S. x uplandicum (2n = 40) with the officinale parent (2n = 48). These hybrids were compared with some other intraspecific hybrids of S. officinale and with other hybrids between S. officinale and S. asperum.
S. asperum 2n = 32 crosses with S. officinale 2n = 40 (male or female) and 2n = 48 (male only). This creates hybrids with 2n = 36."
-'Variation and Hybridization in Some Taxa of the Genus Symphytum' (Variatie en Hybridisatie bij enkele Taxa van het Genus Symphytum) by Th.W.J. Gadella, Department of Population and Evolutionary Biology (Vakgroep Populatie en Evolutiebiologie, Utrecht, The Netherlands); Gorteria: Tijdschrift voor de Floristiek, de Plantenoecologie en het Vegetatie-Onderzoek van Nederland (Journal for Floristics, Plant Ecology and Vegetation Research in the Netherlands), Volume 9, No. 4, pages 88-93, **1978**. All in Dutch except abstract was also in English.
  (Intraspecific means occurring within a species or involving members of one species.)

"According to Wade (1958) the first record of S. x uplandicum in cultivation is in 1827.
Some of these introduced hybrids have been ascribed to (regarded as) **S. peregrinum Ledeb., but it seems that this is simply a more or less central form of the hybrid swarm S. x uplandicum** (Clapham, Tutin and Warburg, 1962)."
-'Comfrey Symphytum spp. as a Forage Crop' by J.C. Forbes, A.D. McKelvie, and P.J.C. Saunders, North of Scotland College of Agriculture, Aberdeen, United Kingdom; Herbage Abstracts, Volume 49, No. 12, pages 523-539, **1979**.

"Cytological and phytochemical studies of the Symphytum officinale polyploid complex, S. asperum and S. peregrinum clearly indicated that S. peregrinum is a distinct taxon.
S. peregrinum differs cytologically, morphologically, chemically and in its distribution from S. officinale and S. asperum. **Therefore S. peregrinum cannot be regarded as a hybrid between these species.**
S. x uplandicum (2n = 36, 40) differs from S. peregrinum (2n = 40) morphologically, chemically and in part cytologically.
The 2n = 40 cytotype of S. officinale differs both from the diploid (2n = 24) and tetraploid (2n = 48) cytotype (of S. officinale) in morphological aspect, but not in chemical respect. It contains the same pyrrolizidine alkaloids as S. officinale (2n = 24, 48). For that reason and also because this cytotype is interfertile (can breed) with the 2n = 48 cytotype, it is regarded as conspecific (same species) with S. officinale (2n = 48).
  The lack of information of the exact identity of the Western 2n = 40 cytotype of S. officinale and the Eastern morphotype of S. tanaicense Stev. does, for the moment, not permit to give a taxonomic recognition and an assignment of the appropriate rank to these taxa."
-'Cyto- and Chemotaxonomical Studies on the Sections Officinalia and Coerulea of the Genus Symphytum' by Th.W.J. Gadella, E. Kliphuis and H.J. Huizing; Botanica Helvetica (now called Alpine Botany), Volume 93, pages 169-192, **1983**.

(Polyploidy is a cell that has more than two paired {homologous} sets of chromosomes. Most species whose cells have nuclei {eukaryotes} are diploid, i.e., they have two sets of chromosomes. One set is inherited from each parent. Polyploidy is common in plants.)

(A cytotype is an individual of a species that has a different chromosomal factor to another.)

(A morphotype is a group of different types of individuals of the same species in a population; a morph. In taxonomy a morphotype is a specimen chosen to illustrate a morphological variation within a species population.)

(Conspecific means belonging to the same species.)

"**The analysis of morphological characters of the taxa studied makes clear that the artifically derived interspecific (between different species) hybrids between S. asperum and S. officinale are intermediate in many aspects.**
Some methods for the analysis of populations suspected of hybridisation were devised by Anderson (*1949). The simplest, albeit (although) somewhat crude method, is the construction of a hybrid index. This is obtained by selecting a number of characters by which two species differ, assigning the score '2' to each of the attributes of the first, '0' to the second species and '1' to the intermediates.

It shows that the artificially produced (and most natural) hybrids between S. asperum and S. officinale are exactly intermediate.

The score of S. peregrinum differs considerably from S. x uplandicum and from S. asperum.

The 2n = 40 cytotype of S. officinale differs from the 2n = 24 and the 2n = 48 cytotype of S. officinale, but not to a large extent.

The 2n = 40 cytotype of S. officinale differs considerably from the S. x uplandicum hybrids.

S. peregrinum and S. asperum:

The differences in chromosome number, morphology, geographical and altitudinal distribution, as well as the differences in steroid and alkaloid **patterns of S. peregrinum and S. asperum are in favour to justify these taxa as being independent.**

There is no doubt that the two species are more closely related to each other than to S. officinale. This close relationship between the former two taxa has been suggested earlier by Popov (**in Komarov 1953).

S. peregrinum and S. x uplandicum:

S. peregrinum and S. x uplandicum are not identical at all. They differ morphologically, partly cytologically, chemically and in their distribution.

For that reason we completely disagree with Bucknall (1913) and with Wickens (1969), who took the description from Bucknall, without having seen native specimens of S. peregrinum.

S. x uplandicum did not arise in the Caucasus, because the parental species are sympatric (same location) in a very restricted area only. Additionally they are bound to different altitudinal zones with this area.

Wade (1958) reported that S. asperum was introduced into the area of S. officinale, where hybridisation between and backcrossing of hybrids with the parental species occurred as well. Usually backcrossing to the S.officinale parent took place.

Two cytotypes of S. officinale (2n = 40, 48) were involved in this hybridisation process. The 2n = 48 cytotype is widely distributed in Europe, but the 2n = 40 cytotype occupies as far as is known up to now, a more restricted area (The Netherlands and northwest Germany).

However, it is highly unlikely that the S. x uplandicum 2n = 36 hybrid originated from west Europe, because the S. x uplandicum hybrids are supposed to have been introduced into west Europe from Leningrad, Russia (Wade, l.c.).

Therefore, we are in the opinion that plants with the morphological characters of the 2n = 40 cytotype most probably have a much wider distribution in Europe.

S. tanaicense Steven:

The west European plants share a number of characters with east European plants of S. tanaicense Steven. However, the latter species has neither been studied cytologically nor chemically. If the material from the low-lying peat lands of The Netherlands is identical with Hungarian and south Russian plants, they should be assigned taxonomically to S. tanaicense.

However, the west European 2n = 40 cytotype crosses readily with S. officinale (2n = 48), producing fertile hybrids (2n = 44).

For that reason **the authors express as their opinion that the crossability and slight morphological differences indicate that 2n = 40 and 2n = 48 cytotypes are conspecific (same species).**

Possibly the differences in ecological requirements may justify treatment at level of subspecies.

At any rate the 2n = 40 cytotype of S. officinale and S. x uplandicum (2n = 40) are not identical at all. They differ morphologically (Gadella & Kliphuis 1973) and chemically (Huizing, Gadella & Kliphuis 1982, Huizing, Malingre, Gadella & Kliphuis, submitted***).

The 2n = 24 and 2n = 48 cytotypes of S. officinale are inseparable on morphological and chemical grounds as far as they have been studied. They do not hybridise in nature and with great difficulty in the experimental garden, giving rise to sterile triploid hybrids (2n = 36). Such hybrids could be produced only between white-flowered parents. The exact taxonomic position of 2n = 56 cytotype is unclear."

-'Cyto- and Chemotaxonomical Studies on the Sections Officinalia and Coerulea of the Genus Symphytum' by Th.W.J. Gadella, E. Kliphuis and H.J. Huizing; Botanica Helvetica (now called Alpine Botany), Volume 93, pages 169-192, **1983.**

(* -'Introgressive Hybridization' by Edgar Anderson, Geneticist, Missouri Botanical Garden, Professor of Botany, Washington University, Saint Louis, Missouri; book printed in New York, 1949.)

(** -'Flora of the U.S.S.R.' edited by V.L. Komarov, Leningrad, Russia, with article by Mikhail Grigorevich Popov, in Volume 19, pages 279-291 Russian version, pages 207-216 English version, 1953.)

(*** The author may be referring to: -'Chemotaxonomical Investigations of the Symphytum Officinale Polyploid Complex and S. Asperum {Boraginaceae}: Phytosterols and Triterpenoids' by H.J. Huizing, Th.M. Malingre, Th.W.J. Gadella, and E. Kliphuis, all from The Netherlands; Plant Systematics and Evolution, Volume 143, pages 285-292, 1983.)
    ('l.c.' is Latin for 'loco citado' which means 'locally cited'. In other words, the author's name was cited in the early part of the book or article with a date and publication name.)

"In a number of cases although we have well established species in a polyploid series, hybridization has resulted in a confused taxonomic situation only clarified by careful cytogenetic studies.
In Symphytum, Symphytum x uplandicum Nyman was originally thought to consist of F1 and backcross hybrids derived from crosses between Symphytum asperum and Symphytum officinale for which chromosome numbers of 2n = 36 and 2n = 42, respectively, were available (*Perring, 1969).
Cytological examination of British plants (Gadella, Kliphuis & Perring, 1974) shows a different and more complex situation. Symphytum officinale has both diploid (2n = 24) and tetraploid cytotypes (2n = 48). Symphytum asperum, a rare relic of cultivation in Britain has 2n = 32.
Symphytum x uplandicum, however, exists as two cytotypes with 2n = 36 and 40.
    The 2n = 36 forms are thought to originate from hybridization of Symphytum asperum with a 2n = 40 cytotype of Symphytum officinale, unknown in Britain, but common in fens (marshes) in Netherlands (Gadella & Kliphuis, 1971).
    The 2n = 40 hybrids are regarded as being derived from a cross of Symphytum asperum with the 2n = 48 cytotype of Symphytum officinale.
Both Symphytum x uplandicum types were almost certainly originally introduced."
-'Cytogenetic Variation in the British Flora: Origins and Significance' by T.T. Elkington, Department of Botany, The University Sheffield, England; New Phytologist: A British Botanical Journal, London, England, Volume 98, pages 101-118, **1984**.
(* -'Network Research, Symphytum Survey' by Franklyn Perring, Biological Records Centre, Monks Wood, Huntingdon, Cambridgeshire County, England; Volume 7, No. 4, 1969. In 'Proceedings of Botanical Society of the British Isles, Volume 7' edited by E.F. Greenwood, London, England, 1967-1969, pages 553-556.)
    (Cytogenetics is study of inheritance in relation to the structure and function of chromosomes.)
    (An F1 hybrid or 'Filial 1' hybrid is the first generation of offspring of distinctly different parental types that produce a new, uniform phenotype {observable properties of an organism} with a combination of characteristics from the parents.)
    (Relict or relic means a surviving species of an otherwise extinct group, or a remnant of a formerly widespread species that lives in isolated areas.)

"The restriction endonuclease fragmentation patterns of cpDNA of **various Symphytum-taxa showed that morphological differentiation and molecular evolution on one hand, and sterility barriers on the other, did not evolve at the same rate. Symphytum officinale (2n = 24, 40, 48) and S. asperum (2n = 32) differ in their cpDNA pattern, but are able to produce fertile hybrids (S. x uplandicum; 2n = 36, 2n = 40).**
Pyrrolizidine alkaloids and isobauerenol can be used as chemotaxonomic markers for the identification of these hybrids."
-'Hybridization in Symphytum: Pattern and Process' by T.W.J. Gadella, Sommerfeltia: The Journal of Natural History Museum and University of Oslo, Norway, Volume 11, pages 79-96, **1990**.
    (cpDNA = chloroplast DNA. A chloroplast in green plant cells is a plastid that has chlorophyll. Plastids are small organelles in the cytoplasm of plant cells that have pigment or food. Cytoplasm is the material in a living cell, excluding the nucleus.)

"**The original cross (of Russian Comfrey) was probably a natural hybridization that took place in the collection of Symphytums gathered by Joseph Busch,** the Hackney (London, England) landscape gardener who laid out the grounds of St. Petersburg Palace for Catherine II (the Great) of Russia.
Plants were sent from there from 1870 onwards (other Symphytums were sent earlier, the first in 1790), and the latest import is reported to have been in 1900."
-Comfrey: Nature's Healing Herb & Health Food by Andrew Hughes. Japan: Sanyusha Publishing Co., Ltd, **1992**, page 136.

"**Symphytum officinale L. and Symphytum asperum L. (Boraginaceae) are allopatric taxa, which are able to intercross and to form interspecific (between different species) hybrids with different chromosome numbers.**
The species differ not only in a number of morphological characters but also ecologically, S. asperum being a species of higher elevations (upper montane zone), S. officinale of lowland and the lower montane zone.
S. asperum is a Caucasian species, which has the sporophytic chromosome number 2n = 32. It was introduced from the Caucasus into Europe as a fodder plant.
S. officinale is variable, containing cytotypes with 2n = 24, 48, 56, 40, and occurs throughout Europe. The most common chromosome number of S. officinale is 2n = 48.
    Scattered diploid populations of 2n = 24 occur in western, central and eastern Europe; they are white-flowered throughout.
    The tetraploids (2n = 48) are white- or purple-flowered in western Europe and purple in eastern Europe.
    Populations in which purple- and white-flowered individuals occur intermingled are very common in western Europe. In eastern Europe, mixed populations are very rare and consist of white-flowered diploid plants with purple-flowered tetraploid plants (Huizing 1985).
The cytotype 2n = 56 has been found at only one location in the high Tatra mountains of Czechoslovakia.

The cytotype 2n = 40 is very common in the low-lying peat regions in The Netherlands.
There exists, however, some controversy whether the latter (2n = 40) represents a true subspecies, S. uplandicum Nym., or a hybrid between S. asperum (2n = 32) and S. officinale (Hills 1976; Huizing 1985)."
-'Symphytum Officinale (Comfrey): In Vitro Culture, Regeneration, and Biogenesis of Pyrrolizidine Alkaloids' by H. J. Huizing and J. H. Sietsma, chapter XXVIII, pages 464-477 in 'Biotechnology in Agriculture and Forestry, Volume 15: Medicinal and Aromatic Plants III' edited by Y.P.S. Bajaj. Cham, Switzerland: Springer, **1991**.
> (Allopatric or geographic speciation occurs when populations of the same species become isolated from each other, and this prevents genetic exchange.)
> (A sporophyte is the diploid multicellular stage in the life cycle of a plant.)

"*Symphytum x uplandicum Nyman, Syll. Fl. Eur.: 80. 1855.*
> *= S. asperum Lepechin x S. officinale L."*

-'Synopsis of Boraginaceae subfam. Boraginoideae Tribe Boragineae in Italy' is the same as 'Boraginaceae in Italy II' by L. Cecchi and F. Selvi, University of Florence, Firenze, Italy; Plant Biosystems: An International Journal Dealing with all Aspects of Plant Biology, Official Journal of Societa Botanica Italiana, Volume 149, No 4, page 630-677, **2015**, Symphytum page 636-646.
End of Hybrid Origin and Classification of Russian Comfrey

Russian Comfrey Distribution (Geographical Location)

"*S. x uplandicum is probably the commonest Symphytum in this country (Britain), being frequent in hedgebanks, beside roads and sometimes in open places in woods.*
It seems to be better suited to conditions in Britain than S. asperum which, though widely distributed, is rare."
-'The Genus Symphytum in Britain' by T.G. Tutin, University College of Leicester, England; Watsonia: Journal of the Botanical Society of the British Isles, Volume 3, pages 280-281, February **1956**. (Watsonia was the journal of the 'Botanical Society of Britain and Ireland' from 1949 to 2010. It is now called 'New Journal of Botany'.)

"*S. x uplandicum Nym. is probably the most common species of Symphytum in this country (England), being frequent in hedgebanks, beside roads, and in clearings in woods; it seems better suited to British conditions than S. asperum (Tutin, 1956; Clapham, Tutin, Warburg, 1962)."*
-'British Medicinal Species of the Genus Symphytum' by K.R. Fell and Janet M. Peck, Pharmacognosy Research Laboratories, University of Bradford, England; Planta Medica: Journal of Medicinal Plant and Natural Product Research, Volume 16, No. 2, pages 208-216, May **1968**.

"*Russian Comfrey hybrids occur throughout the British Isles and have been recorded from all vice-counties with the exception of v.c. 74 (Wigtown, Scotland), 99 (Dunbarton, Scotland), 108 (West Sutherland, Scotland) and 109 (Caithness, Scotland).* **It is the most widespread Symphytum in the British Isles, characteristically appearing in the absence of either parent.**
It occurs on roadsides and river-banks, and generally in more obviously man-made habitats than S. officinale.
The hybrid is also known in Austria, Germany, Netherlands, Scandinavia (Norway, Sweden, Denmark) and the Caucasus."
-Hybridization and the Flora of the British Isles edited by Clive A. Stace, University of Leicester, England in collaboration with 'Botanical Society of the British Isles' now called 'Botanical Society of Britain and Ireland'. London, England: Academic Press, **1975**, page 354.

"*Russian Comfrey spreads mainly through fragmentation and possibly also by seed, although the common garden cultivar is sterile.* **Russian Comfrey was first recorded in the wild in Great Britain in 1884** *in Vice County 7 (Marlborough, North Wiltshire, England).*
Species Status:
Russian Comfrey is widespread and abundant throughout Great Britain, apart from some upland areas, the extreme north-west of Scotland and western Northern Ireland.
Russian Comfrey is regarded as native only in the Caucasus; it is now established in Austria, Belgium, Denmark, Finland, France, Ireland, Luxembourg, Netherlands, Norway and Sweden.
In Great Britain, records would suggest that it initially spread very slowly, colonizing only 200 ten km (10 km =6.2 miles) squares between 1884 and 1986, subsequently record increased to over 2,000 ten km squares by 1999 and nearly 3500 to-date.
> *However caution must be exercised in interpretation of these figures a it is also likely that identification for fieldwork for the year 2000 *'Atlas' was much better than previous recording and part of the apparent expansion involved re-determination of plant previously recorded as Common Comfrey."*

-'Russian Comfrey, Symphytum officinale x asperum = S. x uplandicum' by R.V. Lansdown; NNSS: Great Britain Non-Native Species Secretariat, www.nonnativespecies.org, York, England, **2011**.
(* -'Atlas 2000 and a Problem with Purple Flowered Comfrey' by Eric Chicken, East Yorkshire, England; Botanical Society of Britain and Ireland (BSBI) News, Hertfordshire, England, No. 76, pages 22-23, September 1997.)

"*S. x uplandicum (Section Symphytum), a natural hybrid of S. officinale and S. asperum Lepech.*
It is widespread in Europe and North America; in Central Russia, *it is known only from relatively few localities (Moscow city*

and region).
The species grows in moist meadows, swampy areas, alongside rivers or streams (Pawlowski 1972; *Hege 1972; Frolov 1991; **Tikhomirov et al. 1999)."
-'Pubescence of Vegetative Organs and Trichome Micromorphology in Some Boraginaceae at Different Ontogenetic Stages' by Rimma P. Barykina and Vitaly Y. Alyonkin, Lomonosov Moscow State University, Russia; Wulfenia, Regional Museum of Carinthia, Austria, Volume 23, pages 1-29, **2016**.
(* -Flora Europaea, Volume 3, edited by T.G. Tutin. Cambridge, England: Cambridge University Press, 1972. Article: 'Boraginaceae Juss.' by J.C. Hege, pages 86-120.)
(** -'The Genus Symphytum L. {Boraginaceae} in Central Russia' by V.N. Tikhomirov, S.R. Majorov and D.D. Sokoloff; Novosti Sistematiki Vysshikh Rastenii {News of Higher Plants Systematics}, Moscow and Leningrad {St. Petersburg}, Russia, Volume 31, pages 231-245, 1999. In Russian. If you have the Symphytum pages in English, I would appreciate a copy.)

"Symphytum x uplandicum Nyman *is a perennial fibrous-rooted hemicryptophyte or geophyte with a short rhizome,* **quite widespread in Europe and North America; only very few localities of this species in Central Russia are known (Moscow city and Moscow region).** *The plant grows in wet meadows, marshy areas, along rivers and streams (Pawlowski 1972; Hege 1972; Tikhomirov et al. 1999)."*
-'Syncotyly in Seedlings and Sprouts of Some Boraginaceae: Genesis, Structure and Function of the Cotyledon Tube' by Rimma P. Barykina and Vitaly Y. Alyonkin, Department of Higher Plants, Faculty of Biology, Lomonosov Moscow State University, Russia; Wulfenia, Austria, Volume 24, pages 11-28, **2017**.

Russian Comfrey Description (peregrinum and uplandicum) (in chronological order)

"S. peregrinum is, with the possible exception of S. asperum, **the largest species of the genus,** *attaining a height of 2 meters (6.5 feet) or more, with the stem 3 cm (1.1 inch) in diameter at the base.*
> **From S. officinale it is distinguished** *by its greater size, by the stem, in pure-bred plants, not being winged, by the clear blue flowers, and by the tuberculated nutlets.*
> **From S. asperum, which it greatly resembles, it is distinguished** *by the generally less strongly tubercular-setose stem, by the upper leaves, which are sessile and adnexed to the stem by an uncinate prolongation of the blade of the leaf, not subpetiolate, and especially by the larger calyx with acute segments.*

This applies to typical S. peregrinum, but intermediate forms sometimes occur *in which the stem is more or less tubercular-setose, the calyx larger or smaller with more or less obtuse segments, and the corolla varying in size in relation to the calyx. Such forms are found naturalized in Britain and Sweden, but not, apparently, mixed with the typical plant."*
-'A Revision of the Genus Symphytum, Tourn.' by Cedric Bucknall, (Mus. Bac. Oxon= Bachelor of Music, Oxford University), Journal of the Linnean Society of London, England, Botanical Journal, Volume 41, Issue 284, pages 491-556, December **1913**.

"S. peregrinum Ledeb. (fide C. Bucknall).
When growing on the banks of streams, *is a tall, luxuriant plant, with flowers rose-coloured in bud, then bright blue, the stem without wings, and bearing abundant fruit.*
When growing in dry localities, *the flowers remain rose-coloured or are only partially blue, and the entire plant is not so well developed as when growing in moister situations."*
-'Watson Exchange Club Report: 1912-1913', The Journal of Botany British and Foreign, London, England, Volume 52, page 275, **1914**.   (fide = according to)

"S. peregrinum Ldb.: Calyx parted up to 3/4, 1/3-1/2 as long as corolla, its teeth broader and longer, acute.
Otherwise hardly distinguishable from S. asperum Lepech. *(Talysh). Closely related to S. asperum, probably just a race in it. Stems apparently, lower; leaves usually less hairy beneath, none cordate at base (even the radical cuneate at base), narrower, oblong. Cymes as in S. asperum.*
According to Kuznetsov the main difference *is seen from the calyx; its teeth are dilated below and acute or acutish at the apex; in flower the calyx is 2 to 3 times shorter than the corolla and not 4-5 times. These distinctions do not seem to me to be constant."*
-'Flora of the U.S.S.R.', 30 volumes, 1934-1960, edited by V.L. Komarov; Botanical Institute of the Academy of Sciences of the U.S.S.R, Leningrad, Russia, with article by Mikhail Grigorevich Popov, in Volume 19, pages 279-291 Russian version, pages 207-216 English version, **1953**.   (Talysh Mountains are in southeastern Azerbaijan and northwestern Iran.)

> *"Race: In taxonomic classification, an infraspecific category of uncertain position but occasionally used in floras in place of, or subordinate to, form. The term is also used in lieu of ecotype, implying a category between subspecies and variety covering geographical groupings of plants. Races are often uniform in respect of ecological preference, physiological requirements, and topographical distribution."*
> -Dictionary of Botany, http://botanydictionary.org, 2019. A reference source for anyone with interest in plant science. The dictionary contains some 3000 lucidly written entries, drawn from major fields of pure and applied plant science.

For descriptions of the cultivars of Russian Comfrey of Lawrence D. Hills, see subsection 'Strains (Cultivars) of Russian Comfrey' in this section. He describes them in these books:

-**Russian Comfrey: A Hundred Tons an Acre of Stock or Compost for Farm, Garden or Smallholding**
    by Lawrence D. Hills. London England: Faber and Faber, Limited, **1953**.
-**Comfrey Report: The Story of the World's Fastest Protein Builder and Herbal Healer,** Conservation Gardening
    and Farming Series: Series C by Lawrence D. Hills. England: Henry Doubleday Research Association, 1975.
-**Comfrey: Fodder, Food and Remedy** by Lawrence D. Hills. New York: Rizzoli Universe Books: 1976.

"*S. x uplandicum Nyman (S. asperum x officinale), Blue Comfrey, includes S. peregrinum auct.:*
Similar in general appearance to S. officinale L. and S. asperum Lepech. and showing various combinations of their characters. Stems hispid to scabrid. Upper leaves usually shortly and narrowly decurrent.
Calyx sometimes accrescent, teeth acute or subacute. **Corolla blue or purplish-blue.** Flowers 6-8.
S. peregrinum auct. Ledeb. is more or less a central form of this hybrid swarm.
The commonest Symphytum of roadsides, hedgebanks, woods, etc. but usually absent from the waterside habitats occupied by S. officinale. **Distribution imperfectly known though confusion with S. officinale**, which appears to be much less common."
-Flora of the British Isles, Second Edition by A.R. Clapham, T.G. Tutin and E.F. Warburg. England: Cambridge University Press, **1962**, (first edition 1952).   (auct.= auctorum= various authors)

"**Note on Life Forms:** The Danish botanist Raunkiaer has classified plants, according to the position of the resting buds or persistent stem apices in relation to soil level, into a number of Life Forms, which have in turn been subdivided so as to convey more information about the plants included in the different groups. Life Forms are a convenient method of indicating how a plant passes the unfavourable season (winter), and are also of interest because they show a correlation with climate. The primary classes are:
1. Phanerophytes: woody plants with buds more than 25 cm (9.8 inches), above soil level.
2. Chamaephytes: woody or herbaceous plants with buds above soil surface but below 25 cm.
3. **Hemicryptophytes: herbs (very rarely woody plants) with buds at soil level.**
    Hp: Protohemicryptophytes, with uniformly leafy stems, but the basal leaves usually smaller than the rest.
    Hs: Semi-rosette hemicryptophytes, with leafy stems but the lower leaves larger than the upper ones and the basal internodes shortened.
    Hr: Rosette hemicryptophytes, with leafless flowering stems and a basal rosette of leaves.
4. Geophytes: herbs with buds below the soil surface.
5. Helophytes: marsh plants.
6. Hydrophytes: water plants.
7. Therophytes: plants which pass the unfavourable season as seeds."
-Flora of the British Isles, Second Edition by A.R. Clapham, T.G. Tutin and E.F. Warburg. England: Cambridge University Press, **1962**, (first edition 1952).
    (Christen Christensen Raunkiaer, 1860-1938, was a Danish botanist, who was a pioneer of plant ecology. He discovered plant strategies of how to survive an unfavourable season that he called 'Life Forms', and he showed these strategies corresponded to climate zones. This scheme is still used today and is a forerunner of modern plant strategies such as J. Philip Grime's CSR Triangle theory.)

"In this review, notes are given on the history of **Russian Comfrey and its introduction into Canada as Quaker Comfrey in 1953**. Its agronomy and feeding value are described. Results are given from trials in Canada on yield and chemical composition. **It is persistent, and resistent to insect and disease attack, but does not possess the qualities necessary for use as a forage plant in Quebec.**"
-'Quaker Comfrey' by L. Lachance, Canada Department Agricultural Research Station, Lennoxville, Quebec; Agriculture Montreal, Volume 25, No. 4, pages 29-34, **1968**. (I do not have this report. If you have a copy, could you please send it to me.)

"*S. x uplandicum Nyman,* Syll. 80 (1854) *(S. asperum x officinale).*
Intermediate between the parents. Stem up to 2 meter (6.5 feet). Leaves never cordate, the upper sessile, shortly decurrent or more or less amplexicaul.
Calyx 5-7 mm; lobes usually acute. Corolla 12-18 mm, **pink at first, turning blue, or persistently violet.** Scales as in S. officinale but less wide at the base. **2n = 36.**
Formerly cultivated for fodder, and naturalized; also of spontaneous local origin (Austria, Belgium/Luxembourg, Britain, Denmark, Finland, France, Ireland, Netherlands, Sweden)."
-Flora Europaea, Volume 3: Diapensiaceae to Myoporaceae, editors Tutin, Heywood, Burges, Moore, Valentine, Walters and Webb. United Kingdom: Cambridge University Press, **1972**, page 104. (Symphytum L., pages 103-105 by B. Pawlowski.) (5-volume encyclopedia of plants, published between 1964 and 1993.)

"*S. asperum Lepech. x S. officinale L. = S. x uplandicum Nyman:*
Hybrids form a range of intermediates betweeen the parents in characters of the leaves and flowers. The leaves generally lack the broadly decurent bases of S. officinale but they are never cordate and petiolate as in S. asperum.
The calyx, in bud, is 4-5 mm with acute segments, not 2-3 mm with obtuse segments, or expanding rapidly at fruitings as in S. asperum; the filaments equal or exceed the anthers, never shorter than them as in S. officinale, and the stigma is often bent at the tip. **The flower-colour varies from reddish-purple to violet, but is never sky-blue as in S. asperum or yellowish-white as in some variants of S. officinale.**"

-Hybridization and the Flora of the British Isles edited by Clive A. Stace, University of Leicester, England in collaboration with 'Botanical Society of the British Isles' now called 'Botanical Society of Britain and Ireland'. London, England: Academic Press, **1975**, page 353.

"**S. x uplandicum Nyman (S. officinale x S. asperum)- Russian Comfrey:**
Differs from S. officinale in more bristly stems, leaves and calyx; corolla blue to violet or purplish when open.
Introduced originally as fodder, or possibly arisen anew in a few places; roadsides, rough and damp ground, wood-borders; frequent over most of British Isles.
Fertile and backcrosses to S. officinale, forming a spectrum of intermediates."
-New Flora of the British Isles: Identification of Wild Vascular Plants of the British Isles edited by Clive Stace. Cambridge, England: Cambridge University Press, **1991**, page 647.

"**Symphytum x uplandicum:**
USDA Hardiness Zone: 6-8; *Light: full sun to partial sun; Soil pH: Acid / Neutral / Alakaline.*
Clumping herb. Height: 1 to 4 feet (0.30-1.2 meter); Width: 3 feet (0.9 meter). Growth rate: fast. Native region: Eurasia.
Habitats: disturbed, old fields, edges, other habitats. Areas recently disturbed, whether by humans or natural causes. These plants can be observed as urban or garden 'weeds' or roadside plants. Old fields in the process of becoming forest. Edges are boundaries between two habitats, here generally indicating the boundary between forest and more open fields.
Medicinal: excellent. Some or all parts of the plant are poisonous.
Dynamic accumulator, invertebrate shelter, nectary.
Roots are fibrous & fleshy: *Dividing into a large number of fleshy (thick or swollen) roots immediately upon leaving the crown.*
Nuisance: persistent. Very difficult to remove once planted, often resprouting from even tiny pieces of root."
-Edible Forest Gardens: Volume Two: Design & Practice by Dave Jacke and Eric Toensmeier. White River Junction, Vermont: Chelsea Green, **2005**, page 490.

> (USDA Hardiness Zone 6-8. Zone 6 has a last frost date of May 1st and first frost date of November 1st. Zone 8 has a last frost date of April 1st and first frost date of December 1st.
> The map is the standard by which gardeners and growers can determine which plants are most likely to thrive at a location. The map is based on the average annual minimum winter temperature, divided into 10-degree F zones. United States Department of Agriculture, Agricultural Research Service, https://planthardiness.ars.usda.gov/PHZMWeb/)

"**Symphytum asperum x Symphytum officinale = Symphytum x uplandicum Nyman** is a very rare Comfrey hybrid in New England (eastern United States).
It resembles S. asperum (Prickly Comfrey) in that the **leaves are not decurrent on the stem or infrequently shortly decurrent for a distance of less than 10 mm (0.39 inch).**
However, the hybrid differs in that it has a corolla 13-16 mm (0.51-0.62 inch) long, **purple or pink flower buds,** a calyx 5-7 mm (0.19-0.27 inch) long, and short, broad papillae on the margins of the fornices.
Versus (Prickly Comfrey): corolla 9-14 mm (0.35-0.55 inch) long, red flower buds, calyx 3-5 mm (0.11-0.19 inch) long, and long, narrow papillae on the margins of the fornices.
From S. officinale it (Russian Comfrey) can additionally be distinguished by ascending corolla lobes and dull brown schizocarps (vs. recurved corolla lobes and lustrous black schizocarps).
Two cytotypes of this hybrid are known, depending on which cytotype of S. officinale was involved in the cross.
> Thus far, only the 2n = 40 type of this hybrid has been collected in New England (with a 2n = 48 S. officinale parent).
> This hybrid shows softer stem pubescence, slightly longer leaf base decurrence (on average), and blunter fornices than the 2 n= 36 type (which is known from North America)."
-New England Wild Flower Society's Flora Novae Angliae: A Manual for the Identification of Native and Naturalized Higher Vascular Plants of New England by Arthur Haines. New Haven: Connecticut: Yale University Press, **2011**.

> (A cytotype is an individual of a species that has a different chromosomal factor to another.)
> (Body cells and individuals are described by the number of chromosome sets {ploidy level} they have: monoploid: 1 set; diploid: 2 sets = 2n = 2x, triploid: 3 sets = 2n = 3x, tetraploid: 4 sets = 2n = 4x, pentaploid: 5 sets = 2n = 5x, etc.)

"**S. x uplandicum (Section Symphytum),** a natural hybrid of S. officinale and S. asperum Lepech.
Germination epigeous, hypocotylar. Cotyledon blades oval-elliptic, almost orbicular, 15 to 17 mm long and 10 to 12 mm wide; leaf apex rounded, margins entire, blade base attenuate; indumentum restricted to the adaxial side. Pubescence uniform, scarce, of stiff short hairs.
All trichomes unicellular, more or less similar in size, relatively short, transparent, sessile, subulate (straight and hooked).
> Cotyledon petioles long, 13 to 15 mm, fully (except for basal parts) covered by straight and (or) hooked, short, transparent, upward-directed hairs. Juvenile leaves petiolate, vary from oval-ovate (20 mm long, 11 to 15 mm wide) to wide-lanceolate (40 mm long, 18 to 20 mm wide).

On the first leaf, the indumentum is restricted to the adaxial side, subsequent leaves of all age groups are pubescent on both leaf surfaces.
Only simple trichomes are present: sessile and subtended by rosettes.
> The former are short, one-celled, subulate (straight), thick-walled, with fine-grained surface.
> The latter are long, unicellular, subulate (straight and / or sickle-shaped), thick-walled, with tuberculous surface; brown granular content is present inside them.

Eight to 14 rosette cells thin-walled, much larger than other dermal cells of the leaf."
-'Pubescence of Vegetative Organs and Trichome Micromorphology in Some Boraginaceae at Different Ontogenetic Stages' by Rimma P. Barykina and Vitaly Y. Alyonkin, Lomonosov Moscow State University, Russia; Wulfenia, Regional Museum of Carinthia, Austria, Volume 23, pages 1-29, **2016**.
    (Pubescent means the surface of a leaf or stem is covered with fine short hairs.)
    (Ontogeny is the development of an organism from fertilization of the egg to maturity.)
    (Trichome is a small hair or other outgrowth from the epidermis of a plant. Epidermis is the outer layer.)

"***Symphytum asperum x officinale (S. x uplandicum):***
Ecology: *The habitats of this perennial herb include rough and waste ground, railway banks, roadsides, hedge banks and woodland margins. Generally lowland, but reaching 365 meter (1197 feet) altitude at Alston (Cumberland County, England).*
Status: *Neophyte.*
Trends: *This hybrid was introduced as a forage plant in 1870. It was widely cultivated in Britain in the late 19th and early 20th centuries, and known from the wild by 1884. Two cytotypes occur, with 2n=36 and 2n=40 chromosomes. The map in Perring & Sell* was only provisional; it is therefore difficult to assess changes in distribution, though **the plant is clearly well-established and probably increasing.***
World Distribution: *This hybrid is known from the Caucasus; it has been spread in cultivation and is naturalised in temperate Europe."*
-'Online Atlas of the British and Irish Flora', https://www.brc.ac.uk/plantatlas, Distribution and ecology with photographs, **2018**.
(* -Critical Supplement to the Atlas of the British Flora: Including Distribution Maps of 500 Flowering Plants and Ferns and Explanatory Text by F.H. Perring and P.D. Sell, editors. London, England: Thomas Nelson and Son, 1968.)
    (In botany, a neophyte is a plant species that is not native to a geographical region, and was introduced in recent history. Plants that are long-established in an area are called archaeophytes. In Britain, neophytes are plant species introduced after 1492, when Christopher Columbus arrived in the Americas.)

Determining Hybrid Character from Chromosomes and Chemicals
See subsection 'Hybrids, Hybrid Swarms, Introgression' in section 'Symphytum Species Overview' (Chapter 6).

"**The (*Symphytum officinale* = Common Comfrey) cytotype 2n = 40 is very common in low-lying peat regions in the Netherlands.** *Judging from the descriptions given by Basler (*1972) these plants seem to occur also in Schleswig-Holstein (northern Germany).* **The plants are nearly always purple-flowered, only very exceptionally white.**
We differ from Basler, however, on the taxonomical position of the plants with 2n = 40. Basler refers all plants with 2n = 40 to S. x uplandicum Nym. (Russian Comfrey), *the hybrid between S. asperum (Prickly Comfrey) (2n = 32) and S. officinale (2n = 48).*
A detailed description of the difference between the artificially produced hybrid S. x uplandicum (2n = 40) and the form with the same chromosome number from the low-lying peat lands in the Netherlands has been presented by Gadella and Kliphuis (**1973).
    However, the possibility should not be excluded that the hybrid shows segregation in later generations, producing the forms resembling the plants with 2n = 40.
No segregation of the S. x uplandicum (2n = 40) hybrid, however, occurred in the F1 and F2 generation, all plants retained the characters of S. x uplandicum and did not show the slightest resemblance to the Dutch plants with 2n = 40. **Therefore, Gadella and Kliphuis are of the opinion that the latter plants are conspecific (same species) with S. officinale."**
-'Chemotaxonomical Investigations of the Symphytum Officinale Polyploid Complex and S. Asperum (Boraginaceae): The Pyrrolizidine Alkaloids' by H.J. Huizing, Th.W.J. Gadella, and E. Kliphuis, all from The Netherlands; Plant Systematics and Evolution, Volume 140, pages 279-292, **1982**.
(* -'Cytotaxonomic Studies on the Boraginaceen Genus Symphytum L.: Studies on Predominantly Northern German Plants of the Species S. Asperum Lepech., S. officinale L. and S. x Uplandicum Nym.' by Armin Basler, Botanische Jahrbucher fur Systematik: Pflanzengeschichte und Pflanzengeographie {Botanical Yearbooks for Systematics: Plant History and Plant Geography}, Volume 92, pages 508-553, 1972. All in German. I was unable to find a copy of this report. If you have an English translation, could you please send it to me.)
(** -'Cytotaxonomic Studies in the Genus Symphytum V: Some Notes on W. European Plants with the Chromosome Number 2n=40' by Th.W.J. Gadella and E. Kliphuis, Botanische Jahrbucher fur Systematik {Botanical Yearbooks for Systematics}, Volume 93, pages 530-538, 1973.)
    (An F1 hybrid or 'Filial 1' hybrid is the first generation of offspring of distinctly different parental types that produce a new, uniform phenotype {observable properties of an organism} with a combination of characteristics from the parents. F2 hybrids or second generation hybrids are self or cross pollination of F1 hybrids. They have more variability than F1s, but can have some desirable traits. They are not true to form.)

"***From a comparison of phytosterol and triterpenoid patterns of several Symphytum officinale (Common Comfrey) cytotypes, S. asperum (Prickly Comfrey) and their interspecific hybrids, S. x uplandicum (Russian Comfrey),*** *which were obtained from thin layer chromatography and gas-chromatography (also in combination with mass spectrometry),* **the hybrid character of the latter taxon is clearly shown.**
The specific value of the triterpenoid isobauerenol as a chemotaxonomieal marker within this group is discussed in some detail."

-'Chemotaxonomical Investigations of the Symphytum Officinale Polyploid Complex and S. Asperum (Boraginaceae): Phytosterols and Triterpenoids' by H.J. Huizing, Th.M. Malingre, Th.W.J. Gadella, and E. Kliphuis, all from The Netherlands; Plant Systematics and Evolution, Volume 143, pages 285-292, **1983**.
> (Phytosterols include plant sterols and stanols, which are phytosteroids, similar to cholesterol. Triterpenoid saponins in plants are defensive compounds against pathogenic microbes and animals that eat the plants.)

"For a consideration of the chemotaxonomical value of a specific marker it is necessary to investigate its variation between different individuals from single species.
From our survey of several cytotypes of Symphytum officinale (Common Comfrey) with 2n = 24, 40, and 48, and harvested at different times and locations, only slight qualitative variations in phytosterol and triterpenoid profiles can be demonstrated. The same applies to S. asperum (Prickly Comfrey) and S. x uplandicum (Russian Comfrey).
Because of their ubiquitous (everywhere) occurrence in extracts of the investigated taxa, in general, the phytosterols seem to be of no value as chemotaxonomical markers here.
The triterpenoid isobauerenol, on the other hand, seems useful, for its presence in S. x uplandicum (2n = 36 or 40) readily demonstrates the hybrid character of this taxon which was artificially derived from S. officinale (2n = 40 or 48) plants containing isobauerenol and from S. asperum (2n = 32) plants which lack this typical compound.
However, this specific marker gives no information concerning the question, whether the S. officinale cytotype 2n = 40 belongs to a hybrid swarm between S. asperum and S. officinale (2n = 48), as was suggested by Basler, or has to be considered as conspecific (same species) with S. officinale. For this question one has to use pyrrolizidine alkaloids as markers.
The data we have presented may also be useful for verifying the origin of Comfrey roots used in pharmaceutical preparations or for teas (*Roitman, 1981), especially when pyrrolizidine alkaloid patterns are also included in the phytochemical investigations as was outlined already by us (**Huizing et. al., 1982)."
-'Chemotaxonomical Investigations of the Symphytum Officinale Polyploid Complex and S. Asperum (Boraginaceae): Phytosterols and Triterpenoids' by H.J. Huizing, Th.M. Malingre, Th.W.J. Gadella, and E. Kliphuis, all from The Netherlands; Plant Systematics and Evolution, Volume 143, pages 285-292, **1983**.
(* -'Comfrey and Liver Damage' by James N. Roitman, The Lancet, Volume 317, Issue 8226, page 944, April 25 1981.)
(** -'Chemotaxonomical Investigations of the Symphytum Officinale Polyploid Complex and S. Asperum {Boraginaceae}: The Pyrrolizidine Alkaloids' by H.J. Huizing, Th.W.J. Gadella, and E. Kliphuis, all from The Netherlands; Plant Systematics and Evolution, Volume 140, pages 279-292, 1982.)
> (Chemotaxonomy is classification of plants and animals based on similarities and differences in biochemical composition.)

F1 Hybrid
See subsection 'Hybrids, Hybrid Swarms, Introgression' in section 'Symphytum Species Overview' (Chapter 6).

An F1 hybrid or 'Filial 1' hybrid is the first generation of offspring of distinctly different parental types that produce a new, uniform phenotype (observable properties of an organism) with a combination of characteristics from the parents.
Mules are F1 hybrids between horse and donkey. This happens naturally, and includes hybrids between species, for example, peppermint is an F1 hybrid of watermint and spearmint.
In agronomy (food crop production), these F1 hybrids are usually created by means of controlled pollination, sometimes by hand-pollination. This is not GMO. A Genetically Modified Organism has been altered using genetic engineering techniques in a sophisticated laboratory, frequently using genes from unrelated species.
Russian Comfrey Bocking No. 4 and No. 14 are F1 hybrids.

This F1 hybrid is Henry Doubleday's Russian Comfrey (Quaker Comfrey). The 'Henry Doubleday Research Association' is now known as 'Garden Organic' in Warwickshire, England. It studies and promotes organic gardening and farming.
Russian Comfrey was studied, cataloged and perfected by horticulturist Lawrence D. Hills starting in the 1950s. Henry Doubleday (1810-1902) was a Quaker smallholder in Essex County, England who experimented with different varieties of Comfrey.   (See the 'History' section in Chapters 12 to 16.)

"Symphytum Peregrinum:
The period of the development of this hybrid plant is lost somewhere in the years between 1840-1870, but it has been established that it is a first cross- an F1 Hybrid between asperrimum and officinale.
When peregrinum was introduced into England by Henry Doubleday somewhere around 1870 from St. Petersburg (Russia), with its place of origin established as the Caucasus, it created a sensation by far outproducing both asperrimum (Prickly Comfrey) and officinale (Common Comfrey), which until then had been in fairly common use.
Asperrimum, which had been introduced into England about 100 years before peregrinum, had far outpaced officinale, and was the type mainly being grown for fodder. Peregrinum then proceeded to displace asperrimum, but not completely and much confusion arose."
-Comfrey: Nature's Healing Herb & Health Food by Andrew Hughes. Japan: Sanyusha Publishing Co., Ltd, 1992, page 135-136.
> (The city of Saint Petersburg in Russia was founded in 1703. In 1914 the name was changed to Petrograd.
> In 1924 it was changed to Leningrad. In 1991 it was changed back to Saint Petersburg.)

*"**In spite of its hybrid origin S. x uplandicum is apparently a fixed hybrid** and any variation it shows is due to its back-crossing with S. officinale (Common Comfrey) and more rarely, and perhaps less certainly, with S. asperum (Prickly Comfrey). Populations of S. x uplandicum known to me, unless growing in close proximity to S. officinale, show remarkable uniformity, and plants grown in Cardiff (Wales) from seed collected from a large population in Pembrokeshire (county in southwest Wales) showed no variation from the parent plants.*
It was this uniformity and absence of segregation which led the late C. Bucknall and others to consider it a good species."
-'The History of Symphytum Asperum Lepech. and S. x Uplandicum in Britain' by A.E. Wade, Department of Botany, National Museum of Wales, Watsonia, Volume 4, pages 117-118, **1958**. (Watsonia was the journal of the 'Botanical Society of Britain and Ireland' from 1949 to 2010. It is now called 'New Journal of Botany'.)

    ('Good species' means it is genetically stable. The seeds breed true, meaning the seeds grow into plants that are the same as their parents.)

*"In the F1 hybrid cross, the small triangles that fit together over the stamens around the base of the pistil failed to open to allow the entry of bees to polinate the flowers, unlike the wild species. This meant that the seed was and still is **scarce** and seed that is disseminated only produces similar varieties with the same problem. **The first cross vigor was lost.** Now there is a series of mixtures which, by constant selection and removal of undesirable variations have become **strains**, of which there are a great number."*
-'Comfrey: The Cinderella of Plants' by Maria Wilkes, Herbarist: The Herb Society of America Journal, Volume 33, pages 47-50, **1967**. Also in 'Herbs for Use and for Delight: An Anthology from the Herbarist' by Daniel J. Foley, Herb Society of Ameria. Gloucester, Massachusetts: Peter Smith Publisher Inc., 1974.

    (First cross vigor, hybrid vigor, heterosis, or outbreeding enhancement is the improved function and health of hybrid offspring.)
    (A strain is a cultivar (cultivated plant). Article 28 of the 'International Code of Nomenclature' (ICN) define the rules for cultivars and cultivar groups. *"Cultivar: The basic independent category used for organisms in agriculture, forestry, and horticulture."* A cultivar name is the scientific Latin botanical name followed by a cultivar epithet {descriptive phrase}. This epithet is usually in common language.)

"The greater growth speed came from the vigor of the first cross, or what is now called an F1 hybrid of which there are many among modern vegetables."
-Comfrey Report: The Story of the World's Fastest Protein Builder and Herbal Healer, Conservation Gardening and Farming Series: Series C by Lawrence D. Hills. England: Henry Doubleday Research Association, **1975**, page 9.

*"**The only way to restore the vigor of the hybrid is to cross the two species again, S. assperrimum and S. officinale** (creating an F1 hybrid).*
When this (F1 hybrid) seed was secured and raised, the first cross vigor was lost, and the seedlings sorted out the qualities of their parents carried by the genes. The result was a series of mixtures, which became 'strains' when selected by removing undesirable variations.
Seeds from a hybrid will always produce a further mixture. The variations were of all habits and colors.
The serial numbers with the prefix 'Bocking' are not consecutive for each strain, they are allotted when it becomes possible to be sure that a variation is sufficiently distinct to merit identification."
-Comfrey Report: The Story of the World's Fastest Protein Builder and Herbal Healer, Conservation Gardening and Farming Series: Series C by Lawrence D. Hills. England: Henry Doubleday Research Association, **1975**, page 10.

*"Other historical hybrids (of all genera) have arisen by the stabilization of various F2 or later segregants from an F1 hybrid, of well-defined segments of a hybrid complex, or even of **F1 hybrids themselves.***
In the last case the hybrids are frequently highly or wholly sterile, but have spread vegetatively and come to occupy habitats and geographical areas different from those of the parents.
For instance, Symphytum x uplandicum (S. asperum x S. officinale) is more widespread than either parent, and unlike them, is known to hybridize with S. tuberosum."
-Hybridization and the Flora of the British Isles edited by Clive A. Stace, University of Leicester, England in collaboration with 'Botanical Society of the British Isles' now called 'Botanical Society of Britain and Ireland'. London, England: Academic Press, **1975**, page 4.   (A segregant differs from either parent as a result of segregation or separation.)

*"**Many attempts to repeat the original cross by artificial pollination have been made, but so far without success.** But there is no doubt that peregrinum as it has survived and been carefully retained **exhibits all the vigor and vitality of a first cross**, and has inherited the virtues of its parents, the resistance of its father to disease and insects, and the full medical value of its mother, officinale, plus productive capacity far beyond both."*
--Comfrey: Nature's Healing Herb & Health Food by Andrew Hughes. Japan: Sanyusha Publishing Co., Ltd, **1992**, page 137.

Flowers of Russian Comfrey
For more about Russian Comfrey flowers, see subsection 'Description' in this section.

"The new introduction had an even greater growth speed, reaching 6 feet 8 inches (80 inches = 2 meters) as a maximum when allowed to run to flower.
These flowers were at first blue and changed colour to a magenta pink, and it is this colour that marks the blood of the high-yielding Comfreys."
-Russian Comfrey: A Hundred Tons an Acre of Stock or Compost for Farm, Garden or Smallholding by Lawrence D. Hills. London England: Faber and Faber, Limited, **1953**, page 28.

"The original Russian Comfrey hybrid was sterile, or semi-sterile; the stamens are shut off by a kind of 'false bottom' in the flower *which is (by a freak that may occur perhaps once in every 100,000 flowers) sometimes available for bees to transfer as far as it will go among the neighboring blooms.*
The pistils are, however, always open to receive pollen from our native Common Comfrey *(also found in Europe, including Russia).*
Russian Comfrey may chance to make seed, as it is possible that once in 10,000 chances a bumble bee (not the honey bee, which cannot penetrate to fertilize the flowers) may cross pollinate."
-Comfrey: Nature's Healing Herb & Health Food by Andrew Hughes. Japan: Sanyusha Publishing Co., Ltd, **1992**, pages 137, 159.

(The stamen is the male part of a flower that produces pollen to fertilize the female part.
The pistil is the female part; it produce ovules that develop into fruit and seeds after pollination.)

"Much work on the Symphytum complex was undertaken by the Dutch botanists Th.W.J. Gadella and E. Kliphuis, and together with F.H. Perring they also looked at some British populations.
A relatively recent statement of the accepted position is given by Perring (*1994) in which the ***purple flowered Comfreys are:***
    S. officinale, 2n = 48, with red buds opening to purple flowers. (Common Comfrey)
    S. x uplandicum, 2n = 36, with deep purple buds opening to a colour ranging from purple violet to violet blue. (Russian Comfrey)
A frequent form of S. x uplandicum, 2n = 40, with pink buds opening blue*, and the cream coloured forms of S. officinale, 2n = 24 and 2n = 48 do not pose a problem here, nor does purple flowered S. officinale, 2n = 40 since it is stated to occur only in Holland."*
-'Atlas 2000 and a Problem with Purple Flowered Comfrey' by Eric Chicken, East Yorkshire, England; Botanical Society of Britain and Ireland (BSBI) News, Hertfordshire, England, No. 76, pages 22-23, September **1997**.
(* -The Common Ground of Wild and Cultivated Plants: B.S.B.I. Conference Report No. 22 by A. Roy Perry and R.G. Ellis, {F.H. Perring, pages 64-70}. Cardiff, Wales: Department of Biology, National Museum of Wales, 1994.)

*"****Russian Comfrey Flower:***
Corolla bell-shaped (funnel trumpet-shaped), upper part lightly budded, 12-18 mm (0.5-0.7 inches) long, red, blue or purple, on rare occasions almost white, fused, 5-lobed. Corolla lobes often slightly twisted. Corolla mouth with 5 large, triangular tongue-like scales.
Calyx fused, 5-lobed, 5-7 mm (0.20.3 inches) long, shorter than corolla funnel, hairy. Corolla lobes usually pointed, on rare occasions with roundish tip. Stamens 5. Gynoecium fused, single-styled. Cymes in axils, single-branched or scorpioid."
-'Russian Comfrey: Symphytum x uplandicum' NatureGate, www.luontoportti.com, University of Helsinki, Finland, **2019**. Eija and Jouko Lehmuskallio started documenting nature and developing identification services in the 1990s.

High Yield of Russian Comfrey
See subsection 'Production Per Acre (Yield)' in section 'Productivity and Farm Economics of Comfrey' in Volume 2.

"Russian Comfrey will grow in six weeks what a cabbage takes six months to achieve."
-Compost, Comfrey and Green-Manure: 'Henry Doubleday Research Association' First Gardeners Report by Lawrence D. Hills. Braintree, Essex, England: Henry Doubleday Research Association, **1959**.

"Russian Comfrey grew rapidly after transplanting from glasshouse plants grown from 1-inch (2.5 cm) root cuttings, and yielded 1,815 pounds (823 kg) dry matter per acre in early August.
The leaves were eaten by sheep and heifers; regrowth was good."
-'Russian Comfrey in Southern Alberta' by D.B. Wilson, Forage Notes, Canada, Volume 7, No. 1, pages 38-39, **1961**. (I do not have a copy of this report. If you have it, could you please send it to me.)   (A heifer is a young cow who has never given birth.)

"The plant material used in the trials, identified as Symphytum x uplandicum (S. asperum x S. officinale), *was established in autumn on riverine clay at* ***spacings of 45 x 45, 60 x 60, 75 x 75 and 90 x 90 cm (17.7 x 17.7, 23.6 x 23.6, 29.5 x 29.5, 35.4 x 35.4 inches)*** *and given 30 tons f.y.m. (farm yard manure) each spring and 300 kg (661 pounds) 12-10-18 (NPK) fertilizer after each cut.*
The crop took more than a year to establish, and the highest yield obtained was 86 tons fresh material/ha (hectare) from the closest spacing in 1958.
Average DM (Dry Matter) content was 11.7%. There was 19.5% CP (Crude Protein), which was 38% digestible, and 18% minerals in the DM.

The narrowest spacing gave the highest yields except in very dry weather. The crop appeared to be very sensitive to drought; most roots were in the plough layer, and none penetrated more than 1 meter (3.28 feet) into the soil.
The fresh leaves were not attractive to cattle but were palatable (pleasant taste) to pigs and poultry.
Hay was difficult to make; only 2 of 4 silages were successful, and ensiling losses were 30%.
Harvesting was difficult, dirt tare was high and establishment costs were excessive."
'Report on a Trial with Russian Comfrey in 1953-60' by H. Van Der Zweerde, IBS, Wageningen, Netherlands; Versl 35 Inst Biol scheik Onderz LandbGewass, page 9, **1965**. (I was unable to get this report. If you have a copy, could you please send it to me.)
    (A young, recently planted crop of Comfrey is more sensitive to drought than a well-established crop. An older crop is very drought tolerant.)
    (Tare weight or unladen weight is the weight of an empty vehicle or container.)
    (It is possible to successfully make Comfrey silage. See the subsection 'Comfrey as Silage' in section 'Comfrey Meal, Pellets, Hay and Silage' in Volume 2.)

"The only modern record of a Comfrey yield above 100 tons an acre from an official research station comes from the Agronomy Extension, University of California, Davis, where several large plots of 'Quaker Comfrey' (Russian Comfrey, probably No. 4) were established for pig feeding experiments (1958). The world's record yields were both secured with solid stemmed mixtures.
The second highest yielder over five years at Bocking (England) is Russian Comfrey Bocking No. 14, which is much earlier starting into growth than No. 4. Its value to Vernon Stephenson was to give (feed) early Comfrey for the first foals, and at times he covered plants with cloches to bring them on just a little earlier.
This particular clone is now the most common in South Africa, because of many thousands split from a single plant and given away by my wife before our marriage when she lived in Capetown."
-Comfrey: Fodder, Food and Remedy by Lawrence D. Hills. New York: Rizzoli Universe Books: **1976**, pages 70, 71.
    (Bocking is an area in the town of Braintree, Essex County, England where Lawrence Hills started his Comfrey research station in December 1954.)   (A cloche is a covering to protect plants in cold weather, made of clear glass or plastic.)

Yield of Russian Comfrey No. 4 versus No. 14

"Willey and Knight (*1962) found Russian Comfrey Bocking 14 and Bocking 17 to be relatively high yielding, but obtained low yields from Bocking 4. These selections were neither greatly nor consistently higher yielding than the more variable strains from which they were developed.
McClean (**1964) obtained similar yields from three strains and the selection Bocking 4."
-'Comfrey Symphytum spp. as a Forage Crop' by J.C. Forbes, A.D. McKelvie, and P.J.C. Saunders, North of Scotland College of Agriculture, Aberdeen, United Kingdom; Herbage Abstracts, Volume 49, No. 12, pages 523-539, **1979**.
(* -'Russian Comfrey' by L.A. Willey and R.L. Knight, Journal of the National Institute of Agricultural Botany, Volume 9, No. 2, pages 139-144, 1962. I was unable to get this report. If you have a copy, could you please send it to me.)
(** -'Russian Comfrey' by S.P. McClean, Bridget's Experimental Husbandry Farm, United Kingdom; Experimental Husbandry, No. 10, pages 46-51, 1964. I do not have this report. If you have a copy, could you please send it to me.)

"There are two main strains of Symphytum peregrinum grown in England: Stephenson's and Webster's Giant. These have been classified into some 20 types, natural clones.
The commercial name, Webster's Giant, is the strain that produces the highest quality and yield of Comfrey, and is the type that has been exported to Kenya, Australia, New Zealand and Japan.
There are minor variations in the Webster's Giant strain, but about 60% to 80% of this strain is made up of what has been classified as Bocking No. 4, *and the variations that occur in the balance are not necessarily significant, but need nevertheless to be observed by the specialist."*
-Comfrey: Nature's Healing Herb & Health Food by Andrew Hughes. Japan: Sanyusha Publishing Co., Ltd, **1992**, page 134.   (Russian Comfrey No. 4 came to Japan in April 1958.)

Preferred Soil Type and Roots
See sections 'Care of Plant Overview and How to Propagate' and 'Planting, Soil, Fertilization, Water, Disease' in Volume 2.

*"**The powerful roots of Russian Comfrey** are excellent molecule chosers for gardeners because they take so much from so deep down. Lucerene (alfalfa) roots are as deep but it is not easy in the garden and its thin roots take a very much smaller load."*
-Compost, Comfrey and Green-Manure: 'Henry Doubleday Research Association' First Gardeners Report by Lawrence D. Hills. Braintree, Essex, England: Henry Doubleday Research Association, **1959**.

*"Whereas Common Comfrey 'joyeth in watery places' (ditches) and along the sides of channels and creeks where it grows in England even today, peregrinum does not do well in such places. **Russian Comfrey needs high land, hillsides and mountains to do its best; deep soil where its long roots can go deep down to the subsoil.***
From our experience in Japan it prefers its natural habitat in the mountains, and the best Comfrey so far produced here grows 800 meters (2625 feet) above the sea, a place with a cold winter and hot summer.
The growing habit of Symphytum peregrinum, which, depending on the depth of soil and subsoil layer, puts its roots

down as deep as 3 meters (9.8 feet) in 3-4 years."
-Comfrey: Nature's Healing Herb & Health Food by Andrew Hughes. Japan: Sanyusha Publishing Co., **1992**, pages 139, 151.

Strains (Cultivars) of Russian Comfrey

After a brief overview, this subsection covers the following:

1. Turner Strain also known as Ferne Farm Survival (Bocking No. 1, 16, 17)
2. Stephenson Strain (Bocking No. 2, 14)
3. Webster Strain (Bocking No. 3, 4, 5, 6, 7, 8, 9, 10, 11, Bocking Mixture)
      Gibson Strain (an improved Webster Strain)
4. Other Russian Comfrey Cultivars and Varieties

> *"**A** 'subspecies' is a grouping within a species used to describe geographically isolated variants, a category above 'variety', and is indicated by the abbreviation 'subsp.' in the scientific name.*
> ***A** 'variety' consists of more or less recognizable entities within species that are not genetically isolated from each other, below the level of subspecies, and are indicated by the abbreviation 'var.' in the scientific name.*
> *The words '**variety**' and 'form' are not synonyms for the word cultivars according to the 'International Code of Nomenclature for Cultivated Plants'. The Code considers these terms botanical classifications. The 'Association of Official Seed Certifying Agencies' (AOSCA) considers the terms '**cultivar**' and '**variety**' equivalent."*
> -USDA National Plant Materials Manual, 190-V-NPMM, 4th Edition, Part 542 Acronyms, 542.2 Plant Nomenclature, July 2010.

"In 1955 we planted the Landrover-load (4-wheel-drive vehicle) of Webster strain of Comfrey from Southery (Norfolk, England) that Mrs. Greer brought, the sackfuls of Stephenson strain Vernon fetched down from Little Weighton near Hull (Yorkshire, England).
While Newman Turner sent sacks of black and ugly roots of what I called 'Ferne Farm Survival' (from Shaftestbury, Dorset County, England). These grew behind one of his barns and survived from when this was a famous racing stable between the 1890s and 1930s.
*Ron Suckling drove me over with forks and spades, and **we dug some plants from the headlands of the field at Coggeshall (Essex, England) that had once been part of Henry Doubleday's smallholding (small farm).***
*George Gibson of Guernsey (an island in the English Channel off the coast of Normandy, France), who invented the liquid tomato manure made from Comfrey leaves and stems rotted in water, **sent us his strain** which he had exported to Kenya (east Africa), where John McInnes grew it and secured his record yield. **This had been improved from Webster some time between 1900 and 1914.***
> *So we had a full set as we could of what was being grown as 'Russian Comfrey' in Britain.*
As my plants grew through the summer, I saw that my first conjecture (theory), that the variation was due to neglect or soil differences, was entirely wrong, Comfrey was not a species (Symphytum peregrinum) but a number of mixtures of hybrids resulting from natural self-pollination of the F1 generation seedlings that gave Henry Doubleday his 100 ton an acre yields, and back-crosses of these with the wild S. officinale that John Gerard said 'joyeth in watery ditches'.
As I slowly sorted these, through 1955 and 1956, I gave them 'Bocking Numbers', and found that each of the four mixtures had a dominant variation.
> *The one in the Webster Strain was **Bocking No. 4** which had large, round-ended leaves, thick, solid stems down which the leaves extended in a long triangular fin for about a fifth of the way to the next leaf, and flowers were Bishops Violet 34/3.*
> *The **Stephenson Strain dominant was Bocking No. 14,** with pointed leaves, thin stems to which the leaves fitted without a fin, and Imperial Purple 33/3 flowers.*
> *The **Ferne Farm Survival was dominated by Bocking No. 1**, which had large, pointed leaves that folded in from the edges, thick flower stems down which the leaf fins extended almost from one leaf to the next below it, and deep red flowers that were Indian Lake 826/1.*
> *The **Gibson Strain was very near a Webster** but with more pointed-leaved variations near No. 4, differing only in its flower colours.*
> *The **plants Ron Suckling and I collected at Coggeshall proved to be identical with Bocking No. 14,** but with slightly darker purple flowers."*
-Fighting Like the Flowers: An Autobiography: The Life Story of Britain's Best-Known Organic Gardener by Lawrence D. Hills. Bideford, Devon, England: Green Books, 1989, page 110-114.
> (For more about Vernon Stephenson, see 1942 in the 'History' section in Chapter 16.)
> (For more about F.W. Newman Turner and Mrs. Greer, see 1948 in the 'History' section in Chapter 16.)
> (Lawrence Hills' view is that 'Symphytum peregrinum' and 'Symphytum x uplandicum' are the exact same plant.)

*"**By the second summer (1956) the plants had grown enough to make it clear that the only answer to the problem of about thirty distinct varieties, with only one species name, Symphytum peregrimum, between them was to give them 'cultivar' names,** which all hybrids have to have, whether they are of garden or natural origin from chance crossovers or are bred by a raiser."*

-Comfrey: Fodder, Food and Remedy by Lawrence D. Hills. New York: Rizzoli Universe Books: 1976, page 51.

"There were two commercial strains, the Webster and the Stephenson, and a number of 'Survivals' in hedges as well as the main one at Ferne Farm (Shaftestbury, Dorset County, England, 1954-1955) grown by *F. Newman Turner (Bocking No. 1, 16, 17).* The numbers of each strain are not consecutive because they were allotted when it was clear that a variation was distinct enough to be separated.
Roughly speaking the variations followed either Symphytum officinale with thick stems, many of them solid, and leaves which extend down these stems as a kind of fin (wing), described botanically as 'decurrent'.
Or S. asperrimum (Prickly Comfrey): This started into growth 10 to 14 days earlier, and had thin stems with no wings at all.
All had two habits- the 'fountain of leaves' stage, and the tall flowering stage when the crop was allowed to run past its best in terms of protein production, but ideal for identification."
-Comfrey: Fodder, Food and Remedy by Lawrence D. Hills. New York: Rizzoli Universe Books: **1976**, page 51, and
-Comfrey Report: The Story of the World's Fastest Protein Builder and Herbal Healer, Conservation Gardening and Farming Series: Series C by Lawrence D. Hills. England: Henry Doubleday Research Association, 1975, page 12.
(A plant fin or wing is a green, short, thin, paperlike structure extending along a stem. It is long along the stem but not tall away from the stem.)

Turner Strain (Bocking No. 1, 16, 17) (Ferne Farm, Shaftesbury, Dorset, England)

F. Newman Turner, NDA (National Diploma in Agriculture), NDD (National Diploma in Dairying), FNIMH (Fellow of Britain's National Institute of Medical Herbalists). He lived from 1913-1964. He wrote 'Fertility Farming', 'Fertility Pastures', and 'Herdsmanship', all classics of practical organic husbandry. He was editor of 'The Farmer' (a journal of organic farming) and President of the Henry Doubleday Research Association (1954-1964).

Bocking No. 1

"In 1956 the first Russian Comfrey Bocking number I was sure of was a variation in the Newman Turner strain, with very pointed leaves, a tendency to fold inwards from the sides, and *dark red flowers, and so I called this 'Bocking No. 1'.*
The buds show colour first as a dark red, Indian Lake 826/1 and open to Solferino Purple 26/1, fading to 26/3. The effect is of crimson with a little magenta.
The wings are small and the stem leaves alternate but paired behind the flower cluster."
-Comfrey: Fodder, Food and Remedy by Lawrence D. Hills. NY: Rizzoli Universe Books: **1976**, page 51, 52.
(The color numbers are from the 'Royal Horticultural Society', England, color chart.)

Bocking No. 16 and No. 17

"Bocking No. 16 and No. 17 are also high potash, high yielding, high protein Comfrey varieties as human food or for stock."
-Compost, Comfrey and Green-Manure: 'Henry Doubleday Research Association' First Gardeners Report by Lawrence D. Hills. Braintree, Essex, England: Henry Doubleday Research Association, **1959**.

"Bocking No. 16 is the dominant in the Turner strain. Its flower stems are stout, the wings are broad at the leaf axils and taper sharply. The leaves on the flower stems are alternate and are waved at the edges.
The normal, as distinct from flower stem leaves, are pointed and inclined to fold inwards. The proportion of width measured across the broadest part of the leaf to length measured from end of the stalk to tip is 5 to 12."
The flower colour is mauve (dark purple) 633/1 fading at 633/3 and, especially at the skirt of the bloom, to almost white."
-Comfrey: Fodder, Food and Remedy by Lawrence D. Hills. NY: Rizzoli Universe Books: **1976**, page 52, 53.

Stephenson Strain (Bocking No. 2 and No. 14)

"The Stephenson Strain was started by the late Kenneth Crawley of Lockerbie, Dumfries (Scotland around 1938). It was sold by E.V. Stephenson of Hunsley House Stud, Little Weighton, near Hull (Yorkshire County, England)."
-Comfrey Report: The Story of the World's Fastest Protein Builder and Herbal Healer, Conservation Gardening and Farming Series: Series C by Lawrence D. Hills. England: Henry Doubleday Research Association, 1975, page 12.

"Spring growth commenced earliest in Bocking No. 2, Bocking No. 14 and Bocking No. 17 (one of the Turner strain), while Websters was the last to start growth."
-'Russian Comfrey' by L.A. Willey and R.L. Knight, Journal of the National Institute of Agricultural Botany, Volume 9, No. 2, pages 139-144, **1962**.

Russian Comfrey Bocking No. 14

"Bocking No. 14 is the earliest (to come up in spring) Comfrey and highest in potash (potassium). It averages 7% against 4% *for the large leafed, heavy stemmed mixtures and Bocking No. 4, the kind grown as a vegetable."*
-Compost, Comfrey and Green-Manure: 'Henry Doubleday Research Association' First Gardeners Report by Lawrence D. Hills. Braintree, Essex, England: Henry Doubleday Research Association, **1959**.

"This is the dominant in the Stephenson strain, 80 to 90 percent. The flower stems are slender and frequent and are entirely wingless. The flowers are Imperial Purple 33/3 fading to Lilac Purple 031/3 *(Royal Horticultural Society's colour chart, Great Britain).*
The leaves are pointed, slightly serrated at the edges and vary in proportion from 5 to 12 and 3 to 6.
Bocking No. 14 takes after Symphytum asperrimum (Prickly Comfrey) in flower stem habit. It is early, rust resistant, and high in potash and allantoin, the healing principle."
-Comfrey: Fodder, Food and Remedy by Lawrence D. Hills. New York: Rizzoli Universe Books: **1976**, page 54.
   (I tried to find what those color chart numbers mean but that coding system is no longer in use. A dictionary definition of Imperial Purple is deep purple that is redder and duller than hyacinth violet or petunia violet and redder and less strong than dahlia purple. Lilac is pale violet that is dark mauve or light purple.)

*"This particular variety (Bocking No. 14) is the most often found in the hedges round fields where Comfrey once grew, because, it is the **highest in potash**, and this appears to make it distasteful to rabbits and unattractive to poultry when it is growing.*
It also has the highest resistance to Comfrey Rust (Melampsorella symphyti), the only serious disease of the crop, *which infects the wild S. officinale so badly that this species cannot be grown commercially, except on small scale in herb gardens.*
Bocking No. 14 is highest in allantoin."
-Comfrey: Fodder, Food and Remedy by Lawrence D. Hills. New York: Rizzoli Universe Books: **1976**, page 72.
   (I have fed just-cut Comfrey No. 4 and No. 14 to chickens, ducks, goats, pigeons, pigs and rabbits, and all of them ate it immediately without wilting it first. I have never had rust problems with any of my Comfrey.)

"This variety (Comfrey No. 14) is perhaps the best for garden cultivation, for race and riding horses, for pigs, and for feeding to poultry *if wilted and chaffed (cut in somewhat small pieces).*
Its high potash (potassium) *makes it of special value as 'instant compost', compost material and liquid manure on a garden scale.*
As the basic variation of the Stephenson Strain is Bocking No. 14 (thinner stems), Mr. Stephenson dried better hay on this system than growers with the thick stemmed Webster variations (Bocking No. 4)."
-Comfrey: Fodder, Food and Remedy by Lawrence D. Hills. New York: Rizzoli Universe Books: **1976**, pages 72, 119. (I have cut Comfrey in small pieces for chicks and ducklings. Adults eat whole leaves.)

"The Stephenson strain of peregrinum (clone No. 14) is the oldest established plot of Comfrey growing in Britain (at Hunsley House Stud) and was planted in 1938. *This was grown originally for race horses, and at one time comprised 4 1/2 acres of Comfrey. It is still used mainly for racehorses and pigs, the farm being a stud farm, the Hunsley House Stud at Little Weighton, near Hull, Yorkshire, England."*
-Comfrey: Nature's Healing Herb & Health Food by Andrew Hughes. Japan: Sanyusha Publishing Co., Ltd, **1992**, page 143.

"Russian Comfrey Bocking No. 14 is considered better than No. 4 in terms of medicinal activity and yield."
-'Richters® First Commercial Herb Growing Conference' edited by Rita Berzins and Conrad Richter. Transcripts from October 26 **1996**, Richters: The Herb Specialists, Goodwood, Ontario, Canada.

"A sample analysis of Russian Comfrey concentrate made from 'Bocking 14' contains, per litre (1.05 quart), 79 mg of nitrogen, 26.4 mg phosphorus, and 205 mg potassium."
-Encyclopedia of Organic Gardening: The Complete Guide to Natural & Chemical-Free Gardening by The Henry Doubleday Research Association, edited by Pauline Pears. London, England: DK (Dorling Kindersley), **2001**, page 206.

Webster Strain (Bocking No. 3, 4, 5, 6, 7, 8, 9, 10, 11, Bocking Mixture, Gibson)

"In 1900 Messrs. Webster of Stock, Essex (England) imported it (S. x uplandicum) from Russia, *and they and other nurserymen still carry stocks (as of 1958)."*
-'The History of Symphytum Asperum Lepech. and S. x Uplandicum in Britain' by A.E. Wade, Department of Botany, National Museum of Wales, Watsonia, Volume 4, pages 117-118, 1958. (Watsonia was the journal of the 'Botanical Society of Britain and Ireland' from 1949 to 2010. It is now called 'New Journal of Botany'.)

"This is the most popular commercial strain, stated by the late R.O. Webster to have been imported by his

*father from St. Petersburg in Russia in 1900. **The Gibson strain,** which has now yielded 124 tons an acre in 12 cuts at Nakuru, Kenya (east Africa), is a selection of the plant, has 12 months cutting cycle in that climate and is merely from Webster plants imported from Guernsey before 1914.*
This strain is now the most widely grown in Australia, New Zealand and all parts of Africa."
-Comfrey: Fodder, Food and Remedy by Lawrence D. Hills. New York: Rizzoli Universe Books: 1976, page 57.
    (The city of Saint Petersburg in Russia was founded in 1703. In 1914 the name was changed to Petrograd. In 1924 it was changed to Leningrad. In 1991 it was changed back to Saint Petersburg.)
    (Guernsey is an island in the English Channel off the coast of Normandy, France.)

Russian Comfrey Bocking No. 4

"Bocking No. 14 is the earliest Comfrey and the highest in **potash.** It averages 7% against **4% for the large leafed, heavy stemmed mixtures and Bocking No. 4,** the kind grown as a vegetable."
-Compost, Comfrey and Green-Manure: 'Henry Doubleday Research Association' First Gardeners Report by Lawrence D. Hills. Braintree, Essex, England: Henry Doubleday Research Association, **1959**.

"**Russian Comfrey No. 4 characteristics are large leaves and thick stem, with flower color variations, and some in growth speed. This is the dominant of the Webster strain, about 50 to 60 percent.**
Flower colour is Bishops Violet 34/3 (Royal Horticultural Society's colour chart, Great Britain) when fully open. It has strong stems and small wings. The leaves are broad and round tipped, their proportion is 5 to 10, but they have no incurling, therefore they appear far wider than a No. 1 for example. The edges are unserrated, and the veins are prominent, with bristles thickest on the underside so that they appear smooth. At leafy stage these leaves are very large, recovering rapidly after cutting.
The stems, as in all the variations under trial, are solid. Bocking No. 4 has thick stems like Symphytum officinale, but without the distinctive 'wings'. It is preferred by poultry and for human food, with more protein and less allantoin."
-Comfrey: Fodder, Food and Remedy by Lawrence D. Hills. NY: Rizzoli Universe Books: **1976**, page 57, 58.
    (I tried to find what that color chart number means but that coding system is no longer in use. Bishops Violet is a moderate reddish purple that is bluer, stronger, and slightly lighter than heliotrope, and bluer and duller than eupatorium purple.)

"Here are the words of Thomas Christy, Henry Doubleday's associate (in 1870s):
'**The solid stem variety of peregrinum is far more palatable (tasty),** and in every way it has proved superior to anything grown in this country (England) called Prickly Comfrey (S. asperrimum).'
Their basis for selection was yield, food value, and palatability. **The plants are identified by their solid leaf stems, shape of leaf, no wings continuing down stems from leaves on flower stems, and the color of the flowers.**"
-Comfrey: Nature's Healing Herb & Health Food by Andrew Hughes. Japan: Sanyusha Publishing Co., Ltd, **1992**, page 138.
    (For more about Thomas Christy, see section 'Age of Revolutions and Comfrey 1800s' {Chapter 15}.)

"The Comfrey grown in Australia is an F1 hybrid, a cross between S. officinale and S. asperimum, all from **Webster's Giant, the commercial strain imported into Australia around 1956.**
Careful examination of this strain shows that **the plants are predominantly Bocking No. 4** of the 20 or so classified clones."
-Comfrey: Nature's Healing Herb & Health Food by Andrew Hughes. Japan: Sanyusha Publishing Co., Ltd, **1992**, page 60.

"There are two main strains of Symphytum peregrinum grown in England: Stephenson's and Webster's Giant. These have been classified into some 20 types, natural clones.
**The commercial name, Webster's Giant, is the strain that produces the highest quality and yield of Comfrey, and is the type that has been exported to Kenya, Australia, New Zealand and Japan.
There are minor variations in the Webster's Giant strain, but about 60% to 80% of this strain is made up of what has been classified as Bocking No. 4,** and the variations that occur in the balance are not necessarily significant, but need nevertheless to be observed by the specialist."
-Comfrey: Nature's Healing Herb & Health Food by Andrew Hughes. Japan: Sanyusha Publishing Co., Ltd, **1992**, page 134.   (Comfrey No. 4 came to Japan in April 1958.)

Bocking Mixture

"**The Bocking Mixture: This is the Webster Strain** with the low yielders removed from a stock of 500 plants, established by the writer (Lawrence D. Hills) at Southery Nurseries, Southery, near Downham Market, Norfolk, England, in 1949. This improved selection was supplied to many experimenters, including Mr. I.G. MacDonald and Mrs. P.B. Greer.

The regularity and higher average yield of Bocking No. 4, however, shows up when the 'Mixture' and the pure clone (No. 4) are planted side by side."
-Comfrey: Fodder, Food and Remedy by Lawrence D. Hills. New York: Rizzoli Universe Books: 1976, page 57.
('Bocking Mixture' is Webster Strain of Russian Comfrey before it has been divided into Bocking Numbers.)

Other Russian Comfrey Cultivars and Varieties

A **'variety'** *consists of more or less recognizable entities within species that are not genetically isolated from each other, below the level of subspecies, and are indicated by the abbreviation 'var.' in the scientific name.*
The words **'variety'** *and 'form' are not synonyms for the word cultivars according to the 'International Code of Nomenclature for Cultivated Plants'. The Code considers these terms botanical classifications. The 'Association of Official Seed Certifying Agencies' (AOSCA) considers the terms* **'cultivar' and 'variety'** *equivalent."*
-USDA National Plant Materials Manual, 190-V-NPMM, Fourth Edition, Part 542 Acronyms, 542.2 Plant Nomenclature, July 2010.

*"**Russian Comfrey produces small, tubular flowers in a variety of colours that have been treated as varieties.**"*
-'Russian Comfrey, Symphytum officinale x asperum = S. x uplandicum' by R.V. Lansdown; NNSS: Great Britain Non-Native Species Secretariat, www.nonnativespecies.org, York, England, **2011**.

*"**Russian Comfrey Flower:** Corolla bell-shaped (funnel trumpet-shaped), upper part lightly budded, 12-18 mm (0.5-0.7 inches) long,* **red, blue or purple, on rare occasions almost white,** *fused, 5-lobed. Flowering time: June to August.*
Russian Comfrey, which is a hybrid between Common and Prickly Comfrey, comes in many variations *and grows mainly independently.*
New hybrids are still occurring in the parent species' common habitats, *but Russian Comfrey is far more wide-spread than Prickly Comfrey, which rarely escapes: it grows in roughly the same areas as Common Comfrey."*
-'Russian Comfrey: Symphytum x uplandicum' NatureGate, University of Helsinki, Finland, www.luontoportti.com, **2019**.
Eija and Jouko Lehmuskallio started documenting nature and developing identification services in the 1990s.

Bocking No. 12

*"**Bocking No. 12 was selected from hedgerow remnants of plantings said to have been made prior to 1900, at Abbey Farm, Coggeshall, Essex, England.**"*
-'Russian Comfrey' by L.A. Willey and R.L. Knight, Journal of the National Institute of Agricultural Botany, Volume 9, No. 2, pages 139-144, **1962**.

Symphytum x uplandicum 'Moorland Heather'

"Jennifer Matthews, owner, Moorland Cottage Plants, Pembrokeshire, Wales:
'I also recommend **the variety that originated here- S. x uplandicum 'Moorland Heather'.** *It grows to around 1 meter (3.2 feet), has slowly increasing clumps and* **very dark violet flowers**.'
S. x uplandicum 'Moorland Heather' was a chance seedling found at Moorland Cottage Plants in South Wales. It has striking dark violet flowers from April through to June."
-'Symphytum' by Miranda Kimberley, HorticultureWeek, Twickenham, London, July 9 **2010**. 'Horticulture Week provides senior professionals in the horticulture and related industries with the best possible business-critical information to help them secure success and sustainability for their businesses and estates.'
(Moorland Cottage Plants is out of business so I was unable to contact them for more information.)

Other Moorland Healther descriptions include:
"A lovely form of the native Comfrey which is grown for the stunning magenta-purple flowers, which are on show for most of the summer."   *"A tall Comfrey with flowers in a rich lavender."*

*"**Symphytum x uplandicum 'Moorland Heather'** is a native herb in Britain. The* **dark-purple buds which appear in spring and summer, open to bell-shaped flowers which are magenta.**
This variety is not as tall as some of the others *and is useful as ground-cover between shrubs, making a weed-proof cluster of rough-textured green leaves,* **Height 60 cm (23.6 inch = about 2 feet),** *and spread approximately 1 meter (3.2 feet)."*
-HooksgreenHerbs: Culinary, Medicinal and Scented Pot-Grown Herbs and Seeds, www.hooksgreenherbs.com, Stone, Staffordshire, England, **2019**.

Variegated Varieties of Russian Comfrey
Variegated means leaves that are edged or patterned in a second color, usually white, cream or yellow.

*"**The Symphytum or Comfrey has a most beautiful variegated form-*** *at least, I attribute it to this hybrid plant, supposedly* **S. asperum x S. officinale, with pale mauve (pale purple) flowers.** *The large leaves are a*

*foot (0.3 meters) or more in length, **broadly margined with cream and grey (greyish-green)**. It is a rare plant, slowly increasing, of remarkable beauty."*
-Plants for Ground-Cover by Graham Stuart Thomas, V.M.H., Gardens Advisor to the National Trust. London, England: J.M. Dent & Sons Ltd, **1970**, page 124.

Another variegated Russian Comfrey: **Symphytum x uplandicum 'Axminster Gold'** with yellow along edges of leaves. It has pink or mauve-pink flowers. It is sometimes called Symphytum x uplandicum 'Variegatum'.

"Symphytum x uplandicum 'Variegatum': Hardy perennial: USDA Zones 5-9.
*Flowering time: late spring to early summer. **Flower color: lilac-pink changing to blue.***
Clusters of 1/4 inch (2 cm)-long, tubular flowers are carried on erect stems.
Height: 36 inch (90 cm). Foliage height: 12 inch (30 cm).
*The basal leaves of this plant are big, bold and hairy. **Each gray-green leaf may be more than 12 inch (30 cm) long, and it has a generous, pale cream margin.***
Foliage quality and general growth are best in soil that remains moist throughout the growing season."
-Right Place, Right Plant: Over 1,400 Plants for Every Situation in the Garden by Nicola Ferguson. New York: A Fireside Book, Simon & Schuster, **2005**, page 173.

*"Symphytum x uplandicum 'Axminster Gold' (gold-variegated Russian Comfrey): Vigorous hardy herbaceous perennial with **huge gold-margined leaves. Superb landscape plant for rich moist soil. Scorches in full sun.** Part shade. USDA Zones 4-9.*
Source: Heronswood Nursery, Kingston, Washington, www.heronswood.com"
-'Proceedings of the 24th Annual Horticulture Industries Show' of Oklahoma and Arkansas in Fort Smith, Arkansas, edited by Lynn Brandenberger, Department of Horticulture and Landscape Architecture, Oklahoma State University, Stillwater, Oklahoma, January 14-15 **2005**.

*"For a big bold, ever-present yellow statement in my gardens, I always find room for a specimen or clusters of Yucca filamentosa 'Color Guard', my favorite golden, yellow-centered variegated yucca, and **Symphytum x uplandicum 'Axminster Gold' (yellow variegated Comfrey) with large, coarse, grayish-green leaves artfully edged in vibrant lemon-yellow.***"
-'The HPSO Quarterly: A Publication of the Hardy Plant Society of Oregon', www.hardyplantsociety.org, Portland, Oregon, Volume 4, No. 2, Spring **2016**. Article 'Yellow Ignites the Garden' by Bob Hyland, page 11.

"Symphytum x uplandicum 'Variegatum': USDA Hardiness Zone: 4 to 8.
Height: 3.00 to 4.00 feet (0.9-1.2 meter). Spread: 2.00 to 3.00 feet (0.6-0.9 meter).
Bloom Time: May to June. Bloom Description: Rose aging to purple.
*'**Variegatum' is a variegated-leaf cultivar that grows slightly smaller that S. uplandicum and features striking grayish green leaves with broad white leaf margins.***
Synonymous with Symphytum peregrinum 'Variegatum'.
*Culture: Easily grown in average, medium moisture, well-drained soils in full sun to part shade. Best in moist, organically rich soils. **Best leaf variegation usually occurs with some part afternoon shade in hot summer climates such as the Saint Louis, Missouri area.** Many of the Comfreys, including this hybrid, spread aggressively by creeping rhizomes and can be somewhat invasive in the garden."*
-'Symphytum x Uplandicum Variegatum' by Missouri Botanical Garden, Saint Louis, Missouri, **2018**. Founded in 1859; oldest botanical garden in United States.

End of Strains (Cultivars) of Russian Comfrey

<u>**Russian Comfrey Chromosomes Overview**</u>
See subsection 'Comfrey Chromosomes Overview' in section 'Symphytum Species Overview' (Chapter 6).
See subsection 'Chromosomes' in section 'Details about Symphytum Species: Officinale' (Chapter 8).

Hybrid or crossbreed is the result of combining the qualities of two organisms of different breeds, varieties, species or genera through sexual reproduction.
A hybrid swarm is a group of hybrids that have survived past the first hybrid generation, with interbreeding taking place between hybrid individuals and also backcrossing with its parent type. These plants are highly variable, with genes and phenotypes of individuals ranging widely between two parent types. Hybrid swarms blur the boundary between the parent taxa.
A backcross is a cross of a hybrid with one of its parents or an individual genetically similar to its parent. This creates offspring with genetics closer to that of the parent.

Genotype is the genetic constitution of an organism. Phenotype is the observable properties of an organism.

*"**The whole section (Symphytum) Officinalia seems to be a comparium in the sense of *Danser, to which also S. asperum can be added. These experiments made clear that S. asperum and S. officinale form a very complicated***

species complex, not under natural conditions, but by the influence of man."
-'Cytotaxonomic Studies in the Genus Symphytum III: Some Symphytum Hybrids in Belgium and the Netherlands' by Th.W.J. Gadellai and E. Kliphuis, Biologisch Jaarboek (Biological Yearbook), **1971**.
(* -'Ueber die Begriffe Komparium, Kommiskuum und Konvivium und Ueber die Entstehungsweise der Konvivien' {On the Concepts of the Comparium, the Commiscuum and the Convivium and on the Origin of the Convivia} by B.H. Danser, Amsterdam, Netherlands; Genetica: An International Journal of Genetics and Evolution, Volume 11, Issue 5, pages 399-450, September 1929.)
(Taxonomic ranks: Genus, Subgenus, Section, Subsection, Series, Species, Subspecies, Variety, Subvariety, Form, Subform.)
(Comparium is a group of organisms that can interbreed and are equivalent in scope to a taxonomic genus. For Danser in 1929 comparium is a group comprising all coenospecies which may participate in mutual hybridization, whether direct or indirect. Coenospecies is a collection of related species able to hybridize.)

"Some Hybrids Between S. asperum Lepech. and S. officinale L. in Denmark:
Cytological studies in Danish hybrids of species of Symphytum revealed the occurrence of five cytotypes: 2n = 34, 36, 40, 42 and 44.
These hybrids were compared with other natural and with artificially produced hybrids. **All hybrids are either the result of crossing S. officinale (2n = 40, 48) with S. asperum (2n = 32) or the result of backcrossing to the parental species.**
Summarizing, we are of the opinion that the Danish plants are best interpreted as follows:
    a. Cytotype 2n = 34: *Backcross of S. x uplandicum (2n = 36) with the S. asperum (2n = 32) parent.*
    b. Cytotype 2n = 36:
        In most cases, S. officinale (2n = 40) x S. asperum (2n = 32), or the reciprocal cross; exceptionally, backcross of S. x uplandicum (2n = 40) with the S. asperum (2n = 32) parent.
    c. Cytotype 2n = 42: *S. x uplandicum (2n = 36) x S. officinale (2n = 48), or the reciprocal cross.*
    d. Cytotype 2n = 44: *Backcross of S. x uplandicum (2n = 40) with the S. officinale (2n = 48) parent.*
Local hybrid populations are usually more or less uniform in general appearance and colour of the flower, at least in Great Britain, Belgium and the Netherlands, but the differences between plants from different populations may be considerable."
-'Cytotaxonomic Studies in the Genus Symphytum VII: Some Hybrids Between S. asperum Lepech. and S. officinale L. in Denmark' by Th.W.J. Gadella and E. Kliphuis, Proceedings van de Koninklijke Nederlandse Akademie van Wetenschappen (Proceedings of the Royal Netherlands Academy of Arts and Sciences), Section C, Volume 78, pages 182-188, **1975**.
    (A cytotype is an individual of a species that has a different chromosomal factor to another.)
    (Body cells and individuals are described by the number of chromosome sets {ploidy level} they have: monoploid: 1 set; diploid: 2 sets = 2n = 2x, triploid: 3 sets = 2n = 3x, tetraploid: 4 sets = 2n = 4x, pentaploid: 5 sets = 2n = 5x, etc.)
    (A reciprocal cross shows the role of parental sex on a given inheritance pattern. In one cross, a male expressing the trait is crossed with a female not expressing the trait. In the other, a female expressing the trait is crossed with a male not expressing the trait.)

"S. officinale (2n = 24, white flowers) *does not cross with* **S. asperum (2n = 32, blue flowers).**

S. officinale (2n = 48, white or purple flowers) crossing with **S. asperum (2n = 32, blue flowers)** *creates*
S. x uplandicum (2n = 40, pink-blue flowers).

S. officinale (2n = 40, purple flowers, rarely white) crossing with **S. asperum (2n = 32, blue flowers)** *creates* **S. x uplandicum (2n = 36, blue-purple flowers).**

All North American plants of the genus Symphytum that I have seen can be identified with the aid of this key. "
-'Notes on Symphytum (Boraginaceae) in North America' by T.W.J. Gadella, pages 1061-1067 in book: Annals of the Missouri Botanical Garden, Volume 71, Saint Louis, Missouri. Lawrence, Kansas: Allen Press, **1984**.

    Various Cytotypes of Russian Comfrey Chromosomes

    "Both cytotypes 2n = 40 and 2n = 48 of Symphytum officinale are crossable with Symphytum asperum 2n = 32, resulting in hybrids with the number 2n = 36 and 2n = 40 respectively.
        These hybrids must be regarded as Symphytum x uplandicum, in spite of the fact that they differ slightly from each other and have a different origin.
In view of the fact that tetraploid plants (2n = 48) of Symphytum officinale seem to be the more common in Europe, **it is highly probable that the hybrid Symphytum x uplandicum with the chromosome number 2n = 40 is the commonest type of Symphytum x uplandicum.**
Taxonomically it is not easy to give an interpretation of these results. *If the Symphytum x uplandicum hybrids are fertile, the species Symphytum asperum and Symphytum officinale are fully interfertile.*
The clearcut morphological differences indicate without any doubt that the taxa S. officinale and S. asperum represent distinct morphological species. *The ease with which they hybridize (at least in the experimental garden) indicates that* **they belong to the same biospecies (sensu Mayr)."**
-'Cytotaxonomic Studies in the Genus Symphytum II: Crossing Experiments Between Symphytum Officinale L. and

Symphytum Asperum Lepech' by Th.W.J. Gadella and E. Kliphuis, Botanical Museum and Herbarium of the State University of Utrecht, Netherlands, Acta Botanica Neerlandica, Volume 18, No. 4, pages 544-549, August **1969**.
(Biospecies is a group of interbreeding individuals isolated reproductively from other groups.)

"The so called 'genetic relationship' (sensu Mayr) means, that 2 species are regarded the more related, the more similar are their genotypes (Mayr: 'we do not classify phenotypes but genotypes'). Genetic relationship in this sense is nothing but a phenetic relationship, based on the overall similarity of the genome."
-'Glossary of Phylogenetic Systematics with a Critic of Mainstream Cladism' by Gunter Bechly from Boblingen, Germany at Eberhard Karls University of Tubingen, Germany, 1997.
(Ernst Walter Mayr, born in Germany, 1904-2005, was a respected evolutionary biologist, taxonomist, ornithologist, philosopher, and science historian.)

"Quaker, Russian, or blue Comfrey [Symphytum x uplandicum Nyman (Symphytum peregrinum Ledeb.)] is a hybrid (2n = 36) between a cytotype of Common (2n = 40) and Prickly (2n = 32) Comfrey. The species originated from a natural hybrid, but the hybrid has also been made experimentally*. This hybrid has blue, purple, or purple-red flowers and is taller than either parent.
A pink-flowered cytotype (2n = 40) of Quaker Comfrey is a hybrid of Prickly Comfrey (2n = 32) and a cytotype (2n = 48) of Common Comfrey**.
Both hybrids are fertile and breed true from seed**."
-'Comfrey: A Controversial Crop' by Robert G. Robinson, University of Minnesota, Agricultural Experiment Station, Minnesota Report MR-191, Item No. AD-MR-2210, **1983**, page 2.
(* -'Cytotaxonomic Studies in the Genus Symphytum II: Crossing Experiments Between Symphytum Officinale L. and Symphytum Asperum Lepech' by Th.W.J. Gadella and E. Kliphuis, Botanical Museum and Herbarium of the State University of Utrecht, Netherlands, Acta Botanica Neerlandica, Volume 18, No. 4, pages 544-549, August 1969.)
(** -'Comfrey Symphytum spp. as a Forage Crop' by J.C. Forbes, A.D. McKelvie, and P.J.C. Saunders, Herb Abstr. 49:523-539, 1979.)

"S. x uplandicum= (2n = 36) or (2n = 40)."
-'Chemotaxonomy of the Symphytum Officinale Agg. (Boraginaceae)' by Jaarsma, Lohmanns, Gadella and Malingre, Plant Systematics and Evolution, Volume 167, pages 113-127, **1989**.

"Symphytum asperum x Symphytum officinale = Symphytum x uplandicum Nyman (Russian Comfrey) is a very rare Comfrey hybrid in New England (eastern United States).
Two cytotypes of this hybrid are known, depending on which cytotype of S. officinale was involved in the cross. *Thus far, only the **2n = 40** type of this hybrid has been collected in New England **(with a 2n = 48 S. officinale parent)**. This hybrid shows softer stem pubescence, slightly longer leaf base decurrence (on average), and blunter fornices than the **2n = 36** type (which is known from North America)."*
-New England Wild Flower Society's Flora Novae Angliae: A Manual for the Identification of Native and Naturalized Higher Vascular Plants of New England by Arthur Haines. New Haven: Connecticut: Yale University Press, **2011**.

Russian Comfrey Cytotype 2n = 34

"Some Hybrids Between S. asperum Lepech. and S. officinale L. in Denmark:
Plants with the chromosome number 2n = 34 *have many characters in common with S. asperum (e.g., indumentum, middle and upper stem leaves petiolate, short calyx, colour of the corolla and flower bud) but they differ from this species in chromosome number, length of the stem and in fertility.*
It is highly probable that the Danish plants from Langaa (Denmark) originated as a result of backcrossing S. x uplandicum 2n = 36 to S. asperum (2n = 32). *Not only the chromosome number points in this direction, but also the combination of morphological characters of the hybrid."*
-'Cytotaxonomic Studies in the Genus Symphytum VII: Some Hybrids Between S. asperum Lepech. and S. officinale L. in Denmark' by Th.W.J. Gadella and E. Kliphuis, Proceedings van de Koninklijke Nederlandse Akademie van Wetenschappen (Proceedings of the Royal Netherlands Academy of Arts and Sciences), Section C, Volume 78, pages 182-188, **1975**.

Russian Comfrey Cytotype 2n = 36

"Some Hybrids Between S. asperum Lepech. and S. officinale L. in Denmark:
One type of hybrid 2n = 36:
*The first mentioned hybrid (2n = 36) is the result of **hybridization between S. asperum (2n = 32) and the 2n = 40 form of S. officinale.***
Another type of hybrid 2n = 36:
*The other plants with the chromosome number n = 36 originating from Oksby (Denmark) presented taxonomic problems. **It seems highly probable that these plants (2n = 36) are backcrosses of S. x uplandicum (2n = 40) with S. asperum (2n = 32).***

The most typical differences between these plants and the hybrid S. x uplandicum 2n = 36 are: the length of the stem, the colour of the flowers and the flower buds and the length of the calyx."
-'Cytotaxonomic Studies in the Genus Symphytum VII: Some Hybrids Between S. asperum Lepech. and S. officinale L. in Denmark' by Th.W.J. Gadella and E. Kliphuis, Proceedings van de Koninklijke Nederlandse Akademie van Wetenschappen (Proceedings of the Royal Netherlands Academy of Arts and Sciences), Section C, Volume 78, pages 182-188, **1975**.

"Symphytum L. Chromosome Numbers:
  x uplandicum Nyman   36   Finland   *Vaarama 1948
This book has grown out of the requests by many users of 'Flora Europaea' (Tutin et al., 1964, 1968, 1972, 1976, 1980) for information on the sources of the chromosome numbers cited in that work."
-Flora Europaea: Check-List and Chromosome Index by D.M. Moore. Cambridge, England: Cambridge University Press, **1982**, page 179.
(* -'Chromosome Numbers of Northern Plant Species' book by Askell Love and Doris Love. Reykjavik, Iceland: Ingolfsprent, 1948. Series: Rit Landbunaoardeildar, B-flokkur, Nr. 3. Part by A. Vaarama, Univ. Inst. Appl. Sci. 405 Rep. Dep. Agric, Ser. B 3: 9-131. If you have Symphytum pages in English, I would appreciate a copy.)

Russian Comfrey Cytotype 2n = 40

"Some Hybrids Between S. asperum Lepech. and S. officinale L. in Denmark:
Plants with chromosome number 2n = 40 are hybrids between S. officinale (2n = 48) and S. asperum (2n = 32)
(see also Gadella and Kliphuis, 1973). They have been introduced into western and northwestern Europe for fodder.
The Danish plants (2n = 40), *were compared with other material from Belgium, Great Britain and with artificially produced hybrids that are cultivated in the botanical garden of Utrecht (The Netherlands). (See also Gadella and Kliphuis, 1971, 1974).*
They turned out to be exactly the same and should be assigned to S. x uplandicum Nym."
-'Cytotaxonomic Studies in the Genus Symphytum VII: Some Hybrids Between S. asperum Lepech. and S. officinale L. in Denmark' by Th.W.J. Gadella and E. Kliphuis, Proceedings van de Koninklijke Nederlandse Akademie van Wetenschappen (Proceedings of the Royal Netherlands Academy of Arts and Sciences), Section C, Volume 78, pages 182-188, **1975**.

Russian Comfrey Cytotype 2n = 42

"Some Hybrids Between S. asperum Lepech. and S. officinale L. in Denmark:
The plants with 2n = 42 were compared with artificial hybrids between the tetraploid form of S. officinale (2n = 48) and S. x uplandicum (2n = 36).
These hybrids, that are easily produced, are somewhat variable, depending on the characters of the S. officinale (2n = 48) parent. The flower-colour of the artificially produced 2n = 42 hybrids is somewhat darker if the officinale parent has purple flowers than if this parent has white flowers.
At any rate, we have several artificially produced hybrids (2n = 42) in our collection that exactly match the Danish plants from Lillebraende (Denmark).
The influence of the S. officinale parent is clearly visible in the decurrence of the leaves, which is usually absent in S. x uplandicum hybrids (both 2n = 36 and 2n = 40), and in the fact that the middle and upper stem leaves are not provided with a distinct petiole."
-'Cytotaxonomic Studies in the Genus Symphytum VII: Some Hybrids Between S. asperum Lepech. and S. officinale L. in Denmark' by Th.W.J. Gadella and E. Kliphuis, Proceedings van de Koninklijke Nederlandse Akademie van Wetenschappen (Proceedings of the Royal Netherlands Academy of Arts and Sciences), Section C, Volume 78, pages 182-188, **1975**.

Russian Comfrey Cytotype 2n = 44

*"**The plants with 2n = 44 were presumed to be hybrids between S. officinale (2n = 48) and S. x uplandicum (2n = 40), i.e. backcrosses of S. x uplandicum with the officinale parent.**
In a previous paper (Gadella and Kliphuis 1971), we described some Dutch plants with the chromosome number 2n = 44, collected in the province of Limburg, the Netherlands in the vicinity of Gulpen. These plants were practically sterile in nature and produced very few ripe seeds.
From close comparison of the Danish plants with the Dutch material and with artificially produced backcrosses between S. x uplandicum and S. officinale, **we conclude that the Danish plants perhaps arose also by such backcrosses.**
Study of morphological characters clearly shows S. officinale (2n = 48) traits: the leaves are decurrent but not along the entire internode, and the sepal lobes are very long (one half the corolla length)."
-'Cytotaxonomic Studies in the Genus Symphytum VII: Some Hybrids Between S. asperum Lepech. and S. officinale L. in Denmark' by Th.W.J. Gadella and E. Kliphuis, Proceedings van de Koninklijke Nederlandse Akademie van Wetenschappen (Proceedings of the Royal Netherlands Academy of Arts and Sciences), Section C, Volume 78, pages 182-188, **1975**.

Russian Comfrey (2n = 36) vs (2n = 40)

"Some Hybrids Between S. asperum Lepech. and S. officinale L. in Denmark:
Plants of this cytotype (2n = 40) flower over a considerably shorter time than those with 2n = 36. Another difference is that the latter (2n = 36) prefers moist habitats (ditches, banks of small streams), whereas plants of 2n = 40 seem to prefer drier conditions.
Both plants with 2n = 40 and 2n = 36 always grow in the vicinity of human settlements, and we conclude that all plants studied, originally escaped from cultivation.
The studies by Faegri (*1931) also made clear that two forms of S. x uplandicum (incorrectly assigned by him to S. peregrinum Ledeb.) can be recognized:
    First (2n = 40) has blue flowers and somewhat decurrent stem leaves.
    Second (2n = 36) has purple flowers and nondecurrent leaves.
The first form corresponds to the cytotype 2n = 40, and the second one to the cytotype 2n = 36 of S. x uplandicum."
-'Cytotaxonomic Studies in the Genus Symphytum VII: Some Hybrids Between S. asperum Lepech. and S. officinale L. in Denmark' by Th.W.J. Gadella and E. Kliphuis, Proceedings van de Koninklijke Nederlandse Akademie van Wetenschappen (Proceedings of the Royal Netherlands Academy of Arts and Sciences), Section C, Volume 78, pages 182-188, **1975**.
(* -'Uber in Skandinavien Gefundenen Symphytum-Arten: Nebst einigen Betrachtungen uber das Artproblem Innerhalb der Betreffenden Artgruppe' {About Symphytum Species Found in Scandinavia} by Knut Faegri, Bergens Museum Arbok, Norway, 47 pages, 1931. In German. If anyone has an English translation of it, I would appreciate having it.)

"S. x uplandicum Nyman (2n = 36):
*Stem to 130 cm (51 inches) long, **not decurrent, rough to the touch,** provided with scabrous hairs that are deciduous in older stems; scabrous hairs with a tubercular base; basal leaves elliptic-lanceolate with an acuminate apex and a rounded more or less cordate base; lamina of basal leaves 15-30 cm (5.9-11.8 inch) long and to 6 cm (2.3 inch) wide, petiole to 12 cm (4.7 inch) long; stem leaves smaller, often with winged petiole, the uppermost ones nearly sessile with a cuneate base; indument of the adaxial side of the leaves sometimes as in S. officinale (2n = 40), sometimes as in S. x uplandicum (2n = 40); abaxial side of the leaves with some tubercular based hairs along the veins and otherwise glabrous.*
*Length of the calyx up to 5.5 mm, calyx divided to 3/5 of its length, triangular-lanceolate and obtuse at the apex; calyx lobes with stiff marginal setae with a regular distribution pattern; setae usually without a tubercular base; corolla slightly campanulate 13-16 mm, **purple in bud, blue-purple in flower;** squamae of the corolla 7-7.5 mm long, 2 mm wide at the base, triangular-lanceolate, apex broad and rotundate; papillae acute and densely crowded at the middle of the scale margin; stamens 5-6 mm, shorter than the scales; anthers longer than the filaments, connective not projecting beyond the thecae.*
***Fruit brown and dull,** areolate-granulate, 3-4 mm long and to 3 mm wide at the base.*
Reproduction strictly allogamous (must cross fertilize), plants self-incompatible.
Most plants show a reduced fertility, but some are as fertile as the parental species."
-'Notes on Symphytum (Boraginaceae) in North America' by T.W.J. Gadella, pages 1061-1067 in book: Annals of the Missouri Botanical Garden, Volume 71, Saint Louis, Missouri. Lawrence, Kansas: Allen Press, **1984**.

"S. x uplandicum Nyman (2n = 40):
*Stem to 140 cm (55 inches) **long, soft to the touch,** setae neither prickly, nor scabrid; if the setae have a tubercular base, this is always small and deciduous, never broad and conspicuous as in S. officinale (2n = 40); leaf decurrence sometimes present, but usually not longer than 1 cm (0.3 inch) along the internode; shape and size of leaves as in S. x uplandicum (2n = 36); adaxial side of the leaves with many appressed setae, the majority of which have a small tubercular base; these setae are not deciduous; indumentum of abaxial side of leaf as in S. x uplandicum (2n = 36).*
*Calyx to 4 mm long, divided to 3/5 of its length; calyx lobes triangular-lanceolate and sub-acute or acute; stiff marginal setae irregularly distributed; corolla 13-15 mm long, slightly campanulate, **pink or pinkish blue in flower;** squamae of the corolla with acute papillae, regularly distributed along the scale margin; stamens as in S. x uplandicum (2n = 36).*
Reproduction strictly allogamous; plants self-incompatible. Many plants show a reduced fertility and produce only a few nutlets; sometimes plants as fertile as the parental species."
-'Notes on Symphytum (Boraginaceae) in North America' by T.W.J. Gadella, pages 1061-1067 in book: Annals of the Missouri Botanical Garden, Volume 71, Saint Louis, Missouri. Lawrence, Kansas: Allen Press, **1984**.

*"**S. x uplandicum Nyman**: Hybrids form a range of intermediates between the parents in characters of the leaves and flowers. The leaves generally lack the broadly decurrent leaf bases of S. officinale but they are never cordate and petiolate as in S. asperum.*
The calyx, in bud, is 5-7 mm long with acute segments, not 3-5 mm with obtuse segments expanding rapidly in fruit as in S. asperum. The flower-colour varies from reddish-purple to violet, but is never sky blue as in S. asperum.
The following two forms may be recognised but are only part of the range which occurs:
    a) 2n = 36.

Tall, up to 1.3 meters (4.2 feet), rough and prickly. The leaves are not, or only very slightly and very shortly, decurrent. **The flower-buds are dark purple, changing as they open to purple or violet.** The calyx may become swollen and hispid in fruit. Less variable than the following; often forming large populations on roadsides and spreading vegetatively.
b) 2n = 40.
Tall, up to 1.4 meters (4.5 feet), usually less rough and prickly than 2n = 36 form. The leaves are sometimes decurrent up to half the distance to the next leaf below. **The flower-buds are pink or red-pink, changing on opening to a range of shades from pink to pinkish-blue.**
2n = 40 is probably the commonest Symphytum in this complex: widely naturalised and very variable, back-crossing with the various colour forms of S. officinale with 2n = 48.
Reference: Perring, F. H. (1994). Pages 64-70 in 'The Common Ground of Wild and Cultivated Plants', National Museum of Wales, Cardiff, Wales."
-Plant Crib: Handbook for Field Identification compiled by T.C.G. Rich and M.D.B. Rich with editorial assistance of F.H. Perring for the BSBI Monitoring Scheme. London, England: Botanical Society of the British Isles, 1988, page 74. There is also an updated **1998** edition.
(A species 'complex' is a group of closely related species that are extremely similar in appearance, so much so that the boundaries between them are frequently unclear.)

Russian Comfrey Alkaloids
See the sections 'Alkaloids in Comfrey' in Volume 2.
See subsection 'Alkaloids' in section 'Details about Symphytum Species: Officinale' (Chapter 8).
See subsection 'Alkaloids' in section 'Details about Symphytum Species: Asperum or Asperrimum' (Chapter 7).

"In 1/5 of the diploid plants (2n = 24) of S. officinale echimidine was found. **The amount of echimidine found in S. officinale, however, is significantly lower then in S. x uplandicum and S. asperum where all plants contained echimidine** (data not published)."
-'Chemotaxonomy of the Symphytum Officinale Agg. (Boraginaceae)' by Jaarsma, Lohmanns, Gadella and Malingre, Plant Systematics and Evolution, 167, pages 113-127, **1989**.

"The presence of triterpene and pyrrolizidine alkaloids was employed by Gadella et al. (1983) and Huizing et al. (1982, 1983) to elucidate the hybrid origin of Symphytum x uplandicum from S. asperum and S. officinale. **These authors found that the alkaloid echimidine is present in S. asperum (Prickly Comfrey) and S. x uplandicum (Russian Comfrey) but not in any of the cytotypes of S. officinale (Common Comfrey).**
They also isolated the triterpene isobauerenol from S. officinale and the hybrid, but not from S. asperum.
However, Jaarsma et al.* found that such a sharp chemical differentiation is not always absolute, and there are some anomalies in the chemical profiles within the S. officinale complex. **In fact about 25 percent of the samples of S. officinale were found to contain traces of echimidine.** As many as seven alkaloids have been isolated from S. asperum (**Roitman) and eight from S. x uplandicum (***Culvenor et al., 1980)."
-'Journal of the Arnold Arboretum, Supplementary Series, Volume 1, Harvard University, Cambridge, Massachusetts, **1991**. This volume is a collection of contributions toward a 'Generic Flora of the Southeastern United States'. 'The Genera of Boraginaceae in the Southeastern United States' by Ihsan A. Al-Shehbaz, page 152.
(* -'Chemotaxonomy of the Symphytum Officinale Agg. {Boraginaceae}' by Jaarsma, Lohmanns, Gadella and Malingre, Plant Systematics and Evolution, 167, pages 113-127, 1989.)
(** -'Comfrey and Liver Damage' by James N. Roitman, The Lancet, Volume 317, Issue 8226, page 944, April 25 1981.)
(*** -'Structure and Toxicity of the Alkaloids of Russian Comfrey {Symphytum x uplandicum, Nyman}, a Medicinal Herb and Item of Human Diet' by Claude C.J. Culvenor, M. Clarke, J.A. Edgar, J.L. Frahn, M.V. Jago, J.E. Peterson and L.W. Smith, CSIRO, Division of Animal Health, Victoria, Australia; Experientia, Switzerland, Volume 36, No. 4, pages 337-379, 1980.)

"*Symphytum x uplandicum Nyman (synonym S. peregrinum Ledeb.):*
Russian Comfrey (German: Russischer Beinwell, French: Consoud de Russe), is a hybrid generated from Symphytum officinale L. and S. asperum Lepech.
It originates in Caucasia and has become widespread in Central Europe, England, Canada, United States, Australia, New Zealand, Japan, Kenya and South Africa. Under the designation of Russian Comfrey it is not only cultivated as medicinal plant but also on a large scale as vegetable, fodder and fertilizer plant.
Roots and leaves contain intermedine, lycopsamine, the 7-acetyl derivatives of these alkaloids, uplandicine, symlandine, symviridine, myoscorpine, symphytine, the major alkaloid echimidine, and the N-oxides of these compounds in different concentrations.
A total alkaloid content of about 0.2% was found in the dried aerial parts. Russian Comfrey should no longer be used for medicinal purposes."
-'Medicinal Plants in Europe Containing Pyrrolizidine Alkaloids' by Erhard Thomas Roeder, Pharmazeutisches Institut (Pharmaceutical Institute) der Rheinischen Friedrichs-Wilhelms, University of Bonn, Germany, Pharmazie (Pharmacy) 50, pages 83-98, March **1995**.

"Symphytum x uplandicum is distinguished from S. officinale (Common Comfrey) by containing echimidine which is thought to be the most toxic PA (Pyrrolizidine Alkaloid) in Symphytum spp. (species).
S x uplandicum has a markedly higher overall level of alkaloids which also makes the use of S. officinale preferable."
-'Using Herbs that Contain Pyrrolizidine Alkaloids' by Alison Denham, B.A., MNIMH (National Institute of Medical Herbalists), University of Central Lancashire, England; The European Journal of Herbal Medicine, Volume 2, No. 3, pages 27-38, **1996**.

Creating New Hybrids of Russian Comfrey

"Artificial hybrids have been produced in the Netherlands from Dutch material.
Hybrids with purple flowers and 2n = 36 were made by crossing S. officinale subsp. uliginosum (Kerner) Nyman (2n = 40) and S. asperum (2n = 32).
Other hybrids with pink flowers and 2n = 40 were made by crossing S. officinale (2n = 48) and S. asperum.
Attempts to cross S. officinale (2n = 24) and S. asperum have not been successful."
-Hybridization and the Flora of the British Isles edited by Clive A. Stace, University of Leicester, England in collaboration with 'Botanical Society of the British Isles' now called 'Botanical Society of Britain and Ireland'. London, England: Academic Press, **1975**, page 354.

"As a rough guide to those who are 'roguing' Russian Comfrey mixtures, it should be said that **any Comfrey which cannot average 25 tons an acre is not worth growing**, *and anyone who gets less yield than that should either improve cultivation methods, or get rid of the crop.*
It is unlikely that any variation on the themes of these varieties of Russian Comfrey will turn out to be remarkably higher yield, *and this is all that is likely to happen from cross- or self-pollinating hybrids.*
Real differences in yield can only come from the hybrid vigour of a bi-generic cross, such as the one between S. asperrimum (Prickly Comfrey) and S. officinale (Common Comfrey).
We have tried this on several occasions, and so have our members of Henry Doubleday Research Association, but it needs real skill with a camel hair brush and in breaking the flowers apart. This would involve growing hundreds of seedlings for at least three years to bring them into full production so that their yield could be measured. A 'miracle Comfrey' would also need analysis to see how the change had affected its mineral and vitamin content.
Some years ago a cross between S. caucasium and S. asperrimum was obtained from the U.S.S.R. and grown on the H.D.R.A. (Henry Doubleday Research Association) trial ground, but it proved no more productive than ordinary S. asperrimum, for not all first crosses produce hybrid vigour.
Seed saved from hybrids, whenever it is set, produces variations on the parents."
-Comfrey: Fodder, Food and Remedy by Lawrence D. Hills. New York: Rizzoli Universe Books: **1976**, pages 75-76.
(Roguing is identifying and removing plants with undesirable characteristics so the crop is improved.)

"Knowing the wide variations that Lawrence Hills found in his Russian Comfrey mixtures, **I collected samples** *of as many collections as I could find when I farmed in the north of Scotland prior to moving to Portugal, with a view to trying to find something that might outyield the Nos. 4 and 14 I already had, along with some S. officinale.*
I failed to find one in the collections I acquired. I did not have many years to experiment, up to about 7, but **with only around 100 Comfrey plants it was not difficult to spot low yielders and anything that looked promising. Nothing I had acquired was near the No. 4s and No. 14s for yield."**
-'Comfrey' by Old McDonald in Portugal: Farming, Gardening, Wildlife and Good Food,
http://oldmcdonaldinportugal.blogspot.com, November 23 **2012**.

Russian Comfrey Breeds with Other Symphytum Species

*"**Hybrids vary considerably in fertility**, but British populations are undoubtedly derived from material which was introduced as a crop and has spread into semi-natural habitats vegetatively, as one of its parents, S. asperum, is an extremely rare introduction."*
-Hybridization and the Flora of the British Isles edited by Clive A. Stace, University of Leicester, England in collaboration with 'Botanical Society of the British Isles' now called 'Botanical Society of Britain and Ireland'. London, England: Academic Press, **1975**, page 353.
(Hybrid or crossbreed is the result of combining the qualities of two organisms of different breeds, varieties, species or genera through sexual reproduction.)

*"**The original Russian Comfrey hybrid was sterile, or semi-sterile; the stamens are shut off by a kind of 'false bottom' in the flower** which is (by a freak that may occur perhaps once in every 100,000 flowers) sometimes available for bees to transfer as far as it will go among the neighboring blooms.* **The pistils are, however, always open to receive pollen from our native Common Comfrey** *(also found in Europe, including Russia).*
Russian Comfrey may chance to make seed, as it is possible that once in 10,000 chances a bumble bee (not the honey bee, which cannot penetrate to fertilize the flowers) may cross pollinate."
-Comfrey: Nature's Healing Herb & Health Food by Andrew Hughes. Japan: Sanyusha Publishing Co., Ltd, **1992**, page 137, 159.
(The stamen is the male part of a flower that produces pollen to fertilize the female part. The pistil is the female

part; it produce ovules that develop into fruit and seeds after pollination.)

"Tutin* did find S. x uplandicum growing abundantly and hybridising freely with S. officinale (Common Comfrey).
(backcrossing)
Although there is great variation displayed between one colony of S. x uplandicum and another, none of them shows any variation in the direction of S. asperum (Prickly Comfrey). *So far as my knowledge goes there is no record of S. x uplandicum having been found growing in this country in the company of this parent species.*
In spite of its hybrid origin, S. x uplandicum is apparently a fixed hybrid and any variation it shows is due to its backcrossing with S. officinale and more rarely, and perhaps less certainly, with S. asperum. *Populations of S. x uplandicum known to me, unless growing in close proximity to S. officinale, show remarkable uniformity."*
-'The History of Symphytum Asperum Lepech. and S. x Uplandicum in Britain' by A.E. Wade, Department of Botany, National Museum of Wales, Watsonia, Volume 4, pages 117-118, **1958**. (Watsonia was the journal of the 'Botanical Society of Britain and Ireland' from 1949 to 2010. It is now called 'New Journal of Botany'.)
(* -'The Genus Symphytum in Britain' by T.G. Tutin, University College of Leicester, England; Watsonia: Journal of the Botanical Society of the British Isles, Volume 3, pages 280-281, February 1956.)
    (A backcross is a cross of a hybrid with one of its parents or an individual genetically similar to its parent.
    This creates offspring with genetics closer to that of the parent.)

"Symphytum asperum Lepech., S. uliginosum Kern., S. officinale L.:
These 3 species hybridize and form hybrid swarms consisting of F1 and backcross hybrids *(*,**,***).*
    Symphytum officinale is cytologically heterogenous (diverse):
        $2n = 24$, $2n = 40$, $2n = 48$, *cytotypes occur in various parts of Europe.*
    Symphytum asperum ($2n = 32$) does not hybridize with the diploid ($2n = 24$) form of S. officinale in
    Europe but produces hybrids and backcrosses with the $2n = 40$ and $2n = 48$ cytotypes of S. officinale.
The primary hybrids, with $2n = 36$, or $2n = 40$, are collectively known under the name S. x uplandicum Nyman.
In Europe the parental species are largely allopatric with a very small zone of overlap in the northwest Caucasus (Kusnetosov, 1910). In the zone of overlap, hybridization does not occur because S. officinale and S. asperum grow at different altitudes. Apparently the hybrids arose outside the Caucasus (****Tutin 1956; *****Wade 1958).
In many parts of western, northwestern, or central Europe the hybrid swarms are more common than the parental species."
-'Notes on Symphytum (Boraginaceae) in North America' by T.W.J. Gadella, pages 1061-1067 in book: Annals of the Missouri Botanical Garden, Volume 71, Saint Louis, Missouri. Lawrence, Kansas: Allen Press, **1984**.
(* -'Cytological and Hybridization Studies in the Genus Symphytum' by Th.W.J. Gadella, Symposia Biologica Hungarica {Hungary Biological Symposium}, Vol 12, pages 189-199, 1972. I was unable to find this. If you have it in English, could you send it.)
(** -'Cytotaxonomic Studies in the Genus Symphytum V: Some Notes on W. European Plants with the Chromosome Number 2n=40' by Th.W.J. Gadella and E. Kliphuis, Botanische Jahrbücher fur Systematik {Botanical Yearbooks for Systematics}, Volume 93, pages 530-538, 1973.)
(*** -'Cytotaxonomic Studies on the Genus Symphytum VIII: Chromosome Numbers and Classification of Ten European Species' by Th.W.J. Gadella and E. Kliphuis, Proceedings van de Koninklijke Nederlandse Akademie van Wetenschappen {Proceedings of the Royal Netherlands Academy of Arts and Sciences} Section C, Volume 81, pages 162-172, 1978.)
(**** -'The Genus Symphytum in Britain' by T.G. Tutin, University College of Leicester, England; Watsonia: Journal of the Botanical Society of the British Isles, Volume 3, pages 280-281, February 1956.)
(***** -'The History of Symphytum Asperum Lepech. and S. x Uplandicum in Britain' by A.E. Wade, Department of Botany, National Museum of Wales, Watsonia, Volume 4, pages 117-118, 1958. Watsonia was the journal of the 'Botanical Society of Britain and Ireland' from 1949 to 2010. It is now called 'New Journal of Botany'.)
    (A hybrid swarm is a group of hybrids that have survived past the first hybrid generation, with interbreeding taking place between hybrid individuals and also backcrossing with its parent type. These plants are highly variable, with genes and phenotypes of individuals ranging widely between the two parent types. Hybrid swarms blur the boundary between the parent taxa.)
    (An F1 hybrid or 'Filial 1' hybrid is the first generation of offspring of distinctly different parental types that produce a new, uniform phenotype {observable properties of an organism} with a combination of characteristics from the parents. Mules are F1 hybrids between horse and donkey. This can happen naturally, and includes hybrids between species, for example, peppermint is an F1 hybrid of watermint and spearmint. In agronomy, these F1 hybrids are usually created by means of controlled pollination, sometimes by hand-pollination. This is not GMO. A Genetically Modified Organism has been altered using genetic engineering techniques in a sophisticated laboratory, frequently using genes from unrelated species.)
    (Allopatric or geographic speciation occurs when populations of the same species become isolated from each other, and this prevents genetic exchange.)

"S. x uplandicum is fertile and backcrosses to S. officinale, forming a spectrum of intermediates."
-New Flora of the British Isles: Identification of Wild Vascular Plants of the British Isles edited by Clive Stace. Cambridge, England: Cambridge University Press, **1991**, page 647.

"(S. x uplandicum) x (S. tuberosum): This cross occurs near the parents in very scattered localities in England and Scotland; **it is intermediate in all characters, with yellow corollas tinged with blue or purple,** *and tuberous rhizomes, and is at least*

partially sterile; endemic (native).
Some plants might be S. officinale x S. tuberosum."
-New Flora of the British Isles: Identification of Wild Vascular Plants of the British Isles edited by Clive Stace. Cambridge, England: Cambridge University Press, **1991**, page 647.

"The native and naturalized Symphytum taxa of the group of S. officinale and S. asperum in the urban area of Aachen (North Rhine Westphalia, Germany) are presented and displayed.
Apart from the typical S. uplandicum, two other largely independent taxa are distinguished. Based on morphological features, these two taxa fall between S. asperum and S. officinale and probably arose from backcrossing of S. uplandicum with its parents.
Symphytum uplandicum s.l. (sensu lato) is a very complex hybrid between S. asperum and S. officinale s.st. (sensu stricto) and according to Basler (*1972), in addition to intermediate types, also includes siblings approximated to parents.
In this work, an attempt is made to give an overview of the wealth of forms of S. uplandicum in the Aachen urban area.
Basler (1972) sees a large variability of S. uplandicum, so that backcrossings are hardly distinguishable from S. asperum and in case of doubt make a chromosome counting necessary. According to Basler (1972: 528), Symphytum uplandicum 'is in many cases scarcely distinguishable from backcrossings with one of the parents'.
The morphologically distinguishable clans correspond approximately to the groups resulting from chromosome numbers by Basler (1972): primary hybrids or plants with identical numbers of chromosomes (2n = 40) and backcrossings with the parents. Gadella & Kliphuis (1969) carried out cytological examinations in southern Limburg (southeast Netherlands), 15-20 kilometers (9.3-12.4 miles) west of Aachen, Germany. At Wijlre (in Limburg province) they found **plants with 2n = 43-44 with an intermediate morphology between typical S. uplandicum (2n = 40) and S. officinale***.*
In Epen (in Limburg province) they were able to detect **plants with 2n = 36, whose appearance was intermediate between S. asperum and typical S. uplandicum***. Thus, both backcrossings are cytologically detected in the vicinity of the study area."*
-'Symphytum Bohemicum, S. Officinale s. str. (sensu stricto), S. x Rakosiense und S. Uplandicum s.l. (sensu lato) im Aachener Stadtgebiet' by F. Wolfgang Bomble, Die Veroffentlichungen des Bochumer Botanischen Vereins e. V. (The Publications of the Bochum Botanical Association), Germany, Volume 5, No. 5, pages 44-60, **2013**. In German except abstract is also in English. If you have an English translation, I would appreciate a copy of it.)
(* -'Cytotaxonomic Studies on the Boraginaceen Genus Symphytum L.: Studies on Predominantly Northern German Plants of the Species S. Asperum Lepech., S. officinale L. and S. x Uplandicum Nym.' by Armin Basler, Botanische Jahrbucher fur Systematik: Pflanzengeschichte und Pflanzengeographie {Botanical Yearbooks for Systematics: Plant History and Plant Geography}, Volume 92, pages 508-553, 1972. All in German. I was unable to find this. If you have English translation, could you send it to me.)
(sensu stricto = s.s., s. str., sens. str., sens. strict. In the strict/narrow sense. It is added after a taxon to mean it is being used in the sense of the original author, or without taxa which may otherwise be associated with it.
sensu lato = s.l., sens. lat. In the broad sense. Used in taxonomy to clarify the scope of a taxon when it has been used to define more than one set of lower-level taxons. It includes all its subordinate taxa and/or other taxa that at other times are considered as distinct.)

"S. x uplandicum (Russian Comfrey) sometimes back-crosses with this parent (Symphytum officinale) (*Perring, 1994)."
-'Online Atlas of the British and Irish Flora', https://www.brc.ac.uk/plantatlas, Distribution and ecology with photographs, **2018**.
(* -The Common Ground of Wild and Cultivated Plants: B.S.B.I. Conference Report No. 22 by A. Roy Perry and R.G. Ellis, {F.H. Perring, pages 64-70}. Cardiff, Wales: Department of Biology, National Museum of Wales, 1994.)

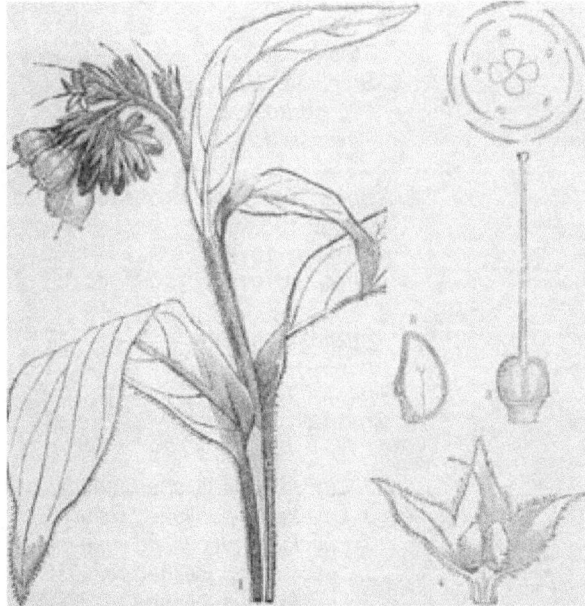

Symphytum Officinale 1846

'The Vegetable Kingdom, Or The Structure, Classification and Uses of Plants' book by John Lindley, Ph.D., F.R.S., Professor of Botany, University of London, England, 1846.

Fig. CCCCXXXIX, Page 655
1. Symphytum officinale
2. A diagram of its flower
3. Its pistil
4. The calyx opened, with two of the nuts remaining
5. A vertical section of a nut

Chapter 11

Details about Symphytum Species Hybrids:
Not Russian Comfrey

This chapter on Symphytum Hybrids:

<u>Hidcote Blue and Pink:</u>
    S. officinale x S. asperum x S. grandiflorum = S. uplandicum x S. grandiflorum

<u>Hybrids with S. officinale</u> (excludes Hidcote and Russian Comfrey)
    S. officinale x S. asperum x S. tuberosum = S. uplandicum x S. tuberosum
    S. officinale x S. bohemicum
    S. officinale x S. cordatum
    S. officinale x S. peregrinum
    S. officinale x S. tuberosum

<u>Other Hybrids:</u>
    S. tuberosum x S. cordatum
    S. tuberosum x S. bulbosum
    S. asperum x S. caucasicum
    S. asperum x S. orientale = S. norvicense
    S. caucasicum x S. orientale

See subsection 'Hybrids, Hybrid Swarms, Introgression' in section 'Symphytum Species Overview' (Chapter 6). For 'Goldsmith' see S. grandiflorum in Chapter 9.

<u>Overview of Symphytum Hybrids</u>

*"**Interspecific (between different species) hybridization appears to be one of the main speciation processes in the evolution of Symphytum.** Therefore a comparative study of maternally inheriting chloroplast DNA together with biparentally inheriting nuclear DNA data, may be a promising route to reconstruct the speciation processes and arrive at more complete conclusions on the evolutionary history of Symphytum species.*
Hybrids known in Symphytum: *In Symphytum a large number of hybrid plants has been described.*
A. Naturally occurring interspecific hybrids:

S. asperum x S. caucasicum	*Popov 1953
S. asperum x S. officinale = S. x uplandicum Nym.	**Smejkal 1978
S. bohemicum x S. officinale = S. x rakosiense (Soo) Penzes	Smejkal 1978
S. bulbosum x S. tuberosum = S. x bicknelli Buckn.	Bucknall 1913
S. cordatum x S. officinale = S. x polonicum Blocki ex Buckn.	***Pawlowski 1961, Bucknall 1913
S. cordatum x S. tuberosum = S. x ullepitschii Wettst.	Smejkal 1978, Bucknall 1913
S. grandiflorum x S. tuberosum	****Stearn 1985
S. grandiflorum x S. x uplandicum = S. 'Hidcote Blue'	Stearn 1985
S. officinale x S. nodosum = S. x foliosum Rehm.	Smejkal 1978
S. officinale x S. orientale = S. x ferrariense Massal.	*****Kurtto 1981
S. officinale x S. orientale = S. floribundum Shuttlew. ex Buckn. (synonym = S. mediterraneum Schulz non Koch.)	Pawloski 1971, Kurtto 1981
S. officinale x S. floribundum = S. x hyerense Pawl.	Buckn1913,Pawl 1971,Kurtto 1981
S. officinale x S. tuberosum = S. x foliosum Rehm.	Popov 1953
S. officinale x S. tuberosum = S. wettsteinii Senholz	Bucknall 1913
S. peregrinum x S. officinale = S. x discolor (synonym: S. x coeruleum Petitmengin)	Bucknall 1912
S. tuberosum x S. officinale = S. x zahlenbruckenzi Sennholz	Bucknall 1913
S. tuberosum x S. x uplandicum	Stearn 1985
S. caucasicum x S. orientale	Stearn 1985
S. creticum x S. circinale	******Runemark 1967

B. Artificial interspecific hybrids from experimental garden:

S. ibericum x S. officinale	J. Luyckx (unpublished results)
S. officinale x S. asperum	J. Luyckx (unpublished results)
S. asperum x S. officinale	J. Luyckx (unpublished results)
S. tuberosum x S. ottomanum	J. Luyckx (unpublished results)
S. ottomanum x S. tuberosum	J. Luyckx (unpublished results)"

-'Phylogenetic Relationships in the Genus Symphytum L. (Boraginaceae)' by J.M. Sandbrink, J. Van Brederode and T.W.J. Gadella, Department of Genome Evolution, University of Utrecht, Netherlands; Proceedings: Koninklijke Nederlandse Akademie van Wetenschappen (Royal Netherlands Academy of Sciences), Series C, Volume 93, No. 3, pages 295-334, **1990**.

(* -'Flora of the U.S.S.R., Volume 19' edited by V.L. Komarov, Leningrad, Russia, 1953. The Borage family part was written by Mikhail Grigorevich Popov, pages 73 to 508 in English version, and pages 97 to 692 in Russian version. Symphytum is pages 207-216 in English, and pages 279-291 in Russian.)

(** -'The Genus Symphytum in Czechoslovakia' {Rod Symphytum v Ceskoslovensku} by M. Smejkal; Zpravy Ceskoslovenske Botanicke Spolecnosti {Czech Botanical Society News}, Prague, Czech Reupublic, Volume 13, pages 145-161, 1978. In Czech. If you have an English translation, I would appreciate a copy.)

(*** -'Observations ad {on} Genus Symphytum L. Pertinentes' by Bogumilus Pawlowski, Fragmenta Floristica et Geobotanica, Poland, Volume 7, No. 2, pages 327-356, 1961. All in Polish. If you have an English translation, I would appreciate a copy.)

(**** -'The Greek Species of Symphytum Boraginaceae' by William T. Stearn, British Museum of Natural History, London, England; Annales Musei Goulandris, Greece, Volume 7, pages 175-220, 1985 or 1986, different sources give different dates, even the pdf has both dates.)

(***** -'Taxonomical Status of Symphytum Floribundum and S. x Ferrariense (Boraginaceae)' by Arto Kurtto, Botanical Museum, University of Helsinki, Finland; Annales Botanici Fennici (Finnish Botanical Annals), Helsinki, Volume 18, No. 1, pages 13-21, 1981.)

(****** -'Studies in the Aegean Flora XI: Procopiana {Boraginaceae} included into Symphytum' by Hans Runemark, Botaniska Notiser {Botanical Notes}, Sweden, Volume 120, pages 84-94, 1967. I was unable to get this report. If you have it, could you please send it to me.)

(Speciation is the formation of new and distinct species in the course of evolution.)

Hidcote Hybrid S. officinale x S. asperum x S. grandiflorum = S. uplandicum x S. grandiflorum
See 'Symphytum grandiflorum' in secton 'Details about Symphytum Species' (Chapter 9).

Current Botanical Nomenclature of Hidcote Hybrid

Symphytum officinale x Symphytum asperum x Symphytum grandiflorum =
Symphytum grandiflorum Hidcote Blue= **Symphytum Hidcote Blue**. (Blue flowers.)

Symphytum officinale x Symphytum asperum x Symphytum grandiflorum =
Symphytum grandiflorum Hidcote Pink= **Symphytum Hidcote Pink (Roseum)**. (Pink flowers.)

*"**Symphytum officinale x asperum x grandiflorum = S. 'Hidcote Blue' hort. ex G. Thomas**.*
Hidcote Comfrey: species hybrid. *Accepted Name.*
Authority: UKSI (United Kingdom Species Inventory, Natural History Museum, London, England).
Synonyms:
Symphytum 'Hidcote Blue'
Symphytum asperum x grandiflorum x officinale
Symphytum 'Hidcote Blue' hort. ex G. Thomas
Symphytum x tauricum
Common Names:
Hidcote Comfrey
Cyfardwf Hidcote (Welsh)"
-National Biodiversity Network®, NBN Atlas, United Kingdom, https://species.nbnatlas.org, **2018**.

('Hort.' or hortulanorum is used for a name that saw significant use in horticultural literature, usually the 19th century and earlier, but was never properly published. This term is used so that non-wild, cultivated plants can be examined by taxonomists so they can be established as species, and published.)

*"**Graham Stuart Thomas** (1909-2003):*
At 17 he became a student at the Cambridge University Botanic Garden, England. After becoming partner at the celebrated Sunningdale Nursery, Thomas developed beautiful planting schemes. This exposed his work to the 'National Trust for Places of Historic Interest or Natural Beauty' (of Great Britain) which had acquired its first garden, Hidcote, and asked him to become a part-time consultant. Thomas saw gardening as an art allied with craft.
Symphytum grandiflorum 'a free-spreading ground cover bearing croziers of cream bells for many weeks, the whole not exceeding a foot in height. **A little taller are the hybrids known as 'Hidcote Blue' and 'Hidcote Pink'. They completely take care of shady banks being weed-proof.*' "*
-'The Man Who Loved Plants: Graham Stuart Thomas' by Dr. David Abbott and Catherine Glass, Brits at Their Best: Sharing the Inheritance (2008). www.britsattheirbest.com

"Boraginaceae **Symphytum x hidcotense** *P.D.Sell (*Fl. Gr. Brit. Ireland 3: 520 (361). 2009)*
= Symphytum grandiflorum DC. x Symphytum uplandicum Nyman"
-'International Plant Names Index'® (IPNI), www.ipni.org, A database of the names and associated basic bibliographical details

of seed plants, ferns and lycophytes, **2018**.
(* -Flora of Great Britain and Ireland by Peter D. Sell and Gina Murrell. England: Cambridge University Press. Updated publications- Volume 1: 2018; Volume 2: 2014; Volume 3: 2009; Volume 4: 2006; Volume 5: 1997. Symphytum is in Volume 3.)

*"**Symphytum x hidcotense** P.D. Sell (synonym: S. 'Hidcote Blue', S. grandiflorum x {S. x uplandicum})."*
-Manual of the Alien Plants of Belgium®, http://alienplantsbelgium.be, Descriptions to the introduced plants that grow wild in Belgium, **2018**.

Description of Hidcote Blue and Pink

My Symphytum Hidcote Blue **blooms earlier than my Common and Russian Comfrey.**

*"**Two hybrids growing at Hidcote Manor, Gloucestershire (England), rather larger in growth, have been named 'Hidcote Pink' and 'Hidcote Blue'.***
These are partially evergreen. Symphytum grandiflorum and its near hybrids or forms, 'Hidcote Blue' and 'Hidcote Pink' make ideal ground-cover.
***'Hidcote Pink' and 'Hidcote Blue' are soft pinkish and pale soft bluish tint respectively**; about twice the height of S. grandiflorum. Plant 2 1/2 feet (0.76 meters) apart."*
-Plants for Ground-Cover by Graham Stuart Thomas, V.M.H., Gardens Advisor to the National Trust. London, England: J.M. Dent & Sons Ltd, **1970**, pages 128, 129, 194, 219, 226.

*"**A garden hybrid of S. grandiflorum DC. (here regarded as including S. ibericum Stev.)** to which Graham Thomas, in 'Plants for Ground Cover' (Dent, 1970), had **given the cultivar name 'Hidcote Blue'**. This new British alien has been ably illustrated by Peter Barnes (Royal Horticultural botanist, England) in the accompanying figure and a description follows:*
    Non-flowering shoots strongly decumbent, vigorous; leaves with lamina up to 19 x 10 cm (7.4-3.9 inches), ovate-oblong, acute above, more or less truncate below, slightly bullate and roughly hairy; petiole long, decurrent as a raised line, but stem not winged.
    Flowering stems up to 3 feet (0.9 meter), erect at first, later decumbent branched above and usually with one or more well-developed, later-flowering axillary shoots; very rough with numerous, rigid, swollen-based hairs throughout and shorter, slender, but rigid hairs especially above; lower leaves petiolate, uppermost more or less sessile.
    *Calyx divided to about 2/3, accrescent in fruit; calyx teeth linear lanceolate, obtuse. Corolla (18-) 20-22 x 10-13 mm, **red in bud** (Royal Horticultural Society Colour Chart, Red 47A, becoming Red 47C/D), whilst at anthesis the tubular basal part of the **corolla is Blue** 101B (sometimes very pale) and the expanded apical bell is white; corolla tube gradually narrowed to the base. Corolla scales about 7-8 mm long, included, not or just exceeding the stamens, with conical teeth densest near the apex. **Some apparently fertile seed is set.***
Commences flowering in March and continues throughout most of the summer (until July) with a peak in April-May.
The combination of habit and flower colour separates Symphytum 'Hidcote Blue' from all other native and alien species (in England).
Although in its habit, early flowering and red buds S. 'Hidcote Blue' shows clear affinities to S. grandiflorum, the identity of the other parent is not so obvious, but it must clearly be a blue-flowered taxon - perhaps a form of S. x uplandicum.
S. grandiflorum differs in its shorter stems, which may be forked at the apex but lack axillary flowering shoots, and in the flowers at anthesis being wholly creamy-white with the corolla tube abruptly contracted at the base.
S. 'Hidcote Blue' is an aggressive plant in cultivation *and is all too likely to become established outside gardens."*
-'A New Alien Symphytum' by A.C. Leslie, Surrey, England; Botanical Society of Britain and Ireland (BSBI) News, Hertfordshire, England, No. 30, pages 15-16, April **1982**.

"S. 'Hidcote Blue' (S. grandiflorum x {S. x uplandicum})- Hidcote Comfrey:
Ascending to erect flowering stems to 50 (100) cm (19.6-39.3 inch) and procumbent to decumbent stolons arising from rhizomes; stem-leaves mostly petiolate, not or scarcely decurrent.
*Calyx divided about 3/5 to 2/3 to base; **corolla blue when open, pink earlier.***
Introduced; grown in gardens, naturalized in hedges and woodland; Surrey County (southwest England), Salop (Shropshire County, west midlands England) and Guernsey (island in the English channel off the coast of Normandy); garden origin."
-New Flora of the British Isles: Identification of Wild Vascular Plants of the British Isles edited by Clive Stace. Cambridge, England: Cambridge University Press, **1991**, page 648.

*"**Stace (*2010) gives as distinguishing features of S. x hidcotense compared to similar S. grandiflorum a branched stem and pink or bluish flowers.***
With naming in the garden trade, caution is appropriate, since many species (not just the genus Symphytum) are under a false name. The pictures of 'Hidcote Blue' in an online field guide (2009) agree with the local plants, while S. grandiflorum is shown (2009) as not branched."
-'Caucasian Comfrey (Symphytum caucasicum M. Bieb.) and Hidcote Comfrey (Symphytum x hidcotense P.D. Sell) in the Region of Aachen' [Kaukasischer Beinwell {Symphytum caucasicum M. Bieb.} und Hidcote-Beinwell {Symphytum x hidcotense P.D. Sell} im Aachener Raum] by F. Wolfgang Bomble and Bruno G.A. Schmitz, Die Veroffentlichungen des Bochumer Botanischen Vereins e. V., Germany, Volume 4, No. 6, pages 50-54, **2012**. (If you have an English translation, could you send it.)

(* -New Flora of the British Isles: Identification of Wild Vascular Plants of the British Isles edited by Clive Stace. Cambridge, England: Cambridge University Press, 1991, 1997, 2010.)

"Symphytum grandiflorum DC. and
S. x hidcotense P.D. Sell (probably S. asperum Lepechin x grandiflorum x officinale L. parentage):
These two cultivated ornamentals were introduced in the key in NF6 (Nouvelle Flore de la Belgique) and full accounts were provided (the former was already briefly cited in NF5).
Both are increasingly seen as escapes and are locally well-established, mostly in Fl. and Brab. (Flanders and Brabant, Belgium) (based on various recent collections in BR and LG, mostly by the authors)."
-'Dumortiera 104', Botanic Garden Meise, Belgium, page 25, **2014**. Publishes articles in English, Dutch or French on flora/vegetation of Belgium and adjacent areas: vascular plants, bryophytes, lichens, algae and fungi. Includes changes in indigenous/non-indigenous flora, revisions of difficult or overlooked groups, keys additions to 'Flora van Belgie / Nouvelle Flore de la Belgique', field surveys. www.plantentuinmeise.be
  (I think 'BR' means BR Herbarium of Botanic Garden Meise, Belgium.
  I think 'LG' means Herbarium, Departement de Botanique, Universite de Liege, Belgium.)

"Symphytum 'Hidcote Blue' (S. asperum x grandiflorum x officinale):
Ecology: *This perennial herb occurs on roadsides and in hedges and woodland. It is a garden escape or throw-out which can occasionally become naturalised. Lowland.*
Status: *Neophyte.*
Trends: This plant, raised at Hidcote Gardens National Trust (East Gloucestershire, England) not long before 1930, grows aggressively in gardens *and is often discarded. It was first noticed in the wild in 1979, and is evidently increasing.*
World Distribution: *A hybrid of garden origin."*
-'Online Atlas of the British and Irish Flora', https://www.brc.ac.uk/plantatlas, Distribution and ecology with photographs, **2018**.
  (In botany, a neophyte is a plant species that is not native to a geographical region, and was introduced in
  recent history. Plants that are long-established in an area are called archaeophytes. In Britain, neophytes are
  plant species introduced after 1492, when Christopher Columbus arrived in the Americas.)

"Symphytum x hidcotense is a popular garden plant of complex, artificial origin.
Many cultivars have been described. The Belgian populations have corollas that are (light) blue and white.
Such plants probably are the result of a crossing between female Symphytum grandiflorum and S. x uplandicum (*Gadella & Perring 2000)."
-Manual of the Alien Plants of Belgium®, http://alienplantsbelgium.be, Descriptions to the introduced plants that grow wild in Belgium, **2018**.
(* -The European Garden Flora: A Manual for the Identification of Plants Cultivated in Europe, Both Out-of-Doors and Under Glass, 6 Volumes, edited by James Cullen, Sabina G. Knees, and H. Suzanne Cubey. England: Cambridge University Press, 2000. (Volume 6: 'Symphytum' by T.W.J. Gadella and F.H. Perring, pages 138-141.)

Hidcote Comfrey Chromosomes

"S. x Hidcote Blue ('c.' means around)

2n=	Cytology Reference	Kew Accession	Collector	Origin	Voucher
c. 52	83-179	1983-242	Hidcote Garden	cultivated	K "

-'New Chromosome Numbers in Petaloid Monocotyledons and in Other Miscellaneous Angiosperms' by Margaret A.T. Johnson and P.E. Brandham, Jodrell Laboratory, Royal Botanic Gardens, Kew, Richmond, Surrey, England; Kew Bulletin, Volume 52, No. 1, pages 121-138, **1997**.

<u>S. officinale x S. asperum x S. tuberosum = S. uplandicum x S. tuberosum</u>

"Symphytum tuberosum x uplandicum.
Vice-county 33, East Gloucester; Woodbridge, near Withington, (England), with both parents, 1955, E. Milne-Redhead, conf. (compare to) A.E. Wade (Hb. Kew = Herbarium, Royal Botanic Gardens, Kew, London, England).
In size and general appearance this hybrid was intermediate between S. tuberosum and S. x uplandicum, with which it was growing. It had the tuberous rootstock of the former species."
-'Symphytum tuberosum x uplandicum' by E. Milne-Redhead, 'Proceedings of the Botanical Society of the British Isles', 'Plant Notes' section, Volume 3, page 46, **1958**.

'Symphytum x uplandicum x tuberosum' by U.K. Duncan, 'Proceedings of the Botanical Society of the British Isles', Volume 3, page 407, **1960**. (I was unable to get a copy of this. If you have it, could you please send it to me.)

"S. asperum Lepech. x S. officinale L. x S. tuberosum L.:
Symphytum x uplandicum (S. asperum x S. officinale) is more widespread than either parent, and unlike them, is known to hybridize with S. tuberosum.

There is no valid binomial or synonym for this hybrid. *Our state of knowledge concerning the nomenclature of hybrids is far less advanced than that concerning species.*
In size and general appearance this hybrid is intermediate between S. tuberosum and S. x uplandicum, its presumed parents. *It has the tuberous rootstock of S. tuberosum. The stems are hispid, not scabrid as in S. x uplandicum; the leaves are 3-4 times as long as broad, not 2-3 times as in S. tuberosum.*
The flowers *are more numerous than in S. tuberosum but resemble them closely except that they are* ***pale mauve (pale purple) in colour;*** *and the calyx segments are lanceolate, the style is slender and the scales considerably exceed the stamens.*
The Arbroath (council area of Angus in Scotland) plant had pollen with a large number of abortive (not fertile) grains. *There has not been any experimental work on artificial hybridization among these 3 species."*
-Hybridization and the Flora of the British Isles edited by Clive A. Stace, University of Leicester, England in collaboration with 'Botanical Society of the British Isles' now called 'Botanical Society of Britain and Ireland'. London, England: Academic Press, **1975**, pages 4, 354.
> (The Swedish botanist Carl Linnaeus developed a system known as 'Linnaean Taxonomy' for categorizing organisms and naming them using binomial nomenclature {two names}.)

"(S. x uplandicum) x (S. tuberosum):
This cross occurs near the parents in very scattered localities in England and Scotland; ***it is intermediate in all characters, with yellow corollas tinged with blue or purple,*** *and tuberous rhizomes, and is at least partially sterile; endemic (native).*
Some plants might be S. officinale x S. tuberosum."
-New Flora of the British Isles: Identification of Wild Vascular Plants of the British Isles edited by Clive Stace. Cambridge, England: Cambridge University Press, **1991**, page 647.

S. officinale x S. bohemicum

"Symphytum bohemicum x officinale: Krizeni nebylo dosud experimentalne prokazano. Zatim jediny doklad vykazujici intermediani znaky pochazi z Vranova u Opocna (PR)."
Online translation:
"Symphytum bohemicum x officinale: Krizeni (hybridization?) has not yet been experimentally proven. *So far, the only document showing intermedian characters comes from Vranova u Opocna, Czechia (PR) (Pardubicky Region)."*
-Kvetena Ceske Republiky (Flora of the Czech Republic), 9 Volumes, edited by B. Slavík, J. Jun. Chrtek and J. Stepankova. Praha (Prague), Czech Republic: Academia- Academy of Sciences of the Czech Republic, 1998-2010. 'Symphytum L. Kostival' article in Boraginaceae Family, Volume 6, pages 202-210, **2000**. (All in Czech. If you have a translation, please let me know.)

S. officinale x S. cordatum

"Hybrid between S. officinale Linn. and S. cordatum Waldst. & Kit.:
x Symphytum Polonicum Blocki. (Symphytum cordatum x officinale)
Habitat: Galicia orientalis (eastern europe)."
-'A Revision of the Genus Symphytum, Tourn.' by Cedric Bucknall, {Mus. Bac. Oxon= Bachelor of Music, Oxford University}, Journal of the Linnean Society of London, England, Botanical Journal, Volume 41, Issue 284, pages 491-556, December **1913**.

"S. x polonicum Blocki ex Buckn. = ***S. cordatum x S. officinale****: Country:* ***Ukraine****, 1908.*
> *Galiciae orientalis. Winniki pr. Leopolim (L'viv, Ukraine). In silva frondosa {In leafy woods}."*

-'Symphytum x polonicum Blocki ex Buckn., Family Boraginaceae', JSTOR® Global Plants, www.jstor.org, **2019**. JSTOR is a digital library for scholars, researchers, and students. It provides access to more than 12 million academic journal articles, books, and primary sources in 75 disciplines.

S. officinale x S. peregrinum

"Hybrids formed by Symphytum officinale L. and Symphytum peregrinum Ledeb.:
> *x S. discolor mihi.*
> *(S. officinale alpha ochroleucum x < peregrinum.)*
> *Habit and stature of S. peregrinum, sparingly fertile or quite sterile. Stem hispid, asperous, narrowly winged; lower leaves rather broadly ovate, upper lanceolate, narrowly decurrent; calyx with acute or acuminate segments, slightly tubercular-setose in fruit; corolla whitish, or more or less tinted with pale rose and blue, turning to pale slaty blue (grayish blue) when dry; anthers a little longer than filaments; nutlets intermediate between those of the parents, faintly areolate, dotted with minute scattered points, shining, slightly constricted above the broad base.*
> *x S. lilacinum mihi.*
> *(S. officinale alpha ochroleucum x beta purpureum x < peregrinum.)*
> *Habit and stature of S. peregrinum, sparingly fertile. Stem asperous, more conspicuously winged than in x S. discolor; lower leaves elliptic-lanceolate, rounded or attenuate at the base, narrower as a rule than in x S. discolor, upper*

lanceolate with broader decurrent base; fruiting-calyx setose; corolla purplish with greenish-yellow tip when in bud, then pale purplish-rose, slaty-blue tinged with purple when dry; anthers as long as the filaments; nutlets as in x S. discolor.
x S. densiflorum mihi.
(S. officinale beta purpureum x > peregrinum.)
Habit of S. officinale, but often taller, fertile or sterile. Stem asperous, narrowly winged; lower leaves narrowly oblong-lanceolate, shortly attenuate at the base, upper narrowly lanceolate, conspicuously decurrent; flowers large, open, crowded; calyx with acute segments, hispid, rarely tubercular-setose in fruit; corolla reddish-violet, dark purple when dry; anthers as long or longer than filaments; nutlets as in x S. discolor, but often olivaceous-black (olive color).
x S. coeruleum Petitmengin.
(S. officinale alpha ochroleucum x peregrinum.)
Habit and stature of S. peregrinum, but almost entirely sterile. Stem hispid, asperous, partially and rather broadly winged; lower leaves oblong, attenuated into the petiole, upper lanceolate, semidecurrent from the rather broad base; calyx with acute segments; corolla rose tipped with green when in bud, then bright blue or rose and blue, the tips of the lobes sometimes yellowish; anthers equal to filaments; nutlets (very seldom produced) closely granulated."
-'Some Hybrids of the Genus Symphytum' by Cedric Bucknall, {Mus. Bac. Oxon= Bachelor of Music, Oxford University}, The Journal of Botany British and Foreign, London, England, Volume 50, pages 332-337, **1912**.
(The symbol '<' means the hybrid shows characters of S. peregrinum in lesser proportions.
The symbol '>' means the hybrid shows characters of S. peregrinum in higher proportions.)

"Symphytum caeruleum Petitmengin (S. officinale alpha ochroleucum x peregrinum):
See Journal of Botany, 1912, p. 335. Cultivated in the University Garden, Bristol, England, June 1913. -J.W. White."
-'The Botanical Exchange Club and Society of the British Isles, Report for 1913' by editor A. Bruce Jackson, Arbroath or Aberbrothock, Scotland, Volume 3, Part 6, October **1914**. Symphytum pages 484-485.

"Hybrids S. officinale L. x S. peregrinum: S. caeruleum Petitmengin:
In Thellung (sub: S. asperum Lep. x S. officinale ? = S. peregrinum, Bot. Mag., t. 6466 non Ledeb.; x S. caeruleum Ptmg., Buckn., Jl. Bot., 1, p. 335, 1912, and (S. officinale alpha ochroleucum x peregrinum).
As is usually the case for the generality of hybrids, these combinations can be distinguished with certainty only when alive, when they are caught growing between the parents."
-'Symphytum Peregrinum Ledeb et ses Hybrides Avec S. Officinale L' {Symphytum Peregrinum Ledeb and Its Hybrids with S. Officinale} by Pierre Senay; Bulletin de la Societe Botanique de France, Paris, France, Volume 87, No. 2, pages 313-322, **1940**. All in French.

"Symphytum x caeruleum Petitmengin ex Thell.:
Petitmengin's binomial was invalid but subsequently validated by Thellung (Vierteljahrschr. Naturf. Ges. Zurich 52: 459, 1907); the author citation was corrected accordingly in NF6 {Nouvelle Flore de la Belgique} (cf. P. Sell & G. Murrell, Fl. Gr. Brit. Irel. 3: 520, 2009)."
-'Dumortiera 104', Botanic Garden Meise, Belgium, page 25, **2014**. Publishes articles in English, Dutch or French on flora/vegetation of Belgium and adjacent areas: vascular plants, bryophytes, lichens, algae and fungi. Includes changes in indigenous/non-indigenous flora, revisions of difficult or overlooked groups, keys additions to 'Flora van Belgie / Nouvelle Flore de la Belgique', field surveys. www.plantentuinmeise.be

S. officinale x S. tuberosum

"Hybrids between S. officinale Linn. and S. tuberosum Linn.:
S. zahlbruckneri Sennholz:
'All leaves, especially the upper ones, elongate lanceolate, broadest at the middle, acuminate. Racemes mostly terminal. Calyx green. Corolla-scales exceeding stamens. Stem 2.6-3.4 dcm. (decimeter) (0.85-1.1 feet) high. Habitat lower Austria: Piesting between Waldegg and Od.'
The above short description does not convey much information as to the exact position of this hybrid, but, the calyx being described as green, while in x S. Wettsteinii it is described as purplish, it may be inferred that x S. Zahlbruckneri = S. officinale alpha ochroleucum x S. tuberosum.
If this inference is correct, a plant gathered by the writer near St. Jean de Luz, France, in 1898, may be placed here. This has the habit of S. tuberosum, but the partially decurrent leaves, the hispidity of the inflorescence and the more numerous flowers, suggest the influence of S. officinale.
x S. wettsteinii Sennholz. (S. officinale beta purpureum x tuberosum):
S. Wettsteinii, G. Sennholz in *Verh. Zool.-Bot. Ges. Wien, xxxviii. (1888), page 69.
Habitat: Lower Austria: Kalksburg, with the parents."
-'A Revision of the Genus Symphytum, Tourn.' by Cedric Bucknall, {Mus. Bac. Oxon= Bachelor of Music, Oxford University}, Journal of the Linnean Society of London, England, Botanical Journal, Volume 41, Issue 284, pages 491-556, December **1913**.
(* -'Verhandlungen der Kaiserlich-Koniglichen Zoologisch-Botanischen Gesellschaft in Wien.' by Zoologisch-Botanische Gesellschaft in Wien., Vienna, Austria, Volume 38, 1888. Symphytum wettsteinii by G. Sennholz, pages 69-70 of Sitzungsberichte {Meeting Reports}. In German.)

"S. officinale L. x S. tuberosum L. = S. x wettsteinii Sennholz has been recorded from Austria and Switzerland."
-Hybridization and the Flora of the British Isles edited by Clive A. Stace, University of Leicester, England in collaboration with 'Botanical Society of the British Isles' now called 'Botanical Society of Britain and Ireland'. London, England: Academic Press, **1975**, page 355.

"Symphytum officinale x S. tuberosum = Symphytum x wettsteinii Sennholz
Verh. Zool.-Bot. Ges. Wien 38: 69, 1888. Kostihoj Wettsteinov.
Hybrid with the shape of the close-up of S. tuberosum with creeping subterranean plants and yellow flowers. Smejkal (1978 page 159).
This may only be a combination of S. officinale x S. angustifolium, or in the vicinity. Hlohovca (town in Slovakia) grows only S. angustifolium A. Kerner non auct. incl. Smejkal.
Up to now we could not find out for this combination corresponding first name. May be possible."
-Flora Slovenska {Flora of Slovakia}, Volume 5/1 (V/1) or 5/2 (V/2), edited by L. Bertova and K. Goliasova. Bratislava, Slovakia: Veda, **1993**. Article: 'Symphytum L.' by J. Majovsky and Z. Hegedusova, pages 76-97. Some sources say 5/2 but I think 5/1 is correct. (This is translated online from Slovak to English. If you have English translation of Symphytum pages, I would like it.)

"Boraginaceae Symphytum wettsteinii Sennholz: (Sitzungsber. Zool.-Bot. Ges. Wien xxxviii. (1888) 69; et in Bot. Centralbl. xxxv. S (1888) 60; G. Beck, Fl. Nied. Oest. ii. II. (1893) 964.)"
-'International Plant Names Index'®. (IPNI), www.ipni.org, A database of the names and associated basic bibliographical details of seed plants, ferns and lycophytes, **2018**.

S. tuberosum x S. cordatum

*"Symphytum ullepitschii = (S. cordatum W.K. x S. tuberosum L.) Wettstein. Conf. (confer/consult) Ullepitsch in *Oest. bot. Zeitschr. 1886, p. 299. Ad confines Hungariae et Galiciae."* {Confines of Hungary and eastern Europe}
-'Schedae ad Floram exsiccatam Austro-Hungaricam opus cura Musei Botanici Universitatis Vindobonensis Conditum, Volume 6' by Anton Joseph Kerner, Vindobonae {Vienna}, Austria, **1893**. Symphytum ullepitschii article by R. Wettstein, page 37.
(* -'Symphytum Cordatum W.K.' by J. Ullepitsch; Oesterreichische Botanische Zeitschrift (Austrian Botanical Journal), Volume 36, pages 298-299, Vienna, Austria, 1886.)

"Hybrid between S. tuberosum Linn. and S. cordatum Waldst. & Kit. = Symphytum x Ullepitschii.
Habitat: Hungary. According to Wettstein, this hybrid does not develop fruit.
The specimens I have seen are mostly nearer to S. tuberosum than to S. cordatum."
-'A Revision of the Genus Symphytum, Tourn.' by Cedric Bucknall, {Mus. Bac. Oxon= Bachelor of Music, Oxford University}, Journal of the Linnean Society of London, England, Botanical Journal, Volume 41, Issue 284, pages 491-556, December **1913**.

"Symphytum tuberosum L. x cordatum Waldst. & Kit. Baen.:
Collector: J. Ullepitsch, #Exsiccatal series: C.G. Baenitz, Herb. Eur., #s.n.. Collected in 1890.
Country: Presovsky; Fl. Hungarica: Lechnitz (Lechnica, **Slovakia**) (Zips region Hungary/Slovakia), unter den Eltern {under the parents}."
-'Symphytum tuberosum L. x cordatum Waldst. & Kit. Baen.', Family Boraginaceae, JSTOR® Global Plants, www.jstor.org, **2019**. JSTOR is a digital library for scholars, researchers, and students. It provides access to more than 12 million academic journal articles, books, and primary sources in 75 disciplines.

S. tuberosum x S. bulbosum

"To the flora of the Riviera, Italy. Of the new states of the surroundings of Genoa, at Bordighera and in the Maritime Alps in the the year 1892 and 1893 by the author collected plants to be emphasized:
   Symphytum bulbosum x tuberosum at Genoa, Italy. by Bornmuller, Weimar, Germany."
-'Botanisches Centralblatt: Refererendes Organ fur das Gesamtgebiet der Botanik des In und Auslandes' {Central Sheet/Source for Botanical Information} by Dr. Oscar Uhlworm and Dr. F.G. Kohl, Cassel {Kassel}, Germany, Volume 62, **1895**. Symphytum page 151. In German. Online translator into English.
   (Maritime Alps are a mountain range in the southwestern Alps. They form the border between the French region of
   Provence-Alpes-Cote d'Azur and the Italian regions of Piedmont and Liguria.)

"Hybrid between S. tuberosum Linn. and S. bulbosum Schimp.
x S. Bicknelli, hybrid nova (new) (S. bulbosum x tuberosum). Habitat: Italy, Liguria."
-'A Revision of the Genus Symphytum, Tourn.' by Cedric Bucknall, {Mus. Bac. Oxon= Bachelor of Music, Oxford University}, Journal of the Linnean Society of London, England, Botanical Journal, Volume 41, Issue 284, pages 491-556, December **1913**.

S. asperum x S. caucasicum

"In the Caucasus there are individuals with densely bristly villous stems without decurrent leaves and with shorter calyx teeth; these may be hybrid forms of S. asperum x S. caucasicum. They are often distinguished by their low habit. Representative specimens: between Kel'ny and Bechenakh (transcaucasia), on gravel in Nakhichevan/Nakhchivan (Azerbaijan) Chai River, flowering 30 V 1947 (A. Grossgeim); between Tskharo and Tabistskhurskoe/Tabatskuri Lake (country of Georgia), meadows in Kuni River valley (Tajikistan), 5-6 VII 1916 (P.N. Krylov and E.I. Shteinberg)."
-'Flora of the U.S.S.R., Volume 19' edited by V.L. Komarov, Leningrad, Russia, **1953**. The Borage family part was written by Mikhail Grigorevich Popov, pages 73 to 508 in English version, and pages 97 to 692 in Russian version. Symphytum is pages 207-216 in English, and pages 279-291 in Russian. Asperum x caucasicum page 214 in English, page 288 in Russian.
 (Plant habit is a plant's overall shape and form. It includes the development, length of stem, branching pattern, density.)

*"A Comfrey found growing in rough grassland at Liberton Dams, Midlothian, **Scotland**, vice county 83 by Richard Learmonth has been identified by the Dutch botanist Professor T.W.J. Gadella and Franklyn Perring as a **hybrid: Symphytum asperum x S. caucasicum.***
 It shares with S. asperum a very short calyx (rare in Symphytum) and short non-decurrent stem leaves, and with S. caucasicum a calyx divided less than 1/2 the length of the tube, leaves not prickly/asperous and stem leaves not decurrent.
One anomaly is the flower colour which is purplish pink whereas both the parents are sky blue."
-'A Symphytum (Comfrey) Hybrid New to Britain' by R.W.C. Learmonth and F.H. Perring, Botanical Society of Britain and Ireland News, Hertfordshire, England, No. 69, page 74, April **1995**.

"Symphytum asperum x caucasicum: Species hybrid: accepted. Native in Scotland."
-NBN Atlas®, https://scotland-species.nbnatlas.org/species/NHMSYS0000464110, National Biodiversity Network®, **2019**. Provides database for the sharing of high quality biological data in the United Kingdom since 2000.

S. asperum x S. orientale = S. norvicense = Norfolk Comfrey

*"**Hybrid Between Symphytum x uplandicum (or other species) with S. orientale:**
For at least the last eight years I have been aware of a very large colony of an unusual Comfrey growing on a road verge (edge/ border) in Intwood, Norfolk, England.
In 2003 I sent some material to Franklyn Perring, late 'Botanical Society of Britain and Ireland' referee for Symphytum, who identified it as the above hybrid.*
 It seems significant to me that the Cambridge, England hybrid colony consists of only a few plants in a very large colony of S. orientale (White Comfrey).
There are three possible origins for such a new taxon; *either it is a new hybrid, or it is a garden cultivar not previously 'gone wild', or it is a taxon from outside the area covered by our usual literature."*
-'Probable Hybrid Between Symphytum x Uplandicum and S. Orientale in Norfolk' by Bob Leaney, Wroxham, Norfolk, England; Botanical Society of Britain and Ireland: BSBI News, Hertfordshire, England, No. 105, pages 6-9, April **2007**.
 (These were some of the initial ideas about Norfolk Comfrey.)

*"**This note describes and formally names a new nothospecies in Symphytum L., its putative (generally regarded as such) parents being Symphytum asperum Lepech. and S. orientale L.***
*This taxon was first recorded from Intwood, East Norfolk v.c. 27, England, in 1999 by R.M. Leaney.
There appears to be no record in the literature of Symphytum asperum x S. orientale, whether occurring naturally or as a result of experimental or horticultural crossing. Therefore an epithet (descriptive phrase) is assigned in order that this taxon may receive full treatment in the forthcoming third edition of the 'New Flora of the British Isles' (Stace 1997).*
Symphytum x norvicense *R.M. Leaney & C.L. O'Reilly* **hybrida nova (new hybrid)** *(Boraginaceae) (putative parentage Symphytum asperum Lepech. 2n = 32 and S. orientale L. 2n = 32, ?64),* **Norfolk Comfrey.** *Flowers May to July.*
The chromosome number of 2n = 48 suggests S. asperum (2n = 32) rather than a cytotype of S. x uplandicum as the other parent taxon.
 Either parent with 2n = 32 may have produced unreduced gametes, or one parent may have been S. orientale with 2n = 64, as unpublished results have shown 2n = 62, 63 for S. orientale, which may represent a miscount of 2n = 64 (T.W.J. Gadella, personal communication, 2009)."
-'A New Nothospecies in Symphytum L. (Boraginaceae)' by C.L. O'Reilly (Northumberland, England) and R.M. Leaney, (Norfolk, England), Watsonia: Journal and Proceedings of the Botanical Society of the British Isles, London, England, Volume 27, pages 372-372, **2009**. (A nothospecies is a hybrid formed by direct hybridization of two species, not with other hybrids.)

*"**The Norfolk Comfrey appears to be endemic to the British Isles, and despite the discovery of new colonies and much increase in recent years, remains confined more or less to Norfolk, England.***
The parentage of the Norfolk Comfrey therefore still remains uncertain (*Stace, 2010).
 That S. orientale is one parent is indicated by *the predominance of fine, soft, uncinate hairs in the indumentum of stem, leaf midrib and calyx; by the broadly ovate upper stem leaves, with widely cuneate to rounded bases; the less*

than half-dissected calyx; and the white sometimes present in the open corolla.
The tall stature (100-150 cm), habit and root type are all much like S. x uplandicum, S. asperum and S. officinale (Common Comfrey), *and there are no other species in the British Isles, or Europe, that can explain this combination of features along with the* **red and blue in the corolla** *(Tutin et al., 1972).*

S. officinale is not a likely parent in view of its carmine or cream flowers and the long and extraordinarily broad winging of the stems, especially from the bases of the very upper leaves. **The leaves of S. x norvicense are not even decurrent.**

The possibility that the Norfolk Comfrey is a previously unrecognized pure species, rather than a hybrid nothospecies of recent origin, also needs consideration. *Sell has suggested that it could be S. savvalense (**Sell, 2009), which occurs in the wild no nearer than Turkey, and does not ever seem to have been taken into cultivation.*

However, examination of a specimen of S. savvalense in BM (British Museum of Natural History) showed very marked and absolute differences in calyx dissection, calyx lobe shape, indumentum, size and height (***O'Reilly & Leaney, 2009).

Since its first discovery, the Norfolk Comfrey has spread quite considerably, *not only vegetatively much like S. x uplandicum, but also, much more than that taxon, by seed, producing numerous seedling plants when conditions are right around the clonal patches and sometimes nearby new populations, or odd new plants, hundreds of yards away.*

Such new populations produced by seed are very constant in character, *with no sign of character segregation.*

**Such a rapid increase in such a short time suggests a very recent origin only a few decades or so ago."*

-'A Further Update on the Norfolk Comfrey (Symphytum x Norvicense) and Another Overlooked Comfrey Hybrid in Norfolk' by Bob Leaney, Wroxham, Norfolk, England; Botanical Society of Britain and Ireland: BSBI News, Broompark, Durham, England, No. 125, pages 21-25, January **2014**.

(* -New Flora of the British Isles: Identification of Wild Vascular Plants of the British Isles edited by Clive Stace. Cambridge, England: Cambridge University Press, 1991, 1997, 2010.)

(** -Flora of Great Britain and Ireland by Peter D. Sell and Gina Murrell. England: Cambridge University Press. Updated publications- Volume 1: 2018; Volume 2: 2014; Volume 3: 2009; Volume 4: 2006; Volume 5: 1997. Symphytum is in Volume 3.)

(*** -'Update on the Identity of the Norfolk Comfrey' by Clare O'Reilly and Bob Leaney, England; Botanical Society of Britain and Ireland: BSBI News, Hertfordshire, England, No. 110, pages 47-48, January 2009.)

"Symphytum asperum x orientale (S. x norvicense) R.M. Leaney & C.L. O'Reilly:
Norfolk Comfrey. Species hybrid: accepted."
-NBN Atlas®, https://species.nbnatlas.org/species/NHMSYS0021109926, National Biodiversity Network®, **2019**. Provides database for the sharing of high quality biological data in the United Kingdom since 2000.

S. caucasicum x S. orientale

"A hybrid between Symphytum caucasicum and Symphytum orientale is described from two sites in Norfolk, and one on the Isle of Wight. The taxon seems to be a new one, certainly for the British Isles and Ireland, and is similar to two previously described hybrids from Norfolk, Symphytum x norvicense (S. asperum x orientale) and Symphytum x uplandicum x caucasicum. Separation of these three taxa is discussed: **all three are characterised by a combination of variegated red, blue, purple or white corolla bell, together with a calyx dissection to less than half way**. This shallowly dissected calyx is the key to recognition of these three entities, which otherwise are easily overlooked as diminutive forms of Symphytum x uplandicum. There seems to be no reasonable doubt that **these plants are two nothomorphs of the hybrid Symphytum caucasium x orientale**. A search of the literature and enquiry to the taxonomist with the most recent interest in Symphytum has not turned up any evidence that this hybrid has been described before. (Wolfgang Bomble, personal communication)"
-'Symphytum caucasicum x S orientale (Boraginaceae) in East Norfolk and Isle of Wight' by Bob Leaney, Wroxham, Norfolk, England; British and Irish Botany, Volume 1, No. 4, pages 327-334, **2019**.

(A nothomorph is any member of a group of different hybrid forms produced by crosses between same 2 parent species.)

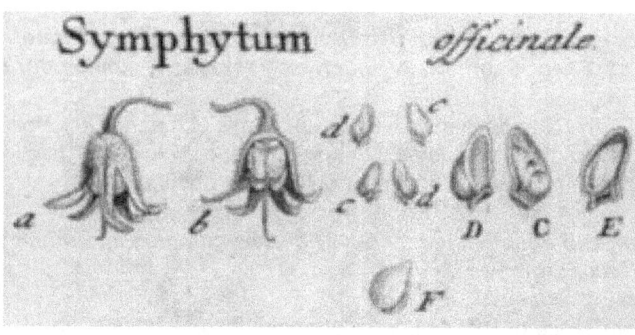

Symphytum Seeds 1788:
Symphytum officinale, Symphytum majus
Symphytum foliis ovato lanceolatis decurrentibus

De Fructibus et Seminibus Plantarum: Volume I (Of Fruits and Seedling Plants)' by Joseph Gaertner, 1788. Page 325, tab. 67. 'De Fructibus' has extremely accurate descriptions of plants with over a thousand species. This book introduced a new era in plant morphology. Gaertner was a German botanist who lived from 1732-1791. He was Professor of Anatomy in Tubingen, Germany, and Professor of Botany at St. Petersburg, Russia.

PART B

HISTORY OF COMFREY

Chapter 12

Prehistory, Ancient Times, Middle Ages and Comfrey

History of the Word 'Comfrey'

The word 'Symphytum' is Latin and comes from the Greek 'symphis', 'symphyo' and 'syumphuo' which mean 'growing together' and from 'phyton' which mean plant.

*"**The bird called chloris** from being yellow beneath, is of the size of the lark, and lays four or five eggs; **it makes its nest of Symphytum, which it pulls up by the root, and lines it with straw, hair, and wool.** The blackbird and jay do the same, and line their nests with the same materials."*
-'History of Animals' or 'Historia Animalium' by Greek philosopher Aristotle, 384-322 BC, ten volumes written in fourth century BC. Book IX: 'Social Behaviour in Animals; Signs of Intelligence in Animals such as Sheep and Birds'. Translated by Richard Cresswell, Oxford England, 1878.

*"**The Comfrey derives its name, according to Dr. Prior, from the Latin word Confirma, from its supposed strengthening qualities.** It is slightly stringent, and was formerly regarded as a steptic and vulnerary.*
***It was known to our fathers by the name of the 'great consound'.** It was also used for 'griefes of the lungs, and possibly with good effect, as the leaves, stems, and the root abound in mucilage'."*
-'English Botany (Sowerby's); or Coloured Figures of British Plants: Volume 7' by John T. Boswell and John Edward Sowerby, London, England, 1880, page 116.

Etymology (origin of words) of Consound:
From Middle English consoude, consowde, consol, consold, consaud, consaude. From Old English consolde. From Old French consolde, consoulde, consoude. From Latin consolida, because it heals.

In the Middle Ages (1275-1325 AD) the word used for Comfrey in Middle English was 'cumfirie' and 'conferye'. In Anglo-French the word was 'cumfirie'. In Old French it was 'confire'. Those words come from the Latin words 'conferva' and 'confervere'.
    'Conferva' (confervae) is a feminine, singular noun. It means a freshwater (aquatic) plant with medicinal power. The word appears in Pliny's "Natural History" (Naturalis Historia), a 77 AD encyclopedia written in Latin by Pliny the Elder, a Roman author and naval commander.
    'Confervere' (confervo) is an intrasitive verb in the third conjugation. It mean to knit broken bones, to grow together, to heal, seethe / boil together. From the "Oxford Latin Dictionary", 1982.

*"**Etymology of Word 'Comfrey':***
***The Comfrey is generally supposed to be the plant described by Dioscorides (40-90 AD) from 'to unite',** on account of the consolidating and vulnerary qualities which were ever attributed to this plant. Hence also the Latin Consolida, Symphytum, and the French Consoude.*
***The term Comfrey was probably derived from the old French word Comfrie or Consyre,** having the same meaning as the foregoing. In rual dialects it has also the names of Consound, Knit-back, Bone-set and Blackwort."*
-'The British Flora Medica or, History of the Medicinal Plants of Great Britain' by Benjamin Barton and Thomas Castle; Revised and Condensed by John R. Jackson; London, England: Chatto and Windus, 1877. See 1838. Symphytum page 117.

"One of the most difficult problems in the study of plant names is the identification of the plant itself, which the very name often helps to obscure.
However, there are cases which present almost insuperable difficulties of identification, and the Old English name of the plant is accordingly difficult or impossible to explain.
    Consider, for example, the term galluc or galloc, which in the Old English Herbarium designates herba confirma ('Comfrey', Symphytum officinale Linn.), whereas in the 'Epinal-Erfurt Glossary' and in the 'Corpus Glossary' it translates the Latin headword galla, 'gall-nut'.

As for the translation of confirma or its synonym Symphytum, Forster and Holthausen consider galloc to be a neologism based on another loan word, gealla ('gall', a blister or painful swelling especially on horses), according to **a normal process in the formation of plant names in which the plant is designated by one of its characteristic features.** *The same process, according to these scholars, lies behind the neologism galloc, that is, 'the plant that cures galls'.* **However, Comfrey is never used in prescriptions for horse galls.**
Yet if the synonyms given in the Latin text of *Pseudo-Apuleius (in the chapter dedicated to the herba confirma) are taken into consideration, it is clear that, in addition to sinfitum, confirma, conserua and pecte, the plant is designated by **alum Gallicum**, *often corrupted to anagallicum, algallicum or anugallicum.* **Galloc, therefore, with the meaning of 'Comfrey', might be another loan word, more recent in its introduction than galloc from Latin galla (galls)."**
-'The Botanical Lexicon of the Old English Herbarium' by Maria Amalia D'Aronco; Anglo-Saxon England, Cambridge, England, Volume 17, pages 15-33, December 1988.
(* -'Pseudo-Apuleius Herbarius' or 'Herbarium Apuleii Platonici', 4th-century {301-400 AD} herbal. This is not the same person as Apuleius of Madaura, 124-170 BC, the Roman poet and philosopher.)

*"***In all western European languages, the name for Comfrey is derived from its application. All the different names focus on uniting and firming.** *For example,* **the Greek term symphyton, (symphytum in Latin), is derived from symphyo:** *'I grow together'. Solidago, derived from solido ('I make firm'), was also a synonym.*
The Latin consolida, *frequently found in historical papers, means 'the one who makes firm'.*
The evolution of the word 'Comfrey' *comprises the middle English comferi, from the old French cumfirie, from vulgar Latin confervia, from confervere ('to boil together').*
The German names, Beinwell and Wallwurz, *are based on the verb 'wallen', which means 'growing together'. 'Bein' originally meant bone, thus Comfrey is an agent that makes bones grow together. Comfrey has also been known as boneset, knitbone, black wort, wall wort, and slippery root."*
-'Comfrey: Ancient and Modern Uses' by C. Staiger, The Pharmaceutical Journal: A Royal Pharmaceutical Society Publication, London, England, Volume 279, pages 22-29, 2008.

*"***Comfrey** *has been used in medicine for many centuries. It was used for healing broken tissue and bone, hence its* **Greek name 'to unite'** *(*Barton and Castle 1877).*
The brief morphology and medicinal uses of two species of Symphytum (S. petraeum and S. alterum) are described in Dioscorides' Herbal (around AD 100) (**Gunther, 1959).
Parkinson (1640) stated that the plant is Symphytum petraeum, whereas Turner (1548) and Matthioli (***1598) used the specific name Symphytum alterum.
The modern name, Comfrey, derives from the mediaeval Latin Comfiria, which in turn replaced the Latin Conferva of Pliny *(****Grigson, 1955).*
It was also known as Consolida maior and, by the Romans, Solidago *(Dodoens, 1586; Salmon, 1610).*
The Romans took the plant to northern Europe (*****Kamm, 1938), and it was used by the Saxons as a vulnerary (******Martindale, 1924). By A.D. 1000 it had appeared in the monastery lists and leech-books (Kamm, 1938).
From the Middle Ages to the middle of the nineteenth century, *there was a belief that both the roots and leaves had the power to heal wounds and bones (Dodoens, 1586, Salmon, 1610). This resulted in the use of names such as* **knitback, bone set, nit-bone, healing blade, bruisewort** *(Kamm, 1938)* **and blackwort** *(Gerard, 1633). It was widely used for quinsy, whooping cough, in poultices for bruises and open wounds and also as a styptic or a pectoral (Kamm, 1938)."*
-'The Anatomy of the Leaf of Symphytum Officinale L.', by J.M. Peck and K.R. Fell, Pharmacognosy Research Laboratory, Bradford Institute of Technology, West Yorkshire, England; The Journal of Pharmacy and Pharmacology, Royal Pharmaceutical Society of Great Britain, Volume 13, pages 154-65, March 1961.
(* -'The British Flora Medica or, History of the Medicinal Plants of Great Britain' by Benjamin Barton and Thomas Castle; Revised and Condensed by John R. Jackson; London, England: Chatto and Windus, 1877. Symphytum page 117.)
(** -The Greek Herbal of Dioscorides by R.T. Gunther. NY: Hafner Publishing Co., 1959. In German. Symphytum page 371, 407.)
(*** -'Petri Andreae Matthioli Senensis, Commentarii, Secundo Aucti' by Pietro Andrea Mattioli, Venetiis {Venice}, Italy, 1560. Also called 'Commentarii Secundo Aucti'. Symphytum page 489.)
(**** -The Englishman's Flora by Geoffrey Grigson. London, England: Phoenix House, 1955. Symphytum page 281.)
(***** -Old Time Herbs for Northern Gardens by Minnie W. Kamm. Boston, Massachusetts: Little, Brown & Co, 1938. Symphytum pages 115-116.)
(***** -'Extra Pharmacopoeia, 18th Edition' by Martindale, Volume 1. London, England: H.K. Lewis & Co., 1924. Symphytum page 862. See 1952.)

Prehistory of Comfrey (Before written records. Some writing began 5,300 years ago.)

*"***As theoretical approaches in Mesolithic (middle stone age) archaeology** *have, over the past fifteen years, made headway in Mesolithic research, it is an objective of this thesis to continue the bridging of the gap between rigorous* **practical analyses of flint assemblages (tools) and the theoretical approaches which seek to illuminate the social factors inherent within the goings-on of hunter-gatherers' lives.**
From his analysis of the **Thatcham Reedbeds (Kennet Valley, Berkshire, England)**, **Holyoak illustrated that whilst*

birch was prominent during the very earliest Mesolithic, it was not long before pine began to dominate the environment (a list of flora from Holyoak's analysis is shown in Table 3.1).
Holyoak's mollusc samples from the Thatcham Reedbeds illustrated that dryland and wetland areas were a congruent (at same time) feature of the valley in the early Mesolithic.
Table 3.1: Comfrey, Symphytum: Pollen evidence from Thatcham Reedbeds."
-'**At the Core of Process: Rethinking the Early Mesolithic Lithic Assemblages from the Kennet Valley, Berkshire (England)**' by Raymond J. Nilson, University of Manchester, England, thesis submitted for degree of Doctor of Philosophy in the Faculty of Humanities, School of Arts Languages and Cultures, 689 pages, 2016.
(* -'Late Pleistocene Sediments and Biostratigraphy of the Kennet Valley, England' by D.T. Holyoak, Unpublished PhD Thesis, University of Reading, England, Volume 1, 1980.)

(Mesolithic {middle stone age} is the time period between the Upper Paleolithic {old stone age} and the Neolithic {new stone age}. It is the final period of hunter-gatherer cultures in Europe and western Asia. In Europe it is 15,000 to 5,000 BP; in southwest Asia it is 20,000 to 8,000 BP. 'BP' means 'Before Present' with standard practice of January 1950 as the beginning date of the age scale. So to convert 15,000 BP to BC if it is the year 2017, then BP + (current year - 1950), or 15,067 BC.)

(The meaning of the above is that during the middle stone age in Europe 5,000 to 15,000 years ago, Symphytum was growing on the land of the hunter-gatherers. Of course, there is no proof but it is possible they were using Comfrey then.)

"**The Hazendonk, the Netherlands, was occupied by humans repeatedly between 4020 and 2410 cal BC and at about 2000 cal BC** (Verbruggen 1992).
The excavations, under direction of Louwe Kooijmans between 1914 and 1976, revealed features such as postholes, pits, hearths, and a palisade. Refuse layers (fossil anthropogenic horizons) along the slopes of the river dune moreover revealed flint, stone, pottery, human remains, bone remains of wild and domestic animals and other organic material.
Human impact resulted in decrease in Quercus (oak), Fraxinus (Ash genus), Alnus glutinosa (alder), Veronica beccabunga-type (speedwell herb), Junxua effusus (rush), Urtica dioica (stinging nettle) and Plantago lanceolata (ribwort plantain), and in a strong increase in dryland anthropogenic indicators (cf. {compare to} Behre 1981) including Cerealia-type (grain), Chenopodium album (lamb's quarters), Solanum nigrum (black nightshade) and Stellaria media (chickweed).
It also shows a moderate increase in ferns, grasses, sedges and wetland taxa, including Sparganium species (bur-reed genus), Filipendula ulmaria (meadowsweet), **Symphytum species**, Ranunculus sceleratus (buttercup) and Rorippa amphibia (yellowcress).
Together, these changes indicate disturbance of the oak vegetation and of the alder carr (wetland), increased presence of open patches, and eutrophication that was probably caused by human dumping of waste."
-'**The Scale of Human Impact at the Hazendonk, the Netherlands, During the Late Neolithic**' by Welmoed Out, Faculty of Archaeology, Leiden University, the Netherlands; Analecta Praehistorica Leidensia: Between Foraging and Farming, 2008.
(Neolithic or 'New Stone Age' is the last part of the Stone Age. It began 12,000 years ago when farming first appeared.)
(Eutrophication or hypertrophication is when water is too enriched with nutrients causing excessive growth of algae.)

"The Jaslo-Sanok Depression is the largest within the whole range of **Polish Carpathian Mountains**. The region borders with Strzyzow-Dynow Foothills to the north, and with Beskid Niski Mountains to the south. The area consists of a series of flat-bottomed valleys separated by series of small hills.
Providing a basis for reconstruction of the **history of vegetation from the Older Dryas to the Boreal period** (in Roz. a) or to the Atlantic period (Roz. b). The pollen diagram from Tarnowiec was divided into twelve pollen assemblage zones. **They covered period from the Older Dryas to Subatlantic period.**
Zone 6 (Roz. a 6, Roz. b 6, Tar. 6):
The sample from the middle part of zone in profile Roz. b is dated at 8670 BP, and two samples from profile Tar. - at 7930 and 5230 BP respectively. The pollen concentration decrease to quite low values. The zone is distinguished by the dominance of Alnus (birch family), Corylus (hazel tree/shrub), Tilia (linden/basswood tree/shrub) and Quercus (oak tree).
Within sporadic species the pollen grains of Rumex acetosella, Plantago maior,
Humulus-Cannabis type, Silene type and **Symphytum are remarkable.**"
-'**Late-Glacial and Holocene History of Vegetation at Roztoki and Tarnowiec Near Jaslo (Jaslo-Sanok Depression)**' by Krystyna Harmata, Institute of Botany of the Jagellonian University, Department of Palaeobotany, Krakow, Poland; Acta Palaeobotanica, Volume 27, No. 1, pages 43-65, 1987.
(The Carpathian Mountains or Carpathians form an arc across central and eastern Europe. It includes parts of Austria, Czech Republic, Hungary, Poland, Serbia, Slovakia, Ukraine and Romania.)
(Sporadic means occurring at irregular intervals or only in a few places.)

"**A Prehistoric Drink Dating Back to the 3rd Millennium B.C. (3000 to 2001 BC):** Human coprolites from Birka, Sweden and Durrnberg, Austria, have been found, dated and palynologically analysed as a part of interdisciplinary studies.
All their pollen spectra are dominated by insect-pollinated taxa well-known as nectar producing flowers, suggesting some consumption of honey. Among those spectra, some show significantly high values of Filipendula ulmaria (meadowsweet) pollen, which was historically used as flavouring in mead production, and which together with other indicators for honey, **suggest that mead was part of the historic and prehistoric human diet both in Birka and Durrnberg.**
Studied human coprolites from the Ferro-Schachtricht site of the Durrnberg salt mine (Hallein, Austria):

Symphytum 0.1% pollen"
-'**Palynological Evidence of Mead: A Prehistoric Drink Dating Back to the 3rd Millennium B.C.**' by Dagfinn Moe (Norway) and Klaus Oeggl (Austria); Vegetation History & Archaeobotany, Volume 23, No. 5, pages 515-526, 2014. Symphytum page 519.
>(Palynology is the study of particulate {dust-like} samples collected from air, water or sediments. These particles, organic and inorganic, give clues to the life, environment, and conditions that produced them.)
>(In paleontology, coprolite is a piece of fossilized dung.)
>(Mead is an alcoholic beverage made by fermenting honey with water and added fruits, spices, grains or hops.)

Symphytum Used in Bronze Age (3000 to 1200 BC):
"*An investigation of pollen, non-pollen palynomorphs and, in some cases, macrobotanical remains found in organic residues in baskets and wooden caskets (box) from Bronze Age burials provided an indication of the remedies placed for the deceased during burial.*
Further evidence on Bronze Age medicines was obtained from organic substances preserved in millstones in the settlement layers.
The medicinal plants used by ancient humans can help to identify diseases that were prevalent in the past. Medicinal herbs from the Bronze Age medicine chest can enrich modern phytotherapy and have a positive effect on development of biomedicine as a whole. Pollen analysis can play an important role in revealing the complex of medicinal plants used by ancient people.
Millstone from Paravani, Georgia burial mound, beginning of third millennium BC (3000-2001 BC):
>*The millstone also uniquely contained pollen grains of Vitis vinifera and that of Symphytum*, Cirsium, Chamaenerium, Brassicaceae, Fagopyrum and Cyperaceae.
Remains of a basket from Bedeni Plateau, Georgia burial mound No. 2:
>*Organic remains were identified as a dark mass with inclusions of plaited, charred, plant branches.* **Herbaceous species included** *Apiaceae, Centaurea, Fabaceae, Artemisia, Polygonum, Trifolium, Heracleum,* **Symphytum**, *Lathyrus, Rumex, Colchicum, Eringium, Astrantia, Fillipendula, etc.*"
-'**Palynological and Palaeobotanical Data about Bronze Age Medicinal Plants from Archaeological Sites in Georgia**' by Eliso Kvavadze, Inga Martkoplishvili, Maia Chichinadze, Luara Rukhadze, Kakhaber Kakhiani, Mindia Jalabadze and Irakli Koridze, National Museum of Georgia, Tbilisi, Georgia; Proceedings of the Georgian National Museum, Natural Sciences and Prehistory Section #5, pages 36-49, January 2013.
>(The Bronze Age was from 3000 to 1200 BC. An ancient civilization is defined as Bronze Age because it produced bronze by smelting copper and alloying it with tin, arsenic or other metals. There were early forms of writing and the beginning of urban civilization.)
>(Georgia is a country in Caucasus region of Eurasia. It is between western Asia and eastern Europe, and bounded to the west by the Black Sea, to the north by Russia, to south by Turkey and Armenia, and to southeast by Azerbaijan.)
>(Paleobotany studies plant remains from geological contexts to reconstruct past environments and determine the evolutionary history of plants.)
>(Palynomorphs are organic-walled microfossils between 5 and 500 micrometers in size.)
>(Macrobotanical means able to be seen with the unaided eye.)
>(Millstones are large, round stones used to grind grain.)

"***It is difficult to find out how prehistoric man discovered the multitude of wound coverings, salves or ointments that were, in all probability, used.*** *It can only be assumed that the selection of these substances occurred by trial and error over a very considerable time. The result was that a number of effective topical (skin) treatments had become available by the time civilizations began to appear about six thousand years ago.*
The first written records containing medical information date from about 2500 BC. *Clay tablets from this time have been discovered in Mesopotamia and the first medical papyri from Egypt are probably some seven hundred years younger, though the Smith papyrus (from about 1650 BC) is thought to be a copy of a much older document.*
Apart from information gleaned from ancient civilizations, much has been learned about the practices of prehistoric peoples from the study of groups of primitive peoples today as well as from the legacy provided by folk medicine.
*****Bergmark (1967) cites the use of a number of plants in wound treatment and it seems possible that these may have been used for many thousands of years.** Many plants have properties useful in wound therapy.*
The Comfrey (Symphytum officinale) contains allantoin *which, apart from being antibacterial, is said to be an excellent healing agent, promoting granulation tissue and being able to promote the healing of fractures (Bergmark 1967).*"
-'**Early History of Wound Treatment**' by Richard D. Forrest MB, Department of Internal Medicine, Centrallasarettet (Central Hospital), Boden, Sweden; Journal of the Royal Society of Medicine, Volume 75, No. 3, pages 198-205, March 1982.
(* -Vallort och Vitlok: Om Folkmedicinens Lakeorter {Comfrey and Garlic: On the Medicinal Herbs in Folk Medicine} by Matts Bergmark. Stockholm, Sweden: Natur och Kultur {Nature and Culture}, 1961, 1967.
This book is only available in Swedish. If anyone has a translation of the Comfrey section, I would appreciate a copy of it.)
>(Mesopotamia is an historical region in western Asia in the Tigris-Euphrates river system. Today it includes most of Iraq, Kuwait, parts of northern Saudi Arabia, eastern parts of Syria, southeastern Turkey, and regions along the Turkish-Syrian and Iran-Iraq borders.)

"**Herb: Comfrey, Symphytum officinale:**
Period of archeological botanical food finds in Britain: **Roman - Anglo - Scandinavia - Medieval.**"
-'**Archeological Botanical Food Finds in Britain**' cataloged by Jennifer Baker, Hodegon NVG (New Varangian Guard),

Melbourne, Australia, from Archaeobotanical Computer Database (ABCD), Dec 2008, http://intarch.ac.uk/journal/issue1/tomlinson/scripts/index-latin.html. New Varangian Guard Inc. is historical re-enactment organisation in Australia.
> **(Roman Britain 44-407 AD, Anglo-Saxon / Scandinavian / Saxo-Norman Britain 850-1150 AD, Middle Ages {Medieval} Britain 597–1485 AD.** So there is archeological evidence that Comfrey has been used in Britain since at least the times of Roman Britain.)

Ancient Times and Comfrey (400 BC- 400 AD)

"The word Symphytum is derived from the Greek word 'a facultate glutandi', i.e. from its glueing properties. The Latins used the word Consolida for the same reason, from consolidare 'to solder, close or glew up.' "
-The Medicinal Uses of Comfrey by Dr. Charles MacAlister, M.D., F.R.C.P., written 1935.

"484-425 B.C. Greek historian <u>Herodotus</u> notes use of Comfrey to staunch severe bleeding."
-National Geographic Guide to Medicinal Herbs: The World's Most Effective Healing Plants by Tieraona Low Dog M.D., Rebecca L. Johnson, Steven Foster and David Kiefer M.D. Washington, DC: National Geographic, 2012.
>(Herodotus, 484-425 B.C., was an ancient Greek historian born in Halicarnassus in the Persian Empire, now Bodrum, Turkey. He wrote the book 'The Histories', a detailed account of the origins of the Greco-Persian Wars, 499-449 B.C.)

<u>Nicander of Colophon</u>, **Greek poet/physician 2nd century BC (200 to 101 BC), wrote about Comfrey in his herbal book 'Alexipharmaca'** of 630 hexameters (poetry) about poisons and their antidotes. In 1856 it was translated into Latin: 'Nicandrea; Theriaca et Alexipharmaca'. He was born at Claros (now Ahmetbeyli, Turkey), near Colophon, where his family held the hereditary priesthood of Apollo. (If anyone has English translation of his writings, let me know what he says about Comfrey.)

<u>Pliny the Elder</u> **(Gaius Plinius Secundus), 23-79 AD, a Roman naturalist and philosopher used Comfrey root medicinally.**
> He wrote: "***The roots be so glutinative that they will solder or glew together meat that is chopt in pieces, seething in a pot, and make into one lump.*** *The same bruysed and layed in the manner of a plaster doth heal all fresh and green wounds."*
> Pliny found Comfrey root to be good as a poultice for binding broken bones together.

Pliny wrote 37 books called **'Naturalis Historia' or 'Natural History'** (77-79 AD). It is one of the largest single works to have survived from the Roman Empire. It includes astronomy, mathematics, geography, ethnography, anthropology, physiology, zoology, botany, agriculture, horticulture, pharmacology, mining, mineralogy, sculpture, painting, and precious stones.
> Books XX to XXIX (20 to 24) are about medicine, especially plants used as drugs. Pliny lists 900 drugs, compared to 600 in Dioscorides's De Materia Medica, 550 in Theophrastus, and 650 in Galen. Diseases and their treatment are covered in Book XXVI (26).

>*"The 'Naturalis Historia' of the Pliny the Elder (23-79 AD) is one of the most important testimonies of ancient phytomedicine. In book 26, chapter 137, Comfrey is mentioned for the first time for the treatment of bruises and sprains, and a syrup of the herb or a decoction of its root are used.*
>*Chapter 148 claims that Comfrey ensures rapid healing of wounds and, in chapter 161, Comfrey is mentioned as an emmenagogue when ground into dark wine."*
>-'Comfrey: Ancient and Modern Uses' by C. Staiger, The Pharmaceutical Journal: A Royal Pharmaceutical Society Publication, London, England, Volume 279, pages 22-29, 2008.
>>(Phytomedicine is herbal traditional medicine using plants for preventive and therapeutic treatments.)
>>(An emmenagogue increases menstrual flow.)

*"I find, also, that various kinds of **aromatites (aromatic wine)** are prepared, differing but very little in their mode of composition from that of the unguents.*
We find mention made of nectarites also, a beverage extracted from a herb known** to some as 'helenion', to others as 'Medica', and to others, again, **as Symphyton (Symphytum officinale of Linnaeus, being all different varieties),** Idaea, Orestion, or nectaria, **the root of which is added in the proportion of forty drachms to six sextarii of 'mnst' (must?), being first similarly placed in a linen cloth."
-'**The Natural History of Pliny, Volume 3**' translated in 1855 by John Bostock and Henry Thomas Riley, London, England, Symphytum page 259.
>(Unguent is a soft greasy or sticky substance used as ointment or for lubrication.)
>(Drachms= 60 grains= 64.79 milligrams)
>(Sextarius is a Roman measure of capacity, 1/6 of a congius, equal to 1 1/4 United States pints.)

*"**Pliny the Younger (nephew of Pliny the Elder) writes only of the Comfrey of the rock, Symphytum Petracum:***
'Singularly good is it for the sides and flanks, the splene, reines, and wings of the belly; for the breast, the lights; for such as reject or cast up blood, and are troubled with the asperity and hoarseness in the throat.
Moreover the chewing of it quencheth thirst and hath a principal virtue to cool the lungs. Being applied outwardly in the

form of a cataplasm it knitteth dislocations, helpeth convulsions, is comfortable to the splene and the bowels if they be fallen by any rupture.' "
-'**The Historie of the World: Commonly called the Natural Historie of C. Plinius Secundus**' (Gaius Plinius Caecilius Secundus = Pliny the Younger, 61-113 AD). Translated by Philemon Holland, Doctor of Physicke. Volume II., page 275, London, 1634.

Greek physician / pharmacologist / botanist Pedanius Dioscorides (40-90 AD) recommended Comfrey for healing wounds, repairing broken bones, and curing lung/stomach problems.
He was born in Anazarbus, Cilicia, Asia Minor (Turkey). He was a doctor for the Roman military.
He authored '**De Materia Medica' (On Medical Material)** in 65 A.D., a 5-volume Greek encyclopedia about 600 herbs and medicinal substances.
    Volume I: Aromatics, Volume II: Animals to Herbs, Volume III: Roots, Seeds and Herbs,
    Volume IV: Roots and Herbs Continued, Volume V: Vines, Wines and Minerals.
    Most plants in his pharmacopeia were found in the Greek-speaking eastern Mediterranean.
During the Middle Ages, 'De Materia Medica' was circulated in the original Greek, as well as in Latin and Arabic translations. 'De Materia Medica' is the prime historical source of information about medicines used by Greeks, Romans, and other people in ancient times. **It is the precursor (forefather) to all modern pharmacopeias. It was used for 1,500 years. It is the most influential and important herbal ever written.** In the 1500s there were 78 different editions of Dioscorides 'De Materia Medica' published (Stannard 1966).

*"In the older botanists Symphytum officinale (Asperifoliaceae), black root; Fraas suggests Symphytum Broehum Bory, Knolly Comfrey, with regard to the fact that D. denotes the flowers white or yellow, while at **Symphytum officinale are very often reddish, purple, or bluish red,** what D. would probably not overlooked.*
By the way, it is questionable if that Symphyton of D. is not one of our different plant, think only to the flower and fruit, which should be like that of Verbascum. Formerly the root radix was Consolidae officinell, it contains a lot of Bassorin, some amylum, tannin, sugar, asparagine, according to Greimer an alkaloid Symphyto-Cynoglossin, which has a paralyzing effect on the central nervous system exercise."
-**The Greek Herbal of Dioscorides** by R.T. Gunther. NY: Hafner Publishing, 1959. In German. Symphytum page 371. (This is an online translation of the German text so some is not clear. If you have a good translation, please send it.)

"The brief morphology and medicinal uses of two species of Symphytum (S. petraeum and S. alterum) are described in Dioscorides' Herbal (around 100 AD) (Gunther, 1959)."
-'**The Anatomy of the Leaf of Symphytum Officinale L.**' by J.M. Peck and K.R. Fell, Pharmacognosy Research Laboratory, Bradford Institute of Technology, West Yorkshire, England; The Journal of Pharmacy and Pharmacology, Royal Pharmaceutical Society of Great Britain, Volume 13, pages 154-65, March 1961.

*"**Compare to Dioscorides, IV. 10,** 'some call it pecte'. Andre (*1985, 253) notes that the **Common Comfrey variety in Greece (and hence Dioscorides) was Symphytum bulbosum Schimp..***
*Pliny 27.41, 'Alum nos vocamus, Graeci Symphyton petraeum' (Alum we call the Greeks Symphyton petraeum) is referring to low pine **(Symphytum tuberosum L.),** for which see **Dioscorides IV. 9.**"*
-**The Alphabet of Galen: Pharmacy from Antiquity to the Middle Ages** by Nicholas Everett. Canada: University of Toronto Press, 2012, pages 30, 343.
(* -'Noms de Plantes Gaulois ou Pretendus Gaulois dans les Textes Grecs et Latins' {Names of Gaulish Plants or Gaulish Pretendus in Greek and Latin Texts} by Jacques Andre, Etudes Celtiques, Volume 22, pages 179-198, 1985.)
(Volume IV: 'Roots and Herbs Continued'.)

*"**Dioscorides De Materia Medica, Book 4, Chapter 10: Sumphuton Allo:***
    Suggested: Symphytum-magnum, Consolida maior (Fuchs), Symphytum consolida major (Bauhin),
    Symphytum officinale (Linnaeus), Comfrey, Knitbone.
Symphyton alterum sends out a stalk *2 feet (0.6 meter) high or more: light, thick, angular, empty, similar to sonchus. Around which comes (from not great distances) rough narrow leaves, somewhat long, similar to those of bugloss.*
The stalk has some extensions of slender leaves adhering to it, stretching along at the corners. **From every wing are yellowish flowers standing up,** *and the seed is around the stalk like verbascum.*
The whole stalk and leaves have a somewhat prickly down that causes itching if touched. **The roots** *are underneath - to the outward appearance black, but within white and slimy- of which use is made.*
    Pounded into small pieces (and taken in a drink), they are good for bloodspitters and hernias. Applied, they
    close up new wounds. Boiled, they join pieces of flesh together. They are smeared on for inflammations- most
    usefully for those in the perineum- with the leaves of senecio.
It is also called pecton, while the Romans call it solidago."
-**Dioscorides De Materia Medica: Being an Herbal with Many Other Medicinal Materials** by Tess Anne Osbaldeston. Johannesburg, South Africa: Ibidis Press, 2000. Symphytum pages 552-553.

*"**Dioscorides and Comfrey:***

In all West-European languages, Comfrey is one of the few medicinal herbs whose name is derived from its application. Its history in European medicine was influenced by Dioscurides' (Dioscorides) 'Materia medica'. A few aspects, including the application of the root in treating rheumatism and gout, were added to its history in the Middle and New Ages (Renaissance?).

It is shown in this article that Dioscurides actually described Symphytum officinale, a fact hitherto under dispute. Further, a surprising continuity of Comfrey's application lasting more than 2000 years is made more vivid.

The 'Materia Medica' by the Greek physician Pedanius Dioscorides is the oldest pharmacology of Europe. The work arose simultaneously and independently of the 'Naturalis Historia' of Pliny. **It has the European and Arabic phytotherapy shaped over almost two millennia (2000 years).**

The 10th chapter of the Fourth Book deals with Comfrey under the heading:

'Symphyton allon' ('The Other Symphytum'). Synonyms are 'Pekte' ('the density', 'the net') and 'Soldago'. **Here is a detailed description of the plant:** 'The stalk is at least two cubits long, light, thick, angular, thistle-like and hollow. On it are in short intervals the hairy, slender, elongated, those of the ox-tongue similar leaves. From every armpit it drives the white or yellow flowers; Flowers and fruit stand around a stem as in Phlomos (Mullein). The whole stalk and leaves have a somewhat rough wooly coating, which at touch causes itching. The roots underneath are black on the outer surface, white on the inside and slimy; these are used.

Finely poked and drunk, they are good for those who have blood spasms and internal ones abscesses suffer, as an envelope they also stick fresh wounds. As cataplasm (poultice) they are used in inflammation, especially at the anus, and indeed with the leaves of the Crosswort (Senecio vulgaris)."

-**'Symphytum Officinale L.: Comfrey in European Pharmacy and Medical History'** by K. Englert, J.G. Mayer and C. Staiger, Institut fur Geschichte der Medizin (Institute for the History of Medicine), Universitat Wurzburg, Germany; Zeitschrift fur Phytotherapie (Journal of Phytotherapy), Volume 26, No. 4, pages 158-168, 2005.

(This report is in German. The first 2 above paragraphs is the abstract it gave in English. After that is from my online translation of the German. If you have an English translation of this report, could you please send it to me.)

(A cubit is an ancient unit of measurement based on the length of the forearm from the elbow to the tip of the middle finger, equal to about 18 inches = 46 centimeters.)

<u>Galen of Pergamon</u> **(Aelius Galenus), a famous Greek physician, surgeon and philosopher (130-200 AD)**, included Comfrey in his medical treatises such as **'Galenou Apanta'**. He influenced the development of anatomy, physiology, pathology, pharmacology and neurology.

"Comfrey (Symphytum officinale L.) (translated from Galen's writings): **Comfrey, which some call 'Gallic garlic', is a well-known plant.** *Its root is oblong, has black bark, and a taste that is slightly bitter and a little sticky. Because of this* **it is administered to stop the coughing up of blood. It vigorously binds and agglutinates (sticks together).**

Translator's interpretation:
According to recent research (as of 2012), Comfrey has analgesic, anti-inflammatory, and anti-exudative actions, and its extract is used in topical (skin) preparations for the treatment of muscle and joint complaints. For Comfrey contains mucilage, tannin, and allantoin, the later of which stimulates tissue regeneraton, hence its use for both external injuries and gastric ulcers. Most of this fits the bill given in 'The Alphabet of Galen'."

-**The Alphabet of Galen: Pharmacy from Antiquity to the Middle Ages** by Nicholas Everett. Canada: University of Toronto Press, 2012, pages 30, 343.

"And Theriac helps those with trouble breathing when thick phlegm builds up in the hollows of the lungs and prevents a man from breathing, cutting and thinning and rendering removable the build-up of sticky fluid. **And Theriac greatly helps those bringing up blood if boiled up with Comfrey and dissolved in water and so administered.** *And it often cures ills of the stomach and makes the man who has lost his appetite and cannot take food turn to it with relish."*
-'On Theriac to Piso, (Treatise) Attributed to Galen: A Critical Edition with Translation and Commentary' by Robert Adam Leigh, University of Exeter, England, thesis for degree of Doctor of Philosophy in Classics, July 2013, page 129. The author translated it from Greek.

(Theriac was a medicinal recipe formulated by Greeks in 1st century AD. It was widely used in ancient world including Persia {Iran}, China and India. It was an antidote to poison and viewed as a panacea or heal-all.)

"When the **Saxon invaders entered Great Britain** *they took with them much knowledge concerning herbal healing.* **It is well known that they made frequent use of the dandelion, Comfrey, nettle, burdock, and other common wayside herbs in treating the sick.** *The Saxon girls were taken into the fields by their parents and taught the names and healing virtues of the plants. And so a knowledge was planted that grew until it became customary to have an 'herb garden' in England."*
-**Back to Eden: The Classic Guide to Herbal Medicine, Natural Foods and Home Remedies** by Jethro Kloss. Wisconsin: Lotus Press, 1939-1999, page 60.

(Starting around 360 AD and after, various Germanic peoples {Alemanni, Saxons, etc.} came to Roman Britain. Saxons were Germanic tribes first mentioned as living near the North Sea coast of what is now Germany in the late Roman Empire around **360-476 AD**.)

Middle Ages (5th- 14th Centuries, 401- 1400 AD)

"**Johnson Papyrus** (London, England; Wellcome Library, MS 5753):
It is a fragment of an early fifth century AD herbal. It is the oldest extant manuscript illustration of a plant.
The papyrus fragment shows a sphere of dark blue-green leaves supported by some small scraggly roots. Below the illustration is a fragment of Greek text. **The illustrated plant has been identified as 'Symphyton' (modern Comfrey), which was an important medicinal plant.**
However the illustration does not closely resemble Comfrey, so that, if the identification is correct, the illustration would have had been of little use as an aid to identification."
-Wikipedia®: The Free Encyclopedia, www.wikipedia.org, 2019.
(Papyrus is similar to thick paper and was used in ancient times to write on. It was made from the pith of the papyrus plant, Cyperus papyrus, a wetland sedge. It also refers to a document written on sheets of this material, joined together, and rolled up into a scroll.)

"**It has to be mentioned that the earliest known illustration of a plant is a drawing of Symphytum or Comfrey (Symphytum officinale L.), from a fragment known as the Johnson Papyrus, made c. (circa) 400 AD.
The action of plant illustration and drawing was a tool of scientific value, related with plant identification, until the beginning of the 20th century** (Lack and Ibanez, 1997; Janick and Paris, 2006; Rhizopoulou, 2007).
It is of greatest interest and importance to search through the methodology, the nomenclature and the etymology (word origin) of scientific names of plants, for the interpretation of plant names used by Dioscorides and his precursors."
-'**The Plant Material of Medicine**' by Sophia Rhizopoulou and Alexandra Katsarou, University of Athens, Greece; Advances in Natural and Applied Sciences: American-Eurasian Network for Scientific Information, Volume 2, No. 2, pages 94-98, 2008.

"**The illustrated herbal is a genre of pharmacological book** known in Graeco-Roman antiquity from at least the first century BCE (Before Common Era = BC). The encyclopaedist Pliny the Elder ('Natural History' 25.8), for example, mentions a number of writers of herbals who provided pictures of plants above descriptions of their medicinal effects.
One of the earliest surviving examples of such illustrated herbals, dating to around 400 CE (Common Era = AD) (to judge from its handwriting), is a lavish, **papyrus codex (manuscript) leaf with full colour illustrations** named 'the Johnson Papyrus' (P. Johnson), after John de Monins Johnson who found it in 1904. He discovered it in the city of Antinoupolis in Middle Egypt, and it is now held in the Wellcome Library in London, England.
On one side of the page, we find a cabbage-like plant with dark, bluish-green leaves bearing the name Symphyton, perhaps to be identified as Comfrey (Symphytum officinale L.).
The surviving caption, written directly underneath the picture, is as follows, though the text may have continued on for several more lines after the break:
'**This plant, when ground down, cures every haemorrhage and agglutinates wounds and severed tendons. It cures coughing up of blood.**' "
-'**Painting Plants in Roman Egypt**' by Dr. David Leith, Exeter, England, specializes in Graeco-Roman medicine; https://recipes.hypotheses.org, The Recipes Project, 'Food, Magic, Art, Science, and Medicine', January 20 2015.

"**This fragment of a leaf from an illustrated herbal from Hellenistic Egypt shows a plant that is possibly Symphytum officinale, or Comfrey.** Found by J. de M. Johnson* in 1904 at Antinoe (Antinopolis), Egypt, while working for the 'Egypt Exploration Fund' of London, England.
The herbal is made of papyrus, a plant that flourished in the valley of the Nile, Egypt, and **the text is in Greek,** the language of science throughout the eastern Mediterranean at this time.
The fragment is probably from a copy of the herbal of Dioscorides of Anazarbus, a first-century Greek physician born in Asia Minor whose work became the foundation text of medieval botany. **The fragment is thought to be the earliest surviving example of an illustrated herbal.**
It is not known for certain whether this fragment originally was part of a roll (the usual format of papyrus manuscripts until the later Roman period) or a codex (the book form with which we are now familiar). The date of the fragment suggests that most likely it was from a codex, as does the fact that it is written and illuminated on both sides, which would have made it difficult to consult in roll form."
-'**Johnson Papyrus', World Digital Library**, www.wdl.org/en/item/3959/, April 2012. A project of the 'United States Library of Congress', with support of the 'United Nations Educational, Cultural and Scientific Organization', and libraries, archives, museums, educational institutions, and international organizations around the world.
(* -'Antinoe and Its Papyri: Excavation by the Graeco-Roman Branch, 1913-14' by J. da M. Johnson; 'Journal of Egyptian Archaeology: Egypt Exploration Society', London, England, Volume 1, No. 3, pages 168-181, July 1914.)

"**On the recto (right side of the codex page) is a cabbage-like plant with heavy leaves of bluish-grey closely applied to each other. It bears the title CYMOYTON.**
There is no figure of Symphyton in the Juliana Anicia manuscript, but **our figure accords well with the Sinfitos of the Leyden (Leiden) Apuleius of the seventh century and better with that of the Cassel (Kassel) Apuleius of the ninth century.** Below the cabbage-like plant, in the papyrus, roots can be seen.
From our examination of this fragment of papyrus, we have inferred that it is from a Greek work similar in form, sub-

stance and illustration to the Latin Apuleius, which was among the most widespread of Latin medical documents during the Dark Ages."
-'The Herbal in Antiquity and Its Transmission to Later Ages' by Charles Singer; The Journal of Hellenic Studies: Society for the Promotion of Hellenic Studies, London, England, Volume 47, Part 1, pages 31-33, 1927.
('Juliana Anicia Codex ', 'Vienna Dioscurides' or 'Vienna Dioscorides' is an early 6th-century Byzantine Greek illuminated manuscript of 'De Materia Medica'.)
(Apuleius Barbarus also known as Apuleius Platonicus and Pseudo-Apuleius. He wrote the 4th-century Latin herbal known as 'Pseudo-Apuleius Herbarius' or 'Herbarium Apuleii Platonici'. Until the twelfth century it was the most influential herbal in Europe. A manuscript of the Apuleius herbal from sixth or early seventh century is in Leiden, Netherlands.)

*"**Archaeological Excavations in Poland:** On the basis of plants found in archaeological excavations in Poland, **natural and semi-natural plant communities most often exploited by man in prehistoric times** are described.*
The opinion is expressed that the present-day synanthropic flora and vegetation reflect local habitat conditions and past and present economic activity of man.
Forest Apophytes in younger phase of the <u>Early Middle Ages</u>: Symphytum officinale: field weeds, 1%.
Percentage of species in relation to the total number of all wild species (not only apophytes) in each period: "
-'History of the **Synanthropic Changes of Flora and Vegetation of Poland**' by Helena Trzcinska-tacik and Krystyna Wasylikowa; Polish Academy of Sciences, Institute of Zoology, Memorabilia Zoologica, Wroclaw, Poland, Vol 37, pages 47-69, 1982.
(Synanthropes are species of wild animals and plants that live near, and benefit from humans' artificial habitats. It does not include domesticated animals.)   (Apophytes are native plants growing in disturbed land.)

*"**In the 6th century AD (501-600 AD), <u>Aetios (Aetius)</u> of Amida (Mesopotamia, now Asian Turkey)** in his medical encyclopedia lists several remedies prescribed for the treatment of non-ulcered cancer:*
Comfrey (Symphytum bulbosum) placed in bread could be applied as a plaster, *and in more serious cases, king's clover with opium poppy seeds (Papaver somniferum) were to be mixed in sweet wine, which had to be boiled."*
-'The Treatment of Cancer in Greek Antiquity' by A.Karpozilosa and N.Pavlidis, University of Ioannina, Greece; European Journal of Cancer, Volume 40, Issue 14, pages 2033-2040, September 2004.
(Aetios of Amida lived around 450-550 AD and wrote 'Sixteen Books on Medicine'. Although his work does not contain much original matter, and is heavily indebted to Galen and Oribasius, it is one of the most valuable medical books of antiquity compiled from authors found in ancient Alexandrian Library of Egypt, whose works are no longer available.)

"<u>**Paulus Aegineta**</u> **Translated from Greek, 625-690 AD:**
Symphytum, Comfrey: *The 'Rock Comfrey' is composed of opposite powers. For it has some incisive powers by which it cleanses the pus in the chest and the kidneys; and it has also some constringency which renders it a suitable remedy for haemoptysis, sprained and ruptured parts, the red flux in women, and intestinal hernia. It contains also some hot humidity, by which it quenches thirst and cures asperities in the trachea.* **The other species, called the Great Comfrey,** *is glutinous and prurient like squills. It is used for the same purposes as the 'Rock'.*
Commentary by Francis Adams, London, England, 1844:
The second species is indisputably the Symphytum officinale, a plant which the Romans, no doubt, naturalized in this country. *The other has been the subject of more controversy. See Parkinson (526) and Matthiolus and Sprengel (Ad Dioscor. iv, 9). We are satisfied that it was the Coris monspeliensis.*
Our author Aegineta manifestly abridges Galen, who borrows from Dioscorides, but improves what he takes. They all agree in commending both as being possessed of great virtues as expectorant and vulnerary medicines.
The Arabians in general seem not to have attached much importance to the Symphytum, *for, after a cursory examination while writing this article, we have not been able to find it in any of the others* **except 'Ebn Baithar', who merely gives extracts from Dioscorides and Galen."*
-'The Seven Books of Paulus Aegineta: Translated from the Greek, with a Commentary Embracing a Complete View of the Knowledge Possessed by the Greeks, Romans, and Arabians on All Subjects Connected with Medicine and Surgery' by Francis Adams, London, England, 1844. Printed in 3 volumes. Symphytum is in Volume 3, Book VII, Section III, 'On the Power of Simples Individually', page 364.
(Paul of Aegina or Paulus Aegineta, 625-690 AD, was a Byzantine Greek physician who wrote the medical encyclopedia 'Medical Compendium in Seven Books'. At the time, this work of Western medical knowledge was unrivaled in its accuracy and completeness. His reputation in the Islamic world was great.)
(Squill is a coastal Mediterranean plant in the lily family.)   (The species Coris monspeliensis is in the Primrose family.)
(Ebn Baithar or Diya Al-Din Abu Muhammad or Ibn al-Baytar, 1197-1248 AD, was an Andalusian {Muslim Spain} Arab pharmacist, botanist, physician and scientist.)

*"**The treatment of rheumatism and gout were added to the indications for Comfrey in the <u>Middle Ages</u>.**"*
-'Comfrey: Ancient and Modern Uses' by C. Staiger, The Pharmaceutical Journal: A Royal Pharmaceutical Society Publication, London, England, Volume 279, pages 22-29, 2008.

Anglo-Saxon Leech Books

In the Middle Ages, a 'Leech' was a physician or surgeon.

A Leechbook was a collection of medical remedies, cures, procedures and recipes.

*"**Saxon Leechdom: While yet the Dark Ages (500-1000 AD) were brooding over Europe, Saxon England was struggling to master the elements of the science of medicine.** In a selection of herbs for the use of the patient minute (detailed) directions were given in order to secure the herb at its best. Bede was a firm believer in the common usages of the Saxon Leech.*
Surgical operations *were confined to the use of the lancet, and to splints; and a number of surgical remedies of a very rude character are preserved in the leech books:*

If a man's bowel be out, pound galluc, *wring through a cloth into milk warm from the cow, wet thy hands therein, and put back the bowel into the man, sew up with silk; then boil him **for nine mornings galluc, that is Comfrey,** except need be for a longer time; feed him with fresh hen's-flesh."*

-'**A Chronology of Medicine: Ancient, Mediaeval, and Modern**' edited by John Morgan Richards, London, England, 1880. Symphytum page 75.

(Saxons or Sachsen were a Germanic people whose name was given in the Early Middle Ages to a large country {Old Saxony} near the North Sea coast of what is now Germany. Early Middle Ages or Dark Ages were from 500 to 1000 AD.)
(Saint Bede or Venerable Bede, 672-735 AD, was an English Benedictine monk at the monasteries of Saint Peter and Saint Paul in Kingdom of Northumbria of the Angles, now Monkwearmouth-Jarrow Abbey in Tyne and Wear, England.)

"In the treatise called 'Book III of the Leech Book of Bald':
*'If a man's bowel be out, pound galluc (Comfrey), wring through a cloth into milk warm from the cow, wet thy hands therein, and put back the bowel into the man, sew up with silk, then **boil him galluc for nine mornings**, except need be for a longer time. Feed him with fresh hens' flesh.'*
Comfrey (Symphytum officinale) had the reputation of possessing great power in uniting wounds or broken parts; and is still used for that purpose in England (1904)."
-'**English Medicine in the Anglo-Saxon Times:** Two Lectures Delivered Before the Royal College of Physicians of London' by Joseph Frank Payne, M.D., Oxford, England, 1904. Lecture II, Part I, page 92.

(The oldest surviving English herbal manuscript is the 'Saxon Leech Book of Bald', also called 'Medicinale Anglicum', written around **900-950 AD**.)

*"My attention had been directed to the fact that from time immemorial the **Symphytum, or Common Comfrey**, had been much recommended by the ancient writers in medicine as a vulnerary and as a dressing for sores and ulcers of various kinds.*
From the days of 'Saxon Leechdom': *(around 1000 AD) to the end of the seventeenth century (late 1600s) there were few authors who did not refer to it, and it was sometimes spoken of as being the 'chief vulnerary herbal'.*
-'**A New Cell Proliferant: Its Clinical Application in the Treatment of Ulcers**' by Charles J. MacAlister, M.D., F.R.C.P., Honorary Physician to the 'Liverpool Royal Southern Hospital' and to 'Royal Liverpool Country Hospital for Children', British Medical Journal, London, England, pages 10-12, January 6 1912. (Vulnerary means used for healing wounds.)

*"**Comfrey** comes from Asia and was known by 400 B.C., when Herodotus recommended it for **mechanically preventing the flow of blood. The Romans likewise knew of this quality and the plant was taken to northern Europe as a valuable curative agent and appears in monastery lists and Saxon Leechbooks of 1000 A.D.**"*
-'**Old Time Herbs for Northern Gardens**' by Minnie W. Kamm. Boston, Massachusetts: Little, Brown & Co, 1938. Symphytum pages 115-116.

*"**The herb Comfrey appears in monastery writings and herbals (Leech-Books) from A.D. 1000.** 'This Wort (plant) strengthens the man,' we read in **the 'Saxon Leechdom', a collection of recipes, prescriptions and remedies for various injuries. Other Saxon herbariums** recommended it for those 'bursten within' (possibly from internal bleeding, ruptures, hernias, etc.) for which purpose, to give one example:*
Comfrey leaves were heated in or over hot, near-ash embers, ground and stirred into honey, and then taken on an empty stomach."
-Comfrey: What You Need to Know by Ben Charles Harris. New Canaan, Connecticut: Keats Publishing, 1982, page 5.

*"**In 1000 AD Comfrey appears in monastic writings such as in the 'Saxon Herbarium'** with one recommendation: 'ground dried Comfrey leaves stirred into honey and take on an empty stomach for internal bleeding and ruptures'.*
Many religious orders are credited with expanding the cultivation of Comfrey throughout Europe in their monastery gardens, *particularly for the treatment of soldiers' wounds."*
-'**Herbal Legacy in the Hedgerows of Thriplow**' by Bernard Meggitt, Thriplow Society: Local History, Environment and Conservation, Thriplow, England, www.thriplow.org.uk, Volume 24, Summer 2015.

"Bald's Leechbook:
*1. For a dry cough again take elecampane and **Comfrey, eat in virgin honey.***
*2. **For foot-ache** take betony, common mallow, fennel, ribwort, equal quantities of each, mix milk with water and bathe the swollen limb from the upper side in case the swelling internalise, **then take boiled Comfrey, lay on.***
*3. **For sudden swelling** for two nights mix together silverweed, flour of malt dregs, cress, the white of an egg, marsh*

mallow (bishop-wort), elecampane, garden radish, parsley and **Common Comfrey and apply.**
4. Salve for an internal wound: wine, oil, Comfrey, honey."
-'**Anglo-Saxon Medicine and Disease: A Semantic Approach,** Volume II: Appendix, Bald's Leechbook' by Conan Doyle, thesis for degree of Doctor of Philosophy, University of Cambridge, England, 2017.

"*Yalluc / Galluc or Comfrey (Symphytum officinale)*: **This wort, which is called 'confirma' (Comfrey), and by another name 'yalluc',** *is produced on moors (fen, heath) and on fields, and also on meadows.*
    1. For wives (womens) flux (discharge), *take this wort confirma, pound it to very small dust, administer it in wine to drink; soon the flux stancheth (stops).*
    2. If one be bursten (ruptured) within, *let him take roots of this wort, let him roast them in hot ashes, then swallow them in honey fasting, he wiill be healed; and it also purges the whole stomach.*
    3. For sore of maw (stomach), *take this same wort, and mingle with honey and with vinegar; thou shalt perceive much advantage.*"
-'**Leechdoms, Wortcunning and Starcraft: Being a Collection of Documents Illustrating the History of Science Before the Norman Conquest**' also called 'Chronicles and Memorials of Great Britain and Ireland During the Middle Ages' edited by Thomas Oswald Cockayne, London, England, 1864. Volume 1, pages 27, 163, 241.
    (The Norman Conquest of England was a 1066 AD invasion and occupation of England by Norman, Breton, Flemish and French soldiers.)
    (Worts are plants used as medicine. Examples include Bellwort, Dragonwort, Honeywort, Lungwort and Saint Johns Wort. Most of these names come from the 1600s or earlier.)

"**Although this Anglo-Saxon herbal about Yalluc is of Roman origin (Apuleius), we see that Symphytum is largely shorn (removed) of the dimensions ascribed to it by Pliny.**
The absence of any mention of its virtues in knitting broken bones is the more striking, as among the ancient Britons a broken limb was one of the 'three dangerous wounds for which the physician received his highest, and an extraordinary fee': one hundred and eighty pence (penny) and his meat, or one pound (240 pence) without his meat."
-'**American Observer Medical Monthly: Devoted to Homoepathic Materia Medica, Surgery, Gynaecology, Obstetrics, Otology, Ophthalmology, Practice of Medicine**', New Series Volume 17, Detroit, Michigan, 1880. Article: 'Materia Medica: The Empirical History of Symphytum Officinale' by Professor S.A. Jones, M.D., Ann Arbor, Michigan, August 1880, pages 384-394.
    (I think by 'Apuleius' he means Apuleius Barbarus also known as Apuleius Platonicus and Pseudo-Apuleius. Pseudo-Apuleius wrote the 4th-century Latin herbal known as Pseudo-Apuleius Herbarius' or 'Herbarium Apuleii Platonici'. Printed versions, rather than the early handmade manuscripts, were first published in the late fifteenth century in Rome.)
    (In early English currency a pound was 20 shillings, and a shilling was 12 pence, so a pound was 240 pence.)

"**The Anglo-Saxon Pharmacopeia (450-1066 AD): The major Saxon medical texts include a total of 327 herbs and spices.** *A total of 191 herbs in what may be called regular use, and of these a group of twenty-five herbs make up almost half the total usage, and a group of fifty-five make up two-thirds.*
This suggests that the Anglo-Saxons exercised some measure of selection and did not use every herb known to them, more or less at random. **The basis of their medicine thus consisted of a nucleus of herbs of proven effectiveness, surrounded by a penumbra (outer region) of herbs of lesser value or more limited effectiveness.**
    This outer group includes *includes feldmore (Pastinaca sativa) which is used 14 times for 13 different ailments,* **Comfrey (Symphytum officinale), used 12 times for 12 different ailments,** *and woad (Isatis tinctoria), used 8 times for 8 different ailments.*
Herbal medicine was still a developing tradition, and the Saxons must have been experimenting with many herbs, using them in different combinations and for different ailments, until **greater experience reduced the number of uses to those for which they proved most effective.**
    This is especially true for herbs such as Comfrey, and to a lesser extent woad, which by the time of sixteenth century (1501-1600 AD) herbalists such as Gerard were used for certain specific conditions: Comfrey for bruises, wounds and broken bones, *and woad for ulcers, inflammation and bleeding.*"
-'**Anglo-Saxon Medicine Within Its Social Context**' by H.M. Cayton, Durham E-Theses for Doctor of Philosophy, Durham University, England, pages 76-82, 1977. (The Anglo-Saxon period in Britain is between 450 and 1066 AD. These dates are after their initial settlement and up to the Norman conquest.)

"**The <u>Benedictines, Cistercians and other religious orders</u> need to be credited with furthering the cultivation of Comfrey plants during long stretches of warfare and unavoidable neglect.
Comfrey became one of the mainstay occupants of monastery gardens during the Middle Ages** *and for good reason: to heal the assortment of wounds and ruptures of returning soldiers.*"
-**Comfrey: What You Need to Know** by Ben Charles Harris. New Canaan, Connecticut: Keats Publishing, Inc., 1982, page 5.
    (Monasteries started forming in western Europe around 340 AD and continue to this day. This section covers only the Middle Ages. The Benedictines or 'Order of Saint Benedict' is a Catholic community of monasteries formed 529 AD in Italy. Cistercians or the 'Cistercian Order', also called Trappists, was formed 1098 AD in France.)

(During the Middle Ages there were no large nation-states like we have now. Instead, there were hundreds of feudal landholders fighting each other. Also soldiers were injured in the Crusades in the 'Holy Land' from 1095-1291 AD. The 'Holy Land' includes Israel, Palestine, Jordan, Lebanon and Syria.)

"In the Middle Ages, the study of medicinal plants was in the hands of monks who in their monasteries planted and experimented on the species described in classic texts. No monastic garden would have been complete without medicinal plants.
The sick went to the monastery, local herbalist, or apothecary to obtain healing herbs. Most monasteries developed herb gardens for use in the production of herbal cures, and these remained a part of folk medicine, as well as were being used by some professional physicians.
Books of herbal remedies were produced by monks as many monks were skilled at producing books and manuscripts and tending both medicinal gardens and the sick.
Lung problems were treated with a medicine made of liquorice and Comfrey."
-'**The Air of History Part II: Medicine in the Middle Ages**' by Rachel Hajar, M.D., Department of Cardiology, Heart Hospital, Hamad Medical Corporation, Doha, Qatar; Heart Views: Official Journal of the Gulf Heart Association, Qatar, Volume 13, Issue 4, October-December 2012.
    (An apothecary is a medical professional who formulates and dispenses medicine.)

"The revival in herbal medicine has been observed since the sixth century AD with the emergence of monasteries, where the gardens in which the medicinal plants were grown were established.
Monastic schools were also established. Monks broadened their skills by studying and copying ancient books, preparing medicines of natural origin according to secret prescriptions used to heal the sick. Although the Middle Ages are often considered the Dark Ages, many achievements of the then phytotherapy have been recorded on the pages of history and man uses them to this day.
Herbs grown in the Middle Ages were divided into utility groups.
    The species used in everyday life included, among others: southernwood, sapling, hops, mullein, stemless carline thistle, juniper, common soapwort, wormwood, cotton thistle and common broom.
    Some healing species are: *St John's wort, red scarlet, valerian, licorice, mallow, marshmallow, tansy and* **Common Comfrey**.
    There were relatively few aromatic plants: iris, blue cupid, true lavender, melissa, verbena, filipendula and tansy. For culinary purposes: small-leaved basil, winter savory, valerian and chives were used."
-'**Medications of Medieval Monastery Medicine**' by Katarzyna Madra-Gackowska, Marcin Gackowski, Emilia Glowczewska-Siedlecka, Zygmunt Siedlecki and Sylwia Ziolkowska, Collegium Medicum of Nicolaus Copernicus University, Bydgoszcz, Poland; Journal of Education, Health and Sport, Voume 8, No. 9, pages 1667-1674, (September 2018).

"It is said that soldiers carried the herb Comfrey with them in medieval times*, and that it was known to the Crusaders as a wound herb."
-**Dictionary of Plant Lore** by Donald Watts, BA, Bath, England. London, England: Academic Press, 2007.
(* -Somerset Folklore, Volume 114 by Ruth L. Tongue. London, England: Folk-Lore Society, 1965.)

"The internal application of Comfrey disrupts the entire order of bodily humors. However, when applied to the skin, it heals ulceration of the limbs."
-**Physica** by <u>Hildegard of Bingen</u> (Hildegard von Bingen, Saint Hildegard, Sibyl of the Rhine) of the '**Order of Saint Benedict**'. She lived from **1098-1179 AD**. She was a German Benedictine Abbess, writer, composer, philosopher, Christian mystic, and polymath. She is considered the founder of scientific natural history in Germany. The Physica contains nine books that describe the scientific and medicinal properties of plants, stones, fish, reptiles, and animals.
    (Humors was a system of medicine of the Ancient Greeks and Romans where four body fluids influence temperament and health. The four humors of Hippocratic medicine are black bile, yellow bile, phlegm, and blood.)
    (A polymath is someone whose expertise spans a large number of different subject areas.)

"If the interior membrane, which encloses the intestines, is cut by some accident, ivy and twice as much Comfrey should be cooked in good wine. After these herbs are cooked, they should be taken from the wine. Then a bit of pulverized zedoary (Curcuma zedoaria), sugar equal to the amount of ivy, and some cooked honey should be added to the wine and brought to a moderate boil. This should be poured through a little sack, making a clear drink. The ill person should drink this frequently, after food, and at night.
*He should also place the herbs which were cooked in the wine over the place where the interior membrane was ruptured. **This draws together the torn places. He should also cut Comfrey root into minute bits and place them in wine, so that it takes their flavor.** He should drink this wine often, until he is healed."*
-**Hildegard von Bingen's Physica: The Complete English Translation of Her Classic Work on Health and Healing** translated from Latin by Priscilla Throop. Rochester, Vermont: Healing Arts Press, 1998.

"Comfrey (consolida) is cold. If a person eats it without any reason, it destroys all the humors that have been correctly established.

But if some member of a person is deficient, ulcerated, or wounded, and the person then eats Comfrey, it quickly heals the bile and the ulcers on the surface of the skin, but not on the inside of the flesh.

Comfrey, eaten immoderately and not in the right way, heals outwardly, but sends all the decay more inwardly. ***If the inner membrane that holds a person's intestines is cut by some fall, let the person cook ivy and twice as much Comfrey in good wine.*** *After the person has cooked these herbs, let the person separate them from the wine. Then mix together a little powder made from zedoary (white turmeric), sugar equal to the ivy, and enough cooked wine and bring to a boil. Then pour this through a little sack and make a pure drink. Let the person drink this after eating food and at night, and do this often.*

But also let the person place the herbs that were cooked in wine over the place where the inner membrane was ruptured. Their warmth pulls the tears together. ***Let the person also cut Comfrey root into little pieces and put them in wine so that it takes their flavor.*** *Always drink the wine until the person is healed."*

-**Hildegard's Healing Plants: From Her Medieval Classic Physica** translated by Bruce W. Hozeski. Massachusetts: Beacon Press, 2001. Comfrey pages 130, 133.

"An elaborate discussion of the physiology of generation and the phenomena of impotence is followed by a collection of remedies for the condition. Concerning a function over which so many fond superstitions still linger in the public mind we may, perhaps, charitably forgive Gilbert for the introduction of **an empirical remedy for sterility, which, he assures us, he has often tried and with invariable success, and which enjoys the double advantage of applicability to either sex.**

   '**Let a man, twenty years of age or more, before the third hour of the vigil of St. John the Baptist, pull up by the roots a specimen of consolida major (Comfrey) and another of consolida minor (healall), repeating thrice (3 times) the Lord's prayer (oratio dominica).**

   Let him speak to no one while either going or returning, say nothing whatever, but in deep silence let him extract the juice from the herbs and with this juice write on as many cards as may be required the following charm: 'Dixit dominus crescite. Uthihoth. multiplicamini. thahechay. et replete terram. amath.' (Increase the owner. Multiply and fill the earth.) If a man wears about his neck a card inscribed with these identical words written in this juice, he will beget a male. Conversely, if a woman, she will conceive a female.' "

-'**Gilbertus Anglicus: Medicine of the Thirteenth Century**' by Henry E. Handerson, A.M., M.D., The Cleveland Library Medical Association, Cleveland, Ohio, 1918.

   (Gilbertus Anglicus or Gilbert of England, 1180-1250, was an English physician. He wrote the encyclopedic 'Compendium of Medicine' {Compendium Medicinae} between **1230 and 1250**. It was a comprehensive overview of medicine and surgery at the time.)

   (Vigil of the Feast of the Nativity {birth} of Saint John the Baptist is a midsummer festival also known as Saint John's Eve or Bonfire Night. Third Hour is 9 a.m., a time of prayer in Christian liturgy. It is the third hour of the day after dawn.)

"**There is a method for healing broken legs or broken arms or any other bone,** which seems to foreshadow the use of starched bandages, though there is an element of white witchcraft in the proceedings. The recipe runs:

   'Take first when thou cometh thereto, and make a cross thereon. In Nomine Patris (in the Name of the Father): and take then a plaster of wheat flour and white of an egg or of the flour of cockel and the seed of woodwax {genista tinctoria}; make the plaster right thick and lay it in a clean linen clout (cloth); take then and lay the plaster upon an even board and lay the broken leg thereon and set it....lap the plaster round about and knit about (bandage it) and **give him to drink the juice of knit-wort (Comfrey).**' "

-'Nova et Vetera: **A Medical Work of the Fourteenth Century (1301-1400)**, Together with a List of Plants Recorded in Contemporary Writings, with their Identifications' by Reverend Professor G. Henslow, The British Medical Journal, page 395-396, translated and published February 17 1900.

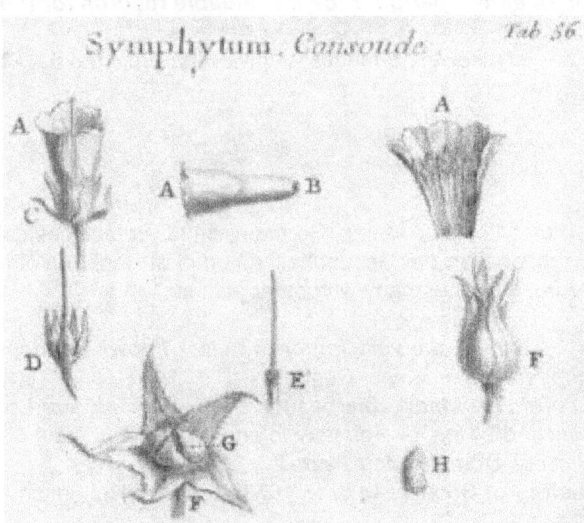

Symphytum Consoude 1700

'Institutiones Rei Herbariae', Volume 2,
by Joseph Pitton de Tournefort, Paris, France, 1700.
Tab. 56, fig. 1. Volume 1, page 138 is description.

Volume 1 is descriptions. Volume 2 and 3 are illustrations.
The 3 volumes include 9000 species in 698 genera,
which directly influenced botanist Carl Linnaeus.

Translated from Latin (Volume 1):
"**A** mono-petalus flower, a funnel-shaped flower
oblong and like bell-shaped: but from the calyx
C & D almost at the base, and furrows the pistil
E from the front of the flower
B fastened like nails and four as if surrounded by embryos,
which then go off in as many seeds   **G** viper's head
H rival, which is in the calyx itself
F when enlarged, they mature."

Chapter 13

The Renaissance and Comfrey

Overview of Renaissance and Comfrey (1401 to 1700)

"Through the Middle Ages new written works on botany were very rare and largely based on those of ancient Greeks. During the Renaissance originality became more of a virtue and the invention of printing in Europe enabled new books to be produced in large numbers.
  The first of these in the field of botany were the herbals, *for in those days botany was virtually synonymous with herbalism, the study of plants in relation to their value to man, particularly as foods and medicines, and herbalists dominated the sixteenth century botanical world."*
-**Plant Taxonomy and Biosystematics,** Second Edition by Clive A. Stace, Professor of Plant Taxonomy, University of Leicester, England. London, England: Edward Arnold, 1989, page 28.

"The earliest species of Symphytum on record are naturally those which are most common and widely distributed in western Europe, viz. (that is) S. officinale Linn. and S. tuberosum Linn.
S. officinale is recorded in Turner's 'Libellus ' (*1538) as 'Symphytum herbarii vocant consolidam majorem vulgus Comfrey (in Latin),' and in L'Obel's 'Historia' (**1576) as Consolida major.
S. tuberosum still retains the name under which it was known before the time of L'Obel, but is described without a name in Gerard's 'Herball' (1597) as follows:
  'There is another kinde of Comfrey that hath leaves like the former (S. officinale) saving that they be lesser; the stalks are rough and tender; the flowers are like the former, but that they be of an overworne yellow colour; the roots are thicke, short, blacke without and tuberous.'
The S. tuberosum minus of Clusius (***Stirp. Pann. Hist. p. 671, 1583) is said to be the plant now known as S. bulbosum Schimp.; but the figure, found also in the 'Historia' of 1601, shows neither the characteristic root nor the exserted corolla-scales of that species. The plant was described by Schimper as S. bulbosum in 1826."
-**'A Revision of the Genus Symphytum, Tourn.'** by Cedric Bucknall, (Mus. Bac. Oxon= Bachelor of Music, Oxford University), Journal of the Linnean Society of London, England, Botanical Journal, Volume 41, Issue 284, pages 491-556, December 1913.
(* -'Libellus de re Herbaria Novus' by William Turner, 1538. Reprinted in Latin with English notes, modern names, list of Turner's writings, and life of Turner. In 1877, London, England by Benjamin Jackson, FLS.)
(** -'Plantarum Seu Stirpium Historia' {Plants or Plants History} by Matthias de L'Obel / Mathias de Lobel / Matthaeus Lobelius, Flemish physician and botanist, 1576. In Latin.)
(*** -'Atrebatis Rariorum Aliquot Stirpium, per Pannoniam, Austriam, & Vicinas Quasdam Provincias observatarum Historia' {Atrebatis Rariorum several plants through Hungary, Austria and neighboring countries, there are some provinces observed History} by Carolus Clusius / Charles de l'Ecluse, Flemish doctor and botanist, 1583. In Latin.)

"Materia medica in medieval Europe had its roots firmly grounded in Greek learning, though usually in Latin translation. The rediscovery of Greek texts, particularly those of Dioscorides, strongly influenced the herbal literature of the fifteenth and sixteenth centuries (1401 to 1600 AD).
With the invention of printing with movable type, this genre flourished and the herbal became a popular form of medical literature. Gradually new plants were added to those described by Dioscorides, but the emphasis on their medicinal value remained.
  This type of literature, together with monastic writings of an earlier period, provide valuable records for the materia medica of the West."
-**'Russian Medical Botany Before the Time of Peter the Great'** by Margery Rowell; Sudhoffs Archiv, Band {Volume} 62, Heft {Issue} 4, 4th Quartal {Quater}, pages 339-358, 1978.

Early Renaissance (15th- 16th Centuries, 1401-1600)

'Herbarius zu Teutsch und Von Aller Hand Kreuttern' published by Peter Schoffer, Mainz, Germany, **1485**. Sometimes called '**Ortus Sanitatis**', 'Hortus Sanitatis', 'Smaller Ortus', '<u>**German Herbarius**</u>' or 'Gart der Gesundheit' (Garden of Health). With 379 illustrations. Copied the same year by Johann (Hans) Schonsperger, Augspurg, Germany with other editions.

  "In 1485, the year following the first appearance of the Latin 'Herbarius', **the very important work known as the German 'Herbarius' or 'Herbarius zu Teutsch',** *made its appearance at Mainz, Germany.*
  Its illustrations, which are executed on a large scale, are often of remarkable beauty. *Dr Payne considered some of them comparable to those of Brunfels (1488-1534) in fidelity of drawing, though very inferior in wood-cutting.*
  Excellent drawings are those of *Winter Cherry, Iris, Lily, Chicory,* **Comfrey** *and Peony."*
-**'Herbals: Their Origin and Evolution: A Chapter in the History of Botany: 1470 to 1670'** by Agnes Robertson Arber,

University College, London; Cambridge Univ. Press, England, 1912, page 162,165. (I was unable to find Comfrey drawing.)

Swiss physician and pharmacologist *Paracelsus (1493-1541 AD) said:
*"To what purpose do you superadde vinegar to the root of Comfrey or bole (a medicinal earth), or suchlike balefull additaments, while **God hath compos'd this simple sufficient to cure the fracture of the bones**? Why add vinegar to Comfrey root, or all sorts of other fatal ingredients when God created it to heal everything by itself, even fractured bones?"*
(* -'Paracelsus: Samtliche Werke' {Complete Works} edited by Karl Sudhoff, Wilhelm Matthiessen and Kurt Goldammer; Munich and Berlin, Germany, 1922-1933. 14 volumes. In German.)

> *"**Observe the following things concerning incarnatives and consolidatives**. They contain in themselves the four grades, while the consolidatives, in the same manner as the laxatives, exclude the elements.*
> **Grade 1: Broken bones are cured by** *Alchimilla, Periwinkle, Perfoliata, Diapensia, Aristolochia rotunda,* **Consolida (Comfrey),** *Serpentina.*
> **Preparation for Wounds:** *Of Litharge, with four times Whitened Vinegar, lb.ss,*
> **An equal quantity:** *Of the juice of the herb Pellitory.* **Of the lesser Comfrey.**
> *Of the round Aristolochy (Birthwort). Make a compound with mucilago lumbricata.*
> **The things which are here enumerated are of a cold nature. Among these those which are produced out of the earth are cold in the first degree: Comfrey."**

-'The Hermetic and Alchemical Writings of Paracelsus, Volume II: Hermetic Medicine and Hermetic Philosophy' by Arthur Edward Waite, London, England, pages 182, 189, 202, **1894**.
(Incarnatives cause new flesh to grow. They are healing and regenerative.)
(Consolidatives cause solidification into a firm mass. It is the power to consolidate.)

*"**If the inaccuracies of herbals, the frauds of herb-importers and the uselessness of apothecary confections made Paracelsus angry, it was because, as a result, herbs were seldom properly used, while their real virtues remained undervalued or unknown by foolish men who thought they could improve on nature:** 'To what purpose do you superadde vinegar to the root of Comfrey...'.*
It is clear both from his writings and from the assurance with which he speaks of them that Paracelsus used herbs knowledgeably and confidently in his own practice.
A strongly-held belief of folk medicine, to which Paracelsus subscribed, was that medicinal plants grew where they were needed, *and that there was no need to travel far to find the remedy for a disease: 'They want medicaments from overseas, and better things grow in their own garden.' "*
-**Green Pharmacy: The History and Evolution of Western Herbal Medicine** by Barbara Griggs. Rochester, Vermont: Healing Arts Press, 1997, pages 50-51. (Also called 'New Green Pharmacy: Story of Western Herbal Medicine' by Barbara Griggs. London, England: Vermillion, 1997.)

*"**Paracelsus (Theophrastus Bombastus of Hohenheim, 1493-1541) brings to 'Consolida', as the Comfrey is called. The root of Consolida is used for wounds and fractures recommended."***
-'**Symphytum Officinale L.: Comfrey in European Pharmacy and Medical History**' by K. Englert, J.G. Mayer and C. Staiger, Institut fur Geschichte der Medizin (Institute for the History of Medicine), Universitat Wurzburg, Germany; Zeitschrift fur Phytotherapie (Journal of Phytotherapy), Volume 26, No. 4, pages 158-168, 2005.
(This report is in German. This is from an online translation. If you have English translation, please send it to me.)

*"**In 1500 Strasbourg's surgeon <u>Hieronymus Brunschwig</u>'s wrote a 'small distillation book'** for many indications.*
For the root of the 'whale root' (Comfrey) Brunschwig uses for the following indications: fresh wounds, cracks on the lips of the mouth, hot gout (as rubbings), that wild fires, 'all tumors', broken bones.
Brunschwig is the first time the external treatment of gout (Podagra) called for the Comfrey root.
Gout designated all forms of rheumatic complaints. Only a few authors followed this indication by Brunschwig."
-'**Symphytum Officinale L.: Comfrey in European Pharmacy and Medical History**' by K. Englert, J.G. Mayer and C. Staiger, Institut fur Geschichte der Medizin (Institute for the History of Medicine), Universitat Wurzburg, Germany; Zeitschrift fur Phytotherapie (Journal of Phytotherapy), Volume 26, No. 4, pages 158-168, 2005. Online translation from German to English.
(If you have an English translation of this report, could you please send it to me.)   (Strasbourg is in France, close to Germany.)
(Hieronymus Brunschwig or Hieronymus Brunschwygk, 1450-1512, was a German surgeon, alchemist and botanist. He was notable for his methods of treating gunshot wounds and his early work on distillation techniques. His most influential book was 'Liber de arte distillandi de simplicibus' {Kleines Destillierbuch} {**Book of the Art of Simple Distillation**}.)

"<u>**Otto Brunfels**</u> (Brunsfels or Braunfels), *1488-1534:*
*First in point of time among the **German botanical reformers of the sixteenth century**, Brunfels is also easily first in rank respecting those educational and literary qualifications which go to the making of what one calls a scholarly book.*
*In this particular his **one botanical treatise, the *'Herbarum Vivae Icones (Eicones)', 1530 and 1536**, is peerless among the several books of botany that appeared in middle Europe within the first half of the sixteenth century.*

Others produced more and better botany; but there are marks of a dignified and conservative erudition (great knowledge) that are characteristically Brunfels' own.

Of such attempted **improvements in classification by appeal to considerations of morphology,** one may come to a fuller appreciation by looking into **Brunfels' way of presenting those many herbs which, in his time, had long been reputed to be good vulneraries (heals wounds), and had therefore passed under the medico-generic name of Consolida, with which Symphytum, Sanicula, Vulneraria, and Solidago were synonymous,** each such name indicative of the property which these plants all had, or were believed to have, of **promoting the closing-up and healing of cuts and wounds.**

Here is a partial list of these plants under their mediaeval names, with their equivalents in modern nomenclature. One thus gains an idea of how great a diversity of plants passed with mediaeval pharmacists and physicians under the generic name Consolida. And the list of Brunfelsian taxonomic betterment:

Medieval	Brunfelsian	Modern
Consolida major	**Consolida major**	**Symphytum officinale**
Consolida media	Consolida media	Ajuga reptans
Consolida minor	Diapensia	Sanicula Europaea
Consolida petraea	**Symphytum petraeum**	**Coris Monspeliensis**
Consolida regalis	Consolida reglais	Delphinium consolida
Consolida rubea	Tormentilla	Tormentilla erecta

One observes that out of 6 Consolida names, 3 have been eliminated, and others brought forward to take their places.
> I say brought forward; for neither Diapensia, nor Symphyton petraeum nor Tormentilla is coined and proposed as new by Brunfels. He picked them up every one out of the ancient and mediaeval synonymy (synonyms) of the vulnerary herbs."

-'**Landmarks of Botanical History: A Study of Certain Epochs in the Development of the Science of Botany**, Part I, Prior to 1562 A.D.' by Edward Lee Greene, first published by The Smithsonian Institution, Washington, DC, 1909. Republished in 1983 by Standford University Press, California, edited by Frank N. Egerton. Part II was published in 1983 from a 1915 handwritten manuscript by Edward Greene. Part II covers 1398 to 1708 A.D. Pages 240, 254.

(* -'Herbarum Vivae Eicones (Icones) ad Naturae Imitationem, summa cum diligentia et artificio effigiatae' {Living Portraits of Plants} by Otto Brunfels {German botanist}, published in Argentorati {Strasbourg}, France, 1530. Symphytum pages 75 to 78. In Latin, 3 volumes {1530 to 1536} with illustrations by Hans Weiditz, German artist.)

> "The first volume of Brunfels's Herbal was published in 1530 under the title 'Herbarum Vivae Eicones', and was followed by a second in 1531, and a third issued posthumously (after death) in 1536. All these were in Latin, and passed through several editions.
>
> **The Herbal of Otto Brunfels forms a link between ancient and modern botany, and may be regarded either as the end of the long line of classical and mediaeval works on medicinal plants or as the beginning of modern taxonomy.**
>
> **The real renascence (Renaissance) of botanical science began in the sixteenth century (1500s),** when the study of living plants gradually replaced the study of what had been written about them by classical authors, and the mediaeval period in botany may therefore be considered as extending to the end of the fifteenth century.
>
> **The text of Brunfels was virtually a compilation of the dicta (statements) of his predecessors, and in this respect his work is purely mediaeval in character.**
>
> The provision of life-like wood-engravings of living plants, however, makes it possible, for the first time in history, to identify a high percentage of the species concerned, so that **modern systematic botany may almost be said to start with Brunfels.**
>
> **Herbarum Vivae Eicones:**
>> 1. **Consolida maior mas, Walwurtz maennlin:** Volume i (editions 1530, 1532) page 75; iii, 6 (text); C.K. (German edition: Contrafayt Kreuterbuoch) p. xii.
>> = Symphytum Consolida major C.B.P. (Caspar Bauhin, Pinax; Latin names not actually occurring in Brunfels, but attributed to him by Bauhin) 259.
>> = **Symphytum officinale L. Sp. 136, var. purpureum** *Pers. Syn. i. 161.
>>
>> 2. **Consolida maior faemina, Walwurtz weiblin:** Volume i (editions 1530, 1532), page 76; iii, 6 (text); C.K. (German edition: Contrafayt Kreuterbuoch) p. xiii.
>> = Symphytum Consolida major C.B.P. (Caspar Bauhin, Pinax) 259.
>> = **Symphytum officinale L. Sp. 136 var. ochroleucum DC.** **Prodr. x, 37. Brunfels (C.K. p. xiv) mentions that there are two kinds of 'Grosz Walwurtz' {Consolida maior} a 'male' kind with brown (i.e., purple) flowers, and a 'female' with white flowers.
>> C. Bauhin also remarks that the purple or purplish-blue flowered form was called the male kind, and the white or yellowish form the female kind.
>> = ***Moretti identified 'Walwurtz weiblin' with **Symphytum tuherosum L.**"

-'**The Herbal of Otto Brunfels**' by T.A. Sprague, B.Sc., F.L.S.; Linnean Society's Journal: Botany, London, England, Volume XLVIII, December 1928, pages 80 to 124.

(* -Pers. Syn. i. 161 = Synopsis Plantarum, Volume 1, page 161 by Christiaan Hendrik Persoon, 1805.)

(** -'Prodromus Systematis Naturalis Regni Vegetabilis, Part 10' by Alphonse Pyramus De Candolle, 1824, 1846. Symphytum pages 36-40.)

(*** -Giuseppe L. Moretti, Italian, 1782-1853. Professor of Botany & Director of botanical gardens, University of Pavia, Italy.)

"**Jean Ruel**, 1474-1537: In his later life he gave the fullest proof of his profound erudition, as well as his field knowledge of many plants, in the composition of **his great classic, the 'De Natura Stirpium', published in the year 1536**.
It is the first volume produced within the period of the Renaissance which from beginning to end carries the implication that botany is botany, and that pharmacy, like agriculture, pomology (fruit growing), and horticulture, is but one of its departments, all of them subsidiary to the philosophy of plant life as a whole.
Ruel was the first botanist to undertake a description of every plant that he mentions. How fully he realizes the importance of descriptive terminology. An example or two taken from the chapter on differences in stems will help us to an understanding of this matter:
'Stems differ in respect to their roughness or smoothness, and as to their being straight or flexous. That of dracunculus is as thick as a walking stick and is marked with variously colored spots and streaks....
The stem of Symphytum is invested with a rough hairiness that stings...' "
-'**Landmarks of Botanical History: A Study of Certain Epochs in the Development of the Science of Botany**, Part I, Prior to 1562 A.D.' by Edward Lee Greene, first published by The Smithsonian Institution, Washington, DC, 1909. Republished in 1983 by Standford University Press, California, edited by Frank N. Egerton. Part II was published in 1983 from a 1915 handwritten manuscript by Edward Greene. Part II covers 1398 to 1708 A.D. Pages 598, 604, 613.
(Jean Ruel, 1474-1537, was a French physician who studied botany and pharmacology. He was Professor at 'Univeristy of Paris', France. In the 3-volume 'De Natura Stirpium' {Nature Plants}, 1536-1537, he describes the habit, habitat, and smell/taste of each plant; in Latin.)

"**Jean Francois Fernel** (1497-1558) was a French physician who introduced the term 'physiology' to describe the study of the body's function. He was **physician to the court of French King Henry II** and wife Catherine de Medici.
His medical works included '**De Naturali Parte Medicinae' (1542)**, 'De vacuandi ratione' (1545), 'De abditis rerum causis' (1548). What has been called his crowning work, 'Universa Medicina', comprises three parts: the Physiologia (developed from the De naturali parte), the Pathologia, and the Therapeutice."
-**Wikipedia®: The Free Encyclopedia,** www.wikipedia.org, 2018.

"*In the Renaissance, the physician Jean Fernel (1497-1558) while accepting the tradition, sought to reform the study of pathology. **He proposed a syrup made from Comfrey (and rose petals, betony, plantain, burnet, scabiosa and tussilage) which had long been prescribed for diarrhea, haemorrhage, cough and phthisis.**"
-**French Wikipedia®: The Free Encyclopedia,** https://fr.wikipedia.org/wiki/Consoude_officinale, 2018.
(Fernel also recommended Comfrey for trauma, fractures and wounds.)

"**Hieronymous Bock**, in writing 'De Stirpium' (Kyber's Edition 1552), clearly recorded the fact that Comfrey was used for both internal and external complaints."
-**About Comfrey: The Forgotten Herb** by G.J. Binding, M.B.E., F.R.H.S. Wellingborough, Northhamptonshire, England: Weatherby Woolnough, 1974, page 29.
(Hieronymus Bock lived from 1498 to 1554. He was a German botanist, physician, and Lutheran minister who began the transition from medieval to modern botany by arranging plants by their resemblance. 'De stirpium maxime earum quae in Germania nostra nascuntur...' {The plants, especially those that are born in Germany...} Written in German. I was unable to find English translation of what he wrote about Comfrey. If you have information, please contact me.)

"A book under the title 'New Kreuterbuch von Vendscheydt, Wurkung' and name of Kreutter, **Hieronymus Bock in Strasbourg 1539 and the 2nd edition 1546. Chapter 79, 'The Common whale-root (Comfrey).'**
Bock separates between inner ones and external application. Inwardly are called: cleansing the chest and lungs (of pus), blood spasms, internal fractions, the chewed root against the thirst.
Externally: fresh wounds (as a plaster) and broken legs. Then follow some formulas for complex remedies."
-'**Symphytum Officinale L.: Comfrey in European Pharmacy and Medical History**' by K. Englert, J.G. Mayer and C. Staiger, Institut fur Geschichte der Medizin (Institute for the History of Medicine), Universitat Wurzburg, Germany; Zeitschrift fur Phytotherapie (Journal of Phytotherapy), Volume 26, No. 4, pages 158-168, 2005. In German.
(This is an online translation. If you have it in English, could you please send it to me.)

'**A New Herball: Wherin are Conteyned the Names of Herbes in Greke, Latin, Englysh, Duch, Frenche**' by **William Turner**, English physician, Part I (**1551**), Part II (**1562**), Part III (**1568**). Part I is Volume 1. Part II and Part III are Volume 2. Volume 2 is titled 'The Seconde Parte of William Turners Herball' and is itself composed of 2 parts.
"**Dioscorides maketh two kindes of Symphytum...in English Comfrey...of Comfrey Symphytum,** the rootes are good if they be broken and dronken (drunk) for them that spitte blood, and are bursten (bursted, broken, ruptured). The same, layd to, are good to glewe (glue) together freshe woundes. They are good to be layd to inflammation, and especially of the fundament (anus), with the leaves of the groundsell."
(William Turner, Master of Arts, lived from 1509 to 1568. He was an English clergyman, physician and natural historian.

He studied medicine in Italy, and was a friend of the Swiss naturalist, Conrad Gessner. He was an early herbalist and ornithologist {studies birds}. 'A New Herball' gave the first clear, systematic survey of English plants that was in a language the average person could understand.)

"Symphytum is of two sortes, the former is called Symphytum petreum (petraeum), and this herbe groweth about Syon (west London), seven myles (miles) above London. It is lyke unto wylde (wild) Mergerum (Marjoram?), but it is neither so hote neither so wel smellyng. It may be called in English unsauery Margeru.
The other kynde (kind) called in Latin Symphytum alterum, is called in Englishe Comfery or Blackewurt, and is Duche (Dutch) walde wurtz or schwart wurtz, in Frenche de la confire."
-'The Names of Herbes' by William Turner, 1548. Reprinted and edited in 1881 by James Britten, F.L.S. of the British Museum, London, England. Symphytum is on page 77.

"Symphytum: Symphytum herbarii uocant consolidam maiorem, uulgus Comfrey."
{They call Symphytum the greater herb, Common Comfrey.}
-'Libellus de re Herbaria Novus by William Turner: Reprinted in Facsimile, with Notes, Modern Names, and Life of Author, Originally Published 1538' by Benjamin Daydon Jackson, FLS London, England, 1877. Symphytum page 6, 52 (C. j. verso)

"William Turner is the ancestor of F. W. Newman Turner who became President of the Henry Doubleday Research Association, England."
-Fighting Like the Flowers: An Autobiography: The Life Story of Britain's Best-Known Organic Gardener by Lawrence D. Hills. Bideford, England: Green Books, 1989, page 98.
(For more about F.W. Newman Turner, see 1948 in section 'The 1900s and Comfrey'.)

'Petri Andreae Matthioli Senensis, Commentarii, Secundo Aucti' by <u>Pietro Andrea Mattioli</u>, Venice, Italy, **1560** (editions from 1544 to 1563). Also called 'Commentarii Secundo Aucti'. Symphytum page 489.
Pietro Andrea Gregorio Mattioli (Matthiolus), 1501-1577, was a doctor and naturalist from Siena, Italy. In 'Discorsi' (Commentaries) on 'Materia Medica' of Dioscorides, he described 100 new plants. Mattioli added descriptions of some plants not in Dioscorides that were of interest botanically but not necessarily medicinal. Illustrations were high quality.

'Stirpium Adversaria Nova' by <u>Pierre Penna and Matthias de L'Obel</u> (Mathias de Lobel / Matthaeus Lobelius), London, England, **1571,** 1576. In Latin. Symphytum page 198.
Includes 1200 plants. Mallet and Jovet (1973) state this book is 'one of the milestones of modern botany'.
Also by L'Obel: **'Plantarum Seu Stirpium Historia'** {Plants or Plants History} by Matthias de L'Obel / Mathias de Lobel / Matthaeus Lobelius, Flemish physician and botanist, 1576. In Latin. Consolida major.

"The rootes of Comfrey pound and drunken are good for them that spitte blood, and healeth all inward woundes and burstings. The same also being brused and layde to in manner of a plaester, do heale all greene and freshy woundes; and are so glutinative that if it be sodde with chopte or minsed meats it will reioyne and bring it all together againe into one masse or lumpe. Rootes of Comfrey boyled and dronken do cleanse the breast from flegmes, and cureth grieffes or hurtes of the Lunges."
-'A Nievve (Niewe or New) Herball, or, Historie of Plantes: Wherein is Contayned the Whole Discourse and Perfect Description of all Sortes of Herbes and Plantes' by **Rembert Dodoens / Dodonaeus** (1517-1585). Translated by Henry Lyte (1529-1607) and Gerard Dewes, London, England, **1578**. Symphytum pages 145-146.
(Rembert Dodoens, Rembert Van Joenckema or Rembertus Dodonaeus, 1517-1585, was a Flemish physician and botanist. The Flemish are a Germanic ethnic group in Flanders, Belgium. His herbal 'Cruydeboeck' was influenced by earlier German botanists, particularly Leonhart Fuchs. Rather than the traditional method of arranging plants in alphabetical order, he divided the plant kingdom into six groups based on properties and affinities. It treated in detail especially the medicinal herbs. This work became one of the most important botanical works of the late 16th century.)

"Dodoens, in this 'Cruydtboeck' (Herb Book) (*Cruydt-Boeck, first published 1554), translated by Lyte (1578), expands Turner's statement, adding that when (Comfrey) 'mengled (mixed) with sugar, syropes (syrups), or honny (honey)...are good to be layde upon all hoate (hot?) tumours.'
*Quite similar statements as to **the value of the Comfrey rhizome (root)** are made in 'Bulleyn's Herbal' (**1562) and in the 'Adversaria' (Stirpium Adversaria Nova) of Pena and Lobelius (1570), and the 'Stirpium Historia' of the latter author (1576), also in 'Camerius' (1586)."*
-The Medicinal Uses of Comfrey by Dr. Charles MacAlister, M.D., F.R.C.P., written 1935.
(In Comfrey: Fodder, Food & Remedy, and Comfrey: An Ancient Medicinal Remedy)
(* -'Cruydt-Boeck Remberti Dodonaei Volghens Sijne Laetste Verbeteringhe' by Rembert Dodoens, Antwerp, Belgium, edited and translated into Dutch by Carolus Clusius, 1644. First published 1554. Symphytum pages 12, 101, 192-198, 243, 498, 551, 557, 979, 983.)
(** -'A Book of Simples: Being a Herbal in the Form of a Dialogue, at the End of Which are the Cuts of Some Plants in Wood' by Dr. William Bulleyn, 1562. Symphyutm pages 'Folio 23, 24'.)

'**De Plantis, Libri XVI**' by <u>Andrea Cesalpino</u> or Andreas Caesalpinus (1519-1603), Florence, Italy, **1583**. Sixteen botany books in Latin with 1500 species. Similar to Theophrastos, he classified plants based on growth habit and fruit/ seed form.
    Consolida: Symphyton and Symphytum petraeum: 'De Suffruticibus and Herbis, Liber Vndecimis, Cap VII' (Book 11, Chapter 7), page 435.
    Hyssopum: Symphytum petraeum: De Suffruticibus & Herbis, Liber Vndecimis, Cap L' (Book 11, Chapter 50), page 462.

"In <u>Sarracenius's</u> (Sarrasin's) version of Dioscorides, published in 1596, descriptions are given of two species of Symphytum: 'De Symphyto Petraeo' and 'De Symphyto Altero'. The roots and their 'vertues' (virtues) are described in the following terms (originally in Latin):
    'The roots of either black outside, inside white glutinous. Can be drunk, the blood of those who are spitting, the fresh wounds of the beaten and broken. They grow, they agglutinate; so that they, too, with whom they are conjungunter (joined together) reigns over the more complicated cases. In addition, inflammation; when leaves are beneficial'. "
-The **Medicinal Uses of Comfrey** by Dr. Charles MacAlister, M.D., F.R.C.P., written 1935.
(* -'Pedakiou Dioskoridou tou Anazarbeos ta Sozomena Hapanta' = 'Pedacii Dioscoridis Anazarbaei Opera quae Extant Omnia' by Dioscorides Pedanius of Anazarbos. Edited by Jean Antoine Sarrasin {Sarracenii, Sarraceni}, Jean Aubry, Claude de Marne and Andreae Wecheli Heredes; Wechel Press, Frankfurt, Germany, 1598.)

<u>John Gerard</u> of England wrote in his 1597 'Herball, or Generall Historie of Plantes': *"Of Comfrey, or Great Consound:*
Symphytum officinale var patens:
    The stalke of this Comfrey is cornered, thicke, and hollow like that of Sow-Thistle: it groweth two cubits (2 times 18 inches or 45 cm = 36 inches or 90 cm) or a yard (3 feet = 0.9 meter) high: the leaves that spring from the root, and those that grow upon the stalkes are long, broad, rough, and pricking withall, something hairie, and being handled make the hands itch; very like in colour and roughnes to those of Borage, but longer, and sharpe pointed, as be the leaves of Elecampane: from out the wings of the stalkes **appear the floures (flowers) orderly placed, long, hollow within, of a light red (or purple) colour:** after them groweth the seed, which is blacke. The root is long and thick, blacke without, white within, having in it a clammy juice, in which root consisteth the vertue (virtue).
Symphytum officinale:
    The Great Comfrey hath rough hairy stalks and long rough leaves much like the garden Buglosse but greater and blacker (darker green), **the floures be round and hollow like little bells of a white colour,** the roote is black without and white within and very slimey. This differeth in no way from the former (Symphytum officinale var patens) but only in the **colour of the floure which is yellowish when the other is reddish or purple.**
Symphytum tuberosum:
    Comfrey with the knobby root: There is another kinde of Comfrey which hath leaves like the former, saving that they be lesser: the stalks are rough and tender: the floures be like the former, but that they be of **an overworn yellow colour:** the roots are thicke, short, blacke without, and tuberous, which in the figure are not expressed so, large and knobby as they ought to have been.
Symphytum pumilum*: Symphytum parvum. Borage-floured Comfrey.
    This pretty plant hath fibrous and blackish roots, from which rise up many leaves like those of Borage, or Comfrey, but much smaller and greener, the stalkes are some eight inches (20 cm) high and on their tops carry pretty floures like those of Borage, but not so sharpe pointed, but of a more pleasing **blew (blue) colour.** This floures in the spring and is kept in some choice Gardens. Lobell calls it Symphytum pumilum repens Borraginis facie, siue Borrago minima Herbariorum.
Comfrey joyeth in watery ditches, in fat and fruiteful meadows, they grow all in my garden. They floure in June and July.
The Temperature:
    The root of Comfrey hath a cold quality, but yet not much: it is also of a clammie and gluing moisture. It causeth no itch at all, neither is it of a sharpe or biting taste, unsavory, and without any qualitie that may be tasted; so far is the tough and gluing moisture from the sharpe clamminesse of the Sea Onion, as that there is no comparison betweene them. The leaves may cause itching not through heate or sharpenesse, but through their ruggednesse, as we have already written, yet lesse than those of the Nettle.
The Vertues (Virtues):
    The rootes of Comfrey stamped and the juice drunke with wine helpeth those that spit blood and healeth all inward woundes and burstings.
    The same bruised and laid to in manner of a plaister, doth heale all fresh and greene woundes, and are **so glutenatiue (glutenous),** that it will sodder or glew together meate that is chopt in peeces (pieces) seething in a pot, and make it in one lumpe.
    The rootes boiled and drunke, doe clense the brest from flegme (phlegm), and cure the griefes of the **lungs,** especially if they be confect with sugar and syrrup; it prevaileth much against ruptures or burstings.
    The slimie substance of the root made in a posset (curdled milk) of ale, and given to drinke against the **paine in the backe,** gotten by any violent motion, as wrastling (wrestling), or overmuch use of women, doth in foure or fiue daies (4 or 5 days) perfectly cure the same: although the involuntary flowing of the seed in men be gotten thereby.
    The roots of Comfrey in number foure (4), Knotgrasse and the leaves of Clarie of each an handfull, being stamped all together, and strained, and a quart of Muscadell (white wine grape) put thereto, the yolkes of three egges, and the powder of three Nutmegs, drunke first and last, is a most excellent medicine against a Gonorrhaea (running of the

reines), and **all paines and consumptions of the backe.**
There is likewise a syrrup made hereof to be used in this case, which staieth (stops) voiding of bloud (blood): tempereth the heate of agues (fever and shivering): allaieth (reduces) the sharpenesse of flowing humors: healeth up ulcers of the lungs, and helpeth the cough: the receit (recipe) whereof is this:

> **Take two ounces of the roots of great Comfrey,** one ounce of Liquorice; two handfulls of Folefoot, roots and all; one ounce and an halfe of Pine-apple kernells; twenty luiubes; two drams or a quarter of an ounce of Mallow seed; one dram of the heads of Poppy; boile all in a sufficient quantitie of water, till one pinte remaine, straine it, and and adde to the liquor strained six ounces of very white sugar, and as much of the best hony (honey), and make thereof a syrrup that must be throughly boiled.

The same syrrup cureth the ulcers of the kidnies (kidneys), though they have been of long continuance; and stoppeth the bloud (blood) that commeth from thence. Moreover, it staieth the overmuch flowing of the monethly sickenesse (menstruation), taken every day for certaine daies (days) together.

It is highly commended for woundes or hurts of all the rest also of the intrailes (entrails) and inward parts, and for burstings or ruptures. The root stamped and applied unto them, taketh away the inflammation of the fundament (anus), and overmuch flowing of the hemorrhoides."

-'The Herball, or, Generall Historie of Plantes Gathered by John Gerarde of London, Master in Chirurgerie (Surgery)' by **Gerard**, 1730 pages, published in London, England in 1633, pages 805-807.
(A copy of this book is in the 'Royal Horticultural Society' Library in London, England. There are 12 plants in the Borage family with the name Buglosse or Bugloss.)
(* I do not know what 'Symphytum pumilum' is in current botanical terminology. It is only mentioned in texts around this time.)
(John Gerard, 1545-1612, was an English botanist in London. His 1,484-page illustrated **'Herball, or Generall Historie of Plantes' was first published in 1597**. It was the most famous botany book in English in the 1600s. Much of Gerard's 'Herbal' is an unacknowledged English translation of the book 'Cruydeboeck' {'Herb Book'} by the Flemish botanist, Rembert Dodoens, published in 1554. Experts have different opinions as to how much of Gerard's 'Herball' was original. Gerard is one of the founders of botany in the English language. As an herbalist and barber-surgeon, he was more interested in the medicinal properties of plants than in botanical theory and technical knowledge.)

"Bloodwort. And I suspect Adders Tongue. Knot Grass. Cheek Weed. **Compherie, with the white Flower.**
May Weed, excellent for the Mother; some of our English Houswives call it Iron Wort, and make a good Unguent (ointment) for old Sores
Great Comfrey (Gerard, page 806), Symphytum officinale, L.: also in the list of garden herbs at page 90."
-'**New England's Rarities: Discovered in Birds, Beasts, Fishes, Serpents and Plants of that Country**' by John Josselyn, Boston, Massachusetts, 1672, page 140.

"**Gerarde (Gerard) gives, as his wont is, a recipe for a syrup, the main compound being Comfrey, for internal use.** Its mucilaginous character ponts to the source of its value in case of Lung, Foot, or Mouth Diseases.
Gerarde recommends Comfrey highly as a salve for external wounds, for internal hemorrhage, and for disorder of the kidneys. -'Herbal', pages 806, 807."
-'**Prickly Comfrey: Its History, Cultivation, Extraordinary Production, and Uses:** A Letter Addressed to His Excellency Sir Hercules Robinson, President of the Agricultural Society of New South Wales' by Arthur T. Holroyd, Sydney, Australia, 1876, page 14.

"**Gerard's Herbal (1597) asserts the efficacy of Comfrey** also in healing up 'ulcers of the lunges' and 'ulcers of the kidneies (kidneys), though they have been of long continuance'. "
-The Medicinal Uses of Comfrey by Dr. Charles MacAlister, M.D., F.R.C.P., written 1935.

"**Gerard gives a long list of virtues of Comfrey**, of which some are a little unexpected:
> 'The slimie sustance of the roote made in a posset of ale (hot drink made with milk curdled with wine or ale) and given to drink against the paine in the backe, gotten by an violent motion, as wrestling, or overmuch use of women, doth in fower (four) or five daies (days) perfectly cure the same, although the involuntarie flowing of the seed in men be gotten thereby.' "

-**Comfrey: Nature's Healing Herb & Health Food** by Andrew Hughes. Japan: Sanyusha Publishing, 1992, page 102.

"**A translation of Dodeons' Latin edition of 1583 was the basis of the most famous of all English herbals, namely that of John Gerard, 'The Herball or General Historie of Plants' in 1597.**
John Gerard (1545-1612) was a surgeon. He was also a very accomplished horticulturalist and a good botanist. His practice was not confined to the use of the knife, splints, suturing, cautery, and forceps.
Being an expert herbalist he used plants and their extracts for conditions ranging from parsley for toothache to **Comfrey for fractures** and yellow henbane for 'inveterate ulcers and burnings'.
The Doctrine of Signatures decreed that the appearance of a plant indicated its use in therapy, and the spotted leaves of Lungwort (Pulmonaria officinalis, Boraginaceae) was used in respiratory disease since it resembles the surface of the lung.
Comfrey root (Symphytum officinale, Boraginaceae) contains mucilage and after being pounded it sets hard and was used to splint fractures, hence the old name of Knitbone."

-'**John Gerard, Physic Gardens and Medicinal Plants**' by Arthur Hollman, Sussex, United Kingdom, Journal of Medical Biography, Volume 19, pages 47-48, May 2011.
(Cautery is the coagulation of blood or destruction of body tissue by heat or caustic substance.)
(The 'Doctrine of Signatures', dating from the time of Dioscorides {40-90 AD} and Galen {129-216 AD}, states herbs resembling parts of the body can be used by herbalists to treat ailments of those body parts.)
(Starting in the Middle Ages, 'physic gardens' are herb gardens with medicinal plants. Botanical or botanic gardens developed from them.)

"*Comfrey:*
1. *The roots stampt (cut or crush into pieces) and drunk are good for spitting of blood and in all inward wounds and ruptures.*
2. *Bruse (bruise) it and apply it to heale fresh wounds wonderfully.*
3. *The root boiled and drank clenseth ye brest, and cureth the griefs of the lungs.*
4. *For all tumors and hot inflammations, especially of the fundament (anus), stamp it with ye leaves of groundswel and apply.*
5. *Ruptures and broken bones, stamp them and apply.*
6. *To stop the flux (diarrhea) and terms (menstruation), seethe (boil) the root in water and drink it.*
7. *Chew thereof to avoid thirst.*
8. *Rost (roast) the roots in the embers and eat them fasting three daies against ruptures and bursting. It helpeth together broken bones, it stoppeth fluxes (discharge) of blood and vomiting blood, the juice being drunk with red wine and a little mustick.*
9. *Hoarsenesse, eat powder of it, and of Elecampana with hony (honey) fasting.*
10. *Bones broken, stamp it with daisies, egrimony and vinegar and apply it till it be whole.*
11. *Rost the roots of Elecampana and Comfrey of each like much, of seethe (boil) them in water to the one-half, then put thereto a third part of clarified hony, and use it morne and even (evening) for chincough (whooping cough).*
12. *Fluxe to stop, apply the juice of Comfrey and both plantins with the burnt clay of an old oven or furnace to the belly.*
13. *Burstings and broken bones, eat the rosted roots with hony.*
14. *Matrix (of mother) rent by travaile of childbirth, put in Comfrey and Cynamon.*
15. *Matrix griefes, drink Comfrey with ale or water three dayes.*
16. *Bloody fluxe, wash the Comfrey root in cold water, and scrape it with a knife of bone or ivorie or of a harts horne (adult male deer horn), and eat thereof two ounces or more; but eat no vinegar that day, for that taketh away the vertue of the root.*
17. *Bursten (break, shatter, burst), put the juice into thy bread, and eat the root either raw or roasted.*
18. *Teethach, seethe it in vinegar and hold thy mouth over it.*
19. *Bruised, drink two spoonfulls of the juice of Comfrey and scabious morn and even, or stampe Comfrey, houseleake, orpin and floure, and apply it.*
20. *Flowers to stop, seethe Comfrey in running water and drink it.*
21. *Spitting blood, drinke Almond and Comfrey with stale ale, nine days and be whole.*
22. *If a man that is wounded, drinke Comfrey and keepe it, he shall live, if not, he shall die.*
23. *Veins broken in the breast, fry the roots with eggs and meate and eat them.*"
-'**The Garden of Health** Conteyning the Sundry Rare and Hidden Vertues and Properties of all Kindes of Simples and Plants' by **William Langham**, Practitioner in Physicke, London, England. First edition **1597**; second edition 1633, Symphytum page 163.
(Found in: 'American Observer Medical Monthly: Devoted to Homoepathic Materia Medica, Surgery, Gynaecology, Obstetrics, Otology, Ophthalmology, Practice of Medicine', New Series Volume 17, Detroit, Michigan, 1880. Article: 'Materia Medica: The Empirical History of Symphytum Officinale' by Professor S.A. Jones, M.D., Ann Arbor, Michigan, August 1880, pages 387-388.)

"**This monograph presents results of research on relics of cultivation and the present vascular flora of <u>sites of medieval fortified settlements and castles in Central Europe</u>.** *Special attention was paid to 109 West Slavic sites located in Poland, northeastern Germany, and the Czech Republic. For comparison, floristic data were collected also at 21 sites of medieval settlements and castles of Baltic tribes, East Slavs, and Teutonic knights.*
Considering the time of cultivation, 3 groups of relics were distinguished:
    (i) relics of medieval cultivation (plants cultivated till the late 15th century).
    (ii) relics of cultivation in modern era (introduced into cultivation in the 16th century or later).
    (iii) relics of medieval-modern cultivation.
Symphytum officinale L.: *(Codes: Sn, G, G5, 16, F, 1.3, 0.85). West Slavic archaeological sites.*
    Non-synanthropic spontaneophytes (native).
    Raunkiaer Life-Form: Geophyte (herbs with buds below the soil surface).
    Meadows *(Molinio-Arrhenatheretea, Plantaginetea majoris: Agropyro-Rumicion crispi). Number of Sites: 16.*
    Frequency Classes of Species Recorded: Frequent, 12.9-27.5%. Mean Abundance: 1.3 (moderately abundant).
    Rarity Coefficient (floristic similarity of the archaeologic sites): 0.85."
-'**Relics of Cultivation in the Vascular Flora of Medieval West Slavic Settlements and Castles**' by Zbigniew Celka, Department of Plant Taxonomy, Adam Mickiewicz University, Umultowska Poznan, Poland; Biodiversity: Research and Conservation, Volume 22, pages 1-110, June 2011.
    (Synanthropes are species of wild animals and plants that live near, and benefit from humans' artificial habitats. It does not include domesticated animals.)

Renaissance (17th Century, 1600s)

"However, other wandering vendors also appeared in the <u>Carpathian Basin in the 17th to 19th centuries</u>. Those coming from the Bihar Mountains (India) were root-dealing healers (radacinari in Romanian).
*They arrived with a load of 40-50 kilograms (88-110 pounds) of roots (Tamus communis, **Symphytum officinale**, Helleborus purpurascens) and stayed for 2-3 weeks."*
-**Pioneers in European Ethnobiology** edited by Ingvar Svanberg and Lukasz Luczaj. Uppsala, Sweden: Uppsala University, October 2014. Chapter: 'Hunting, Gathering and Herding: Bela Gunda (1911-1994) and other Hungarian Ethnobiologists' by Daniel Babai, Zsolt Molnar and Antal File.
    (The Carpathian Basin or Pannonian Basin is in the southeastern part of central Europe. It is surrounded by the Carpathian Mountains and the Alps.)

In the 1622 poem 'Poly-Olbion', Volume 2, by <u>Michael Drayton</u>, part 150, of 'The Thirteenth Song'*:
    "For physick, some again he inwardly applies.
    For comforting the spleen and liver, gets for juice,
    Pale Hore-hound, which he holds of most especial use.
    So Saxifrage is good, and Harts-tongue for the stone,
    With Agrimony, and that herb we call S. John.
    To him that hath a flux, of Shepherd's-purse he gives ;
    And Mouse-ear unto him whom some sharp rupture grieves.
    And for the labouring wretch that's troubled with a cough,
    Or stopping of the breath, by fleagm that's hard and tough,
    Campana here he crops, approved wondrous good:
    As Comfrey unto him that's bruised, spitting blood; (line 228)
(* -'The Complete Works of Michael Drayton, Volume 2 - PolyOlbion' by The Rev. Richard Hooper, London, England, 1876. The 1622 Comfrey poem is on page 150.)
(Michael Drayton {1563 -1631} was an English poet during the Elizabethan era. His poem 'Poly-Olbion', Volume 1 was published in 1612, and Volume 2 in 1622. It is 30 songs in Alexandrine couplets (12-syllable lines) with 15,000 lines of verse. 'Poly-Olbion' is one of the longest poems in English.)

    "The poet Michael Drayton has drawn the portrait of an ancient simpler, and has given a list of the remedies of which he made the most frequent use; the lines are to be found in his 'PolyOlbion', and as they contain examples of herbs selected under the system of the Doctrine of Plant Signatures..."
    -'Plant Lore, Legends, and Lyrics: Embracing the Myths, Traditions, Superstitions, and Folk-Lore of the Plant Kingdom' by Richard Folkard, London, England, 1892, pages 162-163.
    (A simpler is a single medicinal herb rather than a combination of herbs.)

"Herbalists referred to Comfrey as Symphytum majus vulgare (<u>Parkinson</u>, 1640) and Symphytum Consolida major (<u>Bauhin</u>, 1623); other names were associated with colour forms: Consolida major flore purpureo and flore albo (Gerard, 1633).
Several authors described S. tuberosum as Comfrey with knotted roots and white or yellow flowers (Salmon, 1610; Sowerby, 1867; Pharmacopoeia Londinensis Collegarium, 1668; Josselyn, 1672); both species were said to possess healing powers."
-'**British Medicinal Species of the Genus Symphytum**' by K.R. Fell and Janet M. Peck, Pharmacognosy Research Laboratories, University of Bradford, England; Planta Medica: Journal of Medicinal Plant and Natural Product Research, Volume 16, No. 2, pages 208-216, May 1968.

'Theatrum Botanicum: The Theater of Plants Or An Herball of Large Extent' by <u>John Parkinson</u>, London, England, Chapter 24, pages 522-524, **1640**:
"**The rootes of Comfrey, taken fresh,** *beaten small, spread upon leather, and laid upon any place troubled with the gout, doe presently give ease of the paines and applied in the same manner, giveth ease to pained joynts (joints), and profiteth very much for running and moist ulcers, gangrenes, mortifications (death of body tissue), and the like. It helpeth the ulcers of the lungs causing the phlegme that oppressed them to be easily spit forthe.*
Effectual for all those inward griefs and hurts. Being bruised and outwardly applied, helpeth sore, fresh wounds and cuts immediately, and laid thereto, by glueing together their lips and especially good for ruptures and broken bones.
A syrup made thereof *is very effectual for all those inward griefes and hurts; and the distilled water for the same purpose also, and for outward wounds or sores in the fleshy or sinewy parts of the body wheresoever, as also to take away the fits of agues (fevers), and to alay the sharpnesse of humour.*
A decoction of the Comfrey leaves hereof is available to all the purposes, although not so effectual as of the rootes."
    (John Parkinson, 1567-1650, was an English Master Herbalist and apothecary to King James I. There is a marble

statue of him in Sefton Park, Liverpool, England. 'Theatrum Botanicum' is 1,688 pages and describes over 3,800 plants. It was the most complete and beautifully presented English treatise on plants of its time.)
(An apothecary is a medical professional who formulates and dispenses 'Materia Medica' to physicians, surgeons, and patients. Today a pharmacist {chemist in Britain} does the same job. 'Materia Medica' is the study of medicinal drugs.)

*"The title page of the **'Theater of Plants'**, to use its English name, proclaims it as 'An Herball of Large Extent'. It is, for there are 1,755 folio sized pages, over 2,700 woodcuts, and the desciptions of more than 3,800 plants.*
If a reader should happen to drop it on his foot he would be well advised to consult the passages on Comfrey or other plants good for mending broken bones.
Considered to be the last of the great herbals, *the 'Theatrum Botanicum lives up to its reputation, in size as well as in quality.* ***In it Parkinson borrowed from the whole range of writings on materia medica, adding his own considerable knowledge as horticulturist and apothecary,*** *to produce one of the great repositories of herbal literature. His references to older authors and his quotations from them make 'Theatrum' a virtual one-volume herbal library. Should all other herbals be lost, future generations could still sample most of their lore and language through Parkinson."*
-An Illustrated History of the Herbals by Frank J. Anderson. New York, New York: Columbia University Press, 1977.

<u>**North American Colonists:**</u> **"Early settlers raised the herb (Comfrey) in Salem (Massachusetts) and Plymouth (Massachusetts), the Governor planted his on Boston Common (Massachusetts), and homeowners of New Netherlands (Connecticut, Delaware, New Jersey, New York) introduced it around 1649 to their gardens.**
The first frugal New Englanders avoided 'frivolities' such as spending precious minutes to raise ornamental flowers, but **Comfrey had already proven its worth** *as a plaster to external pus-laden sores and as a cough remedy."*
-**Comfrey: What You Need to Know** by Ben Charles Harris. New Canaan, Connecticut: Keats Publishing, Inc., 1982, page 9.
('New Netherland' or 'Novum Belgium' was a 17th-century colony of the Dutch Republic on the east coast of America. It extended from the Delmarva Peninsula to southwestern Cape Cod. Settled areas are now part of New York, New Jersey, Delaware and Connecticut, and small parts of Pennsylvania and Rhode Island.)
(The Colonial period of the United States was from 1607-1775. Jamestown, Virginia was founded in 1607. Plymouth, Massachusetts was founded by Pilgrims in 1620.)

" **'Comferie with white flowers' was taken over to New England (eastern United States) in the 17th century (1601-1700).** *There it escaped and established itself."*
-**Comfrey: Nature's Healing Herb & Health Food** by Andrew Hughes. Japan: Sanyusha Publishing, 1992, page 102.

" **'New England Rarities Discovered'* was first printed in London in 1672 by John Josselyn.** *His visits to Massachusetts (eastern United States) (in 1638 for fifteen months, twenty-five years later for eight years) provided excellent opportunity to observe that 'such garden herbs amongst us as do thrive here'.*
Good examples of the Englishman's herbal transplants *were the well-known plantain, mallow, nettles, dandelion, shepherd's purse, wormwood, knotgrass and the hardy perennial* **'Compherie with the white flower.'** "
-**Comfrey: What You Need to Know** by Ben Charles Harris. New Canaan, Connecticut: Keats Publishing, 1982, page 8.
(* -'New England's Rarities: Discovered in Birds, Beasts, Fishes, Serpents and Plants of that Country' by John Josselyn, Boston, Massachusetts, 1672.)

*"**Our first colonists planted Symphytum officinale at Salem and Governor Winthrop on Boston Common, Massachusetts. It was introduced into 'New Netherlands' before 1649.** Josselyn (1672) says it was in much repute (regard) as a plaster for wounds and as a cough medicine."*
-'**Old Time Herbs for Northern Gardens**' by Minnie W. Kamm. Boston, Massachusetts: Little, Brown & Co, 1938. Symphytum pages 115-116.

*"**Colonial housewives often made apt (good) pupils, from sheer necessity. Many of these dauntless (fearless) women had brought plants and seeds to make their own English physic-gardens in the New World.***
Colonial housewives were pleasantly surprised to find other treasured herbs growing wild and free: clown's woundwart or all-heal (Stachys palustris), much valued in rural England to stop bleeding and heal wounds; the pretty little Maidenhair fern (Adiantum capillus-veneris), so useful in chest troubles, **and other important herbs- marjoram, yarrow, brooklime and Comfrey- were also to be found."**
-**Green Pharmacy: The History and Evolution of Western Herbal Medicine** by Barbara Griggs. Rochester, Vermont: Healing Arts Press, 1997, page 105. (Also called 'New Green Pharmacy: Story of Western Herbal Medicine' by Barbara Griggs. London, England: Vermillion, 1997.)

<u>**"J. (Johann) Bauhin (Swiss botanist, 'Historia Plantarum Universalis',**</u> *Yverdon, Switzerland,***1651) expresses his** *concurrence (agreement) in the views of the sixteenth century herbalists as to* **the curative value of decoctions of Comfrey in all cases of wounds, blood-spitting, or even broken limbs."**
-**The Medicinal Uses of Comfrey** by Dr. Charles MacAlister, M.D., F.R.C.P., written 1935.
(Decoction is boiling herbal or plant material to dissolve the chemicals of the material that can include stems, roots,

bark or rhizomes; for example, making tea.)

"Jean Bauhin's chief botanical work, the 'Histoire Universelle des Plantes', was a most ambitious undertaking. This book is a compilation from all sources, and includes descriptions of 5000 plants. The figures, of which there are more than 3500, are small and badly executed. A large proportion of them are ultimately derived from those of Fuchs. Jean Bauhin's more famous brother, Gaspard or Caspar, was born 1560, and was thus younger by 19 years."
-'Herbals: Their Origin and Evolution: A Chapter in the History of Botany: 1470 to 1670' by Agnes Robertson Arber, University College, London, England; Cambridge University Press, England, pages 93-94, 1912.
(For more about Gaspard Bauhin, see sub-subsection 'Early Botanical History of Symphytum' in subsection 'Symphytum Comfrey Genus' in section 'Borage Family, Symphytum Genus' {Chapter 2}.
See subsections '1596 Bauhin' and '1623 Bauhin' in section 'Symphytum Species Classifications' {Chapter 4}.)

*"**Comfrey roots are full of glutinous and clammy juice**...for all inward hurts...and for outward wounds and sores in fleshy or sinewy parts of the body...It is especially good for ruptures and broken bones. **The great Comfrey** restrains spitting of blood. **The root boiled in water or wine and the decoction drank,** heals inward hurts, bruises, wounds and ulcers of the lungs, and causes the phlegm that oppresses him to be easily spit forth.*
A syrup made there of is very effectual in inward hurts, *and the distilled water for the same purpose also, and for outward wounds or sores in the fleshy or sinewy parts of the body, and to abate the fits of agues (fever with chills) and to allay the sharpness of humours.*
A decoction of the leaves is good for those purposes, but not so effectual as the roots.
The roots being outwardly applied cure fresh wounds or cuts immediately, being bruised and laid thereto; and is specially good for ruptures and broken bones, so powerful to consolidate and knit together that if they be boiled with dissevered (cut) pieces of flesh in a pot, it will join them together again."
-'**Culpeper's English Physician and Complete Herbal** to Which are Now First Added Upwards of One Hundred Additional Herbs' by <u>Nicholas Culpeper</u>, English herbalist, London, England, **1652-1653**, pages 131-132. 'Culpepers Complete Herbal' from 1653 was reprinted in 1816.
(Nicholas Culpeper, 1616-1654, was an English botanist, herbalist, physician and astrologer. His books include: 'The English Physitian' 1652 later known as 'Complete Herbal' 1653, and 'Astrological Judgement of Diseases from the Decumbiture of the Sick' 1655. The systematisation of herbal use by Culpeper was an important development in the evolution of modern pharmaceuticals, most of which originally had herbal origins.)
(According to ancient Greeks, the 4 'Humours' are the major fluids in the body- yellow bile, phlegm {mucous secretions}, black bile and blood- that Greeks believed corresponded to the four elements of fire, water, earth and air.)

*"The **Terms (menstruation)** commonly begin at fourteen years, and then the hair appears on the privities (private parts), the breasts swell, and women begin to be lecherous (sexual desire) and the blood can no longer stay in the veins, but breaks out at the veins of the womb.*
    For provoking (stimulating) the terms: mugwort, pennyroyal, southernwood, savory, thyme, alexander, and anemony.
    For stopping the terms and the whites (leucorrhoea): Comfrey, *mousear, yarrow, mede sweet, adder's tongue, lunaria, trefoil, money-wort, damel, flower gentle, blites, dragon-tree, beech tree, and hasel-nut tree.*
    Stopping the terms (menstruation), as bistort, Comfrey, *tormentil.*
    Stopping the terms, as Comfrey, *houseleek, knotgrass, myrtles, plantane, shepherd's purse, strawberries, and water-lilies.*
    For the mother: mother-wort, feverfew, calamint, burdock, butter-bur, orach, affa foetida, and cow parsnip.
Binding (hold together), *as amomum, agnus caslus, cypress, cinquefoil, **Comfrey**, bawm, fleawort, horsetail, ivy, knotgrass, bay, melilot, myrtles, oak, plantane, purslain, shepherd's purse, sorrel, sengreen, and willow.*
Glutinating (stick together), as birthwort, Comfrey, *daisies, gentian and Solomon's seal."*
-'**Culpeper's English Physician and Complete Herbal** to Which are Now First Added Upwards of One Hundred Additional Herbs' by Nicholas Culpeper, London, England, **1652-1653**, pages 15, 33, 36, 37.

*"**Government and Virtues. This is also an herb of Saturn, and I suppose under the sign Capricorn, cold, dry, and earthy in quality.***
What was spoken of clown's woundwort may be said of this; ***the great Comfrey helpeth those that spit blood, or make a bloody urine; the root boiled in water or wine,*** *and the decoction drunk, helpeth all inward hurts, bruises, and wounds, and the ulcers of the lungs, causing the phlegm that oppresseth them to be easily spit forth.*
It stayeth the defluxions (discharge) of rheum from the head upon the lungs, the fluxes (discharge) of blood or humours by the belly; women's immoderate courses, as well the reds as the whites; and the running of the reins (gonorrhaea), and, happening by what cause soever.
It is good to be applied to women's breasts that grow sore by the abundance of milk coming into them; as also to repress overmuch bleeding of hemorrhoids, to cool the inflammation of the parts thereabout, and to give ease of pains.
The roots of Comfrey taken fresh, beaten small, and spread upon leather, *and laid upon any place troubled with the gout, do presently give ease of the pains; and applied in the same manner, give ease to pained joints, and profit very much for running and moist ulcers, gangrenes, mortifications, and the like, for which it hath by often experience been found helpful."*

-'**Culpeper's English Physician and Complete Herbal** to Which are Now First Added Upwards of One Hundred Additional Herbs' by Nicholas Culpeper, London, England, **1652-1653**, pages 131, 132.

"**Take of the Roots and Tops of Comfry the greater and lesser,** of each three handfuls, red Roses, Betony, Plantane Burnet, Knot-grass, Scabious (Teasel/ Dipsacaceae family), Coltsfoot, of each two handfuls, press the Juyce out of them all being green and bruised, boyl (boil) it, scum it and strain it, ad (add) its weight of sugar to it that it may be **made into a syrup** according to art.
The syrup is excellent for all inward wounds and bruises, excoriations (skin problems), vomitings, spittings, or pissings of blood, it unites broken bones, helps ruptures, and stops the terms in women: you cannot err in taking of it."
-**Pharmacopoeia Londinensis, or, The London Dispensatory** further adorned by the studies and collections of the Fellows, now living of the said colledg' **by Nicholas Culpeper, 1668.**
    (Coltsfoot has toxic ingredients so I do not recommend its use.)

"*Consolida major:* **Comfrey, I do not conceive the leaves to be so virtuous as the roots.**
Temperature of the Roots:
Roots cold in the first degree: Sorrel, Beets white and red, **Comfrey the greater,** Plantain, Rose Root, Madder.
Stop the menses Comfrey: Tormentil, Bistort, etc.
Decoctum Trumaticum:
Take of Agrimony, Mugwort, wild Angelica, St. John's Wort, Wormwod half a handful, Southernwood, Bettony, Bugloss, **Comfrey the greater and lesser roots and all,** Avens, both sorts of Plantain, Sanicle, Tormentil with the roots, the buds of Barberries and Oak, of each a handful, **all these being gathered in May and June and dligently dried.**
Let them be cut and put up in skins or papers against the time of use, then take of the forenamed herbs three handfuls, boil them in four pounds of conduit water and two pounds of White Wine gently till half be consumed, strain it, and a pound of Honey being added to it, let it be scummed and kept for use."
-'**The Complete Herbal; to Which is Now Added Upward of 100 Additional Herbs...to Which are Now Annexed the English Physician Enlarged and Key to Physic by Nicholas Culpeper'**, London, England, update by unknown, pages 54, 235, 257, 259, 294, 1863.

"**Comfrey was also called Consolida, or, in some of the English Herbals, Consound, and was a member of a class of remedies referred to, for instance, in the 'Pharmacopoeia Londinensis Collegarum' (1668 by Culpeper)** among the Herbs, Leaves and Seeds the Consolidae include."
-**The Medicinal Uses of Comfrey** by Dr. Charles MacAlister, M.D., F.R.C.P., written 1935.

"**In the 'Pharmacopoeia Londinensis Collegarum' (1668 by Culpeper), a variety of Comfrey was used by the Turks and Saracens for healing wounds.**"
-**Comfrey and Chlorophyll:** A Report About the Medicinal Value of Comfrey and Chlorophyll as Found in Old and Modern Literature by Vincent Licata. California: Continental Health Research, 1971, page 9.
    (Saracens were nomadic tribes of the deserts between Syria and Arabia.)

"There were probably about nine apothecary gardens in <u>**Moscow, Russia in the seventeenth century**</u>. In fall 1671, the following plants were listed:
    Seeds of 14 plants including Portulaca, Pimpinellay Cochleariay Thymus, Ruta and Nicotiana.
    Flowers of red roses, white bush roses, poppy, and borage (Borago officinalis L.).
    Roots of Petroselinum, Pimpinellay Cichorium and Symphytum officinale.
    And 17 different herbs including marjoram, thyme, sage, mint, fennel, peony, endive, basil, parsley, borage, **Comfrey,** lovage and rosemary - **a list truly reminiscent of the Western medieval monastery garden.**"
-'**Russian Medical Botany Before the Time of Peter the Great**' by Margery Rowell; Sudhoffs Archiv, Band {Volume} 62, Heft {Issue} 4, 4th Quartal {Quater}, pages 339-358, 1978.

"**For Spitting of Blood: Take conserve of Comfrey** and of (rose) hips of each an ounce and a half; conserve of red roses, three ounces; dragon's blood, a drachm; species of hyacinths, two scruples; red coral, a drachm; mix, and with syrup of red poppies make a soft electuary. Take the quantity of a walnut, night and morning."
-Aristotle's Masterpiece: Sex and Midwifery (The Works of Aristotle)' by <u>William Salmon</u>, NY, 1846. **First published 1684.**
    (A conserve is a sweet food made by cooking pieces of fruit with sugar. Dragon's blood is a bright red resin from different species of plant genera: Croton, Dracaena, Daemonorops, Calamus rotang and Pterocarpus. Electuary is an herbal medicine mixed with syrup or honey.)
    ('Aristotle's Masterpiece' was popular in England from late 1600s through 1800s. It was not written by the Greek Aristotle.)

"**Comfrey tis an excellent Wound-Herb,** is Musilaginous and Thickning, and qualifies the Acrimony (ill feeling) of the Humours. (Mollifies the bitterness of body fluids.) Tis used in all Fluxes, especially of the Belly, and for a Consumption.
The Flowers boyl'd (boiled) in Red Wine are very proper for those that make a Bloody Urine.

Outwardly applied, it stops the Blood of Wounds and helps to unite broken Bones, wherefore tis called Boneset."
-John Pechey's 'Compleat Herbal of Physical Plants', English herbalist, 1694
(John Pechey, 1655-1716, was a medical writer, whose name is also spelled Peachey and Peche. He was born in Chichester, West Sussex, England. His methods were those of an apothecary rather than of a physician.)

The book **"A Japanese Herbal in the Wellcome Institute for the History of Medicine: A Contribution to the History of the Transfer of Scientific Knowledge from Europe to Japan"** was written 1997, 2005 by Hartmut Walravens. **It contains a drawing from the 1600s of 'Consolida majoris, Symphyti majoris, Smeer wortel, Waal wortel'.**
Wellcome Institute's excerpt about the book: *"A herbal consisting of hand-coloured illustrations of plants pasted on to Japanese paper and then bound in European fashion. The names of the plants are given in Chinese characters and in Latin, and there are notes in Dutch and transliterations of the Latin and Dutch names into Japanese katakana. The title means 'Herbal extracted from Dodoneas'. Rembert Dodoens or Dodoneas was a Flemish physician best known for his illustrated botanical publications, especially his De stirpium historia (Antwerp, 1554), and Stirpium historiae pemptades sex (Antwerp, 1583)."*

Symphytum Bulbosum 1827

'Iconographia Botanica seu Plantae Criticae', Volume 5,
by Heinrich Gottlieb Ludwig Reichenbach,
Leipzig, Germany, 1827.
He wrote 10 volumes from 1823-1832.

Symphytum bulbosum Schimp., Volume 5,
page 517, t. CCXX (220), fig. 367.

SECTION OF COMFREY WITH SOLID STEM.

Prickly Comfrey Root, Solid Stem 1880

'New Commercial Plants with Directions How to Grow Them to the Best Advantage', No. 3, by Thomas Christy, F.L.S., 1880, London, England, page 13.

Chapter 14

Age of Enlightenment and Comfrey

Age of Enlightenment (18th Century, 1700s)

'Institutiones Rei Herbariae' by French botanist, <u>Joseph Pitton de Tournefort</u>, Paris, France, 3 Volumes, 1700. In Latin. It includes 9000 species in 698 genera, which directly influenced botanist Carl Linnaeus. Tournefort was the first to make a concise definition of the concept of genus for plants.

For more about Tournefort, see sub-subsection 'Early Botanical History of Symphytum' in subsection 'Symphytum Comfrey Genus' in section 'Borage Family, Symphytum Genus' (Chapter 2).
For Carl Linnaeus, see subsection 'Classification of Life' in section 'Taxonomy and Nomenclature of Comfrey' (Chapter 1).

> "*Joseph Pitton de Tournefort (1656-1708), studied medicine in Montpellier, France around 1679 after his father died. Prior to that, he had made his first botanical expedition with Charles Plumier (1646-1706) through Provence (France) and Savoy. Bitten by the 'bug', he also botanised throughout the Languedoc Roussillon region around Montpellier and around Barcelona, Spain in 1681 before coming back to Montpellier at the end of 1681.*
> **The fame of his botanical expertise and that of his herbarium spread** *and he was invited to move to Paris to become became Chief Botanist to Louis XIV and Professor of Botany, in charge of the Royal Garden, the Jardin des Plantes in 1683.*
> **Based on his researches, he published (in Paris, France) his 'Elemens de Botanique ou Methode pour Connoitre les Plantes' (Botanical Elements or Methods to Know Plants) in 1694 which he subsequently translated into Latin as 'Institutiones Rei Herbariae' (1700) to extend its influence throughout Europe.**"
> -'Tournefort' by Botany at the Edward Worth Library, Dublin, Ireland, 2019. This exhibition explores the botanical collections of the library of physician Edward Worth, 1676-1733. https://botany.edwardworthlibrary.ie
>> (Savoy includes the Western Alps between Lake Geneva {between Switzerland and France} in the north and Dauphine {province in southeast France} in the south.)

> "**Comfrey 'Symphitum': The Leaves of the first Species or great Comfrey are insipid (no flavor), glutinous,** *and give a faint Tincture of Red to the blue Paper. The Roots dye it deeper, and abound with a a clammy Juice.*
> **This Plant contains a Salt very like to the Salt of Coral (calcium acetate),** *dissolved in a glutinous Phlegm, in which there is some Sulphur, but very little of Salt-Armoniac (?Sal-Ammoniac= ammonium chloride); for upon a chymical (chemical) Analysis* **the Comfrey yields many acid Liquors, much Earth, very little Sulphur, and no concreted volatile Salt,** *but a small quantity or an urinous (similar to urine) Spirit, and a very moderate quantity of fixed Salt; so that* **its Powers seem principally to depend upon its slimy Mucilage, which the Fire destroys**.
> **The Moderns (modern botanists) deservedly reckon (think) Comfrey among the chief of the vulnerary (wound healing) Plants,** *and they all agree that its Roots incrassate (thickened) or thicken, and blunt or sheath the acrimonious (bitter) Particles of the Humours.*
> *They are used in a Loss or Flux (discharge) of Blood, occasioned by sharp Salts, which render it too fluid; and in a Catarrh or Defluxion (discharge) upon the Breast or Lungs, caused by Salt and corrosive Serosities (serous or watery).*
> **The Roots of the great Comfrey bruised and applied by way of Cataplasm (chopped poultice) to a prick'd Tendon,** *or to the part most sensibly affected with goutith (gout) Pains, give great Ease and Relief, and in the same manner stop running and eating Ulcers, Gangrenes, etc.*
> **Simon Paulli advises not to use the Roots of Comfrey singly for the Cure of goutith Pains,** *least they should too powerfully strike the Humour back, and throw it inwards.*
> **He recommends the following Cataplasm,** *which he learn'd from Sennereus as an incomparable Remedy:*
>> **Take three Ounces of the Roots of the great Comfrey,** *two Ounces of Marsh-mallow Roots, a Handful of Southernwood (southern wormwood) Leaves, two Handfuls of St. John's-wort, three Handfuls of Camomil (Chamomile) Flowers, four Handfuls of Elder Flowers, two Ounces of Fenugreek Seed, three Ounces of Lint Seed (flaxseed); boil all these ingredients, in Elder-Water and make a Cataplasm.*
>> *But this Remedy is too much compounded. I usually mix, says Tournefort, some Drops of fetid Oil with the Comfrey Roots well bruised, and to apply it to the Part affected.*
> *We find among, the 'Observations of Hieronymus Reusuerus' cured a certain person of a malignant ulcer, pronounced to be a cancer by the surgeons, and left by them as incurable, by* **applying twice a day the root of Comfrey bruised,** *having first peeled off the external blackish bark or rind; but the cancer was not above 8 or 10 weeks standing.*"
> -'The Compleat Herbal, or, The Botanical Institutions of Mr. Tournefort with Large Additions from Ray, Gerarde, Parkinson and Others, the Most Celebrated Moderns' by Joseph Pitton de Tournefort, French botanist, London, England, 1719, pages 178-180.
> (Simon Paulli, 1603-1680, was a Danish physician, and professor of anatomy, surgery and botany at University of

Copenhagen, Denmark. He published treatises in medicine and botany such as 'Quadripartitum Botanicum'.)
(Fetid oil is an oily substance that is empyreumatic, i.e., has the odor of burned animal matter.)

"During Tournefort's travels in the East in 1700-1702 he discovered two species which were recorded in his 'Corollarium' in 1703. **One of these, 'Symphytum Constantinopolitanurn...flore albo (white)'** was figured and fully described in the 'Voyage du Levant' (1717), and is **Symphytum orientale Linn.**, as now understood.

The other, 'Symphytum orientale...flore caeruleo (yellow)' may include two or more species. One of these is represented by a specimen in Herb. Mus. Brit. (Herbarium of the Botanical Department of the British Museum of Natural History), marked on the back of the sheet 'Cappadocia, Tournefort.' This has been named 'S. orientale,' but it is the common plant of the Caucasus described as **S. asperum** by Lepechin in 1798, and by Donn as **S. asperrimum** (Bot. Mag. t. 929) in 1806. It was introduced into England in 1801, and there is a specimen in Herb. Mus. Brit. grown at Kew (Gardens, London) in 1803."

-'**A Revision of the Genus Symphytum, Tourn.**' by Cedric Bucknall, (Mus. Bac. Oxon= Bachelor of Music, Oxford University), Journal of the Linnean Society of London, England, Botanical Journal, Volume 41, Issue 284, pages 491-556, December 1913.

(Between 1700 and 1702 Tournefort travelled through Greece, Turkey, Armenia, Georgia and along the Black Sea.)

"<u>**Salmon's Herbal 1710**</u>: *Comfrey: The Preparations. You may have therefrom:*
   1. A Juice of the Leaves or Roots.
   2. An Essence of the same.
   3. A Syrup of the Juice of the Root.
   4. A Decoction of the Root in wine or water.
   5. A Pouder (Powder) of the Root.
   6. A Balsam (resin) of the Juice of the Root.
   7. A Cataplasm (poultice) of the Root.
   8. A Distilled Water from the Leaves, Stalks and Roots.
   9. An Acid Aqueous (water) Tincture.

The Virtues:
1. A Juice of the Leaves or Roots.
Camerarius says, that two ounces of Comfrey being drank at a time, does much good in the Lethargy (tiredness), and Dead Sleep: it is drying and binding in a great measure, and is good for such as spit Blood, bleed at Mouth, or make a bloody Urine: yet it opens Obstructions of the Lungs, and causes easie (easy) Expectoration (removing mucous from lungs).

2. An Essence of the same.
It has all the former Virtues, but is much more effectual to stop any Flux (discharge) of Blood, in any part whatsoever. It prevails against all inward hurts, bruises and wounds, cleanses Ulcers of the Lungs, drys and heals them, and being taken Daily, Morning and Evening, it prevails against Catarrhs (mucous discharge), and stops the defluxion (discharge) of Rheum (watery fluid) from the Head upon the Lungs; fluxes of Blood, or Humors by the Belly, and the immoderate or overflowing of the Courses (menstruation) in Women. It stops also the overflowing of the Whites (leucorrhoea) and universals being premised {introduced} it cures a Gonorrhea, or Running of the Reins in Men, coming from what cause soever. Dose 2 or 3 ounces in Red Port Wine.

3. A Syrup of the Juice of the Root.
It has the Virtues of the essence, but causes a better and more easie Expectoration out of the Lungs; is good against Coughs and Colds, Wheesings (wheezing), and other like Distempers of those Parts. It is said to be good for such as have broken Bones, because it hastens the breeding of the Callous; and for the same Reason, it is said also to be good to cure Ruptures in Children. It is so powerful to Consolidate or Knit together, whatsoever needs knitting, that if the Comfrey Roots be boiled with flesh cut into pieces, or very deeply slash'd, in a Pot, they will join them together again.

4. A Decoction of the Root in Wine or Water.
It is good against inward Bruises and Wounds, inward Bleeding, Spitting, Vomiting, or Pissing Blood, as also the Bloody or Hepatick (hepatic = liver) Flux; and has indeed all the Virtues of the Juice, Essence and Syrup, (but not full out so effectual as they are) being drank to six or eight ounces, Morning, Noon, and Night. It is also good to cleanse, dry, and heal external Wounds, Ulcers, and Running Sores, they being washed therewith once or twice a day.

5. A Pouder (Powder) of the Root.
Being taken inwardly to one dram (drachm= 60 grains= 64.79 milligrams) in a little of the Syrup, it stops inward bleeding, heals Wounds in the Stomach and Thorax (chest), as also Ulcers in the Lungs. If it is applied to green (new) Wounds, as soon as the Wound is made, it conglutinates or joins the Lips thereof together, and causes it speedily to be healed; mixed with the Syrup, and applied to the Hemorrhoids or Piles, it cools the Inflammation, and represses their over much bleeding, and allays the heat of the Parts adjacent, taking away, and easing all the pain.

6. A Balsam (Resin) of the Juice of the Root.
It is a singular Vulnerary, and cures simple green Wounds, generally at one dressing. It is digestive, and cleansing, and dries up and heals running Sores, and old Ulcers, in any part of the Body, but chiefly in those parts which are not depending; resisting Gangrenes, Mortifications (dead tissue), etc.

7. A Cataplasm (Poultice) of the Root.
If it is made of the simple Comfrey Root, beaten into a Mucilage raw, and then spread upon Leather or Linnen Cloth, and applied to parts pained with the Gout; it gives present ease to the pain, and so admirably strengthens the part, as that the Disease never returns any more from the old Cause, and this I have several times proved.

If it is made of the Comfrey Root boiled till it is soft in Water, and then beaten into a Pulp, adding to it the Pouder (powder) of the Root, enough to bring it to the Consistence of a Cataplasm, and it is presently applied to any simple Green Wound, or Cut, it quickly heals it by consolidating, or conjoining the lips thereof together;

Apply'd also upon broken Bones, it facilitates and speeds the Cure, by preventing a flux of Humors, inducing the Callus, and strengthening the Part and applied, is also profitable against Ruptures in Children.

It is good also to be used to Womens Breasts, which swell and grow hard and sore by the abundance of Milk flowing into them, which it does by a repercussive (reducing) Virtue.

It also cools the Inflammation, abates (lessens) the Swelling, and eases the Pain of the Piles, as experience has sufficiently proved. It is also very profitable against moist and running Ulcers, Gangrenes, Sphacelus (gangrenous or dead mass), and the like, in which cases it has been experienced, and found often helpful.

8. A Distilled Water from the Leaves, Stalks, and Roots thin sliced.

It has the Virtues of the Juice and Essence, but very much weaker yet. Authors say, it is good for outward Wounds or Sores, whether in the fleshy or nervous parts of the Body wheresoever; as also to take away the Fits of Agues (fever and shivering), and allay (ease) the sharpness of the Humors; but this it the more effectually does, if it is mixed with equal parts of the Liquid Juice or Essence.

9. An Acid Aqueous (Water) Tincture.

Take Spring Water, a gallon, Oil of Vitriol (sulfuric acid), or Oil of Sulphur per Campanum three ounces, mix them; then put into it of the pouder (powder) of the Root, six ounces; digest in a gentle Sand heat for a Month, shaking the glass three or four times a day; afterwards being well settled, decant the clear Tincture for use. Given inwardly in Wine, or any other proper Vehicle, it stops inward bleedings, and strengthens and restores the Tone of the Stomach; And mixed with Red Port Wine, it makes a good Lotion to cleanse, dry, and heal any old Ulcer, or running Sore, and effectually destroys the Putridity (rotting), if any, therein."

-'**Botanologia, The English Herbal, or History of Plants** with Names, Species, Descriptions, Places, Times, Qualities, Specifications, Preparations, Virtues and Uses' **by William Salmon**, M.D., London, England, **1710**. It is 1,375 pages. Comfrey is on pages 210-213.

(William Salmon, 1644-1713, was an English empiric doctor who called himself a 'Professor of Physick'. Empiric doctor means healing based on experience.)

(Balsam is an aromatic resinous substance, such as balm, exuded by some trees and shrubs and used as a base for some fragrances and medicine. Sulfuric acid causes chemical burns so I do not recommend using it.)

(Sand heat or bath is a container of hot sand in which another container is partially covered.)

"William Salmon in this 'English Herbal' (1710) classifies Comfrey among the agglutinatives or symphitica, 'which is the reason that Comfrey is called Symphytum because of its glewing quality.'
In several Herbals it is spoken of as the chief vulnerary for the same reason."
-The Medicinal Uses of Comfrey by Dr. Charles MacAlister, M.D., F.R.C.P., written 1935.

(Agglutinative means adhesive or uniting.) (Vulnerary means something used for the healing of wounds.)

"*In 1712 there was published in Edinburgh, Scotland a little book entitled 'The Poor Man's Physician, or the Receits (Receipts) of the Famous <u>John Moncrief</u> of Tippermalloch'. The author was Sir John Moncrief, who owned the estate of Tippermalloch in Perthshire (central Scotland) and who was 85 years of age when his book was published.*

Like so many other works of its kind, it is a mere list of diseases with a large choice of remedies. There is no description of the cause or nature of the malady (illness), such as is found in 'Boorde's Breviary'.

Some of the methods of treatment have a modern flavour and may be explained in the light of modern science, such as the use of dried blood for haemorrhage and of watercress for scurvy, but for the most part the 'receits' or prescriptions are of empiric (from experience) nature and the medicaments (medicines) are often disagreeable or even disgusting. A considerable choice, however, is offered to the patient who may dislike heroic or unpleasant doses.

Another strange term is still applied to **pharyngitis** *in some parts of Scotland by the patient who announces 'the pap of the halse has fallen down', meaning his uvula is inflamed and elongated. This diagnosis appears in 'The Poor Man's Physician':*

'If the Pap of the Halse be overlax, first dry it with Decoction of Prune leaves or of Comfrey. *Then divide an egg in two halves, after it has been well boiled, and apply one half to the Crown of the Head."*

-'**Nova et Vetera: The Poor Mans Physician**' by Douglas Guthrie, The British Medical Journal, London, England, page 142, January 26 1946.   (The uvula is a fleshy piece of tissue hanging over the tongue toward the back of the mouth.)

"<u>Johann Wilhelm Weinmann</u> *(1683-1741) was a Regensburg, Germany, apothecary who also produced one of the earliest botanical works to use color-printed mezzotint. He established a small botanical garden in Regensburg.* **In 1723 he began to produce botanical publications. 'Phytanthoza Iconographia' was a monumental work** *of 8 volumes and 1,025 color plates and contained the first published botanical illustrations (unsigned) by famed botanical artist Georg Dionys Ehret (1708-1770).* **The work also included hand-colored etchings by unknown artists, such as Symphytum officinale Linnaeus: Symphytum majus flore atro purpureo** *(Symphytum officinale Linnaeus, Boraginaceae), hand-colored engraving by ?B. Seuter, after an original by an unknown artist (fl. {alive during} 1700s) for Johann Wilhelm Weinmann (1683-1741), 'Phytanthoza Iconographia' (Regensburg, Hieronymus Lentz, 1737-1745, Volume 8, plate 957), Hunt Institute Art accession no. 2223."*

-'**Virtues and Pleasures of Herbs through History: Physic**' by Hunt Institute for Botanical Documentation: A Research Division of Carnegie Mellon University, Pittsburgh, PA, **2019**. Hunt Institute specializes in history of botany and all aspects of plant science.

"Comfrey is also a wound herb. K'Eogh in his 'Irish Herbal' (1735) wrote that it 'heals all inward wounds and ruptures'.

Today, it is still highly regarded for its healing properties."
-**Encyclopedia of Herbal Medicine:** The Definitive Reference to 550 Herbs and Remedies for Common Ailments by Andrew Chevallier, FNIMH. London, England: DK (Dorling Kindersley), 2000.

(**Reverend John K'eogh**, 1681-1754, from Cork, Ireland, was a Doctor of Divinity and naturalist. In 1735 he wrote 'Botanologia Universalis Hibernicaor', or a 'A General Irish Herbal', a book about medicinal plants. This book was republished in 1986 under the title 'An Irish Herbal', edited by Michael Scott.)

"Botanalogia Universalis Hibernica, or, A General Irish Herbal calculated for this kingdom: giving an account of the herbs, shrubs, and trees, naturally produced therein, in English, Irish, and Latin: with a true description of them, and their medicinal virtues and qualities: to which are added, two short treatises."
-Title of Reverend K'eogh's book ('Lus na Cnamh Briste' or Gaelic name for Comfrey means 'plant for broken bones'.)

'Genera Plantarum, Edition 1' by Swedish naturalist <u>Carl Linnaeus</u>, **written 1737**. The first edition of 'Genera Plantarum' contains descriptions of the 935 plant genera known to Linnaeus.
For Carl Linnaeus, see subsection 'Classification of Life' in section 'Taxonomy and Nomenclature of Comfrey' (Chapter 1).

'Species Plantarum, Edition 1' by Swedish naturalist **Carl Linnaeus, written 1753**. The first edition of 'Species Plantarum' lists every species of plant known at the time, classified into genera (plural of genus). It was the first book to consistently apply binomial names.

*"A Salve for a Rupture: Melt a pound of Deer-suet, and put to it a handful of Solomon-Seal, as much **Comfrey-roots**, and as much Mouse-ear, stamp all the Herbs and Roots; strain and boil them with the Suet; when tis well mix'd, and consum'd to half of the quantity, then add four ounces of Adders-spear, and an ounce of fine Bole-Armoniack, well powder'd and sifted; mix all, and keep it for use. When you lay it on the part, bind it hard. For a Rupture in the Throat, lie always in a Neckcloth tied hard."*
-'**A Collection of Above Three Hundred <u>Receipts in Cookery, Physick and Surgery</u>**' by 'Several Hands', London, England, **1746**, page 265-266

(Suet is the hard white fat on the kidneys and loins of cattle, sheep, deer and other animals.)
('Bole-Armoniack' is Armenian bole, also known as bolus armenus or bole armoniac, that is an earthy clay, usually red, native to Armenia. It is red due to the presence of iron oxide.)

<u>Ivan Lepechin</u> was director of the 'Imperial Botanical Gardens' in St. Petersburg, Russia from 1774 to 1802. During this time he became an expert on medicinal plants and published information about 29 new species of flowering plants including **Symphytum asperum**.
For more about Lepechin see sub-subsection 'Early Botanical History of Symphytum (Botanists)' in subsection 'Symphytum Comfrey Genus' in section 'Borage Family, Symphytum Genus' (Chapter 2).

*"In 1771, <u>Joseph Busch</u>, a landscape gardener of Hackney (London borough, England), sold his nursery to <u>Conrad Loddige</u> and set sail with his family to lay out gardens of <u>St. Petersburg Palace</u> for Catherine II (The Great) of Russia. In the 1790s he sent Symphytums back to the Loddiges, as I found from the bound copy of their cataogue for 1836 in Hackney Public Library. One of these species was <u>**Symphytum asperrimum**</u>, with striking cobalt-blue flowers very welcome in the borders of the 18th century."*
-**Fighting Like the Flowers: An Autobiography:** The Life Story of Britain's Best-Known Organic Gardener by Lawrence D. Hills. Bideford, Devon, England: Green Books, 1989, page 100.

(The city of Saint Petersburg in Russia was founded in 1703. In 1914 the name was changed to Petrograd. In 1924 it was changed to Leningrad. In 1991 it was changed back to Saint Petersburg. In 1967 the gardens of Saint Petersburg Palace were called a 'Park of Rest and Culture'.)
(For more about the Loddiges see 1806 in the section 'Age of Revolutions and Comfrey' {Chapter 15}.)

*"**Symphytum Asperrimum (Donn) or Asperum**. This was the earliest type of Comfrey imported into Britain, and its history goes back there to the late 18th century.*
*A nurseryman and gardener named **Joseph Busch** went to St. Petersburg (now Leningrad) to work in the garden of Empress Catherine the Great of Russia.*
__Between 1790 and 1800 he sent back to England several Symphytums for garden plants,__ to be used as showy border plants producing their beautiful bell-like flowers and growing up to 4-5 feet high (120-150 cm)."
-**Comfrey: Nature's Healing Herb & Health Food** by Andrew Hughes. Japan: Sanyusha Publishing, 1992, page 122.

*"**The species (Prickly Comfrey) is believed to have been introduced in Britain from Russia about 1790.***
After the lapse of half a century (50 years) its value as fodder became known. Information is supplied on chemical composition at different periods of growth and on various statements published regarding value of species for stock feed."
-'**Prickly Comfrey**' by J.K. Crawley, Journal of the Land Agents' Society, Volume 41, pages 87-91, 1942.
(A land agent administers a landed estate and its tenancies, or acts as an agent for the sale of land.)
(I could not find this report. If you have it, could you please send it to me.)

*"**In the year 1790, the Symphytum asperrimum, or Caucasian (Prickly) Comfrey, was introduced into England,***

and in distinction from the wild variety indigenous (native) to England it was named Prickly Comfrey. It was described as finding a place in Kew Gardens in 1799.
***Symphytum asperrimum was known, in Russia and Circassia** (a region in the North Caucasus and along the northeast shore of the Black Sea), to be an invaluable plant for medicinal purposes and forage.*
The Russians and Circassians found the Comfrey grew at as high an altitude as 4,000 feet (1219 meter) above the sea level. The roots are sometimes used as food by the inhabitants."
-'**Forage Plants and Their Economic Conservation by the New System of Ensilage:** Part I: Caucasian Prickly Comfrey' **by Thomas Christy,** Jun., FLS (Fellow of Linnean Society), Christy & Co, London, England, 1877, pages 1, 4.

"***Symphytum asperum Lepechin was, according to Aiton* introduced to Britain in 1799 by Conrad Loddiges.***
Loddiges' nursery at Hackney (London borough) was purchased from Joseph Busch, who had been appointed head gardener at the Palace of St. Petersburg (Russia), and who between 1790 and 1801 sent to Conrad Loddiges several species of Symphytum including S. asperum.
The early editions of the sales catalogues issued by 'Conrad Loddiges and Sons' have not been traced, but the 11th edition of 1818 lists the following: Symphytum asperrimum, S. caeruleum, S. coccineum, S. patens, S. tauricum and S. tuberosum.
> *The first is synonymous with S. asperum, of which S.caeruleum may have been a variety.*
> *S.coccineum was presumably the crimson-flowered form of S. officinale, and S. patens may have been the purple-flowered variety of S. officinale or the plant now known as S. x uplandicum."*
-'**The History of Symphytum Asperum Lepech. and S. x Uplandicum in Britain'** by A.E. Wade, Department of Botany, National Museum of Wales, Watsonia, Volume 4, pages 117-118, 1958. (Watsonia was the journal of the 'Botanical Society of Britain and Ireland' from 1949 to 2010. It is now called 'New Journal of Botany'.)
(* -'Hortus Kewensis; or a Catalogue of the Plants Cultivated in The Royal Botanic Garden at Kew: Volume I' by William Townsend Aiton, London,England, 1810.)
> (Kew Gardens is a botanical garden in London, England that houses the 'largest and most diverse botanical and mycological collections in the world'.)

> "***Symphytum. Gen. pl. 245.*** *Corollae limbus tubulato-ventricosus: fauce clausa radiis subulatis.*
> *officinale, tuberosum, orientale, asperrimum."*
-'**Hortus Kewensis**; or a Catalogue of the Plants Cultivated in The Royal Botanic Garden at Kew: Volume I' by William Townsend Aiton, London,England, 1810, pages 293-294.

"***The original cross of Russian Comfrey was probably a natural hybridization that took place in the collection of Symphytums gathered by Joseph Busch,*** *the Hackney (London, England) landscape gardener who laid out the grounds of St. Petersburg Palace for Catherine II (the Great) of Russia.*
*Plants were sent from there from 1870 onwards **(other Symphytums were sent earlier, the first in 1790)**, and the latest import is reported to have been in 1900."*
-**Comfrey: Nature's Healing Herb & Health Food** by Andrew Hughes. Japan: Sanyusha Publishing, 1992, page 136.

"***A rare chance natural hybridization appears to have occurred between the yellow S. officinale (Common Comfrey) found all over Europe including Russia, and the blue S. asperrimum (Prickly Comfrey) planted side by side in the Symphytum border."***
-Fighting Like the Flowers: An Autobiography: The Life Story of Britain's Best-Known Organic Gardener by Lawrence D. Hills. Bideford, Devon, England: Green Books, 1989, page 100.   (This created Russian Comfrey.)

In 1787 English physician <u>Dr. William Withering</u> (1741-1799), and in 1845 naturalist William MacGillivray (1796-1852) wrote about Comfrey in 1801 in "**A Systematic Arrangement of British Plants** with an easy introduction to the study of botany", Volume II. The 4-volumes were published in London, England from 1787 to 1848. It was written so the average person could identify plants. He mentions Comfrey's edibility, but says not all animals seek it as forage.

> "***Symphytum officinale (Common Comfrey):*** *Banks of rivers and wet ditches.*
> ***Symphytum patens.*** *Sibth. Fl. Oxon. who considers it as a distinct species. Frequently found growing with the preceeding, and flowering at the same time.*
> ***Symphytum tuberosum:*** *possibly a variety of S. officinale. Root tuberous. Flowers yellow white."*
-William Withering, M.D., F.R.S. (Fellowship of the Royal Society), Member of Royal Academy of Sciences at Lisbon (Portugal); Fellow of the Linnean Society; Honorary Member of the Royal Medical Society at Edinburgh (England).

<u>'**English Botany (Sowerby's)**</u>**; or Coloured Figures of British Plants with their Essential Characters, Synonyms and Places of Growth' by James Sowerby**, London, England. Sometimes called 'Sowerby's Botany'. The third edition in 1863 was edited by John Boswell Syme with low quality art reproduction. **'English Botany' was a major publication of British plants comprising 36 volumes, issued in 267 monthly parts over 23 years from 1791 to 1814.**
> James Sowerby, 1757-1822, was an English naturalist, illustrator and mineralogist. Contributions to published works, such as 'A Specimen of the Botany of New Holland' or 'English Botany', include his detailed and attractive plates. The

vivid color and accessible texts helped reach a larger audience in works of natural history.
For more about Sowerby see sub-subsection 'Early Botanical History of Symphytum (Botanists)' in subsection 'Symphytum Comfrey Genus' in section 'Borage Family, Symphytum Genus' (Chapter 2).

*"**James Sowerby was one of the 'inner circle' in the great age of botanical illustration in England,** and a central figure in many of the publications of the day. **His illustrations are found in 'Curtis Botanical Magazine', 'Curtis Flora Londinensis' and William Baxter's 'British flowering Plants',** easily recognizable for their characteristic simplicity of line and elegance.*
*Sowerby came from an illustrious English family of naturalists and illustrators. After studying painting at the Royal Academy, London, England, he became one of England's foremost botanical illustrators and engravers, **producing in conjunction with Sir James Smith his great work 'English Botany'**."*
-'James Sowerby Botanical Prints 1791' by Antique Botanical Prints, Rare Antique Prints from Panteek, 2018. https://www.panteek.com/Sowerby/pages/sow1244-251.htm
(The brief, formal descriptions were by the founder of Linnean Society, Sir James Edward Smith. He did not want his name associated with it as he thought his association with an artisan such as Sowerby might degrade his professional standing.)

"Symphytum officinale: As the root of this plant is easily obtained, it may be conveniently substituted for that of althaea (marsh mallow) in all the compositions in which the latter is officially directed or extemporaneously (spur of the moment), for the general purposes of an emollient and demulcent.
This opinion seems also to have the authority of Dr. Cullen who says, 'while mucilaginous matters are retained in our lists, **I do not perceive why both the British Colleges have entirely omitted the Symphytum. It may be of service as alleged in diarrhoeas and dysenteries.'** "
-'**Medical Botany: Containing Systematic and General Descriptions, with Plates of All the Medicinal Plants,**
Comprehended in the Catalogues of the Materia Medica, as Published by the Royal Colleges of Physicians of London, Edinburgh, and Dublin' by **William Woodville**, M.D., F.L.S., London, England, Volume 2, 1832. Symphytum pages 308-309.
(William Woodville, 1752-1805, was an English physician and botanist. He studied medicine at Edinburgh University, Scotland. He was elected a Fellow of the Linnean Society in 1791. 'Medical Botany' is in 3 volumes, written 1790, 1792, and 1793 with updated editions in 1810 and 1832.)
(Emollient is a moisturizer that makes layers of skin softer and increases the skin's hydration. A demulcent forms a soothing film over a mucous membrane such as the throat and intestines. Mucilaginous is a thick, gluey substance.)

*"**Woodville ('Medical Botany', 1794)** writes: 'the mucilaginous matter of Comfrey is the only medicinal principle, and may be used as an emollient and demulcent'. "*
-The Medicinal Uses of Comfrey by Dr. Charles MacAlister, M.D., F.R.C.P., 1935.

*"**William Woodville in 'Medical Botany'** in 1832 wrote of Comfrey's healing properties: 'The dried root, boiled in water, renders a large proportion of the fluid slimy; and the decoction inspissated (thickened), yields a strong flavorless mucilage similar to that obtained from Althea (Marsh Mallow) but somewhat stronger, equal-bodied or more tenacious and somewhat larger quantity, amounting to three quarters the weight of Comfrey.'*
For this reason, he says, it was substituted for Althea, their emollient and demulcent properties being quite equal."
-Comfrey: What You Need to Know by Ben Charles Harris. New Canaan, Connecticut: Keats Publishing, 1982, page 8.

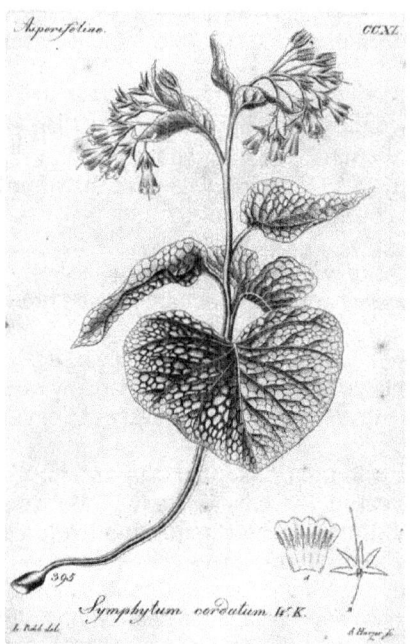

Symphytum Cordatum 1827

'Iconographia Botanica seu Plantae Criticae', Volume 5,
by Heinrich Gottlieb Ludwig Reichenbach,
Leipzig, Germany, 1827.
Reichenbach wrote 10 volumes from 1823-1832.

Symphytum cordatum W.K.,
page 518, t. CCXL (240), fig. 395.

Chapter 15

Age of Revolutions and Comfrey

Age of Revolutions (19th Century, 1800s)

"*From the year 1800, Prickly Comfrey was sold in single plants for shrubberies in England*, as it was found to grow in the shade and reached the height of five or six feet (1.5-1.8 meters).
In 1830 it was introduced as a forage plant, and was found by many people to answer well.
Every chemist (pharmacist) kept a supply of dry leaves, roots and decoctions both for internal and external application; sweetmeats and jams were made from the roots; it was considered a certain cure for all bronchial and chest affections.
"***Symphytum asperrimum (Prickly Comfrey) was brought out in the year 1811 by the Messrs. Loddiges*** (from Hackney, England near London); its graceful pendant, bright blue flowers, and bold foliage.""
-'**Forage Plants and Their Economic Conservation by the New System of Ensilage**: Part I: Caucasian Prickly Comfrey' by **Thomas Christy**, Jun., F.L.S. (Fellow of the Linnean Society), Christy & Co., London, England, 1877, page 1, 4.
    (Sweetmeat is food high in sugar such as candied or crystallized fruit.)

    "*The Prickly Comfrey (Symphytum asperrimum) is a native of the Caucasus, whence it was introduced into England, in 1811, as an ornamental plant, by Messrs. Loddige, of Hackney.*
    It is a perennial plant, with fine reddish-blue flowers."
-'The Journal of the Royal Agricultural Society of England, Second Series, Volume 7', London, England, 1871. Includes article 'XV: **On the Composition and Nutritive Value of the Prickly Comfrey (Symphytum asperrimum)**' by Dr. **Augustus Voelcker**, F.R.S. (Fellow of the Royal Society), pages 387-389.

"*S. asperum Lepechin (S. asperrimum Donn) (Prickly Comfrey) is native in the Caucasus but is now widely naturalised in Europe though very rare in Britain.*
It appears to have been introduced about the beginning of the 19th Century (early 1800s) and to have been described from the gatherings of M. Bieberstein*, for which no precise locality was given.
Donn** records it as having first been cultivated in the Cambridge Botanic Garden (England) in 1801 (not 1811; see ***'English Botany'), and it was grown extensively as fodder a few years later."
-'**The Genus Symphytum in Britain**' by **T.G. Tutin**, University College of Leicester, England; Watsonia: Journal of the Botanical Society of the British Isles, Volume 3, pages 280-281, February 1956. (Watsonia was the journal of the 'Botanical Society of Britain and Ireland' from 1949 to 2010. It is now called 'New Journal of Botany'.)
(* -Baron Friedrich August Marschall von Bieberstein, 1768-1826, was an early explorer of the flora and archaeology of the southern portion of Imperial Russia, including the Caucasus and Novorossiya. He compiled the first comprehensive flora catalogue of the Crimeo-Caucasian region. Baron Bieberstein is recognised as the scientific authority for 1,695 plant taxa. For more about Bieberstein see sub-subsection 'Later Botanical History of Symphytum: Botanists' in subsection 'Symphytum Comfrey Genus' in section 'Borage Family, Symphytum Genus' {Chapter 2}.)
(** -'Hortus Cantabrigiensis: Catalogue of Plants in the Cambridge Botanic Garden' by James Donn, London, England, page 65, 1831. First published in 1796. James Donn from England, 1758-1813. He was a founder member of the Linnean Society, London, England.)
(*** -'English Botany {Sowerby's}; or Coloured Figures of British Plants: Volume 7' by John T. Boswell, James Sowerby, John Edward Sowerby, London, England, 1880.)
    (The Caucasus Mountains are located at the border of Europe and Asia, between the Black Sea and Caspian Sea and occupied by Russia, Georgia, Azerbaijan, and Armenia.)
    (Naturalized plants become established and survive in a region other than their place of origin.)

"*Symphytum, ds.* (= George Don System)

Linnean Name	Flower	English Name	Native	Cult.	Flower Time
officinale, ds. EB.v.12.t.817.	white	Common	Britain		May to July
(*EB= English Botany, Volume 12.)					
purpurascens. pu.	purplish		Britain		
bohemicum,ds.s.bfg.2.L304	crimson	Red Flowered	Bohemia		
tuberosum,ds.FB.v23.t.l502	pale yellow	Tuberous-root	Scotland		May to October
bulbosum, ds.	yellow	Bulbous-root	Levant	1829	May to September
cordatum, ds.	pale yellow	Heart-Leaved	Hungary	1813	
orientale, ds. BM.t.1912.	white	Eastern	Turkey	1752	
bullatum, ds. BM.t.1787.	white	Blistered	Tauria	1806	
tauricum, bm.					
caucasicum, ds. bm.L3182.	blue	Caucasian	Caucasus	1816	

asperrimum, ds. BM.t.929.	blue	Roughest	Caucasus	1799	
echinatum, ds.	purple	Bristly	?	1824	
peregrinum, ds.	white	Oblique-Leaved	Poland	1816	May to July ."

-'**Hortus Cantabrigiensis**: or An Accented Catalogue of Indigenous and Exotic Plants Cultivated in the Cambridge Botanic Garden, Thirteenth Edition' by **James Donn**, Curator, Fellow of the Linnean and Horticultural Societies, London, England, 1845. I could not find the 1831 edition.
(* -'English Botany {Sowerby's}; or Coloured Figures of British Plants: Volume 7' by John T. Boswell, James Sowerby and John Edward Sowerby, London, England, 1880.)
(ds. = Don System = George Don, 'General System of Gardening and Botany: A General History of the Dichlamydeous Plants'. For more about G. Don see sub-subsection 'Later Botanical History of Symphytum: Botanists' in subsection 'Symphytum Comfrey Genus' in section 'Borage Family, Symphytum Genus'.)
(S. officinalis in G. Don: General History, Volume 4, page 312, 1837.)
('Cult.' means when first cultivated in the Cambridge Botanic Garden.)
(Bohemia is the westernmost and largest region of the Czech Republic. Bohemia sometimes refers to the entire Czech territory, including Moravia and Czech Silesia.)
(Levant refers to a large area in Eastern Mediterranean. In its narrow sense, it is Syria. In wide sense, it is all eastern Mediterranean with its islands; i.e., all countries along Eastern Mediterranean shores, from Greece to Cyrenaica.)
(I'm not sure what region 'Tauria' is. Perhaps it means Crimea or the Tauric Peninsula, as it was called in antiquity. Crimea is a peninsula on the northern coast of the Black Sea in Eastern Europe that is almost completely surrounded by the Black Sea and Sea of Azov to the northeast.)

From 1806 to 1836 Loddige Nursery (Hackney, near London, England), sold 7 Symphytums types including Symphytum asperrimum (Prickly Comfrey):
"This species of Symphytum, a native of the Caucasus, is by far the largest of the Genus, growing to the height of five feet (1.5 meter), an ornamental perennial which will thrive in any soil or situation."
-**Curtis's Botanical Magazine, 1806**.
(The Caucasus Mountains are located at the border of Europe and Asia, between the Black Sea and Caspian Sea and occupied by Russia, Georgia, Azerbaijan, and Armenia.)
(**William Curtis**, 1746-1799, was an apothecary and botanist in London, England. He established his own London Botanic Garden in Lambeth district in 1779.
'Curtis's Botanical Magazine' or 'The Botanical Magazine or Flower-Garden Displayed' began in 1787. Natural history illustrators, James Sowerby and Sydenham Edwards both found a start with this magazine. It has been published continuously. From 1984 to 1994 it was called 'The Kew Magazine'. In 1995 the name changed back to 'Curtis's Botanical Magazine'. It is now published by the 'Royal Botanic Gardens, Kew', London, England. It is the longest running botanical magazine in the world. The standard form of abbreviation is 'Curtis's Bot. Mag.' or 'Botanical Magazine' in the citation of botanical literature.)

"*In 1812 the Symphytums were listed and catalogued for sale in England, a total of seven being available* at that time. One of these was identified later by reference to botanical magazines as **Symphytum asperrimum**, the name later being changed and adopted by some as asperum, but not generally used.
The name asperrimum means the roughest, and this type was popularly known as Prickly Comfrey. It was described as native of the Caucasus, the largest of the genus growing to 5 feet high (1.53 meter). Its flowers were vivid blue."
-Comfrey: **Nature's Healing Herb and Health Food** by Andrew Hughes. Japan: Sanyusha Publishing, 1992, page 123.

"**Medicinal Plants of the London (England) and Edinburgh (Scotland) Dispensatories:**
        <u>Symphytum officinale</u>, Comfrey.
Plants According to Period of Flowering:
Symphytum officinale (Common Comfrey) and Symphytum tuberosum (Tuberous Comfrey) flower in June."
-'**A Catalogue of the British Medicinal, Culinary, and Agricultural Plants**, Cultivated in the London Botanic Gardens' by **William Curtis**, London, England, pages 36, 99; 1783.

"<u>Symphytum officinale</u>, Comfrey:
Root: perennial, large, branched, on outside blackish, white within, maukish (mawkish), abounding with a slimy juice.
Stalk: about two feet (0.6 meter) high, upright, branched, round, yet slightly angular, rough; the hairs rigid and bending backwards.
Leaves: alternate, the lower ones standing on footstalks, the upper ones sessile, decurrent, ovate, pointed, seven inches (17.7 cm), or even a foot (30.4 cm) in length, somewhat wrinkly, veined, rough on both sides, the edges slightly waved, and fringed with hairs.
Flowers: a yellowish white colour, rarely purple, drooping, placed on racemi or branches, which usually grow two together, turn spirally inwards, and support many flowers.
Peduncles: both of the racemi and flowers, round and very hairy.
Calyx: a Perianthium of one leaf, hairy, deeply divided into five segments, which are lanceolate, keel'd and upright.
Corolla: funnel-shaped, of a yellowish white colour, deciduous; the tube thick, the length of the calyx, marked

externally at the top with five small depressions. The limb ovate from the gradual widening of the tube, divided into five short roundish segments, which are rolled back; the mouth closed with five long and pointed nectaries, thick at the edge, with numerous teeth-like points, closing at top, shorter than the corolla.

Stamina: *five, lanceolate, white, shortish Filaments; Anthers oblong, bifid both at top and at bottom, of a yellowish colour, upright, hid by the nectaries.*

Pistillum: *Germen divided into four lobes, which are roundish, blunt and green; Style tapering, white, arising from the middle betwixt (between) the lobes, a little longer than the corolla, with a small obliquity at top; Stigma small and blunt.*

Seeds: *four, in the bottom of the calyx, largish, angular, blackish and shining.*

The Comfrey is a very common plant by river sides, on the edges of wet ditches, and in other moist situations. It flowers from June to September; its blossoms are for the most part of a yellowish white colour, but in some parts of England, and abroad, they are more commonly purple.

As a medicinal plant *Comfrey has been held in high estimation, its consolidating virtues have however been carried to a ridiculous excess; the roots, which are full of a glutinous juice, agree in quality with the roots of Marsh-mallow, and hence are recommended, internally, in spittings of blood, purgings, fluxes (diarrhea), and ulcers of the bladder; externally, by way of poultice to fresh wounds, fractured bones, bad ulcers, bruises, gouty swellings, etc.*

It is generally left untouched by cattle. I know of no plant, that on being repeatedly cut down, produces such a quantity of herbage."

-'**Flora Londinensis**, or Plates and Descriptions of such Plants as Grow Wild in the Environs of London', Volume 2, by **William Curtis,** London, England, 1798. Six volumes from 1777-1798.   (Mawkish means having a faint sickly flavor.)

"In the Kew Herbarium, London, is a specimen of <u>S. x uplandicum</u> *collected by Forbes Young at Cobham Lodge, Kent County (southeast England), in 1827, which may have been supplied by Loddiges, and which seems to be the first (official) record of its cultivation in Britain.*

The earliest record of it as a naturalised plant is in 1861 from Marlborough, Wiltshire, England."

-'**The History of Symphytum Asperum Lepech. and S. x Uplandicum in Britain' by A.E. Wade**, Department of Botany, National Museum of Wales, Watsonia, Volume 4, pages 117-118, 1958. (Watsonia was the journal of the 'Botanical Society of Britain and Ireland' from 1949 to 2010. It is now called 'New Journal of Botany'.)

(Kew Gardens is a botanical garden in London, England that houses the 'largest and most diverse botanical and mycological collections in the world'. The library has more than 750,000 volumes, and the illustrations collection has more than 175,000 prints and drawings of plants.)

*"***Comfrey: Symphytum: A common wild plant, of great virtue;** *it is frequent by ditch sides, it grows a foot and half high (0.45 meter): the leaves are large, long, not very broad, rough to the touch, and of a deep disagreeable green: the stalks are green; thick, angulated, and upright.*

The flowers grow along the tops of the branches and are white, sometimes reddish, not very large, and hang often downwards.

The root is thick, black, and irregular; when broken it is found to be white within, and full of a slimy juice. **This root is the part used,** *and it is best fresh, but it may be beat up into a conserve with three times its weight of sugar.*

It is a remedy for that terrible disease the whites (leucorrhoea according to 'Archaic Medical Terms'). It is also good against spitting of blood, bloody fluxes (diarrhea), and purgings, and for inward bruises."

-'**The Family Herbal; or, An Account of all those English Plants**, which are Remarkable for their Virtues' by **Sir John Hill, M.D.**, Suffolk County, England, **1812**, page 85.   (Conserve is a sweet food made by cooking fruit with sugar.)

From 1810 to 1825, nurseryman <u>James Grant</u> near London, England reported yields of 40-60 tons an acre in 5-6 cuttings a year. In 1825 his Comfrey crop is referenced in "The Encyclopaedia of Agriculture" by John Loudon, a Scottish botanist, garden designer, and writer.

"In 1810 James Grant noticed that the herb <u>Prickly Comfrey</u> could grow to 7 feet (2.13 meter) high on good land and grew almost as much underground because of the stimulation created by the extra cutting above ground."
-Comfrey: Symphuo Symphytum A Multi-Purpose Herb by Philip Clarke. Edinburgh, Scotland: Pentland Press, 1997 p. 1.

"A Lewisham (south London, England) nurseryman, **James Grant, saw its agricultural possibilities and advertised it as 'Prickly Comfrey"** *(asperrimum meaning 'the roughest') claiming only 40 to 60 tons an acre for it."*
-**Fighting Like the Flowers: An Autobiography:** The Life Story of Britain's Best-Known Organic Gardener by Lawrence D. Hills. Bideford, Devon, England: Green Books, 1989, page 100.

*"During the next 40-50 years **(1810 to 1850-1860) the plant (Symphytum asperrimum = Prickly Comfrey) became well known and widely used, and largely displaced Symphytum officinale in England**, which had been commonly and widely used before that time, but which had not produced the same quantities of leaf for stockfood."*
-Comfrey: **Nature's Healing Herb & Health Food** by Andrew Hughes. Japan: Sanyusha Publishing, 1992, page 123.

"Symphytum:
1. S. officinale, 2. S. caucasium MB., 3. S. tuberosum Jacq., 4. S. orientale, 5. S. tauricum Willd., 6. S. asperrimum MB., 7. S. bullatum Hornem., 8. S. cordatum Waldst. Kit., 9. S. secundum S.G. Gmel., 10. S. regium S.G. Gmel., 11. S. echinatum Ledeb."
-'**Caroli a Linne Equitis Systema Vegetabilium:** Secundum Classes, Ordines, Genera, Species. Cum Characteribus, Differentiis et Synonymiis', Volume 4 by Johann Jacob Roemer and Joseph August Schultes; Stuttgardtiae (Stuttgart, Germany), **1819**. In Latin. Symphytum pages 64-66.

*"**Comfery, Consolida. Called also Symphytum officinale L.:***
It is a large rough plant, which grows wild in moist grounds; but it is sometimes cultivated in gardens.
The roots are inspissant (thickening agent) and demulcent, having the virtues of marsh mallows.
They correct 'salt sharp serum', heal erosions of the intestines in the diarrhoea and dysentery and prevent spitting of blood.
Externally they are good in ruptures, and to agglutinate wounds and ulcers, but they are but little used in the present practice.
Doses: *Of the 'Comfery' root in powder, a drachm. In decoction, from half an ounce to an ounce, sweetened with sugar."*
-'**The American Herbal; Or, Materia Medica,** Wherein the Virtues of the Mineral, Vegetable, and Animal Productions of North and South America are Laid Open' by Samuel Stearns, L.L.D., Walpole, New Hampshire, **1801**. Symphytum page 111.
    (Drachms= 60 grains= 64.79 milligrams)

*"**On Trottel:** This plant is said, by James Sibbald, Esq. of Paisley, Scotland, to be of the potatoe kind, and cuts of it are planted in the same manner, but in autumn so late as August and September. It grows so rapidly; that its herbage may be gathered the following December and January, and served to all sorts of cattle raw, and to man after being boiled.*
It is not injured by the severest frost, and it bears even in the middle of winter thick leaves, curled and crisp, which are as tender as asparagus, and rather like sea-kale.
In the spring months, when other vegetables are rare, this arrives at maturity, and the roots may then be dug up; there usually are ten or twelve trottels, of eight or ten ounces each, at every plant.
When boiled they are of a yellow colour, and although they are considerably drier than carrots, they have a somewhat similar flavour. It is a most delicious vegetable as well as nutritious, and it keeps well.
It was brought from Labradore, and is now cultivated by the nurserymen near Greenock (Inverclyde Council area in Scotland), and its roots have been sent to Bristol (southwest England) for the purpose of cultivation."
-'**The Farmers Calendar:** Containing the Business Necessary to Be Performed on Various Kinds of Farms During Every Month in the Year, Twelth Edition' by Arthur Young, Esq., F.R.S. (Fellow of the Royal Society), London, England, **1822**, pages 530, 531.
    (In writings after this one there are references to 'Trottel' as being the same as Symphytum asperrimum. This article says it is from 'Labradore'. The current Labrador is part of Canada. I don't know if that is the area they are referring to. Symphytum asperrimum is from Caucasus, not Canada.)

"Symphytum. Tourn. inst. 138. t. 56. Linn. gen. n. 185. Lam. ill. t. 93. Goertn. fruct. 1. p. 325. t. 67. Lehm. asp. p. 343. Endl. gen. n. 3776. Spenn. in ic. fl. germ. fasc. 17."
<u>De Candolle</u> lists these Symphytum species:
    S. officinale (Linn. sp. 195) (alpha ochroleucum, beta purpereum, gamma lanceolatum),
    S. Donii,
    S. peregriunum (Ledeb. ex Spreng, syst. 1. p. 563),
    S. caucasicum (Bieb. fl. taur. n. 326),
    S. asperrimum (Sims bot. mag. t. 929),
    S. tuberosum (Linn. sp. 195),
    S. bulbosum (Schimp. in Flora 8. p. 1. p. 17. Koch syn. 500),
    S. ottomanum (Friv. in Flora 1836. p. 439),
    S. anatolicum (Boiss. diagn. 4. p. 43),
    S. tauricum (Willd. nov. act. nat. cur. berol. 5. p. 120. t. 6. f. 1. ex enum. h. ber. 183),
    S. orientale (Linn. sp. 195),
    S. brachycalyx (Boiss. diagn. 4. p.43),
    S. grandiflorum,
    S. cordatum (Willd. act. soc. berol. 2. p. 121),
    S. echinatum (Ledeb. h. Dorp. suppl. 1811. p. 5),
    S. racemosum (Steph. in Willd. herb. ex Roem. et Seh. syst. 4. p. 752).
-'**Prodromus Systematis Naturalis Regni Vegetabilis,** sive, Enumeratio contracta ordinum generum specierumque plantarum huc usque cognitarium, juxta methodi naturalis, normas digesta, Pars X' (An essay of the system of the natural of the Kingdom of Vegetable, or, An enumeration of the genera and species plant has thus far been contracted cognitarium of the orders, according to the method of the natural, to the norms of being digested, Part 10) by Alphonse Pyramus **De Candolle, 1824,** 1846. In Latin. Symphytum pages 36-40, 587. Standard botanical abbreviation: Prodr. (DC.).
    (Prodromus Systematis Naturalis Regni Vegetabilis is 17 volumes written from 1824 to 1873. Augustin Pyramus De Candolle wrote it as a summary of all known seed plants, including taxonomy, ecology, evolution and biogeography. He authored the first seven volumes. His son, Alphonse Pyramus De Candolle, wrote ten more volumes, with help from other authors. De Candolle further developed his concept of families. This system was published before there were internationally accepted rules for botanical nomenclature.)
    (For more about De Candolle see sub-subsection 'Later Botanical History of Symphytum: Botanists' in subsection

'Symphytum Comfrey Genus' in section 'Borage Family, Symphytum Genus' {Chapter 2}.)

"**To Cure Inward Ulcers: For constant drink, make a beer** of barley malt (one peck = 2 dry gallons), spikenard root (two pounds = 0.9 kg), **Comfrey root (one pound = 0.45 kg)**, burdock roots (two pounds), black spruce boughs (five pounds = 2.2 kg), angelica root (one pound), fennel seed (four ounces =113 grams), for ten gallons (37.8 liters) of beer. Drink one quart (about 1 liter) a day. Let your exercise be light."
-**Dr. John William's Last Legacy, and Useful Family Guide**', O. Taylor & Co., New York, **1826**, page 14.
    (A peck is an imperial and United States unit of dry volume, equivalent to 2 gallons or 8 dry quarts = 9.09 United Kingdom liters or 8.81 United States liters).

"**Symphytum officinale:** Names- Common Comfrey. Fr. Consoude usuelle.
Classification: Nat. Order of Borragines or Asperifolia. Pentandria monogynia L.
History: This plant is a native of Europe, but has been naturalized from New England to Ohio and Virginia, growing spontaeously in thickets, meadows, etc. It blossoms in June and July.
The varieties are:
    1. Purpureum, with purple flowers and spreading calyx. 2. Nigrum, root black. 3. Elatior. 4. Pumilum. 5. Albiflorum.
We have a native American species of this genus, found west of Mississippi, in the prairies and glades, and cultivated at 'Bartram's Garden'. I call it and distinguish as follows:
    Symphytum hirsutum: Whole plant hirsute. Stem erect, somewhat winged, lower leaves petiolate, oblong lanceolate, upper leaves sessile decurrent, oval acuminate; racemes germinate, erect, convolute at the end. Size 4 feet (1.2 meter), lower leaves a foot (0.3 meter) long, flowers white.
Symphytum is much valued in Europe and China, also by our herbalists, but wrongly omitted by all our medical writers, except Schoepf and Cutler. **In China it is called Tihoang (Ti-Hoang), and considered equal to Ginseng in many cases, particularly in preserving health.**
    Pills, lozenges and bolus are made of it, and taken daily in the morning, by people of weak and debilitated habits. In Europe, a conserve and syrup is used. The infusion, decoction, etc. are equally good; the doses need not be very nice (precise), as the effects are mild.
Our herbalists unite it to Burdock and Yarrow to cure the clap (gonorrhea), using at the same time injections of Statice (Sea Lavender) or Tormentil (in rose family).
Boiled in milk, it becomes the best preparation for diseases of the bowels and urinary organs. It may be safely employed in all diseases of debility, relaxation, and overflowing (of menstruation)."
-'**Medical Flora; or Manual of the Medical Botany of the United States of North America,** Volume 2' by Constantine Samuel Rafinesque, A.M., Ph.D., Philadelphia, Pennsylvania, **1830**. Symphytum is on pages 95-96.
    (Bartram's Garden is a Philadelphia, Pennyslvania arboretum. It is the oldest botanic garden in North America.)

    "**Symphytum hirsutum Raf.** (Rafinesque): species doubtful. Med. Fl. 2: 95 (1830)."
-**Global Biodiversity Information Facility®** (GBIF), www.gbif.org, 2019. International network and research infrastructure funded by world's governments and providing open access to data about all types of life on Earth.

"**Ti-Hoang: The Chinese give this name to the root of the large Comfrey,** the best of which is found in Honan (Henan = central China Yellow River Valley), in the neighbourhood of the city Hoai-King. The roots of this plant, when dried, are about the size of a finger, but much longer. The Chinese physicians ascribe (give) to them many salutary (beneficial) properties; and the use of them has become very common in all the provinces of ths empire.
Rich people take pills of Ti-Hoang every morning as people in Europe drink tea, coffee and chocolate. Some cut it into thin slices, and use it in decoction, or when baked in the steam of boiling water. Others pound it, and form it into boluses, which they swallow with warm water.
Five other kinds of plants, or ingredients, are commonly added to it, which are aromatic, cordial, diuretic, acid and a little soporific (creates sleepiness); but the Ti-Hoang is always the basis of these pills."
-'**An Historical, Geographical, and Philosophical View of the Chinese Empire'** by W. Winterbotham, London, England, 1795.
    (An online translation for the Mandarin Chinese word for 'Comfrey' is 'zi cao ke zhi wu' that literally translates as 'Boraginaceae (zi cao ke), Plant (zhi wu)'. That Chinese word includes more plants than just Comfrey. There is very little about Comfrey use in China that is available in English. If you have information, please send it to me.)

"**Symphytum asperrimum**, a native of Siberia, and one of the most rapid growing of herbaceous plants, has been recommended for culture as food for cows. (Gard. Mag., Volume v. page 442) This vegetable is full of mucilage; and **we certainly think it deserves to be tried by every cottager (small farmer) who can procure plants.**
It seeds freely, and propagates most readily by division of the root. **We would strongly recommend gardeners to gather, propagate, and distribute."**
-'**Loudon's Gardeners Magazine'** or 'The Gardeners Magazine and Register of Rural and Domestic Improvement' by John Claudius Loudon, London, England, Volume 6, **1830**. Published from 1826-1844. Symphytum pages 112, 184.

"In the year 1788 when I was in my nineteenth year, my father purchased a piece of land on Onion river, in the state of Vermont. On the 2nd of December, when **I had the misfortune to cut my ancle (ankle) very badly,** which accident prevented me from

doing any labor a long time, and almost deprived me of life.
> **The wound was a very bad one, as it split the joint and laid the bone entirely bare,** *so as to lose the juices of my ancle joint to such a degree as to reduce my strength very much.*

My father sent for a Dr. Cole, of Jericho, Vermont, who ordered sweet appletree bark to be boiled, and the wound to be washed with it, which caused great pain, and made it much worse, so that in eight days my strength was almost exhausted; the flesh on my leg and thigh was mostly gone, and **my life was despaired of**; the doctor said he could do no more for me.

My father was greatly alarmed about me, and said that if Dr. Kittridge, of Walpole, New Hampshire, could be sent for, he thought he might help me; but I told him it would be in vain to send for him, **I could not live so long as it would take to go after him,** without some immediate assistance.

He said he did not know what to do; I told him that there was one thing I had thought of, which I wished to have tried, if it could be obtained, that I thought would help me. He anxiously inquired what it was, and **I told him if he could find some Comfrey root, I would try a plaster made of that and turpentine.**

He immediately went to an old place that was settled before the war, and had the good luck to find some; a **plaster was prepared by my directions and applied to my ancle, the side opposite the wound, and had the desired effect.**

The juices stopped running in about six hours, and I was very much relieved, though the pain continued to be very severe, and the inflammation was great; the juices settled between the skin and bone and caused a suppuration (pus formation), which broke in about three weeks, during which time I did not have three nights' sleep, nor did I eat any thing. This accidental remedy was found through necessity, and was the first time the mother of invention held forth her hand to me. **The success which attended this experiment, and the natural turn of my mind to these things, I think was a principal cause of my continuing to practice the healing art to this time."**

-'**A Narrative of the Life and Medical Discoveries of Samuel Thomson:** Containing an Account of his System of Practice, and the Manner of Curing Disease with Vegetable Medicine' by Samuel Thomson, Columbus, Ohio, **1833-1835**.

> **Samuel Thomson**, 1769-1843, was an American herbalist and botanist, who was the founder of the alternative medicine known as 'Thomsonian Medicine', that was popular in the 1800s. Its goal was the elimination of toxins by physiological processes. He practiced herbalism in Surry, New Hampshire.

"**The Rough Comfrey, Symphytum asperrimum L., a perennial from Siberia, has been brought into notice by D. Grant, a nurseryman at Lewisham (south London, England)**, and tried by a number of cultivators.
> Cattle of every kind are said to be fond of this plant; and so great is its produce on good soil, that Mr. Grant thinks an acre might be made to produce thirty tons of green fodder in one year. He has grown it to the height of seven feet (2.1 meter) as thick as it could stand on the ground.

The plant is of easy propagation by seed or division of the roots. The better way would be to sow in a garden, and transplant when the plants were a year old.

All of the Symphytums are plants of great durability, so that this species, if once established, would probably continue to produce crops for many years; and, in that point of view, **it would seem to be a valuable plant for the cottager who keeps a cow.**"

-'**An Encyclopaedia of Agriculture:** Comprising the Theory and Practice of the Valuation, Transfer, Laying Out, Improvement and Management of Landed Property' by J.C. Loudon, FLGZ, HS, London, England, 1436 pages, **1835**. Symphytum page 870.

"**Symphytum asperimum: Rough or Prickly Comfrey:**
Farther experience tends to prove that cattle become fond of this plant, if not permitted to grow until the leaves and stalks get too hard, although they may refuse it at first. Its great quantity of produce does not seem overrated; and with regard to its duration, what has been there anticipated is correct.
The best mode of propagating it seems to be by dividing the roots, and planting them in a good deep soil, in rows two feet (0.6 meters) apart and at least fifteen inches (38 cm) between the plants.
Seeds seldom ripen in quantity, and seedling plants are long in arriving at maturity."
-'**The Agriculturalists Manual;** Being a Familiar Description of the Agricultural Plants Cultivated in Europe' by Peter Lawson and Son, Seedsmen and Nurserymen to the 'Highland and Agricultural Society of Scotland', Edinburgh, **1836**.

"Doyle states, in 1844, that '**Prickly Comfrey** was introduced into England from Caucasus and Siberia, in 1799, and has been brought into notice by Mr. D. Grant, a nurseryman of Lewisham who has reprinted and extensively circulated some letters on its use from the English paper called the 'Country Times'. **It was brought into Ireland about 6 years ago (1838).**"
-'**Experimental Work of the Agricultural Department of the University of Tennessee** (Report of the Experimental and Other Work of the School of Agriculture and Botany of the University of Tennessee for Session of 1880-1881)' by John M. McBryde, Professor of Agriculture, Horticulture and Botany, Knoxville, Tennessee, 1881.

"**Symphytum maculosum.** See Pulmonaria officinalis (lungwort).
Symphytum minus. See Prunella vulgaris (heal-all).
Symphytum officinale. The Comfrey. Consolida major. This plant, Symphytum officinale- foliis ovatis lanceolatis decurrentibus, is administered where the althaea (marshmallow) cannot be obtained. Its roots abounding with a viscid glutinous juice, whose virtues are similar to those of the althaea.
Symphytum petraeum. See Coris Monspeliensis (in primrose family)."
-'**Lexicon Medicum; or Medical Dictionary,** 7th Edition' by Robert Hooper, M.D., F.L.S., London, England, **1839**.

"**Symphytum officinale, Common Comfrey Root:**
The root is thick, black, and irregular; when broken it is found to be white within and full of a slimy juice. This root is the part used, and it is best fresh, but it may be beat up into a conserve, with three times its weight of sugar. It is a remedy for that terrible disease the whites. It is also good against spitting of blood, bloody fluxes and purgings, and for inward bruises.
The following is the formula, for making the celebrated 'Syrup of Comfrey':
> **Take of Comfrey-root six ounces (170 grams).** Plantain leaves three ounces (85 grams).
> Bruise together in a marble mortar to express the juice; strain the liquid and add an equal quantity of white sugar. This is an excellent remedy for spitting of blood, to be taken in doses of about a wine glass full, it is also good for coughs by adding an ounce or two of Liquorice root."

-'**A New Family Herbal**; Or, a History and Description of All the British and Foreign Plants which are Useful to Man', Sixth Edition, by Richard Brook, London, England, **1840**, pages 40-41.

"Synoptical Table of the Properties and Uses of British Medicinal Plants:
Comfrey, Symphytum officinale Root: Astringent, demulcent, emollient.
> Dysentery; hemorrhages; phthisis; coughs; nephritis; fluor albus; heat of urine."

-'**The British Flora Medica, or, History of the Medicinal plants of Great Britain,** Volume 2' by Benjamin Herbert Barton and Thomas Castle, London, England, **1845**. Symphytum officinale, page 490.   (Synoptic means summary.)

Carl (Karl) Friedrich von Ledebour, 1785-1851, was a German/Swedish botanist. He travelled extensively in Russia and wrote the first comprehensive book on Russian flora: 'Flora Rossica' (Complete Flora of the Russian Empire), **1847**-1849. In this book he classifies Symphytum.
> (For more about Ledebour, see sub-subsection 'Early Botanical History of Symphytum: Botanists' in subsection 'Symphytum Comfrey Genus' in section 'Borage Family, Symphytum Genus' {Chapter 2}.)

"**Symphytum Officinale- Comfrey:** Healing-herb, Gum-plant.
This plant has a whitish, perennial, tapering root, which sends up a rough, erect and branching stem, to the height of four feet (1.2 meters). The leaves have no footstalks, and come off from the stem one above the other on opposite sides. They are rough, of an oblong shape, and diminish to a point.
The flowers are of a yellowish white color, and bloom in June and July. They come out in curved, terminal, nodding bunches.
The whole plant is sometimes used, but it is only the roots which are of much importance, and these should be gathered in the fall, or in early spring. The root has no smell. The taste is slightly acrid (unpleasant) and sweetish, and considerably glutinous.
The most convenient method of using it is in the form of decoction (tea); one or two ounces to a pint and a half (3 cups = 709 ml) of water, boiled down to a pint (2 cups = 473 ml). Dose, a wineglassful, repeated as often as necessity requires.
> In dysentery (bad diarrhea) and diseases of bladder and kidneys it is very useful, as also in scalding urine and piles.

Its principal efficacy (effectiveness) is shown in diseases of the bowels and urinary organs, when taken boiled in milk, in the same proportion as the watery decoction. In families it is much employed as a drink in coughs, colds, and all catarrhal (inflammation of mucous membrane) affections.
Bruised, and applied externally to sprains, wounds and ulcers, it is said to be beneficial."
-'**The Book of Herbs Giving Descriptions of Medical Plants** and Directions for Gathering and Preserving Them' by John B. Newman, M.D., New York, **1847**

"**Comfrey: Symphytum Asperrimum, Rough or Prickly Comfrey, or Trottels,** as it was called by Mr. Young in his 'Farmers Calendar', was introduced into England from Caucasus and Siberia, in 1799, and has been brought into notice by Mr. D. Grant, a nurseryman of Lewisham, who has reprinted and extensively circulated some letters on its use from the English paper called the 'Country Times'.
It was brought into Ireland about six years ago by Dr. Derenzy, who republished, in the 16th number of the 'Irish Farmers Magazine', the letters of Mr. Grant."
-'**A Cyclopaedia of Practical Husbandry and Rural Affairs in General**' by Martin Doyle, London, England, **1851**, page 151.

"**Symphytum asperrimum:**
It was brought into notice as an agricultural plant, by Mr. D. Grant, a nurseryman of Lewisham (south London, England), and was **introduced to Ireland, and recommended to the cottier-farmers (cottage farmers) of that country, about eight years ago (around 1839), by Dr. Derenzy.**
Young mentions it in his 'Farmer's Calendar', under the name of trottles, and appears to think that the whole plant is valuable; but Dr. Derenzy supposes the roots to be useless for either man or beast, and recommends the plant solely for shoots and foliage.
> It is preferred to vetches by pigs; it is not, like clover, dangerous to cows or sheep; and it does not communicate any bad flavor to cow's milk.

In an experiment on the farm of Carnew Castle in Ireland, Symphytum asperrimum amounted to the enormous quantity of **82 tons per Irish acre**, in three cuttings of 28 1/2 tons in the middle of April, 31 tons in the middle of July, and 22 1/2 tons in the middle of September.
> It should be cut about the time of flowering, and never allowed to go to seed.

The tuberous species, Symphytum tuberosum, grows wild in moist shady places in Scotland.

Seven species, besides the very rough (Symphytum asperrimum), have been introduced from foreign countries; all have an ornamental appearance; and several, as well as the two indigenous (native) species, have economical properties similar to those of the very rough, but cannot compete with that species in either productiveness or facility of adaptation."
-The Rural Cyclopaedia, or A General Dictionary of Agriculture: and of the Arts, Sciences, Instruments, and Practice, Necessary to the Farmer, Stockfarmer, Gardener, Forester, Landsteward, Farrier, etc., 4 Volumes, by John M. Wilson, Edinburgh, Scotland. Volume 1: **1852**. Symphytum pages 848-849. The first edition was 1847.

"Symphytum officinale or Common Comfrey:
This plant pushes very early in spring, then producing a great quantity of tender succulent shoots, perfectly free from all noxious (harmful) quality, and freely eaten by cattle, after they are accustomed to it.
Its herbage, even later in the season, is abundant, and evidently a grateful cattle food.
It has therefore been recommended occasionally as a good green crop. It has, however, fallen into disuse, as well as two exotic (not native) species, S. asperrimum and echinatum, both which are more productive than our wild plant (Symphytum officinale).
The Comfreys are of little value, in fact, except in good deep land."
-'A Cyclopedia of Agriculture, Practical and Scientific: In Which the Theory, the Art, and the Business of Farming are Thoroughly and Practically Treated, Volume 2' by John Chalmers Morton, Glasgow, Scotland, **1856**, page 952.
> (John Chalmers Morton, 1821-1888, was a Scottish agriculturist and writer. He was the founding editor in 1844 of the 'Agricultural Gazette' in London, England.)

> *"In 1856 John Morton, editor of the 'Cyclopedia of Agriculture', stated that Symphytum officinale was falling into disuse and that **the new Russian import (Prickly Comfrey) was gaining recognition.**"*
> -**Comfrey: Symphuo Symphytum** by Philip Clarke. Edinburgh, Scotland: Pentland Press, 1997, page 2.

"Diarrhea or Lax (Tsu-ne-squah-lah-tee in Cherokee):
*This disease is characterized by frequent and copious discharges from the bowels, unattended with fever, and has not the appearance of a contagious or catching disease as is the case with Flux. Slippery-elm bark, or the **root of Common Comfrey, forms an excellent drink in this complaint.** Injections (through rectum) of the same are also good. It should be taken in fresh spring water. In many instances a tea of flaxseed, slippery-elm, **Comfrey** or vervain will entirely relieve it in a short time.*
Heart-Burn:
*Slippery-elm bark, powdered and taken in cold water, is an excellent article for this distressing complaint. **Comfrey, either the garden or wild, will generally give speedy relief.** The slippery-elm bark and Comfrey will act as an aperient (mild laxative), and will probably afford the most permanent relief of any of the above-named articles, for costiveness (constipation) should be strictly avoided by persons with heart-burn."*
-'**The Cherokee Physician or Indian Guide to Health** as given by Richard Foreman, A Cherokee Doctor', New York, **1857**.
> (Cherokee Native Americans lived in southeastern United States until many were forced to move to Oklahoma in 1838.)

"**Symphytum L.:**
> **Fornices Inclusi:** *S. officinale, S. cordatum, S. tuberosum.*
> **Fornices Exserti:** *S. bulbosum, S. ottomannum.*"
-'**Icones Florae Germanicae et Helveticae**, Volume XVIII' by Heinrich Gottlieb Ludwig Reichenbach and Heinrich Gustav Reichenbach; Lipsiae {Leipzig}, Germany, **1858**. In Latin. Symphytum pages 57-58.
> (Fornices are small arched scales in the throat of the corolla some flowers.
> Exserted means projected beyond, e.g., stamens beyond the corolla tube.
> Included means enclosed, not protruding, e.g., stamens within the corolla tube.)

'**The Family Companion and Physician**' uses Comfrey as part of an herbal recipe in these diseases:
Remittent fever (page 10), Scarlet Fever (page 14), Hooping Cough (page 23), Inflammation of the Lungs (page 25), Inflammation of the Stomach (page 28), Cholera (page 40), Jaundice (page 46), and Antidote for Poisons (page 59).
> "*Inflammation of the Stomach Symptoms: Burning heat, pain, and swelling, frequently hiccough, vomiting, cold extremities; hard, quick, and tense pulse; pain increased by pressure; thirst and pain increased by drink; restlessness and weakness.*

Treatment: *To ease the pain, give 20 drops of the Cholera Compound, once in fifteen to thirty minutes, until the pain is relieved.* **Drink Slippery Elm Tea freely; Flaxseed, Comfrey, and Marshmallow Root Teas are all good."**
-'**The Family Companion and Physician**: On Reformed Botanical Principles, Treating of the Symptoms and Remedies of Acute Diseases', 21st Edition, by Dr. J. Wilson, Rochester, New York, **1866**.

"There is, however, no need to go to the Caucasus for Comfrey (Symphytum asperrimum), as the wild Symphytum officinale of our rivers and water-courses is equally good, if not better.
> *Symphytum officinale is not so rough in its foliage, and yields a larger crop. Indeed, from a long series of experiments we conclude that our wild plant becomes so much like the foreign form as to be if not identical as a species, at least to be no more than a variety."*
-'**The Gardeners Chronicle and Agricultural Gazette**: A Newspaper of Rural Economy and General News', London, England, **1866**. March 31, page 297.

('The Gardeners' Chronicle' was a British horticulture periodical in publication for almost 150 years with its name changing about every 10 years. It still exists as part of the magazine 'Horticulture Week'. It was founded in 1841 by horticulturists Joseph Paxton, Charles Dilke, John Lindley and the printer William Bradbury.
1841: The Gardeners' Chronicle.
1856: The Gardeners' Chronicle and Agricultural Gazette.
1969: Gardeners' Chronicle & New Horticulturist.
1986: Horticulture Week.)

"*The Gardener's Chronicle, 31st March 1866:* 'Symphytum asperrimum, the Prickly Comfrey:
Some years since we cultivated this to some extent **as a soiling (stall-feeding) crop for cows, and we must say that these animals will eat it greedily,** and it seems very useful.
Its crop is large in May and June; when cut down it speedily throws up a second crop of succulent stems and leaves, but it must be consumed green as no creature could eat so rough a plant in a dry state."
-**Russian Comfrey: A Hundred Tons an Acre** of Stock or Compost for Farm, Garden or Smallholding by Lawrence D. Hills. London England: Faber and Faber, Limited, 1953, page 25.

"*A Codex or Pharmacopoeia does not profess to teach therapeutics.* **The Materia Medica of the 'French Codex' or 'French Pharmacopoeia',** which is very extensive, containing twice as many articles as the 'British Pharmacopoeia'; and the most striking feature in it is the very great number of indigenous (native) plants compared with those present in our own Pharmacopoeia, being at least eight or nine to one.
In running the eye over the individual plants attained in the Materia Medica of the Codex, we are struck by **its strong general resemblance to an herbalist's list**; for in the Labiatae (Lamiaceae family) we have Wood Betony, Bugle, Balm, Catmint, Horehound, Germander, Ground Ivy, Dead Nettle, Marjoram, and Sage.
   In Boraginaceae (Borage or forget-me-not family), there are Borage, Alkanet, Comfrey, Houndstongue, Lungwort, and Bugloss.
   **Comfrey: Used in leucorrhoea and internal hemorrhages (bleeding)."
-'**On the New French Codex**' by J. Birkbeck Nevins, MD, British Medical Journal, London, England, page 766-768, June 29 **1867**.
   (Codex is a manuscript volume, especially a classic work. It is a volume, in book form, of manuscripts of ancient text.)
   (Materia Medica is the scientific study of medicinal drugs and their sources, preparation, and use.)

"*In the course of reading of some of the authors of the seventeenth and earlier portions of the eighteenth centuries- for information, curiosity, and special purpose- I have encountered numerous quaint and absurd medicaments.*
Many trivial medicines which survive amongst the vulgar (uneducated masses of people) date far back to the practice of former physicians, as the compound of oil of almonds and syrup of violets, syrup of buckthorn, dragon's blood, armenial bole, cochineal, sal prunella, and balsam of sulphur.
   In the same category are- I adopt technical language of the period: plants celandine (Chelidonium majus), fenugreek (Famumgr cecum), bistort (Bistortd), burdock {Bardana major), plantain (Plantago), chervil (Chavcfolium) ...turmeric {Curcuma), sweet flag (Acorus), agrimony (Agrimonia), ground ivy (Hedera terrestris), brook lime (Beccabunga), elecampane (Enula campana), angelica,...mallow (Malva), marshmallow (Althaea), **Comfrey (Symphytum)**...
Many of their mixtures, decoctions, infusions, tinctures, pills, plasters, syrups, ointments and other galenicals leave little to desire in way of amendment (change), and have been handed down almost unchanged.
   Some compounds of very ancient origin, however, as the Mithridatium, or Confectio Damocratis, contain fifty ingredients; the Theriaca Andromachi, or Venice treacle, upwards of sixty (Quincy). **It is, in fact, undeniable that very many of the old formulae are ridiculously complicated and overloaded."**
-'**The Practitioner: A Monthly Journal of Therapeutics**, Volume II, January to June' edited by Francis E. Anstie, M.D., F.R.C.P., Westminster Hospital, Baltimore, Maryland, includes 'A Glance at an Obsolete Materia Medica' by W. Boyd Mushet, M.B., M.R.C.P., pages 141-151, **1870**.   (The author had a long list of 'quaint' medicinal herbs. I only included a few here.)
   (Galenical is medicine made of natural rather than synthetic components.)

"*The following account of the medical properties of Symphytum officinale is taken from *Houttuyn's 'Pflanzen-System'*:
   '**The root** is employed as a soothing, sanative (general healer), and astringent remedy in spitting of blood, dysentery, and ulcers in the lungs and urethra. It is also employed externally in emollient and discutient poultices.
   The blossoms also are used in the preparation of a ptisan (tea) for colds and coughs.
   In some parts of the country the blossoms are collected and dried like Cowslips for making Comfrey-wine, which is supposed to have great healing powers.' "
-'**The Gardeners Chronicle and Agricultural Gazette** for 1871' published in Covent Garden, WC, London, England, 1269 pages, January 7 to December 30 **1871**. Article by M.Y.B., August 5 1871, page 1004.
(* -'Des Ritters Carl Von Linne Vollstaendiges Pflanzensystem, Eilfter Theil' by Carl Von Linne and Martinus Houttuyn, Nurnberg {Nuremberg), Germany, 14 Volumes, 1777-1788. In German. 'Pflanzen-System' is German for Plant-System.)
   (Discutient is a medicinal application that disperses morbid matter.)

"**The Prickly Comfrey, although first introduced into England, has not much engaged the attention of English agriculturists, but it appears to be extensively cultivated in several parts of Ireland.**
It was extensively cultivated by the late Bishop of Kildare, Ireland, in a field at Glasnevin (suburb of Dublin, Ireland), where the

plant is still found, growing up as a persistent weed, in spite of every attempt to eradicate it.
Many gentlemen, after the example of the Bishop of Kildare, who was a first-rate dairy-farmer, are reported to cultivate it in their villa or suburban farms around Dublin, and find it a very useful food for their dairy stock."
-'**The Journal of the Royal Agricultural Society of England,** Second Series, Volume 7', London, England, **1871**. Includes article 'XV: On the Composition and Nutritive Value of the Prickly Comfrey (Symphytum asperrimum)' by Dr. Augustus Voelcker, F.R.S. (Fellow of the Royal Society), pages 387-389.

(Born in Germany, Dr. Augustus Voelcker lived from 1822-1884. He was a 'Royal Agricultural Society of England' chemist, living in London. He was known for his methodical and precise analytical practices in agricultural chemistry.)

"**In the 1870's and 1880's Russian Comfrey enjoyed a wave of popularity** mainly through the work of four men whose experiments and writings form part of our basic knowledge of the crop.
The most famous of the four was not a Comfrey grower, he was **Dr. Auguste Voelcker, D.Sc., F.R.S. (Fellow of the Royal Society in Great Britain),** Consulting Agricultural Chemist to the Royal Agricultural Society of England.
His article in the 'Journal' of that body (1871, Volume 7) entitled '**On the Composition and Nutritive Value of the Prickly Comfrey (Symphytum asperrimum)',** gave a very good account of it, including the Carnew Castle (County Wicklow, Ireland) yields."
-Russian Comfrey: A Hundred Tons an Acre of Stock or Compost for Farm, Garden or Smallholding by Lawrence D. Hills. London England: Faber and Faber, Limited, 1953, page 26.

"The Reverend E. Highton, of Bude, Cornwall, England, writes:
'**In November 1875, I obtained about 200 roots of Prickly Comfrey from Messrs. Sutton and Sons, Reading, England.** After digging holes about eighteen inches (45.7 cm) deep, and the same in diameter, I filled them with good half-rotted manure, covering it up with the soil taken out.

Into each of these hillocks, of which there were fifty to the square rod- about 2 1/2 feet (0.76 meters) apart each way- I put a single set (root crown), which was covered with about two inches (5 cm) of soil over the top."
-'**The Journal of the Royal Agricultural Society of England,** Second Series, Vol 18', London, England, 1882. Includes article: 'On Green or Fodder Crops Not Commonly Grown, which Have Been Found Serviceable for Stock-Feeding' by Joseph Darby.

(Messrs. Sutton and Sons sold Prickly Comfrey from 1875-1896.)

(A rod is a linear measure for land equal to 5 1/2 yards {5.029 meters}. A 'square rod' is a square with each side measuring a rod. It is equal to 160th of an acre or 30 1/4 square yards or 25.29 square meters).

"**Prickly Comfrey: This Comfrey (there are several) was introduced into this country (United States) about the year 1875 as a forage plant.** It is described as, 'a deeply rooted plant, independent of weather and climate, affording even in the hottest and dryest seasons several cuttings when other vegetation is either burnt up or at a stand still.' "
-**Experimental Work of the Agricultural Department of the University of Tennessee** (Report of the Experimental and Other Work of the School of Agriculture and Botany of the University of Tennessee for Session of 1880-1881)' by John M. McBryde, Professor of Agriculture, Horticulture and Botany, Knoxville, Tennessee, **1881**.

"The cultivation of Comfrey seems to have become very popular from 1870 onwards, when <u>Henry Doubleday of Coggeshall, Essex, England</u>, imported Symphytum x uplandicum (Russian Comfrey) from Leningrad (then St. Petersburg, Russia).
**The figure (artwork) in 'Curtis Botanical Magazine', plate 6466 (*1879) under the name of Symphytum peregrinum was drawn from plants presented to Kew Gardens in 1875 by Thomas Christy, who was associated with Doubleday in his efforts to popularise its cultivation.
Several seedsmen took up the distribution of the plant, including Messrs. Sutton of Reading, Berkshire, England, who continued to supply it until 1896.**

In 1900 Messrs. Webster of Stock, Essex, England, imported it from Russia, and they and other nurserymen still carry stocks (as of 1958). These and other introductions from Russia and elsewhere are undoubtedly the chief, if not the sole, origin of S. x uplandicum in Britain."
-'The History of Symphytum Asperum Lepech. and S. x Uplandicum in Britain' by A.E. Wade, Department of Botany, National Museum of Wales, Watsonia, Volume 4, pages 117-118, 1958. (Watsonia was the journal of the 'Botanical Society of Britain and Ireland' from 1949 to 2010. It is now called 'New Journal of Botany'.)
(* -'Curtis's Botanical Magazine, Comprising the Plants of the Royal Gardens of Kew, and of Other Botanical Establishments in Great Britain, Volume 105 of Whole Work, Or Volume 35 of Third Series' by Sir Joseph Dalton Hooker, M.D., C.B., K.C.S.I., F.R.S., F.L.S., London, England, 1879.)

"In general, estimates before the 1870's give from 40-60 tons an acre for Comfrey; the greater the number of cuts quoted, the higher the estimated yield.
From 1875 onwards we come to yields of 100 to 120 tons. They spring from Henry Doubleday of Coggeshall, Essex, England, a small farmer but a great experimenter, who imported from St. Petersburg, Russia, what is either Symphytum peregrinum, which grows 4,000 feet (1219 meters) above sea-level near Sawunt in the Talsch Province of the Caucasus, **or a hybrid of S. officinale and S. asperrimum known as S. x uplandicum.
Between them (Doubleday and Thomas Christy) they began the boom in Russian Comfrey,** as they called the new plant to distinguish it from the earlier Prickly Comfrey, S. asperriumum."

-**Russian Comfrey: A Hundred Tons an Acre** of Stock or Compost for Farm, Garden or Smallholding by Lawrence D. Hills. London England: Faber and Faber, Limited, 1953, page 28.
   (The city of Saint Petersburg in Russia was founded in 1703. In 1914 the name was changed to Petrograd. In 1924 it was changed to Leningrad. In 1991 it was changed back to Saint Petersburg.)

Henry Doubleday, scientist and horticulturalist, 1810-1902, was the son of a Quaker grocer from Coggeshall, Essex County, England. He was the first person to improve Comfrey by selecting best plants for propagation.
His records indicated leaf cuttings from second-year plants of 4.08 pounds (1.85 kg) per plant for a first cut on April 30th, and 4.20 pounds (1.90 kg) per plant for a second cut on June 16th. These 2 cuts add up to 31 tons an acre. Doubleday made 6 cuts per year, sometimes even 8 cuts. Maximum production starts in the third year.
He was elected a member of the Royal Horticultural Society, London, England.
His work was recognised by Lawrence D Hills, who named the 'Henry Doubleday Research Association' after him. It is now called 'Garden Organic' and is the largest organic gardening and horticultural organization in Europe.

"**Mr. Henry Doubleday** wrote, 'My experience in feeding on a small scale was quite satisfactory. Lambs and pigs were extremely fond of it and did well; they are also very fond of it in the dry state.
My land is a heavy clay on which the Comfrey grows luxuriantly, the roots striking down deep into the ground. Last year (1875), my first cut was from three to four feet (0.9-1.2 meters) high (plants three feet apart), some producing 10 pounds (4.5 kg) each plant, or 21 1/2 tons per acre, at the first cut 15th May.' "
-'**Prickly Comfrey: Its History, Cultivation, Extraordinary Production, and Uses:** A Letter Addressed to His Excellency Sir Hercules Robinson, President of the Agricultural Society of New South Wales' by Arthur T. Holroyd, Sydney, Australia, 1876, page 7.

"**This new type (Russian Comfrey)** reached 6 feet 8 inches in height (over 2 meters) when allowed to flower, and **was carefully chosen for its various superior characteristics by Henry Doubleday**. It was from this plant that the highest yields of that time were obtained in England, recording 100 tons per acre (25,000 kg per 1/4 acre)."
-**Comfrey: Nature's Healing Herb & Health Food** by Andrew Hughes. Japan: Sanyusha Publishing, 1992, page 124.

"**Henry grew his seedlings and called his F1 hybrid 'Doubleday's Solid Stem Comfrey' or 'Russian Comfrey'** to distinguish it from the earlier introduction.
Henry devoted the last 30 years of his life to research on this crop, leaving the marketing to Thomas Christy, a nurseryman of Sydenham (southeast London, England). It was his dream that his crop would feed a hungry world. We shall never know what he found, for when he died the family said, 'Poor Uncle Henry, he never made any money', and they tidied up all his papers and burned the lot.
Though he was awarded the proud title of 'Fellow of the Royal Society' for his introduction of Russian Comfrey in 1875, he could never afford registration fee of 12 guineas so could not use the letters 'F.R.S' to which he was entitled."
-**Fighting Like the Flowers: An Autobiography:** The Life Story of Britain's Best-Known Organic Gardener by Lawrence D. Hills. Bideford, Devon, England: Green Books, 1989, pages 100, 103.

"**Under the title of 'Forage Plants'*, Mr. Thomas Christy (155 Fenchurch Street, London, England) gives an account of Symphytum asperrimum, a well-known herbaceous plant closely allied to our wild Symphytum officinale, and which is highly recommended as a forage plant** by Professor Buckman and other agricultural authorities.
We have no personal acquaintance with the plant as a forage plant, but if hardiness, rapid growth, abundant foliage, and facility of propagation are requisites, then this plant is likely to be very useful.
We suppose the plant would do best on heavy wet soils and temperate climates. Whether it will be so suitable for hot, dry countries is a matter open to doubt, though its fleshy root-stocks will enable it to with stand the ill effects of drought."
-'**The Gardeners Chronicle: A Weekly Illustrated Journal of Horticulture and Allied Subjects**', New Series, January to June 1877: Volume VII, July to December **1877**: Volume VIII. London, England. This excerpt is from February 3, 1877, page 151.
(* -'Forage Plants and Their Economic Conservation by the New System of Ensilage: Part I: Caucasian Prickly Comfrey' by Thomas Christy, Jun., Fellow of the Linnean Society, Christy & Co., London, England, 1877.)

"**Prickly Comfrey: In England the plant is still uppermost in advertisements of seedsmen, and on the continent of Europe the agricultural experiment stations** have been putting the plant through a course of sprouts to ascertain its growth and value."
-'**Pacific Rural Press**', Volumes 17-18, San Francisco, California, 843 pages, every other week from January 4 to December 27 **1879**. Article: 'Prickly Comfrey' by editors, March 1 1879.

"**Prickly Comfrey from Mr. James J.H. Gregory, the eminent seedsman of Marblehead, Massachussetts.** We copy what follows from Mr. Gregory's catalogue for this year:
   This new forage plant is extensively grown in Europe for feeding of stock. It is a deep rooted plant, and even in the hottest seasons will yield several cuttings of forage. It comes in earlier and lasts longer than almost any forage crop. The method of propagation is by roots only. The cultivation is very simple. In well plowed and well manured ground, plant the cuttings three feet (0.9 meter) apart each way, giving them a liberal dressing of manure the first Winter, and no

further expense is needed.
Prickly Comfrey cuttings by mail, 50 cents/dozen; $2.75/100. Rooted plants by express or freight, $1.50 per pound."
-'**The Farmers Magazine and Kentucky Live-Stock Monthly:** A Comprehensive Monthly Journal for Country Homes' by editor John Duncan, Louisville, Kentucky, 889 pages, November 1877 (Volume 1, No. 1) to August 1880 (Volume 3, No. 10). Article: 'Prickly Comfrey', April **1879**.

"*Mr. John Sagar's Report for farmers from Clitheroe, Lancashire, England of settlement in Canada:*
From a British Farmer in Ontario, Canada to British Farmers in England:
In the improved system of breeding and fattening stock, green fodders are now taking an important place. *The climate is particularly suitable for successive rushes of vegetation during one season. Under liberal treatment (of fertilizers) they can be so arranged as to afford a continuous supply from middle of April to first of November:*

1. Lucern, four cuttings   20 tons per acre.
2. Winter Eye, two cuttings   4 tons per acre.
3. Red Clover, two cuttings   6 tons per acre.
4. Tares (vetch) and Oats, one cutting 3 tons per acre.
5. Millet, two cuttings   4 tons per acre.
6. Maize (corn), one cutting   30 tons per acre.
7. Rape, one cutting   7 tons per acre.
8. The thousand-headed Kale and **Prickley Comfrey have just been introduced with success**."

-'**Reports of Tenant Farmers Delegates on the Dominion of Canada as a Field for Settlement**' by Canada Department of Agriculture, Ottawa, Canada, **1884**, page 133.

"**Symphytum asperrimum (Prickly Comfrey):** *Mr. Mitchell Henry in the (October 5, 1885) 'Times' recommends the use of this plant as a forage crop: 'Five years ago I obtained a small supply of the roots from a London agent, and planted them in a light sandy soil in which they did not do very well.*
The roots were then taken up, divided like Jerusalem Artichokes, and transplanted into reclaimed peat land, receiving a good supply of farmyard manure. Here the Prickly Comfrey has flourished amazingly, and by subdivision now covers several acres.
It has been cut this year already five times, and will be cut again before Christmas (December 25), yielding by careful weighing after the present fifth cutting a total of 40 tons to the acre.
    The plant is uncommonly handsome, and when planted should have intervals for its growth of not less than 2 feet (0.6 meters), and when gathered it should be cut down even with the ground and receive a dose of liquid or other manure.
Cattle eat it greedily, and it is excellent for dairy cows, as it does not flavour the milk. *I have seen it stated that the roughness of the leaves makes it distasteful to cattle, but this is an error.* **It is an invaluable food for pheasants, ducks, and all kinds of fowl,** *and if chopped up for them in that most useful instrument, Starritt's American circular cutter, and mixed with barley-meal or crushed Indian corn, it fattens them rapidly and saves a third of the grain. I have had two of these mincing machines, one large and the other small, both purchased from Gilbertson & Page, Hertford, England.*
Like all broad-leaved plants, which derive much of their food from the air and the rain, **Comfrey grows best wherever Swedes (rutabaga) and Mangels (fodder beets) flourish, and amply repays the expenditure of a fair supply of manure.**
    It has been stated that no manure is wanted, but this, as regards all plants, is nonsense, for in some way or other you must restore to the soil what you have taken out of it, and root crops especially exhaust the soil.
Preserved as ensilage Prickly Comfrey does not seem to have done very well, and product is unusually disagreeable in smell.' "
-'**The Gardeners Chronicle: A Weekly Illustrated Journal: Horticulture and Allied Subjects**', London, England, Oct 10 **1885**.
    ('The Gardeners' Chronicle' was a British horticulture periodical in publication for almost 150 years. It was founded in 1841. It still exists as part of the magazine 'Horticulture Week'.)
    (Silage is fermented, high-moisture fodder that is fed to cattle, goats, sheep and other ruminants. It is fermented in a process called ensilage or silaging, and is usually made from grasses, such as corn, sorghum and other cereals, using all of the green plant. **Comfrey can be successfully made into silage.** See subsection 'Comfrey as Silage' in section 'Comfrey Meal, Pellets, Hay and Silage' in Volume 2.)

"**Symphytum asperrimum (Prickly Comfrey):** *I have been much interested in reading the letter of Mr. Mitchell Henry reproduced in your columns from the 'Times' of October 5 (1885), on the growth of Caucasian Prickly Comfrey (Symphytum asperrimum).* **I have been a grower of it for many years, and can endorse all he says as to its value. Since the introduction of the real Caucasian Comfrey, imported by my friend Thomas Christy, of London, and called by him Russian Comfrey, I have mainly grown this variety, and find it superior in value.**
The last two seasons, during the summer drought, it has been invaluable, my cows keeping in full milk, while others in the pastures failed to give milk.
With regard to ensilage, George Fry's*, of Chobham (Surrey, England), system of sweet ensilage must, I think, be adopted. I have a small silo now filled with it on that principle not yet opened, but it appears to be sweet and promising. All kinds of fowls, partridges, and other game are very fond of roaming over my plantations. They sometimes honeycomb the large lower leaves, but I do not think they damage the crop; indeed, I have a notion they have golden feet, as sheep are said to have.
With regard to the last paragraph **on the healing properties of this plant, I consider the real Caucasian (Russian Comfrey) superior to the Common English Comfrey,** *and I have been recently informed that a poultice made of the*

root is a perfect cure (palliative) for erysipelas. If this is found to be a fact it adds greatly to its value.
by Henry Doubleday, Coggleshall, Essex, England."
-'**The Gardeners Chronicle:** A Weekly Illustrated Journal: Horticulture and Allied Subjects', London, England, October 24 **1885**.
(* -'The Theory and Practice of Sweet Ensilage' by George Fry, FLS. London, England: The Agricultural Press, 66 pages, 1885. It does not mention Comfrey or Symphytum specifically.)
('Golden feet' in sheep means 'rich grasses spring up in their tracks, and useless weeds are eradicated by them'.)
(Erysipelas is an infection with a skin rash caused by group A Streptococcus bacteria.)

"**Advertisement: Russian Comfrey: This pure stock of imported Symphytum asperrimum yields the largest crop of green fodder known.** It will grow luxuriantly on rank clay, and well on light soils, if of some little depth.
For Milch (milk) cows, horses (especially Hunters), and all animals, it is the best of green foods. Its use makes the coats of horses shine, and keeps them hard and in good condition, saving all medicine.
Being a perennial, once planted there is no further expense. Autumn or early winter sown sets (roots) are best, as the leaves will appear before the grass can be fed off in the spring.
For winter keep, it can be preserved by the system of Ensilage or pitting; or it can be made into hay of the finest quality, possessing a magnificent aroma. Russian Comfrey possesses more gum and mucilage than any other variety. It will produce from 100 to 150 tons to the acre the third year on a favourable soil.
-From Thomas Christy and Co., London, England."
-'**New Commercial Plants and Drugs, Issue 8**' by Thomas Christy, F.L.S. (Fellow of the Linnaean Society), London, England, **1885**. Includes Russian Comfrey advertisement, page 95.
(Symphytum asperrimum is not the same as Russian Comfrey, as now understood. In the early days of imported Comfrey in England, these differences were not always clear.)

"**From 1875-1900 Comfrey boomed, though little was done to carefully classify the types,** with the result that many poor types were oversold on the basis of false claims made by get-rich-quick salesmen, somewhat like what happened in **Japan** over the few years after 1960, but with this difference, that the only basic type of Comfrey imported into Japan is peregrinum (Russian Comfrey), so its lesser parents, officinale and asperrimum were not there to compete."
-**Comfrey: Nature's Healing Herb & Health Food** by Andrew Hughes. Japan: Sanyusha Publishing Co., Ltd, 1992, page 124.

"The discovery of the tubercle bacillus, and the way in which it appears to be associated with the progress of the more serious forms of tuberculous (TB) disease, has tended to cast drug treatment rather into the background.
The ancient treatment of phthisis pulmonalis (TB) by demulcent drinks, such as infusion of Comfrey or of mullein, and by such remedies as paregoric tincture and oxymel of squills we now regard with feelings allied to pity, and think we have made some progress in drug healing."
-'**The Use of Drugs in the Treatment of Early Phthisis**' By J.C. Thorowgood, M.D., F.R.C.P., Senior Physician to City of London Hospital for Diseases of the Chest, London, England. Read at Annual Meeting of 'British Medical Association', July **1891**.
(Tubercle bacillus is a rod-shaped aerobic bacterium {Mycobacterium tuberculosis} that causes tuberculosis. Phthisis pulmonalis is a tuberculosis of the lungs with progressive wasting of the body.)
(Paregoric or camphorated tincture of opium is a traditional patent remedy known for its antidiarrheal, antitussive {cough reducer}, and analgesic properties.)
(Oxymel of Squill was used for coughs and was invented by Pythagoras who lived in the sixth century B.C.)

"**We have received a letter from Mr. Thomas Rockford of Bellganny, Nenagh County, Ireland,** who for several years has devoted his entire attention to local prize work in Irish agriculture. Mr. Rockford, **an authority in relation to Comfrey root,** says:
'As its characteristic is to throw out a powerful root system, which penetrates deep into the ground, and may penetrate deep into the subsoil beyond tbe reach of manure, it may be advisable to cut away the older roots after some time; and it may be possible to utilize them as a food for stock in the same way as parsnips or carrots.
The action of our movement since February, 1886, has **induced (persuaded) agriculturists to plant Comfrey sets (roots) cut from plants grown near Nenagh in every county in Ireland**, in the counties of Northumberland, Chester, Derby, Surrey, Gloucester, Northampton, Wilts, Middlesex, Berks and Bedford, **in England**, in the counties of Lanark, Perth, Renfrew, Dumfries, Fife and the Isle of Arran **in Scotland, in Spain, France, Portugal and Belgium, in the States of Iowa and Florida in America, and in New Zealand.** Some of our Comfrey has been sent to San Antonio, Florida, this year where it will be given a thorough trial, and I think results will be satisfactory.' "
-'**The Florida Agriculturist**' by E.O. Painter, Deland, Florida, weekly journal, January to October **1891**. Includes article 'Irish Comfrey' on March 11 1891.

"**The whole Comfrey plant, beaten to a cataplasm (soft moist mass), and applied hot as a poultice, has always been deemed excellent for soothing pain in any tender, inflamed or suppurating (with pus) part.** It was formerly applied to raw indolent (low pain) ulcers as a glutinous astringent, and most useful vulnerary (wound healer).
Pauli (Simon Paulli?) recommended it for broken bones, and externally for wounds of the nerves, tendons, and arteries.
More recently surgeons have declared that **the powdered root** (which, when broken, is white within, and full of a slimy juice), if dissolved in water to a mucilage, is far from contemptible (unpleasant) for bleedings, fractures, and luxations (dislocations), whilst it hastens the callus of bones under repair.

The 'Poet Laureate' tells of: ***'This, the Consound, Whereby the lungs are eased of their grief.'*** *"*
-**'Herbal Simples Approved for Modern Uses of Cure'** by W.T. Fernie, M.D., Bristol, England, **1897**.
    (In 1896 'Poet Laureate' was officially appointed to Alfred Austin, 1835-1913, an English poet. Consound is Comfrey.)

"Plants by their appearance often invite the invalid to cull (harvest) them for his restoration, and assume such shapes as to suggest their curative properties. *For instance, herbs that simulate the shape of the Lungs, as Lungwort, Sage, Hounds-tongue, and* **Comfrey, are all good for pulmonary complaints.**
The Comfrey (Symphytum officinale) plant is demulcent and slightly astringent. **All mucilaginous agents exert an influence on mucous tissues, hence the cure of many pulmonary and other affections** *in which these tissues have been chiefly implicated, by their internal use."*
-**'The Complete Herbalist,** or the People Their Own Physicians by the Use of Natural Remedies' by Dr. O. Phelps Brown, Jersey City, New Jersey, **1897**, pages 16, 74.
    (The 'Doctrine of Signatures', dating from the time of Dioscorides {40-90 AD} and Galen {129-216 AD}, states that herbs resembling various parts of the body can be used by herbalists to treat ailments of those body parts.)

"In **'Squire's Companion to the British Pharmacopoeia'** *(*17th edition, 1899):* **Comfrey as having astringent, mucilaginous and glutinous properties.**
Its author knew a bone-setter who had rendered himself famous by treating fractures with a pulp made of the scraped root spread to the thickness of a crown piece upon cambric or old muslin which was wrapped round the limb and bandaged over. Squire mentions that **Comfrey as officially recognized in the following Pharmacopoeias: in Belgium as 'Radix symphyti", France 'Consonde'; Mexico 'Sinfito'; Portugal 'Consolida major'; Spain 'Sinfito major'."**
-The Medicinal Uses of Comfrey by Dr. Charles MacAlister, M.D., F.R.C.P., written 1935.
(* -'Squires Companion to the Latest Edition of the British Pharmacopoeia: Comparing Strength of Its Various Preparations with those of United States and Other Foreign Pharmacopoeias, 17th edition' by Peter Squire, London, England, 1899, page 619.)
    (An astringent shrinks or constricts body tissue. Mucilaginous means it is a thick, gluey substance. Glutinous comes from the word gluten that means glue in Latin.)
    (A crown piece is a British coin 38 mm across, the same as the United States Silver Dollar- 1.5 inches. So the thickness is the height of these coins that is about 4 mm or 1/8 inch.)

*"**Symphyti Radix (root), Common Comfrey root, Symphytum officinale:**
The black rind (outside of root) is scraped off, and the mucilaginous root is then scraped carefully into a nice even pulp; this spread to the thickness of a crownpiece upon cambric or old muslin, is wrapped round the limb and bandaged over; it soon stiffens, and forms a casing superior to starch, giving great support and strength to the part. The late Author knew a bone-setter who practised more than fifty years ago, and rendered himself famous by treating fractures after this method, which he kept secret, the bandage not being removed until the limb was well."*
-**'Squires Companion to the Latest Edition of the British Pharmacopoeia:** Comparing the Strength of Its Various Preparations with those of the United States and Other Foreign Pharmacopoeias, 17th edition' by Peter Squire, London, England, **1899**, page 619.

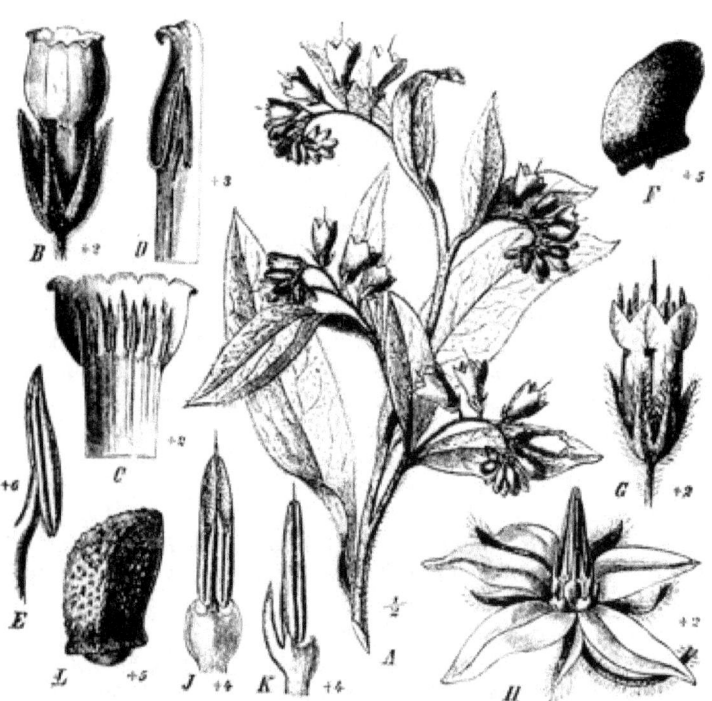

Symphytum and Borrago Flower Parts 1900

'Das Pflanzenreich: Hausschatz des Wissens, Department 5, Volume 7'
(The Plant Kingdom: Treasury of Knowledge)
by Professor Dr. Karl Schumann and Dr. Ernest Gilg; Publisher Neumann, Neudamm, Germany, 1900.
Comfrey art page 758. In German.

Fig. 435
A. Symphytum asperrimum
B-F. Symphytum officinale
G. Symphytum bulbosum
H-L. Borrago officinalis

F and L are Comfrey seeds.

Chapter 16

The 1900s and Comfrey

<u>20th Century (1900s)</u>

"**Comfrey: Symphytum officinale.** Synonym: Nipbone, Knitbone. Part used: Root, leaves.
Action: Demulcent, astringent. Is very highly esteemed as a remedy in all pulmonary (lung) complaints, hemoptysis (hemorrhage of lungs), and consumption (tuberculosis), and forms an ingredient in a large number of herbal preparations. Wherever a mucilaginous medicine is required this may be given.
It has been used of late by the medical profession as a **poultice** to promote healing of obstinate (stubborn) ulcerous wounds. A **decoction** is made by boiling 1/2 to 1 ounce (14-28 gram) of crushed root in 1 quart (0.94 liter) of water or milk. Dose, a wineglassful. The leaves are preferably taken as an infusion prepared in the usual manner (Comfrey tea). Comfrey leaves subdue every kind of inflammation swelling when used as a fomentation (poultice)."
-**Potter's Cyclopedia of Botanical Drugs and Preparations** by Richard Cranfield Wren, London, England, **1900**.

"**In 1900 Messrs. Webster's Nurseries of Stock, Essex County, England, imported a fresh supply of Symphytum peregrinum from Russia.**
They rogued their stock by removing the least vigorous plants and selecting for division those which, when grown on the six cuts a year system, produced the characteristic two-foot (0.60 meter) leaves, between eight and ten inches (20.3-25.4 cm) across, dark green, and bristly on the undersides.
The flower colour was magenta pink varying to purple, but not the dark and dingy colouring of S. officinale var patens, and never including the clear bright blue of S. asperriumum (Prickly Comfrey), **distributed by Messrs. Loddige**."
-**Russian Comfrey: A Hundred Tons an Acre** of Stock or Compost for Farm, Garden or Smallholding by Lawrence D. Hills. London England: Faber and Faber, Limited, 1953, page 37.
(Roguing is identifying and removing plants with undesirable characteristics from agricultural fields. Rogues are removed to preserve the quality of the crop.)
(For more about the Loddige Nursery see 1771, 1806 and 1811 in the 'History' section {Chapters 14 and 15}.)
(The Webster Strain is Russian Comfrey Bocking No. 3, 4, 5, 6, 7, 8, 9, 10, 11, and Bocking Mixture. See Chapter 10.)

"The latest full and favorable account of Russian Comfrey giving yield of eighty tons an acre, is in '**The Complete Grazier**' by **William Fream, published in 1900.**"
-**Russian Comfrey: A Hundred Tons an Acre** of Stock or Compost for Farm, Garden or Smallholding by Lawrence D. Hills. London England: Faber and Faber, Limited, 1953, page 39.
('The Complete Grazier and Farmers and Cattle-Breeders Assistant' was also written by William Youatt. The book was published starting in 1816 with other editions including 1900. William Fream was born and worked in England. He visited Canada and wrote books about their agriculture. He raised sheep and cattle in Lincolnshire, England.)

"**Symphytum Linn.** Gen. Plant. 170: Symphytum foliis ovato- lanceolatis decurrentibus. Syst. Nat.
Symphytum magnum. Raii. Syn. Blossoms yellowish white.
The roots are much used by the common people for sprains. They are glutinous and mucilaginous. The leaves give a grateful flavour to cakes and panadoes."
-'**An Account of Some of the Vegetable Productions, Naturally Growing in This Part of America**' by Reverend Manasseh Cutler, F.A.A., M.S.; Bulletin of the 'Lloyd Library of Botany, Pharmacy and Materia Medica', Bulletin No. 7, Cincinnati, Ohio, **1903**. (Panadoe is a type of bread.)

" '**The Book of Herbs 2**' by Lady Rosalind Northcote, is one of the series of 'Practical Gardening' books edited by Mr. Harry Roberts. The book is divided into 3 parts, the first dealing with the chief herbs used at the present time, the second with herbs chiefly used in the past, and the third with herbs used in decorations, in heraldry, and for ornament and perfumes.
Many herbs have been expunged (removed) from modern Pharmacopoeias. Perhaps we have no use for them now that we in England no longer live in perpetual terror of biting sea-hares, scorpions, or tarantulas, as our forefathers seem to have done! Herbs are dying out among the country poor in favour of quack medicines, though some are still employed, such as wood-sage, dandelions, centaury, and meadow-sweet as bitters. Mallows are likewise used, while elder still keeps its place in the 'British Pharmacopoeia'.
Primrose, poor-man's-friend, and Comfrey are together made into ointment, white flowered Comfrey being used when the ointment is for a woman and red flowered Comfrey for a man."
-'Reviews and Notes: Notes on Books', **The British Medical Journal,** London, England, page 958, April 23 **1904**.
(Lady Northcote, 1873-1950, was born in North Yorkshire, England. 'The Book of Herbs' and another book she published were about the landscape, folklore, flora and fauna of Devonshire, England.)
(Heraldry is the design, display, and study of armorial bearings, ceremony, rank, and pedigree.)

"**The striking results obtainable from the use of coal tar derivatives (drugs) in acute diseases** and in various nervous disorders now so prevalent have given to drugs of this class a preponderating (too much) influence amongst remedies in the treatment of disease, and the supply seems quite equal to the demand, the increase of new drugs being almost wholly from this class. No one constantly using them can possibly deny their great efficacy; they seem suited both to modern diseases and to modern persons.
The contrast between their rapid production and the scanty introduction of remedies from either vegetable or mineral kingdoms is a **phase in pharmacy**.
Our forefatlhers trusted almost entirely to the vegetable kingdom in the treatment of disease. The mothers of families thought it their duty to learn themselves, and to teach their daughters, the use of many simple herbs, which they either kept in their gardens or could gather from the woods or roadsides in the neighbourhood.
The monks of old kept beds of medicinal herbs, or physic-gardens, for use in the treatment of the sick, not only in their own monastery, but amongst the laity (lay people) around.
> It is recorded that the physic-garden of the Cistercian monks at St. Gall, near Lake Constance (Austria, Germany, Switzerland), contained 16 garden beds, each with the name of some herb inscribed on it. Amongst these were: peppermint, rosemary, white lily, sage, rue, **Comfrey**, pennyroyal, fenugreek, rose, watercress or radish, or mustard (sisymbrium) cummin, lovage, fennel, tansy, kidney bean, savory.

The object of this note is to suggest that, **instead of hunting so vigorously for something new and adopting all the new creations of chemists, some effort should be made, by some such company as the Apothecaries' Society, to make reliable preparations from some of the older drugs obtainable from native plants** before they are allowed to become extinct, so that they could be tried as remedies in common ailments.
> The old dandelion still holds its own, and is frequently used in diseases of the liver and stomach. May we not be leaving much that is really valuable to the herbalist? Perhaps the cultivation of these old plants, and **extraction of their alkaloids or other active principles, might give us some useful and powerful drugs of many kinds.**"

-'**A Plea for Some Old Drugs**' by Fred Tresilian, M.D., F.R.C.P. Ed., The British Medical Journal, London, England, pages 922-923, April 16 **1904**.
> (Coal tar is a thick dark liquid which is a by-product of the production of coke and coal gas from coal.
> Coke is a fuel with a high carbon content and few impurities.)

"*Digestible Nutrients and Fertilizing Constituents in 100 Pounds (45 kg):*

	Digestible Nutrients in Pounds				Fertilizing Constituents in Pounds		
	Dry Matter	Protein	Carbohydrates	Ether Extract	Nitrogen	Phosphoric Acid	Potash
Prickly Comfrey	11.6 pound	1.4	4.6	0.2	4.2	1.1	7.5

-'Profitable Stock Feeding: A Book for the Farmer' by Howard Remus Smith, Professor of Animal Husbandry, University of Nebraska, Nebraska Experimental Station, Lincoln, Nebraska, **1905**-1906. Comfrey page 396, 402.

"**Volodymyr Szuchewycz, an Ukrainian ethnographer** and teacher residing in Lviv, Ukraine, observed **extensive use of traditional herbal medicines by Hutsuls** (a neighboring ethnic group of Ruthenian origin to the east of Boykos) and described some of the remedies in the 5th volume of his monograph (*Szuchewycz, **1908), including the widely used Comfrey (Symphytum officinale),** yarrow (Achillea millefolium), marsh woundwort (Stachys palustris), mistletoe (Viscum album), and yellow gentian (Gentiana lutea)."
-'**Botanical Provenance of Traditional Medicines From Carpathian Mountains at the Ukrainian-Polish Border**' by Weronika Kozlowska, Charles Wagner, Erin M. Moore, Adam Matkowski and Slavko Komarnytsky from Wroclaw, Poland and North Carolina; Frontiers in Pharmacology, Lausanne, Switzerland, Volume 9, pages 295-311, April 2018.
(* -'Huculszczyzna' {Hutsulshchyna}, Volume 5, by Wolodymyr Szuchewycz {Volodymyr Shukhevych}; Lviv, Ukraine: The Dzieduszycki Museum, 1908. In Polish.)
> (Hutsuls is an ethnic group living in parts of western Ukraine and Romania. Boykos or Highlanders are a Ukrainian ethnographic group located in the Carpathian Mountains of Ukraine, Slovakia, Hungary, and Poland.)

"**The leading German agricultural periodical, 'Deutsche Landwirtschaftlichen Presse' (German Agricultural Press), has published many articles on 'Beinwell' (Comfrey); one in volume 35, 1908**, by Professor Dr. Werner of Bonn-Poppelsdorf.
> He traces the history from 'den Englander Grant' and gave a rather bad account of Symphytum asperrimum, and was replied to by Dr. Weber in the next issue, who had from his figures and descriptions, got Henry Doubleday's S. peregrinum.

Letters appeared from time to time, including one in volume 59, 1932, page 20, recommending 3-5 kilos (6.6-11.0 pounds) a day as an extra fodder for pigs. In general, opinion of the plant is unfavourable, except for Dr. Weber."
-**Russian Comfrey: A Hundred Tons an Acre** of Stock or Compost for Farm, Garden or Smallholding by Lawrence D. Hills. London England: Faber and Faber, Limited, 1953, page 161.

"*Prickly Comfrey (Symphytum asperrimum) Boraginaceae:*
A perennial forage plant; *stem erect, two to four feet (0.6-1.2 meter); leaves dark rich green, long and narrow, abundant, rough, mucilaginous; flowers purple, in nodding, one-sided clusters.*
It has given greatest success in New York, Michigan and Florida, *in the latter state on waste lands. It is now rarely grown in this country. It is said to be much grown in Europe.*
If cut and fed in the green state, the leaves and stalks make valuable forage. Stock must be trained to like it, as it is somewhat unpalatable (not tasty). It is used for soiling, but is not to be pastured and does not make good hay.

Prickly Comfrey produces an abundance of seeds, *but is nearly always propagated by cuttings of the fleshy roots. The planting distance varies from eighteen to thirty-six inches (45-91 cm) each way. As the plants attain a large size, the greater distance is preferable. A light sandy soil is best; several cuttings may be had each year."*
-'**Cyclopedia of American Agriculture,** Volume II: Crops: A Popular Survey of Agricultural Conditions, Practices and Ideals in the United States and Canada', 4 Volumes, edited by Liberty Hyde Bailey, New York, **1911**, page 309.
    (Soiling is a crop cut green and fed to livestock immediately without curing or processing.)
    (**You can make good hay with Comfrey.** See subsection 'Comfrey as Hay' in section 'Comfrey Meal, Pellets, Hay and Silage' in Volume 2.)

*"****Another pamphlet before us is 'Herbert's Guide to Health Recipe Book: How to Cure Yourself with Herbs'.*** *This bears in different places the names of 'Baldwin and Co., Herbalists and Drug Stores, of Electric Parade, Holloway, London, North' and of 'Herbert's Remedy Co.' at the same address.*
It contains a number of short paragraphs on the preparation of remedies for various complaints, but many of these contain besides the herbs some ingredient which is apparently only to be obtained from the firm in question.
    For example: **Consumption: One packet each Comfrey Leaves,** *Tussilago (Coltsfoot), Mullein, Hyssop, and Liquorice Root. Boil in two quarts (1.89 liter) of water for 20 minutes. Strain, and add 1 pound (0.45 kg) best sugar.*
    Dose: A wineglassful four times a day."
-'**Herbalists and Medical Practice**', no author listed, The British Medical Journal, London, England, page 1274-1277, May 27 **1911**.
    (Consumption is tuberculosis, usually an infection of the lungs.)

Cedric Bucknall, 1849-1921, was an English botanist. In **1912** he wrote the report 'Some Hybrids of the Genus Symphytum' and in 1913 he wrote 'A Revision of the Genus Symphytum, Tournefort'
    (See 'Bucknall 1912' in section 'Borage Family, Symphytum Genus' {Chapter 2}. Also see subsection '1913 Bucknall' in section 'Symphytum Species Classifications' {Chapter 4}. His writings about Comfrey are very important.)

*"****World War 1 United States Army Medical packet of Comfrey for medical display. Commonly carried by Field Medic's for treatment of various symptoms, or conditions.***
    The box says: Comfrey, Symphytum officinale Lin., No. 86, Parke Davis & Co., Detroit, Michigan."
-'United States Medical Items' by Hayes Otoupalik: Military Collectibles and Historical Americana, 1860 to Vietnam, www.hayesotoupalik.com, Missoula, Montana, 2018.   (World War I was from July **1914** to November 1918.)

    "New Cure for Stubborn Wounds Results from Clue Given by Fly:
    The insect that gave the clue to this discovery is one of the flies -in the maggot stage- that gained fame as **a medical aid on World War One battlefields, where an Army doctor found that wounds infested with maggots healed better and faster than wounds without them.** *Since then surgeons all over the world have used maggots in treating deep infections difficult to cure by ordinary surgery.*
    Dr. William Robinson, of the 'Bureau of Entomology and Plant Quarantine', now finds that **allantoin, which is given off by the maggots as they work their way through a wound, is responsible for part of this power.** *Allantoin, Dr. Robinson says, is not a new discovery.*
    Dr. C.J. Macalister, who used it successfully 23 years ago for ulcers, reported that **Europeon peasants had long applied the roots of Comfrey, which contain allantoin to sores.**"
-'**Illinois Medical Journal:** Official Organ of the Illinois State Medical Society, Volume 70' edited by Charles J. Whalen MD and Henry G. Ohls MD, Oak Park, Illinois, July-December 1936.

    Various web sites have statements such as: *"Comfrey was employed as a treatment for wounds infested with maggots during World War I."* and *"Comfrey was used during World War I to help heal maggot-infested wounds."*
    These statements are misleading as shown by the above article. **Both maggots and Comfrey help heal wounds because they contain allantoin.**

In the **1916 "Squire's Companion to the British Pharmacopoedia"** (*19th edition) it recommends an external application on lint changed 3-4 times a day, using .04-.05% of glyoxylic acid **(allantoin)**.
-Russian Comfrey: A Hundred Tons an Acre of Stock or Compost for Farm, Garden or Smallholding by Lawrence D. Hills. London England: Faber and Faber, Limited, 1953, page 20.
(* -'Squires Companion to the Latest Edition of the British Pharmacopoeia: Comparing Strength of Its Various Preparations with those of United States and Other Foreign Pharmacopoeias, 19th edition' by Peter Squire, London, England, 1916, page 1369.)

*"****The root of Symphytum officinale,*** *Linne (Nat. Ord. {Natural Order} Boraginaceae).*
Europe; naturalized in the United States.
Principal Constituents: *Mucilage in quantity, tannin and asparagine (non-essential amino acid).*
Preparation: *Tinctura Symphyti, Tincture of Symphytum-*
    Recent (fresh) root, 8 ounces; Alcohol, 16 fluid ounces. Dose, 1 to 10 drops.
Action and Therapy: *This drug is chiefly mucilaginous and used, therefore, as a demulcent in pulmonary (lung), gastric (stomach), and renal (bladder) irritation and inflammations. With many it is a favorite for irritative cough, with bloody expectoration; and in mucous disorders with a tendency to hemorrhage.*

In ancient days it was lauded as a vulnerary (wound healer), even to promoting the quick healing of fractured bones, a myth that was more recently revived in England because of the discovery of a principle (allantoin) found in the plant."
-'**The Eclectic Materia Medica, Pharmacology and Therapeutics**' by Harvey Wickes Felter, M.D., Eclectic Medical College of Cincinnati, Ohio, **1922**.
(2001 from 'Neo-Eclectic' herbalist: *"From using up to 8 organic solvents and the Lloyd Extractor, 'Specific Medicines' represented the strongest possible concentration of the bioactive aspects of botanicals that would stay in a colloidal solution. Perfected over 4 decades by John Uri Lloyd, each 'Specific Medicine' was prepared according to the nature of that specific plant. A 'Specific Medicine' represented the greatest strength, without degradation, for a particular plant."*)

"Plant collections made by USDA 'Agricultural Explorers' from around the world:
Symphytum officinale L.: Boraginaceae. Comfrey.
*Hardy perennial sometimes grown as a border plant for its foliage. 52885 **from Christiania, Norway.***
Symphytum officinale L.: Boraginaceae. Comfrey.
*A hardy tuberous-rooted perennial 3 feet (0.9 meter) high, with white, yellowish, purple, or rose flowers in drooping cymes. Native to Europe and Asia. 53109 **from Leyden, Netherlands.***
Symphytum asperrimum Donn.: Boraginaceae. Comfrey.
*Coarse European perennial herb with short pricklelike hairs and purple flowers. 53161 **from Upsala, Sweden.**"*
-'**Inventory of Seeds and Plants Imported by the Office of Foreign Seed and Plant Introduction** During the Period from April 1 to June 30 1921' by the United States Department of Agriculture, Bureau of Plant Industry, Washington, DC, **1923**.

"As late as 1923, the herb Comfrey was recognized in the '*British Pharmaceutical Codex'.
Its authors stated that Comfrey root from time immemorial had been employed as a domestic 'simple' for applications to wounds, sores and ulcers of various kinds. The decoction not only is 'applied to such wounds', it is given internally in gastralgia (stomach ache) and gastric ulcer in doses of 1/2 to 2 ounces (14.17-56.69 grams), three times a day."
-**Comfrey: What You Need to Know** by Ben Charles Harris. New Canaan, Connecticut: Keats Publishing, Inc., 1982, page 11.
(* -'The British Pharmaceutical Codex: An Imperial Dispensatory for the Use of Medical Practitioners and Pharmacists' by the Council of the Pharmaceutical Society of Great Britain, London, England, 1923.)

*"A different account is given of the same species (Symphytum asperrimum), with its small blue flowers, by **E. Brandmuller in his article *'Prickly Comfrey as a Fodder Crop', published in 'Farming in South Africa'**, November **1926**.*
He considered its qualities of drought resistance and heavy fodder yields of extreme value on his poor land. A quality of rather surprising value was its frost resistance; it stood up to 15 degrees F (-9 C) of frost in the cold nights."
-**Russian Comfrey: A Hundred Tons an Acre** of Stock or Compost for Farm, Garden or Smallholding by Lawrence D. Hills. London England: Faber and Faber, Limited, 1953, page 40.
(* If you have this Prickly Comfrey {Symphytum asperrimum Donn} article, I would appreciate a copy. Volume 1, pages 275-276.)

Book first published in 1931: 'A Modern Herbal: The Medicinal, Culinary, Cosmetic and Economic Properties, Cultivation and Folk-Lore of Herbs, Grasses, Fungi, Shrubs and Trees with their Scientific Uses' by **Mrs. M. Grieve**, London, England.
She recommends using Comfrey as medicine.

Book first published in 1936: 'Comfrey: An Ancient Medicinal Remedy' by Charles J. MacAlister in England.

*"The first of the **modern Comfrey growers was J. Kenneth Crawley, of Lockerbie, Dumfriesshire (County of Dumfries, England)**. His interest began in **1938**. He carried out a number of experiments with the plant, feeding it mainly to cattle, sheep and pigs.*
*His first article, **'Prickly Comfrey- A Neglected Fodder Crop'**, in the 'Scottish Farmer' for 28th February 1942, was followed by one entitled 'Prickly Comfrey' in the 'Journal of the Land Agents Society' for April 1942.*
*He was, however, not interested in sales so much as in **vindicating (prove good) a crop from the past with great possibilities of helping in the war-time stock-feed shortage, and in experiments to find the best ways of feeding it and conserving it for winter.**"*
-**Russian Comfrey: A Hundred Tons an Acre** of Stock or Compost for Farm, Garden or Smallholding by Lawrence D. Hills. London England: Faber and Faber, Limited, 1953, page 42. (If you have the Prickly Comfrey article, I would appreciate a copy.)

Book first published in 1939: 'Back to Eden: The Classic Guide to Herbal Medicine, Natural Foods and Home Remedies' by Jethro Kloss, Wisconsin. **He recommends Comfrey as medicine.**

"Comfrey tablets were standard issue in the first aid packs given to soldiers during World War II."
-'**Comfrey: Natures First Aid Remedy'** by Dr. Holly Lucille, N.D., R.N.; Better Nutrition Magazine: Healthy Living Guide, Active Interest Media, Healthy Living Group, El Segundo, California, 2014.   (World War II was from **1939** to 1945.)

*"Question from a reader: 'I broke my ankle 18 months ago and doing everything I can to avoid having surgery to put screws in. **I have heard from my uncle who was in the army in World War 2 that it was standard issue for first aid packs to include Comfrey tablets that used to be called Bone Knit.** Just wondering if you think Comfrey tea or something like that would be a good alternative?' "*

-'**Comfrey and Broken Ankle**' by Sally James and Nick Foley, Ask A Naturopath, www.askanaturopath.com, Australia, September 30 2017.

"***Pharmacognosy: Drugs Occurring in European Trade/Commerce:** Most Important, Their Identification, Adulteration, Use: A continuation of Part I dealing with **roots from Symphytum officianale L. (consolida root)**, bryonia, Carlena ncaulis L., curcuma, Carex arenaria L., cynoglossum, Dryopteris Filix-mas Schott, Sainbucus ebulus L., inula, galanga, gelsemium and gentian. Four adulterants are reported for galanga. Eleven illustrations are given. -Franz Berger, Scientia Pharmaceutica, 10 (1939), 71-75.*"
-'**Pharmaceutical Abstracts**' edited by A.G. DuMez, American Pharmaceutical Association, Washington, DC, Volume VI, No. 4, April **1940**, page 149.   (Pharmacognosy is the study of medicinal drugs obtained from plants and other natural sources.)
   (Scientia Pharmaceutica has been published in Switzerland since 1930.)

"***The decline in interest in Comfrey belongs to the period of the development of artificial fertilizers***, and many new discoveries in agricultural science. Comfrey, which is in the nature of a permanent crop, and does not like chemical fertilizers, was forgotten except for special applications such as had to do with the health of the stock.*"
-Comfrey: **Nature's Healing Herb & Health Food** by Andrew Hughes. Japan: Sanyusha Publishing Co., Ltd, 1992, page 124.
   (**In the 1940s in the United States new chemical processes were creating artificial potassium and phosophorus.** Phosphate rock was mined and processed to create superphosphate fertilizers. In later decades, it was replaced by triple superphosphate and ammonium phosphates.
   Nitrogen production started during World War II {1939-1945} because it is one of the main ingredients in explosives. Nitrogen produced took the chemical form of ammonia with the first fertilizer being ammonium nitrate. Anhydrous ammonia fertilizer was developed next.)

"***Vernon Stephenson planted his acre of Russian Comfrey in 1942, buying his plants from the late Kenneth Crawley of Lockerbie, Dumfriesshire County of Dumfries, England,** who grew a different set of variations from those I had at Southery, Norfolk, England- the mainly large-leafed, heavy-stemmed, Webster strain. His had a high proportion of a thin-stemmed, early variety with the same magenta flowers.*
   ***Among my (Lawrence D. Hills) fans was E.V. (Vernon) Stephenson of Hunsley House Stud, Little Weighton, near Hull, Yorkshire, England,** who wanted me to visit him to talk about Comfrey. By the time I visited him, 10 years later (in 1952), he had learnt a great deal about feeding Comfrey to both horses and pigs.*"
-Fighting Like the Flowers: An Autobiography: The Life Story of Britain's Best-Known Organic Gardener by Lawrence D. Hills. Bideford, Devon, England: Green Books, 1989, page 95.
   (The Stephenson strain of Russian Comfrey is Bocking No. 2 and 14. See Chapter 10.)
   (Lawrence Donegan Hills, 1911-1990, was one of Britain's best-known writers on organic gardening. For 8 years he was a gardening writer for the 'Observer', a weekly newspaper, London, England. He wrote for 'Punch' magazine, London, England, and 'The Countryman', Skipton, England. He was Associate Editor of the 'Ecologist and Compost Science' in the United States. His many publications included 'Fertility Without Fertilisers', 'Down to Earth Gardening', 'Organic Gardening' and 'Grow Your Own Fruit and Vegetables'.)

In **1948 Lawrence D. Hills** started researching Comfrey at Convent Lane, town of Bocking and Braintree, Essex County, England. "***I (Lawrence D. Hills) planted the first Comfrey in my life in February, 1949.** Hodson, the best tractor ploughman on Manor Farm, Southery, Norfolk, England, deep-ploughed, cultivated and harrowed the five acres of good, light loam for the new nursery. He, Eric, and I put in the hundred Comfrey offsets (roots) from Websters Nurseries, three feet (0.9 meter) apart along the 320-foot (97 meter) furrow.*
   *The furrow we had planted with three-inch-long (7.6 cm), brown-barked root sections, each with a growing point, had become **a three-foot-high row of inch-thick (2.5 cm) stems and small, purple, hanging bell flowers above clumps of large leaves up to 30 inches (76 cm) long.***
It was now mid-June *and the leys of pedigree Aberystwyth grasses were at their best and, as Ted said, 'If Strawsons' cows will eat Comfrey when they have all the good grass to go for, that'd be a fair test.'*
   *Ted scythed 50 plants at ground level. We loaded the sacks on the cart, and Eric drove us from field to field. Old Mr. Strawson was amazed at the spectacle of his Friesians (Dutch cattle) heaving themselves up to come across for the Comfrey. **A single cow would come over to the cart out of curiosity, and then the others would follow and clear the ration we threw in.***
   The next day the cows were waiting for the cart which we took round at roughly the same time in the morning, and finished all the Comfrey before they went back to the grass. The pigs ate it most greedily."
-Fighting Like the Flowers: An Autobiography: The Life Story of Britain's Best-Known Organic Gardener by Lawrence D. Hills. Bideford, Devon, England: Green Books, 1989, pages 82, 87.
   (Ley farming is growing grass or legumes in rotation with grain or tilled crops for soil conservation.)
   (Scythe is a tool with a long, curving blade fastened at an angle to a handle used to cut grass, grain, etc., by hand.)

   "*Comfrey is unique, not only in its yield or its medicinal qualities, but in the bitter and totally unscientific hatred it can arouse in those who dislike it and the wild enthusiasm it stirs in its supporters.*
      Following Henry Doubleday and Thomas Christy came Kenneth Crawley and Vernon Stephenson; now Comfrey had taken a grip on my life."

-**Fighting Like the Flowers: An Autobiography:** The Life Story of Britain's Best-Known Organic Gardener by Lawrence D. Hills. Bideford, Devon, England: Green Books, 1989, page 103.

In 1949 in "Veterinary Pharmacology, Materia Medica and Therapeutics" by Howard Jay Milks and Alexander Zeissig, Chicago, Illinois, it recommends allantoin skin applications of up to 2% for animals.

"The 1949 'Veterinary Pharmacology' describes glyoxylic acid (allantoin) as a white crystalline powder which dissolves slowly in cold water but easily in hot and can be made synthetically instead of being extracted by herbalists from the dried roots or leaves of Comfrey."
-**Comfrey: Symphuo Symphytum: A Multi-Purpose Herb** by Philip Clarke. Edinb, Scotlandgland: The Pentland Press, 1997, page 2.
(Allantoin is a diureide of glyoxylic acid. A diureide is one of a series of complex nitrogenous substances that has two molecules of urea or their radicals, as uric acid or allantoin.)

"Out of the profits of his organic farm, (Frank) Newman Turner ran 'The Farmer', the first and perhaps the best organic farming publication among the many that sprang up in the post-war years (World War II ended 1945).
It had a section called 'The Gardener' where my (Lawrence D. Hills) first article on Comfrey was published (probably around 1950). This drew enquiries for plants from organic farmers.
I (Lawrence D. Hills) said in an article in 'The Farmer' that I had seen Comfrey grow 'with the muscular, thick-necked fury of the kind of bull who needs watching.'
I got into all the gardening papers, including 'The Smallholder' which took one article on Comfrey that started us digging up more of the original stock plants. Whenever I wrote a Comfrey article my fan mail soared."
-**Fighting Like the Flowers: An Autobiography:** The Life Story of Britain's Best-Known Organic Gardener by Lawrence D. Hills. Bideford, Devon, England: Green Books, 1989, page 91, 92, 95, 97.
(After World War II, Frank Newman Turner bought Goosegreen Farm near Bridgwater, Somerset, England. In 1946 he started the quarterly magazine, 'The Farmer: The Journal of Natural Farming and Living', a magazine about organic farming and healthy living 'published and edited from the farm'.)
('The Smallholder' magazine was founded in 1910 for small farmers, allotment holders and gardeners. It is published monthly by Packet Newspapers in Falmouth, Cornwall, England.)

"The late Mrs. P.B. Greer was one Russian Comfrey customer. She was the first to buy 4,480 plants and put in a whole acre. Others were: Ian G. Macdonald of Goudhurst, Kent, England, who put in a thousand for his pigs. Jim Lackenbry of Middlesborough, Yorkshire, England, also breeding pigs. Paul Weir of Alton, Hants, Hampshire, England. He hoped that Comfrey would provide protein."
-**Fighting Like the Flowers: An Autobiography:** The Life Story of Britain's Best-Known Organic Gardener by Lawrence D. Hills. Bideford, Devon, England: Green Books, 1989, page 91.
(**Mrs. Peggy Greer** was principal of Holmes Chapel Agricultural College, Cheshire County, England. She had 800 acres at Layer-de-la-Haye, near Colchester, Essex county, England. **She started growing Russian Comfrey in 1952.** She used Comfrey to cure sick livestock especially those with diarrhea that she bought at Colchester market. If you know more about her involvement with Comfrey, please let me know.)

"Not only did Mrs. Greer cultivate crops for compost and animal feeds, she revived Comfrey for herbal medicine and kitchen use. On the farm three acres of Russian Comfrey were grown for compost purposes and animal feeds, while half an acre was devoted to a crop for herbal medicine and food.
Comfrey crops were cut, cleaned and dried, and on processing sold in the form of Comfrey tea, tablets, ointment and in powder form."
-**About Comfrey: The Forgotten Herb** by G.J. Binding, M.B.E., F.R.H.S. Wellingborough, Northhamptonshire, England: Weatherby Woolnough, 1974, page 53.

*"The most recent orthodox (established or approved) medical reference to Comfrey is in the 23rd edition of '***The Extra Pharmacopoeia***' (Pharmaceutical Press, London, England, 1952), prescribed as dried root."*
-**Comfrey: The Herbal Healer** by Henry Doubleday Research Association, 1975, page 23. (Included in 'Comfrey Report' book.)

*"**Newman Turner used his Comfrey in his 'alternative veterinary medical treatments' described in his 'Herdsmanship' (book, 1952),** especially for the sterility then common in Jersey cows forced to milk yields beyond their capacity by excessive feeding with concentrates. He used and advised fasting, then Comfrey ad lib (as much as they want).*
His Comfrey plot won the record British yield in 1955 with 67 tons 16 cwt. (hundredweight) an acre, which is still unbeaten."
-**Fighting Like the Flowers: An Autobiography:** The Life Story of Britain's Best-Known Organic Gardener by Lawrence D. Hills. Bideford, Devon, England: Green Books, 1989, page 114.
(Newman Turner grew his Russian Comfrey at Ferne Farm in Shaftesbury, Dorset, England.)
(Herdsmanship: A Guide for the Herd Owner, Herdsman and Cowman by Newman Turner. Austin, Texas: Acres USA, 1952. Excerpts are in the livestock parts of my book.)
(Feeds classified as concentrates include cereal/legume granular feeds, mixed feed concentrates, and animal feeds.)

Thomas Gaskell Tutin, 1908-1987, was Fellow of the 'Royal Society of London' and Professor of Botany at the University of Leicester, England. With A.R. Clapham and E.F. Warburg, he wrote **'Flora of the British Isles'** in 1952. He was editor of **'Flora Europaea'**, a 5-volume encyclopedia of plants, published between 1964 and 1993. **He wrote about the Symphytum genus with works published from 1952 to 1980.**
For more about Tutin, see sub-subsection 'Later Botanical History of Symphytum: Botanists' in subsection 'Symphytum Comfrey Genus' in section 'Borage Family, Symphytum Genus' (Chapter 2).

Book first published in 1952: 'The Complete Herbal Handbook for Farm and Stable' by Juliette de Bairacli Levy in England. She was a pioneer of herbal veterinary medicine with natural and organic cures and farming methods. It is a classic in its field. **She advises using Comfrey to help heal animals.**

"The best botanical library in London, England, is Kew Gardens at Richmond in Surrey county. **It was in the Lindley Library, in October 1952, that I (Lawrence D. Hills) discovered Henry Doubleday (1813-1902) who introduced the crop we know as 'Comfrey' in the 1870s** and in whose honour I named the Association (Henry Doubleday Research Association) in 1954, when I founded it.
I began by plodding through the early numbers of the **'Gardeners Chronicle and Agricultural Gazette'** which was started in 1841. On April 24th of that year, 'Constant Reader' wrote a letter praising Comfrey as a cattle feed, to be followed by others attacking the crop.
This went on year after year, until in a letter on October 24th 1885, **Henry Doubleday of Coggeshall, Essex, England, claimed he had harvested 100-120 tons an acre from between six and eight cuts a season.**"
-**Fighting Like the Flowers: An Autobiography:** The Life Story of Britain's Best-Known Organic Gardener by Lawrence D. Hills. Bideford, Devon, England: Green Books, 1989, page 98.
> (Kew Gardens is a botanical garden in London, England that houses the 'largest and most diverse botanical and mycological collections in the world'. The library has 750,000 volumes, and the illustrations collection has more than 175,000 prints and drawings of plants.)
> (Richmond was part of Surrey county until 1965 when it became part of Greater London.)
> (The Lindley Library is the largest horticultural library in the world. In London at the headquarters of the Royal Horticultural Society. It was started with the book collection of English botanist John Lindley. It has rare books dating to 1514.)
> ('The Gardeners' Chronicle' was a British horticulture periodical in publication for almost 150 years with its name changing about every 10 years. It still exists as part of the magazine 'Horticulture Week'. It was founded in 1841 by horticulturists Joseph Paxton, Charles Dilke, John Lindley and the printer William Bradbury.)

"On the medicinal side I (Lawrence D. Hills) had the help of **Dr. Ronson of the 'Pharmaceutical Society of Great Britain'**, who gave his opinion that 'Comfrey was the best weapon the past had against disease or injury'.
Arrows made clean cuts, but if you had one in the abdomen you would have a far better chance fasting and drinking Comfrey tea than being operated on without antispetics or anaesthetics.
If 'burseten' was a perforating appendix you would have no chance anyway.
The librarian of the 'Royal College of Veterinary Surgeons' was helpful, and the early numbers of their journal were full of references to its use for 'scour' (diarrhea) and as leaf poultices for injuries to horses. Of course, it was all 'anecdotal'."
-**Fighting Like the Flowers: An Autobiography:** The Life Story of Britain's Best-Known Organic Gardener by Lawrence D. Hills. Bideford, Devon, England: Green Books, 1989, page 99.

"**Expansion of Comfrey overseas started in 1953 when the first lot (from England) went to the African continent.**"
-**Comfrey: Nature's Healing Herb & Health Food** by Andrew Hughes. Japan: Sanyusha Publishing Co., Ltd, 1992, page 126.
> (The book was not clear as to the Comfrey species but probably it was Russian Comfrey.)

"**In this review, notes are given on the history of Russian Comfrey and its introduction into Canada as Quaker Comfrey in 1953.** Its agronomy and feeding value are described."
-**'Quaker Comfrey'** by L. Lachance, Canada Department Agricultural Research Station, Lennoxville, Quebec; Agriculture Montreal, Volume 25, No. 4, pages 29-34, 1968.
> (Agronomy is a branch of agriculture about field-crop production and soil management.)

"This hybrid was called Russian or Caucasian Comfrey (Symphytum x uplandicum) in reference to its country of origin. **Cuttings of this hybrid were shipped to Canada in 1954 and it was named Quaker Comfrey, after the religion of Henry Doubleday, the British researcher responsible for promoting Comfrey as a food and forage.**
The majority of Comfrey grown in the United States can be traced to this introduction."
-**'Comfrey'** by Teynor, Putnam, Doll, Kelling, Oelke, Undersander and Oplinger, 'Alternative Field Crops Manual', University of Wisconsin: Cooperative Extension, University of Minnesota: Center for Alternative Plant & Animal Products, and Minnesota Cooperative Extension, February 1992, updated November 1997.

"**In 1954 a few thousand root cuttings of Russian Comfrey were sent to Canada where it was renamed as Quaker Comfrey** to avoid any connections with Russia! Since then, Comfrey has been introduced as a fodder crop in Australia, New Zealand, Africa, Japan and the United States."
-**'Nature Study: Comfrey in the College Farm'** by Gift Siromoney, M.A., M.Sc., Ph.D., F.S.S., India; Madras Christian

College Magazine, Chennai, Tamil Nadu, India, No. xiv, pages 21-26, 1976.

In 1953 Lawrence Donegan Hills wrote 'Russian Comfrey: A Hundred Tons an Acre of Stock or Compost for Farm, Garden or Smallholding', London, England, 167 pages. This book is the primary source about Russian Comfrey. This book is out of print but is available online as a pdf.

> **"Russian Comfrey is another wonder plant, dear to the heart of the compost makers.**
> **If properly handled it can easily produce the hundred tons an acre referred to on the title-page of this enthusiastic book about it.**
> Without these prerequisites (prior conditions) it can lead to reports such as that from the Hannah Dairy Institute, who are cited as stating that 'owing to its very limited value for feeding purposes and the difficulties associated with its management and utilization, it is doubtful whether its cultivation on any considerable scale would be worth while.'
> Equally contradictory reports upon it are quoted from the farming and gardening press for over a century (100 years).
> **This uncertainty is partly accounted for by confusion as to the species referred to, several species exist:**
>> **Symphytum tuberosum which is a tetraploid (chromosome type) and is useless for fodder purposes; S. officinale, the most common British species, is said to be less valuable than Russian Comfrey which seems to be a hybrid, identifiable either with S. peregrinum or S. uplandicum.**
>
> In addition to this it has undoubtedly hybridized further with local species since its introduction to Great Britain towards the end of the eighteenth century (late 1700s). It is thus never certain to what plant the many reports, whether of success or failure, truly relate.
> **By its very nature the plant seems more suitable for the small holder than for extensive cultivation;** it must be reproduced from cuttings and rather carefully tended until established. It is perennial and needs very special systems of weeding to keep it clean. It needs constant cutting if it is to give the high yields quoted for it and, not being a legume, seems to require rather high manuring in order to achieve them.
> In consequence of all this, most research institutions have fought shy of it and **most of the reports about it lack the statistical exactitude demanded by modern science.**
> Comfrey remains therefore largely the subject of speculation and **the chief merit of Mr. Hills's book is in pointing out the sources of confusion in the existing information, explaining the many possible causes of failures in the past which it should now be possible to avoid in future experiments,** and in calling attention, often in very provocative (strong) and picturesque terms, to the undoubted possibilities, at least for certain special purposes, of this unusual plant, which is characterized, we are told, by 'a thick-necked muscular fury- that reminds one of the sort of bull that needs watching.' "

-'Russian Comfrey: A Hundred Tons an Acre' **Book Review** by unknown author, 1953.

> **"This book has been written by a man who obviously has great faith in Russian Comfrey and has gone to considerable trouble in tracing its history in this country (England),** in collecting such scattered information as there is, and committing his experiences and views to paper so that others can benefit from growing the crop if they so wish. There have been many conflicting and often disappointing reports from Comfrey growers and the enthusiasm for the crop, now reflected in the pages of this book, is due basically to the fact that **the author is concerned with a heavy-yielding strain of Symphytum peregrinum having magenta-pink flowers and an absence of wings on the flower stem.**
> This should not be confused with either S. asperrimum (=S. asperum, Rough or Prickly Comfrey) which has clear, bright-blue flowers, or S. officinale (English or Common Comfrey) which is found growing wild in Kent County, southeast England, and has cream or purple flowers.
>> Mr. Hills wrote: 'There is no record of any Comfrey of agricultural value with yellow, cream or white flowers and bright blue is associated with the lower yields.'
>
> **Since the first Symphytum sp. (species) was introduced to England from Russia in about 1770 the plant has been surrounded by an aura of extravagant claims, and this book clearly shows that there have been, and are, plants which will produce consistently high yields of leaf.**
>> Such strains- possibly the result of a natural hybrid- may have a future, particularly on land unsuited to lucerne (alfalfa), but one would have preferred the very high yields quoted to have resulted from well-designed and replicated trials.
>
> **Cultivation, management and the use of Russian Comfrey for stock-feeding, ensiling (making silage) and composting are all adequately and practically dealt with,** and the account given will undoubtedly act as a spur to those who like to try anything once."

-**Book Review** by L. Skidmore, England, 1954, nothing more is known about the reviewer.

> (A leaf wing or plant fin is a green, short, thin, paperlike structure extending along a stem. It is long along the stem but not tall away from the stem.)

"In 1954 Lawrence Hills founded the 'Henry Doubleday Research Association' (HDRA), now named 'Garden Organic', and invited Newman Turner to serve as its first president."
-**Fertility Farming** by Newman Turner. Austin, Texas: Acres USA, 1951, page xii.

> ('Henry Doubleday Research Association' was first based at Bocking near Braintree in Essex, England. In 1985 it moved to Ryton-on-Dunsmore near Coventry in the West Midlands, England. Its working name was changed to

'Garden Organic' in 2005, though the legal name is still the original name. 'Garden Organic' is a United Kingdom charity dedicated to researching and promoting organic gardening, farming and food. It maintains the 'Heritage Seed Library' to preserve vegetable seeds from heritage cultivars. www.gardenorganic.org.uk)

"Then Newman Turner had an idea- the **'International Comfrey Race of 1954'. The contestants were farmers, cutting and weighing their Comfrey yields all over Britain.** The 'race' began in the spring issue of 'The Farmer' in 1953 when Russian Comfrey was first on sale. One of the earliest entrants in the race was **Mrs. P.B. Greer.**
A plot in Kenya (east Africa) won in 1955 with the still unbroken record of 124 tons, 10 cwt. (hundredweight) and 101 pounds in 12 monthly cuts."
-Fighting Like the Flowers: An Autobiography: The Life Story of Britain's Best-Known Organic Gardener by Lawrence D. Hills. Bideford, Devon, England: Green Books, 1989, page 104.

In 1955 Lawrence D. Hills wrote 'Comfrey Report No. 1: The 1954 Research Results', an annual report of the 'Henry Doubleday Research Association', 35 pages.
"**Yield figures are given for individual and total cuts from single plots of Russian Comfrey** (Symphytum peregrinum), 13 of which are located in Great Britain and 5 overseas. The calculated average yield in 1954 from 9 second-year plots in Great Britain was approximately 25 tons green matter per acre.
Accounts are given of exploratory crop-conservation methods suitable for smallholders.
Tables giving the chemical composition of dried and fresh Russian Comfrey show that the plant is relatively high in protein and low in fibre. General information on the agronomy of the crop is also given in the report."
(I do not have this report. If you have a copy, could you please send it to me.)

"**When the full results of the 1954 Comfrey race were in from Britain and overseas, I wrote the very first H.D.R.A. (Henry Doubleday Research Association) publication: 'Russian Comfrey Report Number One'.** This had 38 pages, a plain green cover, and was lavishly illustrated with blocks of my photographs from 'The Farmer' publication. I gave detailed accounts of the competitors in the Comfrey race and how they were using Comfrey on their farms, visiting many of the British plots."
-Fighting Like the Flowers: An Autobiography: The Life Story of Britain's Best-Known Organic Gardener by Lawrence D. Hills. Bideford, Devon, England: Green Books, 1989, page 114.

"**To secure more definite information regarding cultivation of the crop Russian Comfrey, and the methods for conserving and feeding it, a number of trials were carried out in 1954** in Great Britain and overseas under the aegis of the Henry Doubleday Research Association, Bocking, Braintree, Essex, England.
The conclusions of these trials have been published by the Association under the title of **'Russian Comfrey Report No. 1: The 1954 Research Results'** by Lawrence D. Hills.
As with other crops, success depends on good management, and the report includes recommendations for its cultivation, stressing the need for keeping the land clean, well manured, and the crop regularly cut.
The most profitable stock to feed with Comfrey is pigs and poultry, though horses also appear to thrive on it. **The plant might prove useful to the smallholder or poultry keeper.**"
-'Russian Comfrey as a Crop', Nature: Nature Publishing Group, No. 4460, Volume 175, page 711-712, April 23 1955.

"So we had a full set as we could of what was being grown as (Russian) 'Comfrey' in Britain.
As I slowly sorted these, through 1955 and 1956, I gave them 'Bocking Numbers', and found that each of the four mixtures had a dominant variation."
-Fighting Like the Flowers: An Autobiography: The Life Story of Britain's Best-Known Organic Gardener by Lawrence D. Hills. Bideford, Devon, England: Green Books, 1989, page 114.
(For more about Bocking Numbers, see section 'Details about Symphytum Species Hybrids: Russian Comfrey' {Chapter 10}.)

"**Dr. M.L. Peterson, Chairman Department Of Agronomy, University of California, while in England in 1956, checked with English agricultural experiment station workers regarding the merits of Comfrey. Not one of them recommended Quaker Comfrey.** The reasons given were:
 (1) It did not yield the heavy amounts of forage in commercial plantings it was reputed to. Most of the published reports are by private individuals apparently based on calculated per acre yields from a few spaced plants.
 (2) The crop is expensive to propagate.
 (3) The forage is not as palatable as other commonly used grazing and hay species.
 (4) It must be regularly cultivated for weed control.
-'**Quaker Comfrey**' by Milton D. Miller and Lloyd Harwood, Agronomy Progress Report No. 2, Agricultural Experiment Station, Agricultural Extension Service, University of California, Davis, March 1958.

"**Russian Comfrey reached Japan in the late 1950s via our Member Mr. Andrew Hughes of Tokyo, Honorary Secretary of the 'Japanese Comfrey Growers Association'.**
The plant they had is the mixture grown as 'Quaker Comfrey' in Canada and the United States, first imported from Australia where Mr. W.A. Savage of Red Hill South, Victoria, had his stock from our member **Mrs. P.B. Greer** of 'Kerry Cow Dairy Farms', Layer-de-la-Haye, near Colchester, Essex, England, and some imported from her direct.

It is the same stock that has grown 124 tons an acre in New Zealand (North Island) and in Kenya (east Africa), where a longer day length brings a longer growing season.
It is a re-selection of the old 'Webster' strain made by the writer (Lawrence D. Hills) at Southery, Norfolk, England *in 1948, which began modern Comfrey growing."*
-**Comfrey Report No. 3: Feeding Dairy Cattle in Japan'** by Meiji Milk Producing Co., Tokyo, Japan for Henry Doubleday Research Association, Braintree, Essex, England, 31 pages, July 1964.

(Andrew Hughes wrote 'Comfrey: Nature's Healing Herb & Health Food'. Japan: Sanyusha Publishing Co., Ltd, 1992.)

*"**Queensland, Australia Trials on Russian Comfrey: Russian Comfrey, a new fodder crop to Queensland, is being tested under field conditions by the Department of Agriculture and Stock.** Later this year, large-scale trial plantings are planned for the Darling Downs, Calide Valley and Atherton Tableland.*
Three varieties introduced by the C.S.I.R.O. *have been planted by the Department at the Redlands Horticultural Experiment Station and these varieties-* **Bocking 4, Bocking 14 and Bocking 17-** *are being multiplied to provide planting material for larger trials. In proposed bigger plantings at the Hermitage, Biloela and Kairi Regional Experiment Stations,* **Russian Comfrey will be tested in both yield and feeding trials** *during the coming season."*
-'Queensland Agricultural Journal' edited by C.W. Winders, B.Sc.Agr., Queensland Department of Primary Industries, Brisbane, Australia, Volume 83, July to December 1957, No. 7 to No. 12. Article: 'Queensland Trials on Russian Comfrey', Volume 83, No. 11, page 624, November 1 **1957**.

*"**Strains of Russian Comfrey imported from England have been grown under quarantine conditions and are now being multiplied in the field prior to general release in Australia.**"*
-'**Ninth Annual Report of CSIRO** (Commonwealth Scientific and Industrial Research Organisation) for Year 1956-1957' by Parliament of the Commonwealth of Australia, Canberra, Australia, 173 pages, **1957**.

*"**Plant Introduction, Agronomic Trials, Division of Plant Industry, south-eastern Australia:**
Three selected strains of Russian Comfrey tested at Canberra (Australia) showed a marked difference in productivity but withstood summer drought conditions.*
All strains recovered rapidly after cutting at 30-day intervals. **These strains have now been released for further testing throughout Australia.**"
-'**Tenth Annual Report of CSIRO** (Commonwealth Scientific and Industrial Research Organisation) for Year 1957-1958' by Parliament of the Commonwealth of Australia, Canberra, Australia, 173 pages, **1958**.

"**Quaker Comfrey, also called Russian Comfrey (Symphytum peregrinum)** *belongs to a large family of mucilaginous and slightly bitter plants which includes heliotrope, hound's tongue, stickseed, and Gromwell. A similar plant, Prickly Comfrey, received a considerable amount of publicity as a forage plant around the turn of the century (late 1800s).*
Researchers at the USDA (United States Department of Agriculture) and several state experiment stations, after studying this plant, concluded that it did not have a place as a forage plant in the United States. *The green-matter yields were high under favorable conditions, but the quality and dry-weight yields were inferior to those of our common forage plants. Silage was of poor quality because of the watery, gummy nature of the leaves.*
Recently the interest shown in Quaker Comfrey has been scattered. It is reported to be higher yielding than Prickly Comfrey, yielding up to 120 tons of green matter per acre. However, further checking of the literature suggests that these yields were obtained from a few plants grown under garden conditions.
Quaker comfrey is propagated from root cuttings and is usually spaced on three foot (0.9 meter) centers. For optimum yields the crop should be cultivated and grown in soil with a high level of fertility. Cost of planting, disadvantages of grazing spaced plants, and unsuitability for hay or silage would seem to **limit its use to soiling.**
Experimental plantings in the United States have shown that growth during the year of establishment has been modest, and indications are that cattle and swine will have to be induced to eat it.
No recommendations can now be made. Experimental plantings have been made at several locations in the United States; however, they have not been established long enough to warrant definite recommendations.
Dr. Hittle established a small planting of this crop on the 'Agronomy Farm' in May 1956. A small patch (15 plants spaced 3 feet each way) was harvested on June 7, 1957. When expanded to an acre basis, the yield was 26.6 tons of green forage per acre. The plants averaged only 14 percent dry matter; therefore, on a dry-weight basis the yield was 3.73 tons per acre. This was the only harvest made in 1957.
No feeding trial has been conducted and no chemical analysis was made of the crop."
-'**Agronomy News, Crops and Soils: Sorghum Almum and Quaker Comfrey**' by College of Agriculture, University of Illinois, Urbana; 'American Society of Agronomy' (Madison, Wisconsin), 'Crop Science Society of America', and 'Soil Society of America', 1956-1973. Comfrey article is No. 83, May 26, **1958**.
(Based on reports by J.R. Harlan, Oklahoma Experiment Station; E.M. Trew, Texas Extension Service; and Crops Research Division, Agricultural Research Service, U.S. Department of Agriculture.)
(It is possible to successfully make Comfrey silage. See subsection 'Comfrey as Silage' in section 'Comfrey Meal, Pellets, Hay and Silage' in Volume 2.)   (Soiling is a crop cut green and fed to livestock immediately without curing or processing.)

"*A hardy perennial that has produced more than 100 tons per acre annually is the description given* **'Quaker Comfrey' (Russian Comfrey) *by a California concern which is selling the root stock.***

The plant is propagated by root division only, and W.S. Curtis and Associates of Berkeley, California, explain that regulations requiring fumigation, which is deadly to the roots, has hitherto prevented its entry into this country from Canada, where some quantity is available.
Once established, according to the Curtis report, it will live for 20 years and longer, given reasonable care."
-**'The Florida Cattleman and Livestock Journal'** edited by Robert S. Cody, Volume XXII, No. 8, Publisher Aldus M. Cody, Kissimmee, Florida, May **1958**. Article: 'New High Protein Fodder and Silage Crop Reported', pages 13-14.
(See section 'Age of Revolutions and Comfrey, 1800s' {Chapter 15}. 'The Curtis Botanical Magazine or Flower-Garden Displayed' began in 1787. It is the longest running botanical magazine in the world. William Curtis, 1746-1799, was an apothecary and botanist in London, England.' Probably W.S. Curtis and Associates are related to William Curtis.)

*"**Elmer Deetz (1894-1972) of Canby, Oregon**, became the Vernon Stephenson of America, buying his Russian Comfrey stock from **Bodie Seeds Ltd. of Canada**.*
***A great many other growers began to sell Comfrey through the 1960s**. Their advertisements blossomed in 'Organic Gardening and Farming', the one organic movement periodical in the world, with a circulation of 850,000 a month."*
-**Comfrey: Fodder, Food and Remedy** by Lawrence D. Hills. New York: Rizzoli Universe Books: 1976, page 218.

*"Among the letters that poured in was one from **Mr. A.H.A. Lasker of Bodie Seeds Ltd. of Winnipeg, Canada, who wanted to buy 10,000 Russian Comfrey plants**. It was, as far as I know, the only modern Russian Comfrey to reach Canada or the United States.*
***A nurseryman named Elmer Deetz bought a large load from Bodie Seeds in Canada and drove them down to Oregon** over the unguarded frontier.*
Almost all the Comfrey in the United States has grown from that one stock."
-**Fighting Like the Flowers: An Autobiography:** The Life Story of Britain's Best-Known Organic Gardener by Lawrence D. Hills. Bideford, Devon, England: Green Books, 1989, page 104.

*"226543. **Symphytum Officinale L. Boraginaceae. Comfrey.***
***From Canada. Roots presented by Bodie Seeds, Limited, Winnipeg. Received June 17, 1955.**"*
-**'Plant Inventory No 163: Plant Material Introduced January 1 to December 31 1955'** by Eugene Griffith, Botanist, Plant Industry Station, Beltsville, Maryland; United States Department of Agriculture, Washington, DC, June 1964. Comfrey page 136.

*"**Russian Comfrey was brought into Canada in 1953 by Bodie Seeds, Limited, Winnipeg 2, Canada.***
***It was introduced into California shortly thereafter into Sonoma County**. Farm advisor Lloyd Harwood, Agricultural Extension Service, Santa Rosa, California, has been following the progress of plantings in that county."*
-**'Quaker Comfrey'** by Milton D. Miller and Lloyd Harwood, Agronomy Progress Report No. 2, Agricultural Experiment Station, Agricultural Extension Service, University of California, Davis, March 1958.

*"**An observational test was established in 1956 in response to the current interest in Quaker Comfrey. Planting stock is being distributed by the Bodie Seeds, Limited, Winnipeg 2, Canada.***
Promotional literature distributed by Bodie Seeds, Limited, contains reported yields of up to 120 tons of green forage per acre. Quaker Comfrey is propagated by root cuttings. It should be planted on a fertile soil and cultivated to control weed competition. In preliminary palatability studies it was refused by hungry dairy animals and brood sows."
-**'Twentieth Annual Report of Forage (Pasture) Research in the Northeastern United States'** by United States Regional Pasture Research Laboratory, Forage and Range Section, Crops Research Division, State College, Pennsylvania, 1956. Article: 'Observations on Quaker Comfrey {Symphytum Peregrinum}' by A.A. Hanson, page 37.

*"**Our first boom in Japan in Russian Comfrey growing and usage came through 1961**, and continued to move upward thorugh 1962. Comfrey went to both **India and Ceylon (Sri Lanka) from Japan**.*
*So great and growing was the interest that it became necessary to import more offsets (roots) from Australia and England. In the spring of 1962 we obtained further supplies from **Foster Savage of Victoria, Australia**.*
*We also sent for a large quantity of offsets from one original source for all supplies, a champion grower in England, **Mrs. P.B. Greer. Mrs. Greer's farm was in Essex (England)**, the same county where Henry Doubleday had grown his Comfrey and where the experimental research farm of H.D.R.A. (Henry Doubleday Research Association) was then located.*
***It was from this farm that most of the champion stock of Comfrey offsets went overseas, the Australian Comfrey, Kenya Comfrey, New Zealand Comfrey**, all Webster's Giant strain of Symphytum peregrinum."*
-**Comfrey: Nature's Healing Herb & Health Food** by Andrew Hughes. Japan: Sanyusha Publishing Co., 1992, pages 126, 127.
(Webster Strain is Russian Comfrey Bocking No. 3, 4, 5, 6, 7, 8, 9, 10, 11, and Bocking Mixture. There are minor variations in Webster's Giant strain, but 60% to 80% of it is Bocking No. 4.)

*"**Andrew Hughes' (Comfrey: Nature's Healing Herb & Health Food) first knowledge of Comfrey came in 1956, from Foster Savage**: a great advocate of the herb in agriculture, who at that time was farming in Victoria, Australia. **Once Andrew settled in Japan, he arranged, with the Japanese Department of Agriculture, to import 50 plants from Foster Savage**. This was the beginning of a 'great adventure', with Comfrey, in Japan: with extensive trials, testing, and interest, by the Japanese Government.*

When Andrew went to Japan, in 1957, he had little thought that Comfrey was to become a matter of major importance to Japanese agriculture. He did not plan it that way, but circumstances provided the opportunity. He was invited to speak to farmers, schools and research groups. Farmers quickly realised the value of Comfrey, for it added, enormously, to productivity.
Andrew was interviewed by the Yomiuri Shimbum daily press, who saw the potential of Comfrey and with a vision, planted up a vast park-playground-golf course they owned, on the borders of Tokyo.
Once established, they gave away over 1,000,000 Comfrey offsets (root divisions) to schools, agricultural co-operatives and private people, who wanted to grow and test the plant.
Comfrey was not only fed in large quantities to animals producing high-quality table meat, but it was also incorporated into many items of cuisine."
-**How Can I Use Herbs in My Daily Life**: Over 500 Herbs, Spices and Edible Plants: An Australian Practical Guide to Growing Culinary and Medicinal Herbs by Isabell Shipard. New York, New York: David Stewart Publisher / Simon & Schuster, Inc., 2003.

"By 1966 over 11,000 Japanese farmers were cultivating (Russian Comfrey) crops which they used on their farms as a fodder for cattle."
-**About Comfrey: The Forgotten Herb** by G.J. Binding, M.B.E., F.R.H.S. Wellingborough, Northhamptonshire, England: Weatherby Woolnough, 1974, page 53.

"Mr. and Mrs. Foster Savage and their family of 10 at Red Hill South, Victoria, Australia: Mr. Savage said:
'David, our number 10, is the healthiest of all the children, and we feel this is largely due to the fact that my wife drank a glass of Russian Comfrey juice every day during her pregnancy.
The drink is made from leaves of. the Comfrey plant, mixed with water, honey, and lemon juice, it is delicious, and we have no trouble getting the children to drink it.'
Mr. Savage pioneered the production of Comfrey in Australia when he imported the plant from England some years ago. Following its success here, Comfrey was widely grown in Japan after an Australian importer living in Tokyo ordered an initial 200,000 roots from Mr. Savage."
-'**The Australian Womens Weekly**', Sydney, Australia, August 31 1966. Article: 'Australian Family Will Migrate to Spain' by Beverley Cooper, page 27.

Published in 1962, the book 'Silent Spring' by Rachel Carson documented the adverse effects on the environment of the careless use of pesticides. Carson accused the chemical industry of spreading false information and public officials of accepting industry claims without question.

"Rudolph Steiner (1861-1925) and Sir Albert Howard (1873-1947) began the compost and fertility aspects of the organic movement; Sir Robert McCarrison (1878-1960), Weston Price (1870-1948), and Dr. Franklin Bicknell added the nutritional and food additives angle that altered the way we think about food.
Lady Eve Balfour (1898-1990), Newman Turner (1913-1964) and Friend Sykes pioneered organic farming books. Rachel Carson (1907-1964) awoke the world to the dangers of pollution by pesticides, fungicides and all the chemicals that add up to danger in our bodies and through the food chains of the world."
-**Fighting Like the Flowers: An Autobiography:** The Life Story of Britain's Best-Known Organic Gardener by Lawrence D. Hills. Bideford, Devon, England: Green Books, 1989, page 136.
(Friend Sykes is author of 'Humus and the Farmer' written 1946.)
(Lady Eve Balfour wrote a popular book about organic agriculture, 'The Living Soil', in 1943. She also was the founder in 1946 of 'The Soil Association', Britain's leading organic food and farming organization.)

*"**Symphytum officinale** Linnaeus, Sp. Pl. 1: 136. 1753. In Chinese: Ju He Cao.*
This species was introduced in China in 1963 as green forage for livestock."
-**Flora of China,** Volume 16 {Gentianaceae through Boraginaceae}, edited by Z.Y. Wu, P.H. Raven and D.Y. Hong. Science Press {Beijing, China} and Missouri Botanical Garden Press {Saint Louis, Missouri}, 1995. Article: 'Boraginaceae' by G.L. Zhu, H. Riedl and R.V. Kamelin, pages 329-427. Symphytum page 359. With 22 volumes it is the first modern English-language description of 31,000 species of vascular plants of China.

*"**Two pyrrolizidine alkaloids, symphytine, a new compound, and echimidine have been isolated from the dried roots of Symphytum officinale (Common Comfrey)."***
-'**Studies on Constituents of Crude Drugs. I.** Alkaloids of Symphytum Officinale Linn.' by T. Furuya and K. Araki from Kitasato University in Japan; Chemical & Pharmaceutical Bulletin, Volume 16, No. 12, pages 2512-2516, **1968**.
(**This was the first time alkaloids were found in Comfrey.** For more details about this report, see number 2 in the subsection 'Studies Show Dangers of Comfrey' in the section 'Alkaloids' in Volume 2.)

In 1975 Lawrence D. Hills wrote 'Comfrey Report: The Story of the World's Fastest Protein Builder and Herbal Healer', Conservation Gardening and Farming Series: Series C, Reprints, printed by the Henry Doubleday Research Association, England, 139 pages. It includes 'Comfrey: The Herbal Healer'.
*"In 1955 we published **'Comfrey Report No. 1'**, long out of print, like my book 'Russian Comfrey' (Faber and Faber, 1953),*

which began modern interest on this fastest growing perennial fodder crop in the world.
In March, 1963 the work of the 'Henry Doubleday Research Association' made it possible to sum up the progress we had made, in **'Comfrey Report No. 2'**.
In July 1964 in **'Comfrey Report No. 3'** Andrew Hughes translated the Japanese results with dairy cattle that led to a boom in Comfrey growing in Japan.
In August 1966 we issued **'Comfrey Report No. 4'** which dealt mainly with Comfrey for poultry keepers in Britain and overseas.
In August 1969 **'Comfrey Report No. 5'** carried the story further.
This small book (Comfrey Report) is a selection from these reports, with 'Comfrey Report No. 6' printed as an appendix, for it deals with the possibility of a toxic alkaloid in Comfrey, and how that was removed from medicinal aspects of Comfrey by the work of the British Medical Research Council and Dr. Denys Long, Ph.D.

>I am not a doctor of medicine; my knowledge is of Comfrey and how to grow it, feed it, and use it in the garden. Therefore this present works excludes its medicinal aspects, which I trust will some day receive the serious research which they deserve."

(The book 'Comfrey Report' is out of print. If you have Comfrey Report No. 1, No. 4 or No. 5, I would appreciate a copy.)

In 1976 Lawrence D. Hills wrote 'Comfrey: Fodder, Food and Remedy', 253 pages.
From the inside cover: *"**The author is one of the world's leading exponents (promoters) of organic gardening methods and director of the 'Henry Doubleday Research Association'**, which specializes in research on the cultivation of Comfrey. He is a well-known horticultural journalist and the author of many successful books on gardening."* (This book is out of print.)

*"**This is a handy reference covering history, growth, cultivation, analysis, and use of Comfrey (Symphytum officinale and other species and strains) as food and medicine for animals and man.**
It is written by Lawrence D. Hills of Braintree, Essex, England, under the auspices (patronage) of the 'Henry Doubleday Research Association', which the author founded in 1954.
He published an earlier account on Comfrey in 1953 ('Russian Comfrey'), and this book is an update (the author refers to it as still only an interim {provisonal} report).
The chapters dealing with history, growth, and analysis are well written, and several excellent charts and tables are given on documenting such data as yields of plant per acre, protein, amino acid and mineral analysis, various strains used, comparative analysis of Comfreys and other fodder crops, and nutritive data for various animals.
The use of the plant as an inexpensive source of high quality protein for various animals (horses, calves, pigs, etc.) is documented by experiments at various locations around the world."*
-'**Book Review'** of Comfrey: Fodder, Food and Remedy, by Ara Der Marderosian, Philadelphia College of Pharmacy and Science, Pennsylvania; Economic Botany, Volume 33, Issue 1, page 62, January 1979.

*"Comfrey: Fodder, Food and Remedy:
The world's leading authority shows how Comfrey can be cultivated to produce a crop rewarding for the small gardener as well as the largest mechanized farmer, and can be used for the benefit of everyone.
Comfrey is the fastest known builder of vegetable protein. In an age of increasing protein prices, Comfrey's extraordinarily high yield makes it economic for feeding a wide range of livestock. **As a food for cattle,** it has long had a reuptation for producing milk, rich in both quality and quantity.
As a blood purifier, it is famous for its beneficial effects.
And because, in terms of labor, it is probably the cheapest source of Vitamin A, it is being **widely used by poultry farmers,** who save on the purchase of expensive meal without a fall in egg production.
New processes employing yeasts and bacteria are making Comfrey of still more importance to hungry humanity.
Comfrey is not only grown as a kind of 'instant compost' to provide the potassium that every organic gardener needs; it is also the only plant known to take Vitamin B-12 from the soil, and this vital vitamin, missing from an all-vegetable diet, can be concentraed with the protein.
For a thousand years or more, because of its 'gluey' quality, Comfrey was known as 'bone set' or 'knit bone'; it was mentioned in most herbals and in various pharmacopoeias as a remedy for healing wounds.
>Now, in a chapter on the medicinal uses of Comfrey, Dr. Charles MacAlister shows that the allantoin in Comfrey promotes the growth of healthy cells.

A chapter on 'Comfrey for Human Food' includes recipes for preparing Comfrey as a vegetable, in soups, and as a tea; and instructions are given for making Comfrey poultices."*
-**Fertility Without Fertilizers: A Basic Approach to Organic Gardening** by Lawrence D. Hills. New York: Universe Books, 1975.

*"Comfrey is a plant of much folklore, and as such, is particularly relevant to today's search for 'back to nature' experiences.
There are no industrial giants or philanthropic (charitable) organizations concerned with Comfrey study. Many of the experiments discussed in the book are of a 'back yard' variety. The author recognizes this.
However, we are in a period when people have a rekindled interest in natural products.
This book might inspire research in many directions, all of it needed. And the book is enjoyable to read- I recommend it."*

-'**Comfrey: Fodder, Food and Remedy- Book Review**' by Fred Rickson, Oregon State University, Corvallis; The American Biology Teacher, Volume 38, No. 9, page 558, December 1976.

"The Henry Doubleday Research Association (HDRA) grow and market Comfrey (Symphytum officinale) in the United Kingdom.

*Pyrrolizidine alkaloids, which are present in many varieties of plant but particularly those of the Senecio species, have been identified in Symphytum officinale by Furuya and Araki, and in other species of **Comfrey** by Dr. C.C.J. Culvenor and colleagues of the 'Commonwealth Scientific Industrial Research Association' in Melbourne, Australia, and by Pedersen in Denmark.*

__Several pyrrolizidine alkaloids are toxic__ for animals and some predispose to tumour development in animals. In both cases the liver is the principal target organ. Hirono et al recently reported an increased incidence of tumours of the liver in rats fed on diets containing 16-33% Comfrey leaf or 0.5-4.0% Comfrey root.

Before the publication of this report, **HDRA issued a public statement that concluded that until the position is clarified 'no human being or animal should eat, drink, or take Comfrey in any form.'**

An abstract of the HDRA statement appeared in the 'Observer' (newspaper) of 30 July, 1978.

Unless further work shows that for some reason the findings of Hirono et al are misleading, it seems unlikely that Comfrey will ever again be regarded as a plant that can be consumed with complete safety.

The carcinogenic response in animals was seen in response to continuous high dosing over long periods and evidence of liver intoxication preceded liver tumour development.

The consumption of Comfrey by man is generally at a much lower level, *and no examples of liver poisoning have been reported. People who have in the past consumed or used products containing Comfrey have therefore no cause for alarm."*

-'**Letter from Chicago: Henry Doubleday Research Association**', The British Medical Journal, London, England, page 598. (March 3 **1979**).

(These reports are expanded upon and discussed in the Alkaloids section in Volume 2. For dates and information about legal restrictions on Comfrey by governments of various countries, see the 'Alkaloids in Comfrey' sections in Volume 2.)

Symphytum 1777

'Flora Danica: Tab. DXLI-DCCXX' (Tab. 541-720), Vierter Band (Volume 4), by Otto Friedrich Muller, Copenhagen, Denmark, 1777, Tab. DCLXIV (Tab. 664).

"Flora Danica" is 51 volumes and 3 supplements published from 1761-1883. George Christian Oeder at the Botanic Garden in Copenhagen began this atlas of botany. There are 3,240 copper engraved plates.

PART C

GARDEN USES AND NUTRITIONAL VALUE

Chapter 17

Garden Uses of Comfrey: Compost, Fertilizer, Potting Mix

"The aim of this survey was to evaluate the extent to which Comfrey is grown by gardeners and what it is used for. There were 304 people in the United Kingdom taking part in the survey.
Many people had been growing Comfrey for a long period of time.
    64% had grown it for more than 10 years, and 37% for more than 30 years.
Using Comfrey for the purposes as plant feed or improving soil fertility was by far the most popular reason for growing it (97% of respondents). Additionally, 74% stated that they grow it because it is good for attracting bees.
    18% also grow Comfrey as an ornamental.
Very few people grew Comfrey to eat themselves or to feed to animals.
    This represents a change in attitude. Much of the original focus on Comfrey was on its properties as a fodder crop.
Comfrey was most commonly used as a liquid feed (83% of respondents) and also added to compost (81% of respondents). Almost 55% of people also used it as a mulch.
After all the work that has been done in selecting the best varieties of Comfrey, we were interested to know whether gardeners were aware of what type of Comfrey they were growing.
    Encouragingly, 72% of respondents knew what species they were growing. There was most certainty around Russian Comfrey (Symphytum x uplandicum). 55% of people were certain or thought they were certain or probably growing it. Less people thought they were growing Common Comfrey (Symphtum officinale) with 22% certain or thought they were certain or probably growing it.
Encouragingly, 61% said that they knew which cultivar they were growing, with 62% certain or thought they might be growing Bocking 14. Again, this is not surprising, as this is the most commonly named variety sold to gardeners, although there are named ornamental varieties too."
-'Survey of Comfrey Use: Members Experiment 1' by GardenOrganic®, Ryton Gardens, Warwickshire, England, **2018**.
    (GardenOrganic® continues to have these annual Comfrey surveys of the United Kingdom. You can sign up at: www.gardenorganic.org.uk. The results are written up in their magazine 'The Organic Way'.)

For information about fertilizing Comfrey, see subsection 'Fertilization of Comfrey Plant' in section 'Planting, Soil, Fertilization, Water, Disease' in Volume 2.

<u>Comfrey Roots Break Up Hard Soil</u>

"Even Comfrey can often force its roots through a heavy subsoil, if you can keep the soil moist."
-Enchanted Garden: Alan Chadwick's Organic Method of Gardening by Tom Cuthbertson. London, England: Rider & Company / Hutchinson & Co. Publishers Ltd, **1978**, page 169. (Subsoil is layer of soil under topsoil. Topsoil is on the surface of the ground.)

"Ways to break up subsoil: In some conditions, subsoiling, breaking up the subsoil to a depth of several feet (2 feet = 0.6 meter) without bringing any of the subsoil to the surface, can bring about immediate and dramatic improvement. The special tool for accomplishing this is called, obviously enough, a subsoiler.
    Regular use of the mould-board plough is likely to develop an impervious layer called 'plough-sole' (hardpan). Where this has occurred, subsoiling is definitely indicated.
The same result can be brought about more slowly by establishing such deep-rooted plants as alfalfa, Comfrey, and sweet clover."
-'Organic Growing: The Road to Survival' by Haanel Cassidy, Ananda Village, Nevada City, California, **1979**. Practitioner of Biodynamic Agriculture.   (Hardpan is a dense layer of soil, usually below the topsoil layer.)

"Comfrey's fat vigorous taproots push far into the soil and can break up hardpans and heavy clays."
-Gaia's Garden: A Guide to Home-Scale Permaculture by Toby Hemenway. White River Junction, Vermont: Chelsea Green Publishing Company, **2009**, page 127.

*"**Breaks up soil: Comfrey's thick roots break through compacted soil.** It's important to allow plants to perform this function rather than pulling out a spade or shovel and tilling the ground ourselves.*
When we till, we commit a biological genocide with bacteria, fungi, protozoa and micro-arthropods being the victims. Even a slight disturbance in the soil can rip an appendage of a small mite and I have pictures at 400 times magnification to prove this."
-'Why You Should Have Comfrey in Your Garden' by Gabe Garms, Raven's Roots Naturalist School: Wilderness Instruction, http://www.ravensroots.com/blog, Sedro-Woolley, Washington, August 1, **2016**.
 (An arthropod is an invertebrate with exoskeleton, segmented body, and jointed appendages.)

Comfrey as Mulch
 Mulch is a covering such as hay, straw, compost, wood chips, pine needles, paper or plastic sheeting, spread on the ground around plants to reduce evaporation, stop/reduce soil erosion, maintain soil temperature, enrich soil, inhibit weed growth, keep fruit/vegetables clean from soil splashing, etc.

*"**Another system is to spread the cut Comfrey foliage between the rows of tomatoes as a weed-suppressing mulch which will release some potash (potassium) as it is trodden down with picking and trimming.**"*
-Comfrey: Fodder, Food and Remedy by Lawrence D. Hills. New York: Rizzoli Universe Books: **1976**, page 169.

*"**Freshly cut Comfrey leaves make good mulch because they're high in nitrogen, so they don't pull nitrogen from the soil while decomposing**, as high-carbon mulches like straw and leaves do.*
***And Comfrey's high potassium content** makes it especially beneficial for flowers, vegetables (such as tomatoes, peppers and cucumbers), berries and fruit trees.*
*Michele DeFord, owner of 'Crimson Sage Nursery', in Colton, Oregon, has mulched her stock herb plants with Comfrey for 25 years because **it boosts seed yields**. For home gardeners, this means increased flower and fruit production.*
***But using Comfrey to mulch root crops (like carrots) or leafy greens (like lettuce and spinach) may encourage them to go to seed prematurely.**"*
-'Power Plant' by Jean M.A. Nick, Organic Gardening Magazine, Pennyslvania, pages 51-53, June/July **2004**.

*"**Grow your own mulch: Instead of sourcing mulch materials, you can grow your own.***
***Comfrey** - High in nitrogen, potassium and a source of phosphorus, Comfrey is easy to grow.*
 ***Traditionally, it was grown around compost piles/bins** both to capture nutrients and use as a compost starter (it adds enough nitrogen fast enough to get the compost working quickly).*
***Comfrey is considered a permaculture wonder-plant.**"*
-'An Introduction to Permaculture Sheet Mulching: Experimental Strategies and Techniques for the Pacific Northwest Coast' by Harold Waldock, 'Edible Landscape Creations' and 'Vancouver Permaculture Network', British Columbia, Canada, August **2001**.

*"**Cut Comfrey is an excellent mulch, degrading fast and supplying nitrogen and potassium.** I use it a lot.*
Comfrey plants can be cut once or twice a year without having to particularly feed them, but they can be cut four or five times a year if they are fed, for example with urine."
-Creating a Forest Garden: Working with Nature to Grow Edible Crops by Martin Crawford. England: UIT Cambridge, **2010**, p.63.
 (See sub-subsection 'Urine as Fertilizer' in subsection 'Fertilization of Comfrey Plant' in the section 'Planting, Soil, Fertilization, Water, Disease' in Volume 2.)

*"**My grant project 'Alley Cropping in a Hillside Terrace System'** sought to transform the steep slopes of a newly planted fruit orchard in Ohio into a sustainable agriculture system incorporating earth berm terraces, no till annual agriculture, with a cut-and-carry mulch system.*
 ***One hillside plot was successfully populated with Comfrey, stinging nettle, and white clover, another was populated with Comfrey, yarrow, and goumi (cherry silverberry) shrubs,** and the third was left to the existing cover of mostly blackberry, purple knapweed, and goldenrod.*
Cut-and-carry mulch patches:
*Of all the plants used, the best performer by far was Comfrey, and it was the only plant tested that **rebounded quickly enough after the initial cutting to produce enough growth for a significant second cutting that same season.***
 The effort and cost of establishing the mulch patches was further justified when the amount of biomass produced and the quickness of regrowth (in the case of the Comfrey) was compared with the established patch of weeds (which hardly regrew at all). Not only did the test plots provide better coverage per area cut, but they also provided a more easily spread mulch that did not have painful thorns like the blackberry.
In July when plants were probably at about full size, an approximately 8 foot (2.4 meter) wide swath of mulch plants from either of the test plots could provide approximately 2 inch (5 cm) deep mulch over most of 4 foot (1.2 meter) wide bed of same length.
 ***This suppressed weeds entirely for several weeks and then slowed down weed growth** long enough that they could not catch up with squash. The control mulch plot of same size could only be put down roughly 1 inch (2.5 cm) thick.*
*Furthermore, due to the more slender growth of the stalks and stems of the species in the control group compared with the **broad leaves of the Comfrey**, the weeds were not suppressed as well in the control plot. And the best part of establishing these mulch plots is that now they are there and ready to use next year and for years to come."*

-'Alley Cropping in a Hillside Terrace System: Final Report for FNC13-916' by Weston Lombard, Solid Ground Farm, Millfield, Ohio; SARE: Sustainable Agriculture Research and Education: Grants and Education to Advance Innovation in Sustainable Agriculture, https://projects.sare.org, December 31 **2014**.   (Berm is a mound or wall of soil.)

"**Comfrey as mulch: Place a 5 cm (2 inch) deep layer of Comfrey leaves on the soil surface around plants.** *They will slowly rot down and release plant nutrients, particularly potassium.*
An extra layer of grass clippings on top will make the mulch more effective at controlling weeds and also help to speed up decay.
Plants benefitting from a Comfrey mulch include tomato and potato plants and fruit bushes such as blackcurrant and gooseberry.
Do not use around acid-loving plants as decomposing Comfrey produces an alkaline residue. When used regularly in large quantities this may increase the pH of the soil."
-Comfrey for Gardeners: A Garden Organic Guide by Garden Organic. Coventry, England: no date but around **2015**, page 14.

"**Comfrey as Mulch Plant:**
It's significantly more efficient to cover your soil with organic matter than to leave it bare. *One of the principles of permaculture gardening is keeping a mulch layer on top of your soil, mimicking how soil is built in nature, which is never bare. When you leave your soil bare, water is easily evaporated, weeds aren't suppressed, nutrients get quickly leached away, erosion takes place faster, your plants succumb to extreme cold/heat easier, and the good insects and arachnids (8-legged insects) which eat all of our garden pests (so we don't have to use pesticides) won't have any habitat to live in.*
The leaves of Comfrey can get enormous and can easily be chopped and dropped around either itself or other plants growing nearby, covering a lot of bare soil. *The best part about it is that Comfrey can be chopped and dropped numerous times throughout the growing season and provide lots of mulch.*"
-'Why You Should Have Comfrey in Your Garden' by Gabe Garms, Raven's Roots Naturalist School: Wilderness Instruction, http://www.ravensroots.com/blog, Sedro-Woolley, Washington, August 1, **2016**.

"**Comfrey as super-mulch? Preliminary investigation of soil nutrients, microbial communities, and response of an indicator crop.** *Comfrey (Symphytum spp. and varieties) is widely grown as a companion plant or green mulch. With deep roots and a reputation for accumulating potassium and other nutrients in its leaves, Comfrey may serve as a 'nutrient pump', increasing availability of nutrients in topsoil, as has been demonstrated in other species.*
> *This study aims to track nutrients from Comfrey leaf mulch to soil, and then to a fast-growing crop (kale, Brassica oleracea 'Red Russian'), in a preliminary test of how Comfrey mulch might affect soil nutrients, microbial communities, and crop yield and quality.*

Accession: PRJNA354590. Data Type: Raw sequence reads. Scope: Multispecies.
Registration date: November 22 2016. Wellesley College."
-National Center for Biotechnology Information, United States National Library of Medicine, www.ncbi.nlm.nih.gov/bioproject/354590, Bethesda, Maryland, **2018**. (After completing this, it was written about in below 2018 'Comfrey Mulch Enriches Soil'.)

"**Comfrey (Symphytum spp. {species}) is thought to accumulate plant nutrients such as potassium (K) in its leaves and is consequently used widely as a green mulch.**
> *We sought to investigate the efficacy of Comfrey as a soil amendment by measuring its nutritional composition and the effects of mulching with Comfrey on soil nutrients, soil microbial communities, and growth and quality of an indicator crop (kale) over **one growing season** in a small garden plot.*

We found that Comfrey was rich in K and plots mulched with Comfrey had higher concentrations of elemental K, as well as higher concentrations of available nitrogen, compared to plots mulched with paper.
Diversity and composition of soil bacterial communities was similar between Comfrey and paper-mulched plots, but began to show a trend toward divergence by the end of the growing season.
Overall, Comfrey mulch did not enhance the yield or nutritional content of the kale, but perhaps could improve crop performance over a longer period of time or in K-limited soils."
-'Comfrey Mulch Enriches Soil, But Does Not Improve an Indicator Crop within One Season' by Mia M. Howard, Alena A. Plotkin, Amelia R. McClure, Vanja Klepac-Ceraj, Alden B. Griffith, Daniel J. Brabander and Kristina N. Jones, Wellesley College, Massachusetts; International Journal of Plant and Soil Science, Volume 22, No. 2, pages 1-9, **2018**.

Grow-It-Yourself Fertilizer

"**A Comfrey bed is in effect a method of exchanging crude nitrogen for a balanced organic fertilizer**, and so it can have poultry manure straight off the droppings boards, dried sewage sludge, or the mixture of two parts water to one of human urine which is known to H.D.R.A. (Henry Doubleday Research Association) members as **'H.L.A.' or Household Liquid Activator** because it is the best and cheapest of all compost heap activators."
-Comfrey: Fodder, Food and Remedy by Lawrence D. Hills. New York: Rizzoli Universe Books: **1976**, page 163.
> (See sub-subsection 'Urine as Fertilizer' in subsection 'Fertilization of Comfrey Plant' in the section 'Planting, Soil, Fertilization, Water, Disease' in Volume 2.)

"Comfrey has the drawback that with so little fibre it cannot leave much humus in the ground. There are just some black remains

*and none of the accumulation which compost provides, but **it offers organic plant foods as a 'grow-it-yourself' general fertilizer**, to make compost go further, and the art of using it through the season for crops that need most potash (potassium) has become part of **organic gardening**."*
-Comfrey: Fodder, Food and Remedy by Lawrence D. Hills. New York: Rizzoli Universe Books: **1976**, page 160.

*"**Comfrey is not a green manure plant**, for these are annuals to grow in winter, or between other crops in summer, for digging in. **You cannot dig in a Comfrey plant. You cut its folage for many uses in the garden, and it grows on year after year as a kind of fertility mine, bringing up minerals from the subsoil.**"*
-Fertility Gardening: The Organic Way to Make Your Garden Grow by Lawrence D. Hills. London, England: Cameron & Tayleur Books Limited, **1980**.

*"**More-Sustainabile Fertilization:** Each gardener should strive to use less and less fertilizer brought in from outside his or her own garden area. **'Grow' your own fertilizers by raising plants that produce good amounts of compost material, which concentrates the nutrients required in a form that plants can use.***
*Deep-rooting alfalfa (as deep as 125 feet = 38 meter) and **Comfrey (up to 8 feet = 2.4 meter)** also help bring up leached-out and newly released nutrients from the soil strata and rocks below."*
-How to Grow More Vegetables and Fruits, Nuts, Berries, Grains and Other Crops, 6th Edition, by John Jeavons, Ecology Action of Midpeninsula. Berkeley, California: Ten Speed Press, **2002**. First published 1974. A Primer on the Life-Giving Sustainable 'Grow Biointesive'® Method of Organic Horticulture.

*"**Some examples of nutrient sources acceptable for organic certification.***
***Values in pounds/100 square feet** (92.9 square meter) can be converted to pounds per acre by multiplying by **436**.*
Plant Derived Sources:
    Alfalfa meal (3 - 0.5 - 3): 5 to 10.   *Soybean meal (6 - 1.4 - 2): 1 to 5.*
    Cotton seed meal (6 - 2 - 2): 1 to 5.   *Kelp and seaweed (1- 0.2 - 2): 1 to 2.*
    Wood ash (liming action): 1 to 2, compost (may include manure): 5 to 20.
    Dry greens - herbs (nettle, Comfrey, yarrow, etc): 5 to 10 pounds per 100 square feet."
-'Organic Farming Principles and Practices' by John A. Biernbaum, Department of Horticulture, Michigan State University, East Lansing, Michigan, January **2003**.

*"**Fertilizer inputs used by members of 'Garden Organic': Amendments:***
Comfrey (home made). Comfrey liquid, 35% used it, 50.80 liters (53.6 quarts) on average.
Comfrey pellets 10.00 kg (22 pounds) used on average.
*Garden products account for a significant proportion of ecological footprint in 2 areas; those sourced from a long supply chain and those used in large quantities. Improvements are possible by substituting shop bought products with home made ones. **Some amendments like seaweed might be substituted with, for instance home grown and produced Comfrey or liquid worm compost. By trying to close nutrient cycles; e.g. producing amendments at home (e.g. Comfrey),** fixing nitrogen in situ (on site, locally) (e.g., green manures), composting biodegradable materials."*
-'Measuring Your Garden Footprint' by Davies and Schmutz, Garden Organic (formerly Henry Doubleday Research Association), Coventry, United Kingdom, HDRA07/Mem Exp 4, 40 pages, **2007**.
    (Green manure is created by leaving uprooted or sown crop parts to decay on a field as a mulch and soil amendment.
    Cover crops grown for this purpose are ploughed under and incorporated into soil while green or shortly after flowering.)

    In United Kingdom you can buy dried Comfrey pellets to use as a garden fertilizer. It can be used to make Comfrey tea (liquid fertilizer) or be applied directly to soil. It is recommended for beans, peas, tomatoes and other vegetables.

*"My friend Erich discovered that **during the warmer months, a tea made with Comfrey leaf, worm castings, or kelp may prove very useful for watering in seeds, watering plants, or foliar-feeding adult plants, both potted and in the garden.**"*
-The Medicinal Herb Grower: A Guide for Cultivating Plants that Heal: Volume 1 by Richo Cech. Williams, Oregon: Horizon Herbs LLC, **2009**, page 95.

*"Julie van Zevern, a member of ECHO's network, worked with an herbal clinic in Zimbabwe (southern Africa) that uses Comfrey in a medicinal manner. Julie also mentioned that **Comfrey is used as a fertilizer in Zimbabwe by 'Foundations for Farming' (FFF)**. Hazel Edwards from the FFF office in Harare (capital of Zimbabwe):*
    *'We were first introduced to the use of Comfrey as a fertilizer by a local farmer, Rory Maloney, a fairly large market gardener who supplies fresh produce for the Harare vegetable market. He said it **improved the health and disease resistance of his crops.**'*
*Hazel commented that Comfrey is high in potassium (K) and other micronutrients, and 'seems to improve fruiting and disease resistance'. She notes that **Comfrey is most effective when applied to solanaceous (e.g. tomato and potato) and leguminous plants, as they are potassium responsive.** It can be applied every two weeks as a drench or foliar spray.*
*ECHO does not supply Comfrey. **Import from another country of Comfrey can be tricky, because root cuttings can be destroyed by fumigation and other precautionary treatments.** Because of a short historical period of widespread interest in Comfrey, it may already be available in your country."*
-'Russian Comfrey for Fertilizer, Feed and More' by Dawn Berkelaar, ECHO (Educational Concerns for Haiti Organization)

Development Notes: Fighting World Hunger, Issue 123, April **2014**.

(A drench is when the liquid fertilizer is applied to the soil with little or no contact with the plant. This is good for long-term fertilization. A foliar spray is a liquid sprayed on the plant directly. It is good for quick results.)

Minerals in Comfrey Overview
See subsection 'Minerals' in section 'Nutritional Value of Comfrey' (Chapter 19).

"When trees grow, they store away in wood the potash (potassium) and other plant foods which they take in through their roots, going far down to find the molecules which their secretions make available. These are locked away so tightly that they can only be released by fire, or the bacteria like Hutchinson's spirochaete which can break down hard celluloses and hemicelluloses. **Comfrey, however, keeps all its mineral gatherings on current account, ready for instant use. Its roots go down to small-tree depths, and a Comfrey bed is a mineral mine for plants**, just as a large plot on a horse stud farm gathers the calcium for the strong bones of foals, and a field on a pig farm provides the copper which prevents pig anaemia, by giving that 2 to 6 percent in the Comfrey food, instead of artificial nutrients passing over 90 percent out in the dung to build metal toxicity in the soil."
-Comfrey: Fodder, Food and Remedy by Lawrence D. Hills. New York: Rizzoli Universe Books: **1976**, page 157.

Comfrey as Potassium Fertilizer
For more about potassium in Comfrey, see the sub-subsection 'Potassium' in the subsection 'Minerals (Ash)' in the section 'Nutritional Value of Comfrey' (Chapter 19).

"Comfrey is so rich in potassium it has about the same potash balance as a compound chemical potato fertilizer."
-Comfrey Report: The Story of the World's Fastest Protein Builder and Herbal Healer, Conservation Gardening and Farming Series: Series C by Lawrence D. Hills. England: Henry Doubleday Research Association, **1975**, page 40.

"Wilted Comfrey has more than twice as much potash (potassium) as good farmyard manure and about 30 percent more than compost, though the last varies according to what goes into it.
Let us consider the difference between an organic and an inorganic fertilizer, between the potash in Comfrey foliage, in wood ashes, and say, potassium chloride (still called 'Muriate of Potash') from the Strassfurt deposits in Germany or the Dead Sea in Israel.
When you apply 2 ounces (59 ml) a square yard (0.83 square meter) of potassium chloride, 94 percent washes down into the soil, and that 2 or 6 percent is taken up by your plants. That idle potassium in your soil will stop your plants taking up magnesium, another vital plant food, and most gardeners have seen its effect in the **tomato bed**, where the lower leaves turn yellow but the veins stay green.
When you put on compost, or manure, which have all their molecules root-selected, there is no problem- crops manage on that 2 percent passed from 'hand to hand' by soil bacteria and fungi, into plants, and back again to the soil."
-Comfrey: Fodder, Food and Remedy by Lawrence D. Hills. New York: Rizzoli Universe Books: **1976**, pages 156, 160.

Comfrey as Liquid Manure (Liquid Fertilizer)

"Liquid feeds can be made using Comfrey or nettle leaves. Be aware that they both can have a strong smell.
Comfrey leaves are rich in plant foods. The leaves decay rapidly, releasing the goodness they contain. They can also be used as a mulch or compost activator.
Comfrey leaves tend to be slightly alkaline, so the liquid feed should not be used on acid-loving plants.
Comfrey liquid is high in potash (potassium) and has reasonable levels of nitrogen and phosphate (phosphorus); **it is good for fruiting plants**, although its nitrogen levels may not be sufficient for a fully grown hanging basket."
-Encyclopedia of Organic Gardening: The Complete Guide to Natural & Chemical-Free Gardening by The Henry Doubleday Research Association, edited by Pauline Pears. London, England: DK (Dorling Kindersley), **2001**, page 207.

"Greenhouse and field studies were conducted to evaluate manure extracts and their effect on crop yield and selected soil chemical and microbial properties.
For the greenhouse study, Comfrey and tithonia plant materials were each used at rates of 0.0625, 0.0935, 0.125 and 0.156 kg per litre (1.05 quart) of water.
For field studies, 6 combinations of Comfrey and tithonia were used at 0 kg, 0.156 kg Comfrey, 0.156 kg tithonia, 0.0781 kg Comfrey + 0.0781 kg tithonia, 0.117 kg Comfrey + 0.0391 kg tithonia and 0.0391 kg Comfrey + 0.117 kg tithonia per litre of water. 'Kasisi Agricultural Training Centre' and 'Mount Makulu Agricultural Research Station' were used for the field studies, rape (Brassica napus) and tomato were used as test crops.
Application of extracts increased biomass yield of rape to between 2.8 gram Dry Matter (DM) per pot (Comfrey 0.0625 kg) and 12.79 gram (DM) per pot (tithonia 0.125 kg).
Leaf biomass increased with increased biomass of Comfrey, while no particular pattern was observed for tithonia.
Similarly, the application of extracts increased the biomass yield of tomato to between 6.67 gram (DM) per pot (Comfrey 0.0625 kg) and 13.99 gram (DM) per pot (tithonia 0.0938 kg).
Mixing of Comfrey and tithonia residues in the preparation of their extracts resulted in increased tomato yield to

between 5.2 tons per hectare (0.0391 kg Comfrey + 0.117 kg tithonia) and 9.5 tons per hectare (0.156 kg Comfrey), and 32.7 tons per hectare (0.156 kg Comfrey) and 40.2 tons per hectare (0.156 kg tithonia).
Furthermore soil pH, phosphorous, potassium, calcium, magnesium were positively influenced by application of extracts."
-'The Agronomic Effectiveness of Liquid Manure Extracts Derived from Comfrey (Symphytum Officinale) and Tithonia (Tithonia Diversifolia)' by Dina Mambwe, Dissertation for Master of Science in Agronomy, University of Zambia, Africa, **2009**.

"This paper assessed the nutrient content of three organic liquid manures made from Water Hyacinth (Eichhornia Crassipes), Russian Comfrey (Symphytum officinale) and Pig Weed red-root (Amaranthus retroflexus) plants.
The plants were collected and cleaned, and the root section was cut and discarded. Only the leaves and stem were used in the experiment.
In this study a plant to water weight ratio of 1:8 was used to allow the extraction of the highest amount of nutrients. Normal plant to water weight ratios used in both biodynamic and organic farming range between 1:8 to 1:10.
Using an analytical balance 5 kg (11 pounds) of the plant material was weighed, shredded and then placed in a 40 litre (10.5 gallons) container. Forty litres of water was added and the container was sealed to exclude rainwater, prevent evaporation and contain flies and odors.
The material was fermented for 30 days but periodically stirred every 7 days. Samples were analyzed weekly for nitrogen, phosphate and potassium (NPK) and trace elements.
All liquid manures had high K contents, particularly Russian Comfrey (3.90%), hinting against direct foliar application. Liquid manures had NPK contents greater than common solid organic fertilizers such as cattle manure used in Zimbabwe, Africa.
Russian Comfrey Liquid Manure: pH 7.8, TDS (Total Dissolved Solids?) 972.
Percent Nitrogen 2.90, Phosphorus 2.94, Potassium 3.90, Sulfur 1.60, Magnesium 0.08, Calcium 0.06, Zinc 0.04."
-'The Nutrient Content of Organic Liquid Fertilizers in Zimbabwe' by Simbarashe Govere, Benard Madziwa and Precious Mahlatini, Africa; International Journal of Modern Engineering Research, Volume 1, Issue 1, pages 196-202, **2011**.
(Russian Comfrey is Symphytum x uplandicum. Common Comfrey is Symphytum officinale.)

"One of the most popular ways of using Comfrey is to produce an organic liquid feed. This is used for plants in pots and containers, hanging baskets, greenhouse plants and hungry feeders such as tomatoes.
In 1997 the magazine 'Gardening Which?' rated Comfrey as one of the best homemade liquid feeds."
-Comfrey for Gardeners: A Garden Organic Guide by Garden Organic. Coventry, England: no date but around **2015**, page 9.

How to Make Liquid Fertilizer with Water ('Comfrey Tea' for Plants)

"Comfrey liquid manure was invented by an early H.D.R.A. (Henry Doubleday Research Association) member, Mr. George Gibson of Guernsey (island in English Channel off the coast of Normandy, France), who grew a Comfrey bed solely for feeding his greenhouse tomatoes when they began setting their first trusses.
The usual procedure is to fill a water butt (barrel) about quarter full of Comfrey leaves and stems, top up with water and draw off the result. I had a 20 gallon (75.7 liter) fibreglass rainwater butt and propped it up on an old galvanized water tank, so it was well clear of the ground to allow cans to be pushed under the tap for filling. This we filled by hose and then added 14 pounds (6.3 kg) of fresh cut Comfrey leaves.
The Comfrey went black quite quickly and formed a kind of crust on the surface, on which small flies ran. **After a month we tried the first canfuls on the frame tomatoes as they were setting their blossom. More water and Comfrey leaves were added later, for the aim was simply to make a good organic tonic for tomato plants."**
-Comfrey: Fodder, Food and Remedy by Lawrence D. Hills. New York: Rizzoli Universe Books: **1976**, page169, 170.
(The truss is a group or cluster of smaller stems where flowers and fruit develop.)
(A water butt is a large container for collecting and storing liquids.)

"Liquid manure can be made any time of the year, but the autumn brew is useful early in the spring season, and for house-plants and in the greenhouse."
-Organic Gardening by Lawrence D. Hills. Middlesex, England: Penguin Books Ltd., **1977**, page 59.
(You save your autumn Comfrey liquid fertilizer through the winter so it is ready early spring.)

"Comfrey (Symphytum asperum) which is planted as a medicinal herb, for its decorative appearance and to encourage bees, makes good covering material and is suitable for chopped and other types of compost, besides making very good liquid fertilizer!
Liquid Comfrey fertilizer is milder than the stinging-nettle variety, does not have such a penetrating smell, and feeds and promotes the growth and flowering of all pot, tub-grown and outdoor plants.
Comfrey fertilizer is slower acting but is very suitable for mixing with stinging-nettle fertilizer. The two plants can be processed together or the ready-made liquids can be stirred into one another.
Since the main purpose here is to manure the ground, the leaves of the dandelion (Taraxacum officinale), cow parsnip (Heracleum), plantain, yarrow and other roadside weeds may usefully be added to the preparation."
-'Companion Planting: Successful Gardening the Organic Way' by Gertrud Franck; Thorsons Publishers Inc, New York, **1983**. Translated from German.

"Comfrey liquid may be made using a barrel of almost any size, from 5 gallon to 40 or more (18.9 to 151.4 liters).

One method uses 14 pounds (6.3 kg) of Comfrey leaves to 20 gallons (75.7 liters) of water. This ferments in approximately one month and can be fed to tomatoes immediately at concentration 1 part in 20 or stronger as desired.
Whichever method is used the barrels should be completely covered to reduce smell. Drone flies accumulate very quickly around Comfrey liquid. Any unpleasant smells are easily rectified (reduced) with a teaspoonful of Odourcure®."
-Comfrey: Symphuo Symphytum: A Multi-Purpose Herb by Philip Clarke. Edinburgh, England: The Pentland Press, **1997**, page 7
(Eristalis tenax is a hoverfly or drone fly with a wingspan of 15 mm = 0.59 inch. It is stocky and looks like a bee.)

"Recipes for liquid feeds: Steep 3 kg (6 pounds, 12 ounces) Comfrey leaves in 45 litres (12 gallons) of water. Cover with a lid. Use undiluted after four weeks."
-Encyclopedia of Organic Gardening: The Complete Guide to Natural & Chemical-Free Gardening by The Henry Doubleday Research Association, edited by Pauline Pears. London, England: DK (Dorling Kindersley), **2001**, page 207.

"Comfrey Liquid Manure: A quickly available source of potassium for the organic gardener.
One method of extracting it is to put 6 kg (14 pounds) of freshly cut Comfrey leaves into a 90 litre (23.7 gallon) tapped (with a faucet), fibreglass water butt (barrel). Do not use metal as rust will add toxic quantities of iron oxide to the liquid manure. Fill up the butt with rain or tap water, and cover with a lid to exclude the light.
In about 4 weeks a clear liquid can be drawn off from the tap at the bottom. Ideal feed for tomatoes, onions, gooseberries, beans and all potash hungry crops. It can be used as a foliar feed."
-The Complete Book of Vegetables, Herbs & Fruit by Biggs, McVicar and Flowerdew. London, England: Kyle Cathie Limited, **2004**, page 364.   (Foliar feed applies liquid fertilizer directly to the leaves.)

"Liquid Fertilizer: One of the best ways to tap your fertilizer factory is to brew Comfrey tea.
Fill a barrel or trash can about halfway with fresh Comfrey leaves, add water, cover it, and let it steep for 3 to 6 weeks.
The tea may be used full strength or diluted to about half strength- to the color of weak tea.
Use it whenever you water your plants. It's great for water stressed plants to help get them back on track, reports Jerome Osentowski, director of the 'Central Rocky Mountain Permaculture Institute' in Basalt, Colorado."
-'Power Plant' by Jean M.A. Nick, Organic Gardening Magazine, Pennyslvania, pages 51-53, June/July **2004**.

"In England, many forest gardeners grow large patches of Comfrey, nettles and other important accumulators in 'pockets of production' or mulch gardens outside their forest gardens. They then cut these plants for mulch or compost them.
They also make them into fermented 'teas' by submerging a large bunch of cut plants in a barrel of water until the leaves decompose into a slimy liquid, or the water turns a dark color. It also ferments and usually smells pretty bad.
Once the tea is ready, you can water your garden with your own 'green gold'."
-Edible Forest Gardens: Volume Two: Design & Practice by Dave Jacke and Eric Toensmeier. White River Junction, Vermont: Chelsea Green, **2005**, page 20.

"Comfrey Leaf Tea: The most effective and environmentally friendly plant nutrient I have ever discovered is Comfrey leaf tea, a completely home-based, vegetarian, free-of-cost, and nonviolent source of nutrition for plants.
Comfrey leaf has a high protein content (10% to 30% protein, based on the dry leaf matter). By brewing fresh Comfrey leaves in sun, this protein is converted into bioavailable plant nutrients, and it's really good stuff!
First, you get a bucket or barrel, fill it loosely with fresh Comfrey leaves, fill to the brim with water, and then let it steep in the sun.
Every morning, you stir the contents with a stick, first one way and then the other. For starters, this stirring will be difficult, because you have to get all that green Comfrey leaf in motion, but later, as the Comfrey begins to break down, it gets a lot easier to stir and a lot stinkier to smell.
When you stir, you create a vortex. Related to the tornado, the water vortex is extremely powerful magic- as you stir, the tea is oxygenated and enriched with nutrients. Stir first one way, and then break up the vortex by stirring in the opposite direction. Stirring the tea is a repetitive and relaxing task, and lends itself well to the recitation of prayer or mantra. I believe that reciting mantra while stirring serves to energetically potentize the tea.
After 10 days or so, when Comfrey leaves have largely dissolved into the water, the tea is ready, and at this point it is best to use it without delay. Set up a 3-foot (0.9 meter) tripod with a large piece of cheesecloth for a filter. Under the filter, put a bucket to catch filtered tea. Then, dip some raw tea out of the barrel and pour it through the filter. Finished tea will catch in the bucket.
Dilute this tea with water before use. Combine equal parts tea and plain water. Water your plants with the dilute tea, either by spraying as foliar feed or simply by using a watering can, dousing leaves, and allowing the life-giving tea to drip down into the roots.
This is particularly good for nourishing potted plants, plants that are weak or sickly, for watering in new transplants, and for watering field crops that need an extra boost. The plants will be fortified against insect pests, they will grow faster, and they will green up."
-The Medicinal Herb Grower: A Guide for Cultivating Plants that Heal: Volume 1 by Richo Cech. Williams, Oregon: Horizon Herbs LLC®, **2009**, page 134.   (Mantra is a word or sound repeated to aid concentration in meditation.)

How to Make Liquid Fertilizer with No Water (Concentrate)

"There are many ways of making liquid manure from Comfrey.
One of the best is to buy a barrel of the type used as a water butt, 40 gallons (151 liter) is a good sie. Bore a large hole in the

middle of the bottom to take a 1 inch or 1 1/2 inch (2.5-3.8 cm) pipe driven in tight.
> Place a flower-pot over the hole, having chipped the rim a bit so that it does not fit tightly on the bottom but allows the liquid to drain under it. Pour clean gravel round the pot to keep it in place.

Fill the barrel full of cut Comfrey leaves, ramming them in fairly tight. Put on a lid that overlaps round the edge to keep out the water, and prop it up and half an inch (1.2 cm) at one side to run off rain and let in air.

In about three weeks there will be a steady drip of a black fluid from the iron pipe into a jar put to catch it, and as the Comfrey sinks, more is added.

At the end of the season the solid contents can be removed and dug in as compost."

-Grow Your Own Fruit & Vegetables by Lawrence D. Hills. London, England: Faber and Faber Limited: **1971**, page 48.

"Comfrey Liquid Fertiliser (with no water added, i.e., 'concentrate', Russian Comfrey No. 14):
The decomposition of Comfrey leaves in the columns was rapid. Liquid was released from 10 days after setting up the columns and 95% of the total volume produced was released within the next three weeks.

The pH of the Comfrey liquid released from the columns was very high, ranging from pH 8.0 to pH 9.9, and was usually over pH 9.0. (This is alkaline or basic pH.)

There was considerable variation between individual plants but on average plants harvested four times during 1988 yielded 3183 ml (107 fluid ounce) of liquid fertilizer from 6423 grams (14 pounds) of fresh leaf material.

> A residue of 2504 gram/plant (5.5 pounds) was left in the columns after breakdown, and this would be available **for addition to compost heaps or possibly for the production of Comfrey leafmould.**

The concentration of Nitrogen and Phosphorus were significantly lower in the second harvest while the concentration of K (potassium) was significantly higher. Many factors could theoretically have been responsible for these changes in nutrient levels including the repeated cutting of the leaves, natural variation in the nutrient status of the plants, stage in the plant's life cycle, or conditions during decomposition."

-'The Use of Comfrey (Symphytum x uplandicum) as a Source of Plant Nutrients' by P.J.C. Harris, W.F. Bourne, C. Gitsham and M. Lennartsson, 'Henry Doubleday Research Association', Coventry, England and 'Department of Biological Sciences', Coventry Polytechnic, England; January **1990**.

> (pH is a scale from 0 to 14 of how acidic or basic a water-based solution is. Acidic solutions have a lower pH, while basic {alkaline} solutions have a higher pH. pH 7 is neutral.)

"With this method the Comfrey leaves are rammed tightly into the barrel with a weight on top. (No water is added.) Sand or fine mesh can be used to filter the liquid, which will start to drip in approximately 10 days. It can then be kept in an airtight container until required, when it can be diluted from 10 to 1 or 20 to 1 according to the dilution required."

-Comfrey: Symphuo Symphytum: A Multi-Purpose Herb by Philip Clarke. Edinburgh, England: The Pentland Press, **1997**, page 7.

"Comfrey Concentrate:
Comfrey leaves can be made into a concentrate. **Pack cut leaves into a plastic container with a hole in the bottom** (a water barrel, bucket, or drainpipe with an end cap are all good options) and cover with a lid. Place a collection vessel under the hole, and **after 2 to 3 weeks a dark liquid will drip out.**

To use, dilute the concentrate in 10-20 parts water. For example, tomatoes in pots can be fed with concentrate diluted 1:15 with water three times a week once the fruits start to form.

The concentrate can be sotred for a few months in a cool, dry place."

-Rodale's Illustrated Encyclopedia of Organic Gardening edited by Pauline Pears. Emmaus, PA: Rodale Press, **2002**, page 207.

"You can also make liquid fertilizer concentrate by packing fresh-cut Comfrey tops into an old bucket, weighing them down with a big rock or a plastic bag of water, covering tightly, and waiting a few weeks for them to decompose into a lovely thick black goo.

Some gardeners put a hole in the bottom of the bucket and collect the concentrate in another container as it drips out.

Dilute this Comfrey concentrate about 15 to 1 with water, and use as you would Comfrey tea. You can seal this concentrate in plastic jugs until you are ready to use it."

-'Power Plant' by Jean M.A. Nick, Organic Gardening Magazine, Pennyslvania, pages 51-53, June/July **2004**.

"Making Comfrey concentrate differs from the 'ready-to-use' method in that no water is used. The Comfrey leaves decompose to a black liquid that is collected from the base of the container with a faucet.

Store the concentrate in a loosely stoppered, labelled bottle, kept in a cool, dark place. **It may be used immediately or stored for up to one year. Always dilute before use.** An added bonus is that it smells much less than the 'ready-to-use' feed.

There are several variations for making the concentrate, depending on the quantity that you wish to produce.

A. Barrel or bucket method:
> You will need a large plastic bucket or similar container, minimum size about 20 litres (5 gallons). Drill the base to make a hole about 4-6 mm (1/4 inch) wide. A piece of chicken wire to cover hole to prevent blockage. Bricks to stand the container on so that a collecting vessel will fit underneath. A narrow necked collecting vessel about 1 litre (1.05 quart) capacity. **Enough Comfrey leaves to fill the container (packed in).**

B. Plastic drinks bottle method:
> **This method is ideal for small amount of leaves.** You will need 2 litre (2.11 quart) capacity plastic drinks bottle. Collecting vessel such as yoghurt pot, ice cream carton or similar.

Cut off the bottom of the bottle, pack in Comfrey leaves and stand the bottle upside down in a container. Cover the open end with a polythene (plastic) bag. Hold the bag in place with an elastic band, to prevent drying out.
Comfrey liquid will drip out of the bottle into the collecting vessel. Use diluted."
-Comfrey for Gardeners: A Garden Organic Guide by Garden Organic. Coventry, England: no date but around **2015**, page 11.

How to Use Liquid Fertilizer

"**This Comfrey liquid feed could be used for pot (potted) plants and for general garden purposes, with runner beans gaining from the potash almost as much as tomatoes.**
Mr. Gibson considered it was wasteful to compost Comfrey because so much of the **potash** seeped away below the heap, but forking it into the tanks outside every greenhouse preserved it all.
The residue is quite good compost, and as the water butt (barrel) gets low, it can be tipped out and dug in. Some gardeners top off with more water, which reduces the strength of the liquid, but 14 pounds (6.3 kg) to 20 gallons (75.7 liter) is a good nutrient level for tomatoes."
-Comfrey: Fodder, Food and Remedy by Lawrence D. Hills. New York: Rizzoli Universe Books: **1976**, page 170.

"**Analysis of the Comfrey liquid shows about three times as much potash (potassium), almost the same quantity of nitrogen, and about half of the phosphorus as in standard chemical tomato feed, so it is as well to half fill the can with water before topping up with Comfrey liquid manure.**"
-Organic Gardening by Lawrence D. Hills. Middlesex, England: Penguin Books Ltd., **1977**, page 59.

"**How To Use The Liquid Fertilizer:** Definitely with caution. As liquid feed it provides readily available nutrients that could upset the natural balance. It's my understanding for example, that **while Comfrey leaf mould is mainly nitrate nitrogen, Comfrey liquid will contain large amounts of ammonium nitrogen.** Ammonium can have the effect of blocking uptake of other nutrients like potassium and calcium.
Comfrey leaves and Comfrey liquid may not suit acid loving plants.
The big plus is that you get readily available potash for free. This is especially valuable for tomato fruit growing where you might otherwise rely on inorganic sources."
-'Growing and Using Garden Comfrey' by Michael E. J. Scott, www.the-organic-gardener.com, location unknown, **2004**-2017.
    (For more about Comfrey leaf mold, see subsection 'Potting Mixture with Comfrey Leaf Mold' in this section 'Garden Uses of Comfrey' {Chapter 17}.)
    (The 3 common sources of nitrogen used in agriculture are ammonium {NH_4^+}, nitrate {NO_3^-} and urea {CH_4N_2O}. For the best growth, each plant species requires a different ratio of ammonium to nitrate. This ratio varies with temperature, growth stage, soil pH, etc.)

"**To use Comfrey tea or diluted Comfrey extract as a foliar drench or spray, add a few drops of liquid soap (it helps the spray stick to the leaves) and apply to your plants.** You can use a watering can with a fine rose, but you'll get better coverage with a garden sprayer. Be sure to strain your liquid very carefully (let it drip through a large coffee filter) before you put it in your sprayer, or you'll clog up the nozzle before you even get started.
When you spray your plants, don't coat just the top of the leaves; reach under and spray the bottoms, too, at least until the liquid starts to run off."
-'Power Plant' by Jean M.A. Nick, Organic Gardening Magazine, Pennyslvania, pages 51-53, June/July **2004**.

"**Foliar Fertilization: Foliar feeding refers to the application of fertilizers to a plant's leaves.** It is not a substitute for maintaining adequate levels of plant nutrients in the soil but can be beneficial in certain circumstances. Most commonly, it is recommended for alleviating (reducing) specific micronutrient deficiencies. Ideally foliar feeds should be applied in the cooler morning or evening hours. It is not advisable to spray leaves during the heat of the day.
To make herbal or weed teas, pack a five gallon full of herbs like nettle, Comfrey, chamomile or yarrow, or just plain weeds and cover it with water. **After 2 or 3 weeks, strain and use the extract to spray your plants with. Some experimentation might be necessary to determine the best dilution factor.**"
-'Foliar Fertilization' by Dawn Pettinelli, Cooperative Extension Educator, University of Connecticut, College of Agriculture and Natural Resources, Department of Plant Science, Soil Nutrient Analysis Laboratory, Storrs, Connecticut, **2014**.

"**Comfrey is an ideal liquid feed for plants growing in pots, or similarly restricted conditions.**
At 'Garden Organic' Ryton, England, **we use Comfrey feed as a summer feed for the following:**
    Tomatoes and peppers in pots: feed three times weekly.
    Tomatoes and cucumbers in greenhouse beds: feed twice weekly.
    Indoor and outdoor pot plants, hanging baskets: feed once weekly.
Start feeding tomatoes as soon as the first flowers have set fruit. When the Comfrey liquid is thick and black, it is stronger, so dilute to 20:1, when the liquid is thin and brown, dilute to 10:1."
-Comfrey for Gardeners: A Garden Organic Guide by Garden Organic. Coventry, England: no date but around **2015**, page 13.

"*Comfrey supplies extra potassium unavailable to most other plants. The plant's deep roots accumulate potassium from subsoil,*

and it's leaves are high in nitrogen (even more than livestock manure). It has an N:P:K (Nitrogen : Phosphorous : Potassium) ratio of about 8: 2.6: 20.5.
Because Comfrey is high in potassium (K), it promotes the development of flowers and fruit, and so should be applied after the first flowers have set on plants like tomatoes and peppers. Regular feeding may then support better flower and fruit development.
Be advised though that in excess, potassium somewhat stunts growth and coarsens leaves. **Due to its high K concentration, Comfrey tea may be considered too strong for regular use.**
> It can however be adapted by mixing with other garden-made fertilizers like a dilution of 1:19 with worm tea, providing an estimated N:P:K ratio of 2.5: 2.2: 2.5 which can then be diluted with water.

The nitrogen supplied by Comfrey is more likely to occur as ammonium, but the worm tea nitrogen is probably nitrate."
-'Make Your Own Liquid Fertilizers' by Mark Krawczyk, Permaculture and Keyline Educator, Keyline Vermont, Champlain Valley, Vermont, date unknown but on or before **2018**.

Making Compost with Comfrey

Compost is decomposed organic material. It is a soil conditioner rich in nutrients including humus and humic acids. You make compost by piling moist organic products together, and then in a few months this breaks down into humus. Decomposition is speeded up by shredding the plants, adding some water and creating proper aeration (air) by regularly turning it. Aerobic bacteria (uses oxygen) and fungi convert the material into heat, carbon dioxide, and dark decayed organic matter.

"**Comfrey cut for compost means 8 to 10 barrowloads wheeled for every barrow of finished compost.** General garden rubbish runs 5 to 6 barrows to one, but **wilted Comfrey produces three barrows of material of the above analysis for every four cut.** This is a bargain in organic plant foods for space and trouble."
-Comfrey: Fodder, Food and Remedy by Lawrence D. Hills. New York: Rizzoli Universe Books: **1976**, page 160.

How to Make Compost

To make a compost pile, the carbon/nitrogen ratio should be 25 to 30 parts carbon (dry, brown materials) to 1 part nitrogen (fresh, green materials). Finished compost is about 10 parts carbon to 1 part nitrogen.
> **High carbon materials:** corn stalks 60 to 1, dry leaves 60 to 80 to 1, newspaper shredded 175 to 1, nut shells 35 to 1, peat 70 to 1, sawdust 500 to 1, straw/hay 90 to 1, wood chips and twigs 700 to 1.
> **High nitrogen materials:** food scraps 17 to 1, garden debris 30 to 40 to 1, fresh grass clippings 17 to 1, cattle manure 20 to 1, aged chicken manure 7 to 10 to 1, horse manure 25 to 1, pg manure 6 to 1, fresh weeds 20 to 1. **Fresh Comfrey leaves and stems are in the nitrogen category.**

"**Comfrey is sappy greenstuff of the type that the economical straw-basis farm compost heap lacks.**"
-Fertility Farming by Newman Turner. Austin, Texas: Acres USA, **1951**, page 56.

"**For compost Comfrey leaves should wilt for forty-eight hours to lose about 50 percent of the moisture.**"
-Russian Comfrey: A Hundred Tons an Acre of Stock or Compost for Farm, Garden or Smallholding by Lawrence D. Hills. London England: Faber and Faber, Limited, **1953**, page 65.

"**The best small scale system, for those who use lawn mowings and weeds plus kitchen waste as it accumulates and can never get enough to make a big heap in one go, is the New Zealand Box.**
> This is like a deep cold frame with one side removable and no glass, but a temporary roof can be put on in winter with a great improvement of quality.

It can be used with any system of composting, its action is simply to stop heat loss from the sides, and with this and less soil on the roots, there will be fewer weed seeds or none."
-Compost, Comfrey and Green-Manure: 'Henry Doubleday Research Association' First Gardeners Report by Lawrence D. Hills. Braintree, Essex, England: Henry Doubleday Research Association, **1959**.

"Make the heap in a sheltered place. It always rots best away from the prevailing wind. **Build your compost heap** above ground if you can. You can start with brushwood, but in a garden two or three double rows of old bricks or brickends set on their flats to leave a channel down the middle.
Then pile in an eight-inch (20 cm) layer of **garden rubbish,** spread flat, but the material wants to be stemmy, and not all soft weeds. **Then add the 'activator' which supplies the bacterial food, or rather the nitrogen they need for their increase to the numbers required.** This can be horse, cow, pig or any animal manure in a layer about an inch (2.5 cm) thick. If you are using poultry droppings this can be as thick as three inches (7.6 cm).
> **Wilted Comfrey, especially Bocking No. 14, can replace the manure with a three-inch or four-inch (7.6-10.1 cm) layer which is going to add more potash than any of the other activators.** The Comfrey will behave as though it were an organic manure, which, of course, it is, both by laboratory analysis and practical experience.

On top scatter just a little garden soil, to make sure of a good stock of bacteria, but if there are weeds with soil on the roots

this does instead.
On the activator and soil layer spread another eight-inch (20 cm) layer of rubbish, then scatter on about 4 ounces (113 grams) of **lime** to each square yard (0.836 square meter) of surface, enough to make it really white.

      Wood ashes *can replace the lime, for they contain it, but use twice as much, and those who have wood fires can make these layers two inches (5.0 cm) thick.*

Then more rubbish, then more activator and so on. House refuse should always go in the middle of the heap because of rats, and be scattered as it can bind solid. Lawn mowings should also be spread.
In winter or a wet summer, it pays to protect the heap from rain with a propped sheet of corrugated iron or other metal or asbestos. Black polythene (plastic) sheet excludes air, but is a good and cheap winter covering."
-Compost, Comfrey and Green-Manure: 'Henry Doubleday Research Association' First Gardeners Report by Lawrence D. Hills. Braintree, Essex, England: Henry Doubleday Research Association, **1959**.   (I don't recommend using asbestos.)

"**Compost made from straw, with Comfrey as sole activator:**
In August 1979, a straw and Comfrey heap was built in a New Zealand compost box on the Bocking Trial Ground, England. The material consisted of fifteen bales of wheat straw from the previous year's harvest, stacked in the open and partly decayed. And the Comfrey cut from one quarter of an acre (0.6 hectares) grown for plant sales.
It was built into the layers with no added manure, lime or any other activator. The heap was turned after three weeks, when it heated up again.
The result was an excellent compost. Ash 31.55%, Moisture 57.10%, Nitrogen 0.45%, Phosphorus 0.42%, Potash 0.55%."
-Fertility Gardening: The Organic Way to Make Your Garden Grow by Lawrence D. Hills. London, England: Cameron & Tayleur Books Limited, **1980**.
(Bocking is an area in town of Braintree, Essex County, England where Lawrence Hills started his Comfrey research station in 1948.)

Deep Roots Reach Nutrients

"Under every plot of ground, except where bed rock is near the surface, lie reserves of plant foods far greater than any that can possibly be supplied as chemicals. **Composting provides the means of increasing the amount in circulation, and the more deeply rooted the plants the more minerals are transferred to the current account.**
Russian Comfrey provides the best answer we have yet to this problem; **its roots have penetrated as far as eight feet (2.4 meters)**, and its great bulk of readily rotted foliage is as rich in potash as bracken (a large coarse fern), of high phosphorus content, and high in quickly-available carbohydrates.
These last, in a well-made compost heap, not only provide the fuel that can cook weed-seeds like grains of rice, and sterilize the product by the heat that runs up to 180 degrees F (82 C), they also provide additional combined nitrogen."
-Russian Comfrey: A Hundred Tons an Acre of Stock or Compost for Farm, Garden or Smallholding by Lawrence D. Hills. London England: Faber and Faber, Limited, **1953**, page 115.

"Not all soils naturally have all of the nutrients they need for their optimum health and crop productivity. **Deep-rooted crops such as alfalfa and Comfrey can be grown to bring up nutrients from below the range of most roots, then composted and added to the topsoil. However, if the needed nutrients are not in the deeper regions of the soil, they will not be present in the cured compost.**
On the other hand, when cured compost is added to the soil, nutrients that were previously unavailable in the soil may be made available by the biogeologic cycle.
In the biogeologic cycle, humic acid -which is produced from the decomposition process and is contained in the cured compost- along with the carbonic acid developed around the plant's roots, can increase soil microbial activity, decompose larger minerals, and possibly alter soil pH so that previously unavailable nutrients are made available."
-How to Grow More Vegetables and Fruits, Nuts, Berries, Grains and Other Crops, 6th Edition, by John Jeavons, Ecology Action of Midpeninsula. Berkeley, California: Ten Speed Press, **2002**. First published 1974. A Primer on the Life-Giving Sustainable 'Grow Biointesive'® Method of Organic Horticulture.

Mix Comfrey with High Carbon Materials

"**With Comfrey compost it is a good idea to mix the wilted stems with the grass mowings, weeds, and domestic refuse. The thick stems serve to keep open the air spaces and prevent any ingredient clogging up in a solid mass.**"
-Russian Comfrey: A Hundred Tons an Acre of Stock or Compost for Farm, Garden or Smallholding by Lawrence D. Hills. London England: Faber and Faber, Limited, **1953**, page 122, 123.

"**The final cuts of Comfrey for the season** can be dug in as though they were manure in the autumn, but **they are also of value in the compost heap, for they act as the 'paper and sticks' of the bacterial bonfire.**
Use them in two- to four-inch (5-10 cm) layers on top of four to six inches (10-15 cm) of woody refuse such as cut-down herbaceous plants and pea, bean and tomato haulm, and put the activator on them so that the last heaps of summer get away to a flying start with plenty of heat before they slow with the autumn rains."
-Grow Your Own Fruit & Vegetables by Lawrence D. Hills. London, England: Faber and Faber Limited: **1971**, page 49.

(Herbaceous plants have little or no woody tissue. It can be an annual, biennial or perennial. The leaves or entire plant dies back every year when the weather gets cold.)

(Haulm is stems or stalks collectively, such as stalks of grain, peas, beans, potatoes or hops, usually after the crop has been gathered.)

"Another late season use of Comfrey is in autumn compost heaps to add 'paper and kindling' to the rather slow-starting 'bacterial bonfires' of cutdown chrysanthemums, dahlia and Michaelmas daisies (and other garden debris)."
-Fertility Without Fertilizers: A Basic Approach to Organic Gardening by Lawrence D. Hills. New York: Universe Books, **1975**.

"Comfrey is also a kind of 'instant compost', for its foliage has a Carbon/Nitrogen ratio of 9.8 : 1.
It is so rich in protein that it is 'compost' before it goes on the heap.
The few, but stringy, fibers in stems and leaf midribs are mainly cellulose, *because they are grown too fast to build much hemicellulose or lignin, so Comfrey is poor value in the compost pile or as a humus builder, but it is the best mineral mine that will grow in a garden."*
-Fertility Without Fertilizers: A Basic Approach to Organic Gardening by Lawrence D. Hills. New York: Universe Books, **1975**.

(Cellulose and hemicellulose are 2 types of polymers found in plant cell walls. Cellulose is an organic polysaccharide molecule whereas hemicellulose is a matrix of polysaccharides. Cellulose has a strong, crystalline structure and is resistant to hydrolysis. Cellulose is the supporting material in plant cell walls. Hemicellulose has a random structure with little strength. It can be easily hydrolyzed. Hydrolysis is decomposition where a compound is split into other compounds by reacting with water.)

(Lignin is complex organic polymers that form structural materials in tissues of vascular plants. It is important in the formation of cell walls, especially wood and bark, because they give rigidity and do not rot easily.)

*"**Comfrey has so little fibre and so much protein, that its carbon to nitrogen ratio is about 14 to 1,** so in theory it is compost before it goes on the heap.*

*I made my first **all-Comfrey compost** in 1955 with some of the first cuts on the new H.D.R.A. (Henry Doubleday Research Association) trial ground in England, and a whole heap about four feet (1.2 meter) high smashed down to about six inches (15 cm), composed of the fibres and a black and tarry looking mess, that was mainly broken down proteins. It looked like cow dung.*

I learned then that the only way to make good compost with Comfrey was to mix it with other ingredients, just like making silage, gaining the quick heating from the starches and sugars and the extra potash to enrich the heap."
-Comfrey: Fodder, Food and Remedy by Lawrence D. Hills. New York: Rizzoli Universe Books: **1976**, page 157, 158.

Comfrey is a Compost Activator

Compost that breaks down very slowly usually does not have enough nitrogen. Compost activators jump-start the composting process by providing this nitrogen needed by the microbes in the heap of debris.

*"**Ever since I've been including Comfrey leaf in my compost piles, I've discovered that it is a most speedy activator,** a most dependable silent partner, always striving to increase the potency of the compost heap."*
-Comfrey: What You Need to Know by Ben Charles Harris. New Canaan, Connecticut: Keats Publishing, Inc., **1982**, page 91.

*"**Comfrey as a compost activator:** When building your compost heap, alternate layers of Comfrey leaves 5-8 cm (2-3 inch) deep with layers of other garden waste. **Decomposing Comfrey leaves encourage bacterial action causing the heap to heat up, speeding up the composting process.**"*
-Comfrey for Gardeners: A Garden Organic Guide by Garden Organic. Coventry, England: no date but around **2015**, page 15.

Potting Mixture with Leaf Mold

How to Make Non-Comfrey Leaf Mold

*"**Making non-Comfrey leafmould:** When leaves fall from trees in the autumn, they decay on the ground to form a rich, dark material called leafmould, which is an excellent soil conditioner.*

Any leaves fallen from deciduous trees and shrubs can be collected in autumn to make leafmould. *Do not use evergreens such as laurel and holly. An easy way to collect up leaves from a lawn is to run the mower over them with collection bag. The grass and chopped leaf mixture collected by the mower will rot down easily.*

If leaves are dry, soak them well with water. Stuff the leaves into a container or stack them in a corner. Leave them to decay. Make a few air holes if using container.

*It can take anywhere from nine months to two or more years to make a useable batch. **Two-year-old leafmould is much finer and can be used as a soil improver or an ingredient in growing media (potting mixtures).**"*
-Encyclopedia of Organic Gardening: The Complete Guide to Natural & Chemical-Free Gardening by The Henry Doubleday Research Association, edited by Pauline Pears. London, England: DK (Dorling Kindersley), **2001**, page 207.

(Leaf mold or mould is the slow decomposition of deciduous {leaves fall in autumn} shrub and tree leaves. It is a type of compost produced primarily by fungal breakdown. Autumn leaves have very low nitrogen and are usually dry. They are mostly cellulose and lignin so they break down very slowly. Comfrey leaves are not like deciduous leaves.)

"**Leafmold can be use as a mulch, soil conditioner, potting mix or seed compost.** It's a benign fellow, low enough in nutrients so as not to scold tender seedlings but with just the right qualities to dramatically improve soil structure and boost its water retention.
Unlike a compost heap which generates heat and relies on bacteria to break down its contents, leafmold piles are a far more sedate affair. **Fallen leaves are generally broken down by fungi. All this takes place in cool conditions,** so that while compost takes a few months to reach maturity, leafmould usually takes a year - even two.
Not all leaves are equal. The holy trinity of fallen leaves are oak, beech and hornbeam. These three trees produce the best-quality leafmold. Other deciduous tree leaves also work well, though thicker leaves such as chestnut (horse and sweet), walnut and sycamore will take longer to break down.
The easy option involves constructing a wire frame using chicken wire supported in the corners by sturdy posts. **Bigger volumes of leaves work best,** offering a buffer against weather conditions, so ensure your frame is at least a meter (3 feet) wide and deep. Periodically check your leafy investment, adding water if the leaves seem dry. A tarpaulin cover will help to keep the contents more consistently moist.
There are all sorts of leafmold accelerators on the market. Save your money and use what comes naturally - your pee. Fresh urine is sterile, safe, and it works wonders.
The even easier option for making leafmold is to simply scoop leaves up into garbage bags. Fill the bags three-quarters full, tie them closed at the top, then puncture holes into the bottom and sides to allow its contents to breathe. Place the bags out of the way and forget about them for a year or two."
-'How to Make Leafmold: Gardener's Gold' by Benedict Vanheems; GrowVeg®, Growing Interactive Ltd., United Kingdom, July 10 **2011**. www.growveg.com/guides/how-to-make-leafmold-gardeners-gold

Make Potting Mixtures with Leaf Mold (or Peat) and Comfrey

Peat is the accumulation of partially decayed vegetation or organic matter. It is unique to peatlands, bogs, mires, moors or muskegs. Peat is not usually regarded as a renewable source of energy, due to its extraction far exceeding its slow regrowth rate of 1 mm {0.039 inch} per year. Peat regrowth takes place only in 30-40% of peatlands.

"**Potting Medium: The last Comfrey cut of the year before hard frosts begin should be used in preparations for the next year's potting requirements.**
1. Using a plastic bag, polythene sack or plastic dustbin (trash can) **alternate layers of well chopped Comfrey leaves 3-4 inch wilted (7.6-10.1 cm) between 3-4 inch layers peat, leaf mould or coir.** Peat is best for seedlings. A handful of dolomite is optional.
2. **Mixture is left for 2 to 5 months** and is usable when the Comfrey leaves have rotted down and virtually disappeared.
3. **Use as a potting compost** diluted 50/50 with peat or coir for seedlings plus 140 grams (4 ounces) per bushel of calcified seaweed. As a potting mixture 70 grams (2 ounces) of fish, blood and bone can be added per bushel. Comfrey compost is rich in nitrogen and potassium and highly suitable for tomato growing."
-Comfrey: Symphuo Symphytum: A Multi-Purpose Herb by Philip Clarke. Edinburgh, England: The Pentland Press, **1997**, page 9.
(Coir fibers are between the hard, internal shell and the outer coat of a coconut.)
(Dolomite differs from limestone because it has magnesium as well as calcium. It is sourced from rock and is 8-12% magnesium and 18-22% calcium. Limestone is calcium carbonate with up to 40% calcium.)
(A bushel is a unit of weight or mass based on an early measure of dry capacity. The old bushel was equal to 4 pecks or 8 dry gallons {8 dry gallons = 9.3 gallons = 35 liters}, used mostly for agricultural products. Now a bushel refers to mass defined differently for each commodity.)

"**Comfrey leafmould can be used 'neat' (as is) as a potting compost, or as an ingredient of a potting mix. Fill a plastic bucket or dustbin (trash can) with alternate 10 cm (4 inch) layers of damp, two- to three-year old leafmould and chopped Comfrey leaves.** As the Comfrey leaves decompose, the goodness they contain is soaked up by the leafmould. When the Comfrey leaves have disintegrated (usually after two to five months), the leafmould is ready to use.
Plant raising and potting mixes with ratio by volume:
Seed-sowing mixtures: Comfrey leafmould : sand, 4 : 1, will provide sufficient nutrients until potting-on stage.
Potting mixtures: Comfrey leafmould alone, good for flowering and fruiting container-grown plants.
Comfrey leafmould : grit, 4 : 1, to every 35 litres (8 gallons), add 144 grams (5 ounces) general organic fertilizer and 28 grams (1 ounce) seaweed meal."
-Encyclopedia of Organic Gardening: The Complete Guide to Natural & Chemical-Free Gardening by The Henry Doubleday Research Association, edited by Pauline Pears. London, England: DK (Dorling Kindersley), **2001**, pages 117, 207.
('Potting on' or 'potting up' is moving a plant, with its root ball, to a larger pot.)

"**Seed-Sowing Mixtures:**
Comfrey leaf mold : sand, 4 : 1. Will provide sufficient nutrients until transplanting stage.
Potting Mixtures:

Comfrey leaf mold alone. *Good for flowering and fruiting container-grown plants.*
Comfrey leaf mold : sand, *4 : 1. To every 8 gallons (35 liters) add 5 ounces (144 grams) general organic ferltilizer and 1 ounce (28 grams) seaweed meal."*
-Rodale's Illustrated Encyclopedia of Organic Gardening edited by Pauline Pears. Emmaus, PA: Rodale Press, **2002**, page 117.

"**Suggestions for Using Dried Comfrey as Fertilizer:**
Houseplants are a good starting point for suggested use of the dried Comfrey leaves because they need so little by way of fertiliser *provided they are potted into a decent soil in the first place.*
Comfrey can be incorporated into the potting mix.
My recommendation is that 30 grams (1 ounce) is either worked into the top inch (2.5 cm) of soil in the pot with something like an old kitchen fork, or alternatively a shallow trench is made around the pot with the same implement, the dried Comfrey spread around this trench and then covered over with the soil that was removed to make the trench. Continue to water in the usual way and the nutrients in the Comfrey will be released for the plant to use. 30 grams every two months or so whilst the plant is growing is enough for table top or windowsill sized plants.
It is not necessary to add Comfrey if the plant is dormant in the winter - wait until spring growth commences.
Ornamentals Outside:
Ornamentals in outdoor tubs or the ground should be treated in the same way. *The user will need to make a decision as to how much to use, depending upon the size of the plant, but make the Comfrey go a long way.*
4 to 6 week intervals from the beginning of new growth in the spring until early autumn should be sufficient. Stop fertilising before leaf fall or dormancy."
-'Using Comfrey' by Old McDonald in Portugal: Farming, Gardening, Wildlife and Good Food, http://oldmcdonaldinportugal.blogspot.com, November 2 **2014**.

"**Comfrey leafmould potting compost:**
When Comfrey leaves are added to leafmould, they decompose to a nutrient-rich liquid which is absorbed by the leafmould. The end product is known as Comfrey leafmould *which can then be used in seed and potting composts.*
To make Comfrey leafmould:
Fill a dustbin (trash can) or black polythene sack with alternate 7.5-10 cm (3-4 inch) layers of 2-3 year old leafmould and wilted, chopped Comfrey leaves. Firm down gently and add moisture if the leaves are dry. If the leafmould is wet, turn it out to dry off for a few days before making up the mixture.
It is ready to use in 2-5 months or when the Comfrey leaves have virtually disappeared. *Use as a general potting compost or adjust to use as a seed compost by addition of 25% horticultural sharp sand.*
The pH of Comfrey leafmould is usually between 5.8 and 6.2. *If a more alkaline compost is required, ground dolomitic limestone may be added."*
-Comfrey for Gardeners: A Garden Organic Guide by Garden Organic. Coventry, England: no date but around **2015**, page 15.

"**Comfrey potting mixture: originally devised to utilize peat, now environmental awareness has led to a leaf mold based alternative being adopted instead.** *Two-year-old, well decayed leaf mold should be used, this will absorb the nutrient-rich liquid released by the decaying Comfrey.*
In a black plastic sack alternate 7-10 cm (2.8-3.9 inches) layers of leaf mold and chopped Comfrey leaves. *Add a little dolomitic limestone to slightly raise pH. Leave for between 2 to 5 months depending on the season, checking that it does not dry out or become too wet.* **The mixture is ready when the Comfrey leaves have rotted and are no longer visible.**
Use as a general potting compost, although it is too strong for seedlings."
-Wikipedia®: The Free Encyclopedia, www.wikipedia.org, **2018**.

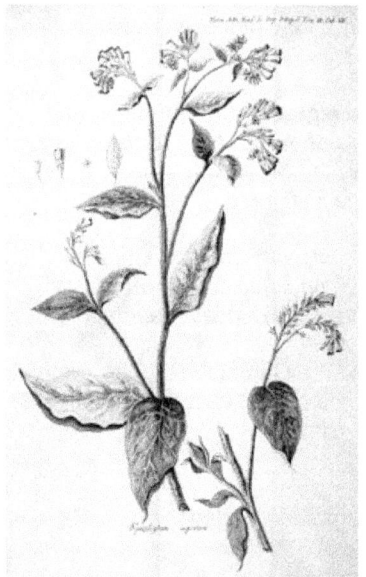

Symphytum Asperum 1802, 1805

'Nova Acta Academiae Scientiarum Imperialis Petropolitanae, Praecedit Historia Ejusdem Academiae, Tomus 14, Pt. 2' (New Journal of the Imperial Academy of Sciences, The History of the Academy, Volume 14, Part 2), by Imperatorskaia Akademia Nauk, Petersburg Imperial Academy of Sciences, Russia, 1797-1798.

Includes article in Latin by Ivan Lepechin about Symphyti Asperi (Boraginaceae Symphytum asperum Lepech.)
Text pages 442-444. Art is tab. 7. Art does not have page number. In the pdf the art is 12 pages after page 826.

The pdf download from BHL, www.biodiversitylibrary.org, dates this volume as 1797-1798.
'Nova Acta' on page 442 has Symphytum asperum date 1802.
IPNI, www.ipni.org/n/120743-1 has date 1805.

Chapter 18

Garden Uses of Comfrey:
Growing Vegetables and Fruit with Comfrey as Fertilizer

How to fertilize Comfrey: subsection 'Fertilization of Comfrey Plant' in section 'Planting, Soil, Fertilization, Water, Disease' in Volume 2.

Growing Potatoes with Comfrey as Fertilizer

"Though growing Comfrey foliage to use as a potato fertilizer would not pay on a farm scale: farmers can make more money feeding the foliage to pigs. Comfrey grows a garden potato crop which will have a chemical fertilizer yield, but with the genuine compost-grown flavour."
-Organic Gardening by Lawrence D. Hills. Middlesex, England: Penguin Books Ltd., **1977**, page 95.

Putting Comfrey Compost in Soil for Potatoes

"Comfrey can have a strong curing effect when it's used as compost, too. We used it for the compost we put in this bed which has potatoes growing in it, because it is good for the prevention of scab and other diseases that potates are susceptible to.
We don't just count on appearance to tell us that potaoes need Comfrey compost, though. **We rely on the research of the 'Henry Doubleday Research Association' on Comfrey.**
We use its compost on the potatoes in particular because it is a dense, almost liquid form of compost that can keep the soil pH down in the acid range far enough to prevent the occurrence of scab. We also use green manuring in the potato bed to make sure the acid level is sufficent.
But you can't quite get the full idea of what the Comfrey compost does for the potato by analyzing the soil pH and studying the admirable work of the 'Henry Doubleday Research Association' on Comfrey.
The result is more than the total achieved by the techniques involved. **You come to know what those potato plants need in a very deep, personal sense that transcends anything that can be learned from books.**
You know in a way that you can only learn if you watch over, nurture, and protect the potato plants throughout their life cycle. You develop a similar feeling for all the plants in the garden. It is based on classroom study, hard field work, and close detailed observation, but it can't be expressed by any other word than love."
-Enchanted Garden: Alan Chadwick's Organic Method of Gardening by Tom Cuthbertson. London, England: Rider & Company / Hutchinson & Co. Publishers Ltd, **1978**, page 97.
(Potato scab is caused by bacterium Streptomyces scabies. It overwinters in soil and fallen leaves. The bacterium survives indefinitely in slightly alkaline soil, but is somewhat scarce in highly acid soil. It is transmitted by infected seed tubers, wind and water.)

Timing Planting of Potatoes with Comfrey Growth

"Most gardeners aim to take their first cutting of Comfrey in time to go into the ground with the early potatoes, but where the Comfrey starts slowly in a savage (very cold) spring, it pays to keep the seed tubers stacked in their sprouting trays waiting for it.
There is no time wasted as long as the potato shoots are growing strong, short, and chunky, even if they stay unplanted till May and follow spring cabbages or early broccoli.
The quick start from Russian Comfrey Bocking No. 14 makes it the most valuable variety for garden use, because most organic gardeners use the bulk of their compost on potatoes, and replacing this with Comfrey saves them more for other crops."
-Fertility Without Fertilizers: A Basic Approach to Organic Gardening by Lawrence D. Hills. New York: Universe Books, **1975**.
(Russian Comfrey Bocking No. 14 is preferred over Bocking No. 4 because No. 14 has thinner stalks so is less likely to start growing into a Comfrey plant. However, all species of Comfrey are good for potatoes.)

"**Potatoes are planted late April through June**, and harvested June through October. Some people plant potatoes late March or early April but there is risk they will rot in cold soil.
Earliest planting is 2-3 weeks before last frost in spring.
In late April plant early/short season potatoes for harvest in June or July.
In late April or May plant mid season and late/storage potatoes for harvest August through October.
In June plant early/short season potatoes for harvest in August or September.
Early/ short season mature in 60-80 days. Mid-season mature in 80-100 days. Storage/late season mature in 100-130 days."
-Western North Carolina Farm and Garden Calendar: The Farming and Gardening Survival Book by Nancy Shirley, **2012**. Good for all eastern states (United States) in hardiness zones 5, 6, 7.

How Much Comfrey Per Row, and Type of Comfrey

"Russian Comfrey Bocking No. 14 is ready much earlier than Bocking No. 4, with a good cut ready for the maincrop potatoes. Because of its proportion of carbon compounds to nitrogen is 9.8 to 1, compared to 10 to 1 for good compost, it is 'compost without a heap'.
Five barrowloads wilt to four for use at the rate of 1 to 1 1/2 pounds (0.45-0.68 kg) per foot (0.3 meter) of potato row. Wilted Comfrey is also balanced just right for potatoes and any potash (potassium) demanding crop, to use dug in or as a surface mulch, or to add to the compost heap after the spring cuts have gone for the potatoes."
-Comfrey Report Number Two: For Gardeners, Farmers and All Comfrey Growers in All Countries, with Analysis, Yields and Cultivation Methods, and the Results of Seven Years More Work Since Our First Report by Henry Doubleday Research Association, Braintree, Essex, England, 45 pages, March **1963**.

"The wilting is because cut Comfrey if left overnight, it is flabby enough to pack neatly in the potato trenches, and it reduces the risk of thick stems growing as cuttings where they are unwanted.
Russian Comfrey Bocking No. 14 is the favourite garden variety partly because its small and slender stems wilt most easily and are least likely to grow in the trench."
-Comfrey: Fodder, Food and Remedy by Lawrence D. Hills. New York: Rizzoli Universe Books: **1976**, page 160.

"The early start which Bocking No. 14 gives is important, but the first cut must always go to the potatoes. We then take out potato trenches to a full digging fork's depth and spread 1 to 1 1/2 pounds (0.45-0.68 kg) of wilted Comfrey to every foot (0.3 meter) of row.
Some gardeners put an inch (2.5 cm) of soil on top of the Comfrey, but we often just set the seed tubers a foot (30 cm) apart for earlies, fifteen inches (38 cm) for maincrop, straight on the Comfrey, then spade the soil back over them.
The potatoes grow in what are, in effect, long, narrow compost heaps which release their plant foods as they are required."
-Comfrey: Fodder, Food and Remedy by Lawrence D. Hills. New York: Rizzoli Universe Books: **1976**, page 164, 165.

"Then in 1956 I tried treating Comfrey foliage as a kind of 'Instant Compost', using it at the rate of 1 1/2 pounds (0.68 kg) a foot (0.3 meter) of row, which doubled the yield of potatoes compared with the 'no-manure' plots."
-Comfrey: Fodder, Food and Remedy by Lawrence D. Hills. New York: Rizzoli Universe Books: **1976**, page 158.

"Spread cut Comfrey out to wilt in sun for 24 hours, ideally turning and leaving it another twenty-four if you are using a thick stemmed variety other than Bocking No. 14 to make sure the stems are really well wilted, to prevent them growing."
-Fertility Gardening: The Organic Way to Make Your Garden Grow by Lawrence D. Hills. London, England: Cameron & Tayleur Books Limited, **1980**.

How to Use Comfrey in Potato Rows or Boxes

"Potato trenches: The Comfrey went in the bottoms, then some soil, and the potaoto seed tubers went 15 inches (38 cm) apart on top. This was to keep the crop well down, to avoid damage by pheasants.
In gardens the trench is usually dug just a bit deeper than normal, and the Comfrey put in like ordinary manure."
-Comfrey Report Number Two: For Gardeners, Farmers and All Comfrey Growers in All Countries, with Analysis, Yields and Cultivation Methods, and the Results of Seven Years More Work Since Our First Report by Henry Doubleday Research Association, Braintree, Essex, England, 45 pages, March **1963**.

"When Comfrey is late in a cold season, or for early potatoes chitted and ready while the Comfrey is still short of a good cut, the seed potatoes can be planted three inches (7.6 cm) deep on the flat and the cut Comfrey spread on the surface as it becomes available, even as late as June.
As the potatoes grow it is raked along the rows, acting as frost protection for earlies in savage (very cold) springs, then soil from the row middles is piled on it, making a smaller than normal earthing-up.
Comfrey has also been spread along the sides of the potato ridges, suppressing seeds and feeding the crop below, but the gain is not as great as with Comfrey in the trenches."
-Grow Your Own Fruit & Vegetables by Lawrence D. Hills. London, England: Faber and Faber Limited: **1971**, page 48.
    (Chitting or greening is pre-sprouting seed potatoes before planting.)

"The cut Comfrey will heat like a long narrow compost heap and the potatoes get away fast, with balanced supply of plant foods just where they are needed. When they are dug, there will only be a little powdery black humus remaining."
-Fertility Without Fertilizers: A Basic Approach to Organic Gardening by Lawrence D. Hills. New York: Universe Books, **1975**.

"Fill in the trenches without treading the soil which could break shoots, and when the shoots come through, draw soil over them with a hoe if frost threatens, for if the growing points are killed, the carefully sprouted seed potatoes must go back to the beginning and start again."

-Organic Gardening by Lawrence D. Hills. Middlesex, England: Penguin Books Ltd., **1977**, page 59.

"Take out potato trenches and lay the wilted Comfrey along them at the rate of 1 to 1 1/2 pounds to the foot of row (1.5 to 2 kg/m); set the potato seed along them, 1 foot (30 cm) between tubers for earlies, 15 inches (40 cm) for maincrops, and spade the soil back.
This is a version of the old trick of filling the trenches with lawn mowings so that **the acidity of these long 'compost heaps' reduces damage by the potato scab fungus, Actinomyces scabies.**
Comfrey has the same effect, but supplies more nitrogen and twice as much potassium."
-Fertility Gardening: The Organic Way to Make Your Garden Grow by Lawrence D. Hills. London, England: Cameron & Tayleur Books Limited, **1980**.

"We have had potatoes growing for 12 years in straw boxes. Near that, you grow a couple of Comfrey plants, because you should always include a Comfrey leaf.
Pick a Comfrey leaf, put your potato in it, wrap it up, put it under the straw, and that is your potash and nutrients."
-'Introduction to Permaculture: Permaculture Design Course Series, Pamphlets 1 to 14' by Bill Mollison at 'The Rural Education Center', Wilton, New Hampshire, **1981**; published by Yankee Permaculture, Sparr, Florida.

Digging Potatoes in the Fall

"When the potatoes are dug, the Comfrey has rotted to a black and powdery humus, but not so much as there would be left from compost or manure, for the fast-grown foliage has not made the harder material that lasts; it cannot replace either, but makes both go further."
-Down to Earth Fruit and Vegetable Growing by Lawrence D. Hills. London, England: Faber & Faber, 1960, page 57.
(Humus is the dark organic material in soils that is produced by the decomposition of vegetable or animal matter.)

Preparing Potato Patch in the Fall

"In 1957 we compared digging the wilted Comfrey under in the autumn with (versus) putting it in the trenches in the spring. **The autumn digging, dug in late and stemmy (to break the green-manuring rules) still gave an increase, but only half as much again, not twice the weight.**
This was probably because the winter rain washed out the potash, but partly because the buried Comfrey heats and warms the potatoes ahead a bit, and **the autumn dug-in foliage had spent its heat."**
-Compost, Comfrey and Green-Manure: 'Henry Doubleday Research Association' First Gardeners Report by Lawrence D. Hills. Braintree, Essex, England: Henry Doubleday Research Association, **1959**.

"I have laid my last Comfrey cut in the autumn on the site reserved for next year's early spuds (potatoes). Over this I have laid a spread of strawy horse manure to remain in situ (in place) till next April. This solves the Comfrey problem for earlies (early potatoes). This policy of spreading the last cut under manure, or grass mowings in September or October for earlies, is popular also for onion beds."
-Comfrey: Fodder, Food and Remedy by Lawrence D. Hills. New York: Rizzoli Universe Books: **1976**, page 166.

Growing Vegetables with Comfrey as Fertilizer

Cabbage, Bean, Fruit (bush), Lettuce, Onion, Pea, Potato, Radish, Sunflower, Tomato

"A Cheshire (northwest England) amateur gardener **used Comfrey dug in like dung and secured an increased yield of outdoor tomatoes and a better lettuce crop with reduced bolting tendency** *than on his normal compost-dressed garden."*
-Compost, Comfrey and Green-Manure: 'Henry Doubleday Research Association' First Gardeners Report by Lawrence D. Hills. Braintree, Essex, England: Henry Doubleday Research Association, **1959**.
(Bolting is the growth of flowers on agricultural crops before the crop is harvested.)

"Green-manuring (mulching) fits the potatoes, then tomatoes outdoors or in a greenhouse, or late peas or lettuce or beans, always giving it to crops that need the potash (potassium) most.
With **potatoes** it gives the real organic flavour, and those who try compost or Comfrey should compare this with some grown with Muriate of Potash (potassium chloride, a non-organic fertilizer).
Lawn mowings are far better value used this way than in a compost heap. An inch (2.5 cm) of Comfrey and two (5 cm) of grass is a fair allowance, but it depends how much you have. With these organic feeds that are released by bacterial action there is no need to think in ounces a square yard."
-Compost, Comfrey and Green-Manure: 'Henry Doubleday Research Association' First Gardeners Report by Lawrence D. Hills. Braintree, Essex, England: Henry Doubleday Research Association, **1959**.

*"**The potato crop takes the first cut from the Comfrey bed.***
***The second cut** is trodden into the trenches, two feet (0.6 meters) apart, in which **pole beans** are sown, with about 4 inches (10 cm) of wilted and trodden Comfrey foliage and four inches of soil on top, which will provide ample potash on the sandiest soil.*
***Another good use is under bush fruits,** spread on the surface and covered with lawn mowings.*
***The last Comfrey cut is best dug in as though it were manure, for the onions,** because it is the best source of cheaply available organic potash fertilizer, quicker acting than any of the rock dust minerals. Onions like firm soil, so do not dig it in for them in the spring, the soil needs the winter to settle down."*
-Comfrey Report: The Story of the World's Fastest Protein Builder and Herbal Healer, Conservation Gardening and Farming Series: Series C by Lawrence D. Hills. England: Henry Doubleday Research Association, **1975**, page 85.

*"**Wilted Comfrey, with rather more than 3 times the potash of farmyard manure and with a similar phosphorus content, has much the same balance of plant foods as a chemical potato and tomato fertilizer,** and one of the first discoveries on the HDRA (Henry Doubleday Research Association) trial ground in England was that it behaved exactly like such a fertilizer.*
***Adding 1 1/2 pounds (0.68 kg) of wilted Comfrey to every foot (30 cm) of potato row doubled the yield** compared with untreated rows and produced a yield roughly the same as the rows treated with the 'complete artificials', but with a superior flavor resembling that of compost-grown potatoes."*
-Fertility Without Fertilizers: A Basic Approach to Organic Gardening by Lawrence D. Hills. New York: Universe Books, **1975**.

*"**The next Comfrey cutting, some five or six weeks later (after the cuttings for potatoes), can go on the surface of the soil between bush fruit, especially gooseberries or tomatoes,** and covered with a 2-inch-thick (5 cm) coat of lawn mowings. This suppresses weeds and makes a sheet of compost, helping the Comfrey to decay and relese its potash.*
***Spread the cut and wilted Comfrey between the tomatoes, but don't bring this or the mowings right up to the stems because of the slug risk.** It should be said that the 'Ruth Stout' and other surface mulch methods used in the United States and Canada work better with their drier climate and lower slug population than in Britain."*
-Fertility Without Fertilizers: A Basic Approach to Organic Gardening by Lawrence D. Hills. New York: Universe Books, **1975**.
    (Ruth Imogen Stout, 1884-1980, was an American author known for her 'No-Work' gardening books and techniques.
    One book is: 'The Ruth Stout No-Work Garden Book: Secrets of the Year-Round Mulch Method' by Rodale Press.)

"Clay soils usually have plenty of potassium. Sandy soils, however, run out fast, as all potassium fertilizers, including wood ashes, are very soluble.
***With potassium shortages tomatoes** have light brown scorching at the edges of the leaves, and the fruit ripens patchily, with green and yellow areas merging into the red surface.*
***Potato leaves** with potassium shortage turn a bronzy green with brown spots between the veins, and the foliage often dies down too soon for a good yield.*
***Beans suffer sorely from a potash shortage**- the chocolate spot fungus of broad beans attacks when this runs low. The symptoms in broad beans are very short joints and dark brown scorching at the leaf edges.*
***The cabbage tribe** show that they are missing their potash by brown leaf margins which curl upwards, and this is also the sign with **lettuce and beet**.*
The best immediate answer is to water with Comfrey liquid manure.
    *Leave this to stand till it is clear, dilute 1 part of liquid with 3 parts of water, and spray on as a foliar feed for tomatoes. With plenty available this can be used to soak potatoes or gooseberries, but the best answer is more compost, or **put cut Comfrey under grass mowings for bush fruit**."*
-Organic Gardening by Lawrence D. Hills. Middlesex, England: Penguin Books Ltd., **1977**, page 181.

*"Methods for the production of **Comfrey (Symphytum x uplandicum = Russian Comfrey Bocking #14) liquid fertilizer** and **Comfrey leafmould** are described.*
Comfrey Liquid Fertilizer:
    Comfrey leaves decomposed rapidly to give a liquid fertilizer. The concentration of nitrogen, phosphorus and potassium in the liquid varied between successive harvests.
    In trials with radish (Raphanus sativus), liquid from the first harvest of Comfrey leaves in spring gave significantly greater yields than did two commercial organic liquid fertilizers.
    In trials with tomato (Lycopersicon esculentum), with the combined liquid of all the harvests for one year, yields with Comfrey liquid did not differ significantly from two commercial organic feeds but were significantly lower than two inorganic liquid fertilizers. The combined Comfrey liquid had higher potassium but lower nitrogen and phosphorus concentrations than the liquid from the first harvest.
Comfrey Leafmould:
    *In separate trials **tomato plants were grown in a potting compost made by mixing Comfrey leaves and leafmould** and the yields compared with those obtained with four commercial organic composts. In 1988 two of the commercial products gave higher yields than Comfrey leafmould, whereas in 1989 only one of the commercial products gave a significantly higher yield.*
    Comfrey leafmould is a good balanced nutrient source for composts and compared favourably with commercial organic potting composts for use by the amateur gardener.
It is concluded that Comfrey gives high yields of leaf material and liquid fertilizer. It is a rich source of nutrients containing particularly high levels of potassium."

-'The Use of Comfrey (Symphytum x uplandicum) as a Source of Plant Nutrients' by P.J.C. Harris, W.F. Bourne, C. Gitsham and M. Lennartsson, 'Henry Doubleday Research Association', Coventry, England and 'Department of Biological Sciences', Coventry Polytechnic, England, January **1990**.

"This research aimed to evaluate the allelopathic effect of the aqueous (water) extracts of the aerial (above ground) part of the carqueja (Baccharis trimera) plants and of the entire plant of Comfrey on the seed germination and seedling development of sunflower.
    The treatment adopted was composed of five concentrations of extract: 0, 50, 100, 150 and 200 grams/liter (7 ounces/ 1.05 quart), of plants in water. The research was carried out in the greenhouse and in the 'Laboratory of Seed Analysis'. In the greenhouse, 10 plastic bags were used to represent an experimental plot. The soil was irrigated with 70 milliliters (2.3 fluid ounce) of the different treatments at sowing and with 35 milliliters (1.18 fluid ounce) two days later. The observations were carried out during 35 days.
An inhibitory effect of Comfrey extract on seeds and seedlings of sunflower in concentrations up to 150 grams/liter was not observed in the laboratory. (In other words, Comfrey extract did not hurt sunflower seeds or seedlings.)
In the greenhouse, this Comfrey extract in all tested concentrations were favorable to seedling development.
The carqueja extract at the concentration 200 grams/liter promoted an emergence delay of seedlings in the greenhouse and the occurrence of abnormal seedlings in the germination tests in the laboratory."
'Allelopathic Effect of the Carqueja (Baccharis trimera) and Comfrey (Symphytum officinale) Extract on Sunflower Seeds -and Seedlings' by A.G. da Silva and R.I.N. de Carvalho, Revista Academica Ciencias Agrarias e Ambientais (Academic Magazine Agricultural and Environmental Sciences), Curitiba, Brazil, Volume 7, No. 1, pages 23-32, January/March **2009**.
    (Allelopathy is a biological phenomenon by which an organism produces biochemicals that influence the growth, survival, and reproduction of other organisms. These biochemicals are known as allelochemicals.)

Fermented Extracts from Comfrey Help Plants Grow

"Ecological agriculture has presented innovations in the productive system, using the products applications into the seeds a routine practice. Plants biostimulant use has shown great potential on the productivity.
This study aimed to evaluate the effect of different concetrations of the Comfrey hydroalcoholic extract in the germination and the early development of maize (corn) seedlings.
    In this study five treatments with different extract concentrations (0, 0.001, 0.01, 0.1 and 1 ml/liter) were used in the processing of corn seeds before planting. The variables analyzed were percentage of seed germination, seedling emergence rate index, shoot length, root length and seedling dry matter. The experimental design was completely randomized design with four replications.
According to the results it was verified that the Comfrey extract concentration of 0.01 ml/liter had a positive effect on root length and dry weight of maize seedlings."
-'Effect of Comfrey Extract on Germination and Vigor of Corn Seeds' by Oliveira et al., Universidade Federal do Ceara, Brazil; Revista Brasileira de Higiene e Sanidade Animal (Brazilian Journal of Animal Hygiene and Health), Volume 7, No. 1, pages 1-8, January-June **2013**. All in Portuguese except abstract is also in English. If you have English translation, could you send it to me.

"**The trial tests of Special Herbal Preparation (SHP) in the capacity of PGR (Plant Growth Regulators) were executed on cucumber, pepper and tomato at a large glasshouse in Orenburg Region of Russian Federation.**
Special preparation was based on wild growing herbs: Comfrey (Symphytum officinale), dandelion (Taraxacum officinale), horsetail (Equisetum arvense), lemon balm (Melissa officinalis), nettle (Urtica dioica), valerian (Valeriana officinalis) and yarrow (Achillea millefolium).
In order to extract the active matter without application of chemical means, an unique production process has been developed. The foundation of this process is based on a special combination of water extraction and biofermentation with continual oxygen enrichment.
    Experiment was executed as per following treatments: 1. Control, 2. Basic fertilization + SHP, 3. Basic fertilization, 4. Basic fertilization + additional fertilization + SHP, 5. Basic fertilization + additional fertilization. Treatments were repeated twice with random allocation on 100 m2 (square meters) sample plots for cucumber and tomato and 50 m2 for pepper.
In order to increase the affinity (attraction) of water towards extraction, tap water was softened with softener and then heated to 40 C (104 F) temperature. **A total of 131 kg (288 pounds) of dried herbs was flooded with 1,160 liters (306 US gallons) of thus prepared water according to the following recipe:**
    Comfrey root (Symphytum officinale) 22%, dandelion leaf (Taraxacum officinale) 10%, horsetail herb (Equisetum arvense) 8.5%, lemon balm leaf (Melissa officinalis) 8.5%, nettle leaf (Urtica dioica) 28%, valerian root (Valeriana officinalis) 15%, yarrow herb (Achillea millefolium) 10%.
The examinations have shown that SHP (Special Herbal Preparation) had a positive impact on growth and development of vegetable crops planted in the glasshouse.
When observed crops were provided with basic fertilization, a higher seasonal yield had always been achieved whenever SHP (Special Herbal Preparation) was added. The same applies when observed crops were provided with basic and additional fertilization. **The results also suggest that SHP has the capacity of an uptake improver."**
-'Influence of Special Herbal Preparation on Growth and Development of Cucumber, Pepper and Tomato' by S. Ljubojevic and V.I. Eliseev, The Independent University of Banja Luka, Faculty for Ecology and Institute for Scientific Research, Bosnia-

Hercegovina; Proceedings of the VII International Scientific Agriculture Symposium, Agrosym 2016, Bosnia and Herzegovina, pages 1745-1754, October 6-9 **2016**.

Tomatoes and Comfrey

"Tomatoes need a liquid manure, beginning when they are setting their first fruit and at weekly intervals through the summer, and this can be supplied effectively with Comfrey liquid manure.
The ideal container for this is a 20-gallon (75.7 liter) fiber glass water tank, propped up on bricks so that a can will go under the tap. Fill this with a hose and stuff with 14 pounds (6.3 kg) of Comfrey, replacing the lid of the tank. The Comfrey will go black rapidly and form a kind of crust on the surface.
    After a month, the liquid from the tap can be used, undiluted, as a tomato feed and the tank refilled for a second but weaker brew in another fortnight (2 weeks).
The residue then can be tipped out for the compost heap and a further charge added.
If the second steeping is drawn off and allowed to settle, it becomes clear enough for use with an automatic feeder, while the crude liquid can be **used for watering runner beans (pole beans) and as a general tonic for vegetable plants.**
The Comfrey liquid manure had nearly three times as much potash (potassium) as nitrogen, and far less phosphorus, which is the ratio that tomatoes need.
This idea was invented by George Gibson of Guernsey, England, in 1955, and he grew a special Comfrey bed to cut for his tomato greenhouse feeding. In his view, it was a waste to compost Comfrey because it broke down so fast and wasted its minerals when they washed into the soil under the compost bin."
-Fertility Without Fertilizers: A Basic Approach to Organic Gardening by Lawrence D. Hills. New York: Universe Books, **1975**.
    (See subsection 'Comfrey as Liquid Manure' in section 'Garden Uses of Comfrey' {Chapter 17}.)

"We give our plants their culinary associates, which have a secondary effect of being weed barriers. When you go for your tomatoes, you get some basil and parsley right in the same basket.
If you want to put a couple of Comfrey plants out there, do it. **A Comfrey leaf under the mulch near the root of your tomato will supply potash.**"
-'Introduction to Permaculture: Permaculture Design Course Series, Pamphlets 1 to 14' by Bill Mollison at 'The Rural Education Center', Wilton, New Hamshire, **1981**; published by Yankee Permaculture, Sparr, Florida.

"Russian Comfrey, Symphytum x uplandicum, was tested as a plant nutrient source in a trial with tomatoes. Tomato plants of the cultivar Aromata planted in a medium of soil : peat (1:1) enriched with organic nutrients and lime, were **fed with Comfrey liquid plus Biofer®, in 3 different concentrations.**
Two different concentrations of BioRika® were used as comparison, along with an unfertilized control group.
    In the '**Comfrey Treatment I**' a total amont of 1 gram (0.03 ounce) Nitrogen,
        0.23 g Phosphorus (P), 0.84 g Potassium (K), was given to the tomatoes.
    In '**Comfrey Treatment II**' they were given 1.3 g N, 0.3 g P, 1 g K.
    In '**Comfrey Treatment III**' they were given 1.7 g N, 0.4 g P, 1.9 g K.
In the 2 treatments with BioRika tomatoes were given 0.23 g N, 0.08 g P, 0.3 g K and 0.44 g N, 0.15 g P, 0.6 g K, respectively.
    The treatments were given at 11 occasions, starting when tomatoes were 34 days and ending when 64 days old.
The treatments gave a significant effect on the fresh weight, water content/dry matter, pH and conductivity, as the unfertilized control group differed considerably from the fertilized groups. No significant differences between the Comfrey treatments and the BioRika treatments appeared during the short experimental period.
The literature study shows that Comfrey liquid is not sufficient as a total fertilizer. The low nitrogen content makes it less suitable for organic transplant production.
However, Comfrey liquid can be used in fruit setting crops as tomatoes, peppers and cucumber, from the time of flowering, as it is a rich source of potassium."
-'Comfrey as Plant Nutrition Source: A Trial and a Literature Study' by Anna Brusling, Swedish University of Agricultural Sciences (SLU), Degree Projects in Horticulture Program, Horticultural Management, Alnarp, Sweden, **2007**. All is in Swedish except abstract is in English. If you have a translation, could you please send it to me.
    (Biofer® is a natural organic fertiliser with nitrogen, phosphorus, potassium and minor nutrients.)

"Suggestions for Using Dried Comfrey:
Tomatoes can have Comfrey added to the planting hole, and then do not use again until the first flowers are open. Other people suggest that you should wait until the first truss is expanding in size. The choice is yours, but I prefer to begin at flowering and cease a while before harvest is complete.
    Comfrey is not as quick acting as soluble chemicals. How much to use is a debatable matter.
Assuming you have planted tomatoes into fertile soil, compost or grow bags, I believe that the tomato plant is capable of producing to its capacity with some little and regular dried Comfrey applications, say 30 grams (1 ounce) every 3 to 4 weeks, worked into the top inch (2.5 cm) of soil.
Cease feeding when there are still a few tomatoes left to harvest. There will be some nutrients left from previous feeds, and there is no point in applying excess that will not be used. **Treat sweet peppers in the same way as tomatoes.**"
-'Using Comfrey' by Old McDonald in Portugal: Farming, Gardening, Wildlife and Good Food,

http://oldmcdonaldinportugal.blogspot.com, November 2 **2014**.
    (A tomato plant truss is a group or cluster of smaller stems where flowers and fruit develop.)

Onions, Garlic and Comfrey

"**The final cut of Comfrey in the autumn has also been wilted and dug into an onion bed as though it were a manure,** which gives the bed time to settle through the winter.
However, Comfrey is not ready early enough in spring to go under onion sets, and it distrubs the crop too much as it sinks, so **the best way to give onions Comfrey potash is as a liquid manure when the bulbs start to plump about May.**"
-Fertility Without Fertilizers: A Basic Approach to Organic Gardening by Lawrence D. Hills. New York: Universe Books, **1975**.
    (This would work with garlic too.)

"**Swedes (Rutabaga), Onions, Garlic, Leeks, Radish, Turnips:**
Comfrey really comes into its own with fruiting and non-brassica species - or brassicas such as swedes (rutabaga) that are grown for the root rather than the leaves.
I like it for alliums in particular.** The dried Comfrey can be worked into the soil once plants have achieved good growth and are beginning to form the parts used for consumption - i.e. **root swelling in swedes, bulbing in onions, clove formation in garlic, and increasing stem circumference for leeks.**
All of them will benefit from applications of Comfrey every few weeks after the growth stages mentioned, and I would suggest 30 grams (1 ounce) per square metre (10.7 square feet) per month is adequate for those plants destined for kitchen use.
Use of Comfrey commences (begins) part way through each plant's growth so the number of applications is limited. **Very short term crops such as radish and white turnips** are used quickly enough that they would only benefit from Comfrey worked into the seed bed. A later application to the soil is not needed."
-'Using Comfrey' by Old McDonald in Portugal: Farming, Gardening, Wildlife and Good Food,
http://oldmcdonaldinportugal.blogspot.com, November 2 **2014**.
    (Brassica is a genus of plants in the mustard family. It includes bok choy, broccoli, Brussels sprouts, cabbage, cauliflower, collard, kale, kohlrabi, mustard, rape, rutabaga/swedes, turnip.)
    (Allium is a genus of monocotyledonous flowering plants that includes onions, garlic, scallions, shallots, leeks, chives.)
    (The Comfrey leaves used can be dry or fresh.)

Beans / Peas and Comfrey

"**The next cut of Comfrey leaves can go in the same type of trench as potatoes but less deep for runner (pole) beans, French beans, or late peas.** The Comfrey below acts as a moisture-holder as well as a lasting potassium-rich feed for the summer. As it decays the soil will sink, allowing room for more watering round the stem."
-Grow Your Own Fruit & Vegetables by Lawrence D. Hills. London, England: Faber and Faber Limited: **1971**, page 48.

"**French beans need potash just as much as broad beans,** and on sandy soils it pays to water them with Comfrey liquid manure. **Cut Comfrey can be spread between the rows,** or trodden into 6-inch (15.2 cm) deep trenches with 3 inches (7.6 cm) of soil on top before sowing on sands where French beans fail. Lawn mowings can be used on the surface or in trenches by those who are short of Comfrey."
-Organic Gardening by Lawrence D. Hills. Middlesex, England: Penguin Books Ltd., **1977**, page 123.

Wheat Seedlings Helped by Comfrey

"**Allelopatic (allelopathic) interactions between plants in nature are caused by physiologically active compounds.** Substances of the plant-donor penetrate into a plant-acceptor, influencing the latters metabolism and thereby the mutual relations with the plant-acceptor.
    In addition they affect soil borne microorganisms including pathogenic fungi. A common pool of organic compounds used by both plants is involved in such exchanges.
It was shown that the weed plants common wormwood (Artemisia absinthium L.), **Comfrey (Symphytum officinale L.),** and cowparsnip (Heracleum sibiricum L.) **influence the growth of wheat seedlings and also the compability of these seedlings to brown sheet rust** (Puccinia recondite f. sp. tritici) under field conditions.
It was established that water extracts from the leaves and the roots of all the weeds used in the work contained great quantities of phenolic compounds. The products of phenols oxidation were **tannins.** The toxic properties of tannins are based on the inhibition of protein synthesis and also on the inactivation of enzymes.
The growth of wheat seedlings was slightly suppressed. There also were some changes in the plant metabolism. Increases in peroxidase activity, in the reaction activity of cell sap and also in the contents of phenols were detected.
These changes resulted in the increase in resistance of the wheat plants to stem rust. Damage due to sheet brown rust was reduced 30%. Inhibitory effects of the weed plants were distinctly observed in a radius up to 30 cm (11.8 inches) after which it weakened."

-'Allelopatic (Allelopathic) Influence of Weeds on Resistance of Wheat Seedlings to Puccinia Recondita f. sp. tritici' by T.P. Yurina, Faculty of Biology, M.V. Lomonosov Moscow State University, Russia; Modern Fungicides and Antifungal Compounds V: 15th International Reinhardsbrunn Symposium, Friedrichroda, Germany, May 6-10 2007, pages 299-301, **2008**.

> (Allelopathy is a biological phenomenon by which an organism produces biochemicals that influence the growth, survival and reproduction of other organisms. These biochemicals are known as allelochemicals.)
> (Puccinia recondita is a plant fungal pathogen. Puccinia triticina causes 'black rust'. Wheat leaf rust affects wheat / barley / rye stems, leaves and grains.)

"**Treatment of healthy wheat (Triticum aestivum L., cultivar Lyubava), with the extracts from Common Comfrey (Symphytum officinale L.), cowparsnip (Heracleum sibiricum L.), and giant knotweed (Reynoutria sashalinensis) stimulated the photosynthetic activity in treated wheat leaves.**
In the third part of our work, we studied the effect of plant extracts on the photosynthetic apparatus of healthy plants.
Spraying of wheat seedlings with fungitoxic extracts from Comfrey (Symphytum officinale L.), and cowparsnip (Heracleum sibiricum L.) caused a slight but well reproducible increase in the chlorophyll content; the same result was obtained for the extract from giant knotweed (Reynoutria sachalinensis).
We also demonstrated an explicit (obvious) increase in the photosynthetic activity (O2 {oxygen} evolution per mg chlorophyll) in the leaves treated with the extracts.
> Biophysical methods revealed the stimulation of photosystem (PS) II activity and the increase in the rate of electron transport between the photosystems in treated leaves."

-'Plant Extracts as the Source of Physiologically Active Compounds Supressing the Development of Pathogenic Fungi' by V.A. Karavaev, et. al., M.V. Lomonosov Moscow State University, Russia; Plant Protection Science, Proceedings of the 6th Conference of European Foundation for Plant Pathology, Prague, Czech Republic, Volume 38, Special Issue 1, pages 200-204, **2002**.

Vegetable Crops Not Helped by Comfrey

"The latest information regarding the trenching of Comfrey according to Fred McPherson, 'Organic Gardening', says that **there are only three crops which Comfrey does not benefit: Brussels sprouts, which are inclined to 'blow' (the sprouts open up like a cabbage) with no increase of yield; carrots and salsify which are inclined to fork in their roots.** As Brussels sprouts do not require too much potassium, this may account for the problem, it having an adverse effect upon their growth."
-Comfrey: Symphuo Symphytum: A Multi-Purpose Herb by Philip Clarke. Edinburgh, England: The Pentland Press, **1997**, page 7.
> (I was unable to find article/book Mr. McPherson in 'Organic Gardening' wrote. If you have a copy, could you send it.)

"**But using Comfrey to mulch root crops (like carrots) or leafy greens (like lettuce and spinach) may encourage them to go to seed prematurely (bolting too soon).**"
-'Power Plant' by Jean M.A. Nick, Organic Gardening Magazine, Pennyslvania, page 51-53, June/July **2004**.

> **An opposite experience with lettuce going to seed:** "A Cheshire (northwest England) amateur gardener **used Comfrey dug in like dung and secured an increased yield of outdoor tomatoes and a better lettuce crop with reduced bolting tendency** than on his normal compost-dressed garden."
> -Compost, Comfrey and Green-Manure: 'Henry Doubleday Research Association' First Gardeners Report by Lawrence D. Hills. Braintree, Essex, England: Henry Doubleday Research Association, **1959**.
> (Perhaps in one situation there was too little potassium, and in the other too much.)

Large Dose of Comfrey Reduces Growth of Lettuce

The following studies do not seem particularly useful to the average gardener. Overdosing a seed or seedling with concentrated Comfrey extract is not what happens in nature.

This study created extracts from plants that contain Pyrrolizidine Alkaloids, and applied them to lettuce to see how the plant reacts.
"**The aim of the present study was to determine whether there is a connection between the total PAs (Pyrrolizidine Alkaloids) content of the mentioned species and their phytotoxicity (plant toxicity).**
Tussilago farfara (coltsfoot), Petasites hybridus (common butterbur), Senecio vernalis (eastern groundsel) and **Symphytum officinale (Comfrey)** are species traditionally used in phytotherapy that besides the therapeutic compounds contain PAs.
The total PAs and the corresponding N-oxides content of the solid plant extracts were measured using Ehrlich's method with senecionine as a standard substance.
> **The highest content of PAs was found in Senecio vernalis extract (424.92 +- 9.81 mg%), followed by Symphytum officinale extract (150.24 +- 10.35 mg%)** and Petasites hybridus extract (0.021 +- 0.00091 mg%). The lowest concentration was found in Tussilago farfara extract (0.0097 +- 0.00072 mg%).

In order to assess the toxicity of the extracts on Lactuca sativa (lettuce), inhibitory concentrations 50% (IC50) of the root elongation were calculated. **The extracts inhibited (reduced) the elongation of the Lactuca sativa radicle, in a dose dependent manner.**
Senecio vernalis extract exhibited the highest toxicity (low IC50), whilst Symphytum officinale extract, having the

second highest concentration of PAs, has the highest IC50. (A high IC50 means less toxicity.)
Further studies should be performed in order to determine whether the inhibitory effect is due to the PAs content and how it is influenced by the other components of the extracts."
-'The Effect of Certain Plant Extracts Containing Pyrroilizidine Alkaloids on Lactuca Sativa Radicle Growth' by O.C. Seremet, D. Balalau and S Negres, Department of Pharmacology and Clinical Pharmacy, Carol Davilla University of Medicine and Pharmacy, Bucharest, Romania; Romanian Journal of Biophysics, Volume 24, No. 1, pages 1-9, **2014**.

> (In botany, the radicle is the first part of a seedling that emerges from the seed during the process of germination.)
> (The IC50 parameter is the concentration that inhibits root elongation 50%, compared to the negative control. The lower the IC50, the more the toxicity.)

"The paper presents results of a large-scale experiment dealing with the impact of various herbal extracts on yield of lettuce (Latuca sativa), conducted at the laboratory of plant physiology.
Plants were grown on hydroponic with half-strength Hoagland's solution. The following characteristics were measured: number of leaves, length of the green parts of plants, root length, total weight, together with visually assessed plant vigor.
Trial included 12 treatments with 30 repetitions each, and that were:
> **Leaf extract of Comfrey (Symphytum officinale),** *leaf and root extract of nettle (Urtica dioica), leaf and root extract of dandelion (Taraxacum officinale), extract of above-ground part of horsetail (Equisetum arvense), flower extract of yarrow (Achillea millefolium), extract of above-ground part of lemon balm (Melissa officinalis), root extract of valerian (Valeriana officinalis), root extract of burdock (Arctium lappa), extract of above-ground part of squirting cucumber (Ecballium elaterium), commercially produced gibberellic acid, commercially produced plant growth regulator (PGR) based on seaweed extract and control group.*

Application of plant extracts did not contribute to a significant increment in the number of leaves on plant compared to the control group.
Treatments with Comfrey and horsetail led to slowing the development of new leaves, and the differences arising in the reporting period were statistically significant.
> Based on these results we can cautiously conclude that the extracts of Comfrey and horsetail act as retardants, when it comes to the number of leaves in the early growth stage of lettuce.

Application of nettle, yarrow, valerian and squirting cucumber extracts contributed to increasing the length of roots in comparison to the control group, however, the observed differences were not statistically significant.
On the contrary to this, **application of Comfrey, dandelion and horsetail extracts showed that root growth in length was significantly slowed compared to the control group.**
The treatment with lemon balm extract did not affect the root's growth in length of those intensity as that they did in case of above ground parts.
Comfrey extract acts as an inhibitor because it slows down the formation of leaves, development of above ground and underground part in length and conclusive increase of fresh biomass."
-'Impact of Various Herbal Extracts on Yield of Lettuce (Lactua Sativa)' by S. Ljubojevic, Faculty for Ecology, Independent University of Banja Luka, Bosnia-Herzegovina; Sixth International Scientific Agricultural Symposium, Agrosym 2015, Jahorina, Bosnia and Herzegovina, pages 1118-1126, October 15-18 **2015**.

> (Hydroponics is a method of growing plants using nutrient solutions in water, without soil.)
> (Hoagland solution is a hydroponic nutrient solution developed by Hoagland and Arnon in 1938. It provides every nutrient necessary for plant growth.)

Fruit and Comfrey (Small Fruiting Plants, Fruit Bushes, Fruit Trees)
See sub-subsection 'Tomatoes and Comfrey' in subsection 'Growing Vegetables and Fruit with Comfrey Fertilizer' in section 'Garden Uses of Comfrey' (Chapter 18).

Small Fruiting Plants and Fruit Bushes

"The late Comfrey leaf cuttings can mulch between bush fruit, especially gooseberries on a sandy soil. **Spread the wilted Comfrey on the surface and put lawn mowings on top of it,** *to rot fast and release the plant foods as well as suppressing weeds.* **This can also be done round tree fruit,** *but the result does not show so quick.*
Lawn mowings are far better value used this way than in a compost heap. An inch (2.5 cm) of Comfrey and two (5 cm) of grass is a fair allowance, but it depends how much you have.
> *With these organic feeds that are released by bacterial action, there is no need to think in ounces a square yard."*

-Compost, Comfrey and Green-Manure: 'Henry Doubleday Research Association' First Gardeners Report by Lawrence D. Hills. Braintree, Essex, England: Henry Doubleday Research Association, **1959**.

*"**All raspberries appreciate** 4 ounces (113 grams) a square yard (0.836 square meter) of wood ashes or seaweed meal each spring, or **cut Comfrey under a two-inch-thick (5 cm) lawn mowings mulch.***
***Put on first in April or May, so that the ground has time to warm up,** but not in the first year because it cools the soil too much for the young canes.*
***A second coat can be added in June,** and through the years these weed-supressing mulches serve to build up the humus."*
-Grow Your Own Fruit & Vegetables by Lawrence D. Hills. London, England: Faber and Faber Limited: **1971**, page 193.

"Pat will randomly assign clusters of Saskatoon berries to receive Comfrey companion planting or no Comfrey as control. She will plant Comfrey no closer than 15 inches (38 cm) from an outside trunk, measured at ground level, of each Saskatoon plant in the treatment groups.
Measurements:
> **May** - woody measurements: height and number of shoots.
> **June** - Brix from leaves and fruit; observe foraging by birds.
> **July** - fruit production (last year the season ended July 11); total amount of fruit picked.
> **October** - woody measurements: height and number of shoots."

-'Comfrey as a Companion Plant for Saskatoon: Protocol' by Farmer-Researcher Patricia Kozowyk from Babalink Farm; Ecological Farmers Association for Ontario, efao.ca/research-library, Research Protocol: Soil Health and Weed Control, Canada, **2017**.
> (For the results of this study, see the following article.)
> (Brix is the sugar content of a water solution. One degree Brix is 1 gram of sucrose in 100 grams of solution, as percentage by mass.)

"Pat collected woody measurements of the Saskatoon berry plants, including height of tallest stem and number of stems in June and October 2017. In future years, she will measure fruit production and Brix from leaves and fruit.
Over their first growing season of the trial, there was measurable growth of some Saskatoon bushes but the growth was not attributable to their Comfrey companion (P=0.37).
> Pat is pleased to see a trend towards greater plant height in the Saskatoons planted with Comfrey companions, even if there's too much variation to know if the difference is a result of the Comfrey."

-'A Comfrey Companion for Saskatoon: Results' by Farmer-Researcher Patricia Kozowyk from Babalink Farm; Ecological Farmers Association for Ontario, efao.ca/research-library, Research Protocol: Soil Health and Weed Control, Canada, **2017**.

"Perennial cover crops:
> They retain mobile nutrients and bring immobile nutrients to the active soil surface.
> They accumulate biomass and help to feed soil microbial activity while moderating soil temperature and moisture.
> Cover crops compete with and suppress weeds without the need for herbicides or tillage, while providing a source of additional nutrients in the case of some legumes and companion plants.

Comfrey Plants by Berry Bushes:
Amelanchier alnifolia (Saskatoon berry) and Ribes nigrum (black currant) are known to be reliable berry producing species in organic production in **southern Ontario**'s climate.
> **Farmer-researchers will plant one Comfrey plant, Symphytum officinale, on either side of each fruit-bearing plant.** This number of Comfrey plants is sufficient for nutrient needs of a black currant or Saskatoon bush, based on Crawford (*2010). The Comfrey is allowed to die naturally and allowed to spread naturally.

Testing May to October:
> Priority measurements include soil tests, woody measurements in May and October, critical phenological dates."

-'Comfrey as a Companion Plant for Saskatoon and Black Currant' by Farmer-Researchers Patricia Kozowyk from Babalink Farm, Ivan Chan from Eden in Season, and Arthur Churchyard from Eramosa Currents; Ecological Farmers Association for Ontario, efao.ca/research-library, Research Protocol: Soil Health and Weed Control, Canada, **2018**.
(* -Creating a Forest Garden: Working with Nature to Grow Edible Crops by Martin Crawford. England: UIT Cambridge, 2010.)
> (For the results of this study, see the following article.)
> (Phenology is study of cycles and seasons of natural phenomena, especially in relation to climate and plant/animal life.)

"In 2018, Pat found no statistical difference between Saskatoon berry shrubs planted with and without Comfrey with respect to height of tallest shrub branch or number of new shoots.
Fruit Production: There was no practical or statistical difference in sugar content (Brix) of the fruits or harvestable yield.
> Average Brix and yields were slightly higher from Saskatoon bushes with Comfrey, but the differences are too small to have enough certainty that they were due to Comfrey.

It is still unclear if Comfrey is beneficial as a perennial cover crop. Pat, Ivan and Arthur will continue to monitor their crops over the next two years to evaluate its impact.
> This year provided insights into how to take practical and meaningful measurements of shrub growth moving forward.
> Pat will include height measurements in 2019 and then focus only on fruit production and quality thereafter.

Leaving Comfrey uncut in its first year may help it establish, and is a practical time savings. Cutting (chop and drop) the Comfrey starting in the second year may be prudent in order to avoid possible caterpillar and mildew disease issues."
-'Does Comfrey Promote Growth and Fruit Production of Saskatoon Berry and Black Currant: Results' by Farmer-Researchers Patricia Kozowyk from Babalink Farm, Ivan Chan from Eden in Season, and Arthur Churchyard from Eramosa Currents; Ecological Farmers Association for Ontario, efao.ca/research-library, Research Protocol: Soil Health and Weed Control, Canada, **2018**.
> (This research is ongoing so check out their website to see the new results.)

"How to Grow More Strawberries: Here in the United Kingdom we have a short fruiting season between frosts so it's crucial to get as much from each plant as we can.
There are many fertilisers on the market and the variation in price is amazing. Some advise a special strawberry fertiliser which is about £9.00 for a litre bottle others go for 'Grow More'®, and so on. **I use my home made Comfrey tea fertiliser, the**

same one I use for my tomatoes, which costs nothing and is very effective.
Once strawberry plants start to produce flowers it's time to fertilise them. Add Comfrey tea, or whatever you decide to use, to a watering can every 7-14 days and continue after plants have finished fruiting for a few weeks."
-'How To Grow More Strawberries' by Steve Jones, Growing Guides, United Kingdom, July 16 **2019**.
https://growing-guides.co.uk/how-to-grow-more-strawberries

Fruit Trees and Other Trees

"3 biostimulant foliar sprays, Lysaplant® (Bugico, Switzerland), Plantali® (Vossen, Netherlands), and Kerry Algae® (Kerry Algae, Ireland), were tested over 5 years in Irish nurseries on a variety of conifer & broadleaf forest tree species. Lysaplant seed coating was found to produce nearly a 2-fold increase of first order root length in 6-month-old Douglas-fir grown in nonfumigated soil.
Lysaplant, Plantali, and Kerry Algae were used in growth experiments in the bareroot and container nurseries over 2 years on 10 species. Plantali and Lysaplant both increased height and diameter growth in most of the species tested, with some height increases greater than 20%.
With the Lysaplant spray, fungicide use in the rooting of cuttings could be reduced to nearly zero with improved rooting (51% with fungicide, 68% with Lysaplant) under stressful rooting conditions.
Lysaplant and Plantali were both effective in promoting growth even when the standard fertilizer rate was halved. Foliar analysis and visual color assessment indicated that the biostimulants improved Nitrogen uptake.

    List of ingredients in Lysaplant® *(taken from German patent DE 38 25 312 C2 29.05.91).*
    Lysaplant was formerly named Elorisan®.
    1. Aluminum nicotinate (natural source: Nicotiana rustica [Tobacco])
    2. Sodium salicylate (natural source: Filipendula ulmaria [Meadow Sweet])
    3. Anthraquinone (natural source: Rheum palmatum [Chinese Rhubarb])
    4. Agininc acid silylester (natural source: Laminaria digitata)
    5. Lithium carbonate
    6. Urea/Saponine (natural source: Carex arenaria [Sedge])
    7. Guanidinium nitrate (natural source: Symphytum officiale [Comfrey])
    8. Potassium-o-ethyl dithiocardamate (natural source: Carnellia sinensis [Green Tea])"
-'Five Years of Irish Trials on Biostimulants: The Conversion of a Skeptic' by Barbara E. Thompson, Nursery Research Consultant, Wicklow, Ireland; National Proceedings: Forest and Conservation Nursery Associations 2003; RMRS-P-33, United States Department of Agriculture, Forest Service, Rocky Mountain Research Station, Fort Collins, Colorado, **2004**.

"Michael Phillips, orchardist and author of 'The Apple Grower', recently started using Comfrey under his apple trees. He says that it creates excellent soil and allows the trees' roots to come to the surface to feed, unlike grass ground covers. The Comfreys produce beautiful flowers and provide overwintering sites for many beneficial insects and spiders."
-Edible Forest Gardens: Volume One: Vision & Theory by Dave Jacke and Eric Toensmeier. White River Junction, Vermont: Chelsea Green, **2005**, page 187.

*"Some of the multiple interactions in a bi-culture (two cultures) of European pear tree (Pyrus communis) and Comfrey (Symphytum officinale): The pear shades and therefore inhibits the Comfrey, reducing its aggressiveness. The Comfrey shades out grasses under the pear, **reducing grass competition**.*
The Comfrey and the pear prefer a similar fungal-bacterial balance in the soil. The Comfrey dynamically accumulates many nutrients from the deep soil and feeds them to the topsoil, aiding the pear.
The two plants share pollinators but flower at different times, so they support each other's pollination needs.
Deep-rooted Comfrey partitions (breaks up) the soil resources with shallow-rooted pear."
-Edible Forest Gardens: Volume One: Vision & Theory by Dave Jacke and Eric Toensmeier. White River Junction, Vermont: Chelsea Green, **2005**, page 140.

"The pear tree is a heavy nutrient feeder with mainly shallow roots that spread wider than the tree canopy.
*Meanwhile, the Comfrey is a potent deep-rooted dynamic accumulator of many nutrients planted closer to the trunk. **Therefore, the Comfrey facilitates the pear when the Comfrey leaves fall to the ground and rot**.*
The Comfrey also facilitates the pear tree by shading out competitive grasses that allelopathically foster the wrong soil biology balance. This is probably good, though we know little about the soil biology of Comfrey.
On the other hand, during drought years, the Comfrey may compete with the pear for water, and the potential exists for it to compete with the pear for nutrients as well if soils are poor.
On balance, however, this would appear to be a favorable combination and pattern of plants, and the experience of forest gardeners indicates that it is."
-Edible Forest Gardens: Volume One: Vision & Theory by Dave Jacke and Eric Toensmeier. White River Junction, Vermont: Chelsea Green, **2005**, page 140.

*"**Comfrey Under Apple Tree:***
***Question: I've been thinking about planting a large patch of Comfrey under a big old apple tree.** I read somewhere that*

Comfrey is beneficial as a companion plant to apples. Last summer, I smothered the grass under the tree with a heavy application of strawy manure. Can you provide advice on whether I should rototill under the tree or plant through the mulch?
Answer: *Comfrey is not usually thought of as a companion plant because it can be invasive and can quickly overtake other plants of similar size.* **Comfrey does very well under and along the drip line of trees.**
Comfrey attracts pollinators and a thick patch of it under apple trees may interfere with the spread of pests.
Also, the deep roots of Comfrey have the effect of 'lifting' nutrients from deep in the ground back up to the surface, *which would over a period of several years gradually release nutrients that surface roots of trees can access. I would be reluctant to rototill under the trees for fear of damaging the roots of the trees. Comfrey is vigourous enough that all you need to do is to plant a few plants around each tree, and you will have a thick patch within a few years."*
-'Comfrey Under Apple Tree' by Conrad Richter; Richters®: The Herb Specialists, www.richters.com, Goodwood, Ontario, Canada, January 24 **2005**.

(A tree 'drip line' or 'dripline' is the outermost circumference of a tree's canopy, where water drips onto the ground. Canopy is the outer layer of leaves of an individual tree or group of trees.)

*"***A quintessential (perfect) permaculture plant, Comfrey (Symphytum officinale) offers a range of uses that is tough to match. It can be grown as a living mulch.**
Ecological orchardists often plant a ring of Comfrey around a fruit tree and periodically practice 'chop and drop' mulching, *which triggers the plant to regrow, converting yet more nutrients from the earth into biomass and then topsoil."*
-Gaia's Garden: A Guide to Home-Scale Permaculture by Toby Hemenway. White River Junction, Vermont: Chelsea Green Publishing Company, **2009**, page 127.

"Comfrey (Symphytum spp. {species}) beneath a **plum tree***:*
The root system of a deep-rooted accumulator such as Comfrey competes very little with that of a fruit tree.*"*
-Creating a Forest Garden: Working with Nature to Grow Edible Crops by Martin Crawford. England:UIT Cambridge, **2010**, page 64.

*"***The marvel of Comfrey from a fruit tree perspective begins with its deep-reaching root system, which effectively mines potassium, calcium, and other untapped minerals.** *Its leaves and stalks are flush with nutrient wealth, producing* **a lush plant that blossoms just after petal fall on apple trees** *in a cascading series of delightful pale purple-pink umbel florets.* **Bumblebees delight in this subsequent nectar source.** *As Comfrey starts to set seed, it becomes carbon-heavy- and thus top-heavy- and soon falls in every random direction as living mulch, thereby* **suppressing grass growth and preventing it from becoming the dominant ground cover.**
A new round of herbal shoots from Comfrey's insistent roots responds to this sunlight opportunity, repeating this same cycle at least two times more in a given year.
The circumference of a Comfrey circle grows as the mother plant expands outward. **The soil here becomes deep brown, even black, brimming with life force.** *Fruit tree feeder roots find this an irresistible invitation, totally unlike the reception provided by a dense sod where high carbon dioxide levels produced by fine grasses (in the process of root transpiration) proves disagreeable.*
Comfrey leaves room in the humus for trees to find full mycorrhizal connection.
And to think, all you had to do was plant Comfrey starts (roots) around the anticipated dripline of the tree to launch this self-renewing orchard plan."
-The Holistic Orchard: Tree Fruits and Berries the Biological Way by Michael Phillips. White River Junction, Vermont: Chelsea Green Publishing, **2011**, page 31.
(An umbel is an inflorescence where the flower stalks, nearly equal in length, spread from a common center.)
(Mycorrhiza is a symbiotic relationship between fungus and roots of a vascular plant. It is important in soil biology / chemistry.)

*"***Comfrey is not planted right up tight against a young tree;** *this would simply be too competitive in those early years when we need to grow wood structure.*
The term 'dripline' refers to the anticipated outer diameter of the tree, *which for free-standing purposes can be 6-8 feet (1.8-2.4 meter) out from the trunk.*
I often don't introduce Comfrey onto a particular tree scene until the fourth or fifth year.*"*
-The Holistic Orchard: Tree Fruits and Berries the Biological Way by Michael Phillips. White River Junction, Vermont: Chelsea Green Publishing, **2011**, page 354.

*"***The mowing timing of an orchard corresponds with when dairy farmers traditionally make their first cutting of hay- sometime between fruit set and as many as three weeks following.** *This wide window allows you to stagger bloom by not necessarily mowing everywhere at once, thereby supporting those bumblebees who are now finished with fruit tree bloom.* **Leave any Comfrey in flower unmowed, as it will indeed fall over soon from becoming top-heavy through its own seed formation.***"*
-The Holistic Orchard: Tree Fruits and Berries the Biological Way by Michael Phillips. White River Junction, Vermont: Chelsea Green Publishing, **2011**, page 110.

*"***Comfrey: Diverse understory species: Comfrey is an incredible plant ally for fruit trees.**
Russian Comfrey propagates by root, rather than by seed. So it doesn't take over the farm. **Russian Comfrey gets flowers right after the apple trees.** *That provides a home for bumblebees and pollinators.*

Deep Roots: *Comfrey has very deep tap roots which draw up nutrition from the subsoil. Grass has shallow but dense roots. All roots give off carbon dioxide, but grass gives off more. This causes feeder roots to dive down rather than up to the top few inches.*
I'll plant Comfrey once a tree is 4 or 5 years old. *Before that I'm trying to grow wood. But then I'll plant it in the edge of the drip line.*
Foliar Sprays: *Herbal foliar sprays including* willow, wintergreen, osage orange, stinging nettle, horsetail (high in silica which increases strength of cuticle), **Comfrey** and garlic, all infused in hot water, stored in cold water for a week where it ferments."
-'Special Supplement on Organic Tree Fruit' by The Natural Farmer, Barre, Massachusetts, Spring **2011**. Includes 'Notes from a Michael Phillips Seminar' by Jack Kittredge, Dorchester, Massachusetts.

"**Herbal Teas and Foliar Sprays:**
Comfrey packs a wallop of calcium in its deep green leaves and this gets included each time with the pure neem.
Making a fermented herb tea is simple. Fill a five gallon (18.9 liter) bucket with fresh herb, lightly packed. Boil a pot of water to pour over the leaves (as opposed to boiling the herb in the water) as this maximizes nutrient extraction. Now fill bucket to brim with un-chlorinated water.
> Let set for a full week, loosely covered to prevent significant evaporation. This fermentation period makes the constituents that much more bioavailable for foliar absorption.

These teas are diluted in that I add the strained tea from each bucket for each herb being used to each 100 gallon batch of spray."
-'Special Supplement on Organic Tree Fruit' by The Natural Farmer, Barre, Massachusetts, Spring **2011**. Includes 'The Holistic Approach: Using Biological Health as an Effective Means for Abating Tree Fruit Disease' by Michael Phillips, Grow Organic Apples: Holistic Orchard Network, GrowOrganicApples.com, Groveton, New Hampshire.

"**I don't recommend planting Comfrey beneath young fruit trees since the dynamic accumulator will steal nitrogen from your tree's roots.** Yellow leaves are a classic sign of nitrogen deficiency.
Before I knew any better, I planted Comfrey in a ring a foot (0.3 meter) or so away from the trunk of a young **nectarine tree**. After a few years, the Comfrey had completely covered the ground beneath the tree's canopy...and **appeared to be outcompeting the nectarine for every important soil nutrient.**
> Specifically, yellowing tree leaves proved that Comfrey was particularly adept at accumulating nitrogen, not from deep in the soil profile but from the same zone where the nectarine roots were hunting this important nutrient.

In the end, I concluded that Comfrey, and perhaps other dynamic accumulators, are only worth planting near fruit trees once the latter are large enough to keep the former in check with heavy shade.
> I recommend letting go of the belief that you can use dynamic accumulators to remineralize your soil. Instead, it's best to either follow the chemical route or to focus on animals and fungi to relieve paucities (deficiencies).
> Then, once your basic deficiencies have been brought into balance, you can start planting cover crops to really boost that moderate soil and turn your growing area into black gold."

-Balancing Soil Nutrients and Acidity: The Real Dirt on Cultivating Crops, Compost, and a Healthier Home (The Ultimate Guide to Soil: Book 3) by Anna Hess. Athens, Ohio: Wetknee Books, **2016**.

"**Companion Plants for Figs: Fig Polycultures:**
I leave our fig trees to branch low to the ground for ease of picking the fruits. This results in a very deep shade cast under the plant and as a result the **Comfrey** and Artichoke produce little biomass in the summer.
Under plantings do grow well before the fig leaves emerge in late spring, **Tuberous Comfrey in particular forms a dense mat and flowers profusely before dying back during the summer when the fig is in full leaf.**
Fertilisation:
> Figs can perform well under soil conditions unsuitable for other crops but recent studies prove that nutrients exert an effect on yield and quality. Drying varieties are more sensitive to nitrogen and adversely affect fruit size and colour. Too much nitrogen also affects the 'breba' crop adversely.
> A good application of compost under a mature tree (20-40 liter = 5.2-10.5 gallon) and young tree (10-20 liter = 2.6-5.2 gallon) **along with regular Comfrey mulching should be sufficient fertiliser for your fig trees.**"

-'Dig the Fig: The Essential Guide to All You Need to Know About Figs- Ficus Carica' by Balkep: The Balkan Ecology Project: A Permaculture-Inspired, Grassroots Project, www.balkep.org, southeastern Europe, Bulgaria, September 17 **2016**.
> (A breba or breva is a fig that develops in the spring on the previous year's shoot growth.)
> In contrast, the main fig crop develops on the current year's shoot growth.)

Fruit Trees, Calcium and Comfrey

"**Comfrey packs a wallop of calcium in its deep green leaves.** This superstar of the orchard understory is known botanically as **Symphytum officinalis**.
The chemical compund allantoin in this herb encourages bone, cartilage, and muscle cells to regenerate in our bodies.
Naturally, Comfrey is a powerful stimulator of all cell multiplication and thus growth in plants as well. I rely on this herb especially as a source of supplementary calcium, a mega-nutrient required for all fruit production.
Calcium translocates (moves) somewhat poorly from leaf to fruit, especially when there are dramatic fluctuations in rainfall patterns or it's a light crop. **This can lead to calcium deficiency in the fruit- even when soil levels appear adequate- causing**

a disease condition known as bitter pit in apples or cork spot in pears.
Commercial growers rely on summer foliar sprays to get more calcium into the fruit to prevent bitter pit.
My homegrown solution to this perennial challenge is the bioavailable calcium found in fermented Comfrey tea, applied every other week after petal fall."
-The Holistic Orchard: Tree Fruits and Berries the Biological Way by Michael Phillips. White River Junction, Vermont: Chelsea Green Publishing, 2011, page 149.
    (For more about 'Comfrey Tea' see subsection 'Comfrey as Liquid Manure' in section 'Garden Uses of Comfrey: Compost, Fertilizer, Potting Mix' {Chapter 17}.)

"**Seasonal Checklist for the Organic Orchard: Those Lazy, Crazy, Hazy Days of Summer:**
Spray foliar calcium at biweekly intervals (every 2 weeks) beginning when the fruit reaches the size of a nickel if bitter pit has been a problem on certain varieties.
Fermented Comfrey tea is a homegrown source for bioavailable calcium and can be included in holistic summer sprays."
-'Special Supplement on Organic Tree Fruit' by The Natural Farmer, Barre, Massachusetts, Spring **2011**. Includes 'Seasonal Checklist for the Organic Orchard' by Michael Phillips. Grow Organic Apples: Holistic Orchard Network, GrowOrganicApples.com, Groveton, New Hampshire.
(United States nickel diameter is 0.835 inches = 2.12 cm. Diameter is a straight line going from side to side through the center.)

Comfrey Heals Other Plants

"**Grown in the garden, Comfrey is beneficial to other plants- a sort of 'plant doctor'.** *Because it is deep-rooting, Comfrey does not rob minerals in the surface soil from other nearby plants. It keeps the surrounding soil rich and moist and gives protective shade and shelter with its large, rough leaves."*
-Roses Love Garlic: Companion Planting and Other Secrets of Flowers by Louise Riotte. Massachusetts: Storey Publishing, **1983**, page 19.

"**Ailing plants, I've found, often respond favorably when a mild Comfrey tea (steeped for two to four days) is sprinkled over the entire plant and soaked into the surrounding soil."**
-'Comfrey' by Pamela D. Jacobsen, Back Home Magazine, The Herb Patch, page 32, November/December **2008**.
    (For more about 'Comfrey Tea' see subsection 'Comfrey as Liquid Manure' in section 'Garden Uses of Comfrey: Compost, Fertilizer, Potting Mix' {Chapter 17}.)

"**The goats clustered around the poor 'Cedar of Lebanon' tree stripping off long strips of slippery bark,** *and then putting back their heads to chew. The strip of bark that had not been chewed was still in one piece, a narrow lifeline between roots and branches. We conjectured that if it could be preserved from drying out, well, the tree might heal itself.*
'Use Comfrey salve!' chimed one of the kids. **Actually, if we smeared Comfrey salve at the sides of the strip of healthy bark, it might serve to keep the bark moist and make it less likely to peel back and crack.** *Perhaps the Comfrey would engender new growth- it certainly worked on humans.*
The children and I spread the waxy green substance all over the naked wound, smearing liberally at the edges. **We tightly wrapped the wound trunk** *with a ripped cotton sheet, being careful to exert plenty of pressure on the bark bridge.*
Over the ensuing (following) days, we anxiously watched the aerial (leafy) parts of our precious Cedar to see if they would droop, and it was a vast relief that **the tree never showed a single sign of stress.**
After a few weeks, I palpitated (touched) the area, and detected the swelling of new bark at the edges.
Sometime the following spring, *the weatherworn bandages on the Cedar tree spontaneously ruptured and fell away of their own accord.* **The tree showed off its new bark, which, to our amazement had completely grown back over the bare wood. The bark of the trunk is, unaccountably, perfect and unblemished. I attribute this to the healing power of Comfrey."**
-The Medicinal Herb Grower: A Guide for Cultivating Plants that Heal: Volume 1 by Richo Cech. Williams, Oregon: Horizon Herbs LLC, **2009**, pages 56-57.

Comfrey Protects Plants from Insects, Slugs and Snails
    See subsection 'Snails as Food' in section 'Livestock / Pet Species and Comfrey' in Volume 2.

"*When young vegetable and bedding plants are transplanted into a newly prepared bed, they become an attractive source of food for hungry slugs.*
A heap of cut Comfrey leaves can help to clear a bed of slugs before it is planted. The leaves, and the slugs feeding on them, are left in place for a few days and removed at night. *Surrounding new plants with a protective ring of cut Comfrey leaves can distract slugs until the plants become more established."*
-Encyclopedia of Organic Gardening: The Complete Guide to Natural & Chemical-Free Gardening by The Henry Doubleday Research Association, edited by Pauline Pears. London, England: DK (Dorling Kindersley), **2001**.

"**The Comfrey method of getting rid of slugs: In the springtime, slugs are particularly fond of wilted Comfrey leaves.**
Comfrey grows strongly in the spring and is readily available.

About a week before planting out, a pile of Comfrey leaves is left in the centre of a plot. The slugs will soon find it and more and more will follow the slime trails. **Don't do anything until about the fifth night** when the pile will be absolutely heaving with slugs. Remove these and dispose of them. On succeeding nights, further slugs can be removed but there will be far fewer. Finally, the pile of leaves is removed and composted. **The plot is planted up, with a continuous ring of Comfrey leaves laid around the edge of the plot.** The Comfrey acts as a decoy to any remaining slugs who may move onto the plot from outside.

After this, only occasional night-time checks are necessary, and renewal of the leaves every week or so. The advantages of the Comfrey method are high-effectiveness and considerable flexibility.

For unknown reasons it is almost entirely ineffective from early July onwards! This disadvantage is a minor one because by July the seedlings are usually big enough to withstand attack, and the hot dry weather inhibits slug activity.

We have never found another plant as good as Comfrey, and nothing that works in high summer. You may still get problems in wet weather. In that case, another method is required."

-'Slugs: What Can You Do About Them?' by Peter Harper, Centre for Alternative Technology Publications, www.cat.org.uk, Machynlleth, Powys, Wales, **2006**.

"**Pest Control: Slugs go for Comfrey, so you could use it to attract slugs away from plants.**
If you really want to go all out against slugs, **grow a ring of Comfrey around your garden**, separating the garden with an electric fence. The Comfrey will attract the slugs from the garden.
Then run pigs in the Comfrey. The pigs will love both the Comfrey and the slugs. And the pig urine and manure will attract in even more slugs, hopefully depleting your local population for a while. In place of the pigs, poultry could be run as well."
-'Comfrey: One Fine Plant' by Douglas Barnes, Tweed, Onatrio, Canada; Permaculture Reflections: Design a Sustainable World, www.permaculturereflections.com, February 15 **2009**.

"*In 2008 the efficacy (effectiveness) of three environmentally friendly substances against apple aphid (Aphis pomi) was tested. The selected substances were cinnamic acid and two plant extracts, namely glycolic extract of Comfrey (Symphytum officinalis) and fluid extract of marigold (Calendula officinalis).*
All of the substances were tested at 0.5, 1 and 5% concentrations. The individuals of apple aphids were collected in the organic orchard of the Biotechnical Faculty in Ljubljana, Slovenia; 10 aphids were then transferred to an apple tree leaf, which was previously sprinkled with water, treated with selected substance and put in a Petri dish. The efficacy was assessed at 15, 20 and 25 C (59, 68, 77 F) with relative humidity being 75%.
The mean corrected mortality rate was determined on the first, second and third day after treatment.
In general all of the tested substances showed aphicidal (kills aphids) properties, and the highest mean corrected mortality rates were determined on the third day at 25 C (77 F).

Only marigold extract exceeded a 50% aphicidal efficacy at 15 C and at 5% concentration of the suspension used, while other treatments showed aphicidal efficacy between 4 +/- 4% (cinnamic acid, 15 C, 0.5%) and 39% +/- 12% (cinnamic acid, 25 C, 1%).
In general, the best mean efficacy (mean corrected mortality rate 15% +/- 2% of the substances tested was found at highest concentration.

The results of our study confirmed the assumption that the substances tested possess a certain aphicidic activity, but their efficacy is much less than that of the synthetic insecticides usually used to control apple aphid (Tuca et al., 2009).

Mortality of apple aphids after application of these substances was generally quite moderate, which could be expected since the substances in question are ecologically friendly and more acceptable (Rajapakse and Van Emdem, 1997).

The study showed that these substances have potential to control the apple aphid, but their application should be optimized before their use could be implemented in the sustainable and ecologically friendly apple production."
-'Efficacy of Three Natural Substances Against Apple Aphid (Aphis pomi De Geer, Aphididae, Homoptera) Under Laboratory Conditions' by Laznik, Cunja, Kac and Trdan, Acta Agriculturae Slovenica (Journal of Agriculture Slovenica), 97-1, pages 19-23, March **2011**.

"**As Comfrey leaves wilt and decompose they become irresistable to slugs and snails.** If you spread them around young plants such as lettuce and other slug-loving plants, this will keep the slugs busy and easy to dispose of."
-Beyond the Pellet: Feeding Rabbits Naturally by Boyd Craven Jr (The Urban Rabbit Project) and Rick Worden (Rise and Shine Rabbitry), **2013**.

"**Comfrey against slugs: Surround crops with a ring of Comfrey leaves.** Renew leaves weekly (more often in dry weather), removing and disposing of any slugs at the same time."
-Comfrey for Gardeners: A Garden Organic Guide by Garden Organic. Coventry, England: no date but around **2015**, page 15.

Comfrey as Plant Antifungal

"**We found that aqueous (water) extracts from the leaves of medicinal Comfrey (Symphytum officinale) and (Siberian) cowparsnip strongly inhibit the germination of Erysiphe graminis conidia (Blumeria graminis called barley powdery mildew or corn mildew) and uredospores of Puccinia graminis (causes stem, black and cereal rusts). The extracts are fungitoxic.*,***

Spraying wheat seedlings with these extracts, in contrast to the irrigation of soil, markedly diminished infection

in plants with powdery mildew.
Antifungal activity in vitro (in laboratory equipment) and protective activity (when plants were sprayed) correlated with the level of phenolic compounds in these extracts.
Experiments with healthy plants have demonstrated that the photosynthetic apparatus of wheat plants is stimulated by extracts. Spraying seedlings with the extracts resulted in an increased rate of O2 (oxygen) evolution calculated per unit of chlorophyll, an increase in the ratio (FM-FT)/FT in the experiments that recorded slow fluorescence induction, an increase in the relative light intensity of band A, and a decrease of relative intensity of band C in experiments with thermoluminescence of wheat leaves.

These results provide evidence that the protective activity of Comfrey and cowparsnip extracts is associated with their action on the pathogenic fungus and with the activation of natural defense reactions of the host plant. Concerning the Comfrey, it is known for its medicinal properties and contains compounds with antiseptic activity *(Valyagina-Malyutina, E.T., 'Medicinal Plants', St. Petersburg, Russia: Spetsial'naya Literatura, pages 79-80, 1996)."*
-'Antifungal Activity of Aqueous Extracts of the Leaves of Cowparsnip and Comfrey' by Karavaev, Solntsev, Yurina, Yurinal, Polyakova, and Kuznetsov, Biology Bulletin of the Russian Academy of Sciences, Moscow, Russia, Volume 28, Issue 4, pages 365-370, July **2001**.
(* -'Protection of Wheat from Powdery Mildew with the Extracts of Symphytum officinale and Heracleum sibiricum' by V.A. Karavaev, Abstracts of Papers, First International Powdery Mildew Conference, Avignon, France, page 48, 1999. If you have this report, could you please send it to me.)
(** -'Changes in the Growth Rate of Phytopathogenic Fungi Induced by Extracts from Leaves of Nontraditional Plants' by Yurina, Yurina, Karavaev et al., International Conference Mycology and Cryptogamic Botany in Russia: Traditions and Current State, St. Petersburg, Russia, pages 281-283, 2000. If you have an English translation of this, I would appreciate a copy.)

"**To use Comfrey tea or diluted Comfrey extract as a foliar drench or spray, add a few drops of liquid soap (it helps the spray stick to the leaves) and apply to your plants.** You can use a watering can with a fine rose, but you'll get better coverage with a garden sprayer. Be sure to strain your liquid very carefully (let it drip through a large coffee filter) before you put it in your sprayer, or you'll clog up the nozzle before you even get started.
When you spray your plants, don't coat just the top of the leaves; reach under and spray the bottoms, too, at least until the liquid starts to run off."
-'Power Plant' by Jean M.A. Nick, Organic Gardening Magazine, Pennyslvania, pages 51-53, June/July **2004**.
> (For more about 'Comfrey Tea' see subsection 'Comfrey as Liquid Manure' in section 'Garden Uses of Comfrey: Compost, Fertilizer, Potting Mix' {Chapter 17}.)

"**Biological control consists of using one organism to attack another that may cause economic damage to crops. Integrated Pest Management (IPM) is a very common strategy.
The white mold produced by Sclerotinia sclerotiorum (Lib.) causes considerable damage to bean crops.** This fungus is a soil inhabitant, the symptoms of which are characterized by water-soaked lesions covered by a white cottony fungal growth on the soil surface and/or the host plant.
> Possible biological control agents taken from plants are being investigated as phytopathogen inhibitors. These are endophytic microorganisms that inhabit the intercellular spaces of vegetal tissues and are often responsible for antimicrobial production.

The objective of present study was to select endophytic fungi isolated from Comfrey (Symphytum officinale L.) leaves with in vitro (in laboratory equipment) antagonist potential against phytopathogenic fungus S. sclerotiorum. Endophytic fungal strains were isolated from Comfrey leaves based on methodology described in Penna* with some modifications.
> Twelve isolates of endophytic fungi and a pathogenic strain of S. sclerotiorum were used in the challenge method. With the aid of this method, four endophytes with the best antagonistic activity against S. sclerotiorum were selected.

Pathogen growth inhibition (reduction) zones were considered indicative of antibiosis (antagonism). The percentages of pathogenic mycelia (fungal) growth were measured both with and without the antagonist, resulting in growth reductions of 46.7% to 50.0% for S. sclerotiorum.
Among 12 endophytic strains tested in the challenge methodology, 4 (Trichophyton sp., Chrysosporium sp., Candida pseudotropicalis and Candida tropicalis) exhibited the best antagonistic activity against the phytopathogen S. sclerotiorum. This was the first report of these 3 genera isolated as endophytes from Symphytum officinale L..
> The growth of phytopathogenic and endophytic strains in the challenge experiments reached maximal rate eight days after inoculation. The isolates with the greatest antagonistic effect against S. sclerotiorum (probably due to antibiosis) were C. pseudotropicalis (50.0% inhibition), Chrysosporium sp. (48.9% inhibition), C. tropicalis (47.8% inhibition) and Trichophyton sp.(46.7% inhibition).
>
> **The data reveal that growth rate of the pathogenic strain in the challenge tests was highly reduced in relation to the control results.**"

-'Selection of Endophytic Fungi from Comfrey (Symphytum Officinale L.) for in Vitro Biological Control of the Phytopathogen Sclerotinia Sclerotiorum' by Rocha, Luz, Engels, Pileggi, Filho, Matiello and Pileggi, Brazilian Journal of Microbiology 40, pages 73-78, **2009**.
(* 'Microrganismos endofiticos em erva-mate {Ilex paraguariensis, St. Hil.} e variabilidade genetica em Phyllosticta sp.' {Endophytic Microorganisms in Yerba Mate and Genetic Variability in Phyllosticta sp.} by E.B. Penna, por Rapd. Curitiba 123 f. Dissertacao:Mestrado em Genetica-Universidade Federal do Parana {Dissertation:Master in Genetics-Federal University of Parana}, Brazil, 2000)
> ('Integrated Pest Management' {IPM}, also known as 'Integrated Pest Control' {IPC} is a broad-based approach that

integrates practices for economic control of pests. It minimizes risk to people and the environment.)
(An endophyte is an endosymbiont, often a bacterium or fungus, that lives within a plant for at least part of its life cycle without causing apparent disease.)
(In biochemistry, an antagonist is a substance that reduces the physiological action of another.)
(Pathogen is a microrganism such as bacteria, virus or fungi that causes disease.)

Permaculture and Forest Gardens

Definition of Permaculture and Forest Gardening

The term permaculture is a contraction of the words 'permanent' and 'agriculture'. **Permaculture is a system of agricultural centered around imitating or directly utilizing the patterns and features observed in natural ecosystems.** Permaculture draws from these disciplines: organic farming, agroforestry, integrated farming, sustainable development, and applied ecology. One concept is the 'Guild' which is a group of species where each provides a unique set of diverse functions that work in conjunction or harmony. Common practices include agroforestry, Hugelkultur, natural building, rainwater harvesting, sheet mulching, intensive rotational grazing, keyline design, fruit tree management, and others.

Forest gardening is a sustainable food production and agroforestry system based on woodland ecosystems, using fruit and nut trees, bushes, herbs, vines and perennial vegetables to produce food, fiber and other needs of people.

"Permaculture is a philosophy of working with, rather than against nature; of protracted and thoughtful observation rather than protracted and thoughtless labour; of looking at plants and animals in all their functions, rather than treating any area as a single-product system."
-Bill Mollison, Australian ecologist and University of Tasmania (Australia) professor who created permaculture in the **1970**'s.
     (Bill Mollison wrote 'Introduction to Permaculture' in 1991, and 'Permaculture: A Designers Manual' in 1997.)

*"**An edible forest garden is a perennial polyculture of multipurpose plants.***
     Most plants regrow every year without replanting: perennials.
     Many species grow together: polyculture.
     Each plant contributes to the success of the whole by fulfilling many functions: multipurpose.
In other words, a forest garden is an edible ecosystem, *a consciously designed community of mutually beneficial plants and animals intended for human food production."*
-Edible Forest Gardens: Volume One: Vision & Theory by Dave Jacke and Eric Toensmeier. White River Junction, Vermont: Chelsea Green, **2005**, page 1.   (Polyculture is the simultaneous cultivation of several crops or kinds of animals.)

Permaculture Polycultures

"Chadwick Garden, University of California, Santa Cruz (UCSC), California.
This **Biointensive garden** was started in 1967 with the direction of Alan Chadwick. While not designed as a food forest, it breaks convention as vegetable, flower, and herb polycultures are grown beneath dwarf fruit trees, and the perimeter of the garden is planted with standard fruit trees, resulting in food forest edges and patches.
Avocadoes, figs, loquats, plums, persimmons, and citrus hang over pineapple guavas, elderberries, and dwarf citrus, peach, apple, and pear trees, while **Comfrey, perennial herbs and flowers, and annual vegetables constitute the lowest layer.** Two or three passionfruit vines in the garden constitute a vertical layer, as well.
Steve Irving, an apprentice for two years now, assures me that this hand-worked garden is much more productive per area than the UCSC Farm, which uses tractors."
-West Coast Food Forestry: A Permaculture Guide by Rain Tenaqiya. Ukiah, California, **2005**.
     (Biointensive agriculture is an organic system for maximum yields from minimum land, while increasing biodiversity and sustaining soil fertility.)

*"**Comfrey is an amazingly productive and multi-functional plant.***
     Great in swales, as living mulch at base of fruit trees, or as garden weed barrier. *Deep roots bring up nutrients.*
     Biomass-producer capable of being cut-back several times a year. Leaves used to heal wounds and as a green.
     Said to be a source of Vitamin B12, protein, Si (silicon), N (nitrogen), Mn (manganese), Ca (calcium), K (potassium), and Fe (iron).
     Edible flowers attract pollinators. Flowers May- June.
     Can grow in partial shade, heavy clay, and poorly drained soil. Drought-tolerant. Fire and deer-resistant.
     Propagated from divisions and pieces of root (can be hard to eradicate).
Comfrey (Symphytum officinale) *is up to 3 feet (0.9 meter) tall.*
Russian Comfrey (S. x uplandicum) is the best species *as it is more productive, less invasive (it rarely produces seed), and accumulates more K.*
S. grandiflorum *is a 4 inch (10 cm), slowly-spreading groundcover."*

-West Coast Food Forestry: A Permaculture Guide by Rain Tenaqiya, Ukiah, California, **2005**.
> (A swale is a low area or basin with gently sloping sides.)

"Comfrey is in the category of 'Fertility / Insectary Understory'. It is a dynamic accumulator, groundcover, general nectary, invertebrate shelter, and medicinal."
-Edible Forest Gardens: Volume One: Vision & Theory by Dave Jacke and Eric Toensmeier. White River Junction, Vermont: Chelsea Green, **2005**, page 45.
> (Nectary is plant gland that secretes nectar. Nectar is sugar-rich liquid that attracts pollinators such as insects & birds.)
> (Invertebrates are animals that do not have a vertebral column {backbone or spine}. This includes all animals except the subphylum Vertebrata. Some examples are insects, crabs, snails, clams, octopuses, and worms.)

"Clumping perennial with clumping shrub:
If the shrub is quite vigorous, they can be planted at the same time. Many shrubs are less vigorous than perennials, though, in which case the shrubs can be planted a year before the perennials. Best if the shrub is larger than the perennial, otherwise it may get swamped. They should have similar light requirements, or the perennial can have slightly more shade tolerance.
> **Example: Chinese indigo (Indigofera incarnata) and large-flowered Comfrey (Symphytum grandiflorum).** *Both species are vigorous, dense, and fast-growing. Chinese indigo fixes nitrogen and grows 18 inches (45 cm) high. Ordinarily the indigo bushes are spaced 3 to 4 feet (1 to 1.3 meters) apart.*

The Comfrey species is an excellent low ground cover, reaching 12 inches (30 cm) high and spaced the same distance apart. You will need to use extra Comfrey plants to fill in the gaps.
This polyculture is a great low-maintenance way to build soil fertility."
-Edible Forest Gardens: Volume Two: Design & Practice by Dave Jacke and Eric Toensmeier. White River Junction, Vermont: Chelsea Green, **2005**, page 296.

"Strata (Layers): The architecture of a temperate Edible Forest Garden typically has between 3 and 7 layers including a combination of a tall tree (e.g. walnut) and short tree (e.g. pear) layer, a tall shrub (e.g. hazelnut) and short shrub or cane (e.g. raspberry) layer, **a herbaceous perennial layer (e.g. Comfrey)**, a groundcover (e.g. strawberry) and a vine layer (e.g. hardy kiwi). Consideration of the below ground layer, the rhizosphere is also extremely important.
Whenever possible, plants with complimentary root morphology should be spaced so that they partition water and nutrients.
An example of the type of guild found in the Michigan State University Edible Forest Garden, centers on a fruit or nut tree (e.g. apple) **with a nutrient aggregating (collecting), mulch producing herbaceous perennial (e.g. Comfrey) planted under the tree canopy to build soil and suppress weeds.**
Surrounding the outside of the mature canopy width of the fruit tree could be found a nut bearing shrub (e.g. hazelnut), a fruit bearing shrub (e.g. beach plum) and a nitrogen fixing shrub (e.g. siberian pea shrub)."
-'Edible Forest Garden Permaculture for the Great Lakes Bioregion: Background, Development and Future Plans for The Michigan State University Student Organic Farm Edible Forest Garden' by Jay Tomczak, East Lancing, Michigan, July **2007**.
> (Temperate climates do not have extremes of temperature and rain/snow. The difference between summer and winter is invigorating but not unbearable. These are the middle latitudes of the Earth. They exclude tropics and polar regions.)

"Courses in permaculture often proffer (propose), as an introduction to students, a basic guild whose dominant member is an apple tree. **The result is a healthier apple tree and a varied ecology.**
In the center of a typical **apple guild,** not too surprisingly, is an apple tree. Under the tree's outermost leaves, at the drip line, is a ring of thickly planted daffodil bulbs.
Inside the bulbs guild area is a broken circle of lush Comfrey plants, each topped with purple blossoms that buzz with bees. Between the Comfrey are two or three robust artichoke plants. Dotted around these grow flowers and herbs: yellow bursts of yarrow, trailing orange nasturtiums, and the airy umbels of dill and fennel.
A closer look shows a number of plants that are normally considered weeds, such as dandelion, chicory, and plantain. Crowding between all the flora is a thick ground cover of clover, and we can spot some fava beans and other legumes growing in the dappled sun beneath the branches."
-Gaia's Garden: A Guide to Home-Scale Permaculture by Toby Hemenway. White River Junction, Vermont: Chelsea Green Publishing Company, **2009**, page 186.

"You'll find that, even with perennials that produce masses of seed, little self-seeding occurs unless there is bare soil or soil with little plant growth in spring.
An example is the **true medicinal Comfrey (Symphytum officinale), which is often regarded as a self-seeding nuisance in conventional gardens. In forest gardens you are likely to get only the occasional plant springing up, and of course you can nearly always leave it to develop where it does, as it is an immensely beneficial plant."**
-Creating a Forest Garden: Working with Nature to Grow Edible Crops by Martin Crawford. England: UIT Cambridge Ltd., **2010**, page 348.

**"The layers or 'horizons' as they are known, can be subdivided from as few as 4 into as many as 9.
Although unreachable by most plants, the subsoil level is probably as important as the topsoil layers as this is where important minerals and trace elements are located.**
> *Only a few plants, such as Comfrey, dandelion and daikon radish, have roots long and strong enough to*

penetrate this level and bring these elements to the surface.
Trace elements often occur in minute (small) amounts and in subtle but crucial balance. They are not readily available in the topsoil layer and need to be accessed from the subsoil clay.
This is where dynamic accumulators such as Comfrey come in. They have the ability to suck up trace elements from the subsoil, making them available to more shallow rooted plants through compost, teas and mulching. Comfrey accumulates N (nitrogen), K (potassium), Ca (calcium), Mg (magnesium), F (flourine), and Si (silicon). *Dandelion accumulates P, K, Ca, Mg, Fe, Cu, and Si. As dynamic accumulators and natural nutrient pumps they definitely deserve a place in any permaculture plan."*
-Permaculture for the Rest of Us: Abundant Living on Less than an Acre by Jenni Blackmore. United Kingdom: New Society Publishers, **2015**, page 19.

(There are seven layers in a food forest, although some include fungi as an eighth layer:
The Canopy: tallest trees in the system. Understory Layer: trees that flourish in dappled light under the canopy. Shrub Layer: woody perennials of limited height. Herbaceous Layer: Plants die back to the ground every winter. Soil Surface: Grow much closer to the ground, grow densely to fill bare patches of soil. Rhizosphere: Root layers in the soil. Vertical layer: climbers or vines, such as runner beans.)

"Use in Design Permaculture: Plant a mulberry bush just outside the chicken yard, Comfrey all around the chicken yard fence, and Siberian pea shrub all through their range- to provide for food that you don't have to carry to the animals."
-'The Principles of Permaculture from Bill Mollison and David Holmgren' by Patricia Allison, earthhaven.org. Source: Organic Growers School, Asheville, North Carolina, organicgrowersschool.org, date unknown but around **2016**.

Dynamic Accumulator and Soil Improver
A dynamic accumulator is a plant that gathers minerals or nutrients from the soil and stores them in a more concentrated and bioavailable form. It is used as fertilizer.

"Plants, especially herbaceous plants, greatly contribute to soil fertility by drawing nutrients out of the soil, storing them, and releasing them as they die. *Some plants do this more actively than others, accumulating nutrients in their tissues to concentrations higher than those found in the soil or than usually found in the average plant.* **We can use these dynamic accumulators to conserve and imporve soil ferilty in our forest gardens.**
The Comfreys (Symphytum spp. {species}) reign as kings of the dynamic accumulators. Comfreys accumulate six different minerals (including Nitrogen, Potassium, Calcium, Phosphorus, Iron, Magnesium, Silica) to higher than average levels, *they produce abundant biomass, and their leaves decompose rapidly. They absorb large amounts of nitrogen and can recycle nutrients from wastewater and human excrement into usable farm products."*
-Edible Forest Gardens: Volume One: Vision & Theory by Dave Jacke and Eric Toensmeier. White River Junction, Vermont: Chelsea Green, **2005**, pages 186-187.

(Herbaceous plants have little or no woody tissue. It can be an annual, biennial or perennial. The leaves or entire plant dies back every year when the weather gets cold.)

"Dynamic Accumulators:

	Macro Primary Nutrients		*Macro Secondary*		*Micro Trace Nutrients*				
	P	K	Ca	Mg	Si	Fe	Mn	Na	Co
Symphytum officinale	242 ppm	1870 ppm	1980	77	1	1.3	0.6	12	129"

-'Nutrient Accumulators Based on James Duke's Research' by Mike H., One Thing Leads to Another, https://portageperennials.wordpress.com, **2013**.

(The part of the plant is not listed in Mike's master chart.
These are the lowest numbers from 'Dr. Duke's Phytochemical and Ethnobotanical Databases': https://phytochem.nal.usda.gov/phytochem/search. Nitrogen is not listed.
In Mike's master chart the lowest possible amount for Phosphorus is listed which is 242 ppm with part of plant not stated. However, in James Duke's database the highest amount is 2,200 ppm. Both numbers are for leaf. Root is not in Duke's database. **So Mike's master chart is misleading.**
The same problem occurs with potassium. In Duke's database the leaf low is 1870 ppm and the high is 17,000 ppm. In Duke's database no low is given for the root. The high is 15,900 ppm.
In Duke's database for calcium in root, no low is listed. The high is 11,300 ppm. For leaf low calcium is 1980 and high is 18,00 ppm. The same problems occur with the other numbers.)

Dr. Duke's Phytochemical and Ethnobotanical Databases:

S. officinale	*Macro Secondary Nutrients*		*Micro Trace Nutrients*				
Low to high	Calcium	Magnesium	Silicon	Iron	Manganese	Sodium	Cobalt
Leaf ppm	1,980 to 18,000	77 to 700	1 to 9	1.3 to 12	0.6 to 5.8	12 to 110	- -
Root ppm	- to 11,300	- to 1,700	- to 35	- to 810	- to 67	- to 3,510	-to 129"

"All of us who have studied permaculture have heard some impressive claims about Comfrey. It is a dynamic nutrient accumulator; it improves the soil; it is 'a slow motion fountain' of nutrients, bringing them up from the subsoil to

improve the topsoil.

*I realized I could just do a **soil test** and compare it to my past years' tests, and then I would have at least one data point to support or refute the claims.*

So far so good, and I think I've made a good case for sheet mulching, but all these figures are blown away by the sample I took this year under the Comfrey plants.

After 5 years of Comfrey, the topsoil in this sample shows a lower pH (we have alkaline clay soil here, so lower is better) and **higher percent organic matter than any of the previous samples, and the nutrient levels are practically off the charts: a 47 to 232% increase over the previously observed highs.**

I did not test for calcium or magnesium either before or after, but **just on the basis of NPK (Nitrogen-Phosphorus-Potassium) the Comfrey is completely vindicated (proved to be good)."**

-'Does Comfrey Really Improve Soil?' by Ben Stallings, Permaculture Research Institute: Education, training and news, https://permaculturenews.org, Geoff and Nadia Lawton, The Channon New South Wales, Australia, March 18 **2014**.

"**My grant project 'Alley Cropping in a Hillside Terrace System'** sought to transform the steep slopes of a newly planted fruit orchard into a sustainable agriculture system incorporating earth berm terraces, no till annual agriculture, with a cut-and-carry mulch system.

One hillside plot was successfully populated with Comfrey, stinging nettle, and white clover, another was populated with Comfrey, yarrow, and goumi (cherry silverberry) shrubs, and the third was left to the existing cover of mostly blackberry, purple knapweed, and goldenrod. The hope was to establish a variety of deep rooted plants designed among them to accumulate all significant micro and macro nutrients.

As to whether the test plots cycled nutrients better or enriched the soil more than the control remains inconclusive. The soil tests exhibited such great variability that no conclusions could be drawn (increases in some nutrients, decreases in others, and overall very random changes that may be the result of sampling errors), and I believe that only a more controlled longitudinal study could answer the question of sustainable nutrient management.

I still believe that this approach may provide a form of closed-loop agriculture that could persist in the face of oil and resource shortages (higher transportation and input costs), or simply for those of limited means.

Growing one's own fertilizer, weed suppression, and moisture retention right next to one's crops results in a system with minimal required energy inputs and thus minimal environmental impact."

-'Alley Cropping in a Hillside Terrace System: Final Report for FNC13-916' by Weston Lombard, Solid Ground Farm, Millfield, Ohio; SARE: Sustainable Agriculture Research and Education: Grants and Education to Advance Innovation in Sustainable Agriculture, https://projects.sare.org, December 31 **2014**.

"**Comfrey: Deep-Rooted Accumulator? As to Comfrey being a dynamic accumulator, it's very hard to find any data.** The difference between plants in their ability to absorb a specific mineral may be due to the nature and the genetics of a plant and/or the conditions under which it's growing.

The point is that roots in the same root zone can, and often do, gather very different amounts of the same mineral or nutrient, and **the shallow roots of Comfrey (as opposed to its deeper roots) may play the primary role in accumulating special nturients and minerals such as calcium, nitrogen, phosphorus and potassium."**

-Understanding Roots: Discover How to Make Your Garden Flourish by Robert Kourik. Occidental, California: Metamorphic Press, **2015**, pages 87-88, 94-98.

"I found excavated root drawings of **Comfrey, Symphytum officinale,** in a rare 1960 book* from Germany.
Comfrey Root Drawing #1:

This illustration of a mature, 32-inch-tall (2.6 feet = 0.8 meter) blooming Comfrey plant.
Soil type plays a large role in root depth. This plant was studied in a 'fresh wet meadow', with the top 12 inches (1 foot = 0.3 meter) of roots growing in a 'humus loam' above a 'clayey loam with a thick structure'. **The 'high-standing, heavy root-resisting gley horizon' at 12 inches prevents deeper roots from forming.** These roots extend 32 inches wide, but barely one foot deep, with no major taproot.

Comfrey Root Drawing #2:

This Comfrey root system, excavated in a wheatfield, has **grown not only a spindly taproot at only 30 inches (2.5 feet = 0.76 meter) deep, but a significant lateral root.**

The top foot of soil was described as 'strong humus, sandy loam, with gauzy-moldy humus'. Below that, the soil was 'sandy, clayey loam, regular with peaty parts'. At the 32-inch depth was a 'fine sandy, groundwater-saturated loam'. The thick horizontal root was presumed to be due to formation **from a stem cutting** displaced diagonally by cultivation, and the single 3-inch (7.6 cm) deep root was thought to follow the descending groundwater level.

Until I see otherwise, I'm staying with the concept that Comfrey roots are fairly shallow, and that their lateral roots are more important than the deeper roots for gathering nutrients and minerals."

-Understanding Roots: Discover How to Make Your Garden Flourish by Robert Kourik. Occidental, California: Metamorphic Press, **2015**, pages 94-98.

(* -Wurzelatlas Mitteleuropaischer Ackerunkrauter und Kulturpflanzen {Root Atlas of Central European Field Weeds & Crops} by Lore Kutschera. Frankfurt, Germany: DLG Verlag, 1960. This book has drawings of roots systems of 2 dug up Comfrey plants.)

(Gley is sticky waterlogged soil low in oxygen, usually gray to blue. This soil is not a good representation of how deep Comfrey roots are.)

(It is possible Comfrey in the wheatfield was disturbed by cultivation throughout the year so never got to grow normally.)

(I would like to know the age of the Comfrey plants that were dug. I could not find any other information stating that Comfrey roots are mostly shallow. Please let me know your experiences with it.)

"Dynamic accumulators are plants that are believed to delve deep into the soil to mine usually inaccessible nutrients, concentrating those found minerals in their leaves and roots.
 Various books recommend that gardeners gather the tops to create high-nutrient compost, or simply site dynamic accumulators next door to more needy crops so the former's organic matter can be allowed to rot into the ground and improve upper layers of the soil for the sake of the latter.
The trouble with this rosy picture is that scientific analyses of the effects of dynamic accumulators are scanty and tend to contradict this common belief.
In 'Understanding Roots'*, for example, Robert Kourik uses data from real-life excavations to prove that most so-called dynamic accumulators merely concentrate nutrients already found in the topsoil rather than digging deep for scanty minerals. To prove this point, he turns to a variety of root maps (2 Comfrey maps from 1960) that show the lack of deep roots in most dynamic accumulator species.
 For example, Comfrey *(a classic dynamic accumulator that is believed to concentrate silica, calcium, nitrogen, magnesium, potassium, and iron)* **actually keeps most of its roots in the top foot (0.3 meter) of soil.**
 Kourik asserts that deeper roots in both Comfrey and in other plants are much less prolific than have been previously believed *and that the few taproots that exist seem to function as anchors and as a way to suck up water during droughts rather than as seekers of plant nutrients."*
-Balancing Soil Nutrients and Acidity: The Real Dirt on Cultivating Crops, Compost, and a Healthier Home (The Ultimate Guide to Soil: Book 3) by Anna Hess. Athens, Ohio: Wetknee Books, **2016**.
(* -Understanding Roots: Discover How to Make Your Garden Flourish' by Robert Kourik. Occidental, California: Metamorphic Press, 2015.)
(I discuss the book 'Understanding Roots' in the excerpt above this one. Mr. Kourik's evidence about Comfrey is not convincing.)

"Is Comfrey a Dynamic Accumulator?:
A dynamic accumulator is a plant that will absorb and retain, in the leaf, at least one nutrient at levels that are at least 10 times higher than the average plant.
 Using this criteria Comfrey is not a dynamic accumulator for NPK (Nitrogen - Phosphorus - Potassium)."
-'Comfrey: Is it a Dynamic Accumulator?' by Robert Pavlis, Master Gardener; Garden Myths: Learn the Truth about Gardening, www.gardenmyths.com, southern Ontario, Canada, September 18 **2016**.
 (This author's conclusion is based on the 'Nutrient Accumulators' master chart by Mike above, which has problems.)

"2018 Soil Analysis Before and After Planting Russian Comfrey Bocking 14:

	pH (KCl)	Nitrogen mg/kg		Phosphorous P2O5	Potassium K2O
		NO3N	NH4N		
March 2015: before planting	5.69	15.4	2.89	16.3 mg/100g	13 mg/100g
November 2016: after cuts	6.67	10.2	3.83	67.1	14.8
March 2017	5.85	12.8	4.5	13.6	16.2
March 2018	6.54	14.6	1.75	47.5	**17.8**

All samples are sent to NAAS (National Agricultural Advisory Service) of the Ministry of Agriculture, Food and Forestry, Bulgaria.
Site Overview: *Shipka Bulgaria, Temperate climate, Latitude 41. Elevation 580 meter (1902 feet).*
 Average annual rainfall: 588 mm (58.8 cm = 23 inch).
 42 plants. *60-70 cm apart (1.9-2.2 feet).* **Bed 13 m2 (square meters) = 140 square feet."**
-'How Much Comfrey Can You Grow on 13 m2? Comfrey Trial Results Year 2: 2017' by Balkep: The Balkan Ecology Project: A Permaculture-Inspired, Grassroots Project, https://balkanecologyproject.blogspot.com, www.balkep.org, southeastern Europe, Bulgaria; October 9 **2017**.

Comfrey as Weed Barrier

"Studies were conducted to evaluate the allelopathic effect of carqueja (Baccharis trimera), Common Comfrey (Symphytum officinale), and yarrow (Achillea millefolium) on nut grass (Cyperus rotundus) (an unwanted weed).
 The different forms of the plants studied were: chopped plants incorporated into the substrate (soil or soil-like medium); chopped plants added to the substrate top; application of crude aqueous (water) extract (Extrato Bruto Aquoso, EBA) to the substrate; and application of EBA residue to the substrate top. The treatments were evaluated for 60 days.
It was concluded that yarrow did not present any allelopathic inhibitory effect, and presented, for the incorporated treatment, stimulus to the emission (growth) of leaves of the nut grass plant.
The carqueja plant for the incorporated treatments, on top and residue of the crude aqueous extract provided less emergence percentage and less leaves per nut grass.
Comfrey plant extract residue and crude aqueous extract treatments reduced the emergence (out of the soil) percentage of nut grass and did not present stimulus to the emission (growth) of leaves of the plant."
-'Allelopathic Effect of the Carqueja, Comfrey and Yarrow on the Development Nutgrass' by L.R.B. Gaziri and R.I.N. de Carvalho, Revista Academica Ciencias Agrarias e Ambientais (Academic Magazine Agricultural and Environmental Sciences), Curitiba,

Brazil, Volume 7, No. 1, pages 33-40, January/March **2009**.
(This report was translated from Portuguese to English so sometimes it is not clear what is being said. It says Comfrey reduced the growth of the invasive weed 'nut grass'. If you have an English translation, I would appreciate a copy.)
    (Nut grass is also called coco-grass, Java grass, purple nut sedge and red nut sedge. It is a perennial that is one of the most invasive weeds known, with a worldwide distribution in tropical and temperate regions, infesting over 50 crops. It is called 'the worlds worst weed'.)
    (Allelopathy is a biological phenomenon by which an organism produces biochemicals that influence the growth, survival, and reproduction of other organisms. These biochemicals are known as allelochemicals.)

"Comfrey is used as a weed barrier because it is able to stop running grasses from spreading.
It needs to be planted in a strip several plants wide, and it's important to use only non-seeding, non-spreading varieties otherwise you will only be replacing one weed problem with another."
-'Comfrey' by Penny Woodward, Edible and Useful Plants, www.pennywoodward.com.au, Australia, November 27 **2012**. Author of books such as 'Herbs for Australian Gardens' and 'Grow Your Own Herbal Remedies'.

"Comfrey can be utilized as a live weed barrier. Comfrey is particularly effective at controlling rhizomatous grasses.
Plant a row of a sterile Comfrey variety alongside crops that you want to protect from encroachment by weeds."
-'Comfrey to the Garden Rescue' by Sheryl Normandeau, Farmers' Almanac: Since 1818, Almanac Publishing Co., www.farmersalmanac.com, Lewiston, Maine, June 23, **2014**.
    (Creeping grasses spread by stems growing from the crown of the plant. New shoots form on underground stems or rhizomes, or on horizontal above ground stems called stolons.)

"If you have problems with running grasses, Comfrey can be a very effective weed barrier. You need to have a stop several plants wide and ensure that you are using non-seeding varieties so that the Comfrey itself does not run riot, but this can be an effective way of keeping grasses out of particular areas, particularly orchards."
-'6 Reasons Why Comfrey is an Amazing Permaculture Plant' by Regnerative: Sustainable Living and Personal Growth, www.regenerative.com, **2017**. The sustainable living and permaculture design school online.

"The idea is to grow a 4 foot (1.2 meter) strip of Comfrey Bocking No. 14 down the side of my garden to create a 'weed barrier'. The idea being that if these plants are established, it'll leave far less room for other plants to take hold. I'll also under sow the plants with clover as a ground cover. I'm not sure how well this idea will work but it seems like a good idea on the surface. It should stop the buttercups creeping their way back in.
On top of this I'll be growing a hugely beneficial plant right next to where I need it. I can use it as a mulch, a plant food or to go in the bottom of trenches to grow beans on top of."
-'Comfrey as a Weed Barrier' by Kevin Alviti, An English Homestead: Self-Reliant Family Living, www.englishhomestead.com, Shropshire, England, May 4 **2017**.

Comfrey as Wind Barrier

"We plant a wind barrier along the opposite sector from the sun, so there is no need to worry about shade from these plants. **It pays to run smaller permanent shelters within the garden, too. I believe the Jerusalem artichoke, the Siberian pea tree, and Comfrey perhaps best meet our criteria.** We want a plant that is soft, easily pruned, nitrogenous, high potash, and preferably alkaline. Given that set of conditions, you will find maybe 50 plants for that barrier.
These plants must be a total barrier, permitting no other vegetation to grow under them, because we want a non-weeding situation. The only weeds that grow are a few dandelions, which we permit, and a couple of bunches of clover, just for teas and salads. We don't have any other weeds in 'Zone One'."
-'Introduction to Permaculture: Permaculture Design Course Series, Pamphlets 1 to 14' by Bill Mollison at 'The Rural Education Center', Wilton, New Hamshire, **1981**; published by Yankee Permaculture, Sparr, Florida.
    (Permaculture 'Zone One' is the most visited areas of your garden. Usually it is the area closest to your house.)

"Wind Break Design: The design is based upon understanding the climate, building soil (humus layer), employing methods to reduce evaporation over precipitation, and increasing biodiversity.
Selecting tree species proficient at producing biomass (leaves) assists in rapid building of soil, and reduces evaporation. Jump starting fungal activity by inoculating each tree at the roots and mulching perimeters with straw increases the ability of trees and shrubs to cope with stressors associated with Bsk/h (semi-arid) climates.
Seeding the alleys (area between trees) with native grasses and forbs, introducing Comfrey, yarrow, worms and other soil biota at stations within Wind Break area aids towards de-compacting the soil."
-'Wind Break Design: Columbia Basin Permaculture, Dry Climate Permaculture' by Lenwood Farms, Connell, Washington, **2015**.
    (Biota is the animal and plant life of a particular region or habitat.)
    (De-compaction increases pore space in soil. Healthy soil is more than 40% pore space.)
    (Forbs are herbaceous flowering plants other than grass.)

Aquaponic Food Production

"Aquaponics-Safe Fertilisers: Plant Based Liquid / Foliar Feed:
A homemade liquid feed is a very good way to boost plant health. It is usually applied as a spray to the leaves, in which case it can be called a foliar feed. Foliar feeds can also help in pest control, both by boosting plant health, and acting as a deterrent to pest organisms.
The best-known liquid feed is made from fermented plants; stinging nettles (Urtica dioica) and Comfrey (Symphytum species) are ideal choices, *though many common 'weeds' may be used. Be careful when picking weeds to use though, as some plants actually have insecticidal properties and can kill beneficial insects as well as pests."*
-'Aquaponics Training Manual' by Lorena Viladomat and Philip Jones with help from Al Basma Centre, Bustan Qaraaqa Project, Operation Blessing Middle East, ELSA Mexico S.A de C.V., September **2011**.
>(Aquaponics is aquaculture {raising fish} and hydroponics {soil-less growing of plants} in one integrated system. The fish waste is food for plants. The plants filter the water for the fish.)

*"**Even balanced aquaponic systems can experience nutrient deficiencies.***
Although fish food pellets are a whole feed for fish, they do not necessarily have the right quantities of nutrients for plants. Generally, fish feeds have low iron, calcium and potassium values. Plant deficiencies can also arise in suboptimal growing conditions, such as cold weather and winter months. ***Thus, supplementary plant fertilizers may be necessary, particularly when growing fruiting vegetables or those with high nutrient demands****.*
Synthetic fertilizers are often too harsh for aquaponics and can upset the balanced ecosystem; instead, ***aquaponics can rely on compost tea for any nutrient supplementation.***
Other Nutrient Teas:
>*In addition to compost, there are many other nutrient-rich organic materials that can be brewed into nutrient tea. One mentioned is to use the solid wastes from the fish tank, captured from the mechanical filter. Brewed in the same way, the solid wastes are completely mineralized and available to add back to the aquaponic system.*
>***Other sources include seaweeds, nettles and Comfrey."***

-'Small-Scale Aquaponic Food Production: Integrated Fish and Plant Farming' by Christopher Somerville (Ireland), Moti Cohen (Israel), Edoardo Pantanella (Italy), Austin Stankus (Italy) and Alessandro Lovatelli (Italy); Food and Agriculture Organization of the United Nations, FAO Fisheries and Aquaculture Technical Paper 589, Rome, Italy, **2014**, pages 141, 143.

Beneficial Insects and Comfrey

*"**The larvae of Panaxia dominula moth have been recorded on many species of plants in the field. There is evidence that preference for Comfrey (Symphytum officinale L.) occurs in Britain,*** *although some colonies of the moth are known in which it is not present. In central Europe the larvae are seldom or never found on Comfrey, so this may be an example of food-plant specialization within a section of an otherwise polyphagous (eats multiple foods) species."*
-'Food-Plant Specialization in the Moth Panaxia dominula L.' by L.M. Cook, Department of Zoology, University of Oxford, England; Evolution: Society for the Study of Evolution, Volume 15, No. 4, pages 478-485, December **1961**.
>(The scarlet tiger moth {Callimorpha dominula, formerly Panaxia dominula} belongs to the subfamily, Arctiinae. It has a wingspan of 45-55 millimeters (1.8-2.2 inches). Adults have various colors. The forewings have a metallic green sheen on blackish areas, with white and yellow/orange markings. Hindwings are red with irregular black markings.)

"Scarce and Threatened Moths of Great Britain:
Order Lepidoptera, Family Ethmiidae, Ethmia funerella (Fabricius, 1787).
Referred to as Anesychia funerella, Fab. by Stainton (1854) and Psecadia funerella by Stainton (1873).
Ethmia funerella is 'Nationally Notable (Scarce) Category A'.
>*Taxa which do not fall within 'Red Data Book' categories but which are none-the-less uncommon in Great Britain and are thought to occur in 30 or fewer 10 kilometers squares (3.86 square miles) of the National Grid or, for less well-recorded groups, within seven or fewer vice-counties.*

In England it had been recorded from North Somerset and East Kent north to Cumberland and Durham but was extremely local. Recent records indicate that this moth is still very local and is now most often found in the fens (marshes) of Cambridgeshire and Huntingdonshire, where larvae can be found in some numbers, with scattered records south to East Kent and north to northwest Yorkshire.
This moth is widely distributed in Europe and is known from Scandinavia, Germany, Poland, Austria, Hungary, Switzerland, Belgium, France, Spain and Italy.
The larva feeds on the underside of a leaf, under a slight web, on Common Comfrey Symphytum officinale, Tuberous Comfrey Symphhytum tuberosum, *lungwort Pulmonaria officinalis, or common gromwell Lithospermum officinale (Emmet 1988) and has also been recorded on wood forget-me-not Myosotis sylvatica (Warren 1980).*
Larvae have been found from August to October. This species overwinters as a pupa in a cocoon spun amongst detritus. The adult occurs from May to July."
-'A Review of the Scarce and Threatened Ethmiine, Stathmopodine and Gelechiid Moths of Great Britain' by M.S. Parsons, United Kingdom Nature Conservation, No. 16, Peterborough, England, **1995**.
>(Ethmia quadrillella = Ethmia funerella = Comfrey Ermel. Larva has orange/yellow dots in a line on top of the back with

gray or black along the sides. Wingspan 15-19 mm. Wings are white with large black marks. Mostly active at night.)

"Comfreys also excel at attracting beneficial insects, providing a preferred egg-laying and overwintering site for many species. Comfreys are also beloved by spiders, hosting as many as 240 per square meter (10.76 square feet) in the soil below them during winter, according to one study*.
> *Spiders are extremely important in ecosystems. Research indicates that excluding spiders from agroecosystms can result in exponential growth of insect pests.*
> No single spider species is critical: generalists predators including the spiders...control prey largely through the asemblage effect: species of varied sizes and habits need to be present throughout the growing season of the pest species to limit the growth of associated pest populuations.

Spiders soft bodies leave them vulnerable to drying out. They like vegetation and shade. **Some spiders live more than one year, so winter headquarters are also important:** they like hollow stems, empty seed capsules, dead flower heads, bark crevices and so on. **Swiss researchers found the highest concentration of hibernating spiders under Comfrey."**
-Edible Forest Gardens: Volume One: Vision & Theory by Dave Jacke and Eric Toensmeier. White River Junction, Vermont: Chelsea Green, **2005**, pages 127, 329.
(* -'Uberwinterung von Arthropoden im Boden und an Ackerunkrautern kunstlich angelegter Achkerkrautstreifen' {Overwintering of Arthropods in the Soil and on Field Weeds of Artificially Created Achkerkraut strips} by H.M. Burki and A. Hausammann. Agraroekologie {Agroecology}, Vol 7, No. 1, page 158, 1992. I was unable to get this. If you have it in English, could you send it.)

"Food and shelter plants for beneficial insects: Symphytym spp. (species):
> *Oviposition (lay eggs): Lacewings.* Oviposition plants provide preferred egg-laying sites.
> *Overwintering: spiders.   Foliage: parasitoid wasps and spiders.*

Insects and spiders prefer overwintering species for hibernation, either in soil under the plant or in the dead stems, flower heads or foliage. Insects and spiders appear to prefer foliage plants as habitat for resting or other purposes."
-Edible Forest Gardens: Volume Two: Design & Practice by Dave Jacke and Eric Toensmeier. White River Junction, Vermont: Chelsea Green, **2005**, page 543.
> (Lacewings, in the order Neuroptera, are insects with a complex network of wing veins that give them a lacy appearance.The most common lacewings are in the green lacewing family, Chrysopidae, and the brown lacewing family, Hemerobiidae. Lacewings are a beneficial general-purpose predator. Adults feed on nectar and pollen. Larvae eat soft-bodied insect pests such as aphids, leafhoppers, mealybugs, red mites, spider mites, thrips and whitefly.)

Bees and Comfrey

See sub-subsection 'Comfrey Flower Buds, Bloom, Nectar and Pollination' in subsection 'Comfrey Flower' in section 'Symphytum Genus Description' (Chapter 5).
See sub-subsection 'Flowers and Pollination' in subsection 'officinale' in section 'Details About Symphtum Species' (Chapter 8).

*"**The inhabitants of our own honeybee hive were particularly attracted to the bluish flowers of balm and Comfrey**, adding yet another dimension of nourishment and flavor."*
-Forest Gardening: Cultivating an Edible Landscape by Robert Hart. White River Junction, Vermont: Chelsea Green Publishing Company, **1991**, page 46.

*"**Sown patches of grass, clover, trefoil and cornflower did not provide significantly more early or late sources of forage.**
This is of importance to both **overwintered bee queens** that emerge from hibernation in April and early May, and males and newly emerged / mated queens still active in September.
It remains vital to ensure that other extensively managed areas of vegetation at field corners and non-cropped parts of farms are left to offer flowers such as Lamium sp., Symphytum officinale (Comfrey) and Salix sp. early in the season.
It is worth considering if some of these components could be incorporated successfully into the sown mixture."*
-'Restoration and Management of Bumblebee Habitat in Agricultural Landscapes' by M.S. Heard (Centre for Ecology and Hydrology), A.F.G. Bourke (University of East Anglia), W.C. Jordan (Institute of Zoology), J.L. Osborne (Rothamsted Research) and C. Carvell (Centre for Ecology and Hydrology), England; Review of Defra's Research Programme on Farmland Conservation and Biodiversity, Volume 1, University of Warwick, England, pages 89-93, November **2004**.
> (The sown patches were Cynosaurus cristatus {Crested Dog-Tails grass}, Festuca rubra {Red Fescue}, Poa pratensis {Kentucky Bluegrass}, Lotus corniculatus {Birds-Foot Trefoil}, Trifolium hybridum {Alsike Clover}, Trifolium pratense {Red Clover} and Centaurea cyanus {Cornflower}.)

"Garden plant recommendations for wild bees of North America:
> *This table contains nearly 200 garden plant genera with species whose flowers are sought by wild bees of North America. It includes: **Symphytum, Borginaceae, Comfrey**.*

Our flower gardens can become valuable cafeterias for local populations of diverse native bees."
-'Gardening for Native Bees in Utah and Beyond' by James H. Cane and Linda Kervin, Utah State University Extension and Utah Plant Pest Diagnostic Laboratory, January **2013**.

Chapter 19

Nutritional Value of Comfrey

Definitions of Nutrient Terms

Ash residue is various types of minerals (inorganic matter). The main components of ash are usually phosphorous and calcium, but it also contains iron, zinc and other minerals. Anything in food that won't burn is counted as ash.

Crude Fat is the crude mixture of fat-soluble material in feed. It is also known as the Ether Extract or the free lipid content. It includes carotene pigments, chlorophylls, diglycerides, fat soluble vitamins, free fatty acids, monoglycerides, phospholipids, steroids, triglycerides, etc. **Ether Extract** substances consist chiefly of fats and fatty acids that are soluble in ether.
Crude Fiber is indigestible portion of plant material. However, some of these substances can be partially digested by microorganisms in rumen (stomach) of ruminants (cow, goat, sheep). The higher the fiber, the lower the energy content of feed.

Crude Protein (CP), also called Total Protein, is the total amount of nitrogen. This includes Non-Protein Nitrogen (NPN) and True Protein Nitrogen.
> **Non-Protein Nitrogen** are feed components such as urea, biuret and ammonia, which are not proteins but can be converted into proteins by microbes in the stomach of ruminants.
> **True Protein** is the total nitrogen minus the NPN. Protein is a class of nitrogenous organic compounds with large molecules of one or more long chains of amino acids.

Nitrogen-Free Extract, N-Free Extract or NFE contains the carbohydrates, sugars and starches and most hemicellulose in feed. The equation to measure NFE starts with 100 and subtracts the sum of the protein, fat, water, ash and fiber.

*"They reveal, what is common knowledge in these days of ley farming and grassland research, that **the best nutritional harvest is in the Comfrey leaf.***
> ***Wide variations in Comfrey analysis depend on the stage and quality of the cut; there is less water and more protein to lower fibre in a crop cut at the leafy stage,** as well as more total weight each season than where it is allowed to run to stem and flower: a fact which also affects palatability (taste).*
*The first analysis confirmed what has never been doubted since, **the high nutritive value of Comfrey**.*"
-Russian Comfrey: A Hundred Tons an Acre of Stock or Compost for Farm, Garden or Smallholding by Lawrence D. Hills. London England: Faber and Faber, Limited, **1953**, page 27.
> (Ley farming is growing grass or legumes in rotation with grain or tilled crops for soil conservation.)

Dry Matter of Comfrey
See subsections 'Production Per Plant or Small Area' and 'Production Per Acre' in section 'Productivity and Farm Economics of Comfrey' in Volume 2.

Dry Matter (DM) is the part of feed that is not water. Percent DM= 100% minus moisture%. The dry matter or dry weight is the mass when completely dried. It is the carbohydrates, fats, proteins, fiber, vitamins, minerals (ash), and other minor constituents.

*"**The composition of Comfrey from budding to fruiting, 29 days, ranged, Dry Matter 11.52 to 18.04 percent.**"*
-'Comfrey: A Valuable Fodder Crop' by E.A. Ziglinskaja, SeveroZapadnyj Nauc- Issled Inst Sel'sko Hoz; Svinovodstvo (Pig Breeding), Soviet Union, No. 9, pages 19-21, **1957**. (The species of Comfrey was not mentioned.)

"*1957 Russian Comfrey Yields of Dry Matter (cwts per acre): Mean 44.08. (4936 pounds = 2238 kg)*
Bocking No. 4	33.3 cwt/acre (3729 pounds/acre = 1691 kg)
Bocking No. 12	53.8
Bocking No. 14	47.4 cwt/acre (5308 pounds/acre = 2407 kg)
Bocking No. 17	45.3
Newman Turner strain	46.0
Stephenson strain	38.3
Webster strain	44.5

Symphytum asperum (Bocking No. 13): 18.0 cwt/acre Dry Matter (not included in total mean).
1957 Russian Comfrey Yields of Dry Matter (percentage of green weight): Mean 11.6%.
Bocking No. 4	11.7%
Bocking No. 12	11.3
Bocking No. 14	11.8
Bocking No. 17	10.7

    Newman Turner strain   11.7
    Stephenson strain   12.1
    Webster strain   11.5
    Symphytum asperum (Bocking No. 13): 12.0% "

-'Russian Comfrey' by L.A. Willey and R.L. Knight, Journal of the National Institute of Agricultural Botany, Cambridge, England, Volume 9, No. 2, pages 139-144, **1962**.

(See subsection 'Strains/Cultivars of Russian Comfrey' in section 'Details about Symphytum Species Hybrid: Russian Comfrey' {Chapter 10}.)

('cwt' is the abbreviation for hundredweight. In the United Kingdom one hundredweight is 112 pounds. In the United States and Canada it is 100 pounds.)

(There are 20 hundredweight in a ton, producing a 'short ton' of 2000 pounds and a 'long ton' of 2240 pounds. A short ton is used in the United States. A long ton is used in Great Britain and countries that used to be territories of it. Metric ton or tonne is the 'International System of Units' that is used worldwide. It is 1,000 kg which equals 2204 pounds.)

(The Dry Matter table provided by the author included Bocking No. 13 which is Symphytum asperum or Prickly Comfrey. The author's 'mean' amount of cwt included S. asperum. I recalcuated the mean with only Russian Comfrey.)

"Percentage composition of Russian Comfrey leaf and stem was Dry Matter (DM) 83.4% (leaf), 80.2% (stem), and in DM crude protein 26.4%."

-'On the Ascorbic Acid Content of Russian Comfrey' by M. Ikeda, S. Uchimura and E. Matsui, Journal of the Faculty of Fisheries and Animal Husbandry, Hiroshima University, Japan, Volume 4, pages 103-109, **1962**.

" **'Chemical Composition of Russian Comfrey'** by Ikeda, Uchimura, and Matsui. 1962. Hiroshima University, Japan, Faculty of Fisheries and Animal Husbandry Journal, Volume 4, pages 103-109.

	Dry Matter
Dried Leaf	83.4%
Dried Stem	80.2
Fresh Leaf	12.0
Fresh Stem	12.1 "

-'Comfrey as Forage' by Lester R. Vough, Extension-Research Agronomist, Oregon State University, Corvallis, Agronomic Crop Science Report, August **1976**, page 4.

" **'Quaker Comfrey (Russian Comfrey) Yields and Protein Content'**, University of California at Davis, 1958, from 'Agronomy Notes', February 28, 1961:

Harvest Date	Yield Per Acre: Green	Yield Per Acre: Dry
June 3	104,544 pounds= 52.2 US tons	7,318 pounds = 3319 kg
July 1	40,656 pounds= 20.3 US tons	4,066 pounds = 1844 kg
August 6	46,464 pounds= 20.2 US tons	5,576 pounds = 2529 kg
November 11	39,688 pounds= 19.8 US tons	7,144 pounds = 3240 kg
Total	231,352 pounds=115.6 US tons	24,104 pounds = 10,933 kg"

-'Comfrey as Forage' by Lester R. Vough, Extension-Research Agronomist, Oregon State University, Corvallis, Agronomic Crop Science Report, August **1976**, page 4.

"Comfrey forage (fresh) averaged about 89% moisture in Minnesota trials.
Moisture content of Comfrey forage is higher than in some legume and grass forages. For example, alfalfa cut in early bloom has 80% moisture and winter rye has 75 to 85% moisture when harvested between the tillering and boot stages."

-'Comfrey' by Teynor, Putnam, Doll, Kelling, Oelke, Undersander and Oplinger, 'Alternative Field Crops Manual', University of Wisconsin: Cooperative Extension, University of Minnesota: Center for Alternative Plant & Animal Products, and Minnesota Cooperative Extension, February **1992**, updated November 1997, page 2.

(Tillers are new grass shoots composed of a growing point, stem, leaves, root nodes, and latent buds. A tiller flowers in good growing conditions. This is called tillering. Boot stage is when the plant begins to concentrate on seed head development rather than the leaf.)

*"Yield of dry biomass of perennial herbs was measured in the mountain zone of Kabardino-Balkarian Republic, Russia in May 2002. Rhizomes of goat's-rue, inula, Comfrey and nettle were planted in 2002 as pure cultures and in a mixture. Yield of dry hay in 2005 was 11.0, 9.9, **8.7** and 16.6 tons/hectare for goat's-rue, inula, **Comfrey** and nettle, respectively."*

-'Effectiveness of Multi-Purpose Grasses in Fodder Production' by A.Ya. Tamakhina, Kabardino-Balkarian State Agrarian Academy, Russia; Kormoproizvodstvo (Plant Breeding), No. 10, pages 2-4, **2009**.

(A hectare is a square 100 meters on each side, also 100 ares {10,000 m2} or 1 square hectometre {hm2}. An acre is 0.405 hectare, and one hectare is 2.47 acres.)

(I'm not sure which type of Comfrey they grew but it is probably Prickly Comfrey or Russian Comfrey, with Prickly Comfrey being more likely. **Their results: Yield dry Comfrey hay = 8.7 tons / hectare. Hectare = 2.47 acres.**
My calculations: 8.7 tons / 2.47 acres = 3.52 tons / acre. Assuming the authors mean the metric ton since they are based in Russia, then the dried Comfrey hay is 7,758 pounds / acre.)

Nitrogen - Phosphorus - Potassium (N-P-K)
Also see in this section the subsections 'Protein (Crude Protein)' and 'Minerals (Ash)'.

"Fertilizing Constituents of Feeding Stuffs: (Symphytum asperrimum, green fodder)

	Water	Ash	Nitrogen	Phosphoric Acid	Potash
Prickly Comfrey	84.36%	2.45	.42	.11	.75 "

-'Yearbook of the United States Department of Agriculture', Washington, DC, 154 pages, **1894**. Includes Report of the Secretary of Agriculture, papers by Bureaus and Divisions for the farmer, statistical tables and references.

"Partial Elemental Composition of First & Second Cuttings Comfrey Forage, Rosemount, Minnesota:

Element	June	July	Average
Potassium	5.89%	5.84%	5.86%
Nitrogen	3.36%	3.70	3.53
Calcium	1.37%	1.51	1.44
Phosphorus	0.49%	0.51	0.50 "

-'Comfrey: A Controversial Crop' by Robert G. Robinson, University of Minnesota, Agricultural Experiment Station, Minnesota Report MR-191, Item No. AD-MR-2210, **1983**, page 5.
    (The author did not say type of Comfrey grown but it may have been Russian Comfrey.)

"I had analyses carried out for two cuts of dried Comfrey leaves and stems for Nitrogen, Phosphorus, Potassium and Calcium, combining the samples for a composite result. The analyses of this mixture were Total Nitrogen (Nitrical and Ammoniacal) 2.2%; Phosphate 0.9%; Potash 6.1% and Calcium 2.2%.
    Phosphorus (P) is low in comparison to the others, but the required level of P in the soil (measured in parts per million or milligrams per kilogram) is also low. Additionally the amount of P removed by crops is normally considerably lower than that of Potassium (K) removed.
The percentages of N-P-K and Ca (Calcium) are lower than in conventional fertilisers, and it is not intended that Comfrey should be the sole source of nutrients for any crop or plant. **Think of it as a booster feed. As with Farmyard Manure (FYM), I am of the opinion that it does more good than its analysis suggests.** I have no scientific explanation for this view but I liken it to having a bowl of soup on a cold day - the effect of the soup far outweighs the nutrients in it."
-'Using Comfrey' by Old McDonald in Portugal: Farming, Gardening, Wildlife and Good Food, http://oldmcdonaldinportugal.blogspot.com, November 2 **2014**.   (He did not give the species of Comfrey.)

"There's a wide range of ideas about the nitrogen-phosphorous-potassium **(NPK) content of Comfrey leaf. One account rates the leaves at a proportion of 1.8 - 0.5 - 5.3. Another study in England found 0.35 - 0.73 - 7.35.** *This is compared to kelp meal with a NPK ratio of 1.0 - 0.5 - 2.5."*
-Understanding Roots: Discover How to Make Your Garden Flourish by Robert Kourik. Occidental, California: Metamorphic Press, **2015**, page 87.

*"In order to understand better what kind of benefits Comfrey has as an organic fertilizer and mineral accumulator I decided to send samples into Maxxam Analytics. **I started by drying some of the last Comfrey leaves in the fall. I then had the sample tested for immediately available NPK and trace elements.***
    The total immediately available NPK of Comfrey is: 0.35 - 0.73 - 7.35. These results represent the NPK that is immediately available to plants in the garden soil.
    This analysis does not account for nutrients that are tied up in larger more complex molecules. As Comfrey is broken down these nutrients are released into the soils nutrient cycle.
Our lab results found total phosphorus and potassium of 5,300 and 70,000 mg/kg and when converted to % molecular weight the total NPK of Comfrey is: 3.7 - 1.21 - 8.43.
Comfrey turns out to have a great NPK both immediately available and long term."
-'Comfrey an Organic Fertilizer and Mineral Accumulator You Can Grow at Home' by Stephen Legaree, Alberta Urban Garden, Edmonton, Canada, February 22 **2015**. We hope to promote organic gardening that is simple, sustainable and does not have to cost a lot. We do this by investigating the science behind gardening, methods, practices and products.
www.albertaurbangarden.cawww.albertaurbangarden.ca
    (He did not give species of Comfrey, though he did recommend Russian Comfrey No. 14 because seeds are sterile.)
    ('Maxxam Analytics'® is now 'Bureau Veritas Laboratories'®. A North American provider of analytical services / solutions to environmental, energy, food, DNA and industrial hygiene industries with extensive laboratory network in Canada.)

*"**Dried Comfrey Leaves Gathered in Fall (mg/kg): Nutrients:***

Available (KCl) **Nitrate plus Nitrite (N)**	3300 mg/kg (1 kg = 2.2 pounds)
Available (KCl) **Nitrite (N)**	7.8
Available (Mod Kel) **Phosphorus (P)**	3,200
Available (Mod Kel) **Potassium (K)**	61,000
Available (CaCl2) **Sulphur (S)**	770

***Physical Properties:** Moisture 9.0% "*

-'Alberta Urban Garden Certificate of Analysis: Revised Report' by Maxxam Analytics®, Canada, January-Februrary **2015**. Includes analysis of soil, compost, Comfrey, rock dust, leaves, biochar and coffee grounds. (Species of Comfrey was not given.)

"Stephen Legaree had some Comfrey analyzed and came up with following dry weight value for total NPK, 3.5 - 1.2 - 8.4. The book, 'Comfrey: Past Present and Future', reports an NPK of 3 - 1 - 4.8 (converted to dry weight)."
-'Comfrey: Is it a Dynamic Accumulator?' by Robert Pavlis, Master Gardener; Garden Myths: Learn the Truth about Gardening, www.gardenmyths.com, southern Ontario, Canada, September 18 **2016**.
> (Stephen Legaree, Professional Biologist, is an Environmental Specialist who is owner of Alberta Urban Garden, Edmonton, Canada that promotes organic gardening.)
> ('Comfrey: Past, Present and Future' is a copy of the book 'Comfrey: Fodder, Food and Remedy' by Lawrence D. Hills.)

"Comfrey: Nitrogen - Phosphorus - Potassium : 1.8 - 0.5 - 5.3."
-'NPK Value of Everything Organic!' by Nigel Davenport, The Nutrient Company: Aquarium and Horticulture Products, https://thenutrientcompany.com, Rochdale, England, January 10 **2019**.   (It did not give the species of Comfrey. This is dried leaf.)

Protein (Crude Protein)
There are more Crude Protein statistics in the subsection 'Fat, Protein, Carbohydrates, Ash, Misc.' in this section.
See subsection 'Comfrey Protein Concentrates' in section 'Humans Eating Comfrey' in Volume 2.

Crude Protein (CP), also called Total Protein, is the total amount of nitrogen. This includes Non-Protein Nitrogen (NPN) and True Protein Nitrogen.
> **Non-Protein Nitrogen** are feed components such as urea, biuret and ammonia, which are not proteins but can be converted into proteins by microbes in the stomach of ruminants.
> **True Protein** is the total nitrogen minus the NPN. Protein is a class of nitrogenous organic compounds with large molecules of one or more long chains of amino acids.

Digestible Crude Protein (DCP) is amount of crude protein absorbed by the animal (crude protein minus protein lost in feces).

"1957 Russian Comfrey Yields of Crude Protein as Percentage of Dry Weight: Mean 25.1%.

Bocking No. 4	**24.8%**
Bocking No. 12	24.8
Bocking No. 14	**24.5**
Bocking No. 17	25.4
Newman Turner strain	25.1
Stephenson strain	25.0
Webster strain	25.6

Symphytum asperum (Bocking No. 13): 25.4%
The average Crude Protein content was highest in the first cut (April/May).
The average yield of Crude Protein was 2.16 cwts per acre."
-'Russian Comfrey' by L.A. Willey and R.L. Knight, Journal of the National Institute of Agricultural Botany, Cambridge, England, Volume 9, No. 2, pages 139-144, **1962**.
> ('cwt' is the abbreviation for hundredweight. In the United Kingdom one hundredweight is 112 pounds. In the United States and Canada it is 100 pounds.)

"On the Nitrogen Content of Russian Comfrey:
The nitrogen content which is soluble in many kinds of solvents, in fresh and dried leaves of Russian Comfrey was investigated. Harvest periods of Russian Comfrey are in the latter part of May and in the early part of September.
Ratios of H2O (water) and 0.3% NaOH (sodium hydroxide) soluble nitrogen are increased and 10% NaCl (sodium chloride) soluble nitrogen decreased in dried leaves compared to fresh leaves. These tendencies are observed in protein nitrogen fraction of dried leaves. The variation of solubility is considered to be caused by denaturation of protein.
NH3-N, NO3-N, NH2-N, Amide-N, and Peptide-N are determined in dried leaves.
Fresh leaves of Russian Comfrey contain 86% of moisture and 0.5% of total Nitrogen. Dried leaves contain about 7% of moisture and 4% of nitrogen, *and a little difference of these elements can be recognized in different growth periods.*
> ***In consideration of the kinds of plant protein, the component of Russian Comfrey is almost uniform although the harvest period is different.***

Russian Comfrey does not wither in a summer of high temperature and little precipitation, and its constituent is almost uniform in both seasons, such as spring and summer. So therefore Russian Comfrey is recongnized to be a suitable feed crop in the region of the south western part of Japan. ***It is well known that peptide is a condensate of amino acid. Peptide nitrogen fraction of Russian comfrey is 1.434 to 1.740%."***
-'On the Nitrogen Content of Russian Comfrey' by Minoru Ikeda, Itaru Kunisaki and Hiroko Matsumura, Department of Animal Husbandry, Fisheries and Animal Husbandry, Hiroshima University, Fukuyama, Japan; Journal of Faculty of Fisheries and Animal Husbandry, Volume 5, No. 1, pages 165-173, **1963**.
> (For the rest of the article about nitrate nitrogen in feed, see sub-subsection 'Problems with Comfrey and Pigs' in subsection 'Pigs' in section 'Livestock / Pet Species and Comfrey' in Volume 2.)

(Peptide is a compound of two or more amino acids linked in a chain.)

"Comfrey (Symphytum perigrinum) harvested monthly from June to November yielded about twice as much Crude Protein as indigenous (native) Indian (India) pasture grasses. Dry matter was also rich in Phosphorus, 0.85%.
The proximate composition was similar to that reported in Poland but only the Polish values are quoted; Japanese values for Russian Comfrey are also given, with no reference."
-'Comfrey: A Forage Rich in Protein and Phosphorus' by J.R. Kaushal, R.S. Gill and S.S. Negi, Indian Veterinary Research Institute, Pradesh, India; Indian Farming, Volume 22, No. 11, pages 37-41, **1973**.
(Proximates are used in the analysis of food into its major constituents. They are approximations. In industry, the standard proximates are: Ash, Moisture, Proteins, Fat and Carbohydrates.)

*"Protein to Fiber Ratio: The especially interesting column in the table is the protein to fiber ratios, for **Comfrey** is far ahead of all the other fodders with 2.27 to 1 compared with 0.88 to 1 for the best pasture grass.*
This is why the pig and poultry keepers are the keenest on Comfrey, and why a high starch equivalent feed must be fed with Comfrey for dairy cattle. Alfalfa is 0.57 to 1; kale is 0.88 to 1; corn is 0.30 to 1."
-Comfrey Report: The Story of the World's Fastest Protein Builder and Herbal Healer, Conservation Gardening and Farming Series: Series C by Lawrence D. Hills. England: Henry Doubleday Research Association, **1975**, pages 33, 34.

"The analysis of Comfrey varies through the season, like that of all fodder crops, especially grass.
Percent protein, dry weight: May 22= 33.68%, June 22= 26.59%, July 26= 23.03%, September 28= 23.58%, October 22= 18.80%. The mineral content is fairly constant through the season."
-Comfrey Report: The Story of the World's Fastest Protein Builder and Herbal Healer, Conservation Gardening and Farming Series: Series C by Lawrence D. Hills. England: Henry Doubleday Research Association, **1975**, pages 39, 41

" 'Quaker Comfrey (Russian Comfrey) Yields and Protein Content', University of California at Davis, 1958, from 'Agronomy Notes', February 28, 1961:

Harvest Date	Yield Per Acre: Green	Yield Per Acre: Dry	Protein Dry Weight
June 3	104,544 pounds= 52.2 US tons	7,318 pounds	21.17%
July 1	40,656 pounds= 20.3 US tons	4,066	24.15%
August 6	46,464 pounds= 20.2 US tons	5,576	22.31%
November 11	39,688 pounds= 19.8 US tons	7,144	15.74%
Total	231,352 pounds=115.6 US tons	24,104 pounds	20.84% average"

-'Comfrey as Forage' by Lester R. Vough, Extension-Research Agronomist, Oregon State University, Corvallis, Agronomic Crop Science Report, August **1976**, page 4.

"What the world needs is a method of extracting the 3 1/2 tons (7840 pounds using British long ton = 3556 kg) a year of pure protein there is in a 100 ton (224,000 pounds = 101,604 kg) an acre crop, compared with the 5 cwt. (560 pounds = 254 kg) there is in a 15 cwt. (1680 pounds = 762 kg) an acre crop of soybeans."
-Comfrey: Fodder, Food and Remedy by Lawrence D. Hills. New York: Rizzoli Universe Books: **1976**, page 177.

*"Crude Protein, **true protein,** ether extract, cell wall constituents (CWC), acid detergent fibre (ADF), and acid detergent lignin were estimated in **plant meals** prepared from lucerne (Medicago sativa), **Comfrey (Symphytum officinale),** amaranthus (Amaranthus hypochondriacus), chenopodium (Chenopodium guinoa), atriplex (Atriplex hortensis) and Sudan grass (Sorghum vulgare sudanese).*
Values are reported for samples frozen immediately after harvest and freeze-dried, for samples wilted for 0.5, 1 and 6 hours before freeze-drying and for air-dried samples. Nitrogen was estimated in CWC and ADF.
In a trial with weanling rats, in which casein and the plant meals each provided half of the dietary protein (16% total Crude Protein), gains were greater for rats given lucerne (alfalfa) or Comfrey than for the other plants."
-'Evaluation of Several Crops as Sources of Leaf Meal: Composition, Effect of Drying Procedure, and Rat Growth Response' by by P.R. Cheeke and R. Carlsson from Department of Animal Science, Oregon State University; Nutrition Reports International (now called Journal of Nutritional Biochemistry), Volume 18, No. 4, pages 465-473, **1978**. (I was unable to get this report. If you have it, could you please send it to me.)

"Much of the interest in Comfrey as a forage crop has centered on its protein content.
The highest single record for Crude Protein is 434 grams/kg DM (Dry Matter) in the leaf blades of S. asperum in spring 1970 in Moscow Province, USSR (Vavilov, Edel'shtein and Solov'eva, 1973).
Normal Crude Protein values in Comfrey are in the range 150-275 grams/kg Dry Matter. This is rather higher than grass and comparable with legumes such as lucerne and red clover. If a typical mean Crude Protein content over the year is 200 grams/kg DM, a Crude Protein yield of 0.8-1.5 tons/hectare can be expected from Comfrey under normal conditions.
Recorded yields are, however, in a somewhat lower range of 0.2-1.2 tons/hectare and are similar to those obtainable from lucerene or other legumes, or from good quality grass.
*Few efforts have been made to determine the true protein content of Comfrey herbage. **Kellner (*1926) estimated that true protein in S. asperum accounted for 60% of the Crude Protein in green herbage, but 80% of that in hay.***
Vavilov, Edel'shtein and Solov'eva (1973) found that on average true protein content of Symphytum asperum was about

78% of Crude Protein. Albumins and globulins made up about 38%, prolamins 4% and glutelins 14% of true protein."
-'Comfrey Symphytum spp. as a Forage Crop' by J.C. Forbes, A.D. McKelvie, and P.J.C. Saunders, North of Scotland College of Agriculture, Aberdeen, United Kingdom; Herbage Abstracts, Volume 49, No. 12, pages 523-539, **1979**.
(* -'The Scientific Feeding of Animals' {Die Ernahrung der Landwirtschaftlichen Nutztiere, 1905} by Professor O. Kellner, German Agricultural Scientist. New York: The Macmillan Company, 1910. Comfrey pages 362, 383. Translated from German by William Goodwin, B.Sc., Ph.D., University of London, England. Also published 1926.)

"**There is little suggestion of any difference in protein content between S. x uplandicum and S. asperum. Medvedev (*1974) found that over eight or nine years, the average Crude Protein contents of these species and of S. officinale, S. caucasicum and S. asperum x caucasicum were all about 170 grams/kg DM (Dry Matter).**
No differences in Crude Protein were found between clonal selections, strains or ecotypes of S. x uplandicum by Willey and Knight (1962), McClean (1964) or Tabin, Berbec and Wrebiakowski (1966). **Seed provenances (different sources) of S. asperum did, however, show considerable differences in Crude Protein content (Medvedev, 1974)."**
-'Comfrey Symphytum spp. as a Forage Crop' by J.C. Forbes, A.D. McKelvie, and P.J.C. Saunders, North of Scotland College of Agriculture, Aberdeen, United Kingdom; Herbage Abstracts, Volume 49, No. 12, pages 523-539, **1979**.
(* -'The Duration of Economic Utilisation and Yields of 5 Species of Comfrey', 1974. In Russian. If you have an English translation, could you please send it to me.)

"The 1950 Comfrey crop was at the rate of 45 tons an acre, so we had enough to put through the shredder and into the grass drier. **Unlike dried grass, which is high in protein and carotene in the spring but falls in July and August, dried Comfrey maintained very nearly the same analysis through the season, though the fibre increased as the flower stems grew.**
Dried grass and meal such as that from sprout stems was sold on the protein level. Though Comfrey meal came out a rather dark brown instead of the bright green the market preferred, Claude, the farm manager, saw a great advantage for it.
If it would produce 21.80% dry protein meal when the grass protein fell, just enough could be added to a poor sample of grass meal to bring it up to 15 to 18 percent- meaning a better price for the whole batch."
-Fighting Like the Flowers: An Autobiography: The Life Story of Britain's Best-Known Organic Gardener by Lawrence D. Hills. Bideford, Devon, England: Green Books, **1989**, page 89.

Amino Acids in Comfrey

Amino acids are organic compounds containing amine and carboxyl functional groups, along with a side chain specific to each amino acid. Nine amino acids are essential for humans because they can not be produced by the body, so they must be in food. They are histidine, isoleucine, leucine, lysine, methionine, phenylalanine, threonine, tryptophan and valine.

"According to the 'Cruzada Nacional Por-Enriquemiento de los Alimentos' (National Crusade for Food Enrichment, **1976**), two-thirds of the population of Mexico are on a bean-maize (corn) diet, which is lacking in three essentail amino acids- methionine, isoleucine and tryptophan.
Comfrey leaf of Bocking 14 is 34.6% protein, 0.58% methionine, 0.64% tryptophan, 1.41% lysine, and 1.15% isoleucine.
    Comfrey is a very good source of tryptophan, rather more than a third better than cashew nuts, more than three times as good as lentils, and more than twice as good as cheese.
    Lentils had slightly more lysine, split peas the same, while the cheeses were well ahead, as all proteins of animal origin can be expected to be, while the nuts were far behind.
Methionine is the most urgent lack in Mexico and among the people of the shanty (shack) towns of Central and South America, and Comfrey is about as good a source as cheese, though Brazil nuts are ahead by two-thirds."
-Comfrey: Fodder, Food and Remedy by Lawrence D. Hills. New York: Rizzoli Universe Books: **1976**, pages 88-90.

"**Shchereva et al. (1965, in Russian) hydrolysed (chemically breakdown) the protein of S. x uplandicum and of lucerne (alfalfa) and compared the amino-acid composition of the two species.**
    They had similar contents of alanine, proline and valine, but **S. x uplandicum had relatively large amounts of aspartic acid, glutamic acid, isoleucine/leucine, lysine, phenylalanine and serine.**
Chekalinskaya and Chekhova (1970, in Russian) found that **S. asperum was rich in lysine, phenylalaine and threonine,** but Egoroa (1973, in Russian) could detect no major differences in amino-acid composition between S. asperum and red clover or lucerne. **Hills (*1976) gives data showing S. x uplandicum to be fairly rich in isoleucine, lysine, methionine and trytophan.** This was confirmed for lysine but not for methionine by Saunders (**1977)."
-'Comfrey Symphytum spp. as a Forage Crop' by J.C. Forbes, A.D. McKelvie, and P.J.C. Saunders, North of Scotland College of Agriculture, Aberdeen, United Kingdom; Herbage Abstracts, Volume 49, No. 12, pages 523-539, **1979**.
(* -Comfrey: Fodder, Food and Remedy by Lawrence D. Hills. New York: Rizzoli Universe Books: 1976.)
(** -'An Investigation Into the Potential of Both Russian Comfrey and a Mixture of Comfrey and Perennial Rye-Grass as a Fodder Crop' by P.J.C. Saunders, B.Sc. Thesis, Dept of Agriculture, University of Aberdeen, Scotland, 1977. I was not able to find this.)

"Among minor constituents of Comfrey are starch and sugar and its asparagin content which lets the herb act as a complementary diuretic. **Asparagin is an amino acid** present also in licorice, marsh mallow, asparagus and leguminous plants."
-Comfrey: What You Need to Know by Ben Charles Harris. New Canaan, Connecticut: Keats Publishing, Inc., **1982**, page 20.

(Asparagin or asparagine is a non-essential amino acid.)   (Diuretics increase the production of urine.)
(Legumes/pulses are in the family Fabaceae/Leguminosae. They include alfalfa, beans, clover, peas, chickpeas, indigo, mimosa, lentils, soybeans, peanuts and tamarind.)

"**Amino Acid Concentrations in the First and Second Cuttings of Comfrey Forage: Oven Dry at Rosemount, Minnesota, Amino Acid Percent in Forage Protein**

	June	July	Average
Glutamic acid	12.25%	12.73%	12.49%
Leucine	10.07	10.01	10.04
Aspartic acid	9.30	9.37	9.34
Valine	6.91	7.13	7.02
Arginine	6.49	6.26	6.37
Phenylalanine	**6.22**	**6.46**	**6.34**
Alanine	5.75	5.83	5.79
Proline	5.47	5.43	5.45
Glycine	5.33	5.41	5.37
Isoleucine	**5.37**	**5.26**	**5.31**
Lysine	**5.60**	**4.81**	**5.21**
Cystine	4.67	—	4.67
Threonine	4.57	4.44	4.51
Tyrosine	3.63	3.84	3.74
Serine	3.10	2.83	2.96
Methionine	**2.34**	**2.59**	**2.46**
Histidine	2.03	2.01	2.02
Tryptophan	**0.91**	—	**0.91**

The total protein and essential amino acid requirements for human nutrition are supplied in 0.75 pound (12 ounces = 0.34 kg) per day of dry Comfrey if digestibility is about 55 percent.*"
-'Comfrey: A Controversial Crop' by Robert G. Robinson, University of Minnesota, Agricultural Experiment Station, Minnesota Report MR-191, Item No. AD-MR-2210, **1983**, pages 4, 5.
(* -'Energy and Protein Requirements', Food and Agricultural Organization of the United Nations {FAO} Nutrition Committee, FAO Nutrition Meetings Report No 52:1-118, 1973.)

"**Amino acid profiles of Russian Comfrey** *during cutting trials, content % Dry Matter, mean of 4 to 12 weeks of growth:*
Alanine 1.15%, Arginine 0.87%, Asparagine 1.84%, Glutamine 2.09%, Glycine 0.99%, Histidine 0.26%, Isoleucine 0.78%, Leucine 1.09%, Lysine 0.59%, Proline 0.94%, Phenylalanine 0.79%, Serine 0.75%, Threonine 0.80%, Tyrosine 0.59%, Valine 1.09%. Assays for Methionine and Tryptophane are not available.
Amino acid content tends to decrease as the plants mature. A comparison of essential amino acid profiles indicates that the protein quality of Comfrey is comparable to that of good quality alfalfa but higher than that of Amaranthus (Amaranth) leaf meal."
-'Potential of Russian Comfrey as an Animal Feedstuff in Uganda' by Bareeba, Odwongo and Mugerwa, Department of Animal Science, Faculty of Agriculture and Forestry, Proceedings of the First Uganda Pasture Network Workshop, Makerere University, Kampala, Uganda, Africa, **1987**.

"**Protein content of Comfrey dry matter (15 to 30%) is about as high as legumes.**
Hart (*1976) mentioned that Comfrey has lower amounts of eight amino acids that are essential for humans than turnip greens or spinach, but more than cabbage.
Comfrey, like most green vegetables, is deficient in methionine and is also low in phenylalanine.
   Three ounces of dried turnip greens or spinach, in comparison to 20 ounces of dried Comfrey, supply adults with the total daily requirement of all essential amino acids, except for methionine."
-'Comfrey' by Teynor, Putnam, Doll, Kelling, Oelke, Undersander and Oplinger, 'Alternative Field Crops Manual', University of Wisconsin: Cooperative Extension, University of Minnesota: Center for Alternative Plant & Animal Products, and Minnesota Cooperative Extension, February **1992**, updated November 1997.
(* -'Comfrey Miracle or Mirage?' {Forage Crops} by R.H. Hart, Crops and Soils: Agronomy for Practicing Professionals, Madison, Wisconsin, Volume 29, No. 1, pages 12-14, 1976. I was not able to find this.)

"**Recent studies with Comfrey leaf have shown that it contains several essential amino acids missing in alfalfa. When it is combined with alfalfa in a 60/40% ratio, it constitutes a 'whole food' for feedlot cattle.**
It is metabolized into harmless proteins, including two essential amino acids missing in alfalfa (lysine and alanine)."
-'Comfrey Leaf: A New Animal Food Supplement' in The Encyclopedia of Alternative Agriculture by Dr. Richard Alan Miller, Agricultural Consultant and Researcher, USA, **1992**.

Fiber (Crude Fiber)
There are more Crude Fiber statistics in the subsection 'Fat, Protein, Carbohydrates, Ash, Misc.' in this section.

Acid Detergent Fiber (ADF) is fiber measurement extracted with acidic detergent to appraise quality of forages. It includes cellulose, lignin, ADIN (Acid Detergent Insoluble Nitrogen) and acid-insoluble ash. The higher ADF, the lower the digestibility.

"The Crude Fibre content was greatest in the second cut (May/June) and the mean content was 12.1%."
-'Russian Comfrey' by L.A. Willey and R.L. Knight, Journal of the National Institute of Agricultural Botany, Volume 9, No. 2, pages 139-144, **1962**.

*"**Fibre: Crude fibre contents of Symphytum spp. (species) range from about 100 to 200 grams/kg DM (Dry Matter)- rather lower than grass or lucerne.** Medvedev (1974, in Russian) found that S. x uplandicum had a consistently lower fibre content than S. asperum. **Crude fibre content increases with age. Within the growing season, fibre content increases with maturity as in other herbage crops.**"*
-'Comfrey Symphytum spp. as a Forage Crop' by J.C. Forbes, A.D. McKelvie, and P.J.C. Saunders, North of Scotland College of Agriculture, Aberdeen, United Kingdom; Herbage Abstracts, Volume 49, No. 12, pages 523-539, **1979**.

<u>Minerals</u> (also called Ash in analysis of feed)
See subsection 'Minerals in Comfrey Overview' in section 'Garden Uses of Comfrey' (Chapter 17).
There are more Mineral statistics in the subsection 'Fat, Protein, Carbohydrates, Ash, Misc.' in this section.

*"**The ash contents were remarkably large being 19.2% in 1957 and 16.6% in 1958.
Potash (potassium) was the main constituent.**"*
-'Russian Comfrey' by L.A. Willey and R.L. Knight, Journal of the National Institute of Agricultural Botany, Volume 9, No. 2, pages 139-144, **1962**.

*"Percentage composition of **Russian Comfrey leaf and stem** was dry matter (DM) 83.4%, 80.2.
In DM, crude protein 26.4% leaf, 10.2% stem. Crude fat 2.8% leaf, 2.1% stem. Nitrogen-free extract 28.2% leaf, 29.3% stem. Crude fibre 11.7% leaf, 28.6% stem. Ash 14.3% leaf, 10.0% stem.
    SiO_2 (Silicon Dioxide) 0.82%, 3.4%.   CaO (Calcium Oxide) 2.1, 0.63.   MgO (Magnesium Oxide) 0.40, 0.14.
    K_2O (Potassium Oxide) 8.0, 8.6. P_2O_5 (Diphosphorus Pentoxide) 0.92, 0.65.
    MnO (Manganese Oxide) 0.03, trace."*
-'On the Ascorbic Acid Content of Russian Comfrey' by M. Ikeda, S. Uchimura and E. Matsui, Journal of the Faculty of Fisheries and Animal Husbandry, Hiroshima University, Japan, Volume 4, pages 103-109, **1962**.

*"**Comfrey was found to be high in both calcium and phosphate and very high in potash with some iron and manganese and a trace of cobalt.**"*
-'Comfrey: The Cinderella of Plants' by Maria Wilkes, Herbarist: The Herb Society of America Journal, Volume 33, pages 47-50, **1967**. Also in 'Herbs for Use and for Delight: An Anthology from the Herbarist' by Daniel J. Foley, Herb Society of Ameria. Gloucester, Massachusetts: Peter Smith Publisher Inc., 1974.

*"**Alfalfa hay is the only feed to come near to Comfrey on calcium, and this richness in minerals** explains why it is a favorite feed for racehorse foals to build the bones that need strength for speed."*
-Comfrey Report: The Story of the World's Fastest Protein Builder and Herbal Healer, Conservation Gardening and Farming Series: Series C by Lawrence D. Hills. England: Henry Doubleday Research Association, **1975**, page 42.

*"**The Bocking clones (Russian Comfrey) are higher in calcium, manganese and iron than the wild Symphytum officinale.**"*
-Comfrey: Fodder, Food and Remedy by Lawrence D. Hills. New York: Rizzoli Universe Books: **1976**, page 80.

*"**Most values for the ash (mineral) content of Comfrey herbage are in the range 125-225 grams/kg DM (Dry Matter)- much higher than the normal range for grasses and legumes.**
This certainly reflects the natural richness of Comfrey in mineral elements, but to some extent it may be a consequence of the liability of Comfrey leaves to become contaminated with soil.
    At one sampling period, Strange (*1959) found that soil accounted for 13.5% of the dry weight of S. x uplandicum herbage, but only 1.7% of the dry weight of lucerne.
The highest recorded ash content for Comfrey is 416 grams/kg DM in S. x uplandicum at the stem elongation stage (**Doring 1959). **Ash content does not vary in any consistent way with growth stage in S. x uplandicum** (Doring, 1959), **S. asperum** (Moiseev et al., 1963; Kharkevich, 1966; Astakhov, 1970; Khrushkova and Odegova, 1971; Vavilov and Kondrat'ev, 1975) **or S. officinale** (Popescu, Pltis and Casanova, 1971). Five species and hybrids of Symphytum in a single trial all had similar ash contents (Medvedev, 1971, in Russian).
Mineral analysis of Symphytum spp. (species) show the crop to be rich in potassium, calcium, phosphorus, iron and copper relative to most herbage species.
    Willey and Knight (1962) found that magnesium content was higher in September than in June, and Saunders (1977) found it to higher in August than in November.
    Sodium, manganese, zinc and copper contents were higher in November than in August.
Ikeda et al. (*1964) found that by supplying the appropriate trace element to the soil, they could enrich the herbage of**

S. x uplandicum in manganese, molybdenum or cobalt but not in zinc, boron or copper."
-'Comfrey Symphytum spp. as a Forage Crop' by J.C. Forbes, A.D. McKelvie, and P.J.C. Saunders, North of Scotland College of Agriculture, Aberdeen, United Kingdom; Herbage Abstracts, Volume 49, No. 12, pages 523-539, **1979**.
(* -'A Comparison Between Russian Comfrey and Lucerne' by Richard Strange, Grassland Research Station, Department of Agriculture, Kenya; East African Agricultural Journal, Volume 24, pages 203-205, 1959.)
(** -'Cultivation of Comfrey' by W. Doring, Inst Acker- u PflBau, University of Halle-Wittenber, East Germany; Deutsche Landwirtschaft (German Agriculture), Volume 10, No. 2, pages 62-66, 1959.)
(*** -'Effect of Trace Elements for Dallisgrass and Russian Comfrey' by M. Ikeda, K. Kurozumi, J. Tsubota and H. Matsumura, Department of Animal Husbandry, University of Hiroshima, Fuku-yama, Japan; Journal of Japanese Society of Grassland Science, Volume 10, Issue 2, pages 100-104, 1964.)

"**Comfrey on a dry weight basis is very high in ash (minerals), averaging 18 percent** at Rosemount, Minnesota and ranging from 13 to 42 percent in other parts of the world*.
The forage is particularly high in potassium and is higher than many other forage crops in calcium, phosphorus, iron, and copper. Carbon, hydrogen, oxygen, nitrogen, and sulfur are the major nonmineral elements.
The combined high nitrogen and high mineral concentrations make Comfrey forage an unusually good material for composting, mulching, and organic fertilization of crops.
Elemental Composition of First & Second Cuttings of Comfrey Forage, Rosemount, Minnesota

Element	June	July	Average
Potassium	5.89%	5.84%	5.86%
Nitrogen	3.36%	3.70	3.53
Calcium	1.37%	1.51	1.44
Phosphorus	0.49%	0.51	0.50
Magnesium	0.27%	0.32	0.30
Aluminum	294 ppm	477 ppm	385 ppm= parts per million
Iron	283 ppm	446	364
Manganese	105 ppm	128	116
Sodium	64 ppm	76	70
Zinc	49 ppm	42	45
Boron	45 ppm	46	45
Copper	9 ppm	11	10
Cobalt	7 ppm	-	-
Lead	5 ppm	7	6
Nickel	2 ppm	2	2
Chromium	1 ppm	2	1
Cadmium	<1 ppm	<1	<1

"
-'Comfrey: A Controversial Crop' by Robert G. Robinson, University of Minnesota, Agricultural Experiment Station, Minnesota Report MR-191, Item No. AD-MR-2210, **1983**, page 5.
(* -'Comfrey Symphytum spp. as a Forage Crop' by J.C. Forbes, A.D. McKelvie, and P.J.C. Saunders, North of Scotland College of Agriculture, Aberdeen, United Kingdom; Herbage Abstracts, Volume 49, No. 12, pages 523-539, 1979.)
(The author did not say which Comfrey type was grown but it may have been Russian Comfrey.)

"**Nutritional constituents of plants: stems, leaves, shoots. Per 100 grams = 3.5 ounces fresh weight:**
Symphytum officinale, Common Comfrey leaves, *Boraginaceae*:
    Copper 1.1 mg, Zinc 3.5 mg, Iron 16.6 mg, Manganese 6.0 mg."
-Traditional Plant Foods of Canadian Indigenous Peoples: Nutrition, Botany and Use by Harriet V. Kuhnlein (School of Dietetics and Human Nutrition, McGill University, Montreal, Quebec, Canada) and Nancy J. Turner (Environmental Studies Program, University of Victoria, British Columbia, Canada). Australia and Canada: Gordon and Breach Publishers, **1991**.

"Comfrey (Symphytum peregrinum= Russian Comfrey) if properly grown, is a deep rooted plant, the main roots going down as far as 3 meters (9.8 feet), depending of course on the soil structure. It is here that the **trace minerals** are found, having leeched through the top soil to the subsoil.
It is this same deep rooting characteristic that gives Alfalfa much of its special value as food, because being deep rooted, it goes down and brings up the micro-elements into the leaves.
The leaf of a mature (3-4 years), deep rooted Comfrey plant is a rich mine of trace minerals."
-Comfrey: Nature's Healing Herb & Health Food by Andrew Hughes. Japan: Sanyusha Publishing Co., Ltd, **1992**, page 51.

"Owing to the hairiness of the leaves, soil contamination was considerably greater than on lucerne (alfalfa). Even without soil, Russian Comfrey had a silica content of nearly 6% which largely accounted for the high mineral content for which Russian Comfrey has been favoured.
Lucerne: 1.72% soil in herage, 0.67% silica. **Russian Comfrey: 13.51% soil in herbage, 5.78% silica."**
-East Africa's Grasses and Fodders: Their Ecology and Husbandry by Joseph G. Boonman. Netherlands: Kluwer Academic Publishers, **1993**, page 302.

"Comfrey Feed Analysis, 100% Dry:
Protein 17.02%, Sodium 0.04%, Phosphorus 0.27%, Potassium 5.54%, Sulphur 0.22%, Calcium 1.75%, Magnesium 0.36%, Copper 15.20 ppm, Iron 551.00 ppm, Manganese 57.30 ppm, Zinc 48.60 ppm, Molybdenum co.50 ppm, Selenium co.20 ppm, Nitrate 0.40%, A.D. (Acid Detergent) Fibre 42.39%."
-'Feasibility of Producing Comfrey (Symphytum spp.) Pellet as a Feed Supplement' by Barl, Gibson, Crerar, Shao, and Sokhansanj, Dept of Plant Sciences and Department of Agricultural & Bioresource Engineering, University of Saskatchewan, Saskatoon, Canada at Soils & Crops Conference, February **1999**. (I do not know what 'co.' means in Molybdenum and Selenium.)

"Horsetail is famous among herbalists as a source of silicon (Si). Comfrey contains from 50-80% of the silicic acid content of horsetail.
*Silicon has no officially recommended dietary intake, but estimates of the requirement for humans range from 5-20 mg per day (Groff et al). Silicon was found to be essential in the 1970s for normal development of the connective tissues, mucopolysaccharides, cartilage, elastin, and bone (Carlisle). It is an important rate-limiting enzyme cofactor in the formation of the collagen matrix of bone, and **its presence facilitates bone repair and the uptake of other minerals into bone.***
Silicic acid, one form in which silicon exists in equisetum (horsetail) and Comfrey (Symphytun spp.), is readily soluble in water, readily absorbed in the digestive tract once dissolved, and readily diffuses to the extracellular fluid reservoir and connective tissues.
> *Thus a small amount of infused (tea) herbal material containing soluble silicic acid may provide more physiologically available silicon than much larger amounts of food in which the silicon is bound by fiber or fails to be extracted into solution in the small volume of fluid in the stomach and intestine."*

-'Equisetum: Silicon in Horsetail and Comfrey' by Paul Bergner, Medical Herbalism: Journal for the Clinical Practitioner, North American Institute of Medical Herbalism Inc., Volume 10, No. 4, page 10, **2001**.
> (Equisetum or horsetail / snake grass / puzzlegrass is the only living genus in Equisetaceae, a family of vascular plants that reproduce by spores rather than seeds.)
> (Silicic Acid is the only bio-available, bio-active silicon molecule for plants and animals. It is water soluble. It is found in plant tissues in amounts similar to macronutrients such as calcium and magnesium. It is very important to health.)

"Germanium: A newcomer to the list of trace minerals, *germanium is now considered to be essential to optimum health. Germanium-rich foods help combat rheumatoid arthritis, food allergies, fungal overgrowth, viral infections and cancer.* **Certain foods will concentrate germanium if it is found in the soil:** *garlic, ginseng, mushrooms, onions and the herbs aloe vera,* **Comfrey** *and suma."*
-Nourishing Traditions: The Cookbook that Challenges Politically Correct Nutrition and the Diet Dictocrats by Sally Fallon and Mary G. Enig, Ph.D. Washington, D.C.: New Trends Publishing Inc, **2001**, page 44.

"Ions (mg/100grams dried crude material)

	Anions -			Cations +			
	Fluoride	Chloride	Nitrate	Sodium	Ammonium	Potassium	Magnesium
S. officinale	13.324	334.684	215.638	16.224	43.449	626.240	2.073 "

-'Investigation of Vasoactive Ion Content of Herbs Used in Hemorrhoid Treatment' by Mahir Gulec, Recai Ogur, Husamettin Gul, Ahmet Korkmaz and Bilal Bakir, Gulhane Medical School, Ankara, Turkey; Pakistan Journal of Pharmaceutical Sciences, Volume 22, No. 2, pages 187-192, April **2009**.
> (Cations {positive charge} and anions {negative charge} are ions. Ions have gained or lost one or more valence electrons giving the ion a net positive or negative charge.)

"In order to understand better what kind of benefits Comfrey has as an organic fertilizer and mineral accumulator **I decided to send dried Comfrey leaf samples into Maxxam Analytics®.** *The second test we had run was the total trace elements*.* **Comfrey contains Boron, significant volumes of Calcium, Iron, Magnesium, Sodium, and Sulphur. Additional elements were reported** *however due to the low levels and the detection limit error rate the reported numbers are not as reliable however they are still present.* **These include Manganese, Molynbdenum, Nickel and Zinc. Other literature sources also found Cobalt and Copper.**
Once the Nitrogen, Phosphorus and Potassium have been added to the total these results represent 15 of the 18 essential and beneficial elements that plants take up from the soil.
Missing from the analysis is Silicon, Selenium, Chlorine.
> Silicon and Chlorine were not tested for as they are abundant and well distributed on the earth's crust.
> Silicon is derived from sands and is the most common mineral on the earth's crust.
> Chlorine is generally found as a part of the common mineral compounds sodium chloride and potassium chloride. It is extremely common on earth's crust at roughly 126 ppm (parts per million).
> Selenium although below the detection limit is an integral part of a number of amino acids and is likely present in the soil. Selenium is a rare earth mineral however it is found commonly as a part of most living things in a number of amino acids. It is likely, although below the detection limit of this test, still in the plant or soils.

I consider this a full complement of essential and beneficial elements."
-'Comfrey an Organic Fertilizer and Mineral Accumulator You Can Grow at Home' by Stephen Legaree, Alberta Urban Garden, Edmonton, Canada, February 22 **2015**. We hope to promote organic gardening that is simple, sustainable and does not have to cost a lot. We do this by investigating the science behind gardening, methods, practices and products.

www.albertaurbangarden.cawww.albertaurbangarden.ca
(* -'Alberta Urban Garden Certificate of Analysis: Revised Report' by Maxxam Analytics®, Canada, January-Februrary 2015. Includes analysis of soil, compost, Comfrey, rock dust, leaves, biochar and coffee grounds.)

"Dried Comfrey Leaves Gathered in Fall (mg/kg):

Total Aluminum (Al)	68 mg/kg (1 kg = 2.2 pounds)
Total Boron (B)	58
Total Calcium (Ca)	20,000
Total Iron (Fe)	200
Total Lithium (Li)	<20
Total Magnesium (Mg)	3,300
Total Manganese (Mn)	91
Total Phosphorus (P)	5,300
Total Potassium (K)	70,000
Total Sodium (Na)	1,600
Total Strontium (Sr)	74
Total Sulphur (S)	2,300
Total Antimony (Sb)	<2.0
Total Arsenic (As)	<2.0
Total Barium (Ba)	90
Total Beryllium (Be)	<0.80
Total Cadmium (Cd)	<0.20
Total Chromium (Cr)	<2.0
Total Cobalt (Co)	<2.0
Total Copper (Cu)	<10
Total Lead (Pb)	<2.0
Total Molybdenum (Mo)	1.8
Total Nickel (Ni)	6.2
Total Selenium (Se)	<1.0
Total Silver (Ag)	<2.0
Total Thallium (Tl)	<0.60
Total Tin (Sn)	<2.0
Total Uranium (U)	<2.0
Total Vanadium (V)	<2.0
Total Zinc (Zn)	29 "

-'Alberta Urban Garden Certificate of Analysis: Revised Report' by Maxxam Analytics®, Canada, January-Februrary **2015**. Includes analysis of soil, compost, Comfrey, rock dust, leaves, biochar and coffee grounds.

"Comfrey, Symphytum officinale, shows up as Silicic Acid in leaf at 40,000 ppm (parts per million).
Horsetail, Equisetum arvense, shows up as Silicic Acid in leaf at 80,000 ppm."
-Understanding Roots: Discover How to Make Your Garden Flourish by Robert Kourik. Occidental, California: Metamorphic Press, **2015**, page 80.
    (These numbers are from Dr. Duke's Phytochemical and Ethnobotanical Databases. See reference below.)

"Minerals per 100 grams (3.5 ounces)	Calcium	Iron	Magnesium	Chromium
Comfrey (Symphytum officinale)	1800 mg	1.2 mg	70 mg	180 mcg (microgram)

from Pedersen's 'Nutritional Herbology'.
Three and a half ounces is a lot of herb, but remember, this is in context of your entire diet. A quart of herbal tea a day, made with an ounce of these or other herbs blended together, can make up for a lot of deficiencies even in a processed-food-heavy diet. It's especially good at filling in the gaps created by modern agricultural methods, which tend to reduce the amount of micronutrients available in vegetables - even organic ones."
-'Practical Phytochemistry: Phytochemistry on the Macro Scale' by Katja Swift and Ryn Midura, CommonWealth Center for Holistic Herbalism, www.commonwealthherbs.com, School and Clinic: local and online, Boston, Massachusetts, **2017**.
(* -'Nutritional Herbology: A Reference Guide to Herbs' by Mark Pederson. Indiana: Wendell W. Whitman Co., 1995.)

Dr. Duke's Phytochemical and Ethnobotanical Databases*:

S. officinale	Macro Secondary Nutrients (Low to High)		Micro Trace Nutrients				
	Calcium	Magnesium	Silicon	Iron	Manganese	Sodium	Cobalt
Leaf ppm	1,980 to 18,000	77 to 700	1 to 9	1.3 to 12	0.6 to 5.8	12 to 110	- -
Root ppm	- to 11,300	- to 1,700	- to 35	- to 810	- to 67	- to 3,510	- to 129"

* -'Dr. Duke's Phytochemical and Ethnobotanical Databases' . For in-depth plant, chemical, bioactivity and ethnobotany searches using scientific or common names. Of interest to pharmaceutical, nutritional, biomedical research, alternative therapies and herbal products. https://phytochem.nal.usda.gov, **2018**.

Potassium in Comfrey
See sub-subsection 'Comfrey as Potassium Fertilizer' in subsection 'Minerals in Comfrey Overview' in section 'Garden Uses of Comfrey' (Chapter 17).

*"**Potassium:** Birch, borage, calamus, carrot leaves, chamomile, coltsfoot, **Comfrey**, couch-grass, cowslip, dandelion, elder, eyebright, fennel, honeysuckle, lady's mantle, meadowsweet, mistletoe, mullein, nettle, oak, peony, peppermint, plantain, primrose, rhubarb, scullcap, toadflax, walnut leaves.*
This mineral promotes general healing of the tissues, tones the bowels, gall bladder and liver. Encourages healing of diseased tissues, relieves pain."
-Herbal Handbook for Farm and Stable by Juliette de Bairacli Levy. London, England: Faber and Faber, Inc., **1952**. First edition that later was called 'The Complete Herbal Handbook for Farm and Stable'.

"When Comfrey goes dormant in winter, it returns all the potash (potassium) in its foliage to store in the roots, which is why it pays to let the last cut die down naturally. This gives a starting stock for the spring."
-Grow Your Own Fruit & Vegetables by Lawrence D. Hills. London, England: Faber and Faber Limited: **1971**, page 50.

*"**The potassium content in Symphytum officinale is only 3.09 percent compared to Russian Comfrey 'Bocking 14' of 7.09 percent.**"*
-The Complete Book of Vegetables, Herbs & Fruit by Biggs, McVicar and Flowerdew. London, England: Kyle Cathie Limited, **2004**, page 363.

Vitamins in Comfrey

"It is possible that Russian Comfrey, cut at the leafy stage, would provide a special poultry meal, with 22.7 percent protein to 10.9 percent fibre, and 38.2 percent carbohydrates (Hannah Dairy Institute figures).
Russian Comfrey with 400 mg per kg carotene (precursor to vitamin A)."
-Russian Comfrey: A Hundred Tons an Acre of Stock or Compost for Farm, Garden or Smallholding by Lawrence D. Hills. London England: Faber and Faber, Limited, **1953**, page 105.
(In 1909 there was an Act of Parliament, United Kingdom, to promote scientific development of agriculture, forestry and fisheries based on state-aided but independent research institutes. The choice of southwest Scotland for the Hannah Dairy Institute was based on size of the local dairy industry. It is now called the Hannah Dairy Research Foundation.)

*"**Vitamin C: Ascorbic acid content of leaves of Symphytum asperrimum (Symphytum asperum, Prickly Comfrey) estimated by indophenol titration was from 323 to 565 mg per 100 gram fresh weight.***
Nitrogen fertilizer increased it.
Deficiency of Nitrogen, Potassium and Phosphorus lowered the ascorbic acid content of the leaves.
***Fertilizer applied with lime or a trace of Cu (Copper) raised the ascorbic acid level.**"*
-'On the Ascorbic Acid Content of Russian Comfrey' by M. Ikeda, S. Uchimura and E. Matsui, Journal of the Faculty of Fisheries and Animal Husbandry, Hiroshima University, Japan, Volume 4, pages 103-109, **1962**.

*"**Comfrey carotene levels are similar to those in grasses and legumes.***
***Carotene and ascorbic acid (vitamin C) contents decline sharply in the course of the growing season in Symphytum asperum**, though carotene contents do so only in the early part of the season (*Edel'shtein and Solov'eva, 1974).*
*Shchereva et al. (1965) found that **Symphytum x uplandicum had a somewhat lower ascorbic acid content in flowering than in vegetative plants.***
***The vitamin content of leaf blades of S. asperum is about ten times that of leaf stalks and stems** (Edel'shtein and Solov'eva, 1974)."*
-'Comfrey Symphytum spp. as a Forage Crop' by J.C. Forbes, A.D. McKelvie, and P.J.C. Saunders, North of Scotland College of Agriculture, Aberdeen, United Kingdom; Herbage Abstracts, Volume 49, No. 12, pages 523-539, **1979**.
(* -'The Carotene and Ascorbic Acid Contents in Silage Crops' by M. Edel'-Shtein and I. Solov'-Eva, Doklady Timiryazevskoi Sel'skokhozyai-stvennoi Akademii {Reports of the Timiryazev Agricultural Academy}, Volume 204, pages 51-54, 1974. In Russian. If you have an English translation of the Comfrey sections, I would appreciate a copy.)

*"**Nutritional constituents of plants: stems, leaves, shoots. Per 100 grams = 3.5 ounces fresh weight:***
***Symphytum officinale, Common Comfrey leaves,** Boraginaceae:*
*   ***Vitamin C: 19.0 mg, Vitamin A: 42 retinol equivalents.**"*
-Traditional Plant Foods of Canadian Indigenous Peoples: Nutrition, Botany and Use by Harriet V. Kuhnlein (School of Dietetics and Human Nutrition, McGill University, Montreal, Quebec, Canada) and Nancy J. Turner (Environmental Studies Program, University of Victoria, British Columbia, Canada). Australia and Canada: Gordon and Breach Publishers, **1991**.
(Vitamin A is expressed in terms of retinol equivalents {RE}. One RE is the biological activity associated with 1 microgram of all-trans retinol. 1 International Unit (IU) retinol = 0.3 micrograms Retinol Equivalents.)

*"**Vitamin Dry Leaf Content of Comfrey per 100 Grams (3.527 ounces)***

Thiamin (Vitamin B1)	0.5 mg
Riboflavin (Vitamin B2)	1.0 mg
Pantothenic Acid (Vitamin B3)	4.2 mg
Nicotinic (Vitamin B5)	5.0 mg
Vitamin B12	0.07 mg
Carotene	.170 parts per million
Vitamin A Equivalent	28,000 International Units
Vitamin C	100 mg
Vitamin E	30 mg
Allantoin	0.18 mg "

-Comfrey: Symphuo Symphytum: A Multi-Purpose Herb by Philip Clarke. Edinburgh, England: Pentland Press, **1997**, page 11.

"The Comfrey study was carried out in Gitathuru Approved School, Nairobi, Kenya. **A sample of 85 children** (8-13 years) whose haemoglobin level was <14 gms/dl were identified from a total of 106 children.
  The subjects were randomly allocated into **an experimental group (CCD) whose diet was supplemented with 10 grams (0.35 ounce) of Comfrey powder/day** and a control group (CFD) whose diet was not supplemented.
The serum beta-carotene and serum retinol levels at the baseline and after 21 days of intervention were determined to **assess the potential of Comfrey as a source of vitamin A.** There were no significant differences in base levels of serum B carotene and retinol concentrations between the two groups of children.
  However, the serum beta-carotene levels of the CCD group (4.99+-1.28 micromol/l) was significantly higher than those of the CFD group post intervention (P<0.05).
On the other hand, post intervention serum retinol levels for CCD group (0.29+-0.18 micromol/l) and those of the CFD group (0.37+-0.17 micromol/l) were not significantly different.
Increasing beta-carotene levels above baseline levels 4.39+-1.28 micromol/l) by supplementation is desirable."
-'Assessment of the Potential of Comfrey (Symphytum Asperrimum) as a Source Of Vitamin A in a School Feeding Programme' by Joyce Violet Chania, Thesis Master of Science in Applied Human Nutrition, Department of Food Technology and Nutrition, University of Nairobi, Kenya, March **1998**.

"**Vitamin A in Comfrey: A beta-carotene rich biscuit (betaCRB) was made using Comfrey** and a relatively low beta-carotene biscuit (betaCLB) without Comfrey was also made and their beta-carotene content determined.
The Comfrey rich biscuit had a significantly higher beta-carotene content (1060 microgram/100 grams) than the Comfrey free biscuit, which had (170 microgram/100 grams).
  There was no significant difference (p>0.05) in serum beta-carotene and retinol levels at baseline (start of the test) but **at post-test, the serum beta-carotene significantly increased** (p<0.05) in the cases from 0.0327 +- 0.069 to 0.096 +- 0.036 than in the control group from 0.050 +- 0.049 to 0.076 +- 0.035.
The difference in post intervention (post test) serum (liquid in blood) beta-carotene levels between the groups was highly significant (p=0.000).
The data strongly indicates that Comfrey (Symphytum peregrinum) can be a source of vitamin A."
-'Assessment of the Potential of Comfrey (Symphytum Peregrinum) as a Source of Vitamin A for Malnourished Children (8-16 Years): A Case Study of Kirigiti Girls Approved School in Kiambu, Kenya' by Wambui Gatigwa, Thesis for Master of Science in Applied Human Nutrition, University of Nairobi, **2002**.

"**Vitamins in Comfrey (Symphytum officinale):**
  Ascorbic-acid root 132 ppm (part per million), Beta-carotene root 660 ppm,
  Carotenes plant 6300 ppm, Niacin root and Riboflavin root 7.2 ppm, Thiamin root 1.2 ppm.
It is almost classed as a vitamin."
-'Natural Anti-Irritant Plants' by Manfred Axterer, Cornelia Muller and Anthony C. Dweck (Symrise GmbH and Co. KG, and Dweck Data) from Dweck Data: Consultants on Natural Products to Cosmetic, Toiletry & Pharmaceutical Industry, United Kingdom, **2017**.

Vitamin B12 in Comfrey
Vitamin B12 is a generic name of a collection of cobalt and corrin ring molecules. All substrate cobalt-corrin molecules from which laboratory B12 is made is synthesized by bacteria. After this synthesis, the body can convert B12 to an active form. Vitamin B12 or cobalamin is a water-soluble vitamin involved in the metabolism of every cell. It is a cofactor in DNA synthesis, and in fatty acid and amino acid metabolism. It is important in the nervous system because it helps in the synthesis of myelin. It helps in the development of red blood cells in bone marrow.

"**Vitamin B12:** The question of vitamin B12 always comes up when vegetarianism is discussed. This is the vitamin of which we need the smallest amount- only three millionths of a gram per day. Vitamin B12 is not found in plant sources. Only the total vegetarian has any difficulty in obtaining enough vitamuin B12, since large amounts are found in milk and eggs.
Comfrey is a source of vitamin B12; twelve Comfrey tablets a day are necessary, however, to fulfill the body's need for this vitamin."
-Back to Eden: The Classic Guide to Herbal Medicine, Natural Foods and Home Remedies by Jethro Kloss. Wisconsin: Lotus Press, **1939**-1999, page 607.

(A vegetarian is someone who does not eat meat, fish, fowl. A vegan is a strict vegetarian who does not eat meat, fish, fowl, eggs, dairy products or any food that comes from animals.)

"**Comfrey is the only known land plant to take up vitamin B12 from soil,** the only other plant doing this being porphyra seaweed as used for laver bread in Wales and dulse in Scotland.
Earthworms produce vitamin B12 for Comfrey to take up and pass on in feed."
-World Protein Resources by Allen Jones. New York: John Wiley & Sons, Inc., **1974**. (Chapter 22: 'Green Leaf Protein', pages 209-216.)

(Porphyra genus is a coldwater seaweed in the red algae Division {Rhodophyta}. Laver bread is usually made with Porphyra umbilicalis seaweed. Dulse or Palmaria palmata is also a red algae in the Rhodphyta Division.)

"**Comfrey is the only land plant known to extract vitamin B12 from the soil. This discovery was made in the summer of 1959 by the late Mr. F. Newman Turner,** MBNOA (Member of the British Naturopathic and Osteopathic Association), MNIMH (Member of the National Institute of Medical Herbalists), NDD, NDA, who was consultant to the 'Society of Medical Herbalists'. Vitamin B12 was found in the course of his work with the **tablets of dried Comfrey** used by the Association (Henry Doubleday Research Association, England) in its medical work for the relief of asthma and bronchial complaints.
His practice included a number of cases of **vitamin B12 deficiency among vegans**, and found that the tablets were producing an effect that indicated the vitamin was present."
-Comfrey Report: The Story of the World's Fastest Protein Builder and Herbal Healer, Conservation Gardening and Farming Series: Series C by Lawrence D. Hills. England: Henry Doubleday Research Association, **1975**, page 43.

"In theory, it would take 3 to 4 pounds (1.3-1.8 kg) of fresh Comfrey or 7 ounces (198 gram) of flour to supply all the B12 an adult needs in a day. Yet the **vegans** whose sore tongues Mr. Newman Turner cured took only **12 Comfrey tablets a day**, when in theory they should have eaten nearer 300. **The evidence is that Comfrey must fit the human digestion rather better than it does that of bacteria (from the laboratory).**
Work continues in this field because of the long term value of a vegetable source of Vitamin B12 for those who cannot afford liver, eggs, milk and all the animal products (liver is the richest source) supplying B12 for well-fed Western men today."
-Comfrey: Fodder, Food and Remedy by Lawrence D. Hills. New York: Rizzoli Universe Books: **1976**, page 101.

"**A number of plankton and seaweeds, of which laver (Porphyra vulgaris) is the best known, can also synthesize this vital vitamin (B12), which depends on a trace of cobalt for its manufacture, and this is present in Comfrey, the only land plant so far known (as of 1976) to take it from the soil.**
One of the symptoms of B12 deficiency is a sore tongue, and though our bodies can store this and other vitamins for long periods, **vegans** who do not take sufficient care over their quite difficult diet often have sore tongues and inflammation round the eyes as their reserves run out.
In 1959 Comfrey tablets cleared up the B12 deficiency symptoms. **In a 1959 test the amount of vitamin B12 present in dried Comfrey Bocking No. 4 leaf was 0.58 micrograms per 100 grams.**
In 1959 two 'Henry Doubleday Research Association' members, Mr. John Beck of Bradford, England, and Mr. Geoffrey Wheeler of Bodmin, England, had discovered that six square inches (38.7 square cm) a day of Comfrey leaf would cure the symptoms of B12 deficiency in chinchillas. These squirrel-like and expensive animals are bred for fur."
-Comfrey: Fodder, Food and Remedy by Lawrence D. Hills. New York: Rizzoli Universe Books: **1976**, pages 97, 98.

(Plankton are small, microscopic organisms floating in the sea or fresh water that include diatoms, protozoans, and small crustaceans.)

"**That Comfrey gets its Vitamin B12 from the soil and stores it in leaf and root was established by HDRA (Henry Doubleday Research Association)** with research done by then Chairman, Dr. F. Newman Turner, who after careful preparation of the experiment, showed under the microscope the red crystals of B12 in leaf and stem, and especially in hairs along both."
-Comfrey: Nature's Healing Herb & Health Food by Andrew Hughes. Japan: Sanyusha Publishing Co., Ltd, **1992**, page 33.

"**This Comfrey B12 was also shown to be more effective** than B12 produced by (laboratory) bacterial action on streptomycin waste, the usual source of the B12 injections in humans. It is injected rather than taken orally because it is less digestable than B12 in calf-liver.
The evidence is that the B12 of Comfrey is more assimilable and suitable than the fungus-grown B12."
-Comfrey: Nature's Healing Herb & Health Food by Andrew Hughes. Japan: Sanyusha Publishing Co., Ltd, **1992**, page 34.

(Vitamin B12 synthesis is confined to a few bacteria and archaea. Its production relies on microbial fermentation. Streptomycin is antibiotic used to treat certain bacterial infections. Streptomycin is synthesized by bacteria Streptomyces griseus. Laboratory production does produce waste though I could not find any information on fungal waste from it.)

"**It is this vitamin in Comfrey that partly accounts for its miraculous effect on the red blood count,** quickly lifting the red corpuscle count in cases of anemia and low vitality.
Vitamin B12 in Comfrey is the same as in calf liver, 0.07 ppm (parts per million)."
-Comfrey: Nature's Healing Herb & Health Food by Andrew Hughes. Japan: Sanyusha Publishing Co., Ltd, **1992**, page 48.

(Anemia is low red blood cell count.)

"Comfrey is solely dependent upon the element cobalt in its manufacture of Vitamin B12.
Cobalt is one of those elusive elements which is provided by such plants as Rosebay Willow herb (Epilobium angustifolium), Bracken (Pteris aquillina), Ribbed Plantain (Plantago lanceolata), Vetches - tares (Vicia sativa) and also by seaweed."
-Comfrey: Symphuo Symphytum: A Multi-Purpose Herb by Philip Clarke. Edinburgh, England: Pentland Press, **1997**, page 11

*"Comfrey has also been touted for its nutritional value; it has been considered **a good source of protein and vitamin B12**, which is unusual for a plant *."*
-'European Medicines Agency- Assessment Report on Symphytum Officinale L., Radix' by European Medicines Agency, Committee on Herbal Medicinal Products (HMPC), London, England, 27 pages, May 5 **2015**.
(* -'Toxicology and Clinical Pharmacology of Herbal Products' by M.J. Cupp, Humana Press: Totowa, New Jersey, pages 203-214, published 2000.)

Doubts about Vitamin B12 in Comfrey

*"**Beliefs that the Comfrey plant (Symphytum officinale) is a natural source of vitamin B12 persist** and are repeated in the current catalogue of at least one firm of horticultural seedsmen and another specialist supplier of herbal products.*
We therefore extracted 12.5 grams of freshly picked Comfrey leaves by boiling in 500 ml acetate buffer (pH 5 0) containing 0.01% sodium cyanide in preparation for assay.
***No vitamin B12 was detected in the extract** using the Euglena gracilis var bacillaris z-strain assay; this implies a vitamin B12 concentration of less than 10 mg/liter of extract. **Thus 1 kg (2.2 lb) of fresh Comfrey leaves could at most have contained 400 mg (0.4 microgram) of vitamin B12. We therefore conclude that Comfrey leaves are not relevant as a source of vitamin B12 in mixed, vegetarian, or vegan diets.**"*
-'Vitamin B12 for Vegans' by Richard W. Payne and Brian F. Savage, Department of Pathology, Worcester Royal Infirmary, England, British Medical Journa, August 13 **1977**.
 (Maybe boiling Comfrey in acetate buffer with sodium cyanide destroyed the B12.)

"On average for 4 samples, fresh Comfrey (Symphytum sp.) had vitamin B-12 in mature leaf 3.7 to 8.4 (mean 6.3), in mature stem 2.4 to 4.2 (mean 2.9), and in root 1.9 to 3.1 (mean 2.6) ng/g (nanogram/gram).
For 5 or 6 samples, commercial Comfrey 'tea' had 2.7 to 4.0 (mean 3.4); tea bags had 1.6 to 3.5 (mean 2.9); and tablets had 0.8 to 1.6 (mean 1.2) ng/g.
*Comfrey has been recommended as a source of vitamin B-12 in vegetarian diet, but **owing to the relatively large amount of Comfrey to be eaten to contribute a significant amount of the vitamin, the potential risk from carcinogens in Comfrey** (Culvenor et al., Australian Journal of Chemistry, 1980, 33, 1105; and Hirono et al., NAR/A 49, 4830) outweigh its benefits."*
-'The Determination of Vitamin B12 in Comfrey and Comfrey Products' by by D.R. Briggs, K.F. Ryan and H.L. Bell, Department of Human Nutrition, Deakin University, Victoria, Australia; Proceedings of the Nutrition Society of Australia, Volume 6, page 148, **1981**.
 (See subsection 'Scientific Studies Showing Dangers of Alkaloids in Comfrey' in section 'Alkaloids in Comfrey' in Volume 2. There is controversy about the dangers of Comfrey.)

*"Comfrey forage is a good source of several vitamins and may be a unique plant source of B-12. **Vitamin B-12 originates in nature from bacteria or fungi that live in the soil or in the intestines of some animals.***
Neither field crops nor animals synthesize B-12 in their tissues. The major sources of B-12 for humans are meat, eggs, and dairy products. Strict vegetarians may develop sore tongues and other symptoms of B-12 deficiency.
Comfrey has been promoted as the only crop that contains B-12.
 In contrast to Comfrey grown in England and Washington*, Comfrey grown in Cheyenne, Wyoming did not contain B-12**. *Comfrey grown at Rosemount, Minnesota showed a B-12 concentration of 0.04 parts per million when tested by the same laboratory that reported none at Cheyenne.*
***Although care was taken in harvesting, contamination of the hairy leaves with soil or with microorganisms that synthesize B-12 is a possibility. Additional research is needed.**"*
-'Comfrey: A Controversial Crop' by Robert G. Robinson, University of Minnesota, Agricultural Experiment Station, Minnesota Report MR-191, Item No. AD-MR-2210, **1983**, page 5.
(* -Comfrey: Fodder, Food and Remedy by Lawrence D Hills. New York: Rizzoli Universe Books: 1976.)
(** -'Forage Yield and Quality of Quaker Comfrey, Alfalfa, and Orchardgrass' by R.H. Hart, A.J. Thompson III, J.H. Elgin Jr., and J.E. McMurtrey III, Agronomy Journal, Volume 73, pages 737-742, 1981.)

*"Comfrey (Symphytum sp. {species}) is a herb of the family Boraginaceae and is eaten as a salad vegetable, tea or tablet. **Tests with Lactobacillus leichmannii showed very low vitamin B-12 activity, 2.50 to 10.87 ng/g (nanogram/ gram) fresh weight, in Comfrey leaf, root, stem, teas and tablets.***
Relatively large amounts of these products would need to be consumed to contribute significantly to dietary vitamin B-12 intake.
 Consumption of such amounts is inadvisable because of potential health hazards associated with Comfrey."

-'Vitamin B12 Acitivity in Comfrey (Symphytum spp.) and Comfrey Products' by D.R. Briggs, K.F. Ryan and H.L. Bell, Department of Human Nutrition, Deakin University, Victoria, Australia; Journal of Plant Foods, Volume 5, No. 3, pages 143-147, **1984**.

(1,000,000 nanograms = 1 milligram. 1 nanogram = 0.000001 milligram. 1 nanogram = 0.000000000001 gram)

"Comfrey B12 content is about 6 ng/gram in fresh mature leaf, so it could not be a reasonable source of this vitamin in view of the large number of leaves which would have to be eaten to reach Recommended Daily Allowance (RDA)."
-'Pharmacists and Comfrey' by Diane Wiesner, B.Pharm, MA, Ph.D., MPS, Principal of the NSW College of Natural Therapies, Sydney, Australia; Australian Journal of Pharmacy, Volume 65, pages 959-963, **1984**.

(United States Food and Drug Adminstration: 'Recommended Dietary/Daily Allowance' of vitamin B12 for adults is 2.4 micrograms. One microgram = 1000 nanograms {ng}. One microgram = .001 milligram.)

Mucilage in Comfrey

Mucilage is a thick, gluey, gummy, gelatinous substance produced by many plants. It helps plants store water/food, improve seed germination, and thicken membranes. It is a polysaccharide that is a carbohydrate such as starch, cellulose or glycogen that is made of sugar molecules bonded together.
It is used in medicine because it relieves irritation of mucous membranes by forming a protective film. Mucous membranes are epithelial tissue that secrete mucus and line body cavities and tubular organs such as the gut, respiratory passages, urogenital tract, eyes and parts of the ear.

"Instead of the starch found in cereals, plants and roots, **Prickly Comfrey contains gum (mucilage)** *which is nearly of the same chemical composition as sugar, and is intermediate between the two (as starch in the germination of grain), or when acted upon chemically is first changed into gum and finally into sugar.*
The same changes occur in the mouths and stomachs of animals eating such food."
-'Forage Plants and Their Economic Conservation by the New System of Ensilage: Part I: Caucasian Prickly Comfrey' by Thomas Christy, Jun., F.L.S. (Fellow of the Linnean Society), Christy & Co., London, England, **1877**, page 11.

('Gum' is a viscous or sticky substance exuded or slowly discharged by certain plants.)

"The root of Symphytum officinale Linne. is mucilaginous. All mucilaginous agents exert an influence on mucous tissues, hence the cure, by their internal use, of many pulmonary and other affections (conditions) in which these tissues have been chiefly implicated. Physicians must not expect a serious disease to yield to remedies which act on mucous membranes only; and to determine the true value of a medicinal agent, they must first ascertain (determine) the true character of the affection, as well as of the tissues involved.
Again, mucilaginous agents are always beneficial in scrofulous ('diseased' or tuberculosis) and anemic ('tired' or not enough red blood cells) habits.

Comfrey root is very useful in diarrhea, dysentery (infection with severe diarrhea), bronchial irritation, coughs, hemoptysis (coughing up blood), other pulmonary affections, leucorrhea, and female debility; these being principally mucous affections."
-'Kings American Dispensatory, Volume 2, 19th Edition' by Harvey Wickes Felter, M.D. and John Uri Lloyd, Phr.M., Ph.D., The Eclectic Medical Institute, Cincinnati, Ohio, **1905**, page 1870.

"The presence of mucilage in Comfrey roots was noted by a number of early workers, its detection is described by both Molisch (1923, page 353, in German) and Tunmann and Rosenthaler (1931, page 947).
Peyer (*1911) believes that the mucilage is a protection against animals (from eating those plants)."
-'Comfrey (Symphytum Officinale L.) Root: Its Anatomy and Its Detection in Admixture with Chicory in Dandelion Coffee' by J.M. Rowson, Museum of The Pharmaceutical Society of Great Britain, Transactions of the Society, Journal Royal Microscopical Society, Great Britain, Volume 75, No. 2, pages 119-128, 1955 and/or January **1956**.
(* -'Biologische Untersuchungen uber Schutzstoffe {Biological Investigations on Protective Substances}' by Willy Peyer, Flora oder Botanische Zeitung: Welche Recensionen, Abhandlungen, Aufsatze, Neuigkeiten und Nachrichten, die Botanik Betreffend, Enthalt {Flora or Botanical Newspaper}; Regensburg, Germany, Volume 103, pages 441-464, Jena, 1911. All in German.)

"Extracted raw mucus % dry weight: Symphytum officinalis (Wurzel = root) 29.0%.
The roots of Symphytum officinalis have a very high content of mucus compared with other root drugs. Thiele (1954) also describes for extracts from Symphytum roots a high viscosity (sticky) yield.
Symphytum officinalis root: Glucose 28.7-30.2%. Fructose 68.8-71.3%.
The mucus polysaccharide from the roots of Symphytum officinalis exists from one or more polysaccharides composed of fructose and glucose.
Boudr (1958) investigated the low molecular weight carbohydrates in Symphytum roots and found oligosaccharides, all of varying proportions of fructose and glucose. With increasing molecular weight the number of fructose molecules per glucose molecule increased, so you see the polysaccharide growth of the oligosaccharide chain can come into existence."
-'Studies on the Mucopolysaccharides of Tussilago Farfara L., Symphytum Officinalis L., Borago Officinalis L. and Viola Tricolor L.' by Von G. Franz; Planta Medica: Society for Medicinal Plant and Natural Product Research, Volume 17, No. 3, pages 217-

220, August **1969**. All in German except for brief Summary in English. The above is an online translation from German.

"Its signaturing abundance of mucilage equals, if not surpasses, that of Elm bark and the Mallows, and indicates its like-cures-like importance: the mucilage has the property of removing mucus or mucilage-like secretions from the bronchial or gastrointestinal linings.
Besides, the jellyish exudation (oozing) becomes a collodian-like (gel-like) first aid remedy and serves as a protective cover for damaged skin.
Comfrey's invaluable effectiveness is pinpointed by its dominant external feature: the stout coarse hairs. They are indicated in painful situations, i.e., the herb is generally employed to ease irritations. True, the herb is listed as an **'anodyne' or pain-reliever**, but it is more important that Comfrey serves as a choice vehicle that soon overcomes the cause of the pain."
-Comfrey: What You Need to Know by Ben Charles Harris. New Canaan, Connecticut: Keats Publishing, **1982**, pages 19, 85.

    (**Like cures like:** 'By similar things a disease is produced, and through the application of the like is cured' from Hippocrates in Greece in 460-377 BC, the 'Father of Medicine'. **'Doctrine of Signatures' of Dioscorides** in 40-90 AD and Galen in 129-200 AD, states that herbs resembling parts of the body can be used to treat ailments of those body parts.)

"Symphyti radix is the fresh or dried root section of Symphytum officinale (synonym: Comfrey) *which is a plant of the Boraginaceae family.* **Mucilaginous substances (up to 30% of the dry weight),** *tannins (4 to 6%) and silicic acid (about 4%) are contained in the Comfrey plant."*
-'European Medicines Agency- Symphyti Radix: Committee for Veterinary Medicinal Products' by European Agency for Evaluation of Medicinal Products, Veterinary Medicines Evaluation Unit, London, England, EMEA/MRL/649/99-FINAL, 5 pages, August **1999**.

*"**Mucilages are long-chain polysaccharides that combine with water to form a slimy, semi-solid mass.** They are not confined to specific plant parts, but may occur in:*
Roots: Comfrey-Symphytum officinalis (Boraginaceae); marshmallow-Althaea officinalis (Malvaceae).
In physical terms mucilages are hydrophilic (they attract water).
In energetic terms mucilages are cooling and sweet or bland. *The primary action is local, by direct contact with the surface of mucous membranes or skin. Here they produce a coating of slime that acts to soothe and protect exposed or irritated surfaces of the gastrointestinal tract: a demulcent action. When this effect occurs on the skin it is referred to as an emollient action. Mucilaginous herbs tend to be ideal for topical (skin) application for a wide variety of conditions ranging from bruises and swellings to irritable dermatological lesions. When applied topically, mucilaginous agents are soothing and emollient.*
Mucilages retain heat due to their hydrophilic properties, a characteristic often utilised in herbal preparations such as warm compresses, as they allow heat to penetrate the tissues progressively. Several herbs, including Comfrey, slippery elm bark, marshmallow and linseed, are used in this way.
Gums and mucilages are invaluable aids in management of irritable digestive disorders, especially where ulceration is a feature."
-The Constituents of Medicinal Plants: An Introduction to the Chemistry and Therapeutics of Herbal Medicine by Andrew Pengelly, BA, ND, DBM, DHom. Crows Nest, NSW, Australia: Allen & Unwin, **2004**, page 124.

*"**Mucilage is a thick, moist carbohydrate substance that coats and moistens the tissues. It also tends to penetrate the tissues and cells to bring out toxins.***
The most important mucilages in Western herbalism include *Chondrus (Irish Moss),* **Symphytum (Comfrey leaf and root),** *Althea (marshmallow), Plantago (psyllium seed, plantain), and Ulmus fulva (slippery elm)."*
-The Practice of Traditional Western Herbalism: Basic Doctrine, Energetics and Classification by Matthew Wood. Berkeley, California: North Atlantic Books, **2004**, page 178.

*"**The constituents of Comfrey root** include 0.6-4.7% allantoin;* **abundant mucilage polysaccharides (about 29%) composed of fructose and glucose units***; *phenolic acids such as rosmarinic acid (up to 0.2%), chlorogenic acid (0.012%) as well as caffeic acid (0.004%) and a-hydroxy caffeic acid."*
-'Comfrey: A Clinical Overview' by Christiane Staiger in Germany, Phytotherapy Research, Volume 26, Issue 10, pages 1441-1448, February **2012**, page 1.
(* -'Studies on the Mucopolysaccharides of Tussilago Farfara L., Symphytum Officinalis L., Borago Officinalis L. and Viola Tricolor L.' by Von G. Franz; Planta Medica: Society for Medicinal Plant and Natural Product Research, Volume 17, No. 3, pages 217-220, August 1969. All in German except for brief Summary in English.)

*"**Sugars and related compounds in Comfrey (Symphytum officinale):***
D-mannose root, Fructose root, Glucose root, Glucuronic-acid root, L-arabinose root, L-rhamnose root, Sucrose root, **Mucilage root 290,000 ppm,** *Mucopolysaccharides root 250,000 - 300,000 ppm, Reducing-sugars root 51,500 ppm, Xylose root."*
-'Natural Anti-Irritant Plants' by Manfred Axterer, Cornelia Muller and Anthony C. Dweck (Symrise GmbH and Co. KG, and Dweck Data) from Dweck Data: Consultants on Natural Products to the Cosmetic, Toiletry & Pharmaceutical Industry, United Kingdom, **2017**.

*"**Long-chain polysaccharides provide the demulcent, moistening effects of mucilaginous herbs.** These include mucilage itself, as well as pectins, gums, and mucopolysaccharides. Honey, glycerin, and syrup preparations all add demulcency to the finished product. It turns out that mucilage per se is not the only constituent that contributes to demulcency.*

Plant	Part	Mucilage	Other Mucilagens
Comfrey (Symphytyum officinale)	root	to 29.0%	gum 5.0 to 10.0%

"

-'Practical Phytochemistry: Phytochemistry on the Macro Scale' by Katja Swift and Ryn Midura, CommonWealth Center for Holistic Herbalism, www.commonwealthherbs.com, School and Clinic: local and online, Boston, Massachusetts, **2017**.

Fat, Protein, Carbohydrates, Fiber, Ash, Miscellaneous

For more about tannins in Comfrey, see subsection 'Leather and Spinning' in section 'Miscellaneous Uses of Comfrey' in Volume 2. For more about carbohydrates and fructans, see subsection 'Photoperiod' in section 'Planting, Soil, Fertilization, Water, Disease' in Volume 2.

"Linnaeus' distinguished pupils (students) Fabricius and Giseke fortunately took notes of his lectures on natural orders; and by the care of the latter, to whom Fabricius communicated what he had likewise preserved, their joint acquisitions (learning or skills) have been given to the public, in an *octavo (8th) volume at Hamburgh (Germany), in 1792.
A great part of the substance of the lectures, published by him, consists of remarks on the genera of each order, as to their mutual distinctions; with numerous botanical and even economical matters.
Order 41: **Asperifolia: These plants were first collected into an order by ***Caesalpinus, and received the above appellation (name) from ***Ray, because of their generally harsh or rough habit.
Their root is fibrous. Cotyledons two. Stem branched; the branches alternate and round. Leaves alternate, simple; neither divided nor compound, for the most part nearly entire, rough with rigid scattered hairs; convolute before they expand.
All the Asperifolia Order are mucilaginous, and act only as such.
The leaves may be eaten as food, by which their small medical use may be estimated. **The root is perennial and mucilaginous.**

> Among the whole (i.e., of all plants in this Order), Symphytum abounds most with mucilage, equalling, in quantity as well as quality, the monadelphous plant Althaea (marshmallow) in this respect.
> Symphytum tuberosum has been recommended in the gout. Possibly its mucilaginous quality may hinder the crystallization of the gouty matter.

Of all plants, the herbs of this Order yield the largest proportion of ashes (minerals).
There is hardly an odoriferous, nor one fragrant, herb in the whole tribe; though Cynoglossum has a somewhat foetid (unpleasant) scent. Their taste is nothing, the great quantity of mucilage involving the stimulating particles."
-'Encyclopaedia Britannica: Supplement to the Fourth, Fifth and Sixth Editions, Volume 2", Edinburgh, Scotland, **1824**, pages 393, 407, 408.
(* -'Praelectiones in Ordines Naturales Plantarum' by Carl Linnaeus, Johann Christian Fabricius, Paul Dietrich Giseke, 1792.)
(** Currently, Symphytum genus is in the Order Boraginales.)
(*** Andrea Cesalpino or Andreas Caesalpinus, 1519-1603, was an Italian physician, philosopher and botanist.
He classified plants according to size, fruit, seed and embryo, distinguishing fourteen kinds of flowering plants and a fifteenth of plants without flowers or fruit. He wrote 'De Plantis' in Florence, Italy, 1583. It is sixteen textbooks of botany in Latin. Volume XVI lists 'Symphytum petraeum'.)
(**** For more about Ray, see sub-subsection 'Early Botanical History of Symphytum' in subsection 'Symphytum Comfrey Genus' in section 'Borage Family, Symphytum Genus' {Chapter 2}.)

"General Composition of Prickly Comfrey (Symphytum asperum):

	Natural State	*Dried at 212 F (100 C)*
Water	90.66%	0.00%
Nitrogenous Organic Compounds	2.72%	29.12%
Non-Nitrogenous Compounds	4.78%	51.28%
Mineral Matter (Ash)	1.84%	19.60%
---	---	---
Oil and Chlorophyll	0.20%	2.20%
Soluble Albuminous Compounds	1.10%	11.81%
Insoluble Albuminous or Nitrogenous	1.62%	17.31%
Gum, Mucilage, and a little Sugar	1.28%	13.65%
Woody Fibre (Cellulose)	3.30%	35.43%
Mineral Saline, soluble in water	1.25%	13.32%
Mineral, insoluble in water	0.59%	6.28%

The preceding figures show that, notwithstanding the large amount of water, the proportion of albuminous compounds (flesh forming matters) in Comfrey is considerable, and that the percentage of cellular fibre is not larger than in similar green food. In comparison with other similar food, I may state that **Comfrey has about the same feeding value as green mustard, or mangold (fodder beet), or turnip-tops, or Italian rye-grass grown on irrigated land.**"
-'The Journal of the Royal Agricultural Society of England, Second Series, Volume 7', London, England, **1871**. Includes article 'XV: On the Composition and Nutritive Value of the Prickly Comfrey (Symphytum asperrimum)' By Dr. Augustus Voelcker, F.R.S. (Fellow of the Royal Society), pages 387-389.

"Fertilizing Constituents of Feeding Stuffs: (Symphytum asperrimum, green fodder)

	Water	*Ash*	*Nitrogen*	*Phosphoric Acid*	*Potash*
Prickly Comfrey	84.36%	2.45	.42	.11	.75 "

-'Yearbook of the United States Department of Agriculture', Washington, DC, 154 pages, **1894**. Includes Report of the Secretary of Agriculture, papers by Bureaus and Divisions for the farmer, statistical tables and references.

"*Composition and Nutritive Value of Feedingstuffs:*

	Dry Matter	Crude Protein	Oil (Ether Extract)	Carbohydrate (N-Free Extract)	Crude Fibre	Ash
Comfrey (S. asperrimum)	11.5%	2.5%	0.3%	5.0%	1.7%	2.0%
Comfrey (Russian)	12.4%	3.4%	0.3%	4.9%	1.5%	2.3%

-'Rations for Livestock' by H.E. Woodman, M.A., Ph.D., D.Sc., School of Agriculture, Cambridge University, England; Bulletin No. 48, Ministry of Agriculture and Fisheries, Her Majesty's Stationery Office, London, England, 12th Edition, August **1952**, pages 117, 128. First published January 1921.   (N-Free = Nitrogen-Free Extractives.)

"*The composition of Comfrey from budding to fruiting, 29 days, ranged, dry matter 11.52 to 18.04 percent, and as percentage dry matter, crude protein 24.35 to 9.80, true protein 22.46 to 8.57, crude fibre 10.52 to 18.25, and **ash (minerals) 12.07 to 15.03 (at flowering). Ash fell to 12.90 at the fruiting stage.***"
-'Comfrey: A Valuable Fodder Crop' by E.A. Ziglinskaja, SeveroZapadnyj Nauc- Issled Inst Sel'sko Hoz; Svinovodstvo, No. 9, pages 19-21, **1957**.   (The species of Comfrey was not mentioned.)

"*Percentage Crude Protein, Fibre, Starch Equivalent, Protein Equivalent Contents of Various Hays (Dry-Matter Basis):*

	Crude Protein	Fibre	Starch Equivalent	Protein Equivalent
Comfrey Hay	24.4%	13.5%	44.1%	12.7% "

-The Conservation of Grass and Forage Crops by Stephen John Watson and Michael J. Nash, University of Edinburgh. Edinburgh, Scotland: Oliver and Boyd, **1960**, page 170. First published 1939.

"*Percentage composition of leaf and stem was dry matter (DM) 83.4%, 80.2, and in DM, crude protein 26.4%, 10.2, crude fat 2.8%, 2.1, Nitrogen-free extract 28.2%, 29.3, crude fibre 11.7%, 28.6, ash 14.3%, 10.0.*
SiO2 (Silicon Dioxide) 0.82, 3.4, CaO (Calcium Oxide) 2.1, 0.63, MgO (Magensium Oxide) 0.40, 0.14, K2O (Potassium Oxide) 8.0, 8.6, P2O5 (Diphosphorus Pentoxide) 0.92, 0.65, MnO (Manganese Oxide) 0.03, trace."
-'On the Ascorbic Acid Content of Russian Comfrey' by M. Ikeda, S. Uchimura and E. Matsui, Journal of the Faculty of Fisheries and Animal Husbandry, Hiroshima University, Japan, Volume 4, pages 103-109, **1962**.

"*In 1960 a New Zealand grower, Mr. F. Stratford of Orotohanga, had a full analysis of the **dried leaves of his Bocking Mixture** carried out by a well-known firm of consulting chemists, on behalf of the Association (Henry Doubleday Research Association): Moisture 13.42%, Fat 2.22%, Protein 22.30%, Carbohydrates 37.62%, Crude Fiber 9.38%, Ash 15.06%.*
Mineral Analysis: *Iron .016%, Manganese .0072%, Calcium 1.7%, Phosphorus .82%.*
Vitamin B Group: *Thiamin (B1) 0.5 mg/100g, Riboflavin (B2) 1.0 mg/100g, Nicotinic Acid (Niacin) 5.0 mg/100g, Pantothenic Acid 4.2 mg/100g, Vitamin B12 .07 mg/100g.*
Other: *Beta Carotene .17 parts per million (equivalent to Vitamin A 28,000 international units/100 g), Vitamin C 100 mg/100g, Vitamin E 30 mg/100g, Allantoin .18 mg/100g.*"
-Comfrey Report: The Story of the World's Fastest Protein Builder and Herbal Healer, Conservation Gardening and Farming Series: Series C by Lawrence D. Hills. England: Henry Doubleday Research Association, **1975**, page 42.
            (100 grams= 3.5 ounces.)

" '***Chemical Composition of Russian Comfrey***' by Ikeda, Uchimura, and Matsui. 1962. Hiroshima University, Faculty of Fisheries and Animal Husbandry Journal, 4:103-109.

	Dry Matter	Moisture	Crude Protein	Crude Fat	N-Free Extract	Crude Fiber	Ash
Dried Leaf	83.4%	16.6%	26.4%	2.8%	28.2%	11.7%	4.3%
Dried Stem	80.2	19.8	10.2	2.1	29.3	28.6	10.0
Fresh Leaf	12.0	88.0	3.8	0.4	4.1	1.1	2.1
Fresh Stem	12.1	87.9	1.5	0.3	4.7	4.1	1.5 "

-'Comfrey as Forage' by Lester R. Vough, Extension-Research Agronomist, Oregon State University, Corvallis, Agronomic Crop Science Report, August **1976**, page 4.

"***Oils and fat:*** *Values for ether extract range from 16 to 54 grams/kg DM (Dry Matter), with a digestibility of 46-85%. **Comfrey is thus similar in oil and fat content to other green forages.***"
-'Comfrey Symphytum spp. as a Forage Crop' by J.C. Forbes, A.D. McKelvie, and P.J.C. Saunders, North of Scotland College of Agriculture, Aberdeen, United Kingdom; Herbage Abstracts, Volume 49, No. 12, pages 523-539, **1979**.

"***Fatty Acids:*** *The distribution of Gamma-Linolenic Acid (GLA) was studied in various parts of the Comfrey (Symphytum officinale L.) plant and at different stages of maturation of the Comfrey seeds.*
Leaves *contained only 1-2%, whereas **stem and roots** contained 12-16% GLA of total fatty acids.*
In seeds, *the 18:3 fatty acids were predominantly of the n-3 type, Alpa-Linolenic Acid (ALA), during early stages of development, whereas a shift was observed towards GLA (n-6 type) during maturation (26% GLA of total fatty acids).*"
-'Distribution of Gamma-Linolenic Acid in the Comfrey (Symphytum Officinale) Plant' by Carl E. Hansen, P. Stoessel and P.

Rossi, Nestle Research Centre, Nestec Ltd, Lausanne, Switzerland; Journal of the Science of Food and Agriculture, Society of Chemical Industry, Volume 54, Issue 2, pages 309-312, **1991**.

"*The constituents of Comfrey root include 0.6-4.7% allantoin (*); abundant mucilage polysaccharides (about 29%) composed of fructose and glucose units (**); phenolic acids such as rosmarinic acid (up to 0.2%), chlorogenic acid (0.012%) as well as caffeic acid (0.004%) and a-hydroxy caffeic acid (***, ****, *****); glycopeptides and amino acids (#); and triterpene saponins in the form of monodesmosidic and bidesmosidic glycosides based on the aglycones hederagenin (e.g. symphytoxide A), oleanolic acid (##) and lithospermic acid (###).*"
-'Comfrey: A Clinical Overview' by Christiane Staiger in Germany, Phytotherapy Research, Volume 26, Issue 10, pages 1441-1448, February **2012**, page 1.
> (This exact same description, word for word, is also in 'Symphyti Radix {Comfrey Root}' by ESCOP Monographs: The Scientific Foundations for Herbal Medicinal Products, European Scientific Cooperative on Phytotherapy, Exeter, United Kingdom, 16 pages, 2012.)

(* -'Studies on Symphytum Species: HPLC Determination of Allantoin' by R. Dennis, C. Dezelak and J. Grime; Acta Pharmaceutica Hungarica, Budapest, Hungary, Volume 57, No. 6, pages 267-274, November 1987.)
(** -'Studies on the Mucopolysaccharides of Tussilago Farfara L., Symphytum Officinalis L., Borago Officinalis L. and Viola Tricolor L.' by Von G. Franz; Planta Medica: Society for Medicinal Plant and Natural Product Research, Volume 17, No. 3, pages 217-220, (August 1969). All in German except for brief Summary in English.)
(*** -Andres RMP. 1991. 'Clinical and instrumental analysis of dermatics with extracts of Symphytum officinale L.', Dissertation, University of Bern, Switzerland.)
(**** -'Phenolic Acids in Symphytum Officinale L.' by B. Grabias and L. Swiatek; Pharmaceutical and Pharmacological Letters, Germany, Volume 8, No. 2, pages 81-83, April 1998.)
(***** -Herbal Drugs and Phytopharmaceuticals: A Handbook for Practice on a Scientific Basis edited by Max Wichtl. Boca Raton, Florida: CRC Press, 2004.)
(# -'Antiphlogistic Glycopeptide from the Roots of Symphytum Officinale' by A. Hiermann and M. Writzel, Pharmaceutical and Pharmacological Letters, Heidelberg, Germany, Volume 8, No. 4, pages 154-157, January 1998.)
(## -Saponins Used in Traditional and Modern Medicine by Manuel F. Balandrin with editors George R. Waller and Kazuo Yamasaki. Part of 'Advances in Experimental Medicine and Biology Series 404'. New York: Springer, 1996. Chapter: Phyto-Pharmacology of Saponins from Symphytum Officinale L. by Khalid Aftab, Fehmeena Shaheen, Faryal Vali Mohammad, Mushtaq Noorwala and Viqar Uddin Ahmad; H.E.J. Research Institute of Chemistry, University of Karachi, Pakistan, pages 429-442.)
(### -Wagner H, Horhammer L, Frank U. 1970. 'Lithospermsaure, the antihormonal principle of action of Lycopus europaeus L. and Symphytum officinale L. (Beinwell): On the ingredients of medicinal plants with hormone and anti-hormone-like effect', Pharm Researchers Drug Res 20: 705-713.)

"**Carbohydrates:** *gum (arabinose, glucuronic acid, mannose, rhamnose, xylose); mucilage (glucose, fructose) ('Herbal Medicines, 3rd Edition' by Barnes, Anderson and Phillipson, Pharmaceutical Press, London, page 188-190, published 2007.)*
Tannins: *pyrocatechol-type, 2.4% (Barnes et al., 2007).*
Triterpenes: *sitosterol and stigmasterol (phytosterols), steroidal saponins, isobauerenol, triterpene saponins symphytoxide A, cauloside D, leontoside A, leontoside B, leontoside D (Barnes et al., 2007; Ahmad et al., 1993; Mohammad et al., 1995)*
Other constituents: *allantoin 0.75-2.55%, caffeic acid, carotene 0.63%, chlorogenic acid 0.037%, caffeic acid 0.035%, choline, lithospermic acid, rosmarinic acid and silicic acid (Barnes et al., 2007; Aftab et al., 1996); vitamins A and B12, calcium, potassium and phosphorus ('The Gale Encyclopedia of Alternative Medicine, Volume 1' edited by J.L. Longe, Thomson Gale, Michigan: Gale Group, pages 526-527, 2005).*"
-'European Medicines Agency- Assessment Report on Symphytum Officinale L., Radix' by European Medicines Agency, Committee on Herbal Medicinal Products (HMPC), London, England, 27 pages, May 5 **2015**.
> (The report did not indicate whether it was Comfrey root or leaf, but probably it is leaf.)

Fructans (Fructose Polymers)

A fructan is a polymer of fructose molecules. A polymer is a moleculer structure with a large number of similar units bonded together. Fructose is fruit sugar, a simple sugar found in food such as honey and fruit.
Fructans with a short length are called fructooligosaccharides. Fructans are in food such as agave, artichoke, asparagus, garlic, jicame, leek, onion, wheat and yacon. It is also found in grass eaten by herbivores such as cattle and goats.

Carbohydrates can be in simple forms such as sugar and in complex forms such as starch and fiber. They contain hydrogen and oxygen in a 2:1 ratio and are broken down by the body to release energy.

"**Roots and Shoots of Symphytum officinale:** *Fructosans are stored and reduced independent of day length.*
Fructosans are consumed during shoot elongation and flowering.
A hydrolysis of starch and fructosans is also caused by low temperatures during the winter.
Starch and fructosans are stored to a different degree in the various subterranean (underground) organs:
> *The most is stored in the shoot-born roots, and the least in the subterranean shoot parts. In the subterranean shoot organs, the carbohydrate content is influenced the most by the growth processes.*"

-'Annual Developmental Cycle of Roots and Shoots in Symphytum Officinale L.' by Karin Staesche, Institut fur Spezielle Botanik und Pharmakognosie der Universitat Tubingen (Institute of Special Botany and Pharmacognosy of the University of Tubingen), Germany; Planta, Berlin, Germany, Volume 71, No. 3, pages 268-282, September **1966**. All in German except this abstract is also in English.

> *"**Staesche (1966) studied the development of the root and shoot system of Symphytum officinale, as well as the storage and consumption of carbohydrates.** The storage and consumption of carbohydrates is dependent upon the day length, as is the development of the cormus.*
> *Plants grown under natural conditions in the field show the following variations of carbohydrate reserves:*
> *In subterranean organs in which fructans are by far the most important reserves, starch is completely absent in winter and spring (**January to May**).*
> *The highest fructan content (up to 50% of the dry weight of roots) is reached in **autumn when the leaves disintegrate.***
> ***From late autumn until spring** the fructans are partially hydrolyzed to oligofructans, sucrose and fructose. The minimal fructan content is reached at **anthesis (flowering period)** and then increases again until late autumn.*
> ***9-Hour Day:*** *Plants grown under 9-hour day condition show a faster fructan storage in the ground axis than plants grown under normal light conditions. Almost no starch is stored in the 9-hour day.*
> ***It seems that starch is used under short-day and stored under long-day conditions, whereas the reverse is true for fructans.****"*
> -Plant Carbohydrates I: Intracellular Carbohydrates edited by F.A. Loewus and Widmar Tanner. Berlin, Heidelberg, Germany: Springer-Verlag, 1982, page 440. (Cormus or corm is a rounded underground storage organ.)

*"**In root-layers of Symphytum officinale,** development as well as storage and consumption of carbohydrates is determined by day length, in a manner similar to that in plants developed from seeds.*
In cultures kept at temperatures of at least +10 C (50 F) fructosans are stored in the young shoot-born roots, while amount of fructosans is reduced in the buds, in the subterraneous shoot parts and in the old root pieces.
The old root piece remains a living part in the root-system of the layer and takes part in the renewed storage of starch just like the primary root of plants developed from seeds."
-'Development of Root-Layers of Symphytum Officinale L.' by Karin Staesche, Institut fur Spezielle Botanik und Pharmakognosie der Universitat Tubingen (Institute of Special Botany and Pharmacognosy of the University of Tubingen), Germany; Planta, Berlin, Germany, Volume 75, No. 4, pages 352-357, **1967**.

*"**The fructans in the rhizomes (roots) of a number of Boraginaceae have been studied** by Bourdu (*1957), who showed that fructans are absent in annual species of this family; in this respect it resembles the Compositae family.*
Three types of perennial species of Boraginaceae can be distinguished:
    1. Species which contain starch only.
    ***2. Species with starch and fructans (eg., Symphytum officinale**, Pulmonaria officinale).*
    3. Species with fructans only in their subterranean parts (e.g., Cynoglossum officinale, Myosotis palustris, Lithospermum officinale, Echium vulgare).
Variations in fructan content *similar to those observed by Binet and Collin (**1974) in Aster tripolium, also occur in the roots of Symphytum officinale (Bourdu 1954, 1957, ***1958)."*
-Plant Carbohydrates I: Intracellular Carbohydrates edited by F.A. Loewus and Widmar Tanner. Berlin, Heidelberg, Germany: Springer-Verlag, **1982**, pages 436, 440.
(* -Contribution a L'etude du Metabolisme Glucidique des Boraginacees. Translated from the French: Contribution to the Study of the Metabolism of Boraginaceae.)
(** -'Cycle Annuel des Fructosanes Chez Aster Tripolium L.' by Paul Binet, Andre Collin and Mlle M. Duyme, Universite de Caen, France; Bulletin de la Societe Botanique de France, Volume 121, No. 9, pages 323-328, 1974.)
(*** -Sur les Glucofructosides de Symphytum Officinale L. et leur Metabolisme. Translated from the French: On the Glucofructosides of Symphytum Officinale L. and their Metabolism.)

> *"**In April,** the fructosans of roots and rhizomes are for the most part used up. The reconstitution of these stores starts from the **end of May** but slows down when the inflorescences (flowers) are forming.*
> *Transitorily, fructosans accumulate in the aerial stems and also, to a lesser extent, in the leaves. The young plants, issuing from the germination of seeds, start forming fructosans from the third month onwards."*
> -'Cycle Annuel des Fructosanes Chez Aster Tripolium L.' by Paul Binet, Andre Collin and Mlle M. Duyme, Universite de Caen, France; Bulletin de la Societe Botanique de France, Volume 121, No. 9, pages 323-328, 1974. All in French except for this summary in English.

"From an extensive literature survey and from analytical data of 130 species from the Sheffield, England flora, the physiological and molecular attributes, and occurrence of fructan are considered.
In view of the contention that fructan may function in low temperature tolerance, estimates are given of their maximum vacuolar concentration.
The association of high fructan concentrations in the shoots of several species with high nuclear 2C DNA values is recorded. It is

*this correlation which may be of significance to several **early-season growing species**.*

By maintaining supplies of fructose and sucrose from vacuoles in tissue undergoing expansion at low temperatures (a feature associated with high DNA values), such species obviate (avoid) the need for transport of carbohydrate over distance as in starch-storing species.

By shortening supply lines at critical cold periods, fructan-rich species may have a considerable advantage over starch-storing, small-celled, transport-dependent species.

Maximum concentration of reserve carbohydrates *in survey of 130 species from the Sheffield flora:*

Symphytum x uplandicum, underground storage organ (rhizome, tuber) in summer:

Fructan:** 74.00 mg/gram fresh weight.* ***Starch: *5.92 mg/gram fresh weight.*

A cellular characteristic of fructan is its relation to cell size.

*On superficial examination, **a number of fructan-containing species in the British flora appear to be soft, rather watery plants with large cells. Examples include** members of the genus Allium, Hyacinthoides non-scripta, Taraxacum spp., Tussilago farfara, **Symphytum spp. (species)** and Monotropa hypopytis.*

DNA Mass:

*In extensive comparative studies, **a correlation has been found between genomic DNA mass and geographical distribution**. Genomic masses of tropical species are comparatively small when compared with the wide range observed among temperate floras (Bennett, 1976; Levin & Funderburg, 1979).*

*Recently, nuclear DNA content has been correlated with shoot phenology; **early-spring growing species have a larger genomic DNA size than the late spring- and summer-growing species** (Grime & Mowforth, 1982; Grime, Shacklock & Band, 1985).*

*In addition, and of immediate relevance, **large nuclear DNA size is associated with large cell size,** consistent with the hypothesis (Grime & Mowforth, 1982) that, **for such species, early-season growth is achieved by cell expansion rather than by division**."*

-'The Ecological Significance of Fructan in a Contemporary Flora' by George Hendry, Unit of Comparative Plant Ecology, Department of Botany, The University, Sheffield, England; New Phytologist: A British Botanical Journal, London, England, Volume 106 (Supplement), pages 201-216, **1987**.

Other Constituents of Comfrey including Phenolic Acids

For allantoin see section 'Comfrey Heals: Allantoin' in Volume 2.

Polyphenols or phytochemicals are chemicals in plants with over 500 known. They include alkaloids, flavonoids, glycosides, phenolics {phenolic acid}, saponins, stilbenes, lignans, tannins, terpenes, anthraquinones, essential oils and steroids.

Phenolic acids or phenolcarboxylic acids are types of aromatic acid compound. Phenolic acids are among the most widely distributed plant non-flavonoid phenolic compounds present in the free, conjugated-soluble and insoluble-bound forms.

Phenolic acids include caffeic acid (a cinnamic acid), chlorogenic, ferulic, gallic acid, gentisic, isovanillic, p-coumaric, p-hydroxybenzoic, p-hydroxycinnamic, protocatechuic, rosmarinic acid, salicylic acid, sinapic acid, syringic, and vanillic. They provide protection against oxidative damage diseases such as coronary heart disease, stroke, and cancer when eaten in fruits and vegetables.

Overview

"Titherley and Coppin (1912) found **0.8 percent of allantoin in the Symphytum officinale (Common Comfrey) rhizome (root)**. Greimer (Arch. Pharm., 1900, 238, 505) found **the poisonous alkaloids,** consolidine and symphytocynoglossine, both of which are depressants to the central nervous system.

Comfrey contains large quantities of mucilage (according to Lewis, more than althea {marshmallow} root), and **some tannin**. It was formerly employed as an application to wounds to stimulate healing (Allantoin)."

-'The Dispensatory of the United States of America, 25th Edition' by Arthur Osol, George E. Farrar Jr., and Editor Horatio C. Wood Jr., Philadelphia, Pennsylvania, **1955**, page 1893. Based on the fifteenth revision (1955) of 'The United States Pharmacopeia', the tenth edition (1955) of 'The National Formulary', the 1953 edition of 'The British Pharmacopoeia', and the first edition (1951) of the 'International Pharmacopeia, Volumes I and II'.

"A bioassay-guided fractionation of the aqueous (water) **extract from Symphytum officinale roots led to the isolation of the antiphlogistic (prevention or reduction of inflammation) principle.**

A glycopeptide was isolated and purified by combination of ultrafiltration, CC on MgO and size exclusion chromatography. The compound with an isoelectric point at pH 4.8 was **found to contain 16 amino acids as well as galactose, fructose, arabinose and glucose.**

The calculated molecular mass was approximately 9000 dalton. On carrageenan-induced rat paw edema (swelling) the isolated glycopeptide exerted a remarkable, dose dependent antiphlogistic (anti-inflammation) effect.

The ED50 was 61 microgram/kg p.o. (control indomethacin 10 mg/kg). Investigations on the release of arachidonic acid, cyclooxygenase and lipoxygenase metabolites as well as on arachidonic acid induced platelet aggregation indicate that **the isolated glyco-peptide inhibits release of prostaglandins and leukotriens** via decreasing expression of phospholipase A2."

-'Antiphlogistic Glycopeptide from the Roots of Symphytum Officinale' by A. Hiermann and M. Writzel, Pharmaceutical and Pharmacological Letters, Heidelberg, Germany, Volume 8, No. 4, pages 154-157, January **1998**.

"Phenolic Acids in Common Comfrey:
The presence of cinnamic acid and 19 phenolic acids in herb and roots of Symphytum officinale L. was discovered. *The content of these compounds was determined by GC (Gas Chromatography).*
In the herb (leaf) *the largest quantity was documented for caffeic acid, followed by p-coumaric and m-hydroxybenzoic acids.*
In the roots *the following acids dominate: m-hydroxybenzoic, salicylic and caffeic.*
The content of rosmarinic acid, was determined by HPLC (High Performance Liquid Chromatography).
Considering the fact, that above mentioned compounds occur in large amounts, may be responsible for the **therapeutic activity of Comfrey, as an antiseptic and anti-inflammatory agent.**"
-'Phenolic Acids in Symphytum Officinale L.' by B. Grabias and L. Swiatek; Pharmaceutical and Pharmacological Letters, Germany, Volume 8, No. 2, pages 81-83, April **1998**.

"**A strategy for screening plants for ecdysteroid content** based on the 'positive tribe' principle is developed and applied, for the first time, to screen the **flora of European North-East Russia** to identify species which accumulate ecdysteroids.
Samples of roots of all the species of the Symphytum genus studied (S. asperum, S. carpaticum, S. caucasicum, S. officinale and S. tanaicense) did not give a positive response for ecdysteroids in the bioassay and RIA (radioimmunoassay).
However, the presence of ecdysteroids was demonstrated in concentrated ethanol extracts of roots for S. asperum, S. caucasicum and S. tanaicense.
Additionally, chromatographic separation of the concentrated extract of roots of S. tanaicense demonstrated RIA- and bioassay-positive material in the region where most ecdysteroids elute and a major peak co-chromatographing with 20-hydroxyecdysone and/or polypodine B.
In future it will be desirable to study these plants at other stages of development to estimate the range of concentrations and to conclude whether this genus contains species with ecologically significant levels of ecdysteroids or not."
-'Screening Plants of European North-East Russia for Ecdysteroids' by V. Volodin, I. Chadin, P. Whiting and L. Dinan; Komi Science Centre, Russian Academy of Sciences, Syktyvkar, Russia, and Department of Biological Sciences, University of Exeter, England; Biochemical Systematics and Ecology, Volume 30, pages 525–578, **2002**.
(Ecdysteroids are arthropod steroid hormones that influence molting, development and reproduction. Phytoecdysteroids are found in many plants to protect against herbivore insects. An arthropod is an invertebrate with an exoskeleton, segmented body, and jointed appendages.) (Elute means to remove an adsorbed substance by washing with a solvent.)

"**Phytochemical study of roots/stems of Symphytum asperum Lepech, (Prickly or Rough Comfrey) was carried out in order to define phenolic constituents.**
Firstly, grinded air-dried S. asperum roots (SAR) and stems (SAS) were fore-extracted exhaustively in a Soxlet apparatus with hexane and chloroform in order to remove lipids, pigments and other nonpolar compounds and afterwards these materials were treated with the aqueous mixtures of different organic solvents. Eight analytical samples were obtained.
Ultrahigh-pressure liquid chromatography coupled with quadrupole time-of-flight tandem mass spectrometry (UHPLC-Q-TOF/MS) analysis of extracts of S. asperum roots/stems was carried out that **revealed the presence of low molecular weight compounds such as caffeic, rosmarinic, chlorogenic acids, salvianolic acid, B/lithospermic acid B and several oligomeric compounds.**
Currently only the brutto-formulas of the oligomeric compounds are known and further investigations are in progress in order to determine exact chemical structures of those. **The obtained results revealed that the Comfrey roots/stems can be used as a source for the isolation of low molecular weight biologically active compounds.**"
'UHPLC-Q-TOF/MS Characterization of Several Compounds from the Roots and Stems Extracts of Symphytum Asperum' by Lela Amiranashvili, Lali Gogilashvili, Sopio Gokadze, Maia Merlani, Vakhtang Barbakadze and Bezhan Chankvetadze from Tbilisi State Medical University and Javakhishvili Tbilisi State University, Tbilisi, Georgia; Bulletin of the Georgian National Academy of Sciences, Volume 10, No. 3, pages 127-133, January **2016**.

"**Seven of the phenolic compounds, in seven plant species within five genera of the Boraginaceae family have been identified** which were (Brunnera orientalis (Schenk) Jonston., Choriantha popoviana H., Cynoglossum creticum Mill, Solenanthus circinnatus Ledeb, Solenanthus stamineus Defed, **Symphytum kurdicum Boiss and Symphytum tuberosum L.**).
The chemical composition of these species were examined for the content of the following phenolic compounds: Caffeic acid, Estragole, 2-6Dimethyl phenol, Coumaric acid, Eugenol, Salicylic acid, and P-Cresol, by using high performance liquid chromatography (HPLC).
The results showed that the most abundant phenolic acids were: Coumaric acid and Salicylic acid which were found in all the studied taxa, followed by Caffeic acid which was absent from Brunnera orientalis and Solenanthus circinnatus.
Eugenol was absent just from Symphytum kurdicum Boiss and Symphytum tuberosum L."
-'Chemotaxonomical Study of the Genera Brunnera (Schenk) Jonston, Choriantha H.Rirdel., Cynoglossum Mill., Solenanthus Ledeb. and Symphytum (Boiss.) L. (Boraginaceae) in Kurdistan Region of Iraq by Using High Performance Liquid Chromatography (HPLC)' by Adel Alzubaidy, Agricultural Technical College, Sulaimani Polytechnic University; Diyala Journal for Pure Sciences, Volume 13, No. 3, pages 88-103, **2017**.

"**Comfrey has pharmacological components that include rosmarinic acid and tannin. Rosmarinic acid is a natural polyphenol antioxidant, and both rosmarinic acid and tannin are considered antiinflammatory agents.**
Tannic acid (TA), an antioxidant, contains anti-mutagenic and anti-carcinogenic properties that exhibit oxygen free radical

trapping activity.
Levanon and Stein suggested that the ability of TA to augment glycosaminoglycan binding to collagen most possibly contributes to the structural reinforcement of synovial articulating (joint) surfaces (Smith et al., 2011)."
-'Role of Symphytum Officinale as an Osteoinducer in Long Bone Fracture Repair in Canine' thesis by Swarop Chandel for Master of Veterinary Science degree, Department of Veterinary Surgery and Radiology, Nagpur Veterinary College, India, **2014**.
> (A synovial joint joins bones with a fibrous capsule with the periosteum of the joined bones. It is the outer boundary of a synovial cavity, and surrounds the bones' articulating surfaces.)

"**The main active constituents responsible for pharmacological activity and mechanisms of action of Comfrey have not been completely elucidated (made clear). It is thought that allantoin and rosmarinic acid are responsible for the main effects** (Horinouchi and Otuki 2013; Staiger 2012).
Rosmarinic acid has been shown to demonstrate anti-inflammatory activity in various tests, although significant absorption through the skin has not been shown. The anti-inflammatory activity of Comfrey root extracts has also been demonstrated in animal studies.
A glycopeptide isolated from Comfrey root dose-dependently inhibited release of prostaglandins and arachidonic acid in rat stomach preparations. An in vitro study found that a 60% ethanolic Comfrey root extract exerted an immunomodulatory effect on elements of the human immune system.
Wound healing effects have also been demonstrated in a test model of fibroblasts in a collagen matrix: 40% ethanolic Comfrey root extract and its high molecular weight fraction both inhibited shrinkage of the collagen matrix (Staiger 2012). THR (Traditional Herbal Registration) products, mainly creams, are now available which contain pyrrolizidine alkaloid (PA)-free extracts."
-'Phytopharmacy: An Evidence-Based Guide to Herbal Medicinal Products' by Sarah E. Edwards, Ines da Costa Rocha, Elizabeth M. Williamson and Michael Heinrich. New York: John Wiley & Sons, Ltd., **2015**, page 116.
> (Ethanolic means derived from ethyl alcohol, i.e., extracted using alcohol. A fibroblast is a cell in connective tissue that produces collagen and other fibers.)

> (The 'Traditional Herbal Registration' Certification Mark was created in 2009 in the United Kingdom by the 'Medicines and Healthcare Products Regulatory Agency'. "Where a herbal medicine carries the Certification Mark this means that the MHRA has assessed the product to ensure that is acceptably safe when used as intended, is manufactured to the quality standards we set, and is accompanied by reliable and accurate product information for the public and patients. The authorised usage and dosage of the medicine is based on evidence of its traditional use. The effectiveness of the product has not been assessed by the MHRA.")

"The qualitative, quantitative and microbial determination of the **aqueous (water) extract of Comfrey (Symphytum officinale) root** was done in this paper. The qualitative and quantitative analyses were done by the UHPLC–DAD–HESI–MS method. **Allantoin, rosmarinic acid and ellagic acid were identified as major bioactive compounds** and their quantification was also done. **The obtained results showed a high content of allantoin, ellagic acid and rosmarinic acid (8.91, 7.4 and 12.8%, respectively)** which indicated that the Comfrey root can be used as a source for the isolation of these three compounds."
-'The Identification and Quantification of Bioactive Compounds from the Aqueous Extract of Comfrey Root by UHPLC–DAD–HESI–MS Method and its Microbial Activity' by Vesna Lj. Savic, Sasa R. Savic, Vesna D. Nikolic, Ljubisa B. Nikolic, Stevo J. Najman, Jelena S. Lazarevic and Aleksandra S. Dordevic, University of Nis, Serbia; Hemijska Industrija (Chemical Industry), Volume 69, No. 1, pages 1-8, **2015**.

"**Comfrey contains many volatile oils,** mucilages, alkaloids (simfitocinoglossine, consollicine, consolidin, coniferidin, cinoglosin, unsaturated necins), allantoin, choline, asparagine, catechic tannins, phenolcarboxylic compounds (caffeic acid, chlorogenic acid) triterpenes, amino acids and carotenoids."
-'Medicinal Herbs as Possible Sources of Anti-Inflammatory Products' by Andreia Corciov, Daniela Matei and Bianca Ivanescu, University of Medicine and Pharmacy, Iasi, Romania; Balneo Research Journal, Bucharest, Romania, Volume 8, No. 4, pages 231-241, December **2017**.

"The Boraginaceae family comprises plants that have important therapeutic and cosmetic applications. Their pharmacological effect is related to the presence of naphthaquinones, flavonoids, terpenoids, phenols, or purine derivative - allantoin.
Average concentration of biologically active compounds in selected species of Boraginaceae family:

	Shoots (mg/gram air-dry matter)				
	Allantoin	PHBA	Hydrocaffeic acid	Rosmarinic acid	Chlorogenic acid
Symphytum cordatum	0.63	nd	nd	12.4	nd
Symphytum officinale	9.38	0.195	0.203	4.5	0.628
	Roots (mg/gram air-dry matter)				
	Allantoin		Hydrocaffeic acid	Rosmarinic acid	
Symphytum cordatum	7.19		0.144	2.8	
Symphytum officinale	25.77		nd	7.1	

-'Comparison of Some Secondary Metabolite Content in the Seventeen Species of the Boraginaceae Family' by Slawomir Dresler, Grazyna Szymczak and Malgorzata Wojcik, Maria Curie-Sklodowska University, Lublin, Poland; Pharmaceutical Biology, Volume 55, No. 1, pages 691-695, (2017).   (Symphytum cordatum Waldst. & Kit ex Willd., and Symphytum officinale L.)
> (PHBA: p-hydroxybenzoic acid. nd = not detectable.)

Rosmarinic Acid in Comfrey (a phenolic acid)

*"It has antioxidant properties. **Rosmarinic acid was first isolated and characterized in 1958** by the Italian chemists M. L. Scarpatti and G. Oriente from **rosemary** (Rosmarinus officinalis). It is found in species used commonly as culinary herbs such as Ocimum basilicum (basil), Ocimum tenuiflorum (holy basil), Melissa officinalis (lemon balm), Rosmarinus officinalis (rosemary), Origanum majorana (marjoram), Salvia officinalis (sage), thyme and peppermint or in plants with medicinal properties such as common self-heal (Prunella vulgaris) or species in the genus Stachys."*
-Wikipedia®: The Free Encyclopedia, 2018.

*"**Rosmarinic acid was isolated from Symphytum officinale L. as the main constituent with anti-inflammatory activity,** and was determined by an HPLC (High-Performance Liquid Chromatography) gradient technique.*
Minor constituents were chlorogenic acid and caffeic acid. The biological activity of the pure compounds was estimated from the inhibition of MDA (Malondialdehyde) formation in human platelets by the TBA (Thiobarbituric Acid) method. For 1 an IC50 of 3.37 mM was found. Chlorogenic acid and caffeic acid do not show significant activity in this model."
-'Isolation of Rosmarinic Acid from Symphytum Officinale L. and its Anti-Inflammatory Activity in an In-Vitro Model (Biochemical-Pharmacological Investigations of Medicinal Agents of Plant Origin, I)' by Lajos Gracza, Heinrich Koch and Eva Loffler in Goppingen, Germany and Vienna, Austria; Archiv der Pharmazie (Weinheim, Germany), Volume 318, pages 1090-1095, **1985**. (All in German except abstract is also in English. If you have an English translation, I would appreciate a copy.)
  (IC50 is the concentration of an inhibitor where the response is reduced by half.)

*"**Rosmarinic acid has been shown to possess anti-inflammatory activity** in various test systems. It inhibits the formation of malondialdehyde in human platelets*, prostaglandin synthesis, and carrageenan- and gelatine-induced erythrocyte (red blood cell) aggregation.**"*
-'Comfrey: A Clinical Overview' by Christiane Staiger in Germany, Phytotherapy Research, Volume 26, Issue 10, pages 1441-1448, February **2012**.
(* -'Isolation of Rosmarinic Acid from Symphytum Officinale and Its Anti-inflammatory Activity in an in Vitro Model: On Biochemical Pharmacological Investigations of Herbal Medicines' by L. Gracza, H. Koch, E. Loffler, Mitt. Archiv der Pharmazie 318: 1090–1095, 1985.)
(** -'Testing the Membrane-Sealing Action of a Phytopharmaceutical and Its Agents' by L. Gracza, Z Phytother 8: 78–81, 1987.)

*"**Rosmarinic Acid (RA) is natural polyphenol antioxidant isolated from Rosmarinus officinalis L. and commonly found in species of the Boraginaceae (e.g. Anchusa officinalis L. and Symphytum officinale L.) and the subfamily Nepetoideae of the Lamiaceae.** RA species of Labiatae are named Salvia officinalis, Rosmarinus officinali.*
*RA exhibits important biological activities that include its **anti-carcinogenic, anti-viral, anti-bacterial anti-microbial, anti-depressant qualities**. RA showed the highest concentrations of antioxidant of all the polyphenols.*
It is a red-orange powder that is slightly soluble in water, but well soluble in most organic solvents.
*RA polyphenolic compounds have been associated with antioxidative action in biological systems, acting as scavengers of singlet oxygen and free radicals. RA protects neurons from oxidative stress significantly attenuated H2O2-induced reactive oxygen species (ROS) generation and apoptotic cell death and could contribute at least in part to neuroprotective effects because **this natural compound exerts neuroprotective and anti-oxidative effects against neurotoxin insult in dopaminergic cells (involving the neurotransmitter dopamine).**"*
-'Phytochemical, Pharmacological and Pharmacokinetics Effects of Rosmarinic Acid: Review' by Rahul Bhatt, Neeraj Mishr and Puneet Kumar Bansal, Department of Pharmaceutics, ISF College of Pharmacy, Punjab, India; Journal of Pharmaceutical and Scientific Innovation, Volume 2, No. 2, pages 28-34, March-April **2013**.
  (Polyphenols or phytochemicals are chemicals in plants with over 500 kn owyn.The include alkaloids, flavonoids, glycosides, phenolics {phenolic acid}, saponins, stilbenes, lignans, tannins, terpenes, anthraquinones, essential oils and steroids.)

*"**The present work evaluated the anti-inflammatory and antinociceptive activity of a Symphytum officinale root extract standardized in rosmarinic acid.** The anti-inflammatory effect of the hydro-glycero-alcoholic extract administered orally was assessed. Quantification of rosmarinic acid, a major compound in S. officinale root, was achieved by a LC/MS/MS method **(74.77 micrograms rosmarinic acid /mL extract)**.*
Our study demonstrated significant anti-inflammatory and antinociceptive effects of S. officinale extract probably mediated by peripheral mechanisms which may justify the traditional use in the treatment of some inflammatory diseases.
Both analgesic and anti-inflammatory effects of S. officinale root extract may be related to the ability of polyphenols such as rosmarinic acid to inhibit the synthesis and release of some pro-inflammatory mediators and cytokines (Erdemoglu et al, 2009) and also to the content in allantoin."
-'Anti-Inflammatory and Antinociceptive Effect of Symphytum Officinale Root' by Oliviu Vostinaru, Simona Conea, Christina Mogosan, Claudia Crina Toma, Corina Claudia Borza and Laurian Vlase in Romania; Romanian Biotechnological Letters, University of Bucharest, 9 pages, **2017**.
(Antinociception is the blocking of detection of painful or injurious stimulus by sensory neurons. It reduces/stops pain sensation.)

Caffeic Acid Polyether (PDPGA) (a phenolic acid)

"Studying the polysaccharide composition of a number of Caucasus flora plants used in folk medicine showed that, unlike the polysaccharides from other plants, the total polysaccharide preparations from the roots of Prickly Comfrey Symphytum asperum and Caucasus Comfrey S. caucasicum (family Boraginaceae) possess a high anticomplementary activity and effectively catch free radicals.
Branched glucofructans, whose structures have been elucidated (shown) by chemical analysis and NMR (Nuclear Magnetic Resonance) spectroscopy, are the main polysaccharide components of these preparations.
Two high-molecular water-soluble preparations with high anticomplementary and antioxidant activity were isolated from the roots of Symphytum asperum and S. caucasicum. **Their main chemical constituent was found to be poly[oxy-1-carboxy-2-(3,4-dihydroxyphenyl)ethylene]** *according to IR (Infrared) and NMR spectroscopy."*
-'Poly[3-(3,4DihydroxyPhenyl)Glyceric Acid] A New Biologically Active Polymer from Two Comfrey Species Symphytum Asperum and S. Caucasicum (Boraginaceae)' by V.V. Barbakadze, E.P. Kemertelidze, I.L. Targamadze, A.S. Shashkov and A.I. Usov, Tbilisi, Georgia and Moscow, Russia; Russian Journal of Bioorganic Chemistry, Volume 28, No. 4, pages 326-330, July **2002**.

(A polysaccharide is a carbohydrate such as starch, cellulose or glycogen made of sugar molecules bonded together.)
(The Caucasus Mountains are located at the border of Europe and Asia, between the Black Sea and Caspian Sea and occupied by Russia, Georgia, Azerbaijan, and Armenia.)
(Anticomplementary means having the capacity to remove or inactivate complement nonspecifically. Capable of reducing or destroying the power of complement.)
(Antioxidants are molecules that reduce the oxidation of other molecules. Oxidation is a chemical reaction that produces free radicals. Free radicals are highly reactive and unstable compounds that damage cell components such as DNA, proteins and lipids.)
(Spectroscopy is the investigation and measurement of spectra produced when matter interacts with or emits electromagnetic radiation. Spectra is the entire range of wavelengths of electromagnetic radiation.)

*"****Comfrey (Symphytum L.)*** *is used to treat bone fractures, tendon injuries, ulcer lesions of gastrointestinal tract. It promotes wound healing, accelerates exudates resorption in lungs and reduces joints' inflammation. In Georgian (country of Georgia) folk medicine, herbal remedies from Comfrey are used to accelerate regeneration processes.*
Comfrey contains hepatotoxic and carcinogenic pyrrolizidine alkaloids, besides **the main active ingredient is poly[3 -(3,4-dihydroxyphenyl) glyceric acid] (PDPGA)**. *The aim of present work was to develop a technology for the substance-poly[3-(3,4dihydroxyphenyl) glyceric acid] from Comfrey stems, free of toxic pyrrolizidine alkaloids."*
-'Development of Technology for the Substance of Poly[3-(3,4-DihydroxyPhenyl) Glyceric Acid] (PDPGA) from Symphytum Asperum' by S.I. Gokadze, V.V. Barbakadze, L.M. Gogilashvili, L.Sh. Amiranashvili and A.D. Bakuridze, Tbilisi State Medical University; Georgian Medical News, Tbilisi, Georgia, No. 5, 218, pages 72-77, May **2013**. All in Russian except for Summary. If you have an English translation, I would appreciate a copy.

"From ancient times extracts, teas and pulps obtained from various Comfrey species (Symphytum L., family Boraginaceae) and amongst them **Caucasian ones (Symphytum asperum Lepech. and S. caucasicum Bieb. widespread in Georgia)** *are known in folk medicine as powerful wound healing and anti-inflammatory remedies.*
It was established that all aforementioned plants contain high molecular constituents namely a caffeic acid derived polyether - Poly[3-(3,4-DihydroxyPhenyl)Glyceric Acid] (PDPGA).
Some of the plants medicinal effects, like its wound healing and anti-inflammatory properties, could be attributed to this polymer. Moreover, the polymer showed antioxidant, immunomodulatory and antitumor activity.
Interestingly, this polymer is a first representative of a previously unknown class of natural occurring biopolymers: phenolic polyethers.*"*
-'Natural Biopolymer-Poly[3-(3,4-DihydroxyPhenyl)Glyceric Acid] from Comfrey and its Synthetic Analogues' by M. Merlani, V. Barbakadze, T. Nakano, L. Amiranashvili, L. Gogilashvili and B. Chankvetadze, Tbilisi, Georgia and Hokkaido, Japan; 4th International Conference on Medicinal Chemistry & Computer Aided Drug Designing, Medicinal Chemistry, Volume 5, Issue 10, page 95, November **2015**. I was only able to get the Abstract.

Choline

Choline is a water-soluble vitamin-like essential nutrient. It is a constituent of lecithin, which is present in many plants and animal organs. Choline is the precursor for the neurotransmitter acetylcholine, which is involved in many functions including memory and muscle control.
(In chemistry, a precursor is a compound that participates in a chemical reaction that produces another compound. In biochemistry, it refers to a chemical compound preceding another in a metabolic pathway.)

Another important constituent of Comfrey is cholin (choline), also known to be a powerful healing agent.*"*
-The Complete Herbal Handbook for Farm and Stable by Juliette de Bairacli Levy. London, England: Faber & Faber, Inc., 1952.

*"****Other constituents:*** *allantoin 0.75-2.55%, caffeic acid, carotene 0.63%, chlorogenic acid 0.037%, caffeic acid 0.035%,*

choline, lithospermic acid, rosmarinic acid and silicic acid (Barnes et al., 2007; Aftab et al., 1996); vitamins A and B12, calcium, potassium and phosphorus (Longe, 2005)."
-'European Medicines Agency- Assessment Report on Symphytum Officinale L., Radix' by European Medicines Agency, Committee on Herbal Medicinal Products (HMPC), London, England, 27 pages, May 5 2015.

Chlorophyll

*"**Good sources of chlorophyll** include green tea; all leafy green vegetables like spinach and chard; **leafy green herbs like Comfrey,** dandelion, borage and lemon grass; young shoots of wheat, rye & barley grasses, and many green and blue-green algae."*
-Viral Immunity: A 10-Step Plan to Enhance Your Immunity Against Viral Disease Using Natural Medicines by J.E. Williams, O.M.D. (Doctor of Oriental Medicine), 2002.

*"**The Chlorophyll molecule bears a striking resemblance to hemoglobin, the red pigment in human blood.** The red blood pigment is a web of carbon, hydrogen, oxygen and nitrogen atoms grouped around a single atom of iron. Nature's green pigment is a similar web of the same atoms, except that its centerpiece is a single atom of magnesium.*
*Chlorophyll (for humans) is mainly used in liquid form, this liquid is not merely green colored water, instead it is the result of a special careful extraction of the green pigment through laboratory processings of the green plants, particularly alfalfa leaves. **Considering the beneficial properties of Chlorophyll and Comfrey in promoting healing of the human tissue internally and externally, it stands to reason that a combination of these two products would be most effective.** The combination could be on a 50% proportion in a suitable liquid. Comfrey and its active ingredient Allantoin has a remarkable quality of promoting healing and the property of cell proliferation of the healthy cells, while **Chlorophyll has a gentle nonirritating antibacteria action, which also promotes healing.**"*
-Comfrey and Chlorophyll: A Report About the Medicinal Value of Comfrey and Chlorophyll as Found in Old and Modern Literature by Vincent Licata. California: Continental Health Research, 1971, pages 23, 24, 26, 30.

Digestibility of Comfrey Leaf

Definitions of Digestibility Terms

Acid Detergent Fiber (ADF) is fiber measurement extracted with acidic detergent to appraise quality of forages. It includes cellulose, lignin, ADIN (Acid Detergent Insoluble Nitrogen) and acid-insoluble ash. The higher ADF, the lower the digestibility.
Ash is the the mineral present in feed. It is measured by burning it at 500 C (932 F) until all organic matter is burned.
Amino Acids (AA) are the builiding blocks of proteins.
Amylase-Neutral Detergent Fiber (aNDF), see Neutral Detergent Fiber.
Average Daily Gain (ADG) is the increase in body weight.

Crude Fiber (CF) is a measure of the quantity of indigestible cellulose, pentosans, lignin, and other similar components.
Crude Protein (CP), also called Total Protein, is total amount of nitrogen. Includes non-protein nitrogen & true protein nitrogen.
Digestible Crude Protein (DCP) or Crude Protein Digestibility (CPD) is the amount of crude protein actually absorbed by the animal (crude protein minus protein lost in feces).
Digestible Dry Matter (DDM) is digestible fiber in a feed. One formula is: %DDM = 88.9- (0.779 x %ADF (dry basis)). Acid Detergent Fiber (ADF) is the least digestible plant components including cellulose and lignin. ADF values are inversely related to digestibility, so forages with low ADF are ususally higher in energy.
Digestible Energy (DE) is the Gross Intake Energy (GIE) minus the fecal energy (DE = GIE - fecal energy). Fecal energy is unused energy that is excreted. Digestible Energy gives the amount of energy the animal has available to use.
Dry Matter (DM) is the part of feed that is not water. Percent DM= 100% - moisture%.

Effective Degradability (ED) is the proportion of feed nutrients contained in the feed that could be degraded in the rumen (of ruminants such as sheep, goats and cattle).
Essential Amino Acids (EAA) are amino acids that can not be sythesized by the animal so must be in the diet.
Ether Extract (EE) is the part of organic material that is soluble in ether and consists chiefly of fats and fatty acids.
In Vitro Digestibility (IVD) is estimated by near infrared reflectance (NIR) analysis or estimated from the acid detergent fiber. In Vitro means in a laboratory setting.
Lignin along with cellulose, forms the cell walls of plants. It is indigestible.

Metabolizable Energy (ME) equals gross feed energy minus energy lost in the feces, urine and gaseous product of digestion.
Metabolisable Protein (MP) is the total amino acids available for digestion for maintenance and growth.
Net Energy (NE) is the energy available to an animal in a feed after removing the energy lost as feces, urine, gas and heat produced during digestion.
Neutral Detergent Fiber (NDF) is a measure of fiber after digesting in a nonacidic, nonalkaline detergent as an aid in determining quality of forages. It has the fibers in ADF, plus hemicellulose. It measures the plant cell wall which consists of lignin, cellulose, and hemicellulose.

Nitrogen-Free Extract (NFE) is carbohydrates, sugars, starches, and most materials classed as hemicellulose. When crude protein, fat, water, ash, and fiber are added and the sum is subtracted from 100, the difference is NFE.

Organic Matter (OM) is carbon-based organic compounds from plants and animals and their waste products.
Starch Equivalent (SE) is the fat-producing capacity of an animal feed expressed as the amount of starch required to produce the same amount of fat. It is basically the same as net energy of feed because both give the productive value of the feed.
Total Digestible Nutrients (TDN) measures available energy of feeds and energy requirements of animals involving a complex formula. It is hard to measure, but is widely used in the United States and Canada. TDN values are quoted as percentages for feeds and as amounts per day for requirements. The simplest and most commonly used formula for estimating: TDN = DE/0.044. One kilogram of TDN is equivalent to 4.4 megacalories of Digestible Energy (DE).

Digestibility Overview

"*Forage crops: The varieties tested were Prickly Comfrey,* kaffir corn (type of sorghum), black eye variety of cow-pea, teosinte (type of Mexican grass), white mustard, Brazilian flour corn. Only the first three seemed to merit further trial.
Total Produce and Digestible Nutrients (Pounds per acre):

	Prickly Comfrey	**Kaffir Corn**	**Cow-Pea**
Total crop per acre	16,500 pounds (7484 kg)	11,000	14,500
Digestible protein per acre	221.0 pounds (100 kg)	119.4	280.9
Digestible fat per acre	30.5 pounds (13.8 kg)	34.2	53.9
Digestible carbohydrates	658.4 pounds (298.6 kg)	941.0	937.0
Total digestible organic matter	909.9 pounds (412.7 kg)	1094.6	1271.8 "

-'Experiment Station Record, Volume 1, September 1889- July 1890', United States Department of Agriculture, Office of Experiment Stations, Washington, D.C., **1889**-1890, page 143.

(Kaffir Corn is a bicolor sorghum that is cultivated in dry regions. It is an important human and animal food. The growth habit and stem are similar to Indian corn. It has sawtooth-edged leaves.)

(Cow-Pea {Vigna unguiculata synonym V. sinensis} is in the legume family related to the bean. It is widely cultivated in the southern United States for forage, green manure, and edible seeds. Also called black-eyed pea and field pea.)

"*Composition of Feeding-Stuffs:*

	Organic Matter	**Crude Protein**	**Crude Fat**	**Nitrogen-Free Extract**	**Crude Fibre**
Comfrey Hay	69%	58%	71%	85%	18%
Comfrey Fresh		2.5%	0.3%	5.0%	1.7%

Digestible Nutrients- Comfrey, Fresh:

Crude Protein	**Crude Fat**	**N-Free Extract**	**Crude Fibre**	**Value**	**%Digestible Protein**	**Starch Equiv**
1.5%	0.2%	3.7%	0.8%	91	0.9%	5.2 /100 lb"

-'The Scientific Feeding of Animals' (Die Ernahrung der Landwirtschaftlichen Nutztiere, 1905) by Professor O. Kellner, German Agricultural Scientist. New York: The Macmillan Company, **1910**, pages 362, 383. Translated from German by William Goodwin, B.Sc., Ph.D., University of London, England.   (Full 'Value' = 100. 'Starch Equiv' = Starch Equivalent per 100 pounds = 45 kg.)

"*Digestible Nutrients- Comfrey (Symphytum asperrimum):*

Crude Protein	**True Protein**	**Oil**	**Carbohydrate**	**Fibre**	**Nutritive Ratio**	**V.**	**Starch Equivalent**
1.5%	0.9%	0.2%	3.7%	0.8%	3	91%	5.2 per 100 pounds

Protein
The percentage of Crude or Total Protein in a feedingstuff is found by determining the percentage of total nigroten using the Kjeldahl method, and multiplying the result by the factor 6.25. Not all nitrogen in feedingstuffs, however, is in the form of True Protein, or as it is sometimes called, pure protein. Some part of it is present as nitrogenous compounds of much simpler chemical nature. These non-protein nitrogenous substances are usually grouped togeher under the heading of 'Amides'.
Nutritive Ratio
Nutritive or nutrient ratio is the ratio of digestible protein, or repair material, to digestible carbohydrate, oil and fibre, or fuel substances. The nutritive ratio thus gives the number of pounds of fuel substances associated with 1 pound of digestible protein in the feedingstuff or ration.
V. Factor
The correction factor 'V' is known as the percentage availability of the feedingstuff. The correction is necessary because the full values of the digestible nutrients do not become available for production on account of the energy that is wasted in the work associated with the mastication (chewing) and digestion of the feedingstuff. The magnitude of the correction naturally increases as the foods under consideration become more fibrous and less digestible.
Starch Equivalent
When, for instance, the starch equivalent of barley is stated to be 71, it is meant that 100 pounds (45.3 kg) of barley when fed to sheep and cattle in conjunction with a maintenance ration, has a productive or fattening value equal to that of 71 pounds (32.2 kg) of starch. The definition implies that the barley would be included in a ration suitably balanced in respect of protein, minerals, and vitamins."
-'Rations for Livestock' by H.E. Woodman, M.A., Ph.D., D.Sc., School of Agriculture, Cambridge University, England; Bulletin No.

48, Ministry of Agriculture and Fisheries, Her Majesty's Stationery Office, London, England, 12th Edition, August **1952**, pages 3-6, 117, 128. First published January 1921.   (Oil is 'Ether Extract'. Carbohydrate is 'Nitrogen-Free Extractives'.)

Nutrition and Digestibility of Leaves/Stems and Maturity

"Of interest are studies on the nutritive value of leaf and stem portions of Russian Comfrey.
Stems were consistently higher than leaves in digestibility at all growth stages (D.N. Mowat, B.R. Christie and E.E. Gamble, 1966, 'In Vitro Digestibility and Protein Content of Russian Comfrey', unpublished data).
With advancing maturity, a decrease in the IVD (In Vitro Digestibility) of both leaves and stems occurred.
Digestibility of whole-plant regrowth would be particularly low due, in part, to a high percentage of leaves.
On the other hand, as usual, leaves were over two times higher than stems in Crude Protein content."
-Forages Economics-Quality, ASA Special Publication Number 13, edited by C.M. Harrison, papers presented at the 'Forage Economics Symposium' and the 'Forage Quality Evaluation Symposium' by the 'American Society of Agronomy', 'Crop Science Society of America' and 'Soil Science Society of America' in November **1967**, published in Wisconsin 1968, pages 88-89.
(I could not find the report: 'D.N. Mowat, B.R. Christie and E.E. Gamble, 1966, 'In Vitro Digestibility and Protein Content of Russian Comfrey'. If you have it, could you please send it to me.
   ('In Vitro' means occurring in a laboratory vessel rather than in a living organism.)

*"Two field samples each of orchardgrass (Dactylis glomerata L.), bromegrass (Bromus inermis Leyss., variety Saratoga), alfalfa (Medicago satiau L., variety Vernal), birdsfoot trefoil (Lotus corniculatus L., variety Empire) and **Russian Comfrey (Symphytum officinale L.)** were collected.*
Russian Comfrey was harvested at two stages of maturity: early (bud) and late (flower). These samples were immediately dried in a forced-air oven at 45 C (113 F), and later separated into leaves and stems. Samples were ground through a hammermill equipped with a 0.8 mm screen.
The total silica content of Russian Comfrey was especially high. However, further analysis showed that approximately 85% of this silica was due to soil contamination. (Silica is a mineral.)
Digestibility decreased with advancing maturity. Furthermore, leaves were higher than stems in digestibility with all forages except Russian Comfrey, where the opposite occurred. Leaves of Russian Comfrey were also higher than stems in lignin content. It is well known that lignin and fiber contents increase as forages mature, whereas cell wall digestibility decreases.
Furthermore, studies have shown that with common grasses and legumes, leaves are lower than stems in lignin and fiber content, and higher in 'in vitro' digestibility from the late vegetative stage onwards. In contrast, with Russian Comfrey, Mowat et al. showed that the 'in vitro' dry matter digestibility of stems was consistently higher than that of leaves.
The relationship of lignin in cell wall constituents to cell wall digestibility was linear in grasses. However, both legumes and Russian Comfrey behaved differently from grasses. They had a relatively low cell wall content which was more lignified than grasses.
The greater lignification was associated with a lower digestion coefficient of cell wall constituents, but this was not as low as would be expected from the amount of lignin present. With legumes and Russian Comfrey different types of lignin may be present, or possibly the lignin is confined to a smaller proportion of the cells or cell walls than with grasses."
-'Lignification and In-Vitro Cell Wall Digestibility of Plant Parts' by D.N. Mowat, M.L. Kwain and J.E. Winch, University of Guelph, Ontario, Canada, Canadian Journal of Plant Science, Volume 49, pages 499-504, **1969**.
   (Lignin forms structural materials to support tissues of vascular plants. They are part of the cell walls, especially in wood and bark. They give rigidity.)

"One aspect of drying Comfrey should be kept in mind. It is that slow drying will reduce the quantity of Beta Carotene for conversion to Vitamin A in the liver. One fresh weight (green leaf) analysis of Comfrey done in England revealed 77 mg per kg of leaf of Beta Carotene. In dried Comfrey leaf the highest figure known was 400 mg per kg. The New Zealand analysis shows 170 mg per kg, which is a good average taken over a year.
Spring growth is highest in Carotene.
Surplus Comfrey should be dried as fast as possible, compressed into bales, and stored in a cool dry dark place to be fed out when needed in the winter."
-Comfrey: Nature's Healing Herb & Health Food by Andrew Hughes. Japan: Sanyusha Publishing Co., Ltd, **1992**, page 161.
   (One kg or kilogram = 2.2 pounds)

"There is evidence that the mineral content falls in the late Autumn, for as Comfrey dies down it returns its minerals to store in the roots."
-Comfrey: Fodder, Food and Remedy by Lawrence D. Hills. New York: Rizzoli Universe Books: **1976**, page 78.

"In summer, the leaves of S. asperum are much richer in Crude Protein than the stems, and in spring and autumn the leaf blades contain more than the petioles (leaf stem) (Vavilov, Edel'shtein and Solov'eva, (1973).
The same is true of the spring leaves of S. x uplandicum (Schereva et al, 1965)."
-'Comfrey Symphytum spp. as a Forage Crop' by J.C. Forbes, A.D. McKelvie, and P.J.C. Saunders, North of Scotland College of Agriculture, Aberdeen, United Kingdom; Herbage Abstracts, Volume 49, No. 12, pages 523-539, **1979**.

Digestibility of Comfrey Protein

"Trials with sheep were made on 8 roughages (fodder) commonly used for winter rations, 16 samples of veld grazing taken at intervals from summer to winter, soya bean seed and mixtures of maize (corn) meal and lucerne (alfalfa) in the proportion 4: 5 or 2: 7.
Total Digestible Nutrients (TDN) in the dry matter of legume and grass hays and Russian Comfrey were from 49 to 58 percent, and Digestible Crude Protein (DCP) from 1.7 to 10.9 percent. Grass hay had the lowest DCP.
Russian Comfrey was high in ash, but this may have been due to contamination.
It is suggested that the interaction of feeds may affect the validity of estimating digestibility by difference."
-'Digestion Trials on Rhodesian Feedstuffs: 2' by R.C. Elliott and A. Croft, Henderson Research Station, Rhodesia (Zimbabwe, Africa); Ministry of Agriculture, Rhodesia Agricultural Journal, Volume 55, pages 40-49, **1958**. (I don't have this report. If you have it, could you please send it to me.)
  (Veld or veldt is a wide open rural landscape in southern Africa. It is a flat area with grass or low scrub.)

"Digestibilities of neem leaves, mangels and Russian Comfrey were estimated with bullocks (castrated bulls). Mangels and Comfrey were low in fibre, and digestibility of DM (Dry Matter) of both was high. Comfrey had 18.7% digestible Crude Protein (CP) in the DM. Comfrey had Ca (Calcium) 2.02% and P (Phosphorus) 0.57%."
-'Nutritive Value of Neem Leaves, Mangolds and Comfrey' by B.M. Patel, P.S. Patel and P.C. Shukla, Institute of Agriculture, Gujarat, Anand, India; Indian Journal of Dairy Science, Volume 15, pages 139-145, **1962**.
  (Mangel is a large coarse yellow-orange to reddish-orange beet grown chiefly as food for cattle and other animals.)

"The swine (pig) digestion trial indicated that the Comfrey used had a Dry Matter content of 12.1% Digestible Crude Protein, and 52.7% TDN (Total Digestible Nutrients) content. *The ratings for Crude Protein, Total Digestible Nutrient content and Nitrogen-free extract were all lower for Comfrey than for the control ration.*
The control ration was ground barley, cottonseed meal (41% of ration's Crude Protein), plus meat and bone meal (50% of ration's Crude Protein)."
-'Comfrey as a Feed for Swine' by Hubert Heitman, Jr. (University of California) and Sergio E. Oyarzun (University of Chile); California Agriculture (Hilgardia: Journal of Agricultural Science by the California Agricultural Experiment Station), Volume 25, No. 1, pages 7-8, January **1971**.

*"In a series of 8 digestibility trials over an 18-month period, 2 **Ossimi (sheep)** wethers (castrated males) were fed **Russian Comfrey plants** cut when the leaves were 50 cm (19.6 inch) long.*
The plants were very palatable (agreeable) and analysis showed that OM (Organic Matter) represented 78.5% of DM (Dry Matter), CP (Crude Protein) 21.7%, CF (Crude Fiber) 10.2%, ether extract 5.9%, NFE (Nitrogen-Free Extract) 40.7%, and ash 21.5%.
DM content was 61.9%, and measured on a DM basis, SE (?) and TDN (Total Digestible Nutrients) were 55.8 and 60.6% respectively.
*Compared with berseem, darawa and maize, **Russian Comfrey had the highest DCP (Digestible Crude Protein) (2.42%) but the lowest digestible CF and digestible NFE.***
SE of Comfrey and berseem were similar (7.96%) but much lower than the values for darawa (11.10%) and maize (11.17%). TDN values of Comfrey and berseem were about 30% lower than those of darawa and maize."
-'Nutritive Value of Russian Comfrey' by A. El-Bassousy, F. El-Sabban, A.M. Makky, M.S. El-Danasoury and M.A. El-Ashry, Animal Production Research Institute, Agriculture Research Center, Cairo, Egypt; Agriculture Research Review, Volume 53, No. 7, pages 59-64, **1975**.

" **'Composition, Digestion Coefficients, and Nutritive Values of Russian Comfrey Leaves and Alfalfa Hay'.**
Russian Comfrey information is from: Elliot and Croft, 1958, Rhodesia Agricultural Journal, 55:40-49.
Alfalfa hay information is from: 'United States-Canadian Tables of Feed Composition', 1969, National Academy of Sciences, National Research Council Publication #1684.
No details on stage of growth, harvest management, etc.

	Dry Matter	Crude Protein	Ether Extract	Crude Fiber	N-Free Extract	Ash
Russian Comfrey Leaves						
Composition	100.0%	15.0%	1.7%	12.3%	43.4%	7.6%
Disgestion Coefficients	67.3	69.0	65.3	46.0	70.9	--
Digestible Crude Protein 10.4%, Total Digestible Nutrients 49.2%						
Alfalfa Hay						
Composition	100.0%	14.4%	1.5%	33.7%	42.0%	8.4%
Disgestion Coefficients	47.6	59.5	33.4	34.9	56.7	--
Digestible Crude Protein 8.6%, Total Digestible Nutrients 45.3%						

-'Comfrey as Forage' by Lester R. Vough, Extension-Research Agronomist, Oregon State University, Corvallis, Agronomic Crop Science Report, August **1976**, page 5.   (Rhodesia is now Zimbabwe, Africa.)

"Hart* pointed out that digestibility of Comfrey protein was only 38 percent in an experiment conducted in the Nether-

lands and only 49 percent in a California experiment. Alfalfa protein, in comparison, is over 70 percent digestible. In South Korea, the total dry matter of Comfrey was 56 percent digestible, while that of orchardgrass and ladino clover was 64 and 73 percent digestible, respectively.

According to Hart, the ash (mineral) content of Comfrey was high in most experiments but much of the ash may have resulted from soil clinging to the hairy leaves of the Comfrey plants. In the Kenya study, such contamination accounted for 14 percent of the uncorrected dry matter yield."

-'Comfrey as Forage' by Lester R. Vough, Extension-Research Agronomist, Oregon State University, Corvallis, Agronomic Crop Science Report, August **1976**, page 2, 3.

(* -'Comfrey Yields and Forage Value' by Richard H. Hart, United States Department of Agriculture, Agricultural Research Service Plant Physiology Institute, Beltsville, Maryland, CA-NE-2, December 1972).

"Estimates of the digestibility of the Crude Protein of Comfrey for ruminants (sheep, goats, cattle) range widely from 38 to 82%. The digestible Crude Protein content normally ranges form 65 to 187 grams/kg DM (Dry Matter), though a rabbit-digestible Crude Protein content of 201 grams/kg DM has been recorded by Han, Kim and Lee (1968).

Lehmann (1944) found that the digestibility of S. x uplandicum for pigs decreased with crop maturity from 73% of total organic matter at the rosette (rose shaped) stage to as low as 55% in the flowering stage. *"*

-'Comfrey Symphytum spp. as a Forage Crop' by J.C. Forbes, A.D. McKelvie, and P.J.C. Saunders, North of Scotland College of Agriculture, Aberdeen, United Kingdom; Herbage Abstracts, Volume 49, No. 12, pages 523-539, **1979**.

"Proximal, mineral and amino acid analysis of Crude Protein Digestibility (CPD), Digestibile Energy (DE) and Metabolizable Energy (ME) were made in 21 feedstuffs for pigs.

Average values expressed on a DM (Dry Matter) basis for CPD % and DE kcal/kg were, respectively...Comfrey hay 45.57 % and 2202 kcal/kg."

-'Chemical Composition and Energy Values of Some Feeds for Pigs' by E.T. Fialho, L.F.T. Albino and E. Blume, Embrapa (Brazilian Agricultural Research Corporation, Ministry of Agriculture Livestock and Food Supply) / CNPSA, SC, Brazil; Pesquisa Agropeduaria Brasilerira, Volume 20, No. 12, pages 1419-1431, **1985**.

(Proximates are used in the analysis of food into its major constituents. They are approximations. In industry, the standard proximates are: Ash, Moisture, Proteins, Fat and Carbohydrates.)

"The aim of this study was to characterize the nutritional quality of nine forage resources:

Comfrey (Symphytum peregrinum, Russian Comfrey), White Mulberry (Mortis alba), Sleeping Hibiscus (Malvaviscus penduliflorus), Nacedero (Trichanthera gigantea), tree marigold (Tithonia diversifolia Hemsl.) Gray, Ramie (Boehmeria nivea L.) Gaud, Arboloco (Montanoa quadrangularis Bipontinus Schultz), Chachafruto (Erythrina edulis Triana ex Michelle) and Andean Alder (Alnus acuminata Kunth), with potential for inclusion in strategic supplementation programs in the highland tropics of Colombia (South America).

The forages that showed the highest levels of PC (Proteina Cruda= Crude Protein) were Comfrey (28.42%), *Chachafruto (26.52%), Arboloco (26.35%) and those that showed the lowest were the Sleeping Hibiscus (15.92%) and Andean Alder (16.88%). The lowest NDF (Natural Detergent Fiber) contents were presented by White Mulberry (33.55%) and Andean Alder (35.79%) while the highest were observed for Chachafruto (49.64%) and Ramie (48.54%).*

The nutrient content of forages evaluated can be included in supplementation programs, due to they can enhance the energy density of the diet and may increase dry matter intake."

-'Nutritional Characterization and Ruminal Degradation Kinetics of Some Forages with Potential for Ruminants Supplementation in the Highland Tropics of Columbia' by J.F. Naranjo and C.A. Cuartas, CIPAV, Cali, Colombia; Revista CES Medicina Veterinaria y Zootechnia (CES Magazine Veterinary Medicine and Zootechnics), Volume 6, No. 1, pages 9-19, **2011**.

"Six sheep fitted with permanent ruminal (rumen) cannulas (thin tubes) were used to determine the degradation characteristics of amino acid (AA) ruminal incubation by nylon-bag technique for 2 forage grasses at different growth stages. The contents of amino acid reduced gradually with the growth of 2 kinds of forages. **The TAA/CP (Total Amino Acids per Crude Protein) of Comfrey showed singer-peak (low-high-low) dynamic pattern.**

Two kinds of grasses were rich in aspartic acid, glutamic acid, alanine, valine and leucine, while the content of sulfur amino acids lower. But the amino acid changed at different growth stages with different rules.

Most amino acids of the two kinds of grasses in the flowering were not easily degradable in the rumen, *showing that amino acids in the rumen of grass flowering had good resistance to degradation, especially proline and alanine.*

In 4 growth periods amino acids were more prone to degradation in the rumen except flowering, while the lysine and arginine were more easily degradable in the rumen."

-'Study on the Changes of Amino Acid Profiles Rumen Degradation for Cluster Leaf Rosinweed and Comfrey at Different Growth Stages' by TaiYu Liu, MengYun Li, Li Zheng and HongXing Qiao, Engineering Technology Research Center of Animal Nutrition and Feed, Zhengzhou, Henan, China; Journal of Northwest A&F University, Natural Science Edition, Volume 39, No. 1, pages 36-42, **2011**.

(I only have access to the abstract. It did not say what species Comfrey. The wording sounds like it was translated from Chinese to English. If you have an English translation of the entire report, could you please send it to me. I understand this report to say that the flowering stage is the least digestible time for the Comfrey plant for ruminants such as sheep and goats.)

Digestibile Dry Matter (DDM)

"*Digestibility of native grasses in Korea was estimated in vitro.*
Digestibility of dry matter (DM) of Russian Comfrey was significantly affected by length of interval between harvests; 25 days was best. Digestibility of ladino clover was the highest among the species tested throughout the experiment. Digestibility of all native grasses in this experiment was less than that of improved pasture grasses."
-'Digestibility of Domestic Feedstuffs Determined In Vitro' by S.H. Park, S.H. Chee and I.K. Han, Livestock Expo Stat, Suwon, Korea; Nong-sa Si-hem Yen-ku Po-ko= Research Report to Office of Rural Development, Suwon, South Korea, Volume 12, No. 4, pages 37-41, **1969**.
(In vitro means in a laboratory vessel or other controlled experimental environment rather than in a living organism.)

" '*In Vitro Digestible Dry Matter (IVDDM) and Crude Protein Content of Russian Comfrey* at Various Stages of Growth (Whole Plant Basis)' by Mowat, Christie, and Gamble, 1966, Canada Deptartment of Agriculture, Forage Crop Division, Forage Notes 12(1):63-64.

Stage	IVDDM	Crude protein
Early	64.8%	21.1%
Medium	66.3	17.0
Late	65.4	15.2
Regrowth	53.2	17.7

"
-'Comfrey as Forage' by Lester R. Vough, Extension-Research Agronomist, Oregon State University, Corvallis, Agronomic Crop Science Report, August **1976**, page 6.
(Digestible Dry Matter {DDM} is digestible part of a feed. 'In Vitro' means done in laboratory equipment as opposed to in/on a living animal that is called 'In Vivo'. Another term is 'In Vitro Dry Matter Digestibility {IVDMD}.)

"**Comfrey was very high in crude protein (Kjeldahl nitrogen percentage x 6.25), and protein percentage increased from first to last harvest.** *Alfalfa cut at one-tenth bloom stage averages about 18 percent protein. Acid Detergent Fiber (ADF) percentage is a measure of digestibility used for grasses and legumes, and ADF data can be converted to percentage Digestible Dry Matter (DDM) by a formula.* **The DDM values for Comfrey using the procedures for grasses and legumes indicate that Comfrey is a forage of high digestibility.**
However, research has not determined whether or not the procedures applicable to grasses and legumes apply to Comfrey. USDA researchers reported that the Tilley and Terry two-stage method of measuring dry matter digestibility gave in vitro dry matter digestibility percentages of 37, 61, and 62 percent, respectively, for Comfrey, orchardgrass, and alfalfa harvested from plots at Beltsville, Maryland.
Crude protein percentage in these USDA trials ranged from 13 to 17 percent for Comfrey and 16 to 17 percent for alfalfa*. Other research indicated that the digestibility of Comfrey ranged from 37 to 77 percent and is similar to or occasionally lower than that of other forages."

"**Yields and Quality of Forage at each Harvest Date from Comfrey**
Cut 4 or 5 times per Year at Rosemount, Minnesota, 1979-81

Cutting Dates	Yield/Acre (pounds)	Moisture %	Protein %	DDM %
First, June 3- June 28	2,130	88	21	65
Second, June 21- July 26	3,190	87	23	60
Third, July 18- August 21	1,910	91	25	64
Fourth, August 14- Oct. 7	1,430	90	26	71
Fifth, September 19	1,470	92	31	74

"
-'Comfrey: A Controversial Crop' by Robert G. Robinson, University of Minnesota, Agricultural Experiment Station, Minnesota Report MR-191, Item No. AD-MR-2210, **1983**, page 4.
(* -'Forage Yield and Quality of Quaker Comfrey, Alfalfa, and Orchardgrass' by R.H. Hart, A.J. Thompson III, J.H. Elgin Jr., and J.E. McMurtrey III, Agronomy Journal, 73:737-742, 1981.)
(The Kjeldahl method determines the nitrogen contained in organic substances plus the nitrogen contained in the inorganic compounds ammonia and ammonium {NH_3/NH_4^+}. Other forms of inorganic nitrogen, for instance nitrate, are not included in this measurement. This was developed by Danish chemist Johan Kjeldahl in 1883.)

Energy: Gross and Metabolizable

Gross Energy (GE) or heat of combustion is the total chemical energy measured from complete combustion of the food in a bomb calorimeter.
Metabolizable Energy (ME) equals gross feed energy minus energy lost in the feces, urine and gaseous products of digestion.

"*All the gross and metabolizable energy values have been calculated from total and digestible nutrients, respectively, using the formulae given by the 'United Kingdom Ministry of Agriculture, Fisheries and Food' (1975).*
Comfrey Gross Energy contents are in the range 14-18 MJ/kg (megajoules per kilogram) DM (Dry Matter), and Metabolizable Energy contents in the range 6.3-12.4 MJ/kg DM.
Taking 10 MJ/kg DM as a standard, Comfrey can be expected to yield 35,000-70,000 MJ/hectare of metabolizable energy, a

sufficient maintenance allowance for 10-20 cattle of 300 kg (661 pounds) body weight for 100 days."
-'Comfrey Symphytum spp. as a Forage Crop' by J.C. Forbes, A.D. McKelvie, and P.J.C. Saunders, North of Scotland College of Agriculture, Aberdeen, United Kingdom; Herbage Abstracts, Volume 49, No. 12, pages 523-539, **1979**.

*"**Energy value of dry biomass of perennial herbs was measured in the mountain zone of Kabardino-Balkarian Republic, Russia in May 2002. Energy value of 1 kilogram of dry biomass was** 16.5, 12.9, **10.7** and 9.1 **MJ (megajoule) for** goat's-rue, inula, **Comfrey** and nettle, respectively.*
Yield and energy value of mixed planting were 13.6 tons/hectare and 12.6 MJ, respectively. Detailed data are presented in a table and a graph."
-'Effectiveness of Multi-Purpose Grasses in Fodder Production' by A.Ya. Tamakhina, Kabardino-Balkarian State Agrarian Academy, Russia; Kormoproizvodstvo (Plant Breeding), No. 10, pages 2-4, **2009**.

Del otro Symphyto 1555
Symphytum Alterum

'De Materia Medica, Acerca de la Materia Medicinal y de los Venenos Mortiferos, Lib. I to VI' by Dioscorides, translated by Andres de Laguna 1555. Symphytum Lib 'IIII' (4), Cap. XI, page 382. Published in Antwerp, Belgium.

During the first century, the Greek doctor and apothecary Pedanius Dioscorides wrote 5 volumes of 'De Materia Medica' (On Medical Material). This book on botany and pharmaceuticals was read throughout Europe and the Middle East for 1,500 years. In the 10th century, the Caliph of Cordova (Abd al-Rahman III) had the work translated into Arabic.

Andres de Laguna (1499-1559) was a Spanish physician, pharmacologist, and botanist. His most important work is the translation of 'De Materia Medica' into Castilian Spanish with many additions and commentaries.

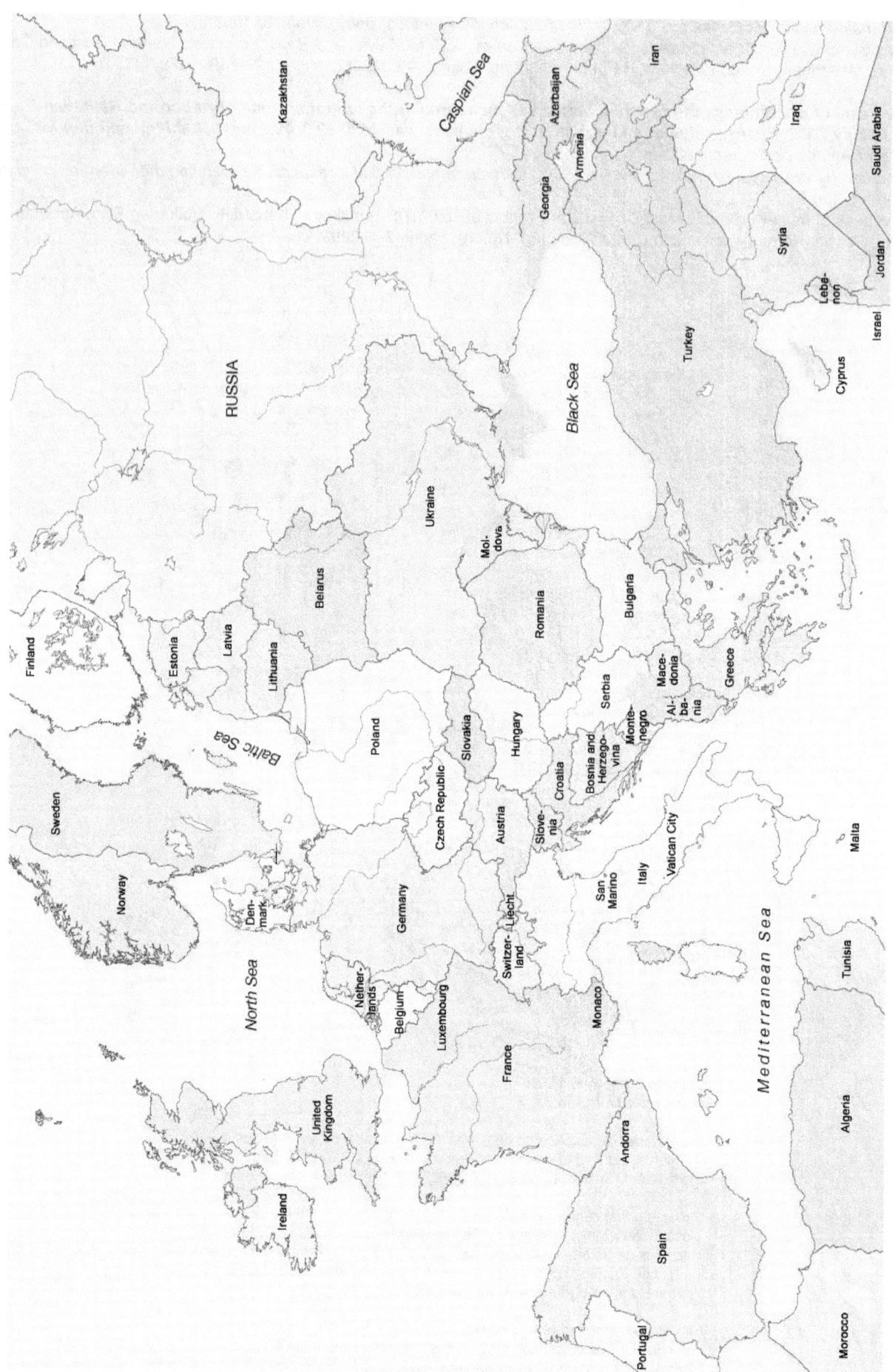